UTILIZATION OF ELECTRIC POWER AND ELECTRIC TRACTION

[Volume-IV of A Course in Electrical Power]

J.B. GUPTA

UTILIZATION OF ELECTRIC POWER AND ELECTRIC TRACTION

For Engineering Students

By

J.B. Gupta

B.Sc. (Elec. Engg.); F.I.E. (India)
C. Engg. (I); M.I.S.T.E.

S.K. KATARIA & SONS®

Publisher of Engineering Books

4885/109, Prakash Mahal, Dr. Subhash Bhargav Lane,
Opposite Delhi Medical Association, Daryaganj, New Delhi-110002 (INDIA)
Phones: +91-11-43551243, Mobile: +91-9871775858
e-mail: katariabook@gmail.com; katariabooks@yahoo.com
Website: www.skkatariaandsons.com
(e-books, subscriptions and bulk buying also available on our website)

Published by :
S.K. Kataria & Sons®
4885/109, Prakash Mahal, Dr. Subhash Bhargav Lane,
Opposite Delhi Medical Association, Daryaganj, New Delhi-110002
Phones: +91-11-43551243, Mobile: +91-9871775858
e-mail: katariabook@gmail.com, katariabooks@yahoo.com
Website: www.skkatariaandsons.com

Edited By:
Rajeev Manglik
B.E. (Elec.) Pb. University
Rohit Manglik
B.E. (Elec.) University of Roorkee

First Edition: July 1968, Tenth Edition: January 2012 Reprint: 2026

Laser Typesetting by : AISA ENTERPRISES

PRINTED AT : AL AZIZ BOOKS (SAHIBABAD)

PREFACE TO THE TENTH EDITION

The author feels great pleasure in presenting the tenth edition of the book with a marked improvement over the previous one. The book has been thoroughly revised and rewritten and so designed as to meet the needs of Degree, A.M.I.E. Sec. B and Diploma students of Electrical Engineering. The subject matter has been presented in an easily comprehensible manner, logical sequence using simple language and to provide a detailed exposition of fundamentals. A large number of the old diagrams have been replaced by new ones and numerous new diagrams have been added to make the subject more illustrative and clear. A large number of worked-out examples and theoretical exercises have been deleted and replaced from latest examination papers of Institution of Engineers (India), different Indian Universities, Boards of Technical Education and other equivalent examining bodies. Objective type questions have also been added at the end of the book. An earnest effort has been made to provide a comprehensive coverage of the course as prescribed by the Institution of Engineers (India), Indian and Foreign Universities, Boards of Technical Education and various agencies conducting competitive examinations.

The book consists of 17 Chapter divided into four sections covering syllabii prescribed by different examining bodies. The book contains adequate material for the papers of *Electric Drives and Control; Electric Drives and Utilization; Electric Drives and Utilization of Electrical Energy; Utilization of Electric Power; Utilization of Electrical Energy; Utilization and Traction; Utilization of Electric Power and Electric Traction; Electric Traction; Illumination Engineering etc.*

The book contains 1134 Objective Questions (700 Multiple Choice Type, 64 Assertion (A)/Reason (R) Type, 20 Matching Type, 200 True (T)/False (F) Type and 150 Fill in The Blank Types) designed to bring out the hidden aspect and also test knowledge, understanding and application aspect of learning. These questions will help the students in improving their performance in their regular as well as competitive examinations.

The solution to illustrative examples covering almost all topics and subtopics of different chapters makes the entire presentation easy to follow. Selected exercises and numerical problems with answers given at the end of each chapter help the students in evaluating their understanding of the subject. Short answer type questions with answers help the students in preparing for viva-voce/practical examinations.

Every care has been taken to eliminate misprints, omissions and errors but it is too much to expect that no inaccuracy/error/misprint has crept in and the author would be grateful to the readers for bringing to his notice any of such errors/misprints/omissions they may come across while going through the book.

The author shall feel satisfied if the book meets the needs of the students for whom it is meant. Suggestions and criticism for the further improvement of the book shall be welcomed.

The author will like to thank the readers (both teachers and students) who sent their valuable suggestions that became the basis for the revision of the book.

The author will like to thank to all the Scientists, Engineers, Authors and Publishers whose works and text have been the source of enlightenment, inspiration and guidance to me in presenting this edition.

Lastly, but not the least, the author is thankful to his publisher for bringing out the book in such a short time and for pricing it moderately despite heavy cost of paper and printing.

January 5, 2012 **J.B. GUPTA**

PREFACE TO THE FIRST EDITION

This book has been written with the aim of presenting a short text-cum-help book covering the syllabii of Degree, A.M.I.E. Sec. B and other equivalent engineering examinations of Indian Universities in the Subject of (*i*) Utilization of Electric Power (*ii*) Electric Traction and (*iii*) Illumination Engineering for self study.

Every effort has been made to deal with each and every topic and problem from the fundamentals in easy and expressive language for the convenience and better understanding to the readers. While writing this book the difficulties and troubles which a common student faces and thrust of knowledge of an intelligent student have been kept in view.

The book has been divided into two parts namely **'Utilization of Electric Power'** and **'Electric Traction'** for convenience. It comprises of 10 chapters covering syllabii prescribed by Institution of Engineers (India) and various Indian Universities. Most of the worked-out examples, unsolved problems and exercises included in the text for illustrations are from examination papers of various Indian Universities, State Boards of Technical Education, Institution of Engineers (India) or of equivalent standard. The solution to each and every problem is step by step and so clear that this helps the readers in understanding the basic theory and creates confidence in them.

Every care has been taken to eliminate omissions and errors but it is too much to expect that no inaccuracy has crept in and the author would be grateful if readers would bring any such error or misprint to his notice.

The author will consider his efforts amply rewarded if the book serves the purpose of the readers for whom it is meant.

Mr. D. D. Sharma deserves my special thanks for his drawing work.

My sincere thanks are due to all of the Scientists, Engineers, Authors and Publishers whose works and text have been the source of enlightenment, inspiration and guidance to me in presenting this small book.

Lastly but not the least I am very thankful to my publishers for bringing out the book in such a short time and for pricing the book so moderately despite of heavy cost of paper and printing.

15th July, 1968 **J.B. GUPTA**

CONTENTS

SECTION-A : ELECTRIC DRIVES AND INDUSTRIAL APPLICATIONS ....1—182

1. ELECTRIC DRIVES .... 3—111

1.1. Introduction 3
1.2. Factors Governing Selection of Electric Motors 4
1.3. Nature of Electric Supply 4
1.4. Types of Drives 5
1.5. Types of Loads 6
1.6. Single Quadrant Operation of Electric Motors 6
1.7. Four Quadrant Operation of Electric Motors 7
1.8. Torque-Speed and Speed-Torque Characteristics 8
1.9. Selection of Electric Drive 8
1.10. Motor Speed-Torque Characteristics 11
1.11. Joint Speed-Torque Characteristic of an Electric Motor and a Driven Unit 11
1.12. Dynamics of Motor-Load Combination 12
1.13. Referred Load Torques and Moments of Inertia 13
1.14. Referring Forces and Masses Having Translational Motion To a Rotating Shaft 14
1.15. Referring Torques and Translating Masses When Speed is Variable 15
1.16. Operating or Running Characteristics of Electric Motors 15
1.17. Starting of Electric Motors 39
1.18. Speed Control of Electric Motors 54
1.19. Electric Braking 67
1.20. Electric Braking With DC Motors 69
1.21. Electrical Braking of Polyphase Induction Motors 76
1.22. Electric Braking of Synchronous Motors 81
1.23. Dynamics of Braking 82
1.24. Types of Enclosures 83
1.25. Bearings 85
1.26. Transmission of Drive 85
1.27. Choice of Drive 87
1.28. Noise 87
1.29. Insulating Materials 88
1.30. Temperature Rise in Electrical Machines 89
1.31. Duty Cycles 90
1.32. Rating of Machines 92
1.33. Choice of Rating of Motors 92
1.34. Methods of Ventilation and Cooling of Machines 98
1.35. Frame Size 99
1.36. Load Equalization 100
1.37. Appearance 102
1.38. Cost 102
Exercises 102
Short Answer Type Questions With Answers 106
Problems 109

2. CONTROL DEVICES FOR INDUSTRIAL MOTORS .... 112—128

2.1. Introduction 112
2.2. Automatic Regulating Equipment 112
2.3. Solid State Devices 118
2.4. Pilot Devices and Accessories 123
2.5. Contactors 125
2.6. Electromagnetic Relays 126
2.7. Protective Devices and Interlocks 127
Exercises 128
Short Answer Type Questions With Answers 128

3. THYRISTOR CONTROL OF ELECTRIC MOTORS .... 129—170

3.1. Introduction 129
3.2. General Configuration of a Motor Drive 130
3.3. Classification of Power Converters 130
3.4. Matching Power Electronic Converter and Motor 132
3.5. DC Motor Control Through Converters 133
3.6. Thyristor Control of Separately Excited DC Motor Drive 133
3.7. Chopper Control of DC Motors 141
3.8. Comparison of Converter Fed and Chopper Fed DC Drives 150
3.9. Special Features of Thyristor Drive Motors 150
3.10. Speed Control of DC Motor Using Feedback Loops 151
3.11. Electric Braking of Rectifier Controlled Separately Excited DC Motor 152
3.12. Thyristor Control of DC Series Motors 152
3.13. Thyristor Control of Three Phase Induction Motors 156
3.14. Thyristor Control of a 3-Phase Synchronous Motor 163
3.15. Speed Control of Single Phase Induction Motor Using Triac (Fan Regulator) 163
3.16. Brushless (or Commutatorless) DC Motor 164
3.17. Switched Reluctance Motor 164
Exercises 166
Short Answer Type Questions With Answers 168
Problems 170

4. MOTORS FOR PARTICULAR SERVICES ... 171—182

4.1. Introduction 171
4.2. Types of Motors Used in Industrial Drives 171
4.3. Applications of Electric Drive 171
Exercises 181

SECTION-B : ELECTRIC POWER UTILIZATION .... 183—298

5. ELECTRIC HEATING .... 185—210

5.1. Introduction 185
5.2. Modes of Transfer of Heat 185
5.3. Classification of Electric Heating Methods 186
5.4. Resistance Heating 187
5.5. Arc Furnaces 193
5.6. Induction Heating 198
5.7. High Frequency Eddy Current Heating 202
5.8. High Frequency Power Supply Sources 203
5.9. Dielectric Heating 204
5.10. Choice of Frequency 207
Exercises 207
Short Answer Type Questions With Answers 208
Problems 209

6. ELECTRIC WELDING .... 211—231

6.1. Introduction 211
6.2. Resistance Welding 212
6.3. Electric Arc Welding 218
6.4. Ultrasonic Welding 221
6.5. Electron Beam Welding 222
6.6. Laser Beam Welding 222
6.7. Requirements of Good Weld 223
6.8. Preparation of Work 224
6.9. Electrodes 224
6.10. Power Supply For Resistance Welding 225
6.11. Machines For Resistance Welding 225
6.12. Electronic Control in Resistance Welding 226
6.13. Power Supply For Arc Welding 228
6.14. Electric Welding Equipment 228
6.15. Arc Welding With DC and AC 229
6.16. Comparison Between Resistance and Arc Welding . 229
Exercises 230
Short Answer Type Questions With Answers 231

7. ILLUMINATION ENGINEERING .... 232—268

7.1. Introduction 232
7.2. Terms Used in Illumination 233
7.3. Laws of Illumination 235
7.4. Polar Curves 239
7.5. Photometry 240
7.6. Integrating Sphere 242
7.7. Measurement of Illumination 242
7.8. Sources of Light 242
7.9. Arc Lamps 242
7.10. Incandescent Lamps 243
7.11. Gaseous Discharge Lamps 245
7.12. Comparison Between Tungsten Filament Lamps and Fluorescent Tubes 253
7.13. Basic Principles of Light Control 255
7.14. Types of Lighting Schemes 255
7.15. Design of Lighting Schemes 256
7.16. Factory Lighting 257
7.17. Methods of Lighting Calculations 258
7.18. Street Lighting 261
7.19. Floodlighting 262
Exercises 264
Short Answer Type Questions With Answers 265
Problems 267

8. ELECTROLYTIC PROCESSES .... 269—281

8.1. Introduction 269
8.2. Electrolysis—Basic Principle 269
8.3. Faraday's Laws of Electrodeposition 269
8.4. Calculation of Current Required For Depositing Given Amount of Metal 270
8.5. Current Efficiency 272
8.6. Voltage 272
8.7. Energy Efficiency 273
8.8. Extraction of Metals 273
8.9. Refining of Metals 273
8.10. Production of Chemicals 274
8.11. Electrodeposition 275
8.12. Electroplating 276
8.13. Deposition of Alloys 278
8.14. Electroforming 278
8.15. Electrotyping 278
8.16. Electrofacing 278
8.17. Electrometallisation 278
8.18. Electrodeposition of Rubber 278
8.19. Anodizing 278
8.20. Electropolishing 278
8.21. Power Supply For Electrolytic Processes 278
Exercises 279
Short Answer Type Questions With Answers 280
Problems 281

9. REFRIGERATION AND AIR CONDITIONING .... 282—298

9.1. Introduction 282
9.2. Terminology 282
9.3. Refrigeration Cycle 282
9.4. Refrigeration Systems 283

9.5. Compound or Multi-Stage Vapour Compression Refrigeration System 285
9.6. Refrigerants 285
9.7. Domestic Refrigerator 285
9.8. Water Cooler 288
9.9. Desert Cooler 289
9.10. Air Conditioning 290
9.11. Comfort Air Conditioning 291
9.12. Industrial Air Conditioning 291
9.13. Effective Temperature 291
9.14. Summer Air Conditioning System 292
9.15. Winter Air Conditioning System 292
9.16. Year-Round Air Conditioning System 292
9.17. Types of Air Conditioning Systems 292
9.18. Room Air Conditioners 293
9.19. Central Air Conditioning Systems 294
9.20. Heating of Buildings 295
9.21. Calculation of Rating of Electrical Equipment 296
Exercises 297
Short Answer Type Question With Answers 298
Problems 298

SECTION-C : ELECTRIC TRACTION .. 299—396

10. TRACTION SYSTEMS .... 301—312

10.1. Introduction 301
10.2. Different Systems of Traction 301
10.3. Systems of Electric Traction 305
10.4. Systems of Track Electrification 308
10.5. Comparison Between DC and AC Systems of Railway Electrification From The Point of View of Main Line and Suburban Line Railway Services 310
Exercises 311
Short Answer Type Questions With Answers 312

11. TRAIN MOVEMENT AND ENERGY CONSUMPTION .... 313—332

11.1. Introduction 313
11.2. Typical Speed-Time Curves 314
11.3. Crest Speed, Average Speed and Schedule Speed ... 314
11.4. Factors Affecting Schedule Speed 314
11.5. Simplified Speed-Time Curves 315
11.6. Mechanics of Train Movement 318
11.7. Tractive Effort For Propulsion of Train 319
11.8. Power Output From The Driving Axles 320
11.9. Energy Output From Driving Axles 322
11.10. Determination of Specific Energy Output Using Simplified Speed-Time Curve 322
11.11. Factors Affecting Specific Energy Consumption of an Electric Train Operating on a Given Schedule Speed 323
11.12. Dead Weight, Accelerating Weight and Adhesive Weight 327
Exercises 329
Short Answer Type Questions With Answers 330
Problems 331

12. ELECTRIC TRACTION MOTORS .... 333—347

12.1. General Features of Traction Motors 333
12.2. Operating Characteristics of DC Motors 334
12.3. DC Series Motor 335
12.4. DC Shunt Motor 337
12.5. AC Series Motor 340
12.6. Three Phase Induction Motor 343
12.7. Linear Induction Motor 343
12.8. Rating and Ventilation 345
Exercises 345
Short Answer Type Questions With Answers 346
Problems 347

13. CONTROL OF TRACTION MOTORS .... 348—362

13.1. Introduction 348
13.2. Starting and Speed Control of DC Traction Motors 348
13.3. Plain Rheostatic Starting (Notching) 350
13.4. Series-Parallel Starting 350
13.5. Transition Methods 352
13.6. Drum Controller 353
13.7. Contactor Type Controller 354
13.8. Buck and Boost Method of Speed Control 357
13.9. Metadyne Control 357
13.10. Thyristor Control of Traction Motors 359
13.11. Speed Control and Starting of Single Phase Series Motors 360
13.12. Starting and Speed Control of 3-Phase Induction Motors 360
Exercises 361
Short Answer Type Questions With Answers 361
Problems 362

14. BRAKING; MECHANICAL CONSIDERATIONS AND CONTROL EQUIPMENT .... 363—375

14.1. Braking 363
14.2. Mechanical Regenerative Braking 369
14.3. Mechanical Braking 369
14.4. Hydraulic Braking 370
14.5. Magnetic Track Brakes 370
14.6. Electromechanical Drum Brakes 370
14.7. Eddy Current Brakes 370
14.8. Mechanical Considerations 370

14.9. Control Equipment 372
14.10. Multiple Unit Control 373
14.11. Auxiliary Equipment 374
Exercises 374
Short Answer Type Questions With Answers 375
Problems 375

15. POWER SUPPLY FOR ELECTRIC TRACTION .... 376—396

15.1. Introduction 376
15.2. Current Collection Systems 377
15.3. Current Collectors For Overhead System 378
15.4. Overhead Construction For Tramways and Trolley Buses and Railways 379
15.5. Sag and Tension Calculation for a Trolley Wire 380
15.6. DC Track Electrification 381
15.7. Low Frequency Single Phase System 383
15.8. AC Track Electrification 383
15.9. Power Supply Arrangement For AC Track Electrification 384
15.10. Block Diagram of an AC Electric Locomotive 386
15.11. Major Substation Equipment 386
15.12. Remote Control Centre 387
15.13. Signalling 387
15.14. Overhead Equipment 388
15.15. Interference in Telecommunication Circuits 390
15.16. Booster Transformer 391
15.17. Negative Booster 392
15.18. Protective Devices 394
Exercises 395
Short Answer Type Questions With Answers 395

SECTION-D : ECONOMIC CONSIDERATIONS .... 397—454

16. ECONOMIC CONSIDERATIONS .... 399—431

16.1. Introduction 399
16.2. Classification of Costs 399
16.3. Cost Analysis of Power Plants 400
16.4. Interest and Depreciation 401
16.5. Methods of Determination of Depreciation 402
16.6. Economics of Power Generation 404
16.7. Significance of Load Factor and Diversity Factor 407
16.8. Capacity Scheduling 407
16.9. Load Sharing Between Base Load and Peak Load Plants 407
16.10. Choice of Size and Number of Generating Units 408
16.11. Load Forecasting 424
Exercises 425
Short Answer Type Questions With Answers 426
Problems 428

17. TARIFFS AND POWER FACTOR IMPROVEMENT .... 432—454

17.1. Tariffs 432
17.2. Types of Tariffs 432
17.3. Spot Pricing 435
17.4. Types of Consumers and Their Tariffs 435
17.5. Power Factor 438
17.6. Disadvantages of Low Power Factor 439
17.7. Causes of Low Power Factor 439
17.8. Methods of Power Factor Improvement 440
17.9. Advantages of Power Factor Improvement 443
17.10. Economics of Power Factor Improvement 446
17.11. Economical Comparison of The Two Methods of Increasing The Power Supplied 448
17.12. Choice of Equipment 449
Exercises 450
Short Answer Type Questions With Answers 451
Problems 453

OBJECTIVE TYPE QUESTIONS .... 455—496

INDEX .... 497—500

SECTION - A

ELECTRIC DRIVES AND INDUSTRIAL APPLICATIONS

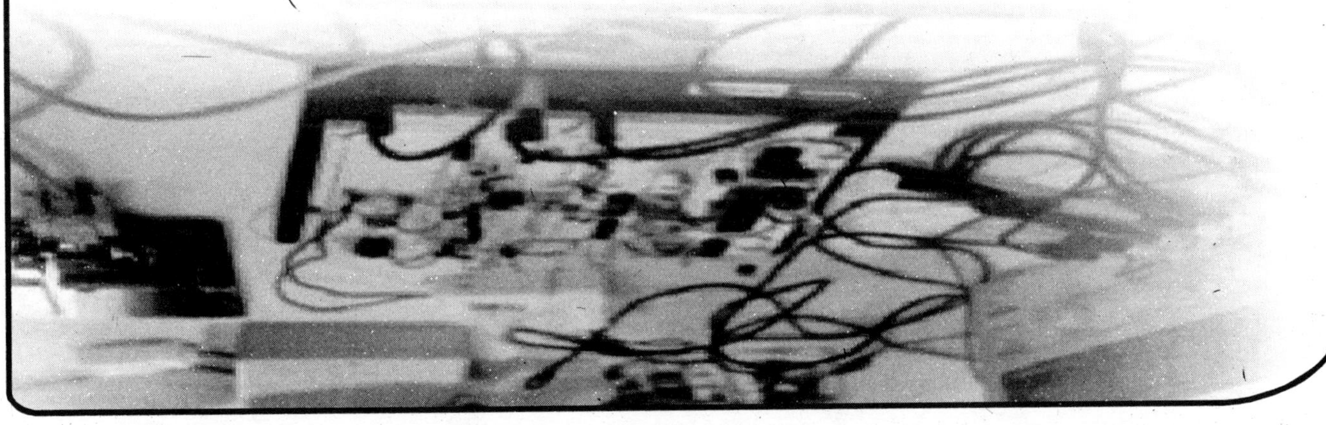

CHAPTER 1

Electric Drives

INSIDE THIS CHAPTER

1.1. Introduction 1.2. Factors Governing Selection of Electric Motors 1.3. Nature of Electric Supply 1.4. Types of Drives 1.5. Types of Loads 1.6. Single Quadrant Operation of Electric Motors 1.7. Four Quadrant Operation of Electric Motors 1.8. Torque-Speed and Speed-Torque Characteristics 1.9. Selection of Electric Drive 1.10. Motor Speed-Torque Characteristics 1.11. Joint Speed-Torque Characteristic of an Electric Motor and a Driven Unit 1.12. Dynamics of Motor-Load Combination 1.13. Referred Load Torques and Moments of Inertia 1.14. Referring Forces and Masses Having Translational Motion to a Rotating Shaft 1.15. Referring Torques and Translating Masses When Speed is Variable 1.16. Operating or Running Characteristics of Electric Motors 1.17. Starting of Electric Motors 1.18. Speed Control of Electric Motors 1.19. Electric Braking 1.20. Electric Braking With DC Motors 1.21. Electrical Braking of Polyphase Induction Motors 1.22. Electric Braking of Synchronous Motors 1.23. Dynamics of Braking 1.24. Types of Enclosures 1.25. Bearings 1.26. Transmission of Drive 1.27. Choice of Drive 1.28. Noise 1.29. Insulating Materials 1.30. Temperature Rise in Electrical Machines 1.31. Duty Cycles 1.32. Rating of Machines 1.33. Choice of Rating of Motors 1.34. Methods of Ventilation and Cooling of Machines 1.35. Frame Size 1.36. Load Equalization 1.37. Appearance 1.38. Cost.

1.1. INTRODUCTION

Every well developed machine may be considered to consist of three important different parts : the prime mover (engine or motor), the transmission system and the working machine proper. The function of the first two parts, the prime mover with its system of control, and the transmission system, which may comprise a set of shafts, pulleys, belts, gears etc., is to impart motion to and operate the working machine. Because of their function of maintaining the working machine in motion the first two parts (prime mover and the transmission system) are embraced by the common term *drive.* The simplest example of a drive is a hand-powered device. Devices of this class are still used in the form of household appliances and some types of agricultural implements, though industrially they are of no importance today. Horse-powered drive which at one time replaced hand-powered drive by using animals for providing motive power also has no industrial importance today. These two means of motive power were eventually replaced by mechanical drives powered by windmills, waterwheels and turbines, steam engines and internal combustion engines.

Nowadays modern water turbines and steam turbines are extensively used in electric power stations generating large amounts of electrical energy for centralised supply of industry, electrified traction facilities, agriculture and domestic needs. In utilising this source of energy, the principal means of driving all kinds of working machines is the electric motor. This makes it the predominate type of drive in modern conditions, the term *electric drive* being applied to it

An electric drive is defined as a form of machine equipment designed to convert electrical energy into mechanical energy and provide electrical control of this process.

The basic elements of an electric drive are the electric motor, the transmission, and the electrical control system. Electric drives often include conversion equipment (frequency changers, rectifier units, motor-generator sets etc.) to attain smoother and more versatile control and make the motor operate on specific speed-torque characteristic.

Electric drive is becoming more and more popular for its simplicity, cleanliness, easy and smooth control, flexibility in layout, compactness, reliability, long life, easy starting, low cost (initial as well as maintenance cost), facility for remote control etc. Modern electric drive with a system of automatic control fulfills several functions other than of imparting motion to the working machine so as to carry out the production processes involved in the most efficient manner. Using automatically controlled electric drives, machinery and production units may be operated with high productivity, ensuring at the same time high quality. Sometimes electric drive is employed although the prime mover is already existing there, such as in diesel electric traction system and for ship propulsion. In diesel electric traction system the electricity is used for driving the locomotive. If the diesel engine had to drive the locomotive directly, the operation would have required a complex gear system, since the diesel engine provides best performance only at one speed. But in diesel electric traction the desired output can be had by controlling the dc motors and running the diesel engine-dc generator set at a constant speed, all the time. In electric ship propulsion the prime mover (steam turbine or diesel engine) drives an ac or dc generator, which supplies ac or dc motors mounted on the propeller shafts. Thus the power which used to be transmitted by the shafts previously is now carried through a cable coming from the engine room and use of long shafts going from engine room to the propellers is avoided.

Its main disadvantage is non-availability of drive on failure of electrical power supply. It also cannot be employed in distant places where electric power supply is not available.

1.2. FACTORS GOVERNING SELECTION OF ELECTRIC MOTORS

Parallel to the growth of machine tools, an equally rapid progress has been made in the field of electric motors. The electric motors of various designs are available to meet the needs of various classes of industry. So the selection of electric motor itself has become an important and tedious process. The conditions under which an electric motor has to operate and the type of load it has to handle, determine its selection. The various factors that are to be considered in the selection of an electric motor for a particular service are :

1. Nature of Electric Supply.
2. Types of Drives.
3. Nature of Load.
4. Electrical Characteristics.
   (*a*) Operating or Running Characteristics.
   (*b*) Starting Characteristics.
   (*c*) Speed Control.
   (*d*) Braking Characteristics.
5. Mechanical Considerations
   (*a*) Types of Enclosures.
   (*b*) Types of Bearings.
   (*c*) Types of Transmission For Drive.
   (*d*) Noise Level.
   (*e*) Heating and Cooling Time Constants of the Motor.
6. Service Capacity and Rating.
   (*a*) Requirements For Continuous, Intermittent or Variable Load Cycle.
   (*b*) Pull-Out Torque and Overload Capacity.
7. Appearance.
8. Cost Considerations.
   (*a*) Capital or Initial Cost.
   (*b*) Running or Operating Cost–power factor, losses, maintenance and depreciation etc.

From the above it is seen that a large number of factors are to be considered in making the choice of an electric motor for a given drive. Although in the list of various considerations the cost has been placed at the end, yet the final choice has to be governed by it. The motor selected must fulfil all the necessary load requirements and at the same time it should not be very costly if it has to be a commercial success. In any particular case some of the factors may be definitely fixed by existing circumstances while others will probably be conflicting and, therefore, for making most satisfactory choice understanding of various possibilities is necessary. The basic problem is of matching the mechanical output of the electric motor with the load requirements, *i.e.*, to provide a motor with the correct speed-torque characteristics as demanded by the load.

The factors mentioned above and their effects on the various types of electric motors available will be individually discussed in this chapter, so that the reader may understand the importance of each.

1.3. NATURE OF ELECTRIC SUPPLY

The electric supply available may be 3-phase ac, single-phase ac or dc.

In case three-phase ac supply is available, polyphase induction motor, squirrel cage type for small ratings and slip-ring type for higher ratings may be used, provided this suits the requirements of the load. In the cases where speed variation is required these cannot be conveniently used, so pole changing motors or motors with stepped pulleys may be used. Where accurate control of speed is required, Schrage motors may be used. Use of single-phase motors is limited to small loads only owing to their limited outputs.

DC motors are not used so widely as ac ones. There are several reasons for this, the most important of which are given below :

1. At present a large percentage (in fact, the whole) of the electrical energy used for domestic and commercial purposes is generated in ac form because of economic and technical reasons, *i.e.*, supply available is usually ac. Additional equipment is, therefore, required for converting existing ac supply into dc supply.
2. DC motors have commutators that are subject to trouble resulting from sparking, brush wear, arc-over and the presence of moisture and destructive fumes in the surrounding air.
3. DC motors are generally more expensive than ac motors for similar working conditions.

On contrary, ac motors have the advantages of possessing few working parts, requiring less maintenance and replacement of spares and providing uninterrupted long service life. Three-phase induction motors have high efficiency and provide high starting and uniform operating torques and balanced loading of supply.

DC motors are nevertheless used in a large number of applications, specially when their excellent torque and speed operating characteristics cannot be duplicated by ac motors. Moreover, the speed control that can be provided for dc motors is far more flexible and satisfactory than that which can be provided for ac motors. These requirements are particularly significant in connection with services such as operation of traction equipment, hoists and elevators. In some cases, such as in electric excavators, steel mills and cranes, the speed control is so important that existing ac supply is converted into dc supply in order to employ dc motors.

With electrical utilities transmitting energy on ac, the usual arrangement used to be a motor-generator set (a dc generator driven by an ac motor) supplying power to a dc motor. However, the advent of solid state diodes and silicon controlled rectifiers (SCRs) eliminated the need for the motor-generator set and made dc motor drives more versatile and more popular than even before.

Nature of dc supply available, to some extent affects the performance, and therefore, size of the motor required. DC supply available from a dc generator will be more or less smooth but the output voltage from thyristor converter consists of a dc component and ac harmonic components. Torque or mechanical power is developed by the dc component of the current where as armature heating is developed by the effective or rms value of current. The form factor for half-wave three-phase thyristors may be taken as 1.2 while for full-wave three-phase thyristors it is 1.1. This increases electrical losses, and, therefore, heating is 5 to 7% more for three-phase full-bridge converters while for three-phase half-bridge convertors it is from 15 to 20%. Thus when a dc motor is driven from thyristor convertor, slightly large size motor will be required for a given kW power requirement.

The commercial ac power is generated at a frequency of 50 Hz but this frequency is not capable of meeting the requirement of certain industries, defence services and railway signalling. As such, there has been recent trend towards development of high frequency power generation. Relative merits and demerits of high-frequency and power-frequency equipment are enumerated below.

Merits of High Frequency. 1. The output of the equipment increases with the increase in frequency or speed, as obvious from the following equations

The output of an alternator per phase,

$$P = 4K_f K_p K_d \phi f T \times I \cos \phi \qquad ...(1.1)$$

where K_f, K_p and K_d are the form factor, pitch factor and distribution factor, ϕ is the useful flux per pole in webers, f is the frequency in hertz, T is the number of turns in series per phase, I is the output current per phase and cos ϕ is the power factor.

The output of a motor is given as

$$P = KD^2LN \qquad ...(1.2)$$

where D is diameter, L is the length and N is the speed of motor in rpm.

For a given power output, high frequency machine may be of one-eighth of the weight of the 50 Hz machine. Because of high power-weight ratio, relative cost of the equipment is reduced with the increase in frequency. Reduction in weight of equipment makes also transportation and handling of equipment easy and convenient at high altitude for defence services.

2. For the generation of high frequency power, along with the small sized generator, the prime mover will also be of smaller size. This will result in reduction of high noise problem.

3. High frequency equipment requires less maintenance and has low operating cost, higher efficiency and long life span requiring minimum repairing cost.

4. More reliable service is provided, in arduous conditions of high altitude, low temperature and high humidity, by high frequency equipment.

The *drawback of high frequency equipment* is that it has a problem of increased power losses due to hysteresis and eddy currents. However, this problem of large power losses can be overcome by using low loss materials, proper flux density and adequate cooling arrangements.

1.4. TYPES OF DRIVES

The various types of electric drives used in industry may be divided into three types : *the group, individual* and *multi-motor* drives.

Group Drive. By group drive is meant a drive in which a single electric motor drives a line shaft by means of which an entire group of working machines may be operated. It is also sometimes called the *line shaft drive.* The line shaft is fitted with multi-stepped pulleys and belts that connect these pulleys and the shafts of the driven machines and serve to vary their speed. This drive is economical in consideration of the first cost of the motors and control gear. A single motor of large capacity costs less than the total cost of a number of small motors of the same total capacity viz a single 100 kW motor costs much less than that of ten motors of 10 kW each. Since all the machines may not be used on full load at the same time, therefore, the kW rating of the motor for group drive can often be less than the aggregate of kW output ratings of all the individual motors, and further causes reduction in cost. The efficiency and power factor of a large group drive motor will be higher, provided it is operated fairly near its rated load. If the machines are liable to short but sharp overloads, group drive is again advantageous, because 100% overload on an individual machine will cause hardly 10% overload when being driven by group drive.

Owing to its following drawbacks and objectionable features, this form of drive has become obsolete nowadays and the modern trend is to employ individual and multimotor drives:

1. In group drive, speed control of individual machine is very cumbersome using stepped pulleys, belts etc.
2. Owing to use of line shafting, pulleys and belts group drive does not provide good appearance and is also less safe to operate.
3. In group drive since machines have to be installed to suit the layout of the line shafting, as such flexibility of layout of the various machines is lost. Also, it is not possible to instal any machine at a distant place.
4. The possibility of installation of additional machines in an existing industry is limited.
5. If at a certain instance all the machines are not in operation, the motor will operate at low capacity and, therefore, operation efficiency will be low.
6. The breakdown of large single motor causes all the operations to be stopped. In some processes, where stoppage of one operation causes stoppage of whole sequence such as in textile industry or flour mills, this drawback is not effective and group drive, therefore, can be used
7. Considerable power loss takes place in the energy transmitting mechanism.
8. The level of the noise produced at the work site is quite large.

Group drive is adopted, when existing factories are changed over from engine drive to electric motor drive simply by replacing the oil or steam engine by an electric motor of corresponding output retaining all the old shafts and belts.

Individual Drive. In an individual drive, a single electric motor is used to drive one individual machine. Example of such a drive are single-spindle drilling machines, various types of electrical hand tools and simple types of metal working machine tools and mechanisms. Though it costs more than group drive, but each operator has complete control of his machine, which enables him to vary its speed if necessary and stop while not is use, thus eliminating no-load losses. The machines can be placed in any desired position and can be moved very easily. The motor and its control unit can be built as an integral part of the machine which results in good appearance, cleanliness and safety. For driving heavy machines such as for lifts, cranes, shapers, lathes etc. and for the purposes where constancy of speed and flexibility of control is required, such as in paper mills and textile industry, individual drive is essential.

For new factories individual drive is preferred, as it causes some saving in the cost of superstructure because of being much lighter and less expensive.

In case of individual drive too, the energy is transmitted to the different parts of the same mechanism by means of mechanical parts (such as gears, pulleys etc.). Thus some power loss occurs in the energy transmission mechanism. This drawback is overcome in the case of *multimotor drives.*

Multimotor Drive consists of several individual drives each of which serves to operate one of many working members or mechanisms in some production unit. For example, in travelling cranes, there are three motors : one for hoisting, another for long travel motion and the third for cross travel motors. Such drive is essential in complicated metal-cutting machine tools, paper making machines, rolling mills, rotary printing machines and similar types of other machinery. The use of multimotor drive is continuously expanding in modern industry as their advantages outweigh the increase in capital cost as compared to the group drives.

The use of individual and multimotor drives has enabled introduction of automation in production processes, which in turn has considerably increased the productivity of various industrial organisations. Complete or partial automation helps to operate various mechanisms at optimum conditions and to increase reliability and safety of operations.

Example 1.1. Compare the total annual cost of a group drive with a motor costing Rs. 60.000 with that of ten individual motors, each costing Rs. 18,750. With group drive annual consumption is 75,000 kWh. With separate drives annual consumption is 60,000 kWh. Electrical energy costs Rs. 4.00 per kWh. Depreciation, maintenance and other fixed charges amount to 12% in the case of group drive and 15% in the other case.

Discuss possible reasons for which you may decide in favour of the drive with a higher capital cost.

Solution : Capital cost on group drive = Rs. 60,000

Annual depreciation, maintenance and fixed charges

$$= \text{Rs. } 60{,}000 \times \frac{12}{100} = \text{Rs. } 7{,}200$$

Energy consumption per annum = 75,000 kWh

Annual energy charges = Rs. 4.00 × 75,000

= Rs. 300,000

Total annual cost in case of group drive

= Rs. 7,200 + Rs. 300,000 = Rs. 307,200

Capital cost on individual drive

= Rs. 18,750 × 10 = Rs. 187,500

Annual depreciation, maintenance and fixed charges

$$= \text{Rs. } 187{,}500 \times \frac{15}{100} = \text{Rs. } 28{,}125.00$$

Annual energy consumption = 60,000 kWh

Annual energy charges = Rs. 4.00 × 60,000 = Rs. 240,000

Total annual cost in case of individual drive

= Rs. 28,125 + Rs. 240,000 = Rs. 268,125

Though the capital cost is higher in case of individual drive, but because of lower annual cost and inherent advantages of individual drive over group drive such as flexibility of control and convenience it is preferred.

1.5. TYPES OF LOADS

There are two types of loads—those which provide active torques and those which provide passive torques. Active torques are due to either gravitational force (as in case of hoists, lifts or elevators and railway locomotive operating on gradients) or deformation in elastic bodies. Such torques are also developed during compression or release of springs. Since functioning of hoisting mechanism, operation of locomotives on gradients and compression/release of springs are all associated with a change in potential energy of the drive, active torques are also closely connected to the potential energy. In upward movement of load/compression of a spring, the stored potential energy increases and the active torque developed opposes the action that takes place *i.e.,* the torque is acted against the upward movement/compression. On the other hand in the downward movement of load/release of a spring the stored potential energy decreases and torque associated with it helps the action. Thus, it can be seen that the active torque continues to act in the same direction even after the direction of the drive has been reversed.

Passive torques are those due to friction or due to shear and deformation in inelastic bodies (lathes, pumps, fans etc.). They always oppose the motion, retarding the rotation of the driven machine. Moreover, with change in direction of motion, the sense of torque also changes. For example, when a weight is raised upward, the friction torque adds to the useful torque but when lowered down it subtracts from the latter.

1.6. SINGLE QUADRANT OPERATION OF ELECTRIC MOTORS

Single quadrant operation means that load and the motor shall be required to run only in one specified direction under the action of developed torque.

When an electric motor is switched to electric supply mains and drives mechanical devices such as line shaft and machine tools, the machine is said to be operating in forward motoring mode or simply motoring mode. In this mode of operation armature current flows in opposition to the emf induced in the armature. The direction of emf induced in the armature is in direct opposition to the applied voltage. Electromagnetic torque developed is in the direction of armature rotation.

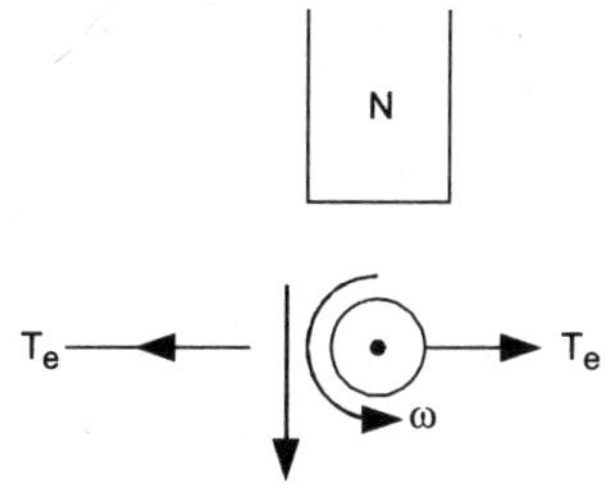

Fig. 1.1. *Single Quadrant Operation of Motor*

1.7. FOUR QUADRANT OPERATION OF ELECTRIC MOTORS

In view of the fact that both active and passive load torques can be present in general, in a drive system, the motor driving the load may operate in different regimes—not only as a motor, but for specific periods, also as a generator and as a brake. Further, in many applications, the motor may be required to run in both directions. Therefore, in sketching the speed-torque characteristics of either the load or the motor, it is preferable to make use of all four quadrants of the speed-torque plane for plotting, rather than to confine, the characteristics to the first quadrant alone.

The conventions used for positive and negative values of speed, motor torque and load torque in a diagram of this type are as follows :

The speed is assumed to have a positive sign, if the direction of rotation is counter-clockwise or is in such a way to cause an 'upward' or 'forward' motion of the drive. In case of reversible drives the positive sign for speed may have to be assigned arbitrarily either to counter-clockwise or clockwise direction of rotation. The motor torque is taken to be positive when it causes an increase in speed in the positive sense. The load torque is assigned a positive sign when it acts against the motor torque.

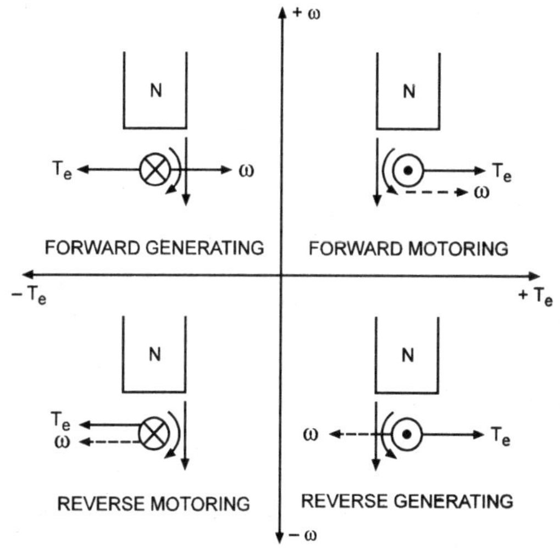

Fig. 1.2. *Four Quadrants of Operation of Motor*

Figure 1.2 illustrates the four different modes of operation of an electric machine. Although the diagram indicates that the field polarity is maintained and the armature current is reversed to obtain negative torque, the same effect is obtained by reversing the field polarity and maintaining the armature

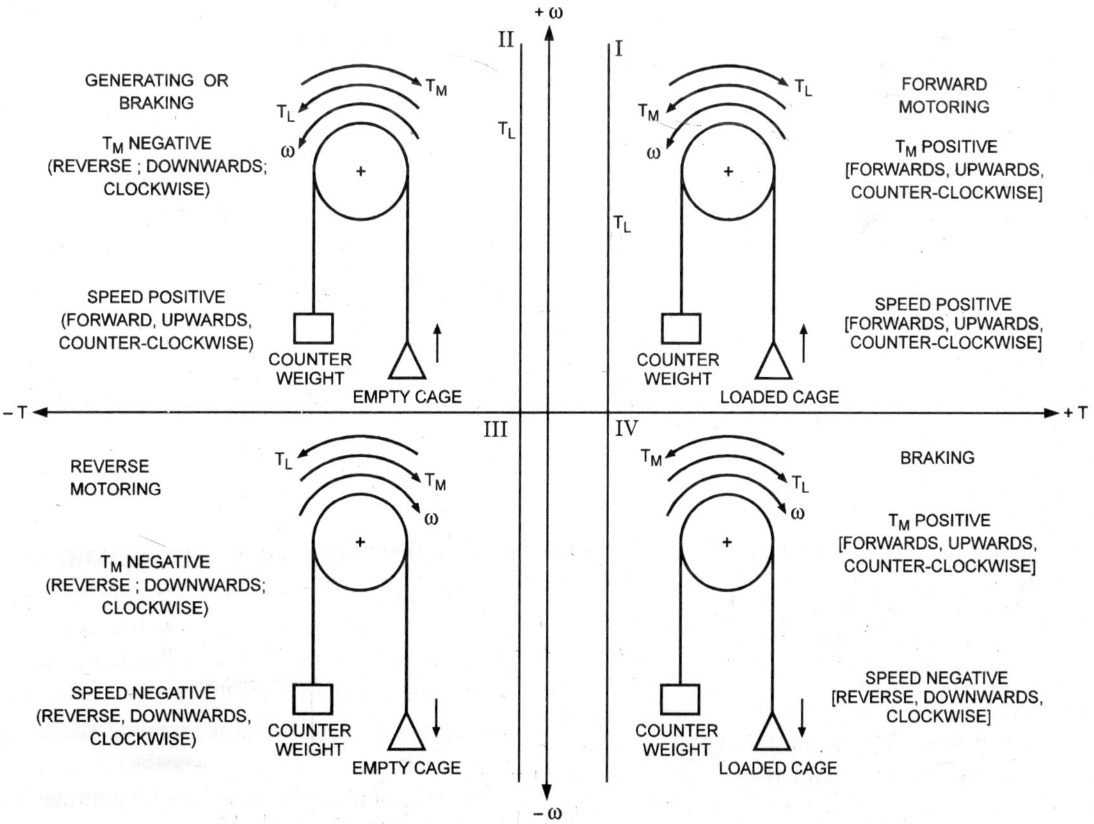

Fig. 1.3. *Four Quadrant Operation of a Motor Driving a Hoist Load*

current direction. Field reversal is necessary with some forms of rectifier control. These four quadrants of operation are feasible for any dc or ac rotating machine but the dc machine is much freer to transfer its operation between quadrants and operates satisfactorily at any point within the envelope.

Figure 1.3 illustrates the four quadrant operation of a motor driving a hoist consisting of a cage with or without load, a rope wound onto a drum to hoist the cage and a balance weight magnitude larger than that of the empty cage but smaller than that of the loaded one. The arrows in this figure indicate the actual directions of motor (or electro-magnetic) torque, load torque and motion in the four quadrants. It can be easily seen that they correspond to the sign conventions mentioned earlier for speed, motor torque and load torque.

The load torque of the hoisting mechanism may be taken to be constant (*i.e.,* independent of speed); forces due to friction and windage being negligible in case of low speed hoists and the torque being primarily due to the gravitational pull on the cage. This torque being an active load torque does not change its sign even when the direction of rotation of the driving motor is reversed. Therefore speed-torque curves of a hoist load can be represented by means of vertical lines passing through two quadrants. The speed-torque characteristic of a loaded hoist is shown in Fig. 1.3 by means of the vertical line passing the first and fourth quadrants. Since the counterweight is assumed to be heavier than the empty cage, the inherent tendency of the load viz the empty cage is to move in an opposite direction to that of load presented by the loaded cage and hence the speed-torque curve of the unloaded hoist is represented by the vertical line passing through second and third quadrants.

In the first quadrant the load torque acts in a direction opposite to that of rotation. Hence, to drive the loaded hoist up, the developed torque in the motor M must act in the same direction as the speed of rotation *i.e.,* T_M should be of positive sign. Since the speed is also positive being an upward motion, the power will also have a positive sign, *i.e.,* the drive is said to be motoring. Quadrant I is arbitrarily and conventionally, thus, designated as 'forward motoring quadrant'.

Since speed and torque are both taken as positive when motoring forwards, upwards or counter-clockwise, the product of E and I is taken as positive under these conditions. It is convenient to assume that both E and I are positive under this condition, *i.e.,* to assume the flux to be positive, so that positive E corresponds to positive speed and positive I to positive torque. If in all circuits the direction of the current in the field winding is assumed to be the same, the flux is automatically assumed to be positive. Negative armature current then indicates negative torque, *i.e.,* the torque in the reverse, downward or clockwise direction. Similarly negative emf indicates negative speed. All that is required to avoid difficulties is to maintain these conventions consistently in the circuit analysis and the signs will indicate the mode of operation. The motor equation V = E + IR must, of course, be used consistently for all cases, and no attempt made to prejudge the behaviour by using the "generator equation" V = E – IR at any time.

The hoisting up of the unloaded cage is represented in the second quadrant. Since the counterweight is heavier than the empty cage, the speed at which the hoist is moved upwards may attain a dangerously high value. In order to avoid this, the motor torque must act in a direction opposite to that of rotation, *i.e.,* the motor should switch over to a braking or generator regime. Note that T_M will have a negative sign and speed still has a positive sign, being forwards, upwards and counter-clockwise, giving power a negative sign, corresponding to the generating or braking operation.

The third quadrant represents the downward motion of the empty cage. The downward journey of the cage opposed by the torque due to the counterweight and friction at the transmitting parts. So for moving the cage downwards, the motor torque must act in the same direction as the motion of the cage. The electrical machine acts as a motor as in the first quadrant, but in the reverse direction. Thus quadrant-III becomes 'reverse motoring'. The motor torque has a negative sign as it causes an increase in speed in the negative sense and the speed also has a negative sign being a downward motion. Power, thus, has a positive sign.

The downward motion of the loaded cage is shown in the IVth quadrant. The motion can take place under the action of load itself, without the use of any motor. But in order to limit the speed of the downward motion of the hoist, the electrical machine must act as a brake. The motor torque has a positive sign as it causes a decrease in speed in the downward motion. The speed, of course, has a negative sign, being a downward journey. The power, thus, acquires a negative sign, corresponding to the braking operation of the motor.

1.8. TORQUE-SPEED AND SPEED-TORQUE CHARACTERISTICS

Machine specialists or designers express the performance of an ac motor by means of a torque-speed characteristic curve, *i.e.,* with speed as abscissa and torque as ordinate. This is because it is much easier to obtain an expression for the torque in terms of speed or slip than the reverse. But in application engineering, it is usually the load torque which is known and the speed at which motor can drive this load is unknown and is to be determined. Thus the torque is independent variable and the speed is dependent variable consequently speed-torque characteristic curves with torque as abscissa are commonly employed in utilization studies. Characteristic curves drawn thus may look a little unfamiliar at first to those accustomed only to the machine designer form of curve.

1.9. SELECTION OF ELECTRIC DRIVE

For selection of reliable and efficient motor it is essential that the conditions of service are well known. It is not sufficient to simply specify the output power in kW and the speed but it is also necessary to know the following additional particulars :

(*i*) Torque at the shaft during running, starting and at different loads.

(*ii*) Accelerating torque and braking torque.

(*iii*) Switching frequency.

(*iv*) Efficiency of motor at different loads.

(*v*) Other working requirements.

In studying the behaviour of a motor selected for a particular driven unit, one of the first problems involved is to determine whether the speed-torque characteristic of the motor suits the requirements imposed by the speed-torque characteristic of the driven unit. Drive behaviour during the transient period of a start-up, braking, or speed change-over also depends upon how the speed-torque characteristics of the motor and the driven unit vary with speed. It is, therefore, imperative to study these characteristics in order to be able to select correctly the motor and obtain an economical drive.

1.9.1. Speed-Torque Characteristics of Machines or Mechanisms

The speed-torque characteristic of a machine or mechanism given by the relation $\omega = f(T_L)$ is defined as the relationship between the speed at which it is operated and the resisting or load torque it develops.

Different kinds of mechanisms and machines exhibit different speed-torque characteristics. However, several general conclusions may be drawn if we use the following empirical equation for the speed-torque characteristic of some driven unit of industrial equipment

$$T_L = T_0 + (T_{rn} - T_0)\left(\frac{\omega}{\omega_n}\right)^x \qquad ...(1.3)$$

where T_L is gross load (or resisting) torque developed by the unit at speed ω, T_0 is the resisting torque developed by the unit due to friction in its moving parts, T_{rn} is the resisting torque developed by the unit when it is driven at its nominal rated speed ω_n and x is the exponential coefficient characterising the change in resisting torque with the change in speed.

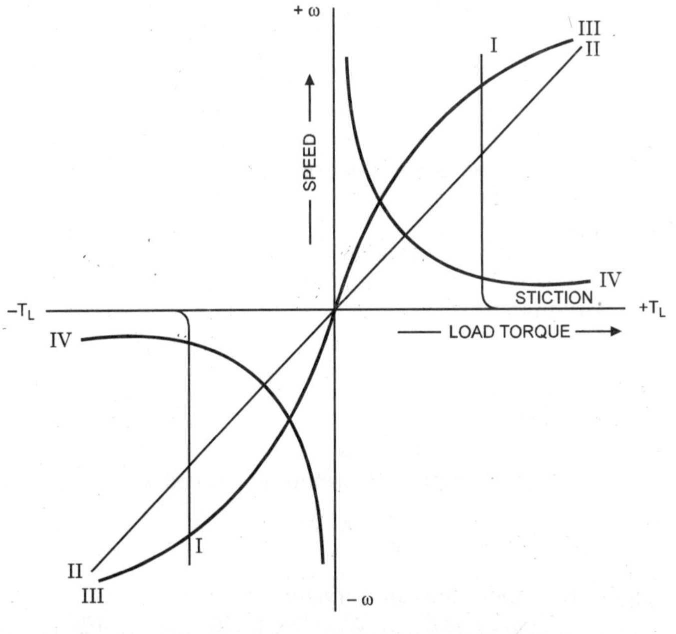

Fig. 1.4. ***Speed-Torque Characteristics of Typical Different Types of Machines and Mechanisms***

The above Eq. (1.3) permits the speed-torque characteristics of different kinds of machines and mechanisms to be roughly divided into the following categories :

1. ***Loads Requiring Constant Torque at All Speeds.*** Such a load presents to the motor a passive torque which is essentially independent of speed. It is characterized also by the requirement of an extra torque at very near zero speed. For this characteristic $x = 0$ and the load torque T_L does not depend upon speed. The speed-torque characteristic for such loads is illustrated by vertical line in Fig. 1.4. Such loads are dry friction, cranes during hoisting, hoist winches, machine tool feed mechanism, piston pumps operating against a constant pressure head, and conveyors handling a constant weight of material per unit time. In power applications it is usually called the breakaway torque and in control systems, it is referred to as *stiction* (derived from sticking friction).

Since it changes sign with reversal of rotation, the dry friction torque characteristic is discontinuous, as illustrated in Fig. 1.4.

2. ***Loads With Linear-Rising Characteristic.*** Such speed-torque characteristics, illustrated by straight line II in Fig. 1.4, are exhibited by calendering machines, eddy current brakes, separately excited dc generators supplying fixed ohmic resistance loads and fluid or viscous friction. In this case $x = 1$ and the load torque T_L rises in direct proportion to the speed.

3. ***Loads With Non-Linear-Rising (Parabolic) Characteristic.*** For such a characteristic $x = 2$ and the load torque T_L is proportional to the square of the speed. Such a characteristic is illustrated by curve III in Fig. 1.4. A load with considerable windage, of which a fan is the extreme example, has a torque which varies almost as the square of the speed. Blowers, centrifugal pumps, propellers in ships or aeroplanes, water wheels, pipe friction, velocity head of pumps etc. also have the same type of speed-torque characteristics.

4. ***Loads With Non-Linear Falling (Hyperbolic) Characteristic (or Constant Power Load).*** For such a characteristic $x = -1$ and the load torque T_L is inversely proportional to the speed, while the power required to drive the given unit remains unchanged. Such a characteristic is illustrated by curve IV in Fig. 1.4. Certain types of lathes, boring machines, milling machines and other kinds of metal-cutting machine tools, steel-mill coilers fall under this category of loads.

The categories of loads listed above do not cover all the cases that may be met with in practice but give a good idea of the characteristics typical of a great many kinds of industrial equipment. In actual practice we may come across loads which are combination of these basic types of loads.

1.9.2. Load Torque-Time Characteristics

Perhaps, the variation of load torque with time is of equal or greater importance in selection of motor. This variation in certain applications, can be periodic and repetitive, one cycle of variation being called a duty cycle. Different types of loads from the point of view of load torque characteristics, may be classified as follows.

(*i*) **Continuous, constant loads** such as paper making machines, centrifugal pumps or fans operating for a long time under the same conditions.

(*ii*) **Continuous, variable loads** such as hoisting winches, metal-cutting lathes, conveyors etc.

(*iii*) **Pulsating loads** such as reciprocating pumps and textile looms and, in general, all machines having crank shaft.

(*iv*) **Impact load** such as rolling mills, shearing machines, presses, forging hammers etc. Apparent, regular and repetitive load peaks or pulses occur in such loads.

(*v*) **Short time loads** such as motor-generator sets for charging of batteries; servo-motors used in remote control of drilling machine clamping rods.

(*vi*) **Short time intermittent loads** such as cranes and hoisting mechanisms, excavators, roll trains etc.

Certain machines (such as ball mills) do not strictly belong to any category mentioned above. If such loads (ball mills, stone crushers etc.) were characterized by frequent impacts of comparatively small peaks, it would be more appropriate to place them in the category of continuous variable loads rather than impact loads. Sometimes, it is quite cumbersome to make distinction between pulsating loads and impact loads, because both of them are periodic in nature.

One and same drive can be represented by a load torque which varies either with speed or with time. The most appropriate example is a fan load whose load torque T_L is proportional to the square of speed, is also a continuous constant load.

1.9.3. Load Torques Varying With Shaft Displacement Angle

In all machines having crankshafts, such as reciprocating pumps and compressors, frame-saws etc. load torque varies with the angular displacement of the shaft or rotor of the motor. For all such machines, the load torque T_L can be resolved into two components—one of constant magnitude T_{av} and the other a variable T_L' which changes periodically in magnitude depending on the angular position of the shaft. Such load torque characteristics, can, for simplicity, be represented by a Fourier series as a sum of oscillations of fundamental and harmonic frequencies, *i.e.*,

$$T_L' = \sum_{r=0}^{m} T_{Lr}' \sin (r\theta + \phi_r) \qquad ...(1.4)$$

where $\theta = \omega t$, ω being the angular speed of the motor shaft driving the compressor.

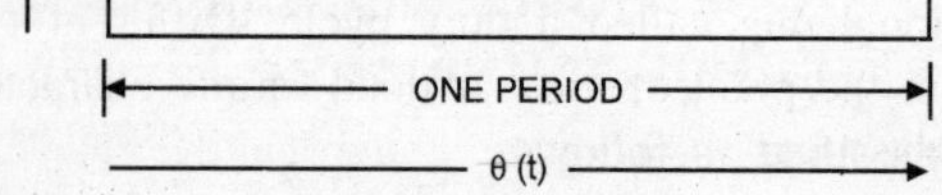

Fig. 1.5

During variations in speed, only small deviations from a fixed value of speed ω_a occur, therefore, displacement can be represented by $\theta = (\omega_a + \Delta\omega)t$. Thus the variable portion of the load torque may be given as

$$T_L' = \sum_{r=0}^{m} T_{Lr}' [\sin (r\,\omega_a t + \phi_r) + r\Delta\omega t] \qquad ...(1.5)$$

The term $r\Delta\omega t$ being very small in magnitude may be neglected. Thus, restricting to small deviations in angle from the equilibrium position, a load torque varying with angular displacement of shaft can be transformed to one which varies periodically w.r.t. time.

1.9.4. Load Torques Depending on Path or Position of Load During Motion

In Art 1.9.1. load torques varying with speed were considered. However, load torques, that depend not only speed but also on the nature of the path traced out by the load during its motion, do exist both in hoisting mechanisms and transport systems. For example, the resistance to motion of a train moving upgradient or taking a turn depends on the magnitude of the gradient or the radius of curvature of the track respectively.

Force due to the gradient is given as

$$F_g = 1{,}000 \text{ W} \sin \theta \text{ kg}. \qquad ...(1.6)$$

where W is the weight of train in tonne.

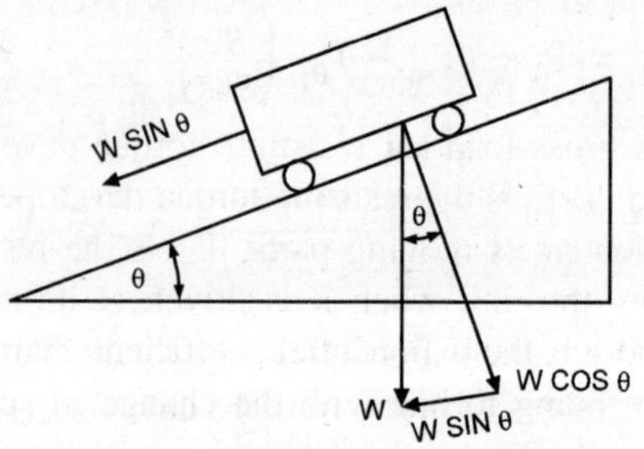

Fig. 1.6. ***Upgradient Motion of Train***

But in railway work gradient is expressed as rise in metres in a track distance of 100 m, and is denoted by *'percentage gradient'* (G%)

i.e., $G = \text{Sin } \theta \times 100$

$$\text{or } \sin \theta = \frac{G}{100}$$

Substituting $\sin \theta = \dfrac{G}{100}$ in Eq. (1.6), we have

$$F_g = 1{,}000 \text{ W} \times \frac{G}{100} = 10 \text{ WG kg}. \qquad ...(1.7)$$

The tractive force required to overcome curvature resistance is given by the empirical formula given below

$$F_c = \frac{700{,}000 \text{ W}}{\text{R}} \text{ kg} \qquad ...(1.8)$$

where R is the radius of curvature in metres.

In hoisting mechanisms in which tail ropes or balancing ropes are not employed (Fig. 1.7), the load torque is not only due to the weight of the unloaded or the loaded cage but also due to that of the lifting ropes or cables, which depends on the position of two cages. When cage 1 is at the bottom most position and is to be lifted upwards, the entire weight of the rope is also to be moved up. When both the cages are at the same height, the weight of the rope to be lifted becomes zero, since the weight of the ropes on both sides balance each other, being equal in length. When cage 1 is at a higher position than cage 2, a portion of the weight of the rope acts in such a way as to assist the upward motion of the cage 1. Ultimately when cage 1 reaches the top most position, the whole weight of the rope assists the upward movement.

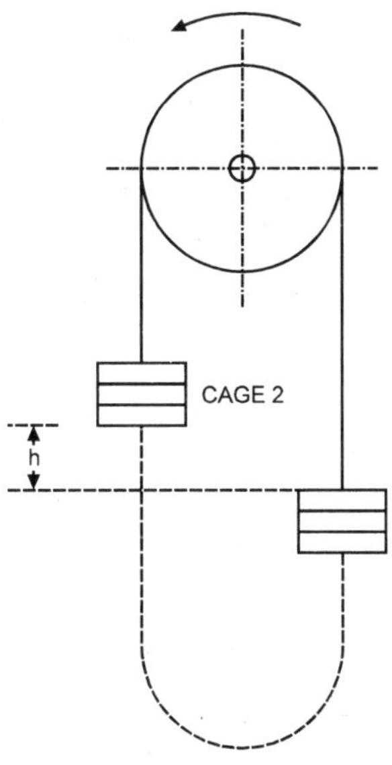

Fig. 1.7. *Hoisting Mechanism*

The force resisting the upward movement of the load, F_r due to varying weight of the rope depending on the position of the load is given as

$$F_r = W_r \left(1 - \frac{2x}{h}\right) \text{ kg} \qquad ...(1.9)$$

where W_r is the total weight of the rope in kg, h is the desired maximum height to which the cage is to be moved upwards, in metres and x is the height of the cage at any arbitrary position from the bottom most position in metres.

For large values of h, the force F_r affects to a large extent, the performance of the drive used in hoisting mechanisms because in such a situation the weight of the rope may be considerably larger than that of the load to be lifted upwards. If we use tail ropes, as illustrated by means of dotted lines in Fig. 1.7, the weight of the connecting rope can be balanced and almost smooth movement of the cages can be had.

1.10. MOTOR SPEED-TORQUE CHARACTERISTICS

A motor speed-torque characteristic is defined as the relationship between the speed at which it operates and the torque it developes, *i.e.*, $\omega = f(T)$.

Practically all electric motors—shunt wound, series wound, compound dc motors, squirrel-cage and slip-ring induction motors, and ac commutator motors, have drooping speed-torque characteristics, *i.e.*, their speed falls off as the load torque increases. However, the degree by which the speed changes with the change in torque differs for various types of motors, it being characterized by the so-called hardness of their speed-torque characteristics.

Electric motor speed-torque characteristics may be classified into three main groups.

1. ***Absolutely hard (flat) speed-torque characteristic,*** a characteristic exhibiting no change in speed with change in load torque. Synchronous motors operate with such a characteristic (horizontal straight line I in Fig. 1.8).

2. ***Hard speed-torque characteristic,*** a characteristic showing a speed which drops only slightly with increase in torque. A hard characteristic is exhibited by a shunt wound dc motor, this also being true of induction motors over the operating region of the speed-torque characteristic (curve II in Fig. 1.8).

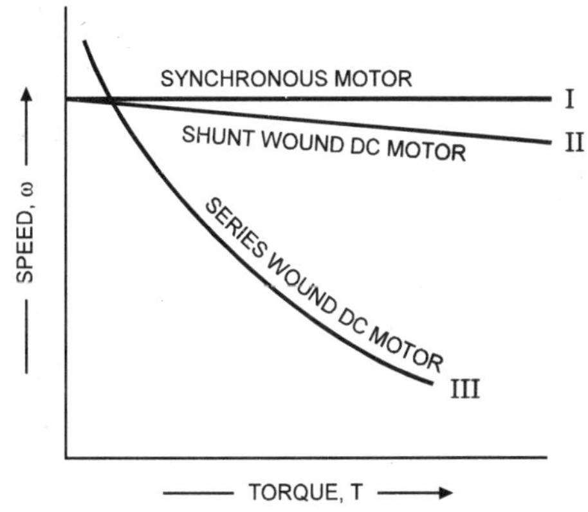

Fig. 1.8. *Speed-Torque Characteristics of Different Types of Motors*

The speed-torque characteristic of an induction motor exhibits a "hardness" that differs according to what point along the characteristic is taken into consideration (Fig. 1.9). Between the points of the maximum torque in motor operation $T_{max\ M}$ and the maximum torque in generator operation $T_{max\ G}$ an induction machine will exhibit a rather hard characteristic.

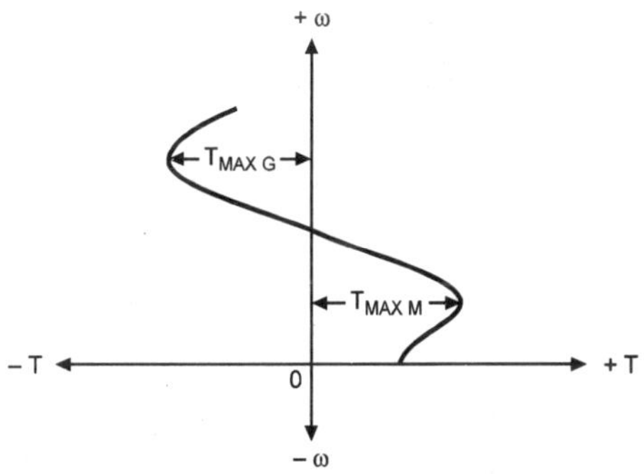

Fig. 1.9. *Speed-Torque Characteristic of an Induction Motor*

3. ***Soft speed characteristic,*** a characteristic showing considerable drop in speed with rise in torque. Series wound dc motor possesses such characteristic, especially along the low-torque portion of the characteristic (curve III in Fig. 1.8). For such motors, the degree of hardness of the characteristic varies all along the curve.

Compound wound dc motors, depending on the degree of hardness their speed-torque characteristics display, may be considered as hard- or soft-characteristic motors.

1.11. JOINT SPEED-TORQUE CHARACTERISTIC OF AN ELECTRIC MOTOR AND A DRIVEN UNIT

The joint operation of an electric motor and the unit it drives, when the speed has a steady value, corresponds to a condition of balance between the driving torque of the motor and the resisting torque developed by the driven unit at a given speed. When the resisting or load torque developed at motor shaft by the driven unit undergoes some change, the speed and torque developed by the motor will change automatically so as to restore stable operation at a new value of speed and load torque.

In case of non-electric prime movers (water turbine, steam turbine or diesel/petrol engine) the balance between the resisting torque and the driving torque is achieved by using respective type of governor to control the inflow of energy to the prime mover by increasing or decreasing the flow of water, steam or fuel. In electric motors, the role of the automatic governor is performed by the emf of the motor. This ability of electric motors to maintain the balance of the drive system on changes in the resisting (load) torque developed by the driven unit is extremely valuable because this torque very often is unstable to a certain degree.

This can be illustrated with the help of Fig. 1.10 which illustrates the speed-torque characteristic (curve III) of a dc shunt motor and the two characteristics I and II of the production unit driven by the motor (for example, a conveyor).*

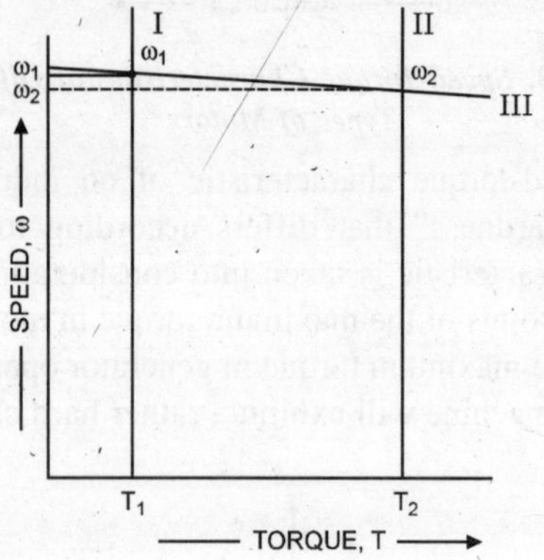

Fig. 1.10. ***Speed-Torque Characteristic of a Shunt Wound DC Motor and Two Characteristics of Driven Unit***

Characteristic I corresponds to the no-load condition of the conveyor unit, while characteristic II corresponds to a greater level of load torque developed by the conveyor when it handles the required flow of material. Initially, at the time when the conveyor operates at no load, the motor torque $T = T_1$ and the motor operates at speed ω_1. As soon as the conveyor starts to maintain a flow of material, the increase in load on the motor acts to brake the motor and reduce its speed. This makes the motor to develop a smaller emf. Consequently the armature current increases and the motor begins to develop a larger driving torque. The motor torque grows until a point of balance is reached at which the torque developed by the motor = resisting torque of the driven unit, *i.e.*, $T = T_2$ (where the speed is ω_2). This new point also constitutes the one that is common both to the speed-torque characteristic II of the conveyor and the speed-torque characteristic III of the motor.

In studying the operation of the motor and the unit it drives, it is sometimes convenient to make use of the so-called joint speed-torque characteristic of the electric drive, a curve representing the algebraic sum of the speed-torque characteristic of the driven unit and that of the drive motor.

Speed-torque characteristics of a fan and the drive motor and the joint speed-torque characteristic of a motor fan unit are represented by curves I, II and III respectively in Fig. 1.11.

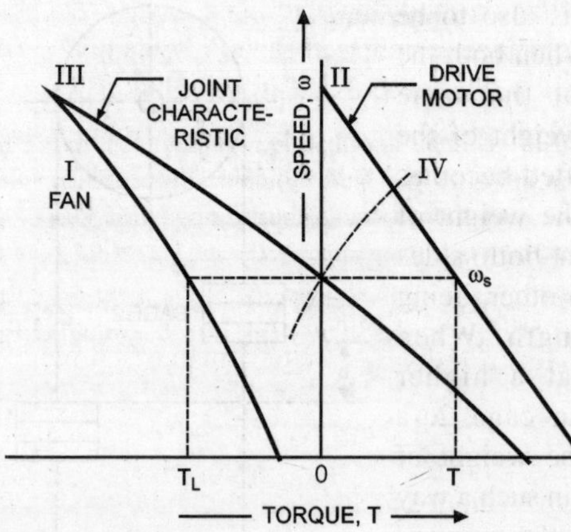

Fig. 1.11. ***Joint Speed-Torque Characteristic of a Motor-Fan Drive***

When the unit attains a steady speed ω_s, the motor operates with the torque T = load torque, T_L. The torque indicated on the joint characteristic will be zero under such a condition. Operation of the unit at the steady speed ω_s will be seen to be stable in this case because any increase in speed leads to a negative change (drop) in torque, while any drop in speed results in a positive change (increase) in torque. Curve III is, therefore, an example of a joint speed-torque characteristic of a drive that will be able to operate with stability. If the joint characteristic had the form of curve IV, operation would not be stable because slight increase in speed gives rise to acceleration as the motor torque exceeds the load torque. On the other hand slight decrease in speed gives rise to retardation because the motor torque becomes less than the load torque.

The conditions of drive operation in the steady state discussed above constitute those required for static stability of the drive, and are only applicable when speed and torque vary slowly. During periods of transient (rapid) change involving the dynamic stability, the conditions of drive stability will be different.

Usually, when an electric drive is designed for a specific drive, its speed-torque characteristic is known before hand. The problem of attaining stable operation in steady-state condition at known speeds and load torques of the driven unit therefore consists in selecting a motor whose speed-torque characteristic will be compatible with that of the driven unit. This may be achieved by first selecting a suitable type of motor and then changing accordingly the electrical parameters of its circuits. Sometimes, in order to provide the required speed-torque characteristics it becomes imperative to set up special power and control circuits for involved switching of the drive motor and the control apparatus.

1.12. DYNAMICS OF MOTOR-LOAD COMBINATION

In translational motion, an active or driving force F_d is counterbalanced by a resisting force F_r developed by the driven

* For convenience, the two speed-torque characteristics of the production unit are drawn in the first quadrant although the resisting torque is actually negative in the sense with respect to the driving torque.

machine and by an inertia force $m\frac{dv}{dt}$ arising from the change in speed. When the body involved has a mass m expressed in kg and a velocity v expressed in m/s, the inertia force, like other forces, will be expressed in newtons (kg-m/s^2).

The equation for force equilibrium in translation of a body may accordingly, be written in the following form.

$$F_d - F_r = m\frac{dv}{dt} \text{ newtons} \quad \text{...(1.10)}$$

The equation for torque equilibrium in translation of a body may, accordingly, be written in the following form.

$$T_M - T_L = J\frac{d\omega}{dt} \text{ Nm} \quad \text{...(1.11)}$$

The above Eq. (1.11) shows that the torque T_M developed by the motor is counterbalanced by a resisting or load torque T_L exerted at its shaft and by an inertia or dynamic torque $J\frac{d\omega}{dt}$. In above Eqs. (1.10) and (1.11) it is assumed that the mass m of the bodies involved and the polar moment of inertia J of the drive remain constant, an assumption which holds good for a large number of industrial machines and mechanisms. In certain drives, it becomes necessary to deal with a variable polar moment of inertia as in the case of crank drives.

From analysis of Eq. (1.11), it is possible to determine the different states at which an electric drive causing rotational motor can remain.

1. When $T_M > T_L$, $\frac{d\omega}{dt} > 0$, *i.e.*, the drive will undergo acceleration, in particular, picking up speed to attain rated speed.

2. When $T_M < T_L$, $\frac{d\omega}{dt} < 0$, *i.e.*, the drive will undergo deceleration and, particularly, coming to rest.
   Deceleration will evidently also occur at negative* values of motor torque. A motor develops a negative torque when it passes over to braking operation.

3. When $T_M = T_L$, $\frac{d\omega}{dt} = 0$, *i.e.*, the drive will run at a steady-state speed.

The above statements, namely, that when $T_M > T_L$ the drive accelerates and that when $T_M < T_L$ the drive decelerates, are valid only when load or restraining torque T_L happens to be a passive torque. The reverse may occur with active torque loads. For example, if a motor is switched on for hoisting up a winch, while it is coming down on its own weight, until the direction of rotation changes, deceleration of the drive takes place when $T_M > T_L$. In case $T_M < T_L$ in the above situation when the motor has been switched on for moving the winch up, the load will continue to come down and the motor will accelerate instead of decelerating.

The inertia or dynamic torque $J\frac{d\omega}{dt}$ appears only during transient conditions, *i.e.*, when the speed of the drive varies. During acceleration of the drive, the inertia torque opposes the drive motion, but during braking it maintains the motion of the drive. The inertia torque, both in magnitude and in sign, is determined as the algebraic sum of the motor torque and the resisting or load torque.

In view of the above, the signs for T_M and T_L in Eq. (1.11) corresponding to motoring operation of the driving machine and to passive load torque (or an active braking torque) respectively. In the general form, the torque equation can be written as

$$\pm T_M \mp T_L = J\frac{d\omega}{dt} \quad \text{...(1.12)}$$

Selection of sign to be placed before each of the torques in above Eq. (1.12) depends upon the operating condition and on the nature of the resisting or load torques. The equation of motion for a drive, makes it possible to determine the dependence of torque, current, speed and path upon the time of operation under transient conditions. All the torques in the equation for motion have to be referred to some given element in the system. Most frequently, both the load torque and the dynamic torque are referred to the motor shaft.

Example 1.2. A motor is coupled to a load having the following characteristics :

Motor : $T_m = 15 - 0.5\omega_m$

Load : $T_l = 0.5\omega_m^2$

Find out the stable operating point for this combination.

[U.P. Technical Univ. Electric Drives 2004-05]

Solution : The stable operation will be obtained when

$$T_m = T_l$$
$$\text{or } 15 - 0.5\omega_m = 0.5\omega_m^2$$
$$\text{or } \omega_m^2 + \omega_m - 30 = 0$$
$$\text{or } \omega_m = 5 \text{ or } -6$$

Discarding minus figure, we have

$$\omega_m = 5 \text{ and } T = 12.5$$

So stable operating point is (12.5, 5) **Ans.**

1.13. REFERRED LOAD TORQUES AND MOMENTS OF INERTIA

A motor generally drives an industrial machine through some transmission system whose individual parts operate at different speeds. In carrying out practical calculations, a necessity arises to refer the torques and masses of individual parts to some convenient element, say, a certain shaft.

Load torques may be referred from one shaft to another on the basis of the power balance of the system. In this case account is taken of the power losses in the intermediate links of the transmission by introducing the respective efficiency values.

Let the speed of the motor shaft be ω_M and the speed of shaft of the given industrial machine ω_L.

On the basis of equality of power flow, we have

$$T_L\omega_L\frac{1}{\eta_T} = T'_L\omega_M$$

or load torque referred to the motor shaft,

* A motor torque is considered to be positive in sense when it acts in the direction necessary for normal operation of the drive. When the motor develops torque opposing this direction of motion, the torque is taken to be negative in sense.

$$T'_L = T_L \frac{\omega_L}{\omega_M} \times \frac{1}{\eta_T} = \frac{T_L}{i\,\eta_T} \quad ...(1.13)$$

where T_L is the load torque, η_T is the efficiency of transmission and i is speed transmission ratio and is equal to ω_M/ω_L.

When there are several stages in transmission between the drive motor and the driven machine, as illustrated schematically in Fig. 1.12, with the transmission ratios i_1, i_2, ..., i_n and the respective transmission efficiencies $\eta_{T_1}, \eta_{T_2}, ..., \eta_{T_n}$, the load torque referred to the motor shaft is given as

$$T'_L = T_L \times \frac{1}{i_1 i_2 \ldots i_n} \times \frac{1}{\eta_{T_1}\eta_{T_2}\ldots\eta_{T_n}} \quad ...(1.14)$$

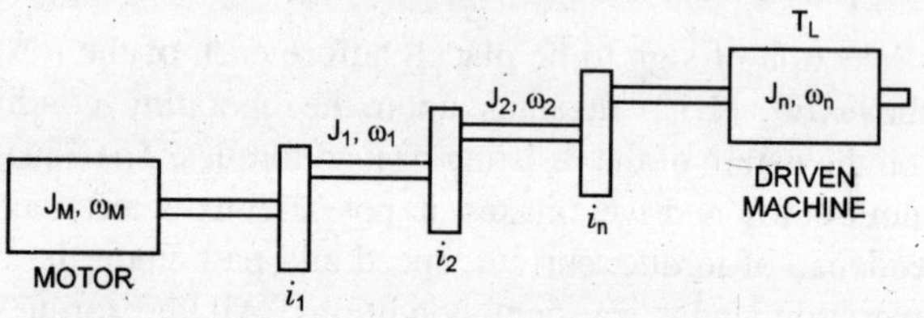

Fig. 1.12. ***Motor and Driven Machine With Gears***

Moments of inertia are referred to a given shaft on the basis that the total amount of kinetic energy stored in the moving parts and referred to the given shaft remains unchanged. With the rotating parts having polar moments of inertia of J_M, J_1, J_2, ..., J_n and angular velocities of ω_M, ω_1, ω_2, ... ω_n (Fig. 1.12), their dynamic action may be replaced by that of a single polar moment of inertia referred, say, to the motor shaft, and we may write the following equations.

$$\frac{J\omega_M^2}{2} = J_M\frac{\omega_M^2}{2} + J_1\frac{\omega_1^2}{2} + J_2\frac{\omega_2^2}{2} \; ... + J_n\frac{\omega_n^2}{2}$$

or the polar moment of inertia referred to the motor shaft

$$J = J_M + J_1\left(\frac{\omega_1}{\omega_M}\right)^2 + J_2\left(\frac{\omega_2}{\omega_M}\right)^2 + \ldots J_n\left(\frac{\omega_n}{\omega_M}\right)^2$$

$$= J_M + \frac{J_1}{i^2} + \frac{J_2}{(i_1 i_2)^2} + \ldots \frac{J_n}{(i_1 . i_2 \ldots i_n)^2} \quad ...(1.15)$$

Example 1.3. A motor drives a rotational load through a reduction gear of teeth ratio a = 0.1 and efficiency as 90%. The load has a moment of inertia of 10 kg-m² and a torque of 50 N-m. The motor has inertia of 0.4 kg-m² and runs at a constant speed of 1,400 rpm. Determine the equivalent inertia of the motor and load combination referred to the motor side and the power developed by the motor. [*U.P. Technical Univ. Electric Drives 2004-05*]

Solution : Motor inertia, J_M = 0.4 kg-m²

Load inertia, J_1 = 10 kg-m²

Teeth ratio of gear, $a = \frac{1}{i} = 0.1$

Total moment of inertia referred to the motor shaft (or side)

$$J = J_M + \frac{J_1}{i^2} = 0.4 + \frac{10}{\left(\frac{1}{0.1}\right)^2}$$

$$= 0.4 + 10 \times (0.1)^2 = 0.5 \text{ kg-m}^2 \quad \textbf{Ans.}$$

Total equivalent torque referred to motor shaft,

$$T'_L = \frac{a_1 T_L}{\eta_T} = \frac{T_L}{i\eta_T} = \frac{50}{\frac{1}{0.1}\times 0.9} = 5.55 \text{ N-m}$$

Power developed by motor

$$= T'_L \times \frac{2\pi N}{60} = 5.55 \times \frac{2\pi \times 1,400}{60} = 815 \text{ W} \quad \textbf{Ans.}$$

1.14. REFERRING FORCES AND MASSES HAVING TRANSLATIONAL MOTION TO A ROTATING SHAFT

In many machines (such as in hoists, cranes, shaping machines etc.) some of the moving parts rotate while others go through a translational motion. A motion of translation will have to be referred to a rotating shaft drive involving such machines.

The resisting force developed by a translating body is referred to a rotating shaft from the condition of unchanged kinetic energy balance in much the same way as torques are referred.

Figure 1.13 illustrates the forces and torques acting on a hoist drive. If a translating body (or moving mass) has a velocity of v m/s and the motor shaft has the angular velocity ω_M rad/second, then

$$F_r\, v \times \frac{1}{\eta_T} = T'_L\,\omega_M \quad ...(1.16)$$

where F_r is the resisting force developed by the given machine due to the translating motion of a load or body of weight W.

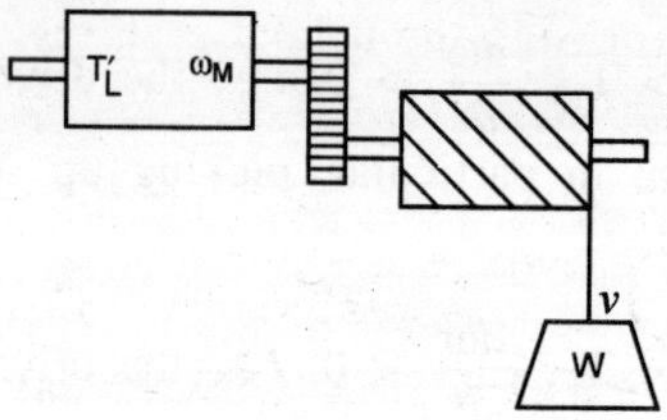

Fig. 1.13. ***Motor-Hoist Drive***

The referred load torque at the motor shaft will be, therefore, given as

$$T'_L = \frac{F_r\, v}{\omega_M \eta_T} \quad ...(1.17)$$

A **translating mass** is referred to a rotating shaft on the basis of equality of kinetic energy

$$\frac{mv^2}{2} = J\frac{\omega_M^2}{2}$$

The polar moment of inertia, referred to the motor shaft, is therefore, given as

$$J = m\left(\frac{v}{\omega_M}\right)^2 = \frac{W}{g}\left(\frac{v}{\omega_M}\right)^2 \quad ...(1.18)$$

Example 1.4. A motor drives two loads one has a rotational motion. It is coupled to the motor through a reduction gear of teeth ratio a = 0.1 and efficiency as 90%. The load has a moment of inertia 10 kg-m² and a torque of 50 N-m. Other load has a translational motion and consists of 600 kg weight to be lifted at a uniform speed of 2.0 m/s. Coupling between this load and the motor has an efficiency of 88%. Motor has inertia of

0.4 kg-m^2 and runs at a constant speed of 1,500 rpm. Determine the equivalent inertia referred to the motor shaft and the power developed by the motor.

Solution : Motor inertia, $J_M = 0.4$ kg-m^2

Load inertia, $J_1 = 10$ kg-m^2

Teeth ratio of gear, $a = 0.1$

$$\text{Motor speed, } \omega_M = \frac{2\pi N}{60} = \frac{2\pi \times 1{,}500}{60}$$

$$= 157 \text{ radians/speed}$$

Uniform speed of weight lifting, $v = 2.0$ m/s

Total moment of inertia referred to the motor shaft,

$$J = J_M + a_1^2 J_1 + M_1 \times \left(\frac{v}{v_m}\right)^2$$

$$= 0.4 + (0.1)^2 \times 10 + 600 \times \left(\frac{2}{157}\right)^2 = 0.6 \text{ kg-m}^2 \quad \textbf{Ans.}$$

Total torque referred to motor shaft

$$T'_L = \frac{a_1 T_L}{\eta_T} + \frac{F_r v}{\eta'_T \omega_M} = \frac{0.1 \times 50}{0.9} + \frac{600 \times 9.81 \times 2}{0.88 \times 157} = 90.75 \text{ N-m}$$

Power developed by motor

$$= T'_L \times \omega_m = 90.75 \times 157 = 14{,}250 \text{ W or } 1.425 \text{ kW} \quad \textbf{Ans.}$$

1.15. REFERRING TORQUES AND TRANSLATING MASSES WHEN SPEED IS VARIABLE

In certain types of machines which convert the motion of rotation into a translational one with the aid of a crank gear, the speed and acceleration of the moving masses vary both in magnitude and in sign throughout each revolution of the crankshaft.

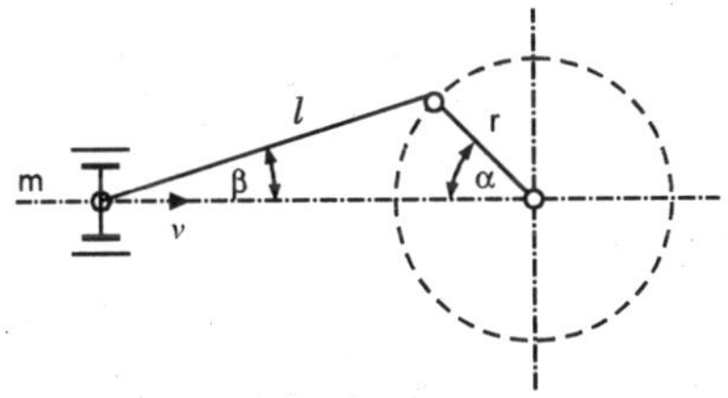

Fig. 1.14. ***Crankshaft Mechanism***

The kinetic energy stored in these masses will, accordingly, change from zero to maximum. The MOI of these masses referred to the crankshaft is given as

$$J = \frac{mv^2}{\omega^2} \qquad \text{...(1.19)}$$

where m is the mass of given translating part moving with velocity v and ω is the angular velocity of the crankshaft.

Using the geometrical relationships of the crank gear shown in Fig. 1.14, after certain transformations, we have the following expression for the velocity of the translating part

$$v = \frac{r\omega \sin(\alpha + \beta)}{\cos \beta} \qquad \text{...(1.20)}$$

where $\beta = \sin^{-1}\left(\frac{r}{l} \sin \alpha\right)$; the value of α depends on the position of the crank pin.

Substituting the value of v from Eq. (1.20) in Eq. (1.19) for moment of inertia, we have

$$J = \frac{mr^2 \sin^2(\alpha + \beta)}{\cos^2 \beta} \qquad \text{...(1.21)}$$

The total moment of inertia will be the sum of the moment of inertia J determined from above Eq. (1.21) and that of all other moving parts referred to the crankshaft. If there is some intermediate stage of transmission between the crankshaft and the driving motor, the moments of inertia referred to the crankshaft should then be referred to the motor shaft by making use of Eq. (1.17).

Thus static load torque referred to the motor shaft,

$$T'_L = \frac{F_r\, r \sin(\alpha + \beta)}{i\, \eta_T \cos \beta} \qquad \text{...(1.22)}$$

where F_r is the resisting force developed by the translating part involved, i is the motor shaft-to-crankshaft transmission ratio, $\left(i.e., \frac{\omega_M}{\omega}\right)$ and η_T is the efficiency of transmission.

With variable moment of inertia as in the case of crank gears, the equation of motion for the electric drive becomes more complex since $J = f(\alpha)$. The kinetic energy stored in the system, at the crankshaft, amounts to

$$(KE)_c = \frac{1}{2} J \omega^2 \qquad \text{...(1.23)}$$

and the dynamic power corresponding to this energy,

$$P_{dynamic} = \frac{d(KE)_c}{dt}$$

$$= J\omega \frac{d\omega}{dt} + \frac{1}{2}\omega^2 \frac{dJ}{d\alpha} \cdot \frac{d\alpha}{dt}$$

$$= J\omega \frac{d\omega}{dt} + \frac{1}{2}\omega^3 \frac{dJ}{d\alpha} \quad \therefore \ \omega = \frac{d\alpha}{dt} \qquad \text{...(1.24)}$$

Taking into account that the dynamic torque $T_{dynamic} = \frac{P_{dynamic}}{\omega}$, the equation of motion becomes

$$T_M - T_L = J\frac{d\omega}{dt} + \frac{1}{2}\omega^2 \frac{dJ}{d\alpha} \qquad \text{...(1.25)}$$

The additional term, that has appeared is obviously owing to the variable moment of inertia.

1.16. OPERATING OR RUNNING CHARACTERISTICS OF ELECTRIC MOTORS

While studying electrical behaviour of a machine under normal operating conditions, the speed-torque characteristic, speed-current characteristic, losses, efficiency, magnetising current and power factor at various loads are kept in view. The last two factors, *i.e.*, magnetising current and power factor are to be considered in case of ac motors only.

Other features, such as temperature rise and insulation strength, are taken care of by Indian Standard Specifications, therefore if a motor is purchased built by a reputed manufacturer in accordance with the appropriate specifications, it will prove satisfactory in these respects.

1.16.1. DC Motors

In all dc machines, the speed and torque are given by expressions

$$N = \frac{V - I_a R_a}{\phi Z} \times \frac{60A}{P} = K_1 \frac{V - I_a R_a}{\phi} = K_1 \frac{E_b}{\phi} \quad ...(1.26)$$

$\because$ Z, A and P are constant for a particular machine

$$T = \frac{\phi}{2\pi} \times Z \times \frac{P}{A} I_a = K_2 \phi I_a \quad ...(1.27)$$

Torque and Speed of a DC Motor. Equations (1.26) and (1.27) reveal that

$$N \propto \frac{E_b}{\Phi} \text{ and } T \propto \Phi I_a$$

The above relations reveal that increase in flux would cause increase in torque and decrease in speed. It cannot be so because torque is the cause of rotation and increase in torque should cause increase in speed rather than decrease. The apparent inconsistency between the above two relations may be reconciled as below.

Let the motor be running at a steady speed and the field strength be weakened, which is possible by inserting an additional resistance in the field circuit in case of a dc shunt motor connected across constant voltage supply. The immediate effect is that, owing to the momentary reduction in back emf, the current flowing through armature circuit increases out of all proportion to the reduction in flux. As a result, inspite of the weakened field, the torque increases momentarily tremendously and will exceed considerably the value corresponding to load. The surplus torque, thus available causes the motor to accelerate, and the back emf to increase. The motor will attain steady speed again when the back emf has attained such a value that the current flowing in the armature develops with the slight weakened flux, only just sufficient torque to drive the load.

For avoiding undue rush of current the field strength should be reduced slowly.

The manner in which the motor slows down on strengthening of the field is also of equally importance. When the field is strengthened, the immediate effect is increase in back emf. Quite a small increase in the flux will normally be sufficient to increase the back emf beyond supply voltage, and in consequence, will not merely reduce the value of armature current, but will cause it to reverse its direction of flow. The machine will, therefore, act temporarily as a generator, and feed energy to the supply system, this energy being available at the expense of the kinetic energy associated with the machine and its load, with the result that a rapid reduction in speed takes place. The decrease in speed will cease when the back emf has fallen to a value such that the motor takes just sufficient current to develop the torque the load requires. By thus strengthening the field the motor speed is reduced by a form of "regenerative braking".

From the above discussion it is obvious that torque developed in a motor is certainly a function of flux and armature current and is independent of speed and the speed depends upon the torque.

(a) DC Shunt Motors

1. ***Speed-Armature Current Characteristic.*** If applied voltage V is kept constant, the field current will remain constant, hence flux will have maximum value on no load but will slightly decrease due to armature reaction as the load increases but for most purposes the flux is considered to be constant, neglecting armature reaction effect.

From speed equation, speed N is directly proportional to back emf E_b or $(V - I_a R_a)$ and inversely proportional to the flux Φ. Since flux is considered to be constant as mentioned above, so with the increase in armature current the speed slightly falls due to increase in voltage drop in armature and the speed-armature current curve coincides with the back emf-armature current curve. Since voltage drop in armature at full load is very small as compared to applied voltage so drop in speed from no load to full load is very small. If demagnetising effect of armature reaction is considered, the drop in speed due to voltage drop in armature is compensated for up to some extent due to decrease in flux with the increase in armature current and speed-armature current characteristic is less drooping, as shown dotted in Fig. 1.15 or even be rising if demagnetisation is high. Thus the drop in speed from no load to full load is very small and for all practical purposes the dc shunt motor is taken as a *constant speed motor*.

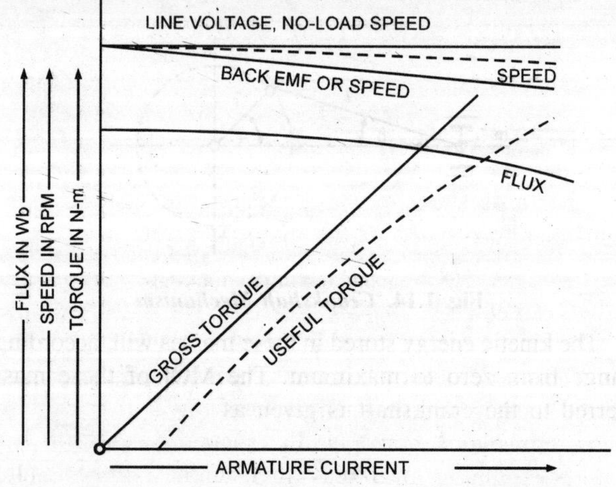

Fig. 1.15

Since there is a slight variation in speed of the shunt motor from no load to full load and this slight variation in speed can be made up by inserting resistance in the shunt field and so reducing the flux. Therefore, shunt motors can be used for the loads which are totally and suddenly thrown off without resulting in excessive speed. Shunt motors being constant speed motors are best suited for driving of line shafts, machine lathes, milling machines, conveyors, fans and for all purposes where constant speed is required. It is not suitable for use with flywheel or with fluctuating loads or for parallel operation due to its constant speed characteristic.

It is also useful where a moderate degree of speed control is required.

2. *Torque-Armature Current Characteristic.* From the expression for the torque of a dc motor, torque is directly proportional to the product of flux per pole ϕ and armature current, I_a. Since in case of a dc shunt motor the flux per pole ϕ is considered to be constant, therefore, torque increases with the increase in load current following linear law, *i.e.,* torque-armature current characteristic is a straight line passing through origin O (Fig. 1.15). But the weakening of field due to armature reaction causes the torque line to droop slightly, and the iron and friction losses cause it to be slightly lower than the line representing the electromagnetic developed torque corresponding to $T = 0.159\phi\, ZP \frac{I_a}{A}$ Nm.

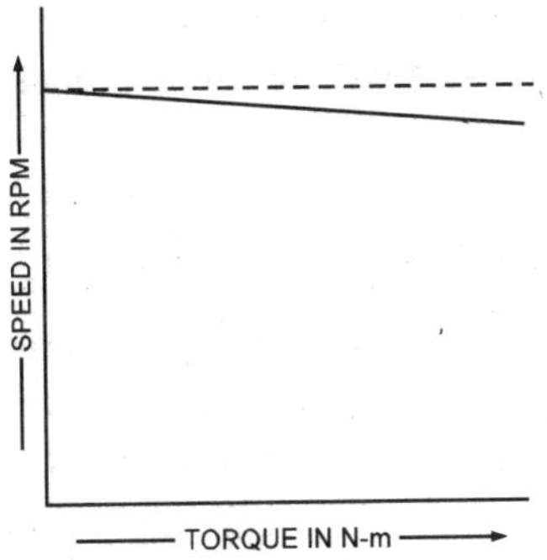

Fig. 1.16. *Speed-Torque Characteristic of DC Shunt Motor*

3. *Speed-Torque Characteristic.* Substituting $I_a = \frac{T}{K_2 \phi}$ from Eq. (1.27) in Eq. (1.26), we have

$$N = K_1 \frac{V}{\phi} - K_1 \frac{\frac{T}{K_2 \phi} R_a}{\phi} = K_1 \frac{V}{\phi} - \frac{K_1}{K_2} \frac{TR_a}{\phi^2} \quad ...(1.28)$$

For a dc shunt motor, ϕ remains more or less constant and, therefore speed of the motor is given as

$$N = C - DT \quad ...(1.29)$$

where C and D are constants for the machine, *i.e.,* the characteristic will be linear in nature, as illustrated in Fig. 1.16.

This type of motor is used in applications requiring medium starting torque such as centrifugal pumps, blowers, fans, conveyors, boring mills, shapers, woodworking machines, spinning and weaving machines, printing presses, machine tools etc.

DC shunt motors should never be started on heavy loads because such loads need heavy starting current.

4. *Break in Field Circuit of a DC Motor.* It is imperative that the field circuit of a dc motor never be opened while it is running. If the field circuit is opened, the flux will drop to practically zero and, therefore, the speed would increase to an extremely high value. The centrifugal force developed at the high speed may cause the windings to be pulled out of the slots or may cause the commutator segments to buckle, resulting in damage to the motor.

Proper use of overload devices would protect the motor. Since back emf would be reduced due to reduction in flux and cause increase in armature current or current drawn from the supply mains by the motor. This excessive current would operate the overload device, disconnect the motor from the supply, and thus protect the motor.

5. *Performance Curves of a DC Shunt Motor.* In Fig. 1.17, four important characteristics of a dc shunt motor, namely, torque, speed, armature current, and efficiency, each plotted against the useful output power, are shown. These curves are also known as *performance curves* of a motor.

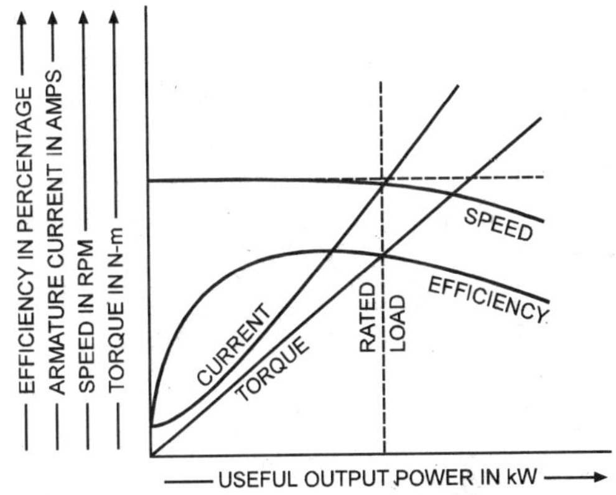

Fig. 1.17. *Performance Curves of a DC Shunt Motor*

From the speed–useful output power curve it is seen that the shunt motor has a definite speed at no load. Hence it does not 'run away' when the load is suddenly thrown off provided that the field circuit remains closed. The fall in speed from no load to full load is small, hence, these motors are used where a substantially constant speed is required as in machine shop drives.

From the efficiency–output power curve it is noted that the efficiency of a shunt motor increases rapidly in the beginning, reaches its maximum value and then falls. In shunt wound motors the field, friction and iron losses are approximately constant but at light loads their percentage is high as compared with the load. The armature copper loss varies as the square of the current so this loss, termed as variable loss, increases rapidly with the increase in load. The efficiency of the shunt machine becomes maximum when variable loss is equal to the constant loss.

From armature current–output power curve it is noted that a certain value of current is required even when the output is zero. The motor input power under no-load conditions is consumed in meeting the various losses occurring within the machine.

A shunt wound motor is said to have a low starting torque as compared to other motors but it does not mean that a shunt motor is incapable of starting a heavy load. A shunt motor is capable of starting heavy load also but it would require more excess of current input over normal values than the series and compound motors. For instance, if twice full-load torque is required at start the shunt motor will draw twice full-load current (since $T \propto I_a$) but a series motor would draw only approximately 1.414 times the full-load current (Since $T \propto I_a^2$). Due to this reason, shunt motors are always started at no load or light load and then the load is applied. When a motor is required to start heavy loads series or compound motor is used.

(b) DC Series Motors

1. *Speed-Current Characteristic.* In case of a dc series motor, the mmf due to the exciting coils increases in direct proportion to the line or armature current, so (neglecting armature reaction effects) the value of flux varies with the load current according to the ordinary magnetization curve. Owing to armature reaction, the actual curve representing the useful flux falls below the open-circuit magnetization curve, as illustrated in Fig. 1.18. With larger currents the magnetic circuit gets saturated and flux Φ tends to approach a constant value.

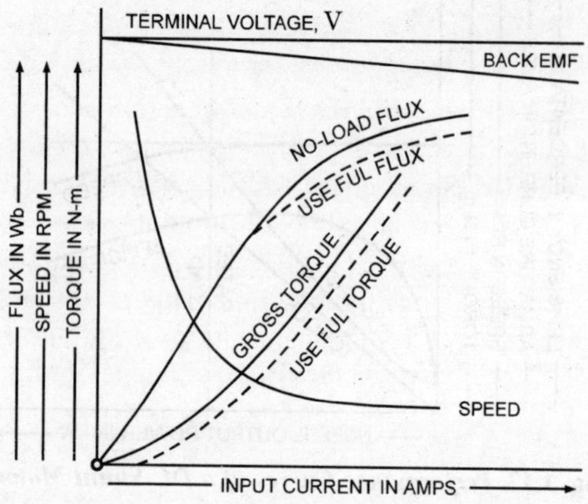

Fig. 1.18

From the speed equation, it is obvious that speed is proportional to back emf E_b and inversely proportional to flux per pole Φ. With the increase in armature current voltage drop in armature circuit and series field $[I(R_a + R_{se})]$ increases and, therefore, back emf E_b decreases, as shown in Fig. 1.18. However, under normal conditions $I(R_a + R_{se})$ drop is quite small and may be neglected. Thus if the applied voltage remains constant, speed N is inversely proportional to flux Φ. If a curve is drawn between speed and input (or line) current I, it will be a rectangular hyperbola before magnetic saturation as up to saturation point the magnetisation curve is a straight line. In this region, the speed decreases abruptly with the increase in input current. After magnetic saturation, the flux Φ tends to become constant and speed-current characteristic becomes a straight line and speed decreases slightly due to voltage drop in armature and series field, as shown in Fig. 1.18. The speed becomes zero when the input current is the normal short-circuit current of the motor *i.e.,* equal to applied voltage divided by the motor resistance $(R_a + R_{se})$. Normally this current is many many times full-load value.

From speed-current characteristic curve it is obvious that the series motor is a *variable speed motor i.e.,* speed varies with the variation in line current. With the decrease in load on a dc series motor, the speed increases and may become dangerously high at very light loads.

Since on no load the speed is dangerously high, the machine may get damaged due to heavy centrifugal forces set up in the rotating parts. *This is the reason that series motors are never started on no load.*

To start a dc series motor, mechanical load is first put and then the motor is started.

Since on no load the series motor attains dangerously high speed, which causes heavy centrifugal force resulting in the damage of machine, series motors are not suitable for the services:

(*i*) where the load may be entirely removed and

(*ii*) for driving by means of belts because mishap to the belt would cause the motor run on no load.

The motors are suitable for gear drive, because gears provide some load on account of the frictional resistance of the gear teeth in case of sudden release of load.

However, very small series motors may be used with belts, since in case of mishap to belt, the comparatively large frictional resistance would represent an appreciable load on it.

The minimum load on a dc series motor should be great enough (not below 15% of full load) to keep the speed of the motor within limits. In case the speed becomes dangerously high the motor must be disconnected from the supply main.

2. *Torque-Current Characteristic.* From the expression of mechanical torque T it is obvious that torque is directly proportional to the product of flux per pole Φ and armature current I_a. Up to saturation point flux is proportional to field current and hence to the armature current, because $I_a = I_f$. Therefore, on light load mechanical torque T is proportional to the square of the armature current *i.e.,* $T \propto I_a^2$ and hence curve drawn between torque and armature current up to saturation point is a parabola, as shown in Fig. 1.18.

After saturation point, flux Φ is almost independent of excitation current and so the torque is proportional to the armature current *i.e.,* $T \propto I_a$. Hence the characteristic becomes a straight line. The useful (or shaft) torque is, of course, less than the total (or gross) torque developed. This is due to torque lost in iron and friction and windage losses.

From the torque-armature current curve it is evident that so long as the field of the motor is not saturated, the series motor exerts a torque proportional to the square of current *i.e.,* starting torque is very high. Hence series motors are used where large starting torque is required for accelerating heavy masses quickly such as in hoists, electric railways, trolleys and electric vehicles.

From torque-current characteristic it is evident that series motor develops large starting torque to accelerate the heavy masses and from speed-current characteristic it is also evident that speed falls as the load increases, so series motor is automatically relieved from heavy excessive load, therefore, series motor is best suited for electric traction work.

3. *Speed-Torque Characteristic.* For low values of load current,

$$I \propto \phi \text{ and } T = K_3\,\phi^2$$

$$\phi = \sqrt{\frac{T}{K_3}} \quad \text{...(1.30)}$$

Substituting $\phi = \sqrt{\dfrac{T}{K_3}}$ in Eq. (1.28) we have

$$N = K_1 \frac{V}{\sqrt{T/K_3}} - \frac{K_1 . K_3}{K_2} \frac{\phi^2 R_a}{\phi^2} = \frac{C_1}{\sqrt{T}} - D_1 \quad ...(1.31)$$

where C_1 and D_1 are two constants, *i.e.*, the shape of the speed-torque characteristic will be hyperbolic.

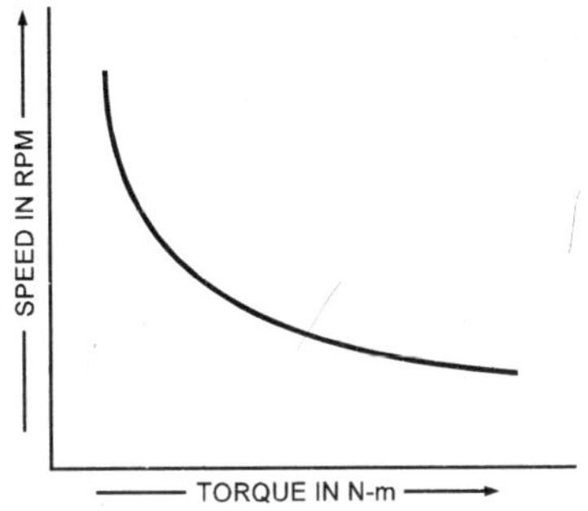

Fig. 1.19. ***Speed-Torque Characteristics of DC Series Motor***

For large values of load current, ϕ remains almost constant and therefore, speed will be given as

$$N = C_2 - D_2 T \quad ...(1.32)$$

where C_2 and D_2 are constants for the machine, *i.e.*, speed-torque characteristic will be linear in nature.

Speed-torque characteristic, is also known as *mechanical characteristic*. Speed sharply falls with the increase in torque for smaller values of load. But at higher loads, the speed drops linearly but slowly with increasing torque.

Hence series motors are best suited for services where the motor is directly coupled to the load such as fans whose speed falls with the increase in load torque.

4. *Performance Curves of a DC Series Motor.* In Fig. 1.20, four important characteristics of a dc series motor, namely torque, speed, current and efficiency each plotted against useful output power are shown.

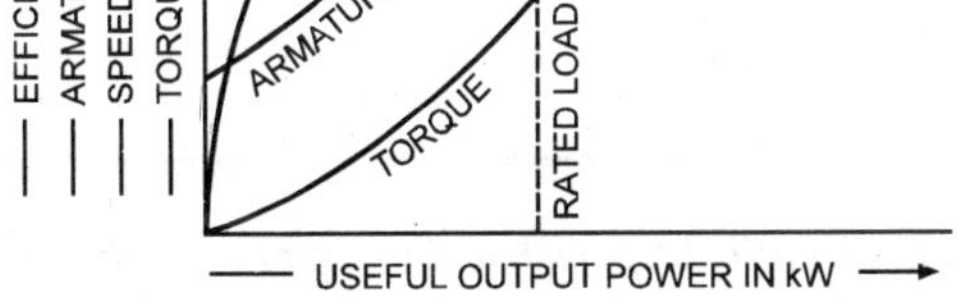

Fig. 1.20. ***Performance Curves of a DC Series Motor***

From the performance curves for a dc series wound motor it is noted that;

(*i*) The speed of series motor falls rapidly with the increase in load, so a series motor is not suitable for services requiring a substantially constant speed.

(*ii*) The efficiency increases rapidly in the beginning, reaches its maximum value and then decreases. This is due to the fact that at light loads the friction and iron losses are large compared with the load and the effect of these losses becomes less with the increase in load. The armature and field copper loss varies as the square of the current, so these losses increase rapidly with the increase in load. The efficiency becomes maximum when friction and iron losses are practically equal to the copper loss.

(*iii*) Series motor develops a starting torque comparatively greater than that developed by a shunt motor for a given current. Hence series motors are best suited where huge starting torque is required *i.e.,* for streetcars, cranes, hoists and locomotives.

(*iv*) In addition to the large starting torque, there is another unique characteristic of series motors which makes them best suited for traction services. For instance let the streetcar ascend a gradient. In case the shunt wound motor is used for driving the above streetcar, it will maintain the speed of the car at approximately the same value that it had on the level ground and, therefore, will tend to draw an excessive current. A series motor, on the other hand, automatically slows down on such a gradient because of increased current demand and, therefore, it develops more torque at reduced speed. This drop in speed allows the motor to develop a large starting torque with but a moderate increase of power. Hence under the same load conditions, the rating of a series motor would be less than for a shunt motor.

(*c*) *Compound Wound DC Motors*

Operating Characteristics of Cumulative Compound Wound DC Motors. The characteristics of cumulative compound wound motor are combination of shunt and series characteristics and lie between those of shunt and series motors.

As the load is increased, the flux due to series field winding increases and causes the torque greater than it would have been with shunt field winding alone for a given machine and for a given current. The increase in flux due to series field winding on account of increase in load cause the speed to fall more rapidly than it would have done in shunt motor. The characteristics are shown in Fig. 1.21.

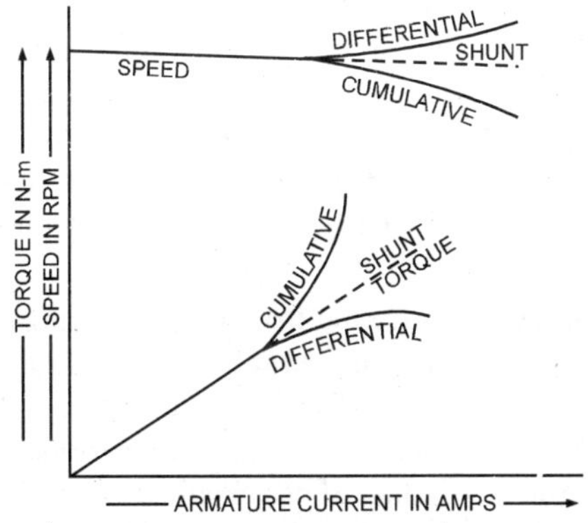

Fig. 1.21. ***Characteristics of Compound Wound DC Motors***

Like the series motor, the cumulative compound wound motor has excellent characteristics for starting heavy loads and yet there is no danger of "running away" at light loads when used without a speed controller.

By varying the proportions of shunt field mmf and series field mmf, compound wound motors may be designed to give speed and torque characteristics anywhere intermediate between those for shunt and series motors. In many cases only enough shunt field is provided to guarantee a safe no-load speed, in which case the motor be called a *stabilised series motor*. Compound wound motors are usually designed so that there will be a drop in speed of about 20% between no load to full load.

Cumulative compound wound motors are used in driving machines which are subject to sudden applications of heavy loads, such as occur in rolling mills, punching and shearing machines, lifts, mine-hoists etc. This type of motor is used also where a large starting torque is required but series motor cannot be employed conveniently such as in cranes and elevators.

Another advantage of the cumulative compound wound motor for suddenly applied loads is that the motor automatically undergoes a substantial drop in speed when the load is applied. Accordingly much of its stored kinetic energy becomes available for supplying a part of the increased load thus reducing the electrical load on the motor as well as the peaks on the power system. The available kinetic energy is frequently increased by the use of a flywheel, particularly with rolling mill motors.

Cumulative compound wound motors are not practical for applications requiring adjustable speed for field control. With a weakened shunt field, the series field becomes dominating hence the changes in load may produce unstable speeds.

1.16.2. Three-Phase AC Motors

(a) Three-Phase Induction Motors. 3-phase induction motors have gained extremely wide application in industry by virtue of the advantages they have over other types of motors. Induction motors are simple in design, rugged in construction and reliable in service, as they have no commutator. Besides this, they have low initial cost, easy operation and simple maintenance, high efficiency and simple control gear for starting and speed control.

The speed-torque characteristics of induction motors are quite important in the selection of an induction motor drive. In addition, the ratio of maximum torque to rated torque, ratio of starting current to rated current, ratio of starting torque to rated torque and the ratio of no-load current to rated current are equally significant. The above characteristics can be conveniently determined by means of the equivalent circuit of the induction motor.

The energy is transferred from primary (stator) winding to secondary (rotor) winding entirely by induction, therefore, induction motor is essentially a transformer. At standstill, the induction motor is actually a static transformer having its secondary (rotor) winding short circuited.

When the motor operates at slip s, the frequency of rotor currents is s times the frequency of the stator currents, therefore, the revolving field produced by the rotor currents revolves with respect to rotor itself at speed

$$\frac{120sf}{P} = \frac{120f}{P} \times s = N_s \times s,$$

Mechanical speed of the motor, $N = N_s(1 - s)$. The speed of the revolving field of the rotor with respect to stator or space is obtained by combining the rotational speed of rotor field with respect to rotor with mechanical speed of the rotor. Hence speed of rotor revolving field with respect to stationary stator or space $= sN_s + N_s(1 - s) = N_s$.

Hence from the point of view of stator, the induction motor still can be considered as a static transformer, even when its rotor is rotating, and it is possible to represent the performance of an induction motor by a transformer phasor diagram. Actually the rotor field does not exist alone but combines with the revolving field of stator to produce a resultant field, just as in the transformer, the resultant field is produced by the combination of primary and secondary ampere-turns.

In the transformer the load on secondary is electrical and in an induction motor the load is mechanical which can be replaced by an equivalent electrical load of resistance R'_L given by

$$R'_L = \frac{R_2}{K^2}\left(\frac{1}{s} - 1\right)$$

where R_2 is the resistance per phase of rotor and K is the turn-ratio of the secondary (rotor) to primary (stator).

If R'_2, the effective rotor resistance referred to stator, is combined with R'_L, the fictitious resistance representing electrically the mechanical power developed in rotor, we have

$$R'_2 + R'_L = R'_2 + R'_2\left(\frac{1}{s} - 1\right) = \frac{R'_2}{s}, \quad \text{...(1.33)}$$

The simplified equivalent circuit for an induction motor is shown in Fig. 1.22 where V is applied voltage per phase, R_1 and X_1 are stator resistance and leakage reactance per phase, R'_2 and X'_2 are rotor resistance and standstill leakage reactance per phase referred to stator, R_0 and X_0 are the resistance and inductive reactance per phase of the magnetizing branch, I'_2 is the rotor current per phase referred to stator and s is the slip.

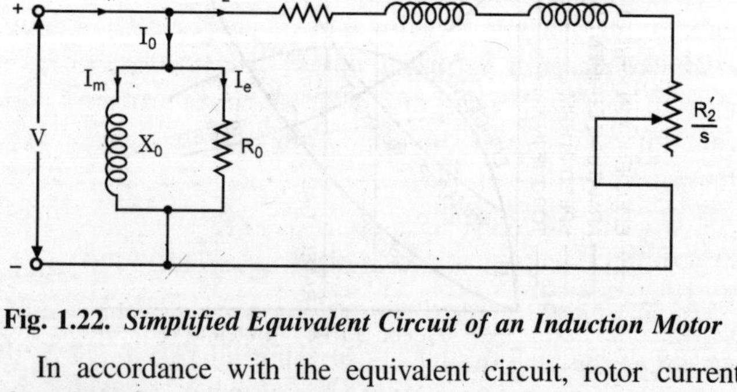

Fig. 1.22. ***Simplified Equivalent Circuit of an Induction Motor***

In accordance with the equivalent circuit, rotor current referred to stator comes out to be

$$I'_2 = \frac{V}{\sqrt{\left(R_1 + \frac{R'_2}{s}\right)^2 + (X_1 + X'_2)^2}} \quad \text{...(1.34)}$$

Using the expression for losses, $\frac{2\pi N_s}{60} Ts = 3(I'_2)^2R'_2$, we have the expression for the torque of the induction motor. The expression for the torque is, therefore, given as

$$T = \frac{3(I'_2)^2 R'_2}{s\frac{2\pi N_s}{60}} \quad ...(1.35)$$

Substituting the value of I'_2, from Eq. (1.34) in Eq. (1.35), we have

$$T = \frac{3V^2 R'_2/s}{\frac{2\pi N_s}{60}\left[\left(R_1 + \frac{R'_2}{s}\right)^2 + (X_1 + X'_2)^2\right]}$$

Newton-metres ...(1.36)

Typical speed-torque curves are illustrated in Fig. 1.23. The two curves are identical; one being drawn for the forward direction of rotation and the other for the reverse. Extension of motoring curve from the first quadrant into the fourth quadrant indicates that the positive torque is developed even though the motor is rotating in the reverse direction. The extension of the same curve into the second quadrant indicates that a negative torque is developed when the machine operates at speeds above synchronous one. Thus regeneration is possible if the system to which the motor is connected is capable of supplying required reactive power for excitation.

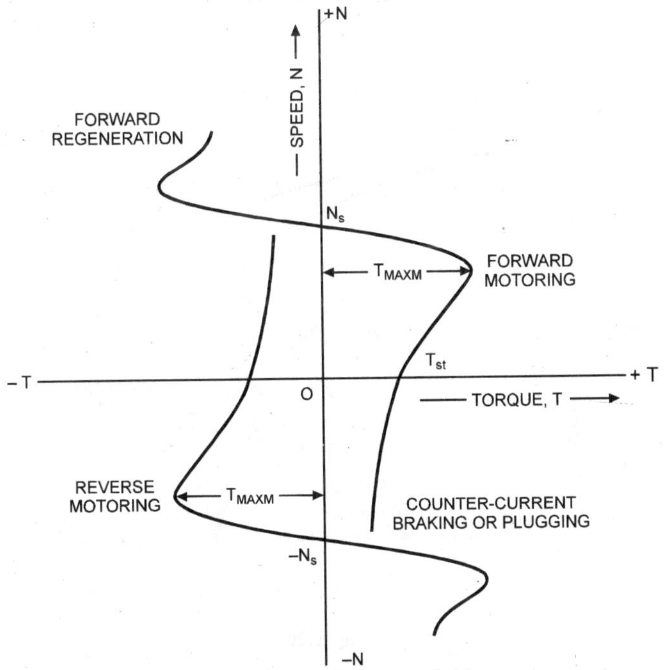

Fig. 1.23. ***Speed-Torque Characteristics of Induction Machine***

By differentiating the torque Eq. (1.36) w.r.t. slip s and equating to zero we get slip corresponding to maximum torque, as

$$s_{max\ T} = \pm \frac{R'_2}{\sqrt{R_1^2 + (X_1 + X'_2)^2}} \quad ...(1.37)$$

Substituting the value of slip corresponding to maximum torque from Eq. (1.37) in Eq. (1.36), the maximum torque is obtained as,

$$T_{max} = \frac{3V^2}{\frac{4\pi N_s}{60}\left[R_1 \pm \sqrt{R_1^2 + (X_1 + X'_2)^2}\right]}$$

Newton-metres ...(1.38)

Plus and minus sign in Eqs. (1.37) and (1.38) corresponds to motoring (or braking) and generating operation respectively.

Dividing Eq. (1.36) by Eq. (1.38), we have

$$\frac{T}{T_{max}} = \frac{2\left[R_1 + \sqrt{R_1^2 + (X_1 + X'_1)^2}\right] \times \frac{R'_2}{s}}{\left(R_1 + \frac{R'_2}{s}\right)^2 + (X_1 + X_2)^2} \quad ...(1.39)$$

For motor operation, R'_2 in the above Eq. (1.39) can be replaced by its value in terms of $s_{max\ T}$ from Eq. (1.37). After simplification, we have

$$\frac{T}{T_{max}} = \frac{1 + \sqrt{\left(\frac{X_1 + X_2}{R_1}\right)^2 + 1}}{1 + \frac{1}{2}\sqrt{\left(\frac{X_1 + X_2}{R_1}\right)^2 + 1}\left(\frac{s}{s_{max\ T}} + \frac{s_{max\ T}}{s}\right)} \quad ...(1.40)$$

Neglecting stator resistance R_1, the above expression becomes more simple and approximate one, which is as

$$\frac{T}{T_{max}} = \frac{2}{\frac{s}{s_{max\ T}} + \frac{s_{max\ T}}{s}} \quad ...(1.41)$$

From above Eq. (1.41) it is obvious that if the maximum torque T_{max} and the slip corresponding to maximum torque, $s_{max\ T}$ are specified, the speed-torque characteristic is almost fixed throughout the entire speed range provided that the motor parameters are constant. The above statement is not true for motors with variable rotor resistance.

Effect of Variation in Supply Voltage on Speed-Torque Characteristics. From the viewpoint of practice, it is of prime importance to note here the influence that variations in supply voltage will have on the speed-torque characteristics of induction motor. As is obvious from Eq. (1.36), for any given value of slip the torque developed by the motor will be proportional to the square of applied voltage. This means that the motor will be sensitive to fluctuations in supply voltage.

Since at full load, slip s is very small, therefore, the terms R_1, X_1 and X'_2 can be neglected in comparison with term $\frac{R'_2}{s}$ and the equation for torque may be written as

$$T = \frac{3V^2 s}{\frac{2\pi N_s}{60} R'_2} \quad ...(1.42)$$

or $T \propto sV^2$

If the supply voltage is changed, it changes the torque under running conditions also. With the decrease in supply voltage, torque under running condition decreases. As the motor now develops a reduced torque, the motor slows down (*i.e.*, slip increases) so as to cause increase in torque to meet the load torque and the motor may come to a standstill (if the load torque exceeds the pull-out torque of the motor corresponding to reduced supply voltage) unless the load is removed. Figure 1.24 shows the variation of torque with slip for three different values of line voltage.

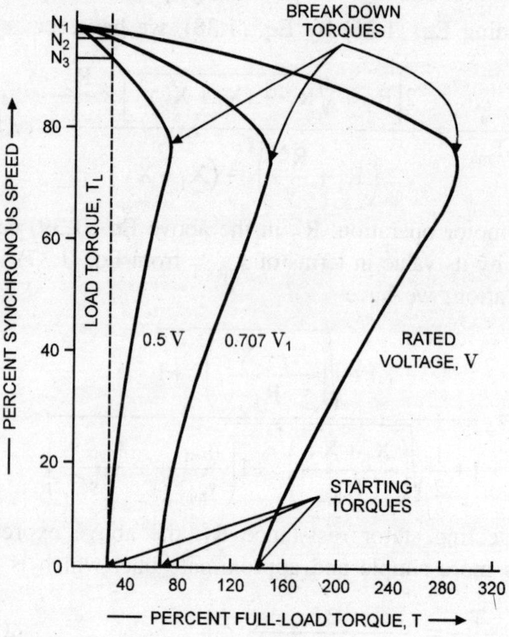

Fig. 1.24

Effects of Varying Supply Voltage and Supply Frequency Together on Speed-Torque Characteristics. This kind of variation can occur for example, if the generator supplying the power to the induction motor (or motors) is subject to speed variations, either due to momentary overloads or because of a noticeable speed regulation. Both output voltage and frequency will vary as the speed if no automatic correction is provided for it. On some large marine drives, the propellor motors are of induction type and are speed-controlled from such a local supply which is provided by synchronous generators coupled to variable speed turbines.

Since emf induced in stator, E_1 is proportional to the product of stator flux ϕ and supply frequency f (*i.e.*, $E_1 \propto \phi f$), a change of frequency will give rise to a change of flux level unless the induced emf is changed in the same ratio. An imbalance in this ratio will cause either an excessive flux and saturation, or diminished flux and reduced torque per ampere of current. Constant flux per pole is approximated if the supply volts per Hz (V/f) is maintained constant because E_1 is approximately equal to V, the supply voltage. However, an idealized variable-frequency control scheme might aim to ensure that any particular torque should be obtained at the same flux.

With this criterion of controlling $\frac{E_1}{N_s}$ to a constant value for a particular torque, the speed-difference $(N_s - N)$ is also fixed for the same torque, since $(N_s - N)$ in turn determines the slip frequency f', the rotor emf E_2, impedance Z_2 and current I_2, which remain unchanged. From this it follows that as the frequency is reduced, the shape of the torque-slip (or torque-speed) curves do not change, but displaced horizontally, as illustrated in Fig. 1.25.

Effect of Rotor Resistance on Speed-Torque Characteristics. From torque Eq. (1.36) it is obvious that for a constant supply voltage when the rotor resistance R_2 is very small, the term $\frac{R_2'}{s}$ is very small as compared to X_2' and the torque for a given slip s is directly proportional to R_2' (neglecting stator impedance). But when the rotor resistance R_2 is very large, the terms R_1, X_1 and X_2' can be neglected and the torque for a given slip s is inversely proportional to R_2.

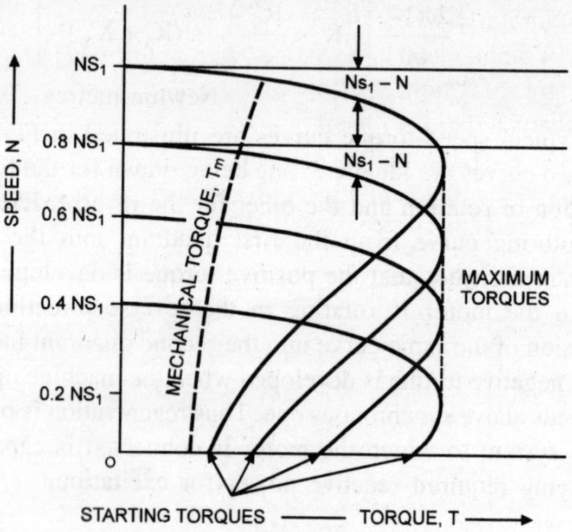

Fig. 1.25

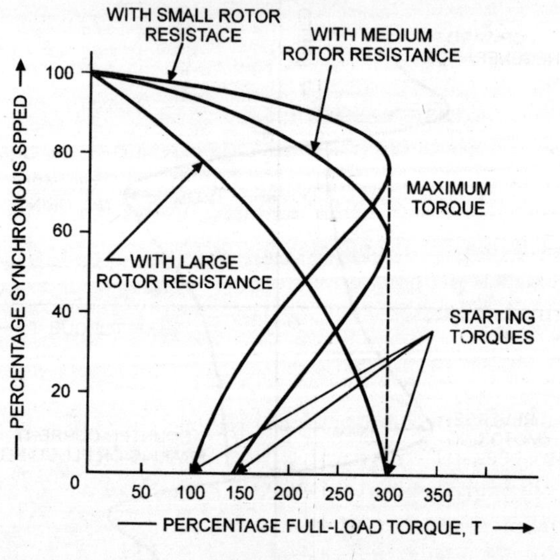

Fig. 1.26

From Fig. 1.26 it is seen that variation of rotor resistance does not change the magnitude of maximum torque T_{max}; merely changes the value of slip at which it occurs. This fact is also obvious from maximum torque Eq. (1.38) and Eq. (1.37) which gives condition for maximum torque. Larger the rotor resistance, the greater is the slip (or smaller is speed) at which the maximum torque occurs.

For a given value of torque T, slip s is proportional to rotor resistance R_2, so addition of external resistance in the rotor circuit does not make the curve narrower but stretches it so that the same torque values occur at lower speeds, as illustrated in Fig. 1.26.

Speed-Torque Curve and Operating Region. The speed-torque curve for an induction motor is given in Fig. 1.27. From Fig. 1.27 it is evident that for any load torque there are two operating points (say B and D). But at position B, the operation is unstable because if there is tendency of speed rise the developed torque also increases than the load torque causing further rise in speed. Thus at position B the operation is unstable. At position D, operation is stable because a tendency to rise in speed will be opposed by the decrease in developed torque. Similarly when there is a tendency to fall in speed, there will be increase in developed torque to bring the motor to operating position D. Thus the region AC of the torque-speed characteristic is *unstable region* of operation while the region CF is stable region of operation.

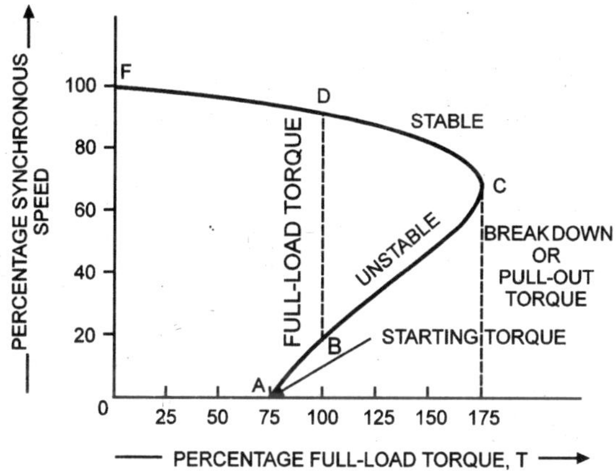

Fig. 1.27

At full load, the motor runs at a speed of N rpm. If the mechanical load applied to the motor is increased, motor speed drops till the torque developed by the motor equals the load torque. As long as the two torques are in balance, the motor will run at a constant speed. However, if the load torque exceeds the breakdown torque of the motor, the motor will immediately stop.

The value of breakdown torque varies with the design of the motor but ranges from 200 to 300 per cent of full-load torque in standard squirrel cage motors.

The speed-torque characteristics of an induction motor in the operating region are quite similar to those of a dc shunt motor. Even at full load the drop in speed from no load does not exceed 5 per cent.

Performance Curves of Induction Motors. Steady-state operating characteristics of an induction motor provides graphically the variation of speed (or slip), power factor, stator current and efficiency with the variation in power output from no load to full load. The performance curves of a 75 kW, 2200 V, 3-phase, 50 Hz squirrel cage induction motor are shown in Fig. 1.28.

1. *Speed- Output Characteristic.* The three-phase, squirrel cage induction motor like a dc shunt motor, operates at substantially constant speed from no load to full load. As the rotor cannot attain the speed of the rotating magnetic field (*i.e.*, synchronous speed), it must at all times operate with a certain amount of slip. Because of the extremely low impedance of the rotor, only a slight decrease in speed is enough to cause a large increase in rotor current to develop the torque necessary to drive the load. The per cent slip at no load is less than 1 per cent while at full load it is usually 2 to 5 per cent. This small percentage change in slip (or speed) from no load to full load likewise indicates why a squirrel cage induction motor is considered a constant speed motor.

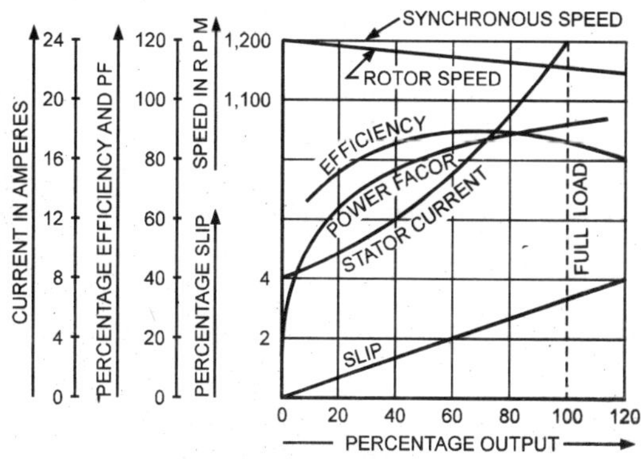

Fig. 1.28. ***Performance Curves of a 75 kW, 2,200 V, 3-phase, 50 Hz Squirrel Cage Motor***

2. *Power Factor Characteristic.* Because of the air gap between the stator and rotor of an induction motor, the reluctance of the magnetic circuit is high. As a result an induction motor draws a large magnetising current (I_m) to produce the required flux in the air gap.

The current drawn by an induction motor operating at no load is largely a magnetising current, so no-load current lags behind the applied voltage by a large angle and thus the power factor of an induction motor at no load is quite low (about 0.1 lagging). As the load is increased, the active or power component of the current increases, but the magnetising component remains practically the same. Consequently, the power factor improves. However, because of the large value of magnetising current, which is present regardless of load, the power factor of an induction motor, even at full load, seldom exceeds 90 per cent.

3. *Efficiency Characteristic.* At all loads, there are certain fixed losses (such as core loss, friction and windage loss) in addition to variable losses that include stator and rotor copper losses varying nearly as the square of the load. At light loads, the efficiency is quite low because the fixed losses are a relatively large part of the input. With the increase in load, the efficiency, at first, increases rapidly and attains a maximum value, the fixed and variable losses being equal at this point. Beyond this point, the copper losses (I^2R losses) become relatively large, causing the efficiency to decrease.

The maximum efficiency occurs at about 80 to 95% of rated output, the higher values being applicable to larger motors.

4. *Stator Current Characteristic.* The no-load current is about 30 to 40 per cent of the rated current, the higher values being applicable to smaller motors. With increase in load, slip increases in a straight line characteristic, the rotor current will likewise increase in practically a direct proportion and cause the stator current to increase accordingly.

Double Cage Rotor Motor. There are three types of 3-phase induction motors namely plain or squirrel cage induction motor, slip-ring or wound rotor induction motor and double cage rotor induction motor.

The simple squirrel cage induction motor is superior to slip-ring induction motor on account of motor output, frame size, initial, maintenance and repair costs, weight, life, reliability, efficiency and power factor. The only drawback of the squirrel cage induction motor is that its speed cannot be controlled by inserting resistance in the rotor circuit as in case of slip-ring induction motor. Maximum torque developed, which is also called the pull-out torque, by the squirrel cage motor, is more than that developed by slip-ring induction motor.

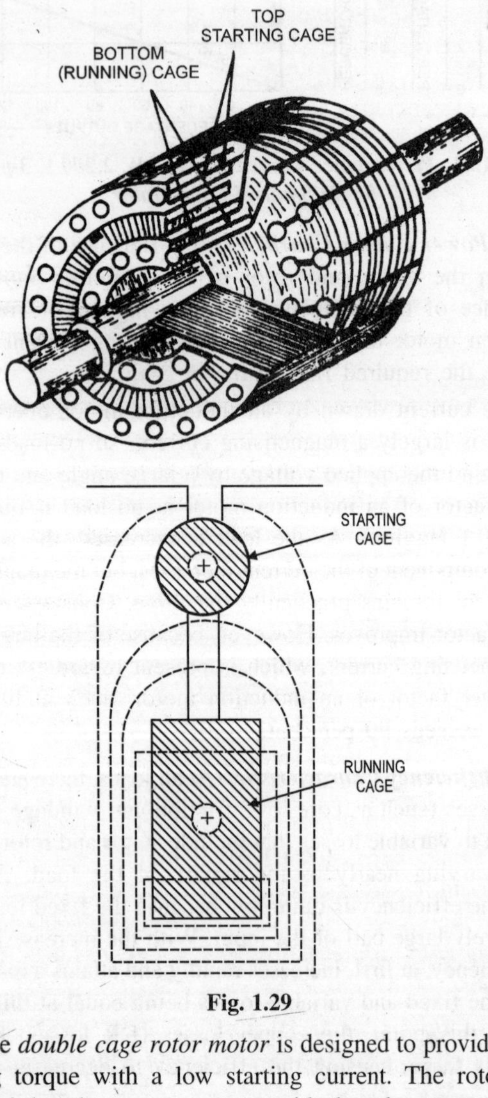

Fig. 1.29

The *double cage rotor motor* is designed to provide a high starting torque with a low starting current. The rotor is so designed that the motor operates with the advantages of high-resistance rotor circuit during starting and a low-resistance rotor circuit under running conditions. The starting torque of a double cage rotor motor ranges from 200 to 250 per cent of full-load torque with a starting current of 400 to 600 per cent of full-load value. The rotor of a double squirrel cage motor carries two squirrel cage windings embedded in two rows of slots (Fig. 1.29). The outer slots contain a high resistance and low leakage reactance winding and the inner slots contain a low resistance and high leakage reactance winding. At starting, the rotor current has the same frequency as the supply current, and the high reactance winding carries very little current, so that the performance approximates to that of high-resistance low-reactance cage alone. At normal speed, however, the rotor current frequency is quite low so that the reactances of the two rotor cages are insignificant compared with the resistances, and the performance approximates to that of the low-resistance cage. The cost of double squirrel motor is 20 to 30% higher than that of an ordinary squirrel cage motor. The speed-torque characteristics of individual cages as well as resultant are shown in Fig. 1.30.

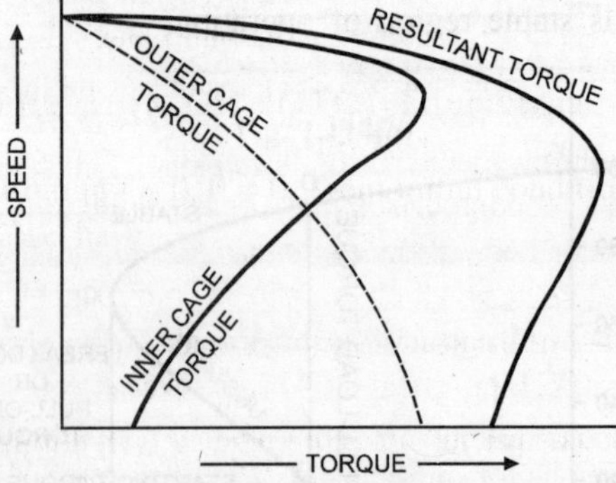

Fig. 1.30. *Speed-Torque Characteristics of Double Cage Induction Motor*

The choice of an induction motor of any type depends on the requirements of starting torque. *Squirrel cage* induction motor is suitable for constant speed industrial drives of small power where speed control is not required and where starting torque requirements are of medium or low value, such as for printing machinery, flour mills and other shaft drives of small power. *Wound rotor* or *slip-ring* induction motors are used for loads requiring severe starting conditions or for load requiring speed control such as for driving line shafts, lifts, pumps, generators, winding machines, cranes, hoists, elevators, compressors, small electric excavators, printing presses, turntables, strokers, large ventilating fans, crushers etc. *Double squirrel cage motor* is particularly suited where both high starting torque and a small slip on full load are required.

(b) Three-Phase Synchronous Motors. The synchronous motor is a constant speed motor whose speed is fixed by the supply frequency and number of poles and, therefore, is independent of load. Since this motor is not self starting, therefore, some special arrangement is required to be made for making it self starting. The special arrangement may be such as embedding of squirrel cage winding in the pole faces (as used in case of plain synchronous motor) or by arranging the field winding in the form of an ordinary induction motor rotor winding (as used in case of synchronous-induction motor). In both cases the motor starts as a plain induction motor and when it attains speed near the synchronous speed the dc excitation is switched on and the motor pulls into synchronism.

On no load synchronous motor draws very little current from the supply to meet the internal losses. With fixed excitation the input current increases with the increase in load. After the input current reaches maximum no further increase in load is possible. If the motor is further loaded, the motor will stop.

It can operate under a wide range of power factor both lagging and leading. This enables the motor to perform phase compensation action in addition to driving the load.

Torque developed by the synchronous motor varies directly as the voltage where as in an induction motor it varies as square of the applied voltage, therefore, synchronous motors are best suited to withstand large voltage variations.

These motors can be constructed with wider air gaps than induction motors, which make them better mechanically. These motors usually operate at higher efficiencies (92 to 96% as against 87 to 90% for induction motors). Such motors can be made to take large overloads.

Synchronous motors are rarely used below 40 kW output in the medium speed range because of their much higher initial cost in comparison to that of induction motors. In addition they need a dc excitation source, and the starting and control devices which are usually more expensive. Where low speeds and high kW outputs are involved the induction motors are no long cheaper. The various classes of services for which synchronous motors are employed may be classified as (*i*) power factor correction (*ii*) voltage regulation and (*iii*) constant-speed, constant-load drives.

Because of the higher efficiency possible with synchronous motors, they can be employed advantageously for the loads where constant speed is required. Typical applications of high-speed synchronous motors (above 500 rpm) are drives such as fans, blowers, dc generators, line shafts, centrifugal pumps and compressors, reciprocating pumps and compressors, constant speed frequency changers, rubber and paper mills etc.

The fields of applications of low-speed synchronous motors (below 500 rpm) are drives such as reciprocating compressors when started unloaded, dc generators, centrifugal and screw type pumps, vacuum pumps, electroplating generators, line shafts, rubber and band mills, ball and tube mills, chippers, metal rolling mills etc. Flywheel is used for pulsating loads.

High power electronic convertors generating very low frequencies made the operation of synchronous motors at ultra low speeds. Synchronous motors in very large sizes (as high as 10 MW size) operating at ultra low speeds are employed to drive crushers, rotary kilns and variable speed ball mills.

(c) Synchronous Induction Motors. The synchronous induction motor, as its name implies, is a machine which is capable of running both as an induction motor and as a synchronous motor, the former being its mode of operation during the starting period, and the latter its mode of operation during normal running.

The secondary winding of a synchronous induction motor consists of a polyphase winding, almost invariably of the three phase type. Since the rotor is uniformly slotted around the whole periphery, there is no saliency, and thus it is a true round rotor machine.

For the synchronous mode of operation it is necessary to supply the rotor winding with direct current. When the rotor windings of a wound rotor induction motor are excited by a direct current, the mmf distribution develops alternate N and S poles in the same way as does a 3-phase current, but the essential difference is that the dc excitation being fixed, the pole axes due to dc excitation are fixed in space and do not shift as in case when the rotor winding carries ac. These fixed rotor poles get magnetically locked with the rotating magnetic field developed by the 3-phase stator windings carrying ac and the motor runs at a constant speed equal to the synchronous speed.

The synchronous induction motor is provided with a large air gap as provided in a plain synchronous motor. A long air gap gives a stiffer machine with a larger overload capacity. The rotor slots are made fewer in number and larger in size. These machines are provided with a heavy rotor winding in order to have a low slip, which facilitates in pulling it into synchronism. Also in order that the induced emf in the field at the starting may not be too high, the field turns provided are few in number and the excitation voltage is kept low. As the exciter winding serves the purpose of damper winding, so there is no need of providing separate damper winding with synchronous induction motors.

The schematic arrangement of starting a synchronous induction motor is shown in Fig. 1.31. The synchronous induction motor is started as a slip-ring induction motor by inserting resistances in the rotor circuit. When the additional resistances are completely cut out and the motor attains the normal induction motor speed, the rotor is disconnected from the starting

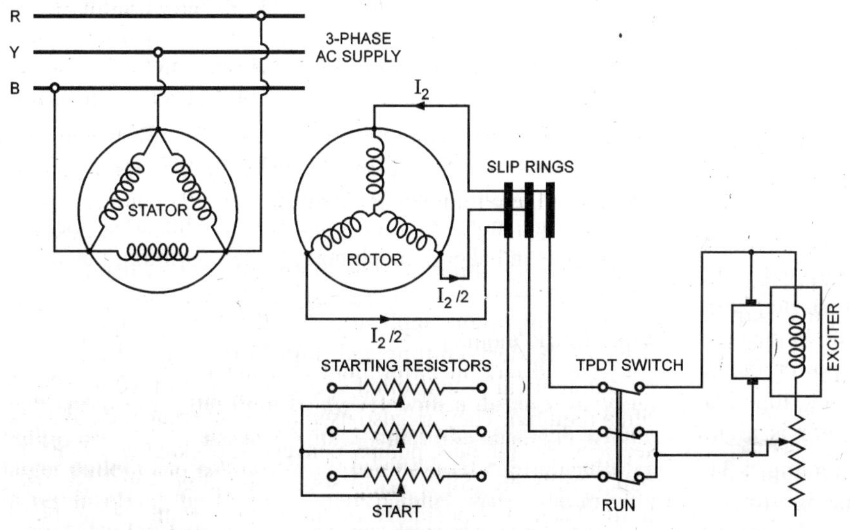

Fig. 1.31

resistances and connected to the exciter which is usually mounted on the same shaft. Motor will now be operating as a synchronous motor.

With secondary rheostats for starting, such a motor gives low starting current and high starting torque of the wound rotor induction motor and an improved power factor under load. Hence the synchronous induction motor is essentially a motor, having the induction motor features like high starting torque with low starting current combined with the synchronous motor features like constant speed and power factor control.

The interesting point about this motor is that its peak torque as an induction motor exceeds that as a synchronous motor. If it is momentarily overloaded as a synchronous motor, it may continue to operate as an induction motor-albeit with considerable pulsations of current and torque due to dc until the load falls and it is able to re-synchronise.

However, the extra cost, low efficiency as compared to that of standard types, the small copper space available with a distributed rotor winding, and the necessary compromise between excessive ring voltage at start, or excessive rotor current in operation are sever handicaps.

The synchronous induction motor is rarely built for ratings below 25 kW because of the relatively higher cost of the exciter. Synchronous induction motors are used where a high starting torque and constant speed operation are required. Typical applications of synchronous induction motors include fans, pumps, blowers, generators, air-compressors, ammonia compressors, machinery and line shafting in industrial works such as cement mills, rolling mills, flour mills, paper mills, rubber works and textile mills.

Synchronous induction motors are very often installed alongwith other induction motors, so that they may improve the overall power factor of the system. Their leading kVAR capacity is designed to offset the lagging kVAR demand from the induction motors. They have been made for ratings as high as 30 MW.

(d) Linear Induction Motor. It will be discussed in chapter 12 (Art. 12.7 page 343).

(e) Variable Speed Commutator Type 3-Phase Induction Motor (Schrage Motor). This is also an improvement of the plain induction motor designed to provide a variable speed. It also consists of a commutator winding placed on the rotor in addition to primary winding, whose emf is collected and injected into the secondary winding, placed on the stator. It is provided with two sets of brushes, as illustrated in Figs. 1.32 and 1.33.

The speed can be controlled over 3 to 1 range by moving the brush sets relatively to one another in opposite directions. The speed which depends upon the total emf induced in the secondary winding, will be affected by the injected emf. The speed will decrease if the injected emf is in phase opposition to the induced emf and will increase if the injected emf is in phase with the induced emf. Some power factor control can be had by moving both the brush sets together in the same direction. The operating power factor is high.

Speed-torque characteristic curves are shown in Fig. 1.34.. The output power available from the machine is directly proportional to the operating speed. Maximum operating torque ranges from 140 to 250% of full-load speed.

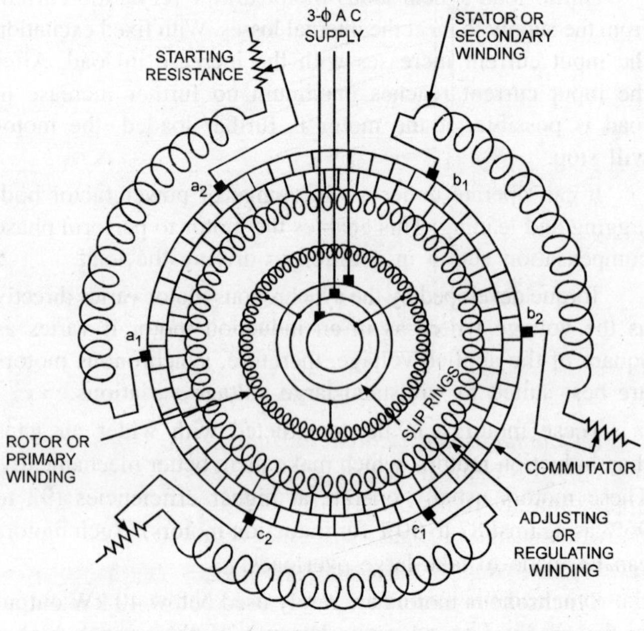

Fig. 1.32. *Connection Diagram of The Schrage Motor*

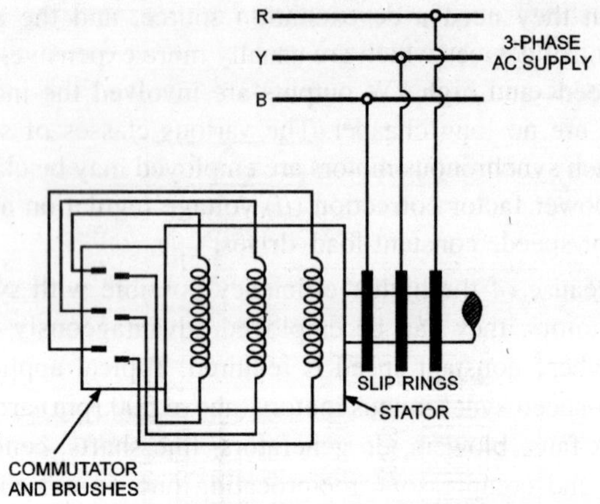

Fig. 1.33. *General Arrangement of The Schrage Motor*

The main advantages of the Schrage motor are (*i*) continuous speed regulation within the required range (*ii*) high power factor for high speed settings (*iii*) high efficiency at all speeds except synchronous speed and very much high at lower speeds.

One major drawback of the Schrage motor is that its operating voltage is limited about to 700 V, because the power is to be supplied to the motor through the slip rings. Other disadvantages are its low power factor at low speed settings and its poor commutation and high cost.

Schrage motor is used only where adjustable speed is required as in bakery, machinery, stokers, printing machines, calenders etc.

One important application of the Schrage motor is to drive the hosiery-knitting and ring-spinning machines, where an automatically controlled speed variation of 10 to 30% is often required. They are also used for a wide range of industrial services, driving fans, pumps, conveyors, packaging machinery, paper mills etc. Sizes up to several hundred kW have been built. They are normally rated on a constant torque basis, the kW output varying in direct proportion to the speed.

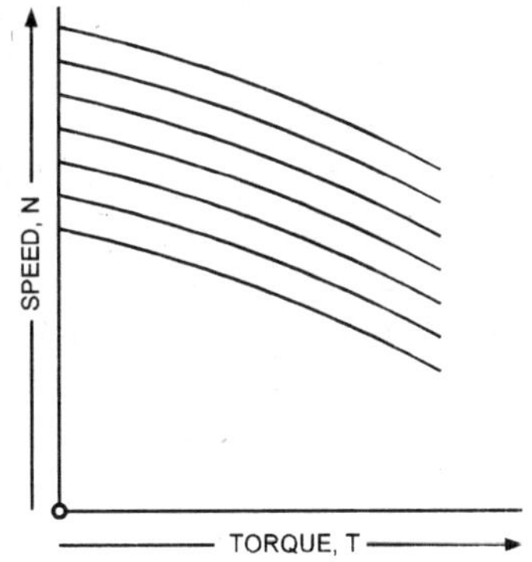

Fig. 1.34

(f) Compensated Induction Motor. This is an improvement of the plain induction motor designed to operate at approximately unity or slightly leading power factor over the whole range of loads. The usual type of motor is *no-lag type* motor, which consists of a primary winding placed on the rotor and the secondary winding on the stator, as shown in Fig. 1.35. The rotor consists of an additional winding, known as *commutator winding,* whose emfs are collected by the brushes from the commutator and injected into the secondary winding in such a way that power factor improvement is accomplished.

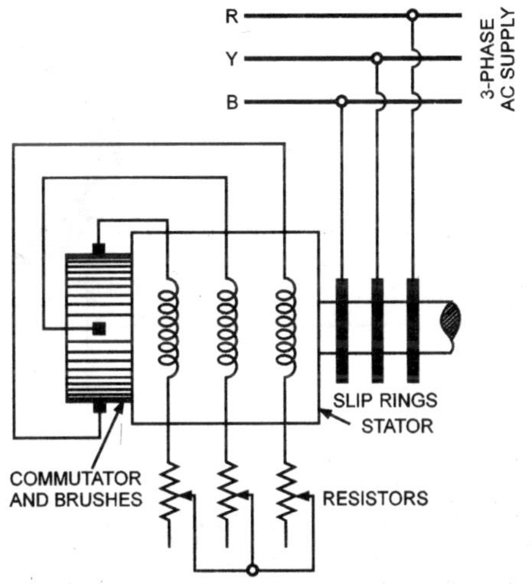

Fig. 1.35. ***General Arrangement of The No-Lag Motor***

No-lag motor was developed earlier and Schrage motor is actually modification form of the no-lag motor. In this motor phase of the emf can be varied but not the magnitude.

(g) Three-Phase AC Series Commutator Motor. This machine has a 3-phase stator winding similar to that of an ordinary induction motor and may be wound for high voltage if required. The rotor carries a winding similar to that of a dc machine and its commutator is provided with three brush sets per pair of poles spaced 120 electrical degrees apart. Connections are shown in Fig. 1.36. In effect, the stator and rotor windings are in series, so that the working flux is dependent on the current providing series speed characteristic. The speed can be controlled by moving the brush gear—the backward movement (*i.e.,* in a direction opposite to that of rotation of the rotor) of brushgear increases the speed and vice versa. For any given position of brushes, the speed falls with the increase in load, as in case of a dc series motor. The power factor is high approaching unity for speeds near and above synchronous speeds. The speed range is limited between 40 and 150 per cent of synchronous speed owing to commutation difficulties. Such a motor is useful when large starting torque is required such as in haulage and hoisting work if only ac supply is available.

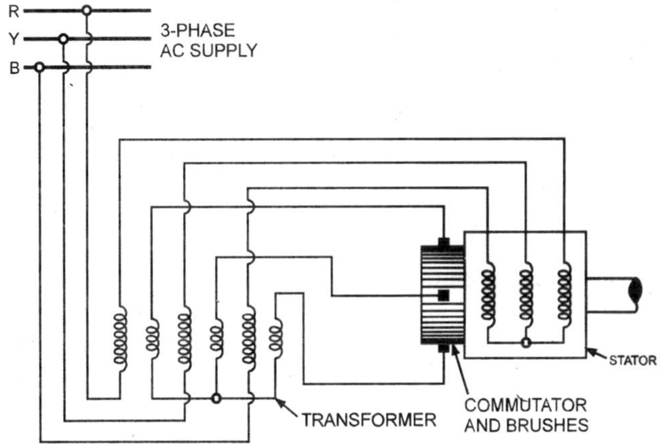

Fig. 1.36. ***Connection Diagram For a 3-Phase AC Series Commutator Motor***

(h) Three-Phase AC Shunt Commutator Motor. The connection diagram of a 3-phase ac shunt commutator motor is shown in Fig. 1.37. In this motor, the stator is connected directly to the supply while the rotor brushes are connected to the supply through a transformer. Speed variation is obtained by changing the tapping points on the transformer. It is to be noted here that brushes are fixed in position in the motor of this type. Such a motor provides more or less constant speed at all loads, for a given tapping on the transformer.

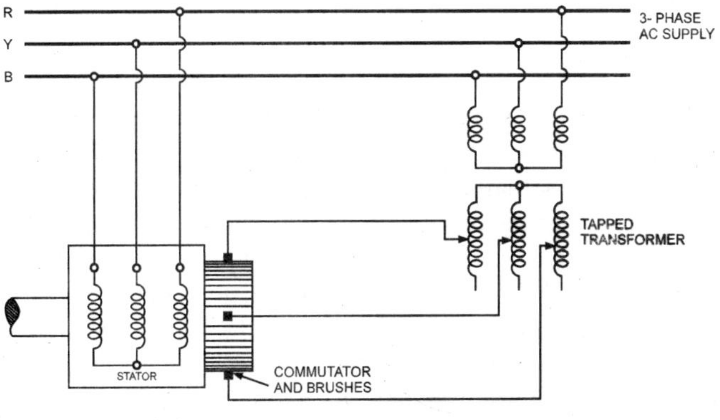

Fig. 1.37. ***Connection Diagram For a 3-Phase AC Shunt Commutator Motor***

1.16.3. Single-Phase AC Motors

Single-phase motors are more widely used than 3-phase motors because of two following basic reasons.

First for reasons of economy, most houses, offices and also rural areas are supplied with single-phase ac and the second factor is the economics of the motor and its branch circuit. Fixed loads requiring not more than 0.5 kW can generally be served most economically with single phase power and a single phase motor. Single-phase motors are simple in construction, reliable, easy to repair and comparatively cheaper in cost and, therefore, find wide use in fans, refrigerators, vacuum cleaners, washing machines, other kitchen equipment, tools, blowers, centrifugal pumps, small farming appliances etc.

Single-phase ac motors may be divided in three general classes namely (*i*) induction motors (*ii*) commutator motors and (*iii*) synchronous motors. Induction motors are further classified as split-phase motors, shaded-pole motors and repulsion-induction motors according to the method of producing starting torque. The commutator motors are the series motors, universal (ac/dc) motors, the repulsion–induction motors, with various modifications and combinations of these types being in use. Single-phase induction motors in very small sizes (1/400 to 1/25 kW) are used in toys, hair dryers, vending machines etc. The universal motor is widely used in portable tools, vacuum cleaners and kitchen equipment. The main drawbacks of single-phase motors are low overload capacity, poor efficiency, low *pf* and low output as compared to that of a 3-phase motor of the given frame size.

(a) Single-Phase Induction Motors. A single-phase induction motor is similar to a 3-phase squirrel cage induction motor in physical appearance. The rotor of a single phase squirrel cage induction motor is essentially the same as that employed in 3-phase induction motors. There is uniform air gap between stator and rotor but no electrical connection between them (stator and rotor). Except for shaded-pole types, the stator core is also very similar. A single-phase motor can be wound for any even number of poles, two, four, and six being most common. Like three-phase machines, adjacent poles have opposite magnetic polarity and synchronous speed equation ($N_s = \frac{120 f}{P}$) also applies.

When the stator winding of a single-phase induction motor is connected to single-phase ac supply, a magnetic field is developed, whose axis is always along the axis of stator coils. With alternating current in the fixed stator coil the mmf wave is stationary in space but pulsates in magnitude and varies sinusoidally with time. Currents are induced in the rotor conductors by transformer action, these currents being in such a direction as to oppose the stator mmf. Thus the axis of the rotor mmf wave coincides with that of the stator field, the torque angle is, therefore, zero, and no torque is developed at starting. However, if the rotor of such a motor is given a push by hand or by another means in either direction, it will pick up the speed and continue to rotate in the same direction developing operating torque. Thus a single-phase induction motor is not self starting and needs special starting means.

Commercial single-phase induction motors employ the principle of 'phase split' and are, therefore, known as *split-phase motors.* Split-phase motors are of the following types.

1. *Resistance–Start Single-Phase Induction Motor.* It is a form of split-phase motor having a high resistance connected in series with the auxiliary or starting winding, as illustrated in Fig. 1.38 (*a*).

A typical torque-speed characteristic is shown is Fig. 1.38 (*b*). The starting torque is 1.5 times to twice the full-load starting torque and starting current is 6 to 8 times full-load current. The speed regulation of a resistance-start single phase induction motor is very good. The per cent slip in most fractional kilowatt split-phase motors is about 4-6 per cent. Such a motor may operate with a power factor of 0.55-0.65 and efficiency of 60-65 per cent.

Owing to their low cost, resistance start split-phase motors are most popular single phase motors. These motors are made in fractional kilowatt $\left(\frac{1}{20} \text{ to } \frac{1}{4} \text{ kW}\right)$ ratings with speed ranging from 2,875 to 700 rpm. Such motors find wide applications for low inertia loads, continuous operating loads and applications requiring moderate starting torque such as for driving washing machines, fans, blowers, centrifugal pumps, domestic refrigerators, duplicating machines, office appliances, food processing machines, woodworking tools, grinders, oil burners etc. Because of low starting torque these motors are seldom employed in sizes larger than $\frac{1}{4}$ kW.

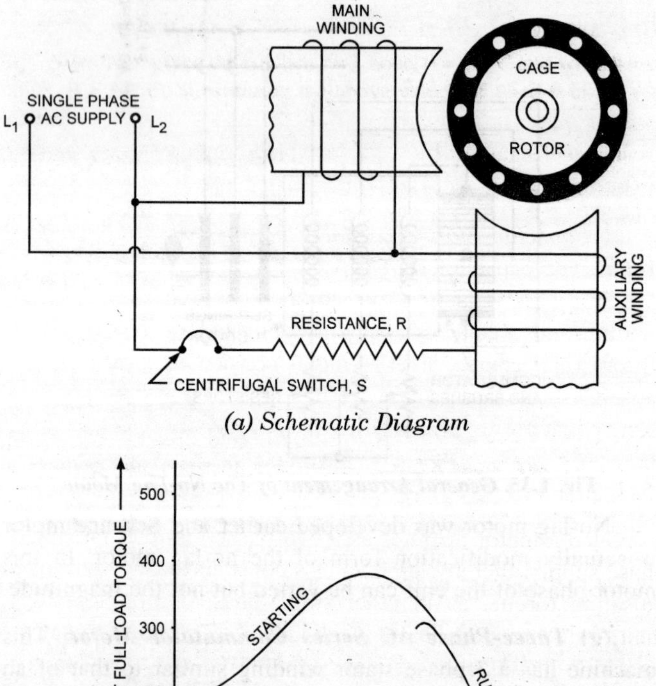

(a) Schematic Diagram

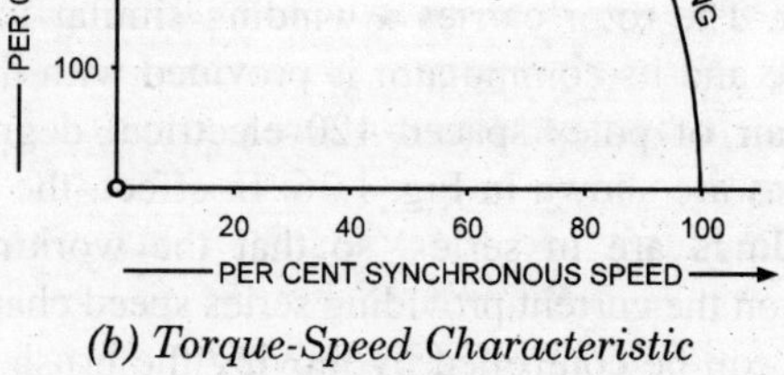

(b) Torque-Speed Characteristic

Fig. 1.38

The direction of rotation of a capacitor-start induction motor may be reversed by inter-changing the connections to the supply of either the main or starting winding. The noteworthy point is that reversal of a split-phase induction motor takes place only when it is at rest and the starting switch is in its normally closed de-energized position.

2. *Capacitor-Start Single-Phase Induction Motor.* Fundamentally this motor is very similar to the resistance-start split-phase motor except that the starting winding has a few more turns and consists of a heavier wire than the starting winding of a resistance-start motor; there is also a large electrolytic capacitor in place of resistor. The capacitor-start motor develops a greater starting torque than does an equally rated resistance-start motor. The line current of a resistance-start motor was found to be two-third higher than line current of a corresponding capacitor-start motor, while the starting torque of the capacitor-start motor was twice of that of resistance-start motor.

It is to be noteworthy that the capacitor-start motor differs from the resistance-start motor only during the starting period, for it develops more starting torque (about 3.5-4.5 times full-load torque) with a lower inrush current. After the machine attains normal speed and the starting winding is cut out, its operating characteristics are identical to those of a corresponding resistance-start motor.

Capacitor-start motors are manufactured in rating ranging from $\frac{1}{10}$ kW to $\frac{3}{4}$ kW, but larger sizes are also available. They are widely used for heavy duty general purpose applications like refrigerator units, air-conditioners, compressors, jet pumps, sump pumps, centrifugal pumps, compressors, conveyors, fans, blowers, farm and home workshop tools, oil burners etc.

3. *Capacitor-Start Capacitor-Run Single-Phase Induction Motor.* The capacitor start capacitor–run motor is similar to capacitor-start, induction-run motor, except that the auxiliary winding and a capacitor remain connected in the circuit at all times. Of course, such as motor consists of two different capacitors–one small value oil impregnated paper continuous rating capacitor C_R and another much larger value electrolytic short-duty capacitor C_S, as illustrated in schematic diagram given in Fig. 1.39 (*a*).

Such a motor operates as a two-phase motor from single phase supply, thereby, producing a constant torque and not a pulsating torque, as in other single-phase motors. Besides their ability to start heavy loads they are extremely quiet in operation, have better efficiency (55-65%) and power factor (0.8-0.95) when loaded and develop up to 25% greater overload capacities. They are indeed splendid machines where the load requirements are severe as in case of compressors, refrigerators and fire stokers, pumps, conveyors etc. The disadvantage of such machines is only high cost. These motors are often employed requiring a quiet operating motor such as in hospitals, studios etc.

In recent years, home air-conditioners have become extremely popular. In many cases, the electrical system in the home may not have the capacity to handle the electrical load of the air-conditioner(s). For this reason, capacitor-start capacitor-run motor have found more wide spread use in the compressors. Both the starting and running currents are low—full-load current may be $\frac{3}{4}$th of that for an equivalent capacitor-start motor.

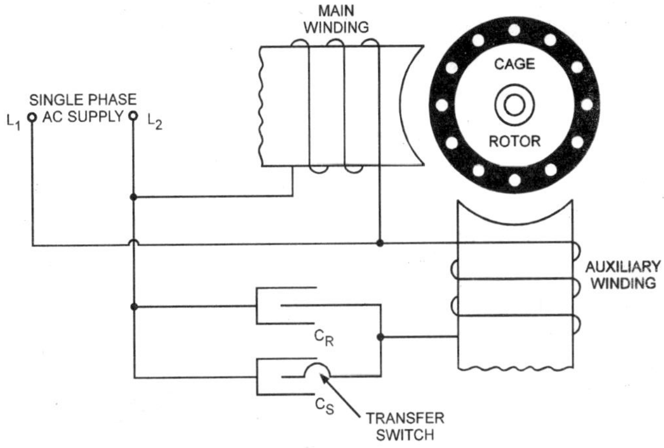

(*a*) *Schematic Diagram*

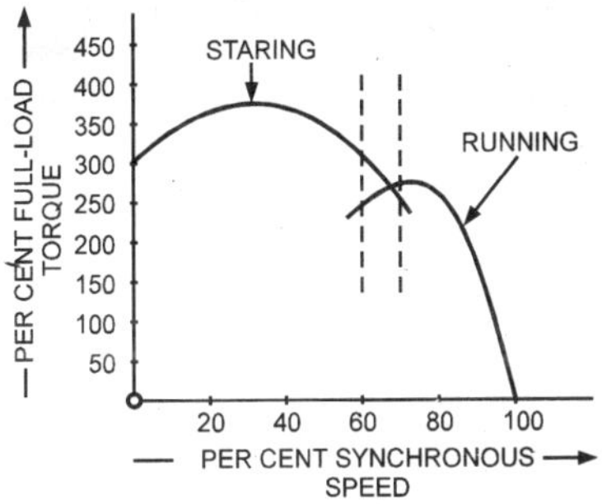

(*b*) *Torque-Speed Characteristic*

Fig. 1.39. ***Capacitor-Start, Capacitor-Run Single-Phase Induction Motor***

Such motors have become increasingly popular and are now manufactured in the larger sized single-phase motors where previously only repulsion-start motors were available. A typical torque-speed characteristic is shown in Fig. 1.39 (*b*). The direction of rotation of motor may be reversed by interchanging the connections to the supply of either the main or auxiliary winding.

4. *Permanent Capacitor Single Phase Induction Motor.* Permanent capacitor single-phase induction motor, also sometimes called the *single-value capacitor–run motor*, has two stator windings placed mutually 90 electrical degrees apart. The main or running winding is connected directly across the single phase ac supply. A capacitor in series with the auxiliary winding is also connected across the supply lines.

The auxiliary winding is always in the motor circuit, and therefore, the operation of the motor when loaded resembles more closely to that of a 2-phase motor. One advantage of such operation is that the rotating field developed by the two stator windings is a more nearly uniform, and therefore, the motor is quieter in operation. The other advantages of such a motor are (*i*) increased pull-out torque (*ii*) a higher power factor (0.8-0.95), (*iii*) higher operation efficiency (55-65 per cent) and (*iv*) smaller full-load line current.

The permanent capacitor induction motor is more expensive than the equivalent resistance-start or capacitor—start induction motor mainly owing to the increased cost of auxiliary winding which is now wound with thicker wire and consists of as many turns as the main winding.

A unique feature of such a motor is that it can be very easily reversed by an external switch, provided the two windings are identical. Because of absence of starting switch it is possible to reverse the leads of either winding while the motor is running.

Such motors are widely used in table and ceiling fans, blowers, oil burners, room coolers, portable tools, other domestic and commercial electrical appliances, where low starting torque is required.

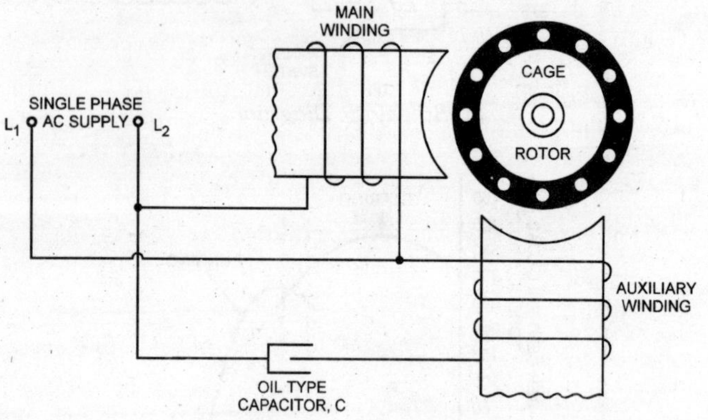

(*a*) *Schematic Diagram*

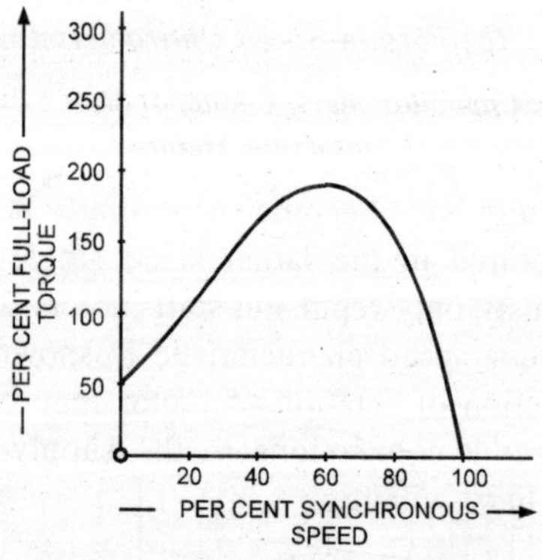

(*b*) *Torque-Speed Characteristic*

Fig. 1.40. ***Permanent Capacitor Single-Phase Induction Motor***

Because of the simplicity of reversal of motor, such motors are often used to operate devices that must be moved back and forth frequently such as induction regulators, rheostats, furnace controls, valves and arc-welding controls.

Because of the absence of starting switch, it is possible to run a permanent (or single value) capacitor-run motor over a wide range of speeds, and such a motor has many applications where adjustable speed is required. Such a motor can be arranged for an adjustable speed by the use of a tapped winding or an auto-transformer regulator. It is known that the torque developed in an induction motor is proportional to the square of the applied voltage. For a given load, therefore, if the applied voltage is reduced, the motor will operate with increased slip to develop the required torque. Thus the speed can be controlled by varying the voltage applied to the motor.

5. *Shaded-Pole Motor.* As illustrated schematically in Fig. 1.41 (*a*) the shaded-pole motor usually has salient poles with one portion of each pole surrounded by a short-circuited turn of copper called a *shading coil.* Induced currents in the shading coil cause the flux in the shaded portion of the pole to lag the flux in the other portion. The result is like a rotating field moving in the direction from the unshaded portion to the shaded portion of the pole, and a low starting torque is developed. A typical torque-speed characteristic is shown in Fig. 1.41 (*b*). Starting torque is very small (about 40-80% of full-load torque). The efficiency is quite low (20 to 50%). Power factor is also very poor ranging from 0.5 to 0.6. It has little overload capacity. Speed falls with the decrease in applied voltage. Speed can be varied by varying the voltage applied across its stator winding (by using an auto-transformer, tapped reactance coil or by using tapped exciting winding). Shaded-pole motors are least expensive type of small power motors.

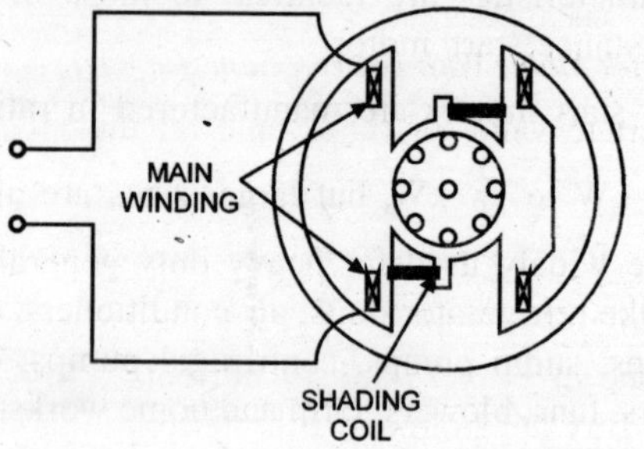

(*a*) *Schematic Diagram*

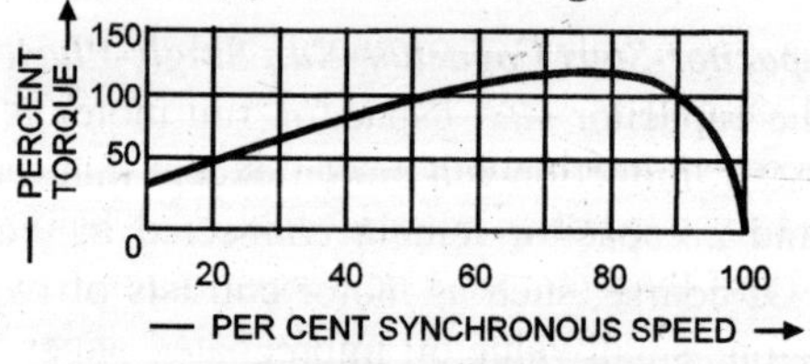

(*b*) *Torque-Speed Characteristic*

Fig. 1.41. ***Shaded Pole Motor***

Because of their poor starting torque, low power factor and poor efficiency, such motors are only suitable for low power applications such as for toys, small fans, electric clocks, hair dryers, time phonographs, slide projectors, humidifiers, small business machines (photocopying machines, vending machines, advertising displays etc.) and other similar applications. They are manufactured in small sizes ranging from below 5 W to about 100–150 W. They are available with built-in gear reducers to give almost any speed (as low as below 1 rpm.)

(b) AC Series Motor. DC shunt or series motor rotates in the same direction regardless polarity of supply, *i.e.,* if the line terminals be reversed, the motor continues to rotate in a direction it was rotating before the line terminals were reversed. From this it seems that any dc motor would operate satisfactorily when connected to single phase ac supply.

However, it is not true. Some modifications are necessary in a dc series motor that is to operate satisfactorily on single phase ac supply. So the construction of an ac series motor is very similar to a dc series motor except that some modifications (such as whole magnetic circuit laminated, series field with as few turns as possible, large number of armature conductors, use of high-resistance carbon brushes, numerous poles with lesser flux per pole, very short air gap etc.) are incorporated. The machine is provided with compensating winding and interpoles to improve commutation. A schematic diagram of a single-phase series motor with interpole and compensating windings is given in Fig. 1.42.

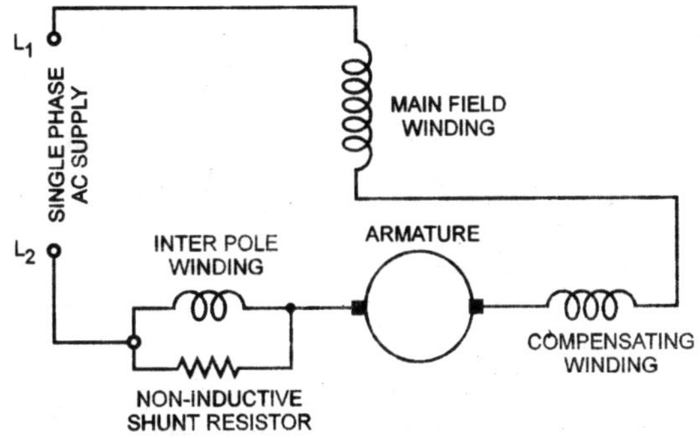

Fig. 1.42. ***Series Motor With Interpole and Compensating Winding***

The average value of the torque on the motor shaft is given as

$$T_{av} = \frac{1}{2\pi\sqrt{2}} Z \frac{P}{A} \phi_{max} I \cos \theta \qquad ...(1.43)$$

where I is the effective value of current, ϕ_{max} is the peak value of the flux per pole and θ is the phase angle between phasors ϕ and I.

For a given value of torque T and applied voltage, the armature current is same but voltage drop in case of an ac series motor is much more than that in case of a dc series motor and so speed of an ac series motor for a given developed torque is less than that of a dc series motor, as illustrated in Fig. 1.43.

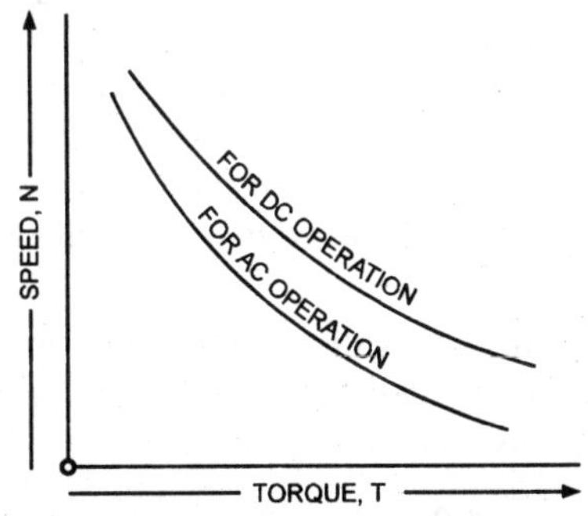

Fig. 1.43. ***Speed-Torque Characteristics of Series Motors***

The single-phase ac series motor has practically the same operating characteristics as the dc series motor. The torque, or tractive effort, varies nearly as the square of the current and the speed varies inversely as the current. This is illustrated in Fig. 1.44.

However, in case of an ac series motor (*i*) power factor is very low at starting and on overloads on account of high inductive nature of the series field and armature circuits (*ii*) efficiency is not as good as in a corresponding dc machine due to eddy current losses and effects of power factor and (*iii*) starting torque is low due to poor power factor at starting.

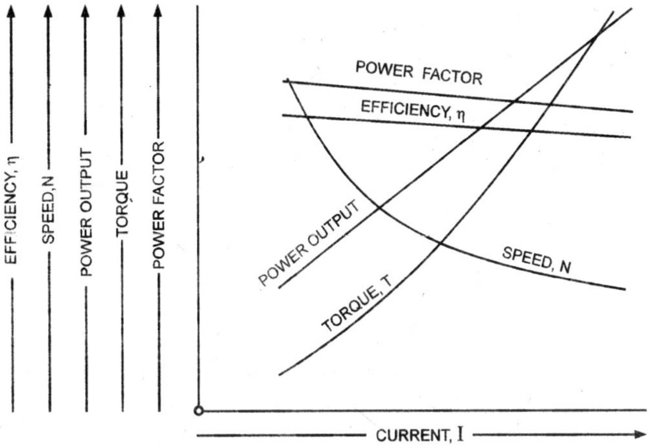

Fig. 1.44. ***Performance Characteristics of AC Series Motors***

For a given kilowatt rating ac series motor is 1.5-2 times in size and weight of the corresponding dc series motor. The construction cost of an ac series motor is much more than that of a dc series motor.

The speed of an ac series motor may be controlled efficiently by taps on a transformer, which is not possible in case of a dc series motor.

The torque-speed characteristic of the single phase series motor is similar to that of the dc series motor *i.e.*, high starting torque and decrease in speed with increase in load making it to have self-relieving property from heavy excessive load, so such a machine is particularly useful for traction services.

(c) Universal Motor. A universal motor is a specifically designed series wound motor, that operates at approximately

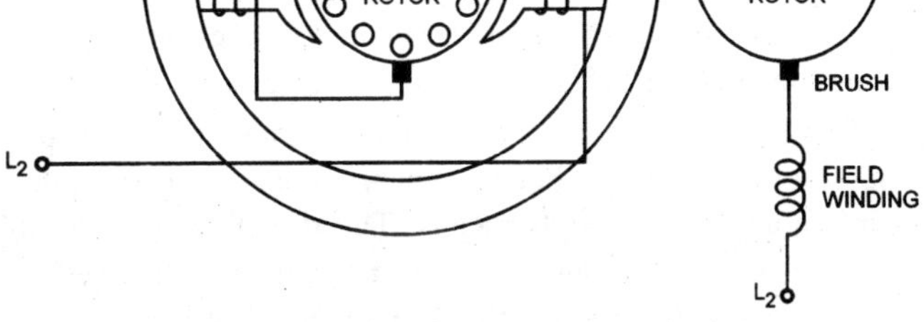

Fig. 1.45. ***Universal Motor***

the same speed and output on either dc or ac of approximately same voltage. Because of the difficulty in obtaining like performance on dc and ac at low speeds, most universal motors are designed to operate at speeds exceeding 3,500 rpm. Motors operating at speeds 8,000 to 10,000 rpm are common. Universal motor is constructed with few series field turns, laminated armature and field circuits, low-reluctance magnetic path, increased armature conductors and commutator segments and using low flux densities so as to minimize the adverse effects caused by high field reactances, eddy current and hysteresis losses.

Universal motors may be either compensated (distributed field) or uncompensated (concentrated field) type, the latter type being used for higher speeds and smaller output ratings (usually not exceeding 200 W) only.

The speed-torque characteristic of a universal motor is quite similar to that of a series wound dc motor, *i.e.,* high starting torque and high no-load speed. Universal motors beings high speed motors, are smaller in size and lighter in weight as compared to other motors of the same output. Full-load pf is high (approximately 0.9) but it is poor at start and on overloads. The direction of rotation of any series motor can be reversed by reversing the direction of flow of current in either the field or armature circuit (but not through both). The speed of universal motor, for any given load, can be changed by changing either flux or applied voltage or both.

Very small power output rating universal motors, which usually do not exceed 5 or 10 watts are employed in such equipment as sewing machines, fans, portable hand tools, hair dryers, motion-picture projectors, and electric shavers. The higher rating (5-500 W) universal motors are used in vacuum cleaners, electric typewriters, food mixers and blenders, motion-picture projectors, cameras, adding machines, and calculating machines.

The small series motors are frequently furnished as motor parts, *i.e.,* consisting of bare stators and rotors (with shaft) but without bearings or feet. They can then be compactly "built in" to the power using devices.

(d) Repulsion Motors. The characteristics of repulsion motors are similar to those of a series motor, *i.e.,* high starting torque and high light-load speed. Its construction is also similar to that of a series motor except that armature is short circuited on itself instead of being connected in series with the stator. Simplified schematic diagram is shown in Fig. 1.46. A repulsion motor develops a torque in the direction in which the brushes are shifted from the field axis.

The torque developed by a repulsion motor should be maximum, theoretically, when the space angle between the pole axis and brush axis is 45°, but in practice the angle of inclination is about 15-25 electrical degrees.

The repulsion motor has better commutation than the series motor at speeds below synchronous speed and poor commutation at very high speeds. The direction of rotation of a repulsion motor can be reversed by shifting the brushes round the commutator, on the other side of the field axis. Speed control can be affected by varying the applied voltage to motor or by mounting the brushes on a rocker which can be rotated by a lever handle mounted on the motor end-shield.

The repulsion motor has high starting torque (about 3-5 times full-load torque) and moderate starting current (about 3-4 times full-load current) but poor speed regulation. Shifting the brushes during operation gives a wide range of speed control, as high as 6:1 ratio, and yet provides a continuous variation. The upper speed is not limited by frequency. The motor is a reversing type, and the direction may be changed during rotation.

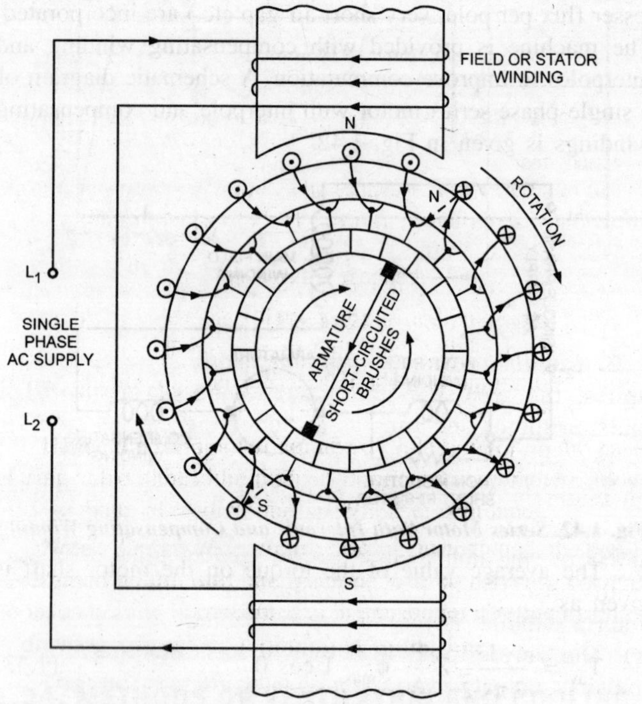

Fig. 1.46. ***Simplified Schematic Diagram of a Repulsion Motor***

The drawbacks of repulsion motors are (*i*) speed variations with the variations in loads-dangerously high at no load (*ii*) low power factor, except at high speeds, (*iii*) tendency to spark at the brushes-sparking at the brushes is negligible at the rated speed which usually occurs near synchronous speed (*iv*) higher cost and (*v*) more attention and maintenance is required.

The repulsion motor has never enjoyed popularity. The motor is used where sturdy motor with large starting torque and adjustable speed is required. Most common use of this type of motor is in the coil winders in which the operator adjusts the speed by shifting the brushes; the motor is equipped with a special lever mechanism, that shifts the brushes when a foot treadle is pressed.

The ratings in which the repulsion motors are built is limited due to commutation problems. The usual rating of the repulsion motor does not exceed 5 kW.

1. *Compensated Repulsion Motor.* It is a modified form of the repulsion motor discussed above. It is provided with an additional winding, called the *compensating winding*, to improve power factor and provide better speed regulation. This winding is much smaller than the stator winding and is usually wound in the inner slots of each main pole and is connected in series with rotor through an additional set of brushes placed midway between the usual short-circuited brushes, as illustrated in Fig. 1.47.

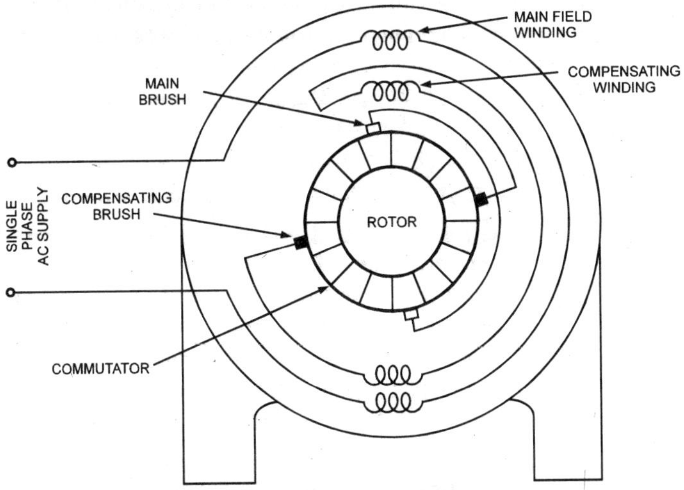

Fig. 1.47. ***Compensated Repulsion Motor***

2. *Repulsion-Start Induction Motor.* As the name implies, the repulsion-start induction motor starts as a repulsion motor and runs as an induction motor.

The general construction of a repulsion-start induction motor is quite similar to that of a repulsion motor (*i.e.*, a laminated stator core complete with one winding similar to the main winding of a split-phase motor and a rotor, also called the *armature,* consisting of slots in which a winding is placed and connected to a commutator). The only difference is that in addition to the basic repulsion-motor construction it is equipped with a centrifugal device which operates at about 75-80 per cent of synchronous speed and short circuits all of the commutator segments.

The repulsion-start induction motor combines the desirable starting characteristics of the repulsion motor (*i.e.*, high starting torque with moderate starting current) with operating characteristic (constant speed running) of the induction motor. A typical performance characteristic of a repulsion-start induction motor is given in Fig. 1.48. The starting torque is 3 to 6 times of full-load torque, and the starting current is approximately 3-4 times of full-load current.

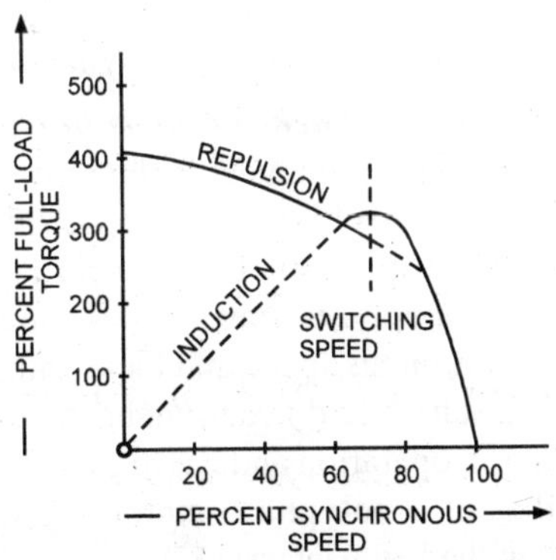

Fig. 1.48. ***Torque-Speed Characteristic For Repulsion-Start Induction Motor***

The direction of rotation of such a motor may be reversed by shifting the brushes, but the brush rigging is not readily accessible, and the method is poorer than with any of the capacitor type motors.

The efficiency and the maximum operating torque of the repulsion-start induction motor are usually less than those of a squirrel cage induction motor of comparative size *i.e.,* a repulsion-start induction motor must be larger than a squirrel cage motor of the same rating to give the same performance.

As compared to capacitor motors, these motors (*i*) need more maintenance because of commutator, brushes and centrifugal switch, (*ii*) are relatively more expensive, (*iii*) cause radio interference during starting, and (*iv*) are noiser, especially when the brushes continue to ride on commutator.

However, despite all the drawbacks, these motors are still used in integral kilowatt ratings because they (*i*) develop higher starting torque with low starting current and (*ii*) can withstand longer starting periods than capacitor type motors.

Such motors are suitable for commercial refrigerators, compressors, pumps, hoists and other constant speed drives, particularly those which have a high inertia and prolonged starting period. The usual rating is from $\frac{1}{3}$ kW to 12 kW but for special applications, ratings as high as 30 kW are available.

3. *Repulsion-Induction Motor.* The repulsion-induction motor has a single phase stator winding, as a respulsion-start

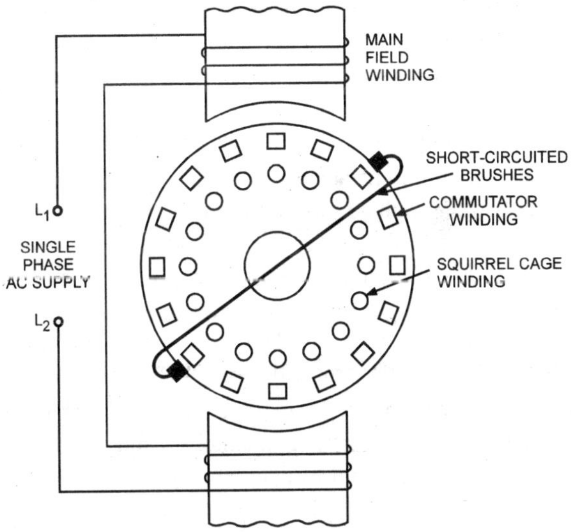

(a) Schematic Diagram

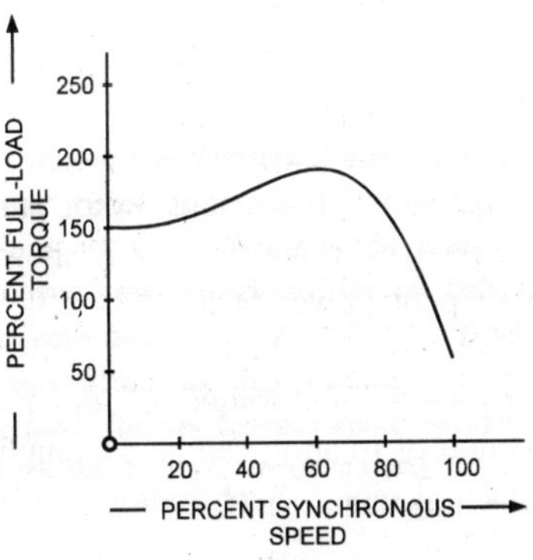

(b)

Fig. 1.49. ***Torque-Speed Characteristic of a Repulsion-Induction Motor***

induction motor has, but it has two separate and independent windings on rotor in common slots. The inner winding is a squirrel cage winding with rotor bars permanently short-circuited while outer winding is a repulsion or commutator winding similar to a dc armature winding. When the motor starts, the squirrel cage winding, due to its high reactance, has no effect and the motor starts as a repulsion motor. As the motor speeds up, the current shifts from the outer winding to the inner winding owing to decreasing reactance of the inner winding with the increasing speed and the squirrel cage winding comes into action. At normal speed both of the windings develop torque, and the output of the motor is the combined output of both of the rotor windings.

The starting torque is about 2.5-3 times the full-load torque and its starting current is about 3-4 times the full-load current. The commutation of the motor is good at all speeds. No-load speed is above synchronous speed but normal speed is somewhat below synchronous one. No-load current is rather high, sometimes greater than the full-load current. The motor besides being costlier, has brushes always on the commutator involving sparking and maintenance problems. This motor is particularly suitable where the load can be removed entirely by the declutching or by loose pulley.

Such a motor is used for applications requiring a high starting torque with essentially a constant running speed. The common ratings are 1/6 to 4 kW although in large ratings are also made. Common applications are household refrigerators, garage air pumps, gasoline pumps, compressors etc.

(e) Synchronous Motors. There are many timing-device applications where small motors having exact constant speed characteristics will be very advantageous. Very small motors with constant speed characteristics have been developed. They operate from a single phase supply. Because of their exact constant speed characteristics, they are called the *single phase synchronous motors.* They do not require dc source of supply for excitation. The principal applications of such single phase synchronous motors are for driving of electric clocks, phonographs, record-player turntables, tape and other timing devices.

The most commonly used types of single-phase synchronous motors are the reluctance motors and the hysteresis motors. The efficiency and torque developing ability of these motors is low. The output of most of the commercial motors available is only a few watts. It is practical to design hysteresis motors up to approximately 125 watts.

Reluctance motor is a split-phase induction motor with properly designed salient shaped poles. It consists of a stator carrying both the main and auxiliary windings for developing a synchronously rotating magnetic field. Rotor punching for a 4-pole reluctance type synchronous motor is shown in Fig. 1.50. Such motors have poor power factor, low efficiency and poor torque. They cannot accelerate high inertia loads to synchronous speed. The pull-in and pull-out torques of such motors are small. Reversal of direction of rotation can be accomplished as in any single phase induction motor. Such motors are widely used for absolute constant speed applications, such as in timing devices, signalling devices, recording instruments, phonograph turntables, control apparatus etc.

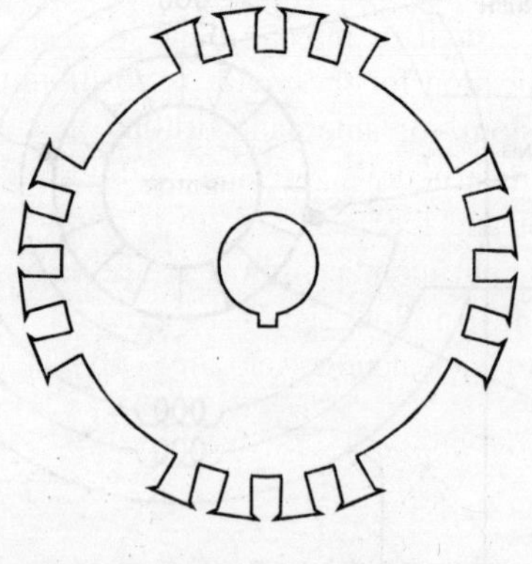

Fig. 1.50. ***Reluctance Motor Rotor***

Hysteresis motor is a synchronous motor with uniform air gap but with no dc excitation. Its operation depends upon the effect of hyteresis. A 2-pole shaded-pole type hysteresis motor employed for driving ordinary clocks is shown in Fig. 1.51. Because of noiseless operation and ability to drive high-inertia loads, hysteresis motors are particularly well suited to drive timing devices, electric clocks, tape decks, turntables and other precision audio-equipment. Commercial motors, being 2-pole motors, run at 3,000 rpm and so for driving an electric clock and other indicating devices, gear train is connected to the motor shaft for reducing the speed. By changing number of stator poles through pole-changing connections, a set of synchronous speeds can be obtained for the motor.

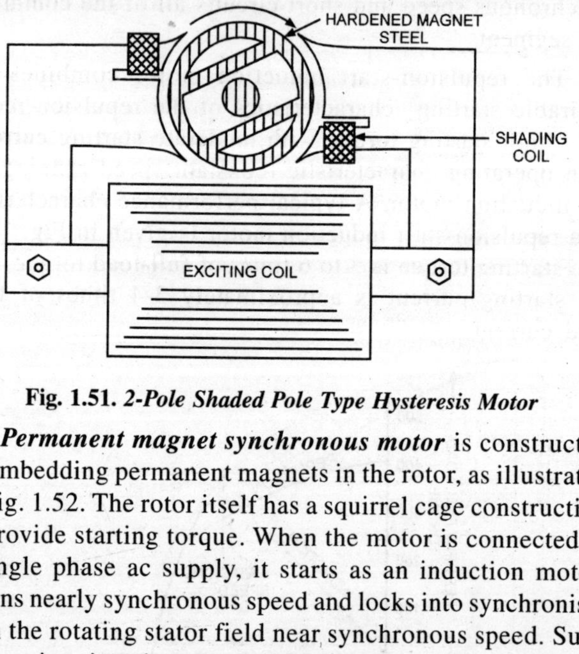

Fig. 1.51. ***2-Pole Shaded Pole Type Hysteresis Motor***

Permanent magnet synchronous motor is constructed by embedding permanent magnets in the rotor, as illustrated in Fig. 1.52. The rotor itself has a squirrel cage construction to provide starting torque. When the motor is connected to a single phase ac supply, it starts as an induction motor, attains nearly synchronous speed and locks into synchronism with the rotating stator field near synchronous speed. Such a motor is quieter in operation and has high pf and efficiency–approaching that of a polyphase motor. It is thus finding more use, even in the low – integral kW ratings (0.5–1.5 kW).

(f) Stepper Motor. The stepper motor* is a form of synchronous motor which is designed to rotate through a

* A stepper motor is an incremental motion machine *i.e.,* the motor which turns in discrete movement (called the steps) is known as the stepper motor. It does not rotate continuously, as a conventional motor does.

specific number of degrees for each electrical pulse received by its control unit. Typical steps are 2, 2.5, 5, 7.5, and 15° per pulse. The stepper motor is used in digital control systems, where the motor receives open-loop commands as a train of pulses to turn a shaft or move a plate by a specific distance. A typical application for the motor is positioning a worktable in two dimensions for automatic drilling in accordance with hole–location instructions on tape. With a stepper motor a position sensor and feedback system is normally not required to make the output member follow input instructions. Stepper motors are built to follow signals as rapid as 1,200 pulses per second and with equivalent power ratings up to several kilowatts.

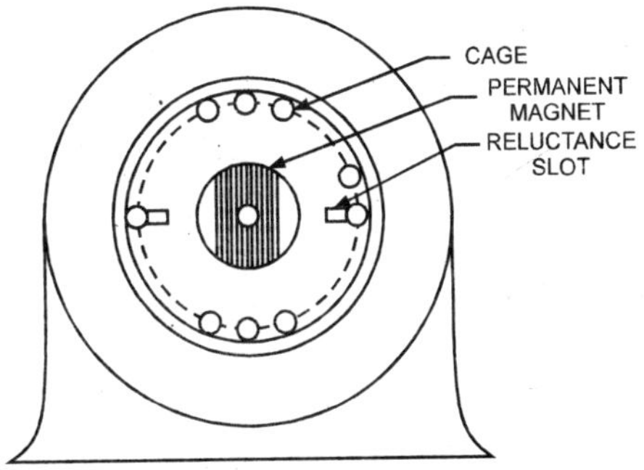

Fig. 1.52. ***Permanent Magnet Synchronous Motor***

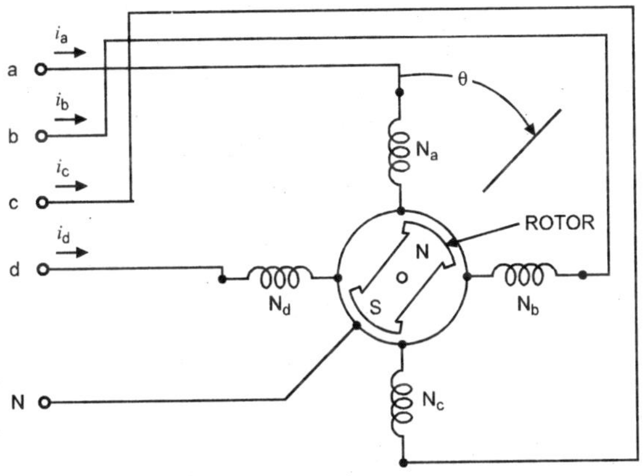

Fig. 1.53. ***Elementary Diagram of a 4-Phase Stepper Motor***

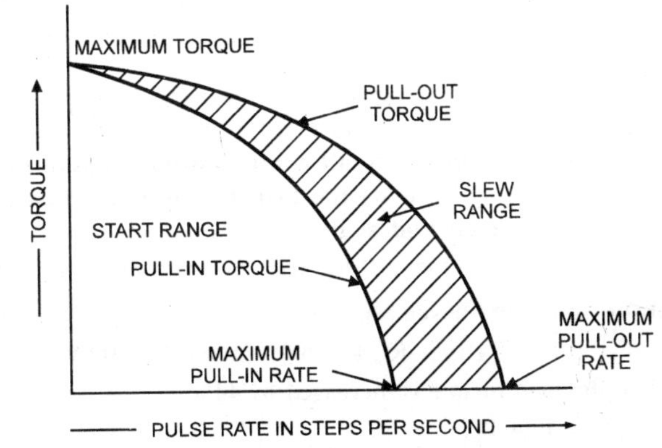

Fig. 1.54. ***Torque Versus Pulse-Rate Characteristics of a Stepper Motor***

Stepper motors are usually designed with a multi-pole, multiphase stator winding that is not unlike the windings of conventional machines. They typically use 3- and 4-phase windings, the number of poles being determined by the desired angular change per input pulse. The rotors are either of the variable reluctance or permanent magnet type. Stepper motors operate with an external drive logic circuit; as a train of pulses is applied to the input of the drive circuit, the circuit delivers appropriate currents to the stator windings of the motor to make the axis of the air-gap field step around in coincidence with the input pulses. Depending upon the pulse rate and the load torque, including inertia effects, the rotor follows the axis of the air-gap magnetic field by virtue of the reluctance torque and/or the permanent-magnet torque.

The elementary operation of a 4-pole stepper motor with a 2-pole, rotor is illustrated in the sequence of Fig. 1.53. The rotor assumes the angles $\theta = 0, 45°, 90°$ ... as the windings are excited in the sequence N_a, $N_a + N_b$, N_b, ... The stepper motor shown in Fig. 1.53 may also be employed for 90° steps by exciting the coils singly. In the latter case, only the permanent-magnet rotor can be employed.

The characteristics of a stepper motor are frequently presented as the torque versus stepping rate of the pulses applied to the drive unit, as illustrated in Fig. 1.54. As the stepping rate is increased, the motor can provide less torque because the rotor has less time to drive the load from one position to the next as the stator-winding-current pattern is shifted.

The stepper motor is essentially a position control device and has the following advantages over a conventional machine :

1. The angular displacement can be precisely controlled without any feedback arrangement.
2. It can be readily interfaced with microprocessor/computer based controller.

Stepper motors have a wide range of applications—paper feed motors in typewriters and teleprinters, positioning of print heads, pens in X-Y graphical plotters, recording heads in computer disk drives and in positioning of worktables and tools in numerically controlled machining equipment.

The stepper motor is also employed to perform many other functions such as metering, mixing, cutting, blending, stirring etc. in many commercial, military and medical applications, usually along, with microprocessor and controlled switches.

Example 1.5. A 220 V, 1,000 rpm, 100 A separately excited dc motor has an armature resistance of 0.1 Ω. The motor is driving a constant torque load equal to the rated torque. Calculate the motor speed if the voltage drops to 200 V.

[U.P. Technical Univ. Electric Drives 2004-05]

Solution : At rated load and rated supply voltage

Supply voltage, $V = 220$ V

Current, $I = 100$ A

Back emf, $E_b = V - IR_a = 220 - 100 \times 0.1 = 210$ V

At reduced voltage, $V' = I = 100$ A $\because$ Load torque $T \propto \phi I_a$ is constant and also ϕ is constant

Back emf, $E'_b = V' - I'R_a = 200 - 100 \times 0.1 = 190$ V

$$\text{Motor speed, } N' = N \times \frac{E'_b}{E_b} = 1{,}000 \times \frac{190}{210} = 905 \text{ rpm}$$ **Ans.**

Example 1.6. A 5 kW, 230 V, shunt motor has an armature resistance of 0.5 Ω and a field resistance of 230 Ω. At no load

the motor runs at a speed of 1,000 rpm and draws a current of 3 A. At full load and rated voltage, the current drawn is 23 A and the armature reaction causes a drop of 2% in flux. Determine (*i*) full-load speed (*ii*) full-load torque.

[*U.P.S.C. I.E.S. Electrical Engineering-II, 1998*]

Solution : Shunt field current, $I_{sh} = \frac{V}{R_{sh}} = \frac{230}{230} = 1$ A

At No Load Line current, $I_{L0} = 3$ A

Armature current, $I_{a0} = I_{L0} - I_{sh} = 3 - 1 = 2$ A

No-load back emf, $E_{b0} = V - I_{a0} R_a$

$= 230 - 2 \times 0.5 = 229$ V

No-load speed, $N_0 = 1{,}000$ rpm

At Full Load

Armature current, $I_{af} = I_{Lf} - I_{sh} = 23 - 1 = 22$ A

Back emf, $E_{bf} = V - I_{af} R_a$

$= 230 - 22 \times 0.5 = 219$ V

Flux $\Phi_f = 0.98\ \Phi_o$

$\because$ Reduction in flux at full load is 2%

(*i*) Speed, $N_f = N_0 \times \frac{E_{bf}}{E_{b0}} \times \frac{\Phi_0}{\Phi_f}$

$= 1{,}000 \times \frac{219}{229} \times \frac{\Phi_0}{0.98\ \Phi_0}$

$= 976$ rpm **Ans.**

(*ii*) Full-load torque, $T_f = \frac{9.55 \times E_{bf} \times I_{af}}{N_f}$

$= \frac{9.55 \times 219 \times 22}{976}$

$= 47.15$ Nm **Ans.**

Example 1.7. A 440 V dc shunt motor has a no-load ($I_0 = 0$) speed of 2,000 rpm. It is running at 1,000 rpm at full-load torque, reduced armature voltage and full field. If load torque is reduced to 50% of rated value with armature voltage and field voltage held constant at previous values, the speed increases to 1,050 rpm. Find the armature voltage drop at full load. Neglect effect of armature reaction. [*GATE 1994*]

Solution : Supply voltage, V = 440 V

No-load speed, $N_0 = 2{,}000$ rpm

No-load back emf, $E_{b0} = V = 440$ V

$\because$ armature current on no load = 0

Full-load speed, $N_f = 1{,}000$ rpm

Half-full load speed, N' = 1,050 rpm

Full-load back emf, $E_{bf} = E_{b0} \times \frac{N_f}{N_0} = 440 \times \frac{1{,}000}{2{,}000} = 220$ V

$\because$ Flux is unchanged

Back emf at half full load, $E_b = E_{b0} \times \frac{N'}{N_0} = 440 \times \frac{1{,}050}{2.000} = 231$ V

Let the reduced armature voltage be V_1 volts and armature voltage drop at full load be v volts.

Since flux remains unchanged, when load torque is reduced to 50%, armature current would reduce to 50%, torque T being proportional to the product of flux per pole and armature current. Since voltage drop in armature is proportional to armature current, armature voltage drop will also reduce to 50% *i.e.* become $\frac{v}{2}$ when load torque is reduced to 50%.

So $E_{bf} = V - v$

and $E'_b = V - \frac{1}{2} v$

or $220 = V - v$ ...(*i*)

and $231 = V - \frac{1}{2} v$ ...(*ii*)

Solving Eqs. (*i*) and (*ii*), we have

Armature voltage drop at full load, $v = 22$ V **Ans.**

Example 1.8. A 230 V dc shunt motor with constant field drives a load whose torque is proportional to its speed. When running at 750 rpm it takes 30 A. Find the speed at which it will run if a 10 Ω resistance is connected in series with its armature. The resistance of armature may be neglected.

[*A.M.I.E. Sec. B. Electrical Drives and Their Control Winter 1993*]

Solution : Armature current, I_a = Line current, $I_L = 30$ A

neglecting shunt field current

Speed, $N_1 = 750$ rpm

Let the speed be reduced to N_2 rpm where $N_2 = KN_1$ after inserting additional resistance of 10 Ω in the armature circuit.

Since with constant field excitation, torque developed is proportional to armature current *i.e.*,

$T \propto I_a$ ...(*i*)

Load torque, $T \propto N$ ...(*ii*) given

From Eqs. (*i*) and (*ii*), we have

$I_a \propto T \propto N$

or $I_{a2} = I_{a1} \times \left(\frac{N_2}{N_1}\right) = K I_{a_1} = 30$ K ...(*iii*)

Since $N \propto [V - I_a(R_a + R)]$ flux φ remaining the same

or $\frac{N_2}{N_1} = \frac{V - I_{a2} R}{V}$ $\because$ armature resistance is negligible

Substituting $\frac{N_2}{N_1} = K$, $I_{a2} = 30$ K from Eq. (*iii*), and R = 10 Ω in above equation, we have

$K = \frac{230 - 30K \times 10}{230}$

or $K = \frac{230}{230 + 300} = 0.434$

New speed, $N_2 = KN_1 = 0.434 \times 750 = 325$ rpm. **Ans.**

Example 1.9. A 200 V dc series motor runs at 700 rpm when operating at its full-load current of 20 A. The motor resistance is 0.5 Ω and the magnetic circuit can be assumed unsaturated. What will be the speed if (*i*) the load torque is increased by 44% (*ii*) the motor current is 10 A ?

[*A.M.I.E. Sec. B. Electric Drives and Their Control Summer 1994*]

Solution : Back emf developed,

$E_{b1} = V - I_1 (R_a + R_{se}) = 200 - 20 \times 0.5 = 190$ V

(*i*) When the load torque is increased by 44%

i.e., $T_2 = 1.44\ T_1$

or $\Phi_2 I_2 = 1.44\ \Phi_1 I_1$ $\because$ $T \propto \Phi I_a$

or $I_2^2 = 1.44\ I_1^2$ $\because$ $\Phi \propto I$

or $I_2 = \sqrt{1.44}\ .\ I_1 = 1.2 \times 20 = 24$ A

Back emf developed,

$E_{b2} = V - I_2 (R_a + R_{se}) = 200 - 24 \times 0.5 = 188$ V

Since $E_b \propto \Phi N$

$$\text{Speed } N_2 = \frac{E_{b2}}{E_{b1}} \times \frac{\Phi_1}{\Phi_2} \times N_1 = \frac{188}{190} \times \frac{20}{24} \times 700$$

$= 577.2$ rpm **Ans.**

(*ii*) **When motor current is 10 A**

Back emf developed,

$$E_{b3} = V - I_3(R_a + R_{se}) = 200 - 10 \times 0.5 = 195 \text{ V}$$

$$\text{Speed, } N_3 = \frac{E_{b3}}{E_{b1}} \times \frac{\Phi_1}{\Phi_3} \times N_1 = \frac{195}{190} \times \frac{20}{10} \times 700$$

$= 1,437$ rpm **Ans.**

Example 1.10. A 230-V dc series motor has an armature resistance of 0.08 Ω and field resistance of 0.05 Ω. The magnetisation characteristic of the machine at 700 rpm is as follows:

Field current	30	60	90	120	150 A
EMF	70	125	180	210	230 V

The motor is run on a 230 V supply. An additional resistance of 1.5 Ω is connected in series with the armature. Determine the torque and speed when the motor draws a current of 90 A. [*A.M.I.E. Sec. B. Electric Drives and Their Control Winter 1994*]

Solution : Total resistance between motor terminals,

$$R_m = R_a + R_{se} = 0.08 + 0.05 = 0.13 \ \Omega$$

Field current, I_f = Line current $I_1 = 90$ A

Additional resistance inserted in armature circuit, R = 1.5 Ω

Back emf developed,

$$E_{b_1} = V - I_1 (R_m + R) = 230 - 90 (0.13 + 1.5) = 83.3 \text{ V}$$

Let speed be N_1

From given datas for magnetisation curve when field current is 90 A

EMF developed, $E_2 = 180$ V

Speed, $N_2 = 700$ rpm

$$\text{Since } N \ \alpha \ \frac{E_b}{\Phi}$$

$$\therefore \frac{N_1}{N_2} = \frac{E_{b1}}{E_2} \times \frac{\Phi_2}{\Phi_1} = \frac{E_{b1}}{E_2}$$

$\because \Phi_2 = \Phi_1$, field current being the same

$$\text{or } N_1 = N_2 \times \frac{E_{b1}}{E_2} = 700 \times \frac{83.3}{180} = 324 \text{ rpm} \quad \textbf{Ans.}$$

$$\text{Torque developed, } T = \frac{9.55 \times E_b I_a}{N} = \frac{9.55 \times 83.3 \times 90}{324}$$

$= 221$ Nm **Ans.**

Example 1.11. A dc series motor has been connected as shown in the figure below, for obtaining different speed-torque characteristics.

(*i*) **Prove that the motor torque is maximum when the speed of the motor is given by**

$$\omega = \frac{RR_1 - RR_2 - R_1R_2}{K_fR} \quad \text{where } K_f = \frac{E_b}{I_f\omega} \text{ is the back}$$

emf constant.

(*ii*) **Derive the expression for the maximum torque in terms of voltage, resistances and the torque constant.**

[*A.M.I.E. Sec. B. Electric Drives and Their Control Summer 1996*]

Solution : From circuit diagram shown in Fig. 1.55, we have

$$V = E_b + RI + R_2I_a = K_f\omega I_f + RI + R_2I_a \quad ...(i)$$

where K_f is the back emf constant.

$$\text{Also } V = R_1I_f + RI \quad ...(ii)$$

Comparing Eqs. (*i*) and (*ii*), we have

$$R_1I_f = K_f\omega I_f + R_2I_a$$

$$\text{or } I_a = \frac{I_f}{R_2}(R_1 - K_f\omega) \quad ...(iii)$$

From Eq. (*ii*), we have

$$V = R_1I_f + R(I_a + I_f)$$

$$= I_f(R + R_1) + \frac{R}{R_2} I_f(R_1 - K_f\omega)$$

$$= I_f\left[R + R_1 + \frac{RR_1}{R_2} - \frac{R}{R_2}K_f\omega\right]$$

$$= \frac{I_f}{R_2}\left[RR_2 + R_1R_2 + RR_1 - RK_f\omega\right]$$

$$= \frac{I_f}{R_2}\left[A - RK_f\omega\right] \quad ...(iv)$$

$$\text{or } I_f = \frac{VR_2}{A - RK_f\omega} \quad ...(v) \quad \text{where } A = RR_1 + R_1R_2 + R_2R$$

Also we know that $T = K\phi I_a = K_tI_aI_f$ where K_t is a torque constant.

Substituting $I_a = \frac{I_f}{R_2}(R_1 - K_f\omega)$ from Eq. (*iii*), we have

$$T = K_t\frac{I_f^2}{R_2}(R_1 - K_f\omega)$$

Substituting $I_f = \frac{VR_2}{A - RK_f\omega}$ from Eq. (*v*) in above equation

we have

$$T = \frac{K_t V^2 R_2 (R_1 - K_f\omega)}{(A - RK_f\omega)^2} \quad ...(vi)$$

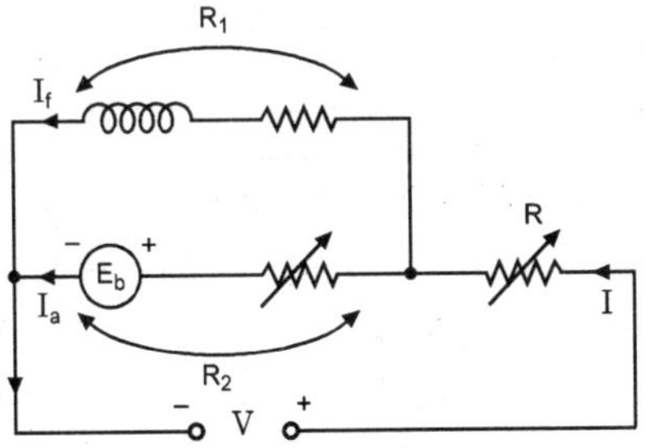

Fig. 1.55

For the torque to be maximum, the differential of T w.r.t. ω is set to zero.

$$\frac{dT}{d\omega} = R_2K_tV^2[(A - RK_f\omega)^2 \times (-K_f) - (R_1 - K_f\omega) \times 2(A - RK_f\omega)(-RK_f)] = 0$$

$$\text{or } (A - RK_f\omega)^2 \times K_f = 2RK_f(A - RK_f\omega)(R_1 - K_f\omega)$$

$$\text{or } RK_f\omega = 2RR_1 - A = 2RR_1 - [RR_1 + R_1R_2 + R_2R]$$

$$= RR_1 - R_1R_2 - R_2R$$

$$\text{or } \omega = \frac{RR_1 - RR_2 - R_1R_2}{K_fR} \quad \textbf{Proved.}$$

Substituting the value of ω in the Eq. (*vi*), we have

$$T = \frac{K_t V^2 R_2 \left[R_1 - \frac{RR_1 - RR_2 - R_1R_2}{R}\right]}{[RR_1 + RR_2 + R_1R_2 - (RR_1 - RR_2 - R_1R_2)]^2}$$

$$= \frac{K_t V^2 R_2 (RR_2 + R_1R_2)}{R[2RR_2 + 2R_1R_2]^2} = \frac{K_t V^2}{4R(R + R_1)} \quad \textbf{Ans.}$$

Example 1.12. A 400 V, 50 Hz, 4 pole, 1,350 rpm star-connected induction motor has the following parameters referred to stator. R_s = 1.5 Ω, R'_r = 4 Ω, X_s = X'_r = 4 Ω, X_m can be neglected for calculation purpose. The motor is controlled using stator voltage controller. When driving a fan load, it runs at a speed of 1,350 rpm at rated voltage. Assuming the load torque to be proportional to speed square for a fan load, calculate the motor torque and terminal voltage at 900 rpm.

[*U.P. Technical Univ. Electric Drives 2004-05*]

Solution : Synchronous speed,

$$N_s = \frac{120f}{P} = \frac{120 \times 50}{4} = 1{,}500 \text{ rpm}$$

Actual speed of rotation,

$$N = 1{,}350 \text{ rpm}$$

$$\text{Slip, } s = \frac{N_s - N}{N_s} = \frac{1{,}500 - 1{,}350}{1{,}500} = 0.1$$

Torque developed by a 3-phase induction motor is given by an expression

$$T = \frac{3V^2 R'_2 / s}{\frac{2\pi N_s}{60}\left[\left(R_1 + \frac{R'_2}{s}\right) + (X_1 + X'_2)^2\right]} \text{ N-m}$$

Refer to Eq. (1.36)

Substituting $V = \frac{400}{\sqrt{3}}$ V, R'_2 = 4 Ω, s = 0.1, N_s = 1,500 rpm, R_1 = 1.5 Ω, X_1 = 4 Ω, X'_2 = 4 Ω in above equation, we have

$$T = \frac{3 \times \left(\frac{400}{\sqrt{3}}\right)^2}{0.1 \times \frac{2\pi \times 1{,}500}{60}\left[\left(1.5 + \frac{4}{0.1}\right)^2 + (4 + 4)^2\right]}$$

$$= 5.7 \text{ N-m } \textbf{Ans.}$$

When speed, N′ = 900 rpm

$$\text{Slip, } s' = \frac{1{,}500 - 900}{1{,}500} = 0.4$$

$$\text{Torque developed, } T' = T \times \left(\frac{N'}{N}\right)^2$$

$$= 5.7 \times \left(\frac{900}{1{,}350}\right)^2 = 2.22 \text{ N-m } \textbf{Ans.} \because T \propto N^2$$

$$\text{Torque develope} \propto \frac{V^2 R'_2}{s\left(R_1 + \frac{R'_2}{s}\right)^2 + (X_1 + X'_2)^2}$$

$$\text{So } \frac{T'}{T} = \frac{(V')^2 \times s\left(R_1 + \frac{R'_2}{s}\right)^2 + (X_1 + X'_2)^2}{V^2 \times s'\left(R_1 + \frac{R'_2}{s'}\right) + (X_1 + X'_2)^2}$$

$$\text{or } V' = V \times \sqrt{\frac{T'}{T} \times \frac{s'\left(R_1 + \frac{R'_2}{s'}\right)^2 + (X_1 + X'_2)^2}{s\left(R_1 + \frac{R'_2}{s}\right)^2 + (X_1 + X'_2)^2}}$$

$$= 400 \times \sqrt{\frac{4}{9} \times \frac{0.4}{0.1} \times \frac{\left(1.5 + \frac{4}{0.4}\right)^2 + (4+4)^2}{\left(1.5 + \frac{4}{0.1}\right)^2 + (4+4)^2}}$$

$$= 400 \times \frac{2}{3} \times 2 \times \sqrt{\frac{(11.5)^2 + (8)^2}{(41.5)^2 + 8^2}} = 176.8 \text{ V } \textbf{Ans.}$$

Example 1.13. A 75 kW, 415 V, 4-pole, 3-φ, 50 Hz induction motor is delivering rated torque at 5% slip driving a pump load. The motor delivers maximum torque at 10% slip. A 3-phase ac regulator controls the rms supply voltage to effect limited speed control. Assume the torque-speed characteristics of the motor in the stable region to be linear. Calculate the rms voltage at which the motor runs at 1,350 rpm.

[*A.M.I.E. Sec. B. Electric Drives and Their Control Summer 1996*]

Solution : Full-load slip, s_f = 5% or 0.05

Full-load speed, N = $N_s(1 - s)$

$$= \frac{120f}{P}(1 - s) = \frac{120 \times 50}{4}(1 - 0.05)$$

$$= 1{,}425 \text{ rpm}$$

Slip corresponding to maximum torque, s_{maxT} = 10% or 0.1

Ratio of rotor resistance to standstill reactance,

$$\frac{R_2}{X_2} = s_{maxT} = 0.1$$

$$\text{or } X_2 = 10 R_2$$

When speed, N′ = 1,350 rpm

$$\text{Full-load slip, } s'_f = \frac{\frac{120f}{P} - N'}{\frac{120f}{P}} = \frac{\frac{120 \times 50}{4} - 1{,}350}{\frac{120 \times 50}{4}} = 0.1$$

For pump load assuming that the torque varies as the speed so torque corresponding to 1,350 rpm.

$$T' = T \times \frac{1{,}350}{1{,}425}$$

$$\text{Torque developed} = \frac{KR_2\, sE_2^2}{R_2^2 + s^2X_2^2}$$

$$\text{So } \frac{T'}{T} = \frac{s'_f (E'_2)^2 \Big/ \left[R_2^2 + (s'_f X_2)^2\right]}{s_f E_2^2 \Big/ R_2^2 + s_f^2 X_2^2}$$

$$= \frac{s'_f (V')^2 \left[R_2^2 + s_f^2 X_2^2\right]}{s_f V^2 \left[R_2^2 + (s'_f X_2)^2\right]} \quad \because E_2 \propto V, \text{ the supply}$$

$$\text{So } V' = V \times \sqrt{\frac{T'}{T} \times \frac{s_f}{s'_f} \times \frac{R_2^2 + (s'_f X_2)^2}{R^2 + (s_f X_2)^2}}$$

$$= 415 \times \sqrt{\frac{1{,}350}{1{,}425} \times \frac{0.05}{0.1} \times \frac{R_2^2 + (0.1 \times 10 R_2)^2}{R_2^2 + (0.05 \times 10 \times R_2)^2}}$$

$$= 415 \times \sqrt{\frac{1{,}350}{1{,}425} \times 0.5 \times \frac{8}{5}} = 361.3 \text{ V } \textbf{Ans.}$$

Example 1.14. An induction motor runs at a slip frequency of 2 Hz when supplied from a 3-phase, 400 V, 50 Hz supply. For the same developed torque, find the slip frequency at which it will run when supplied from a 3-phase, 340 V, 40 Hz system. Slip at which the machine develops maximum torque using 50 Hz supply is 0.1. Neglect stator impedance and assume linear torque-slip characteristic between zero torque and maximum torque in the working region. [*GATE 1992*]

Solution : At 400 V, 50 Hz supply

$$\text{Slip, } s_1 = \frac{f'}{f} = \frac{2}{50} = 0.04$$

Ratio of rotor resistance to standstill reactance,

$$\frac{R_2}{X_2} = s_{\max T} = 0.1$$

$$\text{or } R_2 = 0.1 X_2$$

Torque developed at slip s_1 of 0.04,

$$T_1 = \frac{K s_1 E_2^2}{R_2^2 + s_1^2 X_1^2} = \frac{K_1 s_1 V_1^2}{R_2^2 + s_1^2 X_1^2} \quad \because E_2 \propto V$$

$$= \frac{K_1 \times 0.04 \times (400)^2}{(0.1 X_2)^2 + (0.04 X_2)^2} = \frac{6{,}400 \text{ K}}{0.0116 X_2^2}$$

At 340 V, 40 Hz supply; Let the machine develop torque equal to T_1 at slip s_2

$$\text{So } \frac{K s_2 V_2^2}{R_2^2 + s_2^2 X_2^2} = T_1$$

$$\text{or } \frac{K s_2 \times 340^2}{(0.1 X_2)^2 + s_2^2 X_2^2} = \frac{6{,}400 K}{0.0116 X_2^2}$$

$$\text{or } \frac{1{,}15{,}600 s_2}{0.01 + s_2^2} = \frac{6{,}400}{0.0116}$$

$$\text{or } 6{,}400 s_2^2 - 1{,}340.96 s_2 + 64 = 0$$

$$\text{or } s_2^2 - 0.2095 s_2 + 0.01 = 0$$

$$\text{or } s_2 = \frac{0.2095 \pm \sqrt{(0.2095)^2 - 0.04}}{2}$$

$$= 0.07356 \text{ or } 0.1359$$

Rejecting higher value, slip $s_2 = 0.07356$

$$\text{Slip frequency } f' = s_2 f = 0.07356 \times 40 = 2.94 \text{ Hz } \textbf{Ans.}$$

Example 1.15. A 3-phase, 50 Hz induction motor has $R_s = R'_r = 0.02\ \Omega$. $X_{ls} = X'_{lr} = 0.1\ \Omega$ as the equivalent circuit parameters. The motor is operated at one-half of its rated-voltage and 25 Hz frequency. Calculate the maximum torque at this operating condition in terms of its rated voltage, rated-frequency maximum torque value.

[*A.M.I.E. Sec. B. Electric Drives and Their Control Winter 1998*]

Solution : Maximum torque at rated voltage and frequency,

$$T_{\max} = \frac{3V_1^2}{\frac{4\pi N_s}{60}[R_1 + \sqrt{R_1^2 + (X_1 + X'_2)^2}\,]} \quad ...(i)$$

Refer to Eq. (1.38)

Maximum torque at one-half voltage and 25 Hz frequency (*i.e.*, half the rated frequency)

$$T'_{\max} = \frac{3V_1^2/4}{\frac{4\pi}{60} \times \frac{N_s}{2}\left[R_1 + \sqrt{R_1^2 + \frac{(X_1 + X'_2)^2}{4}}\right]} \quad ...(ii)$$

From Eqs. (*i*) and (*ii*) we have

$$T'_{\max} = T_{\max} \frac{R_1 + \sqrt{R_1^2 + (X_1 + X'_2)^2}}{2\left[R_1 + \sqrt{R_1^2 + \frac{(X_1 + X'_2)^2}{4}}\right]}$$

$$= \frac{0.02 + \sqrt{0.02^2 + (0.1 + 0.1)^2}}{2\left[0.02 + \sqrt{0.02^2 + \frac{(0.1 + 0.1)^2}{4}}\right]} T_{\max}$$

$$= 0.9058\ T_{\max} \textbf{ Ans.}$$

1.17. STARTING OF ELECTRIC MOTORS

The most important processes associated with controlled electric drives are (*i*) starting, (*ii*) speed control (*iii*) braking, (*iv*) reversal of direction of rotation. All these are transient* processes that occur in a drive system. Starting of an electric drive involves a change in its state from rest to a steady-state speed of rotation. The process of starting is the most important phenomenon in the entire operation of the drive.

An electric motor is started by switching on the supply. There are, however, three possibilities which have to be considered :

1. Interference with the supply in the form of an excessive voltage drop which is more than that can be tolerated by other equipment or other consumers connected to the same power supply circuit.
2. The starting currents will add to the motor heating by an amount that depends upon their rms values and the frequency of starting. The excessive currents may cause damage to the motor itself.

   In a dc motor the limitation may be good commutation rather than heating, as dc machine have a certain maximum limit for the current dictated by the commutation process.

* The operation of an equipment between two steady-state conditions is known as transient operation and during this period, for an electric drive, the speed, torque and current undergo change.

3. Damage to the connected load through too rapid acceleration. This may also be taken to include unacceptable discomfort to passengers in lifts and trains.

In some cases none of the above is applicable and full-voltage or direct-on-line (DOL) starting is permissible, but in many cases one or more have to be guarded against, usually at the expense of additional equipment.

Necessity of Starting Equipment. If a motor is switched directly on to the supply it will, except in the small sizes, draw a current much above the permissible limit. The main purpose of the starting equipment is, therefore, to limit the starting current to a safe value without, at the same time, reducing the starting torque to a value less than that required. This limitation of current may be accomplished by reducing the voltage applied to the motor by means of a resistance in series with the motor or by some other suitable means as briefed below.

1. *Direct-on-Line Starting or Full-Voltage Starting.* This involves the application of full line voltage to the motor terminals. Whether or not this method of starting is used, depend, upon the following factors :

(*i*) Size and design of the motor (*ii*) kind of application (*iii*) location of the motor in the distribution system and (*iv*) capacity of the power supply circuit and the rules governing such installations as established by power supply companies.

As a general rule, dc motors up to 2 kW and squirrel-cage induction motors as well as certain small synchronous motors up to 4 or 5 kW are usually started by this method.

2. *Reduced-Voltage Starting.* The starting of a dc motor is usually accomplished by inserting a suitable external resistance in the armature circuit and as the motor speeds up the starting resistance is cut out in steps. For details refer to Art. 1.17.4.

Reduced voltage for starting of 3-phase induction motors is achieved by (*i*) stator resistance starting (*ii*) stator reactor starting (*iii*) star-delta starting and (*iv*) auto-transformer starting. For details refer to Art. 1.17.5. The above methods are applicable equally to the synchronous motors. With reduced voltage starting the transition to full voltage may be done either before or after synchronisation although the former method is usually preferred.

Slip-ring induction motors, though can be started by using the starting methods that are used in case of squirrel-cage induction motors, are usually started with full-line voltage across the stator terminals and by introducing variable resistance in each phase of the rotor circuit. The external resistance introduced in each phase of the rotor circuit not only reduces the current at the starting instant but increases the starting torque also. As the motor accelerates, the external resistance is cut out in steps so that the available electromagnetic torque remains maximum during the accelerating period. Ultimately when the machine attains the normal speed, the rotor winding is short circuited automatically.

3. *Starting By Means of Smooth Variation of Voltage or Frequency.* With ac motor-dc generator sets, dc motors can be started by smooth variation of applied voltage and with variable frequency sources both induction and synchronous motors can be started by smooth variation of supply frequency, simultaneously varying proportionally the voltage applied to the motors.

4. *Starting of AC Commutator Motors.* AC commutator motors may be started by applying a reduced voltage or by shifting the brushes. Both of these methods are also employed for speed control, and in most cases starting can be satisfactorily affected by adjusting the control equipment to the low-speed position and then switching on full-line voltage. In some cases it may be necessary to provide in addition special starting tappings on an auto-transformer or a series resistance.

Special Features of Starting Equipment (Starters). There are a number of features common to starters of motor of different types. Firstly, it is desirable to provide some protection against the flow of excessive currents for a prolonged time duration, even though currents of the same magnitude are permissible for a short period during the initial acceleration. Such 'overcurrent' protection can be had by providing means to return the starter to the 'off' position or by disconnecting the supply by other means. The operation of the device may be achieved electromagnetically by energization of solenoid connected in series with the motor, or thermally by a bimetallic strip heated by the motor current. In the former case a "dashpot" is required to prevent operation during acceleration period; in the latter case, the inherent delay caused by the thermal capacity of the device may be sufficient.

Secondly, it would be pointless to provide starting equipment which remained in the "full-on" position in the event of failure of power supply, since the restoration of supply would cause the very result for the prevention of which the starting equipment was provided. Furthermore, unexpected restarting of the motor and connected load could endanger equipment and personnel. A "no-volt release" is, therefore, provided in all starters, to return them to the "off" position in the event of failure of supply. This applies even to D-O-L starters for the second reason mentioned above.

A third common feature is a matter of definition. Most starters operate in a number of discrete steps and not by continuous smooth variation. It is necessary to define the number of steps, n, as the number of accelerating positions, including "full-on". Since the starter has an "off" position, there are $(n + 1)$ positions in all.

1.17.1. Starting Characteristics of Electric Motors

The study of starting characteristics of a motor is essential to know whether the starting torque that the motor is capable of developing is sufficient to start and accelerate the motor and its load to the rated speed in a reasonable time or not. This feature is very important particularly when the motor is to be selected for the services in which either it is to start against full-load torque (as for driving grinding mills or expellers) or it is to be stopped and started very frequently (as in case of lifts and hoists), *i.e.*, where high starting torque is essential.

The torque developed by the motor at the starting is required. (*i*) to overcome the initial static friction and (*ii*) accelerate the motor and load to the full speed. The torque required to overcome the initial static friction cannot be easily determined and in certain cases as for starting a machine, which has been out of use for some time, it may be much more than the full-load torque.

The torque for accelerating depends upon the load torque. The loads which are usually met with, may be divided according to accelerating torque requirements into the following categories.

(*i*) Load requiring very small accelerating torque in comparison with full-load torque such as when the motor is to be run light.

(*ii*) Load requiring the torque which may increase with speed and it may be proportional to $(\text{speed})^2$ as in case of centrifugal pumps or fans.

(*iii*) Load having constant load torque at all speeds as in case of a hoists, lifts etc.

(*iv*) When the motors have to start and accelerate against full-load torque and in addition to accelerate some heavy moving parts as in case of rolling mills or with motors equipped with a flywheel.

It is obvious that starting of a motor in category (*i*) is quite a different proposition from starting in category (*iv*), even though the full-load torque and speed may be the same in each case; a study of the load characteristic is thus necessary in order to be able to choose the most suitable type of motor and starter.

Another factor of importance is the time taken to run the motor up to speed, since this governs the heating of the motor and control equipment because of the heavy starting currents.

1.17.2. Acceleration Time

The study of behaviour of an electric drive during transient period (*i.e.*, the duration of starting, braking, and transition from one period to another) is of great practical importance, as it helps in determining the requisite control equipment for starting, braking etc. so as to have minimum energy loss during such operation and also to ensure that the machine operates stably after the transients die out.

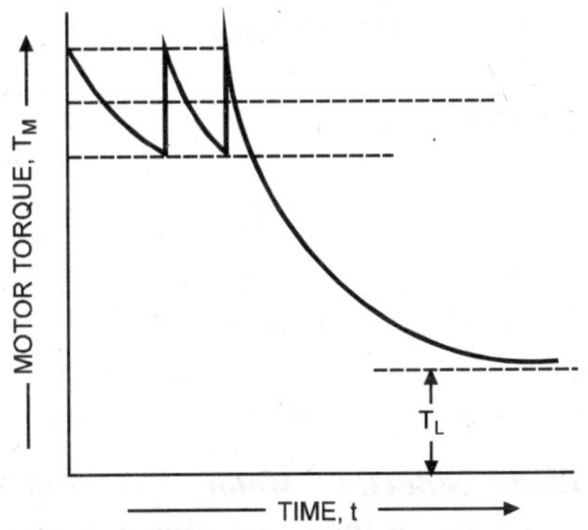

Fig. 1.56. ***Motor Torque During Starting***

Determination of the time during which a transient will continue to exist is based on intergration of the equation of motion for the drive, *i.e.*, Eq. (1.11). Solving the equation for time, we have

$$dt = \frac{J\,d\omega}{T_M - T_L} \quad ...(1.44)$$

The time required to change the speed of drive from ω_1 to ω_2 will amount to

$$t = J\int_{\omega_1}^{\omega_2} \frac{d\omega}{T_M - T_L} \quad ...(1.45)$$

In the above Eq. (1.45) the inertia J is assumed to be constant. In case J varies with time or speed, the equation may be rewritten suitably.

For the case, where ω_1 is zero at t equal to zero and ω_2 is equal to a steady-state speed ω_n attained by the drive, accelerating time is given as

$$t = J\int_{0}^{\omega_n} \frac{d\omega}{T_M - T_L} \quad ...(1.46)$$

The above Eq. (1.46) shows that the acceleration time duration depends upon the area under the curve relating speed ω and $\frac{1}{T_M - T_L}$. Also, as the steady-state speed of the motor is reached, the term $(T_M - T_L)$ tends to be zero and hence $\frac{1}{T_M - T_L}$ tends to be infinite and this would lead to an infinite accelerating time. This difficulty is overcome, in practice, by computing the time required to reach say 95 to 98 per cent of the final steady-state speed. The desired value of accelerating time for any motor-load combination can be achieved by suitably modifying the speed-torque characteristic of the driving motor.

Acceleration Time For Specific Nature of Motor and Load Torques. For determination of accelerating time it is necessary to know the motor torque T_M, load torque T_L, and inertia constant J. We now will derive expressions for the acceleration time for different motor and load torque conditions.

(*i*) Constant Motor and Load Torques. Such a situation may arise, for example, during starting of a hoisting mechanism with a constant load torque. If the motor also develops a constant torque during starting, which is possible with shunted armature connections of a dc series motor or rotor resistance starting of a wound rotor induction motor or a double cage induction motor, the net torque available for acceleration $(T_M - T_L)$ will also be constant, and acceleration time, will therefore, be given as

$$t_1 = \frac{J\omega_n}{T_M - T_L} \text{ seconds} \quad ...(1.47)$$

Also, if the motor is started on no load, *i.e.*, $T_L = 0$, the time required to bring the motor from rest to no-load speed is given as

$$t_1 = \frac{J\omega_n}{T_M} \quad ...(1.48)$$

The time t_1 given by the above Eq. (1.48) is called the mechanical time constant of the motor and denoted by t_m. For dc motors and induction motors rated at 1,000 rpm, t_m varies from 0.4 to 0.6 second.

(*ii*) Linearly Varying Motor Torque and Constant Load Torque. For a particular motor (such as for a slip-ring induction motor with high resistance rotor circuit) let the motor starting torque be T_{st} (*i.e.*, when $\omega = 0$, $T_M = T_{st}$) and be zero when speed is synchronous, ω_s (*i.e.*, when $\omega = \omega_s$; $T_M = 0$). The relationship between motor torque and slip s is expressed as

$$\frac{T_{st}}{1} = \frac{T_M}{s}$$

$$\text{or } s = \frac{T_M}{T_{st}} \quad \text{...(1.49)} \quad \text{where } s \text{ is slip.}$$

$$\text{Also } s = \frac{\omega_s - \omega}{\omega_s}$$

$$\text{or } \omega_s ds = -d\omega \quad \text{...(1.50)}$$

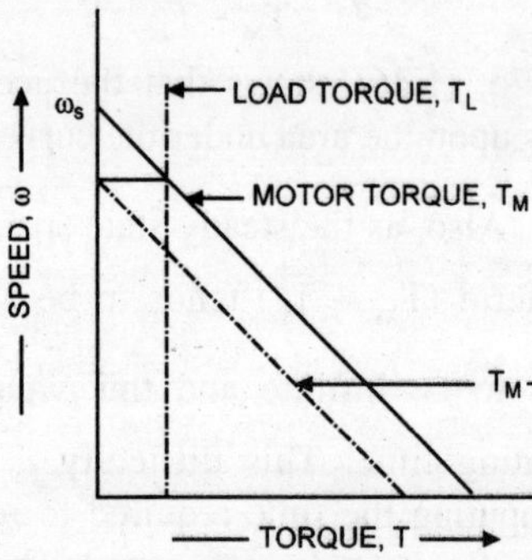

Fig. 1.57. ***Linear Motor Speed-Torque and Constant Load Curves***

Substituting $d\omega = -\omega_s ds$ and $T_M = sT_{st}$ in Eq. (1.46), we have

$$t = J\int \frac{-\omega_s ds}{s\,T_{st} - T_L} = -J\omega_s \int \frac{ds}{s\,T_{st} - T_L} \quad \text{...(1.51)}$$

Similarly the acceleration time duration for different motor torque and load torque conditions can be determined using Eq. (1.46).

1.17.3. Energy Consumption During Starting

An expression for the energy lost in the armature circuit of a *dc shunt motor* can be derived as follows.

The voltage equations of a dc shunt motor is

$$V = E_b + I_a R_a = K\omega + I_a R_a$$

$$\text{or } I_a R_a = V - K\omega \quad \text{...(1.52)}$$

The above equation is also valid for a separately excited dc motor.

The equation of motion on no load will be

$$T_M = K\,I_a = J\frac{d\omega}{dt} \quad \text{...(1.53)}$$

Multiplying Eqs. (1.52) and (1.53), we have

$$I_a^2 R_a = \frac{JV}{K}\frac{d\omega}{dt} - J\omega\frac{d\omega}{dt} \quad \text{...(1.54)}$$

On no load, $I_a^2 R_a$ will be negligibly small and hence $V = K\omega_n$ where ω_n is the no-load speed of the shunt motor.

Substituting $\frac{V}{K} = \omega_n$ in Eqs. (1.52) and (1.54), we have

$$I_a R_a = K(\omega_n - \omega) \quad \text{...(1.55)}$$

$$I_a^2 R_a\, dt = J\omega_n d\omega - J\omega\, d\omega = J(\omega_n - \omega)d\omega \quad \text{...(1.56)}$$

Now energy consumption during starting is given by the expression

$$W_{st} = \int_0^t I_a^2 R_a\, dt = J\int_0^{\omega_n} (\omega_n - \omega)d\omega$$

$$= J\omega_n^2 - \frac{J\omega_n^2}{2} = \frac{J\omega_n^2}{2} \quad \text{...(1.57)}$$

Thus, energy lost in the armature circuit of a dc shunt motor or a separately excited motor during starting on no load will be equal to the kinetic energy stored by the armature and is thus independent of the armature circuit resistance.

If the motor were started with a constant load torque T_L, the energy lost in the armature circuit during starting could be determined as follows.

The equation of motion will, now, be of the form

$$T_M = KI_a = J\frac{d\omega}{dt} + T_L \quad \text{...(1.58)}$$

From Eqs. (1.55) and (1.58), we have

$$I_a^2 R_a = K(\omega_n - \omega)I_a = (\omega_n - \omega)\left[J\frac{d\omega}{dt} + T_L\right] \quad \text{...(1.59)}$$

Now energy consumption during starting is given by the expression

$$W_{st} = \int_0^t I_a^2 R_a\, dt = \int_0^t (\omega_n - \omega)\left[J\frac{d\omega}{dt} + T_L\right]dt$$

$$= J\int_0^{\omega_n} (\omega_n - \omega)d\omega + T_L\int_0^t (\omega_n - \omega)dt \quad \text{...(1.60)}$$

DC Series Motor. Since a dc series motor cannot be started on no load, let us assume that it is started with a constant load torque T_L. The equation of motion will be

$$T_M = KI_a^2 = J\frac{d\omega}{dt} + T_L \quad \text{...(1.61)}$$

Therefore, energy consumed in the armature circuit of a dc series motor is given by the expression

$$W_{st} = \int_0^t I_a^2 R_a\, dt = \frac{JR_a}{K}\int_0^{\omega_n} d\omega + \frac{T_L R_a}{K}t_{st}$$

$$= \frac{JR_a}{K}\omega_n + \frac{T_L R_a}{K}t_{st} \quad \text{...(1.62)}$$

From above Eq. (1.62) it is obvious that unlike in a dc shunt motor, the energy dissipated in the armature circuit of a dc series motor depends upon the armature circuit resistance.

Three-Phase Induction Motor. The mechanical torque developed in a 3-phase induction motor is given as

$$T_M = \frac{3I_2^2 R_2}{\omega_s s} \quad \text{...(1.63)}$$

and the equation of motor under no-load condition, neglecting friction torque becomes

$$T_M = J\frac{d\omega}{dt} = J\omega_s\frac{ds}{dt} \because \omega = \omega_s(1 - s) \quad \text{...(1.64)}$$

The total energy loss in the rotor circuit of an induction motor during starting period is given as

$$W_{st} = \int_0^{t_s} 3I_2^2 R_2 \, dt$$

$$= \int_0^{t_s} s\omega_s T_M \, dt = -J\omega_s^2 \int_1^0 s\, ds$$

$$= \frac{J\omega_s^2}{2} \text{ joules} \qquad ...(1.65)$$

From above Eq. (1.65) it is revealed that the energy lost in the rotor circuit during starting is equal to the kinetic energy stored by the rotor and is independent of rotor circuit resistance. The total energy loss including the loss in the stator circuit, however, depends upon the rotor and stator resistances. An expression for the total energy lost in the motor during a change in speed can be given as

$$W_M = 3\int_{t_1}^{t_2} I_2^2 (R_1 + R_2) dt \quad \because I_2 \simeq I_1 \quad ...(1.66)$$

During starting, the energy lost in the motor will, therefore, be given as

$$(W_M)_{st} = \frac{J\omega_s^2}{2}\left(1 + \frac{R_1}{R_2}\right) \qquad ...(1.67)$$

where J is the moment of inertia of rotor.

The energy loss during starting of a motor can be reduced by adopting the following methods :

(*i*) reduction in moment of inertia of rotor.

(*ii*) smooth variation of applied voltage in case of dc shunt motors.

(*iii*) smooth variation of supply frequency (V/*f* control) in case of induction motors.

1.17.4. Starting of DC Motors

As already stated [refer to Art 1.16.1 Eq. (1.27)], the torque of a dc motor is directly proportional to the product of flux and armature current and is quite independent of speed, *i.e.*, $T \propto \phi I_a$. Hence in order to have high starting torque for a given armature current, the flux must be increased to the maximum value possible.

In case of a *dc shunt motor,* the flux ϕ remains constant as field is connected directly across the constant voltage supply mains and the armature current is controlled by connecting a starting resistance in series with the armature. The torque, which is directly proportional to armature current (as shown in Fig. 1.58) is limited by the maximum starting current–full-load current will develop full-load torque, twice full-load current will provide twice full-load torque, and so on. As the motor speeds up the starting resistance is cut out.

In case of a dc shunt motor reduced-voltage starting method cannot be used because a reduction in supply voltage would reduce the flux proportionately and hence the starting torque.

In the case of a dc series motor as the field winding is connected in series with the armature therefore, the current in the series field winding and the armature is the same. Since up to saturation point the flux is directly proportional to the current flowing through the field and after saturation point the flux is independent of current and remains almost constant, the torque, therefore, varies as the square of the armature current up to saturation point and varies directly as the armature current beyond saturation point. The torque-armature current characteristic is shown in Fig. 1.58.

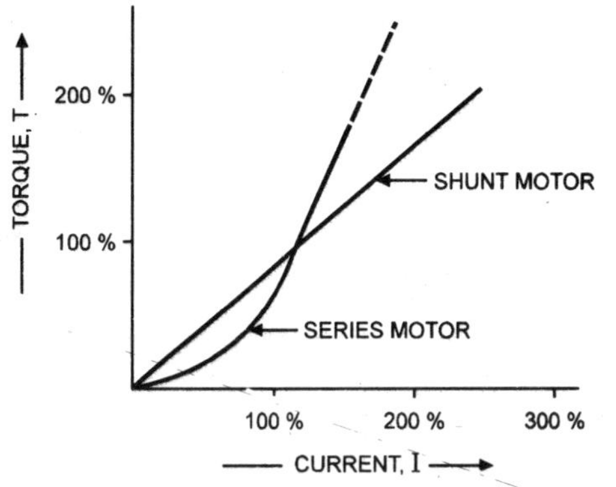

Fig. 1.58. *Torque-Current Curves For DC Motors*

In case of series wound motor since torque is independent of the applied voltage and the speed, hence reduction of voltage applied across the series motor is a very suitable method for its starting, which is achieved by connecting a resistance in series with the motor.

When the series motors operate in pairs, series-parallel method of voltage control (as described in chapter 13) can be employed.

Starters For Shunt and Compound Wound DC Motors. As mentioned above, starters are used to limit the starting current to a safer value and obtain the required starting torque.

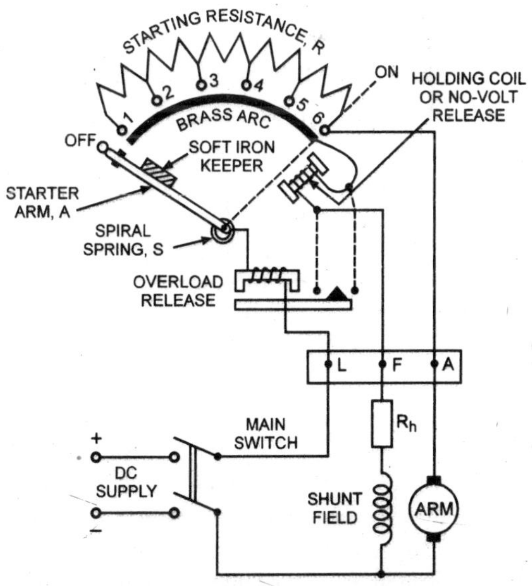

Fig. 1.59. *Three-Point Starter*

In this context it is pertinent to note that while starting dc shunt and compound motors, it is advantageous to keep the field excitation at its maximum value. A large field current, therefore, a higher value of flux will result in a low operating speed and in higher motor torque for a particular value of starting current because motor torque is proportional to the product of flux per pole and armature current. Thus for a given load torque, the motor will accelerate quickly and take less time to reach the lower operating speed from the starting instant. This will result in less heating of the armature during starting. Thus the rheostat, in series with the shunt field winding, should be at zero resistance position at the time of starting of the dc shunt and compound motors.

The starter for a shunt motor (three-point starter) is shown in Fig. 1.59. At start all resistances are in series with the armature, the holding coil (through the starter arm A) is across the line. As the speed builds up and the back emf increases the arm A is moved to the right. This decreases the armature resistance and the armature has greater emf.

At normal speed when the back emf is large enough to limit the armature current, the holding coil and starter arm A make contact. Both the field and armature are directly across the line. Current through the holding coil keeps the arm in this position. If for any reason the line voltage should fail, the magnetic field of the holding coil ceases and under the action of the spring the arm goes back to its off position. This will keep the line voltage open. Because of this action the starter protects the motor. Restarting after a line voltage failure can only be done by using the protective series resistance in the armature circuit. The starter is also provided with a overload release coil to protect the motor against the flow of excessive current due to overload. This coil is connected in series with motor and so carries full-load current. When the motor is overloaded it draws heavy current, which also flows through the coil and magnetises it to such an extent, that it pulls its armature upwards and so short circuit the no-voltage release coil, as shown in Fig. 1.59. The no-volt release coil being short circuited demagnetises and releases the starter arm which goes back to the off position and the motor is automatically disconnected from the supply mains.

In a three-point starter, discussed above, no-volt release coil is connected in series with the shunt field and the field rheostat, and therefore, the current flowing through the field is the same current that flows through the holding coil. Thus if sufficient resistance is cut in by the field rheostat so that the holding coil current is no longer able to create sufficient electromagnetic pull to overcome the spring tension, the starter arm pulls back to the 'off' position. Hence three-point starter is unsuitable for use with speed-controlled motors and that has resulted in the wide spread application of four-point starters. A four-point starter with its internal wiring connected to a long shunt compound motor is shown in Fig. 1.60. In this arrangement 'no-volt release coil' circuit is independent of shunt field circuit and, therefore, it will not be affected by the change of the current in the shunt field circuit.

DC Series Motor Starter. The series motor starter serves the same purpose as the three- and four-point starters employed with shunt and compound wound motors. However, series motor starter has different internal and external connections.

A series motor starter with holding coil and overload release is illustrated in Fig. 1.61. Holding coil protects the motor against "no volt" and "no load" whereas overload release protects the motor against the flow of excessive current.

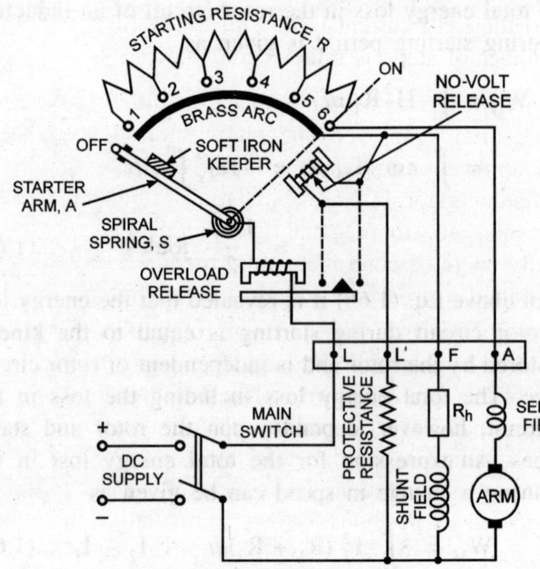

Fig. 1.60. ***Four-Point Starter***

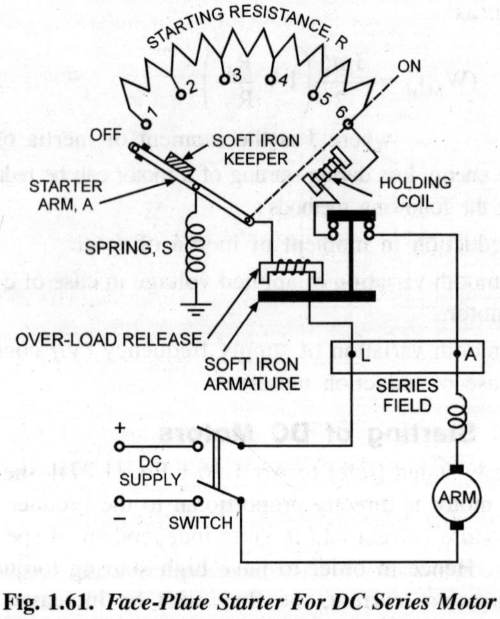

Fig. 1.61. ***Face-Plate Starter For DC Series Motor***

For switching off the dc series motor, the line switch should always be opened rather than throwing back the starting arm, as in case of dc shunt and compound motors. If it is done, then heavy sparking is caused at the stud 1, because here the field circuit is broken and the entire energy stored in the magnetic field is dissipated in the form of heavy spark.

However, there is one significant difference. In case of a dc series motor, the flux does not remain constant but varies with the current because the line current, field current and armature current are the same. Thus the back emf at any given speed varies as the current varies between the upper and lower limits. The practical result of this is that a series motor starter has a smaller number of steps than that required for a starter of a dc shunt motor of the same rating with the same current limits. This is because an increase in current causes increase in back emf and thus the current rise tends to be self-limiting. As a result of the fewer steps, the resistance of each section of the starter resistor is greater than for the shunt motor starter.

Since flux varies during the starting operation and its relation to the field current is non-linear, the determination of the number of steps is rather complicated.

Drum Type Series Motor Speed Controller. DC series motors are often employed on cranes, elevators, tramcars and other applications, where the motor is under the direct control of an operator. In these applications, frequent starting, variations of speed, stopping and reversing may be necessary. A manually operated controller, more rugged than a starting rheostat, called a *drum controller,* is employed.

The controller is in the form of rotating drum having segments which makes contact with the fixed points. Speed controller of this type is shown in Fig. 1.62 in which K is an arc braking coil: EM is a braking electromagnet; FLS is a switch for rotation in one direction and RLS is a switch for rotation in reverse direction. The controller has six positions for forward and six positions for reverse rotation of the motor. The working positions of the controller are shown by vertical dotted lines. The electromagnet EM, is connected in parallel with the motor and releases the motor at starting. When the motor is disconnected it is braked mechanically.

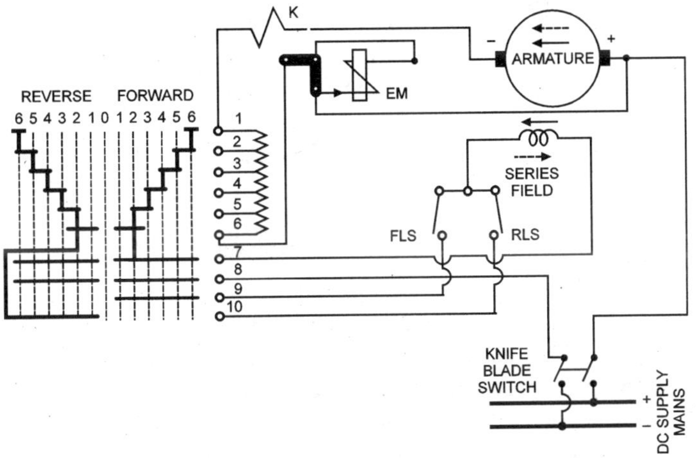

Fig. 1.62. ***Drum Type Series Motor Speed Controller***

Now when the knife blade switch is closed and the controller is set in forward position 1, the connections are along vertical line one. In this position segments make contacts with fixed points 6 and 7, 8 and 9. In this position current flows from +ve bus-bar through armature winding of the motor, arc braking coil K, all the starting resistances 1–6, fixed points 6 and 7 through controller segments, series field winding, the forward limit switch FLS, fixed points 9 and 8 through controller segments and returns back to the negative bus-bar.

In the second forward position, the segments of the controller come in contact with fixed points 5 and 6 thereby bringing some of the starting resistance out of circuit, so speed of the series motor increases. In the subsequent forwarding positions 3rd, 4th and 5th, additional steps of the starting resistances are brought out of the circuit and finally in 6th position all the starting resistances are short circuited and motor attains maximum speed.

In all the six forward positions the direction of current in armature as well as in series field winding is same (from right to left), as shown in Fig. 1.62.

But in first reverse position the current flows from + ve bus-bar through armature winding, arc braking coil K, all the starting resistances 1–6, fixed points 6 and 10 through controller segments, the reverse limit switch RLS, the series field winding, fixed points 7 and 8 through controller segments and returns back to the – ve bus-bar. Hence it is obvious that in reversing positions direction of flow of current in armature winding remains unchanged, while reverses through the series field winding, thereby reversing the direction of rotation of motor.

Automatic Starters. Push-button type of automatic starters are quite common in use in industry. Even an inexperienced operator, with the help of such starters, called the *auto-starters*, can start and stop the motor without any difficulty.

The operation of such starters depends upon, either the time delay or the counter emf developed across the armature terminals. Here only the basic principle of the counter emf starter will be discussed.

Counter-EMF Starter. This type of starter depends upon the build up of the back or counter emf to operate the contactors (a heavy duty relay designed to open or close an electrical power circuit), which in turn causes the starting resistors to be shorted. The connections of a counter-emf starter are shown in Fig. 1.63.

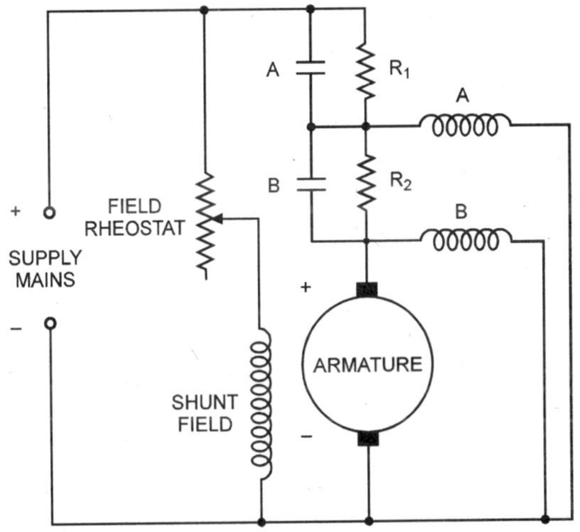

Fig. 1.63

When the motor is switched on, the counter emf developed by the motor is zero and the voltage across contactor coil A or B is insufficient to energize the relays. Consequently the contactors A and B are normally open and, therefore, starting resistances R_1 and R_2 are in series with the armature. As the motor picks up speed and builds up a counter emf the voltage across coils A and B increases until the operating voltage of coil A is reached. Contactor A is then energized, closing the normally open contact A, shorting out the starting resistor R_1. The motor continues to pick up speed building up the counter emf until the operating voltage of coil B is reached. Contactor B is energized, closing normally open contactor B, which in turn shorts out the starting resistor R_2 and places the armature directly across the supply mains.

The drawback of counter-emf automatic starter is that if the motor fails to start, the counter emf remains zero, the voltage sensitive relays cannot operate and so the starting resistances may burn out. Such occurrences can, however, be avoided by using definite time-limit starters.

1.17.5. Starting of 3-Phase Induction Motors

At start, speed of motor is zero and, therefore, slip $s = 1.0$. Substituting $s = 1.0$ in Eqs. (1.34) and (1.36), we have

Starting current per phase,

I_{st} = Rotor current referred to stator,

$$I'_2 = \frac{V}{\sqrt{(R_1+R'_2)^2+(X_1+X'_2)^2}} \text{ amperes} \quad ...(1.68)$$

(neglecting magnetizing current)

and starting torque,

$$T_{st} = \frac{3V^2R'_2}{\frac{2\pi N_s}{60}\left[(R_1+R'_2)^2+(X_1+X'_2)^2\right]} \text{ Nm} \quad ...(1.69)$$

The starting torque becomes maximum when the rotor resistance is made equal to its leakage reactance. Since the rotor resistance is not more than 1 or 2 per cent of its leakage reactance otherwise efficiency will be low therefore, in order to obtain high starting torque resistance must be inserted in the rotor circuit at start and cut out gradually as the motor picks up speed. The additional resistance in the rotor circuit will not only increase the starting torque but will limit the starting current also to a safer value. This method is useful in case of slip-ring motors only, in which the external resistance at the starting instant is introduced in the rotor circuit by taking the rotor winding terminals out to the slip rings mounted on the shaft with brushes resting on them, as shown in Fig. 1.64. When the motor attains the normal speed, the rotor winding is short circuited through the slip rings and brushes are lifted from them automatically.

Since slip-ring induction motor develops high starting torque due to introduction of external resistance in the rotor circuit at the starting instant with reasonable starting current hence such motors can be started under load also. The variation of starting torque and starting current with rotor resistance are depicted in Figs. 1.65 and 1.66 respectively.

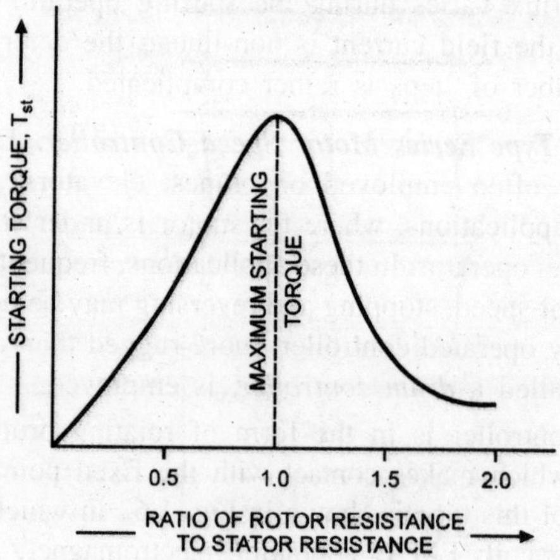

Fig. 1.65

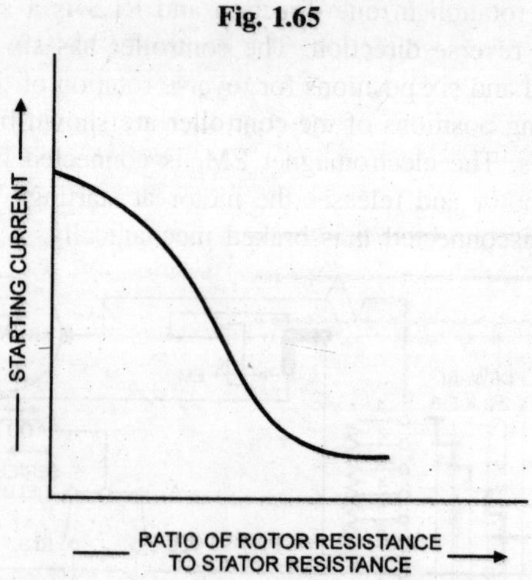

Fig. 1.66

In case of squirrel cage induction motor it is not possible to insert an external resistance in the rotor circuit for starting purpose because of its cage construction and hence such motors will have low starting torque. These motors on account of their robust construction, are more reliable; cheaper and easier to maintain as compared to slip-ring induction motors. These motors are employed where constant torque is required, *e.g.*, in paper mills, textile mills, compressors and conveyers etc.

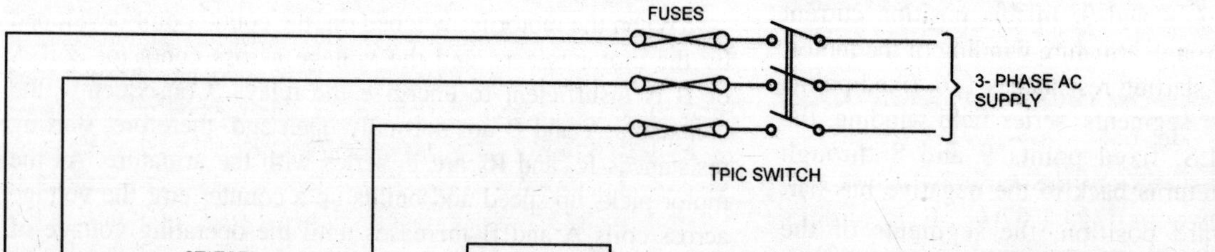

Fig. 1.64. ***Slip-Ring Induction Motor With Starting Rheostat***

Starting of Squirrel-Cage Induction Motors

1. *Direct-on-Line Starting or Full-Voltage Starting.* As already mentioned, squirrel-cage induction motors, as a general rule, up to 4 or 5 kW are started by this method. As the name suggests, this method involves direct switching of polyphase squirrel cage induction motor to supply mains, as illustrated in Fig. 1.67(*a*).

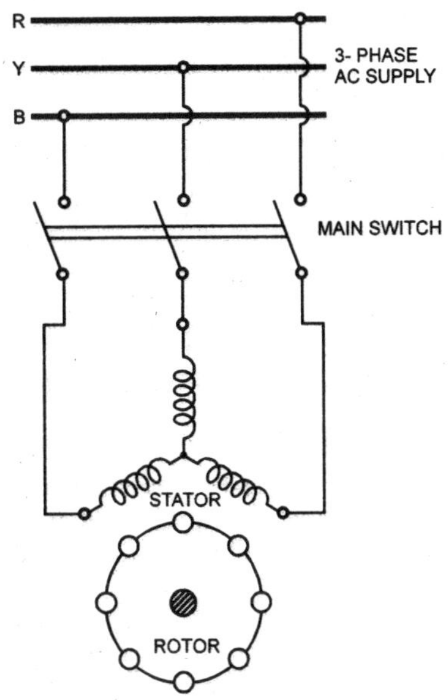

(a)

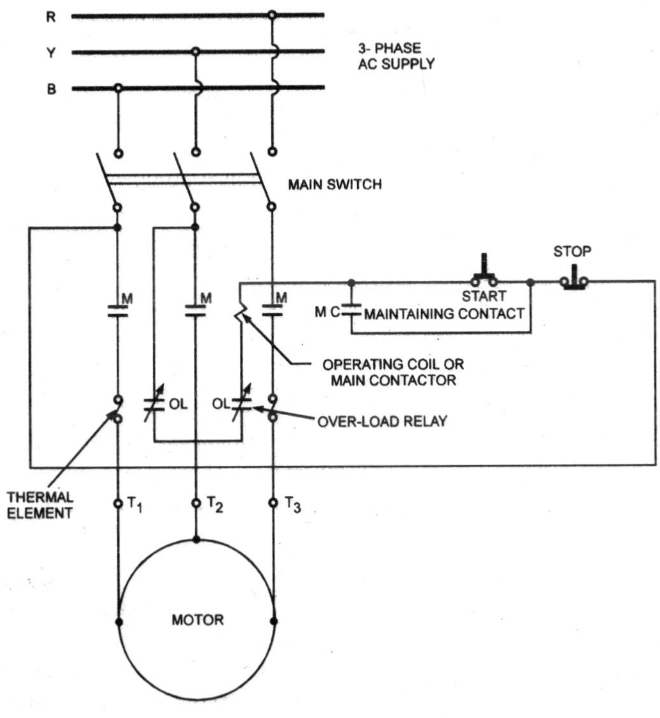

(b) Wiring Diagram of a Direct-on-Line Starter With Protective Devices

Fig. 1.67

The push-button type direct-on-line starter, which is very common in use, is shown in Fig. 1.67(*b*). It is simple, inexpensive and easy to instal and maintain. It consists of a set of 'start' and 'stop' push buttons, a contactor (an electromagnet) with its associated contacts and usually overload and under-voltage protection devices. The start button is momentary contact that is held normally open by a spring. The stop button is held normally closed by a spring. Thermal overload relays are commonly used for motor overload protection.

Under-voltage protection is provided by the contactor (that is controlled by a three-wire control circuit) maintains the interruption of circuit even after the supply is restored.

Torque Developed on Starting of Induction Motor By Direct Switching.

$$\text{Input power to rotor} = T\omega$$

$$\text{Rotor copper loss} = s \times \text{input power to rotor}$$

$$\text{or } 3\, I^2 R_2 = sT\omega$$

where I_2 is the rotor current per phase, R_2 is rotor resistance per phase, T is the torque developed, s is the slip and ω is the supply angular velocity.

$$\text{or Torque developed, } T = \frac{3I^2R_2}{s\omega}$$

$$\textit{i.e. } T \propto \frac{I_2^2}{s} \text{ if rotor resistance is constant}$$

Since rotor current is proportional to stator current I_1, so

$$T \propto \frac{I_1^2}{s}$$

$$\text{or } T = K\frac{I_1^2}{s} \text{ where K is any constant}$$

At starting slip, $s = 1$

So starting torque,

$$T_{st} = K(I_{st})^2 \text{ where } I_{st} \text{ is the starting current}$$

$$\text{Full-load torque, } T_f = \frac{KI_f^2}{s_f}$$

where I_f is the full-load motor current and s_f is the full-load slip

$$\text{So } \frac{T_{st}}{T_f} = \frac{KI_{st}^{\ 2}}{KI_f^2/s_f} = \left(\frac{I_{st}}{I_f}\right)^2 s_f$$

$$\text{or } T_{st} = T_f \times \left(\frac{I_{st}}{I_f}\right)^2 s_f \quad ...(1.70)$$

When the motor is connected directly across the supply mains, the starting current is equal to the short-circuit current, I_{sc}

$$\therefore \text{ Starting torque, } T_{st} = T_f \left(\frac{I_{sc}}{I_f}\right)^2 s_f \quad ...(1.71)$$

2. *Primary Resistor (or Reactor) Starting*. This method of starting of a polyphase cage motor is very simple and provides smooth acceleration of the motor. In this method of starting of a 3-phase squirrel cage induction motor reduced voltage is obtained by means of resistors (or reactors) that are connected in series with each stator lead, as shown in Fig. 1.68, during the starting period. The voltage drop in resistors (or reactors) causes a reduced voltage across the motor terminals. As the motor picks up the speed, the resistors (or reactors) are cut out in steps and finally short circuited when the motor attains the operating speed.

The primary resistor method is quite similar to the dc shunt motor starter with its series armature resistor, because both provide a voltage drop between the voltage supply and the motor terminals. There are, however, some very important differences. In a dc shunt motor the starting torque is directly proportional to the starting armature current whereas in case of

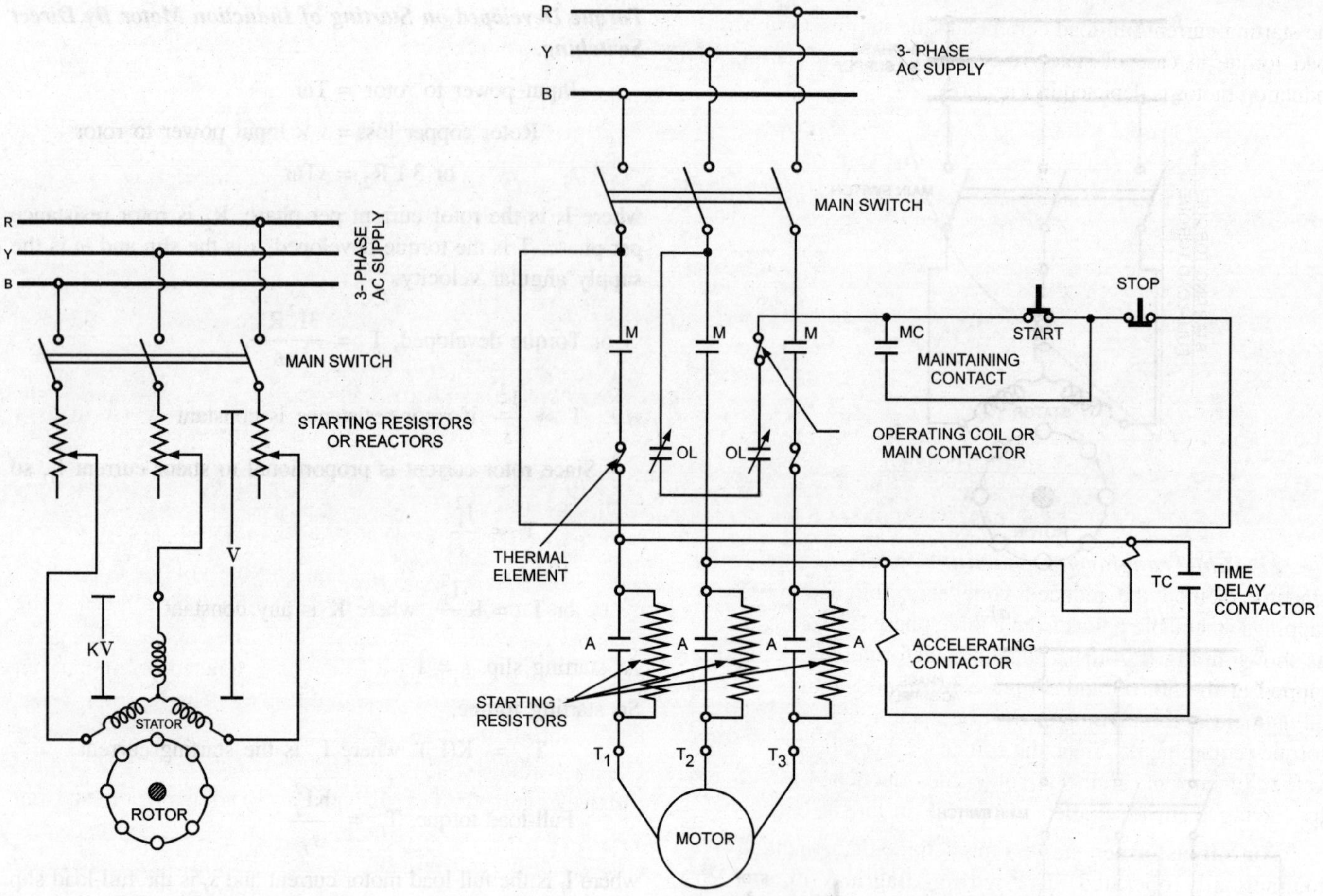

(*a*) (*b*) *Wiring Diagram of a Primary Resistance Starter*

Fig. 1.68

a polyphase induction motor the torque varies as the square of the motor terminal voltage *i.e.*, there is a far greater reduction of starting torque for an equivalent reduction of starting current.

Although the initial cost of reactors is high in comparison to that of resistors, reactor starting is preferred because this method incurs small power losses and is more effective* in reducing the voltage applied to the stator at starting.

The advantages and disadvantages of this method of starting of cage motors are given below:

Advantages. 1. Smóoth acceleration.

2. High power factor during start.
3. Less expensive than auto-transformer starter in lower output ratings.
4. Closed transition starting.
5. Available with as many as 7 accelerating points.

Disadvantages 1. Resistors give off heat.

2. Low torque efficiency.
3. Starting duration usually exceeds 5 seconds, so needs expensive resistors.
4. Starting voltage is difficult to adjust to meet varying conditions.

Circuit diagram of a magnetically operated primary resistance starter is illustrated in Fig. 1.68 (*b*).

Overload and undervoltage protections are provided in the same manner as in the direct-on-line starter.

In the starter described above, the starting resistors are cut out in one step. Starters are also available, in which arrangement is to cut out starting resistances in several steps and thus to give smooth acceleration with less line disturbance.

If the normal supply phase voltage = V volts and by using line resistance starter the voltage is reduced to KV volts, then starting current is also reduced in the same ratio *i.e.*, starting current $I_{st} = K\, I_{sc}$ where I_{sc} is the short-circuit current. Starting torque,

$$T_{st} = T_f \times \left(\frac{I_{st}}{I_f}\right)^2 s_f$$

$$= T_f \times K^2 \left(\frac{I_{sc}}{I_f}\right)^2 s_f \quad \text{...(1.72)}$$

$= K^2 \times$ Torque obtained by switching the motor directly

This method finds its application in some cases where only very low starting torques are required. A curve plotted between

* The power factor of an induction motor is quite low and if reactors are used for reducing voltage across the motor terminals, the voltage drop across the reactors will be nearly in phase with the supply voltage and so the net reduction in the applied voltage to the stator will be more. On the other hand in case of resistance starting, the voltage drop across the resistors will be nearly in quadrature with the applied voltage and so the net reduction in the applied voltage to the stator terminals will be less.

the starting-current/full-load current and the starting-torque/full-load torque in case of stator resistance method of starting of induction motor is depicted in Fig. 1.69.

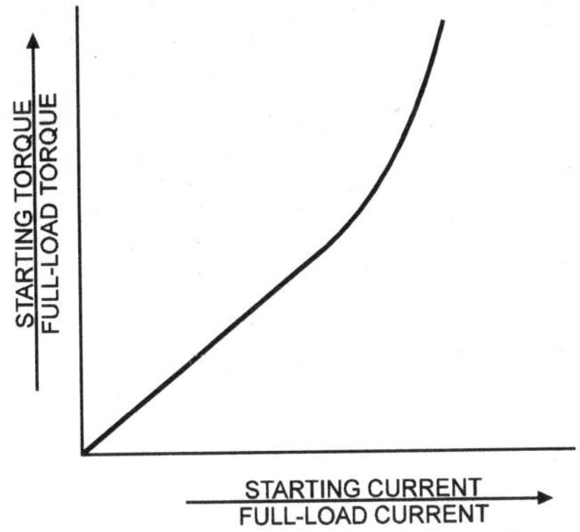

Fig. 1.69

3. *Auto-Transformer Starting*. In auto-transformer starting method the reduced voltage is obtained by taking tappings at suitable points from a three phase auto-transformer, as shown in Fig. 1.70(*a*). The auto-transformers are generally tapped at the 50, 60 and 80 per cent tapping points, so that adjustment at these voltages may be made for proper starting torque requirements. Since the contacts frequently break large values of current, arcing is sometimes quenched effectively by having them assembled to operate in an oil bath.

Auto-transformer starters may be either manually or magnetically operated. The wiring diagram of an auto-transformer starter is shown in Fig. 1.70(*b*).

Let the motor be started by an auto-transformer having transformation ratio, K. If I_{sc} is the starting current when normal voltage is applied, and applied voltage to stator winding at starting = KV

then motor input current,

$$I_{st} = KI_{sc} \qquad ....(1.73)$$

Supply current

$$= \text{Primary current of auto-transformer}$$

$$= K \times \text{secondary current of auto-transformer} = K^2 I_{sc}$$

Starting torque,

$$T_{st} = T_f \left(\frac{I_{st}}{I_f}\right)^2 \times s_f$$

$$= T_f K^2 \left(\frac{I_{sc}}{I_f}\right)^2 \times s_f$$

$$= K^2 \times \text{Torque obtained by direct switching.} \qquad ...(1.74)$$

Hence the line current and the starting torque are reduced in the square ratio.

The advantages of this method of starting of cage induction motors lies in the fact that the voltage is reduced by transformation and not by dropping the voltage in the resistors (or reactors) and, therefore, the current and power drawn from the supply mains are also reduced in comparison to primary resistor (or reactor) starting. The internal losses of the starter itself are small during long starting periods. Other advantages of this method are (*i*) availability of highest torque per ampere of supply current (*ii*) adjustment of starting voltage by selection of proper tap on the auto-transformer (*iii*) suitability for long starting periods (*iv*) closed transition starting and (*v*) motor current larger than supply current.

The drawbacks of such starters are low power factor and higher cost in case of lower output rating motors.

This method can be employed for starting of star-connected as well as delta-connected motors.

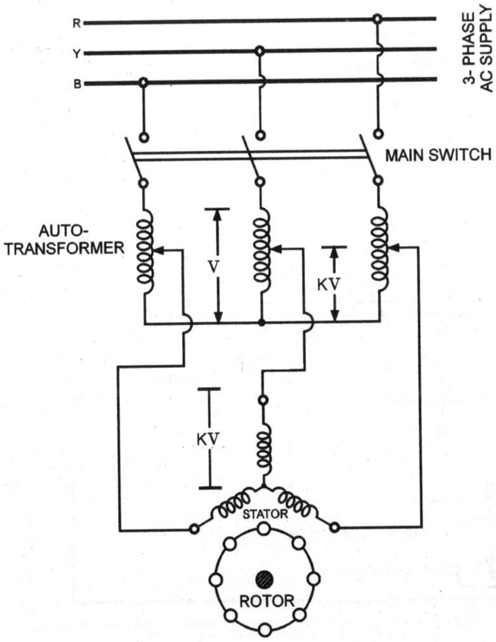

Fig. 1.70(*a*)

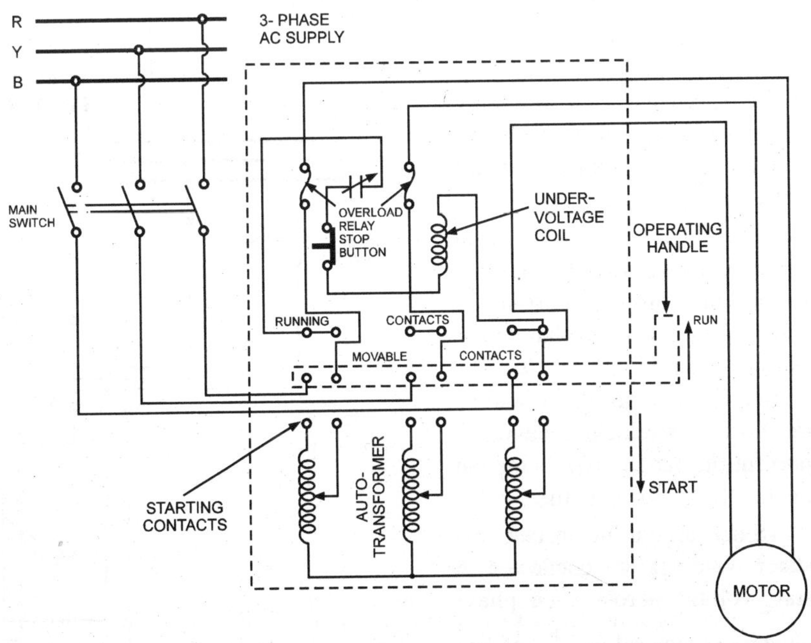

Fig. 1.70(*b*). *Circuit Diagram of an Auto-Transformer Starter*

For starting of large cage motors (of output rating exceeding 20 kW) this method of starting is often used.

The values of starting torque available, together with the corresponding values of the voltage applied to the stator and the motor and line currents expressed as percentages of the normal full-load values, for different tap positions are shown in Fig. 1.70 (*c*).

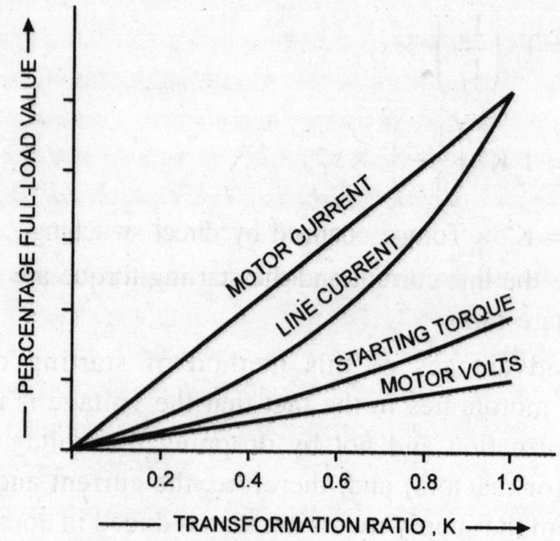

Fig. 1.70(*c*)

4. *Star-Delta Starting*. This method of starting of cage induction motors is based upon the principle that with 3 windings connected in star, the voltage across each winding is $1/\sqrt{3}$ *i.e.* 57.7 % of the line-to-line voltage whereas the same winding connected in delta will have full line-to-line voltage across each.

The star-delta starter connects the three stator windings in star across the rated supply voltage at the starting instant. After the motor attains speed the same windings, through a change-over switch, are reconnected in delta across the same supply voltage.

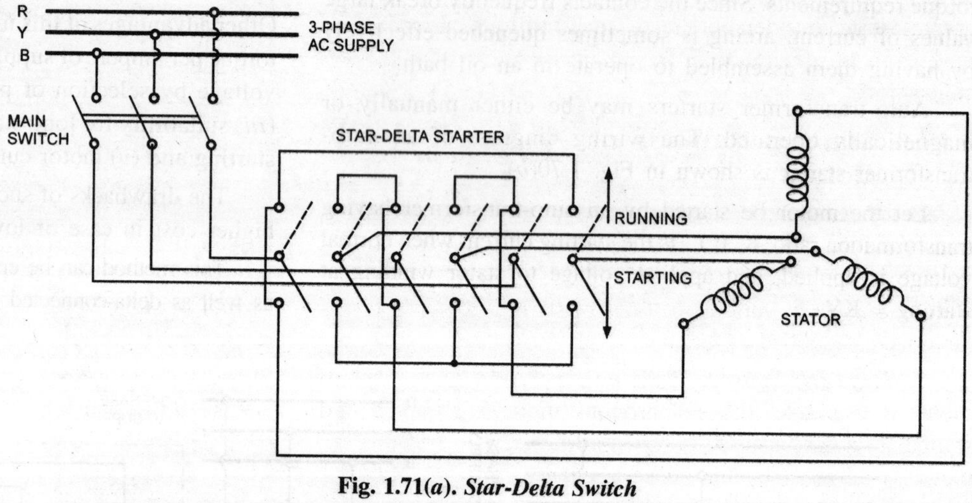

Fig. 1.71(*a*). *Star-Delta Switch*

The basic diagram of connection is shown in Fig. 1.71(*a*). An actual starter incorporates undervoltage and overvoltage coils, as shown in Fig. 1.71 (*b*). The starter is also provided with a mechanical interlocking device to prevent the handle from being put in the 'Run' position first.

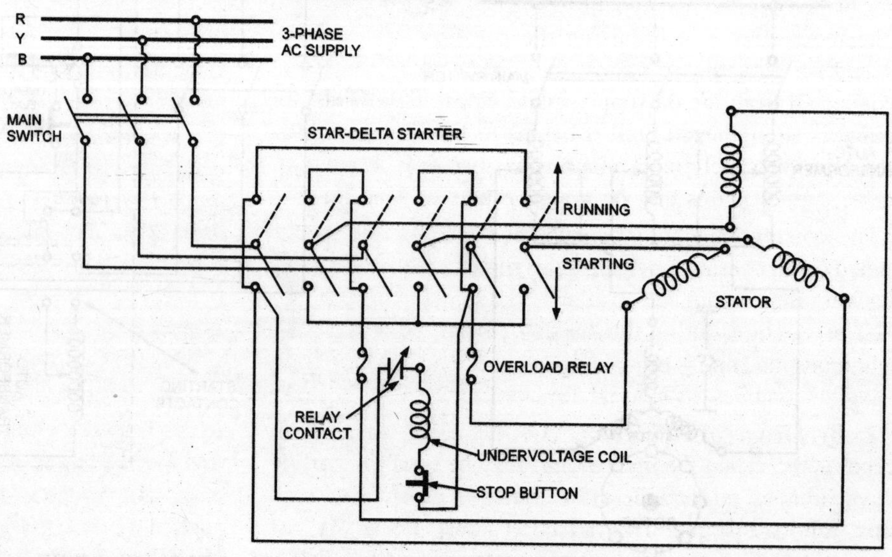

Fig. 1.71(*b*). *Star-Delta Starter*

Since at starting instant, the stator windings are connected in star, voltage across each phase winding is reduced to $\frac{1}{\sqrt{3}}$ of line voltage and, therefore, starting current per phase becomes equal to $I_{sc}/\sqrt{3}$.

Starting line current by connecting the stator windings in star at the starting instant = Starting motor current per phase $= I_{sc}/\sqrt{3}$.

Starting line current by direct switching with stator windings connected in delta $= \sqrt{3}\, I_{sc}$.

$$\therefore \quad \frac{\text{Line current with star-delta starting}}{\text{Line current with direct switching}} = \frac{I_{sc}/\sqrt{3}}{\sqrt{3}\, I_{sc}} = \frac{1}{3}$$

Hence by star-delta starting line current is reduced to one-third of line current with direct switching.

Starting torque,

$$T_{st} = T_f \left(\frac{I_{st}}{I_f}\right)^2 s_f = T_f \left[\frac{I_{sc}/\sqrt{3}}{I_f}\right]^2 s_f$$

$$= \frac{1}{3} T_f \left(\frac{I_{sc}}{I_f}\right)^2 s_f \qquad \text{...(1.75)}$$

Hence with star-delta switching, the starting torque is also reduced to one-third of starting torque obtained with direct switching.

This method of starting of cage motors is simple, cheap, effective and efficient since no power is lost in auxiliary components. This method is also suitable for high inertia and long acceleration loads.

This method needs a motor to be delta-connected for normal operation and all the six terminals of the 3-phase stator windings are to be brought out. The reduction in voltage is fixed and starting torque is also low. So this method is limited to application where high starting torque is not the essential requirement e.g. machine tools, pumps, motor-generator sets etc. This method is unsuitable for line voltage exceeding 3,000 V, because of excess number of stator turns required for delta connection.

Such starters are employed for starting 3-phase squirrel cage induction motors of rating between 4 and 20 kW.

Precaution With Star-Delta Starting. The initial current flowing when the motor is started in star is 57.7% of the short-circuit current in delta together with a transient in each phase. The transient currents decay rapidly but the steady state is not reached until the motor has attained 70 per cent of its synchronous speed. The change-over from star to delta connection should not be made until the motor attains about 90 per cent of synchronous speed, otherwise there will be a current surge considerably greater than full-load current which may even be greater than the standstill current with star-connection.

1.17.6. Starting Methods of Synchronous Motors

As already stated the synchronous motor has got no self starting torque and the motor must first be brought up to synchronous speed by some external means before it is capable of operating with load angle δ. Various methods of starting of synchronous motors are described as follows:

1. *From DC Source.* If the dc supply and dc compound motor is available, the synchronous motor is coupled and started by means of a dc compound motor. The speed of dc motor is adjusted by the speed regulator. The synchronous motor is then excited and synchronised with ac supply mains. At the moment of synchronising the synchronous motor is switched on with the ac mains, and either the dc motor is disconnected from the dc supply mains or the field of the dc machine is strengthened until it begins to function as a generator. Now the synchronous machine is operating, as a motor, from ac supply mains and dc machine acts as load on it.

The synchronous motor can also be started by the exciter mounted on an overhung synchronous motor bracket and shaft extension. Here again, an available dc source operates the exciter as a motor during the starting period; then, after the synchronous machine is brought up to speed and synchronised, the exciter assumes its normal function.

2. *By Means of AC Motor.* A small direct-coupled induction motor, called the *pony motor*, may be used for starting the synchronous motor unless the motor is required to start against full-load torque. The induction motor frequently has two poles less than the synchronous motor and so is capable of raising the speed of the latter to synchronous speed. Before switching the ac supply to the synchronous motor, it must be synchronised with the bus-bars. After normal operation is established, the pony motor is sometimes uncoupled from the synchronous motor. This method is not very satisfactory and not suited to industrial needs. Modern machines are usually of the self starting type and are arranged to start as induction motors.

3. *By Means of Damper Grids in The Pole Faces.* The synchronous motor is made self starting by providing a special winding on the rotor poles, known as *damper winding* or *squirrel cage winding*. The damper winding consists of short-circuited copper bars embedded in the face of the field poles. AC supply given to the stator produces a rotating magnetic field which causes the rotor to rotate, therefore, in the beginning synchronous motor provided with damper winding starts as a *squirrel cage induction motor*. The exciter moves along the rotor. When the motor attains about 95% of synchronous speed, the rotor winding is connected to exciter terminals and the rotor is magnetically locked by the rotating field of the stator and the motor runs as a synchronous motor.

Advantages. When the motor is overloaded it does not stop, continues running as a squirrel cage induction motor. When the speed falls slightly below the synchronous speed due to overloading an emf is induced in the rotor short-circuited winding, called the *damper winding*. Due to this emf a torque is produced which keeps the motor running even on no load.

Disadvantages. Since damper winding resistance is low so at start it takes large current from supply mains.

To avoid large starting current drawn from the supply mains the synchronous motor is started usually as slip-ring induction motor. The rotor winding is connected in series with triplex rheostat at the time of starting by throw-over switch. Rotor circuit is completed and an emf is induced in the rotor circuit as in the rotor of a slip-ring induction motor.

As the motor attains synchronous speed the throw-over switch is put on the side of the exciter terminals and the dc supply passes through the rotor winding. While running, the rotor acts like rotor of a synchronous motor. Hence the motor runs as a synchronous motor. If the starting and excitation is done automatically the motor is known as *auto-synchronous motor*.

Precautions. *In starting a synchronous motor the field winding must be shorted through a suitable resistance so that the induced voltage is distributed throughout the whole winding and no part can be subjected to the high voltage induced in the entire winding.*

Another method of eliminating the hazard of breakdown of insulation of the field winding due to induced high voltage in it is by means of dividing the field winding into several sections during the starting period. While emf is still induced in the field coils, the number of coils in series is reduced, so that the magnitude of emf induced in any section is kept within safe limits. Sectionalizing of field winding may be accomplished by a centrifugally operated switch that keeps the field circuit sectionalized until the rotor revolves at near synchronous speed.

Example 1.16. A 3-phase squirrel-cage induction motor has a starting current of 6 times its full-load current. The motor has a full-load slip of 4 per cent. Calculate (*i*) the starting torque (*ii*) the slip at which maximum torque occurs and (*iii*) the ratio of maximum torque to full-load torque.

[A.M.I.E. Sec. B. Electric Drives and Their Controls Summer 2000]

Solution : Starting current, $I_{st} = 6I_f$

Full-load slip, $s_f = 4\%$, *i.e.*, 0.04

(*i*) Starting torque, $T_{st} = T_f \times \left(\frac{I_{st}}{I_f}\right)^2 \times s_f$

$$= T_f \times (6)^2 \times 0.04 = 1.44\ T_f \quad \textbf{Ans.}$$

From Eq. (1.34), neglecting stator resistance and magnetising current, we have

$$\text{Full-load current, } I_f = \frac{V}{\sqrt{\left(\frac{R_2'}{s_f}\right)^2 + (X_1 + X_2')^2}} \quad ...(i)$$

$$\text{and starting current, } I_{st} = \frac{V}{\sqrt{(R_2')^2 + (X_1 + X_2')^2}} \quad ...(ii)$$

Dividing Eq. (*ii*) by Eq. (*i*), we have

$$\frac{I_{st}}{I_f} = \sqrt{\frac{\left(\frac{R_2'}{s_f}\right)^2 + (X_1 + X_2')^2}{(R_2')^2 + (X_1 + X_2')^2}}$$

$$\text{or} \quad \frac{\left(\frac{R_2'}{0.04}\right)^2 + (X_1 + X_2')^2}{(R_2')^2 + (X_1 + X_2')^2} = \left(\frac{I_{st}}{I_f}\right)^2 = 6^2 = 36$$

$$\text{or } 625(R_2')^2 + (X_1 + X_2')^2 = 36(R_2')^2 + 36(X_1 + X_2')^2$$

$$\text{or} \quad \frac{R_2'}{(X_1 + X_2')} = \sqrt{\frac{36-1}{625-36}} = \sqrt{\frac{35}{589}} = 0.244$$

(*ii*) From Eq. (1.37), slip corresponding to maximum torque

$$s_{max\,T} = \frac{R_2'}{X_1 + X_2'} = 0.244 \text{ or } 24.4\% \quad \textbf{Ans.}$$

Neglecting stator resistance R_1

(*iii*) From Eq. (1.41), we have,

Ratio of maximum torque to full-load torque

$$= \frac{1}{2}\left(\frac{s_f}{s_{max\,T}} + \frac{s_{max\,T}}{s_f}\right)$$

$$= \frac{1}{2}\left[\frac{0.04}{0.244} + \frac{0.244}{0.04}\right] = 3.132. \quad \textbf{Ans.}$$

Example 1.17. A 3-phase squirrel cage induction motor has a starting current 8 times the full-load value. The full-load slip is 4%. Compute the starting torque as a percentage of full-load torque if the motor is started (*i*) direct-on-line (*ii*) by a star-delta starter, (*iii*) using an auto-transformer to limit the starting current to three times the full-load value.

[A.M.I.E. Sec. B. Electric Drives and Their Control Winter 1999]

Solution : Short-circuit current, $I_{sc} = 8\ I_f$

Full-load slip, $s_f = 4\% = 0.04$

(*i*) Starting torque, when started by means of direct switching,

$$T_{st} = \left(\frac{I_{sc}}{I_f}\right)^2 s_f\, T_f = 8^2 \times 0.04\ T_f$$

$$= 2.56 \text{ times full-load torque. } \textbf{Ans.}$$

(*ii*) Starting torque, when started by means of a star-delta starter,

$$T_{st} = \frac{1}{3}\left(\frac{I_{sc}}{I_f}\right)^2 s_f\, T_f = \frac{1}{3} \times 8^2 \times 0.04\ T_f$$

$$= 0.853 \text{ times full-load torque. } \textbf{Ans.}$$

(*iii*) If K is the transformation ratio of auto-transformer, then

Starting current, $I_{st} = K^2 I_{sc}$

$$\text{or Transformation ratio, } K = \sqrt{\left(\frac{I_{st}}{I_{sc}}\right)} = \sqrt{\frac{3I_f}{8I_f}} = \sqrt{\frac{3}{8}}$$

$$\text{Starting torque, } T_{st} = K^2 \left(\frac{I_{sc}}{I_f}\right)^2 s_f\, T_f$$

$$= \frac{3}{8} \times (8)^2 \times 0.04\, T_f$$

$$= 0.96 \text{ times full-load torque. } \textbf{Ans.}$$

Example 1.18. Determine the ratio of starting to full-load current of a 10 hp, 400 V, 3-phase induction motor with star-delta starter. The motor has a full-load power factor of 0.8 lagging and an efficiency of 0.9. The short-circuit current of motor is 7.2 A at 160 V. Neglect the effect of magnetising current.

[Pb. Univ. Electric Drives and Utilization of Electrical Energy August 1984]

Solution : Output = 10 hp = 10 × 0.7355 = 7.355 kW

$$\text{Full-load line current, } I_f = \frac{\text{Output in kW} \times 1{,}000}{\sqrt{3} \times V_L \times \cos\phi \times \eta}$$

$$= \frac{7.355 \times 1{,}000}{\sqrt{3} \times 400 \times 0.8 \times 0.9} = 14.75 \text{ A}$$

Short-circuit current at 160 V

= 7.2 A

Short-circuit current at normal voltage of 400 V,

$$I_{sc} = 7.2 \times \frac{400}{160} = 18.0 \text{ A}$$

Starting current with star-delta starter,

$$I_{st} = \frac{I_{sc}}{3} = \frac{18.0}{3} = 6.0 \text{ A}$$

Ratio of starting current to full-load current

$$= \frac{I_{st}}{I_f} = \frac{6.0}{14.75} = 0.407. \quad \textbf{Ans.}$$

Example 1.19. A 50 kVA, 400 V, 3-phase, 50 Hz squirrel cage induction motor has full-load slip of 5%. Its standstill impedance is 0.866 Ω/phase. It is started using a tapped auto-transformer. If the maximum allowable supply current at the time of starting is 100 A, calculate the tap position and the ratio of starting torque to full-load. *[GATE 1997]*

Solution : Full-load current drawn by a 3-phase induction motor,

$$I_f = \frac{\text{Output in kVA} \times 1{,}000}{\sqrt{3} \times \text{line voltage}} = \frac{50 \times 1{,}000}{\sqrt{3} \times 400} = 72.2 \text{ A}$$

Short-circuit current,

$$I_{sc} = \frac{\text{Supply voltage (line-value)}/\sqrt{3}}{\text{Standstill impedance per phase}} = \frac{400/\sqrt{3}}{0.866} = 266.7 \text{ A}$$

Supply current at start, I_{st} = 100 A

Tap position of auto-transformer,

$$K = \sqrt{\frac{I_{st}}{I_{sc}}} = \sqrt{\frac{100}{266.7}} = 0.6124 \text{ or } 61.24\ \% \textbf{ Ans.}$$

Full-load slip, s = 5 % or 0.05

Ratio of starting torque to full-load torque,

$$\frac{T_{st}}{T_f} = K^2\left[\frac{I_{sc}}{I_f}\right]^2 s_f = 0.6124^2\left[\frac{266.7}{72.2}\right]^2 \times 0.05 = 0.256 \textbf{ Ans.}$$

Note : 3-phase induction motor has been assumed to be star-connected and losses have been neglected.

Example 1.20. A star-connected cage induction motor has the following data : 440 V, 50 Hz, 4-pole, 1460 rpm, $R_s = 2\ \Omega$, $R'_r = 2\ \Omega$, $X'_s = X_r = 3.0\ \Omega$. Calculate starting torque and starting current of this motor at 50 Hz and 10 Hz for V/*f* control.

[A.M.I.E. Sec. B. Electric Drives and Their Control Winter 1996]

Solution : At 50 Hz supply

Supply voltage per phase, $V = \frac{440}{\sqrt{3}} = 254$ V

Equivalent resistance of motor as referred to stator,

$$R_{01} = R_s + R'_r = 2 + 2 = 4\ \Omega,$$

Equivalent reactance of motor as referred to stator,

$$X_{01} = X_s + X'_s = 3 + 3 = 6\ \Omega$$

Starting current,

$$I_{st} = \frac{V}{\sqrt{(R_{01})^2 + (X_{01})^2}} = \frac{254}{\sqrt{4^2 + 6^2}} = 35.23 \text{ A } \textbf{Ans.}$$

Total copper losses in rotor

$$= 3I_{st}^2 \times R'_r = 3 \times (35.23)^2 \times 2 = 7{,}446 \text{ watts}$$

Rotor power input at standstill,

$$P_2 = \text{Total copper losses} = 7{,}446 \text{ watts} \qquad \because s = 1$$

$$\text{Synchronous speed, } N_s = \frac{120f}{P} = \frac{120 \times 50}{4} = 1{,}500 \text{ rpm}$$

$$\text{Starting torque, } T_{st} = \frac{9.55P_2}{N_s} = \frac{9.55 \times 7{,}446}{1{,}500} = 47.41 \text{ N-m } \textbf{Ans.}$$

At 10 Hz supply

$$\text{Supply voltage per phase, } V' = V \times \frac{f'}{f} = 254 \times \frac{10}{50} = 50.8 \text{ V}$$

Equivalent resistance of motor referred to stator,

$R'_{01} = 4\ \Omega$, the same as before

Equivalent reactance of motor as referred to stator,

$$X'_{01} = X_{01} \times \frac{f'}{f} = 6 \times \frac{10}{50} = 1.2\ \Omega$$

Starting current,

$$I'_{st} = \frac{V'}{\sqrt{(R'_{01})^2 + (X_{01})^2}} = \frac{50.8}{\sqrt{4^2 + 1.2^2}} = 12.166 \text{ A } \textbf{Ans.}$$

Power input to rotor at standstill,

$$P'_2 = \text{Total copper losses}$$
$$= 3(I'_{st})^2 R'_r = 3 \times (12.166)^2 \times 2 = 888.1 \text{ watts}$$

$$\text{Synchronous speed, } N'_s = \frac{120 f'}{P} = \frac{120 \times 10}{4} = 300 \text{ rpm}$$

$$\text{Starting torque, } T'_{st} = \frac{P'_2 \times 9.55}{N'_s} = \frac{888.1}{300} \times 9.55 = 28.27 \text{ N-m } \textbf{Ans.}$$

Example 1.21. At particular instant during starting up a drive, following data are recorded.

Motor torque = 100 Nm; Load torque = 30 Nm; Acceleration = 10 revolutions/second². Find the moment of inertia of the drive and specify its units.

[A.M.I.E. Sec. B. Electric Drives and Their Control Winter 1996]

Solution : Motor torque, T_M = 100 Nm

Load torque, T_L = 30 Nm

$$\text{Angular acceleration, } \frac{d\omega}{dt} = 10 \text{ revolutions per second}^2$$
$$= 2\pi \times 10 = 20\pi \text{ rad/s}^2$$

From Eq. (1.44).

Moment of inertia of the drive,

$$J = \frac{T_M - T_L}{d\omega/dt} = \frac{100 - 30}{20\pi} = 1.114 \text{ kg-m}^2 \textbf{ Ans.}$$

Example 1.22. A dc shunt motor delivers rated output of 37.5 kW at 500 rpm. Its moment of inertia is 20 kg-m². Find the time to attain full-load speed when the motor is started from rest against full-load torque. The starting current fluctuates between 1.1 and 1.4 times the full-load current.

[A.M.I.E. Sec. B. Electric Drives and Their Control Summer 1993]

Solution : Full-load torque,

$$T_L = \frac{\text{Output in kW} \times 1{,}000}{2\pi \times \frac{N}{60}} = \frac{37.5 \times 1{,}000 \times 60}{2\pi \times 500} = 716.2 \text{ Nm}$$

The motor is required to be accelerated against full-load. The starting current varies from 1.1 to 1.4 times full-load current, therefore, starting torque fluctuates between 1.1 to 1.4 times full-load torque (in a dc shunt motor torque is proportional to current).

$$\text{Average starting torque, } T_{st} = \frac{1.1 + 1.4}{2} \times 716.2 = 895.25 \text{ Nm}$$

$$\text{Torque available for acceleration, } T_a = T_{st} - T_L$$
$$= 895.25 - 716.2$$
$$= 179.05 \text{ N-m}$$

Time in attaining full-load speed,

$$t_1 = \frac{J\omega_n}{T_M - T_L} \text{ seconds} \qquad \text{Refer to Eq. (1.47)}$$

$$= \frac{J \times 2\pi N/60}{T_a} = \frac{20 \times 2\pi \times 500}{60 \times 179.05} = 5.85 \text{ seconds. } \textbf{Ans.}$$

Example 1.23. A 37.5 kW, 500 rpm shunt motor is used for shop. The stored energy of the machinery averages 660 kg-m per kW. Assuming the load torque during starting to be equal to the full-load torque and the starting current to be twice the full-load current, estimate the time taken to start the motor.

Solution : Load torque during starting,

$$T_L = \text{Full-load torque, } T_f$$
$$= \frac{\text{Output in kW} \times 1{,}000}{2\pi N/60} = \frac{37.5 \times 1{,}000 \times 60}{2\pi \times 500} = 716.2 \text{ N-m}$$

Torque developed by motor during starting,

$$T_M = \frac{\text{Starting current}}{\text{Full-load current}} \times \text{Full-load torque}$$
$$= 2 \times 716.2 = 1{,}432.4 \text{ N-m} \qquad \because \text{shunt motor } T \propto I_a$$

Torque available for acceleration,

$$T_a = T_M - T_L = 1{,}432.4 - 716.2 = 716.2 \text{ N-m}$$

Stored energy of machinery

$$= 37.5 \times 660 \times 9.81 = 242{,}800 \text{ N-m}$$

$$\text{Since kinetic energy stored} = \frac{1}{2} J\omega^2$$

∴ Moment of inertia of the rotating system,

$$J = \frac{\text{KE stored}}{\frac{1}{2}\omega^2} = \frac{242{,}800 \times 2}{\left(\frac{2\pi \times 500}{60}\right)^2} = 177 \text{ kg-m}^2$$

$$\text{Angular acceleration, } \alpha = \frac{d\omega}{dt} = \frac{T_a}{J} = \frac{716.2}{177} = 4.04 \text{ rad/s}^2$$

$$\text{Starting period, } t = \frac{\omega}{\alpha} = \frac{2\pi N/60}{\alpha} = \frac{2\pi \times 500}{60 \times 4.04}$$

$$= 12.95 \text{ seconds } \textbf{Ans.}$$

Example 1.24. A 220 V dc series motor driving a constant load torque runs at 200 radians/second and draws 20 A from the supply. The total resistance of armature and field is 1.0 Ω. The MI of the motor together with load is 5 kg-m². Calculate total energy dissipated in the armature circuit if the motor starts from rest and attains the steady speed of 200 radians/second is 2.5 seconds. *[U.P. Technical Univ. Electric Drives 2008-09]*

Solution : Motor output

$$= V \times I - \text{Power losses in armature and field}$$

$$= 220 \times 20 - 20^2 \times 1.0 = 4{,}000 \text{ W}$$

$$\text{Speed, } \omega_n = 200 \text{ rad/s}$$

$$\text{Load torque, } T_L = \frac{\text{Motor output in watts}}{\omega} = \frac{4{,}000}{200} = 20 \text{ N-m}$$

Starting time, $t_{st} = 2.5$ seconds

$$\text{So } \frac{d\omega}{dt} = \frac{200}{2.5} = 80 \text{ rad/second}^2$$

From Eq. (1.61), we have

$$K = \frac{J\frac{d\omega}{dt} + T_L}{I_a^2} = \frac{5 \times 80 + 20}{20^2} = 1.05$$

From Eq. (1.62), total energy dissipated in the motor circuit

$$W_{st} = \frac{JR_a}{K}\omega_n + \frac{T_L R_a}{K} t_{st}$$

$$= \frac{5 \times 1}{1.05} \times 200 + \frac{20 \times 1}{1.05} \times 2.5 = 1{,}000 \text{ W } \textbf{Ans.}$$

Also, current drawn from supply is constant during starting period load torque being constant, and is equal to 20 A.

So, total energy dissipated,

$$W_{st} = I_{st}^2 R_a \times t_{st} = 20^2 \times 1 \times 2.5 = 1{,}000 \text{ W, same as above.}$$

1.18. SPEED CONTROL OF ELECTRIC MOTORS

Speed control means intentional change of the drive speed to a value required for performing the specific work process. This concept of speed control or adjustment should not be taken to include the natural change in speed which occurs due to change in the load on the drive shaft. The desired change in speed is accomplished by acting accordingly on the drive motor or on the transmission connecting it to the unit it serves to drive. This may be done manually by the operator or by means of some automatic control device.

Any given piece of industrial equipment may have its speed changed or adjusted mechanically by means of stepped pulleys, sets of change gears, variable speed friction clutch mechanism, and other mechanical devices. Historically, this proved to be the first step in transition from non-adjustable speed to adjustable speed drive. The electrical speed control has many economical as well as engineering advantages over mechanical speed control.

The nature of the speed control requirement for an industrial drive depends upon its type. Some drives may require continuous variation of speed for the whole of the range from zero to full speed, or over a portion of this range; while the others may require two or more fixed speeds. Some machines may require creeping speed for adjusting or setting up the work. Such a speed is of the order of few rpm. For most of the drives, however, a control of speed within the range of ± 20% may be suitable.

Some of the important terms related to speed are defined below.

Constant Torque Drive. It is a drive in which the motor shaft torque remains constant over a given speed range. In such a drive, shaft power varies as the speed.

Constant Power Drive. It is a drive in which the motor shaft power, given by the product of shaft torque and speed, remains constant over a given speed range. In a constant power drive, higher torques are available at lower speeds and vice versa. The motor size is always determined by the highest torque requirement at the lowest speed.

$$\text{Percentage speed regulation} = \frac{N_0 - N_f}{N_f} \times 100 \quad ...(1.76)$$

where N_0 and N_f are no-load and full-load speeds respectively.

1.18.1. Speed Control of DC Motors

One of the attractive features the dc motor offers over all other types is the relative ease with which speed control can be achieved and, therefore, dc motors are indispensable for many adjustable speed drives. The various schemes available for speed control can be deduced from the expression of speed for a dc motor which is repeated here with one modification

$$N = K\frac{V - I_a(R + R_a)}{\Phi}$$

The modification involves the inclusion of an external resistance in the armature circuit. The above expression reveals that the speed can be controlled by adjusting any one of the three factors appearing on the right hand side of the expression: (*i*) applied voltage to the armature terminals, V (*ii*) external resistance in the armature circuit, R and (*iii*) flux per pole, Φ. The first two possibilities involve adjustment affecting the armature circuit, whereas the third involves change in the magnetic field. Therefore, speed control methods are broadly classified as *armature control methods* and *field control methods*. Sometimes a combination of the two methods is employed. With armature control the speed decreases as the voltage applied to the armature terminals is reduced, whereas with field control the speed increases as the flux is reduced.

1. *Field Control Methods.* In case of *dc series motors,* the flux can be varied by an one of the following methods.

(i) Field Divertor Method. The field flux can be reduced by shunting a portion of motor current around the series field, thus reducing the excitation mmf and weakening of field. This method is illustrated in Fig. 1.72 (*a*). This method provides speeds above normal because flux is reduced by this method. Lesser the divertor resistance, less the field current, less flux and, therefore, more the speed. This method is convenient as well as economical and provides the speed control range usually not exceeding 2 : 1. This method of speed control is used in electric drives in which the speed should rise sharply as soon as the load falls.

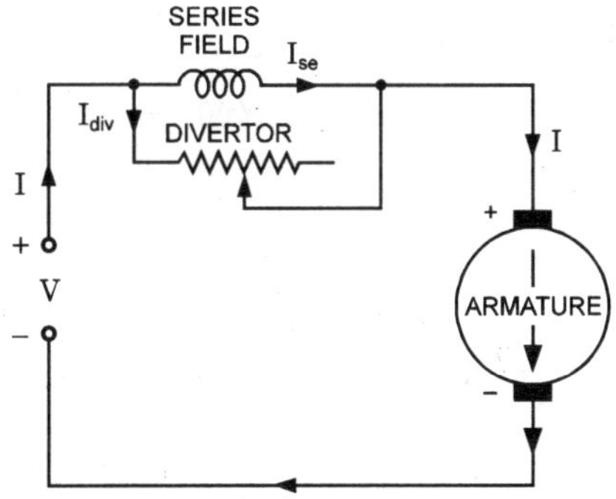

Fig. 1.72 *(a) Field Divertor Method of Speed Control For DC Series Motors*

(ii) Tapped Field Control. This is another method of increasing the speed by reducing the flux and it is accomplised by reducing the number of turns of the field winding through which the current flows. In this method of speed control of dc series motors a number of tappings from the field winding are brought outside, as illustrated in Fig. 1.72(*b*). A number of series field turns can be short circuited according to the requirement. When all field turns are in circuit, the motor runs at the lowest speed and speed increases with cutting out some of the series field turns. This method is usually employed in electric traction.

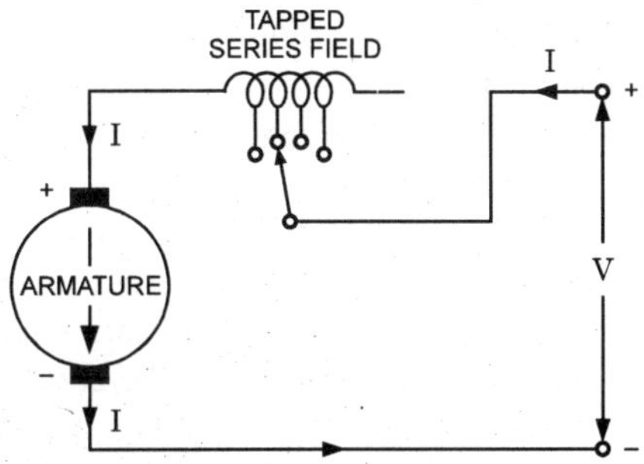

Fig. 1.72 *(b) Tapped Field Control For DC Series Motors*

(iii) Paralleling Field Coils Method. In this method, used in fan motors, several speeds can be obtained by regrouping of field coils, as illustrated in Fig. 1.72 (*c*).

In case of dc **shunt motors** the flux can be varied by inserting a variable resistance in series with the field winding (Refer to Fig. 1.73). Since this resistance has to carry only a small current, so it is made up of slide-wire type of resistor to have continuously variable speed over the range.

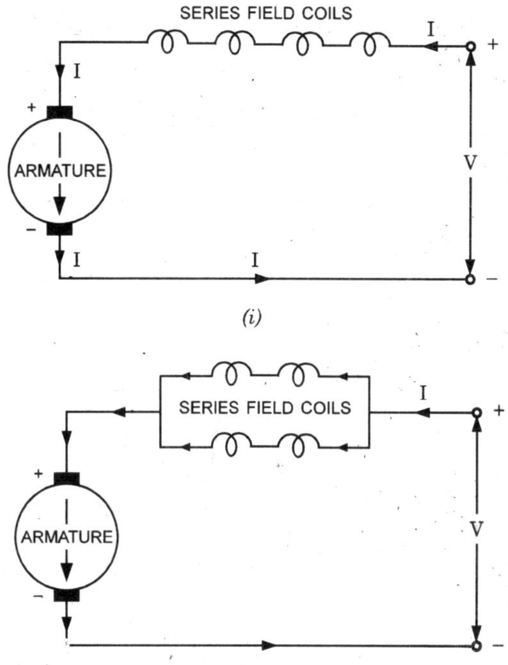

(ii)

Fig. 1.72(*c*)

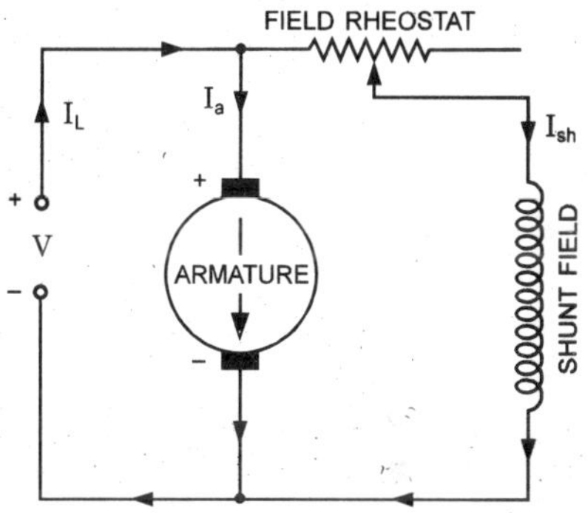

Fig. 1.73. ***Field Rheostat Control For a DC Shunt Motor***

Since the flux can be only decreased (not increased) so the speeds only above normal one can be obtained by this method. The speed is minimum at the maximum value of flux and depends upon the design of the field and its saturation point. The speed is maximum at the minimum value of flux, which is governed by the demagnetizing effect of armature reaction on the field as at higher speeds the motor tends to be unstable and difficulties in commutation arise. The high speed is also restricted due to mechanical considerations as the centrifugal forces are set up at high speeds. Speed variation by this method is limited to ratio of 4 or 5 to 1. Creeping speeds cannot be obtained by this method. The power output being proportional to TN or VI_a remains constant. This method is, therefore, suitable only where power of the load remains constant. This method of speed control is very simple, convenient and most economical and is, therefore, extensively used in electric drives. The power wasted in the controlling resistance is very little, field current being very small. This method of speed control is independent of load on the motor and permits the remote control of speed.

The speed-torque characteristics obtainable with this method are given in Fig. 1.74.

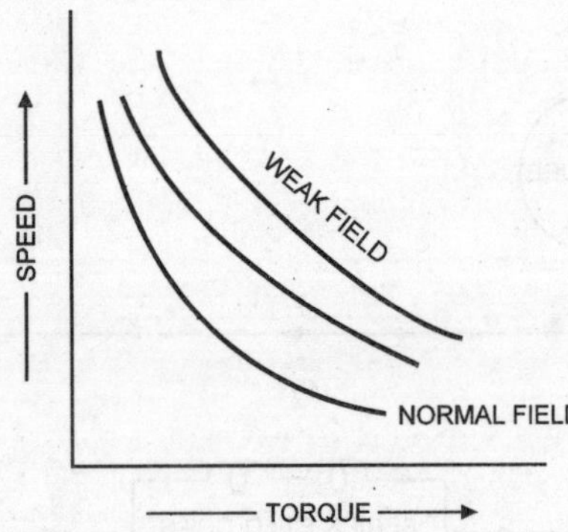

(a) Speed-Torque Characteristics of DC Series Motor With Field Control

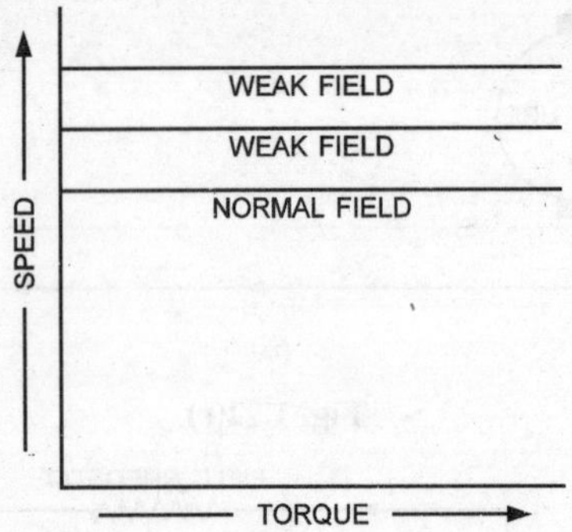

(b) Speed-Torque Characteristics of DC Shunt Motor With Field Control

Fig. 1.74

2. *Armature Resistance Control Methods.* In this method of speed control, reduced speeds are obtained by inserting resistance in the armature circuit. It may be employed with series, shunt and compound wound motors. For the last two types, the series resistance is connected between the shunt field and armature, not between the line and motor.

In case of a dc *shunt motor* an increase in armature circuit resistance will cause more voltage drop in the armature circuit, so the speed will be reduced. Field current will remain unaffected, as the shunt field is directly connected across the supply voltage. For a constant torque load, the armature current remains the same so input to the motor remains the same but the output decreases in proportion to the speeds. Operating costs are, therefore, comparatively high for long time running at reduced speed. In case of fans and centrifugal pumps where the load torque decreases with the decrease in speed, the losses are considerably low and because of its low initial cost and simplicity this method may be quite convenient and economical for short-time or intermittent slowdowns. Wide range of speed (below normal one) can be obtained by this method and at the same time motor will develop any desired torque over its operating range. The main advantage of this method is that speeds below base speed down to creeping speeds of only a few rpm are easily available. This method of speed control, therefore, is employed where speeds lower than rated one are required for a short period only and also occasionally as in printing machines, cranes and hoists where the motor is frequently started and stopped. This method of speed control is also employed where the load drops off rapidly with the decrease in speed, as in fans and blowers.

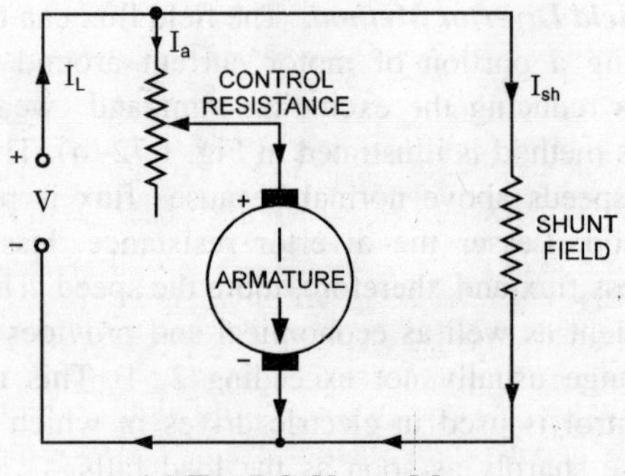

(a) For DC Shunt Motor

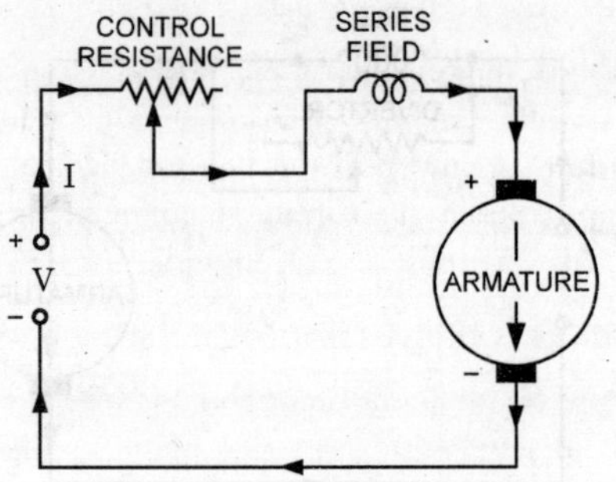

(b) For DC Series Motor

Fig. 1.75

In case of a dc *series motor* an increase in armature circuit resistance will cause more voltage drop in the armature circuit, so speed will be reduced. The torque which is proportional to the product of the flux and armature current will also be reduced with the increase in armature circuit resistance as increase in armature circuit resistance will reduce the armature current and the flux which also depends upon the armature current. In order to obtain different speeds for constant load torque, the armature current is kept constant so the flux. The resistance in the armature circuit is increased to reduce the speed. The drawbacks of armature resistance control for machines with shunt fields are not as important in the speed control of dc series motors.

The poor speed regulation that is inherent in this method has no significance for the control of dc series motors, since the speed characteristic of a dc series motor is a rapidly drooping curve. The power loss in the control resistance for many applications of dc series motors is not too serious, since in these applications the control is utilized for a large portion of time for reducing the speed under light-load conditions and is only employed intermittently when the motor is carrying full load. The speed-torque characteristics of dc series motors with resistance control are shown in Fig. 1.76 (*b*). The maximum range of speed control of about 3 : 1 will be available depending on the load. This method of speed control is most economical for constant torque drives. This method of speed control is employed chiefly for dc series motors driving cranes, hoists, trains etc. because such drives operate on intermittent duty.

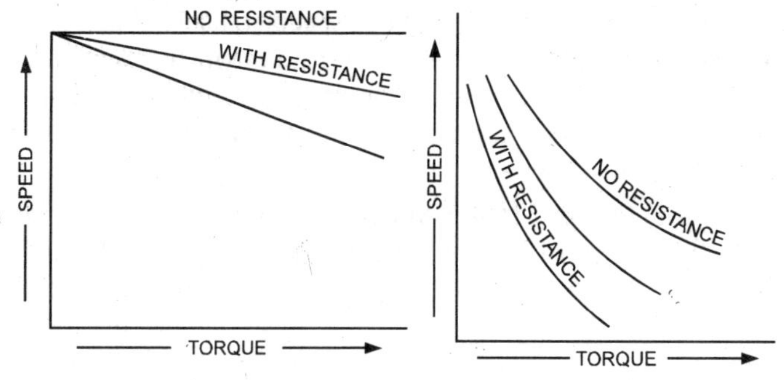

(a) For DC Shunt Motor *(b) For DC Series Motor*

Fig. 1.76. ***Speed-Torque Characteristics With Armature Resistance Control***

The speed-torque characteristics of armature resistance control as applied to dc shunt and series motors are shown in Figs. 1.76 (*a*) and 1.76 (*b*) respectively.

3. *Shunted Armature Control.* The series resistance arrangement described above suffers from a drawback that the speed varies with the load, *i.e.,* the speed increases with the fall in load. This drawback is overcome by shunting the armature with a variable resistance, as shown in Fig. 1.77.

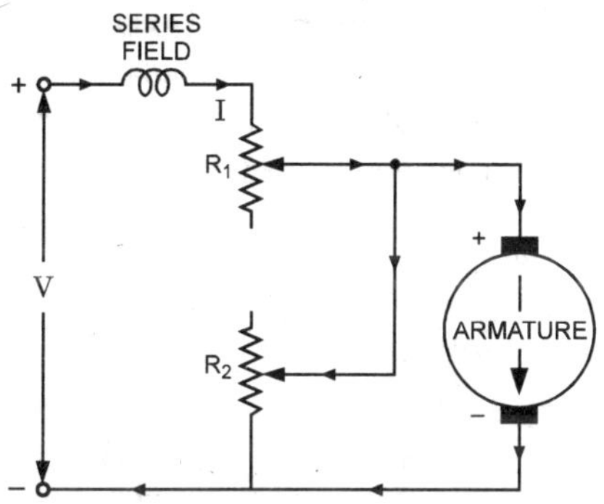

(a) For Series Motor

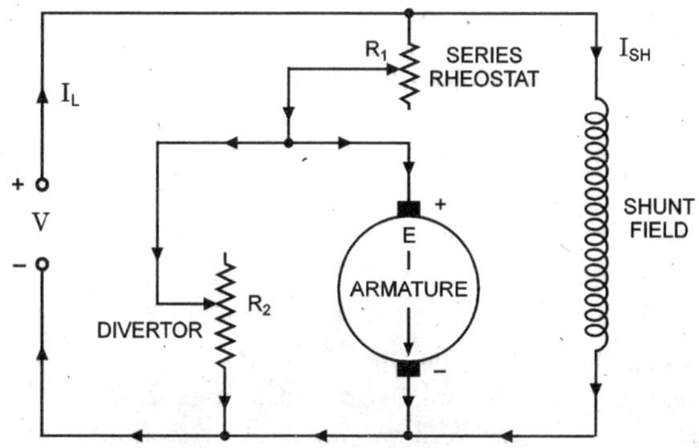

(b) For DC Shunt Motor

Fig. 1.77. ***Shunted Armature Control***

In effect resistors, R_1 and R_2 act as a voltage divider applying a reduced voltage to the armature. Greater flexibility is possible because two resistors may now be adjusted to provide the desired performance. For series motors, no-load speed may be adjusted to a finite, reasonable value, and the scheme is, therefore, applicable to the production of low speeds at light loads. For shunt motors the speed regulation in the low speed range is appreciably improved because the no-load speed is definitely lower than the value with no controlling resistor. The main disadvantages of this method are its low efficiency and the fact that a heavy current is drawn from the supply under certain load conditions.

In Fig. 1.78, natural speed-torque characteristic, speed-torque characteristic with conventional series resistance and shunted armature control are represented by curves I, II and III respectively.

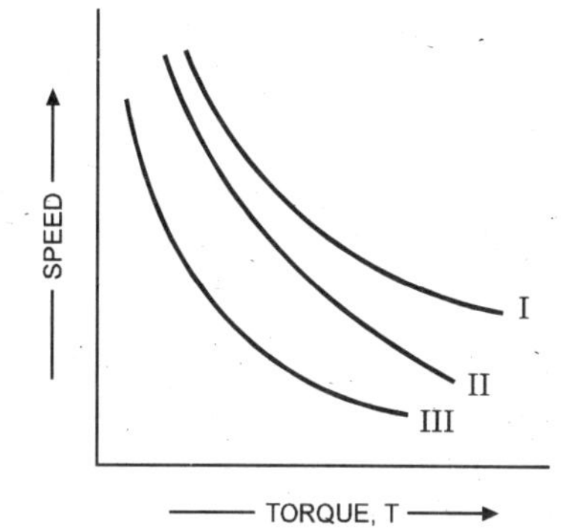

(a) For Series Motor

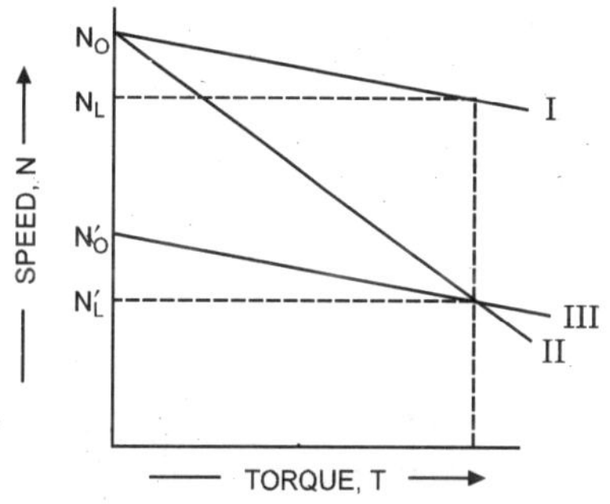

(b) For Shunt Motor

Fig. 1.78. ***Speed-Torque Characteristics***

The shunted armature control is generally employed in drives of comparatively small power rating. This method is also not economical for continuous operation as there is a relatively large power loss in the speed controlling resistances. It is usually employed as a means for sharply reducing the speed so as to achieve accurate stopping.

4. *Armature Voltage Control.* This method of speed control requires a variable source of voltage separate from the source supplying the field current. This method avoids the disadvantages of poor speed regulation and low efficiency which are characteristics of the armature–resistance control method but it is more expensive in initial cost. The adjustable voltage for the armature is obtained from an adjustable voltage generator or from an adjustable electronic rectifier*. This method gives a large speed range with any desired number of speed points. It is essentially a constant–torque system, because the output delivered by the motor decreases with a decrease in applied voltage and a corresponding decrease in speed.

* For most applications dc motors are fed by rectification of the available ac supply. It is usually necessary to insert a transformer immediately ahead of the rectifier.

This particular system has a further advantage that can be employed to provide excellent starting characteristics by bringing the generator voltage gradually up from zero, starting and bringing the motor up to speed with a comparatively slowly increasing voltage. Because of the excellent starting characteristics, this system is used largely for modern high-speed elevators, and on account of the combination of excellent starting characteristics and the wide speed range available, it is used to some extent for reversing planer installations. This method is not applied to any great extent, generally on account of higher initial cost of the generating equipment.

Another drawback of this system is that only the speed below the base speed can be had as the voltage applied to the armature can be reduced from its rated value. If the motor is operated with a voltage across armature terminals higher than the rated value for a long duration of time, the armature winding insulation may get damaged.

5. *Ward-Leonard Method of Speed Control.* The basic adjustable-voltage armature control method of speed control accomplished by means of an adjustable-voltage generator is called the *Ward-Leonard system.* This system consists of simply working the motor with a constant excitation and applying a variable voltage to its armature to provide the required speed. The variable voltage supply is obtained from a motor-generator or convertor set. The scheme is illustrated in Fig. 1.79 (*a*). M_1 is the work motor, powered by the generator G, which is driven by a synchronous or induction motor M_2. The excitation current for the work motor M_1 and the generator G is obtained from the exciter E mounted on the same shaft as the generator. The Ward-Leonard set is started by starting the driving motor. The field rheostat R of the generator is gradually brought out of the circuit as the generator picks up the speed and the work motor begins to rotate. The variable voltage across the terminals of the generator or across the motor is obtained by varying the exciting current of the generator G by means of shunt regulator R. The direction of rotation of motor armature can be reversed by reversing the direction of exciting current of the generator G with the help of reversing switch RS. The converter set runs always in the same direction.

Braking of motor M_1 may be carried out by reducing the generator excitation so that its emf is less than the counter emf of motor M_1. Under these conditions, motor M_1 begins to operate momentarily as a generator, generator G as a motor and ac driving machine M_2 as a generator. As a result kinetic energy of motor M_1 and its load is returned to the supply mains and braking action on the motor M_1 takes place.

Advantages. 1. Very fine speed control over the whole range from zero to normal speed in both directions can be obtained.

2. Uniform acceleration can be obtained.
3. Speed regulation is good.

Disadvantages. 1. Two extra machines are required, so arrangement is costly.

2. Low overall efficiency of the system, especially at light loads.

Applications. This system of speed control is best suited where almost unlimited speed control in either direction of rotation is required such as in steel rolling mills, paper machines, elevators, cranes, mine hoists, diesel-electric propulsion of ships etc.

Ward Leonard Ilgner System, which is a modified form of Ward Leonard system, incorporates a heavy flywheel mounted on the shaft coupling the driving motor M_2 and generator G. In operation with a flywheel, the driving motor has to have a drooping speed-load characteristic *i.e.,* its speed must drop with the increase in load on the shaft. The function of the flywheel is to reduce the fluctuations in the power demand from the supply mains as explained below:

An increase in the load on the shaft causes the work motor to draw more current from the generator so more power is required to drive the later and if there were no flywheel, the driving motor would take all the additional power from the supply line, thus causing sharp fluctuations in it. The heavy flywheel, however, stores a large amount of kinetic energy. When an increase in the generator load causes the driving motor to slow down, some of the kinetic energy of

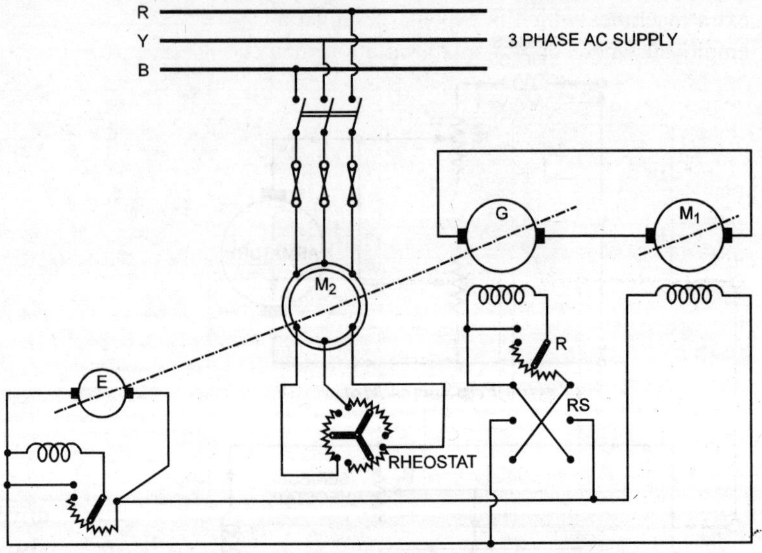

Fig. 1.79 (*a*) *Ward-Leonard Control System*

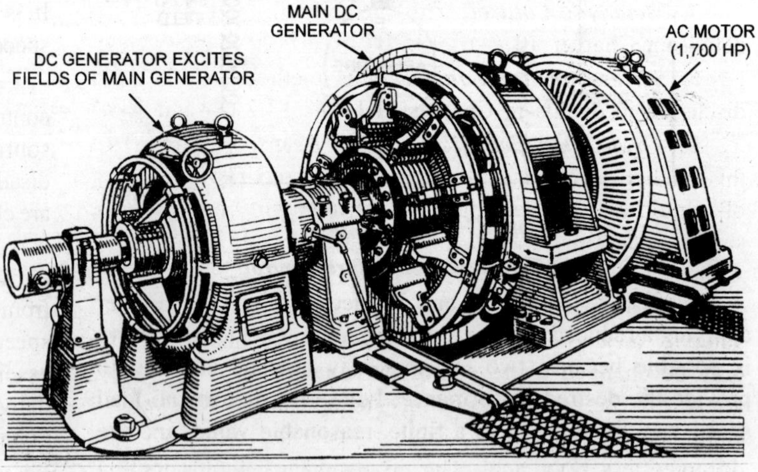

Fig. 1.79 (*b*). *GEC Ward-Leonard Set Used in Mines 1,200 kW Output Generator Controlling a Large Hoist Motor*

the flywheel goes to sustain the peak load on the shaft of the work motor. When the load on the work motor decreases, the driving motor picks up speed and the flywheel stores up kinetic energy.

This system of speed control is employed where the load on the motor shaft sharply varies such as in mine hoists, rolling mills etc.

6. *Booster Speed Control.* A booster in series with the armature of the main motor may be employed for varying the voltage across the driving motor. The booster is a separately excited dc generator, which adds or subtracts the voltage in the motor feeder. The adding (boosting) or subtracting (bucking) depends upon the field current and its direction, which is achieved by parallel resistance arrangement. By this method the speed can be varied over a wide range depending upon the size of the booster. Since booster generator is to provide only the variation in the voltage for boosting and bucking its output, therefore, need not to be as much as that of the main motor inspite of the fact that it has to carry the main motor current. The exact size of the booster however, depends upon the range of speed control required. The chief disadvantages of this method is that it requires two extra machines which is expensive, Generally this method is employed where only a small variation in speed is required.

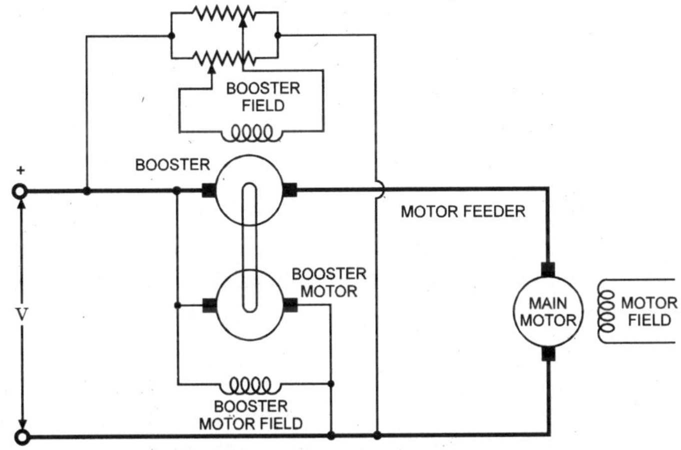

Fig. 1.80. ***Booster Control***

7. *Metadyne Control.* This method will be discussed in details in Chapter 13 (Art. 13.9).

8. *Series-Parallel Control.* This method will be also discussed in details in Chapter 13 [Arts. 13.2 (2) and 13.4]

9. *Grid-Controlled Mercury-Arc Rectifier Method.* In this method, grid-controlled mercury-arc rectifier consisting of a multi-anode steel tank or glass-tube unit or a group of single-anode (thyratron or ignitron) tubes is employed to supply variable voltage to the driving motor.

The output voltage can be varied from zero to maximum value by means of grid control and direction of rotation can only be changed by reversing the connection at the motor armature. The advantage of this method is that the control equipment is static, has a higher operating efficiency and occupies less floor area. The disadvantages of this method are that the power factor on the input side of the rectifier decreases approximately in proportion to speed of the main motor, the control and auxiliary equipment are complicated and it has got no storage capacity so it cannot be employed where large fluctuations of load are involved.

10. *Constant Current System.* For certain services, where there is a possibility of the motors being stalled such as with excavators, ship's windlasses etc. the constant current system is advantageous and is used. In this system a dc generator delivering a constant current by means of a special exciter under all load conditions is employed. The current remains the same through out and the voltage varies with the variation in load. The connections are shown in Fig. 1.81.

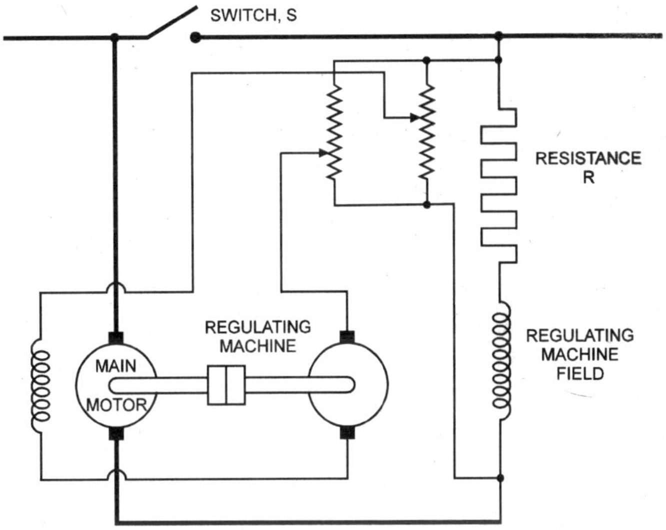

Fig. 1.81. ***Connections For Constant Current Motor***

The field of the main motor is excited from tappings across the resistance R and field winding of the regulating machine so that variable voltage in either direction can be applied to the field system. The regulating machine is mounted on the motor shaft and is so connected in series with the motor field that the emf of the regulating machine opposes the main field current. Any decrease in speed of the motor which may occur due to increase in load, would reduce opposite emf, (emf induced in the regulating machine) and thus field current and torque would increase. The torque would reach maximum when the motor stalls. Speed-torque characteristics for the various positions of the potential divider are shown in Fig. 1.82.

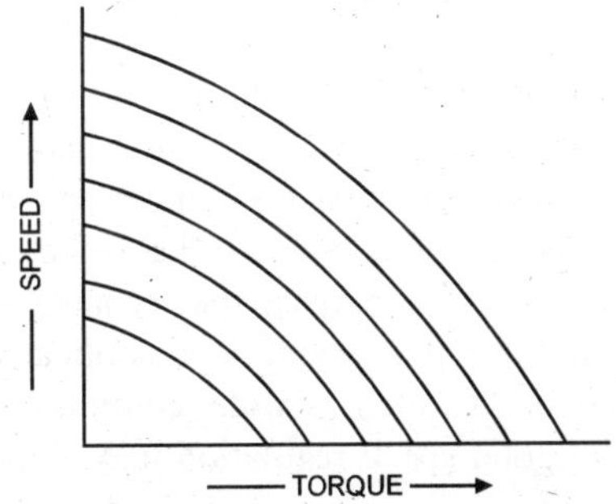

Fig. 1.82. ***Speed-Torque Characteristics of Constant Current Motor***

A full-range of speed control can be had with the maximum torque when the motor stalls without any excessive current.

Any number of motors can be connected in series with each other as the current is independent of load. Thus cable cost and copper losses are reduced. The motor can be stopped simply by closing switch S.

1.18.2. Speed Control of 3-Phase Induction Motors

The problem of speed control of electrical motors in general and of induction motors in particular is of great practical importance.

In a number of industries motors must satisfy very strict speed characteristic requirements, both with respect to the range and smoothness of control, and also with respect to economical operation. From the view point of speed control characteristics, induction motors are inferior to dc motors. The speed of a dc shunt motor can be adjusted between wide range with good efficiency and speed regulation, but in induction motors speed cannot be varied without loosing efficiency and good speed regulation.

The speed of an induction motor is given by the expression $N = \frac{120f}{P}(1-s)$. Thus there are three factors viz, supply frequency f, number of poles P and slip s on which the speed of an induction motor depends. Hence to change the speed of an induction motor it is essential to change at least one of the above three factors.

Methods of speed control are distinguished according to the main action on the motor : (*i*) from the stator side, and (*ii*) from the rotor side.

Various methods of speed control from stator side are (*a*) variation of supply frequency (*b*) variation of applied voltage and (*c*) by changing the number of poles. From the rotor side the speed may be controlled (*a*) by changing the resistance in the rotor circuit and (*b*) by introducing into the rotor circuit an additional emf of the same frequency as the fundamental emf of the rotor.

For the latter method of speed control of induction motors, an additional electrical machine or several such machines are required. A set consisting of a regulated induction motor and one or more additional electrical machines connected to it electrically or mechanically is called *cascade*. Commutator machines are commonly used as the additional machines.

1. *Speed Control By Variation of Supply Frequency.* This method of speed control provides wide speed-control range with gradual variation of the speed throughout this range. The major difficulty with this method is how to get the variable frequency supply. The auxiliary equipment required for this purpose results in a high first cost, increased maintenance and lowering of the overall efficiency. That is why, this method is not employed for general-purpose speed control applications. Inspite of the fact that this scheme is complicated, there are certain applications in which its wide, continuously variable, speed range and good speed regulation makes its use highly desirable.

If an induction motor is to be operated at different frequencies with practically constant values of efficiency, power factor, overload capacity and a constant absolute slip, then, with the iron unsaturated, it is essential that the supply voltage be varied in proportion to supply frequency.

This kind of variation can occur for example, if the supply generator is subjected to the speed variations, either due to momentary overloads or because of a noticeable speed regulation. Both output voltage and frequency vary as the speed if no automatic correction is provided. On some large marine drives, the propeller motors are induction type and are speed controlled from such a local supply which is provided by synchronous generators coupled to variable speed turbines. Even in such cases the range of speed variation is limited, because the efficiency of the prime movers falls rapidly with the change in speed from that for what they are designed.

Typical speed-torque curves for four different frequencies are shown in Fig. 1.25.

2. *Speed Control by Variation of Supply Voltage.* This is a slip-control method with constant frequency variable supply voltage. In this method of speed control of induction motors, the voltage applied to the stator is varied for varying the speed.

This method of speed control is simple, low in first cost and has low maintenance cost but it has limited use because (*i*) the operation at voltages exceeding rated voltage is restricted by magnetic saturation, (*ii*) a large change in voltage is required for a relatively small change in speed (*iii*) the developed torque reduces greatly with the reduction in supply voltage and the motor will come to standstill if the load torque exceeds the pull-out torque of the motor corresponding to reduced supply voltage and (*iv*) the range of speed control is very limited in the downward direction *i.e.*, from rated speed to lower speeds.

In general, this method of speed control is used only on loads where the torque required drops off considerably as the speed is reduced such as with small squirrel cage motors driving fans.

The variable voltage may be obtained by means of either saturable reactors, variac, or tap-changing transformers.

The speed-torque characteristics with reduced voltages are illustrated in Fig. 1.24.

3. *Speed Control By Changing The Number of Poles.* This method is easily applicable to squirrel cage motors because a cage winding automatically reacts to create the same number of poles as the stator. This method of speed control is generally not practicable with wound rotor motors as in such machines this method would involve considerable complications of design and switching, since the interconnections of both primary and secondary would have to be changed simultaneously in a manner to produce the same number of poles in both windings. Otherwise, negative torque will be developed by certain of the rotor conductor belts.

The number of pole pairs in the stator can be changed as follows :

(a) By Using Multiple Stator Windings. In this method of speed control two or more completely independent windings, each wound for different number of poles, are placed in the same stator slots. The number of stator winding poles are in this case in no way interrelated and can be arbitrarily chosen depending on the operating conditions of this motor. For example, a two speed motor may have two stator windings, one wound for 4 poles and another for 6 poles which will give synchronous speeds of 1,500 rpm and 1,000 rpm with a supply frequency of 50 Hz. Motors with four independent stator windings are also used sometimes and they give four different synchronous (and hence running) speeds. Of course, one winding is used at a time, the others being entirely inoperative. Change-over from one speed to the another may be done by a mechanical switch or by contactors. With such an arrangement, the winding or windings which is/are not being used must be kept open circuited by the switch, or at least left in star. Otherwise, because of transformer action, the winding which is connected to the supply would induce voltages in the idle winding(*s*) and cause overheating due to the subsequent circulating currents.

The regulation itself amounts to changing the motor speed in steps as one or another stator winding is connected to the supply mains. At each change, the motor finds itself under conditions essentially analogous to the starting conditions, *i.e.*, a large current and reduced torque. Because of inactive conductors, this method of pole changing requires a large stator than a single-speed motor of the same rating. This method has been used for elevator motors, traction motors and also for small motors driving machine tools.

(b) Consequent Pole Method. The method of pole changing for speed control of 3-phase induction motor described above in (*a*) *i.e.*, multiple stator winding method, is employed only in small motors. In large motors, it is more expedient to employ *consequent pole technique* which eliminates the necessity of having multiple stator windings for changing the number of stator poles. This method of speed control makes use of the entire winding for each speed. Normally, for a single speed motor, successive coil groups are wound so as to create poles of opposite polarity in succession. This is illustrated in Fig. 1.83, where one phase of a four-pole winding is developed.

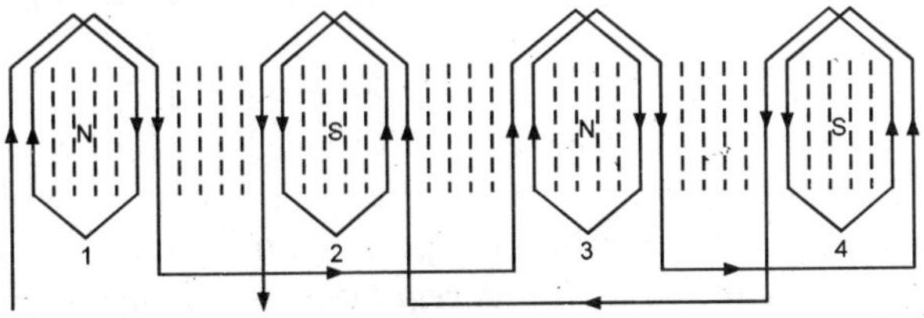

Fig. 1.83. ***Four-Pole Conventional Winding, Series Arrangement***

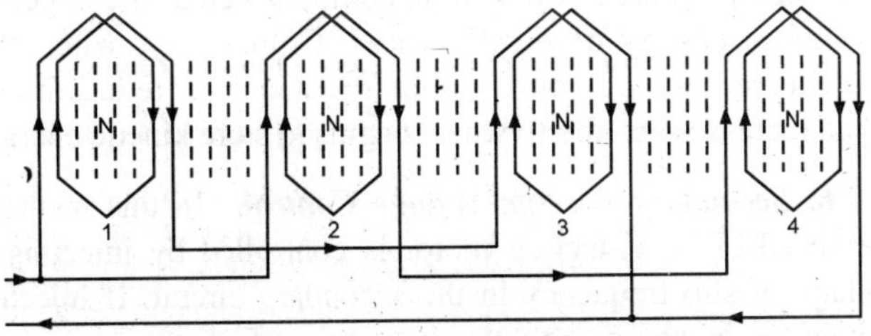

Fig. 1.84. ***Consequent-Pole Method For 8-pole Winding, Series-Parallel Arrangement***

It should be noted that all the coils are in series but that the end connections are made to alternate groups, *i.e.,* group 1 to group 3, and group 4 to group 2. Now the end connections of the groups, with the same coils, can be changed so as to have successive coils of the same polarity, as illustrated in Fig. 1.84. By paralleling the series connections of groups 2-4 with that groups 1-3, the polarity of group 2 and 4 has changed, so that there are now four successive north poles. As a consequence of this south poles are created between each of the north poles, and the motor now has an 8-pole winding. Thus for a 50 Hz supply, the series arrangement gives a synchronous speed of 1,500 rpm while the series-parallel arrangement gives a synchronous speed of 750 rpm. It is also possible to use a series parallel arrangement for the 4-pole winding and then switch to the series arrangement for 8-pole winding. A circuit for switching a three phase winding from 2 to 4 pole winding is shown in Fig. 1.85.

By reconnecting all the coil groups for the same polarity, the stator acts as though it has twice as many poles as pole groups. The additional poles so created are called the *consequent poles,* and the winding is known as a *consequent pole winding.* By an unique switching arrangement, it is possible to have the standard (or conventional) connection on one side of a double-throw switch and the consequent pole connection on the other side of the switch. Thus two speeds

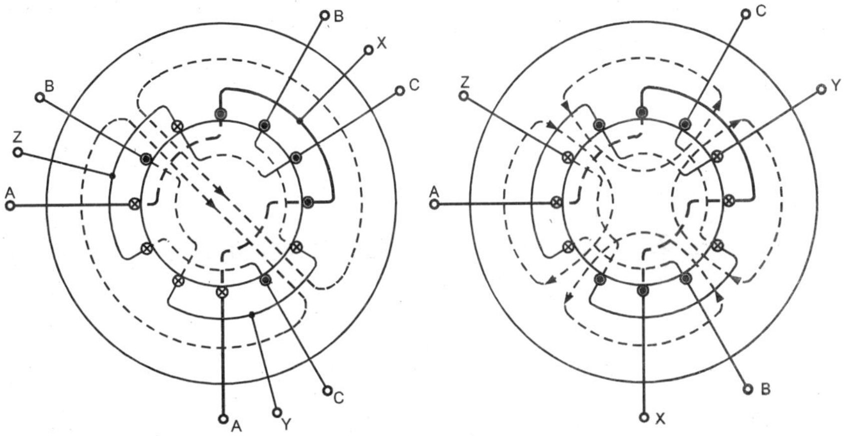

Fig. 1.85. ***Switching Stator Winding From Two to Four Poles***

can be obtained, the higher speed with the conventional connection and half speed with consequent-pole connection.

In the motors using consequent-pole technique for speed control, the best possible design is not usually obtainable at both speeds. That is, desirable characteristics, such as high power factor, are sacrificed at one speed in order that a reasonably good power factor be obtained at the other speed. Sometimes the stator connections are changed from delta to star at the same time that the pole connections are changed. This changes the voltage and makes possible a better motor at each speed.

Two-speed motors are usually made with one winding on the stator, the number of poles being changed in the ratio 1:2. Three- and four-speed motors are provided with two windings on the stators, one or both of which are made with changing the number of poles. For example if it is desired to obtain a motor for four synchronous speeds; 1500, 1,000, 750 and 500 rpm, two windings should be placed on the stator, one of which giving 4 and 8 poles and the other giving 6 and 12 poles.

Employing a series-parallel connection of phase groups of individual phases, the phases can themselves be connected in star/delta resulting in two-speed operation with the three types of torque-speed characteristics viz *constant torque, constant power output* and *variable torque.*

Owing to complications in the design and switching of interconnection of the stator windings, it is not feasible to obtain more than four speeds for any motor through this method of speed control. This method also does not provide gradual speed control. That is why, this method is not employed for general-purpose speed control applications. However, this method is very satisfactory for applications, such as ventilating fans, conveyors, machine tools, or other applications which require operation at only two or four approximately constant speeds which it will provide. This method has the advantage of simplicity, good speed regulation for each setting, high efficiency and moderate initial cost and maintenance.

The number of pole pairs on the stator can also be changed by using *pole amplitude modulation technique*–a suppressed-carrier modulation technique. This method of pole changing has the advantage that the ratio of the two speeds obtainable need not be necessarily 2:1 as in the consequent pole method. A comprehensive discussion of this method of speed control will be beyond the scope of this book. For details, if required, AC Machines by JB Gupta or any other standard book on "AC Machines" be referred.

4. *Stator Impedance Control.* Reduced voltage at the machine terminals can be had by inserting balanced resistors or inductors in series with each stator lead, as illustrated in Fig. 1.68 (*a*). Under such conditions, the motor terminal voltage becomes dependent on motor current and the voltage changes with the acceleration of motor. Typical-speed torque characteristics are illustrated in Fig. 1.86.

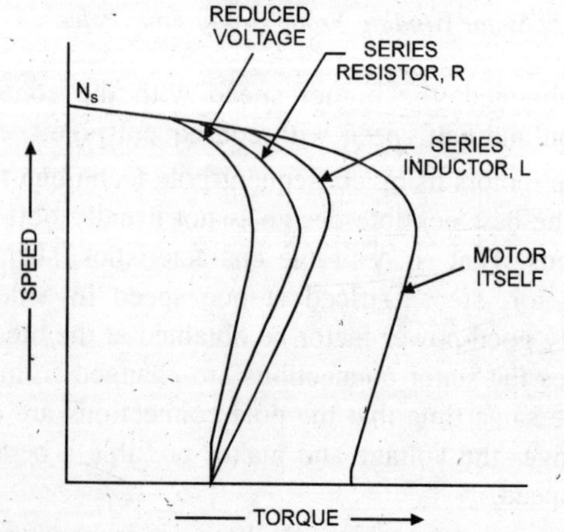

Fig. 1.86. ***Speed-Torque Characteristics with Stator Impedance Control***

If impedance (R or X) is adjusted so as to provide the same starting torque, the speed-torque characteristic in case of addition of inductor would have larger torque than with additional resistance. Besides, both of these characteristics enable us to have torques larger than that obtainable with the characteristic with reduced applied voltage, for same starting torque. Insertion of resistor in the stator circuit improves the power factor, but at the cost of greater energy losses.

5. *Speed Control by Variation of Rotor Resistance (or Rotor Resistance Control).* Wound rotor motors are usually started by connecting starting resistances in the secondary circuit, which are shorted out in steps as the motor speeds up. If the ohmic values of these resistances are properly chosen and if these resistances are designed for continuous operation, they can serve dual purpose, starting and speed control. As obvious this method can be applied to wound rotor motors alone. Smoothness of speed control depends upon the number of steps of resistances that are available. Control is provided in the direction downward from the base speed. The range of speed control is not constant and depends on the value of the load. This method of speed-control has characteristics similar to those of dc shunt motor speed control by means of resistance in series with the armature. Its main disadvantages are lower efficiency and poor speed regulation due to increase in rotor resistance at low-speed settings. For instance, at 50 per cent of synchronous speed the efficiency of rotor is only 50 per cent and the efficiency of the motor would be somewhat less. The speed variation is limited to a range of 50-100% only. Liquid rheostats are preferred over metallic rheostats as they have no inductance and the resistances can be cut out smoothly giving a uniform torque.

The introduction of external resistors in the rotor circuit of a slip-ring induction motor will modify the speed-torque characteristics as illustrated in Fig. 1.26.

This method of speed control as such is, therefore, not suitable for controlling speed at constant torque. But this method is widely used for loads where the torque required drops off considerably as the speed is reduced such as fan loads for which the power input drops noticeably as the speed is reduced which in turn reduces the rotor copper losses. Moreover, this method is not adopted for continuous speed control but is preferred for intermittent (short time) operation. Such a method of speed control is widely used in practice, for low-power motors and in overhead cranes. It is sometimes used however, for speed regulation of rolling mills especially where they are provided with flywheels for reducing the load peaks in the circuit. Here the rheostat, called the *speed regulator,* is automatically switched on when the load increases, and as a result speed decreases and part of the load is compensated at the expense of kinetic energy of the flywheel. Conversely, when the load decreases, the rotor circuit resistance is reduced; the speed increases and the flywheel begins to store kinetic energy.

6. *Secondary Foreign Voltage Control.* In this method, the speed of an induction motor is controlled by injecting a voltage of slip frequency in the secondary circuit. If injected voltage is in phase with the induced emf in the secondary circuit, it is equivalent to decreasing of resistance in the secondary circuit, therefore, slip decreases or speed increases. It is possible to achieve super-synchronous speeds by injecting a voltage of slip frequency in phase with the induced emf in the rotor circuit. If injected voltage is in phase opposition to the induced emf in the secondary circuit, it is equivalent to increasing of resistance in the secondary circuit, hence resulting in increase in slip or reduction in speed. Thus a large range of speed control is possible by this method by including frequency-converting equipment in the rotor circuit.

In this method the disadvantages of lower efficiency and poor speed regulation are overcome, but, since it requires one and usually more auxiliary machines to inject emf of slip frequency in the secondary circuit, therefore, it is costlier and used with motors of very large rating, such as for motors in steel mills.

Several methods have been developed to control speed of induction motors by injecting a voltage of slip frequency in the secondary circuit. The different methods vary in their requirements for auxiliary equipment; from the incorporation of the auxiliary equipment in the design of induction motor itself to a fairly complicated system of auxiliary rotating machines and adjustable ratio transformers. The basic creteria that such a system must fulfill are : (*i*) that the variable foreign voltage have a slip frequency, *i.e.*, frequency equal to s_f and (*ii*) that the energy associated with the foreign voltage be recoverable. A comprehensive discussion of this method of speed control is beyond the scope of this book. The most common systems based on this principle are Kramer system, Leblanc system and Scherbius system of speed control of polyphase induction motors. For details of these speed control systems, AC Machines by J.B. Gupta or any other standard book on AC Machines be referred.

7. *Speed Control by Concatenation.* In this method, two motors are required, at least one of which must have a wound rotor. The two motors may be mechanically coupled together to drive a common load. If the stator winding of one of them, a wound rotor motor, is connected to 3-phase ac supply, and its rotor winding is connected to the stator winding of the second motor, the speed of the combination will be determinated by the sum or difference of the number of poles in the two machines. In practice, it is customary, to connect the rotor output of the first machine to the stator of the second machine in such a way that the revolving fields of both are in the same direction; under this condition, the resulting speed will be determined by the following expression.

$$N = \frac{120f}{P_1 + P_2} \quad ...(1.77)$$

where f is supply frequency, P_1 and P_2 are the number of poles on machines I and II respectively.

For obtaining speed, above normal the torque of the second motor is reversed by simply changing two of the leads of the second motor. This is called the *differential cascading* and under this condition speed will be given by the expression.

$$N = \frac{120f}{P_1 - P_2} \quad ...(1.78)$$

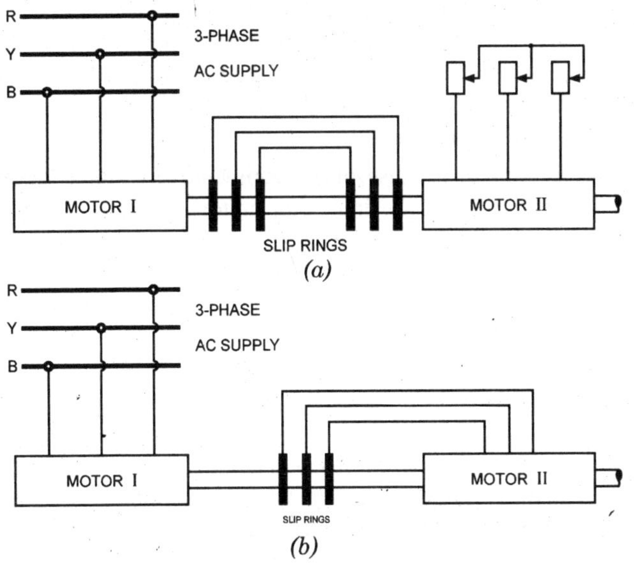

Fig. 1.87. ***Cascade Connections of Induction Motors***

Further speed control, if desired, can be obtained by having second machine also of wound rotor type and inserting control resistance in the rotor circuit of the second machine.

If the two motors have the same number of poles and are connected in cumulative cascade [Eq. (1.77)], they will operate at practically half speed. Since the shaft torque is largely reduced in differential cascade, it is not employed in practice.

The mechanical output of the two machines are approximately in the ratio of $(1 - s_1): s_1$ or $\left(1 - \frac{P_2}{P_1 + P_2}\right) : \frac{P_2}{P_1 + P_2}$ or $P_1 : P_2$, *i.e.*, in the ratio of number of poles on the machines.

1.18.3. Speed Control of Single-Phase Induction Motors

The stator winding of a single-phase induction motor can easily be arranged to give two synchronous speeds, one double the other. In Fig. 1.88 (*a*) the connections for two pole motors

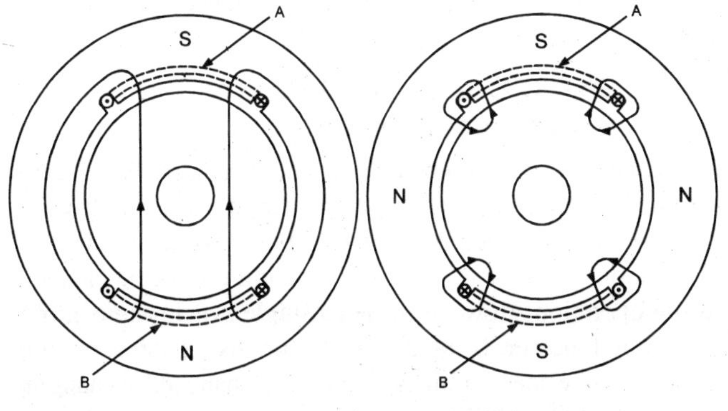

(a) 2-Pole Connection *(b) 4-Pole Connection*

Fig. 1.88

are shown in which two coils A and B have been connected in series with their mmfs helping each other. If the connections to coil B are reversed so that both the coils are again in series but their mmfs opposing each other, as shown in

Fig. 1.88 (*b*), 4 poles are produced and the synchronous speed is reduced to half. Similarly a four-pole motor can be re-connected to give 8 poles. The change from one speed to the other is easily and quickly accomplished by means of a two-pole double-throw switch.

Furthermore, because of the comparatively small power handled by single-phase induction motors, excessive slip to give a slight reduction in speed is not serious, from the stand point of waste of power, as in case of a 3-phase induction motors. For that reason in various cases, such as for fan and blower applications where speeds of 1,100 or 1,300 rpm might be required, it is not unusual to employ 4-pole motor with high slip. This can be accomplished (*i*) by using high rotor resistance, in which case no variation in speed is introduced, but the motor displays a drooping speed curve; or (*ii*) by reducing the voltage applied to the stator winding. For this purpose variable resistance or tapped reactance coil in series with the motor can be employed for speed regulation. This method is inefficient, gives large variations in speed with the variations in load, but it is much used inspite of these draw-backs. Alternatively, the voltage applied to the motor stator winding can be varied either by means of a transformer with taps on its secondary or by means of a variac, which provides the maximum possible number of taps and thus gives finest speed adjustment. This is one of the simplest methods of speed control of fractional kilowatt motors. The speed of the single phase-induction motor can also be controlled by employing tapped exciting winding, so that the constant voltage source may be impressed across the entire winding or some part of it. By this means, a normal speed and a single reduced speed are provided.

Speed control because of its more general use, has brought about the standardization of fan speeds, allowing the use of high slip motors. A squirrel cage motor with 8-10 per cent slip and with low pull-out torque operates satisfactorily not only for constant speed drives, but also for adjustable speed drives through the use of voltage control on the stator winding. With this method of speed control, the torque-speed characteristics resemble with those of a wound rotor motor with different number of external resistances in the rotor circuit.

1.18.4. Speed Control of 3-Phase Synchronous Motors

The speed of a synchronous motor depends upon two factors viz number of poles, P and supply frequency, *f*. The construction of the rotor of a synchronous motor being fixed, the number of poles on the rotor is likewise fixed and cannot be changed during operation. However, when a single synchronous motor is supplied power from an alternator, as in case of ship propulsion, the speed of the motor can be changed by changing the speed of the alternator–the speed of the motor changes exactly in the same proportion as that of the alternator supplying power to it. It is to be noted here that the voltage and frequency are directly proportional to the speed at which the alternator is driven. So no special control is required for maintaining V/*f* ratio constant. Field excitation control, however, is required as voltage applied to the terminals of stator is changed. If not, power factor is greatly affected. Assuming the motor is operating at near unity power factor and normal voltage, increase in speed causes increase in applied voltage across the motor stator terminals and the motor is badly under-excited (*i.e.,* low lagging power factor) for fixed dc excitation. On the other hand, lowering of speed causes overexcitation (*i.e.,* low leading power factor). Thus field excitation control is essential.

1.18.5. Speed Control of AC Commutator Motors

Speed control of ac commutator motors can be had either by varying the applied voltage or by movement of the brushes round the commutator. The former method makes use of tappings on a transformer or, occasionally, a series resistor or inductor so as to provide variation of voltage in steps, while brush shifting provides a gradual control throughout the range.

In case of a **single phase ac series motor,** the speed control can be accomplished, when required, only by voltage variation because it operates with fixed brush position. Transformer tappings provide variation between zero and maximum speed, as depicted in Fig. 1.89, without any appreciable additional loss such as would occur with series resistance. The latter method is employed, only when speed control is required very occasionally, and where capital cost is the prime importance.

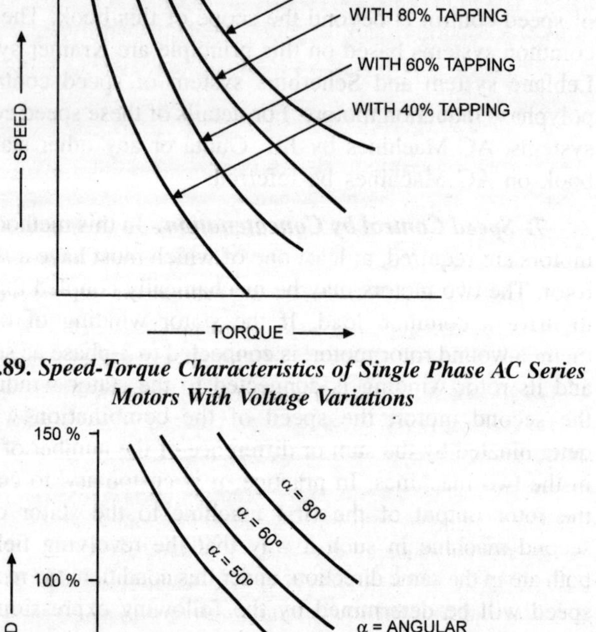

Fig. 1.89. ***Speed-Torque Characteristics of Single Phase AC Series Motors With Voltage Variations***

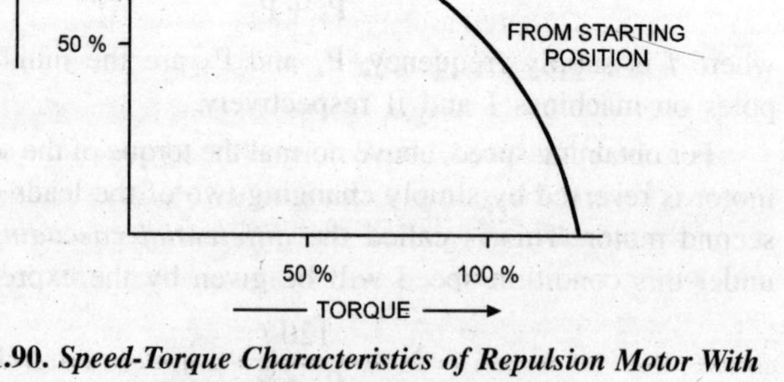

Fig. 1.90. ***Speed-Torque Characteristics of Repulsion Motor With Brush Shifting Speed Control***

In case of *repulsion motors,* brush shifting is the simplest method of speed control for a speed range between about 0.5 and 1.1 times synchronous speed, but beyond this range voltage variation control is preferred due to commutation difficulties. Speed-torque characteristics with different brush positions are shown in Fig. 1.90.

In case of 3-phase series motors, brush shifting is the most economical method of speed control and provides a range of 3 or 4 to 1 with characteristics similar to those of repulsion motors.

Example 1.25. A 220 V dc shunt motor takes 22 A at rated voltage and runs at 1,000 rpm. Its field resistance is 100 Ω and armature circuit resistance is 0.1 Ω. Compute the value of additional resistance required in the armature circuit to reduce the speed to 800 rpm when (*i*) the load torque is proportional to speed (*ii*) the load torque varies as the square of the speed.

[*U.P.S.C. I.E.S. Elec. Engineering II, 1993*]

Solution : In normal conditions

Line current, $I_{L_1} = 22$ A

Shunt field current, $I_{sh} = \dfrac{V}{R_{sh}} = \dfrac{220}{100} = 2.2$ A

Armature current, $I_{a_1} = I_{L_1} - I_{sh} = 22 - 2.2 = 19.8$ A

Back emf, $E_{b_1} = V - I_{a_1} R_a = 220 - 19.8 \times 0.1 = 218.02$ V

Speed, $N_1 = 1{,}000$ rpm

(*i*) Let the additional resistance required in the armature circuit be of R ohms to reduce the speed to 800 rpm when the load torque is proportional to speed.

$$T_2 = T_1 \times \frac{N_2}{N_1} = T_1 \times \frac{800}{1{,}000} = 0.8\, T_1$$

$$\text{or } I_{a_2}\Phi_2 = 0.8\, I_{a_1}\Phi_1 \quad \because T \propto \Phi I_a$$

$$\text{or } I_{a_2} = 0.8 \times 19.8 = 15.84 \text{ A} \quad \because \Phi_2 = \Phi_1$$

$$E_{b_2} = V - I_{a_2}(R + R_a)$$

$$= 220 - 15.84(R + 0.1) = 218.416 - 15.84R$$

$$\text{Since } \frac{E_{b_2}}{E_{b_1}} = \frac{N_2}{N_1} \text{ flux being constant}$$

$$\therefore \frac{218.416 - 15.84R}{218.02} = \frac{800}{1{,}000}$$

or R = 2.778 Ω **Ans.**

(*ii*) Let the additional resistance required in the armature circuit be of R' ohms to reduce the speed to 800 rpm when no load torque varies as the square of the speed

$$T_3 = T_1 \times \left(\frac{N_2}{N_1}\right)^2 = T_1 \times \left(\frac{800}{1{,}000}\right)^2 = 0.64 T_1$$

$$\text{or } I_{a_3}\Phi_3 = 0.64\, I_{a_1}\Phi_1 \quad \because T \propto \Phi I_a$$

$$\text{or } I_{a_3} = 0.64 \times 19.8 = 12.672 \text{ A} \quad \because \Phi_3 = \Phi_1$$

$$E_{b_3} = V - I_{a_3}(R' + R_a)$$

$$= 220 - 12.672\,(R' + 0.1)$$

$$= 218.7328 - 12.672R'$$

$$\text{Since } \frac{E_{b_3}}{E_{b_1}} = \frac{N_3}{N_1} \text{ flux being constant}$$

$$\therefore \frac{218.7328 - 12.672R'}{218.02} = \frac{800}{1{,}000}$$

or R' = 3.497 Ω **Ans.**

Example 1.26. A dc series motor, with unsaturated magnetic circuit and negligible resistance, when running at a certain speed on a given load, takes 50 A at 500 V. If the load torque varies as the cube of the speed, find the resistance to be inserted to reduce the speed by 50%.

[*A.M.I.E. Sec. B. Elec. Drives and Their Control Summer 1993*]

Solution : Motor input current, $I_1 = 50$ A

Speed, $N_1 = N$ (Say)

Speed, $N_2 = 0.5$ N

Back emf, $E_{b_1} = V = 500$ V

Since motor resistance is negligible

Since $T \propto N^3$

$$\therefore \frac{T_2}{T_1} = (0.5)^3 = 0.125$$

and since $T \propto \Phi I$

$$\text{or } \frac{\Phi_2 I_2}{\Phi_1 I_1} = 0.125$$

$$\text{or } \left(\frac{I_2}{I_1}\right)^2 = 0.125 \quad \because \Phi \propto I, \text{ magnetic circuit being unsaturated}$$

$$\text{or } I_2 = I_1\sqrt{0.125} = 50\sqrt{0.125} = 17.678 \text{ A}$$

Let the resistance of R ohms be connected in series with the motor circuit

Back emf, $E_{b_2} = V - I_2R = 500 - 17.678R$

Since $E_b \propto \Phi N$

$$\therefore \frac{E_{b_2}}{E_{b_1}} = \frac{\Phi_2}{\Phi_1} \times \frac{N_2}{N_1}$$

$$\text{or } \frac{500 - 17.678R}{500} = \frac{17.678}{50} \times 0.5$$

or R = 23.283 Ω **Ans.**

Example 1.27. A series motor is required to produce constant full-load torque whilst its speed is raised by 20% above its full-rated value by means of a divertor. Determine the divertor resistance as a percentage of field resistance. Neglect motor losses.

The magnetisation curve of the motor is as follows :

Current	40	60	80	100% of full-load value
Flux per pole	62	80	92	100% of full-load value

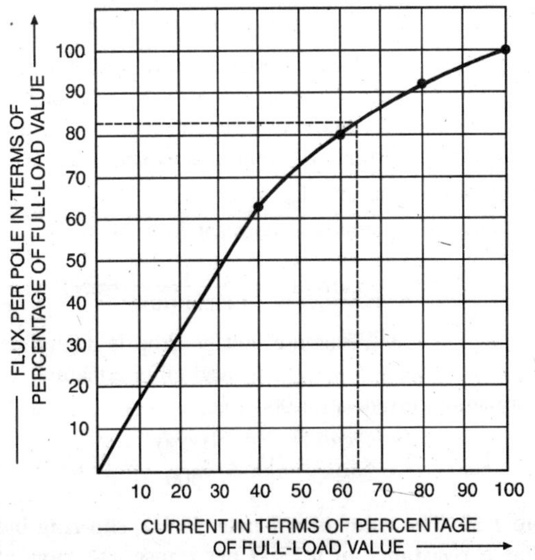

Fig. 1.91

Solution : Let the motor be connected across a supply of V volts and draw current of I amperes while running at normal speed of N rpm and developing full load torque.

Since speed, $N \propto \frac{E_b}{\Phi} \propto \frac{V - IR}{\Phi} \propto \frac{V}{\Phi}$ ∵ losses are negligible

After connecting a divertor, the speed,

$$N_2 = 1.2\ N$$

and since $\frac{N_2}{N_1} = \frac{V/\Phi_2}{V/\Phi_1} = \frac{\Phi_1}{\Phi_2}$

or $\frac{\Phi_1}{\Phi_2} = \frac{1.2\ N}{N} = 1.2$

or $\Phi_2 = \frac{\Phi_1}{1.2} = 83.3\%$ of full load value of flux

From saturation curve the field current corresponding to 83.3% of full-load value of flux is 65% of full-load value of field current.

i.e., $I_{se_2} = 0.65\ I_{se_1} = 0.65\ I_{a_1}$

Since load torque is constant

$$\therefore I_{a_2}\Phi_2 = I_{a_1}\Phi_1$$

or $I_{a_2} = I_{a_1}\frac{\Phi_1}{\Phi_2} = 1.2 I_{a_1}$

Since series field current,

$$I_{se_2} = \frac{R_{div}}{R_{se} + R_{div}} \times I_{a_2}$$

or $0.65\ I_{a_1} = 1.2 I_{a_1} \frac{R_{div}}{R_{se} + R_{div}}$

or $R_{div} = \frac{0.65}{0.55} R_{se} = 1.182\ R_{se}$

i.e., Resistance of divertor is 118.2% of series field resistance. **Ans.**

Example 1.28. Two adjustable speed dc separately excited motors have maximum speeds of 1,000 and 500 rpm. Speed is controlled by variation of field flux. Motor A drives a load requiring constant developed power over the speed range and motor B drives a load requiring constant torque. All losses, including armature resistance and armature reaction may be neglected and magnetic circuit is assumed linear. If the developed power is equal at 500 rpm and armature currents are each 800 A, what will be the armature currents at 1,000 rpm ?

[*A.M.I.E. Sec B. Electric Drives and Their Control Summer 1999*]

Solution : Motor A. Motor operates at developed constant power and, since all losses are negligible, power input to motor A will also be constant. Assuming supply voltage constant, input current and, therefore, armature current will remain unchanged at 800 A. **Ans.**

Motor B. Armature current, $I_{ab} = 800$ A

Since motor B operates at constant load torque,

Armature current $\propto \frac{1}{\text{Field flux}} \propto N$

∵ Armature voltage drop is negligible owing to negligible armature resistance

So armature current at, 1,000 rpm,

$$I'_{ab} = I_{aa} \times \frac{N'}{N} = 800 \times \frac{1{,}000}{500} = 1{,}600 \text{ A} \quad \textbf{Ans.}$$

Example 1.29. The rotor of a 4-pole, 50 Hz, slip-ring induction motor has a resistance of 0.25 Ω per phase and runs at 1,440 rpm at full load. Calculate the external resistance per phase which must be added to lower the speed to 1,200 rpm, the torque being the same as before.

[*A.M.I.E. Sec B. Advanced Elec. Machines Winter 1993; Elec. Machines II Bharathiyar Univ. Nov. 1996; Madurai Kamarajar Univ. Nov. 1996*]

Solution : Synchronous speed of motor,

$$N_s = \frac{120 f}{P} = \frac{120 \times 50}{4} = 1{,}500 \text{ rpm}$$

Full-load speed, N = 1,440 rpm

$$\text{Full-load slip, } s_1 = \frac{N_s - N}{N_s} = \frac{1{,}500 - 1{,}440}{1{,}500} = 0.04$$

After inserting resistance (say R ohms per phase) in the rotor circuit.

Motor speed, $N_2 = 1{,}200$ rpm

$$\text{Slip, } s_2 = \frac{1{,}500 - 1{,}200}{1{,}500} = 0.2$$

$$\text{Since slip, } s = \frac{\text{Rotor copper loss}}{\text{Input power to rotor}} = \frac{3I_2^2 R_2}{\text{Input power to rotor}}$$

For constant power input to rotor and rotor current, $s \propto R_2$

$$\frac{s_2}{s_1} = \frac{R_2 + R}{R_2}$$

∵ for constant load torque power input to rotor and rotor current remains the same

or $\frac{0.2}{0.04} = \frac{0.25 + R}{0.25}$

or R = 1 Ω **Ans.**

Example 1.30. Two 50 Hz, 3-phase induction motors having six and four poles respectively are cumulatively cascaded, the 6-pole motor being connected to the main supply. Determine the frequency of the rotor currents and slips referred to each stator field if the set has a slip of 2%. [*Madras Univ. Elec. Machinery II, 1987*]

Solution : Synchronous speed of the set,

$$N_{sc} = \frac{120 f}{P_1 + P_2} = \frac{120 \times 50}{6 + 4} = 600 \text{ rpm}$$

Actual speed of the set, $N = N_{sc}(1 - s') = 600(1 - 0.02) = 588$ rpm

Synchronous speed of 6-pole motor stator field,

$$N_s = \frac{120 \times f}{P_1} = \frac{120 \times 50}{6} = 1{,}000 \text{ rpm}$$

Slip referred to the stator field,

$$s = \frac{N_s - N}{N_s} = \frac{1{,}000 - 588}{1{,}000} = 0.412 \text{ or } 41.2\% \textbf{ Ans.}$$

The 6-pole motor rotor current frequency or the impressed frequency on 4-pole motor stator field,

$$f' = sf = 0.412 \times 50 = 20.6 \text{ Hz } \textbf{Ans.}$$

The synchronous speed of 4-pole induction motor with frequency f',

$$N'_s = \frac{120 f'}{P_2} = \frac{120 \times 20.6}{4} = 618 \text{ rpm}$$

The slip as referred to the 4-pole induction motor,

$$s'' = \frac{N'_s - N}{N'_s} = \frac{618 - 588}{618} = 0.0485 \text{ or } 4.85\% \textbf{ Ans.}$$

Frequency of rotor current of 4-pole induction motor,

$$f'' = s'' f' = 0.0485 \times 20.6 = 1.0 \text{ Hz } \textbf{Ans.}$$

Example 1.31. A 10 hp, 4-pole, 50 Hz, 220 V induction motor is connected in cascade with another 15 hp, 6-pole, 50 Hz, 220 V motor. What are the possible speeds obtainable with this combination and what is the maximum load in hp which may be delivered without overloading either machine. ? Also determine the ratio of mechanical power outputs of the two machines at this load.

Solution : Synchronous speed of machine I (6-pole machine),

$$N_{s_1} = \frac{120f}{P_1} = \frac{120 \times 50}{6} = 1{,}000 \text{ rpm}$$

Synchronous speed of machine II (4-pole machine),

$$N_{s_2} = \frac{120f}{P_2} = \frac{120 \times 50}{4} = 1{,}500 \text{ rpm}$$

Concantenated speed of the set when cumulatively connected,

$$N_{sc_1} = \frac{120f}{P_1 + P_2} = \frac{120 \times 50}{6+4} = 600 \text{ rpm}$$

Concantenated speed of the set when differentially connected,

$$N_{sc_2} = \frac{120f}{P_1 - P_2} = \frac{120 \times 50}{6-4} = 3{,}000 \text{ rpm}$$

So speeds available are 1,000, 1,500, 600 and 3,000 rpm **Ans.**

In cascade connection the rating of the main motor should be such that it covers the power output of the auxiliary motor also. So the maximum load which can be delivered is 15 hp **Ans.**

The ratio of the mechanical power outputs of the main and auxiliary induction motors are $P_1 : P_2 :: 6 : 4$ or $1.5 : 1$ **Ans.**

Example 1.32. A 3-phase, 440 V, 50 Hz, 110 kW, 24-pole, 245 rpm slip-ring induction motor has both its stator and rotor windings connected in star. The ratio of stator turns to rotor turns is 1.25. The resistance measured between each pair of slip-rings is 0.04 Ω. This motor drives a fan which requires 110 kW at the full speed of the motor. The torque required to drive the fan varies as the square of speed. What resistance should be connected in series with each slip-ring so that the fan will run at 175 rpm. Neglect stator resistance and leakage reactance and rotational losses. *[A.M.I.E. Sec. B. Electrical Drives and Their Control Winter 1993]*

Solution : Synchronous speed of motor,

$$N_s = \frac{120 \times 6}{P} = \frac{120 \times 50}{24} = 250 \text{ rpm}$$

$$\text{Full-load slip, } s_f = \frac{N_s - N}{N_s} = \frac{250 - 245}{250} = 0.02$$

$$\text{Torque at full-load, } T_f = \frac{\text{Output is kW} \times 1{,}000}{2\pi N/60}$$

$$= \frac{110 \times 1{,}000 \times 60}{2\pi \times 245} = 4{,}287.44 \text{ N-m}$$

Rotor resistance per phase,

$$R_2 = \frac{\text{Resistance between a pair of slip-rings}}{2}$$

$$= \frac{0.04}{2} = 0.02\ \Omega$$

Rotor resistance referred to stator,

$$R'_2 = R_2 \times K^2 = 0.02 \times (1.25)^2 = 0.03125\ \Omega$$

Torque developed by rotor is given as

$$T_f = \frac{3}{2\pi N_s/60} \times \frac{V_1^2 R'_2 s_f}{(R'_2)^2 + s_f^2 (X'_2)^2} \qquad \text{Refer to Eq. (1.36)}$$

Substituting $V_1 = \frac{440}{\sqrt{3}}$ V (stator being connected in star), $s_f = 0.02$, $R'_2 = 0.03125\ \Omega$, $T_f = 4{,}287.44$ N-m, and $N_s = 250$ rpm in above equation, we have

$$4{,}287.44 = \frac{3 \times (440)^2 \times 0.03125 \times 0.02 \times 60}{3 \times 2\pi \times 250\left[(0.03125)^2 + (0.02\, X'_2)^2\right]}$$

$$\text{or } (X'_2)^2 = \left[\frac{440^2 \times 0.03125 \times 1.2}{4{,}287.44 \times 500\,\pi} - (0.03125)^2\right] \times \left(\frac{1}{0.02}\right)^2$$

$$= 0.2536$$

$$\text{or } X'_2 = 0.5036\ \Omega$$

$$\text{Slip at 175 rpm, } s = \frac{250 - 175}{250} = 0.3$$

Let the rotor resistance referred to stator per phase corresponding to speed of 175 rpm be R''_2 and torque developed be T

$$\text{So } \frac{T}{T_f} = \left[\frac{N}{N_f}\right]^2 = \left(\frac{175}{245}\right)^2$$

or Torque at 175 rpm,

$$T = T_f \times \left(\frac{175}{245}\right)^2 = 4{,}287.44 \times \left(\frac{175}{245}\right)^2$$

$$= 2{,}187.47 \text{ Nm.}$$

Substituting T = 2,187.47 Nm, $V_1 = \frac{440}{\sqrt{3}}$ V, $s = 0.3$ and $X_2 = 0.5036\ \Omega$ in torque equation, we have

$$2{,}187.47 = \frac{(440)^2 \times R''_2 \times 0.3 \times 60}{2\pi \times 250[(R''_2)^2 + (0.3 \times 0.5036)^2]}$$

$$\text{or } (R''_2)^2 + 0.022825 = 1.0142 R''_2$$

$$\text{or } R''_2 = 0.023\ \Omega \text{ or } 0.991\ \Omega$$

The first value, being lesser than original value, is rejected. So resistance to be added to each slip-ring $= \frac{0.991 - 0.03125}{(1.25)^2}$ $= 0.6142\ \Omega$ **Ans.**

1.19. ELECTRIC BRAKING

If the load is removed from an electric motor and supply to it be disconnected, it will continue to run for some time due to inertia. The time elapsing before it stops will be especially long if the motor is massive and has run at high speed. It is essential, however, in many cases that the motor and its driven machine be stopped quickly (in machine tools, cranes, hoists etc). In fact, quick stopping of a motor is more essential than quick starting. Delay in starting up a motor only causes the machinery to stand idle; a delay in stopping a motor may result in heavy damage to equipment or to the manufactured products and even the loss of human life.

Based on the purpose for which braking is employed, it is of the two types, viz, braking while bringing the drive to rest and braking while lowering loads. In the first type, the device employed for braking absorbs the kinetic energy of the moving parts while in the second one, it absorbs, in addition to the kinetic energy, potential energy, usually gravitational which can drive the system at an excessively high speed.

Braking, while stopping, may be used for any one of the following objectives:

(*i*) Reducing the time taken to stop.

(*ii*) Stopping exactly at specified points, for example in lifts; sometimes such precise stops are necessary for reasons of safety.

(*iii*) Feeding back, atleast a portion of the power, to the supply system.

Braking, while lowering loads, may be employed for any of the following purposes:

(*i*) Controlling the speed at which the load comes down and limiting it to a safe value.

(*ii*) Feeding power back to the supply system.

The main features of a satisfactory brake are that

(*i*) the braking should be quick and reliable in action,

(*ii*) the braking torque must be controllable,

(*iii*) some suitable means must be provided for dissipation of kinetic energy of the moving parts of the motor and its driven machines and

(*iv*) failure of any part of the braking system must result in application of brakes.

Braking torque can be applied either by mechanical (or friction) brakes or electrodynamically. In mechanical braking, the stored energy of rotating parts is dissipated in the form of heat by a brake shoe or band rubbing on a wheel or a brake drum. In electrical (or electrodynamic) braking the stored energy of rotating parts is converted into electrical energy and dissipated in the resistance in the form of heat or returned to the supply. Here we will deal with electric (or electrodynamic) braking only.

Advantages and Disadvantages of Electric Braking Over Mechanical Braking. Before we deal with electric braking it will be desirable to have the knowledge of advantages and limitations of electric braking over mechanical braking.

Advantages. (*i*) Since mechanical braking, due to excessive wear on the brake blocks or brake lining, needs frequent and costly replacement but no such replacement is required in electric braking, hence there is saving on this account.

(*ii*) Frequent adjustments are required to compensate for the wear in mechanical braking thereby making maintenance costly. In electric braking there is no cost by way of maintenance of the brake shoes or lining.

(*iii*) By employing electric braking the capacity of the system can be increased by way of higher speeds and haulage of heavy loads.

(*iv*) In some cases of electric braking, a part of energy is returned to the supply thereby affecting a considerable saving in the operating cost. This is not possible with mechanical braking.

(*v*) In electric braking, heat producd, in no way, is harmful to the braking system where as in mechanical braking heat is produced at brake blocks or brake lining which may result in failure of brake.

(*vi*) Mechanical braking produces metal dust which, if finds ingress to bearings, will cause over-wearing of the bearings. No such dust is produced in electric braking.

(*vii*) Incorrect adjustment of mechanical brakes may result in very harsh and sudden braking being applied which may produce shock loading of ropes and mechanical parts or in case of passenger lifts and trains it may result in the discomfort to the occupants. Electric braking on the other hand is smooth and without shock.

Disadvantages and Limitations. (*i*) In electric braking, the driving motor operates as a generator during the period of braking, and motor ceases to operate as a generator at standstill so that although an electric brake can almost stop a machine or load, but it cannot hold it stationary, therefore, a friction brake is required in addition.

(*ii*) Since the motor has to function as a generator during the braking period, therefore, it must have suitable braking characteristics *i.e.,* choice of motor is limited.

(*iii*) Additional complications, high initial cost, special motors capable of generating electrical energy make electric braking costly.

Economic considerations have a considerable bearing on the use of electric braking.

Types of Electric Braking. There are three types of electric braking, all of which are applicable to the usual types of electric motors, viz plugging (or counter-current braking), dynamic (or rheostatic) braking and regenerative braking.

1. *Plugging or Counter-Current Braking.* This is the simplest type of braking. Plugging or counter-current braking occurs when the motor windings are connected for reverse direction of rotation at a time when the armature is still rotating in the forward direction either under the action of an external torque or due to inertia. Plugging involves reconnection of power supply to the motor so that it tends to drive in the opposite direction. It is obvious that, left to itself, the system will come to rest and then accelerate in the reverse direction. In case, it is required to bring the drive system to rest, it is necessary to include a special device to cut off the supply exactly at the instant when the motor stops.

During the braking period, the energy is drawn from the supply. The energy drawn from the supply and stored or kinetic energy of the rotating parts of motor and its driven machines are dissipated in the series current limiting resistor. Thus this method is wasteful of energy. Other drawbacks of this method of electric braking are the shock caused to the motor and equipment and heavy inrush of current at the time of braking.

This method provides greater braking torque than that provided by rheostatic (or dynamic) braking. This method is used to get either a quick reversal or to get a rapid stop. This method is commonly used in controlling rolling mills, elevators, printing presses, machine tools etc.

This method can be applied to direct current, alternating current induction and synchronous motors.

2. *Dynamic (or Rheostatic) Braking.* In this method of braking the motor is disconnected from the supply and operated as a generator driven by the kinetic energy of the rotating parts of the motor and its driven machines. Thus the kinetic energy of rotation is converted into electrical energy, which is dissipated in the external resistance connected across the motor at the braking instant. By this method, energy required from the supply to brake the motor, has been eliminated as compared to the previous method (plugging) and this is marked advantage. This method of braking can be applied to brake the direct current motors, synchronous motors, and induction motors (with a separate source of dc excitation during braking).

3. *Regenerative Braking.* In the previous two methods of electric braking namely plugging and rheostatic braking stored energy of the rotating parts of the motor and its driven machine is wasted whilst in plugging extra energy is drawn during the braking period and is wasted. In regenerative braking, mechanical energy is converted into electrical energy, part of which is returned to the supply and the rest of the energy is lost as heat in the windings and the bearings of the electrical machines. Regeneration does not, in general, involve any switching operation, unless it is required to change the speed at which it becomes effective. Most of the electrical machines pass smoothly from motoring to generating operation, when overdriven by load.

For regenerative braking it is necessary for (*i*) supply voltage to drop, (*ii*) the motor to be overexcited or (*iii*) the motor to be running at a speed higher than no-load speed. In all of these cases the armature current is reversed and a retarding torque is produced, slowing down the motor until the back emf of the armature becomes equal to the supply voltage. Regenerative braking will not stop the motor. It is effective only for braking hauling loads. They may occur in a crane motor due to action of the lowering load or in the motor of an electric locomotive moving down grade.

1.20. ELECTRIC BRAKING WITH DC MOTORS

1.20.1. Plugging or Counter-Current Braking

In a dc motor a reversed torque is obtained by reversing the current either in the armature or in the field (not both), Polarity reversal of field winding is rarely used because it results in longer braking time due to relatively large inductance of the field winding in comparison to that can be obtained by polarity reversal of armature winding.

The connections for dc series and shunt motors during normal running and braking conditions are shown in Figs. 1.92 and 1.93 respectively. In case of a dc series motor it should be ensured that the direction of flow of current in the field winding remains unchanged when the current flow in the armature winding is reversed.

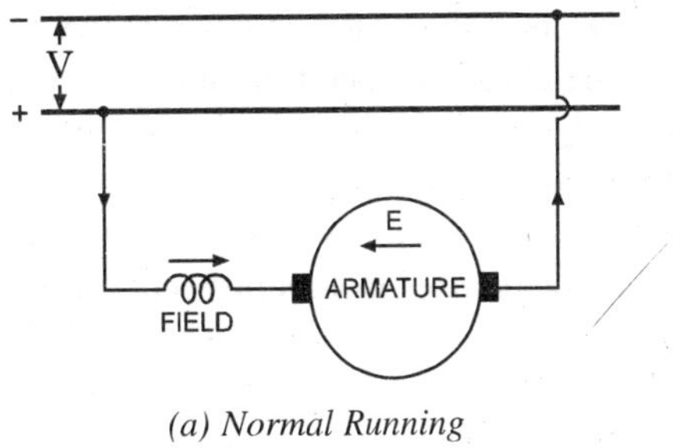

(a) Normal Running

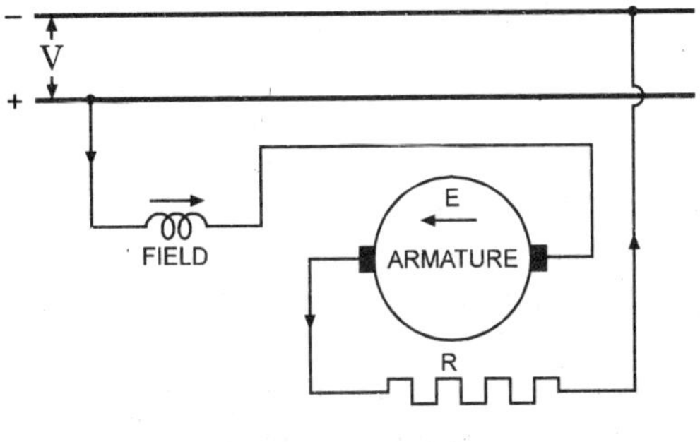

(b) Plugging

Fig. 1.92

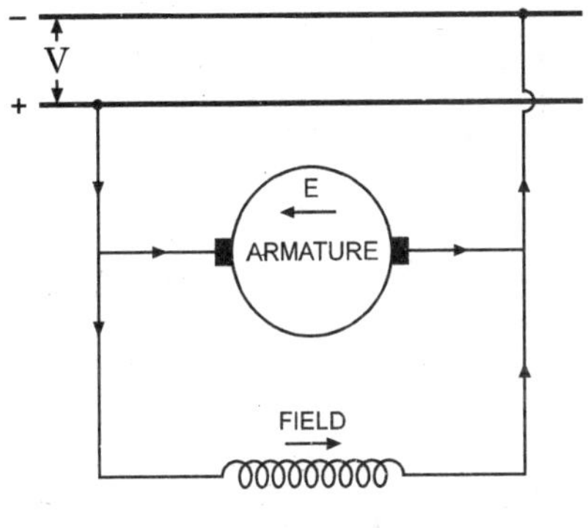

(a) Normal Running

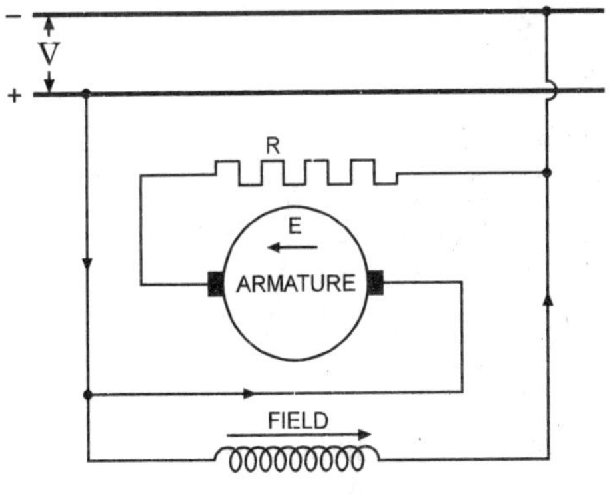

(b) Plugging

Fig. 1.93

In the normal running position the back emf is nearly equal to applied voltage and opposite in direction, so that a small voltage acts across the armature circuit to drive the normal current through a small resistance of the motor. In the plugging position, the induced emf acts in the direction of applied voltage, therefore, at the instant of switching the motor to the plugging position, twice of supply voltage acts across the machine circuit and heavy current would flow (about twice the current drawn by the stationary motor on normal rated voltage). Hence to avoid flow of heavy current and limit it to the safer value, it is necessary that switching performing the plugging operation may also re-insert starting resistance and some additional resistance in series with the armature circuit of the motor.

Figures 1.94 (*a*) and 1.94 (*b*) explain the plugging operation of the dc shunt and series motors respectively on a *quadrantal diagram.*

AB and EF represent the speed-torque characteristics of the motor in normal and reverse direction of rotation without any external resistance, while CL and KG represent the same with external resistances respectively. A is the initial operating point with load, speed N and load torque T_L and the external resistance zero. When plugging is resorted, the motor continues to run in the normal direction but develops torque in reverse direction and point G is the operating point (on characteristic KHG, as resistance is inserted here). The speed falls till it reaches point H (zero) and at this stage supply should be switched off, failing which the motor attains speed in reverse direction under motoring action and operates at point E on switching out external resistance.

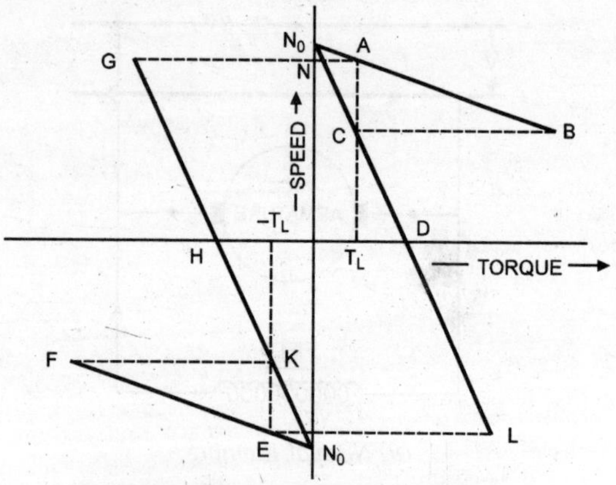

(a) Plugging of a DC Shunt Motor

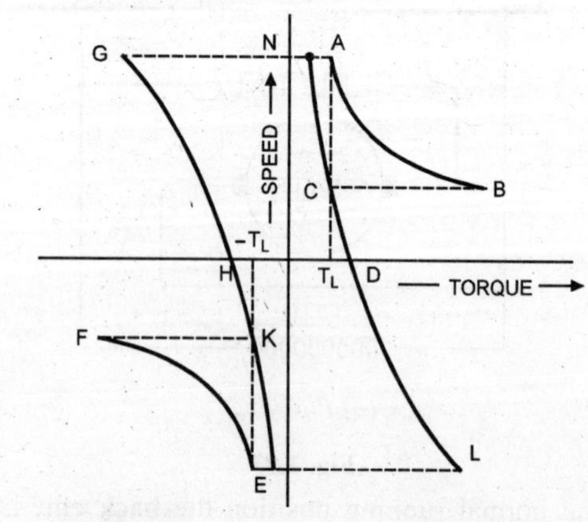

(b) Plugging of a DC Series Motor

Fig. 1.94. ***Quadrantal Diagrams***

1.20.2. Dynamic (or Rheostatic) Braking

1. ***Rheostatic Braking With DC Shunt Motors.*** The armature of a dc shunt motor is disconnected from the supply and connected across a braking resistance, keeping the field connected to the supply, as shown in Fig. 1.95 (*b*). The machine then acts as a generator loaded with a resistance. Since the field current flows as before but the armature current in the reversed direction, therefore, a retarding torque is produced. As the motor speed falls during rheostatic braking, the retarding effect decreases. In order to maintain the retarding torque, some sections of the braking resistance may be then cut out of circuit. If, during rheostatic braking, supply fails, the magnetic flux of the motor will collapse, and there will be no braking effect. This drawback is overcome by providing a series winding in the armature circuit which is connected during the braking period only. The braking effect can be controlled by varying the braking resistance R.

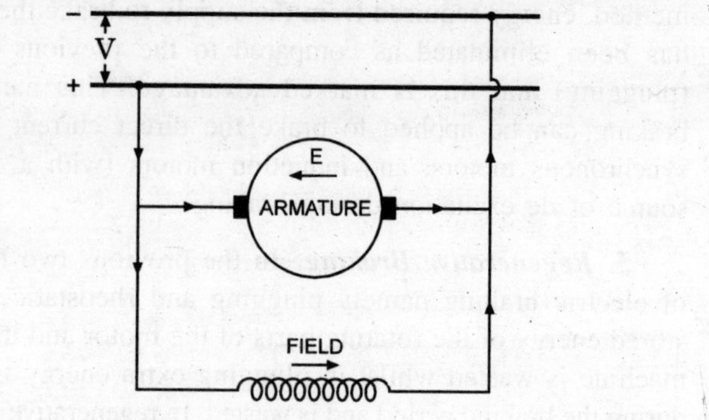

(a) Normal Running

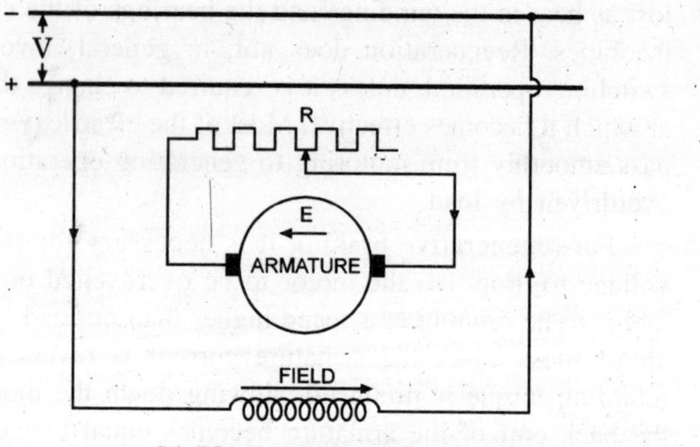

(b) Rheostatic Braking

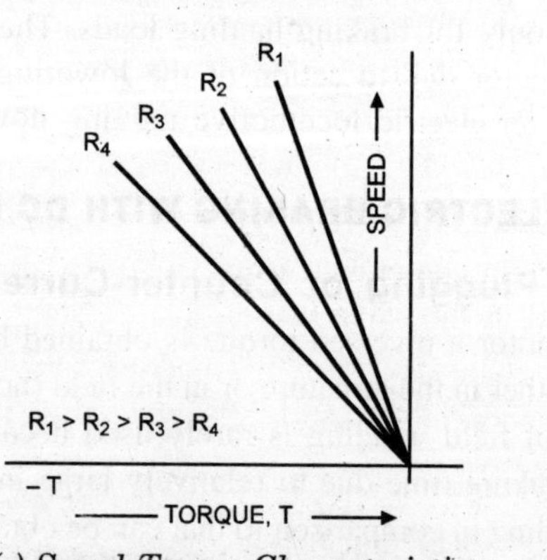

(c) Speed-Torque Characteristics

Fig. 1.95. ***Rheostatic Braking With DC Shunt Motors***

2. ***Rheostatic Braking With DC Series Motors.*** In the case of a series wound direct current motor, when it is disconnected from the supply and is connected across a braking resistance, it is essential that the connections of the field winding are changed over. If this is not done the current through the field will also be reversed, and no braking will occur. To avoid this, the field connections are so changed that the current through the field winding will flow in the same direction as before. The resistance inserted in the circuit must be less than the critical resistance, otherwise the generator will not be self-exciting.

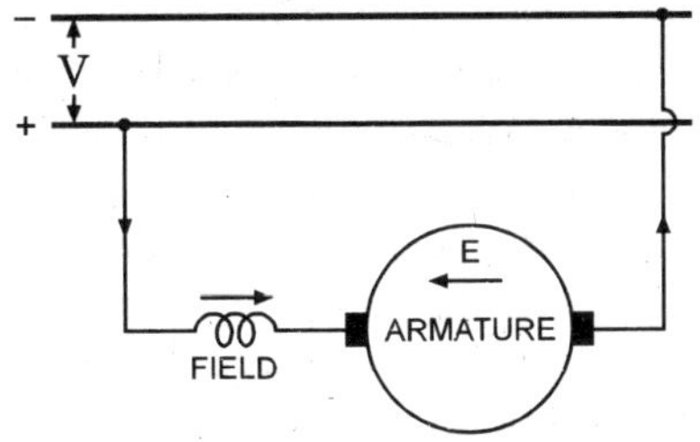

(a) Normal Running

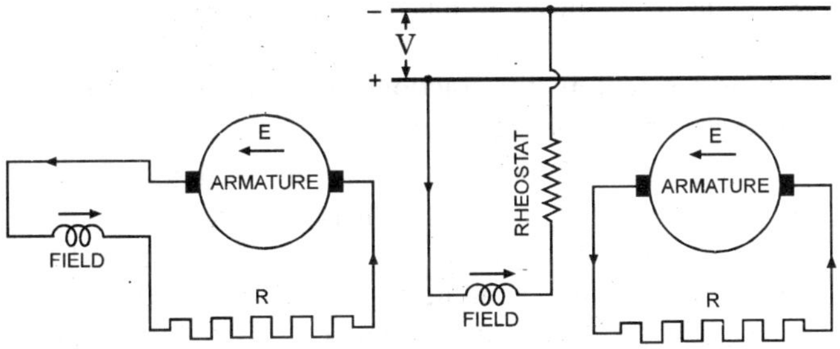

(b) Rheostatic Braking With Self Excitation *(c) Rheostatic Braking With Separate Excitation*

Fig. 1.96

Rheostatic braking of a dc series motor can also be affected by feeding its field winding from an independent source and connecting a resistance across the armature terminals *i.e.,* by reconnecting it as a separately excited machine, as shown in Fig. 1.96 (*c*). An additional resistance is also connected in series with the field winding to limit the field current to a safe value. Since the machine operates as a separately excited generator, the braking characteristics will be same as those for dc shunt motor during dynamic braking.

Dynamic braking with self excitation is normally not used as the field of the machine may build up suddenly and cause the braking torque to jump to high value and the driven unit may experience sudden jerks.

3. *Rheostatic Braking With DC Compound Motors.* For direct current compound wound motors rheostatic braking is performed in the same way as for dc shunt wound motors.

The relation between the braking torque and speed for direct current motors can be determined as below.

$$\text{Electric braking torque} = K_1\Phi I = K_1\Phi.\frac{E}{R}$$

since braking current, $I = \frac{E}{R}$, where E is the induced emf and R is the total resistance in the motor circuit.

$$\text{Again } E \propto \Phi N$$

$$\text{or } E = K_2\Phi N$$

$$\therefore \text{Electric braking torque} = \frac{K_1\Phi}{R} \times K_2\Phi N = K_3\Phi^2 N \quad ...(1.79)$$

In case of series wound motors since flux Φ varies with the variation in current the value of torque therefore, can only be determined graphically from the magnetisation curve.

In case of shunt wound motors since flux remains constant, the electric braking torque, therefore, for any given value of R is directly proportional to speed.

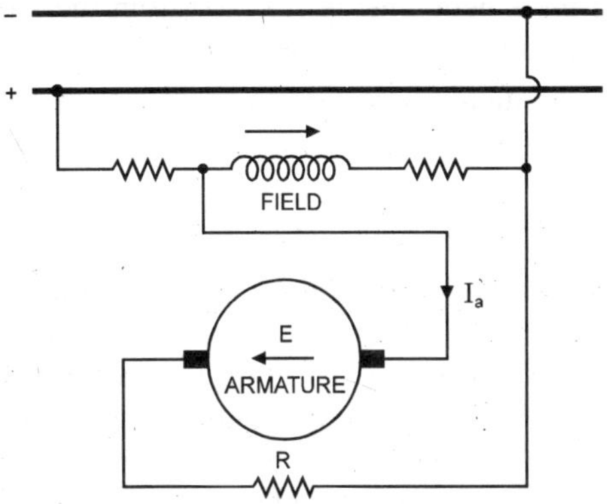

Fig. 1.97

In actual practice, in order to have maximum braking torque resistance in the armature circuit is so reduced with speed that armature current and, therefore, braking torque remains constant till external resistance becomes zero. The braking torque falls linearly with the speed and becomes zero at zero speed. In case of dc series motors with self-excitation speed-torque curve is not linear because magnetic flux changes with the change in armature current which changes with the variations in speed.

In practice rheostatic braking is preferred for non-reversing drives as this simplifies the drive circuit. For reversing drives, plugging is preferred for braking and starting of the motor in the opposite direction in a short time.

Potentiometric Lowering Circuit. Such a circuit is shown in Fig. 1.97 which is just the same as shown in Fig. 1.93 except that the armature carrying reverse current is supplied from a potential divider of which field winding forms a part (not across full supply voltage).

1.20.3. Regenerative Braking

1. *Regenerative Braking With DC Shunt Motors.* Regenerative braking is an inherent characteristic of shunt wound motors and does not require any changing of connections.

Let us consider a shunt motor running in normal condition. Now if due to, say, hauling load, speed increases above normal, due to increase in speed the induced emf will increase and if it exceeds the line voltage, the machine will start supplying current to the line, thus have tendency to prevent any further increase in speed.

Similarly if the field current is increased so that motor emf exceeds the line voltage and it may start supplying current to the line, the motor will slow down to the speed corresponding to this new value of field current.

Regenerative braking can be easily applied to dc shunt motors, particularly in cases where it is required to hold a load at a certain speed for instance lowering a hoist, it is not, however, possible to obtain regenerative braking down to very low speeds, since a sufficient increase in field current cannot be obtained.

Before bringing the motor to rest, the field excitation is increased to the permissible maximum value, as a result of which the speed of the motor is reduced to the minimum value and the kinetic energy released from the motor is fed back to the supply.

Since the motor is requird to operate with a weak field at rated condition, the armature has to be designed to carry a large current to produce the rated torque and hence motor will be larger in size, poor in efficiency and costlier.

DC Shunt Motor in Hoisting Mechanism. Consider a dc shunt motor (or a separately excited motor) used in hoisting mechanism. When it is switched on for lowering a load, the torque developed by the motor and that due to load torque act in unison and, therefore, accelerates the motor. With the increase in speed, induced emf in the armature (*i.e.,* back emf, E) increases till E = applied voltage V. At this moment the speed becomes the ideal no-load speed, the armature current and, therefore, the developed electromagnetic torque is zero, so that the downward motion of the hoist is sustained only by the downward moving load. When the speed exceeds the ideal no-load speed, induced emf E exceeds the applied voltage V and, therefore, armature current becomes negative. The drive, then acts as a generator and provides the braking torque. Under such condition

$$\text{Current } I = \frac{E - V}{R}$$

and electric braking torque,

$$T = K_1 \phi I = K_1 \phi \frac{E - V}{R}$$

$$= K_1 \phi \frac{K_2 N \phi - V}{R}$$

where K_1 and K_2 are constants.

$$\text{or braking torque, } T = \frac{K_1 K_2 \phi^2 N}{R} - \frac{K_1 \phi V}{R}$$

$$= K_3 \phi^2 N - K_4 \phi \qquad \text{...(1.80)}$$

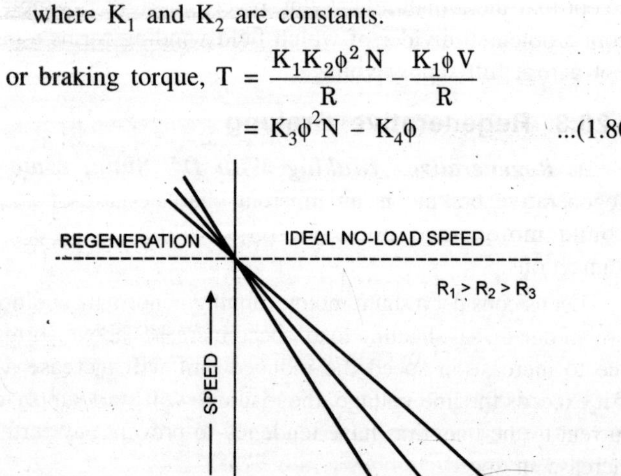

Fig. 1.98. ***Motoring and Regenerative Braking Characteristics of a DC Shunt Motor***

The speed-torque characteristics of motor during motoring and braking operation are shown in Fig. 1.98. From Fig. 1.98 it is obvious that braking or regeneration can occur only at speeds greater than the ideal no-load speed of the motor. From Fig. 1.98 it can also be noted that higher the resistance of the armature circuit, to develop a particular braking torque, higher is the speed required.

2. ***Regenerative Braking With DC Series Motors.*** The dc series motors cannot be used for regenerative braking in an ordinary way. Since the reversal of armature current necessary to produce regeneration would cause a reversal of field, therefore, series field connections must be reversed. But even if the field connections are reversed at the exact moment, this method would still be useless. Because at the instant of reversal, the emf induced in the motor will be small, so current will flow through the field in wrong direction, which will reverse the field and cause the motor emf to help the supply voltage. This will result in short circuit of supply. Due to these complications this method is not used for common industrial purposes. Regenerative braking is, however, used with series motors for traction either by modification of windings or by supplying the machines with separate excitation.

One method of obtaining regenerative braking with series motors is the French method. If there is a single series motor as in case of a trolley buses, tramways, it is provided with a main series field winding and auxiliary field windings connected in parallel with the main series field winding as shown in Fig. 1.99 (*a*). During regeneration (braking period) the auxiliary field windings are put in series with each other and are switched across the supply, as shown in Fig. 1.99 (*b*). The machine acts as a compound generator slightly differentially compounded. Such an arrangement is quite stable. Any change in line voltage causes a change in excitation which produces a corresponding change in the induced emf of the machine so that inherent compensation is provided. For example, if the line voltage increases beyond the emf of the generator the increased voltage across the generator's field will send a large exciting current through it causing the emf of the generator to rise. The reversal of this will happen when the line voltage decreases.

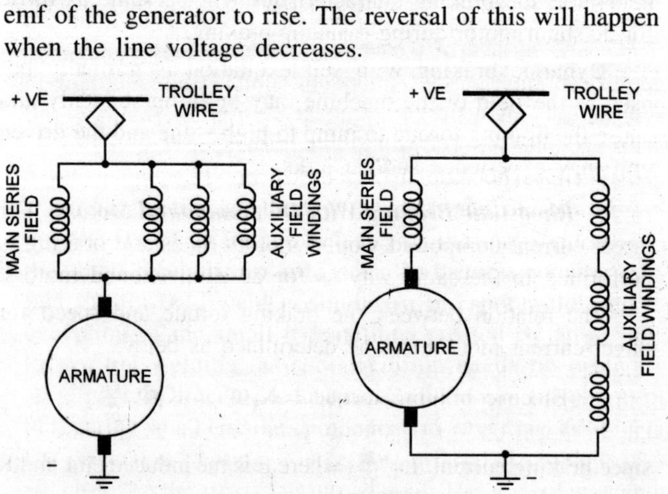

Fig. 1.99. ***French Method of Regenerative Braking***

If there are several motors, we do not require any auxiliary winding. During normal running the motors are connected in parallel with the field winding connected in series with their respective armatures, as shown in Fig. 1.100 (*a*).

But during regeneration the motors are connected, as shown in Fig. 1.100 (*b*), *i.e.*, all armatures are connected in parallel and series field windings of all motors but one are connected in series and placed across the supply. Suitable resistance is also connected in series with the series field windings, as illustrated in the Fig. 1.100 (*b*).

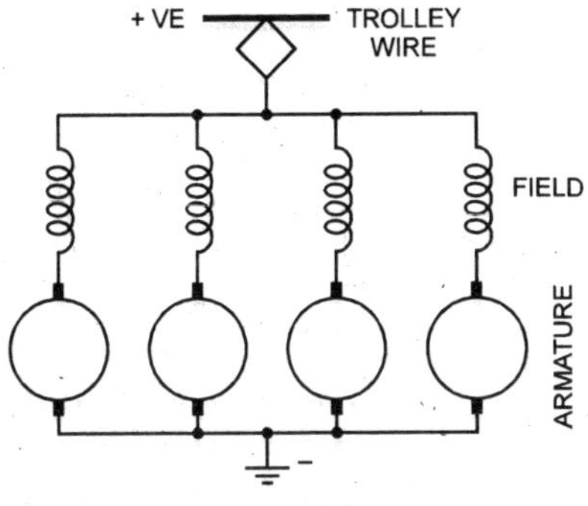

(a)

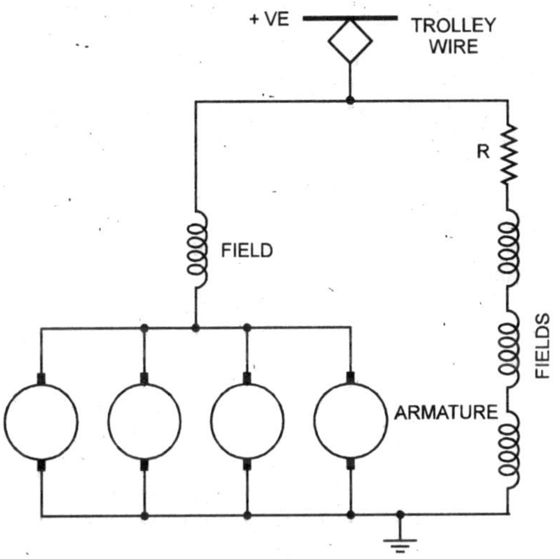

(b)

Fig. 1.100

Another alternative method of regenerative braking, which obviates the losss of power in the resistances in the field circuit, is by using a separate exciter for controlling the excitation of the field windings during regeneration. The exciter may either be axle driven or driven by a motor operated from the auxiliary supply. Many devices have been used to secure compensation against variations in line voltage, most of which make use of some from of differential winding on the exciter. One such method is illustrated in Fig. 1.101 (*a*) in which the exciter is differentially compounded. The shunt field winding of the exciter is separately excited from an auxiliary supply, the series field winding is connected in the main-motor circuit in such a way that it opposes the separately excited winding during regeneration. It can be seen that a rise in line voltage during regeneration, which will tend to reduce the regenerated current, will strengthen the exciter field and counteract the change. The stabilising resistance R also assists in this action.

Another method used for regenerative braking is depicted in Fig. 1.101 (*b*). Here the exciter armature along with the field windings of the series motors are connected across the stabilizing resistance in the main circuit. The current through the stabilizing resistance is the sum of the exciter current and the regenerated current. Any increase in regenerated current, due to fall in line voltage, causes larger voltage drop in the stabilizing resistance and, therefore, less current through the exciter circuit, again counter-acting the effect of line voltage variations.

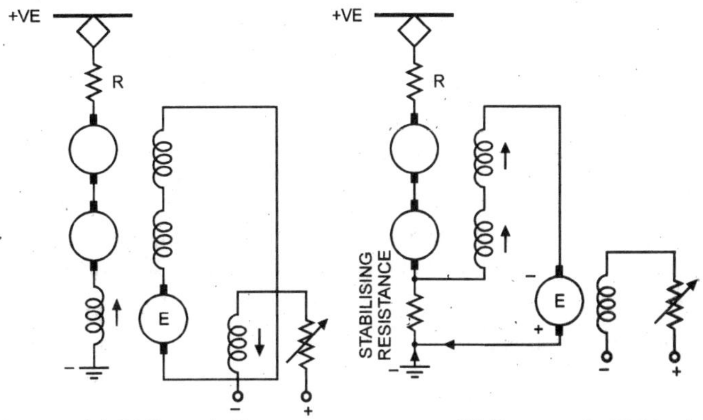

(a) Differentially Compounded Exciter *(b) Separately Excited Exciter*

Fig. 1.101. ***Regenerative Braking With Separate Exciter***

3. *Regenerative Braking With DC Compound Motors*. Compound wound motors can be regeneratively retarded by an hauling load only if their shunt field is stronger than their series field.

Example 1.33. A 220 V separately excited motor running at 1,000 rpm draws 100 A from the source. The motor is braked by plugging. Calculate (*i*) resistance to be inserted in the armature circuit to limit the braking current to twice the full-load current. (*ii*) Initial braking torque. (*iii*) The braking torque when speed has reduced to 500 rpm.

[*U.P. Technical Univ. Electric Drives 2005-06*]

Solution : Full-load motor current,

$$I_f = 100 \text{ A}$$

Supply voltage, $V = 220$ V

$$\text{Full-load torque, } T_f = \frac{V \times I_f}{\frac{2\pi N}{60}} = \frac{220 \times 100 \times 60}{2\pi \times 1000} = 210 \text{ N-m}$$

Neglecting motor losses

Back emf, $E_{b_f} = V = 220$ V Neglecting armature resistance

Voltage across the armature at the instant of braking

$$= V + E_{b_f} = 220 + 220 = 440 \text{ V}$$

(*i*) Resistance to be connected in series with the armature to limit the initial braking current to twice the full-load current (*i.e.* $I_b = 2I_f = 2 \times 100 = 200$ A),

$$R = \frac{V + E_{b_f}}{I_b} - R_a = \frac{440}{200} - 0 = 2.2\ \Omega \quad \textbf{Ans.}$$

(*ii*) Initial braking torque,

$$T_b = T_f \times \frac{I_b}{I_f}$$ $\because T \propto \phi I_a$ and flux in a separately excited motor is assumed to remain constant

$$= 210 \times 2 = 420 \text{ N-m} \quad \textbf{Ans.}$$

(*iii*) When the speed has reduced to 500 rpm, the induced emf falls in the ratio of speed because the flux is assumed to remain constant.

So induced emf, $E'_b = E_{bf} \times \dfrac{N'}{N_f} = 220 \times \dfrac{500}{1{,}000} = 110$ V

Armature braking current,

$$I'_b = \frac{V + E'_b}{R + R_a} = \frac{220 + 110}{2.2} = 150 \text{ A}$$

Electric braking torque,

$$T'_b = T_f \times \frac{I'_b}{I_f} = 210 \times \frac{150}{100} = 315 \text{ N-m } \textbf{Ans.}$$

Example 1.34. A 220 V, 17.6 kW dc shunt motor at its rated speed of 1,200 rpm is to be braked by plugging. The armature resistance is 0.1 Ω and the rated efficiency of the motor is 80%. Calculate : (*i*) the resistance to be connected in series with the armature to limit the initial braking current to twice the rated current, (*ii*) the torque when the speed of the motor falls to 400 rpm. [*U.P. Technical Univ. Electric Drives 2008-09*]

Solution : Full-load (rated) motor current,

$$I_f = \frac{\text{Motor output in kW} \times 1{,}000}{\text{Rated voltage} \times \text{rated efficiency}} = \frac{17.6 \times 1{,}000}{220 \times 0.8} = 100 \text{ A}$$

Full-load torque,

$$T_f = \frac{\text{kW output} \times 1{,}000}{2\pi N / 60} = \frac{17.6 \times 1{,}000}{2\pi \times 1{,}200 / 60} = 140 \text{ N-m}$$

Full-load armature current,

$I_{af} = I_f = 100$ A, neglecting shunt field current

Induced emf in motor,

$E_{bf} = V - I_{af}R_a = 220 - 100 \times 0.1 = 210$ V

Voltage across the armature at the instant of braking

$= V + E_{bf} = 220 + 210 = 430$ V

(*i*) Resistance required to be connected in series with the armature to limit the initial current to twice the rated current (*i.e.* $I_b = 2I_f = 2 \times 100$ = 200 A),

$$R = \frac{V + E_b}{I_b} - R_a = \frac{430}{200} - 0.1 = 2.05 \ \Omega \ \textbf{Ans.}$$

(*ii*) When the speed has fallen to 400 rpm, the induced emf falls in the ratio of speed because in case of a dc shunt motor flux remains constant,

So, induced emf, $E'_b = E_{bf} \times \dfrac{N'}{N} = 210 \times \dfrac{400}{1{,}200} = 70$ V

Armature braking current, $I'_b = \dfrac{V + E'_b}{R + R_a} = \dfrac{220 + 70}{2.05 + 0.1} = 134.9$ A

Electric braking torque, $T'_b = T_f \times \dfrac{I'_b}{I_f} = 140 \times \dfrac{134.9}{100}$

$= 188.84$ N-m **Ans.**

Example 1.35. A 220 V, dc shunt motor having an efficiency of 85% drives a hoist having an efficiency of 80%. Calculate the current drawn from the supply to raise a load of 400 kg at 2.5 m/s. What resistance must be added in the armature circuit in order to lower the load at 2.5 m/s using rheostatic braking, assuming that the efficiency of the motor and load will remain the same. [*U.P. Technical Univ. Electric Drives 2007-08*]

Solution : Work done per second

$= \text{Load in kg} \times g \times v = 400 \times 9.81 \times 2.5 = 9{,}810$ N-m/s or watts

$$\text{Power input} = \frac{\text{Output}}{\eta_m \times \eta_h} = \frac{9{,}810}{0.85 \times 0.8} = 14{,}426.5 \text{ W}$$

Current drawn from supply mains,

$$I = \frac{\text{Power input}}{\text{Rated voltage}} = \frac{14{,}426.5}{220} = 65.6 \text{ A } \textbf{Ans.}$$

When the load is lowered, the machine operates as a generator and the generator output will be dissipated in the resistance connected in the armature circuit.

So, power output of generator,

$P_{out} = \text{Load in kg} \times g \times v \times \eta_m \times \eta_h$

$= 400 \times 9.81 \times 2.5 \times 0.85 \times 0.8 = 6{,}670.8$ watts

If R is the resistance inserted in the armature circuit,

$$\text{Then power dissipated} = \frac{V^2}{R} = 6{,}670.8$$

$$\text{or } R = \frac{V^2}{P_{out}} = \frac{220^2}{6{,}670.8} = 7.255 \ \Omega \ \textbf{Ans.}$$

Example 1.36. The speed of a dc series motor is limited to 250 rpm by application of rheostatic braking system. The total resistance of armature and field of the motor is 1.0 Ω and load exerts a torque of 245 N-m. Calculate the resistance to be connected in the motor circuit, if its magnetisation characteristic at 250 rpm is as follows:

I in Amps	10	20	30	40
E in Volts	90	190	248	295

[*Agra Univ. Traction and Utilization of Elec. Power 1982*]

Solution : Load torque, T = 245 N-m

Input power to machine during rheostatic braking

$$= T \times \frac{2\pi N}{60} = 245 \times 2\pi \times \frac{250}{60}$$

$= 6{,}414$ N-m/s or watts

During rheostatic braking the machine operates as a dc series generator and the electromagnetic power developed by the generator (neglecting rotational losses), *i.e.*, E I should be 6,414 watts.

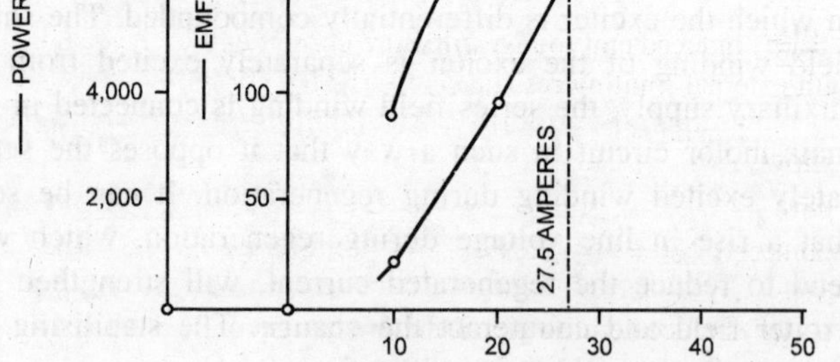

Fig. 1.102

In order to determine E and I separately, plot curves between E and I and power P and I, as shown in Fig. 1.102 from the datas given below :

I in Amps	10	20	30	40
E in Volts	90	190	248	295
Power, P (EI) in watts	900	3,800	7,440	11,800

Current corresponding to power of 6,414 watts is found to be 27.5 A and corresponding induced emf E is 233 V.

Hence total resistance of the circuit $= \frac{E}{I} = \frac{233}{27.5} = 8.47\ \Omega.$

External resistance to be inserted in the motor circuit

$= 8.47 - 1.0 = 7.47\ \Omega$ **Ans.**

Example 1.37. A 500 V, 45 kW, 600 rpm dc shunt motor has a full-load efficiency of 90%. The field resistance is 200 Ω and the armature resistance is 0.2 Ω. Find the speed under each of the following conditions at which the machine will develop an electromagnetic torque equal to rated value—(*i*) regenerative braking: no limiting resistance, (*ii*) plugging: external limiting resistance of 5.5 Ω inserted, (*iii*) dynamic braking: external limiting resistance of 2.6 Ω inserted. The field current is maintained constant and armature reaction and the brush drop may be neglected. [*A.M.I.E. Sec B. Elec. Drives and Their Control Summer 1994*]

Solution : Rated line current,

$$I_{Lf} = \frac{\text{Output in kW} \times 1{,}000}{\text{Supply voltage} \times \text{full-load efficiency}} = \frac{45 \times 1{,}000}{500 \times 0.9} = 100\ A$$

Shunt field current,

$$I_{sh} = \frac{V}{R_{sh}} = \frac{500}{200} = 2.5\ A$$

Armature current on full load, $I_{af} = I_{Lf} - I_{sh} = 100 - 2.5 = 97.5$ A

Induced emf on full load, $E_f = V - I_{af} R_a = 500 - 97.5 \times 0.2 = 480.5$ V

Rated speed, $N_f = 600$ rpm.

Since field current remains constant and the armature reaction and brush drop are negligible so electromagnetic torque developed is directly proportional to armature current and speed is directly proportional to induced emf.

As torque to be developed in each case is equal to rated torque so armature current in each case is equal to armature current at full load *i.e.*, 97.5 A.

(*i*) Induced emf to give armature current of 97.5 A for regenerative braking

$$E_1 = V + I_{af} R_a = 500 + 97.5 \times 0.2 = 519.5\ V$$

and speed, $N_1 = \frac{E_1}{E_f} \times N_f = \frac{519.5}{480.5} \times 600 = 648.7$ rpm **Ans.**

(*ii*) Induced emf to give armature current of 97.5 A for plugging with external limiting resistance of 5.5 Ω

$$E_2 = I_{af} \times (R + R_a) - V = 97.5 \times (5.5 + 0.2) - 500 = 55.75\ V$$

and speed, $N_2 = \frac{E_2}{E_f} \times N_f = \frac{55.75}{480.5} \times 600 = 69.6$ rpm **Ans.**

(*iii*) Induced emf to give armature current of 97.5 A for dynamic braking with external limiting resistance of 2.6 Ω

$$E_3 = I_{af} \times (R + R_a) = 97.5 \times (2.6 + 0.2) = 273\ V$$

and speed, $N_3 = \frac{E_3}{E_f} \times N_f = \frac{273}{480.5} \times 600 = 340.9$ rpm **Ans.**

Example 1.38. A 230 V, 870 rpm, 100 A separately excited dc motor has an armature resistance of 0.05 Ω. It is coupled to an overhauling load with a torque of 400 N-m. Determine the speed at which motor can hold the load by regenerative braking. [*U.P. Technical Univ. Electric Drives 2009-10*]

Solution : Supply voltage, V = 230 V

Armature current, $I_a = 100$ A

Back emf, $E_b = V - I_a R_a = 230 - 100 \times 0.05 = 225$ V

Torque developed by motor, $T = \frac{E_b I_a}{2\pi N / 60} = \frac{225 \times 100 \times 60}{2\pi \times 870} = 247$ N-m

Overhauling load torque, $T_L = 400$ N-m

Assuming field current constant and armature reaction and brush drop negligible electromagnetic torque developed will be directly proportional to armature current and speed will be directly proportional to induced emf

So armature current, $I'_a = I_a \times \frac{T_L}{T} = 100 \times \frac{400}{247} = 162$ A

Induced emf to provide armature current of 162 A for regenerative braking

$$E' = V + I'_a R_a = 230 + 162 \times 0.05 = 238.1\ V$$

and speed, $N' = \frac{E'}{E_b} \times N = \frac{238.1}{230} \times 870 = 900$ rpm **Ans.**

Example 1.39. A 230 V, 500 rpm, 100 A separately excited dc motor has an armature resistance of 0.1 Ω. The motor is driving, under rated conditions, a load whose torque is constant and independent of speed.

Find the amount by which the flux should be reduced to get the motor speed of 800 rpm.

If the above motor is now coupled with a load torque of 800 N-m, find the speed at which the motor can hold the load by regenerative braking. Neglect motor rotational losses.

[*G.B. Technical Univ. Electric Drives 2010-11*]

Solution : Armature current, $I_{a_1} = 100$ A

Back emf, $E_{b_1} = V - I_{a_1} R_a = 230 - 100 \times 0.1 = 220$ V

Speed, $N_1 = 500$ rpm

Now let the flux and armature current be ϕ_2 and I_2 respectively when the speed is raised from $N_1 = 500$ rpm to $N_2 = 800$ rpm

Back emf, $E_{b_2} = V - I_{a_2} R_a = 230 - 0.1 I_{a_2}$ ...(*i*)

Since load torque is constant

$$I_{a_2} \phi_2 = I_{a_1} \phi_1 \qquad \because T \propto \phi I_a$$

or $I_{a_2} = I_{a_1} \times \frac{\phi_1}{\phi_2} = 100 \frac{\phi_1}{\phi_2}$ ...(*ii*)

and $\frac{N_2}{N_1} = \frac{E_{b_2}}{E_{b_1}} \times \frac{\phi_1}{\phi_2}$

or $\frac{800}{500} = \frac{230 - 0.1 I_{a_2}}{220} \times \frac{\phi_1}{\phi_2} \qquad \because N \propto \frac{E_b}{\phi}$

or $352 = \left(230 - 0.1 \times 100 \frac{\phi_1}{\phi_2}\right) \frac{\phi_1}{\phi_2}$

$\because$ From Eq. (*ii*), $I_{a_2} = 100 \times \frac{\phi_1}{\phi_2}$

or $10\left(\frac{\phi_1}{\phi_2}\right)^2 - 230 \frac{\phi_1}{\phi_2} + 352 = 0$

$$\text{or} \quad \frac{\phi_1}{\phi_2} = \frac{230 \pm \sqrt{(230)^2 - 4 \times 10 \times 352}}{2 \times 10}$$

$$= 11.5 \pm 9.85 = 21.35 \text{ or } 1.65$$

$$\text{or } \phi_2 = \frac{\phi_1}{1.65} = 0.6066 \textbf{ Ans.}$$

$$\text{Armature current, } I_{a_2} = 100 \times \frac{\phi_1}{\phi_2} = 100 \times 1.65 = 165 \text{ A}$$

$$\text{Back emf, } E_{b_2} = V - I_{a_2} R_a = 230 - 165 \times 0.1 = 213.5 \text{ V}$$

$$\text{Speed, } N_2 = 800 \text{ rpm}$$

$$\text{Torque developed, } T_2 = \frac{E_{b_2} \times I_{a_2}}{2\pi N_2 / 60} = \frac{213.5 \times 165}{2\pi \times 800} \times 60 = 420 \text{ N-m}$$

When motor is coupled to a load torque T_3 of 800 N-m

$$\text{Armature current, } I_{a_3} = I_{a_2} \times \frac{T_3}{T_2} \quad \text{assuming constant flux}$$

$$= 165 \times \frac{800}{420} = 314.3 \text{ A}$$

Induced emf to provide armature current of 314.3 A for regenerative braking,

$$E_{b_3} = V + I_{a_3} R_a = 230 + 314.3 \times 0.1 = 261.43 \text{ V}$$

$$\text{Speed, } N_3 = \frac{E_{b_3}}{E_{b_2}} \times N_2 = \frac{261.43}{213.5} \times 800 = 980 \text{ rpm } \textbf{Ans.}$$

1.21. ELECTRICAL BRAKING OF POLYPHASE INDUCTION MOTORS

1.21.1. Plugging (or Counter-Current Braking)

Plugging can be achieved in an induction motor merely by reversing two of the three phases which causes a reversal of the direction of the rotating magnetic field. At the instant of switching the motor to the plugging position the motor runs in the opposite direction to that of the field and the relative speed is approximately twice [(2 – *s*) times] of synchronous speed *i.e.*, the slip is very nearly equal to two, being equal to (2 – *s*). So voltage induced in the rotor will be twice of normally induced voltage at standstill and the winding must be provided with the additional insulation to withstand this much voltage.

During plugging period, the motor acts as a brake and it absorbs kinetic energy from the still revolving load causing its speed to fall. The associated power P_{mech} is dissipated as heat in the motor. At the same time, the rotor also continues to draw power from the stator which is also dissipated as heat. Due to increased rotor frequency there are additional iron losses in the rotor. Thus the heat developed in the rotor during braking period are about three times the heat developed during starting period (or under blocked condition of rotor).

The size of the induction motor where plugging is to be applied, is therefore, decided not by loading conditions alone but by braking condition also.

In case of a squirrel cage motor, energy is dissipated wholly within the machine; whereas in case of wound rotor motor this energy is dissipated also in the external resistance added in the rotor circuit for this purpose.

The conditions for braking an induction motor can be studied by considering the torque-slip curves of the motor when extended beyond the point of 100 % slip, as shown in Fig. 1.103(*a*). The ordinate at the point B represents the torque at the instant of plugging and we can see that the torque increases gradually as the motor approaches the standstill, after which, of course, if the supply line is not disconnected it will start up again in the reverse direction in accordance with the ordinary torque-speed curves from A to O. Thus we see that for bringing the motor to rest, it is essential to disconnect the supply to the motor at zero speed.

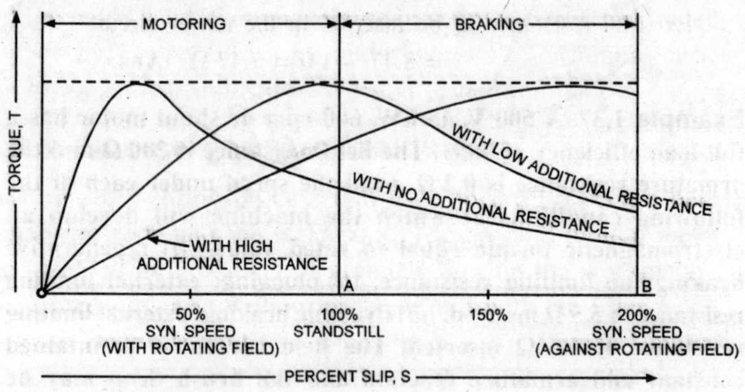

Fig. 1.103(*a*). *Torque-Slip Curves of Polyphase Induction Motor*

From the torque-slip curves shown in Fig. 1.103 (*a*), it is observed that magnitude of the braking torque is very much low as compared to the maximum torque exerted by the motor. This is specially so for low rotor resistances. As more and more resistance is inserted in the rotor circuit, maximum torque occurs at increasing values of the slip. This is the reason that squirrel cage induction motors, which are primarily designed for maximum efficiency and, therefore, for low resistance are not much suitable for such type of braking. In case of wound rotor motors, addition of resistance in the rotor circuit gives rise to more and more braking torque and, therefore, wound rotor motors are more suitable for plugging. If the change in rotor resistance is so adjusted that maximum braking torque is exerted throughout, optimum braking will be achieved in minimum time.

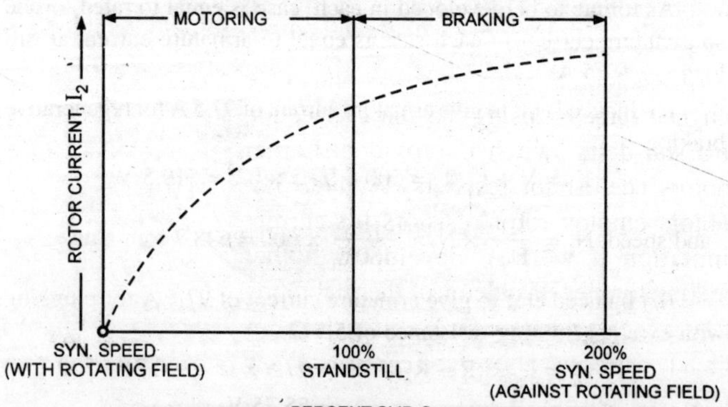

Fig. 1.103 (*b*). *Rotor Current-Slip Curve of Polyphase Induction Motor*

The expression for the braking torque neglecting stator impedance and magnetizing reactance, can be deduced, and is given as

$$T_b = \frac{3 \times 60}{2\pi N_s} \cdot \frac{R_2 s E_2^2}{R_2^2 + s^2 X_2^2} \qquad \text{...(1.81)}$$

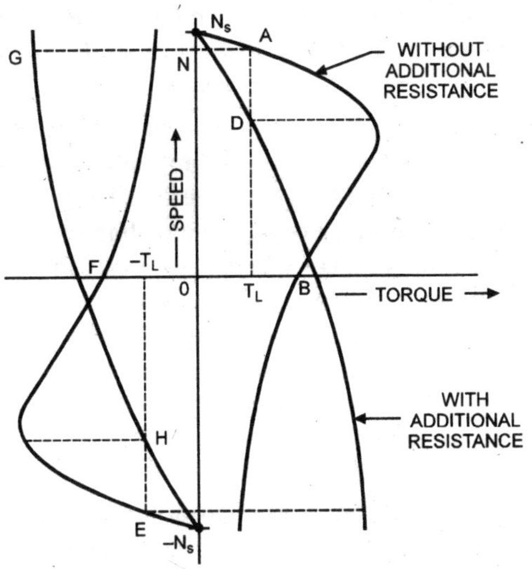

Fig. 1.103(*c*). *Quadrantal Diagram of Induction Motor During Plugging*

The rotor current can be determined during the braking period from the following relation and is plotted, as shown in Fig. 1.103 (*b*).

$$I_2 = \frac{sE_2}{\sqrt{R_2^2 + s^2 X_2^2}} \qquad \ldots(1.82)$$

where E_2 is the emf induced in each phase of the rotor at standstill, R_2 is the rotor resistance per phase and X_2 is the rotor standstill reactance per phase and s is the slip.

During plugging period, since the induced emf in the rotor is very high, as already stated above, so the rotor current and thereby stator current are very high. However, braking current can be reduced by inserting external resistance in the rotor circuit. From this point also, the wound rotor motors are beneficial as compared to squirrel cage rotor motors.

Cage motors of about 20 kW output are plugged directly, using the star connection if a star-delta switch is provided. Larger motors need stator resistors. Wound rotor motors employ rotor resistors for current limitation as well as developing higher braking torques. Problems of thermal rating may arise for frequent braking duty, as already explained above, and contactors may need frequent servicing.

Simultaneous reversal of stator connections and inclusion of rotor resistance in the wound rotor motors is quite practicable, but the same effect can be achieved by including saturistors in the rotor windings. Alternatively a deep bar rotor cage may give rise in effective rotor resistance following the plugging operation as the slip frequency is suddenly increased from sf to $(2 - s)f$ where f is the supply frequency.

In practice, for reversing drives where braking and starting up of induction motor in reverse direction comprises stages of the same continuous process, plugging is advantageously employed.

The plugging operation can also be explained with the help of quadrantal diagram shown in Fig. 1.103 (*c*). Speed-torque characteristics shown in Fig. 1.103 (*c*) are actually extension of the motor characteristics in the second and fourth quadrant. The explanation of Fig. 1.103 (*c*) is exactly identical to one for Fig. 1.94 (*a*) or 1.94 (*b*).

1.21.2. Dynamic (or Rheostatic) Braking

The rheostatic braking with a polyphase induction motor can be obtained by disconnecting the stator winding from the ac supply and exciting it from a dc source to produce a stationary dc field. In rheostatic braking, the stator winding is employed

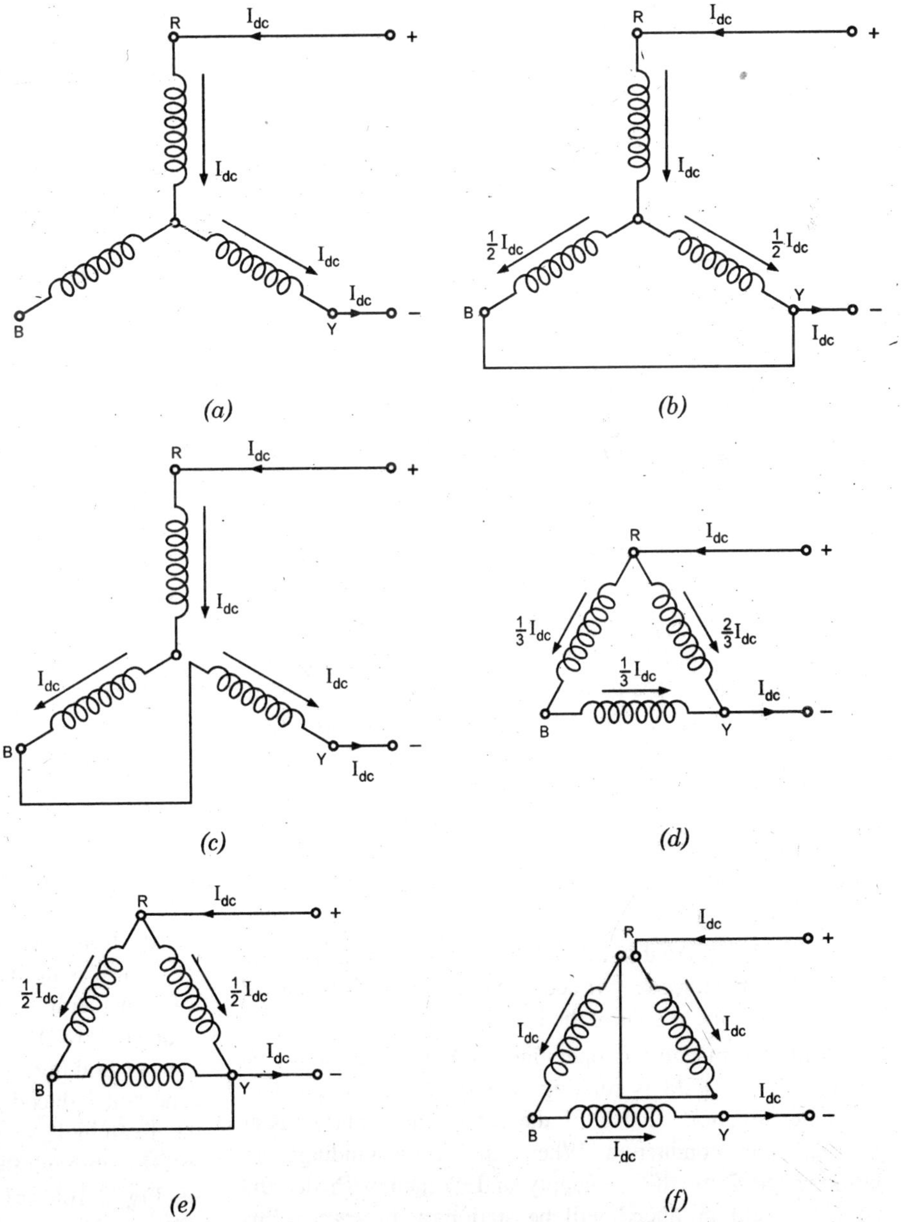

Fig. 1.104. *Connection of Stator Windings of 3-Phase Induction Motors For Dynamic Braking*

as a dc field winding and the rotor winding as an armature winding. With a wound rotor machine, external resistors can be inserted into the rotor circuit to provide a load. With squirrel cage machines, however, the rotor winding itself has to form the load.

The source of excitation may be provided either by an independent dc source or from the ac mains through a transformer rectifier set, as shown in Fig. 1.105.

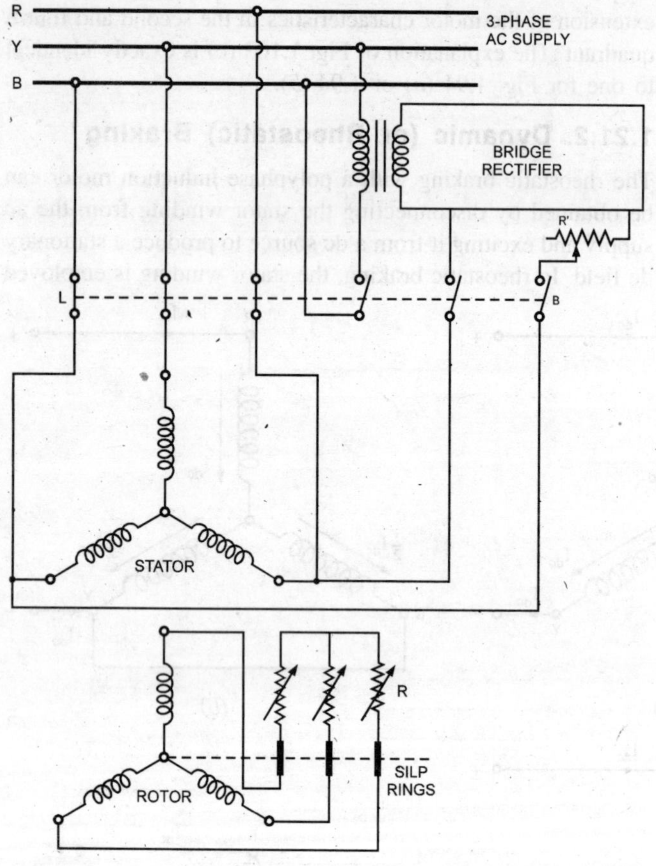

Fig. 1.105. ***DC Rheostatic Braking of a 3-Phase Induction Motor***

Various methods of connecting the stator winding to a dc source are shown in Fig. 1.104. Though connections shown in Figs. 1.104 (*c*) and 1.104 (*f*) give uniform current loading of all the three phases but are complicated in switching operation. Being more simple in this respect, the connections shown in Figs. 1.104. (*a*) and (*b*) are usually employed.

A typical connection diagram is given in Fig. 1.105; the machine operates as a motor with contactors L closed, while with L open and B closed a direct current is supplied through two stator phases, the third being left open circuited. The resistance R′ is inserted in the stator circuit to limit its current. In case of wound rotor motors additional resistance R is inserted in the rotor circuit to control the braking effect.

While the machine is operating normally as a motor, its stator magnetic field is rotating at a synchronous speed in the same direction as that of the rotor, but slightly faster than the rotor conductors. When the stator windings are disconnected from the ac supply and excited with dc, the magnetic field produced will be stationary in space, thus making the rotor conductors move past the field with a speed $(1 - s)N_s$ or SN_s. The currents induced in the rotor conductors will be opposite in direction to that corresponding to motoring operation, producing a braking torque.

Although the air-gap flux is stationary, there will be the same number of poles as when the machine is excited with ac and the rotor currents will, therefore, have a frequency which is near to the frequency corresponding to synchronous speed (*i.e.*, N_s) initially, but which decreases with rotor speed and becomes zero at standstill *i.e.*, rotor current frequency can be expressed as S*f*, where *f* is the frequency of ac supply. Likewise, the emf induced in the rotor decreases from a maximum when the rotor is running, to zero at standstill. In other words, the emf induced in the rotor will be given by SE_2 where E_2 is the magnitude of the emf induced in the rotor when it rotates past the field at speed N_s. It is, thus seen that the conditions in rotor during dc rheostatic braking with the speed falling from synchronous to standstill are very much the same as when the motor accelerates in the normal manner. Hence the equivalent circuit of the rotor can be represented as shown in Fig. 1.106 (*a*). Dividing the voltage and the impedance by S, we have the circuit shown in Fig. 1.106 (*b*) carrying the same current I_2.

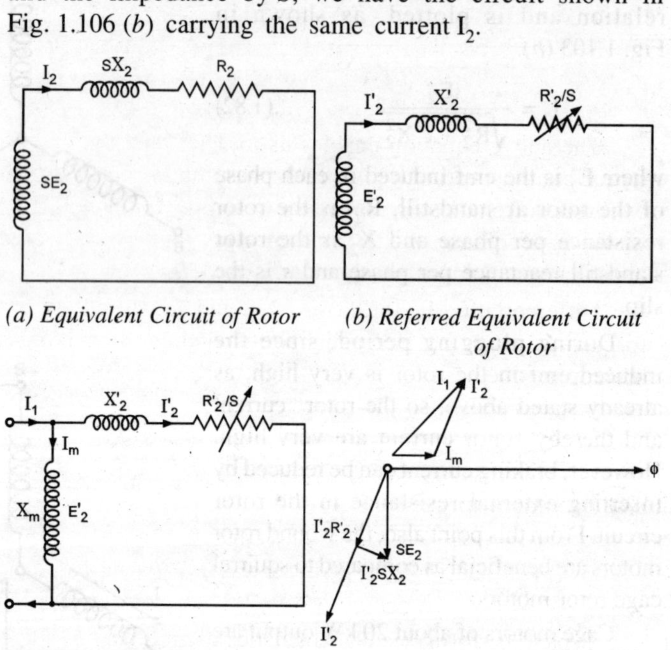

(*a*) *Equivalent Circuit of Rotor* (*b*) *Referred Equivalent Circuit of Rotor*

(*c*) *Referred Equivalent Circuit For Braking Operation* (*d*) *Phasor Diagram*

Fig. 1.106

Since the stator winding carries only direct current, the inductance of the stator has no effect under steady-state operation. The dc voltage applied across the stator winding is fixed only by the stator winding resistance. There will be no iron loss in the stator core. But the rotor core loss will be of considerable magnitude and hence the rotor resistance value referred to stator, which determines the torque developed during braking, must be corrected to take this into account. The emf induced in the rotor, at synchronous speed, E'_2 will be given by $I_m X_m$ where I_m represents the magnetizing current passing through the magnetizing reactance. The circuit shown in Fig. 1.106 (*b*) can be modified into a circuit shown in Fig. 1.106 (*c*). The phasor diagram corresponding to this modified circuit is shown in Fig. 1.106 (*d*).

It may be noted that the rotor current is alternating in nature although it is produced by a constant flux, caused by a magnetizing mmf $I_m N_1$, which remains stationary in space. However, when viewed from the rotor, both the magnetizing mmf and the resultant flux in the stator will appear as alternating. Also, since the rotor ampere-turns $I_2 N_2$ ($= I'_2 N_1$) must be balanced by the stator ampere-turns so as to maintain the magnetizing ampere-turns $I_m N_1$, the stator ampere-turns $I_1 N_1$ as seen from the rotor, represents the phasor sum of $I_m N_1$ and $I'_2 N_1$. The effective alternating current I_1 depends on the magnitude of direct current carried by the stator and the nature of the stator winding connections, as explained above.

From above discussion, it can be concluded that the operation of the motor during dc rheostatic braking can be analysed by using the equivalent circuit (the type of which we are already familiar) shown in Fig. 1.106 (*c*).

The braking torque can be determined from the relation

$$T_b = \frac{60}{2\pi N_s} \cdot 3(I'_2)^2 \frac{R'_2}{S} \text{ N-m}$$

$$= \frac{60}{2\pi N_s} \cdot 3I_1^2 \frac{X_m^2 \left(\frac{R'_2}{S}\right)}{\left(\frac{R'_2}{S}\right)^2 + (X_m + X'_2)^2} \text{ N-m} \quad ...(1.83)$$

The above equation enables us to determine the braking torque at any speed of a given induction motor, when excited by direct current, I_1 corresponding to an equivalent ac, which would have produced the same mmf as that by the actual direct current carried by the stator windings.

Differentiating Eq. (1.83) w.r.t. S and equating it to zero for a maximum we have slip corresponding to maximum braking torque

$$S_{\max T} = \frac{R'_2}{X_m + X'_2} \quad ...(1.84)$$

and maximum torque,

$$T_{\max} = \frac{60}{2\pi N_s} \cdot 3 I_1^2 \frac{X_m^2}{2(X_m + X'_2)} \quad ...(1.85)$$

Typical performance characteristics of an induction motor are illustrated in Fig. 1.107. The magnitude of the braking torque developed by the motor depends upon the excitation (strength of the field developed by the stator winding), the rotor circuit resistance, and the speed of the motor. Effect of variation of excitation I_1 and rotor resistance R_2 are also illustrated in Fig. 1.107 (*a*). For design purposes, the torque-rotor resistance curves illustrated in Fig. 1.107 (*b*) are sometimes used. With the increase in rotor resistance R_2, the speed at which maximum braking torque occurs also increases. But, the maximum torque would not increase in proportion to the square of current (I_1^2), as pointed out by Eq. (1.85), because magnetising reactance X_m decreases due to saturation caused by increase in current I_1.

Stable braking operation is on the steeper portion of torque-speed curve [*i.e.,* on the left hand side of the maximum torque position in Fig. 1.107 (*a*)]. If inadvertently the speed rises above that at which the torque is a maximum, further rise of speed is caused by a decrease in braking torque, which may lead to a further rise in speed in the case of an overhauling load. At the start of applying braking, therefore, we require high resistance in the rotor circuit, which has to be gradually reduced with decrease in speed. Braking torque can be controlled by any or both of the following methods: (*i*) By controlling the dc excitation (*ii*) By varying the rotor resistance.

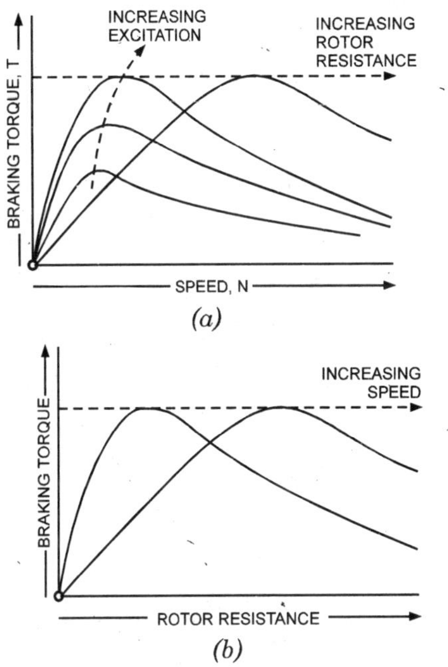

Fig. 1.107. ***DC Rheostatic Braking Characteristics***

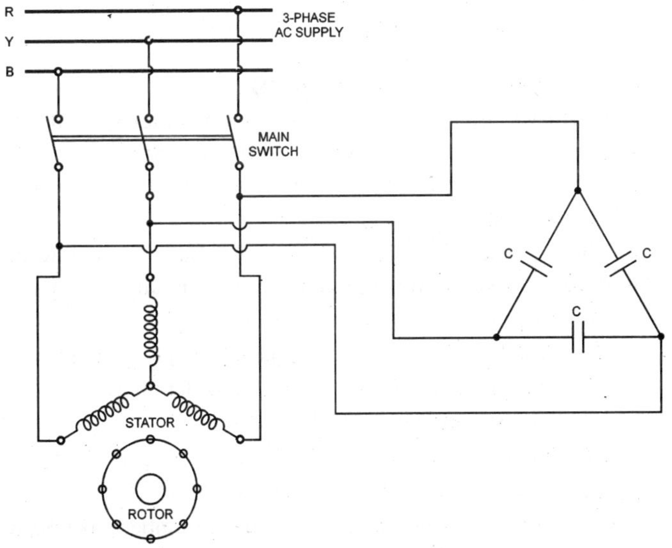

Fig. 1.108

AC dynamic braking may be obtained with self excitation by connecting the stator windings to a bank of static capacitors as shown in Fig. 1.108. In this case the induction machine operates as an induction generator receiving its ac excitation from the static capacitors. The generated electrical energy is dissipated as heat in the rotor circuit. AC dynamic braking is not popular owing to relatively high cost of capacitor banks.

DC dynamic braking takes little power from the supply and provides smooth braking torque, useful for mine winders and high inertia loads. Its advantages over plugging are the absence of the reverse rotation air-gap field [and, therefore, no tendency for the machine to run backwards] and a lower rotor I^2R loss. By increasing rotor resistance, the braking torque can be made more effective at higher speeds. This method is, therefore, very much suitable for retarding over-hauling loads. In case of a squirrel cage motor, since rotor resistance is small, it will apply braking torque at very low speeds and, therefore, useful only for bringing the load to rest.

The advent of automatic control of dynamic braking of induction motors employing closed loop systems, has made the induction motors more popular than dc motors, especially for the drives employing in mine hoists.

Figure 1.109 shows the quadrantal characteristics of induction motor under dynamic braking condition.

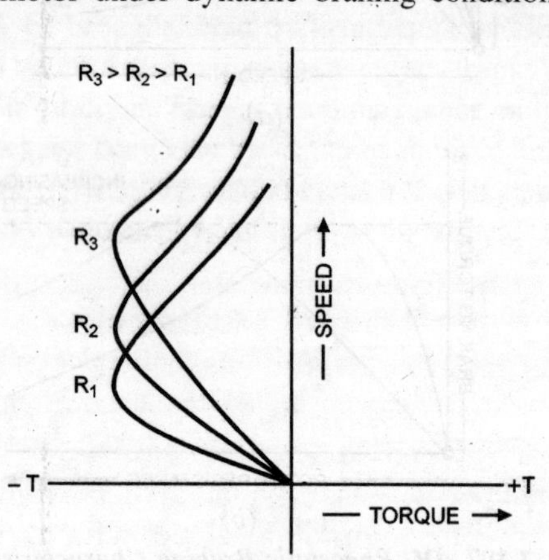

Fig. 1.109. ***Quadrantal Diagram of Induction Motor During Dynamic Braking***

1.21.3. Regenerative Braking

Regenerative braking is an inherent characteristic of an induction motor, since it operates as an induction generator when it runs at speed above synchronous and it feeds power back to the supply line.

The 3-phase induction motor can be made to operate at speed above synchronous speed by employing any one of the following processes.

(*i*) Switching over to a low frequency supply in frequency controlled induction motors in order to reduce the speed of operation of the drive.

(*ii*) Downward motion of a loaded hoisting mechanism such as cranes, hoists, excavators etc.

(*iii*) Switching over to a larger pole number operation from a smaller one in multi-speed squirrel cage motors.

In all the above processes, the slip and torque developed become negative, as shown in Fig. 1.110, and thus the machine acts as a generator, receiving mechanical energy and giving it back to the supply system in the form of electrical energy.

If the load drives the motor above synchronous speed, no switching operation is required. Once the machine is driven above synchronous speed, the braking operation automatically starts. The operating point will depend upon the magnitude of load torque and the nature of torque-speed characteristic of the machine during generating operation. By varying the resistance in the rotor circuit, it is possible to operate at any speed above synchronous speed during braking. In case the driving torque of the load exceeds the maximum braking torque, of which the machine is capable, the system will become unstable and the speed will rise further, probably to a disastrous value, since, the faster the machine runs, the lesser will be the braking torque developed.

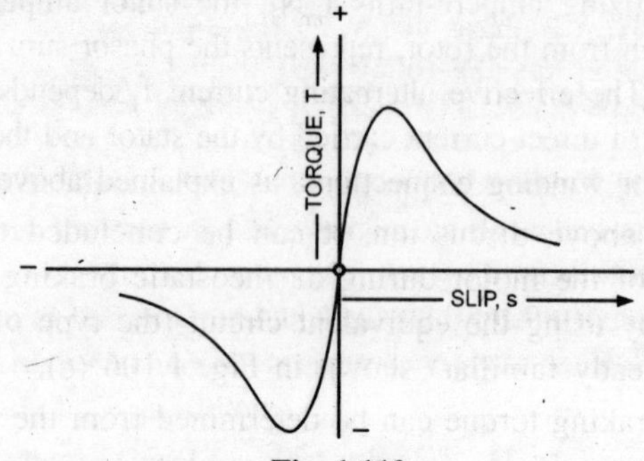

Fig. 1.110

In the case of a squirrel cage induction motor, stable speed is obtained at a speed considerably in excess of the synchronous speed and the regenerative braking cannot be applied unless the motor is specially designed to withstand the excessive speed.

Regenerative braking has the disadvantage of the possibility of braking only at super-synchronous speeds and, therefore, is seldome used for braking. This method can be used only in hoisting type of mechanism or with a multi-speed squirrel cage motor. It is advantageous on mountain railways too. It returns about 20% of the total energy on certain railway runs and saves a great deal of brake shoe wear.

Example 1.40. A 30 kW, 400 V, 3-phase, 4-pole, 50 Hz induction motor has full-load slip of 5%. If the ratio of standstill reactance to resistance per rotor phase is 4, estimate the plugging torque at full speed.

[Pb. Technical Univ. Electric Drives and Utilization December 2008]

Solution : Synchronous speed, $N_s = \dfrac{120f}{P} = \dfrac{120 \times 50}{4} = 1{,}500$ rpm

Full-load slip, $s_f = 5\%$ or 0.05

Full-load speed, $N_f = N_s(1 - s_f) = 1{,}500\,(1 - 0.05) = 1{,}425$ rpm

$$\text{Full-load torque, } T_f = \frac{\text{Power developed in kW} \times 1{,}000}{2\pi N/60} = \frac{30 \times 1{,}000}{2\pi \times 1{,}425/60} = 201.04 \text{ N-m}$$

Since torque developed,

$$T \;\alpha\; \frac{sR_2E_2^2}{R_2^2 + (sX_2)^2}$$

$$\text{So} \quad \frac{T_2}{T_1} = \frac{s_2R_2E_2^2/(R_2^2 + s_2^2X_2^2)}{s_1R_2E_2^2/(R_2^2 + s_1^2X_2^2)} = \frac{s_2(R_2^2 + s_1^2X_2^2)}{s_1(R_2^2 + s_2^2X_2^2)}$$

$$= \frac{s_2}{s_1}\,\frac{1 + s_1^2\left(\dfrac{X_2}{R_2}\right)^2}{1 + s_2^2\left(\dfrac{X_2}{R_2}\right)^2} = \frac{s_2}{s_1}\,\frac{1 + 16s_1^2}{1 + 16s_2^2} \qquad \because \frac{X_2}{R_2} = 4$$

Slip, $s_1 = s_f = 0.05$

Slip, $s_2 =$ Slip at plugging $= 2 - s_f = 2 - 0.05 = 1.95$

So plugging torque, $T_P = \dfrac{1.95}{0.05} \times \dfrac{1+16(0.05)^2}{1+16(1.95)^2} \times T_f$

$$= 39 \times \frac{1.04}{61.84} \times 201.04 = 131.86 \text{ N-m } \textbf{Ans.}$$

Example 1.41. A 3-phase, 400 V, 50 Hz, 8-pole, star-connected induction motor has the following parameters: $R_1 = 0.1\ \Omega$, $R_2 = 0.5\ \Omega$, $X_1 = 0.6\ \Omega$, $X_2 = 2.4\ \Omega$, $s_f = 0.05$ and the ratio of effective stator to rotor turns $\frac{1}{2}$. The motor is to be braked at rated speed and an external resistance of 2 Ω per phase (referred to stator) has been inserted into the rotor circuit. Determine the initial braking torque in (*i*) plugging and (*ii*) dc dynamic braking.

Solution : Rotor resistance referred to stator,

$$R_2' = 0.5 \times \left(\frac{1}{2}\right)^2 = 0.125\ \Omega$$

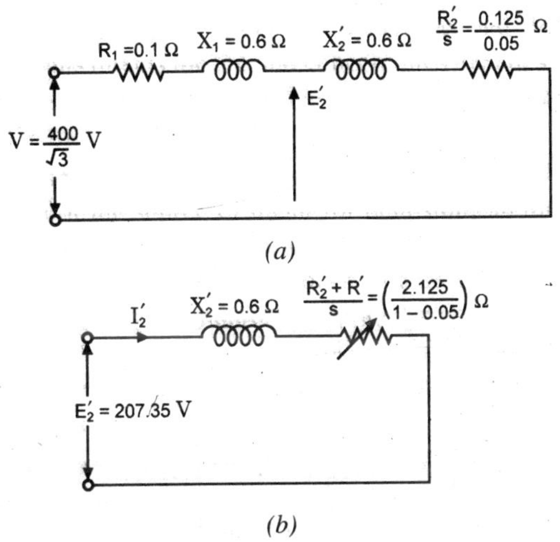

Fig. 1.111

Rotor leakage reactance referred to stator,

$$X_2' = 2.4 \times \left(\frac{1}{2}\right)^2 = 0.6\ \Omega$$

(*i*) During plugging

Slip, $s = 2 - sf = 2 - 0.05 = 1.95$

Synchronous speed, $N_s = \dfrac{120f}{P} = \dfrac{120 \times 50}{8} = 750$ rpm

Initial rotor current during plugging,

$$I'_2 = \frac{V}{\sqrt{\left[R_1 + \dfrac{(R_2' + R_L')}{s}\right]^2 + (X_1 + X_2')^2}}$$

$$= \frac{400/\sqrt{3}}{\sqrt{\left(0.1 + \dfrac{0.125+2}{1.95}\right)^2 + (0.6+0.6)^2}}$$

$= 136.7$ A

Initial braking torque, $T_b = \dfrac{3 \times 60}{2\pi N_s} \times (I_2')^2 \times \dfrac{R_2' + R'}{s}$

$$= \frac{3 \times 60}{2\pi \times 750} \times (136.7)^2 \times \frac{2.125}{1.95}$$

$= 777$ Nm **Ans.**

(*ii*) *During dc dynamic braking.* The equivalent circuit of the motor corresponding to slip of 0.05 and the equivalent-circuit of the machine during dynamic braking at rated speed are given in Figs. 1.111 (*a*) and 1.111 (*b*) respectively,

$$E'_2 = \sqrt{\frac{\left(\dfrac{0.125}{0.05}\right)^2 + 0.6^2}{\left(\dfrac{0.125}{0.05} + 0.1\right)^2 + (1.2)^2}} \times \frac{400}{\sqrt{3}} \text{ V} = 207.35 \text{ V}$$

During dc dynamic braking, S = 1 − 0.05 = 0.95

$$I'_2 = \frac{207.35}{\sqrt{0.6^2 + \left(\dfrac{2.125}{0.95}\right)^2}} = 89.53 \text{ A}$$

Initial braking torque,

$$T_b = \frac{3 \times 60}{2\pi N_s} \times (I_2')^2 \times \frac{R_2' + R'}{S} = \frac{3 \times 60}{2\pi \times 750} \times (89.53)^2 \times \frac{2.125}{0.95}$$

$= 684.9$ Nm **Ans.**

1.22. ELECTRIC BRAKING OF SYNCHRONOUS MOTORS

In principle, a synchronous motor may be stopped by either plugging or rheostatic braking. But the former method has certain limitations, as explained below.

If the direction of dc excitation is reversed, dc and ac fields will rotate in opposite directions and no braking torque will be developed in case of plain synchronous motors (*i.e.*, synchronous motors provided with no damper windings). The braking effect will be produced, however, due to eddy currents induced in damper windings, if provided for starting purposes. Plugging as applied to synchronous motors provided with damper windings, does not prove satisfactory. Because of usually large size of such motors it results in heavy surges of stator current causing line disturbances beyond permissible limits and also it calls for additional control equipment to switch off the motor when the speed falls to zero. Also, since the plugging torque arises from the damper windings and unless it is specially designed for the purpose, the braking torque developed during plugging are of smaller magnitudes. Hence, *rheostatic or dynamic braking* is almost exclusively used for braking synchronous motors.

As already mentioned above, synchronous motors are usually braked with rheostatic braking. In rheostatic braking, the stator of synchronous motor is disconnected from the ac supply and is short circuited through a set of balanced 3-phase resistors, keeping the field excitation unchanged. In case field excitation is increased, greater braking torques will be developed. During dynamic braking, the machine acts as an alternator and dissipates the stored energy in the resistors. The terminal voltage and frequency drop with the fall in speed and both become zero at standstill. The maximum torque occurs at approximately that speed for which the machine reactance becomes equal to the braking resistance. The appearance of high torque immediately prior to stopping has a very positive effect on the braking process.

1.23. DYNAMICS OF BRAKING

Let electrical and mechanical torque in N-m be T_E and T_F respectively.

Electrical torque,

$T_E \propto$ speed of rotation if excitation is constant.

Braking torque,

$$T_B = T_E + T_F = K\omega + T_F$$

Angular retardation,

$$\beta = \frac{T_B}{J} = \frac{K\omega + T_F}{J} \text{ rad/s}^2; \ \beta \text{ is negative being retardation.}$$

Time required to bring the motor from an initial speed ω_1 to final speed ω in the same direction is given as

$$t = \int_{\omega_1}^{\omega} \frac{d\omega}{\beta} = -J\int_{\omega_1}^{\omega} \frac{d\omega}{K\omega + T_F} = \frac{-J}{K}\left[\log_e (K\omega + T_F)\right]_{\omega_1}^{\omega}$$

$$\text{or } t = \frac{-J}{K}\left[\log_e (K\omega + T_F) - \log_e (K\omega_1 + T_F)\right]$$

$$= \frac{J}{K}\log_e \frac{K\omega_1 + T_F}{K\omega + T_F} \quad \text{...(1.86)}$$

Time required to bring the motor to standstill is obtained by substituting $\omega = 0$ in above equation,

$$\textit{i.e.,}\ t = \frac{J}{K}\log_e \frac{K\omega_1 + T_F}{T_F} \quad \text{...(1.87)}$$

From Eq. (1.86)

$$K\omega + T_F = (K\omega_1 + T_F)\, e^{\frac{-Kt}{J}}$$

$$\text{or } \omega = \frac{1}{K}\left[(K\omega_1 + T_F)\, e^{\frac{-Kt}{J}} - T_F\right] \quad \text{...(1.88)}$$

Number of revolutions made,

$$n = \frac{1}{2\pi}\int \omega\, dt = \frac{1}{2\pi}\int_0^t \left[(K\omega_1 + T_F)\, e^{\frac{-Kt}{J}} - T_F\right] dt$$

$$= \frac{1}{2\pi K}\left[\frac{(K\omega_1 + T_F)\, e^{-Kt/J}}{\frac{-K}{J}} - T_F t\right]_0^t$$

$$\text{or } n = \frac{1}{2\pi K}\left\{\left[\frac{K\omega_1 + T_F}{\frac{-K}{J}}\, e^{-Kt/J} - T_F t\right] - \left[\frac{K\omega_1 + T_F}{\frac{-K}{J}} - 0\right]\right\}$$

$$= \frac{1}{2\pi K}\left[\frac{J}{K}(K\omega_1 + T_F)(1 - e^{-Kt/J}) - T_F t\right] \quad \text{...(1.89)}$$

Example 1.42. A 400 V, 8-pole, 3-phase synchronous motor and its load have a total moment of inertia of 630 kg-m². Determine time taken and the number of revolutions made for it to come to standstill if (*i*) it is braked by plugging, which produces a constant braking torque of 165 kg-m. (*ii*) rheostatic braking is employed which gives an initial electric braking torque of 690 kg-m. Assume a frictional torque 1.4 kg-m and supply frequency 50 Hz.

Solution : Total moment of inertia of rotating parts,

$$J = 630 \text{ kg-m}^2$$

Frictional torque, $T_F = 1.4 \times 9.81 = 13.734$ N-m

(*i*) Electrical torque produced by plugging,

$$T_E = 165 \times 9.81 = 1{,}618.65 \text{ N-m}$$

Total braking torque, $T_B = T_E + T_F = 1{,}618.65 + 13.734 = 1{,}632.4$

$$\text{Angular retardation, } \beta = \frac{T_B}{J} = \frac{1{,}632.4}{630} = 2.591 \text{ rad/s}^2$$

Synchronous speed of motor,

$$N_s = \frac{120 f}{P} = \frac{120 \times 50}{8} = 750 \text{ rpm}$$

$$\omega_1 = \frac{2\pi \times 750}{60} = 78.54 \text{ radians/second}$$

Time taken to stop the motor,

$$t = \frac{\omega_1}{\beta} = \frac{78.54}{2.591} = 30.3 \text{ seconds. } \textbf{Ans.}$$

$$\text{Now } t = -\int \frac{d\omega}{\beta}$$

$$\text{or } dt = -\frac{d\omega}{\beta}$$

and number of revolutions made,

$$n = \frac{1}{2\pi}\int \omega\, dt = \frac{-1}{2\pi}\int_{\omega_1}^{0} \frac{\omega\, d\omega}{\beta} = \frac{-1}{2\pi\beta}\left[\frac{\omega^2}{2}\right]_{\omega_1}^{0}$$

$$= \frac{1}{2\pi \times 2.591}\left[\frac{78.54^2}{2}\right] = 189.5 \textbf{ Ans.}$$

(*ii*) Initial braking torque,

$$K\omega_1 = 690 \times 9.81 = 6{,}769 \text{ N-m}$$

$$\text{and } K = \frac{6{,}769}{78.54} = 86.2$$

Time taken to bring the motor to rest,

$$t = \frac{J}{K}\log_e \frac{K\omega_1 + T_F}{T_F} = \frac{630}{86.2}\log_e \frac{6{,}769 + 13.734}{13.734}$$

$$= 45.3 \text{ seconds } \textbf{Ans.}$$

Number of revolutions made,

$$n = \frac{1}{2\pi K}\left[\frac{J}{K}(K\omega_1 + T_F)(1 - e^{-Kt/J}) - T_F t\right]$$

$$= \frac{1}{2\pi \times 86.2}\left[\frac{630}{86.2}(6{,}769 + 13.734)(1 - e^{-86.2 \times 45.3/630}) - 13.734 \times 45.3\right]$$

$$= \frac{1}{541.8}\left[49{,}555\,(1 - 0.00203) - 622\right] = 90 \text{ revolutions } \textbf{Ans.}$$

Example 1.43. A 37.5 kW, 400 V, 750 rpm synchronous motor having a total moment of inertia of 20 kg-m² employs rheostatic braking for obtaining rapid stopping in cases of emergency. When the motor is operating on full load, estimate the time taken and the number of revolutions made before the motor is stopped after switching over to a star-connected braking resistor of 2.5 Ω per phase. Assume full-load efficiency of 90% and full-load power factor to be 0.85.

Solution : Full-load torque,

$$T_f = \frac{\text{Output in kW} \times 1{,}000}{2\pi N/60} = \frac{37.5 \times 1{,}000 \times 60}{2\pi \times 750} = 477.46 \text{ N m}$$

Full-load current,

$$I_L = \frac{\text{Output in kW} \times 1{,}000}{\sqrt{3}\, V_L \times \cos\phi \times \eta} = \frac{37.5 \times 1{,}000}{\sqrt{3} \times 400 \times 0.85 \times 0.9} = 70.75 \text{ A}$$

Approximate line current at the start of braking period (neglecting impedance drop)

$$I_b = \frac{400}{\sqrt{3} \times 2.5} = 92.376 \text{ A}$$

Assuming the excitation to remain constant

Initial electric braking torque

$$= 477.46 \times \frac{92.376}{70.75} = 623.4 \text{ Nm}$$

Total initial braking torque = 477.46 + 623.4 = 1,100.86 Nm

If T_F is the mechanical braking torque due to load (assumed to be constant) and T_E is the electric braking torque, we have

Total braking torque, $T_B = T_F + T_E = T_F + K\omega$

$$\text{Now } \omega = \frac{2\pi N}{60} = \frac{2\pi \times 750}{60} = 78.54 \text{ rad/s}$$

$$T_E = 623.4 \text{ N/m}$$

$$\text{and, therefore, } K = \frac{623.4}{78.54} = 7.94$$

$$\text{Thus, retardation, } \beta = \frac{T_B}{J} = \frac{T_F + K\omega}{J}$$

Time taken to bring the motor to standstill,

$$t = \int_0^{\omega_1} \frac{d\omega}{\beta} = J\int_0^{\omega_1} \frac{d\omega}{K\omega + T_F} = \left[\frac{J}{K}\log_e(T_F + K\omega)\right]_0^{\omega_1}$$

$$\text{or } t = \frac{J}{K}\log_e \frac{T_F + K\omega_1}{T_F} = \frac{20}{7.94}\log_e \frac{1,100.86}{477.46} = 2.1 \text{ seconds } \textbf{Ans.}$$

Number of revolutions made,

$$n = \frac{1}{2\pi K}\left[\frac{J}{K}(K\omega_1 + T_F)(1 - e^{-Kt/J}) - T_F t\right]$$

$$= \frac{1}{2\pi \times 7.94}\left[\frac{20}{7.94} \times 1,100.86\left(1 - e^{\frac{-7.94 \times 2.1}{20}}\right) - 477.46 \times 2.1\right]$$

$$= \frac{1}{49.8885}[2,772.95\ (1 - 0.43444) - 1,002.666]$$

= 11.34 revolutions.

Example 1.44. Determine the time taken by a dc shunt motor to fall in speed from 1,500 rpm to 600 rpm with the excitation maintained constant. The armature is connected across a resistance of 15 Ω. The emf induced in the armature at 1,500 rpm is 240 V. The moment of inertia of the armature is 20 kg-m². Neglect the effect of the friction in motor. Also determine the time taken for the same fall in speed if there existed a frictional torque of 1.5 kg-m.

Solution : The emf induced in the armature at 1,500 rpm,

$$E = 240 \text{ V}$$

Power supplied to the resistance connected across the armature.

$$P = \frac{E^2}{R} = \frac{240^2}{15} = 3,840 \text{ W}$$

Electric braking torque,

$$T_B = \frac{P}{2\pi N/60} = \frac{3,840 \times 60}{2\pi \times 1,500} = 24.45 \text{ N-m}$$

Since rheostatic braking is being used keeping field excitation constant and connecting the armature to a resistance of 15 Ω. At any state the braking effect will be due to the current supplied which in turn depends upon the emf induced in the armature at that speed. The electric braking torque is thus proportional to rotational speed or angular velocity.

$$T_E = K\omega = 24.45 \text{ N-m}$$

$$\omega_1 = \frac{2\pi N}{60} = \frac{2\pi \times 1,500}{60} = 157 \text{ radians/second}$$

$$K = \frac{T_E}{\omega_1} = \frac{24.45}{157} = 0.1557$$

Time taken to bring the motor speed from 1,500 rpm to 600 rpm (*i.e.*, 157 rad/s to 62.832 rad/s),

$$t = J\int_{\omega_2}^{\omega_1} \frac{d\omega}{K\omega + T_F} = \left[\frac{J}{K}\log_e(T_F + K\omega)\right]_{\omega_2}^{\omega_1}$$

$$= \frac{J}{K}\log_e \frac{T_F + K\omega_1}{T_F + K\omega_2}$$

When frictional torque $T_F = 0$

$$t = \frac{J}{K}\log_e \frac{K\omega_1}{K\omega_2} = \frac{20}{0.1557}\log_e \frac{157}{62.832}$$

= 111.7 seconds **Ans.**

When frictional torque, T_F = 1.5 kg-m = 1.5 × 9.81 = 14.715 Nm

$$\text{So } t = \frac{20}{0.1557}\log_e \frac{14.715 + 24.45}{14.715 + 24.45 \times \frac{600}{1,500}}$$

$$= \frac{20}{0.1557}\log_e \frac{39.165}{24.495} = 60.3 \text{ seconds } \textbf{Ans.}$$

1.24. TYPES OF ENCLOSURES

In order to protect machines against the ingress of dirt and dust or larger foreign bodies like spanners, vermin etc. into it, it is desirable that some suitable enclosure be provided. But the enclosure provided for this protection will interfere with the free entry of cool air into the machine with the result that an enclosed machine does not give the same output as does an open type machine of the same dimensions for a given temperature rise.

The purpose of providing enclosure is two fold—firstly providing protection to persons against contact with live or moving parts inside the enclosure and protection to machines against the ingress of solid foreign bodies, and secondly providing protection to machines against harmful ingress of water. Various types of enclosures giving various degrees of protection and cooling facilities for use in different working conditions can be built. ISI 4691 provides an ingress protection code which consists of the letter IP followed by two numbers—the first characteristic numeral designates the extent of protection of persons against contact and of the machine against ingress of solid foreign bodies, while the second characteristic numeral indicates the extent of protection to the machine against harmful ingress of water. Further addition of letter S or M indicates that the degree of protection against ingress of water is applicable only while the machine is stationary or rotating respectively. Absence of both the letters means that protection is provided under both the conditions. For example enclosure marked IP 22 will provide protection against ingress bodies of diameter more than 12 mm and against ingress of water falling up to 15° from vertical under both the conditions. Similarly enclosures marked IP 44 and IP 56 will provide protection against ingress of bodies of diameter more than 1 mm and against water splashed under both the conditions and protection against ingress of bodies in powder form and water projected by a nozzle under both the conditions respectively. For further details refer to ISI code 4691.

The various types of enclosures, as specified by relevant ISI are as follows:

1. *Open Type.* The bearings are supported on simple pedestals or end brackets. This type of enclosure does not protect the machines against dust, dirt and larger foreign bodies but facilitates in cooling. The machines of this type are usually large dc generators and dc motors, which are installed in a neat and clean separate room such as in case of power stations, substations, collieries, rolling mills. Because all the moving parts are exposed, hand rail and other suitable rail guards are provided to prevent unauthorized persons coming into contact specially with moving parts.

2. *Protected Type.* Protected type machines are such in which access to live and rotating parts is so limited as to prevent accidental contact with such parts or ingress of foreign bodies while ventilation is not materially obstructed. Protected type motors are used mostly for industrial purposes.

3. *Screen Protected Type.* In this type of machines openings provided for ventilation are covered with wire screen, expanded metal or other suitable perforated covers having apertures not exceeding 3.25 cm^2 in area but not less than 0.6 cm^2 in area. Though such an enclosure does not protect the motor against dirt and dust but larger bodies and big insects cannot enter into the machine . Such motors are very widely employed for most of industrial applications where dry clean conditions prevail and gas or chemical fumes are not present. Fig. 1.112 (*a*) illustrates a screen-protected motor in which no attempt has been made to exclude dust or fluff from the windings. It is suitable for most forms of electric drive and permits ample ventilation.

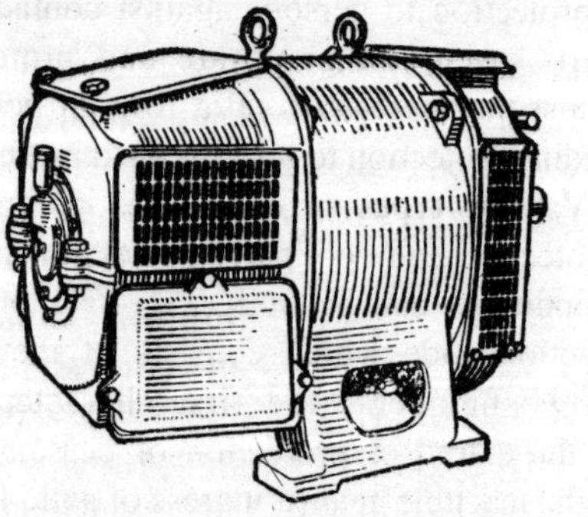

(a) Screen-Protected Motor

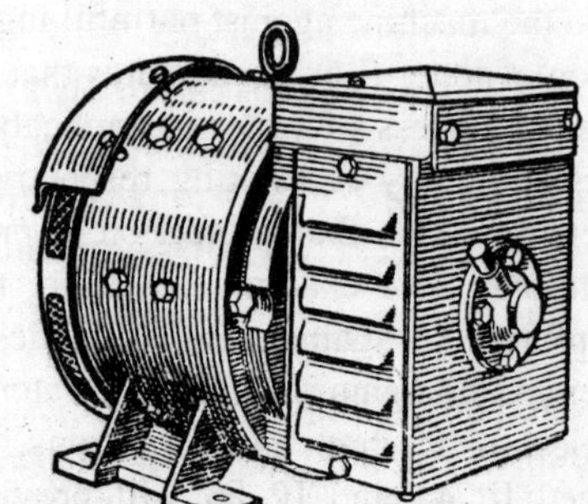

(b) Drip-Proof Motor

Fig. 1.112

4. *Drip Proof Type.* In such machines openings provided for ventilation are so protected by overhanging cowls that liquid drops or dust particles falling on the machine vertically cannot enter the machine. Such motors are used in damp situations such as for pumping sets. Fig. 1.112 (*b*) illustrates a drip-proof motor, similar to the screen-protected type, but having protection against dripping moisture.

5. *Splash Proof Type.* In such machines the ventilating openings are so constructed that liquid drops or dust particles falling on the machine or coming towards it in a straight line at any angle not greater than 100° from the vertical cannot enter the machine. Such motor can be used in rains.

6. *Totally Enclosed Type.* Such machines have solid frames and end shields but no opening for ventilation. They get cooled by surface radiation only. In such machines no dirt or foreign matter can enter the motor and clog the air passage. These machines are used in situations where dust, water, oil, chemicals, gases or other injurious substances are likely to enter the machine but natural cooling is sufficient such as for driving coal handling plant, saw mills etc. In these machines due to cooling difficulties the output obtainable is only about 50 to 70% of that obtainable from a protected type of same dimensions and for same temperature rise. Totally enclosed motors are seldom furnished in sizes above 1.5 or 2 kW because higher rating motors require frames of much larger sizes than fan-cooled motors of corresponding rating.

7. *Totally Enclosed, Fan Cooled Type.* In order to assist in heat dissipation, most of the manufacturers have developed totally enclosed fan cooled motors in which an outer casing is built around the motor and air is blown by means of a fan mounted on the shaft between the casing and external surface of the motor. To increase cooling rate, motor is provided with external fans. Such motors are used in saw mills, flour mills, cement works etc. These motors are not water proof and should not be used in places likely to be submerged. Fig. 1.112 (*c*) illustrates a totally enclosed motor in which the main means of cooling is by radiation and internal air circulation.

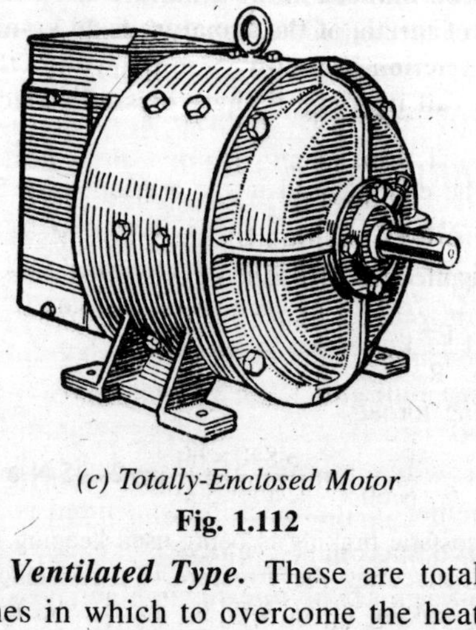

(c) Totally-Enclosed Motor

Fig. 1.112

8. *Pipe Ventilated Type.* These are totally enclosed type machines in which to overcome the heat dissipation difficulties, a duct or pipe is provided through which clean air is led to the motor from outside the dust laden building or area. The cost of the piping may be overweighed by the saving in cost of motor, as motor of smaller size can be used with same output.

Pipe ventilated motor is shown in Fig. 1.112 (*d*). Fresh clean air is brought to the motor by piping and drawn in through a big inlet, as illustrated in (*i*), and air is then expelled through louvres or taken away by pipe, as illustrated in (*ii*).

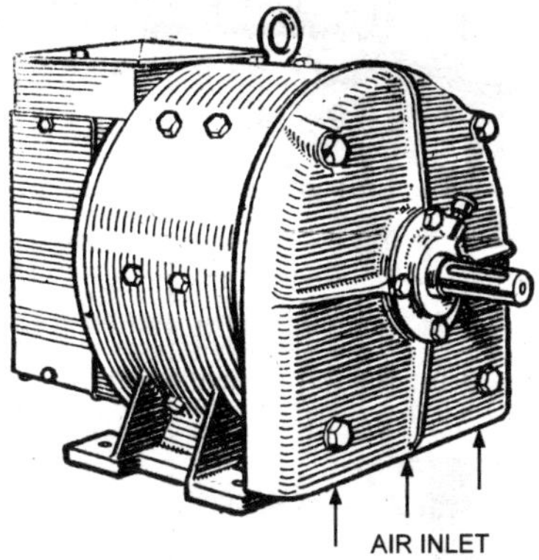

(*i*)

(*ii*)

(*d*) *Pipe-Ventilated Motor*

Fig. 1.112

9. *Flame Proof Type.* The flame proof type machine is provided enclosure designed and constructed to withstand an explosion of the gas or vapour which may occur within it and to prevent the ignition of the gas or vapour surrounding the machine. The ventilating openings in the machine are made long enough for the flame to be contained within the frame. Flame proof type motors are used in collieries, petroleum stations or chemical plants.

1.25. BEARINGS

Bearings are those parts of a machine which house supports and restrain the rotating parts. The moving part of the bearing is known as *journal* and the stationary part that supports the moving part is known as *bearing*.

The main functions of a bearing, besides supporting the rotating parts, are (*i*) to permit free rotation of the moving components with minimum friction and (*ii*) to maintain the rotating member of the machine in a fixed physical location relative to the stationary member.

The bearings are of two types, namely (*i*) ball or roller bearings and (*ii*) sleeve or bush bearings.

1. *Ball or Roller Bearings.* Ball or roller bearings are those bearings in which shafts are supported by balls or rollers which themselves are movable. The action in ball or roller bearings is not that of rubbing but of rolling between the balls and or rollers and their races, the inner surface being fixed to the shaft while the outer surface is fixed to the housing supporting the bearing. Such bearings are supplied and filled with grease. The period between the renewal of grease depends upon the speed of rotation of the motor. These are also known as *antifriction bearings*.

Up to about 75 kW use of ball or roller bearings is made. Very small motors are usually provided with ball bearings. In vertical shaft drives the ball bearings are essential as they are capable of taking an axial thrust.

Though the initial cost of bearings of this type is high but their life is longer, occupy less space, friction loss is smaller and their maintenance cost is low. These have made use of induction motors with small air gap possible as the roller bearing does not wear appreciably. However, these cannot be recommended where noise is to be avoided.

2. *Sleeve or Bush Bearings.* Sleeve or bush bearings are those bearings in which the rotating shaft is supported by a bearing component rigidly fixed to the frame of the machine. They may be (*i*) journal bearing offering support to the shafts at right angles to the shaft axes (*ii*) foot-step bearings supporting shafts parallel to the shaft axes and (*iii*) thrust or collar bearings offering support to shafts subjected to end or axial thrust.

Sleeve bearings are normally of bronze but for universal and fan-cooled motors, sintered powder or porous metal bearings may be used because of their self lubricating properties due to capillary action.

Sleeve bearings on very small motors are lubricated with an oil wick pressed lightly against the shaft by a spring. The wick must be kept saturated with a fairly light oil and will remain moist through several weeks of heavy use. Large sleeve bearings are ring lubricated which consists of a ring freely rotating on the shaft, carrying oil to the bearings. Lubricating oil, which is free from acid and resin, is renewed every six months. Temperature of bearing (either sleeve or roller) is the indication of the bearing condition. ISS 325 provides for maximum temperature rise at sleeve or ball bearings to be 50°C.

Bearings of this type are necessary to reduce the noise level for certain types of couplings and for the positions subject to external vibrations. They are also used where there is a danger of individual balls or rollers indenting the race during long rest periods.

Motors with sleeve bearings should always be used with shaft horizontal and fixed coupling between motor and the driven machine. Because of larger wear of the bearings, relatively large air gap in the induction motor has to be introduced.

1.26. TRANSMISSION OF DRIVE

Mechanical power available at the motor shaft has to be transmitted to the driven machine or machines. There are various methods of doing so. Among these may be mentioned the following :

1. *Direct Drive.* In this type of drive the driving member is connected direct to the driven member, without any interposed gearing, by means of solid or flexible coupling. It is the simplest method, space required is less than with belt drive and efficiency is 100 per cent. Solid coupling requires very accurate aligning otherwise there is possibility of damage to shafts. Another shortcoming of solid coupling is that sudden jerks of load are transmitted to the motor. By employing flexible coupling both of the above shortcomings are overcome to some extent. Flexible coupling can accommodate some angular, lateral or vertical misalignment. The other disadvantages of direct drive are that (*i*) end thrust is exerted on the motor by the load if it happens to be a centrifugal pump and (*ii*) the driving and driven members must run at the same speed, which is usually not desirable. This type of drive is employed where there is a possibility of arranging the driving member in line with the driven member and where speed of driven member is same as that of driving member.

2. *Belt Drives.* These are the least expensive drives. These are employed where a speed change is desired in the transmission of power and where it is not absolutely essential to maintain a fixed speed ratio between the driving and driven shaft. The belt drives are of two types namely (*i*) flat belt drive and (*ii*) V-belt drive.

(i) Flat Belt Drive. It is a long distance drive with a limited degree of slip. The belts used for drive are made of hair, woven cotton, rubberized cotton ply, nylon, terylene and leather. The leather belt drive is the most economical type of drive. Vertical belt drive should be avoided to the possible extent. In horizontal belt drive the slack side of the belt should be kept preferably on the top, so that the natural sag increases the arc of contact on two pulleys.

Two motors of 15 kW and 10 kW arranged for flat belt drive are illustrated in Fig. 1.113 (*a*).

The advantages, disadvantages and limitations of flat belt drive are :

Advantages. (*i*) Greater flexibility in the original design of a plant is possible and new arrangements of old apparatus can be made at any time.

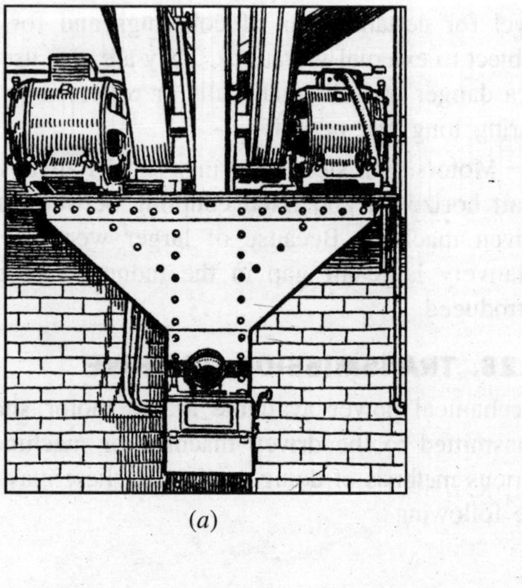

(*a*)

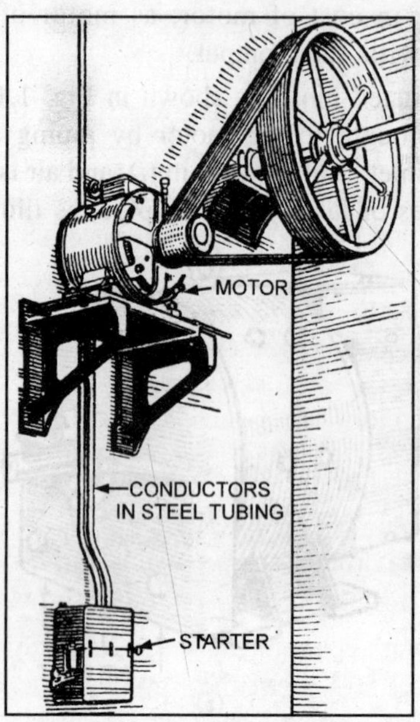

(*b*)

(*c*)

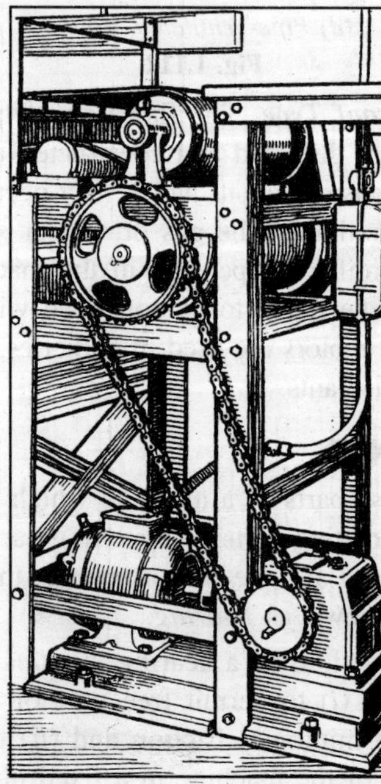

(*d*)

Fig. 1.113. ***Transmission of Drive***

(*ii*) It provides conveniently any desired speed ratio and permits the use of high-speed motors, which are cheaper than low-speed motors.

(*iii*) The tendency of slipping, especially under heavy overloads, is often beneficial in that it will absorb a portion of the shock of suddenly applied loads, and thus protect, to some extent, both the driving and driven machines.

Disadvantages and Limitations. (*i*) Considerable space is required and action is not positive. Distance between drive and driven pulley should not be less than four times the diameter of larger pulley. Typical figures for the minimum distance between pulley centres for motors of 22 to 30 kW are about 6 m for a 6 : 1 ratio and about 1.5 m for a $\frac{3}{2}$: 1 ratio.

(*ii*) Belts exert a side pull on the bearings which results in wear, also loss of power by friction.

(*iii*) The maximum power which can be conveniently transmitted by means of belts is about 150 to 220 kW.

(*iv*) The most economical speed has been found to be from 1,000 to 1,500 metres per minute, though practical limitations such as speed of shafting and pulley diameters, often limit belt speeds to much lower values.

Makers of belts and some motor makers publish tables from which the recommended pulley sizes, driving centres and section of belts or ropes can be conveniently calculated. In arriving at the final speed of the driven machine it must not be forgotten that there is always some slip (3 or 4 per cent) in a belt drive.

(ii) V-Belt Drive. It is essentially a short centre, silent drive with a negligible slip (say about 1 per cent). Such type of drive is suitable for motors of ratings up to 450 kW. The power transmitted depends upon the size of the belt, the speed of the belt (limited between 2.5 to 25 m/s), the size of the sheaves, (the larger the sheave, the greater the kW power transmitted) and the arc of contact on the smaller sheave.

The drives should be so designed that the arc of contact on the smaller V-pulley is not less than 90° and more than 130°, as far as possible 12 kW motor with multiple V-belt drive is shown in Fig. 1.113 (*b*).

3. *Rope Drive.* Rope drive, the drive by means of ropes running over pulleys having a number of grooves, is used for power ranges beyond the limit of V-belt drive. It is a long centre drive. Main advantages of rope drive are negligible slip and ability of taking sudden loads. Where the speed ratio is very high rope drive is indispensable. This type of drive is coming into use on small and medium sized machines, nowadays.

Coupled motors driving through multiple ropes working in grooved pulleys is shown in Fig. 1.113 (*c*).

4. *Chain Drive.* Though chain drive is more costlier than belt and rope drives but it is more efficient, can be used for high speed ratio (the limit being 7:1 with a distance between pulley centres of about 1.5 to 2 times the diameter of the larger pulley) and has no slip. It can be used conveniently for drives involving up to 12 or more parallel shafts; the driving centres can be shorter than with any drive except the direct or gear drive. It is particularly suitable for damp and dirty conditions because the chain case protects the drive from outside influences. With chain drive, it must be ensured that shafts are parallel and chains run centrally on their sprockets at right angle to the shafts to give long life and avoid end thrust on the motor. This method is suitable for speeds up to 900 m/minute needing positive drives without any slipping. Chain drive, through reduction gearing, from a 2.5 kW motor is shown in Fig. 1.113 (*d*).

5. *Gear Drive.* It is a short centre positive drive. It is very important to have proper alignment in the absence of which, otherwise, we may have to pay in the form of bent motor shafts. It was the first form of power transmission in the early steam driven mills. Modern straight cut gears, and the particular variation found in the worm reduction drive, are common features in many heavy machines such as cloth calenders. A typical 35 kW worm gear with a speed reduction of 15 : 1 will have a mechanical efficiency of about 98 per cent.

6. *Vertical Drive.* In this type of drive motor is arranged with its axis vertical. This arrangement frequently proves to be convenient.

1.27. CHOICE OF DRIVE

Choice of drive is governed by the speed of driving and driven machines, convenience, space available, clutching arrangement required and cost. Choice of speed of motor is most important factor as it not only affects the performance of the motor but also overall cost. The dimensions and, therefore, the first cost of a motor for a given output are approximately inversely proportional to the speed, so for the same output kW the cost of high-speed motor is less than that of a low-speed motor. The efficiency and, in case of induction motors, power factor also decreases with the decrease in speed. Thus for a low-speed drive high-speed motor using a reduction gear is usually found cheaper than a low-speed direct coupled motor. These days there is a trend to make motors for low-speed drives by incorporating gear in the motor frame, the motor being designed for a high speed. The overall cost of such a motor including the gear is less than that of the corresponding low-speed motor; gear efficiency ranging from about 96 per cent; for a 5 : 1 ratio to about 80 per cent for a 50 : 1 ratio can be obtained.

1.28. NOISE

A feature of importance in choosing a motor, as well as any other apparatus, is that of noise; extreme silence is, of course, essential in motors for domestic purposes, use in hospitals, theatres, another similar institutions. Even for ordinary industrial purposes it is essential to keep motor noise to a minimum because it causes fatigue to the workers and, therefore, affects the overall efficiency and output of the factory adversely. The increasing use of steel-framed buildings, which transmit noise readily, tends to make the problem of greater importance than hitherto.

The level of sound is measured in bels (B), which is defined as

Difference in loudness

$$= \log_{10} \frac{W_2}{W_1} = \log_{10} \frac{P_2^2}{P_1^2} \text{ bels} \qquad \because W \propto P^2$$

$$= 2 \log_{10} \frac{P_2}{P_1} \qquad \text{...(1.90)}$$

where W_1 and W_2 are the powers of sound waves under consideration and P_1 and P_2 are the pressures corresponding to the sound waves concerned.

In practice the decibel (one-tenth of a bel) is more convenient unit and represents approximately the smallest difference in sound which the ear can detect.

Thus the difference in loudness between two sounds

$$= 10 \log_{10} \frac{W_2}{W_1} \text{ dB} = 20 \log_{10} \frac{P_2}{P_1} \text{ dB} \qquad \text{...(1.91)}$$

Noise may be due to purely mechanical causes, such as resonance or vibration of laminations, or it may be due to magnetic pulsations caused by an incorrect choice of the number of slots. The elimination of such features is a matter of the manufacturer, but their effects can be minimized by preventing transmission of vibrations to other parts of the building. This can be accomplished by resilient or spring mounting the motor. Resilient mounting makes use of cushioning material such as felt mat, cork and rubber slabs which is effective in the audible frequency range. Spring mounting on the other hand can absorb all vibration frequencies. To avoid transmission of vibrations electrical connections to the motor are made by means of flexible pipe rather than by conduit pipe. It is, however, impossible to exterminate the noise.

1.29. INSULATING MATERIALS

Insulating materials are employed for providing electrical insulation between parts at different potentials. These materials begin to deteriorate at relatively small temperatures. For reliable operation, it is essential that the temperature rise in electrical machines and equipments do not exceed the permissible maximum temperature of the insulating materials used therein.

The three fundamental electrical properties of insulating materials of utmost importance in the operation of electrical machines or equipment are resistivity (insulation resistance), dielectric strength and dielectric loss angle. In addition to the above, other important properties of insulating materials are mechanical strength, heat resistance, hygroscopicity.

An insulating material should have high resistivity, high dielectric strength, low dielectric loss, good heat conductivity, sufficient mechanical strength to withstand vibrations etc., capability of withstanding a repeated heat cycle without deterioration and should be non-hygroscopic.

The materials used for insulation in electrical machines and apparatus are numerous (such as insulating varnishes, mineral waxes, synthetic waxes, varnish, cloth, silk, cotton, rubber, vulcanized rubber, hard rubber, bakelite, paper, wood, mica, asbestos, glass, porcelain, marble, slate, insulating oils etc.). Application of some of the most important insulating materials are given below:

Cotton (cotton tape, empire cloth etc.) is mainly employed for covering bare wire used in winding of small magnet coils, armature windings of small machines, chokes and coils of small transformers. **Silk** is coated with a waxy form of lubricant having good insulating properties and is mainly used for winding of small machines. In electrical machines, **glass** is mainly used as a glass tape and varnished glass cloth. **Glass tapes** are used for insulating the windings and winding ends of electrical machines operating at high temperatures. **Varnished glass cloth** is used mainly for insulating the stator windings in hv electrical machines. **Impregnating varnishes** serve to treat porous fibrous or solid insulators used in electrical machines and apparatus. **Bakelite** is used as an insulating cylinder in transformers. **Mica** is largely used in commutators for separating its segments.

1.29.1. Classification of Insulating Materials

The temperature rise which electric motor can safely withstand is determined by the thermal stability (limiting temperature) of the insulating materials used in them. By using materials of higher thermal stability, motors of the same physical size may be rated for larger power output.

Different insulating materials have different limiting temperatures. According to their limiting temperatures all insulating materials employed in electrical machines are classified as per IS 1271 as below.

1. **Class Y-** Cotton yarn and fabrics, paper and fibrous materials of cellulose or silk neither impregnated with nor immersed in liquid dielectric. The limiting temperature is 90° C.
2. **Class A-** The same materials as mentioned above, but impregnated with or immersed in liquid dielectric. Limiting temperature is 105° C.
3. **Class E-** Certain synthetic organic films and other materials having the same thermal stability. Limiting temperature is 120° C.
4. **Class B-** Mica, asbestos or glass-fibre base materials combined with an organic binding agent, The limiting temperature is 130° C.
5. **Class F-** Mica, asbestos or glass-fibre base materials combined with suitable synthetic binding and impregnating agents. The limiting temperature is 155° C.
6. **Class H-** Mica, asbestos or glass fibre combined with silicon binding and impregnating agents. The limiting temperature is 180° C.
7. **Class C-** Mica, ceramic materials, glass, or quartz used without or with inorganic binding agents. For this class of insulating materials, the limiting temperature is above 180° C and is restricted only by the specific physical, chemical or electrical properties of the material concerned.

The limiting temperatures are never exceeded in practice, otherwise it will result in deterioration and breakdown of the insulation and will shorten the life of the motors.

1.30. TEMPERATURE RISE IN ELECTRICAL MACHINES

Heat is developed in all electrical machines due to the losses in the various parts, principally copper loss (I^2R loss) in conductors, eddy current and hysteresis loss in iron and mechanical losses (in rotating machines only) due to friction of the bearing, air friction or windage causing the temperature of that part to rise. This temperature rise continues until all the heat generated is dissipated to the surroundings by one or more of the natural modes of heat transfer, viz. conduction, convection and radiation.

Ultimately then, under steady load, each part achieves a final temperature, the magnitude of which depends on the balance between the rate at which heat is developed in that part (or received by conduction from a hotter part), and the rate at which the heat can be dissipated which is determined by effectiveness of the cooling method.

The temperature rise depends upon (*i*) the amount of heat produced and (*ii*) the amount of heat dissipated per 1°C rise of the surface of a machine.

According to Newton's law of cooling the rate of loss of energy of a hot body is proportional to the difference in temperature between that body and its surroundings. This law is approximately true for moderate temperature difference (up to 100° C) and for bodies dissipating heat by radiation and natural convection. It means that the amount of heat dissipated per 1° C rise of the surface of a machine depends on the surface area of cooling.

The size of motor for any service is governed by the maximum temperature rise when operating under the given load conditions and the maximum torque required. The former is more important because if the motor operates satisfactorily for maximum temperature rise, it will usually provide the required maximum torque, except in special cases where the load consists of heavy peaks followed by relatively long intervals of no load. Electrical machines are, therefore, designed for a limited temperature rise. In fact, the continuous rating of a machine is that rating for which the final temperature rise is equal to or just below the permissible value of temperature rise for the insulating material used in protection of motor windings. When the machine is overloaded for such a long time that its final temperature rise exceeds the permissible limit, it is likely to be damaged. In worst cases, it will result in an immediate thermal breakdown of the insulating material which will cause a short circuit in the motor, thus putting a stop to its functioning. The short circuit may also lead to a fire. In less severe cases immediate thermal breakdown of the insulating material may not occur, but the quality of the insulation will deteriorate such that thermal breakdown with future overloads or even normal loads might soon occur, thus shortening the useful life of the machine.

The temperature rise to which a motor be allowed to rise is limited by the insulation. The maximum temperature rises which should not be exceeded by different types of motor are fully set out in the relevant ISS. Since temperature rise is one of the chief features in fixing the size of a motor, its calculations become a matter of importance and may be carried out as detailed below.

Heating-Time Curve. For determination of an expression for the temperature rise of an electrical machine after time *t* seconds from the instant of switching it on, let

Power converted into heat = P joules/s or watts

Mass of active parts of machine = *m* kg

Specific heat of material = C_p joules/kg/°C

Surface area of cooling = S metres^2

Coefficient of cooling = α in watts per metre^2 of surface per °C of difference between surface and ambient cooling temperatures.

Assumptions Made. (*i*) Losses or heat produced remains constant during the temperature rise.

(*ii*) Temperature of the cooling medium remains unchanged.

and (*iii*) Heat dissipated is directly proportional to the difference in temperature of the motor and the cooling medium.

Suppose a machine attains a temperature rise of θ°C above ambient temperature after *t* seconds of switching on the machine and further rise of temperature by $d\theta$ in very small time dt.

Energy converted into heat = Pdt joules

Heat absorbed = $mC_p d\theta$ joules

Heat dissipated = $S\theta\alpha dt$ joules

Since energy converted into heat

= Heat absorbed + heat dissipated

$$\therefore \; Pdt = mC_p d\theta + S\theta\alpha dt \quad ...(1.92)$$

$$\text{or } (P - S\alpha\theta)dt = mC_p \, d\theta$$

$$\text{or } \frac{dt}{mC_p/S\alpha} = \frac{d\theta}{\frac{P}{S\alpha} - \theta} \quad ...(1.93)$$

When final temperature is reached, there is no absorption of heat. Whatever heat is generated, has to be dissipated.

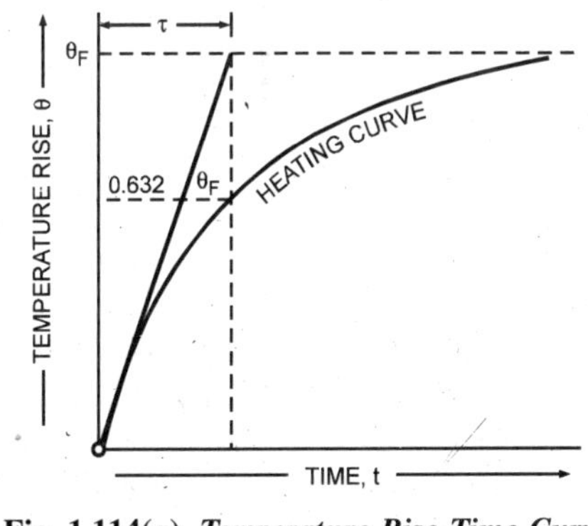

Fig. 1.114(*a*). *Temperature Rise-Time Curve*

$$\therefore \; Pdt = S\theta_F\alpha \, dt$$

or Final temperature rise,

$$\theta_F = \frac{P}{S\alpha} \quad ...(1.94)$$

Substituting $\frac{P}{S\alpha} = \theta_F$ in Eq. (1.93), we get

$$\frac{S\alpha dt}{mC_p} = \frac{d\theta}{\theta_F - \theta}$$

Integrating both sides, we get

$$\frac{S\alpha}{mC_p}t = -\log_e(\theta_F - \theta) + K_1 \qquad ...(1.95)$$

where K_1 is a constant of integration.

Substituting $t = 0, \theta = \theta_1$, initial temperature rise from initial conditions, we have

$$0 = -\log_e(\theta_F - \theta_1) + K_1$$

or $K_1 = \log_e(\theta_F - \theta_1)$

Substituting $K_1 = \log_e(\theta_F - \theta_1)$ in Eq. (1.95), we get

$$\frac{S\alpha}{mC_p}t = -\log_e(\theta_F - \theta) + \log_e(\theta_F - \theta_1)$$

$$= \log_e\left(\frac{\theta_F - \theta_1}{\theta_F - \theta}\right)$$

or $e^{\frac{S\alpha t}{mC_p}} = \frac{\theta_F - \theta_1}{\theta_F - \theta}$

or $\theta_F - \theta = (\theta_F - \theta_1)\, e^{-S\alpha t/m\,C_p}$

or $\theta = \theta_F - (\theta_F - \theta_1)\, e^{-S\alpha t/m\,C_p}$

$$= \theta_F - (\theta_F - \theta_1)\, e^{-t/\tau} \qquad ...(1.96)$$

where $\tau = \frac{mC_p}{S\alpha}$ and is known as *heating time constant.*

If motor is started from ambient temperature, $\theta_1 = 0$, we have

$$\theta = \theta_F - \theta_F^{-t/\tau} = \theta_F(1 - e^{-t/\tau}) \qquad ...(1.97)$$

Heating time constant is defined as the time taken by the motor in attaining the final steady value if the initial rate of rise of temperature were maintained throughout the operation. Substituting $t = \tau$ in Eq. (1.97), we have

$$\theta = \theta_F(1 - e^{-1}) = 0.632\,\theta_F \qquad ...(1.98)$$

So *heating time constant may also be defined as the time duration during which the machine will attain 63.2% of its final temperature rise above ambient temperature.*

From Eq. (1.94) it is obvious that θ_F is directly proportional to the power losses and inversely proportional to the surface area S and specific heat dissipation α.

For poorly ventilated machines, it will attain a higher final temperature rise.

Heating time constant τ, being equal to $\frac{mC_p}{S\alpha}$, has small value for well ventilated machines and large value for poorly ventilated machines. Large size machines have large heating time constant because with the increase in size of machine, the volume and hence the mass increases in proportion to the third power of linear dimensions and surface area S in proportion to second power. Typical values of heating time constant lie between about $1\frac{1}{2}$ hours for small motors (7.5 kW to 15 kW) up to about 5 hours for motors of several hundred kW.

Cooling-Time Curve. Let the machine be switched off after reaching final steady-temperature rise of θ_F. When the machine is switched off, no heat is produced, therefore,

0 = Heat absorbed + heat dissipated

or $0 = mC_p d\theta + S\theta\alpha' dt$ where α' is the rate of heat dissipation during cooling

or $dt = \frac{-mC_p\, d\theta}{S\alpha'\theta}$

Integrating both sides, we have

$$t = \frac{-mC_p}{S\alpha'}\log_e\theta + K_2 \qquad ...(i)$$

where K_2 is a constant of integration

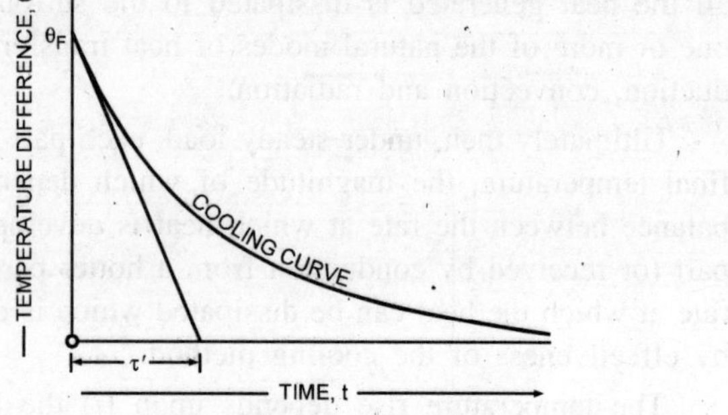

Fig. 1.114(*b*)

Substituting $t = 0$, $\theta = \theta_F$ from initial conditions in above equation, we have

$$K_2 = \frac{mC_p}{S\alpha'}\log_e\theta_F$$

Substituting $K_2 = \frac{mC_p}{S\alpha'}\log_e\theta_F$ in Eq. (*i*), we have

$$t = \frac{-mC_p}{S\alpha'}\log_e\theta + \frac{mC_p}{S\alpha'}\log_e\theta_F = \frac{-mC_p}{S\alpha'}\log_e\frac{\theta}{\theta_F}$$

or $\theta = \theta_F\, e^{-S\alpha' t/mC_p} = \theta_F\, e^{-t/\tau'} \qquad ...(1.99)$

where $\tau' = \frac{mC_p}{S\alpha'}$, called the cooling time constant.

Substituting $t = \tau'$ in Eq. (1.99), we have

$$\theta = \theta_F e^{-1} = 0.368\,\theta_F \qquad ...(1.100)$$

i.e., cooling time constant may also be defined as the time required to cool the machine down to 0.368 times the initial temperature rise above ambient temperature.

The heating and cooling curves follow an exponential law. Heating time constant and cooling time constant may be different for the same machine, as the ventilation conditions in the two cases may not be the same. The cooling time constant of a rotating machine is usually larger than its heating time constant owing to poor ventilation conditions when the machine cools. In self-cooled rotating machines the cooling time constant is about 2-3 times greater than its heating time constant because cooling conditions are worse at standstill.

1.31. DUTY CYCLES

The nominal duty of a drive motor is the duty corresponding to the service conditions and performance marked on its name plate.

There are three types of duties viz continuous duty, short-time duty and intermittent duty.

The heating and cooling curves for *continuous-duty motor* are shown in Fig. 1.115 (*a*). Continuous duty is that duty when the on-period is so long that the motor attains a steady-state temperature rise.

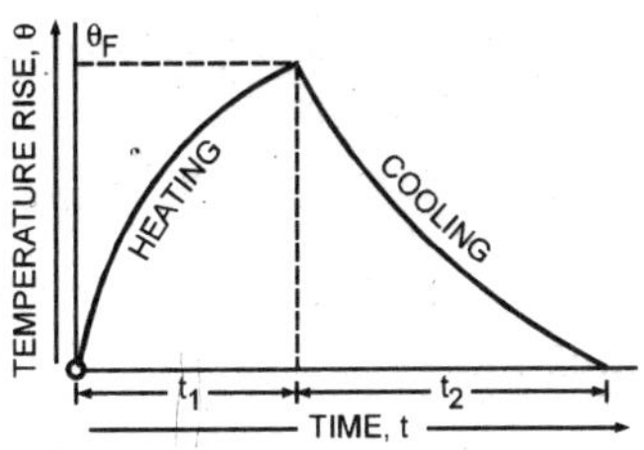

(a) For Continuous Duty Motor

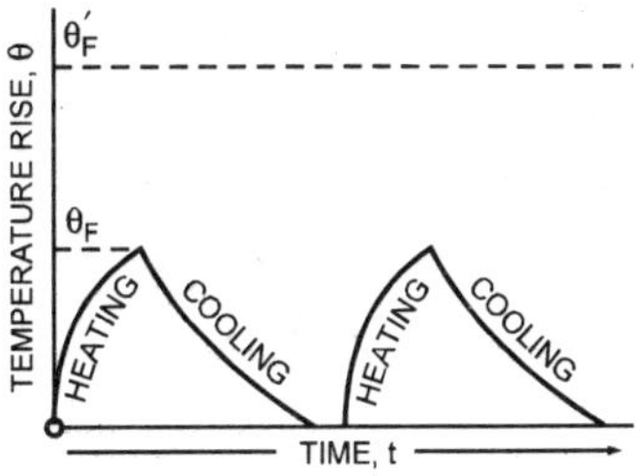

(b) For Short-time Duty Motor

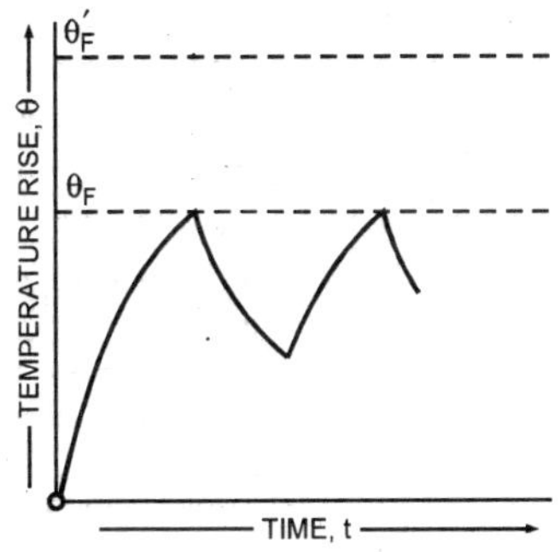

(c) For Intermittent Periodic Duty Motor

Fig. 1.115. *Heating and Cooling Curves*

The heating and cooling curves for *short-time duty motor* are given in Fig. 1.115 (*b*). The short-time duty motor operates at a constant load for some specified periods which is then followed by a period of rest. The period of run (or load) is so short that machine cannot attain its steady temperature rise while the period of rest is too long that the motor temperature drops to the ambient temperature.

The heating and cooling curves for *intermittent periodic duty motors* are illustrated in Fig. 1.115 (*c*). On intermittent duty the periods of constant load and rest with machine de-energized alternate. The loading periods are too short to allow the motor to attain its final steady-state value while periods of rest are too small to allow the motor to cool down to the ambient temperature. *Intermittent rating of a machine is defined as the load which is applied during a certain fraction of time of a load cycle and the temperature rise limit is not exceeded.*

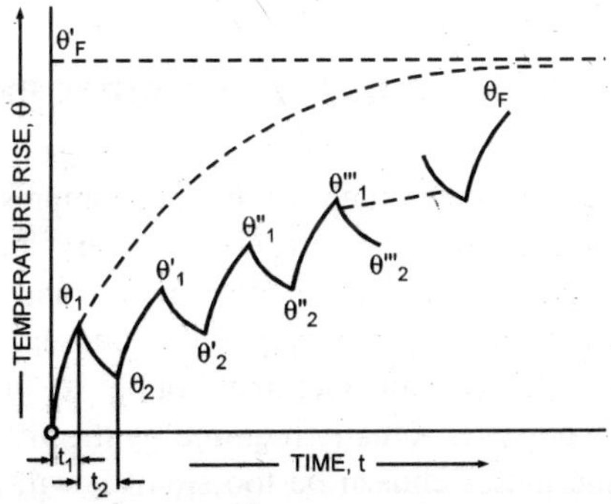

Fig. 1.116

When a machine is intermittently loaded, it will cool down during the time it is off, and temperature will rise when it is on, as illustrated in Fig. 1.116.

Continuous duty motors are employed to drive fans, compressors, generators etc. They may be in operation for many hours and even days in succession. *Short-time duty motors* are used in navigation-lock gates, railway turntables, bascule bridges, and the like. *Intermittent-duty motors* are employed in cranes, hoists, lifts, rolling mills, some metal-working machines, etc.

The duty cycles for various motor duties are shown in Fig. 1.117.

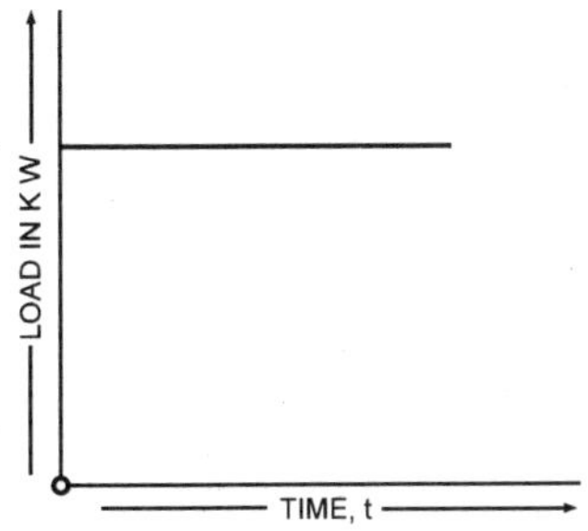

(a) Continuous Duty Cycle

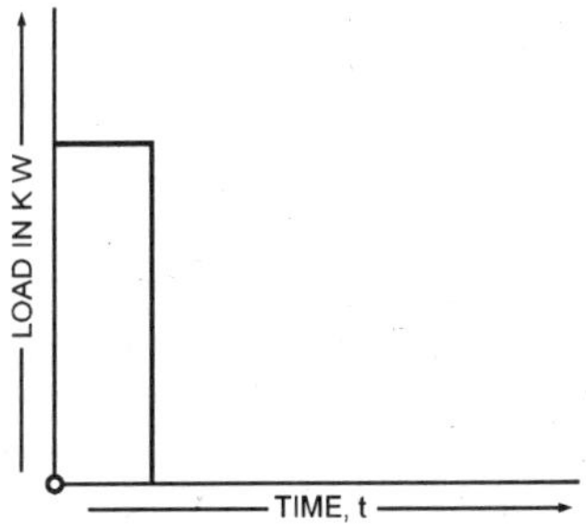

(b) Short-Time Duty Cycle

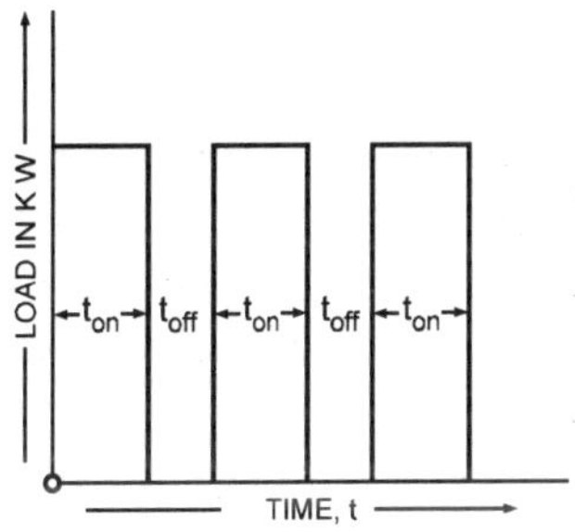

(c) Intermittent Duty Cycle

Fig. 1.117

1.31.1. Maximum Temperature Attained With Intermittent Loading

When a machine is intermittently loaded it will cool down during the time duration it is off and its temperature will rise when it is on, as illustrated in Fig. 1.116. Let θ_1, θ'_1, θ''_1, be θ'''_1 be the temperature rise after heating and θ_2, θ'_2, θ''_2, θ'''_3 be the temperature rise after cooling, as illustrated in Fig. 1.116.

$$\text{Now } \theta_1 = \theta'_F(1 - e^{-t_1/\tau}) = \theta'_F(1 - e^x) \text{ where } x = \frac{-t_1}{\tau}$$

$$\text{and } \theta_2 = \theta_1 e^{-t_2/\tau'} = \theta_1 e^y = \theta'_F(1 - e^x)e^y \text{ where } y = \frac{-t_2}{\tau'}$$

$$\begin{aligned}
\theta'_1 &= \theta'_F - (\theta'_F - \theta_2)e^x \\
&= \theta'_F(1 - e^x) + \theta_2 e^x \\
&= \theta'_F(1 - e^x) + \theta'_F(1 - e^x)\, e^y e^x \\
&= \theta'_F(1 - e^x)\,(1 + e^x e^y) \\
\theta'_2 &= \theta'_1 e^y = \theta'_F(1 - e^x)\,(1 + e^{x+y}).e^y \\
\theta''_1 &= \theta'_F(1 - e^x) + \theta'_2 e^x \\
&= \theta'_F(1 - e^x) + \theta'_F e^x e^y (1 - e^x)\,(1 + e^x e^y) \\
&= \theta'_F(1 - e^x)\,[1 + e^x e^y + e^{2x}.e^{2y}]
\end{aligned}$$

For n times intermittency, we have

$$\theta_1^{1-n} = \theta'_F(1 - e^x)[1 + e^x e^y + e^{2x} e^{2y} + \ldots e^{(n-1)x} \times e^{(n-1)y}]$$

$$= \theta'_F(1 - e^x)\left[\frac{1 - e^{nx} e^{ny}}{1 - e^x e^y}\right]$$

When $n = \infty$ both e^{nx} and e^{ny} will be zero as x and y are negative. If θ_F is the maximum temperature rise with intermittent load then

$$\theta_F = \theta_F' \left(\frac{1-e^x}{1-e^x e^y} \right) = \theta_F' \frac{1-e^x}{1-e^{x+y}} = \theta_F' \frac{1-e^{-t_1/\tau}}{1-e^{-(t_1/\tau+t_2/\tau')}} \quad ...(1.101)$$

1.32. RATING OF MACHINES

A name plate fixed to the outside frame of an electrical machine records the data pertaining to its rating. A machine rating specifies the voltage, current, speed, excitation, power factor, efficiency, power output etc.

The rating of a machine should give all information necessary so that, if the machine is operated within the limits of all factors specified in its rating, the machine will operate satisfactorily and safely and will give reasonable length of service. Therefore, the rating of a machine must give the necessary information to safeguard the application of the machine from conditions of operation which (*i*) would result in unsafe mechanical or electrical strains upon any part of its structure or (*ii*) would result in excessive deterioration of the mechanical or electrical characteristics of the materials of which the machine is constructed, To give this information, the rating of an electric machine should include the output, voltage, speed, and any other information that may necessary for the proper operation of the machine.

Generators and motors are rated in terms of kW output at a given speed and voltage. The size and rating of an electric machine for any service is mainly governed by the factor 'temperature rise'. The maximum temperature, to which an electric machine is allowed to reach, is limited by the type of insulation used.

The maximum temperature rise permissible with insulation A and insulation B are 40ºC and 50ºC respectively. Overloads are generally permissible for short period of time but when machines are required to carry greater loads than those specified, they must be kept under inspection to see that the temperature does not rise too much and that severe sparking at the commutator does not occur.

The type of service to which a machine is subjected is of great importance. The machines operating continuously at rated (or near rated) load are physically larger than those working at intermittent loads. Also electric machines that are not enclosed and are, in addition, well cooled by fans are likely to have higher ratings than covered up machines or those machines located where air does not circulate freely through and over them. High-speed machines and machines employing mica, glass tapes, and the new silicons as insulation can generally be physically smaller, in given ratings, than the low-speed machines employing standard insulations. The output of dc machines is also limited by the factor 'commutation sparking'. This factor often limits the output, even though heating may not have proceeded to permissible values.

The machines according to ISI specifications are classified as follows.

Continuous Rating. This is an output, which a machine delivers continuously without exceeding the permissible temperature rise. It can deliver 25 per cent overload for two hours.

Continuous Maximum Rating. Similar to continuous rating but not allowing any overload. It is used for motors of capacity larger than 1.84 kW (2.5 hp) per rpm.

Short-Time Rating. This is an output which an electric machine can deliver for a specified period (say 1 hr., $\frac{1}{2}$ hr., $\frac{1}{4}$ hr. etc.) without exceeding the specified temperature rise.

1.32.1. Effect of Altitude on Rated Output

At high altitudes, the density of cooling air decreases. As a result, cooling capacity is reduced. The effect is negligible for elevations not exceeding 1,000 m. The output is reduced as stated below in tabular form.

Altitude above sea level in metres	*Permissible output in percentage of the rated value for motor speeds*	
	up to 1,000 rpm	*above 1,000 rpm*
0–1,000	100	100
2,000	95	92
3,000	90	85
4,000	87	80
5,000	85	77

1.32.2. Overload Capacity of Induction Motors

Motors are designed to carry overloads, without any adverse effect, as mentioned below.

Sustained overloads	*Time duration*	*Motor kW output per 1,000 rpm*
25% overload in torque	2 hours	7.5 and above
	30 minutes	3 to 6.8
	15 minutes	0.75 to 3
Momentary overloads		
50% overload in torque	1 minute for all sizes	
100% overload in torque	15 seconds for all sizes	

Single phase induction motors are usually not designed to carry overload. However, 20-25% overload in torque intermittently may be permitted in split-phase ac motors and 40-45% overload in torque intermittently for capacitor-start motors.

1.33. CHOICE OF RATING OF MOTORS

The choice of ratings of motors for the services requiring fairly constant power such a fans, air-compressors, blowers, pumps, motor-generator sets etc. is every simple. In such cases the power required is determined and motor having continuous rating of power required is chosen. It is, however, worthwhile to take considerab e trouble to find out exactly how much power is actually required by the driven machinery, since if the motor chosen be too small it will be overloaded and will, therefore, overheat and deteriorate, while if it is too large the efficiency and power factor will probably be less than those of a motor of the correct size. If there is no possibility of overloading then a motor with a continuous maximum rating may be selected, which will be comparatively smaller in size and cheaper in cost. Such a motor may operate at a higher temperature (55°C in place of 40°C).

For the constant torque loads, the size of the motor can be determined as follows :

In case of rotary motion, rating of motor required is given as

$$P = T \times \frac{2\pi N}{60} \times \frac{1}{\eta} \times \frac{1}{1{,}000} \text{ kW} = \frac{TN}{9{,}550\eta} \text{ kW} \quad ...(1.102)$$

where T is the load torque in N-m, N is the speed in rpm and η is the product of the efficiency of the driven load and that of the transmitting device. Efficiency varies considerably with the type of drive, bearings etc. and its value can best be dictated by experience in applications.

In case of linear motion, rating of the motor required is given as

$$P = \frac{F \times 9.81 \times v}{\eta} \times \frac{1}{1{,}000} \text{ kW} = \frac{F \times v}{102\,\eta} \text{ kW} \quad ...(1.103)$$

where F is the force caused by the load in kg and v is the velocity of motion of load in m/s.

The above expression [Eq. (1.103)] can be applied in case of hoisting mechanism, lifts or elevators with slight modification and that is due to counterweight (balancing the weight of the cage or car as well as one-half of the useful load). The modified expression is

$$P = \frac{Fv}{2 \times 102\,\eta} \text{ kW} \quad ...(1.104)$$

v in case of normal passenger lift cabins lies in the range of 0.5–1.5 m/s.

In case of pumps, the rating of the motor required,

$$P = \frac{Q \times \rho \times h}{102\,\eta} \text{ kW} \quad ...(1.105)$$

where Q is the delivery of pumps in m^3/s, ρ is the density of liquid pumped in kg/m^3 and h is the gross head (static head + friction head) in m. Efficiency η lies in the range of 80 to 90% in case of reciprocating pumps and 40 to 80% in case of centrifugal pumps.

Similarly, the rating of a fan motor is given as

$$P = \frac{Qh}{102\eta} \text{ kW} \quad ...(1.106)$$

where Q is the volume of air or any other gas in m^3/s and h is the pressure in mm of water or in kg/m^2. Efficiency may be taken to be 0.8 for larger power fans and 0.6 in case of smaller power fans.

The selection of size of motor is also simple for the services where the motor is run for a short time so that the temperature rise is less than the maximum permissible value and then it is allowed to cool down to the ambient temperature. For such services a motor rated at the required output power for the given period is chosen.

In cases where the load fluctuates over a given cycle, as in rolling mills or colliery winders, the size of the motor is determined accurately by finding the heating and cooling curves of motors under consideration and plotting heating curve for each motor, when working on the given cycle. The smallest motor which delivers the load according to given cycle without exceeding the specified temperature rise may be choosen.

Other methods for determination of rating of a motor for continuous duty and variable load are given below.

1. *Method of Average Losses.* In this method it is assumed that the temperature rise attained by the motor with fluctuating loading conditions over a certain period of time will be the same as that attained by the motor with a certain load of constant magnitude. This will be true provided that the average losses in the motor for the same time are the same for both the conditions of operation.

This method consists of determining average losses in the motor when it operates according to the given load diagram and then comparing with the losses corresponding to the continuous duty of the machine when operated at its nominal rating.

2. *Equivalent Current Method.* This method is based upon the assumption that the actual variable current may be replaced by an equivalent current I_{eq} which produces the same losses in the motor as the actual current.

$$\therefore \quad I_{eq} = \sqrt{\frac{I_1^2 t_1 + I_2^2 t_2 + I_3^2 t_3 + \ldots + I_n^2 t_n}{t_1 + t_2 + t_3 + \ldots + t_n}} \quad ...(1.107)$$

The heating and cooling conditions in self ventilated machines depend upon its speed. At low speeds the cooling conditions are poorer than at normal speeds. Therefore, if the work cycle involves slow-speed operation, it must be taken into consideration when using Eq. (1.107).

The equivalent current is compared with the rated current of the motor selected. The equivalent current may be less than or equal to the rated current of the machine.

3. *Equivalent Torque Method.* This method is based upon the assumption that the motor current is proportional to the torque and heating is proportional to the square of current (*i.e.*, heating is proportional to the square of torque).

The equivalent torque is determined in the same manner as the equivalent current *i.e.*,

$$T_{eq} = \sqrt{\frac{T_1^2 t_1 + T_2^2 t_2 + T_3^2 t_3 + \ldots + T_n^2 t_n}{t_1 + t_2 + t_3 + \ldots + t_n}} \quad ...(1.108)$$

A motor having torque nearly equal to T_{eq} is selected.

4. *Equivalent Power Method.* At constant speed or where the variations in speed are small, the equivalent power is given by the relationship,

$$P_{eq} = \sqrt{\frac{P_1^2 t_1 + P_2^2 t_2 + P_3^2 t_3 + \ldots + P_n^2 t_n}{t_1 + t_2 + t_3 + \ldots + t_n}} \quad ...(1.109)$$

A motor having power rating P_{eq} is selected.

The equivalent current method is the most accurate out of the four methods discussed above. This method may be employed for determining the capacity for all applications except where it is necessary to take into account the variations in so called 'constant losses'.

The equivalent torque method cannot be employed for applications where equivalent current method cannot be applied or in cases in which flux does not remain constant like dc series motors.

The equivalent power method cannot be used for motors whose speed varies considerably under load, especially when dealing with starting and braking conditions.

Example 1.45. The outside surface of a 12 hp (metric) motor is equivalent to a cylinder of 65 cm and 1 metre length. The motor weighs 400 kg and has a specific heat of 700 J/kg/°C. The outer surface is capable of heat dissipation of 12 W per metre square per degree centigrade. Find the final temperature rise and thermal constant of motor when operating at full load with an efficiency of 90 %. *[J.N. University Hyderabad April 2003]*

Solution : Power output of machine = 12 × 735.5 = 8,826 watts

$$\text{Power input} = \frac{\text{Power output}}{\eta} = \frac{8,826}{0.9} = 9,807 \text{ watts}$$

Power losses, P = Power input – power output = 9,807 – 8,826 = 981 watts

Mass of active parts of machine, m = 400 kg

Specific heat of material, C_p = 700 J/kg/°C

Coefficient of cooling, α = 12 watts/m^2/°C

Surface area of cooling = $\pi dl = \pi \times 0.65 \times 1 = 2.042$ m^2

$$\text{Final temperature rise, } \theta_m = \frac{P}{S\alpha} = \frac{981}{2.042 \times 12} = 40^\circ \text{ C } \textbf{Ans.}$$

Heating time constant,

$$\tau = \frac{mC_p}{S\alpha} = \frac{400 \times 700}{2.042 \times 12}$$

= 11,427 seconds or 3 hours 11 minutes **Ans.**

Example 1.46. The mean temperature rise of a machine to run on full load is :

Time in hours	0.2	0.4	0.6	1.0	1.5	2.0	3.0
Temperature rise in °C	7.0	13.5	19.3	29.1	39.1	46.7	57.3

Estimate the final steady-state temperature and heating time constant. *[Agra Univ. Traction & Utilization of Elec. Power 1979]*

Solution : For determination of steady-state temperature plot temperature-rise against time from the given datas, as shown in Fig. 1.118. Now plot increments of temperature for equal intervals of time against temperature rise to the left of Y-axis, as shown in Fig. 1.118.

Straight line drawn through the plotted points cuts the Y-axis, where the rate of change of temperature rise is zero. Therefore the point of intersection of the line drawn, as mentioned above, with the Y-axis gives steady-state temperature. From Fig. 1.118, θ_F = 74°C. **Ans.**

From definition, heating time constant is the time at which temperature rise is 0.632 times of final steady state temperature, *i.e.*, 0.632 × 74 = 46.7°C.

From Fig. 1.118, time corresponding to temperature rise of 46.7°C is 2 hours. Hence heating time constant is 2 hours. **Ans.**

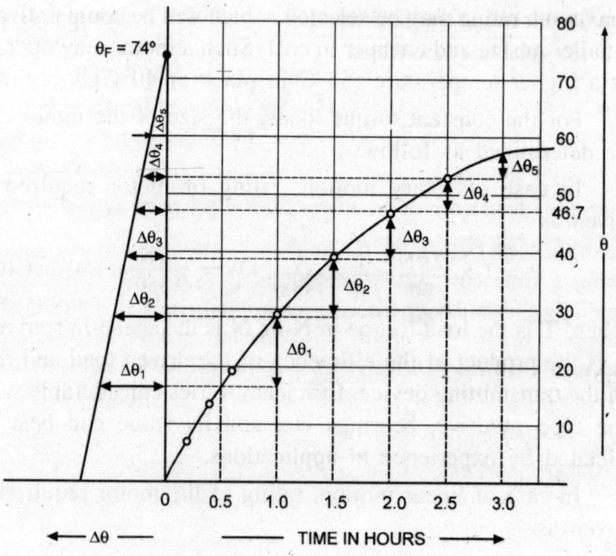

Fig. 1.118

Example 1.47. The full-load temperature rise of totally enclosed motor is 20°C after one hour and 34°C after two hours. Find (*i*) final steady-temperature rise on full load, (*ii*) the heating time constant, (*iii*) one hour rating of the motor for the same final temperature rise. Take no-load losses = copper losses at rated continuous load.

Solution : From Eq. (1.97)

$$\theta = \theta_F(1 - e^{-t/\tau}) \quad ...(i)$$

Now from given datas when t = 1 hour, θ = 20° C and when t = 2 hours, θ = 34° C. Substituting the above values of t and θ in Eq. (*i*) we get

$$20 = \theta_F(1 - e^{-1/\tau}) \quad ...(ii)$$

$$\text{and } 34 = \theta_F(1 - e^{-2/\tau}) \quad ...(iii)$$

Dividing Eq. (*iii*) by Eq. (*ii*) we get

$$\frac{34}{20} = \frac{1 - e^{-2/\tau}}{1 - e^{-1/\tau}}$$

$$\text{or } 1 + e^{-1/\tau} = 1.7$$

$$\text{or } e^{-1/\tau} = 0.7$$

$$\text{or Heating time constant, } \tau = \frac{-1}{\log_e 0.7} = 2.8 \text{ hours } \textbf{Ans.}$$

Substituting τ = 2.8 hours in Eq. (*ii*) we get

$$\text{Final steady temperature rise, } \theta_F = \frac{20}{1 - e^{-1/2.8}} = 66.67^\circ \text{ C } \textbf{Ans.}$$

If temperature rise after one hour is 66.67°, final temperature rise θ_F will be given as,

$$\theta'_F = \frac{\theta_F}{1 - e^{-t/\tau}} = \frac{66.67}{1 - e^{-1/2.8}} = 222^\circ \text{ C}$$

Total losses at full load

= Iron loss + copper loss = 2 × copper loss

∵ Iron loss = no-load loss = full-load copper loss

= 2P_c where P_c is full-load copper loss

Let one hour rating be x times full load then

Total losses = Iron losses + x^2 P_c

= $P_c + x^2 P_c = (1 + x^2) P_c$

$$\therefore \quad \frac{\text{Total losses at one-hour rating}}{\text{Total loss at full-load}} = \frac{\theta'_F}{\theta_F}$$

$$\text{or} \quad \frac{(1+x^2)P_c}{2P_c} = \frac{222}{66.67}$$

$$\text{or} \quad x = 2.38$$

So one-hour rating of the motor is 2.38 times full-load rating. **Ans.**

Example 1.48. Determine the half-hour rating of a 25 kW motor having a time constant of 1.5 hours. Assume that the motor cools down completely between each load period.

[U.P. Technical Univ. Electric Drives 2005-06]

Solution : Heating time constant, $\tau = 1.5$ hours

Let P kW be the half-hour rating, θ_F be the final temperature rise in case of continuous rating and W be full-load copper loss.

Final temperature rise in case of half-hour loading

$$\theta'_F = \frac{\theta_F}{1-e^{-t/\tau}} = \frac{\theta_F}{1-e^{-0.5/1.5}} = \frac{\theta_F}{1-e^{-1/3}} = 3.528\,\theta_F$$

Losses at 25 kW

$= 0.9W + W = 1.9$ W assuming iron loss to be 90 per cent of copper loss at full load

$$\text{Losses at P kW} = 0.9\text{ W} + \left(\frac{P}{25}\right)^2 \text{W} = \text{W}\left(0.9 + \frac{P^2}{625}\right)$$

Since copper loss varies as the square of the load

$$\frac{\text{Total losses at P kW}}{\text{Total losses at 25 kW}} = \frac{\theta'_F}{\theta_F}$$

$$\text{or} \quad \frac{\text{W}\left(0.9 + \frac{P^2}{625}\right)}{1.9\text{ W}} = 3.528$$

or $P = \sqrt{(3.528 \times 1.9 - 0.9) \times 625} = 60.2$ kW say, 60 kW **Ans.**

Example 1.49. A motor running continuously on full load has a temperature rise of 20 °C. The heating time constant is 60 minutes. How long the motor should be run at twice the continuously rated output without overheating ? The maximum efficiency of motor at full load is 80%.

[Pb. Technical Univ. Electric Drives and Utilization May 2008]

Solution : Heating time constant, $\tau = 60$ minutes

Final temperature rise, $\theta_F = 20°\text{C}$ in case of continuous rating

Let P kW be the continuous rating, and t be the time duration in minutes for which motor can be run at twice the continuously rated output without overheating.

Since maximum efficiency occurs 80% of full-load *i.e.*, 0.8 P, therefore at 0.8 P load copper loss is equal to constant loss, let each be equal to W watts.

$$\text{Total losses at P kW} = \text{W} + \left(\frac{P}{0.8P}\right)^2 \text{W} = 2.5625\text{ W}$$

$$\text{Total losses at 2P kW} = \text{W} + \left(\frac{2P}{0.8P}\right)^2 \text{W} = 7.25\text{ W}$$

Since copper loss varies as the square of load

$$\frac{\text{Total losses at 2P kW}}{\text{Total losses at P kW}} = \frac{\theta'_F}{\theta_F}$$

$$\text{or} \quad \theta'_F = \theta_F \times \frac{\text{Total losses at 2P kW}}{\text{Total losses at P kW}}$$

$$= 20 \times \frac{7.25}{2.5625} = 56.585\,°\text{C}$$

$$\text{Also} \quad \theta'_F = \frac{\theta_F}{1-e^{-t/\tau}}$$

$$\text{or } t = \tau \log_e \frac{1}{1-\frac{\theta_F}{\theta'_F}} = 60 \log_e \frac{1}{1-\frac{20}{56.585}}$$

= 26.17 minutes say, 26 minutes **Ans.**

Example 1.50. If one hour rating of a machine is $\sqrt{3}$ times the continuous rating and if the steady temperature rise for one hour rating is twice for normal load, find the ratio of iron losses to I^2R losses at full load.

Solution : Losses on continuous rating $= P_c + P_i$

Losses on one-hour rating $= x^2 P_c + P_i = 3P_c + P_i$

$\because$ one hour rating is $\sqrt{3}$ times the continuous rating *i.e.*, $x = \sqrt{3}$

Ratio of final steady temperature rises with 1 hour and continuous rating is given as

$$\frac{\theta'_F}{\theta_F} = \frac{P_i + x^2 P_c}{P_i + P_c}$$

$$\text{or } 2 = \frac{3P_c + P_i}{P_c + P_i}$$

$$\text{or } P_i = P_c$$

Thus iron losses are equal to full-load copper losses at full load. **Ans.**

Example 1.51. The 15 minute rating of a motor is 400 watts. The heating time constant is 60 minutes. Determine the continuous rating of the motor if the maximum efficiency occurs at 80 per cent of full load.

[A.M.I.E. Sec. B. Electric Drives and Their Control Winter 1996]

Solution : Since maximum efficiency occurs at 80% of full load so at 80% of full-load, iron loss is equal to copper loss, let each equal to P_c watts

$$\text{Losses on full load} = P_c + \left(\frac{1}{0.8}\right)^2 P_c = 2.5625\,P_c$$

$$\text{Losses at load of 400 watts} = P_c + \left(\frac{400}{0.8 \times \text{full load}}\right)^2 P_c$$

Ratio of final temperature rises with 15 minute rating and continuous rating is given as,

$$\frac{\theta'_F}{\theta_F} = \frac{1}{1-e^{-t/\tau}} = \frac{\text{Total loss on 15 - minute rating}}{\text{Total loss on full load}}$$

$$\text{or} \quad \frac{1}{1-e^{-15/60}} = \frac{P_c + \left(\frac{400}{0.8 \times \text{full load}}\right)^2 P_c}{2.5625\,P_c}$$

or Continuous rating = 153.7 watts **Ans.**

Example 1.52. A motor has a thermal time constant of 45 minutes. When the motor runs continuously on full load, its final temperature rise is 80° C. (*i*) What is the temperature rise after 1 hour if the motor runs continuously on full load? (*ii*) If the temperature on one-hour rating is 80° C, find the maximum steady-state temperature at this rating. (*iii*) How much time does the motor take for its temperature to rise from 50° C to 80° C if it is working at its 1 hour rating ?

[A.M.I.E. Sec. B. Electric Drives and Their Control Winter 1993]

Solution : (*i*) From given datas :

Final temperature rise, $\theta_F = 80°$ C

Thermal time constant, τ = 45 minutes = 0.75 hour

Temperature rise after 1 hour (*i.e* when t = 1) can be had by substituting θ_F = 80° C; τ = 0.75 hour and t = 1 hour in Eq. (1.97)

$$i.e., \ \theta = \theta_F (1 - e^{-t/\tau}) = 80 (1 - e^{-1/0.75}) = 58.9° \text{ C } \textbf{Ans.}$$

(*ii*) Let the maximum steady-temperature rise at one hour rating be θ'_F

Now from given datas τ = 0.75 hour, θ = 80° C when t = 1 hour

$$\therefore \theta'_F = \frac{\theta_F}{1 - e^{-t/\tau}} = \frac{80}{1 - e^{-1/0.75}} = 108.64° \text{ C } \textbf{Ans.}$$

(*iii*) Time taken to attain temperature of 80° C (assuming initial temperature as 0° C) in one hour or 60 minutes,

Final temperature rise, θ'_F = 108.64° C

Now let the time taken to attain temperature of 50° C from 0° C be t hours then

$$50 = 108.64 (1 - e^{-t/0.75})$$

$$\text{or } t = 0.4625 \text{ hour or } 27.75 \text{ minutes}$$

Hence time taken to increase temperature from 50° C to 80° C = 60 – 27.75 = 32.25 minutes **Ans.**

Example 1.53. A motor has a continuous rating of 100 kW. The heating and cooling time constants are 50 and 70 minutes respectively. The motor has a maximum efficiency at 80% of full-load and is employed in an intermittent periodic load cycle consisting of a load period of 10 minutes followed by no-load period of 10 minutes. Calculate the value of load in kW during the load period. [*U.P. Technical Univ. Electric Drives 2009-10*]

Solution : Heating time constant, τ = 50 minutes

Cooling time constant, τ' = 70 minutes

On-load period, t_1 = 10 minutes

Off-load period, t_2 = 10 minutes

From Eq. (1.101)

$$\frac{\theta'_F}{\theta_F} = \frac{1 - e^{-\left(\frac{t_1}{\tau} + \frac{t_2}{\tau'}\right)}}{1 - e^{-t_1/\tau}} = \frac{1 - e^{-\left(\frac{10}{50} + \frac{10}{70}\right)}}{1 - e^{-10/50}} = \frac{1 - 0.70974}{1 - 0.81873} = 1.6$$

Let P be rating of motor when subjected to a duty cycle of 10 minutes on certain load followed by 10 minutes on no load.

Since maximum efficiency occurs at 80 per cent of full load *i.e.* at a load of 80 kW, therefore, at 80 kW load copper loss is equal to constant loss, let each be equal to W watts.

$$\text{Losses at 100 kW load} = W + \left(\frac{100}{80}\right)^2 W = 2.5625 \text{ W}$$

$$\text{Losses at P kW load} = W + \left(\frac{P}{80}\right)^2 W = W\left(1 + \frac{P^2}{6{,}400}\right)$$

$$\text{Thus } \frac{\text{Total losses at P kW output}}{\text{Total losses at 100 kW output}} = \frac{\theta'_F}{\theta_F}$$

$$\text{So } \frac{W\left(1 + \frac{P^2}{6{,}400}\right)}{2.5625 \text{ W}} = 1.6$$

$$\text{or } P = \sqrt{(2.5625 \times 1.6 - 1) \times 6{,}400} = 140.85 \text{ kW } \textbf{Ans.}$$

Example 1.54. The temperature rise of a motor after operating for 30 minutes on full load is 20°C. After another 30 minutes on the same load the temperature rise becomes 30°C. Assuming that the temperature increases according to an exponential law, determine the final temperature rise and the heating time constant. [*U.P. Technical Univ. Electric Drives 2007-08*]

Solution : From Eq. (1.97)

$$\theta = \theta_F (1 - e^{-t/\tau})$$

Now from given datas when t_1 = 30 minutes, θ_1 = 20°C and when t_2 = 30 + 30 = 60 minutes, θ_2 = 30°C. Substituting the values for t and θ in above equation, we have

$$20 = \theta_F (1 - e^{-30/\tau}) \quad ...(i)$$

$$30 = \theta_F (1 - e^{-60/\tau}) \quad ...(ii)$$

Dividing Eq. (*ii*) by Eq. (*i*), we have

$$\frac{1 - e^{-60/\tau}}{1 - e^{-30/\tau}} = \frac{30}{20} = 1.5$$

$$\text{or } 1 + e^{-30/\tau} = 1.5$$

$$\text{or Heating time constant, } \tau = \frac{-30}{\log_e 0.5} = 43.28 \text{ minutes}$$

Substituting τ = 43.28 minutes in Eq. (*i*), we have

$$\theta_F = \frac{20}{1 - e^{-30/43.28}} = 40 \text{ °C } \textbf{Ans.}$$

Example 1.55. Calculate the maximum overload that can be carried by a 25 kW motor if the temperature rise is not to exceed 50°C after one hour on overload. The temperature rise on full load after one hour is 30°C and after 2 hours is 40°C. The losses vary as square of the load.

Solution : The expression for temperature rise is given as

$$\theta = \theta_F(1 - e^{-t/\tau}) \quad ...(i)$$

From given datas when t = 1 hour, θ = 30°C and when t = 2 hours, θ = 40°C.

Substituting these values of t and θ above expression (*i*), we have

$$30 = \theta_F(1 - e^{-1/\tau}) \quad ...(ii)$$

$$\text{and } 40 = \theta_F(1 - e^{-2/\tau}) \quad ...(iii)$$

Solving above Eqs. (*ii*) and (*iii*), we have

$$e^{-1/\tau} = \frac{40}{30} - 1 = \frac{1}{3}$$

Substituting $e^{-1/\tau} = \frac{1}{3}$ in Eq. (*ii*), we have

$$\theta_F = \frac{30}{1 - \frac{1}{3}} = 45°\text{C}$$

If temperature rise after 1 hour is not to exceed 50°, final temperature rise, θ'_F will be given as

$$\theta'_F = \frac{50}{1 - e^{-1/\tau}} = \frac{50}{1 - \frac{1}{3}} = 75°\text{C} \quad \because t = 1 \text{ and } e^{-1/\tau} = \frac{1}{3}$$

Let the maximum overload that can be carried by a 25 kW motor be P kW.

Since losses vary as the square of load

$$\frac{\theta'_F}{\theta_F} = \frac{\text{Losses at P kW}}{\text{Original losses}}$$

$$\frac{75}{45} = \left(\frac{P}{25}\right)^2$$

$$\text{or } P = 25\sqrt{75/45} = 32.3 \text{ kW } \textbf{Ans.}$$

Example 1.56. A totally enclosed motor has a temperature rise of 20°C after half an hour and 35°C after 1 hour on full load. Determine the temperature rise after 2 hours on load. If a fan is fitted driving cool air over the external surface of the machine which reduces the final steady-temperature rise of the machine to 80 per cent of the original value, determine the temperature rise from cold after 1 hour on full load.

Solution : From Eq. (1.97)

$$\theta = \theta_F (1 - e^{-t/\tau}) \quad ...(i)$$

Now from given datas when $t = \frac{1}{2}$ hour, $\theta = 20°$ C and when $t = 1$ hour, $\theta = 35°$ C

Substituting the above values for t and θ in Eq. (*i*) we get

$$20 = \theta_F (1 - e^{-0.5/\tau}) \quad ...(ii)$$

$$35 = \theta_F (1 - e^{-1/\tau}) \quad ...(iii)$$

Dividing Eq. (*iii*) by Eq. (*ii*) we get

$$\frac{1 - e^{-1/\tau}}{1 - e^{-0.5/\tau}} = \frac{35}{20} = 1.75$$

$$\text{or } 1 + e^{-0.5/\tau} = 1.75$$

$$\text{or } \tau = \frac{-0.5}{\log_e 0.75} = 1.738 \text{ hours}$$

Substituting $e^{-0.5/\tau} = 1.75 - 1 = 0.75$ in Eq. (*ii*) we get

$$\theta_F = \frac{20}{1 - 0.75} = 80°\text{C}$$

Temperature rise after 2 hours,

$$\theta' = \theta_F (1 - e^{-t/\tau}) = 80(1 - e^{-2/1.738}) = 54.7° \text{ C } \textbf{Ans.}$$

When a fan is fitted driving cooling air over the external surface of the machine, coefficient of cooling is increased resulting in decrease in steady final temperature rise θ_F and heating time constant τ. Let the new coefficient of cooling, steady final temperature rise and heating time constant be α_1, θ_{F_1} and τ_1 respectively.

$$\theta_{F_1} = 0.8\theta_F$$

$$\text{Since } \theta_{F_1} = \frac{P}{S\alpha}$$

$$\text{So } \frac{\alpha_1}{\alpha} = \frac{\theta_F}{\theta_{F_1}}$$

$$\text{or } \alpha_1 = \frac{\theta_F}{\theta_{F_1}} \alpha = \frac{\alpha}{0.8} = 1.25\,\alpha$$

Heating time constant is given as

$$\tau = \frac{mC_p}{S\alpha}$$

$$\text{So } \frac{\tau_1}{\tau} = \frac{\alpha}{\alpha_1}$$

$$\text{or } \tau_1 = \frac{\alpha}{1.25\alpha} \tau = 0.8\,\tau = 0.8 \times 1.738 = 1.39 \text{ hours}$$

$$\theta_{F_1} = 0.8\,\theta_F = 0.8 \times 80 = 64° \text{ C}$$

Temperature rise from cold after 1 hour at full load

$$\theta_I = \theta_{F_1} (1 - e^{-t/\tau_1}) = 64(1 - e^{-1/1.39}) = 32.8° \text{ C } \textbf{Ans.}$$

Example 1.57. A certain motor has to perform the following duty cycle: 100 kW for 10 minutes, no load for 5 minutes, 50 kW for 8 minutes and no load for 4 minutes. The duty cycle is repeated indefinitely. Draw the curve for the load cycle. Assuming that the heating is proportional to the square of the load, determine the suitable size of a continuously rated motor.

Solution : The load cycle is illustrated in Fig. 1.119.

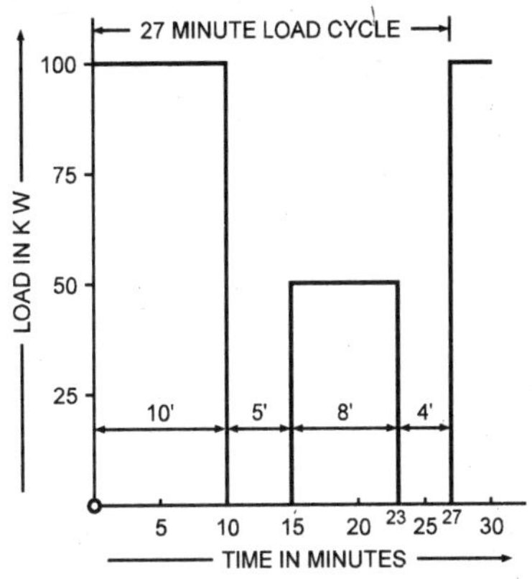

Fig. 1.119

Since heating is proportional to the square of the load,

$$\text{Rating of continuously rated motor} = \text{RMS kW}$$

$$= \sqrt{\frac{(100)^2 \times 10 + (50)^2 \times 8}{27}}$$

$$= 66.67 \text{ kW}$$

i.e. A motor of 70 kW will be adequate.

Example 1.58. An electric motor has load variation as given below: Torque 240 N-m for 20 minutes, 140 N-m for 10 minutes, 300 N-m for 10 minutes, 200 N-m for 20 minutes. If the speed of the motor is 720 rpm, find the power rating of motor.

[*A.M.I.E. Sec. B. Electric Drives and Their Control Winter 1993*]

Solution : Equivalent load torque, from Eq. (1.108),

$$T_{eq} = \sqrt{\frac{T_1^2 t_1 + T_2^2 t_2 + T_3^2 t_3 + T_n^2 t_n}{t_1 + t_2 + t_3 + t_n}}$$

$$= \sqrt{\frac{(240)^2 \times 20 + (140)^2 \times 10 + (300)^2 \times 10 + (200)^2 \times 20}{20 + 10 + 10 + 20}}$$

$$= 225.39 \text{ Nm}$$

Speed of motor N = 720 rpm

Power rating of motor,

$$P = T_{eq} \times \frac{2\pi N}{60} = 225.39 \times 2\pi \times \frac{720}{60}$$

$$= 16{,}994 \text{ Watts or 17 kW } \textbf{Ans.}$$

Example 1.59. Based on the rms torque, estimate the kW rating of a 750 rpm motor used for driving an equipment having the following load torque curve :

(*i*) For first 10 seconds, the torque is constant at 40 kg-m;

(*ii*) For next 30 seconds the torque varies linearly with time from 35 kg-m to 15 kg-m;

(*iii*) For the last 50 seconds, the torque is constant and equal to 10 kg-m.

[*A.M.I.E. Sec B. Electric Drives and Their Control Summer 2000*]

Solution : The load cycle is shown in Fig. 1.120.

$$(T_{eq})^2 \times \text{Time of cycle} = \int T^2\,dt \text{ over the cycle}$$

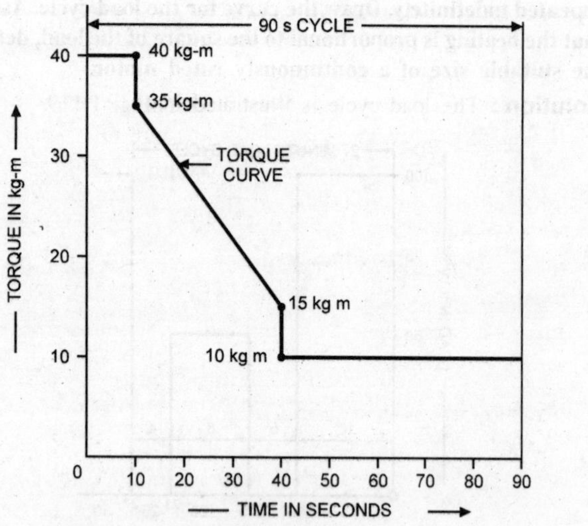

Fig. 1.120

$$= \int_0^{10} 40^2\, dt + \int_{10}^{40}\left[35 - \frac{(35-15)}{30}(t-10)\right]^2 dt + \int_{40}^{90} 10^2\, dt$$

$$= \int_0^{10} 1{,}600\, dt + \int_{10}^{40}\left[1{,}225 - \frac{140}{3}(t-10) + \frac{4}{9}(t-10)^2\right]dt + \int_{40}^{90} 100\, dt$$

$$= \left[1{,}600t\right]_0^{10} + \left[1{,}225t - \frac{140}{6}(t-10)^2 + \frac{4}{27}(t-10)^3\right]_{10}^{40} + \left[100t\right]_{40}^{90}$$

$$= 16{,}000 + 1{,}225 \times 30 - \frac{70}{3} \times 900 + \frac{4}{27} \times 27{,}000 + 5{,}000$$

$$= 16{,}000 + 36{,}750 - 21{,}000 + 4{,}000 + 5{,}000 = 40{,}750$$

or $T_{eq} = \sqrt{\dfrac{40{,}750}{90}} = 21.278$ kg-m

Power rating of motor,

$$P = T_{eq} \text{ in kg-m} \times 9.81 \times \frac{2\pi N}{60} = 21.278 \times 9.81 \times \frac{2\pi \times 750}{60}$$

$$= 16{,}400 \text{ W or } 16.4 \text{ kW } \textbf{Ans.}$$

Example 1.60. A motor driving a colliery winder has the following duty cycle :

(*i*) Acceleration period 0–15 s – Load rising uniformly from zero to 1,050 hp

(*ii*) Full-speed period 15–85 s – Load constant at 600 hp

(*iii*) Deceleration period 85–95 s – HP returned 200 to zero uniformly

(*iv*) Idle period 95–120 s – Motor stationary.

Estimate continuous rating of the motor for this load cycle.

[*U.P. Technical Univ. Electric Drives 2008-09*]

Solution : The load cycle is shown in Fig. 1.121.

Assuming losses proportional to $(hp)^2$, we have

$(\text{HP rating of motor})^2 \times$ cycle duration

$$= \int (\text{HP})^2\, dt \text{ over the complete cycle}$$

$$= \int_0^{15}\left(\frac{1{,}050}{15}t\right)^2 dt + \int_{15}^{85} 600^2\, dt + \int_{85}^{95}\left[\frac{200}{95-85}(t-85)\right]^2 dt$$

$$= \left[\frac{4{,}900t^3}{3}\right]_0^{15} + \left[360{,}000t\right]_{15}^{85} + \left[\frac{400(t-85)^3}{3}\right]_{85}^{95}$$

$$= 5{,}512{,}500 + 25{,}200{,}000 + 133{,}333 = 30{,}845{,}833$$

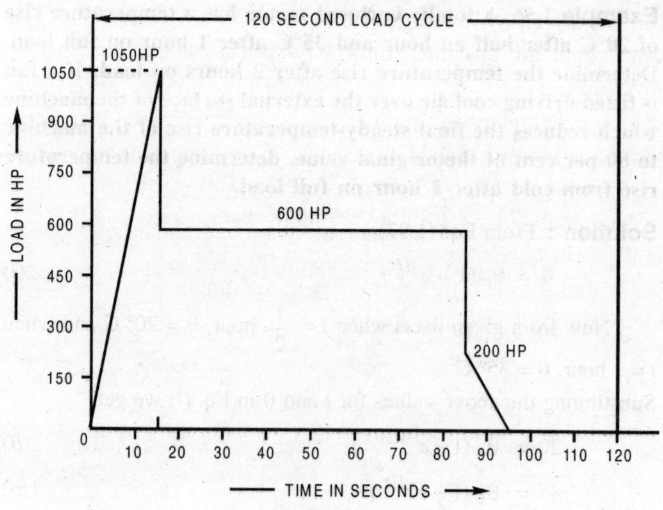

Fig. 1.121

or HP rating of motor $= \sqrt{\dfrac{30{,}845{,}833}{120}} = 507$ hp **Ans.**

Hence the rating of continuously rated motor on the basis of rms value should be 507 hp in order to supply the above load without exceeding the specified temperature.

Note : During regenerative braking, even though the power is returned to the line, the machine will be carrying current. So far as heating is concerned, it is immaterial whether machine is drawing current or returning it to the line.

1.34. METHODS OF VENTILATION AND COOLING OF MACHINES

It is necessary to provide suitable ventilation and cooling for the machines so that the temperature rise at any part of the machine does not exceed the permissible limit governed by the type of insulation employed in the construction. The cooling of electrical machine by means of an air stream is called *ventilation of the machine.*

The circulation of air in the machine can be arranged by the use of a built-in fan or fan with a separate drive. Either a centrifugal type fan or an axial type fan is employed for rotating electrical machines. A centrifugal fan forces air from the centre to flow outwards and operates the same way irrespective of the direction of rotation of the machine. An axial or propeller type fan moves the air in the opposite direction when the direction of rotation of the machine reverses.

According to IS : 4722-1968, the cooling system for rotating electrical machines are classified into three types depending up the origin of cooling.

1. *Natural Cooling.* The machine is cooled by air movements in the machine due to its rotation or due to temperature difference between inside parts and the outside temperature of the air. The machine thus is cooled without the use of a fan, by the movement of air and radiation. This method of cooling is employed only in small fractional kW machines with rating within several hundred watts, since the condition of cooling are comparatively favourable.

2. *Self Cooling.* The machine is cooled by air, blown by fan integrally built with the motor or mounted on the shaft.

3. ***Separate Cooling.*** The machine is cooled either by a fan not driven by its shaft, or it is cooled by a cooling medium other than air put into motion by means not belonging to machine.

According to the manner of cooling the ventilation systems are classified as

1. ***Open-Circuit Ventilation.*** The heat is given up directly to the cooling air flowing through the machine which is being continuously replaced. The open-circuit ventilation can be further divided into two types in accordance with how the air is brought into the machine.

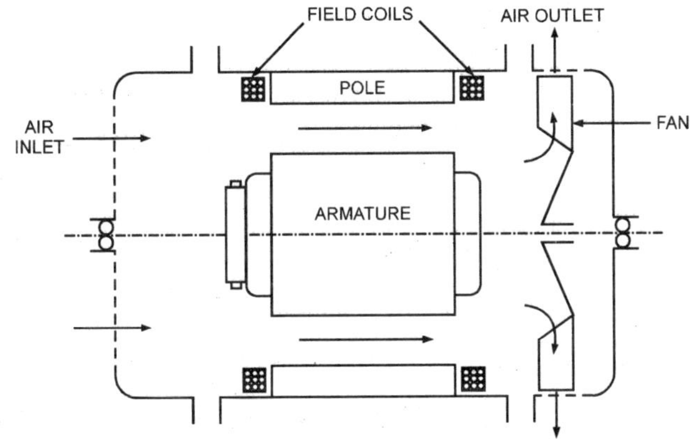

Fig. 1.122. ***Induced Ventilation in Electrical Machine***

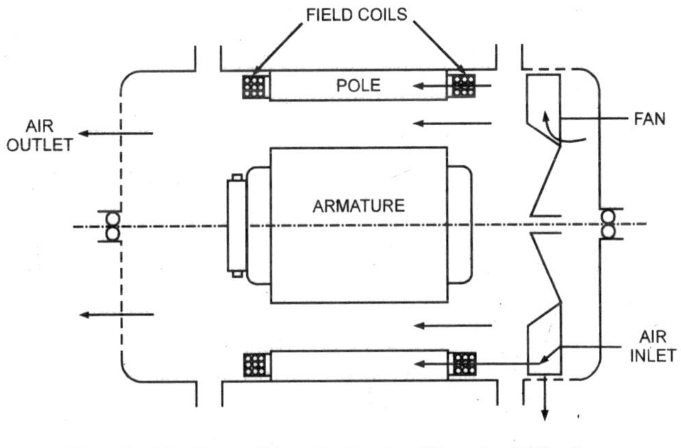

Fig. 1.123. ***Forced Ventilation in Electrical Machine***

(a) Induced Ventilation. Induced ventilation method used in an electrical machine is illustrated in Fig. 1.122. In this arrangement, a fan produces a reduced pressure of air inside the machine and air from outside is sucked into the machine owing to the atmospheric pressure. The air is circulated through the machine and then pushed out by the fan into the atmosphere. The fan is normally arranged inside the machine, as illustrated in the figure. The fan can be provided externally also.

(b) Forced Ventilation. In this method of ventilation, illustrated in Fig. 1.123, the fan sucks the air from outside and forces it into the machine and finally pushed out into the atmosphere. For forced ventilation, the pressure of air inside the machine is greater than the atmospheric pressure outside. The fan may be arranged internally, as illustrated in the figure or externally,

2. ***Closed Circuit Ventilation.*** In this system of cooling, heat is transferred to the cooling medium such as air or hydrogen through an intermediate cooling medium circulating in a closed circuit between the interior of the machine and the cooling device, *e.g.*, a water cooler can be used to cool the cooling medium such as air.

For increasing the surface of contact for cooler, stator and rotor cores of electrical machines are provided with ducts. These ducts may be radial or axial type depending on the direction of air flow through them.

3. Surface Ventilation. The heat is transferred from inside of the machine, by cooling medium, to the external surface of a totally enclosed machine. The external surface is being cooled by natural means or mainly by air blown by fan.

1.35. FRAME SIZE

A frame is the mechanical structure required to house a stator along with its bearings, end covers and terminal box. The standardization of frame size leads to economy. All modern motors of small size are built with standard frame sizes as specified in IEC 72 which lists a coherent range of main structural dimensions with centre heights between 56 and 1,000 mm.

A frame size is designated by a number which is its centre height (H) expressed in mm. Thus frame designated 132 has a centre height-(H) of 132 mm. The frame sizes with heights between 56 and 1,000 mm as per recommendations of IEC are 56, 63, 71, 80, 90, 100, 112, 132, 160, 180, 200, 225, 250, 280, 315, 355, 400, 450, 500, 630, 710, 800, 900, 1,000 mm.

The outputs from the standard frames are periodically assessed to take into account the latest technological advancements.

The frames may be diecast or fabricated. Machines of ratings up to 50 kW have their frames diecast in a strong silicon aluminium alloy and in some cases with the stator core cast in. The frames of larger machines are fabricated by welding steel plates. The advantage of fabrication is its adaptability to new designs and modifications.

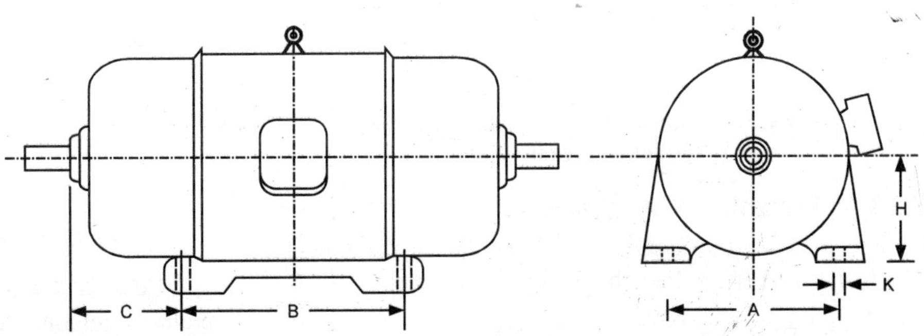

Fig. 1.124. ***Structural Dimensions of Standard Frames***

1.36. LOAD EQUALIZATION

In many industrial drives, such as in rolling mills, planning machines, electric hammers, presses, reciprocating pumps, the load fluctuates over a wide range in the space of a few seconds. It is desirable to smoothen out the fluctuations in load, otherwise during interval of peak load, it will draw a heavy current from the supply either producing large voltage drop in the distribution system or requiring cables and wires of heavy section. The process of smoothing out of fluctuations in the load is known as *load equalization*. In this process, energy is stored during the interval of light load and given out during the interval of peak load. Thus the power drawn from the supply mains remains almost constant.

The most common method of load equalization is by use of flywheel. During the light load period the flywheel accelerates and stores the excessive energy drawn from the supply and during peak load period the flywheel decelerates and supplies some of its stored energy to the load in addition to the energy supplied from the supply. Thus the load demand is reduced.

The motors used for such load should have drooping speed-torque characteristics, so that the speed may fall with the increase in load and enables the flywheel to give up its stored energy. For the load in which the motor have to run in the same direction and is not to be stopped and started frequently, flywheel may be mounted on the motor shaft. For a reversing drive, such as for colliery winder, the Ward Leonard control system is generally used for reversing and speed control, so flywheel can be mounted on the shaft of the motor-generator set. The load torque required and motor torque developed as well as speed variations with time are shown in Fig. 1.125.

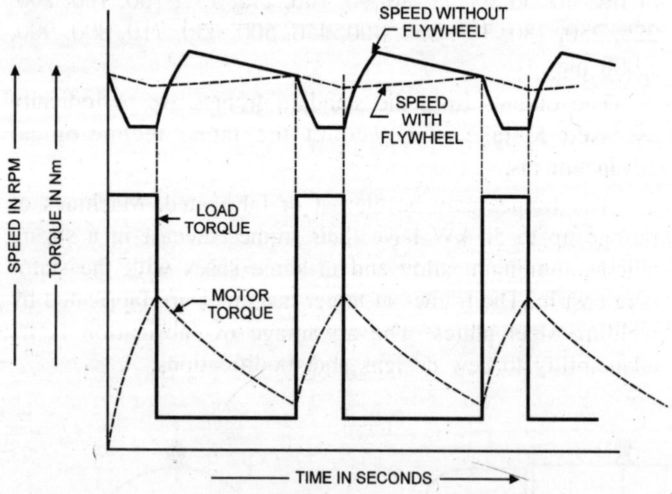

Fig. 1.125. ***Variations of Speed, Load Torque and Motor Torque Against Time***

1.36.1. Flywheel Calculations

Let the load torque (assumed constant during the time for which the load is applied) be T_L in newton-metres.

Flywheel torque $= T_F$ N-m

No-load torque $= T_0$ N-m (say, due to friction and windage etc.)

Motor torque at any instant $= T_m$ N-m

Motor speed on no load $= \omega_0$ radians/s

Motor speed at any instant $= \omega$ radians/s

Motor slip, $s = \omega_0 - \omega$

Moment of inertia of flywheel $=$ J kg-m^2

Time $= t$ seconds.

Case 1. ***Load Increasing.*** During this period the flywheel decelerates and gives up a part of stored energy in it. The torque required to be supplied by the motor, $T_m = T_L - T_F$.

Energy given out by flywheel when its speed is reduced from ω_0 to ω

$$= \frac{1}{2} J(\omega_0^2 - \omega^2) = J\frac{\omega_0 + \omega}{2}(\omega_0 - \omega)$$

but $\omega_0 - \omega = s$ and $\frac{\omega_0 + \omega}{2}$ is the mean speed and may be taken equal to ω, since drop of speed is assumed not to exceed 10 per cent.

$\therefore$ Energy given out $= J\omega s$

The power given out by flywheel

$$= \text{Rate of energy given up}$$

$$= \frac{d}{dt}(J\omega s) = J\omega\frac{ds}{dt}$$

Flywheel torque, $T_F = \frac{\text{Power}}{\omega} = J\frac{ds}{dt}$

Expression for the torque thus becomes

$$T_m = T_L - J\frac{ds}{dt} \quad \text{...(1.110)}$$

Since for value of slip up to 10 per cent, the slip is proportional to the torque,

$$\textit{i.e.,}\ s = KT_m$$

$$\text{So } T_m = T_L - JK\frac{dT_m}{dt}$$

$$\text{or } T_L - T_m = KJ\frac{dT_m}{dt}$$

$$\text{or } \frac{dT_m}{T_L - T_m} = \frac{1}{KJ}dt$$

$$\text{or } -\log_e(T_L - T_m) = \frac{1}{KJ}t + C_1 \quad \text{...(i)}$$

From initial conditions when $t = 0$, $T_m = T_0$

Therefore from Eq. (*i*) $C_1 = -\log_e(T_L - T_0)$

$$\text{and Eq. (i) becomes } \frac{-1}{KJ}t = \log_e\frac{T_L - T_m}{T_L - T_0}$$

$$\text{or } \frac{T_L - T_m}{T_L - T_0} = e^{-t/KJ}$$

$$\text{or Motor torque, } T_m = T_L - (T_L - T_0)e^{-t/KJ} \quad \text{...(1.111)}$$

Case 2. ***Load Decreasing.*** While load decreases, the motor accelerates the flywheel to its normal speed, slip is therefore decreased and $\frac{ds}{dt}$ is –ve.

Motor torque, $T_m = T_0 + T_F$

$$\text{or } T_m = T_0 - J\frac{ds}{dt} \text{ since } \frac{ds}{dt} \text{ is negative}$$

By solving the above expression, we get

$$T_m = T_0 + (T'_m - T_0)e^{-t/KJ} \quad ...(1.112)$$

where T'_m is the motor torque at the instant when the load is removed.

Example 1.61. A 4-pole. 50 Hz induction motor has a flywheel on its shaft. Total inertia at the motor shaft is 1,000 kg-m². Load torque is 100 kg-m for 10 seconds followed by a no-load period long enough for the flywheel to regain its full speed. Motor has a slip of 6% at a torque of 50 kg-m. Calculate the speed at the end of deceleration period. Assume motor speed-torque characteristic to be a straight line in the region of interest and neglect friction and windage.

Solution : Load torque, T_L = 100 kg-m = 100 × 9.81 = 981 N-m

Load period, t = 10 seconds

Moment of inertia, J = 1,000 kg-m²

No-load torque, $T_0 = 0$

$$\text{Synchronous speed, } N_s = \frac{120f}{P} = \frac{120 \times 50}{4} = 1{,}500 \text{ rpm}$$

$$\text{Slip speed} = 0.06 \times 1{,}500 = 90 \text{ rpm or } 2\pi \times \frac{90}{60} = 9.42 \text{ rad/s}$$

$$K = \frac{\text{Slip speed}}{\text{Torque}} = \frac{9.42}{50 \times 9.81} = 0.0192$$

$$\therefore T_m = T_L - (T_L - T_0)\, e^{-t/KJ}$$

$$= 981 - (981 - 0)\, e^{\frac{-10}{0.0192 \times 1{,}000}} \quad \because T_0 = 0$$

$$\text{or } T_m = 981(1 - e^{-0.52}) = 981\,(1 - 0.594) = 398 \text{ N-m}$$

$$\text{Slip speed} = K \times T_m = 0.0192 \times 398 \text{ rad/s}$$

$$= \frac{0.0192 \times 398}{2\pi} \times 60 = 73 \text{ rpm}$$

∴ Speed at the end of deceleration period

= 1,500 – 73 = 1,427 rpm **Ans.**

Example 1.62. The following data refers to a 500 hp rolling mill induction motor equipped with a flywheel.

No-load speed – 40 rpm; Slip at full load – 12%; load torque during actual rolling – 41,500 kg-m. Duration of each rolling period – 10 seconds.

Determine the inertia of the flywheel required in the above case to limit the motor torque to twice its full-load value. Neglect the no-load losses and assume that the rolling mill torque falls to zero between each rolling period. Assume that the motor slip is directly proportional to the torque.

[*Pb. Univ. Elec. Drives and Utilization of Elec. Energy Nov. 1980*]

Solution : Output of motor = 500 hp = 500 × 735.5 W

No-load speed, N_0 = 40 rpm

Full-load slip, s_f = 12% or 0.12

Full-load speed, N_f = 40(1 – 0.12) = 35.2 rpm

$$\text{Full-load torque, } T_f = \frac{500 \times 735.5}{2\pi \times \frac{35.2}{60}} = 99{,}766 \text{ N-m}$$

Motor torque, $T_m = 2 \times T_f$

= 2 × 99,766 = 199,532 N-m

Load torque, T_L = 41,500 × 9.81 = 407,115 N-m

No-load torque, T_0 = 0

Load period, t = 10 seconds

Slip speed = Full-load slip × no-load speed

= 0.12 × 40 = 4.8 rpm

$$\text{or } \frac{2\pi \times 4.8}{60} = 0.5026 \text{ rad/s}$$

$$K = \frac{\text{Slip speed}}{\text{Torque}} = \frac{0.5026}{99{,}776} = 5.0383 \times 10^{-6}$$

$\because s \propto$ torque

$$\because T_m = T_L - (T_L - T_0)\, e^{-t/KJ} = T_L(1 - e^{-t/KJ})$$

$\because T_0 = 0$

$$\therefore J = \frac{-t}{K \log_e\left(1 - \frac{T_m}{T_L}\right)}$$

$$= \frac{-10}{5.0383 \times 10^{-6} \log_e\left(1 - \frac{199{,}532}{407{,}115}\right)}$$

$$= 2.947 \times 10^6 \text{ kg-m}^2 \textbf{ Ans.}$$

Example 1.63. A 50 kW, 3-phase, 6-pole, 50 Hz induction motor fitted with a flywheel supplies a constant load torque of 300 N-m and at wide intervals an additional load torque of 1,500 N-m for 10 seconds. The full-load slip of the motor is 4%. Assuming that the stator losses are negligible and the motor operates on the linear portion of the torque-slip curve, determine :

(*i*) Weight of the flywheel if the radius of gyration is 0.9 m and the motor torque is not to exceed twice rated torque.

(*ii*) Time taken after removal of extra load before motor torque is 700 N-m.

Solution : Output of motor = 50 kW

$$\text{Synchronous speed, } N_s = \frac{120f}{P} = \frac{120 \times 50}{6} = 1{,}000 \text{ rpm}$$

Full-load speed of motor,

$$N_f = N_s(1 - s_f) = 1{,}000(1 - 0.04) = 960 \text{ rpm}$$

$$\text{Full-load torque, } T_f = \frac{50 \times 1{,}000}{2\pi \times \frac{960}{60}} = 497.36 \text{ N-m}$$

Motor torque, $T_m = 2 \times T_f$ = 2 × 497.36 = 994.72 N-m

Load torque, T_L = 300 + 1,500 = 1,800 N-m

No-load torque T_0 = 300 N-m

Load period, t = 10 seconds

$$\text{Slip speed} = \frac{0.04 \times 1{,}000 \times 2\pi}{60} = 4.1888 \text{ rad/s}$$

$$K = \frac{\text{Slip speed}}{T_f} = \frac{4.1888}{497.36} = 8.422 \times 10^{-3}$$

$$\because T_m = T_L - (T_L - T_0)\, e^{-t/KJ}$$

$$\therefore e^{-t/KJ} = \frac{T_L - T_m}{T_L - T_0} = \frac{1{,}800 - 994.72}{1{,}800 - 300} = 0.5368$$

$$\text{or } J = \frac{-t}{K \log_e 0.5368}$$

$$= \frac{-10}{8.422 \times 10^{-3} \log_e 0.5368} = 1{,}909 \text{ kg-m}^2$$

$$\text{Weight of flywheel} = \frac{J}{r^2} = \frac{1{,}909}{(0.9)^2} = 2{,}357 \text{ kg } \textbf{Ans.}$$

(*ii*) Let the time taken be t_1 seconds.

The expression for the motor torque is given as

$$T_m = T_0 + (T'_m - T_0)e^{-t_1/KJ}$$

$$\text{or } t_1 = KJ \log_e \frac{T'_m - T_0}{T_m - T_0}$$

$$= 8.422 \times 10^{-3} \times 1{,}909 \log_e \left(\frac{994.72 - 300}{700 - 300}\right)$$

$$= 8.876 \text{ seconds. } \textbf{Ans.}$$

∵ Motor torque at peak speed,

$T'_m = 994.72$ Nm and $T_m = 700$ N-m given

Example 1.64. A motor equipped with a flywheel has to supply a load torque of 600 N-m for 10 seconds followed by a no-load period long enough for the flywheel to regain its full speed. It is desired to limit the motor torque to 450 N-m. What should be the moment of inertia of the flywheel ? The no-load speed of the motor is 600 rpm and it has a slip of 8% at torque 400 N-m. Assume the motor speed-torque characteristic to be a straight line in the range of operation. Motor has an inertia of 10 kg-m².

[*U.P. Technical Univ. Electric Drives 2009-10*]

Solution : Full-load torque, $T_L = 600$ N-m

No-load torque, $T_0 = 0$

Motor torque, $T_m = 450$ N-m

$$\text{No-load speed, } \omega_0 = \frac{2\pi N}{60} = \frac{2\pi \times 600}{60} = 20\pi$$

$$\text{Slip speed} = 0.08 \times 20\pi = 1.6\pi$$

$$K = \frac{\text{Slip speed}}{\text{Torque}} = \frac{1.6\pi}{450} = 0.01117$$

From Eq. (1.111)

$$\text{Motor torque, } T_m = T_L - (T_L - T_0)\, e^{-t/KJ}$$

Substituting the values, we have

$$450 = 600 - (600 - 0)\, e^{\frac{-10}{0.01117 J}}$$

$$\text{or } e^{\frac{-895}{J}} = 0.25$$

$$J = \frac{-895}{\log_e 0.25} = \frac{-895}{-1.3863} = 645.6 \text{ kg-m}^2 \textbf{ Ans.}$$

Moment of inertia of flywheel,

$$J_F = J - J_M = 645.6 - 10 = 635.6 \text{ kg-m}^2 \textbf{ Ans.}$$

Example 1.65. A 3-phase 100 kW, 6-pole, 960 rpm induction motor has a constant load of 600 N-m at wide intervals and is subjected to additional torque of 2,400 N-m for 10 seconds. Calculate the moment of inertia of the flywheel used for load equalization, if motor torque is not to exceed twice the rated torque. The moment of inertia of the motor is 200 kg-m².

[*U.P. Technical Univ. Electric Drives 2005-06*]

Solution : Output of motor, P = 100 kW

$$\text{Synchronous speed, } N_s = \frac{120 f}{P} = \frac{120 \times 50}{6} = 1{,}000 \text{ rpm}$$

Full-load speed, N = 960 rpm

$$\text{Slip speed} = N_s - N = 1{,}000 - 960$$

$$= 40 \text{ rpm} = 40 \times \frac{2\pi}{60} = 4.1888 \text{ rad/s}$$

$$\text{Full-load torque, } T_F = \frac{100 \times 1{,}000}{2\pi \times \frac{960}{60}} = 994.72 \text{ N-m}$$

$$\text{Motor torque, } T_m = 2 \times 994.72 = 1{,}989.44 \text{ N-m}$$

$$\text{Load torque, } T_L = 600 + 2{,}400 = 3{,}000 \text{ N-m}$$

$$\text{No-load torque, } T_0 = 600 \text{ N-m}$$

$$K = \frac{\text{Slip speed}}{T_f} = \frac{4.1888}{994.72} = 0.004211$$

$$\text{From Eq. (1.111), } T_m = T_L - (T_L - T_0)e^{-t/KJ}$$

$$\text{or } J = \frac{-t}{K \log_e \frac{T_L - T_M}{T_L - T_0}}$$

$$= \frac{-10}{0.04211 \log_e \frac{3{,}000 - 1{,}989.44}{3{,}000 - 600}} = 2{,}745.5 \text{ kg-m}^2$$

Moment of inertia of flywheel,

$$J_F = J - J_M = 2{,}745.5 - 200 = 2{,}545.5 \text{ kg-m}^2 \textbf{ Ans.}$$

1.37. APPEARANCE

Pleasing appearance is a good aspect of design, which can be obtained by harmonious blending of the motor finish, mounting and dimensions in line with those of the driven machine.

1.38. COST

Although the cost has been placed as last factor in the above list, but it is the major factor for the choice of motor. Sometimes we have to choose a motor which may not be an ideal choice from engineering consideration but cheap in cost—because special features are not worth the additional investment. While considering the cost, the capital as well as running cost (cost of losses and cost due to maintenance, interest and depreciation) of a motor should be considered. In most of the cases, we find that equipment very cheap in first cost is costlier in running cost due to poor efficiency, low power factor and short life. In a nutshell the motor choosen should give the desired service at minimum overall cost.

As compared to plain squirrel cage 3-phase induction motors, which are the cheapest, the cost of dc motors range from 1.5 to 3 times its cost. The cost of synchronous motors, 3-phase slip-ring induction motors and commutator motors range from 1.5 to 4 times the cost of a plain squirrel cage induction motor of the same output. For single phase motors it is 1.5 to 2.5 times the cost of plain squirrel cage induction motor of the same size.

Exercises

1. What is an electric drive ? What are its advantages ?
[G.B. Technical Univ. Electric Drives 2010-11]
2. Discuss criteria for selection of electric motor for a particular duty. **[U.P. Technical Univ. Electric Drives 2005-06]**
3. Discuss various factors which are required to be considered for selecting a motor.
[Pb. Technical Univ. Electric Drives & Utilization Dec. 2006]
4. What are the disadvantages of dc drives due to which the three-phase induction motor drive is fastly replacing it ?
[G.B. Technical Univ. Electric Drives 2010-11]
5. Enlist the classification of electric drives and explain.
[G.B. Technical Univ. Electric Drives 2010-11]

6. Differentiate bewteen group drive and individual drive and give their merits.
[U.P. Technical Univ. Electric Drives 2005-06]

7. What is meant by "group drive" and "individual drive" ? Discuss their merits and demerits.
[U.P. Technical Univ. Electric Drives 2008-09]

8. Write advantages and disadvantages of group drives.
[Pb. Technical Univ. Electric Drives & Utilization Dec. 2006]

9. Explain the various types of loads. What are the components of the load torque ?
[G.B. Technical Univ. Electric Drives 2010-11]

10. What is meant by nature of load torque ? Explain difference between "passive load" and "active load" and give two examples of each category.
[U.P. Technical Univ. Electric Drives 2008-09]

11. Explain active load and passive load giving two examples for each. **[U.P. Technical Univ. Electric Drives 2005-06]**

12. How do you define passive load torques and active load torques ? What are the differences between the two ?
[U.P. Technical Univ. Electric Drives 2009-10]

13. What are the components of load torque in a drive ? How each component is dependent on speed ?
[U.P. Technical Univ. Electric Drives 2007-08]

14. Discuss the motoring and braking operations of an electric drive in both directions, forward and reverse taking the hoist as an example. **[A.M.I.E. Sec. B. Electric Drives & Their Control Summer 1999]**

15. Explain four quadrant operation of an electric drive.
[G.B. Technical Univ. Electric Drives 2009-10, 2010-11]

16. Describe four quadrant operation of a motor with illustrative figures for each. Show clearly the direction of motor torque, load torque and speed for each quadrant.
[U.P. Technical Univ. Electric Drives 2007-08]

17. A motor is driving a hoist load. Discuss four quadrant operation of the motor and show that in the speed-torque plane. Indicate the polarity of torque, speed and power in each quadrant. **[U.P. Technical Univ. Electric Drives 2004-05]**

18. Discuss factors which decide selection of electric drive for a particular industrial application.
[U.P. Technical Univ. Electric Drives 2008-09]

19. Classify various mechanical loads on the basis of their offered torque-speed characteristics.
[U.P. Technical Univ. Electric Drives 2005-06]

20. Give typical example for following types of loads encountered in a drive.
    (*i*) Continuous constant load.
    (*ii*) Continuous variable load.
    (*iii*) Pulsating load.
    (*iv*) Impact load.
    (*v*) Short time intermittent loads.
[U.P. Technical Univ. Electric Drives 2007-08]

21. What are constant torque and constant power operations of an electric drive ?
[G.B. Technical Univ. Electric Drives 2010-11]

22. What is meant by "constant torque drive" and "constant power drive" ? Explain them and mention one application of each such drive. **[U.P. Technical Univ. Electric Drives 2005-06]**

23. Derive equation of motion and explain under what conditions motor will be accelerating, decelerating or at steady state. **[U.P. Technical Univ. Electric Drives 2008-09]**

24. Draw the torque-speed characteristics, giving examples with justification, of the following types of loads:
(*i*) Constant torque (*ii*) Constant power (*iii*) Torque directly proportional to speed.
[A.M.I.E. Sec. B. Electric Drives & Their Control Winter 1996]

25. A motor running at a speed of N rpm is driving a rotational load L_1 directly coupled to its shaft and another rotational load L_2 through a gear to reduce the speed by a factor K the inertia of motor, L_1 and L_2 are respectively J_m, J_1 and J_2. The load torque of L_1 and L_2 are τ_1 and τ_2 respectively. Derive an expression for effective inertia and load torque on the same motor in terms of the above parameters.
[U.P. Technical Univ. Electric Drives 2007-08]

26. Derive the necessary relationship between the variation of electromagnetic torque and load torque with respect to speed for a stable operation of an electric drive. Comment on the stability of the equilibrium points A, B, C and D in Fig. 1.
[G.B. Technical Univ. Electric Drives 2010-11]

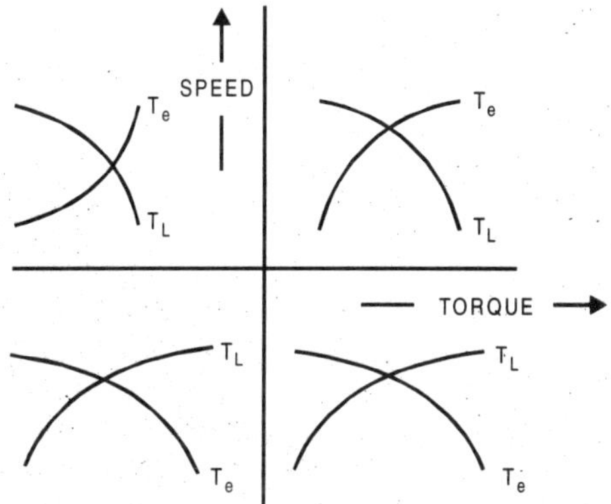

Fig. 1

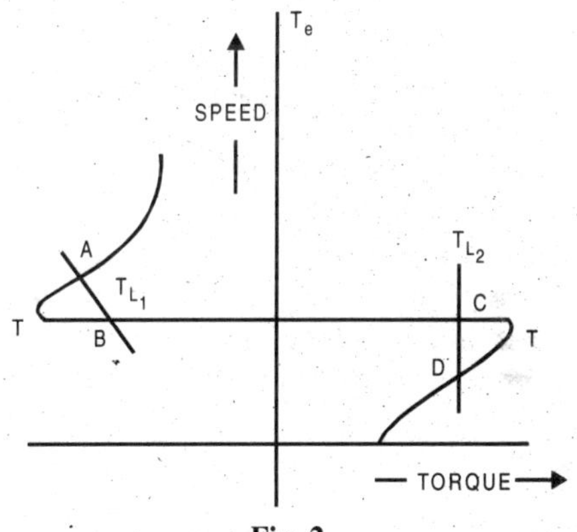

Fig. 2

27. Figure 2 shows plots of speed vs motor and load torques. Comment on the stability of operating points A, B, C and D (Fig. 2). **[U.P. Technical Univ. Electric Drives 2009-10]**

28. Give the characteristics of dc shunt motors. Why are such motors not suitable for traction purposes ?

29. Draw speed-torque characteristics of a dc series motor. What modifications are expected in above characteristics when armature reaction is considered ?
[A.M.I.E. Sec. B. Electric Drives & Their Control Summer 1998]

30. Explain that the characteristic of a dc series motor is suitable for traction application.
[A.M.I.E. Sec. B. Electric Drives & Their Control Winter 1993]

31. Define steady state stability of electric drive. Derive a condition for steady state stability. Using this condition

determine steady state operation point on torque-slip characteristics of three phase induction motor supplying constant load torque. **[U.P. Technical Electric Drives 2005-06]**

32. Discuss, with the help of curves, the characteristics and applications of the following electric motors:

(*a*) 3-phase, cage type induction motor (*b*) 3-phase wound rotor induction motor (*c*) single phase induction motor (*d*) synchronous motor (*e*) dc shunt motor.

[A.M.I.E. Sec. B. Electric Drives & Their Control Summer 1996]

33. What do you understand by "series" and "shunt" speed-torque characteristics of electric motors used in industrial drives ? Name (at least one of each type) ac and dc motor with such characteristics.

Name the characteristics and applications of the following type of motors :

(*a*) DC compound wound motor.

(*b*) Synchronous-induction motor.

(*c*) Double-squirrel cage induction motor

(*d*) Schrage motor.

[Gorakhpur Univ. Utilization of Electrical Power & Traction Sept. 1983]

34. Derive an expression for accelerating time of three-phase induction motor having pure inertia load, started from a constant voltage supply. Mention the factors which affect the accelerating time.

[U.P. Technical Univ. Electric Drives 2005-06]

35. Explain how the accelerating time and energy losses are calculated during a starting of a three-phase induction motor. Also enlist the methods of reducing the energy loss during starting. **[G.B. Technical Univ. Electric Drives 2010-11]**

36. Explain the dynamics during starting of a three-phase induction motor. What are the methods of reducing energy loss during starting ? Explain.

[U.P. Technical Univ. Electric Drives 2009-10]

37. Draw and explain the starting characteristics of various types of dc motors.

[Pb. Technical Univ. Electric Drives & Utilization Dec. 2009]

38. What is the essential difference in the principle of operation of a starter for slip-ring and squirrel cage motor ?

[Pb. Technical Univ. Electric Drives & Utilization May 2007]

39. Describe with suitable diagrams the various methods of starting 3-phase induction motors. Give relative merits and demerits of each method.

[Pb. Univ. Electric Drives & Utilization of Electrical Energy August, 1984]

40. Discuss, with circuit diagram, the star-delta starter for a squirrel cage induction motor. Explain factors which decide the starter-switch change-over from star-to-delta position.

[A.M.I.E. Sec. B. Electric Drives & Their Control Winter 1998]

41. What is the drawback of a star/delta starter for an induction motor ? How is it overcome ?

For a three-phase induction motor with negligible stator resistance and no-load current, show that the ratio of starting current to its stator current at any slip can be expressed as

$$\frac{I_{st}}{I} = \sqrt{\frac{s^2 + s_{max}^2}{s^2 (1 + s_{max}^2)}}$$

where I_{st} = starting current, I = stator input current at any slip s and s_{max} = slip at which maximum torque occurs.

[A.M.I.E. Sec. B. Electric Drives & Their Control Summer 1999]

42. How do you start a synchronous motor ?

[A.M.I.E. Sec. B. Electric Drives & Their Control Winter 1999]

43. Explain why field (or flux) control is employed for obtaining speeds higher than rated value and armature voltage control for speeds lower than rated in the case of dc separately excited motor.

[A.M.I.E. Sec. B. Electric Drives & Their Control Summer 1999]

44. Explain which properties make series motors more suitable for heavy torque applications. Also discuss various methods for controlling speed of dc series motors.

[Pb. Technical Univ. Electric Drives & Utilization May 2007]

45. Draw the speed-torque characteristics for dc series motor. Show the effect of increase in series resistance on above characteristics. What is base speed in dc motors? How these motors can be run above the base speed ?

[A.M.I.E. Sec. B. Electric Drives & Their Control Summer 2000]

46. Describe with the help of connection diagram the Ward-Leonard method of speed control and explain its working. What are the merits and demerits of this method ? State its field of application.

47. Name the different methods of speed control of a 3-phase induction motor. Describe in brief speed control of induction motor by pole changing. **[Rajasthan Univ. 2008]**

48. Describe the variable voltage, variable-frequency method of speed control of induction motors. Explain, with reasons, the voltage-frequency relationship for full-range speed control.

[A.M.I.E. Sec. B. Electric Drives & Their Control Winter 1998]

49. Derive the induction motor torque-speed characteristics under V/*f* control. Explain why E/*f* control is superior to V/*f* control. **[A.M.I.E. Sec. B. Electric Drives and Their Control Winter 1993]**

50. What is electric braking ? What are the advantages and disadvantages of electric braking.

[Rajasthan Univ. Utilization of Electric Power 2003, 2004, 2005, 2007, 2008]

51. What do you understand by electric braking? Why is it necessary to have mechanical brakes in electric drives? What are the advantages of electric brakes over mechanical brakes? Describe in brief various methods of electric braking and mention their relative merits and demerits.

52. Explain various methods of electric braking. State the conditions to be fulfilled for each method of braking.

[M.D. Univ. Utilization of Elec. Power & Traction Dec. 2008]

53. What are the advantages and disadvantages of electric braking ?

[Pb. Technical Univ. Electric Drives & Utilization May 2007]

54. Explain the types of braking of dc motors. Which one is usually employed and why ?

[G.B. Technical Univ. Electric Drives 2010-11]

55. Explain the various types of braking as used in a dc motor. What necessary modifications are required while braking a dc series motor ?

[U.P. Technical Univ. Electric Drives 2009-10]

56. How can plugging be applied to a separately excited dc motor ? **[U.P. Technical Univ. Electric Drives 2008-09]**

57. Describe different methods of braking used for dc shunt and 3-phase squirrel cage induction motor.

[U.P. Technical Univ. Electric Drives 2007-08]

58. Explain dc dynamic braking of 3-phase slip-ring induction motor and derive an expression for braking torque. How can the magnitude and the point of occurrence of the maximum braking torque be varied ?

[U.P. Technical Univ. Electric Drives, 2008-09]

59. Explain dc dynamic braking of 3-phase induction motor. Discuss effects of excitation and rotor resistance on the torque-slip characteristics of the motor.

[U.P. Technical Univ. Electric Drives 2005-06]

60. Explain in 4-quadrant torque-speed plane the reversal of direction of a loaded dc shunt motor by plugging using a single resistor in the motor armature circuit.

[A.M.I.E. Sec. B. Electric Drives & Their Control Summer 1993]

61. Explain, with diagram, the Ward-Leonard dc drive using rotating machines. How is the regenerative braking achieved in this drive ? **[A.M.I.E. Sec. B. Electric Drives and Their Control Summer 1993]**

62. What is regenerative braking? Why it cannot be accomplished with a series wound dc motor without modifications in the circuit diagram. Give its merits and demerits.

[Gorakhpur Univ. Utilization of Electric Power and Traction Sept. 1984]

63. Specify the different types of enclosures of motors, highlighting the special feature of each type. Mention the area of application of each type.

[A.M.I.E. Sec. B. Utilization of Elec. Power Summer 1985]

64. Discuss the different types of drives required for transmission of power from the driving machines to the loads.

[A.M.I.E. Sec. B. Utilization of Elec. Power Winter 1983]

65. Discuss the following features pertaining to the selection of a motor:

Types of enclosures, bearings, transmission of drive and noise.

[A.M.I.E. Sec. B. Utilization of Elec. Power Summer 1982]

66. What are the mechanical features to be considered in the selection of a motor for a particular service ?

What do you understand by the type of enclosure indicated as IP 44 ?

List the advantages of using roller or ball bearings for an induction motor.

State the factors that contribute to mechanical vibrations and noise.

Describe (*i*) flat belt drive (*ii*) V-belt drive (*iii*) rope drive (*iv*) chain drive (*v*) gear drive and (*vi*) direct drive.

[A.M.I.E. Sec B. Utilization of Elec. Power Summer 1980]

67. What are the limiting temperatures of motors when the class of insulation used is :

(*i*) Y, (*ii*) A, (*iii*) E, (*iv*) B, (*v*) F, (*vi*) H and (*vii*) C ?

Give typical examples of loads demanding continuous duty, intermittent duty and short-time duty.

68. From basic principles of heat transfer, develop an expression for temperature rise of an electric motor delivering fixed rated power at a fixed (rated) speed.

[U.P. Technical Univ. Electric Drives 2007-08]

69. What is heating time constant ? Explain how the rating of a motor is affected by the temperature rise.

[Pb. Technical Univ. Electric Drives & Utilization Dec. 2008]

70. Explain what do you understand by 'continuous', 'short time', intermittent rating of an electric machine.

[Pb. Technical Univ. Electric Drives & Utilization May 2009]

71. Define continuous duty, short time duty and periodic intermittent duty. Draw typical variations of output power losses, temperature rise with time in each case.

[U.P. Technical Univ. Electric Drives 2007-08]

72. Derive an expression for the temperature rise of an electric motor. State the assumptions made.

[A.M.I.E. Sec. B. Electric Drives & Their Control Winter 1994]

73. A motor of smaller rating can be selected for a short time duty. Why ? **[U.P. Technical Univ. Electric Drives 2009-10]**

74. Discuss different classes of motor duty cycles. Give industrial application for each class of duty cycle.

Discuss the equivalent power method for determining the rating of an electric motor which is subjected to intermittent periodic duty.

[A.M.I.E. Sec. B. Electric Drives & Their Control Summer 1993]

75. Draw the pattern of temperature rise characteristic under steady-state for (*i*) short-time duty (*ii*) intermittent duty.

What is the equivalent current method of estimating motor rating ? Explain. Can this method be used if motor speed varies during a load cycle ?

[A.M.I.E. Sec. B. Electric Drives & Their Control Winter 1996]

76. Explain the load diagram. Why is it important to know for selection of drive rating ?

[A.M.I.E. Sec. B. Electric Drives & Their Control Summer 1998]

77. Discuss 'rms current', 'rms torque' and 'rms power' methods for evaluation and rating of motor for continuous duty, variable load conditions. Give their merits and limitations.

[U.P. Technical Univ. Electric Drives 2005-06]

78. What are the problems faced in case a motor of wrong rating is chosen ?

[A.M.I.E. Sec. B. Electric Drives & Their Control Winter 1999]

79. What is load equalization ? How it is achieved ?

[A.M.I.E. Sec. B. Electric Drives & Their Control Summer 2000]

80. What are the reasons for load equalization in electric drive ?

[U.P. Technical Univ. Electric Drives 2007-08]

81. Derive expression of torque equation in case of load increasing or flywheel releasing energy.

[U.P. Technical Univ. Electric Drives 2009-10]

82. Derive an expression for the motor torque driving a rolling mill when equipped with flywheel.

[Pb. Technical Univ. Electric Drives & Utilization May 2009]

83. What do you understand by load equalization ? Where do you find its applications ? Derive the expression for instantaneous motor torque, MOI of the flywheel, and the motor slip. State any assumptions made.

[Pb. Univ. Elec. Drives and Utilization of Electric Energy, August 1984]

84. Explain the following statements.

(*a*) Pole changing method of speed control is used only in squirrel cage induction motors.

(*b*) Ward-Leonard method of speed control facilitates speed variation below the base speed, and is suited for constant torque drives.

(*c*) Motors used in domestic fans should not have low rotor resistance.

(*d*) By feeding energy to the rotor of an induction motor the speed variation can be achieved above the synchronous speed.

[A.M.I.E. Sec. B. Utilization of Elec. Power Winter 1984]

85. Justify the following statements.

(*i*) DC series motors are best suited where high starting torque is required.

(*ii*) Starting torque of a dc series motor is proportional to the square of its armature current.

(*iii*) DC series motors should never be started on no load.

(*iv*) DC series motor is able to produce a far greater torque compared to a dc shunt motor.
[Gorakhpur Univ. Sept. 1984]

(*v*) A differential compound motor tends to run in reverse direction when started and is found to have a tendency to race on overload.

(*vi*) 3-phase induction motors are most commonly used.

(*vii*) The speed of the rotor cannot be equal to that of the revolving field.

(*viii*) Schrage motor can run above and below synchronous speed.

(*ix*) Synchronous motors are not self-starting.

(*x*) Synchronous motors will operate at constant speed irrespective of load.

(*xi*) In case of fans and centrifugal pumps where the load torque decreases with the fall in speed armature control is quite convenient and economical.
[Gorakhpur Univ. Sept. 1984]

(*xii*) Ward-Leonard system of speed control is used where fine speed adjustment and smooth acceleration is required. **[Gorakhpur Univ. 1982]**

(*xiii*) Quick stopping of motor is more essential than quick starting. **[Gorakhpur Univ. 1982]**

(*xiv*) Friction brake is required in addition to electric braking system. **[Gorakhpur Univ. Sept. 1984]**

86. Write short notes on the following :

(*i*) Group and individual drives. **[Pb. Univ. April 1979]**

(*ii*) Ward-Leonard method for speed control.
[Rajasthan Univ. 2005]

(*iii*) Cascade connections of induction motors.
[Pb. Univ. April 1974]

(*iv*) Different methods of electric braking.
[M.D. Univ. May 2009]

(*v*) Regenerative braking as applied to traction system.

(*vi*) Types of enclosures for electrical machines.
[Pb. Univ. April 1981]

(*vii*) Rating of electric motors for different use with examples.

(*viii*) Load equalization.
[Pb. Univ. 1983; Pb. Univ. Nov. 1983; April 1981]

Short Answer Type Questions With Answers

Q. 1. What is an electric drive ?

Ans. An electric drive is defined as a form of machine equipment designed to convert electrical energy into mechanical energy and provide electrical control of this process.

The basic elements of an electric drive are the electric motor, the transmission, and the electrical control system. Electric drives often include conversion equipment (frequency changers, rectifier units, motor-generator sets etc.) to attain smoother and more versatile control and make the motor operate on specific speed-torque characteristic.

Q. 2. Why electric drive is preferred over mechanical drive ?
[Gorakhpur Univ. 1982]

Ans. Electric drive is preferred over mechanical drive because of its simplicity, cleanliness, easy and smooth control, flexibility in layout, compactness, reliability, long life, easy starting, low cost (initial as well as maintenance cost), facility for remote control etc. Modern electric drive with a system of automatic control fulfills several functions other than imparting motion to the working machine so as to carry out the production process involved in the most efficient manner. Using automatically controlled electric drives, machinery and production units may be operated with high productivity, ensuring at the same time high quality.

Q. 3. Enumerate the factors governing selection of electric motors.

Ans. The various factors that govern the selection of electric motor for a particular service are :

1. Nature of electric supply available—dc or ac.
2. Types of drives—individual, group or multimotor.
3. Nature of load—active or passive.
4. Electrical characteristics (operating, starting, speed control and braking).
5. Mechanical considerations (Types of enclosures, types of bearings, types of transmission for drive, noise level etc.)
6. Service capacity rating
7. Appearance and
8. Cost

Q. 4. Write classification of electric drives. **[P.T.U. December 2006]**

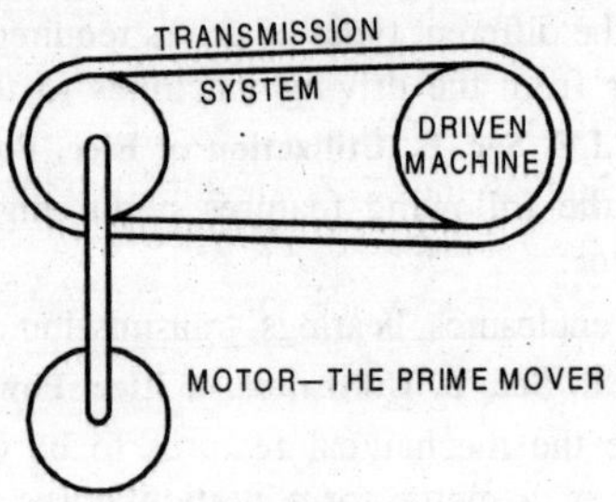

Fig. 1

Ans. The various types of electric drives used in industry may be divided into three types : The group drive, the individual drive and multimotor drive.

Q. 5. What is group drive ?

Ans. Group drive, also sometimes called the line shaft drive, means a drive in which a single electric motor drives a line shaft by means of which an entire group of working machines may be operated.

Q. 6. What is individual drive ?

Ans. In an individual drive, a single motor is used to drive one individual machine. Example of such drives are single-spindle drilling machines, various types of electrical hand tools and simple types of metal working machine tools and mechanisms.

Q. 7. What is multimotor drive ?

Ans. Multimotor drive consists of several individual drives each of which serves to operate one of many working members or mechanisms in some production unit.

Q. 8. Why individual drive is preferred over group drive nowadays ? **[Gorakhpur Univ. 1981]**

Ans. Though the individual drive costs more than group drive, but each operator has complete control of his machine, which enables him to vary its speed if necessary and stop while not is use, thus eliminating no-load losses. The machines can be placed in any desired position and can be moved very easily. The motor and its control unit can be built as an integral part of the machine which results in good appearance, cleanliness and safety. For driving heavy machines such as for lifts, cranes, shapers, lathes etc. and for the purposes where constancy of speed and flexibility of control is required, such as in paper mills and textile industry, individual drive is essential.

For new factories individual drive is preferred, as it causes some saving in the cost of superstructure because of being much lighter and less expensive.

Q. 9. How does active torque differ from passive torque ?

Ans. Active torques are due to either gravitational force or deformation in elastic bodies whereas passive torques are due to friction or due to shear and deformation in inelastic bodies. Active torque continues to act in the same direction even after the direction of drive has been reversed while passive torque always opposes the motion, retarding the rotation of the driven machine.

Q. 10. What is meant by two-quadrant operation of a dc motor ?

Ans. The operation of an electric motor in forward motoring mode and in regenerating (or braking mode) is called the two-quadrant operation of the motor.

Q. 11. What factors govern selection of motor for a drive ? **[P.T.U. December 2007]**

Ans. For selection of reliable and efficient motor it is essential that the conditions of service are well known. It is not sufficient to simply specify the output power in kW and the speed but it is also necessary to know the following additional particulars :

(*i*) Torque at the shaft during running, starting and at different loads.

(*ii*) Accelerating torque and braking torque.

(*iii*) Switching frequency.

(*iv*) Efficiency of motor at different loads.

(*v*) Other working requirements.

In studying the behaviour of a motor selected for a particular driven unit, one of the first problems involved is to determine whether the speed-torque characteristic of the motor suits the requirements imposed by the speed-torque characteristic of the driven unit. Drive behaviour during the transient period of a start-up, braking, or speed change-over also depends upon how the speed-torque characteristics of the motor and the driven unit vary with speed.

Q. 12. Why speed-torque (not torque-speed) characteristic curves are commonly used in utilization studies ?

Ans. In application engineering, it is usually the load torque which is known and the speed at which the motor can drive this load is unknown and is to be determined. Thus the torque is independent variable and speed is dependent variable. Consequently speed-torque characteristic curves with torque as abscissa are commonly used in utilization studies.

Q. 13. What will be the effect on direction of rotation of a dc motor if the supply terminals are reversed ?

Ans. No effect. Motor will rotate in the same direction, as before the reversal of the line terminals.

Q. 14. Why should a dc series motor not be run without load ?

Ans. On no load dc series motor would attain tremendously high speed and the motor may get damaged due to heavy centrifugal forces set up in rotating parts.

Q. 15. Why is series motor preferred for traction purposes ?

Ans. Because dc series motor develops large starting torque and slows down with the increase in load and so gets automatically relieved from heavy excessive load.

Q. 16. Why the dc motors are mostly gear motors ?

Ans. Because the low-speed motors are much more expensive than high-speed motors of the same kW output rating.

Q. 17. Why are induction motors called asynchronous motors ?

Ans. Because the rotor does not turn in synchronous with the rotating field developed by the stator currents.

Q. 18. Is the maximum torque of a 3-phase induction motor dependent on the rotor resistance ?

Ans. No, it is independent of rotor resistance.

Q. 19. Why the power factor of an induction motor is low at starting ?

Ans. The rotor frequency and rotor reactance are high under starting conditions and, therefore, rotor currents lag the rotor emf by a large angle. This results in low power factor at starting.

Q. 20. The starting torque of a squirrel cage induction motor cannot be altered, when the applied voltage is constant, why ? **[M.K. Univ. Nov. 1996]**

Ans. The starting torque of a squirrel cage induction motor cannot be increased as there is no provision for inserting resistance in the rotor circuit.

Q. 21. How will you improve the starting torque of a 3-phase induction motor ? **[Madras Univ. Elec. Machines II October 1995]**

Ans. The starting torque of a 3-phase induction motor can be improved by inserting external resistance in each phase of the rotor circuit, which is possible only in case of wound rotor motors.

Q. 22. What are the main advantages of a cage motor ? **[M.S. Univ. April 1996]**

Ans. The main advantages of a cage motor are simple and rugged construction, low initial as well as maintenance cost, better efficiency and power factor, better cooling, more pull-out torque, greater maximum power output, nearly constant speed and simple starting arrangement.

Q. 23. What are the main disadvantages of a cage motor ? **[M.S. Univ. April 1996]**

Ans. The main drawbacks of a cage motor are low starting torque with large starting current and no possibility of a speed control.

Q. 24. What are the advantages and disadvantages of wound rotor over squirrel cage rotor in case of induction motor ? **[P.T.U. December 2002]**

Ans. In wound rotor external resistance can be inserted and, therefore,

(*i*) high starting torque can be obtained by inserting an external resistance of proper value in each phase of the rotor circuit and

(*ii*) adjustable speed can be obtained simply by adding external resistors in the rotor circuit.

The drawback of wound rotor over squirrel cage rotor are initial and maintenance costs are more because wound rotor needs slip rings, brushes, short-circuiting devices etc.

Q. 25. Why in some induction motors double cage rotors are provided ? **[A.M.I.E. Sec. B. Summer 1994]**

Ans. In some induction motors double cage rotors are provided so as to get high starting torque with a low starting current as well as high operating efficiency. Such motors operate with the advantages of high-resistance rotor circuit during starting and low-resistance rotor circuit under running conditions.

Q. 26. State the advantages of squirrel cage induction motors over dc motors. **[Rajasthan Univ. 2008]**

Ans. The important advantages of squirrel cage induction motors over dc motors are: (*i*) Ruggedness (*ii*) Lower maintenance (*iii*) Better reliability, (*iv*) Lower cost, weight, volume and interia, (*v*) Higher efficiency, (*vi*) Ability to operate satisfactorily with sharp voltage fluctuations and in dirty environment.

Q. 27. Why the *V*/*f* is kept constant while controlling the speed of a 3-phase induction motor ?

Ans. In order to maintain flux level constant.

Q. 28. How the power factor of a synchronous motor is changed keeping the shaft load undisturbed ?

Ans. By varying the field excitation.

Q. 29. For which type of loads, induction motor with voltage controller is suitable and why ? **[A.M.I.E. Sec. B. Summer 1996]**

Ans. Induction motor with voltage controller is suitable only for loads where the torque required drops off considerably with the drop in speed such as for driving fans because a large change in voltage is required for a relatively small change in speed and developed torque reduces greatly with the reduction in supply voltage.

Q. 30. How can the speed of a linear induction motor be controlled ? **[Bharathiyar Univ. Nov. 1995]**

Ans. By varying both frequency and voltage simultaneously.

Q. 31. Is the torque developed by a single phase induction motor at synchronous speed zero ?

Ans. No, it is negative.

Q. 32. Why single phase induction motors are usually set on rubber spring mounts ?

Ans. In a single phase induction motor, pulsating torque is developed as a direct consequence of the pulsating power in a single phase circuit.

Q. 33. Does a shaded-pole motor have large starting torque ?

Ans. No, it has very little starting torque.

Q. 34. What is a shaded-pole motor ?

Ans. Shaded-pole motor is a salient-pole, split-phase type, squirrel cage rotor single-phase induction motor provided with one or more auxiliary short-circuited stator windings displaced in magnetic position from the main winding.

Q. 35. State the applications of shaded-pole single phase induction motors. **[Madras Univ. April 1996]**

Ans. Shaded-pole motor develops very little starting torque and, therefore, it is used only for small fans, electric clocks, hair dryers, time phonographs, advertising displays, toys, instruments and such other applications.

Q. 36. Why is the normal full-load slip of a single phase induction motor higher than that of a three-phase motor ? **[A.M.I.E. Sec. B. Summer 1993]**

Ans. The normal full-load slip of a single phase induction motor is higher than that of a 3-ϕ motor, owing to development of backward rotating field. The power is to be delivered to the backward field from the power converted into mechanical power by the forward field.

Q. 37. What are the applications of electric drives ? **[P.T.U. December 2009]**

Ans. Squirrel cage induction motor is suitable for constant speed industrial drives of small power where speed control is not required and where starting torque requirements are of medium or low value, such as for printing machinery, flour mills and other shaft drives of small power.

Wound rotor (or slip-ring) induction motors are used for loads requiring severe starting conditions or for loads requiring speed control such as for driving line shafts, lifts, pumps, generators, winding machines, cranes, hoists, elevators, compressors, small electric excavators, printing presses, turn tables, strokers, large ventilating fans, crushers etc.

Single phase induction motors find wide use in fans, refrigerators, air conditioners, vacuum cleaners, washing machines, other kitchen equipment, record players, small farming appliances etc.

Q. 38. State whether repulsion motor has series speed-torque characteristics, or shunt speed-torque characteristics.

Ans. Series speed-torque characteristics.

Q. 39. What is a stepper motor ?

Ans. A motor which turns in discrete movement is called a stepper motor.

Q. 40. For stopping the motor, the line switch should always be opened rather than throwing back the starting arm. Why ?

Ans. If the starter arm is thrown back the field circuit is broken at the last contact button and owing to inductive nature of field, this will cause a hot arc and burn the contact.

Q. 41. Why do you require starters for starting a 3-phase induction motor ?

Ans. Starter is required to limit the initial rush of current to a predetermined value at the starting instant.

Q. 42. Which is the cheapest method of starting a 3-phase induction motor ?

Ans. Direct-on-line (D-O-L) starting.

Q. 43. What type of protection is provided in the starters meant for 3-phase induction motors ?

Ans. Overload and undervoltage protection.

Q. 44. Which method we must adopt to control the speed of a dc shunt motor above the base speed ?

Ans. Field control method.

Q. 45. While controlling the speed of a dc shunt motor what should be done to achieve a constant torque drive ? **[A.M.I.E. Sec. B. Summer 1995]**

Ans. Applied voltage should be maintained constant so as to maintain field strength constant.

Q. 46. If the number of poles on a motor is increased to lower the speed, how will the power factor be affected ?

Ans. Power factor will be reduced.

Q. 47. How is speed control affected in a 3-phase ac series commutator motor ?

Ans. In a 3-phase ac series commutator motor, the speed control is affected by moving all brushes simultaneously round the commutator by means of a hand-wheel.

Q. 48. How is speed control affected in a 3-phase ac shunt commutator motor ?

Ans. In a 3-phase ac shunt commutator motor, the speed is adjusted by changing the tapping points on a transformer inserted between the supply mains and the brushes, which are in fixed position.

Q. 49. How speed variation is affected in a repulsion motor ?

Ans. Speed variation in a repulsion motor is affected either by varying the impressed voltage or by changing the position of brushes.

Q. 50. What is meant by plugging ?

Ans. Plugging means stopping of an electric motor by instantaneously reversing its torque till it stops.

Q. 51. How plugging of a dc motor is done ?

Ans. Plugging is obtained in a dc motor by reversing the supply terminals to the armature of the motor.

Q. 52. What is the most economical method of electric braking ?

Ans. Regenerative braking.

Q. 53. Electric braking is accomplished in an induction motor by interchanging two stator terminals. Name the method.

[A.M.I.E. Sec. B. Winter 1995]

Ans. Plugging or counter-current braking.

Q. 54. When it is required to control the induction motor speed during deceleration what kind of braking is preferable to employ ? **[A.M.I.E. Sec. B. Winter 1996]**

Ans. DC dynamic braking is highly suitable for controlling the speed of an induction motor during deceleration by controlling the dc excitation.

Q. 55. How is choice of drive governed ?

Ans. Choice of drive is governed by the speed of driving and driven machines, convenience, space available, including clutching arrangement required and cost.

Q. 56. What is meant by "rating of motor" ? **[P.T.U. December 2009]**

Ans. A name plate fixed to the outside frame of an electrical machine records the data pertaining to its rating. A machine rating specifies the voltage, current, speed, excitation, power factor, efficiency, power output etc.

The rating of a machine should give all information necessary so that, if the machine is operated within the limits of all factors specified in its rating, the machine will operate satisfactorily and safely and will give reasonable length of service. Therefore, the rating of a machine must give the necessary information to safeguard the application of the machine from conditions of operation which (*i*) would result in unsafe mechanical or electrical strains upon any part of its structure or (*ii*) would result in excessive deterioration of the mechanical or electrical characteristics of the materials of which the machine is constructed, To give this information, the rating of an electric machine should include the output, voltage, speed, and any other information that may necessary for the proper operation of the machine.

Q. 57. On what factors does the rate of rise of temperature depend ?

Ans. The rate of temperature rise depends upon (*i*) amount of heat produced and (*ii*) the amount of heat dissipated per °C rise of the surface of the machine.

Q. 58. Why the cooling time constant of a rotating machine is usually larger than its heating time constant ?

[A.M.I.E. Sec. B. Winter 1996]

Ans. The cooling time constant of a rotating machine is usually larger than its heating time constant owing to poor ventilation conditions when the machine cools.

Q. 59. What is meant by load equalization ?

Ans. The process of smoothing out of fluctuations in load is known as load equalization.

Q. 60. How is load equalization achieved ?

Ans. The most common method of load equalization is by use of flywheel. During the light load period the flywheel accelerates and stores the excessive energy drawn from the supply and during peak load period the flywheel decelerates and supplies some of its stored energy to the load in addition to the energy supplied from the supply. Thus the load demand is reduced.

Problems

1. Compare the total annual cost of a group drive with a motor costing Rs. 65,000 with that of 8 individual motors, each costing Rs. 25,000. With group drive annual consumption is 60,000 kWh. With individual drive the consumption is 45,000 kWh. Electrical energy costs Rs. 1.20 per kWh. Annual interest, depreciation and maintenance charges may be taken as 18% in either case.

   [**Ans.** Rs. 83,700 per annum in case of group drive and Rs. 90,000 per annum in case of individual drive]

2. A drive has the following equations for motor and load torques

   $T = 1 + 2\omega_m$ and $T_L = 3\sqrt{\omega_m}$.

   Obtain equilibrium points and determine their steady state stability. **[U.P. Technical Univ. Electric Drives 2008-09]**

3. A 220 V dc series motor is running at a speed of 800 rpm and draws 100 A. Calculate at what speed the motor will run when developing half the torque. Total resistance of the armature and field is 0.1 Ω. Assume that the magnetic circuit is unsaturated. [**Ans.** 1,147 rpm]

4. A series motor takes 20 A at 400 V and runs at 250 rpm. The armature resistance is 0.6 Ω and the field resistance 0.4 Ω. The torque required by the device varies as the square of the speed. Find the applied voltage and current to run the device at 350 rpm. [**Ans.** 772.8 V; 28 A]

5. A 3-phase induction motor at rated voltage and frequency exerts a maximum torque of 275 per cent and a starting torque of 125% of full-load torque. Determine the slip (*i*) at maximum torque and (*ii*) at full-load. Assume constant rotor resistance and neglect stator resistance and rotational losses.

   [**Ans.** (*i*) 24% (*ii*) 4.52%]

6. A 3-phase, 50 Hz, 8-pole induction motor has a full-load slip 4%. The rotor resistance is 0.001 Ω per phase and standstill reactance 0.005 Ω per phase. Find the ratio of maximum to full-load torque and the speed at which the maximum torque occurs.

   [Madras Univ. Elec. Machines II, April 1997]

   [**Ans.** 2.6 ; 600 rpm]

7. A 3-phase star-connected 6.5 kV, 20-pole, 50 Hz induction motor has rotor resistance of 0.12 Ω and standstill reactance of 1.12 Ω. The motor has speed of 292.5 rpm at full load. Calculate slip at maximum torque and ratio of maximum torque to full load torque. **[Rajasthan Univ. Utilization of Elec. Power 2008]**

   [**Ans.** 0.107; 9.68]

8. The star-connected stator winding of a 3-phase, 6-pole induction motor is supplied from a 240 V, 50 Hz source. The rotor resistance and standstill reactance are 0.12 Ω and 0.85 Ω per phase respectively. The stator to rotor turn-ratio is 1.8. Find the pull-out torque and the corresponding speed.

   [**Ans.** 99.86 Nm; 859 rpm]

9. Determine approximately the starting torque of an induction motor in terms of its full-load torque when started by means of (*i*) star-delta starter and (*ii*) by an auto-transformer starter with 50 % taps. The motor draws 5 times full-load current when switched directly on line and has a full-load slip of 4 %.

   [**Ans.** One-third and one-fourth of full-load torque]

10. A squirrel cage induction motor takes 5 times full-load current and develops 2.5 times full-load torque when switched directly

on the line. Determine the torque developed and the initial value of the current drawn from the supply mains when started through an auto-transformer with 60 % tapping.

[**Ans.** 1.8 times of full-load current; 81% of full load torque]

11. A 250 V dc shunt motor has an armature resistance of 0.5 Ω and a field resistance of 250 Ω. When driving a load of constant torque at 600 rpm, the armature current is 20 A. If it is desired to raise the speed from 600 to 800 rpm, what resistance should be inserted in the shunt field circuit? Assume that the magnetic circuit is unsaturated. [**Ans.** 88 Ω]

12. A 220 V shunt motor has an armature resistance of 0.5 Ω and takes an armature current of 40 A on a certain load. By how much the main flux be reduced to raise the speed by 50% if the developed torque is constant ? Neglect saturation and armature reaction. [**Ans.** 37.3%]

13. A 200 V dc shunt motor running at 1,000 rpm takes an armature current of 17.5 A. It is required to reduce the speed to 600 rpm. What must be the value of resistance to be inserted in the armature circuit if the original armature resistance is 0.4 Ω ? Take armature current to be constant during this process.

[**Ans.** 4.41 Ω]

14. A 250 V shunt motor with armature resistance of 0.5 Ω runs at 600 rpm on full load and takes an armature current of 20 A. If resistance of 1.0 Ω is placed in the armature circuit, find the speed at (*i*) full load torque (*ii*) half-full load torque.

[**Ans.** (*i*) 550 rpm (*ii*) 587.5 rpm]

15. A 230 V dc shunt motor takes an armature current of 20 A on a certain load. Resistance of armature is 0.5 Ω. Find the resistance required in series with the armature to halve the speed if (*i*) the load torque is constant (*ii*) the load torque is proportional to the square of the speed. [**Ans.** (*i*) 5.5 Ω (*ii*) 23.5 Ω]

16. A 250 V dc series motor takes 40 A when developing full-load torque at 1,500 rpm. Its total resistance is 0.5 ohm. If the load torque varies as the square of the speed, determine the resistance to be connected in series with the armature to reduce the speed to 1,200 rpm. Assume a straight line magnetisation curve.

[**Ans.** 2.7125 ohms]

17. A series motor working on a 500 V dc supply mains at a speed of 800 rpm when the load is such that the current is 120 A. The motor armature resistance is 0.11 ohm and field resistance is 0.04 ohm. Assuming an unsaturated field calculate the speed of the motor when the torque is one half of the full-load value and the field winding is connected in parallel with divertor resistance of 0.08 ohm. [**Ans.** 1,395 rpm]

18. A 4-pole, 250 V dc series motor takes 20 A and runs at 900 rpm. Each field coil has resistance of 0.025 Ω and the resistance of the armature is 0.1 Ω. At what speed will the motor run developing the same torque if (*i*) a divertor of 0.2 Ω is connected in parallel with the series field (*ii*) rearranging the field coils in two series and parallel groups. Assume unsaturated magnetic operation.

[Nagpur Univ. Electric Drives & Their Control 1993]

[**Ans.** (*i*) 1,101.6 rpm (*ii*) 1,275.4 rpm]

19. The rotor of a 6 pole, 50 Hz slip-ring induction motor has a resistance of 0.2 Ω per phase and runs at 960 rpm on full load. Calculate the approximate resistance per phase of a rotor rheostat such that speed is reduced to 800 rpm for full-load torque.

[**Ans.** 0.8 Ω/phase]

20. The voltage across two slip rings of a stationary star-connected 100 hp, 415 V induction motor is 273 V with full voltage across stator. If full-load slip is 2 per cent, find the resistance to be added in series with each slip ring to reduce the speed by 20 %. Assume load torque to be constant and rotor reactance negligible.

[Pb. Univ. Elec Machines II Dec. 1993]

[**Ans.** 0.196 Ω]

21. What synchronous speeds can be obtained from the combination of 50 Hz induction motors having 8 poles and 4 poles.

[**Ans.** 1,500 rpm; 750 rpm; 500 rpm]

22. A cascade set consists of two motors A and B with 4 poles and 6 poles respectively. The motor A is connected to 50 Hz supply. Determine (*i*) the speed of the set (*ii*) the electric power transferred to motor B when the input to motor A is 25 kW. Losses may be neglected. [**Ans.** (*i*) 600 rpm (*ii*) 10 kW]

23. A 50 hp, 440 V dc shunt motor is braked by plugging. Calculate the value of the resistance to be placed in series with the armature circuit to limit the initial braking current to a double of the full-load current. Armature resistance is 0.1 Ω. Full-load armature current is 100 A. [**Ans.** 4.25 Ω]

24. A 25 hp, 220 V dc shunt motor with a full-load speed of 600 rpm is to be braked by plugging. Estimate the value of the resistance which should be placed in series with it to limit the current to 130 A. What should be the initial value of the electric braking torque and value when the speed has fallen to half its full load value ? Armature resistance of motor is 0.1 Ω. Full-load armature current is 95 A.

[**Ans.** 3.211 Ω ; 400.5 Nm; 302.57 Nm]

25. A 400-V,25 hp, 450 rpm dc shunt motor is braked by plugging when running on full-load. Determine the braking resistance necessary if the maximum braking current is not to exceed twice the full-load current. Determine also the maximum braking torque and the braking torque when the motor is just reaching zero speed. The efficiency of the motor is 74.6% and the armature resistance is 0.2 Ω. [**Ans.** 6.1 Ω; 1,028 Nm; 522 Nm]

[**Hint :** Maximum braking torque will be developed initially when motor speed is maximum *i.e.* at a speed of 450 rpm]

26. A 250 V dc shunt motor has an armature resistance of 0.1 Ω and with full field has an emf of 240 V at a speed of 1,200 rpm. The motor is driving an overhauling load with a torque of 180 Nm.

Calculate the minimum speed at which the motor can hold the load by means of regenerative braking. [**Ans.** 1,297 rpm.]

27. Rheostatic braking is used to limit the speed of a series type dc crane motor to 500 rpm when the supply is cut off. The motor has a resistance of 1.35 Ω and the descending load exerts a torque of 442 N-m. Determine the resistance to be connected across the motor terminals.

The magnetisation characteristic of the machine at 500 rpm is as follows:

Current (A)	20	30	40	50
EMF (V)	360	440	490	520

State assumptions made if any.

[Gorakhpur Univ. Utilization of Elec. Power and Traction Sept. 1984] [**Ans.** 10.005 Ω]

28. A 400 V, 50 Hz, 4-pole cage type induction motor is braked by plugging, while working at full load delivering 50 hp; the full-load slip being 4%. The ratio of rotor resistance to standstill reactance = 1:5.

Calculate the initial braking torque.

[Gorakhpur Univ. Utilization of Electric Power and Traction 1983] [**Ans.** 128 N-m]

29. A 40 hp, 400 V, 3-ϕ, 4-pole, 50 Hz induction motor has full-load slip of 5%. If the ratio of standstill reactance to resistance per motor phase is 4, estimate the plugging torque at full speed.

[Pb. Univ. Elec. Drives and Utilization of Elec. Energy April 1981] [**Ans.** 129.3 N-m]

30. A 50 kW, 415 V, 3-phase, 4-pole, 50 Hz induction motor has a full-load slip of 5%. If the ratio of standstill reactance to resistance per phase of rotor is 5, calculate the torque developed at the start of reverse current braking at rated speed. Neglect stator impedance and magnetizing current.

[U.P. Technical Univ. Electric Drives 2007-08]

[**Ans.** 144.5 N-m]

31. The outside of a 12.5 kW dc motor is equivalent to a cylinder of a 100 cm diameter and 140 cm in length. The machine may be considered to be a homogenous material weighing 650 kg and of specific heat 700 J/kg/°C. Assuming that the outer cylindrical surface dissipates heat at 0.0012 W/cm^2/°C, estimate the thermal time constant of the machine and its final temperature rise when operating on full load with an efficiency of 85%.

[**Ans.** 2.394 Hours; 41.8° C]

32. In a temperature rise test on a dc machine at full load, the following temperature rises were observed; 16° C after one hour and 28° C after 2 hours. Find (*i*) final steady temperature rise and the time constant (*ii*) the steady temperature rise after one hour at 50% overload from cold. Assume that the final temperature rise on 50% overload is 92° C.

[**Hint :** Heating time constant of a motor is independent of rate of heat production so the heating time constant at 50% overload is the same as on full load.]

[**Ans.** (*i*) 64° C; 3.481 hours (*ii*) 123° C]

33. The heating time constant of a 22 kW motor is 90 minutes. If the motor is to be run continuously on full load, it attains a maximum steady temperature rise of 40° C. Calculate the half-hour rating of the motor for this temperature rise. Assume that losses are proportional to the square of the load and the motor cools down completely between each load cycle.

[BHU Electric Drives & Power Electronics 1978]

[**Ans.** 41.32 kW]

34. A motor driving some haulage equipment for a mine has to deliver a load which follows the following duty cycle: 100 kW for 10 minutes, no load for 4 minutes, 50 kW for 10 minutes, no load for 6 minutes. The cycle being repeated indefinitely. Estimate the suitable size of continuously rated motor for the purpose.

[**Ans.** 65 kW]

35. A machine on a short test run gave the following mean temperature rise :

Time hours	0	0.25	0.5	0.75	1.0	1.25	1.5
Temp. rise °C	42.5	45.0	47.4	49.5	51.4	53.0	54.4

Estimate graphically the final temperature rise and the time constant. Justify the procedure adopted by you.

[**Ans.** 62°C, 1.6 hours]

36. The temperature rise of a motor when operating for 25 minutes on full load is 25°C and becomes 40°C when operates for another 25 minutes on the same load. Determine the heating time constant and steady-state temperature rise.

[A.M.I.E. Sec. B. Electric Drives & Their Control Winter 1998]

[**Ans.** 48.94 minutes; 62.5°C]

37. A motor driving some haulage equipment for a mine has to deliver a load which follows the following duty cycle :

100 hp for 10 minutes, no load for 4 minutes, 50 hp for 10 minutes, no load for 6 minutes.

The cycle being repeated indefinitely. Estimate the suitable size of continuously rated motor for the purpose.

[Agra Univ. Traction and Utilization of Elec. Power 1982]

[**Ans.** 65 hp]

38. A constant speed drive has the following duty cycle :

Load rising from 0 to 400 kW : 5 min

Uniform load of 500 kW : 5 min

Regenerative power of 400 kW returned to supply : 4 min

Remains idle for : 2 min

Estimate the power rating of the motor. Assume losses to be proportional to (power)^2.

[A.M.I.E. Sec. B. Electric Drives & Their Control Summer 1998]

[**Ans.** 330 kW]

39. A 500 rpm synchronous motor and its load have a total moment of inertia of 420 kg-m^2 and a frictional torque of 140 N-m. The motor is braked by rheostatic method, field current and the braking resistance being kept constant at value which gives an initial electric braking torque of 5,400 N-m. Find the time taken to bring the motor to test. [**Ans.** 15 seconds]

40. A motor fitted with a flywheel supplies a load torque of 100 N-m for 20 seconds. During the no-load period, the flywheel regains its original speed. The motor torque is required to be limited to 500 N-m. Determine the moment of inertia of the flywheel. The no-load speed of the motor is 500 rpm and it has a slip of 10% on full load.

[Pb. Univ. Elec. Drives & Utilization of Elec. Energy April 1980]

[**Ans.** 2,755 kg-m^2]

41. A dc shunt motor 30 hp, 400 V, 500 rpm draws a current of 40 A when running at full load. The moment of inertia of the rotating system is 12.4 kg-m^2. If the current during the starting period varies from 1.05 to 1.25 times of full-load current, calculate (*i*) the full-load torque, and (*ii*) the time required for the motor to attain the rated speed against full load.

[**Ans.** (*i*) 421.4 N-m (*ii*) 10.27 seconds]

42. A 6-pole, 50 Hz induction motor has a flywheel of 1,200 kg-m^2 moment of inertia. Load torque is 100 kg-m for 10 seconds. No-load period is long enough for the flywheel to regain its full-speed. Motor has a slip of 6% at a torque of 50 kg-m. Calculate (*i*) maximum torque exerted by motor (*ii*) speed at the end of deceleration period.

[Pb. Univ. Utilization of Elec. Energy 1973 Supp.]

[**Ans.** (*i*) 469 N-m (*ii*) 925 rpm]

❑❑❑

CHAPTER 2

Control Devices For Industrial Motors

INSIDE THIS CHAPTER

2.1. Introduction 2.2. Automatic Regulating Equipment 2.3. Solid State Devices 2.4. Pilot Devices and Accessories 2.5. Contactors 2.6. Electromagnetic Relays 2.7. Protective Devices and Interlocks.

2.1. INTRODUCTION

The apparatus such as suitable switches required for opening and closing the necessary circuits and for operating rheostats required for starting, reversing, speed control and braking of electric motors together with the protective equipment for guarding against faults or careless usage is known as the *control gear.*

The simplest method of motor control is manual control, which is accomplished by plain knife switches, rotary switches, starting and speed control rheostats, push buttons and controllers. The use of manual control devices involves a certain additional expenditure of time for accomplishing control and hence lowers the productivity of the equipment concerned, especially when its operation calls for frequent starting or speed control. The use of manual control devices also precludes the possibility of remote control. This makes it altogether unsuitable for many kinds of modern automatic equipment. In large capacity electric drives, manual control is difficult to provide, and in some cases is altogether impracticable because the operators would be unable to exert the effort required for switching the apparatus.

The need to eliminate the shortcoming of manual control has led to the development of semi-automatic and fully-automatic control equipment. Semi-automatic control makes use of control apparatus which is energized by pressure on a push button. In the case of automatic control, the driven machinery itself may start up the drive motor, or this can be done by suitable controllers.

Automatic electric drive control is of great importance to the national economy because it permits the productivity of labour to be raised, makes for better working conditions, improves the quality of the products being produced, reduces the expenditure of electrical energy, and raises the reliability of operation of the equipment.

Automation of the drive simplifies the operation of the equipment and allows it to be remotely controlled. This is particularly valuable where local control of drive motors is inconvenient by reason of position or impossible because of the nature of the production process.

The above advantages are tending to increase the applications of fully-automatic control equipment, inspite of the fact that it costs about three times as much as the corresponding manual control equipment.

2.2. AUTOMATIC REGULATING EQUIPMENT

In industrial processes a certain quantity, such as the speed of a motor, the temperature of an oven, the tension in a wire, output voltage of the generator or output current is required to be held at some specified value, usually constant, although in certain cases a predetermined cycle of variation may be required.

Electrical devices used for providing such regulation operate on the closed cycle or feedback principle. Such an arrangement requires essentially (*i*) an actuating device for providing a control signal dependent on the deviation of the controlled quantity from its specified value (*ii*) an operating device for controlling the power supply feeding the apparatus to be regulated in such a way as to counteract any deviation and (*iii*) a connecting link for transferring the actuating signal in an appropriate manner to the operating device.

The connecting link used for this purpose should be of high power gain and of a very rapid response so that the smallest departure from the specified value will be sufficient to initiate the correcting effect of the operating device and there is no time delay between the occurrence of the actuating signal and the correcting effect.

The thermionic valves, usually of gas filled type, rotating amplifiers (amplidynes) and magnetic amplifiers are extensively used as connecting link.

2.2.1. Thyratron

Thyratron is a hot cathode gas filled triode. Normally it is filled with a small amount of inert gas such as argon, hydrogen, neon or mercury vapour. The fast acting devices do not use mercury vapour as the mercury vapour has the disadvantage of characteristics varying with temperature and it takes time in deionization. The characteristics of hydrogen thyratron are stable and the devices are fast. Plate is a disc of graphite. The special construction oxide coated cathode is used so that its surface is not damaged due to bombardment of positively charged particles. The control grid has a metal cylinder surrounding the cathode with one or more perforated discs known as *grid baffles* near the centre. Thus the control grid acts as an electrostatic shield between cathode and anode except for the holes in the grid baffles.

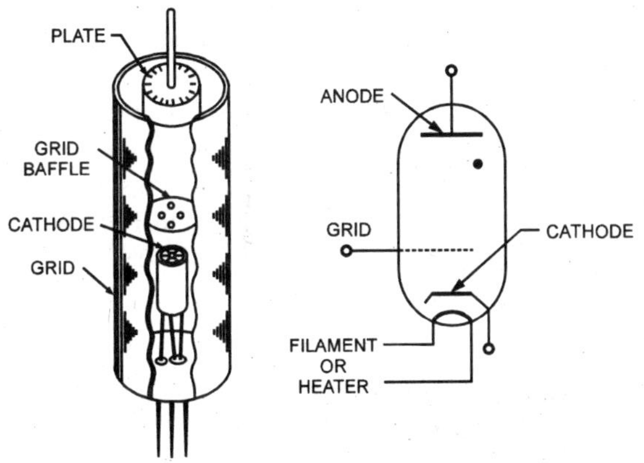

(a) Construction *(b) Symbolic Representation*

Fig. 2.1. *Thyratron*

Before the tube fires, its characteristics are similar to those of a vacuum triode *i.e.,* the plate current (small in magnitude) can be varied by variation of potential applied to the grid. After firing of the tube the grid loses control over plate current completely and plate current is limited by the external resistance in the plate circuit *i.e.,* the thyratron behaves like a diode. The introduction of grid in between the plate and cathode makes it possible to control the conditions under which the discharge should start.

Since the only function of the grid is to delay or prevent the starting of the current, the grid is naturally designed to achieve its purpose with as little grid voltage as possible. The grid, therefore, is quite an extensive structure. The critical grid voltage varies with the plate voltage to some extent, and also with the design of the grid.

Plate Voltage-Plate Current (or Plate) Characteristics. These characteristics are drawn for various grid voltages.

For a given grid voltage when the plate voltage is increased from zero, initially there is no plate current but when the plate voltage is further increased, a voltage is reached, called the *breakdown voltage*, the current at once shoots up and is limited only by the external resistance in the plate circuit. If the grid is made more negative, the tube will break down at higher plate voltage, as illustrated in Fig. 2.2.

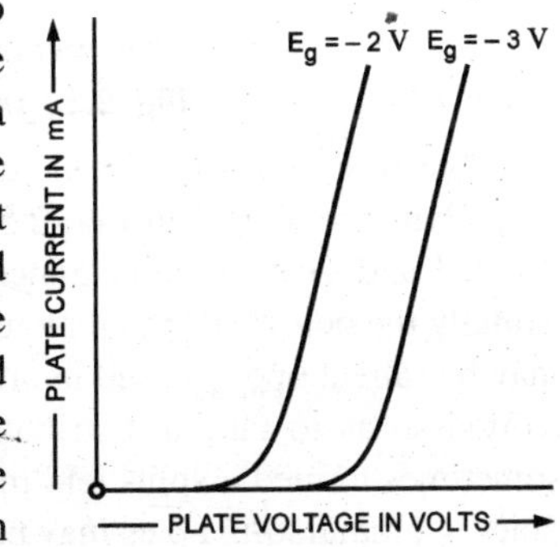

Fig. 2.2

Grid Control Characteristics. The curve drawn between plate voltage and critical grid voltage is called *the grid control characteristic or starting characteristic or breakdown characteristic*. This characteristic can be determined by determining the value of critical grid voltage for different values of plate voltage and plotting them, as shown in Fig. 2.3. The slope of this characteristic is known as *control ratio*.

Applications. Thyratron has two main fields of applications :

(*i*) as an electronic switch to handle heavy currents and (*ii*) as a grid controlled rectifier when voltage control is required. The thyratron, when used as electronic switch, keeps the circuit open from the plate to cathode so long as the grid is kept at high negative potential. As soon as the grid potential is brought to such a value that the tube fires, the circuit is closed.

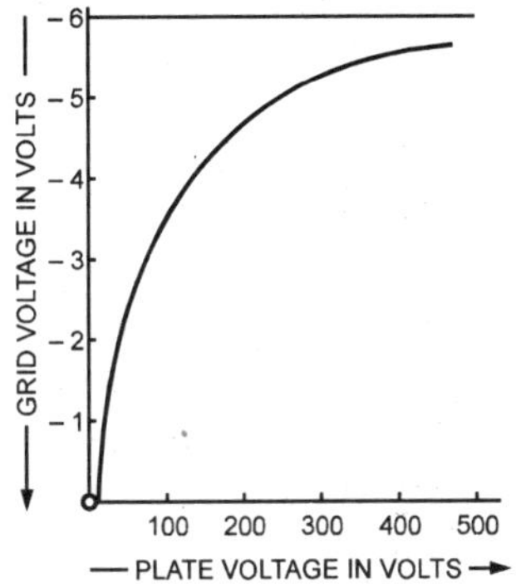

Fig. 2.3

The thyratron switch has two great advantages over mechanical switches : (*i*) it is electrically operated and requires only a minute amount of energy to close it, so that it leads itself readily to automatic operation; (*ii*) operation is very fast. However, it has got some disadvantages also such as (*i*) voltage drop of 8 to 18 volts in the circuit; (*ii*) it cannot open dc circuit; (*iii*) a single thyratron will pass only the positive half waves. A second thyratron connected in parallel with the first one but with opposite polarity may be used to pass the negative half waves. A thyratron can open ac circuit, with a delay that never exceeds one half cycle, so long as frequency is not so high that the time of one half cycle is less than the deionizing time.

The thyratron finds wide use in timing and other control applications.

The fact that its time of firing may be controlled makes the thyratron adaptable to heat control in resistance welding. By delaying or advancing the time during the cycle at which the tube fires, the average amount of welding current allowed to flow per cycle may be increased or decreased. In addition to this, the number of cycles that the welding current is allowed to flow may be controlled very accurately.

Thyratron rectifiers are also used in electronic methods of electric motor control for both small and large powers. The main advantages of electronic motor control are that the control can be made to depend on very minute signals such as those obtained from a photocell or even a thermocouple and that where necessary an extremely precise control can be obtained.

Control may be affected in several ways such as (*i*) with direct voltage applied to the plate as well as to the grid (*ii*) with alternating voltage to the plate circuit and direct voltage to the grid and (*iii*) alternating voltage applied on both the grid and plate. In the first case the tube will fire and conduct whenever the grid voltage is made less negative than critical grid voltage and will again stop to conduct when the plate voltage is reduced below striking voltage or the plate circuit is opened. In the second case the grid will regain control when the plate voltage becomes negative, hence conduction will be stopped every half cycle and the tube breakdown will occur each half cycle as the plate voltage becomes positive, the point at which the tube fires depends upon the grid voltage. The average current flow can be controlled, therefore, by controlling the length of time during the half cycle that the tube is conducting *i.e.,* by controlling the breakdown point. In the last case breakdown point is determined by the phase relation of the two voltages as well as by their magnitudes. There will thus be a point in each positive half cycle of the cathode–plate path when the tube will break down, if the grid voltage has a small negative value than the critical grid voltage with reference to the plate voltage, which will result in ionization and conduction. By using a phase shifting device to control the phase of grid voltage with respect to the cathode-plate voltage wave the duration of current flow may be controlled from zero to point represented by practically all the positive half cycles of plate voltage.

The thyratron cannot be employed as an amplifier like a vacuum triode because the grid voltage has no control over the magnitude of plate current after firing of the tube.

2.2.2. Ignitron

Ignitron is a gaseous tube and consists of carbon anode, boron carbide ignitor tube and mercury pool cathode. It is a cold cathode tube and does not have any heating element or grid control as in thyratron. Therefore, electrons are to be made available before the tube starts conducting. This is achieved by the ignitor which remains dipped in the mercury pool but being made of boron carbide does not get wet and, therefore, gives high resistance, say 10 to 500 Ω, between the mercury pool and ignitor. On closing of switch S full voltage is impressed between ignitor and mercury pool. This causes intense electric field which pulls out electrons from mercury pool to start with. The electrons of the arc ionize the mercury vapours if the ignitor current is of the magnitude between 20 and 40 amperes. Positive ions thus released are attracted towards cathode mercury pool and momentum of these mercury ions releases large number of electrons from the cathode pool by secondary emission. These electrons are attracted towards the carbon anode. Now resistance of main path of the current is very low as compared to that of ignitor path and, therefore, ignitor current stops. This process is to be repeated every time anode becomes positive in ac voltage cycle.

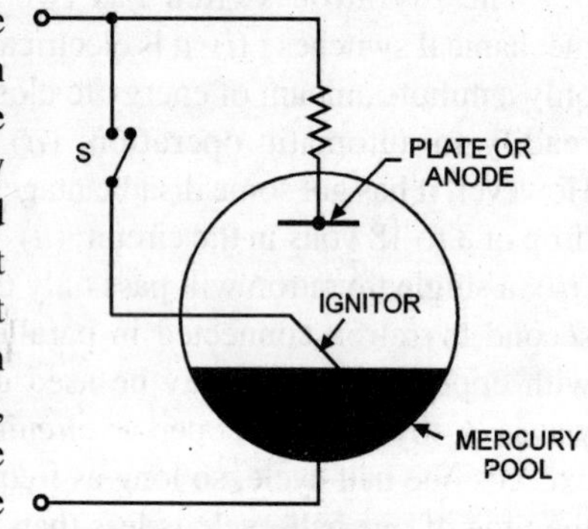

Fig. 2.4

Ignitrons can handle current from 40 to 10,000 A. It may also be noted that electronic current flow from ignitor to cathode, spoils the ignitor and, therefore, should not be permitted.

The power output can be controlled by timing the firing by means of the ignitor, usually by shifting the phase of the pulse to the ignitron.

The single units may be connected in 3-phase, 6-phase, or 12-phase. In smaller ratings, sealed-in glass tubes are used; in the intermediate ratings, sealed-in water-cooled metal tanks are used. In the larger power units, pumped all steel water-cooled tanks are used.

2.2.3. Amplidyne (Rotary Amplifier)

The amplidyne is a rotary amplifier in which a small amount of power supplied to excite the field is amplified in the generator output. Though the ordinary separately excited generator operated at low flux densities can be used as a power amplifier since the power required for exciting the field is only 1 to 3 per cent of the rated output of the generator and a small change in the power to the field can be made to produce a large proportionate change in the power output. However, a much larger power amplification can be had by using the amplidyne, since in this machine power is amplified in two stages. Rotary amplifiers are generally built for outputs up to 100 kW.

This machine uses the cross-field principle and a single two-pole armature winding.

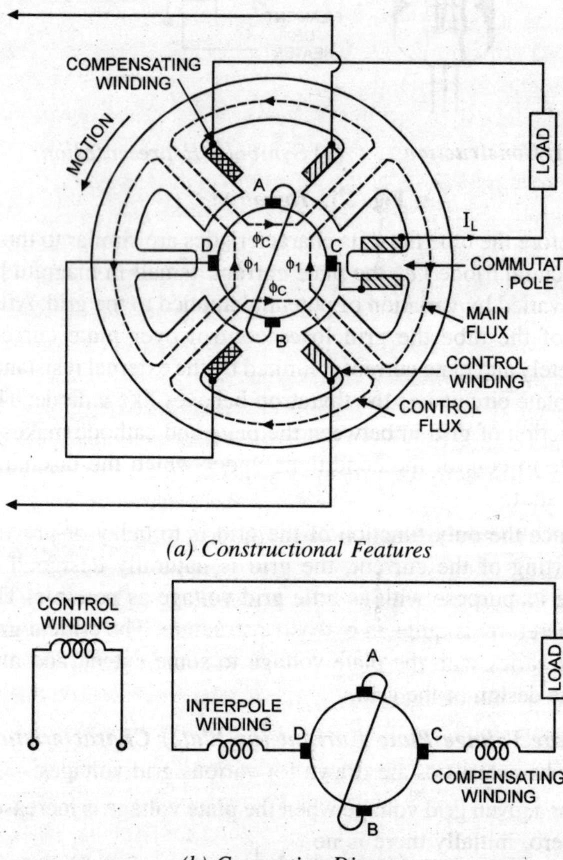

(a) Constructional Features

(b) Connection Diagram

Fig. 2.5. ***Amplidyne***

The basic constructional features for a 2-pole armature amplidyne are shown in Fig. 2.5 (*a*). The poles, however, are divided and give the appearance of a four-pole field though actually the poles are in pairs with the same polarity. The poles may be salient ones as used in ordinary dc machines or may be slotted so as to take a distributed winding. The machine is sometimes termed a split-field machine because it has divided poles. Commutating poles may be arranged in all the interpolar gaps. Control windings and compensating windings are arranged, as illustrated in Fig. 2.5 (*a*). Compensating winding in the form of a concentrated coil wound around the field cores, as shown in Fig. 2.5 (*a*) is used in many cases but such a winding does not completely neutralize the armature load current mmf and some distortion of the flux, therefore, exist. For better compensation a distributed type of compensating winding, properly located in slots in the pole faces must be used. The magnetic circuit may then consist of slotted punchings. This form does not provide as much space for a variety of field windings as the form with polar projections, and the various control inputs are more usually combined in an electronic or magnetic amplifier providing the control excitations instead of in the machine itself. Magnetic circuit should be of laminated construction throughout.

The action of the machine is now explained. The control windings produce a small control flux Φ_c in the horizontal axis. When the machine is driven, a voltage is developed between the short-circuited brushes A and B and, therefore, current flows between them (the control winding is provided with few ampere-turns in order to limit the current through the short-circuited armature winding). Now the current I_1 through the short-circuited armature winding produces armature reaction flux at right angles to the control flux Φ_c which in turn develops emf between the output brushes C and D. In the figure separate flux paths have been shown. Although the fluxes are superimposed in the poles and parts of the armature and yoke, in the absence of saturation they are mutually independent.

When the amplidyne is loaded, load current I_L flows which in turn sets up the flux opposing the control flux Φ_c. It is essential, therefore, for the armature reaction mmf set up by the output current I_L to be neutralised if the machine is to operate as a high-gain amplifier. A compensating winding is, therefore, connected in series with the power output brush terminals. The machine operated in this manner is called the amplidyne. This device is virtually two machines in one. Current through the short-circuited brushes which form the load for the first stage, acts also as the field excitation for the second stage. Mechanical input at the shaft is converted into an electrical output, modulated by the control flux. With full compensation, gains of the order of 20,000 : 1 are possible and the power supplied to the control winding may be fraction of a watt even in a 10-20 kW rated machine. Hence the amplidyne can be fed directly from an electronic amplifier in an automatic control scheme.

The basic machine, described above, had only one control winding, which was separately excited. Actual amplidynes are provided with additional field windings separately excited as well as self excited to provide desirable characteristics. Separately excited field windings are called the control windings since the energy or signals, as they are generally called, which are supplied to the control windings control the machine output. Since these windings are of very small power capacity, a small change of input power to a control field winding will develop a large change in the output voltage across brushes C and D and, therefore, a large change in power output. It is feasible to construct amplidynes with four separate control field windings so that the output of the machine may be controlled from several independent signals. Commutating poles are fitted at the brush axis to assist commutation, but it is not usual to fit them at the cross axis, except in large machines owing to the small value of I_1, the exciting current.

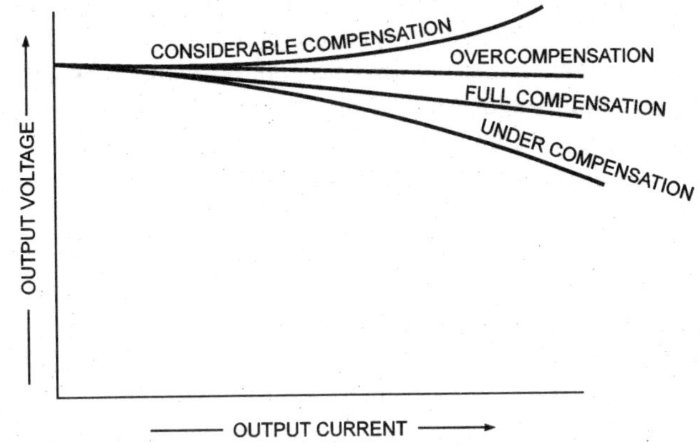

(a) For Various Degrees of Compensation Amplidyne External Characteristics

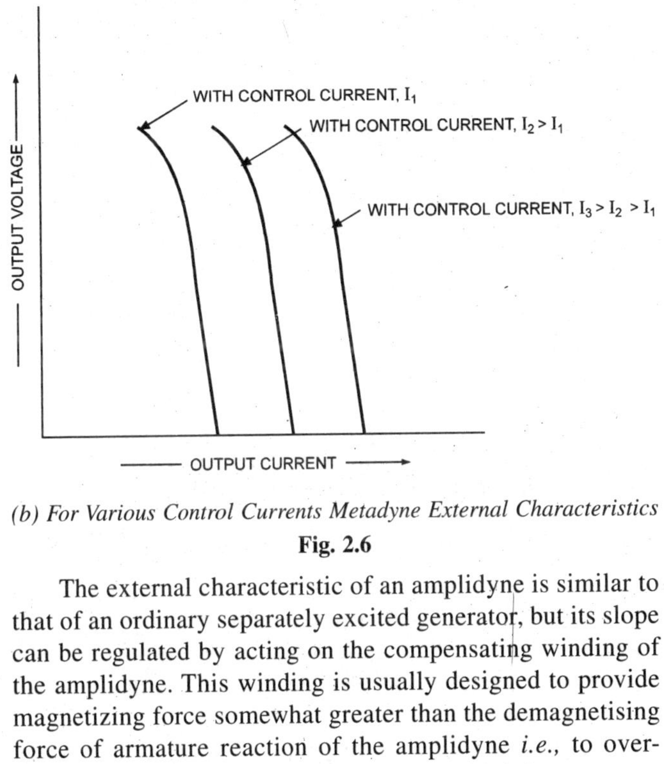

(b) For Various Control Currents Metadyne External Characteristics

Fig. 2.6

The external characteristic of an amplidyne is similar to that of an ordinary separately excited generator, but its slope can be regulated by acting on the compensating winding of the amplidyne. This winding is usually designed to provide magnetizing force somewhat greater than the demagnetising force of armature reaction of the amplidyne *i.e.,* to over-compensate for the armature reaction, which is necessary under certain amplidyne operating conditions. But usually the compensating winding is shunted by a rheostat to reduce the magnetising force. This allows the slope of the external characteristic to be regulated. In the case of under compensation, the voltage drop, with increasing load, will be due not only to the voltage drop in the armature, but also to armature reaction. In the case of full compensation of armature reaction a fall in the amplidyne voltage with increasing load will be due only to voltage drop in the armature. With slight overcompensation, an increase in flux with the load can be made to compensate exactly for the increase in voltage drop in the amplidyne armature, in which case the voltage of the amplidyne will not vary with the load. However, in the case of extreme overcompensation, undesirable self excitation of the amplidyne *i.e.*, spontaneous rise in voltage and load current at constant control current (or with the control circuit open) can occur. Such self-excitation (similar to that of generator) can continue until saturation of the amplidyne magnetic system becomes effective. Self excitation can cause considerable overloading of the amplidyne and totally disrupt operation of the system containing the amplidyne.

In partial compensated machine the opposing flux produced by the output current I_L would neutralize most of the control flux. In control engineering terms, it is a negative current feedback tending to suppress any increase of output current. The action is sufficient to maintain the output current almost constant regardless of change in load resistance. The machine operated in this way is called the Metadyne. In such cases electrical input and output powers are the same, neglecting copper losses and the only mechanical power that is required to overcome the mechanical losses. The output current is controlled by the current through control windings. The output characteristics of Metadyne with different control currents are shown in Fig. 2.6(*b*).

Note : From the characteristics shown in Figs. 2.6(*a*) and 2.6(*b*) for amplidyne and metadyne respectively it is obvious that an amplidyne should never be operated on short circuit and a metadyne on open circuit.

Metadyne Generator Rating. The rated output power of a metadyne is appreciably less than that obtainable from the same armature in a normal generator. First; the output current has to be reduced by about 10 per cent, owing to the heating caused by the quadrature current I_1 flowing in the armature. Secondly the provision of the commutating zones for two sets of brushes reduces the effective pole-face area, and hence the total flux and terminal voltage by about one-third. The rated output power of the armature is thus about $0.9 \times \frac{2}{3} = 0.6$ of that of the same armature in an ordinary dc generator. For this reason it is not economical to use metadynes for large output power, and, above 15 to 20 kW separately excited dc generators are employed with metadynes as exciters to provide similar results as from a metadyne generator.

Applications of Amplidynes. An amplidyne is widely used for various purposes, particularly in feedback control systems where the regulated power requirement is in the range of 1 to 50 kW. It may sometimes be found economical to employ it as a voltage regulator in the excitation system of a large alternator to buck or boost voltage in series with the field winding of the main exciter. It can also be employed as a generator in a Ward-Leonard system of speed or position control. In modern control systems these functions of an amplidyne have been taken over by SCR circuitry.

Amplidynes are employed in installations for controlling reversing rolling mills, cold rolling mills, mine hoists, metal cutting and paper machines. However, amplidynes posses a number of shortcomings: poor commutation, which sometimes disrupts the operating conditions of the installation, and a low efficiency of operation. There is a trend in industry to replace amplidynes by static devices, such as magnetic amplifiers operating in conjunction with transistor amplifiers.

2.2.4. Magnetic Amplifiers

In a magnetic amplifier, the basic component is a steel-cored coil with an additional winding energized by direct current. It is intended for controlling alternating current of relatively large power by means of a low power dc.

The simplest practical magnetic amplifier has two identical steel cores with identical ac windings, called the *load winding,* on each. In addition they have a common dc winding called the *control winding.* The ac windings are connected together in series or in parallel, and in series with them there is a load. The series connection is used when a short-time response and a high voltage are required, usually in controlling loads requiring low power. The parallel connection, on the other hand, has a slow response of approximate 1 to 3 seconds, but permits control of large current. The control winding is energized by dc.

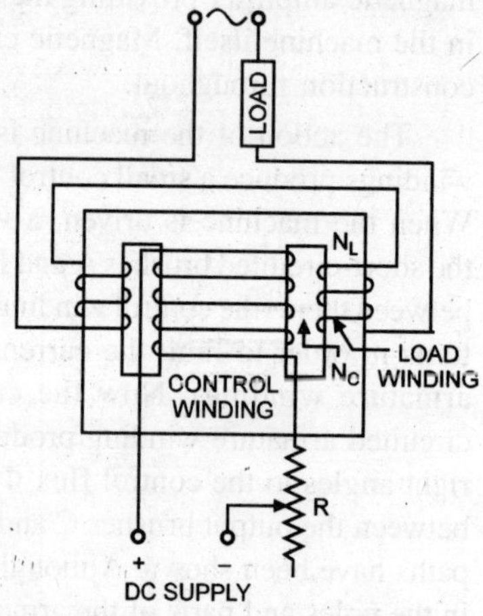

Fig. 2.7. ***Simplest Magnetic Amplifier Circuit***

With no current in control winding, the ac winding has a large inductance, and a high reactance is offered to an ac potential. As a result alternating current to the load is limited to a low value by the inductive reactance, and the load voltage is small. When a dc voltage is now applied to control winding the dc magnetic flux passes through the core and the core approaches magnetic saturation, even when a small value of dc voltage is applied. This causes the effective inductance of the ac winding to decrease, so that the ac winding impedance drops. Consequently the alternating current flowing through the ac winding (and the load voltage) increases as the flow of direct current through the control winding increases. Thus, varying the magnitude of control current over a small range causes the load voltage to vary over a wide range.

If a magnetic amplifier consisted only of one core with two windings (ac and control), the load current would be noticeably non-sinusoidal. This is one of the reasons why the simplest practical magnetic amplifier is made double. A second reason is that for one core, the control winding would be linked with an alternating magnetic flux, and an alternating emf would be induced in it. This emf could become very significant for a large value of the ratio $\frac{N_C}{N_L}$. With two cores, the total flux linking the control winding does not have an alternating component and, hence, no alternating emf. is induced in it. At present there are a large number of different magnetic amplifier circuits and types of construction.

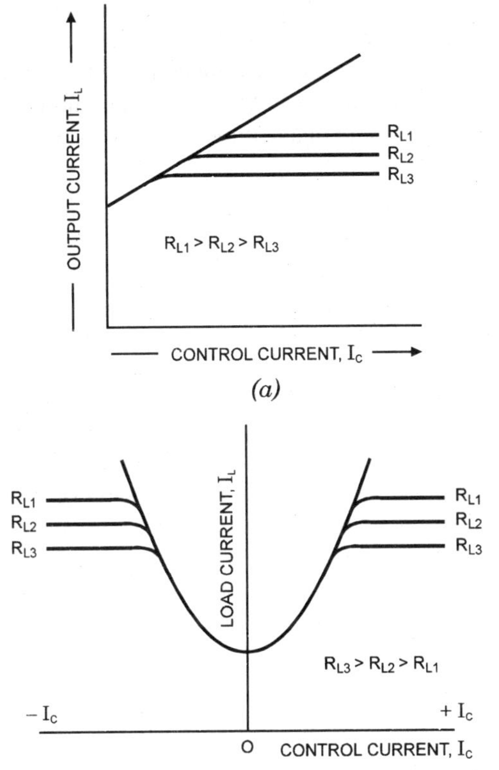

(a)

(b)

Fig. 2.8. ***Dynamic Characteristic of Non-Polarised Magnetic Amplifier***

The output or transfer characteristic curves of a magnetic amplifier are shown in Fig. 2.8 for various load resistances. From the curves it is obvious that with the exception of a small region near the origin the load current is directly proportional to the control current up to a value determined by the load resistance, and secondly that the characteristic curve does not begin from the origin but intersects the ordinate axis above the origin. This is due to the current flowing in ac winding which is determined by the applied voltage and the inductive reactance of the winding. Thus for zero control current, load current, I_L will be equal to $\frac{E}{X_L}$, known as *quiescent current.*

Magnetic amplifiers have got following advantages over electronic ones :

1. Mechanically sturdy and inherently shockproof.
2. Relatively inexpensive amplifiers can be designed to have very high gain, with low input power requirements, and to be compact and light in weight.
3. It does not have any filament to be heated as is in the case of vacuum tube amplifiers.
4. It can be designed for a wide range of input and output impedances and since the control and load circuits are not conductively coupled, they provide somewhat more flexibility than vacuum tube circuits.

On the other hand magnetic amplifiers have limited frequency response and produce an output that is a modulated ac carrier. They have a much lower input impedance than vacuum tubes and provide only a finite power gain where as vacuum tube may theoretically at least, give an infinite power gain. Other disadvantage is that magnetic amplifier, in general, is required to be designed specifically for each application, since both the supply voltage and the load impedance critically affect the operation.

Magnetic amplifiers are used in electrical systems of control, measurement and automatic regulation as power amplifiers (for example for automatic control of ac and dc motors or in smooth regulation of lighting).

The schematic diagram of a magnetic amplifier control circuit for squirrel cage motor starting as a function of voltage is shown in Fig. 2.9.

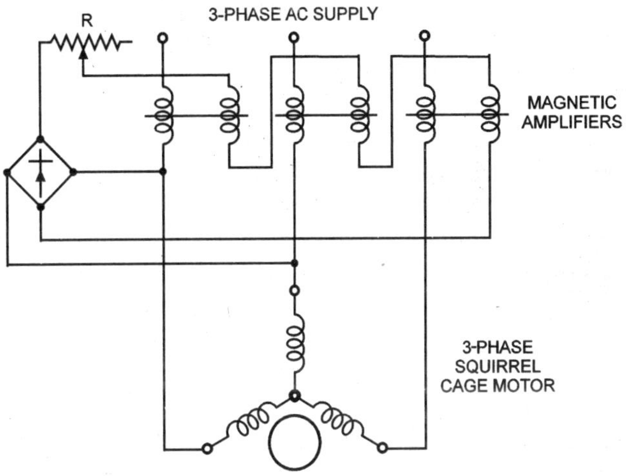

Fig. 2.9. ***Magnetic Amplifier Control Circuit For Squirrel Cage Motor Starting***

When the ac supply is applied to the motor for starting, the load windings of the magnetic amplifier have a high inductive reactance and only a reduced voltage acts across the stator terminals of the motor. As the motor begins to gain speed, the starting current drops and allows the voltage at the stator terminals to rise. Two of the stator winding terminals are connected to a semiconductor bridge rectifier which serves to excite the control windings of all the three magnetic amplifiers. As the voltage across the stator terminals rises during starting, the direct current through the control windings also rises. As a result, the inductive reactance of the ac windings is reduced and the voltage at the stator terminals rises. The control winding current, due to rise in stator terminal voltage, increases until it finally saturates the amplifiers. By this time the inductive reactance of the ac winding drops practically to zero and the motor operates at full voltage.

The character of motor current and torque change with time during starting may be adjusted by setting up a suitable ratio of the electromagnetic time constant of the magnetic amplifier control winding circuit to the electromechanical time constant of the drive.

2.3. SOLID STATE DEVICES*

Solid state devices find more and more applications in industry and, therefore, it becomes imperative to have preliminary knowledge of such devices. The main devices usually used for motor control are diode, thyristor or silicon controlled rectifier, triac, diac, power transistors etc. Every device has certain maximum voltage, current and time response (speed) capabilities. These can very rarely be achieved simultaneously in a single device. Only one of these is realized in one device.

2.3.1. Diode

'The term *diode* denotes a two-electrode device. A *semiconductor diode* is simply a P-N junction with connecting leads or terminals on the two sides of the P-N junction. A diode is a unidirectional device permitting the easy flow of current in one direction but restraining the flow in opposite direction. A major application of diodes is in *rectification i.e.,* in conversion of ac into dc. Semiconductor diode is becoming more and more popular these days due to its smaller size, cheapness, robustness and higher operating efficiency.

The symbol for a semiconductor diode is shown in Fig. 2.10. The arrow in the symbol indicates the direction of conventional current flow when the diode is forward biased *i.e.,* from the positive terminal through the device to the negative terminal. The P-side of the diode is always the positive terminal for forward bias and is designated the *anode*. The N-side is called the *cathode* and is the negative terminal when the device is forward biased.

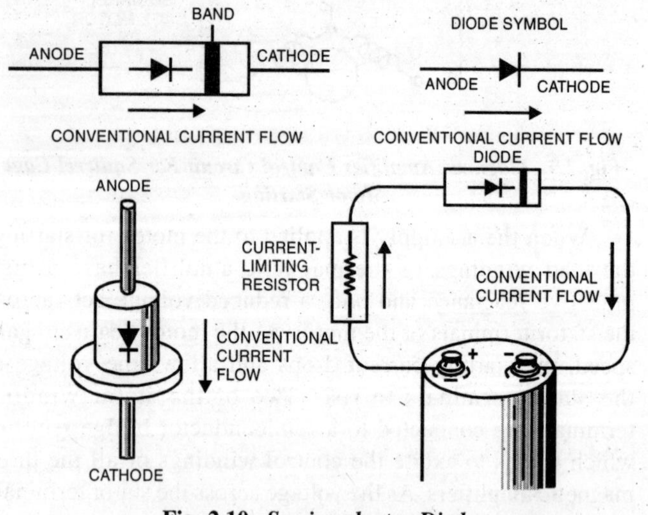

Fig. 2.10. ***Semiconductor Diode***

The semiconductor diode may be either silicon one or germanium one. The silicon junction diode is similar in appearance to the germanium diode but differs in properties. Silicon diodes have, in general, higher peak inverse voltage (PIV) and current ratings and wider temperature ranges than germanium diode.

The volt-ampere characteristic of a diode are illustrated in Fig. 2.11. When the junction is in forward biased very little current, called the *forward current,* flows until the forward voltage exceeds the junction barrier potential (0.3 V for Ge and 0.7 V for Si). The characteristic follows an exponential law. With the increase in forward voltage forward current increases almost linearly and the P-N junction starts behaving as a resistor and when the applied voltage exceeds a certain value, extremely large current would flow and the P-N junction may get damaged due to overheating. A practical diode has a very low forward resistance (of the order of 75 Ω in case of Ge and 150 Ω in case of silicon) and has a voltage drop of 0.3 to 0.7 V at all current levels in forward direction.

When an ordinary P-N junction diode is reverse biased, normally only very small reverse saturation current flows. It is almost independent of the voltage applied. However, if the reverse bias is increased, a point is reached when the junction breaks down and the reverse current increases abruptly, as shown in Fig. 2.11. This current could be large enough to destroy the junction. If the reverse current is limited by means of a suitable series resistor, the power dissipation at the junction will not be excessive, and the device may be operated continuously in its breakdown region to its normal (reverse saturation) level. It is found that for a suitably designed diode, the breakdown voltage is very stable over a wide range of reverse currents. This quality gives the *breakdown diode* many useful applications as a *voltage reference source.*

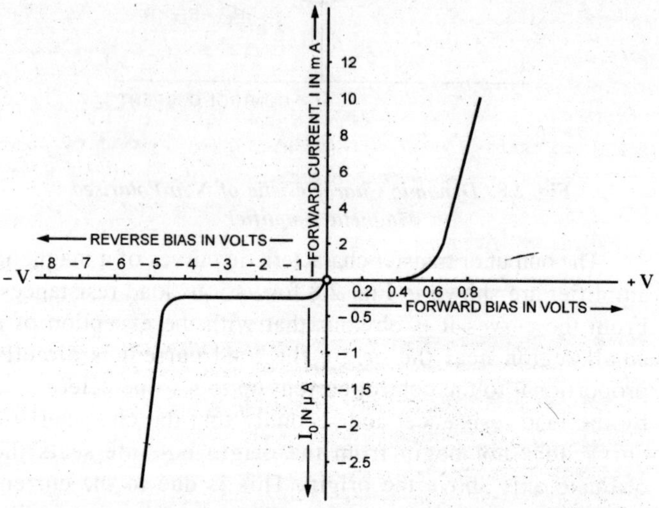

Fig. 2.11. ***Volt-Ampere Characteristics of Semiconductor Diode***

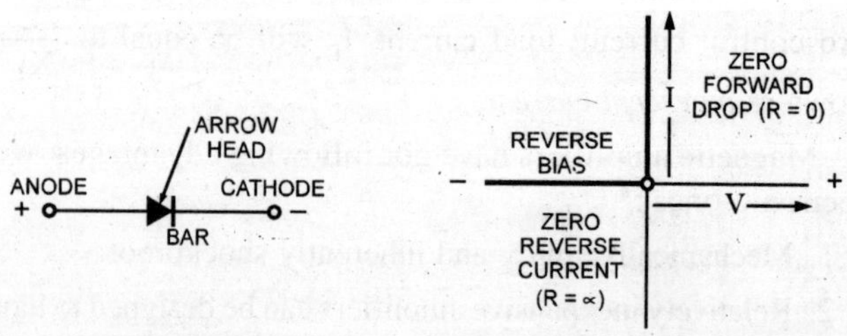

(a) Ideal Diode *(b) Ideal Diode Characteristics*

Fig. 2.12

An *ideal diode* presents zero impedance to current flow in one direction and infinite impedance in opposite direction. The symbol and V-I characteristic of an ideal diode are shown in Fig. 2.12. An ideal diode acts like an *automatic switch.*

* For details refer Electronic Devices and Circuits by J.B. Gupta or any other standard book on the subject.

The switch is closed when the diode is forward biased and is opened when reverse biased.

The main parameters of a diode are blocking or peak inverse voltage or PIV (about 1,000 V for Si and 400 V for Ge), average forward current and maximum operating junction temperature (nearly 125°C for Si and 60° for Ge). Protection is necessarily provided against voltage surges, overcurrents and excessive temperature rise.

A "freewheeling" diode is commonly employed in motor control systems to provide an alternate path for continuity of current in the inductive circuit following the switching off of some power device between the motor and the supply source.

Zener Diode. Zener diode, also sometimes called the *breakdown diode,* is a P-N junction diode specially designed for operation in the breakdown region in reverse bias condition. The breakdown diode may be silicon or germanium one but silicon is preferred over germanium because of higher operating temperature and current capability. The knee point is also more sharp in case of a silicon diode. These diodes are analogous to gas diodes in which large current appears on reaching the breakdown potential.

The symbol of a zener diode is given in Fig. 2.13. This is similar to that of a normal diode except that the line representing the cathode is bent at both ends *i.e.,* the bar is turned into Z-shape; stands for zener.

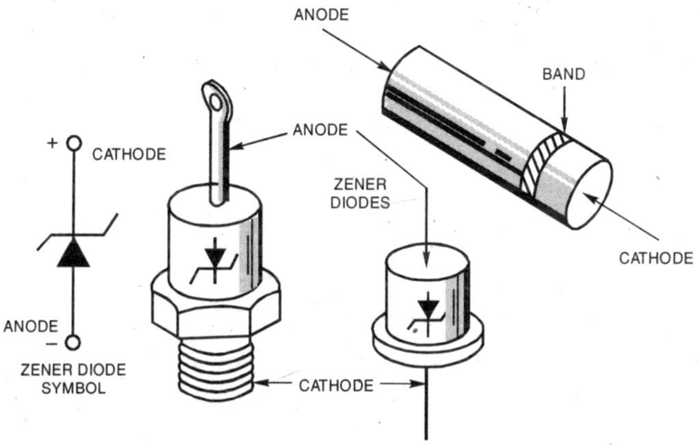

Fig. 2.13. ***Zener Diode***

The zener diode, though not a true power controlling device, is quite often used as a voltage control and sensing device in many motor controllers.

Power Diode. Diodes find many applications in electronics and electrical engineering circuits. Power diodes play a significant role in power electronics for conversion of electric power. A diode acts as a switch to perform various functions, such as switching in rectifiers, freewheeling in switching regulators, charge reversal of capacitor and energy transfer between components, voltage isolation, energy feedback from load to the power source, and trapped energy recovery.

Power diodes can be assumed as ideal switches for most applications but practical diodes differ from the ideal characteristics and have certain limitations. The power diodes are similar to P-N junction signal or low-power diodes. However, the power diodes have large power–, voltage–, and current-handling capabilities than those of ordinary diodes. The frequency response (or switching speed) is low in comparison to that of ordinary diodes.

Power diodes are of three types: General purpose, high speed (or fast-recovery), and Schottky. General-purpose diodes are available up to 6000 V, 4500 A, and the rating of fast-recovery diodes can go up to 6000 V, 1100 A. The reverse recovery time varies between 0.1 and 5 μs. The fast-recovery diodes are essentially required for high-frequency switching of power converters. Schottky diodes have low on-state voltage and very small recovery time, typically nanoseconds. The leakage current increases with the voltage rating and their ratings are limited to 100 V, 300 A. A diode conducts when its anode voltage is higher than that of the cathode; and the forward voltage drop of a power diode is very small, typically 0.5 and 1.2 V.

2.3.2. Thyristor

Thyristor is the general name given to a family of semiconductor devices having four layers with a control mechanism, although this term is most commonly applied to the SCR (silicon controlled rectifier). This term is derived from *thyratron* and *transistor* because the device combines the rectification action of thyratron and control action of transistor.

Since its inception the thyristor has come to stay as a basic building block in many industrial and power system applications. Its ability to be controlled, compactness, fast response, high reliability, better efficiency, large power handling capacity, high voltage and current ratings, good trigger sensitivity, static operation, large power gain, sturdy construction, long life, very little maintenance and low cost of fabrication—due to advancement in the field of fabrication—have given the thyristor a colourful reception in every field. Today thyristors are finding applications in the control of dc/ac motors; for the improvement of power factor; and as switching devices. Thyristors have helped in further cost reduction and in the development of drive system by changing the emphasis from dc motors to ac motors. With cycle converters and inverters, the speed of an ac motor can also be controlled with ease and reliability. Apart from these main applications it finds use as a switching device, particularly in the improvement of power factor of transmission lines and mains. Thyristor can be used as a power switching device with a power handling capacity ranging from a few watts to as high as 4 MW (2,500 A at 1,600 V). Some thyristors have a rating as high as 35,000 A, 10,000 V for use in hv dc transmission lines.

Thyristors, with its large number of advantages and tremendous control capabilities, have numerous applications and have completely replaced the electromagnetic control systems. Thyristor basically serves two functions viz electronic switching and electronic control. Some of the applications of thyristors are listed below:

1. Speed control of dc and ac motors.
2. As rectifier for conversion of ac into dc.
3. As inverter for conversion of dc into ac.
4. As dc chopper or dc to dc converter for converting dc at one level to dc at another level.
5. As cycloconverter for converting ac of one frequency into ac of another frequency.
6. Control of temperature, level, position and illumination.
7. Power switches (dc and ac circuit breakers).

8. As static switches.
9. Control of induction heating.
10. Relay control.
11. Phase control.

SCRs and triacs having high voltage and current ratings are widely employed for power control applications whereas other members of thyristor family are employed for small power applications and for switching in control and digital circuits.

2.3.3. Silicon Controlled Rectifier (SCR)

The SCR is a four-layer, three-junction and a three-terminal device and is shown in Fig. 2.14 (*a*). The end P-region is the anode, the end N-region is the cathode and the inner P-region is the gate. The anode to cathode is connected in series with the load circuit. Essentially the device is a switch. Ideally it remains off (voltage blocking state), or appears to have an infinite impedance until both the anode and gate terminals have suitable positive voltages with respect to the cathode terminal. The SCR then *switches on* and current flows and continues to conduct without further gate signals. Ideally the SCR has zero impedance in conduction state. For switching off or reverting to the blocking state, there must be no gate signal and the anode current must be reduced to zero. Current can flow only in one direction. Now SCRs of voltage rating 10 kV and an rms current rating of 35,000 A with speed of 1 μs corresponding to a power-handling capacity of 30 MW are available. It can be switched on by a low voltage supply of about 1 A and 10 W. Obviously it has tremendous power amplification (of the order of 3×10^6). Because of its compactness, high reliability and low loss, the SCR has almost replaced the earlier power switching devices—thyratron and a magnetic amplifier.

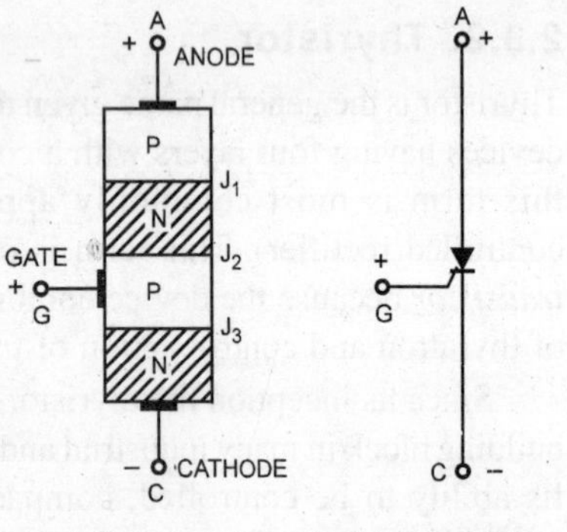

(*a*) Schematic Diagram (*b*) Symbolic Representation

Fig. 2.14. ***Silicon Controlled Rectifier***

Volt-Ampere Characteristics. As mentioned above, the SCR is a four layer device with three terminals, namely, the anode, the cathode and the gate. When the anode is made positive with respect to the cathode, junctions J_1 and J_3 are forward biased and junction J_2 is reverse biased and only the leakage current will flow through the device. The SCR is then said to be in the *forward blocking state* or in the *forward mode* or *off state*. But when the cathode is made positive with respect to the anode, junctions J_1 and J_3 are reverse biased, a small reverse leakage current will flow through the SCR and the SCR is said to be in the *reverse blocking state* or *in reverse mode*.

When the anode is positive with respect to cathode *i.e.,* when the SCR is in forward mode, the SCR does not conduct unless the forward voltage exceeds certain value, called the *forward breakover voltage*, V_{FBO}. In non-conducting state, the current through the SCR is the leakage current which is very small and is negligible. If a positive gate current is supplied, the SCR can become conducting at a voltage much lesser than forward breakover voltage. The larger the gate current, lower the breakover voltage. [Fig. 2.15(*c*)] With sufficiently large gate

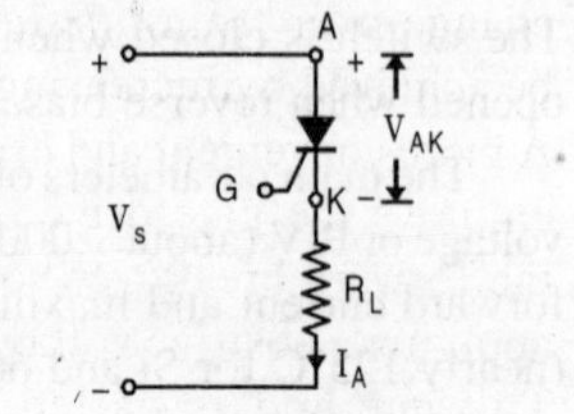

(*a*) Circuit Diagram

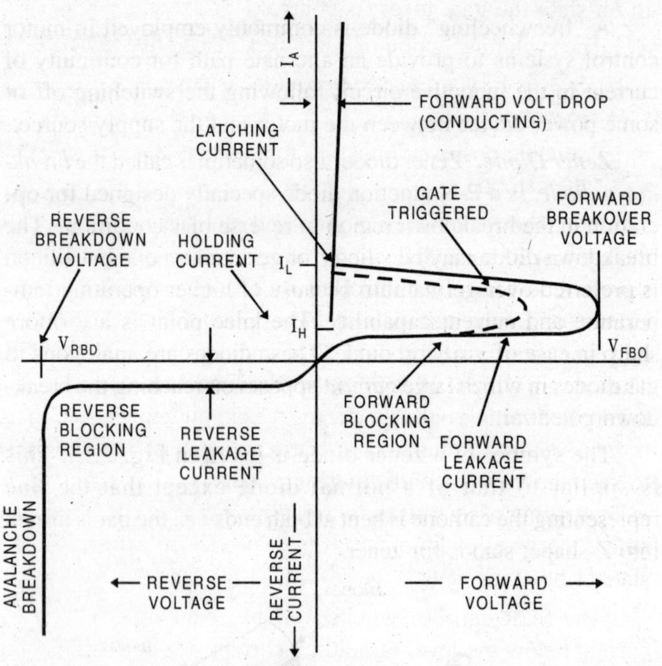

(*b*) V-I Characteristics

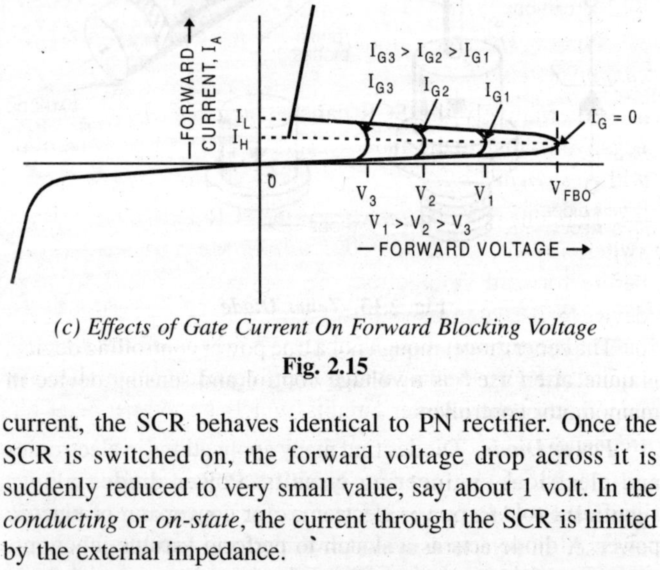

(*c*) Effects of Gate Current On Forward Blocking Voltage

Fig. 2.15

current, the SCR behaves identical to PN rectifier. Once the SCR is switched on, the forward voltage drop across it is suddenly reduced to very small value, say about 1 volt. In the *conducting* or *on-state*, the current through the SCR is limited by the external impedance.

When the anode is negative with respect to cathode *i.e.,* when the SCR is in reverse mode or in blocking state no current flows through the SCR except very small leakage current of the order of few microamperes. But if the reverse voltage is increased beyond a certain value, called the *reverse breakover voltage*, V_{RBO} *avalanche breakdown* takes place. Forward breakover voltage V_{FBO} is usually higher than reverse breakover voltage, V_{RBO}.

From the foregoing discussion, it can be seen that the SCR has two stable and reversible operating states. The change over from *off-state* to *on-state*, called *turn-on*, can be achieved by increasing the forward voltage beyond V_{FBO}. A more convenient and useful method of turning on the device employs the gate drive. If the forward voltage is less than the forward breakover voltage, V_{FBO}, it can be turned on by applying a positive voltage between the gate and the cathode. This method is called the *gate control*. Another very important feature of the gate is that once the SCR is triggered to on-state the gate loses its control.

The switching action of gate takes place only when

(*i*) SCR is forward biased *i.e.*, anode is positive with respect to cathode, and

(*ii*) suitable positive voltage is applied between the gate and the cathode.

Once the SCR has been switched on, it has no control on the amount of current flowing through it. The current through the SCR is entirely controlled by the external impedance connected in the circuit and the applied voltage. There is, however, a very small, about 1 V, potential drop across the SCR. The forward current through the SCR can be reduced by reducing the applied voltage or by increasing the circuit impedance. There is, however, a minimum forward current that must be maintained to keep the SCR in conducting state. This is called the *holding current* rating of SCR. If the current through the SCR is reduced below the level of holding current, the device returns to off-state or blocking state.

The SCR can be switched off by reducing the forward current below the level of holding current which may be done either by reducing the applied voltage or by increasing the circuit impedance.

Note : *The gate can only trigger or switch on the SCR, it cannot switch off.*

Alternatively the SCR can be *switched off* by applying negative voltage to the anode (reverse mode), the SCR naturally will be *switched off*.

Here one point is *noteworthy*, the SCR takes certain time to switch off. The time, called the *turn-off time*, must be allowed before forward voltage may be applied again otherwise the device will switch on with forward voltage without any gate pulse. The turn-off time is about 15 microseconds, which is immaterial when dealing with power frequency, but this becomes important in the inverter circuits, which are to operate at high frequency.

There are mainly two types of SCRs for use in motor controllers—*inverter grade* and *converter grade*. Former ones are used in inverters, cycloconverters, and brushless dc motor systems while the latters are used in choppers, phase-controlled rectifiers, ac regulators etc. These two grades vary in their time responses. Inverter grade SCRs are costlier than the converter grade.

Typical SCR Parameters. Commercially available SCRs have a wide range of voltage and current ratings, turn-on and turn-off times and dynamic resistances in cut-off and in saturation regions. Typical values of these parameters for SCRs are given below in tabular form.

Table 2.1. Typical SCR Parameters

S. No.	Parameters	Typical values
1	Forward Breakover voltage, V_{FBO}	50 to 500 volts
2	Maximum on-state voltage	About 1.5 V
3	Peak reverse voltage, PRV	Up to 2.5 kV
4	Holding voltage, V_H	0.5 to 20 volts
5	Forward breakover current	Less than a few hundred μA
6	Peak forward current	30 A to over 100 A
7	Holding current	A few mA to few hundred mA
8	Turn-on and turn-off times	A few tenths of μs for fast acting SCRs
		A few μs for slow acting SCRs
9	Dynamic resistance in cut-off region	A few MΩ to a few hundred MΩ
10	Dynamic resistance in saturation region	Lesser than 1 Ω for currents of several amperes Lesser than 10 Ω for large currents

2.3.4. Triac

The triac is another three-terminal ac switch that is triggered into conduction when a low-energy signal is applied to its gate terminal. Unlike the SCR, the triac conducts in either direction when turned on. The triac also differs from the SCR in that either a positive or negative gate signal triggers it into conduction. Thus the triac is a three terminal, four layer bidirectional semiconductor device that controls ac power whereas an SCR controls dc power or forward biased half cycles of ac in a load. Because of its bidirectional conduction property, the triac is widely used in the field of power electronics for control purposes. Triacs of 16 kW rating are readily available in the market.

"Triac" is an abbreviation for three terminal ac switch. 'Tri'- indicates that the device has three terminals and 'ac' indicates that the device controls alternating current or can conduct in either direction.

Though the triac can be turned *on* without any gate current provided the supply voltage becomes equal to the breakover voltage of the triac but the normal way to turn on the triac is by applying a proper gate current. As in case of SCR, here too, the larger the gate current, the smaller the supply voltage at which the triac is turned *on*.

Triac can conduct current irrespective of the voltage polarities of terminals MT_1 and MT_2 with respect to each other and that of gate and terminal MT_2.

Typical V-I characteristic of a triac is shown in Fig. 2.16 (*b*). The triac has on- and off-state characteristics similar to SCR but now the characteristic is applicable to both positive and negative voltages. This is expected because triac consists of two SCRs connected in parallel but opposite in directions.

Next to SCR, the triac is the most widely used member of the thyristor family. In fact, in many of control applications, it has replaced SCR by virtue of its bidirectional conductivity. Motor speed regulation, temperature control, illumination control, liquid level control, phase control circuits, power switches etc. are some of its main applications.

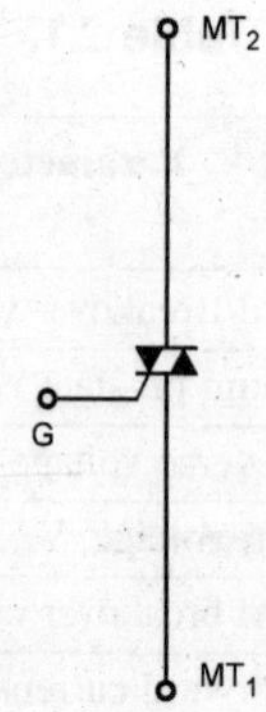

(a) *Schematic Symbol*

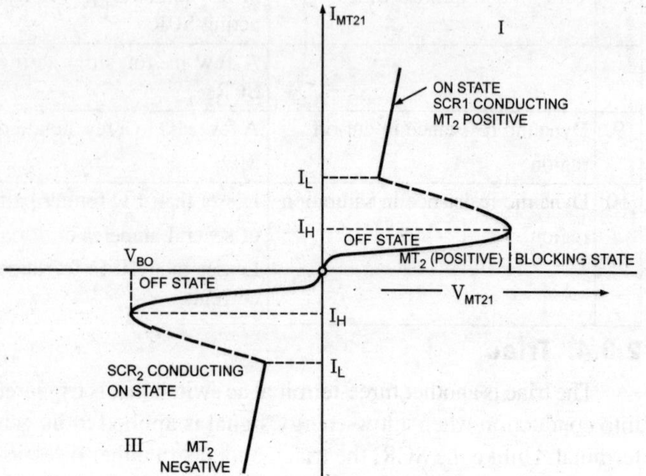

Fig. 2.16. *Triac*

2.3.5. Diac

A diac is an important member of the thyristor family and is usually employed for triggering triacs. A diac is a two-electrode bidirectional avalanche diode which can be switched from off-state to the on-state for either polarity of the applied voltage. This is just like a triac without gate terminal. Its equivalent circuit is a pair of inverted four layer diodes. Two schematic symbols are shown in Fig. 2.17 (*a*). Again the terminal designations are arbitrary since the diac, like triac, is also a bilateral device. The switching from off-state to on-state is achieved by simply exceeding the avalanche breakdown voltage in either direction. Volt-ampere characteristic is shown in Fig. 2.17 (*b*).

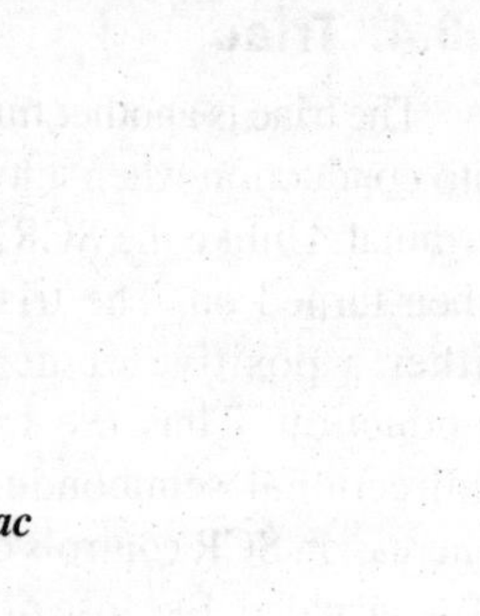

(a) *Schematic Symbols*

The diacs, because of their symmetrical bidirectional switching characteristics, are widely used as triggering devices in triac phase control circuits used in heat control, universal motor speed control etc.

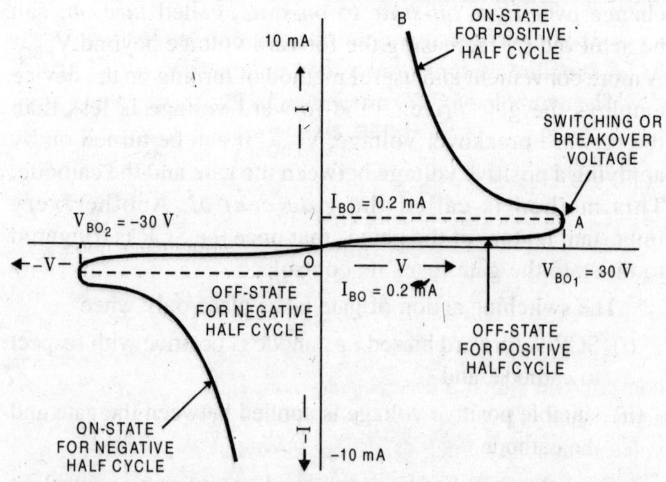

(b) *V-I Characteristics*

Fig. 2.17. *Diac*

2.3.6. Gate Turn-Off (GTO) Thyristors

Conventional thyristors are used as ideal switches in power electronic applications. They have the capability of blocking several thousand volts in the *off*-state and conducting several thousand amperes in the *on*-state with only small *on*-state voltage drop. The only drawback of conventional thyristor is that when the gate turns into *on*-state, the gate loses the control over the device and, therefore, it cannot be turned *off* without the technique called commutation. The circuitry required for commutation makes the device costlier and bulky. This can be avoided by using a device of thyristor family, known as the gate turn-off thyristor (GTO).

A GTO like an SCR can be *turned on* by applying a positive gate signal. However, a GTO can be *turned off* by a negative gate signal. A GTO is a non-latching device and can be built with current and voltage ratings similar to those of an SCR. The device incorporates the character of both of a thyristor and a transistor.

The GTOs have the advantages over SCRs : (*i*) elimination of commutating components in forced commutation, resulting in reduction in cost, weight and volume (*ii*) faster turn-off permitting high switching frequencies (*iii*) improved efficiency of converters (*iv*) reduction in accoustic and electromagnetic noise due to elimination of commutation chokes (*v*) high blocking voltage and large current capability.

In low-power applications, GTOs have the following advantages over BJTs :

1. Higher blocking voltage capability.
2. High ratio of peak controllable current to average current.
3. High ratio of peak surge current to average current, typically 10 : 1.
4. High on-state gain (anode current and gate current) typically 600.
5. Pulsed gate signal of short duration.

Under surge conditions, a GTO goes into deeper saturation due to regenerative action. On the other hand, a BJT tends to come out of saturation.

However, the GTO has the following disadvantages :

1. High latching and holding currents.
2. More on-state voltage drop and associated loss.

3. Higher triggering gate current due to multi-cathode structure as compared to conventional thyristor.

The symbols of GTO are shown in Fig. 2.18.

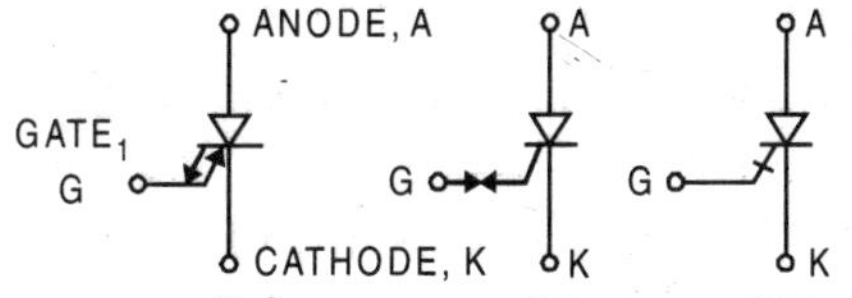

Fig. 2.18. ***Circuit Symbols For GTO***

Applications : GTOs are mostly used in voltage-source converters in which a fast recovery antiparallel diode is required across each GTO. Thus, GTOs normally do not require reverse voltage capability. Such GTOs are known as *asymmetric* GTOs. This is achieved by a so-called buffer layer, a heavily doped N^+ layer at the end of the N layer. Asymmetric GTOs have lower voltage drop and higher voltage and current ratings.

GTO devices are now used in (*i*) high performance drive systems, such as the field-oriented control scheme used in rolling mills, robotics and machine tools, (*ii*) traction purposes because of their lighter weight and (*iii*) adjustable-frequency inverter drives. At present, GTOs with ratings up to 5,000 V and 3,000 A are available.

2.3.7. Power Transistors

The transistor is a two junction, three terminal, current controlled solid state device. This device is a 3-layer P-N-P (or N-P-N) structure as illustrated in Fig. 2.19. The transistor is capable of amplification and in most respect it is analogous to a vacuum triode. The transistor is only six decade old, yet it has replaced vacuum tubes in almost all applications. The reasons are obviously its advantages over vacuum tubes such as compact size, light weight, rugged construction more resistive to shocks and vibrations, instantaneous operation (no heating required), low operating voltage, high operating efficiency and long life with essentially no ageing effect if operated within permissible limits of temperature and frequency.

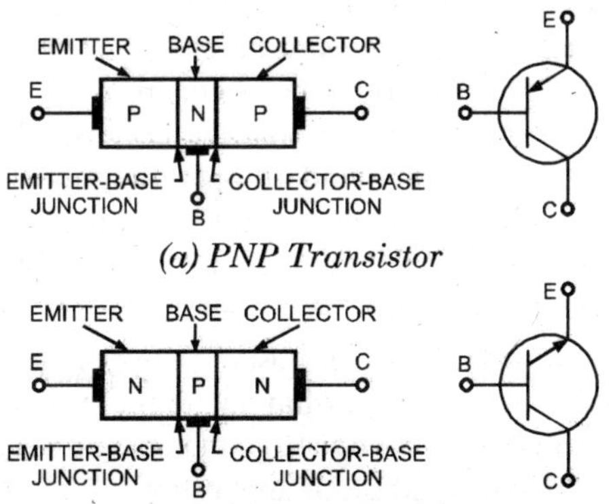

(*a*) *PNP Transistor*

(*b*) *NPN Transistor*

Fig. 2.19

In normal operation the emitter-base junction is forward biased and collector-base junction is reverse biased. The collector current is almost equal but slightly less than the emitter current. The collector current can be controlled by base current. Except for small motors, transistors are used in the switching (on/off) mode. The common-emitter configuration is normally employed as it has high power gain.

Power transistors are now available in ratings suitable for motor control. The common-emitter saturation voltage $V_{CE\,(Sat)}$ for typical power transistors varies from 0.3 to 0.7 V. This is sufficiently less than the on-state anode-cathode voltage drop of an SCR, and therefore, the average power loss in a transistor is less than that of an SCR of an equivalent power rating.

The switching of a power transistor is controlled through the base current. A continuous base current is required to keep the transistor in the saturated state, *i.e.*, the on-state. In a high-current device, a base current of several amperes is required to keep the device in the on-state owing to its low current gain (about 10) and, therefore, the power loss in the base circuit may be appreciable.

Power transistors have no surge current capacity and can withstand only a low rate of change of current. The switching operations of power transistors are generally faster than those of SCRs, and the commutation problems of SCR are not present in transistors. Power transistors are becoming more and more popular in low-to-medium-power applications, where they compete successfully with thyristors and GTOs. Power transistor applications are in the range of a few kW to several hundred kWs size in voltage-fed choppers and inverters with switching frequency up to 10-15 kHz.

2.4. PILOT DEVICES AND ACCESSORIES

Motor control devices may be classified as primary control devices and pilot control devices. Pilot devices control the primary control devices which in turn controls (connects or disconnects) the motor or load. For example in D-O-L (direct-on-line) starter main contactors and tripple pole switch, which connect the motor to the line are classified as *primary control devices* where as start or stop push buttons which energizes or de-energizes the operating coil are called the *pilot devices.* Pilot devices usually have sensing element to sense temperature, liquid level, fluid pressure or mechanical pressure applied etc. Pilot devices convert what they sense into the control of primary control devices. The most important of pilot devices are push buttons, master switches, float switches, thermostats, pressure regulators, plugging switches, limit switches.

1. *Push Button.* This is the most simple pilot device and provides control by pressing a button. The push-button switches are used in remotely operated electromagnetic apparatus and signalling circuits. Push buttons are usually of two types namely *maintained contact type* in which on pushing of the button the contacts close and remain closed until stop button is pushed and the *momentary contact type* in which contacts remain closed till push button is held pushed. The contacts of push buttons are usually double break, actuated by plastic buttons. Two sets of contacts are usually provided namely normally open contacts 'NOC' and normally closed contacts 'NCC.' so that when the button is

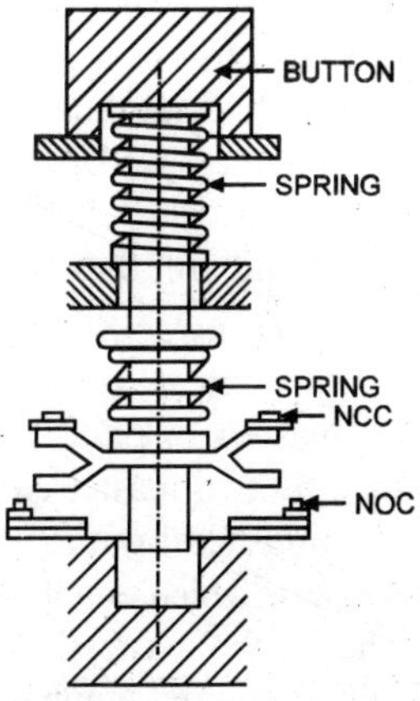

Fig. 2.20. ***Push Button***

pressed one set is opened and the other set is closed. The push button station usually consists of a green button to start and red button to stop. Start push button is provided with normally open contact (NOC) and stop push button is provided with normally closed contact (NCC). The contacts are made silver tipped to withstand numerous of switching operations. Since push buttons are used in control circuits, not used in power circuits, current rating of push button station is, therefore, usually very small (2 A, 500 V for ac working). On pressing start push-button contactor coil is energized, which closes the main contactor in the circuit.

2. *Limit Switches.* Limit switches are ordinarily used as pilot devices in control circuits of magnetic starters to govern starting, stopping, or reversing of electric motors. These may be used either as control devices for regular operation or as emergency switches to prevent improper functioning of machinery. These are actuated by the travel of the driven machinery and automatically disconnect the machine after it has travelled through its entire path. These switches can be actuated by cams, levers, push rods and the like, and also through gears. Lever type actuators may be either of normal roller type or angular type. The switch is either stationary and actuated by some moving member or it moves with the machine and is actuated by a stationary member. These are usually provided with one NOC (normally open contact) and one NCC (normally closed contact). These switches are usually air break type. These are also used for opening and closing of doors, cinema curtains, lifts and elevators etc.

3. *Master Controllers.* A device or a group of devices that governs in a predetermined manner the supply of electric power to an electric apparatus with a view to accelerate, decelerate, start, stop or reverse it according to requirement is called the *controller.* The device performing the above functions in respect of many units at a time is called the *master controller.* Master controllers can employ different methods for closing or opening the contactors namely all-electric method, electropneumatic method and the camshaft method. In the all-electric method, the contactors are operated by solenoids and the drawback of this method is that supply variations may cause uncertainty of action of the contactors. In the electropneumatic method solenoids operate the valves that admit air into cylinders in which movable piston close or open the contactors and in the cam shaft method, the contactors are operated by cams mounted on a shaft driven by an electric or a pneumatic motor. In electropneumatic or in cam shaft method large mechanical forces can be put on the contactors, so that vibration can cause no harm : moreover the systems are electrically simple and mechanically very robust and are, therefore, favoured.

4. *Transfer Switches.* Sometimes operation and control of an electrical equipment becomes necessary from either of two masters of pilot devices, such as control of multiple unit stock by two master controllers at the ends of a train. This is accomplished by making use of switches, called the *transfer switches,* which enable electrical apparatus to be connected to either of the masters. These are merely double throw switches having required number of poles.

5. *Speed Governors.* These are employed for protection of rotating machinery against overspeed, under-speed or to perform some functions at a given speed of the machine. These are actuated by centrifugal force.

6. *Float Switches.* This is a type of limit switch and is actuated by a float through a chain, one end of which is connected to the float and the other to the counterweight. Stops connected to the chain move a lever which turns and tilts the mercury tube forming a switch contact. These are designed and used for maintaining a certain level of liquid. The raising or lowering of a float mechanically attached to electrical contacts may start motor driven pump to empty or to fill tanks, whichever is desired. They are also used for opening or closing piping valves for the control of fluids.

7. *Pressure Switches and Regulators.* The pressure switches and regulators are widely used in industry. These are designed to cover the wide range of requirements encountered in the control of pneumatic or hydraulic machines. These controls are most commonly used in the control circuits of welding equipment, high pressure lubricating system, machine tools and motor driven pumps and compressors. These are available in various pressure ranges. A diaphragm actuated device is designed for applications requiring sensitive response to small pressure changes at low pressure ranges. A seamless metal-bellows actuated control is used for pressures as high as 1.5 kg/mm^2 and the piston operated hydraulic switches are used for pressures up to 10 kg/mm^2.

Gauge type pressures regulators provide accurate control of pressure or vacuum conditions on the systems on which they are used. When used as pilot control devices with magnetic starters, they govern operation of pump or compressor motor similar to pressure switches. Reverse action regulators may also be employed on pressure system interlocks to prevent an operation until the pressure in that system has increased to the point desired.

Pressure regulators consist of a Bourdon type pressure gauge and control relay. Delicate contacts on the gauge energise the relay, causing it to be closed or opened, as the case may be. Thus the relay contacts are used to control a large motor starter, avoiding possible burning of the gauge contacts. Regulators for standard operation open the circuit at high pressure and close the circuit at low pressure. Special regulators operate on reverse operation to close the circuit at high pressure and to open the circuit at low pressure.

8. *Thermostats.* The thermostat is a two or three (or more) wire control pilot device that is temperature actuated. Thermostats indirectly control large motors in air-conditioning systems and in many industrial applications.

General industrial temperature controllers are recommended for heating applications where the temperature to be controlled is higher than the normal or ambient temperature. In general, the application would be in connection with the controlled liquids rather than gases because of the relatively greater conductivity between bulb and liquid when compared to conductivity between bulb and gas. Generally, where air or gas temperature is to be controlled, the sensitivity will decrease and the difference between "on and off" will widen.

Gas vapour or liquid pressure expands a metal bellows against the force of a spring to give a motion to an operating pin, which snaps a precision switch to its operating position when the setting is reached. This is very similar to industrial pressure switches. The operating pressure is generated in proportion to the temperature of liquid in a closed bulb. The precision switch snap back to its normal position when the pressure in the bellows drops enough to allow the main spring to compress the bellows. Thermostats are of many different designs and are used in control circuits to operate heaters, blowers, fans, pumps, solenoid, valves and other devices.

9. *Plugging Switches*. Plugging switches, also called *zero-speed switches* or speed relays, were primarily designed for incorporation into control circuits as pilot devices to provide quick automatic stopping of machines driven by squirrel cage motors. Properly adjusted plugging switches are capable of preventing reversal of rotation of the controlled drive after it has been brought to standstill by the reversal of motor connections. These are often used on machine tools that must come to sudden stop at some point in their cycle of operation to prevent inaccuracies in the work or damage to the machine. Another application is found in processes where the machine must come to a complete stop before the next step of work is started, here the reduced stopping time saves time for a greater total output.

The shaft of the plugging switch, is mechanically coupled to the motor shaft or to a suitable shaft on the driven machine. Motion of the motor or driven machine is transmitted to the switch contacts by a centifugal mechanism or a magnetic induction (eddy current disk) arrangement within the switch. The contacts are wired to the reversing starters which control the motor. Acting as a link between the motor and reversing starter, the switch allows the starter to apply just enough 'reverse' power to bring the motor to a quick stop.

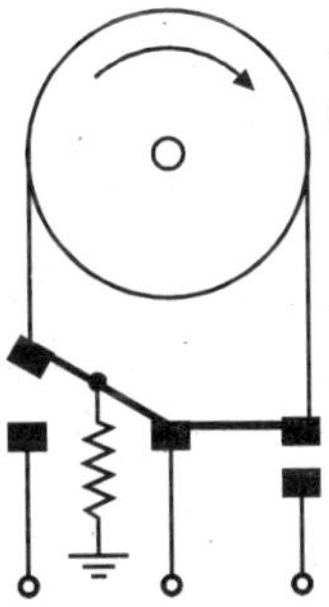

Fig. 2.21. ***Plugging Switch***

2.5. CONTACTORS

Magnetic contactors are electromagnetically operated switches that provide a safe and convenient means for repeatedly making and breaking branch circuits. The principal difference between a contactor and a motor starter is that contactor does not contain overload relays. Contactors are used to switch, with pilot control device, such loads as lighting, heating and controlling electric motors where overload protection is provided separately. The larger sizes are used extensively for remote control of relatively high current circuits where expense prohibits running the power leads to the remote controlling location. This is one of the main advantages of electromagnetic control over manual control. Pilot devices such as push buttons, pressure switches, float switches, limit switches or thermostats are used to provide the necessary control for operating contactors.

The contactor consists of an electromagnet (having an armature, a core and a pick-up coil connected in its control circuit), main contacts (one stationary and the other movable) that make and break the power circuit, electrical interlocks that are used for blocking and control of automation and signal circuits, a blowout structure (a chute with magnetic blow-out or with special grids) that provides quick arc suppression when breaking a circuit.

Contactors are manufactured for ac and dc circuits and may have one or several poles. Besides they may have normally open and (or) normally closed contacts that are switched instantaneously or with some delay. DC contactors have long pick-up coils with a small diameter and a large outside surface, which enhances cooling and enables larger currents to be carried. AC contactors are manufactured for rated voltages of 220, 380 and 500 volts and for rated currents from 20 A to 600 A. They can have from one to five main contacts and from two to four interlocks. AC contactors consist of the same elements as dc contactors. Their magnetic circuit, however, is built up from varnished electrical steel laminations to reduce eddy current losses. Three-pole ac contactors with normally open contacts are the most widespread.

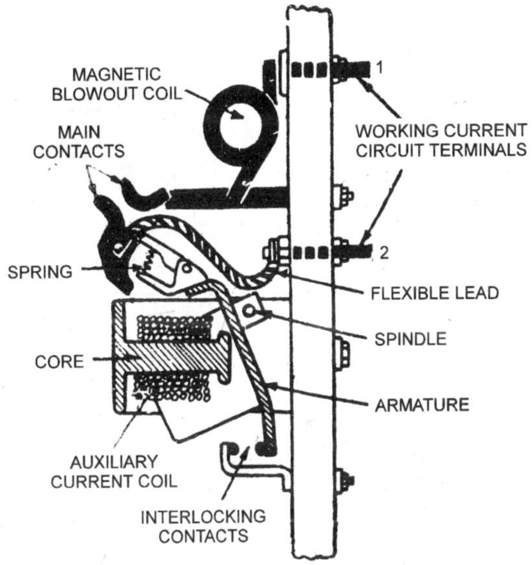

Fig. 2.22 (*a*) *DC Electromagnetic Contactor*

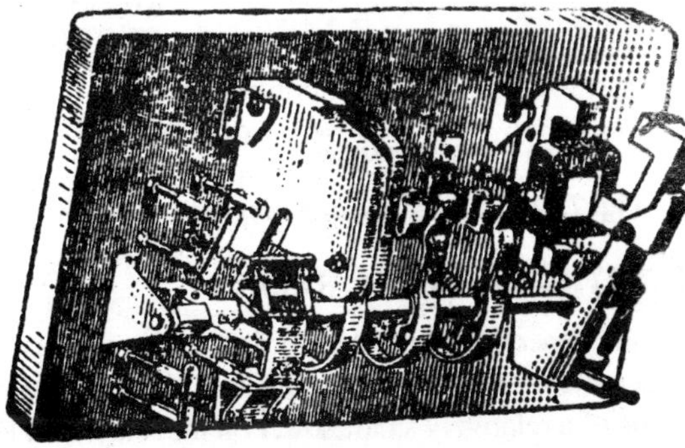
Fig. 2.22 (*b*) *AC Contactor*

Figure 2.22 (*a*) shows the construction of a dc electromagnetic contactor. The auxiliary current passes through the contactor coil and attracts a steel armature to its core; the armature rotates about an axis and closes the main contacts in the working-current circuit. The spring ensures proper pressure of the movable contact. The main circuit is connected to the terminals 1 and 2; its current passes through the magnetic blowout coil, the main contacts and the flexible connecting wire. This contactor closes the contacts in the auxiliary current circuit so-called interlocking contacts which serve to perform the auxiliary control operations in addition to the main circuit. Figure 2.22(*b*) shows a diagram of an ac contactor.

2.5.1. Magnetic Starters

Magnetic starters are special forms of magnetic contactors designed to include thermal overload relays and electrical interlocks. Like contactors, magnetic starters are also used for remote control of electric motors. However, unlike contactors they will disconnect the motor from supply when overload arise. A magnetic starter, shown in Fig. 2.23, essentially comprises a contactor and a set of thermal relays. Each relay consists of two closely bonded strips of different alloys with different temperature coefficients of expansion. When the current flowing through such a relay exceeds its setting, it heats the bimetal element and makes it bend because the two strips increase in length by different amounts. By bending, the bimetal element opens the holding coil circuit and thus trips open the starter. The connections of a usual three-pole magnetic starter are shown in Fig. 2.24. The starter in such cases usually incorporates two built-in thermal overload relays and one electrical interlock.

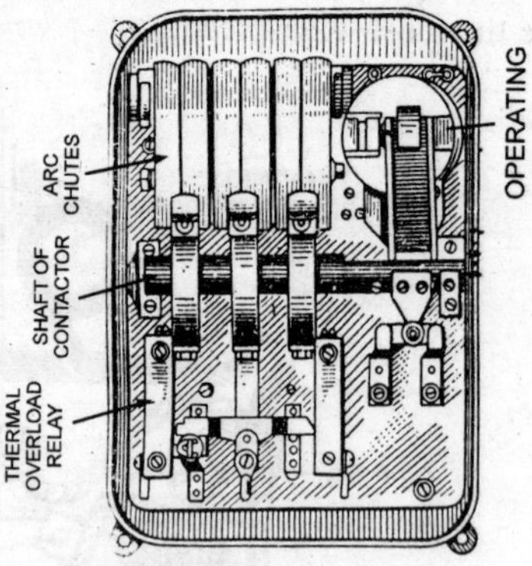

Fig. 2.23. ***Magnetic Starter***

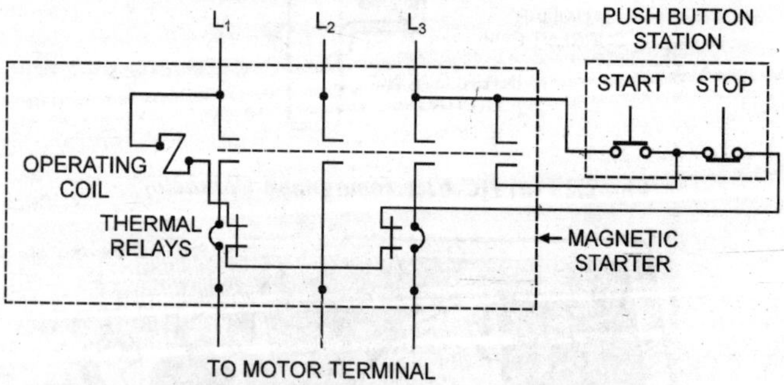

Fig. 2.24. ***Connections of a Magnetic Starter***

2.6. ELECTROMAGNETIC RELAYS

A relay, in the general sense of the word, is any apparatus which serves to actuate or control the regime of a high power system by the action of a relatively small power on the relay. An electric relay is a mechanism which actuates auxiliary electric circuits or mechanical devices, when electrical factors act on it. The relay mechanism consists of a sensing member and an operating member; the former "perceives" the action of the electrical factors and actuates the operating member, which performs some operation.

Relays are widely used for protection against circuit overload, keying remote switching, reversing, control etc. The uses of relays in industrial control circuits are numerous and it will not be possible to describe all of them here. It can generally be said, however, that the relay is used in any industrial circuit in which a large amount of load power is to be controlled by a small amount of control power.

Basic Construction. Figure 2.25 (*a*) shows the basic construction of a typical electromagnetic relay. The electromagnetic relay essentially consists of a coil to which a voltage is applied, a core upon which the coil is wound, relay contacts, and a movable steel armature which is held against one of the contacts by means of the spring. Before the voltage is applied to the coil the relay is said to be in its *normal* or *de-energized* position. Its action depends on the interaction of the magnetic field set up by the coil carrying the current and the movable steel armature.

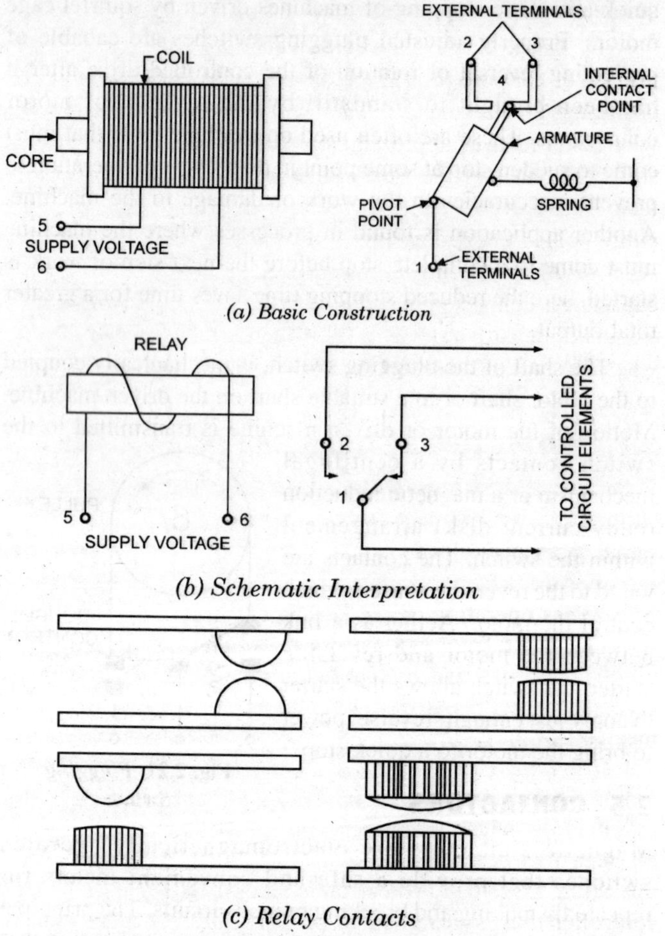

(*a*) *Basic Construction*

(*b*) *Schematic Interpretation*

(*c*) *Relay Contacts*

Fig. 2.25. ***Electromagnetic Relay***

Operation. In the relay shown in Fig. 2.25(*a*) any external circuit connected between terminals 1 and 3 is in a closed position at this instant and the external circuit connected between terminals 1 and 2 is in a open position. When the voltage is applied to the coil, the relay is energized and the magnetic field set up around the coil makes the armature to be attracted toward it. Contact 4 breaks with terminal 3 but makes with terminal 2, and any external (controlled) circuit connected between terminals 1 and 2 is now completed. The relay remains in the position till the coil remains energized.

The schematic interpretation of this basic relay is illustrated in Fig. 2.25 (*b*). Terminals 2 and 3 are the fixed contacts, and terminal 1 is connected to the movable armature (in some relays the armature remains isolated from all contacts). The contacts of the basic relay acts as single pole double throw (SPDT) switch. Although switch contacts are manually controlled, but relay contacts are remotely controlled.

Relay Contacts. Relay contacts are made of metals and alloys of low resistance in order to prevent overheating. In addition, the metal used in the manufacture of the contacts must be able to resist both the welding action that occurs when the contacts close and the arcing that forms when the contacts open. The shapes of the contacts should be such as to resist the pitty action. Some common shapes are shown in Fig. 2.25 (*c*). Metal used in the manufacture of relay contacts include silver, tungsten, palladium and certain alloys. The current at which any given pair of contacts will interrupt successfully (without arcing) is generally smaller for dc than ac voltages, and is smaller at higher voltages.

AC Relay. Such relays have a *shading coil* placed over a portion of the pole face at the air gap to reduce vibration and chattering of the contacts. The shading coil (a single turn of wire fastened over approximately half of the pole face area) shifts the phase of the flux in the shaded portion of the pole face. This prevents the total force of the flux (taken over the entire pole) from decreasing to zero at any point in the cycle. As a result, chattering is minimized. In certain ac relays, a capacitor is connected in series with the relay coil, producing a frequency-sensitive device. The relay is then energized only when the applied ac signal is at or near the resonant frequency of the coil-capacitor combination.

2.7. PROTECTIVE DEVICES AND INTERLOCKS

Protective devices must be incorporated in the control circuit (or circuits) to interrupt the flow of current to the motor to guard against overloads or other conditions dangerous to the motor or the driven machine, and also to guard against faulty or careless operation of the equipment itself.

The most frequent causes of abnormal operation of electric motors are overloads, open-field, open-phase, reversed phase, overspeed, overheating of bearings or other mechanical parts, low voltage, or voltage failure.

Overload Protection. An overload is an increase in motor current above its safe limit and may be caused by an overload on driven machinery, by a low line voltage, or by an open line polyphase system that results in single phase operation. Overload protection is an essential feature of any controller that is designed to properly protect a motor and still obtain its maximum available power under a range of overload and temperature conditions.

An excess current lasting for only a second or two will not normally cause any serious damage hence protective device should not operate on overloads of short duration and limited magnitude. On the other hand, the protective device must respond to overloads of long duration and of high magnitude which are dangerous to the motor in that they may burn the windings and insulation. So whatever overload protective device is employed, it should possess a time lag to ensure that it operates only for a sustained overload.

Apparatus having an inverse time lag will trip the motor out after a very short interval on a heavy overload but after a relatively long interval on a small overload. Such protection is obviously more satisfactory for most purposes, because the motor temperature will rise more quickly on a heavy overload. This inverse-time characteristic is possessed by certain special types of relays, by a fuse, and also by an ordinary solenoid relay provided with an oil dashpot.

Overload or overcurrent protection may be affected by fuses, electromagnetic and thermal overload relays. Overload relays or fuses need only be placed in two of the three phases of a 3-phase machine if the neutral of the system is not earthed, but should be placed in all three phases if it is earthed. Although an overload relay is more expensive than a fuse but it has the advantage that it automatically resets when the switch is reclosed. The selection of a particular protective device depends on the kW rating, type and purpose of the motor, starting duty and nature of overload.

Open-Field Protection. DC motors of large size may attain dangerously high speed with the loss of field excitation while those of small size may not race due to friction etc. Field loss relays are available to protect dc shunt or compound wound motors against loss of field excitation. Different arrangements exist with starting equipment for dc and ac synchronous motors.

Open-Phase Protection. This condition may be caused by a blown fuse, an open connection, or a broken line. If the motor is at standstill at the time of occurring phase failure, stator current will rise to and remain at a very high value, but the motor will remain stationary and thus the windings may get damaged due to poor ventilation. Dangerous conditions may exist also while the motor is running.

Reversed Phase Protection. In case two phases of the supply of a 3-phase induction motor are interchanged, the direction of rotation of the motor will reverse. This is known as *phase reversal.* In elevator operation and industrial applications, this might result in serious damage. Phase failure and phase reversal relays are available. These protect the motor, machines, and personnel from the hazards of open-phase or reversed phase sequence conditions.

Overspeed Protection. Excessive speeds are possible with certain motors that might damage a driven machine, materials in the industrial process, or the motor. Overspeed protection may involve a fine selection and use of control equipment for such applications as paper and printing plants, steel mills, processing plants and the textile industry.

A centrifugal device operates at a certain predetermined speed, usually about 20 per cent above normal, and causes the motor to be shut down.

Protection Against Overheating of Bearings. Protection against overheating of bearings or other mechanical parts may be provided by placing thermocouples in the desired situations and arranging so that when the temperature reaches a certain value, relays are operated and the motor is caused to shut down or alarm bells are caused to be rung.

Low-Voltage Protection. This protection is necessary, partly to avoid damage to the motor and its driven machine due to heavy currents on switching on full rated voltage after a temporary shutdown, and partly to protect the operator of the equipment. The low-voltage release consists usually of a simple solenoid held up by the voltage of the supply and which causes the starting equipment to return to the "off" position on the failure of the supply.

In addition to the features mentioned above to guard against faults on the motor or driven machine, protection must also be provided against careless manipulation of hand-operated gear or faulty operation of automatic equipment owing to sticking of relays or contactors or to dirty contacts.

The main danger with hand-operated gear is the too fast movement of the starting arm and it can be avoided by providing a mechanical ratchet to prevent the arm being moved more than one step at a time. The other possibility is starting of dc machines with a weak field resulting in an unnecessarily heavy current to provide a given torque. For protection against such possibility mechanical or electrical interlocks may be provided. The mechanical interlock may consist of a metal plunger, operated by the field rheostat arm, that prevents the movement of the starting arm except when the rheostat is in the full-field position. In an electrical interlock, the plunger is controlled by a solenoid, which is energized only when the field is fully on. Similar interlocks are to be provided in the control equipment for ac commutator motors used in speed control by brush shifting; the main switch should be prevented from closing unless the brushes are in the low-speed position.

With fully automatic equipment, there is always possibility that one or more of the accelerating contactors remain closed after being de-energised, due to dirt having clogged their movement. However, this can be avoided by placing auxiliary contacts on the moving arm which are closed when the contactor is open. These contacts are connected in series with the start button, so that the latter is inoperative if any of the contactors have remain closed.

Exercises

1. What is meant by control gear ? Give the advantages of automatic control over manual control.
2. What is thyratron? How does it differ in construction and properties from triode ? Give it uses too.
3. What is an ignitron ? What is its usual application ?
4. How does an amplidyne differ from an ordinary dc generator ? Give its construction, principle of operation and uses.
5. Describe drawing neat sketches the principle of working and construction of magnetic amplifier. Why ac windings are provided in pair ? Mention a few applications of magnetic amplifiers.
   [A.M.I.E. Sec B. Utilization of Elec. Power Summer 1983]
6. Explain with diagram the principle of operation of magnetic amplifiers, rotating amplifiers and thyristors.
   [Pb. Univ. Elec. Drives and Utilization of Elec. Energy April, 1983]
7. Why are solid state devices becoming more and more popular in industry nowadays ? What is diode? Give its V-I characteristic and uses.
8. What is Zener diode ? Give its schematic symbol and uses.
9. What do you understand by the word "Thyristor" ? Give its applications.
10. What is an SCR ? Give its volt-ampere characteristic and typical parameters.
11. With a neat diagram explain the constructional features of a triac. Sketch its V-I characteristics. How does it differ from an SCR ? Mention its applications.
12. What is diac ? Give its schematic symbol, V-I characteristic and uses.
13. What is GTO ? Give its schematic symbol and its advantages over SCR.
14. Describe the construction of a dc electromagnetic contactor.
15. What is a relay ? What are their applications ?
16. What are the differences between an amplidyne and an ordinary dc generator ?
    How does a saturable reactor differ from an ordinary reactor ? Explain.
    Distinguish clearly the primary control devices and pilot control devices of a motor.
    Where do (*i*) limit switch (*ii*) pressure switch (*iii*) plugging switch and (*iv*) transfer switch find applications ? Explain.
    [A.M.I.E. Sec B. Utilization of Elec. Power Winter 1981]
17. Explain the action of (*i*) push buttons (*ii*) limit switches (*iii*) float switches (*iv*) pressure switches (*v*) thermostats (*vi*) speed governors (*vii*) plugging switches and (*viii*) master controllers.
    [A.M.I.E. Sec B. Utilization of Elec. Power Summer 1980]
18. Write short notes on the following :
    (*i*) Amplidyne
    (*ii*) Magnetic amplifier
    (*iii*) Thyratron and its applications
    (*iv*) SCR
    (*v*) Triac, diac and GTO switch.

Short Answer Type Questions With Answers

Q. 1. What are the two main applications of a thyratron ?
Ans. Two main applications of thyratron are (*i*) switching and (*ii*) rectification.

Q. 2. Why is series compensating winding employed in amplidyne ?
Ans. Compensating winding is connected in series with the power output brush terminals of the amplidyne to neutralize the armature reaction set up by the load current and increase amplification factor.

Q. 3. Why are amplidynes provided with split-poles ?
Ans. Amplidynes are provided with split-poles to provide space for interpoles.

Q. 4. What is the effect of temperature on the reverse current of a P-N junction ?
Ans. Reverse current of a P-N junction increases with the increase in junction temperature. It has been found that reverse current of a P-N junction approximately doubles for each 10°C rise in temperature in case of germanium and for each 8°C in temperature in case of silicon.

Q. 5. What is Zener diode ?
Ans. Zener diode is a P-N junction diode specially designed for operation in its breakdown region in reverse bias condition.

Q. 6. How is current limited in conducting state of an SCR ?
Ans. The current in conducting state of an SCR is controlled by an external impedance.

Q. 7. What is false triggering ?
Ans. False triggering is unintended turn-on of an SCR either through gate due to noise pick-up or excessive anode voltage.

Q. 8. Why is it necessary to keep supply voltage much less than breakover voltage of an SCR ?
Ans. The supply voltage is kept much less than breakover voltage of an SCR otherwise the device will get damaged.

Q. 9. Where do you recommend the use of freewheeling diode ?
Ans. Use of freewheeling diode is recommended in motor control circuits to provide an alternate path for continuity of current in the inductive circuit following the switching-off of some power device between the motor and the supply source.

Q. 10. Where are float switches used ?
Ans. Float switches are used for opening or closing pipe valves for the control of fluids.

□□□

CHAPTER 3

Thyristor Control of Electric Motors

INSIDE THIS CHAPTER

3.1. Introduction **3.2.** General Configuration of a Motor Drive **3.3.** Classification of Power Converters **3.4.** Matching Power Electronic Converter and Motor **3.5.** DC Motor Control Through Converters **3.6.** Thyristor Control of Separately Excited DC Motor Drive **3.7.** Chopper Control of DC Motors **3.8.** Comparison of Converter Fed and Chopper Fed DC Drives **3.9.** Special Features of Thyristor Drive Motors **3.10.** Speed Control of DC Motor Using Feedback Loops **3.11.** Electric Braking of Rectifier Controlled Separately Excited DC Motor **3.12.** Thyristor Control of DC Series Motors **3.13.** Thyristor Control of Three-Phase Induction Motors **3.14.** Thyristor Control of a 3-Phase Synchronous Motor **3.15.** Speed Control of Single-Phase Induction Motor Using Triac (Fan Regulator) **3.16.** Brushless (or Commutatorless) DC Motor **3.17.** Switched Reluctance Motor.

3.1. INTRODUCTION

Electric motors are employed in a very large power range, from a few watts to thousands of kilowatts. Many applications require very precise position adjustments (as in robotics). In many applications optimum performance and efficiency are the main concern. Variable speed drive (VSD) systems help in optimisation of process so as to reduce investment costs, operational costs and maintenance costs. Energy saving is another big advantage of variable speed drives.

The advent of thyristors capable of handling large currents has revolutionized the field of electric power control. Thyratrons, ignitrons, mercury-arc rectifiers, magnetic amplifiers motor-generator sets have all been replaced by solid state circuits employing semiconductor diodes and thyristors. Thyristor controlled drives employing both dc and ac motors find wide application in industry as variable speed drives. In the 1960's, ac power used to be converted to dc power for direct control of drive motors with solid state devices (high power silicon diodes and silicon controlled rectifiers). Initially saturable reactors were employed in conjunction with high power silicon rectifiers for dc drives. Nowadays, thyristors are extensively used for ac-to-dc conversion.

Thyristorized control has the following advantages and disadvantages:

Advantages. 1. The response of the control device is faster as it eliminates the time lag introduced by the inductances of the generator field and the armature.

2. Due to low voltage drop across the thyristor, the efficiency of the control system is high.
3. The control device is smaller in size, lighter in weight, cheaper in cost, requiring less space and minimal maintenance.
4. Simple and reliable operation.

Disadvantages. 1. Because of higher ripple content in the converter output, motor heating and commutation problems are serious.

2. Due to switching action of thyristors and non-sinusoidal nature of current, there is more possibility of interference with the communication networks.

In all variable speed drive systems power electronic converter acts as an interface which accepts electric power from the existing source and converts it in a controlled manner into a suitable form compatible with the particular load or the process for which it is employed. The main sources for electric power are: Single or three-phase 50 Hz ac from utility systems and dc from storage batteries or solar cells. The four basic forms of power conversion required are: ac to dc, dc to dc, dc to ac and ac to ac. The modern converters are compact, cheap, reliable, durable, flexible, completely controllable. They need reduced maintenance too. They are suitable for all the four basic forms of power conversion mentioned above through rectifiers (ac/dc), choppers (dc/dc), inverters (dc/ac) and cycloconverters or ac regulators (ac/ac).

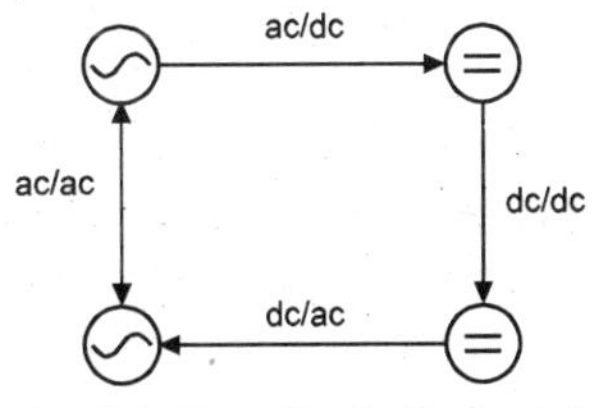

Fig. 3.1. ***Four Basic Forms of Power Conversion***

For dc motor control, controlled dc power from a constant voltage ac supply is obtained by means of *controlled rectifiers* (usually termed as *converters*) using thyristors and diodes. The control of dc voltage is achieved by varying the phase angle at which the thyristors are fired relative to the applied alternating voltage waveform. This scheme of control is known as the *phase control*. In another control system, known as the *integral cycle control*, the current is gated to flow from the ac supply for a number of complete cycles and is then quenched for a further few cycles, the process being repeated continuously. Control is applied by adjusting the ratio of *on* and *off* durations. This method is suitable for the control of fractional kW output dc motors.

Phase controlled converters are simple to operate and are less expensive as they do not require additional circuitry for commutation process. In such converters natural commutation is achieved, *i.e.,* when an incoming thyristor is turned on, it immediately reverse biases the outgoing thyristor and turns it off.

Phase control and integral cycle control methods are also applicable for ac motors wherein the converter circuitry is not required.

Control of dc motors supplied from dc supply is achieved by means of a thyristor switching circuit called the *chopper*. In chopper circuits, the control of the average voltage is achieved by changing the on-to-off time duration ratio for which the dc supply voltage is applied to the motor. This provides an efficient and stepless control of dc motors. The motor can also be made to operate in the regenerative braking mode. Instead of converter circuits, it is possible to use an uncontrolled rectifier, which provides a constant direct voltage, followed by a chopper to provide a variable mean direct output voltage. The chopper controller needs forced commutation of the thyristor.

For the control of ac motors supplied from dc supply, thyristor, transistor or MOSFET based inverters are used. Such switching circuits transfer energy from dc supply to ac load of variable frequency and/or variable voltage. Because of switching operation, the ac voltage waveforms are stepped, the harmonic contents of which are filtered out. As the power supply is normally ac, the complete scheme for obtaining variable-voltage and frequency power involves the use of both an inverter and a converter.

A *cycloconverter* is a control unit for providing variable-voltage and frequency power directly from a fixed frequency supply without the necessity of an intermediate dc stage. The mechanism of voltage and frequency control is a combination of those employed in the phase-controlled converter and the pulse-modulated inverters. Cycloconverters inspite of their attraction of direct ac to ac conversion, suffer from certain drawbacks because of which they did not become popular. Some of these drawbacks are : cycloconverters can generate only a sub-frequency output; they produce output with large harmonic content and have a low input power factor. Cycloconverters are employed for low-speed drives and for controlling linear motors in high-speed transportation systems.

3.2. GENERAL CONFIGURATION OF A MOTOR DRIVE

Figure 3.2 shows a block diagram illustrating the control of a motor drive. The main constituents are power electronic converter, motor, processes, process control computer and controller.

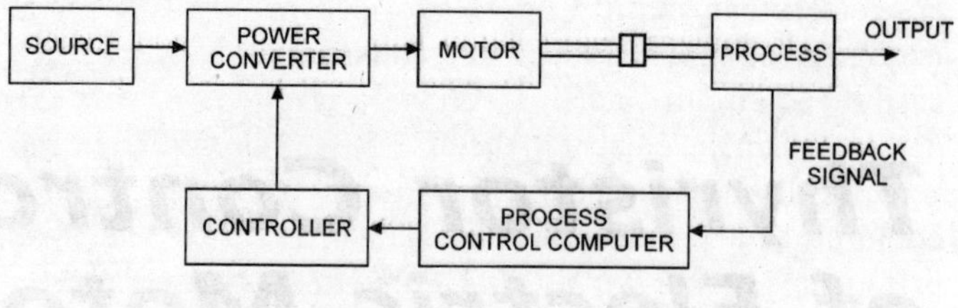

Fig. 3.2. ***Block Diagram of a Motor Drive***

The required characteristics of the motor drive are determined by the process. The process may need a variable-speed drive or a servo drive etc. The motor is selected on the basis of these requirements. The power electronic converter converts ac single phase or 3-phase input into a supply suitable to provide desired characteristics of the motor. The process control computer receives a feedback from the process as regards the extent to which the requirements are met. This feedback may be of speed, position etc. The computer feeds an error signal to the controller which, in turn, does the corrective action. The corrective action may be, say, the change in the firing angle of thyristors of the power electronic converter. In some situations the accuracy and response time of the motor to adjust the new speed is of utmost importance. In some other situations the requirements may not be so critical.

3.3. CLASSIFICATION OF POWER CONVERTERS

Solid state power converters are employed for obtaining the appropriate form of electrical energy such as direct current or adjustable-frequency alternating current (required to operate most electronic circuits and motor drives) from fixed-frequency alternating current.

According to functions, the static power converters may be classified in the four basic categories namely ac/dc converters, ac/ac converters, dc/dc converters and dc/ac converters, already discussed in Art. 3.1 [refer to Fig. 3.1].

The inverters (dc/ac converters) may be further categorised as per their source nature as voltage source and current source inverters. A voltage source inverter (VSI) is fed by a constant voltage source system while a current source inverter (CSI) is supplied from a constant current source. A voltage source inverter is one in which dc source has small or negligible impedance. Because of a low internal impedance, the terminal voltage of a VSI remains almost constant with variations in load. It is, therefore, equally suitable to single motor and multimotor drives. On the other hand, CSI is supplied with a controlled current from a dc source of high impedance. Because of a large impedance, the terminal voltage of a CSI varies substantially with a change in load. CSIs are not suitable for multimotor drives because a change in load on any motor affects other motors.

The static power converters may also be categorised into two categories viz. transistor converters and thyristor converters. The transistor converters, which do not require additional commutation circuitry as needed in thyristor converters, have almost completely replaced thyristor converters below 75 kW. The availability of low-cost high-power BJTs has also made possible the use of pulse width modulation (PWM) for improving the motor current waveform.

According to the method of commutation the thyristor converters can mainly be categorised in two types viz. naturally commutated and force commutated converters. In case of ac circuits, ac line voltage is available across the device. When the current in the SCR passes through a natural zero, the device is turned off. This process is known as natural commutation process and the converters based on this principle are known as *line commutated* or *naturally commutated converters*. AC/DC converters, cycloconverters, and ac/ac voltage regulators using SCRs are of the naturally commutated type.

In case of dc circuits, since the supply voltage does not pass through the zero point, some external source is required to commutate the device. This process is known as forced commutation process and the inverters based on this principle are known as force commutated inverters. As the device is to be commutated forcefully, such inverters need complicated commutation circuitries.

Converters may also be grouped into one-quadrant, two-quadrant and four-quadrant converters. A semiconverter is a one-quadrant converter and it has one polarity of output voltage and current. A full converter is a two-quadrant converter and polarity of its output voltage can be either positive or negative. However, its output current has one polarity only. A dual converter is a four-quadrant converter and both its output voltage and current can be either positive or negative.

3.3.1. AC/DC Converters

Rectification is a process of converting alternating current or voltage into a direct current or voltage. This conversion can be achieved by a variety of circuits based on and using switching devices. The widely used switching devices are diodes, thyristors, power transistors, power MOS etc. The rectifier circuits using only diodes as the rectifying elements are known as *uncontrolled rectifiers* because for a fixed value of ac input voltage, the output dc voltage is fixed and cannot be controlled. For some low-power devices where control of output voltage is not required, rectifier circuits using only diodes may be employed. Such circuits have the advantages of low cost and simple circuitry.

Rectifier circuits using thyristors are known as *controlled rectifiers* or *ac-dc converters*. The basic principle of a controlled rectifier is to control the point in time at which the thyristors are allowed to conduct during each cycle. Thus, it is possible to select the time segments of the ac voltage waves which appear at the dc terminals, and the mean output voltage is controlled continuously. This process of control is called the phase control and the controlled rectifiers are, therefore, also called the phase-controlled rectifiers. Since thyristors are available in high voltage and large current ratings, controlled rectifiers using thyristors are employed in many high-power devices. A controlled rectifier may be a single-phase or 3-phase depending on whether the input is single-phase or 3-phase ac.

Controlled rectifiers are of two types namely fully-controlled and half-controlled. The fully-controlled rectifier uses thyristors as the rectifying elements and the dc output voltage is a function of the amplitude of the ac supply voltage and the point-on-wave at which the thyristors are triggered (called firing angle α). The half-controlled rectifiers consist of a mixture of diodes and thyristors, allowing a more limited control over the dc output voltage level than the fully-controlled rectifier. The half-controlled rectifier is cheaper than a fully-controlled rectifier of the same ratings but has operational limitations.

Uncontrolled and half-controlled rectifiers permit power to flow only from the ac system to the dc load and are, therefore, referred to as unidirectional, one-quadrant or semiconverters. However, with a fully-controlled rectifier it is possible, by control of the point-on-wave at which switching occurs, to allow power to be transferred from the dc side of the converter back into the ac system. When this occurs, operation is said to be in the inverting mode. The fully-controlled converters may, therefore, be referred to as bidirectional or two-quadrant converters. Thus it is possible for the phase-controlled converters to provide either a one-quadrant, two-quadrant or four-quadrant operation at its dc terminals. In a four-quadrant operation, two fully-controlled converters, connected back-to-back with one another, are used. Four-quadrant converters or dual converters provide the facility for bidirectional current flow through the load.

The converter type depends on the power to be handled and how much voltage ripple is tolerable. For low powers (below 20 kW), single-phase circuits are adequate but they themselves can take different forms. For high powers (above 20 kW), three-phase circuits are invariably employed.

3.3.2. AC/AC Converters

Thyristor-based ac voltage controllers using the phase control principle convert fixed alternating voltage directly to variable alternating voltage without a change in frequency. Applications of ac voltage controllers or regulators include domestic and industrial heating, transformer tap changing, lighting control, speed control of single-phase and three-phase induction motors etc. Earlier, the devices used for such applications were auto-transformers, tap-changing transformers, magnetic amplifiers, saturable reactors etc. But these devices are now being replaced by ac regulators using thyristors and triacs because of their high efficiency, flexibility in control, compact size and less maintenance. However, ac regulators using thyristors and triacs introduce objectionable harmonics in the circuit. AC regulators are classified as single phase or three phase. Each of these may be half-wave (unidirectional) or full-wave (bidirectional). Because of ac input, ac regulators are always line or naturally commutated and as such no complex commutation circuitry is required in these controllers.

Cycloconverter is defined as a power controller which converts input power at one frequency to output power at a different frequency with one-state conversion. Thus it is an ac/ac converter without any intermediate dc link. Cycloconverters can be classified as single phase-to-single phase, three phase-to-single phase, 3-phase-to-three phase devices. They may also be classified as step-down and step-up. In step-down cycloconverters, the output frequency is lower than the supply or input frequency. In step-up cycloconverters, output frequency is higher than supply frequency. Step-down cycloconverters use line or natural commutation while step-up cycloconverters need forced commutation. Generally step-down cycloconverters are in much use. The desired amplitude of output voltage can be had by changing firing angle of thyristors used in the circuit. With the advent of high-power thyristors cycloconverters are becoming more and more popular. At present, the applications of cycloconverters include speed control of high-power ac drives, induction heating, static VAR compensation, conversion of variable-speed-alternator voltage to constant frequency output voltage for use as power supply in aircraft or shipboards.

3.3.3. DC/DC Converters

A dc chopper, also known as dc-dc converter, is a static device (switch) used to obtain variable dc voltage from a source of constant dc voltage. Thus, chopper may be thought of as dc equivalent of an ac transformer as it behaves in an identical manner. Besides the saving in power, the dc chopper offers greater efficiency, fast response, lower maintenance, small size, smooth control, and, for many applications, lower cost than motor-generator sets or gas tubes approaches.

Solid state choppers due to their inherent advantages are widely employed in trolley cars, battery-operated vehicles, traction motor control, control of a large number of dc motors from a common dc bus, marine hoists, forklift trucks, mine haulers etc. They can also be used for regenerative braking in electric traction.

3.3.4. DC/AC Converters

The dc to ac power converters are known as inverters. In other words, an inverter is a circuit which converts a dc power into an ac power at desired output voltage and frequency. The ac output voltage could be fixed at a fixed or variable frequency. The dc input supply may come from a battery, solar cells, fuel cells etc.

The conversion can be achieved either by transistors or thyristors. For low and medium power outputs transistorised inverters are used but for high power outputs, thyristors should be used. For low power outputs, self-oscillating transistorised inverters are suitable but for high power outputs, driven inverters are more common than self-oscillating ones. Moreover for multiphase ac output, there is no alternative other than driven inverters. The driven inverters have better frequency stability because a separate master oscillator is used for this purpose. For the applications in inverters, transistors have some edge over thyristors regarding the switching speed, simplicity in control circuits, higher efficiency and greater reliability. This is mainly due to the fact that thyristor inverters need extra circuits to turn thyristors off, moreover, additional complex logic circuits may be required to avoid false triggering and provide proper commutation timings. Thyristors can handle much higher load currents than transistors and consequently, for high power outputs, thyristors become more desirable than transistors.

The output voltage waveforms of an ideal inverter should be sinusoidal, but the output voltage waveforms of practical inverters are non-sinusoidal and contain certain harmonics. Square wave or quasi-square wave voltages may be acceptable for low and medium power applications, but for high power applications low-distorted sinusoidal waveforms are required. The output frequency of an inverter is determined by the rate at which the semiconductor devices are switched on and off by the inverter control circuitry and consequently, an adjustable frequency ac output is readily provided. The harmonic contents of output voltage can be reduced significantly by switching techniques of available high speed power semiconductor devices. The filtering of harmonics is not feasible when the output frequency varies over a wide range, and the generation of ac waveform with low harmonic content is important. When the ac output voltage of an inverter is supplied to a transformer or ac motor, this output voltage must be varied in conjunction with frequency to maintain the proper magnetic conditions. Some of the important applications of inverters are: variable speed ac motor drives, induction heating, aircraft power supplies uninterruptable power supplies (UPS), high-voltage dc transmission lines etc.

3.4. MATCHING POWER ELECTRONIC CONVERTER AND MOTOR

The motor has to be selected to match the load requirements. The power electronic converter has to match the motor. Some important considerations in matching power electronic converter and motor are as under.

1. *Current Rating.* The torque developed by an electric motor depends on the current drawn by it. During most of the operating duration a motor is required to deliver its rated torque, and therefore, the current drawn is also equal to its rated current. But a motor is capable of developing much higher torque (as compared to its rated torque) provided the duration of the torque is small as compared to thermal constant of the motor. Since torque depends on the current, a motor for developing a large torque, requires a large current which has to be supplied by the converter. However, the overcurrent capability of electronic devices is very small. A large current leads to high junction temperature which may destroy the electronic converter. The thermal time constants of the power electronic devices are very small in comparison to those of motors. The current of the converter must be selected properly on the basis of average, rms and peak currents of the motor.

2. *Voltage Rating.* Every motor develops a back emf e_b that opposes the applied voltage v. The rate of change of torque depends on the rate of change of current which, in turn, can be expressed as

$$\frac{di}{dt} = \frac{v - e_b}{L} \qquad ...(3.1)$$

where $\frac{di}{dt}$ is the rate of change of current and L is the inductance of the motor. For a reasonably large value of $\frac{di}{dt}$, the output voltage v of converter should be much larger than e_b. Since e_b depends on the motor speed, it is necessary that the voltage rating of the converter is selected on the basis of maximum speed of the motor.

3. *Switching Frequency and Motor Inductance.* The steady-state ripple content in the motor current depends on motor inductance L. Large ripple content leads to large motor losses. So the steady-state ripple content in the motor current should be small. This is possible if inductance L is large. But large inductance means that the motor cannot respond quickly to the change in load demand. To meet these opposing requirements, the switching frequency of converter should be high (so that the steady-state ripple content is small). However, a high switching frequency leads to large power losses in the converter. Therefore, a proper selection of switching frequency and inductance is essential.

3.5. DC MOTOR CONTROL THROUGH CONVERTERS

DC motors are widely used in adjustable-speed drives and position control applications. Their speeds below base speed can be controlled by armature voltage control while speeds above base speed are obtained by field flux control. DC motors are preferred over ac motors because these are inherently most suitable for smooth, efficient, wide-range speed control and quick reversals.

Phase-controlled converters provide an adjustable dc output voltage from a fixed ac input voltage. DC choppers also provide adjustable dc output voltage from a fixed dc input voltage. When available supply is ac, an ac-dc converter (phase-controlled converter) is ideally suited to convert the constant ac voltage into a variable dc voltage. Phase-controlled converters are simple to operate and are less expensive as they do not require additional circuitry for commutation process. In such converters natural commutation is achieved, *i.e.,* when an incoming thyristor is turned on, it immediately reverse biases the outgoing thyristor and turns it off. The power conversion efficiency of these converters is more than 95% because of low losses in thyristors. Thus, dc motors can be controlled conveniently and efficiently through phase-controlled converters.

The dc motors used in conjunction with power converters are dc separately excited motors and dc series motors. The power converters are used in speed control of fractional kW dc motors as well as in large sized motors used in variable-speed reversing drives for rolling mills with motor ratings as large as several MWs.

3.6. THYRISTOR CONTROL OF SEPARATELY EXCITED DC MOTOR DRIVE

3.6.1. Single Phase Drives

The single phase ac is converted into dc by a controlled rectifier or converter and the dc is supplied to the dc motor. By varying the firing angle the voltage applied to the motor can be varied and thus the speed of the motor can be controlled. Single phase drives can be further classified as

1. Single phase half-wave converter drives.
2. Single phase semiconverter drives.
3. Single phase full converter drives.
4. Single phase dual converter drives.

1. *Single Phase Half-Wave Converter Drives.* Figure 3.3 (*a*) shows a single phase half-wave converter for controlling a separately excited dc motor. It needs a single thyristor and a freewheeling diode (D_{FW}). Freewheeling diode, also sometimes known as *bypass diode* or *commutating diode*, is used to improve the waveshape of load current and power factor. Freewheeling diode is connected across the motor terminals to allow for dissipation of energy stored in the motor inductance and to provide for continuity of motor current when the thyristors are blocked. It also provides protection against transient overvoltages. Separate converters are employed for the armature and field circuits. It is desirable that the supply to the field winding is provided through a semiconverter or full converter. If half-wave converter is used for supply to field circuit, the high ripple content in the field circuit would cause increase in the iron losses of the machine. It is one quadrant drive, as illustrated in Fig. 3.3 (*b*).

Because of inductance of the field and armature, the thyristor would not turn off at $\omega t = \pi$. Therefore it is desirable to have freewheeling diode (D_{FW}), as shown.

Average value of voltage applied across armature is given by

$$V_a = \frac{1}{2\pi}\left[\int_{\alpha_a}^{\pi} V_{max} \sin \omega t \, d(\omega t) + 0\right]$$

$$= \frac{V_{max}}{2\pi}(1 + \cos \alpha_a) \quad \text{for } 0 \le \alpha_a \le \pi \qquad ...(3.2)$$

where α_a is the firing angle of converter in the armature circuit.

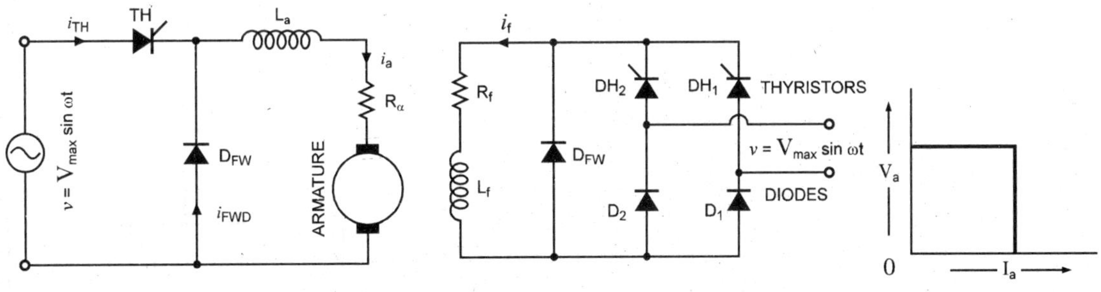

(*a*) Circuit Diagram (*b*) Quadrant of Operation

Fig. 3.3. *Separately Excited DC Motor Fed From Single Phase AC Supply Through Half-Wave Converter*

In this circuit the motor current is always discontinuous, resulting in poor performance. This type of drive is used only for small dc motors of rating up to 500 W or so.

2. *Single Phase Semiconverter Drives.* Figure 3.4 (*a*) shows the circuit of a semiconverter feeding a separately excited dc motor. Both armature and field circuits are supplied from single phase ac supply through semiconverters. Because of inductance in the armature and field circuits, freewheeling diodes

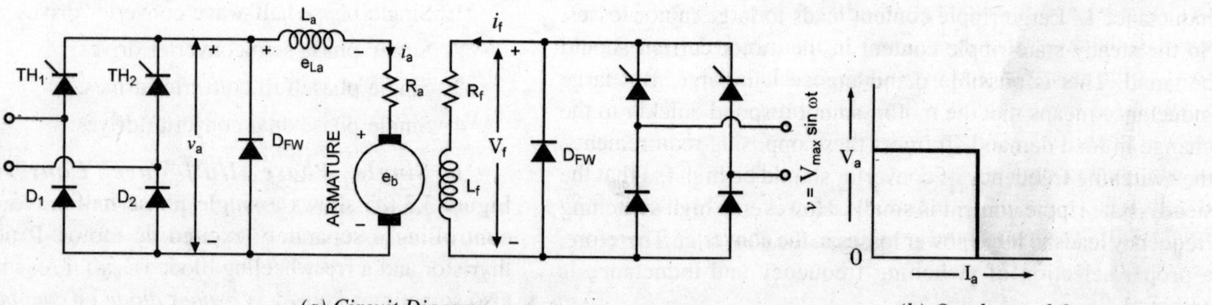

(a) Circuit Diagram (b) Quadrant of Operation

Fig. 3.4. ***Separately Excited DC Motor Fed From Single Phase AC Supply Through Semiconverters***

are required in both the circuits. The voltages applied to the armature and field are

$$V_a = \frac{V_{max}}{\pi}(1 + \cos \alpha_a) \text{ for } 0 \le \alpha_a \le \pi \quad \ldots(3.3)$$

$$V_f = \frac{V_{max}}{\pi}(1 + \cos \alpha_f) \text{ for } 0 \le \alpha_f \le \pi \quad \ldots(3.4)$$

where α_a and α_f are the firing angles of converters in the armature and field circuits respectively.

This is also a one quadrant drive [Fig. 3.4 (*b*)] which provides voltage and current of one polarity at dc terminals. It, therefore, does not provide for regenerative braking, *i.e.*, power flow from dc motor to the ac supply. Where regeneration is not required, this converter is used for reasons of economy. This drive is used for motors up to 15 kW rating.

Typical steady-state voltage and current waveforms are illustrated in Fig. 3.5. The thyristor TH_1 is fired at angle α_a while TH_2 at angle $\pi + \alpha_a$ with respect to supply voltage v and process is repeated continuously.

Under steady conditions, as thyristor TH_1 is fired ($\omega t = \alpha_a$), TH_1 and diode D_2 conduct and the motor gets supply, *i.e.*, v_a. At $\omega t = \pi$, v_a tends to become negative as the input voltage reverses in polarity. This makes freewheeling diode to become forward biased and armature current flowing through $TH_1 D_2$ is transferred to D_{FW} for the freewheeling period $\pi < \omega t < \pi + \alpha_a$ providing for continuity of the armature current during this period when the motor remains disconnected from the supply. At $\omega t = \pi + \alpha_a$, the thyristor TH_2 is fired and TH_2D_1 conduct, causing D_{FW} to become reverse biased, and therefore, open circuited. The motor is once again connected positively to the supply for the next period of $\pi + \alpha_a < \omega t < 2\pi$. This process repeats continuously.

Different voltage and current waveforms of a separately excited dc motor supplied through a semiconverter are illustrated in Fig. 3.5. Though the voltage across motor terminals [Fig. 3.5 (*c*)] contains harmonics over and above a steady dc value, it is rightly assumed here that the motor does not respond to these harmonics and thus runs at constant speed of *n rps* and has constant induced back emf e_b. As TH_1 fires at $\omega t = \alpha_a$, the motor current is given as

$$\frac{1}{L_a}\int_{\alpha_a}^{\omega t}(v - e_b)\, d(\omega t)\,;$$

assuming R_a negligibly small

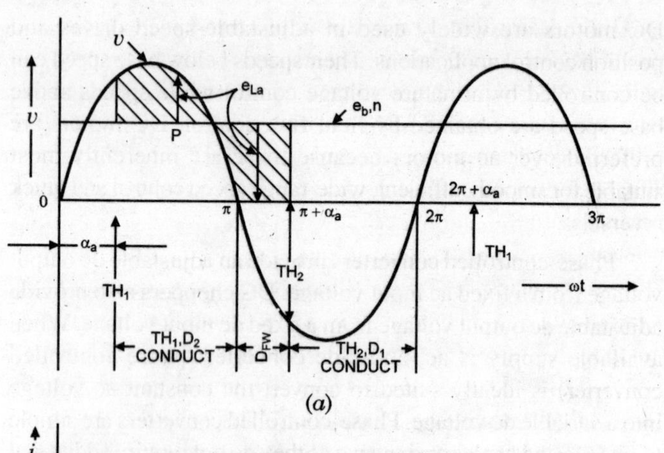

Fig. 3.5. ***Voltage and Current Waveforms of Semiconverter Feeding Separately Excited DC Motor; Continuous Armature Current***

up to the point P illustrated in Fig. 3.5 (*a*) ; $v > e_b$ so that the motor current increases. So does the motor back emf e_b. During this period, apart from energy being delivered to the load, energy is also being stored in the motor armature inductance L_a. Beyond the point P, voltage v becomes lesser than induced back emf e_b and the motor current starts to drop. This also implies the reversal of voltage across motor armature inductance which now feeds energy into the system. During the freewheeling period ($\pi < \omega t < \pi + \alpha_a$), the diode continues to be forward biased by the reversal of the inductive voltage. During this period a part of the energy stored in armature inductance is consumed in supplying the mechanical load. The motor current, speed, and emf, therefore, all decrease. This process then repeats over the next period ($\pi + \alpha_a < \omega t < 2\pi + \alpha_a$) through $TH_2 D_1$, and later through D_{FW}. The current drawn from the supply shown in Fig. 3.5 (*d*) is that part of the armature current which flows over the periods (α_a, π), ($\pi + \alpha_a$, 2π), when the motor is connected to the supply. It is not necessary to employ the freewheeling diode. In its absence at $\omega t = \pi$, diode D_1 becomes forward biased so that freewheeling occurs through $TH_1 D_1$ till TH_2 is fired. At $\omega t = 2\pi$ freewheeling occurs through $TH_2 D_2$ and so on.

Discontinuous Armature Current. The armature current becomes discontinuous for large values of the firing angle, high speed and low values of torque. Discontinuous armature current results in deterioration of motor performance. The ratio of peak to average and rms to average value of armature current increases. Thus it is desirable to operate the motor in the continuous current mode. This can be achieved by using an external armature circuit choke, which reduces the rate of decay of current during the freewheeling operation.

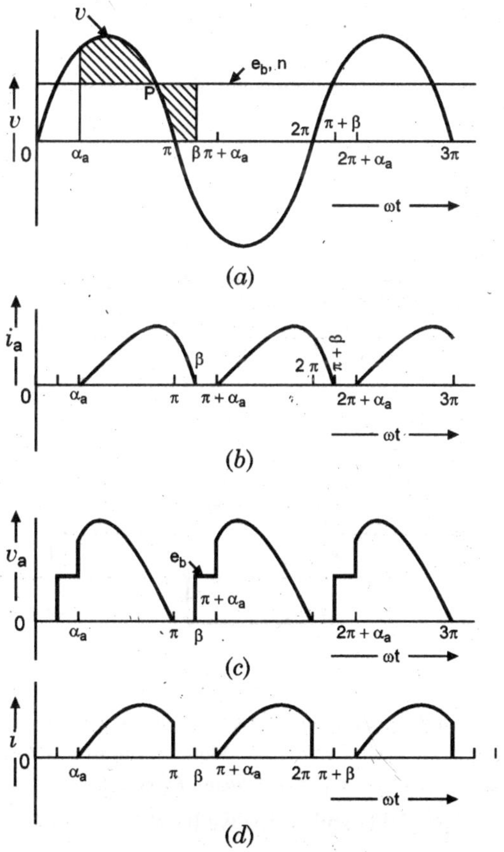

Fig. 3.6. ***Voltage and Current Waveforms of Semiconverter Feeding Separately Excited DC Motor: Discontinuous Armature Current***

The voltage and current waveforms for semiconverter with discontinuous current are illustrated in Fig. 3.6. The motor is connected to the supply through TH_1 and D_2 for the period $\alpha_a < \omega t < \pi$. Beyond π, the motor is shorted through the freewheeling diode D_{FW}. The armature current falls to zero at angle β (extinction angle) $< \pi + \alpha_a$ *i.e.*, before the thyristor TH_2 is fired, thereby making the armature current discontinuous, During α_a to π, the conduction period through TH_1 and D_2, the motor terminal voltage is the same as the input voltage. During π to β, the motor terminal voltage is zero, motor terminals being shorted by the freewheeling diode D_{FW}. From β to $\pi + \alpha_a$, the motor coasts and, therefore, its terminal voltage is the same as its induced emf. It should be noted here that the fundamental of the current drawn from the mains lags behind the voltage by an angle ϕ, ($< \alpha_a$).

Torque-Speed Characteristics. The emf induced in armature conductors of a motor, known as back emf, is given by

$$E_b = k_e \phi N \qquad ...(3.5)$$

where K_e is emf constant and is equal to $\frac{PZ}{60A}$.

In a separately excited dc motor, the field winding is excited from a separate source. So voltage applied to the armature is given by

$$V_a = E_b + I_a R_a \qquad ...(3.6)$$

where I_a is armature current in amperes and R_a is the armature resistance in ohms.

Thus, from Eqs. (3.5) and (3.6), we have

$$\text{Back emf, } E_b = V - I_a R_a = k_e N \phi$$

$$\text{or speed of motor, } N = \frac{V - I_a R_a}{k_e \phi} \qquad ...(3.7)$$

Since field is excited separately, field current I_f and flux ϕ are constant. Moreover voltage drop in armature, $I_a R_a$ is negligible. So motor speed N is constant if voltage applied to the armature is constant, and speed varies directly in proportion to the applied voltage *i.e.*, $N \propto V_a$.

Electromagnetic torque developed in a dc motor is given by

$$T_e = k_T \phi I_a \qquad ...(3.8)$$

where k_T is torque constant and is equal to $\frac{1}{2\pi}\frac{PZ}{A}$

$$\text{and torque constant, } k_T = \frac{1}{2\pi}\frac{PZ}{A} = \frac{60}{2\pi} \times \frac{PZ}{60\,A} = \frac{60}{2\pi} k_e \qquad ...(3.9)$$

From Eq. (3.8), we have

$$\text{Armature current, } I_a = \frac{T_e}{k_T \phi} \qquad ...(3.10)$$

From Eqs. (3.7) and (3.10), we have

$$N = \frac{V_a}{k_e \phi} - \frac{T_e R_a}{k_T k_e \phi^2} \qquad ...(3.11)$$

With a semiconverter,

$$V_a = \frac{V_{max}}{\pi}(1 + \cos \alpha_a) \qquad \text{... Refer to Eq. (3.3)}$$

$$\text{So } N = \frac{V_{max}(1 + \cos \alpha_a)}{\pi k_e \phi} - \frac{T_e R_a}{k_T k_e \phi^2} \qquad ...(3.12)$$

The first term of RHS of Eq. (3.12) represents the theoretical speed and the second term represents the speed drop due to voltage drop in armature circuit.

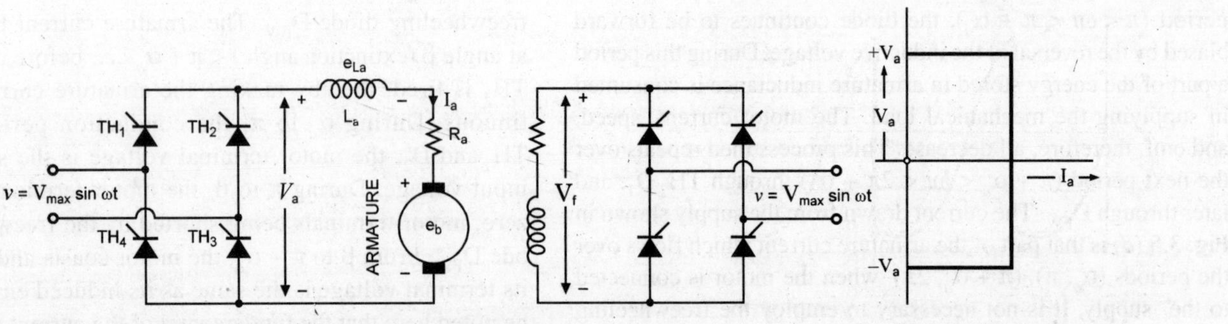

(a) Circuit Diagram (b) Quadrant of Operation

Fig. 3.7. ***Separately Excited DC Motor Supplied From Single Phase AC Supply Through Full Converters***

The theoretical no-load speed can be controlled by varying the firing angle α_a.

3. ***Single Phase Full-Converter Drives.*** The circuit diagram is illustrated in Fig. 3.7 (*a*). A fullconverter needs four thyristors but no freewheeling diode. Full converters are employed for both the armature and field supply. As illustrated in Fig. 3.7 (*b*), two quadrant operation is possible. The converter in the armature circuit provides $+V_a$ or $-V_a$ thus allowing operation in first and fourth quadrants. Current remains unidirectional because of the unidirectional thyristors. When operating in the fourth quadrant regenerative braking is possible and motor feeds back energy to the source. If α_a and α_f are the firing angles for the converters used in the armature and field circuits, the armature voltage V_a and field voltage V_f are given as

$$V_a = \frac{2V_{max}}{\pi} \cos \alpha_a \text{ for } 0 \le \alpha_a \le \pi \qquad ...(3.13)$$

$$V_f = \frac{2V_{max}}{\pi} \cos \alpha_f \text{ for } 0 \le \alpha_f \le \pi \qquad ...(3.14)$$

This drive is also employed for motors of rating up to 15 kW.

The voltage and current waveforms are illustrated in Fig. 3.8; armature current i_a being assumed to be almost constant. During the time interval $\alpha_a < \omega t < \pi + \alpha_a$, thyristors TH_1 and TH_3 conduct and connect the motor to supply. At $\pi + \alpha_a$, thyristors TH_2 and TH_4 are triggered. Immediately the supply voltage appears in reverse bias across thyristors TH_1 and TH_3 and turns them off. This is *natural* or *line commutation*. The motor current is transferred from thyristors TH_1 and TH_3 to thyristors TH_2 and TH_4. During the time interval α_a to π energy is supplied to the motor (both *v* and *i* are positive and, therefore, are v_a and i_a). But during time interval π to $\pi + \alpha_a$, some of the motor energy is fed back to the supply (*v* and *i* and so the v_a and i_a being of opposite polarity). The noteworthy point is that the fundamental of the current drawn from the supply mains lags behind the applied voltage by angle $\phi_1 = \alpha_a$.

The voltage and current waveforms for $\alpha_a > 90°$ are illustrated in Fig. 3.9. The average motor terminal voltage is now negative. On reversing of the motor terminals, it will operate as a generator supplying power back to the ac supply. This is the *inversion operation* of the converter and is employed in regenerative braking of the motor. The noteworthy point here is that during the conduction period of either TH_1 TH_3 or TH_2 TH_4 as the supply voltage becomes negative, the armature current begins to drop, causing the inductance polarity to reverse and thus conducting thyristors continue to be forward biased.

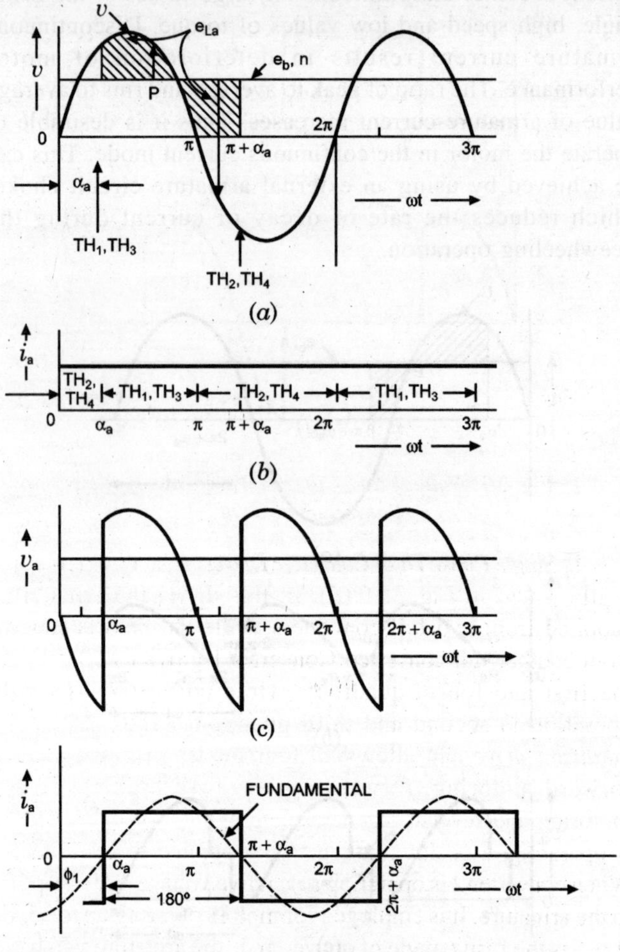

Fig. 3.8. ***Voltage and Current Waveforms of Full converter Supplying Separately Excited DC Motor; Continuous Armature Current (Almost Constant): α < 90°***

From Eqs. (3.11) and (3.13) we have

$$\text{Speed } N = \frac{2V_{max} \cos \alpha_a}{\pi k_e \phi} - \frac{T_e R_a}{k_T k_e \phi^2} \qquad ...(3.15)$$

In the case of discontinuous current, the average voltage at motor terminals depends upon the extinction angle β which itself depends on the average motor speed N, average armature current I_a and firing angle α_a. We need not to go in its analytical treatment.

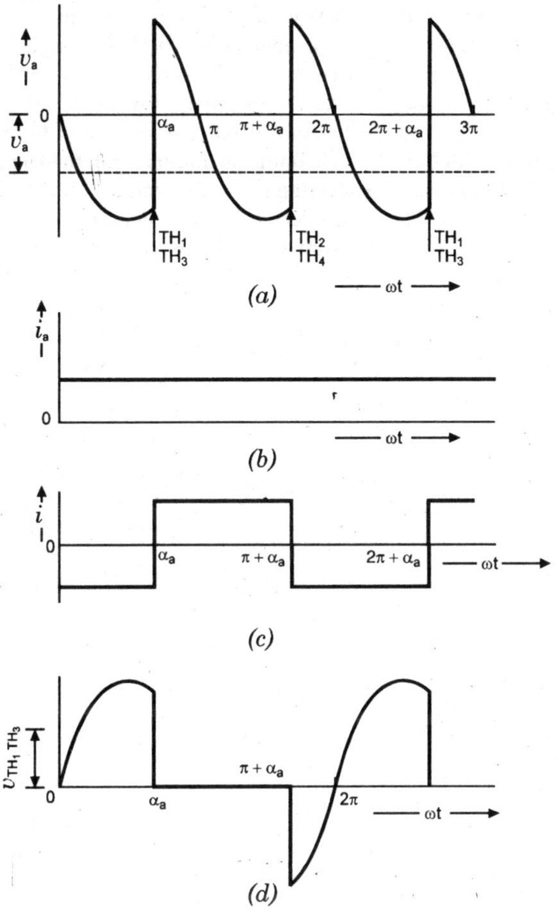

Fig. 3.9. ***Voltage and Current Waveforms For Full Converter:*** $\alpha_a > 90°$

4. ***Single Phase Dual Converter Drives.*** The circuit diagram is illustrated in Fig. 3.10 (a). In this drive, the armature is supplied from two full converters while the field is supplied from another full converter. Converter 1 provides operation in the first and fourth quadrants while converter 2 provides operation in second and third quadrants. Thus it is a *four quadrant drive* and allows all four modes of operation, *i.e.,* forward motoring, forward regenerative braking, reverse motoring and reverse regenerative braking. When converter 1 is operating, a positive voltage V_a is applied to the armature. When converter 2 is operating a negative voltage $-V_a$ is applied to the armature. It is employed for motors of rating up to 15 kW. If α_{a_1} is the firing angle of converter 1, the armature voltage V_a is given as

$$V_a = \frac{2V_{max}}{\pi}\cos\alpha_{a_1}, \text{ for } 0 \le \alpha_{a_1} \le \pi \qquad ...(3.16)$$

The firing angle α_{a_2} of converter 2 will be equal to $(\pi - \alpha_{a_1})$ and when converter 2 is operating, armature voltage is given as

$$V_a = \frac{2V_{max}}{\pi}\cos\alpha_{a_2} \quad \text{for } 0 \le \alpha_{a_2} \le \pi \qquad ...(3.17)$$

The voltage applied to the field is given as

$$V_f = \frac{2V_{max}}{\pi}\cos\alpha_f \quad \text{for } 0 \le \alpha_f \le \pi \qquad ...(3.18)$$

where α_f is the firing angle of converter supplying the field. Figure 3.10 (b) illustrates the four quadrants of operation.

Example 3.1. A single-phase half-wave converter is operated from a 120 V, 60 Hz supply. If the load is resistive, R = 10 Ω and the delay angle $\alpha = \frac{\pi}{3}$, determine:

(*i*) the efficiency (*ii*) the form factor (*iii*) the ripple factor (*iv*) the transformer utilisation factor ; and (*v*) peak inverse voltage of the thyristor.

[*A.M.I.E. Sec. B. Electric Drives and Their Control Summer 2000*]

Solution : In a half-wave converter, negative half waves are eliminated. The average voltage across load is

$$V_{dc} = \frac{1}{2\pi}\int_\alpha^\pi V_{max}\sin\omega t\, d(\omega t) = \frac{V_{max}}{2\pi}(1+\cos\alpha)$$

$$= \frac{120\sqrt{2}}{2\pi}\left(1+\cos\frac{\pi}{3}\right)$$

$$= \frac{120\sqrt{2}}{2\pi} \times 1.5 = 40.5 \text{ V}$$

$$I_{dc} = \frac{V_{dc}}{R} = \frac{40.5}{10} = 4.05 \text{ A}$$

RMS voltage across the load,

$$V_{rms} = \left[\frac{1}{2\pi}\int_\alpha^\pi (V_{max}\sin\omega t)^2\, d(\omega t)\right]^{1/2}$$

$$= V_{max}\left[\frac{1}{4\pi}\int_\alpha^\pi (1-\cos 2\omega t)\, d(\omega t)\right]^{1/2}$$

$$= V_{max}\left[\frac{1}{4\pi}\left|\left(\omega t - \frac{\sin 2\omega t}{2}\right)\right|_\alpha^\pi\right]^{1/2}$$

$$= V_{max}\left[\frac{\pi-\alpha}{4\pi} + \frac{\sin 2\alpha}{8\pi}\right]^{1/2}$$

$$= 120\sqrt{2}\left[\frac{\pi-\pi/3}{4\pi} + \frac{\sin 2\pi/3}{8\pi}\right]^{1/2} = 76.1 \text{ V}$$

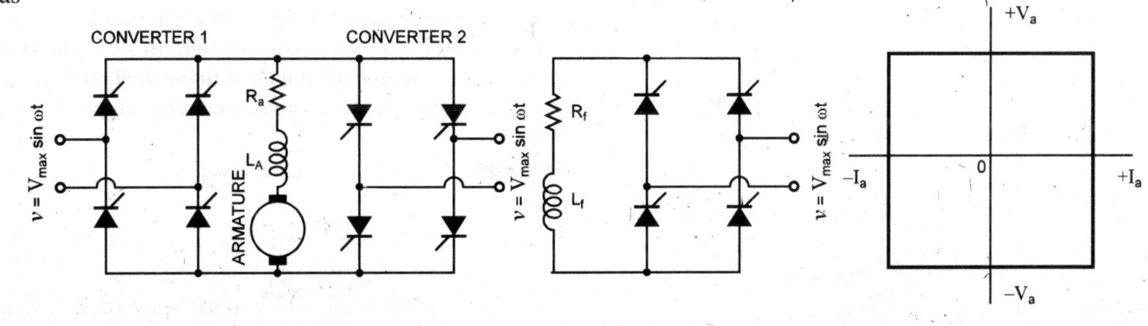

(*a*) *Circuit Diagram* (*b*) *Quadrant of Operation*

Fig. 3.10. ***Separately Excited DC Motor Supplied From Single Phase AC Supply Through Dual Converters***

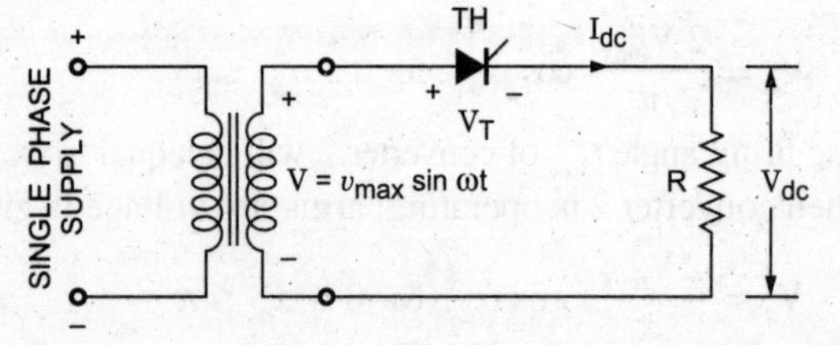

(a) Circuit Diagram

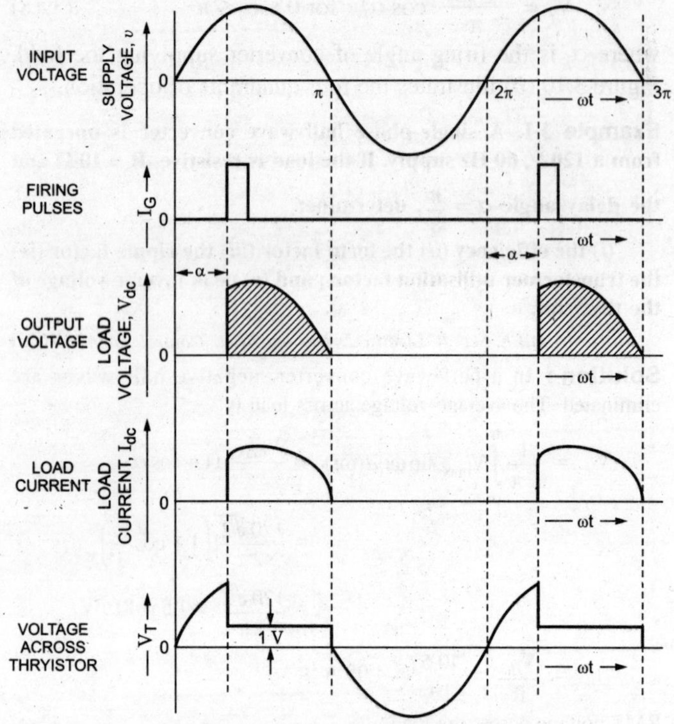

(b) Waveforms of Input Voltage, Firing Pulses, Output Voltage, Current and Voltage Drop Across Thyristor

Fig. 3.11. ***Single Phase Half-Wave Converter***

$$I_{rms} = \frac{V_{rms}}{R} = \frac{76.1}{10} = 7.61 \text{ A}$$

DC output, $P_{dc} = V_{dc} I_{dc} = 40.5 \times 4.05 = 164$ W

AC input, $P_{ac} = V_{rms} I_{rms} = 76.1 \times 7.61 = 580$ W

(*i*) Rectification efficiency, $\eta = \frac{P_{dc}}{P_{ac}} = \frac{164}{580} = 0.283$ **Ans.**

(*ii*) Form factor, $K_f = \frac{\text{RMS voltage}}{\text{Average voltage}} = \frac{76.1}{40.5}$

$= 1.879$ **Ans.**

(*iii*) Ripple factor, $\gamma = \sqrt{(K_f^2 - 1)} = \sqrt{1.879^2 - 1}$

$= 1.59$ **Ans.**

Volt-ampere rating of transformer $= 120 \times 7.61 = 913.2$

(*iv*) Transformer utilisation factor $= \frac{P_{dc}}{\text{VA rating of transformer}}$

$= \frac{164}{913.2} = 0.18$ **Ans.**

(*v*) Peak inverse voltage, PIV $= V_{max} = 120\sqrt{2} = 169.7$ V **Ans.**

Example 3.2. A single phase, 220 V, 50 Hz supply feeds a separately excited dc motor through two single phase semiconverters, one for the field and another for armature. The firing angle for the field semiconverter is zero. The field resistance is 250 Ω and armature resistance is 0.2 Ω. The load torque is 50 N-m at 1,000 rpm. The voltage constant is 0.8 V/A-rad/s and the torque constant is 0.8 N-m/A². Assuming armature and field currents to be continuous and neglecting losses determine (*i*) field current (*ii*) firing angle of the converter in the armature circuit and (*iii*) power factor of the converter of the armature circuit.

Solution : Field circuit voltage,

$$V_f = \frac{V_{max}}{\pi}(1 + \cos\alpha_f) = \frac{220\sqrt{2}}{\pi} \times (1 + \cos 0°) = 198 \text{ V}$$

(*i*) Field current, $I_f = \frac{V_f}{R_f} = \frac{198}{250} = 0.792$ A **Ans.**

(*ii*) Since torque developed, $T_d = K_f I_f I_a$ where K_f is torque constant

So armature current, $I_a = \frac{T_d}{K_f I_f} = \frac{50}{0.8 \times 0.792} = 78.9$ A

$$\omega = \frac{2\pi N}{60} = \frac{2\pi \times 1{,}000}{60} = 104.72 \text{ rad/s}$$

Back emf, E_b = Voltage constant × ω × I_f

$= 0.8 \times 104.72 \times 0.792 = 66.35$ V

Armature voltage, $V_a = E_b + I_a R_a = 66.35 + 78.9 \times 0.2 = 82.13$ V

Since from Eq. (3.3),

$$V_a = \frac{V_{max}}{\pi}(1 + \cos\alpha_a)$$

So $82.13 = \frac{220\sqrt{2}}{\pi}(1 + \cos\alpha_a)$

or $\alpha_a = 99.83°$ **Ans.**

(*iii*) Power output of converter in the armature circuit,

$= V_a I_a = 82.13 \times 78.9 = 6{,}480$ W

$$\text{Input current} = \left[\frac{2}{2\pi}\int_{\alpha_a}^{\pi} I_a^2 \, d(\omega t)\right]^{1/2}$$

$$= I_a\left[\frac{\pi - \alpha_a}{\pi}\right]^{1/2} = 78.9\left[\frac{180 - 99.83}{180}\right]^{1/2}$$

$= 52.66$ A

Input volt-amperes $= 220 \times 52.66 = 11{,}585.2$

Input power factor $= \frac{6{,}480}{11{,}585.2} = 0.56$ lagging **Ans.**

Example 3.3. A single phase full converter is used to control the speed of 5 kW, 220 V, 1,000 rpm separately excited dc motor. The armature resistance is 0.2 Ω and the rated armature current is 30 A. The ac supply voltage is 230 V. The motor voltage constant is $K_a\phi$ = 1.668 volt/radian/second. Determine (*i*) the speed of the motor and (*ii*) motor torque for a firing angle α = 45° and rated motor current. Assume continuous and ripple free armature current.

Solution : Voltage across the armature,

$$V_a = \frac{2V_{max}}{\pi}\cos\alpha_a \qquad \text{Refer to Eq. (3.13)}$$

$$= \frac{2 \times 230\sqrt{2}}{\pi}\cos 45° = 146.42 \text{ V}$$

Back emf, $E_b = V_a - I_a R_a = 146.42 - 30 \times 0.2 = 140.42$ V (*i*) So speed, $\omega = \frac{E_b}{K_a\phi} = \frac{140.42}{1.668} = 84.186$ radians/second

or $N = \frac{\omega}{2\pi} \times 60 = \frac{84.186}{2\pi} \times 60 = 804$ rpm **Ans.**

(*ii*) Motor torque, $T = K_a\phi I_a = 1.668 \times 30 = 50.04$ Nm **Ans.**

Example 3.4. A 200 V, 875 rpm, 150 A separately excited dc motor has an armature resistance of 0.06 Ω. It is fed from a single phase fully controlled rectifier with an ac source voltage of 220 V, 50 Hz. Assuming continuous conduction calculate the firing angle for rated motor torque at 750 rpm.

[*A.M.I.E. Sec. B. Electric Drives and Their Control Winter 1998*]

Solution : Back emf of motor when operated from 200 V dc supply and runs at 875 rpm,

$$E_{b_1} = V_{dc} - I_a R_a = 200 - 150 \times 0.06 = 191 \text{ V}$$

Back emf of motor when fed from 220 V 50 Hz single phase supply through fully controlled rectifier and runs at 750 rpm (assuming flux ϕ to remain constant),

$$E_{b_2} = E_{b_1} \times \frac{N_2}{N_1} = 191 \times \frac{750}{875} = 163.71 \text{ V}$$

Armature current $I'_a = I_a = 150$ A

∵ torque remains the same and flux is assumed to remain constant

Armature voltage, $V_a = E_b + I_a R_a = 163.71 + 150 \times 0.06 = 172.71$ V

Since for a fully controlled single phase rectifier voltage across armature is given as

$$V_a = \frac{2V_{max}}{\pi} \cos \alpha_a$$

So $172.71 = \frac{2 \times 220\sqrt{2}}{\pi} \cos \alpha_a$

or $\alpha_a = 29.31°$ **Ans.**

Example 3.5. A single phase fully controlled thyristor converter is connected to a load comprising 2 Ω resistance and 0.3 H inductance. The supply voltage is 230 V at 50 Hz. Estimate the average load voltage, average load current and input power factor for triggering angle of 30°. Assume continuous and ripple free load current. [*U.P. Technical Univ. Energy Conversion 2005-06*]

Solution : Supply voltage $V_{(rms\ value)} = 230$ V

$$V_{max} = 230\sqrt{2} \text{ V}$$

Average load voltage, $V_{L\,av} = \frac{2V_{max}}{\pi} \cos \alpha = \frac{2 \times \sqrt{2} \times 230 \cos 30°}{\pi} = 179.33$ V **Ans.**

Average load current, $I_{L\,av} = \frac{V_{L\,av}}{R} = \frac{179.33}{2} = 89.665$ A **Ans.**

As load current is ripple free,

RMS value of load current, $I_{rms} = I_{L\,av} = 89.665$ A

Active power of load, $P = V_{L\,av} \times I_{L\,av} = 179.33 \times 89.665 = 16{,}080$ W

Reactive power of load, $Q = \frac{2V_{max}}{\pi} I_{L\,av} \sin \alpha = \frac{2 \times \sqrt{2} \times 230}{\pi} \times 89.665 \sin 30° = 9{,}283.6$ VAR

Input power factor

$$= \cos \tan^{-1} \frac{Q}{P} = \cos \tan^{-1} \frac{9283.6}{16{,}080} = \cos 30° = 0.866 \text{ (lagging)}$$ **Ans.**

Example 3.6. A separately excited dc motor is controlled by varying its armature voltage using a single phase full-converter bridge as shown in Fig. 3.12. The field current is kept constant at the rated value. The motor has an armature resistance of 0.2 Ω, and the motor voltage constant is 2.5 V/rad/sec. The motor is driving a mechanical load having a constant torque of 140 Nm. The triggering angle of the converter is 60°. The armature current can be assumed to be continuous and ripple free.

(*a*) **Calculate the motor armature current.**

(*b*) **Evaluate the motor speed in rad/sec.** [*GATE 2001*]

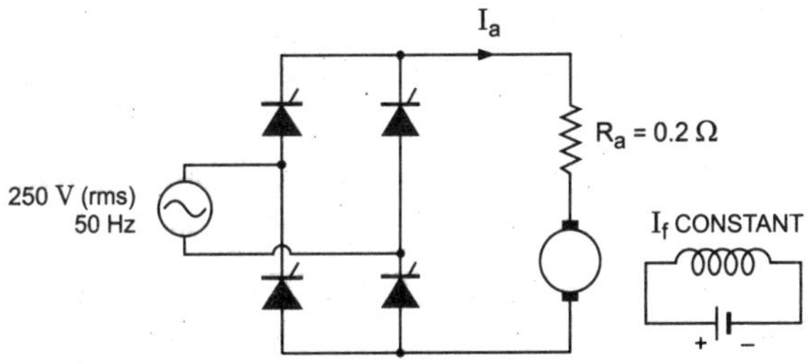

Fig. 3.12

Solution : (*a*) Voltage across the motor armature

$$V_a = \frac{2V_{max}}{\pi} \cos \alpha = \frac{2 \times 250\sqrt{2}}{\pi} \cos 60° = 112.5 \text{ V}$$

∵ $V_{max} = 250\sqrt{2}$ V and $\alpha = 60°$

Now $T\omega = E_b I_a = K_a \omega I_a$ ∵ $E_b = K_a \omega$ ...(*i*)

or Armature current, $I_a = \frac{T}{K_a} = \frac{140}{2.5} = 56$ A **Ans.**

(*b*) Back emf, $E_b = V - I_a R_a = 112.5 - 56 \times 0.2 = 101.34$ V

From Eq. (*i*) Speed, $\omega = \frac{E_b I_a}{T} = \frac{101.34 \times 56}{140} = 40.536$ radians/second **Ans.**

Example 3.7. A 220 V, 1500 rpm, 10 A separately excited dc motor is fed from a single phase fully controlled rectifier with an ac source voltage of 230 V, 50 Hz, R_a = 2 Ω. Assuming continuous conduction, find the firing angles for :

(*i*) **Half the rated torque 500 rpm**

(*ii*) **Rated motor torque and (–1000) rpm.**

[*U.P. Technical Univ. Electric Drives 2009-10*]

Solution : Back emf of motor when operated from 220 V dc supply and runs at 1,500 rpm

$$E_{b_1} = V_{dc} - I_{a_1} R_a = 220 - 10 \times 2 = 200 \text{ V}$$

(*i*) Back emf of motor when fed from 230 V, 50 Hz single phase supply through fully controlled rectifier and runs at 500 rpm (assuming flux ϕ to remain constant)

$$E_{b_2} = E_{b_1} \times \frac{N_2}{N_1} = 200 \times \frac{500}{1{,}500} = 66.67 \text{ V}$$

Armature current, $I_{a_2} = \frac{1}{2} I_{a_1} = \frac{1}{2} \times 10 = 5$ A

∵ Torque is half and flux is assumed to remain constant

Armature voltage, $V_{a_1} = E_{b_2} + I_{a_2}R_a = 66.67 + 5 \times 2 = 76.67$ V

Since for a fully controlled single phase rectifier voltage across armature is given as

$$V_a = \frac{2V_{max}}{\pi}\cos\alpha_a$$

$$\text{or } 76.67 = \frac{2\times\sqrt{2}\times 230}{\pi}\cos\alpha_{a_1}$$

$$\text{or } \cos\alpha_{a_1} = \frac{76.67\times\pi}{2\sqrt{2}\times 230} = 0.37$$

$$\text{or } \alpha_{a_1} = \cos^{-1} 0.37 = 68.27° \quad \textbf{Ans.}$$

(*ii*) Back emf of motor when fed from 230 V, 50 Hz single phase supply through fully controlled rectifier and runs at –1,000 rpm (assuming flux *f* to remain constant)

$$E_{b_3} = E_{b_1} \times \frac{N_3}{N_1} = 200 \times \frac{-1{,}000}{1{,}500} = -133.33 \text{ V}$$

Armature current, $I_{a_3} = I_{a_1} = 10$ A ∵ Torque remains the same and flux is assumed to remain constant

Armature voltage, $V_{a_2} = E_{b_3} + I_{a_3}R_a = -133.33 + 10\times 2 = -113.33$ V

$$\text{and } V_{a_2} = \frac{2V_{max}}{\pi}\cos\alpha_{a_2}$$

$$\text{So } -113.33 = \frac{2\times\sqrt{2}\times 230}{\pi}\cos\alpha_{a_2}$$

$$\text{or } \cos\alpha_{a_2} = \cos^{-1}(-0.5473) = 123.18° \quad \textbf{Ans.}$$

Example 3.8. A 220 V single 50 Hz supply feeds the armature and field circuits of a separately excited dc motor through two full converters. The firing angle of field circuit converter is zero. The armature and field resistances are 0.2 Ω and 200 Ω respectively. The torque and voltage constants are 1.12. The firing angle of the armature circuit converter is 30° and armature current is 40 A. Find torque developed and motor speed. Assume that field and armature currents are continuous and constant and neglect brush contact drop.

Solution : Field circuit voltage,

$$V_f = \frac{2V_{max}}{\pi}\cos\alpha_f = \frac{2\times 220\sqrt{2}}{\pi}\cos 0° = 198 \text{ V}$$

Field circuit current,

$$I_f = \frac{V_f}{R_f} = \frac{198}{200} = 0.99 \text{ A}$$

Armature circuit voltage,

$$V_a = \frac{2V_{max}\cos\alpha_a}{\pi} = \frac{2\times 220\sqrt{2}}{\pi}\cos 30° = 171.53 \text{ V}$$

Back emf, $E_b = V_a - I_aR_a = 171.53 - 40\times 0.2 = 163.53$ V

Torque developed,

$$T_d = K_t I_a I_f = 1.12\times 40\times 0.99 = 44.352 \text{ N-m} \quad \textbf{Ans.}$$

$$\text{Speed, } N = \frac{60\omega}{2\pi} = \frac{60}{2\pi}\times\frac{\text{back emf}}{K_t I_f} = \frac{60}{2\pi}\times\frac{163.53}{1.12\times 0.99}$$

$$= 1{,}408 \text{ rpm} \quad \textbf{Ans.}$$

3.6.2. Three Phase Drives

For large sized (rating exceeding 15 kW) motors 3-phase drive is employed. The output voltage of a 3-phase converter has less ripple contents than that of single phase converter. So the filtering requirements for smoothing out the armature current are less and the current is mostly continuous. Three-phase converters could be half-wave, semiconverter, full-converter and dual converter. The 3-phase half-wave converter drives are only of theoretical importance and are generally not employed in industrial applications. This is because of presence of dc component in supply currents. Semi- and full-converters are widely used. A dual converter is employed in case of reversible drives with a power rating up to several MW.

1. *Three-Phase Semiconverter Drives.* It is a one quadrant drive and is employed for motors of rating up to about 100 kW. The field converter may be 3-phase or single-phase. The armature voltage V_a is given as

$$V_a = \frac{3\sqrt{3}V_{max}}{2\pi}(1+\cos\alpha_a) \text{ for } 0 \le \alpha_a \le \pi \quad \text{...(3.19)}$$

where V_{max} is the peak value of line-to-neutral voltage and α_a is the firing angle of the converter in the armature circuit.

The circuit diagram for a separately excited dc motor supplied from 3-phase ac supply through a three-phase semiconverter is shown in Fig. 3.13.

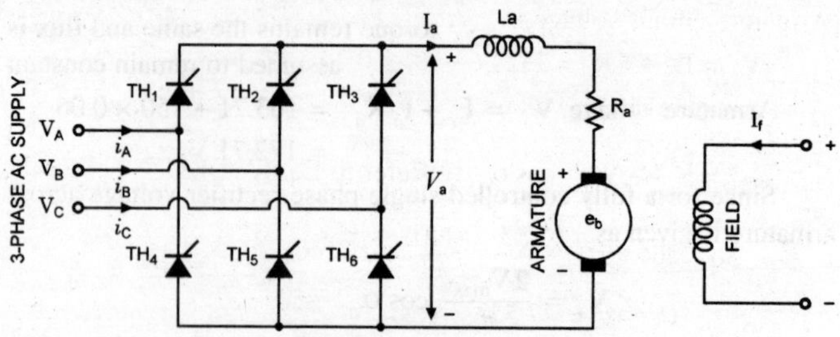

Fig. 3.13. *Three-Phase Semiconverter Supplying Separately Excited DC Motor*

2. *Three-Phase Full Converter Drives.* Three-phase full converter drive is a two quadrant drive and is employed up to 1,500 kW motors. The converter in the field circuit may be single-phase or 3-phase. When 3-phase full converters are used for both armature and field, the armature and field voltages are given as

$$V_a = \frac{3\sqrt{3}V_{max}}{\pi}\cos\alpha_a \text{ for } 0 \le \alpha_a \le \pi \quad \text{...(3.20)}$$

$$V_f = \frac{3\sqrt{3}V_{max}}{\pi}\cos\alpha_f \text{ for } 0 \le \alpha_f \le \pi \quad \text{...(3.21)}$$

3. *Three-Phase Dual Converter Drives.* 3-phase dual converter consists of two converters and can be operated in all the four quadrants. Either converter 1 is operated to supply armature voltage V_a or converter 2 is operated to supply armature voltage $-V_a$. It is used for motors of rating up to 1,500 kW. The armature voltage supplied by converter 1 is given as

$$V_a = \frac{3\sqrt{3}V_{max}}{\pi}\cos\alpha_{a_1}, \text{ for } 0 \le \alpha_{a_1} \le \pi \quad \text{...(3.22)}$$

where α_{a_1} is the firing angle for converter 1.

For converter 2, the voltage V_a is given as

$$V_a = \frac{3\sqrt{3}V_{max}}{\pi}\cos\alpha_{a_2} \text{ for } 0 \le \alpha_{a_2} \le \pi \quad \text{...(3.23)}$$

where $\alpha_{a_2} = \pi - \alpha_{a_1}$.

If the field is also supplied by a 3-phase full converter,

$$V_f = \frac{3\sqrt{3}\, V_{max}}{\pi} \cos \alpha_f \text{ for } 0 \le \alpha_f < \pi \qquad ...(3.24)$$

Example 3.9. A 400 V, 3-phase ac supply feeds a separately excited dc motor through two 3-phase full converters, one for armature and the other for the field. The firing angle of the field circuit converter is zero. The armature and field resistances are 0.2 Ω and 250 Ω respectively. The armature current is 60 A. The motor voltage constant is 1.3 V/A-rad/second. The speed is 1,200 rpm. Determine the firing angle of the armature circuit converter.

Solution : Field circuit voltage,

$$V_f = \frac{3\sqrt{3}}{\pi} V_{max} \cos \alpha_f = \frac{3\sqrt{3}}{\pi} \times \frac{400}{\sqrt{3}} \times \sqrt{2} \cos 0° = 540.2 \text{ V}$$

...Refer to Eq. (3.21)

Field circuit current,

$$I_f = \frac{V_f}{R_f} = \frac{540.2}{250} = 2.16 \text{ A}$$

Back emf developed,

$$E_b = KI_f\omega = 1.3 \times 2.16 \times \frac{2\pi \times 1{,}200}{60} = 352.86 \text{ V}$$

Armature circuit voltage,

$$V_a = E_b + I_aR_a = 352.86 + 60 \times 0.2 = 364.86 \text{ V}$$

Also armature circuit voltage

$$= \frac{3\sqrt{3}}{\pi} V_{max} \cos \alpha_a \quad \text{...Refer to Eq. (3.20)}$$

$$\text{So} \quad 364.86 = \frac{3\sqrt{3}}{\pi} \times \frac{400}{\sqrt{3}} \times \sqrt{2} \cos \alpha_a$$

or $\alpha_a = 47.52°$ **Ans.**

Example 3.10. A 3-phase half-controlled thyristor bridge with 400 V, 3-phase, 50 Hz supply is feeding a separately excited dc motor. Armature resistance is 0.2 Ω, armature rated current is 100 A and back emf constant is 0.25 V/rpm. Determine no-load speed if no-load armature current is 5 A and firing angle is 45°. Also determine firing angle to obtain a speed of 1,500 rpm at rated current.

Solution : Armature circuit voltage at no load,

$$V_{a_0} = \frac{3\sqrt{3}}{2\pi} V_{max} (1 + \cos \alpha_a) \qquad \text{...Refer to Eq. (3.19)}$$

$$= \frac{3\sqrt{3}}{2\pi} \times \frac{400\sqrt{2}}{\sqrt{3}} (1 + \cos 45°) = 461 \text{ V}$$

Back emf at no load, $E_{b_0} = V_{a_0} - I_{a_0} R_a = 461 - 5 \times 0.2 = 460$ V

Speed at no load, $N_0 = \frac{460}{0.25} = 1{,}840$ rpm **Ans.**

Back emf developed at a speed of 1,500 rpm,

$$E_b = 1{,}500 \times 0.25 = 375 \text{ V}$$

Input voltage to motor, $V_a = E_b + I_aR_a = 375 + 100 \times 0.2 = 395$ V

$$\text{Also} \quad V_a = \frac{3\sqrt{3}}{2\pi} V_{max} (1 + \cos \alpha_a)$$

$$\therefore \quad 395 = \frac{3\sqrt{3}}{2\pi} \times \frac{400\sqrt{2}}{\sqrt{3}} (1 + \cos \alpha_a)$$

or $\alpha_a = 62.45°$ **Ans.**

Example 3.11. A separately excited dc motor is supplied from 3-phase, 400 V, 50 Hz supply through half controlled 3-phase bridge rectifier. The motor parameters are inductance 0.012 H, resistance 0.8 Ω, armature constant 2 V/rad/s (N-m/A). Neglecting rectifier losses and supply impedance, determine speed-torque characteristics up to 80 N-m at fixed firing angle of 90°.

Solution : Because of armature inductance, the armature current can be assumed to remain constant.

Armature circuit voltage,

$$V_a = \frac{3\sqrt{3}}{2\pi} V_{max} (1 + \cos \alpha_a)$$

$$= \frac{3\sqrt{3}}{2\pi} \times \frac{400}{\sqrt{3}} \times \sqrt{2} \; (1 + \cos 90°) = 270 \text{ V}$$

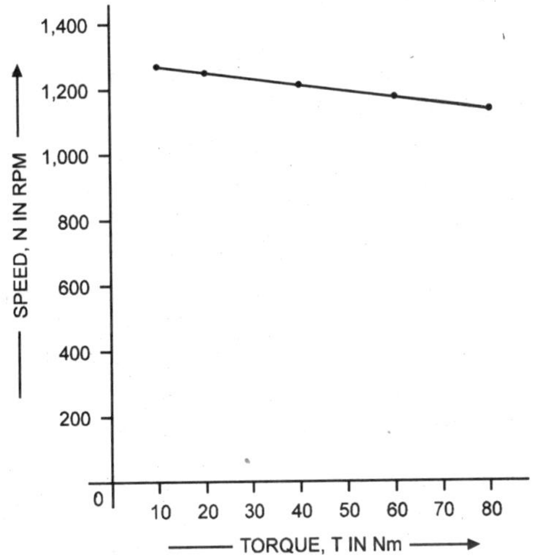

Fig. 3.14

Let us determine torque and speed for armature current of 5, 10, 20, 30, 40 A, as tabulated below :

Armature Current in Amperes, I_a	*Torque in N-m* $= K_aI_a$ $= 2 \times I_a$	*Back emf,* $E_b = V_a - I_aR_a$ $= 270 - I_a \times 0.8$ *in volts*	*Speed, N in RPM* $= \frac{E_b}{K_t} \times \frac{60}{2\pi} = \frac{E_b}{2} \times \frac{30}{\pi}$
5	$2 \times 5 = 10$	$270 - 5 \times 0.8 = 266$	$\frac{266}{2} \times \frac{30}{\pi} = 1{,}270$
10	$2 \times 10 = 20$	$270 - 10 \times 0.8 = 262$	$\frac{262}{2} \times \frac{30}{\pi} = 1{,}250$
20	$2 \times 20 = 40$	$270 - 20 \times 0.8 = 254$	$\frac{254}{2} \times \frac{30}{\pi} = 1{,}213$
30	$2 \times 30 = 60$	$270 - 30 \times 0.8 = 246$	$\frac{246}{2} \times \frac{30}{\pi} = 1{,}175$
40	$2 \times 40 = 80$	$270 - 40 \times 0.8 = 238$	$\frac{238}{2} \times \frac{30}{\pi} = 1{,}136$

Speed-torque characteristic is shown in Fig. 3.14.

3.7. CHOPPER CONTROL OF DC MOTORS

In many cases, conversion of dc supply source voltage to different levels is required. For example, subway cars, trolley buses, or battery operated vehicles are fed from a fixed voltage dc supply

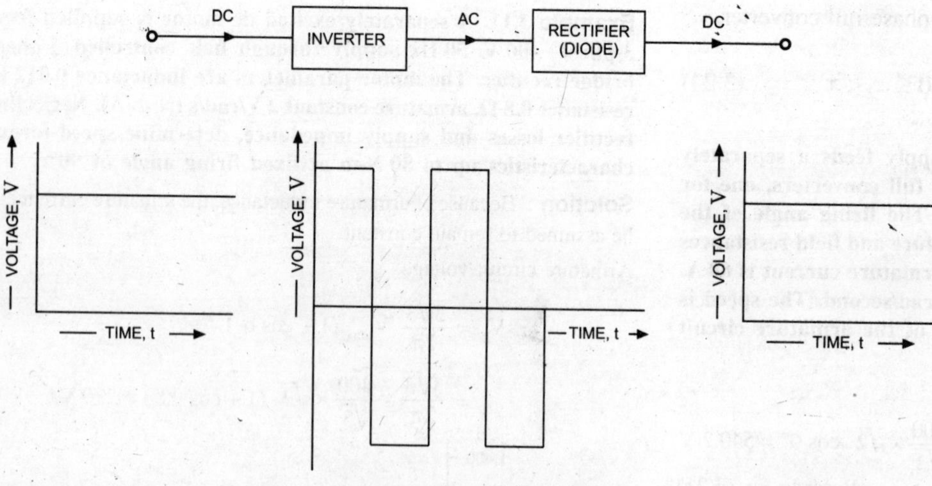

Fig. 3.15. *AC-Link Chopper*

source. But for their speed control we need conversion of fixed voltage dc source to a variable voltage dc source. Conventional methods for obtaining the variable dc voltage from a fixed dc voltage source are given below:

1. Resistance Control. In this method, a variable resistance is introduced between the load and the source. The drawbacks of this method are wastage of lot of energy and for a given output voltage, requirement of resistances of different values for different values of load current. This method is still employed for older traction installations.

2. Motor-Generator Set (M-G Set). Separately excited dc generator provides a voltage which can be varied from zero to rated value with either polarity. It has disadvantages of higher cost, large bulk, low efficiency and sluggish response because of the generator field time constant.

3. AC Link Chopper (Inverter-Rectifier). In this method, the dc supply is first converted to ac supply by an inverter (dc to ac converter). The obtained ac supply is then stepped up or down by a transformer and then rectified back to dc by a rectifier. Because of two stages of conversion, the set-up is costly, bulky and less efficient.

4. DC Chopper (DC to DC Power Converter). As already mentioned in Art. 3.3.3, a dc chopper is a static switch to provide variable dc voltage from a source of constant dc voltage. Choppers are now being used all over the world for rapid transit systems. These are also employed in trolley cars, marine hoists, forklift trucks and mine haulers. The future electric automobiles are likely to use choppers for their speed control and braking. Chopper systems offer smooth control, high efficiency, fast response and regeneration.

3.7.1. Principle of Operation of Chopper

A *dc chopper* is a high speed on/off semiconductor switch that connects the load to and disconnects it from the supply and produces a chopped load voltage from a constant input supply voltage. As it is an on/off switch between source and load, it can be symbolically represented by a switch S inside a dotted rectangle, as shown in Fig. 3.16. Actually, a chopper consists of main power semiconductor device together with their turn-on and turn-off mechanisms. In low-power chopper circuits, power transistors, GTOs etc. are used widely. In high-power chopper circuits, however, thyristors are in common use.

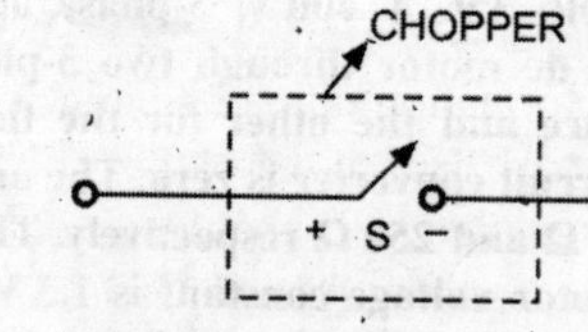

Fig. 3.16. *Symbol of Chopper*

Figure 3.17 illustrates the principle of operation of a chopper. The switch may be closed or opened by supplying on/off pulses to it. Here the mechanism of sending pulses to

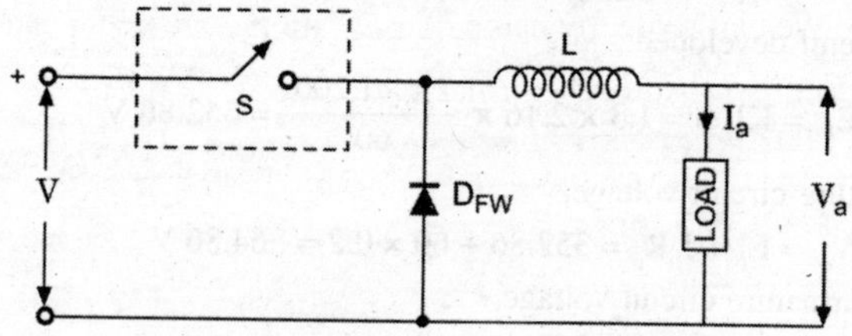

Fig. 3.17. *Elementary Chopper Circuit*

a switch has not been shown. A shunting diode D_{FW} is provided across the load for freewheeling the load current when the switch S is off. When the switch S is closed, the source voltage V, assuming that voltage drop across switch S in on-state is negligible, appears across the load for the duration the switch is closed. This duration is designated T_{on}. After the pulse to switch is removed the load is disconnected from the source. The energy stored in the inductor circulates a current through the freewheeling diode D_{FW}. As a result the load is short circuited and voltage across it falls to zero but the current through the load falls exponentially, as illustrated in Fig. 3.18. The duration for which switch is off is called off time T_{off}. The total time of a pulse is, therefore, given by $T = T_{on} + T_{off}$

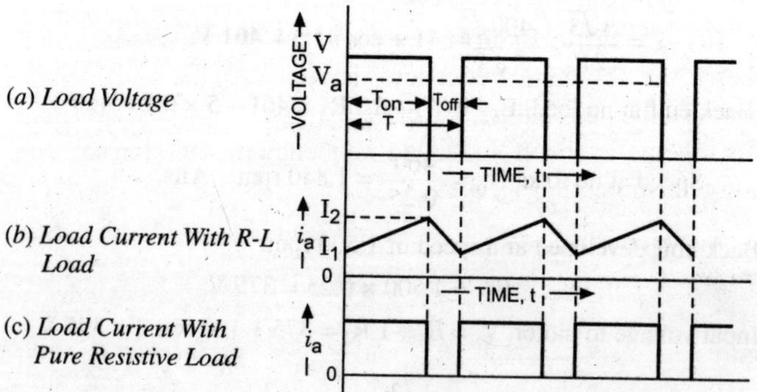

Fig. 3.18. *Voltage and Current Waveforms*

Thus, a dc chopped voltage is produced at the load terminals [Fig. 3.18 (*a*)] and the load current is continuous as shown in Figs. 3.18 (*b*) and (*c*).

From Fig. 3.18 (*a*), the average load voltage V_a is given by

$$V_a = V \frac{T_{on}}{T_{on} + T_{off}} = V \frac{T_{on}}{T} = \alpha V \qquad ...(3.25)$$

where $\alpha = \frac{T_{on}}{T}$ is called *duty cycle* of the circuit. It is analogous to transformation ratio of ac transformer. The value of α ranges from 0 to 1. Hence the output voltage varies between zero and supply voltage V. In the present case the output voltage is less than the supply voltage. It is, therefore, called **step-down** chopper. In choppers, called step-up choppers, the output voltage is more than supply voltage.

The load current and voltage waveforms, as shown in Figs. 3.18 (*b*) and (*a*) are for R-L load. However, in case, resistive load is used in the output, the current and voltage waveforms have similar variations as illustrated in Fig. 3.18 (*c*) and (*a*).

In case of R-L load, the inductor absorbs energy from the source during on-period while same is released by it during off-period. We can write, by using conservation of energy principle;

$$\text{Energy absorbed} = \text{Energy released}$$

$$\text{or} \quad VI = V_a I_a$$

$$\text{or} \quad I = \frac{V_a}{V} I_a = \alpha I_a$$

$$\text{or} \quad \frac{I}{I_a} = \alpha = \frac{V_a}{V} \qquad ...(3.26)$$

Above relation is similar to that for an ac transformer. Here α, the duty cycle, corresponds to transformation ratio of transformer. The output of a chopper, as can be observed from Eqs. (3.25) and (3.26), can be controlled by a variation in α, the duty cycle.

Control Strategies. It is seen from Eq. (3.25), that, average value of output voltage, V_a can be controlled by periodic opening and closing of the semiconductor switch. There are two types of control strategies for operating the switches that can be used in dc choppers. These control strategies are time-ratio control (TRC) and current limit control.

1. Time-Ratio Control. In this control strategy, the value of $\frac{T_{on}}{T}$ is varied. This is affected in two ways namely variable frequency control and constant frequency control.

(*a*) **Variable Frequency Control (FM).** Under such scheme the chopping frequency *f* is kept variable and for this purpose either (*a*) *on* time, T_{on} or (*b*) *off* time, T_{off} is kept constant. This scheme is, therefore, called frequency modulation scheme and is illustrated in Fig. 3.19.

In Fig. 3.19 (*a*) and (*b*) *on* time T_{on} is kept constant while *off* time T_{off} of the switch is variable. The duty cycles in two figures are $\alpha_1 = \frac{T_{on}}{T_{on} + T_{off}} = \frac{3}{4}$ and $\alpha_2 = \frac{1}{2}$ respectively.

Figure 3.19 (*c*) and (*d*) are for variable on time T_{on} and constant off time T_{off}. The duty cycle in the two cases are respectively $\frac{1}{4}$ and $\frac{2}{5}$. This technique is suitable for switches which require forced commutation to turn off.

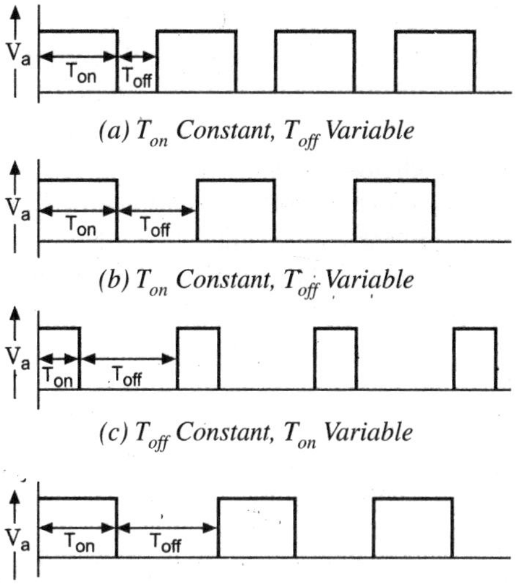

(*a*) T_{on} *Constant*, T_{off} *Variable*

(*b*) T_{on} *Constant*, T_{off} *Variable*

(*c*) T_{off} *Constant*, T_{on} *Variable*

(*d*) T_{off} *Constant*, T_{on} *Variable*

Fig. 3.19. ***Output Voltage Waveforms For Variable Frequency Control***

(*b*) **Constant Frequency Control.** Under this control scheme the chopping period T remains constant but the pulse width T_{on} is varied to control the output. This is, therefore, called pulse width modulation (PWM) control. The scheme is illustrated in Fig. 3.20. In the figure, time $T = T_{on} + T_{off}$ is constant but T_{on} and T_{off} both are variable. The duty cycles of α = 0.75 and 0.50 are shown, respectively, in Figs. 3.20 (*a*) and (*b*).

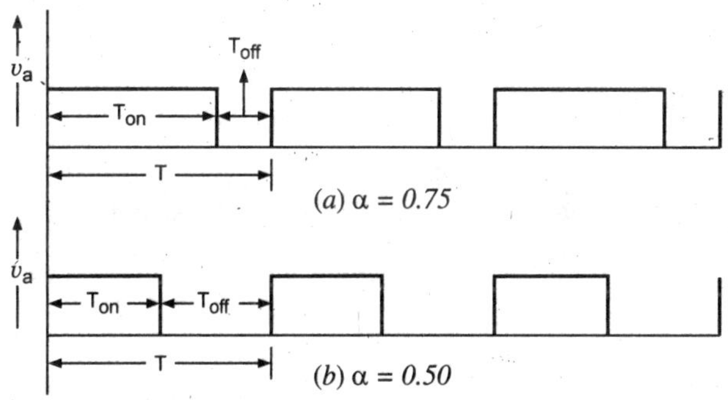

Fig. 3.20. ***Constant Frequency Control***

Variable frequency control has some disadvantage as compared to constant frequency control method which are:

(*i*) Chopping frequency is required to be varied over a large range to control the output. Thus the output has wide range of harmonics. The design of a filter to remove such a wide range of harmonics is quite difficult.

(*ii*) For the control of duty cycle the off period may be so large that load current falls to zero between two consecutive pulses. This leads to discontinuity in the load current which is not desirable.

(*iii*) For the control of duty cycle α, frequency variations would be wide. As such, there is a possibility of interference with signalling and telephone lines in frequency modulation control.

For above reasons PWM control is preferred over FM control in chopper circuits.

2. Current Limit Control. Under this scheme the chopper is switched *on* and *off* between upper and lower limits of load current. As the load current exceeds upper limit the chopper is turned *off*. The load current now freewheels until it reaches the lower value prescribed. As the load current attains lower limit, the chopper is switched *on*. As the freewheeling operation is involved in this scheme, a storage element is a necessity in the load. As the load current varies between two prescribed limits, this scheme is applicable only either at constant frequency or at constant turn *on* time T_{on}. The waveforms for current limit control are shown in Fig. 3.21. In this case the load current is continuous and difference between upper and lower values of current decides the switching frequency. At smaller values of difference in current, the switching frequency is higher. As a result the ripples in the output are reduced. As an example the switching frequencies of the chopper for $I'_{a\,max}$ and $I_{a\,max}$ as upper limits of the current are, respectively $\frac{1}{T'}$ and $\frac{1}{T}$ where $I'_{a\,max} < I_{a\,max}$ and $T' < T$. By further reducing the difference between upper and lower limits of the current the switching frequency can be increased correspondingly which leads to reduction in the ripple content in the output.

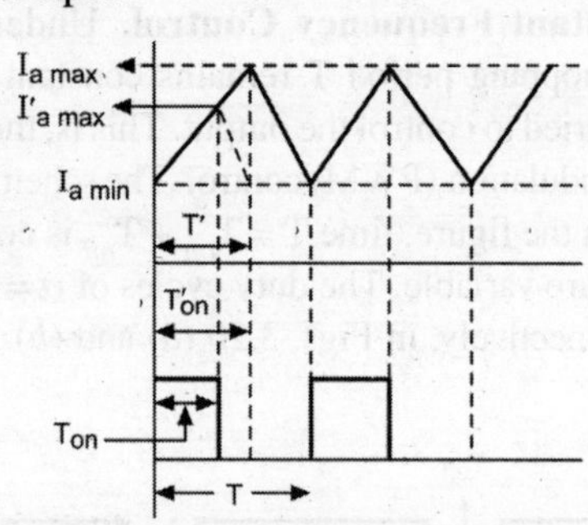

Fig. 3.21. ***Current Limit Control***

Example 3.12. For a dc to dc converter shown in Fig. 3.17, the source voltage is 200 volts, load resistance 10 Ω and duty cycle α = 0.4. If load is resistive, calculate the following (*a*) Average load voltage (*b*) Thyristor average current (*c*) Average and rms freewheeling diode currents (*d*) Effective input resistance of chopper.

Solution : For resistive load, the load voltage and load current are in phase:

(*a*) Average load voltage, $V_a = \alpha V_{dc} = 0.4 \times 200 = 80$ volts **Ans.**

(*b*) Average thyristor current $= V_d/R = \frac{80}{10} = 8$ Amps

(*c*) Diode currents are zero in case of resistive load. **Ans.**

(*d*) Effective input resistance $= \frac{\text{DC source voltage}}{\text{Average source current}}$

$$= \frac{V}{\alpha \frac{V}{R}} = \frac{R}{\alpha} = \frac{10}{0.4} = 25\ \Omega \quad \textbf{Ans.}$$

3.7.2. Step-up Choppers

In a step-down chopper the output voltage lies between zero and supply voltage. However, there are techniques by which the output of a chopper obtained is higher than the supply voltage. Such types of choppers are called step-up choppers. The circuit diagram of a step-up chopper is shown in Fig. 3.22 (*a*). In figure, the chopper is represented by a switch S.

In this chopper, a large inductor in series with the supply voltage source is essential, as shown in Fig. 3.22 (*a*). When the chopper is on, the current flows through inductance L, chopper S and power supply source, as shown in Fig. 3.22 (*b*). During this on-period T_{on}, the inductor L stores energy.

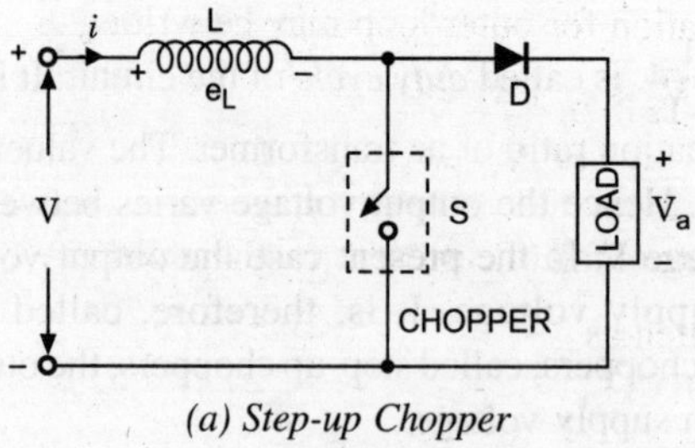

(a) Step-up Chopper

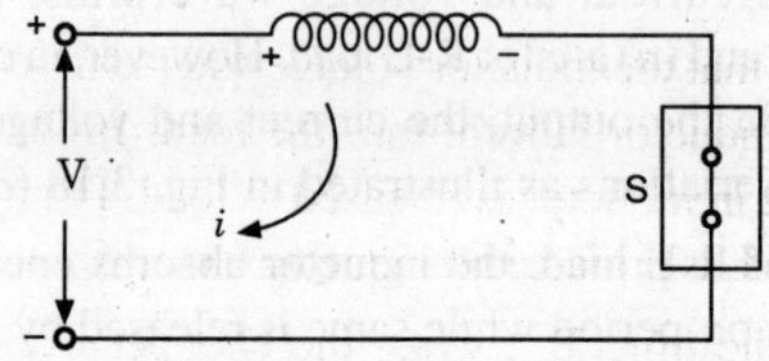

(b) Equivalent Circuit During On-Period

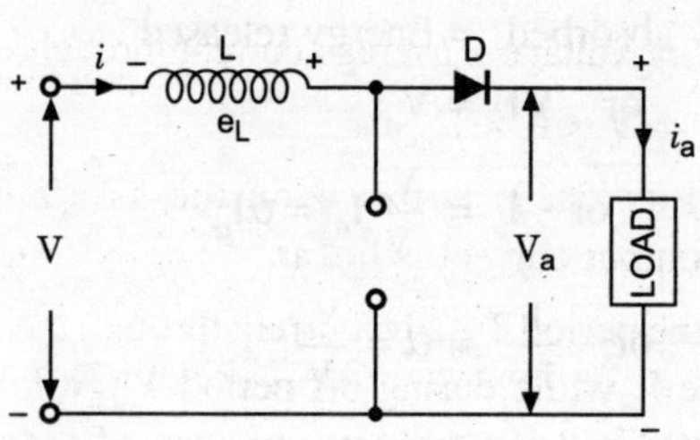

(c) Equivalent Circuit During off-Period

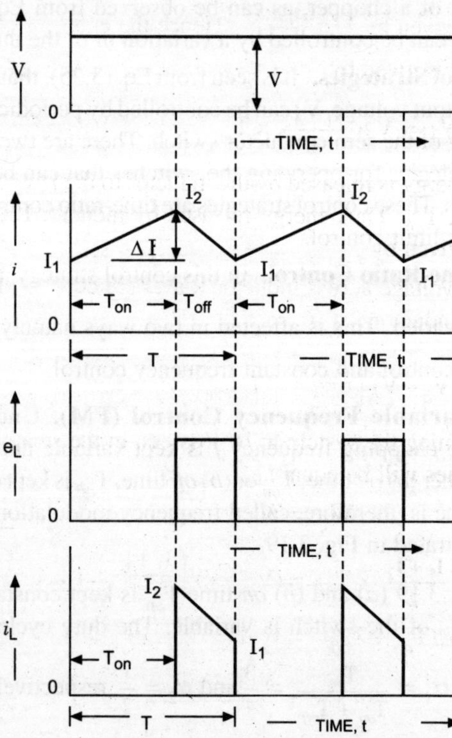

(d) Voltage and Current Waveforms

Fig. 3.22. ***Step-Up Chopper***

When the chopper is off, the inductor current is forced to flow through the diode and load for a period T_{off} because the inductor current cannot die instantaneously. As the current tends to decrease, polarity of the emf induced in the inductor is reversed, as illustrated in Fig. 3.22 (*c*).

Assuming negligible voltage drop across diode D, the voltage equation for outer loop may be written as

$$-V_a + e_L + V = 0$$

$$\text{or } V_a = V + e_L \qquad ...(3.27)$$

If the rate of fall of current through inductor is $\frac{di}{dt}$, Eq. (3.27) may be written as

$$V_a = V + L\frac{di}{dt} \qquad ...(3.28)$$

It means that the inductor voltage adds to the source voltage to force the inductor current into the load. In this manner, the energy stored in inductor is released to load. Here, higher value of inductance L is preferred for getting lesser ripple in the output.

During the time T_{on}, when the chopper is on, the energy stored in the inductor is given by

$$W_1 = \text{Voltage} \times \text{average current through L} \times T_{on}$$

$$= V \times I_a \times T_{on} \qquad ...(3.29)$$

Above Eq. (3.29) is based on the assumption that the variation of output current is linear.

During the period T_{on}, the current through the inductor rises to peak value I_2 while during off period T_{off} it falls to value I_1 and rises again to I_2 in the next on state of the chopper. It is shown in Fig. 3.22 (*d*). Thus, in a cycle the average value of current is given by $I_a = \frac{I_1 + I_2}{2}$. Hence the energy stored in the inductor L is given by

$$W_1 = V \times \frac{I_1 + I_2}{2} \times T_{on} \qquad ...(3.30)$$

and this energy is released by the inductor to the load during the off-period, T_{off}. Thus energy released by inductor L during off period is given by

$$W_0 = \text{Voltage across inductor} \times \text{average current through inductor} \times T_{off}$$

$$= (V_a - V)\left(\frac{I_1 + I_2}{2}\right) T_{off} \qquad ...(3.31)$$

Assuming the system to be lossless, in the steady-state the two energies will be equal *i.e.*,

$$W_1 = W_0$$

$$\text{or } V \times \frac{I_1 + I_2}{2} \times T_{on} = (V_a - V)\left(\frac{I_1 + I_2}{2}\right) T_{off}$$

$$\text{or } V(T_{on} + T_{off}) = V_a T_{off}$$

$$\text{or } V_a = \frac{V(T_{on} + T_{off})}{T_{off}}$$

$$= \frac{V \times T}{T - T_{on}} = \frac{V}{1 - \frac{T_{on}}{T}} = \frac{V}{1 - \alpha} \qquad ...(3.32)$$

The value of α, the duty cycle lies between zero and unity. Corresponding average values of load voltage V_a lies between V and infinity. Thus the output voltage of step-up chopper is never less than the supply voltage V *i.e.*, for $0 \le \alpha \le 1$ the output voltage of step-up chopper lies in the range $V \le V_a \le \alpha$.

The step-up chopper can be employed for regenerative braking of dc motors even at low operating speeds. Let V represent the motor armature voltage and V_a the dc source voltage in Fig. 3.22 (*a*). Regenerative braking takes place when $\left(V + L\frac{di}{dt}\right)$ exceeds V_a. Even at decreasing motor speeds, duty cycle α can be so adjusted that $\left(V + L\frac{di}{dt}\right)$ exceeds fixed supply voltage V_a.

Example 3.13. A step-up chopper shown in Fig. 3.22 (*a*) is fed by a dc source of 220 volts. The inductance has value of 30 mH. If load voltage is 250 volts across a resistance of 10 ohms, calculate average current and duty cycle.

Solution : Average value of load voltage is

$$V_a = \frac{V}{1 - \alpha} = 250 \text{ volts}$$

$$\text{or } 250 = \frac{220}{1 - \alpha}$$

$$\text{or } \alpha = 1 - \frac{220}{250} = 0.12 \textbf{ Ans.}$$

$$\text{Average load current} = \frac{V_a}{R} = \frac{250}{10} = 25 \text{ A } \textbf{Ans.}$$

3.7.3. Step-up/down Chopper

A chopper can also be used for stepping up or down the supply voltage by varying its duty cycle. The principle of operation is illustrated in Fig. 3.23. As shown, the output voltage polarity is opposite to that of input voltage V.

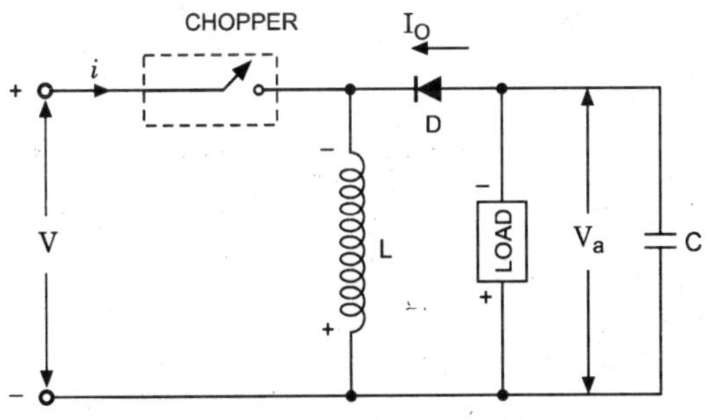

Fig. 3.23. ***Step-Up/Down Chopper***

When the chopper is on, the supply current flows through the path V_+ – chopper – inductor L – V_-. As a result energy is stored in the inductor. When the chopper is made off, the current through the inductor L decreases. As a result an emf is induced in the inductor with lower end positive and the energy stored in the inductor is released through the load and diode and the voltage across the load is negative as illustrated. It is thus used to reverse the polarity of load voltage w.r.t. source voltage.

The energy stored in inductor L during on-period T_{on} is given by

$$W_1 = V \times \text{average current} \times T_{on}$$

But the average current through the inductor, as in case of step-up chopper discussed in Art. 3.7.2, is

$$I_a = \frac{I_1 + I_2}{2} \text{ giving}$$

$$W_1 = V \times \frac{I_1 + I_2}{2} T_{on} \quad \text{...(3.33)}$$

During *off* period T_{off}, the energy released by inductor to the load is

$$W_0 = V_a \times \frac{I_1 + I_2}{2} T_{off} \quad \text{...(3.34)}$$

Under steady-state condition, assuming no loss of power.

Energy stored in inductor = Energy released to load

$$\text{or } V \times \frac{I_1 + I_2}{2} T_{on} = V_a \times \frac{I_1 + I_2}{2} T_{off}$$

$$\text{or } V_a = V \times \frac{T_{on}}{T_{off}} = \frac{V \times T_{on}}{T - T_{on}} = \frac{V}{\frac{T}{T_{on}} - \frac{T_{on}}{T_{on}}}$$

Substituting $\frac{T_{on}}{T} = \alpha$ in above equation, we have

$$V_a = \frac{V}{\frac{1}{\alpha} - 1} = \frac{\alpha}{1 - \alpha} V \quad \text{...(3.35)}$$

Equation (3.35) can be used for step-up and step-down operation of chopper. In the range $0 < \alpha < 0.5$ the circuit operates in step-down mode while for duty cycle in the range $0.5 < \alpha < 1$ it acts as a step-up chopper.

Example 3.14. A step-up/step-down chopper has an input voltage of 125 V and output voltage of 200 V. If the period of conduction is 1 ms, determine the frequency of switching pulse.

Solution : Average output voltage, $V_a = 200$ V

Supply voltage, $V = 125$ V

Period of conduction, $T_{on} = 1 \text{ ms} = 0.001$ s

Since output voltage of step-up/step-down chopper is given by

$$V_a = \frac{\alpha V}{1 - \alpha},$$

$$200 = \frac{\alpha \times 125}{1 - \alpha}$$

$$\text{or } \alpha = \frac{8}{13}$$

Frequency of switching pulse,

$$f = \frac{1}{T} = \frac{1}{T_{on}/\alpha} = \frac{\alpha}{T_{on}} = \frac{8/13}{1 \times 10^{-3}} = \frac{8{,}000}{13} \text{ Hz} = 615.4 \text{ Hz} \quad \textbf{Ans.}$$

3.7.4. Chopper Control of Separately Excited DC Motor Drive

Chopper drives can be operated in the following modes :

1. Motoring mode.
2. Regenerative braking mode.
3. Rheostatic braking mode.

1. *Motoring Mode.* Figure 3.24 (*a*) illustrates this mode of operation with a one quadrant chopper feeding the motor from dc supply of voltage V, the field being excited from a separate source. Figure 3.24 (*b*) shows the quadrant of operation. When the chopper conducts, voltage across the motor armature is supply voltage V, and when it blocks, voltage across the motor armature is zero. Therefore, both average voltage and current are positive and power can flow from supply source to the load, and thus, represents *motoring operation*. The voltage across the motor armature is given as

$$V_a = \alpha V \quad \text{...(3.36)}$$

where α is the duty cycle of the chopper.

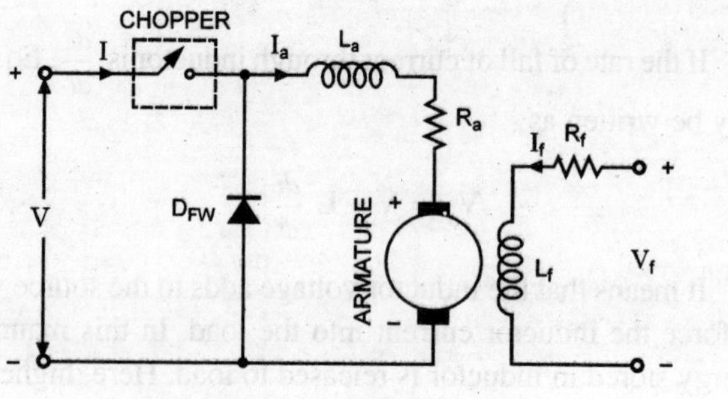

(*a*) *Circuit Diagram*

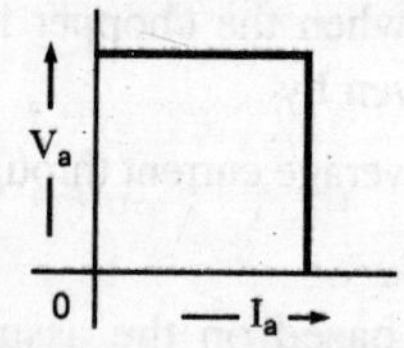

(*b*) *Quadrant of Operation*

Fig. 3.24. ***Separately Excited DC Motor Supplied by a Chopper***

Freewheeling diode D_{FW} is necessary so that the armature current may flow when the chopper is off. The power supplied to the motor is given as

$$P = V_a I_a = \alpha V I_a \quad \text{...(3.37)}$$

The current drawn from the supply is

$$I = \alpha I_a \quad \text{...(3.38)}$$

By varying the duty cycle α, the speed of the motor can be controlled.

2. *Regenerative Braking Mode.* When the motor is to be stopped, brakes are to be applied. In regenerative braking the motor is made to operate as a generator and the kinetic energy of the motor is converted into electrical energy and returned to the supply. Fig. 3.25 (*a*) shows the circuit diagram for this mode of operation. During regenerative braking mode, the chopper is gated so that the motor armature is short circuited through the chopper. Kinetic energy of the motor is partly dissipated in armature resistance R_a and partly stored in inductance L_a. When chopper is turned off, freewheeling diode D_{FW} is turned on and the energy stored in inductance L_a is returned to the supply source (provided the source is receptive).

The operation is in second quadrant, as illustrated in Fig. 3.25 (*b*).

The average voltage V_{ch} is given as

$$V_{ch} = (1 - \alpha) V \quad \text{...(3.39)}$$

Regenerated power P_g is given as

$$P_g = V_{ch} \times I_a = (1 - \alpha) V I_a \quad \text{...(3.40)}$$

where I_a is the armature current

Generated voltage E_g is given as

$$E_g = V_{ch} + I_a R_a = (1 - \alpha) V + I_a R_a \quad \text{...(3.41)}$$

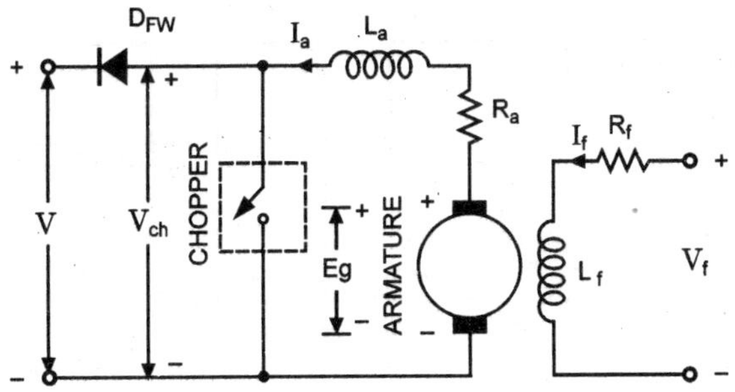

(a) Circuit Diagram

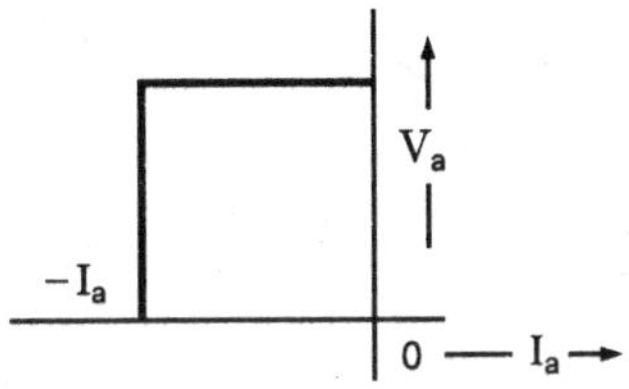

(b) Quadrant of Operation

Fig. 3.25. ***Regenerative Braking of Separately Excited DC Motor***

3. ***Rheostatic Braking Mode.*** Figure 3.26 (*a*) illustrates the circuit for this mode of operation. In this mode also the motor is made to operate as a generator and the kinetic energy of the motor is converted into electrical energy. This electrical energy is dissipated in braking resistance R_b connected to the terminals, as illustrated in the figure. When this mode is to be used, chopper is turned on by a gate pulse.

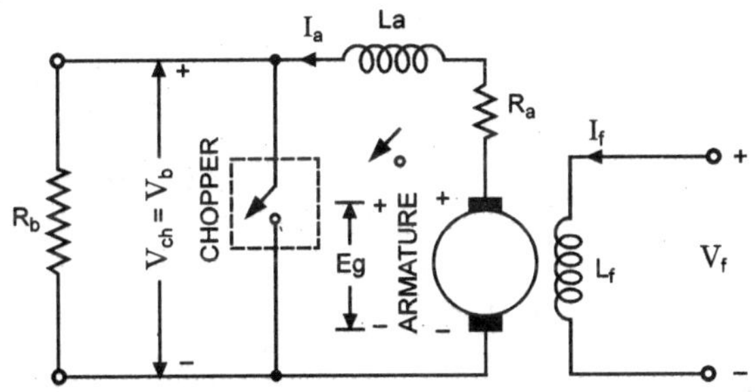

(*a*) Circuit Diagram

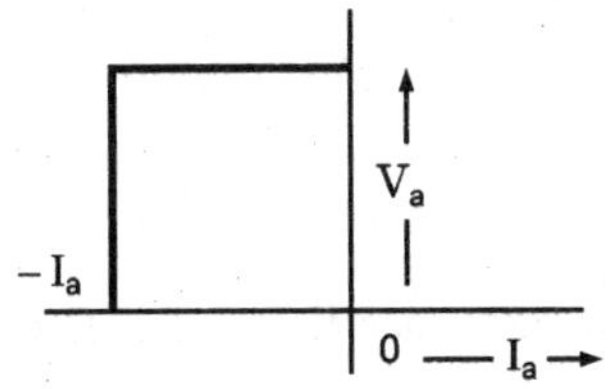

(*b*) Quadrant of Operation

Fig. 3.26. ***Rheostatic Braking of Separately Excited DC Motor***

The kinetic energy is partly dissipated in armature resistance R_a and partly stored in armature inductance L_a. When chopper is turned off, the energy stored in inductance L_a is transferred to braking resistance R_b and dissipated as heat.

The average current in braking resistance R_b is given as

$I_b = (1 - \alpha)\, I_a$ ...(3.42) where I_a is the armature current.

Voltage across braking resistance R_b,

$$V_b = I_b R_b = R_b I_a (1 - \alpha) \qquad ...(3.43)$$

Power dissipated in braking resistance R_b,

$$P_b = I_a^2 R_b (1 - \alpha)^2 \qquad ...(3.44)$$

In this mode also the operation is in second quadrant, as illustrated in Fig. 3.26 (*b*).

It is also possible to combine the regenerative and rheostatic braking modes. Regenerative braking is used when the supply system is receptive. During remaining period the rheostatic braking is used.

3.7.5. Step-Down Chopper With R-L-E_b Load

The choppers are widely employed for speed control of dc motors in industrial and traction drives. Figure 3.27 (*a*) shows the basic

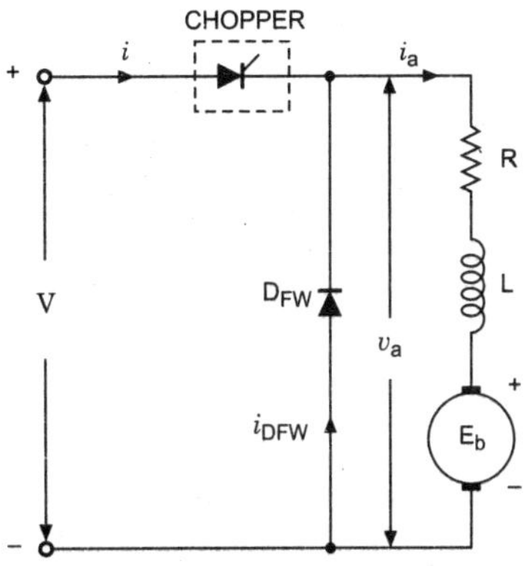

(*a*) Circuit Diagram

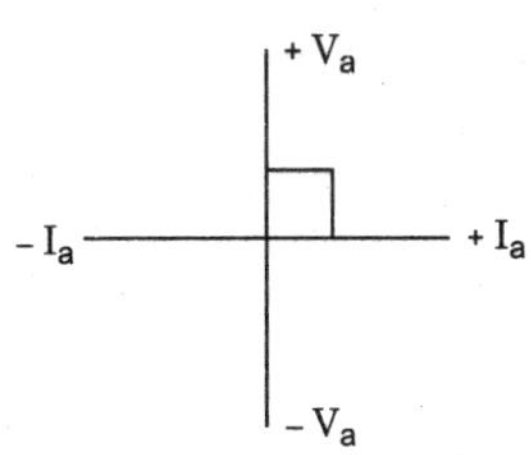

(*b*) Quadrant Diagram

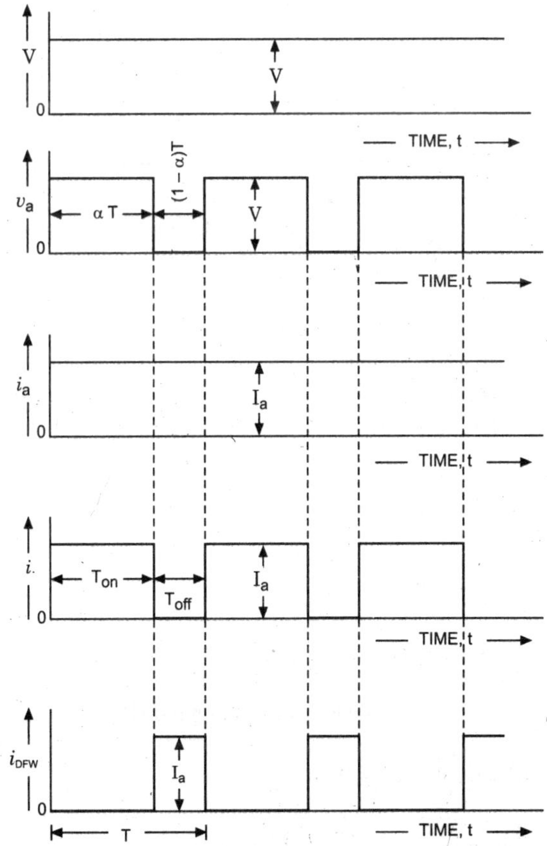

(*c*) Waveforms

Fig. 3.27. ***Step-Down Chopper With R-L-E_b Load***

chopper circuit for the control of a dc series motor. The chopper is shown to consist of a force-commutated thyristor, it could well be a transistor switch. It offers one-quadrant drive [Fig. 3.27 (*b*)]. Armature current is assumed continuous and ripple free. The waveforms for the source voltage V, armature terminal voltage v_a, armature current i_a, supply current i and freewheeling-diode current i_{DFW} are shown in Fig. 3.27 (*c*). From these waveforms we have

Average value of voltage across motor,

$$V_a = \frac{T_{on}}{T} V = \alpha V = f T_{on} V \quad \text{...Refer to Eq. (3.25)}$$

where α = Duty cycle = $\frac{T_{on}}{T}$ and f = Chopping frequency = $\frac{1}{T}$

Power delivered to load (motor)

= Average value of voltage across motor × average motor current

$= V_a I_a = \alpha V I_a$

Average supply current

$$= \frac{T_{on}}{T} I_a = \alpha I_a$$

Input power to chopper

= Average supply voltage × average source current = $V\alpha I_a$

For the motor armature circuit,

$$V_a = \alpha V = E_b + I_a (R_a + R_{se}) = k_e \omega_m + I_a (R_a + R_{se})$$

or Motor speed,

$$\omega_m = \frac{\alpha V - I_a (R_a + R_{se})}{k_e} \quad \text{...(3.45)}$$

From above Eq. (3.45) it is obvious that the armature terminal voltage and therefore, speed of the dc motor can be controlled by varying the duty cycle α of the chopper.

So far, armature current i_a has been assumed ripple free and accordingly waveforms in Fig. 3.27 (*c*) are drawn. But in practice, the motor armature current increases during chopper on period and decreases during chopper off period, as illustrated in Fig. 3.28.

For R-L-E_b type of load, E_b is load voltage (back emf of motor). When the chopper is switched on for a period T_{on} and power is supplied to R-L-E_b load. By using Kirchhoff's voltage law (KVL) in the loop consisting of V, chopper and R-L-E_b load, we have

$$V - E_b - iR - L\frac{di}{dt} = 0$$

$$\text{or } V = L\frac{di}{dt} + iR + E_b \text{ for } 0 \le t \le T_{on} \quad \text{...(3.46)}$$

When chopper is off, the load current continues flowing through the freewheeling diode due to energy stored in the inductor. The KVL for the loop containing R-L-E_b and D_{FW} yields

$$-iR - L\frac{di}{dt} - E_b = 0$$

$$\text{or } L\frac{di}{dt} + iR + E_b = 0 \text{ for } T_{on} \le t \le T \quad \text{...(3.47)}$$

The above Eqs. (3.46) and (3.47) can be solved by using Laplace transform. It is seen from Fig. 3.28 that initial value of load current is I_{min} for Eq. (3.46) and I_{max} for Eq. (3.47).

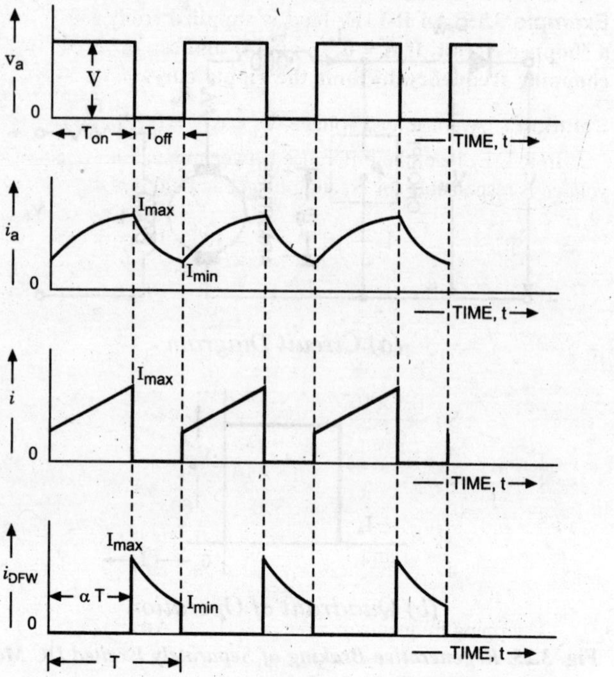

Fig. 3.28. ***Waveforms For DC Chopper Drive of Fig. 3.27 (a)***

So Laplace transform of Eqs. (3.46) and (3.47) we have

$$RI(s) + L[s\,I(s) - I_{min}] = \frac{V - E_b}{s} \quad \text{...(3.48)}$$

$$\text{and } RI(s) + L[s\,I(s) - I_{max}] = \frac{-E_b}{s} \quad \text{...(3.49)}$$

From Eq. (3.48), we have

$$I(s) = \frac{V - E_b}{s(R + Ls)} + \frac{LI_{min}}{R + sL} = \frac{V - E_b}{sL\left(s + \frac{R}{L}\right)} + \frac{I_{min}}{s + \frac{R}{L}}$$

Laplace inverse of above equation is

$$i(t) = \frac{V - E_b}{R}\left(1 - e^{\frac{-R}{L}t}\right) + I_{min}\, e^{\frac{-R}{L}t} \quad \text{...(3.50)}$$

Similarly the time-domain expression for current from Eq. (3.49) is

$$i(t') = -\frac{E_b}{R}\left(1 - e^{\frac{-R}{L}t'}\right) + I_{max}\, e^{\frac{-R}{L}t'} \quad \text{...(3.51)}$$

Here R = Armature resistance R_a + series field resistance R_{se}

and L = Armature inductance + series field inductance L_{se}.

The values of maximum current I_{max} and minimum current I_{min} can be determined by using the fact that $i(t) = I_{min}$ when $t = 0$ and $i(t') = I_{max}$ when $t' = 0$. The values are found to be

$$I_{max} = \frac{V}{R}\left[\frac{1 - e^{\frac{-R\alpha T}{L}}}{1 - e^{\frac{-RT}{L}}}\right] - \frac{E_b}{R} \quad \text{...(3.52)}$$

$$\text{and } I_{min} = \frac{V}{R}\left[\frac{e^{\alpha TR/L} - 1}{e^{RT/L} - 1}\right] - \frac{E_b}{R} \quad \text{...(3.53)}$$

Example 3.15. An R-L-E_b load is supplied from 400 V through a chopper circuit. If R = 0, L = 0.5 H and α = 0.25, calculate the chopping frequency to limit the ripple current to 10 A.

Solution : Average load voltage, $V_a = \alpha V = 0.25 \times 400 = 100$ V

In R-L-E_b load, the difference between input voltage and average voltage is responsible for creating magnetic field in the coil. So

$$L\frac{di}{dt} = V - V_a = 400 - 100 = 300 \text{ V}$$

$$L = 0.5 \text{ H}$$

$$\text{and } \frac{di}{dt} = \frac{10\text{ A}}{T_{on}}$$

$$\text{So } \frac{0.5 \times 10}{T_{on}} = 300$$

$$\text{or } T_{on} = \frac{0.5 \times 10}{300} = \frac{1}{60} \text{ second}$$

$$\text{and } T = \frac{T_{on}}{\alpha} = \frac{1/60}{0.25} = \frac{1}{15} \text{ second}$$

$$\text{Chopping frequency } f = \frac{1}{T} = 15 \text{ Hz.} \quad \textbf{Ans.}$$

Example 3.16. The speed of a separately excited dc motor is controlled by a chopper. The supply voltage is 120 V, armature circuit resistance = 0.5 Ω, armature circuit inductance = 20 mH and motor constant 0.05 V/rpm. The motor drives a constant load torque requiring an average current of 20 A. Assume motor current is continuous. Calculate (*a*) the range of speed control (*b*) the range of duty cycle. [*Pb. Univ., Nov. 1990*]

Solution : The motor will operate at maximum speed (or base speed) when the ratio of time-on (T_{on}) to cycle (T), α = 1, In this case

$$\text{Voltage across the motor armature, } V_a = V = 120 \text{ V}$$

$$\text{Back emf, } E_b = V_a - I_a R_a = 120 - 20 \times 0.5 = 110 \text{ V}$$

$$\text{Motor speed, } N = \frac{E_b}{\text{Motor constant}} = \frac{110}{0.05} = 2{,}200 \text{ rpm}$$

The minimum speed is zero and at zero speed

$$\text{Back emf, } E_{b_0} = 0$$

Voltage across the motor armature,

$$V_a = E_b + I_a R_a = 0 + 20 \times 0.5 = 10 \text{ V}$$

$$\text{Since, } V_a = V \times \frac{T_{on}}{T}$$

$$\therefore \text{ Duty cycle, } \alpha = \frac{T_{on}}{T} = \frac{V_a}{V} = \frac{10}{120} = \frac{1}{12}$$

Hence (*a*) Range of speed control is from 0 to 2,200 rpm **Ans.**

(*b*) Range of duty cycle is from $\frac{1}{12}$ to 1 **Ans.**

Example 3.17. A 200 V, 940 rpm, 100 A separately excited dc motor has an armature resistance of 0.02 Ω. The motor is fed from a single quadrant step-down chopper for motoring operation. The input to the chopper is 200 V dc. Assuming continuous load current, calculate the duty cycle of the chopper for motoring at rated torque and 500 rpm.

[*U.P. Technical Univ. Electric Drives 2004-05*]

Solution : Back emf induced in the motor armature when it is drawing a current of 100 A from 200 V supply and running at 940 rpm

$$E_{b_1} = V - I_a R_a = 200 - 100 \times 0.02 = 198 \text{ V}$$

Back emf induced at a speed of 500 rpm

$$E_{b_2} = E_{b_1} \times \frac{N_2}{N_1} = 198 \times \frac{500}{940} = 105.32 \text{ V}$$

Voltage across armature when running at 500 rpm

$$V_a = E_{b_2} + I_a R_a = 105.32 + 100 \times 0.02$$
$$= 107.32 \text{ V} \qquad \because \text{ Torque is rated one}$$

$$\text{Duty ratio, } \alpha = \frac{T_{on}}{T} = \frac{V_a}{V} = \frac{107.32}{200} = 0.5366 \quad \textbf{Ans.}$$

Example 3.18. A dc chopper fed from 400 V supply runs a separately excited motor. The armature resistance R_a = 0.1 Ω. The motor voltage constant is 4 Vs/radian, the average armature current I_a = 150 A. The armature current is continuous and has negligible ripple. Determine (*i*) the input power, (*ii*) the motor speed and (*iii*) the developed torque for a duty cycle of 60%. Neglect the losses in chopper. If the duty cycle of the chopper varies between 10% and 90%, find the minimum and maximum speeds. [*A.M.I.E. Sec. B. Electric Drives and Their Control Summer 1999*]

Solution : (*i*) Power input $= \dfrac{\alpha \times V \times I_a}{1{,}000} = \dfrac{0.6 \times 400 \times 150}{1{,}000}$

$= 36$ kW **Ans.**

$$\text{Voltage across armature circuit, } V_a = \alpha V = 0.6 \times 400 = 240 \text{ V}$$

$$\text{Back emf developed, } E_b = V_a - I_a R_a$$
$$= 240 - 150 \times 0.1 = 225 \text{ V}$$

$$(ii) \text{ Speed, } N = \frac{60}{2\pi}\,\omega = \frac{60}{2\pi} \times \frac{E_b}{\text{voltage constant}} = \frac{60}{2\pi} \times \frac{225}{4}$$
$$= 537 \text{ rpm} \quad \textbf{Ans.}$$

$$(iii) \text{ Torque, } T = \frac{E_b I_a}{2\pi N/60} = \frac{225 \times 150 \times 60}{2\pi \times 537} = 600 \text{ Nm} \quad \textbf{Ans.}$$

For duty cycle of 10%,

$$E_b = \alpha V - I_a R_a = 0.1 \times 400 - 150 \times 0.1 = 25 \text{ V}$$

$$\text{Speed, } N = \frac{60}{2\pi} \times \frac{25}{4} = 60 \text{ rpm}$$

For duty cycle of 90%,

$$E_b = \alpha V - I_a R_a = 0.9 \times 400 - 150 \times 0.1 = 345 \text{ V}$$

$$\text{Speed, } N = \frac{60}{2\pi} \times \frac{345}{4} = 824 \text{ rpm}$$

Minimum speed will be at 10% duty and will be equal to 60 rpm while the maximum speed will be at 90% duty and will be equal to 824 rpm. **Ans.**

Example 3.19. A one quadrant chopper is used for rheostatic braking of a separately excited dc motor. Armature resistance is 0.1 Ω, braking resistance is 7.5 Ω, voltage constant 1.6 V/A–rad/s, armature current 150 A and field current is 1.5 A. The duty cycle of the chopper is 0.4. Determine (*i*) voltage across the chopper (*ii*) power dissipated in the braking resistance and (*iii*) motor speed.

Solution : (*i*) Average voltage across chopper,

$$V_{ch} = V_b = (1 - \alpha) R_b I_a \qquad \text{... Refer to Eq. (3.43)}$$
$$= (1 - 0.4) \times 7.5 \times 150 = 675 \text{ V} \quad \textbf{Ans.}$$

(*ii*) Power dissipated in braking resistance,

$$P_b = I_a^2 R_b (1 - \alpha)^2 \qquad \text{... Refer to Eq. (3.44)}$$
$$= 150^2 \times 7.5 (1 - 0.4)^2 = 60{,}750 \text{ W or } 60.75 \text{ kW} \quad \textbf{Ans.}$$

Generated emf, $E_g = V_a + I_aR_a = 675 + 150 \times 0.1 = 690$ V

(iii) Speed, $N = \dfrac{60}{2\pi} \times \omega$

$$= \frac{60}{2\pi} \times \frac{E_g}{\text{voltage constant} \times \text{field current}}$$

$$= \frac{60}{2\pi} \times \frac{690}{1.6 \times 1.5} = 2{,}745 \text{ rpm} \quad \textbf{Ans.}$$

Example 3.20. The motor given in Example 3.8 is to be employed for regenerative braking by reversing the direction of field current. Find the firing angle of the armature circuit converter for the given value of armature current. Also determine the power supplied back to the supply source.

Solution : When the field current is reversed, the direction of back emf is reversed. The motor is now acting as a generator.

Generated emf, $E_g = -163.53$ V

Voltage across armature circuit,

$$V_a = E_g + I_a R_a = -163.53 + 40 \times 0.2 = -155.53 \text{ V}$$

Since $V_a = \dfrac{2V_{max}}{\pi} \cos \alpha_a$

$$\therefore \quad -155.53 = \frac{2 \times 220 \times \sqrt{2}}{\pi} \cos \alpha_a$$

or $\alpha_a = 141.74°$ **Ans.**

Power supplied back to the source

$$= V_aI_a = 155.53 \times 40$$

$$= 6{,}221.2 \text{ W or } 6.2212 \text{ kW} \quad \textbf{Ans.}$$

Example 3.21. A constant frequency system is used to control the speed of a dc traction motor from 220 V dc supply. The combined armature and field resistance is 0.2 Ω. The average current in the circuit is 25 A and chopper frequency is 200 Hz. Determine the pulse width, if the average value of back emf is 100 V. [*A.M.I.E. Sec B. Electric Drives and Their Control Winter 1995*]

Solution : Average voltage across chopper,

$$V_{ch} = V_a = E_b + I_a(R_a + R_{se}) = 100 + 25 \times 0.2 = 105 \text{ V}$$

Also $V_a = V \times \dfrac{T_{on}}{T} = V \times T_{on} \times f$

So $105 = 220 \times T_{on} \times 200$

or $T_{on} = \dfrac{105}{220 \times 200} = 0.00239$ second or 2.39 ms **Ans.**

3.8. COMPARISON OF CONVERTER FED AND CHOPPER FED DC DRIVES

For speed control of a dc motor it is necessary that it is supplied with an adjustable voltage which can be obtained either from a controlled converter or from a chopper. The relative advantages and disadvantages of these two are given below :

1. When available supply is ac, an ac-dc converter is ideally suited to convert the constant ac voltage into a variable dc voltage. If we use chopper, which converts a constant dc voltage into a variable dc voltage, diode rectifier will be required to convert constant ac voltage into a constant dc voltage for feeding it to chopper.
2. The ripple in the output of a chopper is much less than that in the output of a converter. Less ripple means lesser losses.
3. The chances of discontinuous operation in a chopper drive are lesser than that in a converter drive. So speed regulation and transient response of chopper drive are better than that in converter drive.
4. For reducing ripple in the output a converter with higher pulse number is required and a converter with higher pulse number has lower utilisation factor and more cost.

3.9. SPECIAL FEATURES OF THYRISTOR DRIVE MOTORS

The most commonly used dc motors for thyristor drive are separately excited and series excited dc motors. The thyristor drive motors usually differ from conventional dc motors in construction. Before discussing the special features incorporated in thyristor drive motors, discussion on the effects of thyristor power supply on performance of a dc motor is necessary.

3.9.1. Effects of Thyristor Power Supply on The Performance of a DC Motor

1. The output voltage from thyristor converter consists of a dc component and ac harmonic components.
2. The output voltage can change very rapidly in comparison to that of a motor-generator set owing to the absence of field time constants associated with the generator.
3. An abnormally high value of armature current may rise in the event of thyristor fault when operating in the inverter mode.
4. The magnitude of harmonics decreases with the increase in frequency of harmonic for all converters. The magnitude of the harmonics increase as the firing angle is increased and dc voltage is reduced for all bidirectional converters. The higher the armature inductance, the lesser will be the harmonic currents.
5. Torque is developed by the dc component of the current whereas heating is developed by the effective (or rms) value of current. The form factor (ratio of effective value to average value) for half-wave three-phase thyristors may be taken as 1.2 while for full-wave three-phase thyristors it is 1.1. This increases electrical losses and, therefore, heating is 5 to 7 per cent more for three-phase full-bridge converters while for three-phase half-bridge converters it is from 15 to 20 per cent.
6. The commutating ability is seriously affected by the presence of harmonic currents. The peak value of the current is increased, interpole flux will be reduced in magnitude and a time lag will be introduced between the interpole current and the flux due to eddy currents generated in the iron path of the inter polar flux.

The other effects of thyristor power supply on motor performance are heating of interpole winding, saturation of interpole magnetic circuit, transformer voltage at the brushes and increase in voltage commutator segment.

3.9.2. Special Features of Thyristor Drive Motors

The thyristor drive dc motors are designed with the following special features in order to improve their performance.

1. The thyristor drive dc motors are made with larger diameter armature and larger size poles of reduced height.
2. The commutators are made larger in order to provide extra insulation to withstand larger and rapid voltage fluctuations.
3. The yoke as well as the main and commutating poles are laminated to reduce the eddy current effects.
4. Low inertia armatures are employed for improving the response.
5. Compensating windings are used in large motors to reduce armature reaction effect.
6. Split brushes of good commutating quality are used for reducing the effect of transformer voltage in the coils undergoing commutation.
7. The use of a laminated yoke instead of a solid yoke improves the commutation to a greater extent.
8. The use of a large number of commutator bars reduces the voltage between commutator segments and improves the commutation.
9. The use of an octagonal, rather than circular, shape for the frame accommodates more material and gives a larger rating for the same frame size.
10. The current densities used for the armature and interpole winding are reduced compared to conventional dc motors of the same frame size and rating in order to reduce the effect of heating armature and interpoles.
11. A better class of insulation (class F materials as standard insulation) is used to allow higher temperature rise and dissipation of more losses from a given frame.
12. The ratio pole arc/pole pitch is reduced in order to reduce the ratio of commutating zone to neutral zone.
13. Armature inductance is increased to reduce the ripple current. It may also increase reactance voltage. The number of turns per armature coil is kept at the minimum since the reactance voltage is proportional to the square of the turns per coil. Use of dummy coils is avoided.
14. Forced cooling by an auxiliary ac motor is widely used for improving cooling of the motor at reduced speeds.
15. Great care is required to be taken for accurate construction : spacing of brushes, poles and manufacture of commutators.

3.10. SPEED CONTROL OF DC MOTOR USING FEEDBACK LOOPS

The systems of speed control discussed in Art. 3.6 are open-loop systems. An open-loop system can be converted into a closed-loop system by employing a feedback loop, as illustrated in Fig. 3.29. The tacho-generator is a permanent magnet dc generator whose output emf is directly proportional to the speed. The difference of input speed setting and speed feedback provides the error signal. The error signal is amplified by the amplifier and supplied to the firing circuit of the converter. Thus the firing of converter thyristors is advanced or delayed depending on whether the actual speed is less than or more than the desired speed.

Let the desired speed of motor be 1,200 rpm, error signal be 0.1 V, converter output be 240 V and output of tacho-generator be 10 V per 1,000 rpm. If the input setting is 12.1 V, the converter output will be 240 V and motor speed will be 1,200 rpm. If the load torque causes the decrease in speed and let the speed become 1,190 rpm, the output of tacho-generator will be 11.9 V increasing the error signal to 0.2 V. The converter output voltage will increase causing the speed to increase to 1,200 rpm. If the speed tends to increase beyond 1,200 rpm, the error signal will become less than 0.1 V and the output voltage of the converter will decrease, thus reducing the motor speed. Thus negative feedback is employed for control of motor speed.

A change in motor speed causes a temporary increase in motor and thyristor currents. Excessive current, however, can be avoided by an automatic current limiting circuit using another feedback loop, as shown in Fig. 3.30. The dc current transformer (dc CT), measures the converter output current and develops a voltage proportional to current. This signal is fed back, as illustrated at a position between the two amplifiers. If the converter current tends to increase, the output of dc CT increases thus reducing the current error. Thus the output current of the converter is reduced to the safe value.

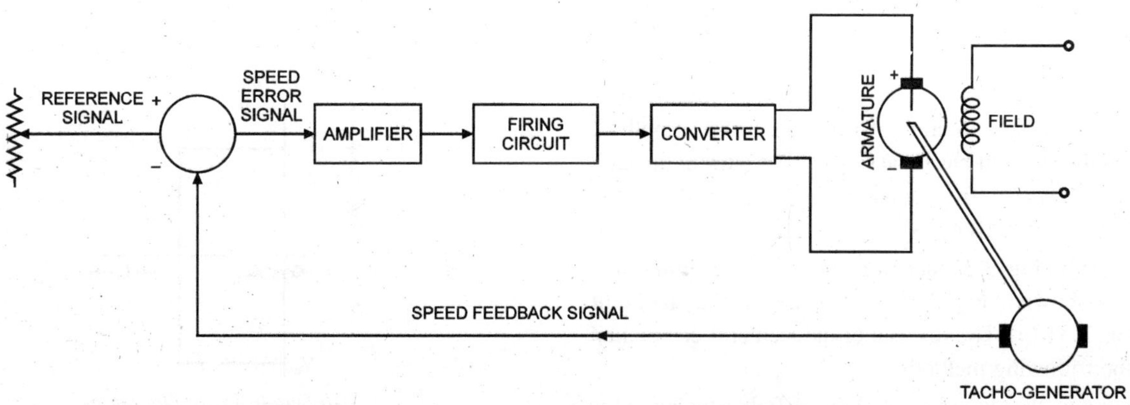

Fig. 3.29. *Speed Control of DC Motor With Negative Feedback*

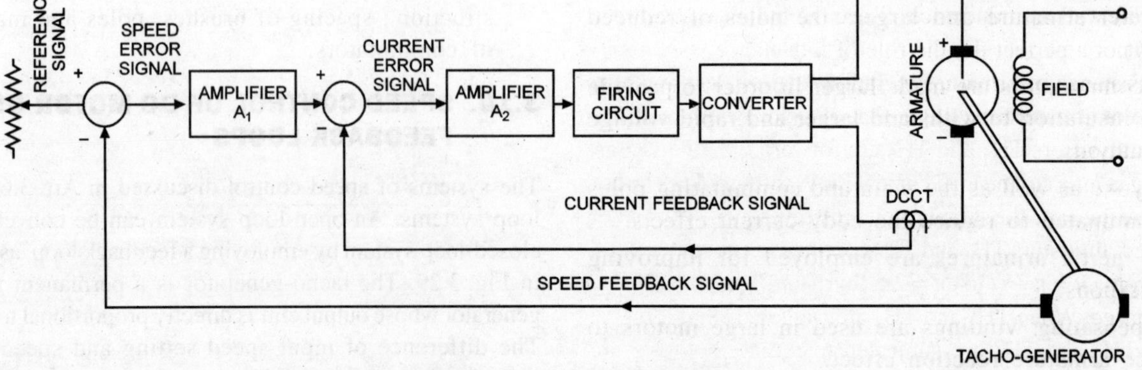

Fig. 3.30. *Speed Control of DC Motor With Current Limiting*

3.11. ELECTRIC BRAKING OF RECTIFIER CONTROLLED SEPARATELY EXCITED DC MOTOR

Figure 3.31 (*a*) illustrates a simplified representation of a single phase fully controlled rectifier supplying a separately excited dc motor. The polarities of voltages and directions of currents are also marked. The firing angle of the rectifier is in the range of $0 \le \alpha \le 90°$.

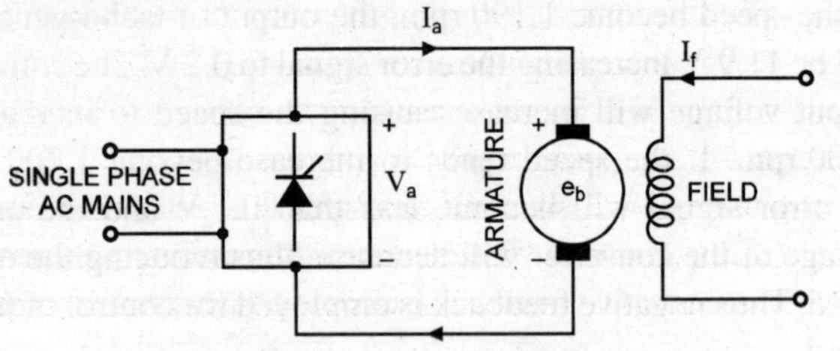

(a) Motor Action $0 \le \alpha \le 90°, V_a > E_b$

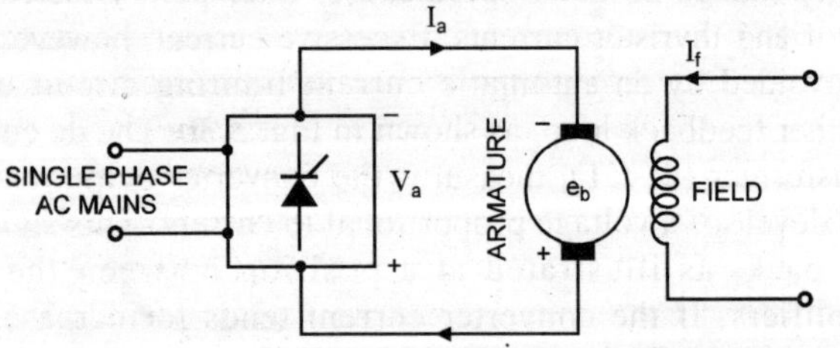

(b) Generator Action $90° \le \alpha \le 180°, E_b > V_a$

Fig. 3.31. *Electric Braking of Separately Excited DC Motor*

For regenerative braking, the direction of power flow is to be reversed. The motor has to operate as a generator and supply power to converter which now operates as an inverter and feeds back power to the supply mains. The circuit diagram for this operation is illustrated in Fig. 3.31 (*b*). Since the inverter output is connected to the supply mains, natural commutation is possible. It is seen that direction of flow of current in both of the circuits is the same. It has to be so since the current through a thyristor can flow only from anode to cathode. For this mode of operation $90° \le \alpha \le 180°$.

It is seen from Fig. 3.31 that direction of induced emf (back emf) in circuit shown in Fig. 3.31 (*b*) is reverse of that in circuit shown in Fig. 3.31 (*a*). The reversal of induced emf can be had by any of the following methods :

1. By reversing the direction of flow of field current. This provides forward regeneration.
2. By reversing the armature terminals (keeping direction of rotation unchanged). This also provides forward regeneration.
3. By driving the motor by an active load coupled to it in reverse direction. No change in connection is required. This provides reverse regeneration.

If the motor is supplied by a semiconverter regenerative braking is not possible because a semiconverter operates only in a first quadrant (*i.e.*, cannot operate as an inverter). However, plugging is possible with both fully controlled and semiconverter. In this case it is necessary to reverse direction of induced emf by any of the above three methods. However, plugging needs an external resistor to be connected to the machine (to dissipate energy) and results in wastage of energy. Therefore, plugging is generally not employed in converter fed dc motors.

3.12. THYRISTOR CONTROL OF DC SERIES MOTORS

3.12.1. Fully Controlled Rectifier

Figure 3.32 (*a*) illustrates the circuit diagram for a dc series motor supplied from single-phase ac mains through a fully

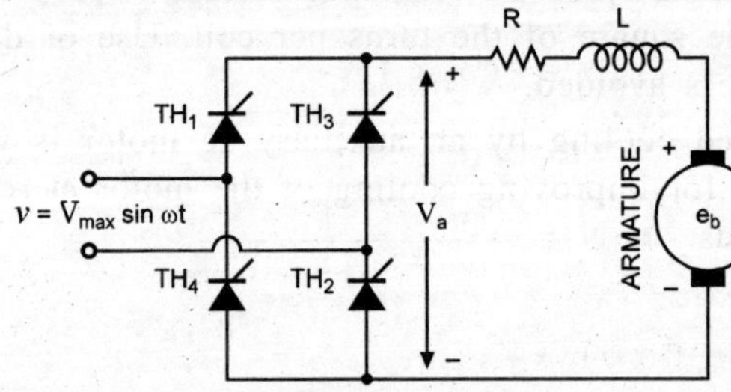

(a) Circuit Diagram

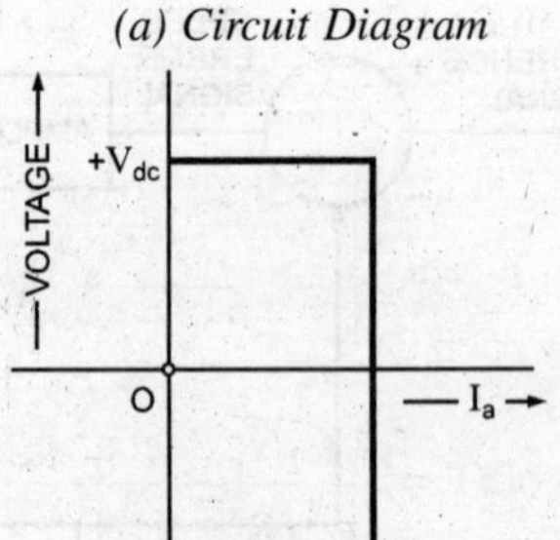

(b) Quadrants of Operation

Fig. 3.32

controlled rectifier. The motor armature and field windings have resistance as well as inductance. Since the output current of a rectifier is not a perfect dc, the role of inductance also comes in. R is resistance of armature including that of field and L is inductance of armature including that of field. During positive half cycle, thyristors TH_1 and TH_2 are forward biased and start conducting at $\omega t = \alpha$. The load current flows through TH_1, motor and TH_2. At $\omega t = \pi$, the supply voltages reverses. Because of inductance L thyristors TH_1 and TH_2 continue to conduct beyond $\omega t = \pi$. From $\omega t = \pi + \alpha$ to $\omega t = 2\pi$, thyristors TH_3 and TH_4 are forward biased. When TH_3 and TH_4 are triggered at $\omega t = \pi + \alpha$, thyristors TH_1 and TH_2 are subjected to reverse bias and are turned off by natural commutation. Load current is transferred from TH_1 and TH_2 to TH_3 and TH_4. This mode of operation continues till TH_1 and TH_2 are triggered in the next positive half cycle. The quadrants of operation and waveshapes are shown in Figs. 3.32 (*b*) and 3.32 (*c*). If inductance L is sufficiently large, the motor current is more or less constant. The value of α should be such that when thyristors are triggered the instantaneous value of ac input voltage $V_{max} \sin \omega t$ is more than back emf E_b. This sets a lower limit on firing angle α.

Average output voltage of converter,

$$V_a = \frac{2}{2\pi} \int_{\alpha}^{\pi+\alpha} V_{max} \sin \omega t \, d(\omega t)$$

$$= \frac{2\,V_{max}}{2\pi} \Big| -\cos \omega t \Big|_{\alpha}^{\pi+\alpha} = \frac{2\,V_{max}}{\pi} \cos \alpha$$

For continuous motor current I_a, we can write

$$V_a = \frac{2V_{max}}{\pi} \cos \alpha = RI_a + E_b$$

Since the field current is also equal to I_a, the back emf E_b can be written as $K_1 I_a N$ where K_1 is a constant. In writing E_b as $K_1 I_a N$ we have neglected the residual magnetism in the motor.

$$\text{So } \frac{2V_{max}}{\pi} \cos \alpha = RI_a + K_1 I_a N$$

$$\text{or } N = \frac{\frac{2V_{max}}{\pi} \cos \alpha - RI_a}{K_1 I_a} \quad ...(3.54)$$

and torque, $T \propto \phi I_a = K_2 I_a^2$ ...(3.55)

Substituting value of I_a from Eq. (3.54) in Eq. (3.55), we have

$$T = K_2 \left[\frac{\frac{2\,V_{max} \cos\alpha}{\pi}}{R + K_1 N} \right]^2$$

$$\text{or } T = \frac{4K_2 V_{max}^2 \cos^2 \alpha}{\pi^2 (R + K_1 N)^2} \quad ...(3.56)$$

Using above Eq. (3.56) speed-torque characteristics for different values of α can be plotted, as illustrated in Fig. 3.32 (*d*).

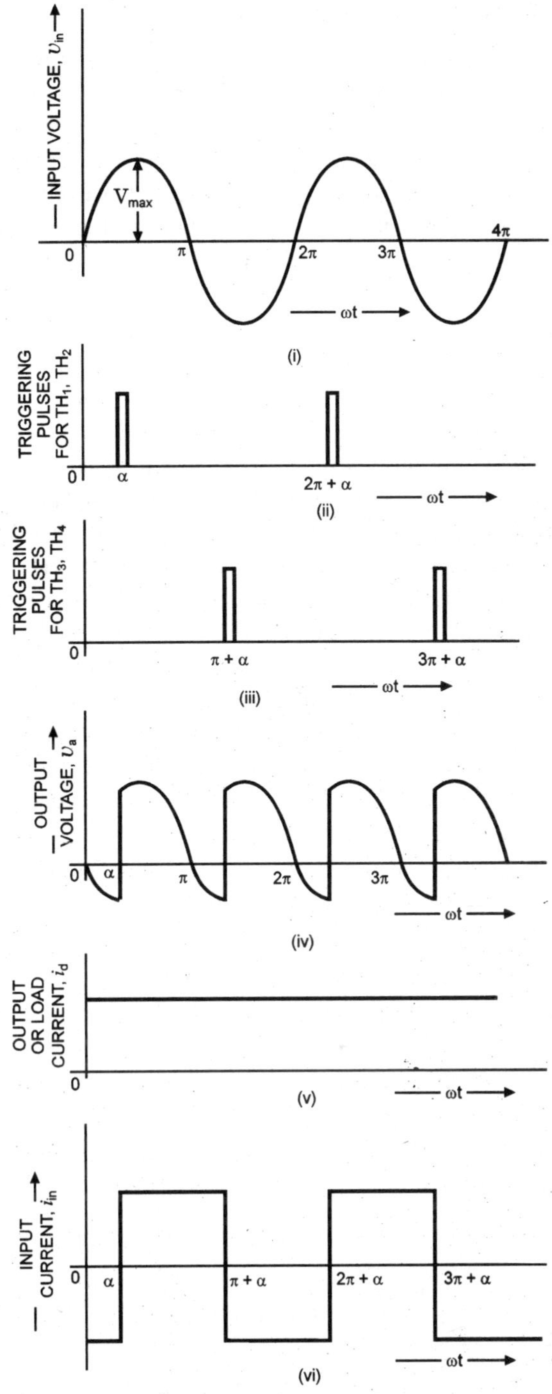

(*c*) *Waveshape*

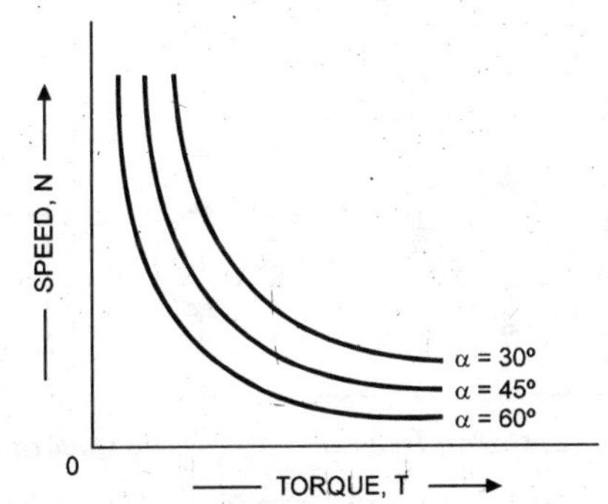

(*d*) *Speed-Torque Characteristics*

Fig. 3.32. ***DC Series Motor Supplied From Single Phase AC Supply Through Fully Controlled Rectifier***

3.12.2. Semicontrolled Rectifier

A dc series motor supplied from single phase ac supply mains through a semicontrolled rectifier is shown in Fig. 3.33 (*a*). The rectifier has two thyristors TH_1 and TH_2 and two diodes D_1 and D_2. The freewheeling diode D_{FW} helps in conduction of current when thyristor is not conducting. R is the resistance of armature and field and L is the inductance of armature and field windings.

During positive half cycle thyristor TH_1 is triggered at $\omega t = \alpha$ and starts conducting. The value of α should be such that $V_{max} \sin \alpha > E_b$. The current flows through TH_1, motor and diode D_1 from $\omega t = \alpha$ to $\omega t = \pi$. At $\omega t = \pi$ the input voltage becomes negative and freewheeling diode D_{FW} is forward biased. Thus at $\omega t = \pi$, thyristor TH_1 and diode D_1 stop conducting and current is transferred to D_{FW}. During negative half cycle thyristor TH_2 is forward biased and when it is triggered at $\omega t = \pi + \alpha$, the freewheeling diode D_{FW} stops conducting and current is transferred to $TH_2 - D_2$ combination. TH_2 and D_2 conduct from $\omega t = \pi + \alpha$ to $\omega t = 2\pi$. At $\omega t = 2\pi$, thyristor TH_2 and diode D_2 get turned off and current freewheels through D_{FW} from $\omega t = 2\pi$ to $\omega t = 2\pi + \alpha$. At $\omega t = 2\pi + \alpha$, thyristor TH_1 is triggered again and next cycle of operation starts. Thus the circuit operation is: D_{FW} conducts for $0 \le \omega t \le \alpha$; TH_1 and D_1 conduct for $\alpha \le \omega t \le \pi$; D_{FW} conducts for $\pi \le \omega t \le \pi + \alpha$; and TH_2 and D_2 conduct for $\pi + \alpha \le \omega t \le 2\pi$.

The quadrant of operation and waveshapes are shown in Figs. 3.33 (*b*) and 3.33 (*c*) respectively.

The average output voltage of semiconverter is given as

$$V_a = \frac{2}{2\pi}\int_{\alpha}^{\pi} V_{max} \sin \omega t \, d(\omega t)$$

$$= \frac{2V_{max}}{2\pi}\Big|-\cos\omega\Big|_{\alpha}^{\pi} = \frac{V_{max}}{\pi}(1+\cos\alpha)$$

The back emf E_b can be written as $K_1 I_a N$, neglecting residual magnetism

$$\text{So } V_a = \frac{V_{max}}{\pi}(1+\cos\alpha) = RI_a + K_1 I_a N$$

$$\text{or } N = \frac{\frac{V_{max}}{\pi}(1+\cos\alpha) - RI_a}{K_1 I_a} \qquad ...(3.57)$$

and torque $T \propto \phi I_a = K_2 I_a^2$

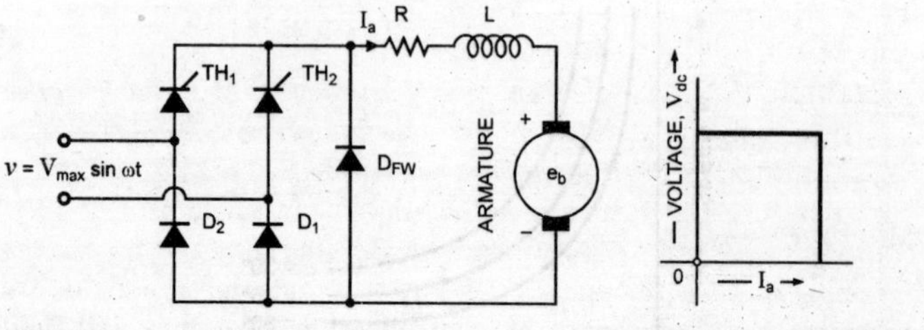

(*a*) *Circuit Diagram* (*b*) *Quadrant of Operation*

Fig. 3.33

Substituting the value of I_a from Eq. (3.57) in above equation, we have

$$T = \frac{K_2 V_{max}^2 (1+\cos\alpha)^2}{\pi^2 (R + K_1 N)^2} \qquad ...(3.58)$$

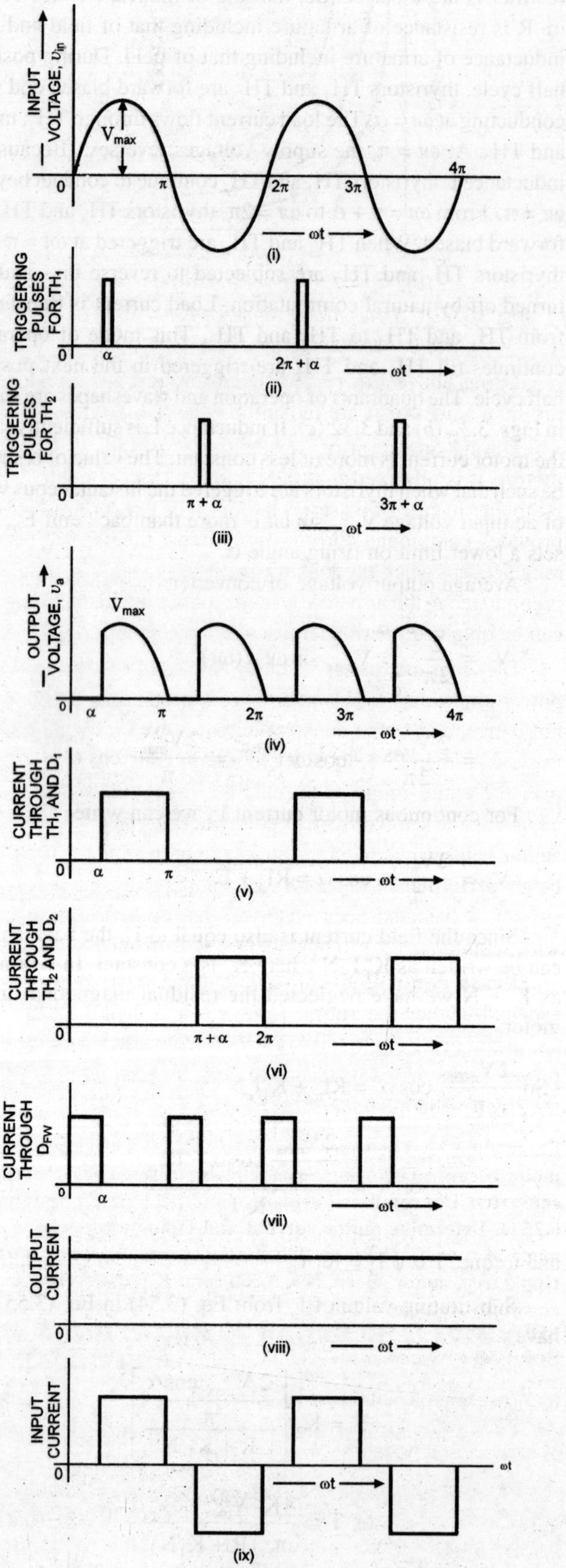

(*c*) *Waveshapes*

Fig. 3.33

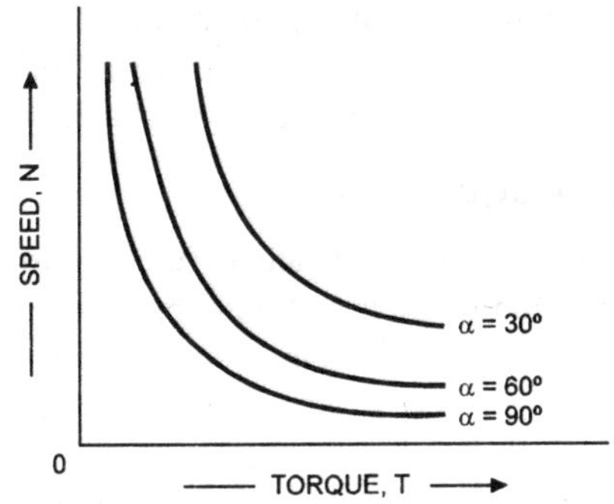

(d) Speed-Torque Characteristics

Fig. 3.33. ***DC Series Motor Supplied From Single Phase AC Supply Through Single Phase Semiconverter***

Using above Eq. (3.58) speed-torque characteristics for different values of α are plotted, as illustrated in Fig. 3.33 (*d*).

During operation of dc separately excited and dc series motors, it is observed that the region of discontinuous current is smaller in case of dc series motors. The use of semiconverter provides continuous current over almost the entire operating range except when the load torque is very small. If inductance is employed in the armature circuit, continuous armature current can be obtained over the entire range of operation.

DC series motors are normally employed for constant output power applications. However, speed-torque characteristics do not correspond to the constant output power for a given firing angle and, therefore, in order to obtain constant output power over the entire range of speeds, the firing angle α has to be adjusted. The semiconverter system, because of its freewheeling action, helps in maintaining continuous current and thus provides better performance to motors in comparison to full-converter system. It has also been observed that dc series motor along with semiconverter provides better performance.

Phase controlled converters have poor power factor especially when the output voltage is less than the maximum, *i.e.*, when firing angle α is large. Semiconverters provide better power factor in comparison to full converters even though the improvement is marginal.

Example 3.22. The speed of a 10 kW, 220 V, 1000 rpm dc series motor is controlled using a single phase (*i*) semiconverter (*ii*) full converter. The combined armature and field circuit resistance is 0.25 Ω. Determine motor current and motor torque for a firing angle of α = 30°. Assume that motor current is continuous and ripple free, motor speed, N = 1,000 rpm, K = 0.025 N-m/A², and ac supply voltage = 240 V.

Solution : Speed, $\omega = \dfrac{2\pi N}{60} = \dfrac{2\pi \times 1{,}000}{60} = 104.72$ radians/s

(*i*) When controlled through semiconverter

$$\text{Voltage across armature, } V_a = \frac{V_{max}}{\pi}(1 + \cos\alpha)$$

$$= \frac{\sqrt{2} \times 240}{\pi}(1 + \cos 30°) = 201.6 \text{ V}$$

$$\text{Armature current, } I_a = \frac{V_a}{R + K\omega} = \frac{201.6}{0.25 + 0.025 \times 104.72}$$

$$= 70.3 \text{ A} \quad \textbf{Ans.}$$

$$\text{Motor torque, } T = KI_a^2 = 0.025 \times 70.3^2 = 123.53 \text{ N-m} \quad \textbf{Ans.}$$

(*ii*) When controlled through full converter

$$\text{Voltage across armature, } V_a = \frac{2V_{max}}{\pi}\cos\alpha = \frac{2 \times \sqrt{2} \times 240}{\pi}\cos 30°$$

$$= 187.13 \text{ V}$$

$$\text{Armature current, } I_a = \frac{V_a}{R + K\omega} = \frac{187.13}{0.25 + 0.025 \times 104.72}$$

$$= 65.25 \text{ A} \quad \textbf{Ans.}$$

$$\text{Motor torque, } T = KI_a^2 = 0.025 \times 65.25^2 = 106.43 \text{ N-m} \quad \textbf{Ans.}$$

Example 3.23. A dc series motor is supplied from a single phase, 230 V, 50 Hz supply through a full-wave diode rectifier. The armature and field resistance is 1.0 Ω and torque constant K_t is 0.3 Nm/A². If load torque is 30 N-m, find average motor current and speed. Assume that motor current is constant and emf constant is 0.3 V/A-rad/second.

Solution : DC output voltage,

$$V_{dc} = \frac{1}{\pi}\int_0^{\pi} V_{max}\sin\omega t \, d(\omega t) = \frac{\sqrt{2} \times 230}{\pi}\int_0^{\pi}\sin\omega t \, d(\omega t)$$

$$= \frac{\sqrt{2} \times 230}{\pi} \times 2 = 207 \text{ V}$$

Applying Kirchhoff's second or voltage law to the motor circuit, we have

$$RI_a + (L_a + L_f)\frac{dI_a}{dt} + K_e\omega I_a = V_{max}\sin\omega t$$

where I_a is the armature current, L_a is the armature winding inductance, L_f is the field winding inductance, R is the combined resistance of field and armature windings, ω is the motor angular speed and K_e is the emf constant.

$$\text{Since armature current } I_a \text{ is constant, } \frac{dI_a}{dt} = 0$$

$$\text{So} \quad 1.0 \times I_a + 0.3\,\omega I_a = 230\sqrt{2}\sin\omega t$$

Integrating above equation over one half cycle, we have

$$I_a + 0.3\omega I_a = 207 \qquad \ldots(i)$$

$$\text{Moreover, load torque, } T_L = K_t I_a^2$$

$$\text{So } 30 = 0.3\,I_a^2$$

$$\text{or } I_a = \sqrt{\frac{30}{0.3}} = 10 \text{ A} \quad \textbf{Ans.}$$

Substituting I_a = 10 A in Eq. (*i*), we have

$$\omega = \frac{207 - I_a}{0.3\,I_a} = \frac{207 - 10}{0.3 \times 10} = 65.667 \text{ rad/second}$$

$$\text{and speed, } N = \frac{\omega \times 60}{2\pi} = \frac{65.667 \times 60}{2\pi} = 627 \text{ rpm} \quad \textbf{Ans.}$$

Example 3.24. A dc series motor is coupled to a load whose torque varies as square of the speed and is supplied from a transformer through a semiconverter. The secondary voltage of the transformer is 675 V. When the firing angle is 100°, the motor speed and armature current are 400 rpm and 36 A. It is desired to raise the speed to 600 rpm. Determine the armature current and firing angle of semiconverter. Armature and field resistance is 0.4 Ω. Assume that magnetic circuit is unsaturated.

Solution : Torque developed by a dc motor is given as

$$T = K_t\,\phi\,I_a$$

or $T \;\alpha\; I_a^2$ ∵ magnetic circuit is unsaturated and, therefore, $\phi \;\alpha\; I_a$

Also load torque, $T \propto N^2$

Under steady-state conditions

$I_a^2 \propto N^2$

or $I_a \propto N$

So $I_{a_2} = I_{a_1} \times \frac{N_2}{N_1} = 36 \times \frac{600}{400} = 54$ A **Ans.**

Voltage across armature when firing angle $\alpha_1 = 100°$

$$V_{a_1} = \frac{V_{max}}{\pi}(1 + \cos\alpha_1) = \frac{675\sqrt{2}}{\pi}(1 + \cos 100°) = 251.09 \text{ V}$$

Back emf, $E_{b_1} = V_{a_1} - I_{a_1}(R_a + R_f) = 251.09 - 36 \times 0.4 = 236.69$ V

Let V_{a_2} be the output voltage of semi-converter when speed is 600 rpm

Back emf, $E_{b_2} = V_{a_2} - I_{a_2}(R_a + R_f) = V_{a_2} - 54 \times 0.4 = (V_{a_2} - 21.6)$ volts

Since $E_b \propto \phi N \propto I_a N$

$$\text{So } E_{b_2} = E_{b_1} \times \frac{I_{a_2} N_2}{I_{a_1} N_1} = \frac{236.69 \times 54 \times 600}{36 \times 400} = 532.56 \text{ V}$$

Also $E_{b_2} = V_{a_2} - 21.6$

or $V_{a_2} = E_{b_2} + 21.6 = 532.56 + 21.6 = 554.16$ V

Also $V_{a_2} = \frac{V_{max}}{\pi}(1 + \cos\alpha)$

or $554.16 = \frac{675\sqrt{2}}{\pi}(1 + \cos\alpha)$

or $\alpha = 34.54°$ **Ans.**

3.13. THYRISTOR CONTROL OF THREE PHASE INDUCTION MOTORS

As already stated in Art. 1.18.2, the speed of an induction motor is given as

$$N = \frac{120 f}{P}(1 - s)$$

Thus the speed of an induction motor with fixed number of poles depends upon the supply frequency f and the slip s which in turn depends upon the voltage or current supplied to the motor. Following methods are available for speed control of 3-phase induction motors using thyristors.

1. Stator voltage control or variable voltage constant frequency control.
2. Variable voltage and variable frequency control.
3. Rotor resistance control.
4. Secondary foreign voltage control.

Stator voltage variation is accomplished by means of *ac regulators*, which control the rms value of the ac voltage applied to the motor by introducing thyristors connected back-to-back in each supply line. Variable frequency power is obtained by means of a *cycloconverter*, which directly converts fixed frequency ac to variable frequency ac or by an inverter which converts dc to ac. The effective value of an external resistance introduced into the rotor circuit can be controlled by connecting a high frequency chopper across the resistance and varying the time for which the chopper is *on* during the cycle. Static frequency converters are being employed for replacing the auxiliary machines in the Scherbius system. The Kramer scheme has also been changed by making use of a diode bridge rectifier in place of rotor converter, but a dc motor is still required for converting the rectified slip power to mechanical power.

3.13.1. AC Regulators

An ac regulator converts a constant ac voltage into a variable ac voltage of the same frequency. No doubt, ac voltage level can be changed by auto-transformer, tap-changing transformer, saturable reactors etc. These devices have been in use for long time and still are in use. But ac regulators using thyristors and triacs are becoming more and more popular because of their high efficiency, fast control and compact size. However, ac regulators using thyristors and triacs introduce objectionable harmonics in the circuits. AC regulators are classified as single-phase or three-phase. Each of these may be half-wave (*i.e.*, unidirectional) or full-wave (*i.e.*, bidirectional).

Since the input to an ac regulator is ac, it is always line commutated. Therefore, forced commutation is not required. As such the circuits of ac regulators are quite simple. Two types of controls are used in ac regulators. These are known as *integral cycle control* and *phase control*.

In an *integral cycle control*, also known as *on-off control*, the thyristors are used as switches to connect the motor to the supply source for a certain number of cycles of the source voltage and then to disconnect it for another certain number of cycles. Each of the on- and off-times consists of an integral number of cycles. Thyristors are switched on by gate pulses at zero voltage crossing of input voltage.

In *phase control*, the thyristors are employed as switches for connecting the motor to the supply for a certain portion of each cycle of the supply voltage. Most ac regulators use phase control. The power circuit configurations for integral cycle control and phase control do not differ in any manner.

Smooth variation of three-phase ac voltage can be realized by means of different configurations of the power circuit.

The three-phase regulators, as already said, may be half-wave or full-wave. Circuit of *three-phase half-wave regulator* for delta-connected motors or star-connected motors in which neutral point is not accessible is shown in Fig. 3.34. This circuit uses three thyristors and three diodes. Though half-wave ac regulator illustrated in Fig. 3.34 affects a saving in the cost of semiconductor devices and does not give rise to dc components in any part of the system but it introduces more harmonics into the line current than does the full-wave regulator. Half-wave circuit is not used in actual practice.

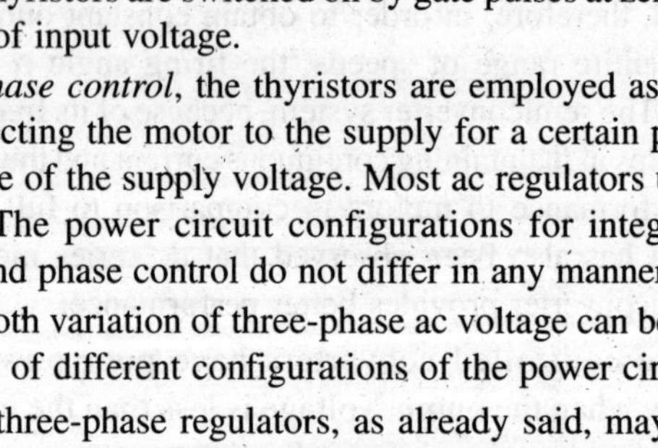

Fig. 3.34. ***Three-Phase Half-Wave AC Regulator***

Figure 3.35 shows a 3-phase full-wave regulator. It uses 6 thyristors, 2 for each phase. The input transformer may or may not be employed. As regards the heating of motor windings are concerned, star-connected motor supplied through a full-wave ac regulator is preferred over a delta-connected motor supplied through a full-wave ac regulator. This is so, because any third harmonic voltages generated by the motor back emf can cause circulating currents in case of delta-connected motor.

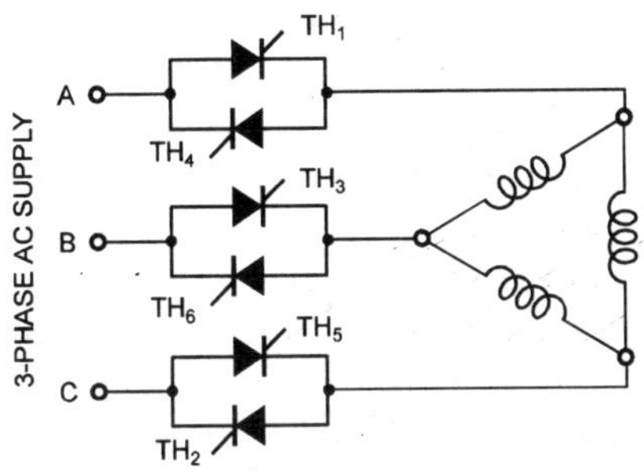

Fig. 3.35. *Three-Phase Full-Wave AC Regulator*

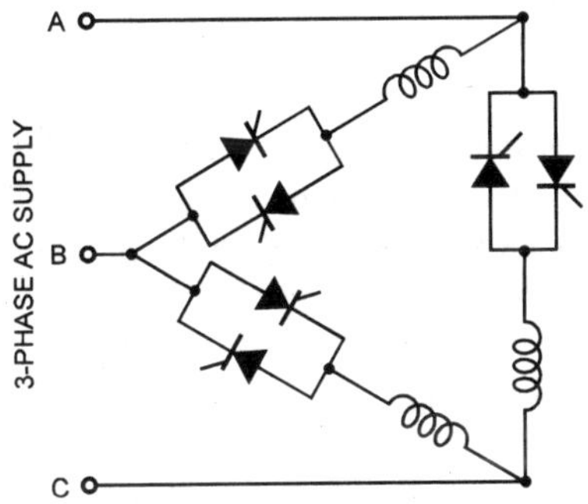

Fig. 3.36. *Delta-Connected Speed Controller*

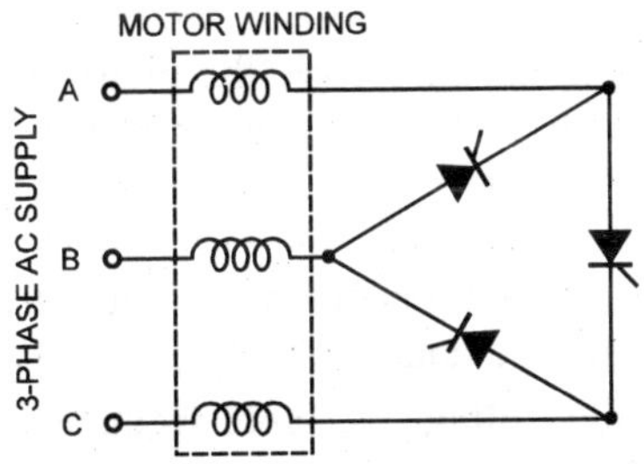

Fig. 3.37. *Neutral Connected Regulator*

For delta-connected load circuits in which each end of each phase is accessible, the arrangement illustrated in Fig. 3.36 is employed. This arrangement has the advantage of reducing the current of the device as it has to carry now $1/\sqrt{3}$ of the current if they were connected in the line of the delta winding. Once the phase current wave is known, the line current wave can be constructed by superposition.

For star-connected load circuits in which neutral point is accessible and can be opened, the arrangement illustrated in Fig. 3.37 may be used. In this arrangement, the number of thyristors required reduces to three and control circuitry becomes considerably simplified. The power consumption of the motor may be as high 100% greater than that with sine wave voltage control especially at reduced speeds.

3.13.2. Variable Voltage and Variable Frequency Control

If only frequency is changed and stator voltage is kept constant, the stator flux will not be at its rated value. The operation with flux below or above the rated value is not desirable. For constant flux operation, it is necessary that the induced emf increases or decreases linearly with applied frequency. At higher voltages and at high frequency operation stator drops are very small and thus constant flux operation is obtained by keeping V/f ratio constant. Speed-torque characteristics of a 3-phase induction motor with variable voltage variable frequency (constant V/f control) are illustrated in Fig. 1.25. It may be seen from the Fig. 1.25 that the maximum torque does not alter because the flux per pole remains constant. This mode of operation ensures the ability of developing the highest possible torque per unit stator current and, therefore, results in the best possible utilisation of the available current capability of the drive.

The variable stator voltage and frequency can be obtained from the systems illustrated in Fig. 3.38 or in Fig. 3.41, known as square-wave inverter and pulse width modulated (PWM) inverter respectively.

A square-wave inverter power circuit is illustrated in Fig. 3.38. The three-phase ac supply is converted into dc by a controlled rectifier. The output of the rectifier is supplied to the filter circuit to remove the harmonics. The dc output from filter is fed to a controlled inverter which provides variable voltage variable frequency output. This supply is fed to the stator of the 3-phase induction motor whose speed is to be controlled.

Figure 3.39 shows the phase voltage waveforms V_{AN}, V_{BN}, V_{CN} and line- to-line voltage waveforms V_{AB}, V_{BC} and V_{CA}. Each line voltage waveform is displaced in time phase by 120 electrical degrees from one another and is of quasi-square wave with 120° width. The noteworthy point is that the inverter thyristors are force commutated because

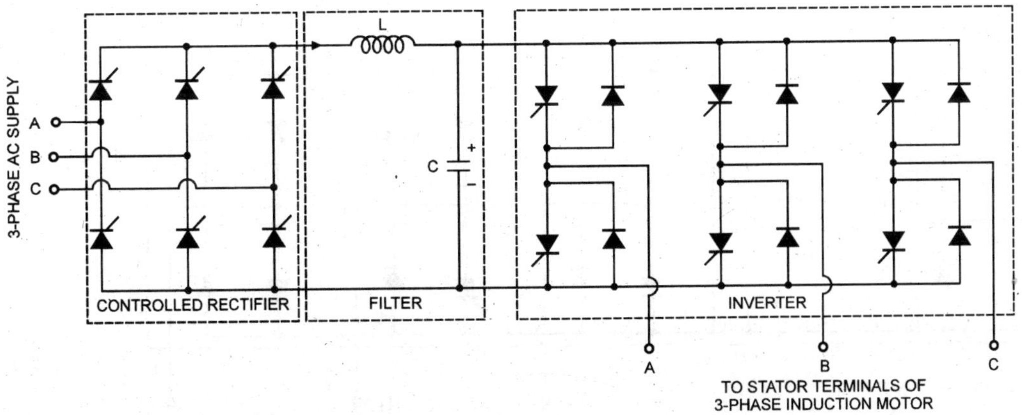

Fig. 3.38. *Variable Voltage Variable Frequency Control of 3-Phase Induction Motor (Square-Wave Inverter Power Circuit)*

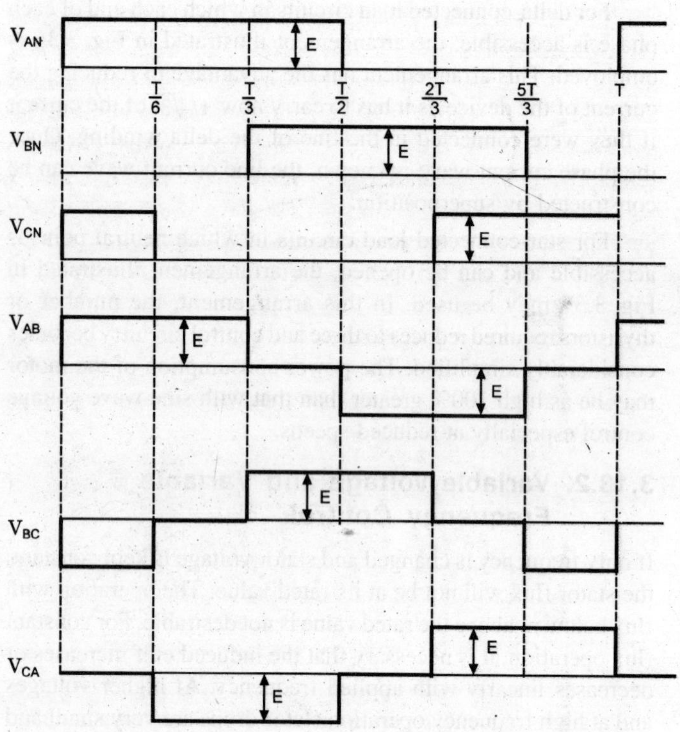

Fig. 3.39. ***Phase and Line Voltage Waveforms at the Motor Terminals***

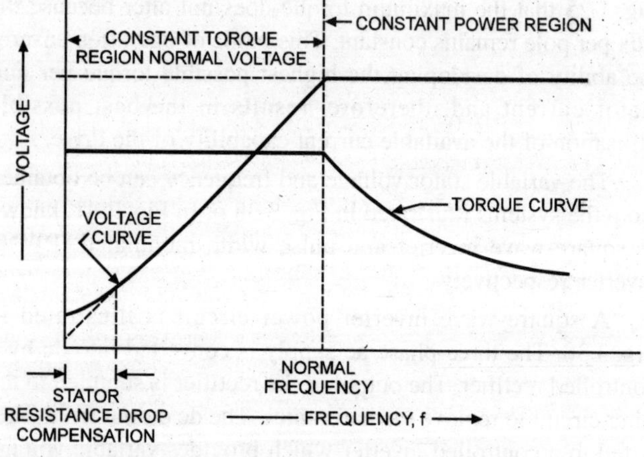

Fig. 3.40. ***Voltage-Frequency Relationship***

induction motor is a lagging power factor load. The feedback diodes help in circulation of load reactive power with filter capacitor and maintain the output voltage waves fixed to the level of dc link voltage.

Desired voltage-frequency relationship of induction motor is illustrated in Fig. 3.40. When the frequency is less than the normal frequency, the voltage is reduced by the same proportion so as to keep V/*f* constant. At very low frequencies, as the reactance drop becomes smaller in comparison to stator resistance drop ($\omega L < R$), additional voltage is required to be applied to compensate this effect. It means a higher V/*f* ratio. When the frequency exceeds the normal frequency, the torque decreases with the decrease in air-gap flux and the motor now operates in the constant power region, as illustrated in Fig. 3.40. This is equivalent to field weakening mode of speed control of a dc motor.

When operating at reduced voltage the converter voltage decreases and, therefore, the commutation capability of the capacitor also decreases. So the inverter is usually provided with an auxiliary constant voltage dc supply for commutation purposes.

The inverter mentioned above cannot return power back to the ac supply lines unless another phase controlled rectifier is added to form a reversing system. This method of speed control is used in low and medium size 3-phase induction motors where the speed ratio is usually restricted to 10 : 1.

The noteworthy point is that electrical machine is designed corresponding to near saturation point on the magnetization curve (or B-H curve). This is done from the point of view of full utilisation of the core. If the stator frequency is reduced, keeping the stator voltage constant, motor will operate in the saturation region and therefore, the motor will draw large magnetising current causing increase in core and stator copper losses and therefore, decrease in motor efficiency. However, if only supply frequency is increased keeping the stator voltage constant, the motor will operate at low flux density and thus motor capacity will be under utilised.

The variable voltage variable frequency pulse width modulation (PWM) control circuitry for an induction motor is illustrated in Fig. 3.41. This is recent method and supersedes the square-wave inverter circuit discussed above.

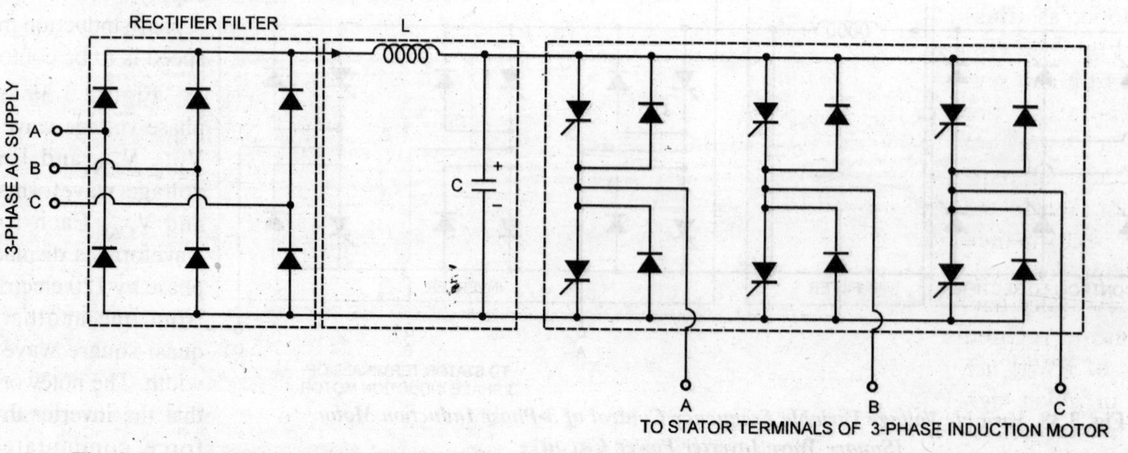

Fig. 3.41. ***Variable Voltage Variable Frequency Control of 3-Phase Induction Motor (PWM Control)***

Pulse width modulated (PWM) inverters use the chopping or pulsing technique for control of alternating voltage output of a static inverter. The dc link voltage is kept uncontrolled by a diode rectifier. The square or step-wave output voltage is rapidly switched on and off several times during each half cycle so that a number of pulses of equal amplitude are formed. Each pulse has the amplitude of the inverter input voltage V_{dc}. The magnitude of the fundamental output voltage is controlled by variation of the total on-time during a half cycle. By commutating one side of the bridge several times during a half cycle, the output voltage of the waveform shown in Fig. 3.42 (*a*) can be obtained. Simple PWM inverters can be easily made to produce a waveform with only two pulses per half cycle within a six step envelope, as illustrated in Fig. 3.42 (*b*). Significant fifth and seventh harmonics are present in such waveform and they cause appreciable deterioration in the low-speed performance of the ac motor. For elimination of low order harmonics, more refined techniques of PWM are employed, in which high frequency pulsing occurs throughout the half cycle.

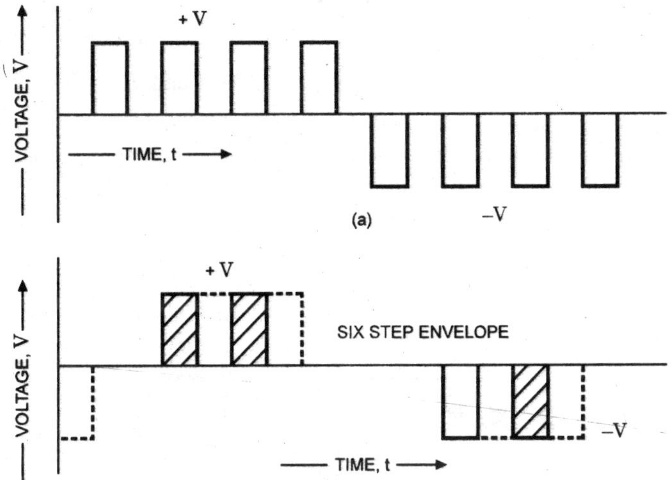

Fig. 3.42. ***Basic Waveform of PWM***

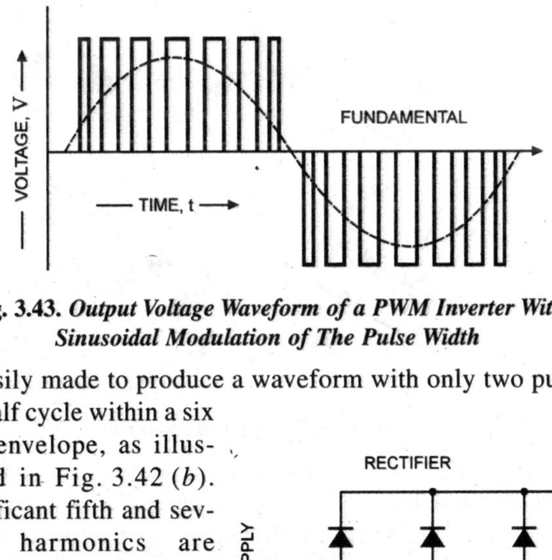

Fig. 3.43. ***Output Voltage Waveform of a PWM Inverter With Sinusoidal Modulation of The Pulse Width***

In sophisticated PWM systems, the pulse width is varied throughout the half cycle in a sinusoidal manner, as illustrated in Fig. 3.43. Actually, the pulses should be regularly spaced and the pulse width at a particular position should be proportional to the area under the sine wave at that position. In the PWM waveform, the lowest harmonic frequency is at the pulse repetition frequency and, if this is much higher than the fundamental frequency, adequate filtering is provided by the inductance of the machine. Such waveforms are usually produced by means of a control circuit in which a high frequency triangular waveform is mixed with a sinusoidal waveform of the desired frequency. Voltage control is obtained by varying the widths of all pulses without affecting the sinusoidal relationship.

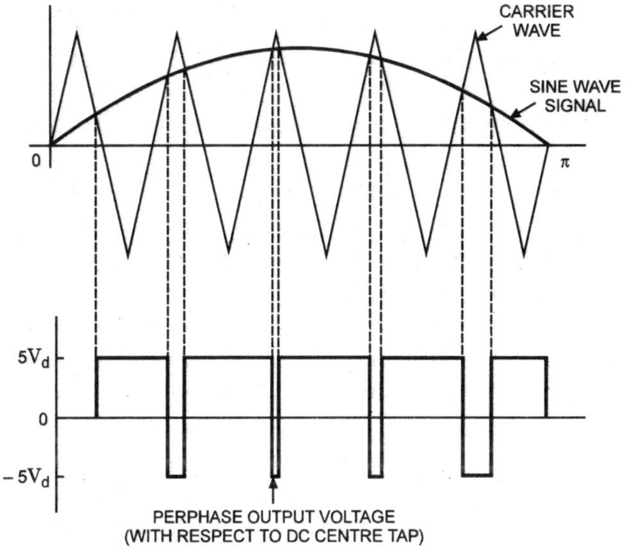

Fig. 3.44. ***Principle of Pulse Width Modulation (PWM)***

Figure 3.44 illustrates sinusoidal PWM technique where an isosceles triangular wave is compared with the sinusoidal wave signal and the points of commutation are determined by the crossover points. If modulation index is found to be less than unity, only carrier frequency harmonics with fundamental frequency related side bands appears at the output. Such waveform generates less harmonic heating and torque pulsation in comparison to that of the square wave. When the modulation index exceeds unity, maximum voltage is

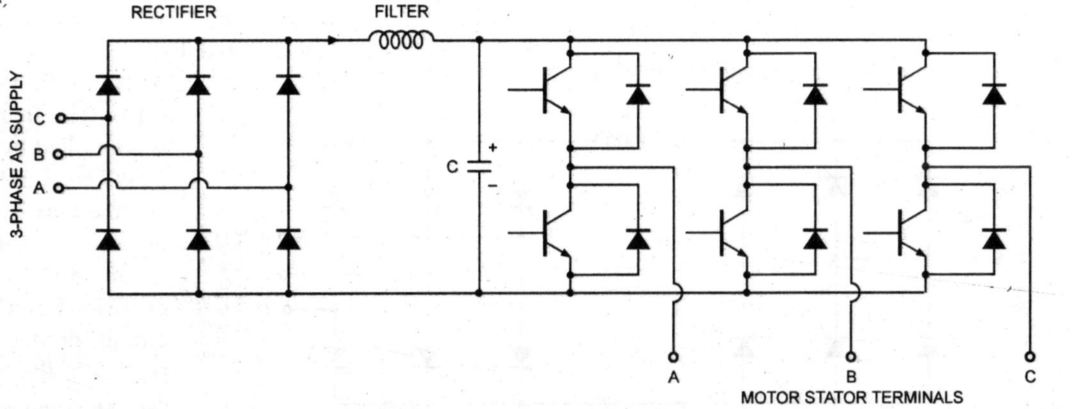

Fig. 3.45. ***Transistorized PWM Control***

obtained in the square-wave mode. Thus, PWM mode is applicable in the constant torque region while in the constant power region the operation is similar to that of the square-wave mode.

The transistorized PWM control, shown in Fig. 3.45, is used for control of low to medium size motors. No doubt power transistors cost much more than that of the thyristors of the same capability but savings due to elimination of commutation circuit and corresponding commutation losses, the circuit proves to be more economical and efficient. Also, the transistors operate faster, PWM is possible at higher frequency. This cuts down further the machine losses.

3.13.3. Variable Current Variable Frequency Control

A variable current variable frequency control circuit for an induction motor is illustrated in Fig. 3.46. Variable dc voltage provided by a phase controlled rectifier is converted to a current source by connecting a large inductor in series. The large inductance maintains the current constant. The voltage available at the stator terminals of the 3-phase induction motor is almost sinusoidal with superimposed voltage spikes due to commutation. The converter employed is a line commutated while the inverter is forced commutated because induction motor operates at lagging power factor. The phase controlled converter can be replaced by a diode rectifier followed by a dc chopper.

The circuit has the following ***advantages*** :

(*i*) Since the input current is constant, misfiring of devices and short circuits do not pose any problem.

(*ii*) Less number of components in inverter circuit and less commutation loss.

(*iii*) Rugged and reliable power circuit.

(*iv*) Simpler and more reliable control circuit. This is because only 6 thyristors are to be controlled.

(*v*) Peak current of devices is limited.

(*vi*) It can handle reactive or regenerative loads without freewheeling diodes.

Disadvantages : (*i*) Somewhat sluggish response of the drive.

(*ii*) Somewhat bulky and expensive inverter. This is due to large size of the inductance and the commutation capacitors.

(*iii*) Low frequency range of the inverter.

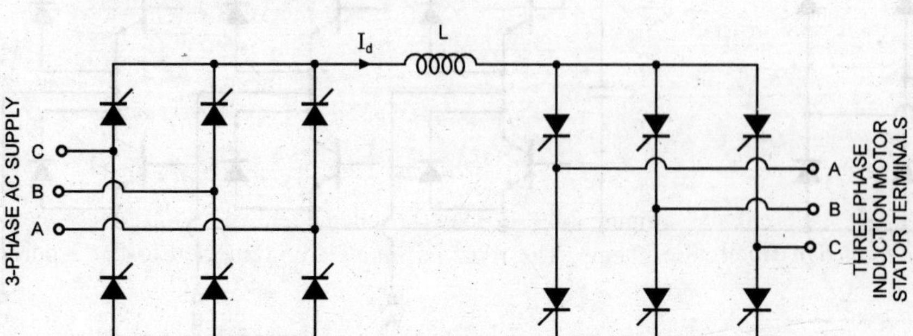

Fig. 3.46. ***Variable Current Variable Frequency Control***

(*iv*) It cannot operate under no-load condition. This is because some minimum load current is necessary for the satisfactory commutation of the inverter.

3.13.4. Cycloconverters

A cycloconverter converts ac at one frequency to an ac of another frequency. Cycloconverters can be classified as single-phase to single-phase, three-phase to single-phase and three-phase to 3-phase devices. They can also be classed as step-up and step-down cycloconverters. A step-up cycloconverter provides an output whose frequency is higher than that of input while a step-down cycloconverter provides an output of frequency lower than that of input. Step-down cycloconverter uses line or natural commutation. Cycloconverters were developed initially for electric traction systems operating at 25 Hz and 16 $\frac{2}{3}$ Hz. In their early stages of use, the cycloconverters used mercury-arc rectifiers. With the development of thyristors the applications of cycloconverters have increased.

The basic power circuit scheme of a 3-phase cycloconverter is shown in Fig. 3.47.

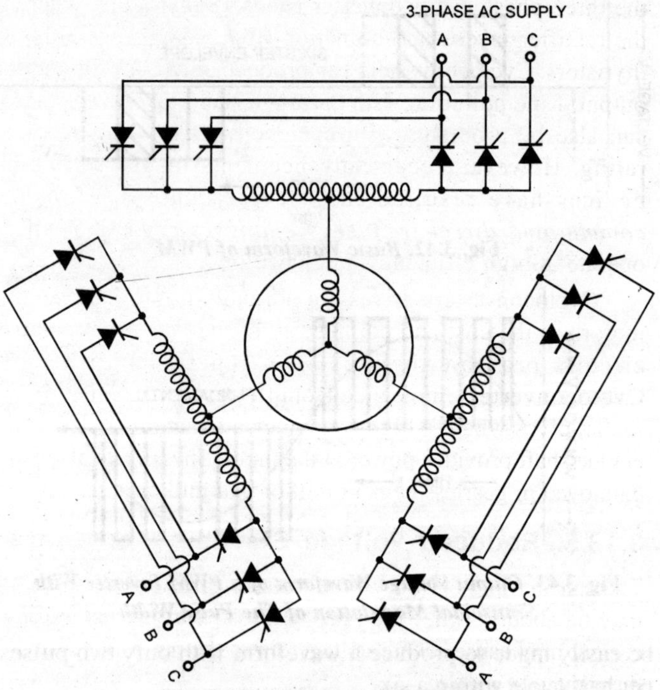

Fig. 3.47. ***3-Phase Cycloconverter***

Independent control of output frequency and voltage is obtained with only one parameter variation, *i.e.*, viz., by variation of the firing points of the controlled rectifiers. The frequency of the output voltage is controlled by the rate at which the firing points are varied about the quiescent point and the output voltage is controlled by the maximum excursion of the firing points from the quiescent point. The cycloconverter, with its associated firing circuit, produces an output voltage that is replica of the reference voltage.

The operation of cycloconverter is characterized by several features. They are, generally, employed as step-down frequency converters. There is no fixed minimum ratio of input to output frequency; however, the output frequency is restricted typically to one-third or one-half of the input or line frequency. Below these ratios, the efficiencies of both the cycloconverters and motors supplied by them start falling significantly.

Reversibility is another feature of cycloconverter drive systems. A cycloconverter fed ac motor drive will respond to a change in polarity of the input signals by changing the direction of rotation of the motor without the use of contactors to reverse phase sequence.

The ability of cycloconverter to handle power flow in either direction is another important feature. This, together with the above mentioned reversibility feature, provides an induction motor drive capable of operating in any of the four quadrants of the motor's speed-torque curve.

While the cycloconverter has many attractive features from a theoretical point of view, there are several limitations because of which they have not gained popularity. It needs more power semiconductors than an inverter. For example, the three-phase cycloconverter needs 18 thyristors, whereas the rectifier-inverter combination (Fig. 3.38) needs only 12 thyristors. Cycloconverters can produce only a sub-frequency output. Line pollution with harmonics and low power factor can also be problems with cycloconverters of high power rating. However, recent advancements in fast switching devices have resulted into devices known as *forced commutated direct frequency changers* (FCDFC) which operate at high efficiency and have low harmonic content.

Cycloconverters drives are normally used for large size motors as the cost and complexity of the power and control circuits prohibit their use for general applications. Cycloconverters have been employed in diesel electric locomotives where a high frequency alternator coupled to the engine shaft provides power at the input. These have also been employed in gearless cement mill or ball mill drives.

3.13.5. Reduced Voltage Starting (Soft Start)

The starting line current at full voltage of an induction motor may be about 6 times the rated full-load current. Such a high current may cause severe voltage dip in the network supplying the induction motor.

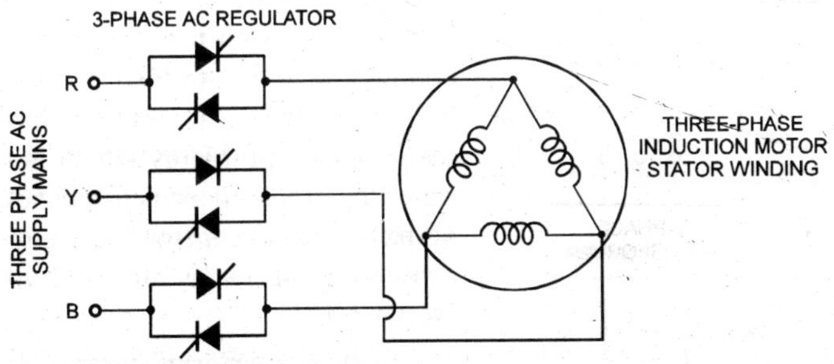

Fig. 3.48. ***Reduced Voltage Starting of a 3-Phase Induction Motor***

The circuit illustrated in Fig. 3.48 can be used for feeding a reduced voltage at the start. As seen it is a 3-phase ac regulator. By proper control of firing angle the regulator provides a low output voltage which is supplied to the induction motor. When the motor attains the full or rated speed, the regulator can be short circuited by mechanical contactor so that the motor operates normally at rated voltage. Moreover, if the motor is employed for a constant speed drive it is possible to operate the motor at reduced voltage when the mechanical load is small. Operation at reduced voltage causes decrease in power losses in the motor and thus results in energy saving.

3.13.6. Rotor Resistance Control

The conventional method of connecting resistances across the slip rings of a wound rotor induction motor is a form of rotor voltage control [Art. 1.18.2 (5)]. The main drawback of this method of speed control is its poor efficiency because of power wasted in the external resistors.

Figure 3.49 (*a*) shows a 3-phase diode rectifier and gate-turn-off thyristor (GTO)* connected in the rotor circuit of a wound rotor induction motor. The GTO operating as a chopper varies the resistance R as per the duty cycle α. The effective resistance R_e is given by

$$R_e = R(1 - \alpha) \quad \text{...(3.59)}$$

Thus the speed of a wound rotor induction motor is controlled by varying the ratio of on-time to off-time.

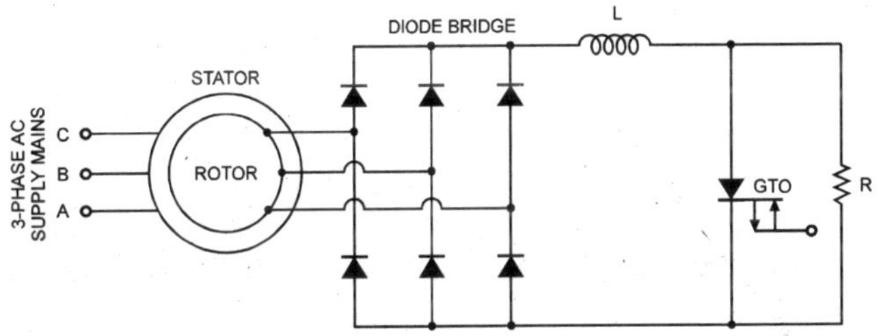

Fig. 3.49 (*a*). ***Rotor Voltage Control of a 3-Phase Wound Rotor Induction Motor (Chopper Control)***

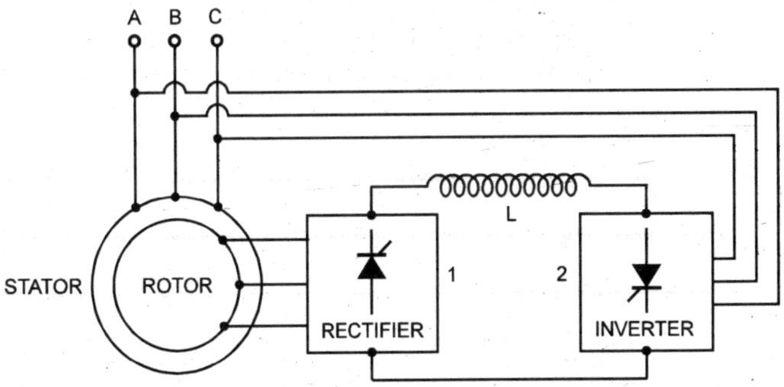

Fig. 3.49 (*b*). ***Rotor Voltage Control of a 3-Phase Wound Rotor Induction Motor (Slip-Power Recovery Scheme)***

3.13.7. Slip-Power Recovery Scheme

Figure 3.49 (*b*) shows a scheme known as *slip-power* recovery scheme. The rotor terminals are connected to the 3-phase ac

* A gate turn-off thyristor (GTO) is a special thyristor which can be turned on by a positive gate signal and can be turned off by a negative gate signal. Evidently the use of a GTO in a power electronics circuit eliminates the need of forced commutation circuit because turn-off is achieved by applying a negative gate signal.

supply mains through two fully controlled thyristor bridges. Bridge 1 acts as a rectifier (or converter) and bridge 2 acts as an inverter. Power output from rotor can be supplied back to the supply source. Since the frequency of rotor currents is slip frequency, this method is known as *slip-power recovery scheme.* By controlling the firing angles of the two bridges, the rotor power output can be varied. Thus the motor slip and speed (for the same torque) will also change. However, the drawback of this scheme is that both the bridges draw reactive power from the supply mains. Therefore, the overall power factor of the motor is poor. If speeds only below synchronous are desired, bridge 1 may be uncontrolled and, thus, may consists of diodes. If both bridges are controlled, the operation of the two bridges can be reversed also to get speeds above synchronous one.

Infact, the slip-power is either returned to the supply network as in Scherbius scheme or used to drive an auxiliary motor which is mechanically coupled to the induction motor shaft, as in Kramer scheme.

1. ***Static Scherbius Drive.*** The static Scherbius drive also uses the principle of slip-power recovery. The schematic diagram is shown in Fig. 3.50. For achieving both sub-sychronous and supersynchronous speed control, converters 1 and 2 must be fully-controlled thyristor bridges, one functioning at slip frequency as a rectifier or inverter while the other operating at supply system frequency as an inverter or rectifier. The cost of converters is quite appreciable and a slip-frequency gating circuit is also required. Moreover, at speeds near synchronous, when the slip-frequency emfs are quite small for natural commutation, special connections for forced commutation methods are required. If converter 1 is taken uncontrolled one (diode bridge) the converter cascade and the control unit would become economical and simple, but only sub-synchronous speed control would then be available.

The 3-phase transformer between the supply source and inverter 2 is meant to bring the rotor circuit voltage up to the value corresponding to the voltage of the supply source. The main drawback of sub-synchronous cascade drive is its poor power factor particularly at reduced speeds.

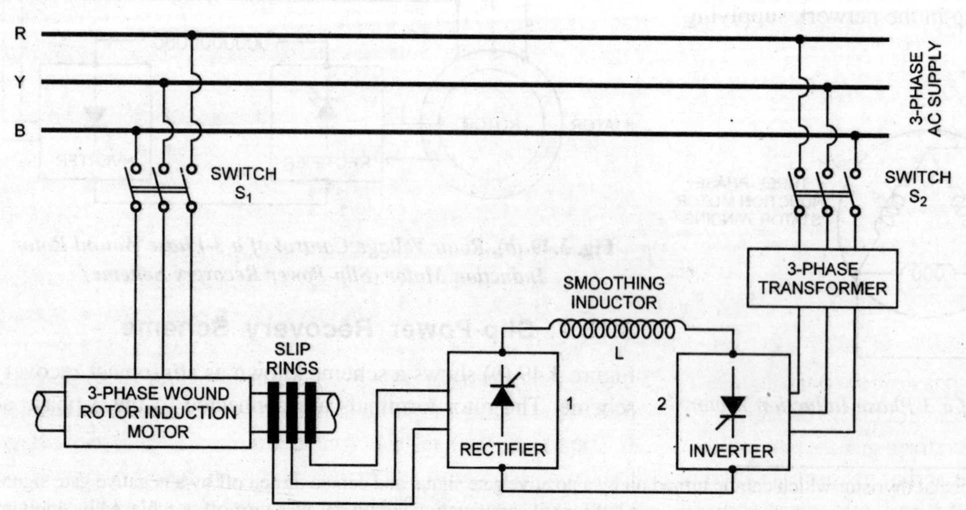

Fig. 3.50. ***Static Scherbius Drive***

This drive has applications in large power fan and pump drives which need speed control in a narrow range only. Power ratings of diode bridge, inverter and transformer is just maximum slip times the power rating of the motor resulting in a low cost drive. This drive provides a constant torque control. Constant power control can be obtained from Kramer drive discussed below.

2. ***Static Kramer Drive.*** Figure 3.51 shows the schematic diagram of the Kramer cascade with static converter. The rotor circuit of a slip-ring induction motor feeds the slip power, rectified by a diode bridge, to the armature of a separately excited dc motor mechanically coupled to the induction motor. Speed control is achieved by varying the field current of the dc motor. An emf proportional to the back emf of the dc motor may be considered to be injected into the rotor circuit of the induction motor to cause variation in the speed of the system.

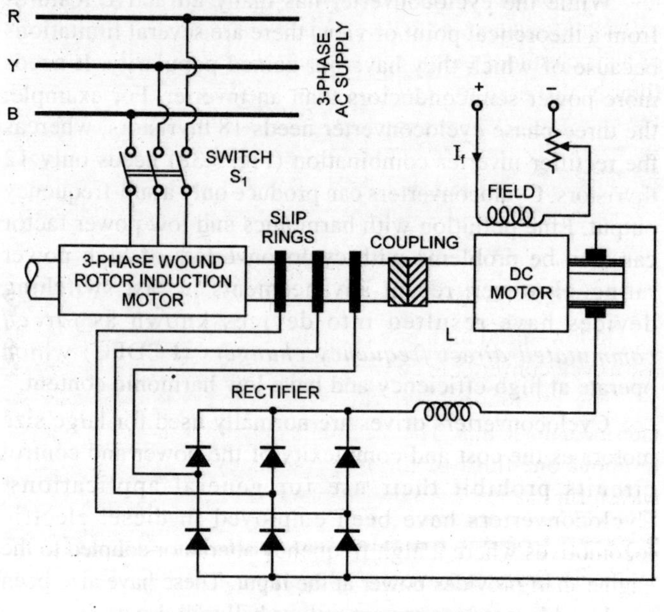

Fig. 3.51. ***Static Kramer Drive***

For achieving larger speed range, diode bridge will need replacement by a thyristor bridge. With thyristor bridge speed can be controlled up to standstill.

Static Kramer drive possesses no line-commutated inverter, it takes less reactive power and introduces smaller harmonic contents of currents than a static Scherbius drive. However, it has maintenance problems which are posed by commutator and brushes of the auxiliary dc motor. It has also drawback of large moment of inertia.

Static Kramer drive systems are used in large power pumps and compressor type loads where speed control is within narrow range and below synchronous speed.

3.14. THYRISTOR CONTROL OF A 3-PHASE SYNCHRONOUS MOTOR

The synchronous motor is a constant speed motor and it develops torque only at synchronous speed which is directly proportional to supply frequency. Therefore, variation of frequency of ac supply is a convenient method to control the speed of a synchronous motor. Before the advent of power electronic devices change of frequency was a very difficult task. However, the power electronic circuits provide a simple method of changing the frequency. The 50 Hz ac is converted into dc by a controlled rectifier. A variable frequency inverter converts dc into variable frequency ac. To keep the flux density in the motor constant voltage and frequency must be varied in the same ratio so that voltage/frequency ratio is constant.

Figure 3.52 shows a system of speed control of a synchronous motor. The 3-phase 50 Hz ac is rectified by a controlled 3-phase rectifier. The filter removes harmonics from the output of the rectifier. The variable frequency inverter feeds a variable frequency ac to the synchronous motor. Thus the speed of a synchronous motor can be controlled. The dc output from filter is also used to excite the rotor winding.

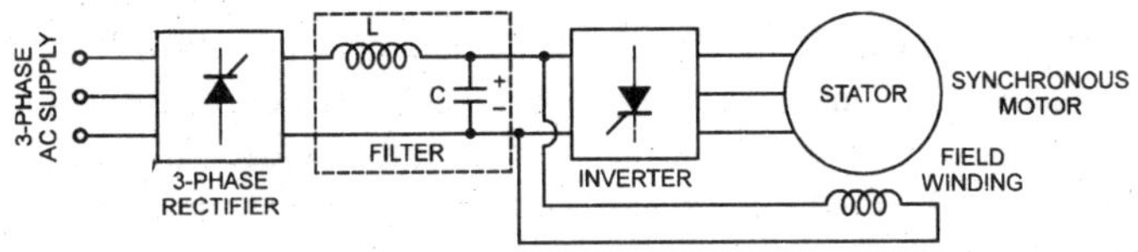

Fig. 3.52. *Speed Control of a Synchronous Motor*

In low-speed high power applications it is also possible to use a cycloconverter to convert 50 Hz ac into variable frequency ac. The cycloconverter does the conversion without an intermediate dc link. The maximum output frequency is limited to about one-third of the line frequency so that the output waveform has low harmonic content.

3.14.1. Load Commutated Inverter Drive For Synchronous Motor

For large sized motors (motors having output rating of more than 750 kW), load commutated synchronous motor drive becomes competitive with induction motor drive in adjustable speed requirements. Figure 3.53 shows the circuit for a load

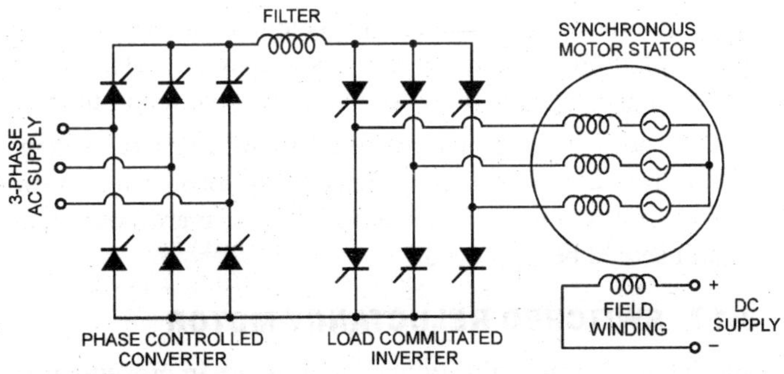

Fig. 3.53. *Load Commutated Inverter Drive*

commutated inverter drive. The three-phase ac supply is rectified by phase controlled 3-phase bridge converter. The dc supply is passed through a filter to improve waveform and then fed to the load commutated inverter. Each phase of the synchronous motor is represented by an internal induced emf in series with an inductance of stator winding. The frequency and phase of stator currents are synchronised to the rotor position. The current commutation in the load commutated inverter to supply currents to the stator phases in an appropriate sequence is provided by the stator induced emfs. The presence of 3-phase induced emf in the motor stator winding causes the commutation of the thyristor. A control of firing angle of converter controls its dc output voltage and, therefore, the current.

A complete circuit of load commutated inverter drive controller is shown in Fig. 3.54. The input voltage to stator of a synchronous motor is measured to calculate the rotor field position as function of time. The measured voltage is rectified to provide a dc signal proportional to instantaneous speed of synchronous motor. The turn-off time T_{off} available to thyristor in the inverter is kept constant. Keeping field current and T_{off} constant, the actual speed is compared with the reference speed. The error signal is amplified to provide reference I_d. If actual current I_d is less than reference, the rectifier increases the voltage supplied to inverter, thus increasing I_d and the motor torque. Based on current I_d and measured voltage, the fire pulses to the gates of thyristors of inverter are provided to keep T_{off} constant.

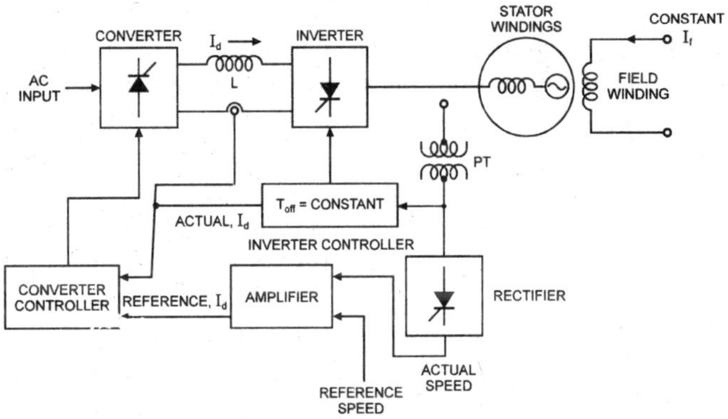

Fig. 3.54. *Load Commutated Inverter Drive For a Synchronous Motor*

3.15. SPEED CONTROL OF SINGLE PHASE INDUCTION MOTOR USING TRIAC (FAN REGULATOR)

The conventional speed regulator for a fan uses a resistance regulator. The regulator resistance is in series with the fan motor. The speed of the fan is reduced, whenever desired, by increasing the regulator resistance. Thus the voltage drop across the regulator resistance increases and, therefore, voltage applied to the motor is reduced and so the speed. The resistance regulator causes loss of energy. This loss of energy becomes significant at low speeds.

Figure 3.55 shows a speed control circuit using a diac-triac pair. The diac is meant to trigger the triac into conduction. It is preferable to use a matched diac-triac pair. The R-C circuit forms the triggering network. By adjusting the resistor R, the voltage across capacitor C can be adjusted. When voltage across capacitor C exceeds breakdown voltage of diac, it is triggered into conduction and sends a triggering signal to the triac gate. When triac is turned on the motor starts. By varying resistance R, The firing angle of triac can be altered and thus the voltage applied to the motor is changed and consequently the motor speed is changed. Since resistance R, carries very small current, the loss of energy is very small. Sometimes an R-C snubber circuit is added in parallel with the triac to protect it from high voltage transients and high $\frac{dv}{dt}$

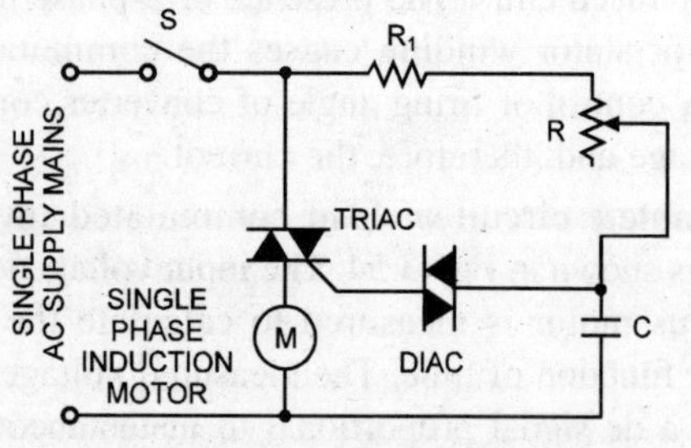

Fig. 3.55. *Speed Control of Single Phase Induction Motor Using Triac*

3.16. BRUSHLESS (OR COMMUTATORLESS) DC MOTOR

A dc motor can be thought of as an ac synchronous machine in which the field is stationary and the armature with its multiphase ac winding is rotating. The armature receives ac supply from a dc source through brushes and commutators. The brushes and commutator constitutes an inverter sensitive to the shaft position. In a similar way, a synchronous motor may be considered to operate as a dc motor. In a synchronous machine the field is rotating whereas the armature is stationary but it should be supplied by an inverter controlled by shaft-position sensing signals. The static inverter with the shaft-position-sensitive controller can very well be regarded as an *electronic commutator* serving the same function as does the mechanical commutator. This facilitates the operation of a synchronous motor as a versatile variable-speed drive like a dc motor but having no mechanical commutator and brushes. This is, no doubt, a great advantage.

No doubt, the brushless dc motors are generally more expensive for the same kW rating that conventional (commutator and brush) motors but possess certain advantages too such as little or no maintenance, much longer operating life, no risk of explosion or possibility of RF radiation due to arcing, higher efficiency, rapid response etc. These motors produce no brush or commutator particles or gases as by- products of operation. These motors are capable of operation submerged in fluids, combustible gases and may even be hermetically sealed. They provide a fairly linear output torque vs input current characteristic, which lends itself to servo applications.

The stator winding of a brushless dc motor may be conventional three-phase winding of a synchronous motor or the conventional armature winding of a dc motor. However, in both cases, the stator winding has to be supplied from a static inverter triggered by shaft-position sensitive signals so that the supply frequency is proportional to the shaft speed. The dc field winding is placed on the rotor and supplied from a static dc source through slip rings mounted on the motor shaft. The equivalent block diagram of a brushless dc motor is shown in Fig. 3.56.

It has already been explained that static inverter together with the shaft-position sensitive trigger circuit is equivalent to the mechanical commutator of a dc machine. A 3-phase inverter is equivalent to a three-segment commutator. If the stator winding is similar to the armature winding of an ordinary dc machine, six symmetrical tappings from the winding can be taken out and the stator winding may be supplied from a six-phase SCR bridge inverter. This is equivalent to six segment commutator. There may be many other variations of the electronic commutation arrangement. For instance, as individual stator coils may be controlled by series SCR switches. However, three segment commutation with a three-phase static inverter is widely used for drives.

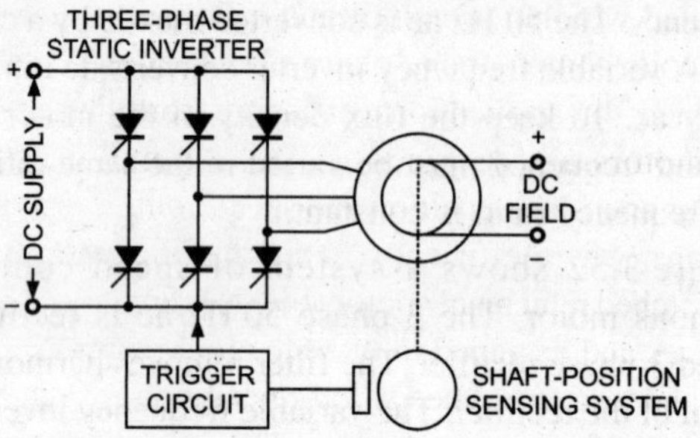

Fig. 3.56. *Equivalent Block Diagram of Brushless DC Motor*

It is, therefore, obvious that a 3-phase synchronous motor when supplied by a 3-phase inverter behaves like a simple dc motor but the SCRs of the inverter should be triggered in proper sequence and instant proportional to the position of the rotor shaft. The SCRs may be turned off naturally owing to the nature of the load (synchronous motor). The synchronous motor may be operated at any power factor (leading, lagging behind or unity) by variation of the dc field excitation. When the motor is operated at leading power factor, the inverter can be operated with load commutation *i.e.*, the SCRs are turned off naturally. Otherwise, a force commutated inverter should be employed.

Since the system behaves like a conventional separately excited dc motor, the speed can be controlled by the variation of the dc supply to the inverter or to the field. The speed is inversely proportional to the field strength. The torque-speed characteristics are similar to those of a separately excited dc motor but are slightly more drooping in this case.

The shaft-position sensing unit normally consists of a toothed disc on a free motor shaft end and two sets of three position detecting sensors. The detecting sensor may be an oscillator triggered into action for a definite period by the toothed disc or may be an optical unit using phototransistor, light sources producing pulses at definite intervals or may be a Hall device for sensing a magnetic pulse for the required period of time.

The brushless dc motor may be controlled either by a voltage fed inverter, a current fed inverter or a cycloconverter. Open-loop control is simple. To achieve higher accuracy closed-loop control should be employed.

3.17. SWITCHED RELUCTANCE MOTOR

The increasing availability and progressive improvement of power semiconductor devices as well as intensive computer aided study of the geometry of conventional reluctance motors for optimum torque production have led to the introduction of yet another type of drive called the *switched reluctance motor drive.* It is becoming a serious competitor to converter supplied dc and ac variable speed drive systems. The switched reluctance motor is being considered, nowadays, for applications ranging from low power servomotors to high power traction drives. Motors of power ratings varying from 4 to 22 kW are commercially available at present for many applications.

Construction. The switched reluctance motor (SRM) has both salient pole stator and rotor like a variable reluctance stepper motor. Figure 3.57 shows a one phase winding of a 4-phase switched reluctance motor having 8 poles on stator and 6 poles on rotor. While the rotor has no windings, each stator pole has a concentrated winding around it and each pair of diametrically opposite coils comprise one phase of the motor. These motors are designed for applications different from those for which stepper motors are designed. A stepper motor is designed suitable for open-loop position and speed control in small power applications, where efficiency is not significant. On the other hand a switched reluctance motor (SRM) is employed in variable speed drives and designed to operate efficiently for wide range of speed and torque and needs rotor position sensing. It is, also, quite different from a standard synchronous reluctance motor, described in Art. 1.16.3 (*e*), in two aspects. A synchronous reluctance motor has the same number of poles on stator and rotor but the rotor of a SRM has lesser poles than the stator, which is an essential feature to provide self-starting capability and bidirectional control. They are also different in stator construction—the synchronous reluctance motor has a cylindrical stator with distributed winding, while the SRM has a salient pole stator with concentrated coils like a dc motor.

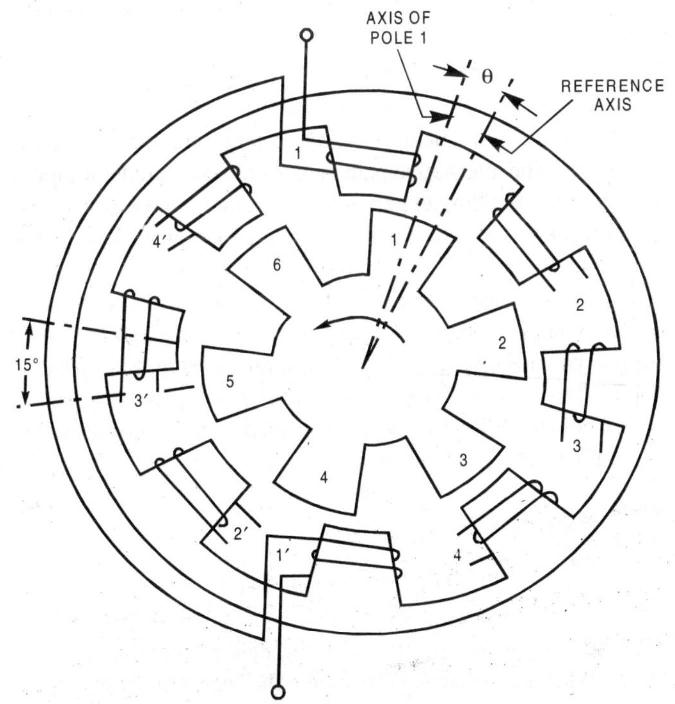

Fig. 3.57. One Phase Winding of a 4-Phase 8/6 Pole Switched Reluctance Motor

Though various combinations of stator and rotor pole numbes are possible in SRMs, the commonly used are 8/6 and 6/4. As already mentioned, the stator has concentrated coils and diametrically opposite coils are connected in series or parallel to provide one phase. Thus, motors with pole numbers 6/4 and 8/6 will have three and four phases respectively.

Operation. Reluctance torque is produced when a stator phase is excited by means of unidirectinal currents. This results in the magnetic attraction of an adjacent rotor pole as it tends to align into a position of minimum reluctance. When the number of stator and rotor poles differ, the sequential switching of the excitation from one set of stator poles to the next, in synchronism with the rotor position, produces an almost constant torque resulting in an uniform rotation of rotor. The synchronisation of the switching on the excitation with rotor position can be accomplished with simple rotor position feedback.

Neglecting nonlinearity of the magnetic circuit, the instantaneous torque produced in such machines may be given as

$$T = \frac{1}{2} i^2 \frac{dL}{d\theta} \qquad ...(3.60)$$

where i is the instantaneous current in the exciting winding and L is the self inductance of that winding varying as a function of the angular position of the rotor. It may be noted that the torque developed is independent of the direction of current flow in the windings, so that unidirectional currents can be used to control the motor. The direction of rotation can be reversed by exciting the phases in the reverse sequence. The motor can also provide regenerative braking. If a phase is excited after the rotor has crossed the position of minimum reluctance, the rotor will experience a torque in opposition to its motion, it will decelerate, mechanical energy drawn from it will be converted into electrical energy and supplied to the source. In fact, the possibility of operating in all of the four quadrants of the speed-torque plane and obtaining flexible speed-torque characteristics simply by appropriate switching of current pulses makes the motor very versatile.

Figure 3.58 (*a*) and (*b*) illustrates the ideal variation of inductance of the exciting winding with respect to the angular position of the rotor over a periphery of one rotor pole pitch and the corresponding torque developed, for an assumed value of constant current [using Eq. (3.60)].

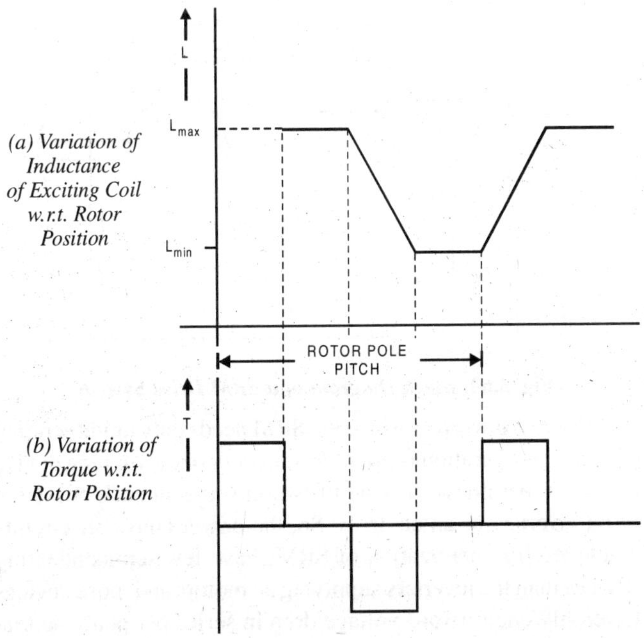

Fig. 3.58

Modes of Operation of Motor. There are two distinct modes of operation corresponding to low or high speed. Monitoring of exciting current during low speed operation is essential because of long duration of each phase period and needs chopping of energization to restrict each phase current within the semiconductor ratings. Moreover, the developed torque is controlled by varying the average phase current. Hence, accurate monitoring of the exciting current is required for obtaining high degree of controllability possible. During high speed operation, current control is not essential because the inductance of the winding and the motional back emf induced restrict the excitation to single pulses of current. Torque is controlled by optimal positioning of these pulses rather than the current level. Current monitoring, however, is retained for the sake of protection.

Typical phase current waveforms of the two modes (low speed and high speed) of operation are shown in Figs. 3.59 (*a*) and (*b*).

The complete drive system, comprising a SRM coupled to a load, a power converter and a control system involving rotor position transducer and current sensor is shown in Fig. 3.60.

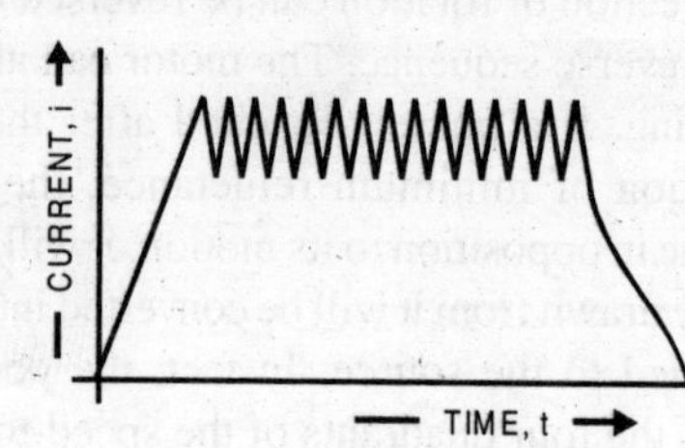

(*a*) *Low Speed Operation*

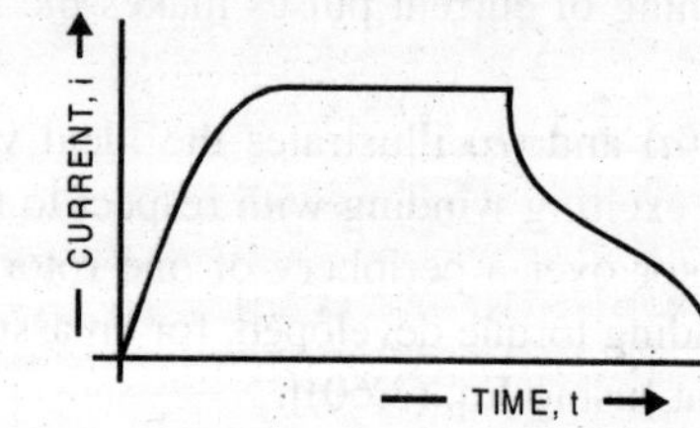

(*b*) *High Speed Operation*

Fig. 3.59. ***Current Waveforms***

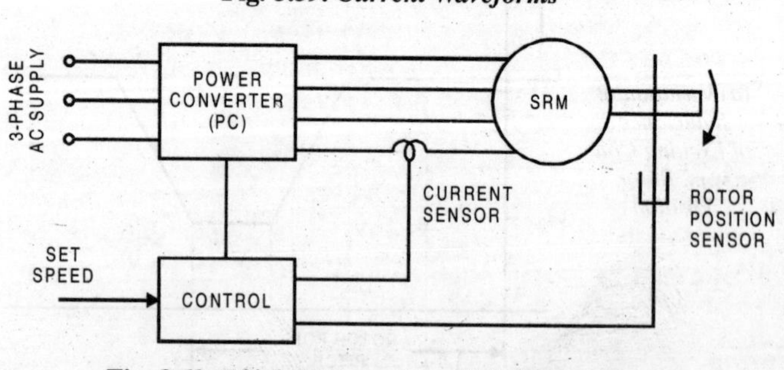

Fig. 3.60. ***Block Diagram of a SRM Drive System***

Power Converter. Since the SRM needs only unidirectional currents, its operation is possible with only one switching device in service per phase, instead of two in series in each phase leg of an inverter for an ac drive. So, the power converter circuits employed for energization of SRMs have few semiconducting devices than the inverters supplying ac motors and those devices have only one forward voltage drop in series per phase so that the power losses may, in general, be lower than in conventional inverters. Because of these facts, other factors being the common, results in reduction in the physical size of converter and increase in its reliability.

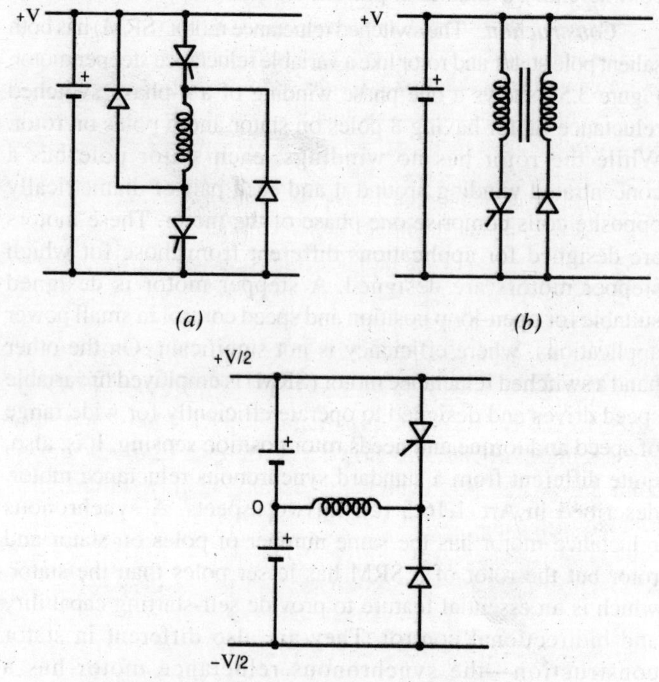

Fig. 3.61. ***Different Power Circuit Configurations***

Different circuit configurations are available for the power converters for SRM drive. All of these have two essential elements—a controlled switch or switches to connect the dc voltage source to the exciting winding to build up current and an alternative path for the current to take when the switch is turned off. The alternative path is provided by a diode or diodes such that the winding experiences a reverse voltage to collapse the current. Three alternatives for one phase of a SRM power converter are depicted in Fig. 3.61. A flexible circuit using two switches per phase is shown in Fig. 3.61 (*a*), whilst the use of bifilar wound motor (as in stepper motors) or a centre tapped supply permits the use of only a single switch per phase as depicted in Figs. 3.61 (*b*) and (*c*). The correct choice of power circuit configuration will depend on the power level of the drive, the supply voltage and the application.

Exercises

1. Why thyristor controlled drives employing both dc and ac motors find wide application in industry ? Explain.
2. Give the classification of static power converters.
3. What are the factors that govern the choice of current rating, voltage rating and switching frequency of the power electronic converter for electric motor drive ?
4. Explain working principle of single phase half controlled bridge converter fed separately excited dc motor and draw waveforms of motor terminal voltage, motor current and supply current.

   Compare its merits and demerits with respect to when fed by fully controlled bridge converter.

   [U.P. Technical Univ. Electric Drives 2005-06]

5. Explain how the speed control of a separately excited dc motor can be achieved by a fully controlled single phase thyristor bridge converter.

[Madurai Kamaraj Univ. November 1996]

6. Draw the complete drive circuit of a single phase fully controlled rectifier controlling a separately excited dc motor. The circuit should include the single phase supply, controlled rectifier (including the thyristors) and the motor. Draw typical waveforms of supply voltage, armature voltage and current at a fixed speed under continuous conduction. Derive an expression for motor speed in terms of supply voltage, firing angle and motor constants.

[U.P. Technical Univ. Electric Drives 2007-08]

7. Compare and contrast a half-controlled converter with a fully controlled converter for speed control of a separately excited dc motor. In which case supply power factor is better and why ? **[A.M.I.E. Sec. B. Electric Drives & Their Control Summer 1996]**

8. Compare the performance of a separately excited dc motor when fed from a 1-phase fully controlled bridge converter with that of the same motor fed from a 1-phase half controlled bridge. Draw the relevant waveforms in each case and sketch the variation of average output voltage *vs* firing angle. **[A.M.I.E. Sec. B. Electric Drives & Their Control Summer 1994]**

9. Describe the operation of a single phase half controlled bridge converter fed separately excited dc motor and draw waveforms of motor terminal voltage, armature and supply currents assuming continuous current conduction. List its merits and demerits in comparison to that of with fully controlled bridge rectifier.

[U.P. Technical Univ. Electric Drives 2008-09]

10. How do you change from motoring to braking operation in forward direction of a dc separately excited motor when two fully controlled thyristor converters are used in back-to-back connection? Explain in detail.

[A.M.I.E. Sec. B. Electric Drives.& Their Control Summer 1999]

11. Draw the circuit arrangements for a dual converter and explain its operation. **[Madras Univ. April 1997]**

12. Describe, with power circuit diagram, the dual converter. What is the relationship between firing angles of the two full converters of the dual converter under continuous current conduction mode ? Explain Why ?

[A.M.I.E. Sec. B. Electric Drives & Their Control Summer 1993]

13. Explain two modes in which a dual converter is used to control the dc motor speed. Which of the two methods is better and why ? Explain.

[U.P. Technical Univ. Electric Drives 2009-10]

14. A three phase fully controlled bridge converter is feeding a separately excited dc motor. Assuming continuous load current, derive the equation relating speed and torque of a motor as a function of triggering delay 'α'.

[U.P. Technical Unv. Electric Drives 2004-05]

15. Enumerate the different methods of obtaining variable dc voltage from a fixed dc voltage source. Give their advantages and disadvantages.

16. What is a dc chopper ? Explain the principle of operation of a chopper giving its circuit diagram and output voltage waveforms.

17. Explain the time ratio control (TRC) and current limit control used for choppers.

18. What is the analogy between a chopper and transformer ? Describe the working and obtain an expression for average load voltage of step-up chopper.

19. Discuss in full, the chopper control of a dc separately excited motor.

[Bharathiyar Univ. Electrical Drives and Control Oct. 1996]

20. Draw the block diagram for chopper fed dc motor drive. Explain the working. **[A.M.I.E. Sec. B. Electrical Drives & Their Control Summer 2000]**

21. Draw the power circuit of a single quadrant dc chopper and explain its working.

[A.M.I.E. Sec. B. Electric Drives & Their Control Winter 1998]

22. Draw the circuit for the chopper control of a separately excited dc motor and derive an expression for speed in terms of motor constants, supply voltage, duty ratio and load torque. **[U.P. Technical Univ. Electric Drives 2007-08]**

23. Regenerative braking is energy efficient braking, explain in view of chopper drives.

[A.M.I.E. Sec. B. Electric Drives & Their Control Summer 1998]

24. Discuss a chopper circuit for regenerative braking of a separately excited dc motor. Explain the operating principle.

[U.P. Technical Univ. Electric Drives 2004-05]

25. State the advantages of chopper controlled dc motors.

[A.M.I.E. Sec. B. Electric Drives & Their Control Winter 1994]

26. How is the performance of a dc motor affected by thyristor power supply ? Explain.

27. What are the special problems associated with the operation of dc drives where speed is controlled by phase controlled converters and choppers ?

[A.M.I.E. Sec. B. Electric Drives & Their Control Summer 1993]

28. Explain the speed control of a dc series motor using thyristors.

29. Explain in brief, on the importance of supply harmonics, power factor and ripples in the motor current, in the performance of a converter fed dc motor drive.

[G.B. Technical Univ. Electric Drives 2010-11]

30. Explain the chopper control techniques for each of a separately excited dc motor and dc series motor.

[G.B. Technical Univ. Electric Drives 2010-11]

31. How ac motor speed control is achieved by using voltage source inverter ?

[A.M.I.E. Sec. B. Electric Drives & Their Control Summer 2000]

32. Explain, giving block diagram, the operation of a dc link voltage source inverter for speed control of 3-phase induction motors.

[A.M.I.E. Sec. B. Electric Drives & Their Control Winter 1998]

33. What is 'pulse width modulation (PWM)' ? Explain the purpose. **[A.M.I.E. Sec. B. Electric Drives & Their Control Summer 2000]**

34. State the advantages of ac voltage control by PWM technique. Explain the sinusoidal PWM technique together with relevant waveforms.

[A.M.I.E. Sec. B. Electric Drives & Their Control Winter 1998]

35. What is meant by VVVF control of induction motors ? Explain with a block diagram and with reference to both constant torque and constant power modes of operation.

[A.M.I.E. Sec. B. Electric Drives & Their Control Winter 1999]

36. Draw the power circuit of a 3-phase current source inverter and explain its working. State also its inherent advantages over voltage source inverter for speed control of ac motor.

[A.M.I.E. Sec. B. Electric Drives & Their Control Winter 1998]

37. Distinguish between voltage source and current source inverters. Compare the two inverters for speed control of 3-phase induction motor.
[A.M.I.E. Sec. B. Electric Drives & Their Control Summer 1996]

38. Discuss the advantages and disadvantages of cycloconverter for speed control of 3-phase induction motors.
[A.M.I.E. Sec. B. Electric Drives & Their Control Summer 1996]

39. What is a cycloconverter ? Enumerate some of its industrial applications.
[A.M.I.E. Sec. B. Electric Drives & Their Control Summer 1995]

40. What is meant by "soft start" ? State and explain the soft start schemes (including electronic means, if any) employed for induction motors. **[A.M.I.E. Sec. B. Electric Drives and Their Control Winter 1999]**

41. Draw neat diagram for static rotor resistance control scheme using diode bridge and switch controlled resistor.
[A.M.I.E. Sec. B. Electric Drives & Their Control Summer 1998]

42. With neat circuit diagram, explain the speed control of a 3-phase slip-ring induction motor by static control of rotor resistance through chopper.
[A.M.I.E. Sec. B. Electric Drives & Their Control Summer 1996]

43. Discuss a static slip-power recovery scheme for speed control of slip-ring induction motor.
[A.M.I.E. Sec. B. Electric Drives & Their Control Winter 1998]

44. Explain the use of chopper for rotor resistance control in induction motors. Compare this scheme with slip-power recovery scheme, specially at lower speeds, stating the merits and demerits. Include circuits as needed.
[A.M.I.E. Sec. B. Electric Drives & Their Control Summer 1999]

45. Explain thyristor equivalent Kramer system of controlling speed of a slip-ring induction motor and regenerative operation. **[A.M.I.E. Sec. B. Electric Drives & Their Control Winter 1993]**

46. Explain any one slip power recovery scheme for the control of three-phase induction motor drive. Can supersynchronous speeds be achieved through it ? Explain.
[G.B. Technical Univ. Electric Drives 2010-11]

47. Explain the slip power recovery scheme of a three-phase induction motor control. What are its advantages and disadvantages ? **[U.P. Technical Univ. Electric Drives 2009-10]**

48. Explain working of static slip power recovery control scheme returning power to supply mains for speed control of three phase slip-ring induction motor. Why is supply power factor poor and how can it be improved ?
[U.P. Technical Univ. Electric Drives 2008-09]

49. Discuss static Scherbius scheme for speed control of a slip-ring induction motor. Draw a neat circuit diagram of the complete scheme. Mention one main advantage of this scheme compared to rotor resistance control.
[U.P. Technical Univ. Electric Drives 2004-05]

50. Discuss with the help of a suitable schematic diagram, the operation of a load commutated-fed synchronous motor drive. Why is this drive suitable for high speed and high power applications?
[A.M.I.E. Sec. B. Electric Drives & Their Control Winter 1998]

51. Explain the self controlled scheme for the synchronous motor control. **[G.B. Technical Univ. Electric Drives 2010-11]**

52. Describe self-controlled and load commutated inverter controlled synchronous motor drives in detail and compare them. **[U.P. Technical Univ. Electric Drives 2009-10]**

53. Explain static speed control of a single phase induction motor. **[U.P. Technical Univ. Electric Drives 2005-06]**

54. What are the similarities between a brushless dc motor and a conventional dc motor? Give the advantages of brushless dc motor over conventional dc motor.

55. Briefly explain the construction and principle of operation of a switched reluctance motor.
[G.B. Technical Univ. Electric Drives 2010-11]

56. Describe the construction and principle of operation of a switched reluctance motor.
[U.P. Technical Univ. Electric Drives 2009-10]

57. Describe working principle of switched reluctance motor. Discuss any one converter circuit used. What are its advantages over other ac motor drives.
[U.P. Technical Univ. Electric Drives 2005-06

58. Write short notes on the following :
(*i*) Half controlled converters.
[Madurai Kamaraj Univ. Nov. 1996]
(*ii*) Full controlled converters.
(*iii*) Dual converter drives.
(*iv*) Chopper drives.
(*v*) Comparison of converter fed and chopper fed dc drives.
(*vi*) Thyristor control of dc series motors.
(*vii*) Control circuits for ac drives.
[Bharathidasan Univ. Nov. 1996]
(*viii*) Cycloconverters.
(*ix*) Slip-power recovery scheme for speed control of slip-ring induction motor.
(*x*) Load commutated inverter drive for synchronous motor.
(*xi*) Brushless dc motor.
(*xii*) Switched reluctance motor.

Short Answer Type Questions With Answers

Q. 1. Why the thyristor control is preferred over Ward Leonard system of speed control ?

Ans. The thyristor control has some distinct advantages over Ward Leonard system of speed control such as low installation cost, ease of maintenance, higher accuracy, greater reliability, quick response, higher efficiency of operation, lesser floor space requirements, operation at a wide range of temperature etc.

Q. 2. What are freewheeling diodes ?
[Bharathiyar Univ. April 1997]

Ans. Freewheeling diode is a bypass or commutating diode. It is connected across motor terminals to allow for dissipation of energy stored in motor inductance and to provide for continuity of motor current when the thyristors are blocked. It also provides protection against transient overvoltages.

Q. 3. Why half-wave converter is not used for supply to the field circuit of a dc motor ?

Ans. If half-wave converter is used for supply to the field circuit of a dc motor, the high ripple content in the field circuit would cause increase in iron losses of the machine.

Q. 4. What is the relationship of V_{dc} and V_{ac} in terms of firing angle α for half-controlled and full-controlled converters?
[Madurai Kamaraj Univ. Nov. 1996]

Ans. For half-controlled converter,

$$V_{dc} = \frac{\sqrt{2}\, V_{ac}}{\pi} (1 + \cos \alpha) \text{ for } 0 \le \alpha \le \pi.$$

For full-controlled converter,

$$V_{dc} = \frac{2\sqrt{2}\, V_{ac}}{\pi} \cos \alpha \text{ for } 0 \le \alpha \le \pi.$$

Q. 5. What are the drawbacks of half-wave rectifier circuits ?

[Bharathiyar Univ. April 1997]

Ans. The output voltage wave of a half-wave rectifier is not a perfect dc. It is a discontinuous current and contains various harmonics. So half-wave converter drives have poor performance and their use is limited to small drives (rating up to 500 W).

Q. 6. What is the necessity of dc choke coil and freewheeling diode in a converter circuit ?

[Madurai Kamaraj Univ. Nov. 1996]

Ans. Freewheeling diode allows for dissipation of energy stored in the motor inductance and dc choke coil which reduces the rate of decay of current during the freewheeling operation when thyristors are blocked and thus motor current is made continuous. Discontinuous armature current results in deterioration of motor performance.

Q. 7. Why three-phase half-wave drives are generally not employed in industrial applications ?

Ans. Because of presence of dc component in supply currents.

Q. 8. What is a chopper ?

Ans. A chopper is a static switch to provide variable dc voltage from a source of constant dc voltage source.

Q. 9. Give the applications of dc choppers.

Ans. DC choppers are employed in trolley cars, marine hoists, forklift trucks, mine haulers and electric traction.

Q. 10. What is meant by duty ratio of a dc chopper ?

[Madras Univ. April 1997]

Ans. The ratio of on-period to the time period is known as duty ratio, *i.e.*,

$$\text{Duty ratio, } \alpha = \frac{T_{on}}{T} \text{ or } T_{on} \times \text{frequency or } \frac{T_{on}}{T_{on} + T_{off}}.$$

Q. 11. What are the advantages of chopper fed drive over converter fed drive ?

Ans. The advantages of chopper fed drive over converter fed drive are:

1. The ripple in the output of chopper is much less than that in the output of a converter. Lesser ripple means lesser losses.
2. The chances of discontinuous operation in chopper drive are lesser than that in converter drive. So speed regulation and transient response of chopper drive are better than that in converter drive.

Q. 12. Enumerate the control strategies used in dc choppers.

Ans. There are two types of control strategies for operating the switches that can be used in dc choppers. These control strategies are time ratio control (TRC) and current limit control.

Q. 13. What is TRC ? In how many ways can it be affected ?

Ans. TRC (time-ratio control) is a control strategy in which the value of $\frac{T_{on}}{T}$ is varied. This is affected in two ways namely variable frequency control (FM control) and constant frequency control (PWM).

Q. 14. What quantity is varied in PWM and frequency modulation ?

Ans. In pulse width modulation (PWM), the chopping period T is kept constant but the pulse width T_{on} is varied to control the output.

In frequency modulation the chopping frequency is varied and for this purpose either T_{on} or T_{off} is varied.

Q. 15. What is a step-up chopper ?

Ans. A step-up chopper, like a step-up transformer, raises the level of input dc voltage.

Q. 16. On what principle does a step-up chopper operate ?

Ans. A step-up chopper operates on the principle of storing energy in an inductance to provide output voltage higher than source voltage.

Q. 17. What is steady-state ripple ?

Ans. When chopper feeds R-L or R-L-E_b load, the current varies between I_{max} and I_{min}. The difference of I_{max} and I_{min} is known as steady-state ripple.

Q. 18. When is steady-state ripple maximum.

Ans. When $\alpha = 0.5$, steady-state ripple is maximum.

Q. 19. What factors can make the current discontinuous in a chopper feeding an R-L-E_b load ?

Ans. Higher L/R ratio, low duty cycle and higher back emf can make the current discontinuous.

Q. 20. The commutators of the thyristor controlled dc motors are made larger than those for ordinary dc motors. Why ?

Ans. The commutators of the thyristor controlled dc motors are made larger in order to provide extra insulation to withstand larger and rapid voltage fluctuations.

Q. 21. Semiconverter system is preferred for dc series motor drive. Why ?

Ans. The semiconverter system, because of its freewheeling action helps in maintaining continuous current and thus provides better performance to dc series motors in comparison to full converter system.

Q. 22. Why are ac regulators preferred over conventional methods of providing ac variable voltage of constant frequency for ac motor drives ?

Ans. AC regulators using thyristors are becoming popular because of high efficiency, fast control and compact size.

Q. 23. What is difference between on-off control and phase control in the case of ac voltage controller ?

[Madras Univ. October 1996]

Ans. In on-off control, also known as integral cycle control, the thyristors are used to connect the motor to the supply source for a certain number of cycles and then to disconnect it for another certain number of cycles; each of the on and off times consist of an integral number of cycles. On the other hand in phase control, the thyristors are employed as switches for connecting the motor to the supply for a certain portion of each cycle of the supply voltage.

Q. 24. What is the significance of maintaining the ratio of supply voltage to frequency constant in variable frequency drives of induction motors. **[U.P.T.U. Electric Drives 2007-08]**

Ans. If only frequency is changed and stator voltage is kept constant, the stator flux will not be at its rated value. The operation with flux below or above the rated value is not desirable.

The noteworthy point is that electrical machine is designed corresponding to near saturation point on the magnetization curve (or B-H curve). This is done from the point of view of full utilisation of the core. If the stator frequency is reduced, keeping the stator voltage constant, motor will operate in the saturation region and therefore, the motor will draw large magnetising current causing increase in stator core and copper losses and therefore, decrease in motor efficiency. However, if only supply frequency is increased keeping the stator voltage constant, the motor will operate at low flux density and thus motor capacity will be under utilised.

For constant flux operation, it is necessary that the induced emf varies linearly with applied frequency. At higher voltages and at high frequency operation stator voltage drops are very small and constant flux operation may be obtained by keeping V/f ratio constant.

Q. 25. State two advantages of current source inverter.

[Madurai Kamaraj Univ. Nov. 1996]

Ans. Current source inverter has the following two main advantages:

1. It can handle reactive or regenerative loads without freewheeling diodes.
2. Misfiring of devices and short circuits, because of constant input current, do not pose any problem.

Q. 26. Mention any two drawbacks of cycloconverters.

[Bharathiyar Univ. April 1997]

Ans. The two drawbacks of cycloconverters are :

1. It needs more power semiconductors than an inverter.
2. They can produce only a sub-frequency output.

Q. 27. Give some applications of cycloconverter.

[Madurai Kamaraj Univ. Nov. 1996]

Ans. Cycloconverter drives are normally used for large sized motors. They have also been employed in gearless cement mill or ball mill drives.

Q. 28. What is meant by GTO ? **[Madras Univ. April 1997]**

Ans. Gate-turn-off switch is a special thyristor which can be turned on by a positive gate signal and can be turned-off by a negative gate signal.

Problems

1. A single phase 230 V, 50 Hz supply feeds a separately excited dc motor through single-phase semiconverters. The firing angle of the field semiconverter is zero. The armature and field resistances are 0.1 Ω and 150 Ω respectively. The load torque is 150 N-m at 750 rpm. The voltage constant is 0.9 V/A radian/s and torque constant is 0.9 N-m/A^2. Determine (*i*) field current (*ii*) firing angle of converter in armature circuit and (*iii*) power factor of armature circuit converter. Assume continuous current operation and neglect losses. **[Ans.** (*i*) 1.38 A (*ii*) 96.6° (*iii*) 0.585 lagging**]**

2. A 230 V single phase ac supply is used to supply a separately excited dc motor through two full converters. The firing angle of converter in the field circuit is kept at zero. The field and armature resistances are 150 Ω and 0.15 Ω respectively. The firing angle of converter in the armature circuit is 30° and armature current is 80 A. The torque and voltage constants are 1.2 N-m/A^2 and 1.2 V/A-radian/second respectively. Determine torque developed and motor speed. **[Ans.** 132.48 N-m; 964.7 rpm**]**

3. A 220 V, 1500 rpm, 10 A, separately excited dc motor with an armature resistance of 3 Ω is fed from a single phase fully controlled rectifier connected to an ac source voltage of 230 V, 50 Hz. Assuming continuous conduction, calculate the firing angle for half the motor torque at 600 rpm.

[U.P. Technical Univ. Electric Drives 2007-08] [Ans. 63.93°**]**

4. A 400 V 3-phase supply feeds a separately excited dc motor through two three-phase full converters, one for armature and the other for field. The firing angle of the field converter is zero. The armature and field resistances are 0.2 Ω and 200 Ω respectively. The armature current is equal to rated value of 60 A. The motor voltage constant is 1.05 V/A-rad/second. The speed is 1,000 rpm. Find firing angle of armature circuit converter. **[Ans.** 56.65°**]**

5. A 3-phase fully controlled bridge converter is used to control the speed of a 40 kW, 400 V, 1,000 rpm separately excited dc motor. The converter is fed from a 3-phase, 400 V, 50 Hz ac power supply. The induced emf constant is 0.3 V/rpm and the armature resistance is 0.5 Ω. Determine (*i*) rated torque (*ii*) firing angle when the motor operates at full load and rated speed. (*iii*) firing angle when the motor should operate at half load and at 500 rpm.

[U.P. Technical Univ. Electric Drives 2008-09]

[Ans. (*i*) 382 N-m (*ii*) 42.23° (*iii*) 68.27°**]**

6. A 230 V ac supply feeds the armature and field circuits of a separately excited dc motor through two full converters. The firing angle of the field converter is zero. The armature and field resistances are 0.5 Ω and 300 Ω respectively. The torque and voltage constants are 1.35 each. Determine motor speed and torque for armature current of 60 A and firing angle of armature circuit converter of 45°. **[Ans.** 1,173 rpm; 55.89 N-m**]**

7. A dc chopper is supplying a resistive load. If the input and output voltages are 200 V and 128 V and chopping frequency is 800 Hz, find the periods of conduction and blocking in each cycle. **[Ans.** 0.8×10^{-3} s and 0.45×10^{-3} s**]**

8. A chopper circuit is operating on TRC principle at a frequency of 4 kHz on 200 V dc supply. If the load voltage is 160 V, compute the conducting and blocking period of thyristor in each cycle. **[Ans.** 0.2 ms and 0.05 ms**]**

9. A step-up chopper has an ac input voltage of 100 V and an output voltage of 250 V. The period of blocking in each cycle is 0.6 ms. Find period of conduction and chopping frequency. **[Ans.** 0.36 ms; 1.6667 kHz**]**

10. A step-up chopper has input voltage of 200 V and output voltage of 600 V. If the conducting period of chopper is 50 ms, compute the pulse width of output voltage. In case output voltage pulse width is halved for constant frequency operation, determine the average value of new output voltage. **[Ans.** 25 ms; 1,200 V**]**

11. A separately excited dc motor is supplied from a 500 V source through a chopper. Armature current is 200 A and armature resistance is 0.1 Ω. The voltage and torque constants are 1.4 V/A rad/s and 1.4 Nm/A^2. The field current is 2.5 A. The duty cycle of the chopper is 0.5. Determine (*i*) input power (*ii*) speed and (*iii*) torque.

[Ans. (*i*) 50 kW (*ii*) 627.5 rpm (*iii*) 700 N-m**]**

12. The chopper used for control of a dc separately excited motor has supply voltage of 220 V, on-period is 10 ms and off-period is 15 ms. Determine the average load current when the speed of the motor is 1,200 rpm and has a voltage constant of 0.5 V/rad per second. The armature resistance is 2.5 Ω. **[Ans.** 10.07 A**]**

13. A one quadrant chopper is used for rheostatic braking of a separately excited dc motor. Armature and braking resistances are 0.2 Ω and 6 Ω respectively. Armature and field currents are 160 A and 1.9 A respectively. Duty cycle of chopper is 0.4 and voltage constant is 1.1 V/A-rad/second. Determine (*i*) average voltage across chopper (*ii*) speed of motor and (*iii*) power dissipated in braking resistance. **[Ans.** (*i*) 576 V (*ii*) 2,778 rpm (*iii*) 92.16 kW**]**

14. A dc series motor is coupled to a load whose torque varies as the square of speed. The motor is supplied through a semiconverter with input voltage of 700 V. For firing angle of 90° the armature current is 25 A and motor speed is 400 rpm. Determine the firing angle and armature current if the motor speed is to be raised to 500 rpm. Assume unsaturated magnetic circuit and the combined resistance of armature and field to be 0.4 Ω. **[Ans.** 56.46°; 31.25 A**]**

❑❑❑

CHAPTER 4

Motors For Particular Services

INSIDE THIS CHAPTER

4.1. Introduction **4.2.** Types of Motors used in Industrial Drives **4.3.** Applications of Electric Drive.

4.1. INTRODUCTION

We have so far discussed dynamics of electrical drives, electrical characteristics (operating or running characteristics, starting characteristics, speed control and braking characteristics), mechanical features, service capacity and ratings of different types of electrical motors available in the market. In this chapter we will discuss the requirement of individual application and suggest possible choice of motor for particular application, keeping in mind its requirement. Space does not permit of a discussion of more than a few of applications, so only the important applications will be included here for discussion.

4.2. TYPES OF MOTORS USED IN INDUSTRIAL DRIVES

While selecting a motor for a particular service the following points must be known about it.

(*i*) Speed-torque characteristics—whether they are series, shunt or constant.

(*ii*) The power factor at which the motor normally operates.

(*iii*) Speed control range.

(*iv*) Starting torque and starting current.

(*v*) Maintenance cost—whether low, medium or high.

(*vi*) Initial cost.

The characteristics including all the above six points and applications of various types of motors in common use are given in Table 4.1.

4.3. APPLICATIONS OF ELECTRIC DRIVE

1. ***Domestic Uses.*** Requirements are low torque, nearly constant speed, frequent start and comparatively small load duration. Small universal motors, usually series type, are used for domestic uses such as for driving sewing machines, vacuum cleaners, mixers, electric shavers, hair dryers, small fans, cloth-washing machines etc. Single phase capacitor-start induction motor with direct-on-line start is used for domestic refrigerators, window air-conditioners etc.

2. ***Machine Tool Drives.*** Many machine tools require a simple more or less constant speed drive. For this squirrel-cage induction motors together with simple manual controls are adequate. For transmission of power to cutting tools or work pieces at different finite number of speeds, we use gear boxes and stepped pulleys. The use of gear mechanism gives rise to vibrations and noise, and thus affects the machining accuracy. Stepless speed control, if used, will provide better machine timings and surface finishes. Electro-hydraulic and electromagnetic controls are used for providing desired smooth speed control. Though thyristor controlled drives have replaced these methods up to some extent. At present, in our country, the use of thyristorized drives is limited to special machine tools only, as the cost of such drives is sufficiently high.

In some cases pole changing is preferred over gear changing. If more than two speeds are required, two pole changing windings may be used in the same stator slots, *e.g.*, 4/8 and 6/12 poles providing speeds in the ratio of 1 : 1.5 : 2 : 3. Such motor is convenient for small vertical drilling machines.

In some cases, speeds in excess of 3,000 rpm are required as for example in woodworking machinery. A direct drive is preferred over a geared drive, but with a 50 Hz supply this can be achieved only by commutator machines. If there are several such drives, it is worthwhile to install an induction type frequency-changer to provide high frequencies (100 Hz and/or 150 Hz). This permits the relatively cheap and robust squirrel cage motors to be employed for speeds approaching 6,000 and 9,000 rpm respectively.

Other drives need a variable speed. For a wider speed range the Ward-Leonard system is supreme but now it has been replaced by a cheaper alternative such as a rectifier-dc motor combination. For a smaller speed range, field-controlled

Table 4.1

Type of Motor	*Characteristics and Costs*	*Applications*
1. DC Motors		
(*a*) Series wound	(*i*) Very high starting torque (3 times full-load torque with twice full-load current at start). (*ii*) Torque varies nearly as the square of the current. (*iii*) Varying speed; dangerously high at no load. (*iv*) Self adjustable speed–decrease in speed with increase in load. (*v*) Speed control possible, maximum range of 3 : 1. (*vi*) Maintenance cost — medium. (*vii*) Relative cost — 1.5.	For heavy duty applications such as electric railways, rolling mills; metallurgical works, mine hoists, continuous conveyors, cranes, valve operation etc.
(*b*) Shunt Wound	(*i*) Nearly constant speed. (*ii*) Torque varies nearly as the current. (*iii*) Medium starting torque (twice full-load torque with twice full-load current at start). (*iv*) (*a*) Continuous speed range of 4 : 1 maximum. (*b*) Continuous speed range of 100 : 1 and more by means of Ward Leonard control. (*v*) Maintenance cost — medium. (*vi*) Relative cost — 1.5 to 3.0.	For driving constant speed line shafts, lathes, vacuum cleaners, pressure blowers, constant head centrifugal pumps, compressors, reciprocating pumps, fans, wood-working machines such as circular saws, wood planners, laundry washing machines, milling machines, passenger elevators, continuous conveyors, grinders, polishers, small printing presses, paper making machines, metal cutting machines etc.
(*c*) Cumulatively Compound Wound	(*i*) High starting torque (2 to 3 times full-load torque with twice full-load current at start). (*ii*) Varying speed but within limits. (*iii*) Self adjustable speed with changing load. (*iv*) Speed control possible 2 to 1 range. (*v*) Maintenance cost — medium. (*vi*) Relative cost — 1.5 to 3.0.	For driving compressors, stamping machines, pressure blowers, variable head centrifugal pumps, door lifts, rotary presses, flat bed presses, circular saws, metal planner, shearing machines, punch presses, rolling mills, freight elevators, passenger elevators, continuous conveyors, hydro-extractors etc.
(*d*) Differentially Compound Wound	(*i*) Low starting torque. (*ii*) Constant speed but dangerously high at no load. (*iii*) Maintenance cost — medium. (*iv*) Relative cost — 1.5 to 3.0	Seldom used.
2. AC Motors (*a*) *Three-Phase Motors* (*i*) *Synchronous Motors*		
Plain Synchronous Motor	(*i*) Constant speed at varying load. (*ii*) No self-starting torque. (*iii*) Speed control—not possible. (*iv*) Capacity of working over wide range of power factor both leading and lagging. (*v*) Maintenance cost — medium. (*vi*) Relative cost — 2.	(*i*) For driving continuously operating equipment at constant speed such as ammonia and air compressors, motor-generator sets etc. (*ii*) For main drives of continuous rolling mills, medium and large drives in the paper and cement industries. (*iii*) For voltage control at the end of transmission line. (*iv*) For improving power factor like capacitors in large power systems in which considerable lagging power factor equipment is installed.
Synchronous Induction Motor.	(*i*) Constant speed. (*ii*) Twice full-load torque with twice full-load current at start. (*iii*) Capability of working over wide range of power factor both leading and lagging. (*iv*) Speed control — not possible. (*v*) Maintenance cost – medium. (*vi*) Relative cost – 2.	It is used when some starting torque is to be developed but load torque remains fairly steady, such as for large fans, compressors, line shafts, pumps etc.

Type of Motor	Characteristics and Costs	Applications
(ii) Induction Motors		
Squirrel Cage Type	(*i*) Nearly constant speed. (*ii*) Poor starting torque (full-load torque with five times full-load current at start). (*iii*) High overload capacity. (*iv*) Simple starting arrangement. (*v*) Speed control — not possible. (*vi*) Power factor — 0.75 to 0.9 (lagging). (*vii*) Maintenance cost — very low. (*viii*) Relative cost —unity.	For low and medium power drives where speed control is not required such as for water pumps, tube wells, lathes, drills, line shafts, circular saws, grinders, polishers, wood planers, compressors, laundry washing machines, fans, blowers etc.
Double Squirrel Cage Type	(*i*) Nearly constant speed. (*ii*) High starting torque with low starting current (2 to 2.5 times of full-load torque with a starting current of 4 to 6 times of full-load current). (*iii*) High overload capacity. (*iv*) No need of starting equipment. (*v*) Speed control — not possible. (*vi*) Power factor — 0.75 to 0.9 (lagging). (*vii*) Maintenance cost — very low. (*viii*) Relative cost — 1.5.	For driving loads requiring high starting torque such as for compressors, pumps, large refrigerators, crushers, conveyor equipment, boring mills, woodworking equipment, textile machinery, cranes, punches, lathes etc.
Multiple-Speed Motor	(*i*) Nearly constant speed for a given setting. (*ii*) Poor starting torque (full-load torque with a starting current of 5 times of full-load current). (*iii*) Power factor — 0.75 to 0.9 (lagging). (*iv*) Speed control by changing number of poles of stator winding; speed range 6 : 1; no control between steps; maximum 1/6th of top speed. (*v*) Maintenance cost — very low. (*vi*) Relative cost — 1.5 to 3.0.	For applications requiring several definite speeds but not necessarily a continuously adjustable speed such as for ventilating fans, conveyors, machine tools, elevators, traction etc.
Slip-Ring Type or Wound Rotor Type	(*i*) Nearly constant speed. (*ii*) High starting torque (twice full-load torque with a starting current of twice full-load current). (*iii*) High overload capacity. (*iv*) Speed control can be affected as to 50-75%. (*v*) Lower efficiency. (*vi*) Power factor — 0.75 to 0.9 (lagging). (*vii*) Maintenance cost — low. (*viii*) Relative cost — 1.5.	For industrial drives of high power where high starting torque and speed control are required such as for driving line shafts, lifts, pumps, generators, winding machines, auxiliary fans, smoke exhausters, printing presses, elevators, compressors etc.
(iii) Commutator Motors		
3-Phase Series Motor	(*i*) Speed torque characteristic—series. (*ii*) High starting torque (2.5 times full-load torque with a starting current of twice full-load current). (*iii*) Power factor — 0.9 to 1.0 (lagging). (*iv*) Speed control over the full range. (*v*) Maintenance cost — high. (*vi*) Relative cost — 2 to 3.	For drives where load increases with the speed such as fans, centrifugal pumps and compressors or where large starting torque is required such as for haulage and hoist work.
Schrage Motor	(*i*) Nearly constant speed for a setting. (*ii*) High starting torque (2.5 times full-load torque with a starting current of twice full-load current). (*iii*) Power factor — 0.7 to 1.0. (*iv*) Speed control range 3:1. (*v*) Maintenance cost — high. (*vi*) Relative cost — 3 to 4.	For drives where adjustable speed is required such as for driving paper making machines, printing presses, calenders and textile works. It is also employed for lifts, pumps, machine tools, belt conveyors etc. because of its high starting torque and convenient speed control.

Type of Motor	*Characteristics and Costs*	*Applications*
(b) Single Phase Motors		
(i) Synchronous Motor	(*i*) Constant speed. (*ii*) Power factor — 0.9 to 1.0. (*iii*) Speed control — not possible. (*iv*) Maintenance cost — medium. (*v*) Relative cost — 2.5.	For applications requiring constant speed but very low torque such as teleprinters, clocks, control apparatus, all kinds of timing devices, recording instruments, sound recording and reproducing apparatus.
(ii) Commutator Motors		
Single Phase Series Motor	(*i*) High starting torque (3 times full-load torque with a starting current of twice full-load current). (*ii*) Torque varies nearly as the square of the current. (*iii*) The speed varies inversely as the current nearly. (*iv*) Speed control — full range with transformer. (*v*) Power factor — 0.8 to 0.9 (lagging). (*vi*) Cost of maintenance — high. (*vii*) Relative cost — 2.5.	In small size it is used in driving small domestic appliances like vacuum cleaners, refrigerators etc. In large size it is employed for traction work.
Repulsion Motor	(*i*) High starting torque (3 times full-load torque with a starting current of twice full-load current). (*ii*) High speed at light load. (*iii*) Speed control possible by shifting of brushes–full range. (*iv*) Power factor–0.8 to 0.9 (lagging). (*v*) Maintenance cost–high. (*vi*) Relative cost–2.5.	For services requiring large starting torque and adjustable but constant speed as in coil winder.
Repulsion Induction Motor	(*i*) Constant speed. (*ii*) High starting torque (2.5 to 3 times full-load torque with a starting current of 3 to 4 times full-load current). (*iii*) Power factor — 0.85 to 0.9 (lagging). (*iv*) Speed control — not possible. (*v*) Maintenance cost — medium. (*vi*) Relative cost — 2.	For applications requiring high starting torque with an essential constant running speed such as for driving compressors, lifts etc.
(iii) Induction Motors		
Capacitor-Start Type	(*i*) High starting torque (1.5 times full-load torque with a starting current of twice full-load current). (*ii*) Constant running speed. (*iii*) Power factor — 0.7 to 0.9 (lagging). (*iv*) Speed control — not possible. (*v*) Maintenance cost — low. (*vi*) Relative cost — 1.5.	For compressors, refrigerators, and other applications involving high starting torque such as in small portable hoists.
Capacitor-Start, Capacitor-Run Type	(*i*) High starting torque (1.5 times full-load torque with a starting current of twice full-load current). (*ii*) Constant speed. (*iii*) Quiet in operation. (*iv*) Speed control — not possible. (*v*) High efficiency. (*vi*) Power factor — 0.9 to 1.0 (lagging). (*vii*) Maintenance cost — low. (*viii*) Relative cost — 2.0.	For applications requiring quiet operating motor and replacing repulsion start motor.
Repulsion-Start Type	(*i*) High starting torque (2.5 to 4.0 times of full-load torque with a starting current of 3.5 times full-load current). (*ii*) Constant running speed. (*iii*) Speed control — not possible. (*iv*) Power factor — 0.8 to 0.9 (lagging). (*v*) Maintenance cost — medium. (*vi*) Relative cost — 2.	For refrigerators, compressors, and other drives particularly those which have a high inertia and prolonged starting period.

dc shunt motor is adequate, and if dc supply is not available a free-firing rectifier may be employed. AC commutator motors with induction regulators or brush-shifting control of speed may also be employed for variable-speed drives.

For use such as internal grinding spindles for horological applications, drilling of printed circuit boards etc., high frequency, three-phase induction motors operating up to speeds of 180,000 rpm and providing fairly high output powers in small sizes are used. Typical ratings are 1.2 kW at 120,000 rpm for high-speed grinder drives and 0.5 kW at 150,000 rpm for high precision horological applications.

Some applications need braking torques. Almost all the methods described in chapter 1 are applicable. Many motors associated with machine-tool drives are of sufficiently small rating to allow capacitors to be employed for braking purposes without the size and cost being prohibitive. Where considerable inertia is involved, as in planing-machine drives, the regenerative braking of the Ward-Leonard set is a further advantage in addition to its others.

3. *Cranes and Hoist Work.* Essential requirement of a crane or hoist motor is that it should develop high starting torque and should withstand large number of switching operations. DC series or compound wound motors are preferred for cranes on account of their high starting torque and smooth speed control. The motors used for lifting, travelling and reversing, and for conveying and hoisting are dc compound wound, dc series wound and ac slip-ring induction motors (with rotor resistance control) respectively.

Hoist motors are provided with special electromechanical brakes, which by means of springs, are made to hold the load as soon as supply fails. On restoration of supply, solenoid connected across the terminals of motor is energized and brakes are released. Such motors are half-, or one-hour rated, developing starting torque twice of full-load torque, and robust to withstand severe strains to which these are exposed.

Drum-type controllers are used, operated by contactors, which are controlled by press buttons.

4. *Lifts.* High smooth accelerating torque (twice full-load torque at start), high overload capacity and pull-out torque, high degree of silence and moderate speed are essential requirements. Motors are normally one-hour rated for duty cycle of 150 to 200 starts per hour. DC compound wound and ac slip-ring induction motors are mostly used. Induction repulsion and variable speed ac commutator motors are also used. In single phase installation a shunt type commutator motor is also used. In case of dc motors speed control is affected by shunt field variation or by variable voltage control (Ward Leonard). In case of polyphase slip-ring induction motor the speed is controlled by rotor resistance and involves copper loss (I^2R loss).

5. *Lathes, Milling and Grinding Machines.* Lathes are usually driven by constant speed squirrel cage induction motors. Multi-speed ac motors or adjustable speed dc motors are sometimes used.

Milling machines are usually driven by constant speed squirrel cage induction motors. Larger machines such as planer type milling machines have separate motors for each milling head and separate motors for each of feed motions. The general practice is to use adjustable speed motors for headstock drive and either constant or adjustable-speed motor for the wheel drive. Separate motors are often used to feed the wheel with respect to the work.

Grinding machinery varies considerably in the design of individual machines. The general practice is to employ adjustable speed dc motors for the headstock drive, constant speed squirrel cage or adjustable speed dc motors for wheel drive and constant speed squirrel cage motors for traverse drive.

Constant speed squirrel cage induction motors with the grinding wheels mounted directly on the motor shaft extension are used for bench, pedestal and centreless grinders.

6. *Planers.* Planer consists of a bed having attached to it a platen which moves forwards and backwards on the bed. The job is clamped to the platen and is planed by a stationary clamped tool; at the end of the stroke the tool moves slightly in order to make a new cut. Cutting stroke is slow and return stroke is quick, therefore adjustable speed, reversing motors are required. In case of dc supply shunt or compound wound motor and in case of ac supply slip-ring induction motor is used.

7. *Punches, Presses and Shears.* Special features are high peak loads and heavy starting torque. These are usually equipped with flywheels to supply momentarily demand of power. Motors used for these machines should have drooping speed-torque characteristics to allow the motor to slow down when heavy loads are encountered. High-slip squirrel cage induction motors, slip-ring induction motors or dc cumulative compound motors are used for such applications.

8. *Shapers and Slotters.* Constant speed squirrel cage induction motors are used for shapers and slotters.

9. *Blowers and Fans.* Squirrel cage induction motors and synchronous motors are used for driving blowers and fans.

10. *Frequency Converters.* Squirrel cage induction motors and synchronous motors are used for driving converters. Nowadays owing to the recent advances in the field of silicon controlled rectifiers, static frequency changers are being increasingly used.

11. *Drilling Machines.* Constant speed squirrel cage or dc shunt or multi-speed ac motors are used for this purpose.

12. *Air Compressors.* Slip-ring induction motors and synchronous motors are usually employed in large size units. The latter type can be used for power factor correction in addition to driving air compressors if so designed. Squirrel cage induction motors are used for small compressors only.

13. *Electric Traction.* DC series motors, which are simple and robust in construction, provide high starting torque and smooth speed control, are suitable for all types of services but more particularly for suburban services where high rate of acceleration is essential.

Single phase ac compensated series motors, which have speed-current and speed-torque characteristics similar to those of dc series motors, are extensively used for the main line work. Due to poor starting torque these motors are not suitable for suburban services where stops are frequent.

14. *Pumps.* Drip proof or totally enclosed surface cooled polyphase induction motors are usually employed, often mounted on a common bed plate and direct coupled to the pump. Sometimes motor is flange mounted. In case the speed does not come within the range of the fixed motor speed, V-belt drive is employed. Squirrel cage motors with reduced voltage starters are used for centrifugal pumps, where starting torque required is about 40 to 50 per cent of full-load torque. Slip-ring induction motors are used for reciprocating pumps in which starting torque required may be 100 to 200% of full-load torque.

15. *Refrigeration and Air Conditioning.* In vapour compression system of refrigeration, motor employed to drive the compressor is thermostat controlled. On re-starting the motor has to drive compressor against high head pressure. Motor used, therefore, should be capable of developing starting torque 2 to 2.5 times full-load torque.

Single phase, 230 V, capacitor-type induction motors with D-O-L starters are usually employed for small units. For large installations high torque squirrel cage induction motors or slip-ring induction motors are employed. For very large plants, synchronous motor, driving turbo-compressor, may suit specially if power factor correction is also required.

16. *Breweries.* Totally enclosed surface cooled type motors are used because working atmosphere is damped one.

17. *Excavators.* Totally enclosed motors are used because of arduous and rugged work.

18. *Belt Conveyors.* Belt conveyors are used for moving sand and gravel. Heavy loads are required to be accelerated. Normal starting current and high starting torque double cage induction motor with direct-on-line starter or wound rotor induction motors are used. These motors must be totally enclosed, surface cooled type as grit and dust remain present in the atmosphere.

19. *Flour Mills.* Pipe-ventilated or totally enclosed type motors are used. Speed of the motor varies from 600-1,000 rpm.

20. *Quarrying.* Totally enclosed surface cooled type motors are used. Such motors must be of robust construction.

21. *Woodworking Machinery.* Screen-protected motors, except where there are chances of motor being burried in saw dust are usually used. In places where there are chances of motor being buried in sawdust totally enclosed surface cooled motor is used. Requirement of speed above 3,000 rpm, if required, can be met with induction motors in conjunction with induction type frequency converter or by using commutator motors.

22. *Printing Machinery.* In printing machinery for constant speed work such as guillotines, driving platens or other small machines squirrel cage induction motors are used. Where line capacity and moderate starting torque permits single phase capacitor motors may be used. In case of rolling presses where a slow forward stroke and a quick return stroke is required, two squirrel cage motors–one for each stroke, are used. For other purposes; where variable speed motors are required such as for rotary presses, dc compound or three-phase slip-ring induction motors with rotor resistance control or ac commutator motors may be used. For obtaining the steady crawling speed for inching the press, a pony or barring motor is used.

23. *Petrochemical Industries.* In petroleum and chemical industries, fluid handling equipment is extensively used. Pumps have been driven by induction motors with flow control, when desired, accomplished by throttling valves. As per estimates 30% or more of electrical energy used by a typical refinery is wasted due to throttling effect of control valves. However, adjustable-speed pumping utilizing multi-speed induction motors or inverter-fed induction motors are quite economical because of reduced energy consumption and maintenance costs.

Other important tasks within the production processes of the petrochemical industry are gas liquefaction, compression, refrigeration and heat recovery. Compressor drive systems constitute important components in plants for such processes. The majority of the drives operate at a constant speed, employing a 4- or 6-pole motor with step-up gear. For large ratings, synchronous motors with cylindrical rotors having both excitation and starting windings (damper bars) are preferred. Such motors are specially designed considering the oscillating torques developed during asynchronous acceleration owing to the magnetic and electrical anisotropy in the rotor and special cooling conditions.

24. *Sugar Mills.* In sugar mills, a centrifuge is employed for separating out crystalized sugar from the syrup obtained from steam evaporator and for drying it out by the action of centrifugal force. The duty cycle of a centrifuge motor is illustrated in Fig. 4.1. The functions involved are charging, intermediate spinning, spinning, regenerative and reverse current braking and plugging for ploughing. All these functions are to be performed at different speeds, variation of which may be as much as 1 : 30. The motors used for this purpose are usually four-speed, pole-changing motors having two sets of stator windings that enable us to obtain synchronous speeds 1,500/750/214/107 rpm or 1,000/500/214/107 rpm. They not only are capable of providing the desired fixed speeds of operation, but also of returning a

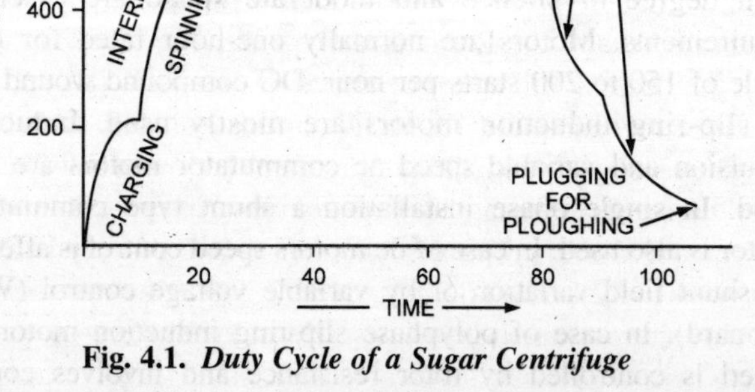

Fig. 4.1. ***Duty Cycle of a Sugar Centrifuge***

portion of energy back to the supply mains during regenerative braking accomplished by switching over to a higher pole operation from a lower pole one.

In order to satisfy the duly cycle illustrated in Fig. 4.1, first 28 pole winding is switched on to provide a speed of approximately 200 rpm. During charging supply is cut off and then intermediate spin speed of about 450 rpm is obtained by energizing the 12 pole winding. The final spin speed of about 950 rpm is achieved by switching over to the 6 pole winding. After centrifuging operation is complete, supersynchronous regenerative braking is applied by connecting the motor for 12, 28 and 56 poles successively which brings the speed down to 500, 214 and 107 rpm respectively. Ploughing speed of about 50 rpm is obtained by the application of reverse current braking. Automatic control of the entire duty cycle is achieved by feeding the control equipment from the output of a tachogenerator directly coupled to centrifuge motor.

Motors employed for driving centrifuges have their own special features in construction. They are vertically mounted so as to be coupled with the centrifuge shaft. The motors may be made with larger air gap so as to take care of the possible rotor oscillations about the vertical axis. Insulation used in motors must be humid proof so that it can operate successfully in a humid environment. For protection of motor against overheating, thermal elements are embedded in the windings. These elements, which are also called sensotherms by trade name, operate few degrees below the maximum permissible temperature of the winding. Operation of thermal elements will either directly trip the motor or give visible and audible warning signals so that the particular duty cycle may be completed. Following this, next cycle cannot be commenced until temperature of the motor has come down to normal.

25. *Cement Mills.* The driving motors used in cement works can be broadly classified as (*i*) raw mill and cement mill drives (*ii*) kiln drives (*iii*) crusher drives (*iv*) waste gas fan drives and (*v*) compressor drives etc.

The starting torque for the mill motors for large cement plants is limited to 125% of the full-load torque and the pull-out torque is restricted to nearly 240% of the full-load torque. Slip-ring induction motors of 6.6 kV with liquid resistance starters are usually employed. Gearboxes are employed in order to get the desired mill speed of about 15 rpm and high voltage capacitors of adequate reliability and automatic capacitor control switchgear and circuit breakers are used for power factor correction. Owing to large ratings (above 3,000 kW) required for the raw and cement mill drives and due to the limitations in the availability of large size gearboxes and motors, twin drives are used–the two motors employed for twin drives have to be more or less identical to each other and so also their liquid resistance starters.

The rotary kiln is an indispensable part of the cement plant. The rating of the motors employed for driving the kilns vary from 100–1,000 kW. The maximum speed of the kiln is about 1 rpm and speed range required is of the order of 1 : 10. The starting torque required may be between 200% and 250% of rated torque. In early years, variable speed ac commutator motors to be employed for kiln drives, which has been superseded by Ward-Leonard drives as ac commutator motors proved expensive and had maintenance problems. Nowadays use of dc motors with static supply is made. To cope up with increasing kiln capacities, the modern trend is to employ twin motor dc drives for kiln application.

Usually, the starting torque for crusher drives is limited to 160% of the full-load torque and pull-out torque is limited to 200–250% of the full-load torque. The slip-ring induction motors are usually used in crushers. Normally the motors are designed to withstand locked rotor current during running, without any external resistance inserted in the rotor circuit, for one minute–very important feature owing to frequent chances of getting crushers jammed.

The starting torque required for fan drives is about 120% of the full-load torque and the pull-out torque is from 200% to 250% of the full-load torque. The slip-ring induction motors with a speed variation are used for such drives. For starting and controlling the speed, the cast iron grid resistance controllers are normally used. The slip-rings and brush gears are totally enclosed and kept external to the motor enclosures for convenience in maintenance.

The air compressor motors are usually of smaller ratings, say 300-450 kW. Squirrel cage or wound rotor induction motors can be used as per power needs. The enclosures are usually of TEFC type and the speeds vary between 1,000 and 750 rpm.

26. *Mining Work.* The electric motors employed in coal mines can be classified into two categories namely the auxiliary motors, used for driving auxiliaries such as compressors, fans, conveyors, pumps, hoists etc., and the mine face motors, used on continuous miners, drills, shuttle cars, cutting machines, loaders etc. The motors of the two categories differ considerably from each other. The auxiliary motors are normally modifications of general purpose industrial motors, while the face motors are specially designed motors as per specific needs. The duty of the auxiliary motors is, often, well defined and steady, while that of face motors consists of random loading and has a number of high-shock loads.

The motors employed for coal drills are usually short-time rated induction motors of rating 1 kW, 125 V, 2860 rpm, 50 Hz with a flame proof enclosure. Sometimes, high frequency motors (150 or 200 Hz) with the help of a frequency changer housed in the drill panel are also used.

Slip-ring induction motors with water rheostats are, usually employed for driving the colliery winder used for raising the coal, raising or lowering the persons and other loads. DC rheostatic braking is used initially and mechanical braking is employed only towards the fag end of the winding period. For higher capacities dc motors supplied from a Ward-Leonard generator are used for driving the mine winders as precise control of drum speed and position is necessary for accurate decking. The recent trends are to use thyristor convertors for supplying variable dc voltage supply.

The motors employed for haulage (mine-winches) work are subjected to frequent starting and stopping and also to braking and reversing. For this work the loads, often, require different speeds at different states of haulage. Hence slip-ring induction motors with drum resistance controllers are used. The drum controller and the rotor resistances are separately mounted in flame proof enclosures.

There are two types of ventilation fans used in mines namely axial flow type and radial flow type. Although squirrel cage induction motors having large number of poles can be used directly but high speed motors having 4 to 6 poles with a reduction gear are usually employed.

There are two types of colliery pumps–triple ram type driven by a high speed motor with a gear and the centrifugal type directly coupled to the driving motor. The ram type pump requires a starting torque of about 200% of rated torque, and, therefore wound rotor motors are used. For driving centrifugal pump, which requires only about 40% of rated torque squirrel cage induction motors are used.

Working surroundings is the chief factor which distinguishes electrical motors used in mining from those used in general industry. The motors employed in mining work must be flame proof, and all commutators and slip-rings must be in flame proof chambers with an inch metal to metal flange.

27. *Textile Mills.* In early years the textile mills were usually equipped with diesel generators, steam turbines. The aspect of boiler, engine etc. made it essential to consider the electric drive. Then the ac power supply was preferred. This was on account of power cost to drive the mill. Thereafter, the group drive of machines were designed, but with the recent developments, the need for individual drive was recognised.

The standard motors in open type or in totally enclosed type cannot be used in a textile mills but specially designed motors are required on account of environmental (presence of fluff and dust fly, high humidity, restricted ventilation etc.), operating (wide voltage fluctuations, frequent starting and stopping, intermittent operation resulting in rapid change in torque and power requirement) and drive (restricted space, special mountings and special shaft extension requirement) conditions.

The special features of textile motors are :

Electrical : Extra high or moderate torque with smooth acceleration characteristics.

Thermal : Normally high thermal withstand capacity for frequent starting, prolonged starting time or restricted ventilation or inching.

Environmental : Ability to withstand high humidity, temperature, dust and fluff.

Mounting : Foot, Cradle and Flat Base.

Construction : Totally enclosed with axial or circumferential ribs with steam-lined shapes for smooth floor for fluff.

Cooling : Surface cooled or fan cooled.

Miscellaneous : Terminal box or loose leads and special shapes for restricted overall dimensions, special shaft extensions.

Loom Motors. In order to achieve a short pick-up time, the starting torque of the loom motor should be high (2–2.5 times of the rated torque). The loom machine operates on a reciprocating mechanism, which causes both torque and current pulsations. Also the loom motor is subjected to frequent starts and stops. This results in a high temperature rise and is taken care by having a liberal thermal dissipation capacity of the motor. These motors are totally enclosed surface cooled. A high degree of fluff in the atmosphere calls for a smooth surface finish of the housing and shields so that the fluff do not get accumulated on the motor surface. The insulation of the motor must be able to withstand high moisture content. These motors have normally a higher slip or lower speed than that of a standard motor in order to give flywheel effect to smoothen the current pulsations. The ratings of the motors employed for driving looms for light fabrics such as silk, rayon, cotton, nylon etc. are 0.37, 0.55, 0.75, 1.1 and 1.5 kW; while those of the motors used for making heavy fabrics (wool and canvas) are 2.2 and 3.7 kW. The motors are usually of 6 or 8 poles.

Carding Motors. The inertia of the carding drum is high, as such the starting of such motors have to be designed to withstand prolonged starting time. Once the drum is started, the operation is continuous and uninterrupting. The commonly used motors are totally enclosed and totally enclosed fan cooled, 3-ϕ, high torque squirrel cage induction motors of rating 1.1 and 1.5 kW for light fabrics and 2.2, 3, 3.7 and 5.5 kW for heavy fabrics. Here again, the motors are usually of 6 or 8 poles.

Spinning Motors. In spinning, the moderate starting torque and the smooth acceleration is essential. This calls for a low dip in the speed-torque curve of the motor. Totally enclosed fan cooled motors of ratings 5–30 kW are used for spinning frame operation. In general, three types of drives are employed for spinning frame operation namely single-speed motor, two-speed motor and two-motor drives. Usually a 4 pole or 6 pole squirrel cage induction motor is employed for single-speed drive. Two-speed motors (4/6 or 6/8 poles) are used in order to give maximum production with minimum breakage. The increased production may compensate for the additional outlay owing to larger size and higher cost. In case of two-motor drive, two separate motors are employed to drive the common pulley of the ring frame. Although this drive is costlier and requires more space, it has several advantages such as (*i*) availability of any desired speed by adjustment of speed ratios, (*iii*) facility of adjustment of yarn tension independently, (*ii*) continued production even in case of failure of one of the motors.

The most common electrical method of controlled-torque starting involve the use of standard squirrel cage induction motors and different methods of applying reduced voltage to the motors during a selected starting period. One of the most effective methods of controlling the stator voltage of motors employed in textile mills is that of using series reactors. This method provides stepless, closed transition increase to approximately full speed. Variable iron-core inductors offer almost infinite choices of starting torques. Reactors with fixed tappings may be used to provide variety of starting torques. With the ever increasing use of solid state devices, nowadays ac regulators are being used to provide control of starting torque.

28. *Woollen Mills.* Loom motors are totally enclosed type, capable of developing 3 times full-load torque at start if direct driven or alternatively started light and controlled through clutch. Motors including switchgear in dyeing section should be of totally enclosed type, and should be treated with acid proof paint as atmosphere is laden with fumes.

29. *Paper Mills.* There are two main processes, that take place in a paper mill namely pulp making and paper making. The drives used are quite different for each one of these processes.

There are two methods of pulp making–purely by mechanical means and by both mechanical and chemical processing. The former method involves grinding logs of wood of about a metre length on large grind-stones. Grinders operate at almost constant speeds of 200–300 rpm and can be started under light load conditions. Hence constant speed synchronous motors and geared drives are most suitable. Usually, pulp making by purely mechanical means consumes more than half of the total power requirement of a paper mill, therefore, large size grinders driven by 3,000–4,000 kW motors are, usually, considered as economical. Pulp can also be made by cutting wood logs into chips of several cms length and treating them with alkalies alongwith other raw materials such as grass, rags etc. During the chemical treatment the material is continuously beaten. Wood choppers have random load characteristics and their inertia is large, depending upon the size of the disc, on which the knives of the chopper are mounted. Beaters, usually are required to start with large load. Depending upon the size of the mill, the ratings of motor used for chipping, beating, refining and storing range from several hundreds to thousand killowatts. Except for beaters, synchronous motors are employed in these drives. Since beaters, very often, require speeds less than 200 rpm and large starting torque, wound rotor induction motors are, therefore, most suitable for such drives.

The paper making machine has to perform the job of forming sheets, removing water from sheets, drying of sheets, pressing of sheets and reeling up of sheets.

The basic requirements of paper making machine drive are :

(*i*) For the actual formation and production of sheets and from the point of view of economy, constancy of the speed of the drive is necessary.

(*ii*) For the paper machine to be multipurpose, its speed should be adjustable over a range as large as 10 : 1.

(*iii*) In the calendar section and reel section, variations in tension in sheet can occur even if the correct relative speeds are maintained due to uneven drying and other factors. It is, therefore, imperative to augment the speed control circuit by an overriding tension control.

(*iv*) While cleaning the wire, it has to be moved forward a few cms at a time for its proper cleaning and inspection. Hence the motor employed should be capable of running at inching speeds of 10-25 m/minute as long as a particular button in pressed.

(*v*) Each section should be able to run at the crawling speed of 10–25 m/minute for running in felts, wire and heating up of dryer cylinders.

(*vi*) Smooth and quick starting of the sections, without causing excessive starting current, are required.

(*vii*) Control system employed should be flexible in nature.

There are two types of drives used for making paper from pulp, namely *line shaft drive* and *sectional drive.* In line shaft drive, the various sections of the paper machine are driven from a line shaft running the full length of paper machine. Cone pulleys and belt-combinations drive the various sections of the machine from line shaft through right angled gear reductions. Electric motors are usually employed for driving the transmission shaft. Both ac and dc drives can be employed for practically loss-less speed control. In ac drive, only the ac commutator motor with shunt characteristic is suitable to provide an economic speed control system. However, the speed of an ac commutator motor depends on load and, therefore, its use as a paper making machine drive with stringent requirements of constant speed is not advisable.

Also, its speed range (usually of the order of 1 : 3) as well as the power required greatly affects the size of the motor. The open-loop speed control of the ac commutator motor is sluggish in comparison with a dc drive, as speed is varied by adjusting an induction regulator and shifting the brush rocker. In dc drives, the speed of the paper making machine is controlled by varying the armature voltage of a separately excited dc motor. The variable dc voltage is obtained from ac by means of either rotary converters or static converters.

In *sectional drive* each section of the paper making machine has its own electrical motor. The speed of the paper making machine can be controlled by varying the supply voltage. By adjusting the field excitation of any motor, it is possible to vary the speed of that particular motor with respect to other motors.

The requirements of a line shaft drive are not as stringent as those of the sectional drive. It has a number of major limitations as compared to the sectional drive system. The lower cost of the electrical equipment of a line shaft drive system is offset significantly by the additional cost of the mechanical equipment required and its maintenance.

30. *Ship-Propulsion.* Except for small vessels, electric drive for ship-propulsion is universal. In electric ship-propulsion the prime mover (steam turbine or diesel engine) drives an ac or dc generator, which feeds power to ac or dc motor mounted on the propeller shafts. For large ships requiring power above 2,500 kW steam turbines with ac equipment and for smaller vessels diesel engines with dc equipment are generally employed.

In case of ac system 3-phase induction or synchronous motors are used. Speed is controlled partly by variation of applied voltage and partly by frequency variation. Speed of induction motors can also be changed by pole-changing method which gives two economical speeds. Reversal is achieved by changing the phase sequence of the supply. Power on the ship is generated by turbo-alternators at 2,200 or 6,600 volts.

In case of dc system, the speed control and reversing are carried out by varying field excitation of generator field in Ward-Leonard system. Power on the ship is generated by diesel-electric sets comparatively at low voltage of 650 volts.

Instead of constant voltage system, constant current system can be used to feed propeller motor armatures while their fields are supplied from constant voltage supply. Such a supply system is inherently proof against overloads, provides maximum safe torque without any risk of interruption of current. Motor is controlled through field control. This is quite suitable for small vessels requiring maximum manoeuvrability in restricted waters, such as tugs and ferries.

In case of very small vessels, propeller shafts are directly coupled to diesel engines through slip couplings. Speed of propeller shaft can be controlled by varying dc excitation, and hence the slip, of the coupling.

31. *Rolling Mills*. Rolling of steel, the process during which the cross section of the metal gets reduced, while increasing its length proportionately, is the major function of steel mills. Rolling mills are usually classified according to the material produced by them. The principal types of mills, classified according to their products rolled are :

Blooming mills (the mills used for producing blooms–any length of rolled metal having almost a square cross section), *slabbing mills* (the mills used to produce slab—any length of metal having a rectangular cross section), *billet mills* (the mills used for producing only a limited range of sizes which are further reduced in finishing mills), *plate mills* (the mills used to produce plates from slabs previously rolled by a blooming or slabbing mill), *structural* mills (the mills used for rolling beams, heavy angles, channels etc. from blooms or billets), *rail mills* (the mills used for rolling rails only from blooms or billets), *merchant mills* (the mills used for rolling small angles, channels, rounds, squares, etc. from billets), *rod mills* (the mills used for rolling small rounds, which are later drawn into wire), *hot-strip mills* (the mills used for rolling strips from slabs), *cold-strip mills* (the mills used for cold-rolling strips previously rolled on hot-strip mills to obtain sheets of high quality or to obtain gauges thinner than those produced on a hot mill), *sheet mills* (the mills used for rolling sheet bars or packs prepared from strips rolled on hot strip mills), *skinpass or temper mills* (the mills used to produce strips, previously rolled on a cold mill, of the desired temper, flatness, surface, and luster by using rolling pressure or a combination of pressure and tension and *tube mills* (the mills used for producing tubes, *e.g.*, pipe and conduit, by either the seamless or the butt or lap-welded processes).

Direct drive is preferred whenever the speed of the mill permits. In general, it is not desirable to use direct drive (direct-connected motor) for speed less than 150 rpm and outputs less than 750 kW. In cases, the speed of the mills is not suitable for direct drive, gear drive is employed. It is common practice to use machine-cut, double helical gears for such work, although sometimes single helical gears are used, and provision is made to carry the thrust that is produced by such gears. Spur gearing is used only for the lower speed drives where the pitch line speeds are 450 metres per minute or less. The use of gears introduces losses and offsets the increased efficiency of high-speed motors, but in many cases the total cost of installation is less, and there are other advantages, *e.g.*, higher power factor, with induction motors and sometimes a reduction of space required.

The flywheel where necessary, should be mounted between the gear and the mill, if the speed of the mill permits, so that gear does not have to transmit the peak loads which are carried by the flywheel. Where the mill shaft runs at a speed too low to permit economic flywheel design, the flywheel should be mounted on the pinion shaft of the gear unit in which case the peripheral speed of the wheel can be relatively high, and the weight correspondingly less, but the gears must be large enough to transmit the peak loads to the mill.

The dc motors, because of their inherent characteristics, are best suited for the rolling mills. Motors for reversing mills must have high starting torque, wide speed range, precise speed control, be capable to withstand overload and pull-out torque which may be as three times the rated value and have good commutation throughout. Acceleration from zero to base speed and then to top speed and subsequent reversal from top speed forward to top speed backward must be achieved in a couple of seconds. The moment of inertia of the armature must be as small as possible. Requirement of higher constant torque at low speed needed for affecting heavy reduction in first few passes can be achieved by varying the applied voltage to the armature, keeping field constant. On the other hand low torque and higher speed required for rolling operation in final passes can be achieved by weakening the field. The motors employed for steel mill duty are, therefore, of totally enclosed, forced ventilated, higher class of insulation, lesser diameter and larger length. In general, the use of dc motors increases the initial cost and introduces losses owing to the necessity of providing converting equipment, but the advantages obtainable by use of such motors greatly outweigh these additional losses and costs. The supply to dc motors for rolling mill drive can be made by means of grid controlled mercury-arc rectifier, silicon controlled rectifier or Ward-Leonard system. The main advantages of grid-controlled mercury-arc rectifier are higher efficiency, higher overload capacity, stepless voltage control from 0–100% and quick response to very rapid changes in operating variables but suffers from the drawback of causing higher current peaks in the power system accompanied by wide voltage fluctuations. The silicon controlled rectifier has the advantage of small size but suffers from the drawback that it does not have overload capacity. Hence grid controlled mercury-arc rectifiers or silicon controlled rectifiers are used for supplying the rolling mills where ac supply network is sufficiently strong. The main advantages of employing Ward-Leonard set are possibility of having speed variation within ± 100% of rated speed, possibility of electric braking by working the mill motor as a generator and of reducing the fluctuations in the power demand from the supply system. Ward-Leonard system, however, suffers from having low efficiency, longer response time and requires more maintenance.

Synchronous motors are used for driving constant speed mills and are advantageous where power factor correction is desirable or where slow-speed motors are necessary. These are particularly adaptable to continuous mills where the load is maintained for relatively long periods or for drives where the length of the material is such that the load exists for such long periods that a flywheel cannot be economically used and consequently the motor must carry the entire load. For slow speeds, synchronous motors can be built more economically than induction motors.

Slip-ring type induction motors are suitable for roughing and re-rolling mills where very precise speed control is not required. Their efficiency is low because of wastage of energy taking place in the rotor resistance. There is also abrupt rise in the motor speed when the material leaves the rolling stands. The disadvantage of the slip-ring induction motor with variable rotor resistance can be overcome by employing cascaded induction motor. The low frequency rotor output is rectified by silicon rectifier and the rectified power is made to operate against an adjustable back emf provided by a separately excited dc motor mounted on the same shaft, which converts the rectified power into mechanical power. The slip-ring induction motor now has a shunt speed-torque characteristic and speed control is affected without power loss in the secondary resistances. The starting torque is about 4 times of the rated torque when the motor is started with full dc excitation on. The cascading arrangement becomes cheaper as compared to dc motors where speed variation desired is not more than 25%.

Exercises

1. Give comparative costs and applications of various types of electric motors.
2. State with reasons, the type of ac motor suitable for following loads:
   (*i*) Table fan (*ii*) Domestic mixie.
   [A.M.I.E. Sec. B. Electric Drives & Their Control Summer 1993]
3. Explain the selection of electric motors for machine tool application regarding (*i*) electrical features (*ii*) mechanical features. (*iii*) How are frequent starting and reversing requirements are handled ? (*iv*) Methods of braking used for machine tool motors.
   [A.M.I.E. Sec B. Electric Drives and Their Control Winter 1993]
4. What are the desirable characteristics of the drives for use in cranes and hoists ? Make a comparative study of dc and ac motors for crane applications in the light of the above points.
   [A.M.I.E. Sec B. Electrical Drives and Their Control Winter 1999]
5. Explain the factors that must be taken into account while selecting a motor for a particular work. On the basis of these factors suggest and justify motors for the following :
   (*a*) Paper mill drive (*b*) Cranes
   (*c*) Marine drive (*d*) Textile mill drive.
   [Pb. Univ. Elec. Drives & Utilisation of Elec. Energy Nov. 1983]
6. State, giving reasons, what type of motor you would suggest for the following drives :
   (*i*) Rolling mills (*ii*) Centrifugal pumps
   (*iii*) Textile machinery (*iv*) Passenger lifts.
   [Gorakhpur Univ. Utilisation of Elec. Power & Traction Sept. 1984]
7. State, giving reasons, what type of motor you would consider suitable for the following drives :
   (*a*) Air compressor (*b*) Reciprocating pump
   (*c*) Paper mill drive and (*d*) Elec. traction.
   [Gorakhpur Univ. Utilisation of Elec. Power & Traction Sept. 1983]
8. Give an account of the special conditions imposed upon the electrical equipment of rolling mills in a steel works. Describe fully how these conditions are fulfilled in the case selected. Assume an ac supply, the driving motors being either direct or alternating.
   [Pb. Univ. Elec. Drives & Utilisation of Elec. Energy August 1984; April 1983]
9. What type of drives do you suggest for paper mill and steel rolling mill ? Explain in brief.
   [A.M.I.E. Sec. B. Electric Drives & Their Control Summer 2000]
10. Suggest suitable type of motors with reasons for spinning mills and cement mills.
    [Pb. Technical Univ. Electric Drives & Utilization Dec. 2008]
11. Suggest suitable type of motors with reasons for textile mill and paper mill drives.
    [Pb. Technical Univ. Electric Drives & Utilization May 2009]
12. Write typical motors and controllers for
    (*i*) Pump drives, (*ii*) Celing fan, (*iii*) Kiln drive of cement mills, (*iv*) Rolling mill drives and (*v*) Electric traction.

❑❑❑

SECTION - B

ELECTRIC POWER UTILIZATION

CHAPTER 5

Electric Heating

INSIDE THIS CHAPTER

5.1. Introduction **5.2.** Modes of Transfer of Heat **5.3.** Classification of Electric Heating Methods **5.4.** Resistance Heating **5.5.** Arc Furnaces **5.6.** Induction Heating **5.7.** High Frequency Eddy Current Heating **5.8.** High Frequency Power Supply Sources **5.9.** Dielectric Heating **5.10.** Choice of Frequency.

5.1. INTRODUCTION

Heating is required for domestic purposes such as cooking and heating of buildings, as well as for industrial purposes such as melting of metals, hardening and tempering, case-hardening, drying and welding. Practically all the heating requirements can be met by some form of electric heating equipment. The main advantages of electric heating over other systems of heating (coal, oil or gas heating) are given below :

1. *Economical.* Electric heating is economical as electric furnaces are cheaper in initial cost as well as maintenance cost. It does not require any attention so their is a considerable saving in labour cost over other systems of heating. Electrical energy is also very cheap as it is being produced on large scale.

2. *Cleanliness.* Since dust and ash are completely eliminated in electric heating system, it is a clean system and cleaning costs are rendered to a minimum.

3. *Absence of Flue Gases.* Since no flue gas is produced in this system, there is no risk of atmosphere or objects being heated and operation is, therefore, hygienic.

4. *Ease of Control.* Simple, accurate and reliable temperature control can be had either by hand operated or by fully automatic switches. Desired temperature or temperature cycle can be had accurately in electric heating system, which is not convenient in other heating systems.

5. *Automatic Protection.* Automatic protection against overcurrents or overheating can be provided through suitable switchgears in the electric heating system.

6. *Upper Limit of Temperature.* There is no upper limit to the temperature obtainable except the ability of the material to withstand heat.

7. *Special Heating Requirements.* Certain requirements of heating such as uniform heating of material or heating of one particular portion of the job without effecting others, heating of non-conducting materials, heating with no oxidation, can be met only in electric heating system.

8. *High Efficiency of Utilization.* The overall efficiency of electric heating is comparatively higher, since in this system of heating the source can be brought directly to the point where heat is required, thereby reducing the losses. Further there is no product of combustion in which heat losses are involved.

It has been practically ascertained that 75 to 100% of heat produced by electric heating can be successfully utilised whereas in cases of gas, solid fuel and oil heating the efficiencies are 60%, 30% and 60% respectively.

9. *Better Working Conditions.* Electric heating system produces no irritating noise and also the radiating losses are low. Thus working with electric furnaces is convenient and cool.

10. *Safety.* Electric heating is quite safe and responds quickly.

5.2. MODES OF TRANSFER OF HEAT

A heated substance will give off heat to another substance at a lower temperature. The different modes by which heat is transferred are conduction, convection and radiation.

1. *Conduction.* In this mode of transfer of heat, one molecule of the substance gets heated and transfer the heat to the adjacent one and so on. Thus heat is transferred through a substance from one part to another, or between two substances in contact.

The rate of conduction of heat along a substance depends upon the temperature gradient. It may be expressed in MJ per hour per square metre per metre or in watts per square centimetre per centimetre while dealing electric heating.

In a plate of thickness t metres, having x-sectional area of its two parallel faces A square metres and temperature of its two faces T_1 and T_2 °C absolute the quantity of heat passed through it during T hours is given by

$$Q = \frac{kA}{t}(T_1 - T_2)\,T \qquad ...(5.1)$$

where k is the coefficient of thermal conductivity for the material in MJ/m²/m/°C/hour.

2. *Convection.* Heat is transferred by convection in case of immersion type water heater or in case of low temperature heating equipment for buildings. The air in contact with a heated radiator element in a room receives heat from contact with the element. The heated air expands and rises, cold air flowing into take its place. Thus there is a constant flow of air upwards across the heating element. This process is called *convection.* These convection currents give up some of their heat to the colder parts of the room. The room and its contents are thus gradually heated by this means. A similar action takes place in an electric water heater, a continuous flow of water passing upwards across the immersed heater element, with the result that the whole of the water in the tank becomes hot.

The quantity of heat absorbed from the heater by convection depends mainly upon the temperature of the heating element above the surroundings and upon the size of the surface of the heater. It also depends partly on the position of the heater. Heat dissipation is given by the following expression.

$$\text{Heat dissipation, } H = a(T_1 - T_2)^b \text{ W/m}^2 \quad \text{...(5.2)}$$

where a and b are constants, whose value depend upon the heating surface facilities for heating etc. T_1 and T_2 are the temperatures of the heating surface and the fluid in °C absolute respectively.

For vertical surfaces in air, the above expression becomes

$$\text{Heat dissipation, } H = 3.875\,(T_1 - T_2)^{1.25} \text{ watts/m}^2 \quad \text{...(5.3)}$$

In furnaces heat transfer by convection is not to any great extent.

3. *Radiation.* In this mode of heat transfer the heat reaches the substance to be heated from the source of heat without heating the medium in between. Rate of heat radiation is given by Stefan's law, according to which

Heat dissipation,

$$H = 5.72 \times 10^4\, k\, e \left[\left(\frac{T_1}{1,000}\right)^4 - \left(\frac{T_2}{1,000}\right)^4\right] \text{W/m}^2 \quad \text{...(5.4)}$$

where T_1 is temperature of source of heat in °C absolute, T_2 is temperature of substance to be heated in °C absolute, k is a constant known as *radiant efficiency,* k is unity for a single element and between 0.5 and 0.8 for several elements placed side by side and e is emissivity, which is unity for black body and is equal to 0.9 for resistance heating elements.

Since radiation is proportional to the difference of fourth powers of temperatures, we can have very efficient heating at high temperatures.

5.3. CLASSIFICATION OF ELECTRIC HEATING METHODS

Electric heating can be broadly classified as (*i*) power frequency heating and (*ii*) high frequency heating.

Power frequency heating can be further classified as (*i*) resistance heating and (*ii*) arc heating.

Resistance heating can be further classified as (*i*) direct resistance heating, (*ii*) indirect resistance heating and (*iii*) infrared or radiant heating.

Similarly arc heating can be further classified as (*i*) direct arc heating and (*ii*) indirect arc heating.

High frequency heating can be classified into (*i*) induction heating and (*ii*) dielectric heating. Induction heating can further be classified as (*i*) direct induction heating and (*ii*) indirect induction heating.

1. *Direct Resistance Heating.* Electric current is made to pass through the body to be heated. This principle of heating is employed in resistance welding and electrode boiler for heating water.

2. *Indirect Resistance Heating.* Electric current is made to pass through a wire or other high resistance material forming a *heating element;* heat so developed is transferred from the heating element to the body by the agency of radiation or convection. Normally this method is used in immersion heaters, resistance ovens, domestic and commercial cooking and heat treatment of metals.

3. *Infrared or Radiant Heating.* Heat energy from an incandescent lamp is focused upon the body to be heated up in the form of electromagnetic radiations. This is employed to dry the wet paints on an object.

4. *Arc Heating.* The arc drawn between two electrodes develops high temperature (about 3,000–3,500°C) depending upon the electrode material. The electric arc may be used in the following different ways:

(*i*) By striking the arc between the charge and the electrode or electrodes. In this method the heat is directly conducted and taken by the charge. The furnaces operating on this principle are known as *direct arc furnaces.*

(*ii*) By striking the arc between the two electrodes. In this method the heat is transferred to the charge by radiation. The furnaces operating on this principle are known as *indirect arc furnaces.*

(*iii*) By striking an arc between an electrode and the two metallic pieces to be joined, as in arc welding.

5. *Direct Induction Heating.* In this method of heating the currents are induced by electromagnetic action in the body to be heated. The induced currents when flowing through the resistance of the body to be heated develop the heat and thus raise the temperature. In induction furnace heat is used to melt the charge and eddy current heaters used for heat treatment of metals are other forms of direct induction heating.

6. *Indirect Induction Heating.* In this method of electric heating the eddy currents are induced in the heating element by electromagnetic action. Eddy currents set up in the heating element produce the heat which is transferred to the body to be heated up, by radiation and convection. Certain types of induction ovens used for heat treatment of metals operate on this principle.

7. *Dielectric Heating.* In this method of electric heating use of dielectric losses is made to heat the non-metallic materials. Non-metallic material to be heated is placed between two metal electrodes across which a high voltage having high frequency is applied, the heat is developed owing to the dielectric losses taking place.

5.4. RESISTANCE HEATING

Resistance heating is based upon the I^2R effect. This method of heating has wide applications such as heat treatment of metals (annealing, hardening etc.), drying and baking of potteries, stoving of enamelled ware and commercial and domestic cooking. Temperature up to about 1,000°C can be obtained in ovens employing wire resistances for heating elements. There are two methods of resistance heating.

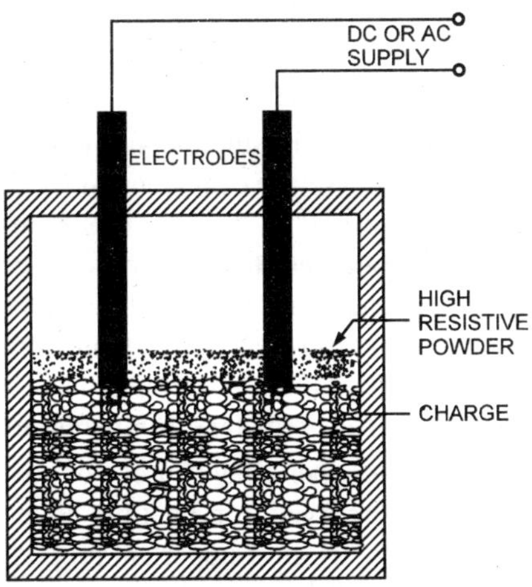

Fig. 5.1. *Direct Resistance Heating*

1. *Direct Resistance Heating.* In this method of heating, the material or charge to be heated is taken as resistance and current is passed through it. The charge may be in the form of powder, pieces or a liquid. Two electrodes are immersed in the charge and connected to the supply in case of availability of direct current or single phase ac supply and three electrodes are immersed in the charge and connected to supply in case of availability of 3-ϕ ac supply. When some pieces of metals are to be heated some highly resistive powder is sprinkled over the surface of pieces to avoid direct short circuit. The current flows through the charge and heat is produced. This method has high efficiency since heat is produced in the charge itself. As the current in this case is not easily variable, therefore, automatic temperature control is not possible. However, uniform and high temperature can be obtained. This method of heating is used in salt bath furnaces and in the electrode boiler for heating water.

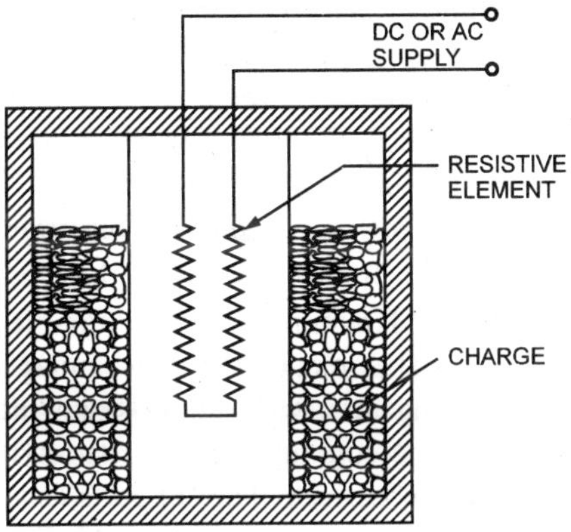

Fig. 5.2. *Indirect Resistance Heating*

2. *Indirect Resistance Heating.* In this method of heating, the current is passed through a wire or other high resistance material forming a heating element. The heat proportional to I^2R loss produced in the heating element is delivered to the charge by one or more of the modes of transfer of heat viz. conduction, convection and radiation. If the heat transfer is by conduction the resistor must be in contact with the charge. An enclosure, known as *heating chamber,* is required for heat transfer by radiation and convection for the charge. For industrial purposes, where a large amount of charge is to be heated, the heating element is kept in a cylinder surrounded by jacket containing the charge.

This arrangement provides a uniform temperature. Automatic temperature control can be provided in this case. This method of heating is used in room heaters, immersion water heaters and in various types of resistance ovens employed in domestic and commercial cooking, and salt bath furnaces.

Resistance Ovens and Furnaces. Resistance furnaces may be classified according to their operating temperature. Low temperature heating chamber with provision for ventilation is termed as *oven.* Resistance ovens are used for drying and baking potteries, drying varnish coatings, vulcanising and hardening of synthetic materials and commercial and domestic heating. They are also employed for tempering hardened steel pieces. *Medium temperature furnaces* having operating temperatures between 300°C and 1,050°C are employed for annealing, normalising of steel and non-ferrous metals, melting of non-ferrous metals and stove enamelling. *High temperature furnaces having* operating temperatures between 1,050°C and 1,350°C are employed for hardening applications. Typical furnace using heating elements and having provision of carrying out heating in particular atmosphere is depicted in Fig. 5.3.

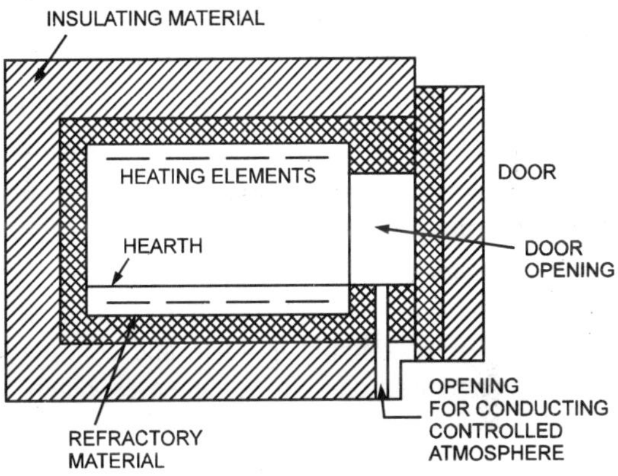

Fig. 5.3. *Resistance Furnace*

The resistance oven is constructed of firebricks or other heat insulating material supported on a metal framework. The heating elements are mounted on the top, sides or bottom of the oven, according to circumstances.

The resistance furnace is an enclosure with a refractory lining : a surrounding layer of heat insulation and outer casing of steel plate, bricks or tiles. The inside proportions of a heating chamber are made to suit the character of the charge and type of furnace or oven. The nature of material required for the insulation is determined by the maximum temperature of the inner face of the layer of insulation of a heating chamber. The heating elements are mounted on top, sides or bottom of the oven as the circumstances permit.

Heating Chambers. An enclosure of a charge to be heated by radiation or convection or by the joint effect of these modes of heat transfer is known as *heating chamber.* Apart from mechanical considerations the design of the chamber is related primarily to temperature and the major mode of heat transfer to be used. The essential features are : the enclosing chambers, the resistors, and the atmosphere within the chamber. The functions of the heating chambers are :

(*a*) to control the distribution of heat within the chamber;

(*b*) to control the cooling rate of charge, if required;

(*c*) to confine the atmosphere around the charge; and

(*d*) to store as much of the heat supplied as may be practicable and economical.

A cylinder with an inside diameter equal to its height represents an ideal shape and proportions of a heating chamber, keeping only the utilisation of material and rate of heat loss from the exterior surfaces in view. This ideal is seldom attained because of many practical considerations in heating service. Electric heating permits wide latitude in the proportions and shape of a heating chamber, and factory planning takes this into account in developing the sequence of movements of material through manufacturing operations.

Heating chambers may be of 'batch' or 'continuous' type. In the former type, the charge remains stationary during the heat application. The cycle may include cooling the charge in the chamber. In case of continuous type chambers, the charge is heated as it moves through the chamber. In some cases the chamber is extended for more or less cooling of the charge before it leaves the chamber.

The application of the batch type heating chamber is defined by the following conditions.

(*i*) Intermittent production.

(*ii*) Long period of heating and in some cases slow cooling.

(*iii*) For material that cannot be handled on conveyors.

(*iv*) Heating service beyond the range of the capacity of conveying mechanisms.

(*v*) Supplementary heating service.

A continuous type heating chamber is used where the flow of material is reasonably uniform and continuous, *i.e.*, mass production conditions. In some cases batch heating chambers with automatic charging and discharging are essentially of the continuous type.

5.4.1. Heating Element Materials

A material which is used as a heating element must possess the following properties :

(*i*) ***High resistivity.*** The material to be used for heating element should be of high specific resistance so that a small length of wire may be sufficient to produce the required amount of heat.

(*ii*) ***High melting point.*** Melting point of material to be used for heating element should be high so that charge can be heated to a high temperature.

(*iii*) ***Low temperature coefficient.*** The material for heating element should have low temperature coefficient so that resistance may not vary with the change in temperature otherwise starting current would be high.

(*iv*) ***Free from oxidation.*** The material for heating element should be such that it may withstand the required temperature without getting oxidised, otherwise it would have to be replaced frequently.

The materials commonly used for heating elements for low and medium temperature services are either alloy of nickel and chromium (Nickel 80% and chromium 20%) or alloy of nickel, chromium and iron (Ni 65%, Cr 15% and Fe 20%). The addition of iron of the alloy reduces the temperature at which oxidation takes place but the cost of the product is also reduced.

Ni-Cr alloy is suitable for temperatures up to 1,150°C and for work in severe conditions Ni-Cr-Fe alloy is recommended for use up to 950°C. Nickel-chromium (Nichrome) alloy has not only good resistance to oxidation up to 1,150°C but at the same time it has sufficient strength. Ni-Cr-Fe alloy is the cheapest, and most economical and stronger for temperatures up to 950°C. Alloys containing iron, chromium, cobalt and aluminium can withstand temperatures as high as 1,350°C.

Nickel-copper alloy (Eureka or constantan by trade name) is often employed for heating elements operating at low temperatures. Its most important property is that it has virtually zero resistance temperature coefficient. Iron-chromium-aluminium alloy (Kanthal by trade name) has very good resistance to oxidation at high temperatures but are weaker in mechanical strength and, therefore, such elements need more support.

Resistors for operating temperatures above 1,150°C are made of silicon carbide, molybdenum, tungsten and graphite.

Silicon carbide is the basis of a resistor material for operating in air for temperatures up to about 1,500°C. The material is formed into rods of diameters and lengths for combination into circuits of the required electrical rating. The range between the cold and hot resistances is small enough to permit single voltage operation. The resistance gradually increases with use, and an auto-transformer with taps to compensate for this increase of resistance is desirable.

Molybdenum resistors are suitable for temperatures up to 1,650°C. This metal is ductile enough at room temperature for drawing into wire for resistor windings. The wide range between the cold and hot resistances makes multiple voltage operation necessary. The supports (insulators) of the winding should be magnesia or zirconia. Molybdenum resistors cannot be operated in air and also must be protected against reactions with silicon and carbon. The metal is immune from reactions with sulphur, nitrogen, hydrogen, and water vapour. A hydrogen atmosphere is usually used for the protection of these resistors. Molybdenum is not suitable for resistors of vacuum furnaces because of its high vapour pressure.

Tungsten resistors can be used for temperatures up to 2,000°C. The maximum temperature is limited by the refractory supports of the resistor. The metal must be heated for drawing into shapes. The low vapour pressure of tungsten makes it useful for resistors of vacuum furnaces.

Graphite resistors are suitable for any temperature that can be used. The resistors require protection against oxidation above about 600°C. Also because of the chemical activity of carbon, special consideration must be given to the surrounding atmosphere. The material is available in a wide range of shapes and dimensions.

5.4.2. Maximum Operating Voltage

Maximum operating voltage is limited, by electrical insulation at high temperatures and from safety consideration, to 600 volts. This value may exceed in special cases. Resistor winding may be in one, two or more circuits and may be connected either to single or to polyphase supply system.

5.4.3. Design of Heating Element

Knowing the electrical input and its voltage the size and length of the wire required as the heating element to produce the given temperature can be calculated. The wire employed may be circular or rectangular like a ribbon. The latter permit the use of higher wattage per unit area and are used in ovens, vulcanizers, toasters etc. The heating element on reaching a steady temperature will dissipate the heat from its surface equivalent to electrical input. Since generally the heat will be dissipated from the heating elements at high temperatures, it is reasonable to assume that the whole of the heat energy is dissipated solely by radiation.

Heat dissipated according to Stefan's law,

$$H = 5.72ke\left[\left(\frac{T_1}{100}\right)^4 - \left(\frac{T_2}{100}\right)^4\right] \text{ watts/m}^2 \qquad ...(5.5)$$

$$\text{Electrical input, } P = \frac{V^2}{R}$$

where V is the supply voltage and R is resistance of heating element and is given by the expression,

$$R = \rho\frac{l}{a} = \frac{\rho l}{\frac{\pi d^2}{4}} = \frac{4\rho l}{\pi d^2}$$

for a circular wire of diameter d, length l and resistivity ρ.

Electrical power input,

$$P = \frac{V^2}{R} = \frac{V^2}{4\rho l/\pi d^2} = \frac{\pi d^2 V^2}{4\rho l}$$

$$\text{or } \frac{l}{d^2} = \frac{\pi V^2}{4\rho P} \qquad ...(5.6)$$

Surface area, $S = \pi dl$

Heat dissipated $= \pi dl\, H$

Since at steady temperature

Power input = Heat dissipated

$$\therefore \quad P = \pi dl\, H$$

$$\text{or } \frac{\pi d^2 V^2}{4\rho l} = \pi dl\, H$$

$$\text{or } \frac{d}{l^2} = \frac{4\rho H}{V^2} \qquad ...(5.7)$$

Solving Eqs. (5.6) and (5.7) length and diameter of wire can be determined.

For ribbon type of conductor let w be the width and t be the thickness.

Electrical power input,

$$P = \frac{V^2}{R} = \frac{V^2}{\rho l/a} = \frac{V^2}{\rho l/wt} = \frac{V^2 wt}{\rho l}$$

$$\text{or } \frac{l}{wt} = \frac{V^2}{\rho P} \qquad ...(5.8)$$

Since at steady temperature

Power input = Heat dissipated

$\therefore$ $P = 2wlH$ neglecting the side area $2tl$ as the thickness is negligible

$$\text{or } \frac{V^2 wt}{\rho l} = 2wlH$$

$$\text{or } \frac{t}{l^2} = \frac{2H\rho}{V^2} \qquad ...(5.9)$$

So by solving the two Eqs. (5.8) and (5.9), length l and width w for a ribbon of thickness t will be evaluated.

5.4.4. Efficiency and Losses

The heat produced in the heating elements is also to overcome the losses occurring due to

(*i*) heat used in raising the temperature of oven or furnace;

(*ii*) heat used in raising the temperature of the containers or carriers;

(*iii*) heat conducted through the walls;

(*iv*) escapement of heat due to opening of door; in addition to heat required to raise the temperature of the charge to the required value.

The efficiency of the oven

$$= \frac{\text{Heat required to raise the temperature of the charge to the required value}}{\text{Heat required to raise the temperature of the charge to the required value + losses}}$$

The heat required to raise the temperature of the charge to the required value, W = Mass of charge in kg × specific heat of charge in J/kg/ °C × temperature rise = *mst* Joules.

The heat used for raising the temperature of the oven or furnace can be calculated in a similar way by knowing the weight of the refractory material and its specific heat. In case the oven is used continuously this loss becomes negligible.

Heat used for raising the temperature of container is calculated in exactly the same way as for oven or furnace. The container usually has to be heated up afresh for each charge.

Since the heat is continuously conducted through the walls, this source of heat loss is most important. It can be determined knowing the mean value of the inside and outside surface areas A in square metres; the thickness of walls t in metre; inside and outside temperatures T_1 and T_2 in °C and thermal conductivity of walls k in MJ per hour per square metre, per metre, from the following expression

Heat loss by conduction through walls

$$= \frac{kA}{t}(T_1 - T_2) \text{ MJ/hour} \qquad ...(5.10)$$

Though there is no specific formulae for determination of loss occurring due to opening of door for inspection of the charge, however, this loss may be taken as 0.575 to 1.15 MJ/m^2 of the door area if the door is opened for a period of 20 to 30 seconds.

The efficiency defined above lies between 60 and 80 per cent. A more convenient figure for expressing the behaviour of an oven, however, is the energy required per tonne of charge treated. Typical values for different processes are given in Table 5.1.

Table 5.1.

Process	*Energy Consumption in kWh per Tonne of Charge*
Baking of bread	50-100
Annealing copper	100-200
Annealing steel	200-250
Carburising	250-500
Vitreous enamelling of sheet steel	400-700

5.4.5. Temperature Control of Resistance Ovens/Furnaces

Temperature control is necessary in resistance ovens/furnaces–temperature may have to be kept constant or varied according to requirements. Control may be manual or automatic.

In resistance ovens/furnaces heat developed depends upon $I^2 R\, t$ or $\frac{V^2}{R}\, t$. So there are three ways in which the temperature can be controlled. Firstly, by varying the applied voltage to the elements or current flowing through the element; secondly, by varying the resistance of elements and thirdly, by varying the ratio of on and off times of supply.

Voltage across the oven can be controlled by changing the transformer tappings. This is economical and most suitable if the transformer is to be used for stepping down the voltage for supply to ovens or furnaces, but such conditions do not arise usually. Auto-transformer or induction regulator can also be used for variable voltage supply. In case of large furnace, there may be an independent generating set for the supply to the furnace and in that case variable voltage supply can be provided.

Alternatively voltage across the oven or furnace can be controlled by varying the impedance connected in series with the circuit. But this method is not economical as power is continuously wasted in the controlling resistance. Its use is, therefore, limited to small furnaces.

Temperature can also be controlled by switching the various combinations of groups of resistances used in the ovens or furnaces in the following ways.

1. *Use of Variable Number of Elements.* In this method, the number of heating elements in working are changed; so total power input or heat developed is changed. This method does not provide uniform heating unless the number of heating elements in the circuit at any particular instant are distributed over the surface area, which requires complicated wiring.

2. *Change of Connections.* In this method the elements are arranged to be connected either all in series or all in parallel or combination of both or in star or in delta by means of switches at different instants according to the requirements. This is the simplest and most commonly used method of control.

An *on-off switch* can also be employed for temperature control but its use is restricted to small ovens. The time duration for which the oven is connected to the supply and the time duration for which it remains cut off from supply will determine the temperature. Here an oven is supplied through a thermostat switch which makes and breaks the supply connections at particular temperature. The ratio of $\frac{\text{Time duration during which supply remains on}}{\text{Total time duration of an on-off cycle}}$ is an indication of temperature. The higher the ratio, the larger will be the temperature of the oven. Advantage of this method is that it is more efficient than series impedance method.

Normally the temperature is required to be maintained between limits. This necessitates some type of automatic control of the thermostat type. When the temperature of the oven exceeds or falls below certain predetermined values, the thermostat connected in the circuit operates a relay which varies the current in the elements by any one of the methods given above. The relay may even switch the oven out of service if so required. The actual switching is carried out by contactors.

For large capacity ovens, some protection is to be provided against overloading, excessive temperatures, and in some cases against the possibility of workmen loading or unloading the furnace with the power on.

The only possibility of overload is due to a part of a metallic charge falling against and short circuiting the elements. An instantaneous overload relay is provided and set to trip the circuit at 10 or 15 per cent, above normal current. This adequately guards against damage due to such a contingency. In addition to above, fuses are also provided either in the main oven circuit or in the hold-on coil of the energizing contactors to provide protection in case of failure of automatic control system. For the protection of nickel-chrome wire in this way a gold fuse is employed, because it has a sharp melting point of 1,060°C; and does not oxidise. Fuses melting at a lower temperature can be had, but although of some service are not completely reliable.

5.4.6. Causes of Failure of Heating Elements

The chief causes of failure of heating elements are given below.

1. *Formation of Hot Spots.* Hot spots are the points in the heating element which are at a higher temperature than the main body of the element. One of the reasons of formation of hot spot in the heating element is high rate of local oxidation that may reduce the element wire x-section, thereby increasing the resistance at that spot and producing more heat locally and causing breakdown of the element. Another reason may be due to shielding of element by supports etc. that will reduce the local heat loss by radiation and cause an increase of temperature of shielded portion of the element. However, it can be minimized by using minimum number of supports without producing distortion of the element. Other causes are due to too high element temperature, insufficient support for the element or selection of wrong fuse material, may result in sagging and warping of the material. The sagging and warping may cause uneven spacing of sections, thereby developing hot spots where sections are closest together or even there may be actual shorting of adjacent sections of an element.

2. *Oxidation and Intermittency of Operation.* Continuous and tenacious oxide scale is formed on the surface of the heating element when it attains high temperature. This oxide layer is so strong that it prevents further oxidation of the inner metal of the heating element. But in case of frequent use of the element, the scale is subjected to thermal stresses developed owing to frequent heating and cooling. The oxide scale, therefore, cracks and flakes off, exposing further fresh metal to oxidation and thus resulting in increased local oxidation and producing hot spots.

3. *Embrittlement Due to Grain Growth.* All heating alloys containing iron tend to form large brittle grains at high temperatures. When cold, the heating elements are very brittle and liable to rupture easily on slighest handling and jerks.

4. *Contamination and Corrosion.* Gases of the controlled atmosphere prevailing in annealing furnaces or fumes from flux used in brazing furnaces, or oil fumes caused by heat treatment of components contaminated with lubricant contaminate the elements and produce dry corrosion.

Of all the conditions stated above, which determine the element life, most critical are the temperature of the hottest point and ratio of intermittent to continuous working.

5.4.7. Special Types of Resistance Furnaces

The most important applications of resistance ovens are air-circulation oven and bright annealing furnace.

1. *Air-Circulation or Convection Type Furnaces.* The charge may not get uniformly heated in an ordinary furnace in which heat is transferred to the charge almost entirely by radiation. If uniform distribution of heat is necessary and the temperature is not to exceed about 600°C, such furnaces are used. The common applications are drawing and hardening of steel, heat treatment of aluminium and light metals. In such furnaces, air is passed over the heating elements and this heated air is circulated over or through the charge imparting its heat uniformly to all parts of the charge by conduction. The direction of flow in air-circulation furnaces is sometimes reversed periodically in order to make the distribution of heat more uniform. If this were not done, the temperature of that part of the charge which comes across the hot air first will be higher than that which comes across it afterwards. However, this difference can be minimised by circulating large volume of air. Blowers and fans are used for circulating the gas or air. The advantage of the circulating air or gas is to impart all the parts of the charge a uniform application of the atmosphere. Generally the charge is to be cooled in the furnace itself and to achieve this gases are circulated after being cooled. Gases are let through cooling pipes during the cooling process. Nitriding is an important application. In this process the steel after being heated by hot air, is exposed to ammonia. The steel absorbs nitrogen and gets hardened. An important item in the design of such an oven is the characteristic of the fan. A centrifugal fan is used, but it must be of such a design that a change in the resistance of the air path due to charges of different size and shape does not appreciably affect the volume of air delivered or kW required by the motor.

2. *Resistance Furnace For Bright Annealing.* The process of cooling a heated job slowly for removing its brittleness is known as *annealing*. In the usual process of annealing the job is left covered with a scale of oxide which must be removed in order to have bright finish. The formation of oxide scale can be prevented if annealing process is carried out in an atmosphere free from oxygen, water vapour and carbon dioxide. Such a condition can be achieved by heating the job in an air tight furnace to which is connected a non-return air valve. During the period of heating air and any oil absorbed by the job are expelled through the non-return valve due to their expansion. During cooling the non-return valve closes, thereby preventing access of oxygen or other gases to the job. Thus the surface of the job under annealing process is maintained bright.

5.4.8. Salt-bath Heating

Salt-bath furnaces are used for the purposes of tempering, quenching and hardening of steel tools. The characteristics of salt-bath heating are rapidity and uniformity and selective localised heating combined with protection from oxidation.

This type of furnace consists of a bath of some salt such as molten sodium chloride and two electrodes immersed in it. When the current between electrodes immersed in the salt is passed, heat is developed and the temperature of the salt bath may be in the range between 1,000°C and 1,500°C depending upon the type of salt used. In this bath the material to be heated is dipped and the necessary heat treatment, as required for tool steel, is given to it. The heat transfer in this case is by conduction and by better contact of the body with the heated salt, which is in molten state.

As dc would cause electrolysis of the salt, therefore, alternating current is used. The voltages up to 20 volts and currents up to 3,000 amperes depending on the size of the furnace are used. Voltage on the secondary side is changed by tap changing gear included in the primary of the transformer.

The resistance of the salt decreases with the increase of temperature of the salt therefore, in order to maintain constant power input it is necessary that the salt furnaces are started with highest tap and as the temperature rises, this is gradually brought to lower taps. Control of power input is also affected by varying the depth of immersion and distance between electrodes. It must be ensured that the current flows through the salt and not through the job being heated.

5.4.9. Infrared or Radiant Heating

In the ordinary resistance furnaces the heat is transferred from heating elements to the charge partly by radiation and partly by convection, the latter predominating at low and medium temperatures but in this form of heating suitable for low and medium temperature heating purely radiant or infra-red heating is used. In this method of heating, heating elements consist of tungsten filament lamps together with reflectors to direct the whole of the heat emitted on to the charge. The lamps are operated at 2,300°C instead of 3,000°C giving greater proportion of infrared radiation and a longer life. The reflectors are plated with rhodium which, being harder, has a longer life and is easier to maintain. The lamps employed are usually of rating between 250 and 1,000 watts operating at 115 volts. The operation at low voltage results in a robust filament.

Plant sizes range from a single lamp to chambers containing several hundred kW of lamps. With this arrangement charge temperature between 200ºC and 300ºC can be obtained. Heat emission intensities up to 7,000 watts/square metre of the chamber surface can be obtained. These are much higher than those obtained with ordinary resistance ovens (1,500 watts/m^2). In radiant heating heat absorption remains practically constant whatever the charge temperature is where as it falls off rapidly as the temperature of charge rises in the ordinary resistance furnace.

In this process the lamps with their reflectors are grouped close together to form the walls and possibly the top of a heating chamber; the lamps being mounted on a light angle-iron framework, no heat insulation is required as all heat is directed inwards. Individual reflectors for each lamp or trough reflectors may be used, the latter enabling a greater number of lamps to be installed in the given area. The precise shape of the chamber depends on the size and shape of the charge to be heated and the framework can easily the made adjustable. Continuous processes can be had, by carrying the articles to be heated through the chamber on a conveyor belt.

Infrared heating possesses the advantages of (*i*) rapid heating, (*ii*) compactness of heating units, (*iii*) flexibility and (*iv*) safety and finds application in (*i*) paint stoving (*ii*) drying of wood furniture (*iii*) preheating of plastics prior to moulding (*iv*) softening of thermoplastic sheets and (*v*) drying of pottery, paper, textiles etc. where moisture content is not large.

For obtaining best results, the infrared lamps should be located at a distance of 25-30 cm from the object to be heated.

Example 5.1. Six resistances each of 80 Ω are used in resistance heating furnace. How much power is drawn with following connections and supply voltages ?

(*a*) 400 V single phase (*i*) Six groups in parallel, each of one resistance.

(*ii*) Three groups in parallel, each of two resistances in series.

(*iii*) Two groups in parallel, each of three resistances in series.

(*b*) 400 V three-phase star-connections (*i*) Two elements in parallel in each phase.

(*ii*) Two elements in series in each phase.

(*c*) 400 V three-phase delta-connections (*i*) Two elements in parallel in each branch.

(*ii*) Two elements in series in each branch.

(*d*) Same as in case of (*b*) and (*c*) but with 50% tapping on auto-transformer.

[Rajasthan Univ. Utilization of Elec. Power 2000]

Solution : (*a*) (*i*) *Six groups in parallel, each of one resistance*

Equivalent resistance,

$$R_1 = \frac{1}{\frac{1}{80}+\frac{1}{80}+\frac{1}{80}+\frac{1}{80}+\frac{1}{80}+\frac{1}{80}} = \frac{80}{6} \text{ or } \frac{40}{3}\ \Omega$$

$$\text{Power drawn, } P_1 = \frac{V^2}{R_1} = \frac{400^2}{\frac{40}{3}} = 12{,}000 \text{ W or 12 kW } \textbf{Ans.}$$

(*ii*) *Three groups in parallel, each of two resistances in series*

Equivalent resistance,

$$R_2 = \frac{1}{\frac{1}{80+80}+\frac{1}{80+80}+\frac{1}{80+80}} = \frac{160}{3}\ \Omega$$

$$\text{Power drawn, } P_2 = \frac{V^2}{R_2} = \frac{400^2}{\frac{160}{3}} = 3{,}000 \text{ W or 3 kW } \textbf{Ans.}$$

(*iii*) *Two groups in parallel, each of three resistances in series*

Equivalent resistance,

$$R_3 = \frac{1}{\frac{1}{80+80+80}+\frac{1}{80+80+80}} = 120\ \Omega$$

$$\text{Power drawn, } P_3 = \frac{V^2}{R_3} = \frac{400^2}{120} = 1{,}333 \text{ W or 1.333 kW } \textbf{Ans.}$$

(*b*) (*i*) *Two elements in parallel in each phase*

Equivalent resistance,

$$R_4 = \frac{1}{\frac{1}{80}+\frac{1}{80}} = 40\ \Omega$$

$$\text{Phase voltage, } V_P = \frac{400}{\sqrt{3}} \text{ V}$$

Power drawn per phase,

$$P_4 = \frac{V_P^2}{R_4} = \frac{\left(\frac{400}{\sqrt{3}}\right)^2}{40} = \frac{4{,}000}{3} \text{ W or } \frac{4}{3} \text{ kW}$$

$$\text{Total power drawn} = 3P_4 = 3\times\frac{4}{3} = 4 \text{ kW } \textbf{Ans.}$$

(*ii*) Two elements in series in each phase

Equivalent resistance, R_5 = 80 + 80 = 160 Ω

$$\text{Total power drawn, } P_5 = 3\times\frac{\left(\frac{400}{\sqrt{3}}\right)^2}{160} = 1{,}000 \text{ W or 1 kW } \textbf{Ans.}$$

(*c*) (*i*) This is the same case as (*a*) (*i*),

Power drawn = 12 kW **Ans.**

(*ii*) This is the same case as (*a*) (*ii*), Power drawn = 3 kW **Ans.**

or Power drawn will be three times that have been drawn under (*b*) (*i*) and (*b*) (*ii*).

(*d*) Power drawn will be of one-fourth of the values obtained under (*b*) (*i*), (*b*) (*ii*), (*c*) (*i*) and (*c*)(*ii*) when applied voltage is 50 per cent.

Example 5.2. A 2.5 kW, 240 V, single-phase resistance oven is to have nichrome wire heating elements. If the wire temperature is to be 1,500°C and that of the charge 450°C, estimate the diameter and length of wire. The resistivity of nichrome alloy is 42.5 μΩ-cm. Assume the radiating efficiency and the emissivity of the element as 1.0 and 0.9 respectively.

[G.B. Technical Univ. Utilization of Electrical Energy and Electric Traction 2010-11]

Solution : Power supplied, P = 2.5 kW = 2,500 W

Supply voltage, V = 240 V

Radiating efficiency, K = 1.0

Emissivity, *e* = 0.9

Resistivity, ρ = 42.5 × 10^{-6} Ω-cm = 0.425 × 10^{-6} Ω-m

From Eq. (5.6), $\frac{l}{d^2} = \frac{\pi V^2}{4\rho P} = \frac{\pi \times 240^2}{4 \times 0.425 \times 10^{-6} \times 2{,}500}$

$= 42.5778 \times 10^6$ ...(*i*)

From Eq. (5.5),

$$H = 5.72\, ke\left[\left(\frac{T_1}{100}\right)^4 - \left(\frac{T_2}{100}\right)^4\right] \text{ W/m}^2$$

$$= 5.72 \times 1 \times 0.9\left[\left(\frac{1{,}500 + 273}{100}\right)^4 - \left(\frac{450 + 273}{100}\right)^4\right] \text{ W/m}^2$$

$= 5.72 \times 1 \times 0.9\ [(17.73)^4 - (7.23)^4]$ W/m^2 = 494,647

Since heat dissipated = Electrical power input

$\therefore\ 494{,}647 \times \pi \times dl = 2{,}500$

or $d^2 l^2 = \left(\frac{2{,}500}{\pi \times 494{,}647}\right)^2 = 0.000002588$ ...(*ii*)

Multiplying Eqs. (*i*) and (*ii*), we have

$l^3 = 110.2$ m^3

or $l = 4.79$ m **Ans.**

and diameter of wire, $d = \frac{\sqrt{0.000002588}}{4.79}$

$= 0.000336$ m or 0.336 mm **Ans.**

Example 5.3. In a three-phase 440 V, 50 Hz, star-connected 20 kW oven, the temperature of wire is 1,200°C and that of charge is 700°C. If radiating efficiency is 0.6 and emissivity is 0.9, design the heating element. A strip of thickness 0.025 mm having resistivity 1.05 × 10^{-6} Ω-m is used.

[*U.P. Technical Univ. Utilization of Electrical Energy and Electric Traction 2006-07*]

Solution : Power per phase, $P = \frac{20 \times 1{,}000}{3} = 6{,}667$ W

Phase voltage, $V_P = \frac{440}{\sqrt{3}}$ V = 254.034 V

Resistance of strip, $R = \frac{V_P^2}{P} = \frac{254.034^2}{6{,}667} = 9.68\ \Omega$

If l is the length of the strip and w is the width of strip, then

Resistance of the strip, $R = \frac{\rho \times l}{w \times t} = \frac{1.05 \times 10^{-6} \times l}{w \times 0.025 \times 10^{-3}}$

or $\frac{l}{w} = \frac{9.68 \times 0.025 \times 10^{-3}}{1.05 \times 10^{-6}} = 230.476$ ...(*i*)

Heat dissipated from the surface,

$$H = 5.72\, ke\left[\left(\frac{T_1}{100}\right)^4 - \left(\frac{T_2}{100}\right)^4\right] \text{ W/m}^2$$

$$= 5.72 \times 0.6 \times 0.9\left[\left(\frac{1{,}200 + 273}{100}\right)^4 - \left(\frac{700 + 273}{100}\right)^4\right]$$

$= 117{,}727.4$ W/m^2

Surface area available for heat radiation = $2wl$ m^2

Since heat dissipated = Power input

$117{,}727.4 \times 2wl = 6{,}667$

or $wl = \frac{6{,}667}{2 \times 117{,}727.4} = 0.0283$ m^2 ...(*ii*)

Dividing Eq. (*i*) by Eq. (*ii*), we have

$w = \sqrt{\frac{0.0283}{230.476}} = 0.011084$ m = 11.084 mm **Ans.**

Length of strip, $l = 230.476 \times 11.084 \times 10^{-3} = 2.555$ m **Ans.**

Example 5.4. A cubic water tank has surface area of 6.0 m^2 and is filled to 90% capacity six times daily. The water is heated from 20 °C to 65 °C. The losses per square metre of tank surface per 1 °C temperature difference is 6.3 watts. Find the loading in kW and the efficiency of the tank. Assume specific heat of water = 4,200 J/kg/°C and one kWh = 3.6 MJ.

Solution : Let the one side of the tank be l metres

Surface area of the tank, $6\, l^2 = 6.0$ m^2

or $l = 1.0$ m

Volume of the tank = $1.0 \times 1.0 \times 1.0 = 1.0$ m^3

Water to be heated daily = 6×0.9 m^3 = 5,400 kg.

$\because$ 1 cubic metre water weights 1,000 kg

Heat required to raise the temperature of water

$= ms\,(t_2 - t_1) = 5{,}400 \times 4{,}200 \times (65 - 20)$

$= 1{,}020$ MJ $= \frac{1{,}020}{3.6} = 283.5$ kWh

Losses from the surface of the tank

$= \frac{6.3 \times 6.0 \times (65 - 20) \times 24}{1{,}000}$ kWh = 40.8 kWh

Energy supplied = 283.5 + 40.8 = 324.3 kWh

Loading in kW $= \frac{324.3}{24} = 13.5$ kW **Ans.**

Efficiency of the tank $= \frac{\text{Output}}{\text{Input}} \times 100 = \frac{283.5}{324.3} \times 100$

= 87.4% **Ans.**

5.5. ARC FURNACES

When a high voltage is applied across an air gap, the air in the gap gets ionised under the influence of electrostatic forces and becomes conducting medium. Current flows in the form of a continuous spark, called the *arc*. It is to be noted that a very high voltage is required to establish an arc across an air gap but to maintain an arc small voltage may be sufficient. Alternatively an arc can also be produced by short circuiting the two electrodes momentarily and then withdrawing them back. In this method of striking an arc, high voltage is not required. Arc drawn between two electrodes produces heat and has a temperature between 1,000 °C and 3,500 °C depending on the material of the electrodes used. *The use of this principle is made in electric arc furnaces.*

Usually arc furnaces are of cylindrical shape but recently conical shaped shells have been used. Even in conical shapes the horizontal cross section is cylindrical. The conical shape has the advantage of a large surface area per unit bath volume. It consumes less power and the radiation losses are reduced. Melting time is also reduced. Bottom, side walls and roof of the furnace are lined with firebricks, magnesite bricks and silica bricks. The bottom is lined with magnesite mix or ground ganister mix depending upon whether the basic or acidic lining is required. The electrodes project through the top or sides of the chamber and are arranged for easy replacement and adjustment. Each furnace is provided with charging door and tap hole for introducing the charge and taking out the molten metal. These doors and tap holes are also to be suitably lined. To minimize heat losses the size of the doors/tap holes and the opening time has to be reduced. The furnace rests on platform which can be titled for pouring the molten metal through the door opening in the side of the shell. Tilting mechanism consists of a motor which drives a pinion which meshes with a semicircular toothed rack at the bottom of the platform.

The furnaces may be *door-charge type* or the *top-charge type.* Large furnace installations are often equipped with side door charging. Top charging is growing in favour for medium size furnaces. In this method for charging the roof of the furnace is removed, and a complete charge is placed in the chamber by a drop bucket handled by an overhead crane. This saves labour and time.

The electrodes used in arc furnaces are of three types namely carbon electrodes, graphite electrodes and self-baking electrodes. Material of the electrodes namely carbon and graphite has been selected on account of their electrical conductivity, insolubility, infusibility, chemical inertness, mechanical strength and resistance to thermal shock. The size of these electrodes may be 18 cm to 27 cm in diameter. With graphite or carbon electrode, the temperature obtainable from the arc is between 3,000 °C and 3,500 °C. The trend is toward the general use of graphite electrodes. Carbon electrodes are used with small furnaces for manufacture of ferro-alloys, aluminium, calcium carbide, phosphorous etc. Self-baking electrodes are employed in ferro-alloys and electrochemical furnaces and in electrolytic production of aluminium. Owing to lower resistivity of graphite (one fourth that of carbon), graphite is required half in size for the same current resulting in easy replacement. Graphite begins to oxidise at about 600 °C where as carbon at about 400 °C. Under average conditions the consumption of graphite electrodes is about one half that of carbon electrodes. Electrode consumption for steel-melting furnaces varies between about 4.5 and 9 kg of electrodes per tonne of steel with carbon electrodes, and between about 2.27 and 6.8 kg for graphite electrodes. Carbon electrodes are very cheap and cost less than one half as much for same weight as graphite electrodes. The large area of carbon electrodes allow for more uniform heating. However, the arc has to be brought nearer the surface of the refractory lining and, therefore, the life of the refractory is shortened. Carbon electrodes are made of anthracite coal and coke where as graphite electrodes are obtained by heating the carbon electrodes to a very high temperature. Self-baking electrodes are made of special paste, whose composition depends upon the type of process for which it is used, contained in thin steel cylinder. The flow of current produces heat and the paste is baked and formed into an electrode.

5.5.1. Types of Arc Furnaces

Arc furnaces are of two types namely (*i*) direct arc furnaces and (*ii*) indirect arc furnaces. In a direct arc furnace, shown in Fig. 5.4, the arc is formed between the electrodes and the charge, while in the indirect arc furnace the arc is formed between the electrodes, as shown in Fig. 5.5, and heat is transmitted to the charge solely by radiation.

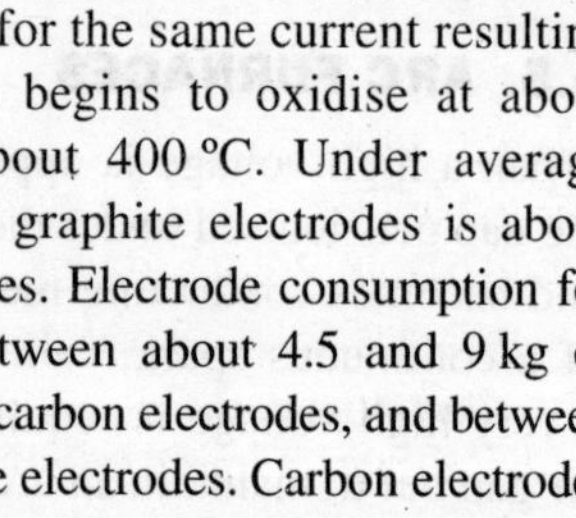

Fig. 5.4. ***Direct Arc Furnace***

1. *Direct Arc Furnace.* In a direct arc furnace charge acts as one of the electrodes and the charge is heated by producing arc between the electrodes and the charge. Since in a direct arc furnace, the arc is in direct contact with the charge and heat is also produced by flow of current through the charge itself, the charge can be, therefore, heated to highest temperature. In case of a single phase arc furnace two electrodes are taken vertically downward through the roof of the furnace to the surface of the charge and in a 3-phase furnace three electrodes put at the corners of an equilateral triangle, project on the charge through the roof and three arcs are formed. The current passing through the charge develops electromagnetic field and necessary stirring action is automatically obtained by it. Thus uniform heating is obtained.

It is commonly used for production of steel. The usually size of such a furnace is between 5 and 10 tonnes, though 50 and 100 tonne arc furnaces have also been developed. The main advantage of direct arc furnace over cupola method for production of steel is that purer production is obtained and the composition can be exactly controlled during refining process. Another advantage is that arc furnace can operate on 100% steel scrap which is cheaper than pig iron where as the cupola requires a proportion of pig iron in cupola charge. This is the reason, that direct arc furnaces even being costlier in initial as well as operating costs are preferred. Though such furnace is employed both for melting and refining, but due to its higher cost it is used for refining than melting. The power factor is about 0.8 lagging. For 1 tonne furnace power required is about 200 kW and energy consumed is 1,000 kWh/tonne.

2. *Submerged Arc Furnace.* These are cylindrical furnaces in which arc is formed between the carbon electrodes (from the top) and hearth electrodes. The hearth lining is of magnesite which becomes comparatively good electrical conductor when hot. It is also mixed with coke or graphite. Sometimes a conducting hearth is used as electrode. The number of electrodes, taken from the roof depends on the type of supply. One for single phase, two or four for two phase and three for three phase supply, bottom conductor being connected to the neutral. Usually 3-phase supply is used for dealing with large power, usually required. The current from the top electrode passes through the arc to the charge and returns through the electrode at the bottom of the charge. Better distributed resistance heating is obtained since charge behaves as the resistance. Similarly better mixing of charge takes place. The current under short circuit is limited due to charge, which otherwise in indirect furnace is very high. Power is controlled by varying distance between the electrodes or by varying the voltage applied to the electrodes. In a direct arc furnace the length of arc is varied for this purpose. These furnaces are used for the manufacture of ferro-alloys like ferro-chrome and ferro-manganese. The power factor is about 0.8 lagging. The stirring action is inherent due to electromagnetic forces set up by the current resulting in uniform heating of charge.

3. *Indirect Arc Furnace.* In this case arc is formed between two electrodes above the charge, as shown in Fig. 5.5

and heat is transmitted to the charge solely by radiation. In this case the temperature of the charge is, therefore, lower than that in case of direct arc furnace. Since in this furnace current does not flow through the charge, there is no stirring action and the furnace is required to be rocked mechanically. That is why the furnace is made of cylindrical shape, with the electrodes projecting through the chamber from each end and along the horizontal axis. By rocking action there is thorough mixing of the charge. The life of the refractory lining also increases since the molten metal comes in contact with the lining and takes away some of its heat thus preventing it from attaining excessive temperature. The efficiency is increased because the charge is heated not only by radiation from the arc but by conduction from the heated refractory during rocking action. Its construction limits the number of electrodes to two, so single-phase supply is required. The size of the furnace is thus limited by the amount of single-phase load, which can be taken from one point. The arc is produced by bringing the electrodes into solid contact and then withdrawing them. Power input is regulated by adjusting the arc length by moving the electrodes.

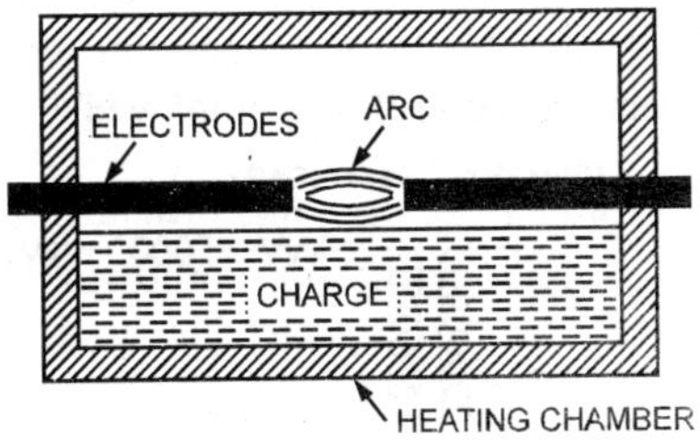

Fig. 5.5. ***Indirect Arc Furnace***

An electric motor is employed for operating suitable grinders and rollers to provide rocking action to the furnace. At start, the rocking action is carried through an angle of 15 to 20° and as the melt proceeds the angle is increased to about 200° at a frequency of about 2 cycles per minute.

Such furnaces have got following advantages :

(*i*) ***Low Metal Losses.*** Since furnace chamber is closed and a reducing atmosphere above the metal is produced due to carbon arc, therefore, metal losses due to oxidation and volatilisation are quite low.

(*ii*) ***Economical.*** Melting is rapid and takes place in a completely closed chamber resulting in small heat losses and low power consumption. Higher production rate gives low labour costs resulting in lower overall production cost per tonne of molten metal.

(*iii*) ***Sound Castings.*** The agitation produced by the rocking action of the furnace and the absence of combustion gases ensure products free from blow holes, inclusion and segregations. Higher pouring temperature may readily be attained. Thus sound castings in thin and intricate designs can be produced.

(*iv*) ***Flexible.*** Single furnace is capable of handling small or large heat of widely differing analysis. Interchangeable furnace shells can be used for different alloys and can be substituted in a few minutes.

Due to indirect heating the furnace is suitable for comparatively low melting point. So these furnaces are used for melting non-ferrous metals. These are also used to a considerable extent in iron foundries where relatively small quantity of metal is required intermittently. The capacity of furnace varies from 0.25 tonne to 3 tonnes. The power factor is about 0.85. The energy consumption is about 275 kWh/tonne for brass; 325 kWh/tonne for bronze and 400 kWh per tonne for copper.

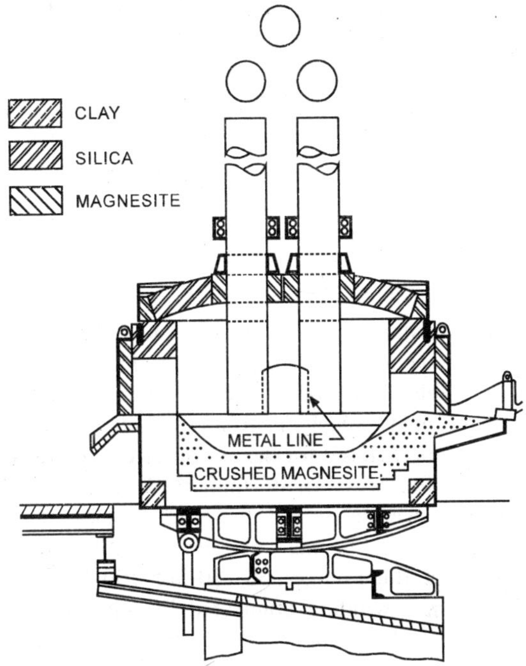

Fig. 5.6. ***Three-Phase Arc Furnace***

Even though Ajax Wyatt induction furnace is its chief rival in the field of the non-ferrous metals, owing to its higher efficiency, it is more suitable in situations where the charge has to be varied frequently or where heating is intermittent.

The three-phase arc furnace having a steel bowl chamber with a refractory lining is shown in Fig. 5.6. Single-phase rocking arc furnace is shown in Fig. 5.7.

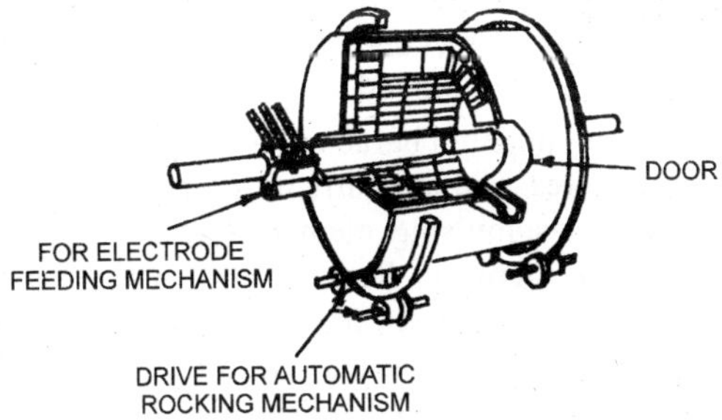

Fig. 5.7. ***Single-Phase Rocking Arc Furnace***

5.5.2. Power Supply and Control

An arc furnace used for melting and refining of steel requires power of about 500 kW per tonne for small furnaces (say for 1/2 tonne capacity) and of about 200 kW per tonne for very large furnaces (say for 50–100 tonne capacity). The energy required is usually between 600 and 800 kWh per tonne. For the arc furnaces used for refining only electrical energy consumption is about 100–120 kWh/tonne. Thus it is seen that the power consumption of the arc furnace is very high. The arc voltage lies between 50 and 150 volts, the current required to give the above mentioned power is, therefore, of the order of several hundred or thousand amperes. The reasons for employing low voltage high current power supply for arc furnaces are as follows :

(*i*) Heating effect is proportional to the square of the current, therefore, to achieve higher temperatures heavy currents are essential.

(*ii*) The maximum secondary voltage is also limited to 275 V (line-to-line open-circuit voltage) because of insulation and safety considerations.

(*iii*) By using low voltage and high current the electrodes are kept very near to the charge as the arc is of small length. Thus arc remains away from the roof and, therefore, life of the roof refractory is increased.

(*iv*) Higher voltage causes higher voltage gradient between the electrode and the charge causing nitrogen of furnace atmosphere ionised and absorbed by the charge, which produces embrittlement.

Thus a transformer having low voltage and large current on the secondary and of a special design having mechanical rigidity to enable the windings and the electrical insulation to withstand the heavy mechanical stresses set up by the high current surges is required for the arc furnace. The transformer used with arc furnace is of oil-immersed water-cooled type. A typical specification for a 3-phase arc furnace transformer includes an extended primary winding with taps there in for the secondary voltage range 235-220-205-190-175-160 volts, with the primary winding connected in delta. This voltage range is extended by changing the connections of the primary windings from delta to star giving 58% voltage from each tap. Both core type and shell type transformers, can be used, but the latter type is preferred because it facilitates the bringing out and bracing of the heavy current leads.

It is desirable to arrange the furnace and transformer in such a way that the current leads are of shorter length. Also in order to keep inductance of these leads low they should be placed as close together as possible and they must be of such a shape as to reduce skin effect to a minimum. Since the leads from the transformer to the furnace are to carry heavy currents, so leads consisting of rectangular copper strips spaced a few mm apart are mostly employed as current carrying conductors.

The voltage-ampere characteristic of the arc is negative, *i.e.*, with the increase in arc current, the resistance of the arc decreases, and a stabilising element (*reactor*) is, therefore, required for the stability. Reactor also serves as a safety device by limiting the current in the circuit when an electrode touches the charge. The inherent reactance in the circuit of a large furnace 5,000 kVA and larger may be and usually is sufficient for the purpose but the smaller installations require more or less supplemental reactance.

The equivalent circuit of an electric arc furnace is shown in Fig. 5.8 where R_T and X_T are the equivalent resistance and reactance of the furnace transformer referred to the secondary side. R_L and X_L are the resistance and reactance of leads. R_E is the resistance of electrodes and $E_A = IR_A$, the voltage drop across the arc. The arc current depends upon the resistance and reactance of transformer (R_T and X_T), resistance and reactance of the leads (R_L and X_L), resistance of the electrodes (R_E) and the arc resistance R_A.

The power input can be controlled by raising or lowering the electrodes which results in variation of the arc resistances R_A. The power input can be also controlled by changing the tapping of transformer which results in variation of voltage across the furnace. For complete control of furnace temperature and to achieve best operating conditions both voltage and electrode controls are employed.

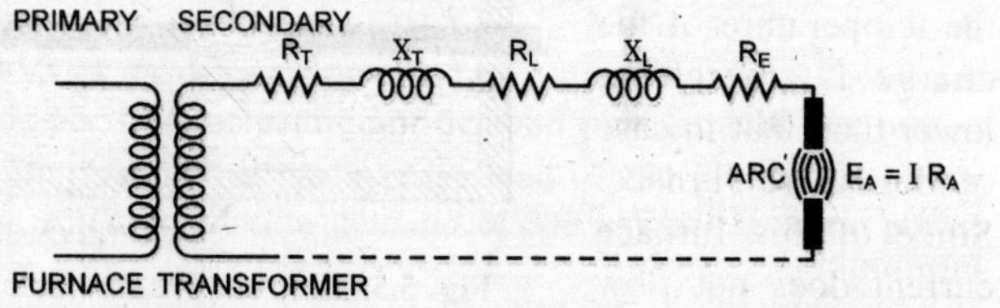

Fig. 5.8. ***Equivalent Circuit of An Arc Furnace***

Figure 5.9 illustrates the performance characteristics of a typical arc furnace. Initially when the electrodes are touching the charge, *i.e.*, arc resistance is zero, the current is limited by the impedance of the furnace circuit and the power input to the furnace is almost zero (copper losses in the transformer windings). Again when the electrodes are raised so that the arc is extinguished, the input to the furnace is zero. In between two these two limits there is a particular loading when the power input to the furnace is maximum.

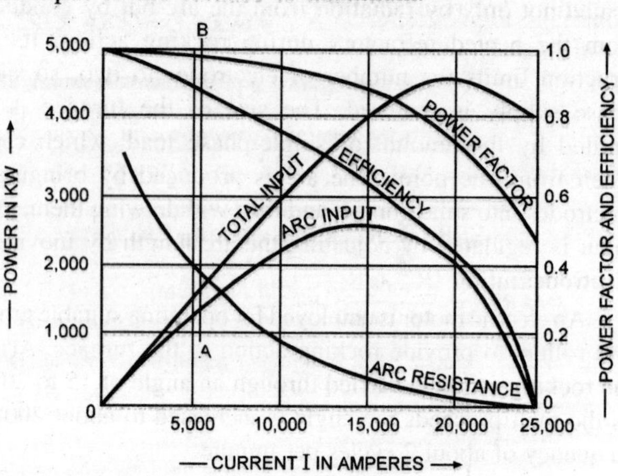

Fig. 5.9. ***Performance Characteristics of a Typical Arc Furnace***

It can be seen, from the performance characteristic curves shown in Fig. 5.9, that for maximum power input to the furnace, the power factor is about 0.87 and the efficiency is less than 70%. The furnace is normally operated at a point in the neighbourhood of line AB so that reasonable efficiency and power factor alongwith a reasonable service from the furnace can be had. The efficiency stated above is the electrical efficiency (taking into account only the ohmic losses occurring in the furnace circuit), overall efficiency is about 50-60 per cent taking into account all other losses such as used in heating up the refractory material and furnace chamber, losses through walls, losses through escaping gases, losses through water cooling. Water cooling is applied to the openings in the roof through which electrodes are placed, the roof ring and the door frame.

Condition For Maximum Output

$$\text{Arc current, } I = \frac{V}{\sqrt{(R_T + R_L + R_A)^2 + (X_T + X_L)^2}}$$

$$\text{Power loss in the arc} = I^2 R_A$$

$$= \frac{V^2}{(R_T + R_L + R_A)^2 + (X_T + X_L)^2} \times R_A$$

$$= \frac{V^2 R_A}{R_A^2 + 2R_A(R_T+R_L) + (R_T+R_L)^2 + (X_T+X_L)^2}$$

$$= \frac{V^2}{R_A + 2(R_T+R_L) + \frac{(R_T+R_L)^2}{R_A} + \frac{(X_T+X_L)^2}{R_A}}$$

Power loss will be maximum when denominator will be minimum

$$\therefore \quad \frac{d}{dR_A}\left[R_A + 2(R_T+R_L) + \frac{(R_T+R_L)^2 + (X_T+X_L)^2}{R_A}\right] = 0$$

$$\text{or } 1 + 0 - \frac{(R_T+R_L)^2 + (X_T+X_L)^2}{R_A^2} = 0$$

$$\text{or } R_A = \sqrt{(R_T+R_L)^2 + (X_T+X_L)^2}$$

i.e., Power loss in the furnace will be maximum when the arc resistance R_A will be numerically equal to the impedance of the whole electrical circuit referred to the secondary excluding arc resistance R_A.

Power factor at maximum power loss,

$$\text{Cos } \phi = \frac{R_A+R_T+R_L}{\sqrt{(R_A+R_T+R_L)^2 + (X_T+X_L)^2}}$$

$$= \frac{R_A+R_T+R_L}{\sqrt{(R_A^2 + 2R_A(R_T+R_L)) + (R_T+R_L)^2 + (X_T+X_L)^2}}$$

Substituting $(R_T + R_L)^2 + (X_T + X_L)^2 = R_A^2$, we get

$$\text{Cos } \phi = \frac{R_A+R_T+R_L}{\sqrt{2R_A^2 + 2R_A(R_T+R_L)}}$$

$$= \sqrt{\frac{R_A+R_T+R_L}{2R_A}}$$

$$= \frac{1}{\sqrt{2}}\sqrt{1+\frac{R_T+R_L}{R_A}} = \frac{1}{\sqrt{2}} = 0.707$$

if we neglect $R_T + R_L$ in comparison with R_A. However, it is not economical to operate an arc furnace with primary side power factor under 0.8.

Example 5.5. A 3-phase arc furnace is used to melt 4.3 tonnes of steel in one hour. Determine the average kW and kVA input to the furnace, arc voltage, arc resistance and pf of the current drawn from the supply. Given:

Specific heat of steel = 444 J/kg/°C

Latent heat of steel = 37.25 kJ/kg

Melting point of steel = 1,370°C

Initial temperature of steel = 19.1°C

Overall efficiency of furnace = 50% = 0.5

Input current = 5,700 A

Resistance of transformer referred to secondary and reactance of the transformer referred to secondary are 0.008 and 0.014 ohm respectively. [*Pb. Technical Univ. Electric Drives and Utilization May 2009*]

Solution : Energy required to melt 4.3 tonnes of steel

$= ms(t_2 - t_1) + mL$

$= m[s(t_2 - t_1) + L]$

$= 4.3 \times 1{,}000[444(1{,}370 - 19.1) + 37.25 \times 1000]$

$= 2{,}739{,}313{,}280$ J

$$= \frac{2{,}739{,}313{,}280}{3.6 \times 10^6} = 760.92 \text{ kWh}$$

Time taken in melting of steel = 1 hour

$$\text{Average output} = \frac{\text{Total energy required in kWh}}{\text{Time of melting in hours}}$$

$$= \frac{760.92}{1} = 760.92 \text{ kW}$$

$$\text{Average input} = \frac{\text{Average output}}{\text{Overall efficiency}} = \frac{760.92}{0.5} = 1{,}521.84 \text{ kW} \quad \textbf{Ans.}$$

Current input, I = 5,700 A

Voltage drop due to resistance of furnace leads (including transformer)

$= 5{,}700 \times 0.008 = 45.6$ V

Voltage drop due to reactance of furnace leads (including transformer)

$= 5{,}700 \times 0.014 = 79.8$ V

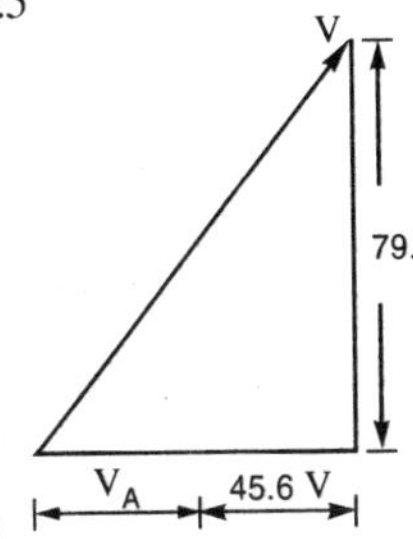

Fig. 5.10

Let the arc drop be V_A volts, resistive in nature

Open-circuit phase voltage of trans-former secondary

$$= \sqrt{(V_A + 45.6)^2 + (79.8)^2}$$

$$\text{Power factor, cos } \phi = \frac{V_A + 45.6}{\sqrt{(V_A + 45.6)^2 + 79.8^2}}$$

Total power input = 3 × power drawn per phase

= 3 × current drawn per phase × secondary phase voltage × pf

$$\text{or } 1{,}521.84 \times 1{,}000 = 3 \times 5{,}700 \times \sqrt{(V_A + 45.6)^2 + (79.8)^2} \times \frac{V_A + 45.6}{\sqrt{(V_A + 45.6)^2 + (79.8)^2}}$$

$$\text{or } V_A = \frac{1{,}521.84 \times 1{,}000}{3 \times 5{,}700} - 45.6 = 43.4 \text{ V} \quad \textbf{Ans.}$$

$$\text{Arc resistance, } R_A = \frac{\text{Arc voltage}}{\text{Arc current}} = \frac{43.4}{5{,}700} = 0.0076\ \Omega \quad \textbf{Ans.}$$

Open-circuit phase voltage of transformer (secondary)

$$= \sqrt{(V_A + 45.6)^2 + (79.8)^2}$$

$$= \sqrt{(43.4 + 45.6)^2 + (79.8)^2} = 119.54 \text{ V}$$

Total kVA drawn from the supply

= 3 × current drawn per phase × secondary phase voltage × 10^{-3}

= 3 × 5,700 × 119.54 × 10^{-3}

= 2,044 kVA **Ans.**

Power factor of current drawn from supply

$$= \frac{V_A + 45.6}{\sqrt{(V_A + 45.6)^2 + (79.8)^2}}$$

$$= \frac{43.4 + 45.6}{\sqrt{(43.4 + 45.6)^2 + (79.8)^2}} = 0.7445 \text{ (lagging)} \quad \textbf{Ans.}$$

Example 5.6. A 3-phase arc furnace is used to melt 4.3 tonnes of steel in one hour. Determine the average kW and kVA input to the furnace, arc voltage, arc resistance and pf of the current drawn from the supply. Given:

Specific heat of steel = 0.12 kcal/kg/°C

Latent heat of steel = 8.89 kcal/kg/°C
Melting point of steel = 1,370°C
Initial temperature of steel = 19.1°C
Overall efficiency of furnace = 50%
Input current = 5,700 A

Resistance of transformer referred to secondary and reactance of the transformer referred to secondary are 0.008 and 0.014 ohm respectively. [*Pb. Technical Univ. Electric Drives and Utilization Dec. 2008*]

Solution : Energy required to melt 4.3 tonnes of steel

$= m[s(t_2 - t_1) + L]$

$= 4.3 \times 1000[0.12(1{,}370 - 19.1) + 8.89]$

$= 735{,}291.4$ kcals

$= \frac{735{,}291.4}{860} = 855$ kWh $\because$ 1 kWh = 860 kcals

Average output $= \frac{855}{1} = 855$ kW

Average input $= \frac{855}{0.5} = 1710$ kW **Ans.**

Current input, I = 5,700 A

Voltage drop due to resistance of furnace leads (including transformer)

$= 5{,}700 \times 0.008 = 45.6$ V

Voltatge drop due to reactance of furnace leads (including transformer)

$= 5{,}700 \times 0.014 = 79.8$ V

Let the arc drop be V_A volts, resistive in nature

Open-circuit phase voltage of transformer secondary

$$= \sqrt{(V_A + 45.6)^2 + (79.8)^2}$$

$$\text{Power factor, } \cos\phi = \frac{V_A + 45.6}{\sqrt{(V_A + 45.6) + (79.8)^2}}$$

Total power input = 3 × I × secondary phase voltage × pf

$$\text{or } 1{,}710 \times 1{,}000 = 3 \times 5{,}700 \times \sqrt{(V_A + 45.6)^2 + (79.8)^2} \times \frac{V_A + 45.6}{\sqrt{(V_A + 45.6)^2 + (79.8)^2}}$$

$$\text{or Arc voltage, } V_A = \frac{1{,}710 \times 1{,}000}{3 \times 5{,}700} - 45.6 = 54.4 \text{ V} \quad \textbf{Ans.}$$

$$\text{Arc resistance, } R_A = \frac{VA}{I} = \frac{54.4}{5{,}700} = 0.00954\ \Omega$$

Open-circuit voltage of transformer secondary

$$= \sqrt{(V_A + 45.6)^2 + (79.8)^2}$$

$$= \sqrt{(54.4 + 45.6)^2 + (79.8)^2} = 127.94 \text{ V}$$

Total kVA drawn $= 3 \times 5{,}700 \times 127.94 \times 10^{-3} = 2{,}187.7$ kVA **Ans.**

$$\text{Power factor, } \cos\phi = \frac{\text{kW}}{\text{kVA}} = \frac{1{,}710}{2{,}187.7} = 0.7816 \text{ (lagging)} \quad \textbf{Ans.}$$

5.6. INDUCTION HEATING

In induction heating effect of currents induced by electromagnetic action in the charge is employed. The heat developed depends on the power drawn by the charge. The power consequently depends upon the voltage and resistance of the charge, because power drawn is equal to $\frac{V^2}{R}$. So to develop heat sufficient to melt the charge, the resistance of the charge must be low, which is possible only with metals, and voltage must be higher, which is obtained by employing higher flux and higher frequency. Magnetic materials, therefore, can be easily treated than non-magnetic materials because of their higher permeability.

The various types of induction furnaces are described as follows.

5.6.1. Core Type Furnaces

The core type furnace is just like a transformer having primary connected to the supply and the charge to be heated as secondary, as shown in Fig. 5.11.

1. *Direct Core Type Induction Furnace.* Direct core type induction furnace is shown in Fig. 5.11 which consists of an iron core, crucible of some insulating material and primary winding connected to an ac supply. The charge is kept in the crucible, which forms a single turn short-circuited secondary circuit. The current in the charge is very high, of the order of several thousand amperes.

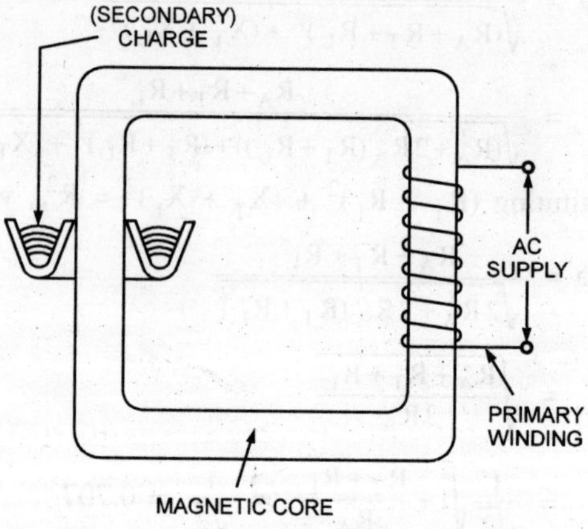

Fig. 5.11. ***Direct Core Type Induction Furnace***

This type of furnace has following drawbacks :

1. As magnetic coupling between the primary and secondary circuit is poor, therefore, leakage reactance is high and power factor is low. This difficulty, however, is overcome by employing supply of frequencies as low as 10 Hz for operation of such furnaces. For obtaining low frequency supply motor-generator set or frequency changer is required, which involves extra cost.
2. If normal frequency supply is employed for operation of such furnaces, the electromagnetic forces cause severe stirring action in the molten metal. Low frequency supply, therefore, is also necessary from this point of view.
3. If the current density exceeds about 5 amps per mm^2 the *pinch effect* (formation of bubbles and voids etc.) due to electromagnetic forces may cause complete interruption of the secondary circuit and so of supply.
4. The crucible for the charge is of odd shape and inconvenient from the metallurgical point of view.
5. For functioning of the furnace the closing of the secondary circuit is essential which necessitates the formation of

complete ring of the charge around the core. For starting the furnace, either molten metal is poured into the crucible or sufficient molten metal is allowed to remain in the crucible from the previous operation. This is required otherwise, the secondary will remain open circuited and no current will circulate and no heating will take place. However, once the melting starts additional metal can be added and the furnace tapped periodically. Also in order to close the secondary circuit, an iron ring may be placed in the crucible or the lining may be of graphite.

Such furnaces are not suitable for intermittent services or where different types of charges are to be melted.

On account of the drawbacks mentioned above such furnaces have become obsolete nowadays.

2. *Vertical Core Type Induction Furnace.* This is an improved form of furnace described above. The furnace of this type, known as Ajax Wyatt vertical core type furnace (shown in Fig. 5.12) employs a vertical channel instead of horizontal one for the charge. The convection currents keep the circulation of molten metal round the V portion. As V-channel is narrow, so even a small quantity of charge is sufficient to keep the secondary circuit closed. Hence the chances of discontinuity of the circuit are less. Due to pinch effect the adjoining molecules carrying current in same direction will try to repel each other, but because of the weight of the charge they will remain in contact and chances of interruption will be reduced.

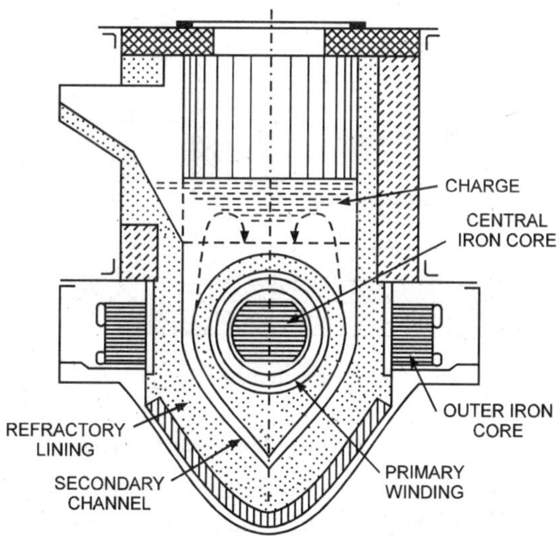

Fig. 5.12. ***Ajax Wyatt Vertical Core Furnace***

The output of the furnace depends upon the type and dimensions of the channels used. In certain furnaces instead V-shaped channels U-shaped channels or rectangular channels are employed.

The inside of the furnace is lined depending upon the charge. Clay lining is used for yellow brass. For red brass and bronze an alloy of magnesia and alumina or corrundum having high contents of alumina is employed.

The shell of the furnace is of heavy steel. The top of the furnace is covered with an insulated cover which can be removed for charging. Necessary hydraulic arrangements are usually made for tilting the furnace to take out the molten metal.

Advantages. 1. Highly efficient heat, low operating costs and improved production.

2. Accurate temperature control, uniform castings, reduced metal losses and reduction of rejects.
3. Absence of crucibles.
4. Consistent performance and simple control.
5. Ideal working conditions in a cool atmosphere with no dirt, noise or fuel.
6. Absence of combustion gases resulting in elimination of the most common source of metal contamination.
7. Comparatively high power factor (the average power factor between 0.8 and 0.85) with normal supply frequency since primary and secondary are both on the same central core.

However, it is to be noted that the Vee must be kept full of charge in order to maintain continuity of the secondary circuit. For this reason, this type of furnace is suitable only for continuous operation. These furnaces are widely used for melting and refining of brass and other heavy non-ferrous metals. Its efficiency is about 75 per cent. Standard sizes of these furnaces range 60 to 300 kW, all single phase, 50 Hz in this country for standard voltage up to 600 V.

3. *Tama Furnace.* The main drawback of Ajax Wyatt furnace is that it cannot be used for melting aluminium and its alloys. This is mainly because of tendency of ahderence, of any aluminium oxide formed, to the walls of V channel. This makes the operation of furnace impossible any longer. This difficulty, however, can be overcome by using a ring type furnace, known as the *Tama furnace,* depicted in Fig. 5.13.

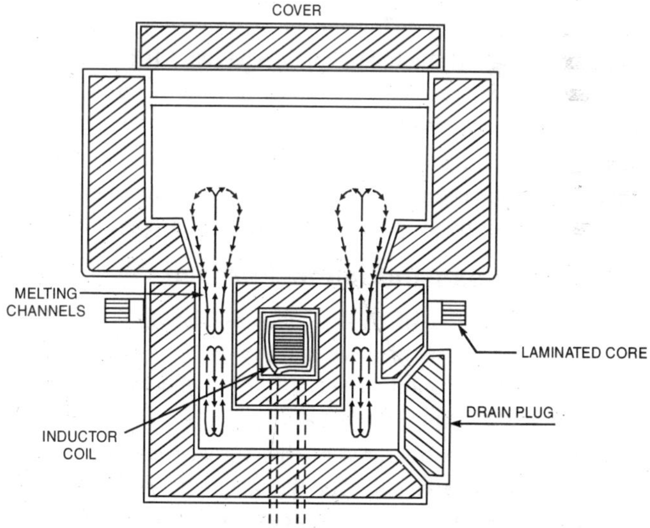

Fig. 5.13. ***Tama Furnace***

In this furnace inductor channels are built vertical which facilitates removal of the deposits with a simple and cheap type cleaning tool. Vertical channels can be cleaned without emptying the furnace and without taking long time. The horizontal channel through the base of furnace does not require frequent cleaning. No doubt, it needs emptying of furnace before removal of drain plug but the operation is speedy. The pf of the Tama furnace is reduced because of its straight sided channels and therefore installation of additional capacitors is required which makes the Tama furnace costlier than the Ajax Wyatt furnace. Stirring action is also set up due to electromagnetic forces in this furnace like Ajax Wyatt furnace.

4. ***Indirect Core Type Induction Furnace.*** In such a furnace an inductively heated element is made to transfer its heat to the charge by radiation. In this type of furnace the principle of induction has been utilised for providing heat treatment of metallic and other charges in addition to its use for melting metals. Such a furnace is shown in Fig. 5.14.

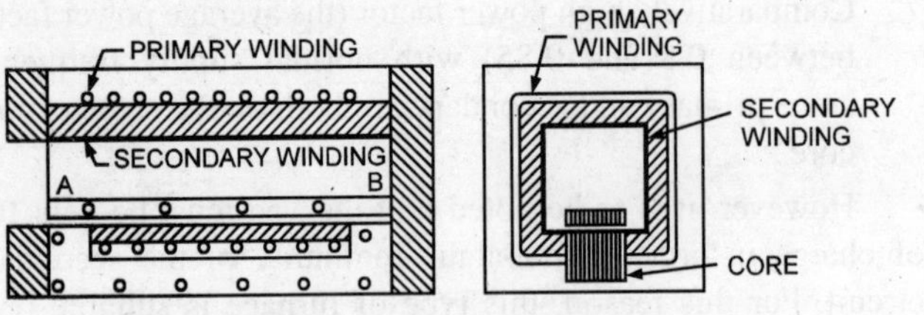

Fig. 5.14. ***Indirect Core Type Induction Furnace***

It consists of an iron core linking with the primary winding and secondary also. In this case secondary consists of a metal container forming the walls of the oven proper. Primary winding is connected to the ac supply, inducing currents and heating the metal container. Heat is transmitted to the charge by radiation. It is advantageous in respect of temperature control without use of external control equipment. It consists of part AB of the magnetic circuit situated in the oven chamber and consisting of a special alloy which looses its magnetic properties at a particular temperature and regains them when cooled to the same temperature. As soon as the oven attains the critical temperature, the reluctance of the magnetic circuit increases many times and the inductive effect correspondingly decreases, thereby cutting off the heat supply. The bar AB is detachable type and can be replaced by others having different critical temperatures between 400 °C and 1,000 °C, according to needs.

From the mode of transmission of heat it will be seen that this furnace is directly in competition with resistance oven; but has comparatively poor power factor (0.8 approximately).

5.6.2. Coreless Induction Furnace

The general design of this furnace is shown in Fig. 5.15. It essentially consists of three main parts (*i*) the primary coil (*ii*) the refractory container and (*iii*) the frame which includes supports and a tilting mechanism.

The distinctive features of this furnace are the absence of a continuous iron path for the magnetic flux and small quantity of refractory material in comparison with other types of melting furnaces in construction.

Standard preformed crucibles are employed for small furnaces (up to about 200 kg holding capacity). The base and the wall around the crucible are made by ramming granular refractory material. The top of the wall is sealed with refractory cement. The containers of the large furnaces are made in place, the procedure being same as for the smaller furnaces except that a hollow collapsible form is substituted for the crucible to form the receptacle. Acid or basic materials are used as per requirements.

For floor level mounting, the electrical connections are made by knife contacts in order to make the handling of the furnace as a ladle for pouring. For platform mounting the electrical connections are made with flexible cables. This is done as with this arrangement, power can be left on the furnace while pouring—a feature often desirable when a number of castings are poured from one heat. A variation is the lift-coil furnace made in the smaller sizes. The primary coil is lowered and raised over the load crucible.

The charge is put into the crucible and primary winding coil is connected to high frequency ac supply. The flux created by primary winding sets up eddy currents in the charge which tend to flow concentrically with those in the inductor. These eddy currents heat up the charge to its melting point and also set up electromagnetic forces producing stirring action which is essential for obtaining uniform quality of metal. Because of high frequency employed, which is necessary to induce the required voltage in the secondary, the skin effect produces heat in the primary winding coils. The primary winding coils are, therefore, made from hollow tube and are cooled by circulation of water through it. Insulated supporting structure is employed for such furnaces, otherwise stray magnetic field outside the primary will set up emf in it, which will result in circulation of eddy currents in it and so reduction of efficiency.

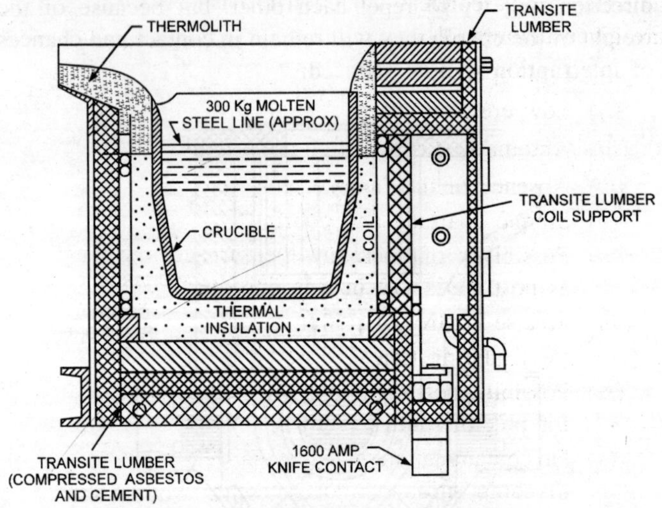

Fig. 5.15. ***Coreless Induction Furnace***

Standard sizes of coreless induction furnaces for melting non-ferrous metals and alloys range from 50 kg to 500 kg holding capacity.

The choice of frequency of operation plays vital role. It is governed by the factors, material to be heated and thickness of cylinder layer at the outside edge of the crucible. The frequency of the primary current can be ascertained by using penetration formula. Accordingly,

$$t = \frac{1}{2\pi}\sqrt{\frac{\rho \times 10^7}{\mu f}} \text{ m} \quad ...(5.11)$$

where ρ is the resistivity of molten metal in Ω-m, f is supply frequency in Hz and μ is relative permeability (for molten steel μ = 1).

The exact theory also shows that for efficient operation the ratio of radius of piece of material in the charge to the thickness should be greater than 3. If it is taken as 4, then the expression for the frequency, for efficient operation, becomes

$$f = \frac{1}{4\pi^2} \times \frac{\rho \times 10^7 \times 16}{\mu \times r^2} \quad ...(5.12) \quad \because \frac{r}{t} = 4$$

From above Eq. (5.12) it can be seen that by increasing r efficient operation can be obtained with lower frequencies. In most of the modern coreless induction furnaces the frequencies in the range of 500 to 1,000 Hz are used. However, for smaller units for melting small quantities of finely divided metal frequencies up to 100 kHz or even 1,000 kHz are used.

The refractory container makes necessary a large air gap (loose coupling) with consequent low power factor (between 0.1 and 0.3). Static capacitors are, therefore, invariably employed in parallel with such a furnace in order to improve the pf. Since in case of coreless induction furnace operation pf does not remain constant, capacitance in the circuit during heat cycle is varied to maintain power factor approximately unity.

The choice of frequency is influenced also by the cost of the capacitors, used for power factor improvement, which decreases with increase of frequency, and the cost of the converting apparatus, which increases with frequency.

Advantages of Coreless Induction Furnaces. The advantages of a coreless induction furnace over other types are given below :

(*i*) Low operating cost.
(*ii*) Low erection cost.
(*iii*) Automatic stirring action produced by eddy currents.
(*iv*) Absence of dirt, smoke, noise etc.
(*v*) Simple charging and pouring.
(*vi*) Possibility of operating the furnace intermittently, as no time is lost in warming up.
(*vii*) Precise control of power.
(*viii*) Less melting time.
(*ix*) Possibility of employing vacuum heating necessary for precious metal melting.
(*x*) No contamination of charge and very accurate control of composition so most suitable for production of high grade alloy steels.

The coreless induction furnace is mainly employed as a metal melting unit. An important application of this furnace is the production of carbon free ferrous alloys. Various special uses are : vacuum melting, duplexing steel, heating of charges of non-conducting materials (with or without melting) by the use of conducting crucibles etc. The energy consumption is 600 to 1,000 kWh per tonne of steel.

Example 5.7. Estimate the rating of an induction furnace to melt two tonnes of zinc in one hour if it operates at an efficiency of 80%. Specific heat of zinc = 0.1, latent heat of fusion of zinc = 26.67 kcals per kg. Melting point is 555°C. Assume initial temperature to be 35°C.

[*G.B. Technical Univ. Utilization of Electrical Energy and Electrical Traction 2010-11*]

Solution : Quantity of zinc to be melt, m = 2 tonnes = 2,000 kg

Initial temperature, t_1 = 35°C

Melting temperature, t_2 = 555°C

Specific heat of zinc = 0.1 kcal/kg/°C

Latent heat of fusion, L = 26.67 kcals/kg

Heat energy required to melt 2 tonnes of zinc

$= m[s(t_2 - t_1) + L]$

$= 2{,}000[0.1(555 - 35) + 26.67]$

$= 157{,}340$ kcals

$= \frac{157{,}340}{860}$ kWh = 182.95 kWh

$$\text{Energy input} = \frac{\text{Energy required}}{\eta} = \frac{182.95}{0.8} = 228.7 \text{ kWh}$$

$$\text{Rating of furnace} = \frac{\text{Energy input}}{\text{Time of operation in hours}}$$

$$= \frac{228.7}{1} = 228.7 \text{ kW} \quad \textbf{Ans.}$$

Example 5.8. Estimate the efficiency of a high frequency induction furnace which takes 10 minutes to melt 1.8 kg of aluminium. The input to the furnace being 5 kW and initial temperature 15°C.

Given : Specific heat of aluminium = 880 J/kg/°C

Melting point of aluminium = 660°C

Latent heat of fusion of aluminium = 32 kJ/kg;

1 J = 2.78 × 10^{-7} kWh

[*Pb. Univ. Electric Drives and Utilization of Electric Energy August 1984*]

Solution : Quantity of aluminium to be melt, m = 1.8 kg

Initial temperature, t_1 = 15°C

Melting temperature, t_2 = 660°C

Specific heat of aluminium, s = 880 J/kg/°C

Latent heat of aluminium, L = 32,000 J/kg

Heat required to melt 1.8 kg of aluminium

$= m\,[s\,(t_1 - t_1) + L]$

$= 1.8\,[880\,(660 - 15) + 32{,}000 = 1{,}079{,}280$ J

$= \frac{1{,}079{,}280}{3.6 \times 10^6} = 0.3$ kWh $\quad \because$ 1 kWh = 3.6 MJ

$$\text{Energy input} = 5 \times \frac{10}{60} = 0.8333 \text{ kWh}$$

$$\text{Efficiency of furnace, } \eta = \frac{\text{Output}}{\text{Input}} \times 100 = \frac{0.3}{0.8333} \times 100 = 36\% \textbf{ Ans.}$$

Example 5.9. A low frequency induction furnace operating at 10 volts in secondary circuit takes 500 kW at 0.5 pf when the hearth is full. If the secondary voltage be maintained at 10 volts, estimate the power absorbed and the power factor when the hearth is half-full. Assume the resistance of the secondary circuit to be thereby doubled and the reactance to remain the same.

[*U.P. Technical Univ. Utilization of Electrical Energy and Traction, 2009-10*]

Solution : Secondary voltage, V = 10 volts

Power drawn, P = 500 kW

$$\text{Secondary current, I} = \frac{P}{V\cos\phi} = \frac{500 \times 1{,}000}{10 \times 0.5} = 100{,}000 \text{ A}$$

Impedance of the secondary circuit when hearth is full

$$Z = \frac{V}{I} = \frac{10}{100{,}000} = 1 \times 10^{-4}\,\Omega$$

Resistance of secondary circuit when hearth is full

$R = Z \cos\phi = 1 \times 10^{-4} \times 0.5 = 0.5 \times 10^{-4}\ \Omega$

Reactance of secondary circuit when hearth is full

$X = Z \sin\phi = 1 \times 10^{-4} \times 0.866 = 0.866 \times 10^{-4}\ \Omega$

When the hearth is half-full

Resistance of secondary circuit, $R' = 2R = 2 \times 0.5 \times 10^{-4}\ \Omega$
$= 1 \times 10^{-4}\ \Omega$

Reactance of secondary circuit, $X' = X = 0.866 \times 10^{-4}\ \Omega$

Impedance of secondary circuit,

$$Z' = \sqrt{(R')^2 + (X')^2} = \sqrt{(1\times10^{-4})^2 + (0.866\times10^{-4})^2} = 1.323 \times 10^{-4}\ \Omega$$

$$\text{Power factor} = \frac{R'}{Z'} = \frac{1\times10^{-4}}{1.323\times10^{-4}} = 0.756 \textbf{ Ans.}$$

$$\text{Secondary current, } I' = \frac{V}{Z'} = \frac{10}{1.323\times10^{-4}} = 75{,}593\ \text{A}$$

$$\text{Power drawn, } P = (I')^2R = (75{,}593)^2 \times 1.0 \times 10^{-4} = 571{,}428\ \text{W or } 571.428\ \text{kW} \textbf{ Ans.}$$

Example 5.10. A low frequency induction furnace, whose secondary voltage is maintained constant at 10 volts, takes 400 kW at 0.6 power factor when the hearth is full. Assuming the resistance of the secondary circuit to vary inversely as the height of the charge and reactance to remain constant, find the height up to which hearth should be filled to obtain maximum heat.

[*Gorakhpur Univ. Utilization of Electrical Power & Traction Sept., 1983*]

Solution : Secondary voltage, V = 10 volts

Power drawn, P = 400 kW = 400,000 watts

$$\text{Secondary current, } I = \frac{P}{V\cos\phi} = \frac{400{,}000}{10\times0.6} = 6.667 \times 10^4\ \text{A}$$

Impedance of secondary circuit when hearth is full

$$Z = \frac{V}{I} = \frac{10}{6.667\times10^4} = 1.5 \times 10^{-4}\ \Omega$$

Resistance of secondary circuit when hearth is full,

$R = Z\cos\phi = 1.5 \times 10^{-4} \times 0.6 = 0.9 \times 10^{-4}\ \Omega$

Reactance of secondary circuit, X

$= Z \sin\phi = 1.5 \times 10^{-4} \times 0.8 = 1.2 \times 10^{-4}\ \Omega$

Let the height of charge be x times of full hearth, *i.e.*, $h = x\text{H}$

Since resistance varies inversely as the height of the charge

∴ Resistance of the charge with height of charge as

$$x\text{H} = \frac{R}{x} = \frac{0.9\times10^{-4}}{x}\ \Omega$$

Power drawn or heat produced will be maximum when resistance of secondary circuit will be equal to reactance of secondary circuit

$$\textit{i.e.,}\quad \frac{0.9\times10^{-4}}{x} = 1.2 \times 10^{-4}$$

$$\text{or } x = \frac{0.9}{1.2} = \frac{3}{4}$$

i.e., Maximum heat will be obtained with the height of charge as three-fourth of height of hearth. **Ans.**

5.7. HIGH FREQUENCY EDDY CURRENT HEATING

The articles to be heated are placed within a high frequency current carrying coil, alternating magnetic field is set up, eddy currents are induced in the article and heating is, therefore, affected (Fig. 5.16). The high frequency current carrying coil is known as *heater coil* or *work coil*, the material which is to be heated is known as the *charge* or *load* and the process employed is referred to as *high frequency eddy current heating.*

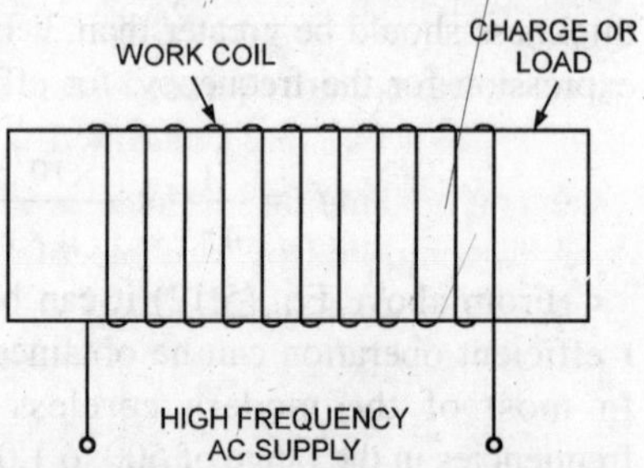

Fig. 5.16. ***High Frequency Eddy Current Heating***

The eddy current loss is primarily responsible for the production of heat, but hysteresis loss also contribute to it, though to a little extent, in the case of magnetic materials.

Since the eddy current loss is proportional to the square of the product of supply frequency and flux density, therefore, by controlling the frequency and the flux density the amount of heat can be controlled.

However, it is observed that higher the frequency employed in induction heating, the greater will be the tendency of the induced heating currents to remain at the surface of the material being heated. This property is called *skin effect.* This has a strong effect upon the uses and limitations of high frequency eddy current heating.

Induced eddy current is of greatest magnitude at the surface of the material to be heated and its value decreases with the distance normal to the surface as we go inside the material.

Since the depth of penetration of the eddy currents into the charge is inversely proportional to the under root of supply frequency $\left(\delta = \frac{1}{2\pi}\sqrt{\frac{\rho\times10^7}{\mu f}}\ \text{m}\right)$, therefore eddy current heating can be restricted to any desired depth of the material to be heated by judicious selection of frequency of the heating current. The supply frequency is usually employed between 10 to 400 kHz.

To be effective the coil must be so shaped to provide as tight a magnetic coupling as possible between itself and the work (or load), the turns of the coil being parallel to the paths in which it is desired that the eddy currents shall flow. With the higher frequencies it is possible to achieve power inputs to the charge of 1.6 kW or more per square centimetre of surface.

Advantages of Eddy Current Heating. (*i*) It is quick, clean and convenient method.

(*ii*) There is little wastage of heat, as heat is produced in the body to be heated up directly.

(*iii*) It can easily take place in vacuum or other special atmosphere, where as other conventional types of heating are not possible in such places.

(*iv*) The control of temperature is very easy.

(*v*) The heat can be made to penetrate into the metal surface to any desired depth.

(*vi*) Unskilled labour can also operate the equipment.

(*vii*) The area of surface over which heat is produced can be accurately controlled.

(*viii*) The amount of heat produced can be accurately controlled by suitable timing devices.

(*ix*) The work coils are not required to fit closely around the object being treated. It enables the same coil to heat many different objects of different shapes and sizes and in fact often obviates the necessity of requiring odd shaped coils to heat irregular shaped objects.

From the economic point of view, the generation of heat is costly, efficiency of equipment is quite low (less than 50%) and initial cost of equipment is also high.

Applications of High Frequency Eddy Current Heating. The important applications of high frequency eddy current heating are :

(*a*) *Surface Hardening.* It is very important application of high frequency eddy current heating. The bar whose surface is to be hardened, by heat treatment, is surrounded by the work coil. An alternating current of high frequency is made to pass through the work coil. The desired depth of penetration is obtained by judicious selection of the frequency of the heating current in the work coil. After about a few seconds, when the surface has reached the proper temperature, the ac supply is cut off and the bar is at once dipped in water. In the case of hardening done by other methods, the heat treatment of the inner metal is disturbed which produces scales on the surface and also warps a long piece. This is avoided in this method of surface hardening. Such treatment due to induction heating reduces the cost, labour and time considerably.

(*b*) *Annealing.* In conventional method of annealing the process takes long time, which results in scaling of metal—a great shortcoming. But in eddy current heating method, the time taken in the process is less, thus, there is no scale formation. By this method a temperature of the order 750°C can be attained in one minute (approximately) up to depth of 25 mm.

(*c*) *Soldering.* For soldering it is necessary that required amount of heat is produced at the soldering point and the remaining portion of the solder may remain cold. Eddy current heating can be economically employed for soldering precisely for high temperature soldering, Silver, copper and their alloys are used as solders. By induction heating heat develops very rapidly melting the solder which runs into the joint sealing it properly.

Other applications are welding, drying of paints, melting of precious metals, sterilization of surgical instruments and forging of bolt-heads and rivet heads.

Example 5.11. The depth of penetration, in the case hardening of a steel pulley, required is 1.5 mm. The relative permeability is unity and the specific resistance for steel is 5 × 10^{-7} Ω-m. Determine the frequency required.

Solution : Specific resistance, $\rho = 5 \times 10^{-7}\ \Omega\text{-m}$

Relative permeability, $\mu = 1$

Depth of penetration, $\delta = 1.5\text{ mm} = 0.0015\text{ m}$

$$\text{Since } \delta = \frac{1}{2\pi}\sqrt{\frac{\rho \times 10^7}{\mu f}}$$

Frequency required,

$$f = \frac{\rho \times 10^7}{\mu \delta^2 \times 4\pi^2} = \frac{5 \times 10^{-7} \times 10^7}{1 \times (0.0015)^2 \times 4\pi^2}$$

$= 56{,}290$ Hz or 56.3 kHz **Ans.**

5.8. HIGH FREQUENCY POWER SUPPLY SOURCES

The power supply for coreless induction furnace is usually obtained from ordinary supply system and its frequency is converted to higher value either by means of motor-generator set with salient pole alternator (suitable for frequencies up to about 1,000 Hz and for any output required) or by means of motor-generator set with inductor alternator (suitable for frequencies up to about 10,000 Hz and output of 1 tonne) or by means of valve oscillators (suitable for very small furnaces which necessitates frequencies up to 1 million Hz). The oscillator which is employed in induction heating differs from the oscillator used in radio transmitters. The induction heaters are rugged, compact, portable and are almost automatic in operation. They are to be handled by unskilled operators and in comparatively dirty atmosphere. They are also subjected to overload and vibration. Hence while designing an oscillator for high frequency current supply for induction heating, the above factors should be given a proper consideration.

The round about value of frequency used for induction heating is 400 kHz, the circuit being used is shown in Fig. 5.17.

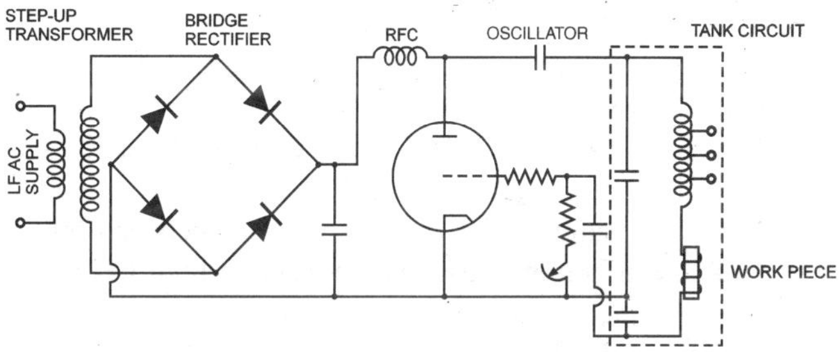

Fig. 5.17. ***Simplified Circuit of a Vacuum Tube Oscillator***

The ac supply is stepped up by the transformer initially, the stepped up voltage is rectified by using a bridge rectifier circuit, the rectified voltage is applied to the oscillator to obtain high frequency currents. The capacitances and inductances (including that of work piece) decide the frequency of supply to work piece. The vacuum tube oscillators are less efficient as compared to SCR, the voltage drop across the SCR being very small (of the order of 1 V). The operating efficiency of SCR is about 90%. The SCRs are fired by pulses of gate current produced by a UJT. The circuit is shown in Fig. 5.18. The frequency of supply to work piece depends upon the value of R and C. The smaller the product of R and C, higher is the frequency across the work piece.

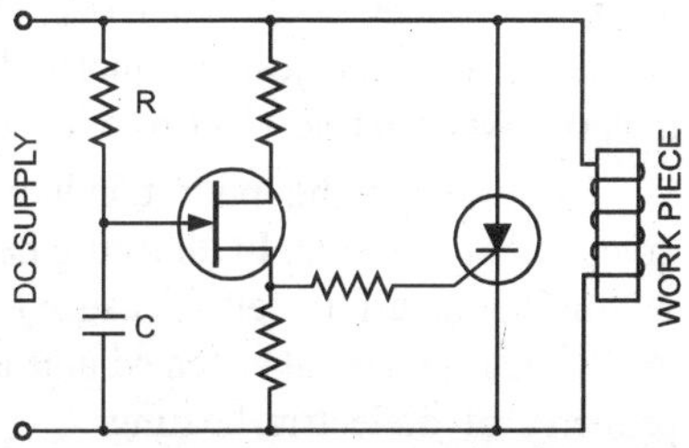

Fig. 5.18. ***SCR Inverter Circuit***

Sometimes spark gap oscillators are also employed to provide high frequency supply. The basic principle of operation of a spark gap converter is the alternate charging and discharging of a capacitor.

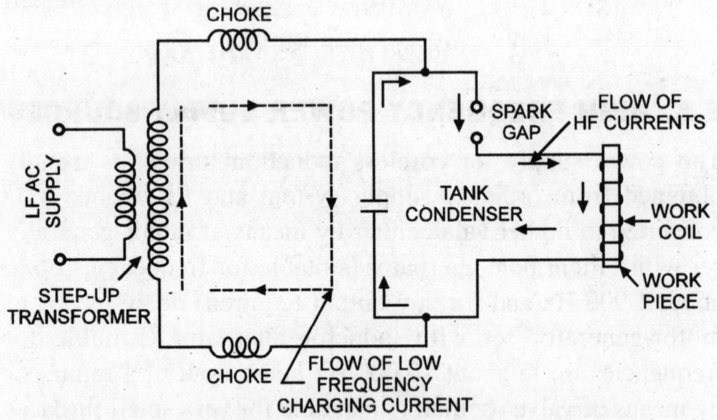

Fig. 5.19. ***Basic Spark-Gap Converter Circuit***

The transformer step ups the voltage, say to about 500 V, the tank condenser gets a voltage build-up till the spark gap breaks down and flashover takes place. The value of inductance and capacitance in the discharge circuit decides the frequency of current through the gap. Frequencies up to 1 MHz can be obtained.

5.9. DIELECTRIC HEATING

When non-metallic parts such as wood, plastics, bones are subjected to an alternating electrostatic field dielectric loss occurs. In dielectric heating use of these losses is made. The material to be heated is placed as a slab between metallic plates or electrodes connected to high frequency ac supply. For producing sufficient heat frequency between 10 and 30 MHz is used. Even though voltage up to 20 kV have been used but from personnel safety point of view voltages between 600 V and 3 kV are in common use. The necessary high-frequency supply is obtained from a valve oscillator, as in the case of high frequency eddy current heating, explained in Art. 5.8. An overall efficiency is about 50%.

The current drawn by the capacitor, when an ac supply voltage is applied across its two plates, does not lead the supply voltage by exactly 90° and there is always an in-phase component of the current. Due to this in-phase (or active) component of current, heat is always produced in the dielectric material placed in between the two plates of the capacitor. The electric energy dissipated in the form of heat energy in the dielectric material is known as *dielectric loss.* The dielectric loss is directly proportional to the frequency of ac supply given to the two plates of the capacitor. The physical conception of the dielectric loss is just as a molecular friction in the dielectric material when an ac electrostatic field is applied to it.

Insulators being poor conductors cannot be heated up quickly from outside. In dielectric heating the heat is produced within the material itself. Because heat generation is uniform, the dielectric material is heated uniformly. This is the important property of dielectric heating.

In insulators or non-conducting materials, the amount of heat produced by dielectric heating can be calculated as follows :

The material to be heated may be considered as the imperfect dielectric of a condenser and may be, therefore, represented as a capacitance placed in parallel with a resistance, as shown in Fig. 5.20 (*b*). The phasor diagram of the circuit is shown in Fig. 5.20 (*c*). If V is supply voltage in volts, *f* is supply frequency in Hz, C is the capacitance of the condenser in farads and cos ϕ is power factor of the load or charge,

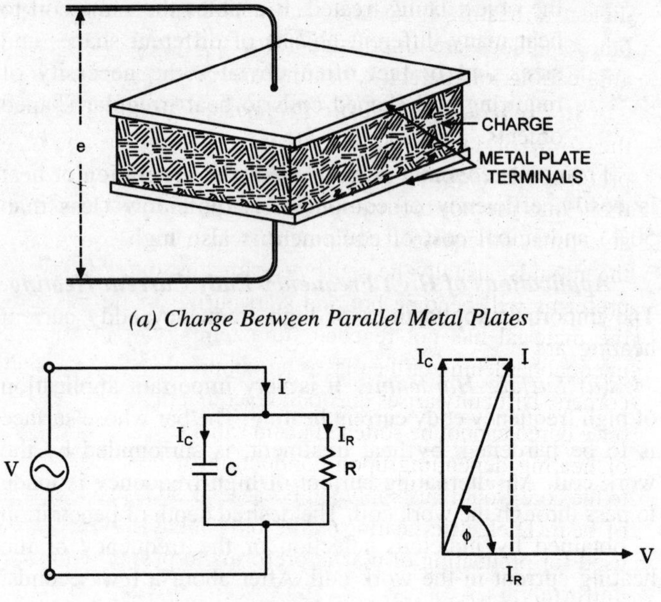

(a) Charge Between Parallel Metal Plates

(b) Equivalent Circuit *(c) Phasor Diagram*

Fig. 5.20. ***Dielectric Heating***

Current through the capacitor,

$$I_C = \frac{V}{X_C} = \frac{V}{1/2\pi fC} = 2\pi fCV \text{ amperes}$$

where C is in farads and V is in volts.

The current drawn from supply,

$$I \simeq I_C = 2\pi fCV \text{ amperes}$$

Power produced, $P = VI \cos\phi = V \times 2\pi fCV \times \cos\phi$ watts

$$= 2\pi fCV^2 \cos\phi \text{ watts} \quad ...(5.13)$$

The value of power factor for a particular non-conducting material is constant, and the capacity is determined from the dimensions of plates, the charge which serves as the dielectric medium and the dielectric constant. Therefore, the dielectric heating depends upon the values of the frequency and the voltage. By varying one of these two quantities, the rate of dielectric heating can be varied. The insulation problem limits the voltage to be used; hence to achieve more heat, high frequency is used. However, there is a limit also, on the high frequency, imposed by cost involved in getting a circuit for obtaining very high frequencies and other difficulties.

The capacity of the condenser can be calculated from the following relation

$$C = \frac{\epsilon_r \epsilon_0 A}{t} \text{ farads} \quad ...(5.14)$$

where ϵ_r is relative permittivity of dielectric, ϵ_0 is absolute permittivity of vacuum and equals 8.854×10^{-12} F/m, *t* is the thickness of dielectric in metres and A is the surface area of plates in m^2.

Advantages : (*i*) If the material to be heated is homogeneous, and the alternating (or varying) electric field is uniform, heat is developed uniformly and simultaneously throughout the entire mass of the charge.

(*ii*) As materials heated by this process are non-conducting, so by other methods heat cannot be conducted to inside so easily.

Applications. The cost of the equipment required for dielectric heating is so high that it is employed only where other methods are impracticable or too slow. Some of the applications of dielectric heating are given below.

1. *Preheating of Plastic Preforms.* The raw material in the form of tablets or biscuits, commonly called plastic preforms, is required to be heated uniformly before putting them into the hot moulds so that whole mass becomes fluid at a time, otherwise if the raw material is put directly into the moulds, usually heated by steam, the outer skin of the preforms will become hot and start curing while the core of the material has not reached fluid temperature resulting in unequal hardening of the plastic and improper filling of moulds corners. Difficulty arises due to the fact that plastic raw material once cured cannot be softened again satisfactorily. Any method of heating depending upon conduction of heat from surface to the core would miserably fail because plastic is bad conductor of heat. Dielectric heating is the only method which can be used for preheating of plastic preforms to proper temperature uniformly.

2. *Gluing of Wood.* Dielectric heating is most commonly used for gluing of wooden sheets or boards as in this method of gluing the moisture contents of the wooden sheets remain unaltered. It is due to the fact that heat can be applied to the desired surface. Main difficulty in using animal glues is of long curing time and that parts to be joined are to be kept under mechanical pressure after application of glue for a period of about 24 hours. Mechanical pressure may be applied in gluing of wood by dielectric heating in order to secure better adhesion. Because of higher loss factor of glue as compared to that of wood most of the heat developed goes into the glue and very little heat is wasted.

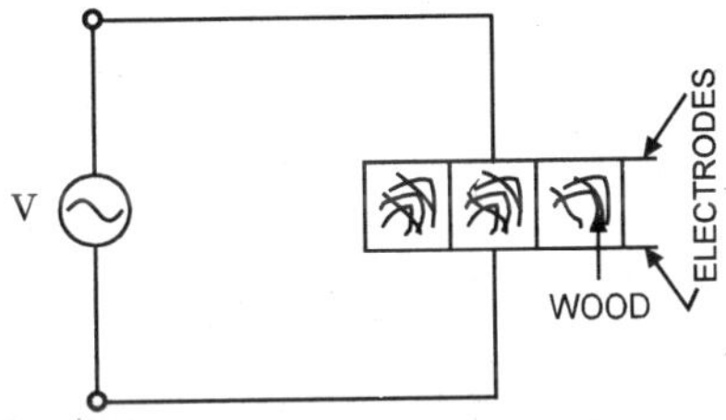

Fig. 5.21

High frequency dielectric heating is very economical for obtaining curved wood sections such as furniture. The curves obtained by this method are stable.

3. *Baking of Foundary Cores.* In foundaries resin type thermosetting binders are employed as they set almost instantaneously when brought to polymerizing temperature. The dielectric heating evaporates water rapidly from the core mix and at the same time raises the temperature of the core material to polymerization point. Hence dielectric heating is most suitable for baking foundary cores mixed with thermosetting resin type core binders.

4. *Diathermy.* Dielectric heating is also employed for heating tissues and bones of the body required for the treatment of certain types of pains and diseases.

5. *Sterlization.* The dielectric heating is quite suitable for sterilization of bandages, absorbent cotton, sterile gauge, instruments etc.

6. *Textile Industry.* In textile industry the dielectric heating is employed for drying purposes.

7. *Electronic Sewing.* Nowadays rain coats, umbrellas, food containers, medicine containers etc. are made from plastic film materials. In case of ordinary stitching by thread they will be no longer water tight and also become weak at seam. With adhesives curing times will be longer. In case of electronic sewing the films to be stitched are rolled in between cold rollers to which radio-frequency voltage is applied. The heat produced in the material seals it all along the line where mechanical pressure is applied. The cold rollers prevent the outer surface of the films from being softened.

8. *Food Processing.* The use of dielectric heating for food processing is one of the most modern method. It has brought many advantages for the food processing industry and has set forth such processes which are outside the realm of cooking. The dielectric heating can be appreciably employed for the following purposes :

(*a*) Pasteurising of milk and beer inside bottles.

(*b*) Dehydrating of fruits, milk, cream, vegetables and eggs etc.

(*c*) Cooking of foods without removing the outer shells.

(*d*) Defrosting of frozen foods such as meat and vegetables.

(*e*) Germicidal heating–In dielectric heating process the products do not loose flavour.

Disinfestation of cereals and grains–Dielectric heating may be used to destroy the eggs of insects and pests saving grains and cereals from damage and also to prevent these eggs, larva or moth etc. to pass on to cereal processing equipment and develop into dangerous worms later on under favourable conditions of temperature and humidity.

Example 5.12. A laminated wooden board 30 cm × 16 cm × 25 cm thick is to be heated from 20°C to 170°C in 10 minutes by dielectric heating using 30 MCPS (30 MHz) supply source. Specific heat of wood is 0.35 and density 0.55 gm/cc, relative permittivity = 5, pf = 0.05. Determine the voltage across the workpiece and current during heating. Assume loss of energy by conduction, connection and radiation as 15%.

[*Rajasthan Univ. Utilization of Elec. Power 2006*]

Solution : Volume of plywood to be heated

$$= 30 \times 10^{-2} \times 16 \times 10^{-2} \times 25 \times 10^{-2} = 0.012 \text{ m}^3$$

$$\text{Density of plywood, } d = 0.55 \text{ gm/cc} = \frac{0.55}{1{,}000} \times 10^6 = 550 \text{ kg/m}^3$$

Weight of plywood $= 550 \times 0.012 = 6.6$ kg

Specific heat of wood = 0.35 calorie per gm per °C

$= 0.35 \times 4.18 \times 1{,}000$ joule per kg per °C

$= 1{,}463$ J/kg/°C

Heat required to raise the temperature of plywood from 20°C to 170°C

$$= 1{,}463 \times 6.6 \times (170 - 20) = 1{,}448{,}370 \text{ joules}$$

$$= \frac{1{,}448{,}370}{3{,}600} \text{ kWh} = 402{,}325 \text{ Wh}$$

Since it is to be done in 10 minutes or 1/6 hour

$$\text{Power utilised} = \frac{402.325}{1/6} = 402.325 \times 6 = 2,413.95 \text{ watts}$$

$$\text{Power input required, P} = \frac{2,413.95}{(100-15)} \times 100 = 2,840 \text{ W}$$

Power factor, $\cos\phi = 0.05$

Capacitance of capacitor formed,

$$C = \frac{\varepsilon_0 \varepsilon_r A}{t} = \frac{8.854 \times 10^{-12} \times 5 \times 30 \times 10^{-2} \times 16 \times 10^{-2}}{25 \times 10^{-2}}$$

$$= 8.5 \times 10^{-12} \text{ F}$$

Voltage required,

$$V = \sqrt{\frac{P}{2\pi f C \cos\phi}} = \sqrt{\frac{2,840}{2\pi \times 30 \times 10^6 \times 8.5 \times 10^{-12} \times 0.05}}$$

$$= 5,954 \text{ V} \quad \textbf{Ans.}$$

Current flowing through the material,

$$I = \frac{P}{V\cos\phi} = \frac{2,840}{5,954 \times 0.05} = 9.54 \text{ A} \quad \textbf{Ans.}$$

Example 5.13. A slab of insulating material 150 cm² in area and 1 cm thick is to be heated by dielectric heating. The power required is 400 W at 30 MHz. Material has relative permittivity of 5 and pf 0.05. Absolute permittivity = 8.854 × 10⁻¹² F/m. Determine the necessary voltage.

[M.D. Univ. Utilization of Elec. Power and Traction December 2009]

Solution : Power absorbed, P = 400 W

Power factor, $\cos\phi = 0.05$

Supply frequency, $f = 30 \text{ MHz} = 30 \times 10^6$ Hz

Capacitance of capacitor formed,

$$C = \frac{\varepsilon_0 \varepsilon_r A}{t} = \frac{8.854 \times 10^{-12} \times 5 \times 150 \times 10^{-4}}{1 \times 10^{-2}}$$

$$= 66.405 \times 10^{-2} \text{ F}$$

Voltage required,

$$V = \sqrt{\frac{P}{2\pi f C \cos\phi}} \qquad \because P = 2\pi f C V^2 \cos\phi$$

$$= \sqrt{\frac{400}{2\pi \times 30 \times 10^6 \times 66.405 \times 10^{-12} \times 0.05}} = 800 \text{ W} \textbf{ Ans.}$$

Example 5.14. The power required for dielectric heating of a slab of resin 150 cm² in area and 2 cm thick is 200 watts, frequency of 30 MHz. The material has a relative permittivity of 5 and pf of 0.05. Determine the voltage necessary and current flowing through the material. If the voltage is limited to 600 V, what will be the value of the frequency to obtain the same heating ?

[Gorakhpur Univ. Utilization of Electrical Power With Traction 1980]

Solution : Power absorbed, P = 200 watts

Power factor, $\cos\phi = 0.05$

Supply frequency, $f = 30 \times 10^6$ Hz

Capacitance of the capacitor formed,

$$C = \frac{\epsilon_0 \epsilon_r A}{t} = \frac{8.854 \times 10^{-12} \times 5 \times 150 \times 10^{-4}}{0.02} = 33.2 \times 10^{-12} \text{ F}$$

∴ Voltage required,

$$V = \sqrt{\frac{P}{2\pi f C \cos\phi}} = \sqrt{\frac{200}{2\pi \times 30 \times 10^6 \times 33.2 \times 10^{-12} \times 0.05}}$$

$$= 800 \text{ V} \textbf{ Ans.}$$

Current flowing through the material,

$$I = \frac{P}{V\cos\phi} = \frac{200}{800 \times 0.05} = 5.0 \text{ A} \textbf{ Ans.}$$

Since heat produced $\alpha\, f V^2$

$$\therefore f_2 V_2^2 = f_1 V_1^2$$

or Frequency required with 600 V supply to obtain the same heat, *i.e.*, of 200 watts,

$$f_2 = f_1 \left(\frac{V_1}{V_2}\right)^2 = 30 \times 10^6 \times \left(\frac{800}{600}\right)^2 = 53.3 \text{ MHz} \textbf{ Ans.}$$

Example 5.15. A small piece of plywood having dimensions : length 5 cm, width 2 cm and thickness 1 cm is placed in between two electrodes having dimensions, length 25 cm, width 2 cm and with 2 cm distance between them, Electric current at frequency of 10 MHz is passed through electrodes for heating of wood by the process of dielectric heating. If the power consumption for heating the piece of wood is 1 kW, determine (*i*) voltage across the electrodes and (*ii*) current through it during the process of heating. Assume that the plywood is an imperfect dielectric and has relative permittivity of 5 and power factor 0.04.

[A.M.I.E. Sec. B. Utilization of Electric Power Summer 1983]

Solution : The arrangement of the heating material is shown in Fig. 5.22. With this configuration the capacitance of the parallel plate condenser formed with two dielectrics, plywood and air is given by the expression

$$C = \epsilon_0 \left[\frac{A_1 \epsilon_{r2}}{t} + \frac{A_2}{\frac{t_1}{\epsilon_{r1}} + \frac{t_2}{\epsilon_{r2}}} \right] \text{ Farads}$$

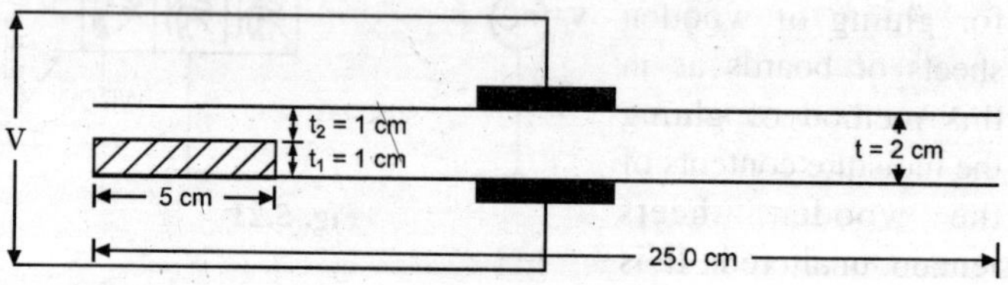

Fig. 5.22

where $A_1 = (25 - 5) \times 2 = 40 \text{ cm}^2 = 0.004 \text{ m}^2$

$A_2 = 5 \times 2 = 10 \text{ cm}^2 = 0.001 \text{ m}^2$

$t = 2 \text{ cm} = 0.02 \text{ m}$

$t_1 = 1 \text{ cm} = 0.01 \text{ m}$

$t_2 = 2 - 1 = 1 \text{ cm} = 0.01 \text{ m}$

$\epsilon_0 = 8.854 \times 10^{-12}$ F/m

ϵ_{r1} = Relative permittivity of plywood = 5

ϵ_{r2} = Relative permittivity of air = 1

∴ Capacitance of the capacitor,

$$C = 8.854 \times 10^{-12} \left[\frac{0.004 \times 1}{0.02} + \frac{0.001}{\frac{0.01}{5} + \frac{0.01}{1}} \right]$$

$$= 2.509 \times 10^{-12} \text{ Farads}$$

(*i*) Voltage across the electrodes,

$$V = \sqrt{\frac{P}{2\pi f C \cos\phi}} = \sqrt{\frac{1,000}{2\pi \times 10 \times 10^6 \times 2.509 \times 10^{-12} \times 0.04}}$$

$$= 12,594 \text{ V} \textbf{ Ans.}$$

(*ii*) Current through plywood,

$$I = \frac{P}{V\cos\phi} = \frac{1,000}{12,594 \times 0.04} = 1.985 \text{ A} \textbf{ Ans.}$$

5.10. CHOICE OF FREQUENCY

The selection of frequency for heating purposes is an important factor. Although, the selection of frequency has a great bearing on the work to be heated and the method of heating to be employed (induction heating or dielectric heating). Power frequency (50 Hz) furnaces can be of 1 MW capacity where as medium frequency (500 Hz to 1,000 Hz) furnaces have capacity of 500 kW and high frequency (100 kHz to 2 MHz) furnaces have capacity from 200 kW to 500 kW.

Induction Heating. For making a choice of frequency for induction heating the following factors are to be considered.

1. The thickness of surfaces to be heated. Higher the frequency, thinner the surface will get heated.
2. The time of continuous heating. Longer the duration, deeper the penetration of heat owing to conduction.
3. The temperature to be obtained, and duration of heating. If higher temperatures are to be obtained in shorter time, higher capacities of generator should be selected.

Dielectric Heating. The rate of heat produced in dielectric heating is given by the expression $P = 2\pi f CV^2 \cos\phi$ watts where V is the supply voltage, f is the supply frequency and C is the capacitance of the condenser formed, which depends upon the relative permittivity $\in_r$ of the material.

Thus the rate of heat production $\alpha\ V^2 \times f \times \cos\phi \times \in_r$

The voltage across any specimen is limited by its thickness, *i.e.*, potential gradient, breakdown voltage, insulation and safety consideration. The voltages used for dielectric heating are usually from 600 V to 3,000 V, however voltage up to 20,000 volts is also sometimes used. Alternatively higher rate of production of heat can be obtained by employing higher frequency but this is also limited due to the following considerations.

(*i*) Necessity of incorporating a special matching circuit at higher frequencies due to the fact that maximum power output is obtained from the supply oscillator only when the impedance of the oscillator match with that of load.

(*ii*) Possibility of existing of a standing wave across the surface of electrodes having wavelength approximately equal to or more than one quarter of the frequency wavelength at any time.
It causes a variation in the voltage across the electrodes which results into uneven heating. This is avoided by imposing limits on the length of the electrode at a particular frequency.

(*iii*) At very high frequency it is difficult for tuning inductance to resonate with the charge capacitance.

(*iv*) At higher frequencies uniform voltage distribution cannot be obtained.

(*v*) The higher frequencies may disturb nearby radio station services due to radiation. So special care should be taken that there is no radiation etc.

Exercises

1. What are the advantages of electric heating ? Give classification of various electric heating methods alongwith brief account of their working principles.
[Gorakhpur Univ. Utilization of Elec. Power & Traction Sept., 1984]
2. Write the Stefan's law of heat radiation and label all constants.
[A.M.I.E. Sec B. Utilization of Elec. Power Winter 1983]
3. Discuss various methods of electric heating and explain any one of them.
[Pb. Technical Univ. Electric Drives and Utilization December 2009]
4. Explain the different methods of electric heating and their field of applications.
[Rajasthan Univ. Utilization of Electrical Power 2007]
5. What are the desirable properties the materials for heating element should have ? Name a few materials.
Discuss the methods of temperature control of resistance ovens.
Explain the design procedure of the heating elements when the power and voltage of the oven are known.
[Gorakhpur Univ. Utilization of Elec. Power & Traction 1983]
6. What are the essential properties of resistance heating elements ? How will you design such elements in case of resistance heating.
[Pb. Technical Univ. Electric Drives and Utilization May 2008]
7. What are the requirements of a good heating material. Describe different type of heating materials. Also describe causes of failure of heating elements.
[Rajasthan Technical Univ. Utilization of Elec. Power 2010]
8. Describe with the help of neat sketches salient features of (*i*) air circulation ovens (*ii*) bright annealing (*iii*) radiant heating and (*iv*) salt bath heating and give application of each.
[A.M.I.E. Sec B. Utilization of Elec. Power Summer 1983]
9. What are the different methods of heating ? Describe briefly the methods of direct and indirect resistance heating. List the applications of these two methods.
[U.P. Technical Univ. Utilization of Electrical Energy and Electric Traction, 2005-06]
10. Describe various methods of controlling the temperature in resistance ovens.
[West Bengal Univ. of Technology Utilization of Electric Power 2009]
11. Describe the construction and operation of direct type arc furnace.
[U.P. Technical Univ. Utilization of Electrical Energy and Electric Traction 2006-07]
12. Describe advantages of electrical heating and functioning of indirect arc furnace.
[U.P. Technical Univ. Utilization of Elect. Energy and Traction 2009-10]
13. Describe the construction and working of a 3-phase arc furnace.
[A.M.I.E. Sec B. Utilization of Electric Power Winter 1982]
14. With neat sketches, describe the construction, principle of operation, applications and control methods of direct and indirect arc furnaces.
[M.D. Univ. Utilization of Elec. Power and Traction May 2009]
15. Explain construction, operation, performance and application of arc furnace.
[Rajasthan Univ. Utilization of Electrical Power 2007]
16. Explain Principle of induction heating.
[U.P. Technical Univ. Utilization of Elect. Energy and Traction 2009-10]
17. Explain different methods of induction heating. Give some applications of induction heating.
[Rajasthan Univ. Utilization of Elec. Power 2008]

18. With neat sketches, describe the construction, principle of operation, applications and control methods of Ajax-Wyatt vertical core furnace.
[G.B. Technical Univ. Utilization of Elect. Energy and Traction 2010-11]

19. Explain with neat sketches and describe the construction, principle of operation and application of Ajax Wyatt furnace. **[Rajasthan Univ. Utilization of Elec. Power Dec. 2008]**

20. Explain the working of Tama furnace with the help of neat diagram. **[U.P. Technical Univ. Utilization of Elect. Energy and Traction 2009-10]**

21. With the help of diagram describe the working of a coreless type induction furnace. Mention advantages of coreless type furnace.
[Pb. Technical Univ. Electric Drives & Utilization Dec. 2008]

22. Describe a coreless induction furnace.
[West Bengal Univ. of Technology Utilization of Elec. Power 2009]

23. Explain the method of induction heating and describe coreless type of induction furnace.
[M.D. Univ. Utilization of Electric Power and Traction Dec. 2009]

24. Describe the construction and working of an induction furnace suitable for melting non-ferrous metals. What are the various sources of generating high frequency energy for induction heating ? Describe in brief.
[U.P. Technical Univ. Utilization of Elect. Energy and Traction 2005-06]

25. Explain the principle of high frequency induction heating. Discuss the factors affecting the depth of penetration of heat.
[M.D. Univ. Utilization of Electric Power and Traction Dec. 2008; May 2009]

26. Describe with the help of a neat sketch, the working of a vertical core type induction furnace. Give the applications of high frequency eddy current heating.
[Rajasthan Technical Univ. Utilization of Electrical Power 2011]

27. Describe the principle of induction heating at high frequency and highlight applications of the eddy current heating.
[M.D. Univ. Utilization of Electric Power & Traction May 2010]

28. What is dielectric heating ? Explain the factors on which the dielectric loss in a dielectric material depend.
[D.E.I. Agra Utilization of Electric Power 1989]

29. Explain dielectric heating. What is the rate of rise of temperature in dielectric heating ? Discuss the advantages of dielectric heating.
[Pb. Technical Univ. Electric Drives and Utilization May 2007]

30. State the advantages of electric heating. Explain the principle of induction heating. What are the applications of induction heating ?
What is dielectric heating and dielectric loss ?
[West Bengal Univ. of Technology Utilization of Electric Power 2009]

31. Explain the principle of 'dielectric heating'. Derive the mathematical expression of power consumed in such process. State important applications of dielectric heating.
[R.T.U. Utilization of Electrical Power 2010]

32. What is dielectric heating ? Explain the factors on which the dielectric loss in a dielectric material depends.
[M.D. Univ. Utilization of Electric Power and Traction Dec. 2008; Dec. 2009]

33. What is dielectric heating ? Explain why very high frequency should not be used for dielectric heating.
[M.D. Univ. Utilization of Electric Power & Traction May 2009]

34. Compare induction heating and dielectric heating. Discuss different applications where dielectric heating and induction heating are used. **[U.P. Technical Univ. Utilization of Elect. Energy and Traction 2006-07]**

35. Discuss various applications of dielectric heating.
[U.P. Technical Univ. Utilization of Elect. Energy and Traction 2009-10]

36. Write short notes on the following :
(*i*) Electric heating. **[M.D. Univ. May 2010]**
(*ii*) Resistance heating. **[G.B.T.U. 2010-11]**
(*iii*) Arc furnace.
(*iv*) Induction heating. **[G.B.T.U. 2010-11]**
(*v*) Coreless induction furnace.
(*vi*) Dielectric heating. **[W.B. Univ. of Technology 2009]**

Short Answer Type Questions With Answers

Q. 1. Why electric heating is preferred over other forms of heating ?

Ans. Electric heating is preferred over other forms of heating because of its numerous inherent advantages like economy, cleanliness, absence of flux gases, easy of control, automatic protection, upper limit of temperature, special heating needs.

Q. 2. What is an oven ?

Ans. Low temperature heating chamber with provision for ventilation is termed as oven.

Q. 3. Write and explain Joules law of electric heating.
[P.T.U. December 2006]

Ans. According to Joule's law of electric heating, the energy expended or heat generated in joules is given as

Work done or heat generated

$$= I^2Rt = \frac{V^2}{R}t = VIt = Wt \text{ joules.}$$

i.e., According to Joules law of heating, heat produced in an electric circuit of resistance R and carrying current I for *t* seconds is

(*i*) proportional to the square of the current *i.e.*, $H \propto I^2$

(*ii*) proportional to the resistance of the circuit *i.e.*, $H \propto R$, and

(*iii*) proportional to the time duration for which the current flows through the circuit *i.e.*, $H \propto t$

Heat produced in kcals,

$$H = \frac{I^2 Rt}{4{,}180} \simeq \frac{I^2 Rt}{4{,}200} \qquad [\because 1 \text{ kcal} = 4{,}180 \text{ J} \simeq 4{,}200 \text{ J}]$$

Q. 4. What properties are considered for selecting material for heating element ?

Ans. The properties to be considered in selection of material for heating elements are resistivity, melting point, temperature coefficient of resistance, and freedom from oxidation.

Q. 5. What are the essential properties of resistance heating elements ? **[P.T.U. Dec. 2008]**

Ans. The essential properties of resistance heating elements are : High resistivity, high melting point, low temperature coefficient of resistance and freedom from oxidation.

Q. 6. Name few materials, that are widely used as heating element materials.

Ans. The materials widely used as materials for heating elements are nickel-chromium (Nichrome) alloy, Ni-Cr-Fe alloy, nickel-copper alloy (Eureka or constantan), iron-chromium-aluminium alloy (Kanthal) and alloys containing iron, chromium, cobalt and aluminium.

Q. 7. What are the causes of failure of heating elements ?

Ans. The chief causes of failure of heating elements are formation of hot spots, oxidation and intermittency of operation, embrittlement due to grain growth, contamination and corosion.

Q. 8. How can you control the temperature inside the resistance furnace ? **[P.T.U. Dec. 2006]**

Ans. Refer to Art. 5.4.5.

Q. 9. Name few applications of infrared heating.

Ans. Infrared heating finds applications in (*i*) paint stoving (*ii*) drying of radio-cabinets and wood furniture (*iii*) preheating of plastics prior to moulding (*iv*) softening of thermo-plastic sheets, (*v*) drying of pottery, paper, textiles etc. where moisture content is not large.

Q. 10. Name the types of electrodes used in arc furnaces.

Ans. The electrodes used in arc furnaces are of three types namely carbon, graphite and self-baking electrodes.

Q. 11. Why are indirect arc furnaces not built in large sizes ? **[P.T.U. May 2009]**

Ans. The arc is produced between two electrodes and heat is transferred to the charge by radiation only. Such construction limits the use of single phase supply. The size of the furnace is thus limited by the amount of single-phase load, which can be taken from one point.

Q. 12. How is power input to submerged arc furnace controlled ?

Ans. The power input to submerged arc furnace is controlled by varying distance between electrodes or by varying the voltage applied to the electrodes.

Q. 13. How is control of power affected in electric arc furnaces ?

Ans. The power input to arc furnaces can be controlled by raising or lowering the electrodes or by varying the voltage applied to the arc furnace (by changing of tapping on supply transformer); preferably both voltage and electrode controls.

Q. 14. At what power factor would you like to operate the electric arc furnace ?

Ans. At power factor of about 0.9, but not below 0.8 in any case.

Q. 15. What type of supply is used for direct core type induction furnace in our country ?

Ans. Single-phase, 50 Hz, 600 V supply is used for direct core type induction furnaces.

Q. 16. How amount of heating is controlled in high frequency eddy current heating ?

Ans. The amount of heating is controlled by controlling the supply frequency and the flux density in high frequency eddy current heating.

Q. 17. What are the applications of high frequency eddy current heating ?

Ans. The applications of high frequency eddy current heating are surface hardening, annealing, soldering, welding, drying of paints, melting of precious metals, sterilization of surgical instruments and forging of bolt-heads and rivet heads.

Q. 18. How can the rate of dielectric heating be varied ?

Ans. Rate of dielectric heating can be varied by varying either supply voltage or supply frequency.

Q. 19. Why dielectric heating is employed only where other methods are impracticable or too slow ?

Ans. Because of very high cost of equipment required for dielectric heating.

Problems

1. A 4.5 kW, 200 V, 1-phase resistance oven is to have nichrome wire heating elements. If the wire temperature is to be 1,000°C and that of charge 500°C, estimate the diameter and length of wire. Resistivity of nichrome alloy = 42.5 μΩ-cm, radiant efficiency = 1.0 and emissivity of the element = 0.9.
[M.D. Univ. Utilization of Electric Power & Traction May 2009]
[**Ans.** 0.9072 mm; 13.5176 m]

2. Determine the length and diameter of the nichrome wire in the resistance element of a single phase electric furnace rated at 20 kW and 220 V. The wire temperature is not to exceed 1,170°C and the temperature of the charge is to be 500 °C. Take k = 0.57, *e* = 0.95 and ρ = 10 μΩ-cm.
[Pb. Univ. Drives and Utilization of Elec. Energy April 1981]
[**Ans.** 16.70 m; 3.1 mm]

3. A 40 kW, 3-phase, 400 V resistance oven is to employ Ni-Cr strip of 0.3 mm thickness. The heating elements are star-connected. If the wire temperature is to be 1,127°C and that of the charge is to be 727°C, estimate the suitable width and length of wire required. Take radiation efficiency = 60%; specific resistance of Ni-Cr = 1.03×10^{-6} Ω-m; emissivity = 0.9.
[Gorakhpur Univ. Utilization of Elec. Power & Traction 1982]
[**Ans.** 8,075 mm, 9.4 m]

4. A 30 kW, 3-phase, 400 V resistance oven is to employ nickel-chrome strip 0.025 cm thick for the 3-phase star-connected heating elements. If the wire temperature is to be 1,100 °C and that of charge is to be 700°C, estimate a suitable width for the strip. Assume radiating efficiency as 0.6 and emissivity as 0.9. The specific resistance of the nichrome-alloy is 1.03×10^{-6} Ω-m. State any assumption made.
[A.M.I.E. Sec B. Utilisation of Elec. Power Summer 1982]
[**Ans.** 6.86 mm]

5. In a 3-phase arc furnace to melt 10 tonne steel in 2 hours, estimate the average input to the furnace, if overall efficiency is 50%. If the current input is 9,000 A with the above kW input and the resistance and reactance of furnace leads (including transformer) are 0.003 Ω and 0.005 Ω respectively, estimate the arc voltage and total kVA taken from the supply.
Specific heat of steel = 444 J kg^{-1} $°C^{-1}$
Latent heat of fusion of steel = 37.25 kJ/kg
Melting point of steel = 1,370 °C
[Pb. Univ. Elec. Drives & Utilization of Electric Energy Nov., 1983]
[**Ans.** 38.5 V; 2,146 kVA]

6. Estimate the energy required to melt 0.5 tonne of brass in a single-phase induction furnace. If the melt is to be carried out in 0.5 hour, what must be the average power input in the furnace. **[West Bengal Univ. of Technology Utilization of Electric Power 2009]**
[**Hint:** Assume room temperature to be 20°C, specific heat of brass 393.6 J/kg/°C, latent heat of fusion 163,000 J/kg, melting point of brass 920° and furnace efficiency 70%]
[**Ans.** 102.625 kWh; 205.25 kW]

7. Estimate the efficiency of high frequency induction furnace which takes 12 minutes to melt 2.5 kg of aluminium, the input to the furnace being 3.75 kW. The initial temperature is 20°C, specific heat of aluminium 880 J kg^{-1} C^{-1}, melting point 650°C, latent heat of fusion 32 kJ/kg.
[B.H.U. Traction & Utilization 1978] [**Ans.** 54.3%]

8. A low frequency induction furnace operating at 12 V in the secondary circuit takes 48 kW at 0.5 pf when the hearth

is full. If the secondary voltage be maintained at 12 V estimate the power factor and power absorbed when hearth is half-full. Assume the resistance of the secondary circuit to be thereby doubled and the reactance to remain the same.

[M.D. Univ. Utilization of Elec. Power and Traction May 2010]

[Ans. 0.756 (lagging); 548.584 kW]

9. A low frequency induction furnace, whose secondary voltage is maintained constant at 10 volts, takes 400 kW at 0.6 power factor when the hearth is full. Assuming the resistance of the secondary circuit to vary inversely as the height of the charge and the reactance to remain constant, determine the height up to which the hearth should be filled to obtain maximum heat.

[Gorakhpur Univ. Utilization of Electrical Power & Traction Sept., 1981]

[Ans. Three-fourth of height of hearth]

10. An insulating slab 2 cm thick and 150 cm^2 in area is to be heated by dielectric heating. The power required is 300 watts at a frequency of 30 MHz. The material has a relative permittivity of 5 and pf of 0.05. Determine the voltage necessary and current that will flow through the material. If the voltage is limited to 600 V, calculate the frequency that will produce the same loss.

[Pb. Univ. Electric Drives & Utilization of Electric Energy Nov., 1980]

[Ans. 980 V; 6.12 A; 80 MHz]

11. A piece of plastic material of size 4 × 2 × 1 cm is heated by being placed between two electrodes, each having an area of 20 × 2 cms and the distance of separation between two being 1.6 cm. The frequency of voltage impressed across the electrodes is 20 MHz. If the power consumed is 80 watts, find the voltage applied across the electrodes and current through the material. Assume relative permittivity as 5 and power factor 0.05.

[Gorakhpur Univ. Utilization of Electrical Power & Traction 1981]

[Ans. 2,189 V; 0.73 A]

12. A 3-phase electric arc furnace has the following data :

Current drawn = 5,000 A

Arc voltage = 50 V

Resistance of transformer referred to secondary = 0.002 Ω

Reactance of transformer referred to secondary = 0.004 Ω

(*i*) Calculate the pf and the kW drawn from the supply.

(*ii*) If the overall efficiency of the furnace is 65% find the time required to melt 2 tonnes of steel when latent heat of steel = 8.99 kcal/kg; specific heat of steel = 0.12, melting point of steel = 1,370ºC and initial temperature of steel = 20ºC.

[A.M.I.E. Sec. B. Utilization of Elec. Power Winter 1976]

[Ans. 0.9487; 900 kW; 40 minutes 46 seconds]

13. A 3-phase arc furnace takes 7,500 A when the arc voltage is 75 V. The circuit comprising the transformer and the electrodes excluding the arc has a reactance of 0.003 Ω and a resistance of 0.002 Ω. The steel has a specific heat of 278 J/kg/ºC; a latent heat of fusion 37,000 J/kg and melting point 1,370ºC.

Assuming furnace overall efficiency as 50% and initial temperature of steel as 15ºC, determine

(*i*) the time taken to melt 5 tonnes of steel, (*ii*) the kVA taken from the supply and (*iii*) the electrical efficiency.

[Ans. (*i*) 2,040 seconds (*ii*) 2,088 kVA (*iii*) 83.3%]

14. Determine the efficiency of a high frequency induction furnace, which takes 10 minutes to melt 1.815 kg of aluminium, the input to the furnace being 5 kW and the initial temperature is 15ºC.

Specific heat of Al = 0.212

Melting point of Al = 660ºC

Latent heat of Al = 76.8 kcal/kg. **[Ans.** 54%]

15. A slab of insulating material 130 cm^2 in area and 1 cm thick is to be heated by dielectric heating. The power required is 380 W at 30 MHz. Material has a relative permittivity of 5 and pf of 0.05. Absolute permittivity = 8.854×10^{-12} F/m. Determine the necessary voltage.

[Rajasthan Univ. Utilization of Electrical Power 2008]

[Ans. 837 V]

❑❑❑

CHAPTER 6

Electric Welding

INSIDE THIS CHAPTER

6.1. Introduction **6.2.** Resistance Welding **6.3.** Electric Arc Welding **6.4.** Ultrasonic Welding **6.5.** Electron Beam Welding **6.6.** Laser Beam Welding **6.7.** Requirements of Good Weld **6.8.** Preparation of Work **6.9.** Electrodes **6.10.** Power Supply For Resistance Welding **6.11.** Machines For Resistance Welding **6.12.** Electronic Control in Resistance Welding **6.13.** Power Supply For Arc Welding **6.14.** Electric Welding Equipment **6.15.** Arc Welding With DC and AC **6.16.** Comparison Between Resistance and Arc Welding.

6.1. INTRODUCTION

Welding is a materials joining process that produces coalescence* of materials by heating them to the welding temperature with or without the application of pressure or by the application of pressure alone, and with or without the use of filler metal. It is used to make welds. A weld is "a localized coalescence of metals or non-metals produced either by heating materials to the welding temperature, with or without the application of pressure, or by the application of pressure alone and with or without the use of filler metal."

All metals can be joined by one welding process or another. There is a saying, "If it's metal, weld it," and it is certainly true. This should be qualified by stating that all metals commercially employed for structural or strength parts are weldable. Some metals are easy to weld, and others are difficult to weld. The metals that are easily weldable can be welded in thickness from the very thinnest, about the thickness of this paper page, to the thickest or heaviest produced. The difficult-to-weld metals require special procedures and techniques that must be developed for specific applications. Some metals may never be welded or joined. For example, mercury is a liquid at room temperature and cannot be welded, while sodium and potassium melt just below the temperature of boiling water and are of no use as a strength member and cannot be welded. In general, metals that have a low melting point or low strength would not be welded. Some metals are so scarce or expensive that they would not be used where welding would be required. The physical and mechanical properties (such as melting temperature, density, thermal conductivity, tensile strength, ductility etc.), availability, and price all help in determining if a metal will be used in applications where welding is required.

6.1.1. Importance of Welding

Welding is the most efficient way to join metals. It is the only way to join two or more pieces of metal to make them act as one piece. Welding is widely employed to manufacture or repair all products made of metal. Look around, almost every thing made of metal is welded; the world's tallest building, moon rocket engines, nuclear reactors, home appliances, and automobiles barely start the list.

The use of welding is still increasing. If a joint is welded, it is permanent joint. Obviously, if the joint must be disassembled occasionally, it should not be welded. Thus we should change our statement to "welding is the most economical method to permanently join metal parts." To join two members by bolting or riveting requires holes in the parts to accommodate the bolts or rivets. These holes reduce the x-sectional area of the members to be joined by up to 10 per cent. The joint may also require the use of one or two gusset plates, thus increasing the weight of material required and the cost. This expense can be eliminated by the use of a weld. The greatest economy of a welded design will be obtained if the x-sectional area of the entire structural member is reduced by the amount of the bolt holes. This can be done since the entire cross section of a member of a welded design is utilized to carry the load. The amount of material required is reduced as well as its cost. This same design concept applies to joining plates used to build a ship or a container. In view of this material savings, ships and storage tanks are no longer riveted.

Pipes joined by welding offer similar economies. The wall thickness of a pipe should be heavy enough to carry the required load. However, if the pipe is joined by screw threads a heavier wall thickness is used to allow for cutting away a portion of the thickness for the threads. A thinner pipe wall thickness is used for the entire welded pipe system. This reduces the amount of metal required and the cost. The inside surface of the welded joint is smoother. Larger diameter pipes are no longer connected together with screw threads and pipe fittings.

Converting castings to weldments allows the designer to reduce weight by reducing metal thickness. Welding is a design concept which allows freedom and flexibility not possible with

* Coalescence means the growing together or growth into one body of the materials being welded.

cast construction. Heavy plates can be used where strength is required and thin ones can be used where possible. The uniform thickness rule and minimum thickness required for foundry practice are not necessary for weldments. Additionally, high-strength materials can be employed in specific areas, while normal-strength materials are used where required.

Welding is the best way to protect and conserve materials by protecting their surface with special metal overlays. Corrosion and wear of metals account for losses running into billions of dollars annually. Together they are responsible for an untold loss of lives. Waste from both of these destructive forces can be largely reduced by welding. Special alloys are weld deposited on base metals to provide corrosion resistance surfaces. Hard surfacing overlays can be made by welding to provide special alloys with wear resistant surfaces. A typical application is the resurfacing of a cinder crusher roll with hard weld metal. Weld surfacing is used to reduce the costly abrasive and corrosive wear of machinery.

Thus in short we can say, "today, the science and art of welding encompasses a wide range of processes and procedures applicable to materials of any thickness and shape—from tiny electronic components to mammoth machines and structures. Welding has opened up prospects for radical improvements in the manufacture of a large variety of machines, devices and structures. It has promoted mechanization and automation in their manufacture. The use of automatic, semi-automatic and mechanized welding equipment has relieved a large number of workers for employment elsewhere in the national economy. Welding is nowadays extensively used in automobile industry, aircraft machine frames, structural work, tanks, machine repair work, shipbuilding, pipeline fabrication in thermal power plants and refineries, fabrication of metal structures."

6.1.2. Advantages and Disadvantages of Welding

Some of the advantages of welding are given below :

1. Welding is the lowest-cost joining method.
2. It affords lighter weight through better utilization of materials.
3. It joins all commercial metals.
4. It can be used anywhere.
5. It provides design flexibility.

The limitations of welding are :

1. Some welding depends on the human factor.
2. It often needs internal inspection.

Most of these limitations can be overcome by means of good controls and supervision.

Welding is also an economical manufacturing method wherever there is a need to join parts permanently. In the high-volume production industries it is common to see welding operations intermixed with bending, machining, forming, assembly and so on. Welding is an important manufacturing process taking its place with other metal working operations to help bring us good-quality products at economical prices.

6.1.3. Personnel Protection and Safety Rules

There are a number of safety and health problems associated with welding. When correct precautionary measures are followed, welding is a safe occupation. The hazards that are more or less peculiar to welding are electrical shock, arc radiation, air contamination, fire and explosion, compressed gases, welding cleaning and other hazards related to specific processes or occupations.

Combustible materials must not be allowed to collect in or near the welding workplace. Good housekeeping practices should always be employed in the welding shop. Adequate safety devices should be provided, such as fire extinguishers, life-saving and support equipment, first-aid kits, and so on, plus the training of personnel to utilize this equipment properly. Only approved equipment be used, and it must be properly installed and maintained in good working conditions.

6.1.4. Classification of Welding Processes

There are many ways to make a weld and there are many different kinds of welds. The welder behind the hood making sparks is using one of the more popular welding processes, known as *arc welding.* Some welding processes do not cause sparks; in some cases electricity is not used, and in some cases there is not even extra heat. Welding has become complex and technical. It needs considerable knowledge to select the proper welding process for critical work.

Modern methods of welding may be classified under two broad headings—plastic welding and fusion welding. They are also called pressure welding, and non-pressure welding respectively. In the *plastic welding* or *pressure welding*, the pieces of metal to be joined are heated to a plastic state and then forced together by external pressure. This includes forge welding, resistance welding, thermit welding and gas welding, in which pressure is required. In the *fusion welding* or *non-pressure* welding, the material at the joint is heated to a molten state and allowed to solidify. This includes gas welding, arc welding and thermit welding without pressure.

The various welding processes used in general engineering are given below :

1. Gas Welding — (*i*) Oxyacetylene (*ii*) Air-acetylene and (*iii*) Oxy-hydrogen.
2. Resistance Welding — (*i*) Butt (*ii*) Spot (*iii*) Projection (*iv*) Seam and (*v*) Percussion.
3. Arc Welding — (*i*) Carbon arc (*ii*) Metal arc (*iii*) Gas metal arc (*iv*) Gas tungsten arc (*v*) Atomic hydrogen arc (*vi*) Plasma arc (*vii*) Submerged arc (*viii*) Flux-cored arc and (*ix*) Electro-slag.
4. Thermit Welding.
5. Solid State Welding — (*i*) Friction (*ii*) Ultrasonic (*iii*) Diffusion and (*iv*) Explosive.
6. Newer Welding — (*i*) Electron beam and (*ii*) Laser.

Here we will discuss only about electric welding (Resistance welding and arc welding) and some introduction to other modern welding techniques.

6.2. RESISTANCE WELDING

By definition, *resistance welding is that process in which a sufficiently strong electric current is sent through the two metal pieces in contact to be welded which melts the metals by the resistance they offer to the flow of the electric current.*

Resistance welding includes butt welding, spot welding, projection welding, seam welding and percussion welding. All are alike in the principle of resistance heating but differ in the details of application.

In resistance welding a heavy current (above 100 A) at a low voltage is passed directly through the workpiece and heat developed by the resistance to the flow of current, given by the expression I^2Rt (where I is the current in amperes, R is the resistance in ohms and t is the time or duration of flow of current in seconds) is utilised. The heat developed at the contact area between the pieces to be welded reduces the metal to a plastic state; the pieces are then pressed together to complete the weld. In this process, preferably two copper electrodes are incorporated in a circuit of low resistance and the metals to be welded are pressed between the electrodes. The electrical voltage required ranges from 4 to 12 volts depending on the composition, area, thickness etc. of the metal pieces to be welded. The amount of power supplied to the weld usually ranges from about 60 watts to 180 watts for each sq. mm of area. *Alternating current is found to be most suitable for resistance welding as it can provide any desired combination of current and voltage by means of a suitable transformer.*

In order to avoid the surface distortion, the portion of the metal adjacent to the weld or joint should not be allowed to be overheated.

Resistance to the flow of current is made of (*i*) resistance of the current path in the work (*ii*) resistance between the contact surfaces of the parts being welded and (*iii*) resistance between the electrodes and the surface of the parts being welded. In order to develop higher temperature between the interfaces of the work to be welded rather than at the surface of the work in contact with the electrodes it is necessary to keep the resistance between the electrodes and the surface of the body being welded to minimum.

In order to obtain a good weld it is necessary to maintain the contact resistance uniform which depends upon the surface condition.

For welding thin materials the resistance of the current path in the work is kept minimum. For welding thick materials of low conductivity the resistance of the current path have a comparatively greater value and the control of contact resistance is not necessary. For welding thick materials of high conductivity either reduced pressure or high resistance electrodes having melting point higher than that of metal to be welded, can be used. For welding two dissimilar metals having different conductivity, low conductivity electrodes on high conductivity metal side and vice versa are used in order to prevent overheating on the low conductivity metal and to develop sufficient heat to melt high conductivity metal side.

The pressure which is to be applied on the weld is also an important factor. At high pressure, low temperature plastic welds can be obtained and where as if the pressure is lowered the resistance to the welding current is to be increased. There is a limit up to which the resistance can be increased and after that there will be surface burning. The pressure necessary to effect the weld varies from 2.5–5.5 kgf/mm^2.

The magnitude of current is controlled by varying either the primary voltage of the welding transformer (by using auto–transformer between supply and the welding transformer) or changing the primary turns of the welding transformer. Alternative method of controlling the current to weld is to vary the magnitude and wave of the primary as well as secondary current by using Thyratron or Ignitron tubes in the primary circuit.

In resistance welding, the time for which current flows is very important. Usually automatic arrangements are devised which switch off the supply after a predetermined time from applying of pressure (starting of weld). The pressure may be applied manually, by air pressure, by springs or by hydraulic means. After switching off the supply, the pressure is maintained on the electrodes until the weld cools. In machines which are operated continuously, the electrodes are cooled by water circulating through hollow electrodes.

Electrical circuit diagram for resistance welding is shown in Fig. 6.1. The machine employed for resistance welding contains a transformer provided with necessary taps, a clamping device for holding the metal pieces, and a mechanical means for forcing the pieces, to be welded, together to complete the weld.

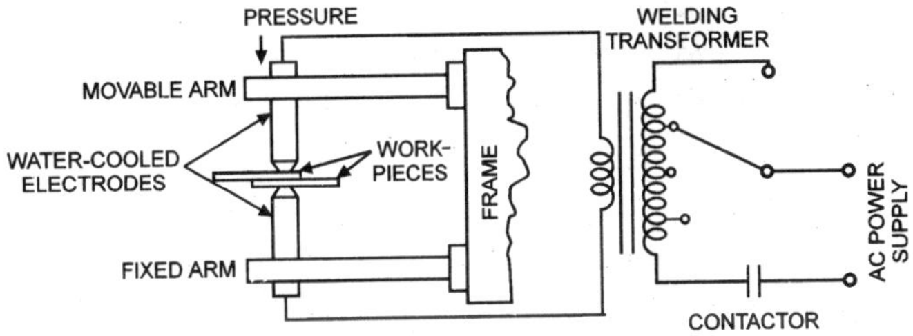

Fig. 6.1. ***Electrical Circuit For Resistance Welding***

Resistance welding has the advantage of producing a large volume of work at high speeds that are reproducible with high quality. Resistance welds are made very quickly; however, each process has its own time cycle. Resistance welding operations are automatic. Good quality welds do not depend on welding operator skill but more on proper set-up and adjustment of the equipment and adherence to weld schedules.

Resistance welding is employed mainly for mass production. It is easily adapted to those components which can be moved to the machine and are light. The operation is extremely rapid and simple. This is the only process where heat can be controlled and which permits a pressure action at the weld. Metals of medium and high resistance, such as steel, stainless steel, monel metal and silicon bronze are easy to weld. Special control gear is required, however, in the case of high-carbon steels and special equipment providing very high current impulses (stored energy welding) is used in case of materials of low electrical resistance.

The automotive industry is the major user followed by the appliance industry. It is used by many industries manufacturing a variety of products made of thinner-gauge metals and for manufacturing pipes, tubing, and smaller structural sections.

When specifying the material intended to be resistance welded, consideration must be given to the state in which this to be supplied to the welding shop. Whilst slight rust, mill scale etc., on the material may not affect the efficiency of arc welds to a considerable degree, lack of cleanliness will be fatal to resistance welded joints. Pickling or shot-blasting immediately prior to the resistance welding operation is essential for making this latter method a success.

Material up to 5 mm thickness which is to be used on resistance-welded jobs is usually purchased in a pickled and slightly oiled condition, and should be carefully stored in order to keep it clean. It can then be used without removing the oil film provided that the oil is clean. Material above 5 mm thickness should be shot-blasted before being taken to the resistance welding machines. No long delay should occur between the shot blasting and the welding, in order to avoid new corrosion which might eliminate the advantage gained by the former. Sand-blasting is not recommended, as particles of the siliceous material may be embedded in the steel surface and influence its electrical resistance.

High frequency resistance welding is done with 400 to 450 kHz current usually supplied by an oscillator. The high frequency current readily breaks through oxide film barriers and produces a thin heat-affected zone because it travels on the surface of the material.

However, resistance welding have got some limitations and drawbacks also, as enumerated below :

(*i*) The initial cost of equipment required is high.

(*ii*) Skilled persons are required for the maintenance of equipment and its controls.

(*iii*) In some materials, special surface preparation is required.

(*iv*) Certain resistance welding processes are limited to lap joints. A lap joint has an inherent device between the two metal pieces, which cause stress concentration in applications where fatigue is present. The device may also cause trouble when corrosion is present.

6.2.1. Butt Welding

Welding of two pieces of metal together with a butt weld can be carried out by different resistance welding processes. The butt weld consists of joining of two pieces of metal together either on face or on edge. It is of two types namely (*a*) upset butt welding and flash butt welding.

(*a*) *Upset Butt Welding*. It is a resistance welding process that produces coalescence over the entire area of faying surfaces or progressively along a butt joint by the heat obtained from the resistance to the flow of welding current through the area where those surfaces are in contact. It can be used only if the parts to be welded are equal in cross-sectional area.

End to end welds, lap welds, and butt welds are included. The faces of the metal pieces to be joined are prepared for even contact. The metal pieces to be joined are fixed in clamps and butted squarely against each other, as shown in Fig. 6.2 and heavy current is passed through them. In this case heating is obtained by the contact resistance of the metals to be welded. When sufficient heat is developed, the pieces are rammed by force to complete the weld. Pressure is applied either manually or with toggle mechanism. The pressure applied upsets or forges the parts together. The joint is cooled under pressure. After cooling, the force is released and the weld is completed. Upset butt welding, which has the advantage of creating no flash spatter and a smooth and symmetrical upset, is used principally on non-ferrous materials for welding bars, rods, wire, tubing, formed parts etc. The voltage required for welding is 2–8 volts and current varies from 50 A to several hundred amperes, depending upon the material and the area to be welded at a time. The current, pressure and time may be manual or automatic controlled. The latter method is preferred in production machines.

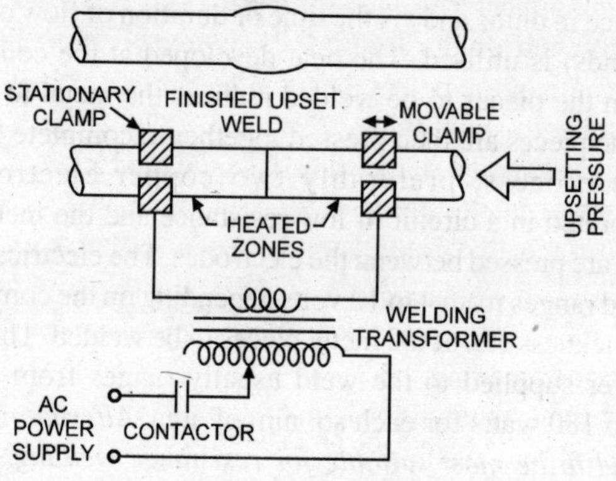

Fig. 6.2. *Upset Welding Process*

If intimate contact is not obtained because of improper joint preparation, the weld will be defective.

(*b*) *Flash Butt Welding*. Even though it is not a resistance welding process, it is desirable to consider it here, as in this method the two metal pieces are butted together and a current is passed through the joint which heats the ends to red hot. Flash welding may be defined as a resistance welding process that produces coalescence at the faying surfaces of a butt joint by flashing action and by application of pressure after heating is substantially completed (Fig. 6.3).

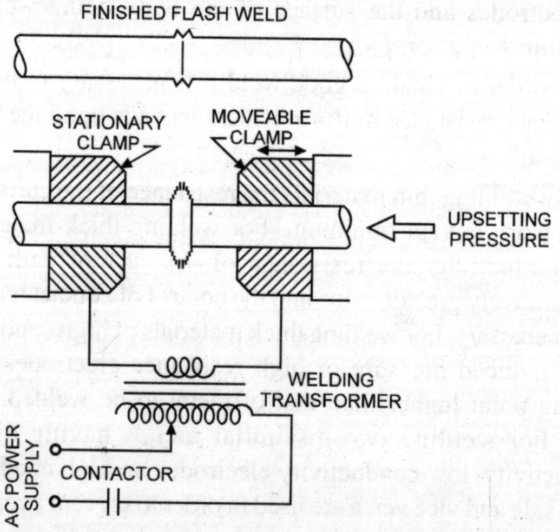

Fig. 6.3. *Flash Butt Welding Process*

In this process of welding no special preparation of the faces to be welded is necessary. In flash butt welding the voltage to the metal parts to be joined together is applied before putting them together. Protruding small parts of the faces to be welded are brought into contact; the heat developed owing to high resistance of the contact areas raises the temperature of the material at these points and causes them to melt. The molten metal is blown out, and a small arc is formed which raises the temperature of adjoining parts of the abutting surfaces. The ejection of molten metal from high spots continues until the whole of the surfaces are heated to the welding temperature. Some additional heat may be created by the combustion of molten particles, the forceable action of which is probably due to the pressure of metallic vapour. The flashing action is aided by an oscillating movement of one of the components. As soon as the whole of the surfaces to be welded reach forging temperature, the pressure is applied, and the power supply is cut off.

Flash butt welding creates a joint which has practically the full strength of the parent metal under static loading conditions and a strength slightly less than that of the parent metal under dynamic loads. The shape and area of the two components must be identical at the point to be joined, and a suitable upset allowance must be made.

Flash butt welding can be used on most metals (low carbon steels, tool steels, stainless steels, aluminium alloys with thickness exceeding 1.25 mm, copper alloys with high zinc content, magnesium alloys, molybdenum alloys, nickel alloys, titanium alloys, medium strength and high strength low alloy steels) leaving cast iron, lead, zinc, antimony alloys and bismuth. Flash butt welding is considered superior to upset butt welding on the following accounts.

(*i*) Many dissimilar metals with different melting temperatures can be flash welded.

(*ii*) Flash weld provides a smaller upset.

(*iii*) Flash welding is faster than other methods.

(*iv*) All the foreign metals appearing on the butting surfaces is burnt due to first few flashes preceding application of welding pressure. Thus clean and pure weld is obtained.

(*v*) Power requirement is less.

(*vi*) Flash welding can be used for the welding of those highly alloyed steels which cannot be welded satisfactorily by any other process. This is because, in flash welding, under correctly controlled conditions, the heating is not only even but extremely local; so that the cooling stresses are maintained at a minimum; this avoids hardening and cracking in highly alloyed steels.

Flash welding have some limitations and drawbacks also, as enumerated below :

(*i*) The most undesirable feature of flash welding is the flashing operation during which it is impossible to protect the welding machine and the surrounding area from these particles, which can burn into sideway bearings, insulation etc. This necessitates more frequent maintenance.

(*ii*) The process presents a considerable fire hazards. Operators need to be protected from flying particles.

(*iii*) Metal is lost during flashing and upsetting.

(*iv*) Concentricity and straightness of the workpieces during welding process is often difficult to maintain.

Applications of Flash Welding. (*i*) Flash welding is applied primarily in the butt welding of metal sheets, tubing, bars, rods, forgings, fittings etc.

(*ii*) Flash welding finds applications in automotive and aircraft products, household appliances, refrigerators and farm implements.

(*iii*) The process is also used for welding the band-saw blades into continuous loops, and joining of tool steel drill, tap and reamer bodies to low carbon steel and alloy steel shanks.

(*iv*) Flash welding is used to produce assemblies that otherwise would require more costly forgings or castings.

Difference Between Flash Welding and Upset Butt Welding. (*i*) In upset butt welding no arcing (hence flashing) occurs between the surfaces being joined. Heat is produced solely by the electrical resistance at the abutting surfaces to the passage of an electric current.

(*ii*) In flash welding, unlike upset butt welding, the movable platen keeps on moving constantly toward the stationary platen.

(*iii*) Flash welding consumes much less welding current than consumed by upset butt welding process; the time allowed for weld to be completed is, however, more.

(*iv*) In flash welding heat application precedes the pressure where as in upset butt welding constant pressure is applied during the heating process which eliminates flashing.

6.2.2. Spot Welding

By definition, *spot welding is a form of resistance welding in which the parts or pieces are joined in spots, accompanied by heated relatively small sections of the parts or pieces between suitable electrodes under pressure.*

Spot welding is perhaps the most widely used form of resistance welding. It is characterized by low cost, speed, and dependability making it a common electrical resistance welding process.

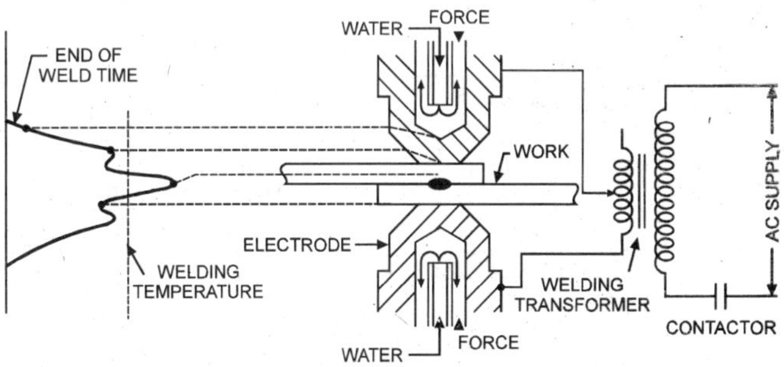

Fig. 6.4. *Spot Welding*

Spot welding is usually used for joining or fabricating sheet metal structure. This type of welding only provides mechanical strength and is neither airtight nor watertight. For obtaining good welds having strength, and to avoid the excessive heating effects due to foreign filmson sheet surfaces, it is desirable to clean the sheets thoroughly before welding. In this type of welding the workpieces to be welded are pressed together by mechanical pressure exerted through electrodes (made of pure copper or alloy of copper) as shown in Fig. 6.4 and then supply is switched on for a particular period. The mechanical pressure may be developed by a foot lever or by air pressure or by hydraulic cylinders. The metal in the zone of pressure gets up heated to fusion and the joint thus made gets cooled under pressure. The pressure remains applied till the weld cools and regains sufficient strength. Current required is above 5,000 A and the voltage between the electrodes is usually less than 2 volts. The open-circuit voltage is less than 12 volts. The time period of flow of current varies widely depending upon the thickness of sheets, kinds of metal etc. For thin sheets this period is about 1/50 second for each 0.25 mm of total thickness of the two sheets to be joined.

In order to avoid electrode sticking it is necessary to avoid heating of junction faces between electrodes and work. This is achieved (*i*) by using water-cooled electrodes and (*ii*) by employing electrodes made of material having high electrical as well as thermal conductivity. In order to minimise the wear and tear at the tip of the electrodes it is necessary to employ electrodes made of material having high mechanical strength.

Efficient spot welds are possible only by a correct combination of current, pressure, and welding time. The heat developed being proportional to the product of the welding time and the square of the current, it is possible to obtain good welds by employing either low currents for a longer duration or higher currents for a shorter duration. Longer welding time usually produces stronger welds but this involves lot of distortion of workpieces, electrode maintenance and high energy expenditure. The welding pressure required is determined on the basis of a specific electrode pressure of 7 kg per sq. mm over the electrode tip area. In view of the fact, however, that plates will not always be perfectly flat, a certain additional pressure is usually recommended, which will overcome slight distortion of the plates without the resultant necessary welding pressure.

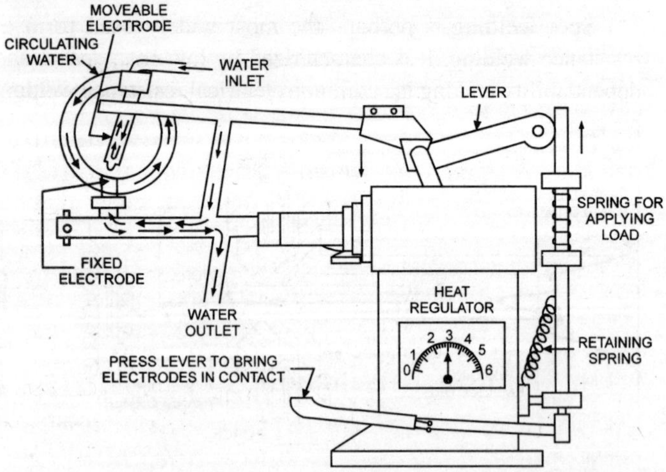

Fig. 6.5. ***Spot Welding Machine***

For obtaining good joints spot welding machine (Fig. 6.5), which makes the welding operation independent of the operator by automatically controlling welding current, pressure, and timing, once the same have been set correctly, may be employed. By toggle mechanism electrodes are pressed on the work and according to the requirement the current can be turned on and off. In certain machines the electrodes of spot welders are equipped with water cooling arrangement. In such cases the electrodes are made hollow and the water is circulated through them. Spot welding machines are of capacities ranging from 10 to 150 kVA.

Spot welding is applicable for joining components made from plate material, the plate thickness being limited (usually to 10 or 12 mm) by the pressure and current capacity of the available spot welding machine. It has an advantage over projection welding in cases where the number of welds required is relatively small, where the size of the job is too large for the capacity of projection welders, and where the design allows the necessary spacing and arrangement of the welds to be obtained. The lower cost of the equipment makes spot welding economical for small quantities.

6.2.3. Projection Welding

Projection welding is a resistance welding process that produces coalescence by the heat obtained from the resistance to the flow of the welding current. The resulting welds are localized at predetermined points by projections, embossments or intersections.

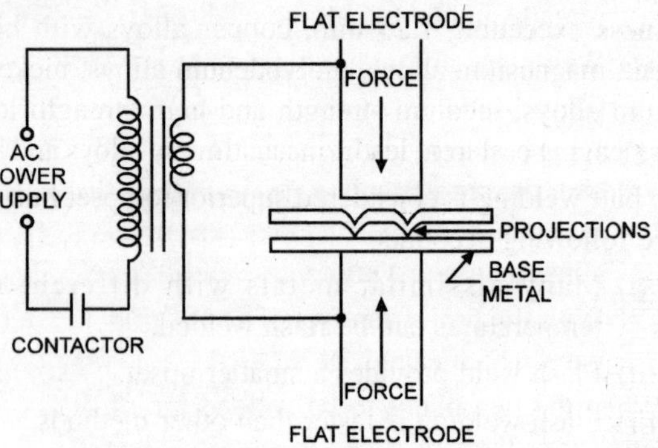

Fig. 6.6. ***Projection Welding***

Projection welding uses the same equipment as spot welding. The only difference is that the electrodes used are flat on the ends and larger in diameter than the flat electrodes used in spot welding. Successful projection welding depends greatly on the surface preparation of the pieces to be welded. Projections, small deformations that will touch the surface of the material to be welded, are made on the weld areas. One of the main advantages of these projected points is that welding area can be located easily, which makes projection welding a high-production welding technique. As the current flows through the two parts to be welded, the projected points are the main contact area. Whenever a point or projection touches the metal, a weld nugget will start. These weld points soon reach the plastic state, and the force applied by the electrodes finishes the weld nugget. The cycle time is the same as the spot-weld time. As the points reach their plastic state, the metal is compressed so that the finished weld is similar to the spot weld with the exception of the small indentations created by projections. Projection welding, as spot welding, needs no protective atmosphere in order to produce successful results. It is also useful in securing small circular component, to larger pieces of metal. For projection welding, a large flat-surfaced electrode is used, which covers the areas of the projections. The areas of projection, one or more, and the spots where they touch on the other piece of metal will be the only places where fusion takes place, even with a large electrode. Projection welding has been used by auto manufacturers for many years. Auto bodies have many areas that are projection welded. Another area that uses projection welding is stud welding, which has also been utilized for car bodies.

Projection welding reduces the amount of current and pressure required in order to form a good bond between two surfaces. With this reduction there is less chance of shrinkage and distortion in the areas surrounding the weld zone, one reason why it has been incorporated into many manufacturing processes.

Projection welding has the following advantages over spot welding :

(*i*) Projection makes the welding process simple.

(*ii*) It is possible to join several welding points in projection welding.

(*iii*) More output is obtained since more than one weld are done at a time.

(*iv*) Electrode life is increased due to use of low current density and low pressure.

(*v*) Good finished appearance is obtained as surface remains unidented by the electrodes.

(*vi*) Welds are automatically located by the position of projections.

(*vii*) As the electrodes are flat, therefore the contact area over the projection is sufficient. It facilitates the easier maintenance of electrodes, alignments of parts and construction of various jigs and fixtures.

(*viii*) With the projection welding, it is easy to weld certain parts which cannot be welded by spot welding.

When the two plates to be welded are of different cross-sectional area then in order to obtain desirable strength it becomes necessary to have the projections on the thicker plate.

When the two plates to be welded are of different material then the projections should be made on the high conductivity materials.

The thickness of plates to be jointed is limited to 2 × 6 mm in case of steel, 2 × 4 mm in case of aluminium, and 2 × 2 mm in case of copper. This reduction in maximum thickness of different metals is due to increase in conductivity.

The initial cost of the equipment required for projection welding is high as press type machines are required for this type of welding.

This type of welding is usually employed on punched, formed or stamped parts, where the projections automatically exist.

Small fasteners, nuts etc. can be welded to larger components.

Projection welding is particularly applicable to mass production work, *e.g.*, the welding of refrigerator condensers, crossed-wire welding, refrigerator racks, gratings, grills etc.

6.2.4. Seam Welding

Seam welding can be defined as series of continuous spot welds. This process is employed for making a continuous joint between two overlapping pieces of sheet metal. In this process the work to be welded is placed between the two wheels which apply sufficient mechanical pressure and also carry sufficient current for producing continuous welds. The mechanical pressure applied is kept constant and the current is regulated by a timer (the current is kept on for a definite time and then off for another fixed interval). As the pressure is applied, the drive is started, the welding current switched on and simultaneously the overlapping surfaces of the metal are forced together as fast as they are heated. The heat is generated due to flow of current through the resistance in the welding circuit. The heat generated can be controlled by either varying the current or pressure between the sheets which varies the contact resistance. If the rate of heat generation is high then the speed of the rollers is increased thereby reducing the welding time and vice versa. The electrodes are made of copper alloys and are water circulated in order to dissipate heat from them. In high speed seam welding using continuous current the frequency of the current acts as an interrupter. In order to localise the current and pressure to the welding point, the contact area of the electrodes is kept small. Usually number of spots obtained per metre of the joint are between 200 and 400 depending upon the nature of the joint required. Welding speed ranges from 0.25 metre to 10 metres per minute. This type of welding is very important as it provides pressure tight or leak proof joint.

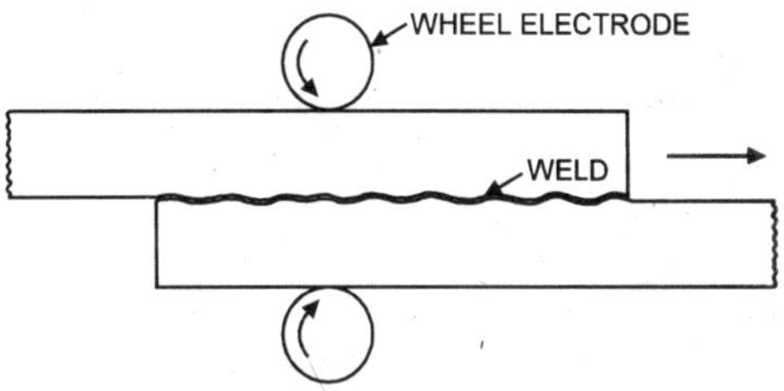

Fig. 6.7. ***Seam Welding***

It is very essential in case of seam welding that the surfaces to be welded are cleaned and made dust and dirt free. This is not necessary in other forms of resistance welding, *e.g.*, in spot welding the detrimental effect of scale and surface dirt can be compensated for by adjustment of time, pressure, current or by other means but this is not true of seam welding. Materials which can be seam welded after they are cleaned include high-carbon, stainless and coated steels and alloys of aluminium, nickel and magnesium. Seam welding of copper and high copper alloys is not recommended. Seam welding is employed for welding pipes, conduits, tanks, transformers, refrigerators, gasoline tanks, aircrafts and various types of containers. This process is best adopted for metal thickness ranging from 0.25 mm to 3 mm.

6.2.5. Percussion Welding

It is a recent development in welding. It is a resistance welding process where in coalescence is produced simultaneously over the entire area of abutting surfaces by heat obtained from an arc produced by a rapid discharge of stored electrical energy. This is a self-timing spot welding method.

In this process, the pieces to be welded are held apart, one in a stationary holder and the other in a clamp mounted in a slide and backed up against heavy spring pressure. When the movable clamp is released, it moves rapidly carrying with it the piece to be welded. When the two pieces are about 1.5 mm apart, there is a sudden discharge of electrical energy, causing intense arcing between the two surfaces and heating them to high temperature. As pieces come in contact with each other under heavy pressure the arc is extinguished due to the percussion blow of the two parts and the force between them effects the weld.

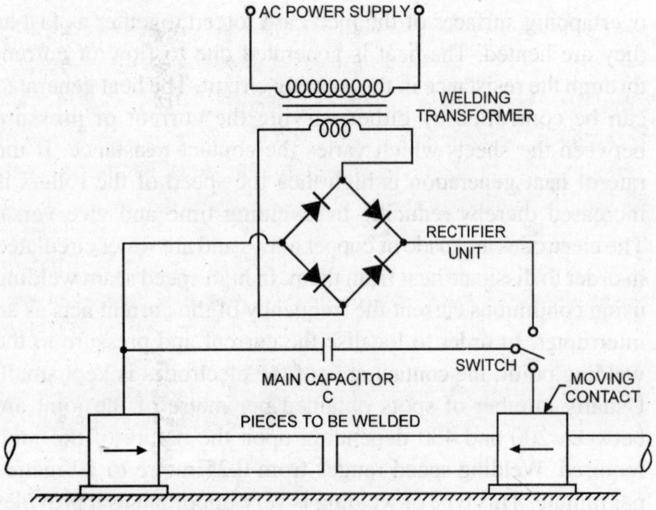

Fig. 6.8. ***Percussion Welding***

This process is quite similar to flash welding and upset welding, but is limited to parts of the same geometry and cross section. It is more complex than the other two processes. The advantage of the process is that there is an extremely shallow depth of heating and the time cycle is very short. This process is so fast (takes about 0.1 *s*) that there is little heating effect in the material adjacent to the weld. It may be noted that this process is limited to only small areas (up to 3.2 cm^2) of nearly regular sections. Thin sheets of equivalent area cannot be joined by this process. It can be used for welding a large number of dissimilar metals. It is used for very specialized applications and the process is entirely automatic. Metals which can be percussion welded include, copper alloys, aluminium alloys, nickel alloys, low-carbon steel, medium-carbon steel and stainless steels. This process is used for welding satellite tips to tools, copper to aluminium or stainless steel, silver contact tips to copper, cast iron to steel, lead-in-wires on electric lamps and zinc to steel. Gold, silver, copper-tungsten, silver-tungsten and silver-cadmium oxide percussion welded to copper alloys for commonly used assemblies for electric contacts.

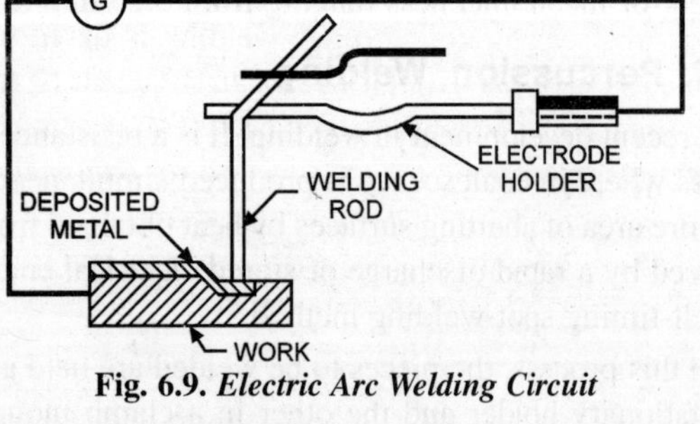

Fig. 6.9. ***Electric Arc Welding Circuit***

6.3. ELECTRIC ARC WELDING

By definition, *arc welding is that process in which the pieces of metal to be welded are brought to the proper welding temperature at point of contact by the heat liberated at the arc terminals and in the arc stream so that the metals are completely fused into each other, forming a single solid homogeneous mass, after it solidifies.*

In this process an electric arc is produced by bringing two conductors (electrode and metal piece) connected to a suitable source of electric current, momentarily in contact and then separating by a small distance. The current continues to flow across the small gap and gives intense heat. The heat developed is utilised to melt the part of the workpiece, and the filler metal and thus forms the joint. So arc welded joint is a union of metal parts made by localised heating without any pressure. That is why sometimes this type of welding is known as the *non-pressure welding*. The heat developed by arc is also used for cutting of metal.

The voltage necessary to strike the arc is higher than that required to maintain it. The temperature is of the order of 3,600°C, at which mechanical pressure is not required for jointing. According to BSS 638 maximum voltage specified for welding is about 100 volts for the safety of the operator. The arc voltage varies from 20 to 40 volts and current from 50 A in sheet metalwork to 1,000 A in heavy automatic welding.

The welding circuit for an electric arc welding process is shown in Fig. 6.9. The welding circuit consists of a welding machine, two leads, an electrode holder, an electrode and the work itself.

The electric arc welding is widely used for joining of metal parts, the repair of fractured castings and the fillings by the deposition of new metal on worn-out parts.

Arc welding is again subdivided into five groups namely carbon arc welding, metal arc welding, atomic hydrogen arc welding, inert gas metal arc welding and submerged arc welding.

6.3.1. Carbon Arc Welding

Carbon arc welding differs from metal arc welding in that it is a puddling process and is somewhat similar to the gas welding process. The carbon arc is very stable and easy to maintain. The length of the arc can be varied over wide limits without causing the arc to go out. There is no tendency for the electrode to freeze or stick, as in the case of the metallic electrode. Accordingly the arc can be struck without difficulty at any point, and rapidly moved over the surface of the work to the point where the weld is to be made. High melting speed and efficient welds are the main advantages claimed.

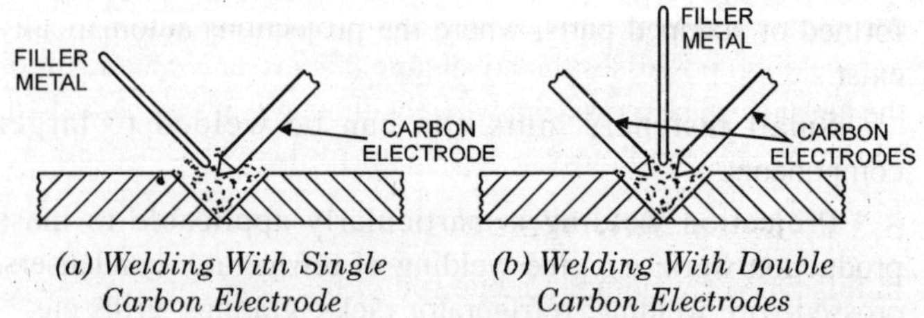

(a) Welding With Single Carbon Electrode *(b) Welding With Double Carbon Electrodes*

Fig. 6.10. ***Carbon Arc Welding***

In this process, a carbon or graphite rod is used as a negative electrode and the work being welded as positive. Mostly graphite electrodes are used as they yield longer life and have low resistance and thus capable of conducting more current. The arc produced between the two electrodes heats the metal to the melting temperature. This is about 3,200°C on the negative electrode, and 3,900°C on the positive electrode. The reason to use carbon rod as negative electrode is that less heat will be generated at the electrode tip than that at the workpiece, and carbon from the electrode will not fuse and mix up with the job. If this so happens the resultant weld will be rich in carbon, and consequently very much brittle and unsound. For this type of welding only dc can be used. The use of ac is not recommended because no fixed polarity can be maintained. This process is normally adopted where addition of filler material is not required, *e.g.*, flange or edge joints, but if filler metal is required, it is provided by the welding rod made of the similar metal to that of the metal to be welded. Some protection for the molten weld metal may be provided by using a long arc which produces a carbon monoxide gas envelope. There are two methods of carbon arc welding. In one method no flux is used and in the other method flux either in the form of powder or paste is used to prevent the weld from oxidation. Former method is confined to non-ferrous metals and the later method is usually employed for ferrous metals.

Electrodes up to 25 mm in diameter with currents up to 600 or 800 A are used, so that the carbon arc process is well suited to rapid work, or cases such as arise in the repair of castings, where large quantities of filler metals have to be deposited.

Carbon arc welding is employed for welding sheet steel, copper alloys, brass, bronze, and aluminium. It is not suitable for vertical and overhead welding.

In another process of carbon arc welding, known as *twin carbon arc welding,* two carbon electrodes are used between which the arc is formed, as illustrated in Fig. 6.10 (*b*). In this way the arc generates heat which is applied close to the parts to be welded, causing the edges to fuse. In heavy welds, a filler rod is melted by the arc and deposited in the weld. An ac supply is recommended for twin carbon arc welding. In case dc supply is used, the positive electrode will disintegrate and consume at a much faster rate as compared to the –ve electrode because two-thirds of the total heat is generated at the positive pole. This will produce an unstable arc and need more frequent adjustment of the electrodes. The electrodes used for twin carbon arc welding are approximately of the same diameter as the thickness of workpiece. The magnitude of arc current required depends upon both, electrode diameter and plate thickness. Twin carbon arc welding, though, more complex than single carbon arc welding, possesses the advantage that arc is independent of the job and can be moved anywhere without getting extinguished. Moreover, the workpiece is not a part of the electrical circuit.

6.3.2. Metal Arc Welding

In this type of welding, a metal rod of the same metal as being welded forms one of the electrode and also serves as a filler and no filler rod is used separately. The arc struck between the work being welded and the electrode causes the melted rod to flow across the arc into the metal pool of the parent metal. This deposition of metal is accomplished by contact made between the molten metal and the globules formed on the end of the electrode filler rod. The temperature produced is about 2,400°C and 2,600°C on the negative electrode and positive electrode respectively. The concentration of heat energy at the terminals of the electrode causes a small part of the work being welded to melt almost instantaneously and an intermittent flow of metal across the arc stream. The metal in the arc stream is in both the liquid and gaseous forms, the liquid metal being transferred across the arc by molecular attraction, adhesion, cohesion, surface tension or a combination of these. For this type of welding both ac and dc can be used. For dc supply 50-60 volts and for ac supply 70-100 volts are used for welding. For current above 750 A ac equipment is preferred as it has high efficiency, negligible loss at no load and minimum maintenance and initial cost. Welding with dc also gives rise to "arc blow", owing to the magnetic forces created by the current, a trouble which does not exist with ac welding as each magnetic force is immediately followed by one in opposite direction and the arc-deflecting forces are automatically cancelled. Metal in the molten state has an affinity for foreign elements such as oxygen, nitrogen etc.; and in order to prevent impurities from entering the weld modern arc welding electrodes are covered with a coating which, under the welding heat, generates a gas shield round the arc, and also forms a slag which is deposited on and around the molten metal, protecting it against foreign matter during the cooling stages. A metal arc welding circuit is shown in Fig. 6.11.

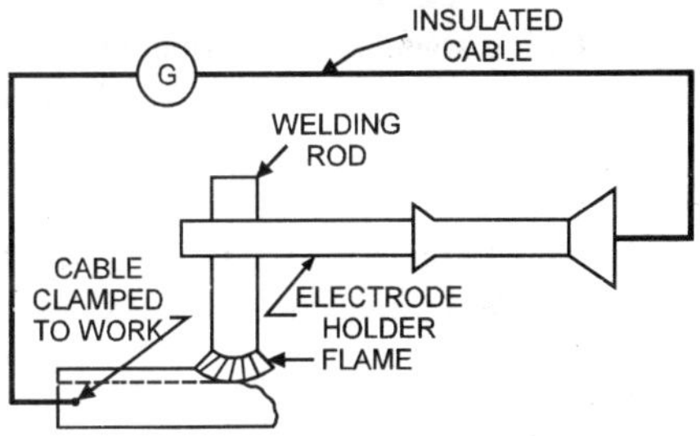

Fig. 6.11. *Metal Arc Welding Circuit*

The correct welding current, voltage and welding speed are very important. Various welding currents can have a deciding effect on the forming of proper beads. A welding shows bead characteristics under different conditions : (*i*) welding current too low excessive piling of the metal (*ii*) welding current too high causing excessive spatter, (*iii*) voltage too high bead too small (*iv*) welding speed too low excessive piling up of weld metal. Proper current and timing create a smooth, regular, well-formed bead. Undercutting is also a result of too much current. No enough current results in overlapping and a lack of fusion with the metal.

6.3.3. Atomic Hydrogen Arc Welding

The principle of atomic hydrogen welding is based on (*i*) the possibility of obtaining atomic hydrogen by means of an electric arc between the two tungsten electrodes, in an atmosphere of hydrogen at atmospheric pressure, (*ii*) the very high temperature produced by the recombination of the atoms, which occurs in the cooler regions immediately outside the arc, and (*iii*) the very great heat conductivity of hydrogen at high temperatures, owing to the dissociation of the hydrogen molecules into the atomic state, and resulting in an extremely high rate of delivery of heat to the surfaces to be welded–roughly twice that for the oxyacetylene welding flame.

In this method of arc welding an ac arc is maintained between the two non-consumable tungsten electrodes while a stream of hydrogen gas, under a pressure of about 0.5 kg/cm^2, is passed through the arc and around the electrodes. AC supply is used in order to obtain equal consumption of the electrodes. The hydrogen is usually supplied from steel cylinders and serves two-fold purpose—firstly as protective screen for the arc, secondly as cooling agent for the glowing tungsten electrode points. As the molecules of hydrogen pass through the electric arc, they are changed into the atomic state and thus absorb a considerable amount of energy—resulting in cooling of glowing tungsten electrodes. But when the atoms of hydrogen recombine into molecules just outside the arc, a large amount of heat is liberated. This extra heat, added to the intense heat of the arc itself, produces a temperature of about 4,000°C, as compared to 2,000°C produced by the combination of normal hydrogen and oxygen. This heat is used in making fusion welds. When additional metal is required, filler rods are melted into the joint. Due to the decarbonizing character of the welding flame (with the formation of hydrocarbons) the filler rod material should have a carbon content higher than that of the material to be welded, if the finished weld metal is to have the same carbon content as the parent material. Owing to the concentrated heat, high welding speeds may be obtained keeping distortion small. Further more hydrogen excludes all oxygen and other gases which might combine with the molten metal to form oxides and other impurities. It also removes oxides from the surface of the work. Thus, this process is capable of producing smooth, uniform, strong, and ductile welds. Automatic equipment for this process has been developed, the hydrogen being fed through a water-cooled nozzle block.

Arc currents up to 150 A can be used and the power supply arrangements are similar to those for ordinary ac arc welding except that the transformer voltages are higher, being about 300 V on open circuit for striking the arc and between 80 and 100 V while operating.

This method is very flexible—practically any metal or alloy, ferrous and non-ferrous, can be welded. Its greatest usefulness is for the fusion welding of certain special ferrous alloys, such as chrome nickel steels, aluminium, and duralumin, but it must use flux to weld some of these materials. Welding of thin sheets, production of tubing, and the repairing of expensive tools and dies are some of the common uses of this method.

6.3.4. Inert Gas Metal Arc Welding

It is a gas shielded, metal arc welding process which uses the intense heat of an electric arc between a continuously fed, consumable electrode wire and the material to be welded. A consumable bare electrode wire producing the filler metal is automatically fed from a reel on to the job through a welding gun which also carries a nozzle. Through this nozzle helium or argon is blown around the arc and on to the weld. Because of low arc voltage for a given welding current argon is suitable for thin materials as tendency to burn through is reduced. Because of low arc voltage and low power for given welding current in case of argon gas welding tendency for metal fusion and metal sag or run is reduced and, therefore, it is preferred for inposition welding. On the other hand, because of high arc voltage and high power in case of helium gas welding, it is used for welding of thick materials and metals with high heat conductivity. Mixture of the two gases is useful to get intermediate characteristics. AC and dc both can be used. This type of welding has the following advantages :

(*a*) In this method concentration of heat is possible and thus distortion is minimised.

(*b*) In this method no flux is required since atmosphere is inert and air does not come in contact with the molten metal.

This process is particularly employed for welding light alloys, stainless steel and non-ferrous metals such as copper, aluminium and their alloys. For welding aluminium and aluminium alloys ac supply and argon gas is used where as for welding magnesium and magnesium alloys either ac supply and argon gas or dc supply, reversed polarity and either argon or helium gas is used. For the welding of stainless steel, mild steel, copper and copper alloys, dc and either argon or helium or ac high frequency stabilized with argon or helium used.

6.3.5. Submerged Arc Welding

It is an arc welding process that uses an arc between a bare metal electrode and the weld pool. The arc and molten metal are shielded by a blanket of granular flux on the workpieces. The process is used without pressure and with the filler metal from an electrode, and sometimes from a supplemental source such as a welding rod, flux, or flux with metal granules.

It is an automatic process developed primarily for the production of high quality butt welds in thicker steel plate than is normally suited to other manual arc welding processes.

As in open arc welding, the source of heat in the submerged arc welding process is an electric arc or arcs between a metal electrode, or electrodes and the work. The welding zone is shielded by a blanket of flux, so that the arc is not visible. Hence the name "*submerged arc welding*". The arc melts the parent metal, the electrode, and the flux. The fused flux produces a blanket of liquid slag which forms a protective envelope both around the arc and the welding zone. Right beneath the electrode tip a weld pool is formed, holding an amount of molten metal. On melting, the electrode metal forms globules which go over into the puddle and mix with the molten parent metal. The pressure of the arc forces the mix out of the puddle so that it forms the weld (Fig. 6.12). This action continues as long as the arc advances. The shape and size of the puddle have a strong effect on those of the weld.

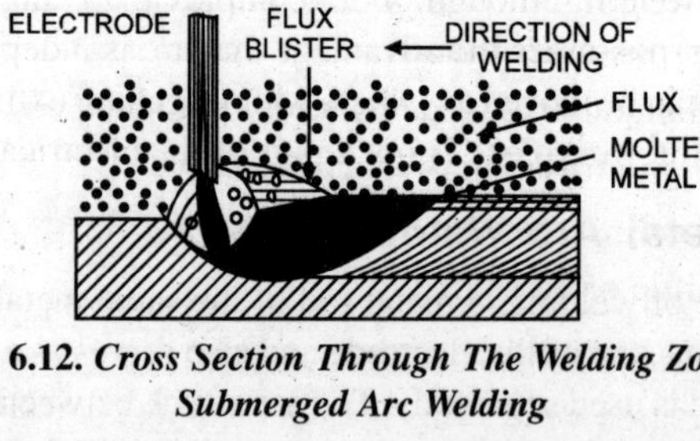

Fig. 6.12. ***Cross Section Through The Welding Zone in Submerged Arc Welding***

The method of making a weld by submerged arc welding is illustrated diagrammatically in Fig. 6.13. Flux is fed continuously onto the work just ahead of the advancing arc. Since the arc is not visible, the operator can wear goggles with ordinary glasses, using neither a hand screen nor a helmet. Since molten flux is lighter than molten metal, it floats to the surface and solidifies as a crust of brittle slag which is readily removable from the weld surface. Much of the flux spread on the work remains unfused. It is recovered by suction into the hopper for re-use.

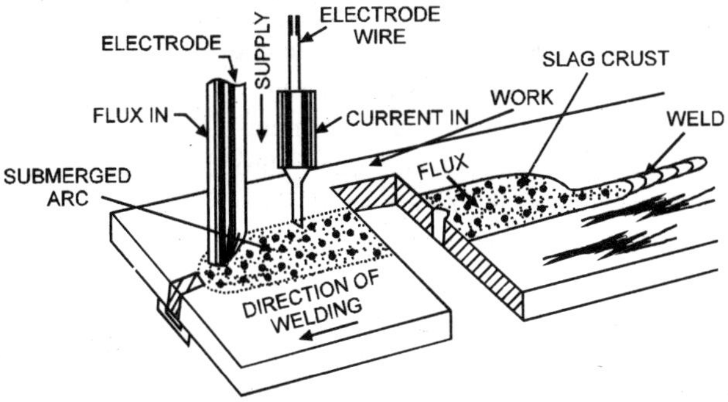

Fig. 6.13. ***Submerged Arc Welding***

The flux may be made of silica, metal oxides, and other compounds fused together and then crushed to proper size. Another group of fluxes is made of similar material "bonded" and formed into *granules.*

The bare electrode is fed from a reel down through the gun or nozzle. The operator can move it slowly, a little at a time for the start of the arc and then set the proper feed rate on the control box. Wire of several alloys for steel, stainless steel, copper etc. is available in diameters from 3 to 6 mm. Voltage employed is from 25 to 40 volts. Current employed depends considerably on workpiece thickness. Normally dc is used employing 600 to 1,000 A current for welding alloy and stainless steel, although ac is preferable particularly for low carbon steel. The current for ac is usually 2,000 A.

The advantages of submerged arc welding are :

(*i*) Partly because it is often automated, it is much faster than regular arc welding. Speeds up to 3,800 mm/minute are possible on 3 mm thick steel at 100% efficiency.

(*ii*) Deep penetration with high quality weld is possible.

(*iii*) Less distortion occurs from high speed and uniform heat input, specially when automated.

(*iv*) No edge preparation is required.

(*v*) Operator is not exposed to usual spatter and can work more easily without helmet and other safety equipment.

(*vi*) The welds obtained have good ductility, impact strength, uniformity, low nitrogen content and high corrosion resistance.

(*vii*) Manipulative skills normally not involved.

(*viii*) High utilisation of electrode wire.

(*ix*) Easily automated for high operator factor.

(*x*) Little or no smoke.

The submerged arc welding is widely used in heavy steel plate fabrication work. This includes the welding of structural shapes, the longitudinal seam of larger diameter pipe, the manufacture of machine components for all types of heavy industry, the manufacture of vessels and tanks for pressure and storage use. It is widely used in the ship building industry for splicing and fabricating subassemblies, and by many other industries where steels are used in medium to heavy thickness.

It is also used for surfacing and build-up work, maintenance, and repair.

6.4. ULTRASONIC WELDING

Ultrasonic welding is a solid state welding process that produces coalescence of materials by the local application of high-frequency (20-60 kHz) vibratory energy as the workpieces are held together under pressure. No flux or filler metals are used, no electrical current flows through the weld metal, and usually no heat is applied. Since vibratory action in ultrasonic welding breaks up and disperses moisture, oxide and other (*e.g.*, insulation) coatings, normally, degreasing may be the only cleaning required before welding (especially with aluminium).

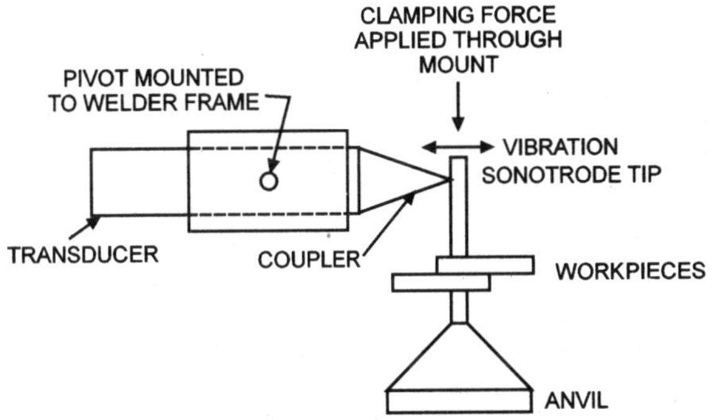

Fig. 6.14. ***Ultrasonic Welding***

Welding occurs when the ultrasonic tip or electrode, the energy coupling device, is clamped against the workpieces and is made to oscillate in a plane parallel to the weld interface. The combined clamping pressure and oscillating forces introduce dynamic stresses in the base metal. This produces minute deformations which create a moderate temperature rise in the base metal at the weld zone. This coupled with the clamping pressure provides for coalescence across the interface to produce the weld.* Ultrasonic energy will aid in cleaning the weld area by breaking up oxide films and causing them to be carried away. The vibratory energy that produces the minute deformation comes from a transducer which converts high-frequency alternating electrical energy into mechanical energy. The transducer is coupled to the work by various types of tooling which can range from tips similar to resistance welding tips to resistance roll welding electrode wheels.

The temperature at the weld is not raised to the melting point and, therefore, there is no nugget similar to resistance welding. Weld strength is equal to the strength of the base metal. Most ductile metals can be welded together and there are many combinations of dissimilar metals that can be welded. The process is restricted to relatively thin materials normally in the foil or extremely thin gauge thicknesses. The maximum thickness by these processes ultrasonically may vary from 0.38 to 2.5 mm depending upon the metal.

This process is used extensively in the electronics, aerospace, and instrument industries. It is also used for producing packages and containers and for sealing them. The process can also be employed for joining plastics and is finding wider use in this field than in joining metals.

* The coupling system and welding tip form a unit called a sonotrode.

6.5. ELECTRON BEAM WELDING

Basically, electron beam welding in vacuum utilises the kinetic energy of electrons travelling with high velocity in a high vacuum (10^{-3} to 10^{-5} mm Hg). When the electrons strike the surface of the metal, they give up the bulk of their energy as heat, and this goes to melt the metal. If the work is done in a high vacuum, no electrodes, gases, or filler metals need contaminate it and pure welds can be made. Moreover, high vacuum is necessary around the filament so that it will not burn up and will also produce and focus a stable beam (Fig. 6.15).

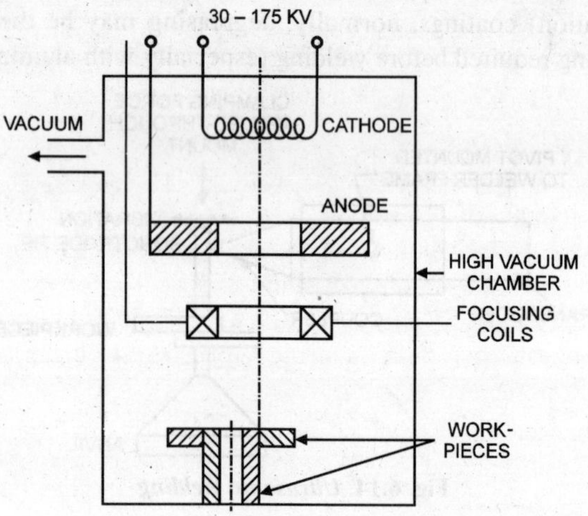

Fig. 6.15. *Electron Beam Welding*

In all types of electron beam machines, a tungsten filament which serves as a cathode emits a mass of electrons that are accelerated and focused to a 0.25–1 mm diameter beam of high energy density up to 0.5–10 kW/mm^2. The temperature produced is about 2,500°C. This is sufficient to melt and vaporize the workpiece material and thus fills a narrow weld gap even without a filler rod (a'though filler rods may be employed).

The speed of the beam is stepped to one-half to two-third of the speed of light by passing it through a high-voltage electrostatic field. An electromagnetic lens is used to obtain correct focusing of the beam.

A similar process, known as *non-vacuum electron beam welding* is rapidly coming into use today, *i.e.*, many welds are now made without elaborate arrangement required for vacuum electron beam welding. In this case, the vacuum chamber in which the beam is created is evacuated to a lower pressure. In situation where contamination of the workpiece must be held to a minimum, the beam should be passed through argon or helium. To make the chamber high vacuum, it takes about 5 to 30 minutes to evacuate the air, depending on the size of the chamber. A medium size electron beam welder operates below 60 kV. The welding head or the work is moved by numerical control or by hand.

The advantages of electron beam welding are that the welds are clean, with no porosity since there is no air ; no shielding gas is required ; and as the energy input is in a narrow, concentrated beam, distortion is almost eliminated. The speed may be as fast as 2,500 mm per minute, and it will weld or cut any metal or ceramic, diamond, sometimes as thick as 150 mm.

The major advantage of electron beam welding is its tremendous penetration, which occurs when the highly accelerated electron hits the base metal. It will penetrate slightly below the surface and at that point release the bulk of its kinetic energy, which turns to heat energy. This brings about a tremendous temperature rise at the point of impact. The succession of electrons striking the same place causes melting and then evaporation of the base metal. This creates metal vapours, but the electron beam travels through the vapour much easier than solid metal. This causes the beam to penetrate deeper. The depth-to-width ratio can exceed 20 : 1. With the increase in power density, penetration is increased. Penetration also depends on the beam current. With the increase in beam current, penetration is increased. The other variable, travel speed, also affects penetration. As travel speed is increased, penetration is reduced.

Since the electron beam has tremendous penetrating characteristics, with the lower heat input, the heat affected zone is much smaller than of any arc welding process. In addition, because of almost parallel sides of the weld nugget, distortion is greatly reduced. The cooling rate is much higher, and for many metals this is advantageous; however, for high carbon steel this is a drawback and cracking may occur.

Almost all metals can be welded with the electron beam welding process. The metals that are most often welded are super-alloys, the refractory metals, the reactive metals, and the stainless steels. Many combinations of dissimilar metals can also be welded.

One of the disadvantages of the electron beam process is its high capital cost. The price of equipment is very high, and it is expensive to operate, due to the requirement for vacuum pumps. In addition, fitup must be precise and locating the parts with respect to the beam must be perfect.

Today automobile, airplane, aeroscape, farm and other types of equipments including ball-bearing over 100 mm are being welded by the electron beam process.

6.6. LASER BEAM WELDING

Laser beam welding is a welding process that uses the heat from a laser beam impinging on the joint. The process is used without a shielding gas and without the application of pressure. The laser is a device that produces a concentrated coherent light beam by stimulating electronic or molecular transitions to lower energy levels. The word *laser* is an acronym for "*light amplification by stimulated emission of radiation*". The laser beam is a highly concentrated source of energy which has many applications. It can be used for welding, for cutting metals and non-metals, for surface heat treating of metals, and for cladding by fusing powders to base materials. It can also be used for brazing and soldering, and for drilling, machining, and marking. It is also used in other fields : medicine, communications, marking and surveying.

There are two basic types of lasers employed in metal working. The original types are the solid state lasers, which use a solid sodium. The second type are the gas lasers, which normally use a mixture of helium, nitrogen, and CO_2 gas in a tube. In either case, when the medium is excited, it emits photons, which become the laser beam.

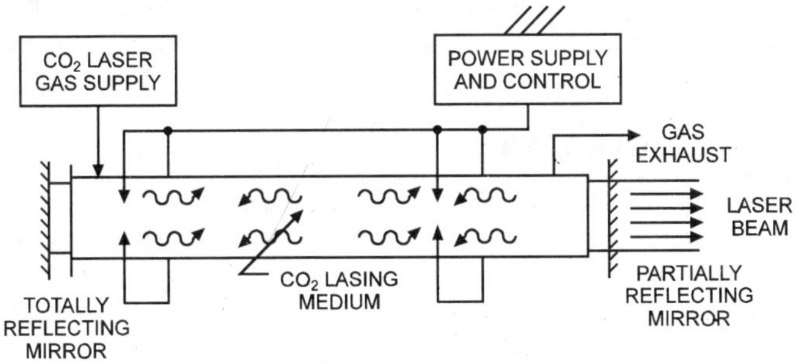

Fig. 6.16. *Basic CO_2 Laser System*

A gas laser beam is shown in Fig. 6.16. Excitation of gas laser is by means of high-voltage low-current electric power. The electrical discharge excites the CO_2 molecules, which on returning to their original state, emit photons. Mirrors are placed on both ends of the tube, one entirely reflective and the other with small partially transmissive area to permit the beam to exit. This forms a cavity in which photons builds up. The freed photons travel between the mirrors and excite the CO_2 molecules, starting a chain reaction of photon emissions. A stream of photons, the laser beam, exits through the unsilvered section of the one mirror. The wavelength of the laser beam is 10.6 μm. This wavelength is longer and does not have the safety problems of the short wavelength beam. The CO_2 gas lasers can be operated in the continuous wave mode or in the pulsed mode.

There are several types of CO_2 lasers. The early or low powered type used a "sealed tube" which has a power output of from 3 to 100 W. The axial flow type, which is more complex method of producing laser beams, will produce from 50 to 2,000 W average output power. Another type, known as the transverse flow type, has an average output power of from 2 to 15 kW. The laser power supply system is called an oscillator. These different types of CO_2 lasers involve different designs of the beam-producing apparatus. The size of the apparatus and the efficiency of the system are different.

The laser beam is very intense and unidirectional but can be focused and reflected in the same way as an ordinary light beam. The focus size is controlled by the choice of lenses and mirrors and the distance to the workpiece. The spot size can be made as small as 1 mm to up to 12 mm. The smaller focus spot size is used for cutting and welding, and the larger spot size is used for heat treating. The laser beam can be used in open air and can be transmitted long distances with only minimal loss of power.

A block diagram of a laser beam welding system is shown in Fig. 6.17. This shows the major components.

When using a laser beam for welding, the beam impinges on the surface of the base metal with such a concentration of energy that the surface is melted and volatilized. When the metal is raised to its melting temperature, the surface conditions have a minor effect on reflecting the beam. The distance from the optical cavity to the workpiece has little effect on the laser, because it can be focused to the proper spot size at the work with the same amount of energy available whether it is close or far away.

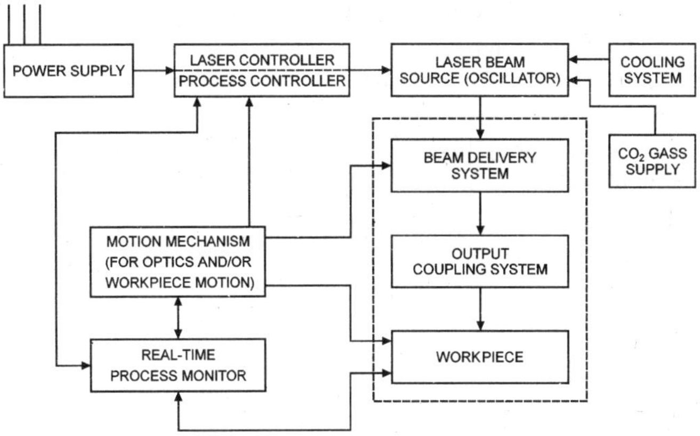

Fig. 6.17. *Block Diagram of CO_2 Laser System*

With laser beam welding the molten metal takes on a radial configuration similar to plasma arc welding*, known as "melt-in" welding. When the power density rises above a certain threshold level, keyholing occurs, the same as with plasma arc or electron beam welding. Keyholing provides for extremely deep penetration, which provides the weld a high depth-to-width ratio. For some applications, inert gas is used to shield the molten metal from the atmosphere.

The welding characteristics of the laser are similar to those of the electron beam. The laser can weld the same joint types. The concentration of energy by both beams is similar, with the laser having a power density of the order of 10^6 W/cm^2. The power density of electron beam is slightly greater.

The location of the focal point of the beam with respect to the surface of the workpiece is important. Maximum penetration occurs when the beam is focused slightly below the surface. Penetration is less when the beam is focused on the surface or deep within the surface. As the power is increased, the depth of penetration increases.

Heating and cooling rates are much higher in laser beam welding than in arc welding, and the heat-affected zones are much smaller. Rapid cooling rates can create problems, such a cracking in high-carbon steels.

The laser beam has been used to weld carbon steels, high-strength low-alloy steels, aluminium, stainless steel, and titanium. Laser welds are similar in quality to welds made by electron beam process. Filler metal is used to weld metals that tend to show porosity when welded with either electron beam or laser beam welding.

Materials 12 mm thick are being welded at a speed of 250 mm per minute. The efficiency of laser beam welding equipment is 5 to 10%, but as new equipment becomes available this will increase. The cost of the equipment will decrease in the future. More welding applications for laser welding are being found daily, and its use will increase.

6.7. REQUIREMENTS OF GOOD WELD

The weld is said to be good weld if it is free from external defects such as irregular width and heights of beads, deviation of the weld from the prescribed dimensions, unfilled craters on

* Plasma arc welding is an arc welding process that uses a constricted arc between a non-consumable electrode and the weld pool (transferred arc) or between the electrode and the constricting nozzle (non- transferred arc).

the surface of welds, undercuts, slag on the surface of welds, porosity in the outer layer of the welds, visible cracks in and around the weld and internal defects such as poor fusion, hidden porosity in the deposited metal, hidden cracks in the weld and parent metal, poor fusion between layers in multiple layer welding, and entrapment of slag and oxides.

Poor fusion, which is among the most dangerous defects, is lack of through and complete union between the deposited and parent metal, or between individual layers in multiple layer welding. Lack of fusion impairs the strength of welds and makes welded structures unreliable. In order to get sound weld it is essential that absolute fusion of parent metal and electrode metal is obtained.

Porosity is entirely ruled out in weldments where tight seams are critical (pressure vessels, storage tanks, pipelines etc.). Porosity may be brought about by insufficient puddling, as it does not give time for the trapped gas to escape from the molten metal before it solidifies. The gas remaining in the molten metal forms gas pockets. They may be scattered about or link up into strings of pores, giving the welded metal a spongy appearance. Porosity makes welds leaky, pervious to liquids and gases and weak.

Cracks in welds may arise from locked-up stresses up by non-uniform heating and cooling, excess sulphur or phosphorus in the weld metal, and some other causes. Cracks most often occur in butt welds and may be either longitudinal or transverse. Apart from visible cracks, there may be hair cracks, usually appearing at the line of fusion between the parent and deposited metal. They may be brought about by lack of fusion, large gas pockets, or slag inclusions where thermal stresses concentrate, giving rise to hair cracks which may develop into large cracks. That is why hair cracks are intolerable.

6.8. PREPARATION OF WORK

There are several factors which must be considered while preparing work for welding in order to obtain good results.

For obtaining good results it is desirable that provision may be made for expansion and contraction wherever possible. The strength of the weld will depend on the correct bevelling and spacing of parts to be welded. Uniform fusion is directly dependent on the proper bevelling and spacing.

The cleaning of the surfaces is another important factor. The cleaning can be done by wire brushing, machining or sand blasting. Impurities if present tend to make the joint weaker as the welded portion is filled with gas and slag inclusions and metal becomes brittle and the cohesion between the metals is, therefore, poor.

Another fundamental of good preparation or joint design is to keep the cross section of the added metal as small as practicable, so as to localise the effective heat in adjacent sections.

Another important point in welding is to overcome the oxidation at high temperatures. The oxides of metals formed at high temperatures encountered in welding have serious weakening effects. This influence is counteracted by using a flux which removes the oxides and permits perfect cohesion of the metals. In arc welding the flux is coated on the electrodes. The flux forms a protective coating of slag over the weld metal and creates a non-oxidising atmosphere.

6.9. ELECTRODES

An electrode is a piece of wire or rod (of a metal or alloy), with or without flux covering, which carries current for welding. At one end it is gripped in a holder and an arc is set up at the other end.

Electrodes are made of materials having high electrical conductivity as well as thermal conductivity so that heat developed at junction faces between electrodes and work is minimum in the first instance and secondly, whatever heat is produced is conducted away efficiently. In addition to high electrical and thermal conductivity, electrode material should be of very high mechanical strength so that the wear and tear at the electrode tip is minimum.

The materials used for electrodes are : Hard drawn copper (softening temperature 150°C and for welding of aluminium alloys); Cadmium copper (softening temperature 250°C and for joining MS sheets); Chromium copper (softening temperature 500°C and for joining of steel) and Tungsten copper (softening temperature 1,000°C and for welding of steel) and copper alloys.

The diameter of the electrode is approximately given by the expression $d = 5\sqrt{t}$

Where d is the diameter of tip in mm and t the thickness of sheet in mm.

Either non-consumable or consumable electrodes may be employed in arc welding.

Non-consumable electrodes may be of carbon, graphite or tungsten which do not consume during the welding operation. Consumable electrodes may be made of various metals depending upon their purpose and the chemical composition of the metals to be welded.

In general, the electrodes can be divided into three categories depending upon outer aspects as well as their technological properties.

Bare electrodes are most commonly used in automatic and semi-automatic welding. In using the plain or bare electrodes, as the globules of the metal pass from the electrode to the work, they (globules) are exposed to the oxygen and nitrogen in the surrounding air. This causes the formation of some non-metallic constituents which are trapped in the rapidly solidifying weld metal and thereby reduces the strength and ductility of the metal.

Lightly coated (or washed) electrodes have a coating layer several tenths of a millimetre thick. This coating usually consists of lime mixed with soluble glass which serves as a binder. The primary purpose of the light coating is to increase the arc stability, so they are also called the stabilising coatings. No attempt is made to prevent oxidation and no slag is formed on the weld, nor are the mechanical properties of the weld metal improved. For this reason lightly coated electrodes may only be used for welding non-essential jobs.

Heavy coated electrodes, sometimes referred to as *shielded arc electrodes*, are used to obtain a weld metal of high quality, comparable with, and even superior to, the parent metal in terms of mechanical properties. Heavy coatings employed are composed of ionising (chalk), deoxidising (aluminium, ferro-manganese etc.), gas generating (starch), slag forming (kaolin), alloying and binding materials. These electrodes have the following distinct advantages :

(*i*) The physical and metallurgical properties of the weld can be influenced by adding alloying components to the covering.

(*ii*) Basic salts of silicon, magnesium and calcium in the covering form slag which floats on the surface of metal and prevents rapid cooling of the weld. Thus weld does not become brittle.

(*iii*) The weld metal is protected from oxidising action of atmospheric oxygen and nitrifying action of nitrogen of air due to gases formed from the covering material.

(*iv*) In case of ac supply arc cools at zero current and there is a tendency of deionizing the arc path. Covering gases keep the arc space ionized.

(*v*) During welding the covering extends beyond the core wire. This directs the arc and concentrates the arc stream, reduces thermal losses and causes increase in temperature of electrode tip.

There may also be *powder-cored electrodes* which have a good portion of the source of metal for depositing in the joint located as powdered iron mixed in with the flux coating. This makes the electrodes much larger on the outside for the same diameter core wire as standard flux coated electrodes. These electrodes are well suited to down-hand welding and are said to deposite more metal in a given period time than standard flux covered electrodes.

Both bare and coated electrodes, for manual arc welding, are made in the shape of rods up to 12 mm in diameter and 450 mm long. Semi–automatic and automatic welding use electrode wire in coils.

6.10. POWER SUPPLY FOR RESISTANCE WELDING

AC supply is used for resistance welding because of the ease and covenience with which the required high current at a low voltage can be obtained by means of a transformer. The kVA required for resistance welding, when actually making a weld, ranges from a few kVA to as much as 1 MVA. The power factor will be about 0.25 or 0.3 lagging. The power factor is low mainly due to the high ratio of reactance to resistance of the loop formed by the jaws of the welding machine. Such heavy intermittent single-phase loads may cause serious voltage drop difficulties in the supply network. Such problems can be overcome to some extent by connecting capacitors of suitable capacity in parallel with the welding transformer so as to improve the power factor. But with this arrangement the power factor will become leading when welding current is not being drawn. This problem can be avoided by connecting the capacitors in series with the welding transformer to neutralize the reactance drop in the supply circuit.

6.11. MACHINES FOR RESISTANCE WELDING

The machine for resistance welding incorporates a transformer, suitable electrodes for supplying current to the weld and arrangement for controlling the mechanical pressure, and finally, means for controlling the duration of weld current flow. The mechanical pressure may be exerted through levers and clutch by an electric motor or by compressed air. The magnitude of pressure required depends upon the type of work and may vary from a few kg for thin sheets or wires up to a tonne or more for heavy work.

In the older type of welding machines, the electrodes were brought on to the work and the electrical circuit closed by the operation of a pedal. Thus application of pressure and the time duration of current flow used to be controlled by the operator and for this operator needs to be experienced and skilled. The modern practice is to pass heavy currents for shorter time durations (ranging from 10 ms to 100 ms). The equipment used for this purpose may be *constant time, current-actuated,* or *energy-actuated* types.

Constant time equipment is employed in high speed production where the work has a consistently clean surface. Constant time equipment may be provided with mechanical control or electrical control. In mechanical control providing up to 300 welds per minute, the device employed is a cam-operated switch, connected in the primary circuit of a welding transformer, driven from the welding machine. For a large number of welds per minute the mechanical arrangement becomes unsuitable because it is not capable of providing consistently accurate timing, due to wear of the cam and operating mechanism, arcing and burning of the contacts and irregularities caused by closing of switch at different instants in the cycle.

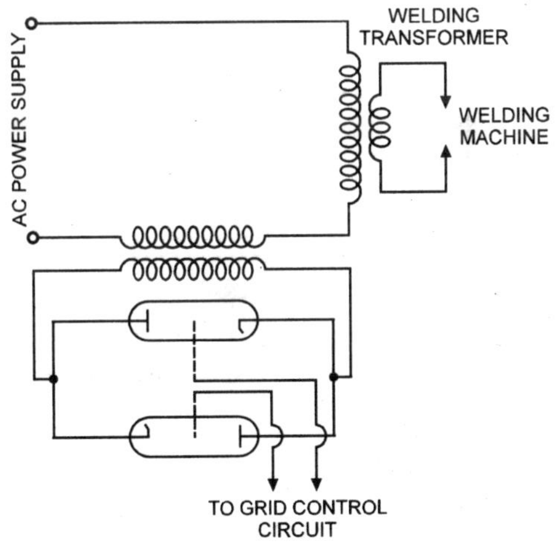

Fig. 6.18. ***Valve Controlled Resistance Welding Circuit***

An alternative arrangement is to control the timing through grid controlled ignitrons or thyratrons. It is easier to build tubes for a high voltage and small current than with a low voltage and high current. An arrangement using valves in the secondary circuit of a series transformer is shown in Fig. 6.18. When the tubes conduct, the series transformer secondary is almost short-circuited, and whole of the supply voltage is available across the welding transformer primary. But when the tubes are not conducting, the series transformer primary winding offers a high impedance in the circuit of the welding transformer and the current is reduced to a negligible value. Auxiliary valves are employed for controlling the timing of the negative potential applied to the grids of the main tubes.

Constant-time method of control does not yield consistently good results when there may be variations in the conditions under which successive welds are made, due to variations in supply voltage or mechanical pressure, wearing of electrodes, surface irregularities etc. The energy-actuated control, wherein definite amount of energy is supplied to the weld, is used. Constant-time method of control did not prove to be successful, particularly with modern high-speed welding. The energy-actuated control, which permits the current to flow until a predetermined amount of energy has been supplied to the weld, is theoretically an ideal method. However, the control equipment is quite complicated.

6.12. ELECTRONIC CONTROL IN RESISTANCE WELDING

As there are several factors, *i.e.*, current, pressure, heat, time, to be considered, manual control does not yield good results in case of resistance welding. For the precise control of these factors electronic control welding circuits are used. Some of the electronic control circuits are given here.

6.12.1. Ignitron Contactor

Theory of ignitron has already been dealt in Art. 2.2.2. We will now discuss its use as a contactor for controlling heavy currents.

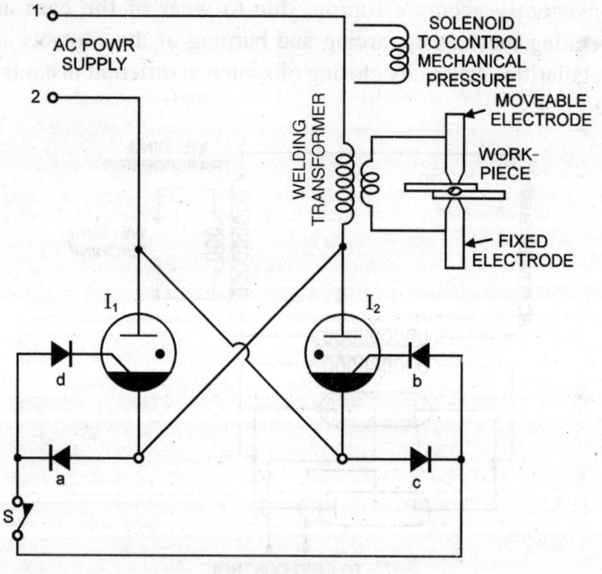

Fig. 6.19

A simple line contactor using two ignitrons is shown in Fig. 6.19. If the switch S is closed at the instant the line 1 is positive, current will rush through the primary of welding transformer, rectifier *a*, switch S, rectifier *b*, the ignitron I_2 and back to the line 2. The current will strike an arc in ignitron I_2 and the tube starts conducting. Now voltage drops across I_2 to a low value, causing a voltage drop in the ignitron circuit. Therefore, the ignitron will conduct just long enough to strike an arc. Similarly, during next half cycle, the line 2 will be positive and the current will flow from line 2 through rectifier *c*, switch S, rectifier *d* to ignitron I_1. During this half cycle, as the anode of ignitron I_2 becomes – ve, it stops conducting. Metallic rectifiers are used in this circuit. They conduct the current in proper direction, thus, preventing the application of negative voltages to the electrodes and saving the ignitrons from inverse current damages. A solenoid is used to apply suitable pressure through the upper movable electrode. Manual control of the contactor switch is possible only if the welds are of long duration. But for the accurate time control of the welds of short duration, thyratrons are used to ignite the ignitrons, as shown in Fig. 6.20. The grid circuits of the thyratrons are controlled by a suitable timing control circuit.

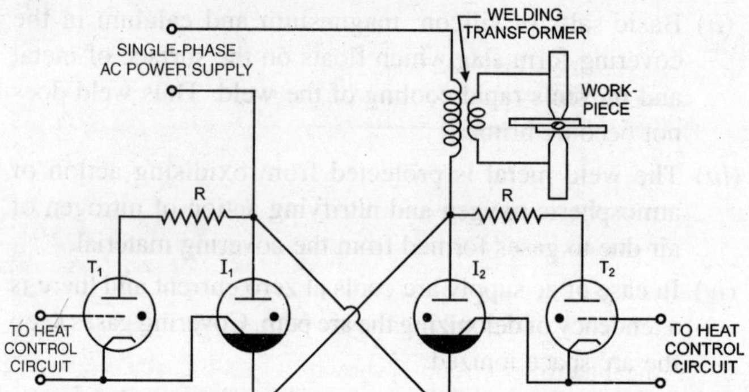

Fig. 6.20

Since very heavy current flows through the ignitrons, say 1,000 amperes, and arc drop be taken constant at 10 volts, therefore, losses to the tune of 10 kW will take place. So the ignitrons are always water cooled. In case of very heavy load, temperature of water becomes too high, normally closed thermostat contacts open and the ignitrons stop conducting.

6.12.2. Heat Control Unit

It is an electronic circuit which helps the delay in the firing of the ignitrons by a definite, predetermined angle in each cycle and operates in conjunction with the line contactor. A typical circuit employed for the heat control is shown in Fig. 6.21. This is essentially a phase shift control circuit and delays the firing of the ignitrons, thus reducing the magnitude of welding current as per requirements.

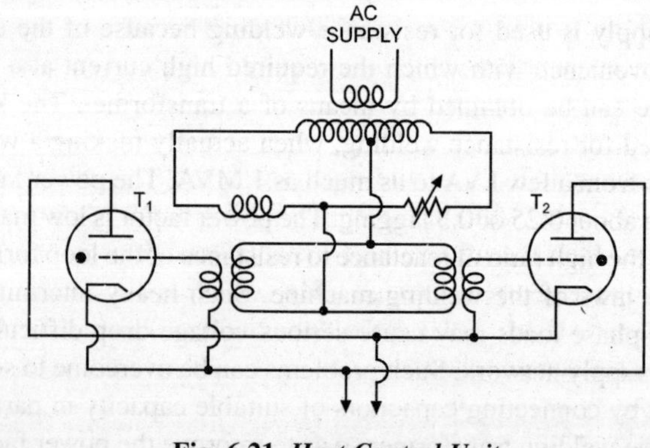

Fig. 6.21. *Heat Control Unit*

6.12.3. AC Timer Circuit

When the condenser C is discharged through a resistor R the voltage across the condenser falls exponentially as given by the expression

$$v = Ve^{-t/CR}$$

$$\text{or} \quad t = RC \log_e \frac{V}{v}$$

From the above expression it is clear that greater the capacity of condenser and value of resistor, greater will be the time required for voltage to fall by given amount. In all timer circuits provision is, therefore, made for charging a condenser to a particular value of voltage and then discharge by short-circuiting switch till condenser is discharged to a particular value when relay will operate and particular contact will open or close.

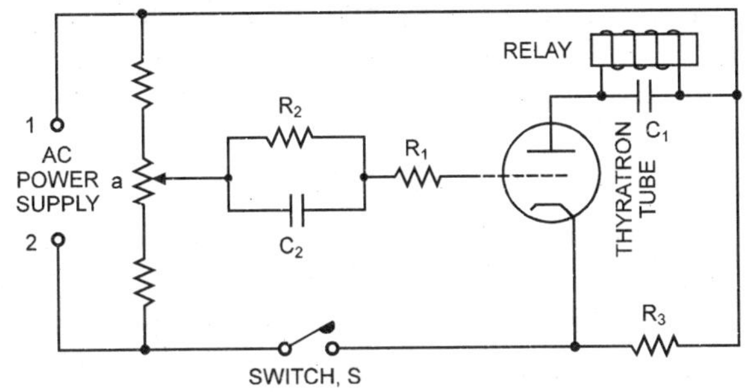

Fig. 6.22. *AC Timer Circuit*

Typical ac timer circuit is shown in Fig. 6.22. Such a timer circuit is used to control the number of cycles for which power may be supplied into the weld. The action of such a timer circuit is explained below.

When switch S is open and supply terminal 1 is positive w.r.t. terminal 2, the cathode and anode of thyratron are at the same potential, and grid is – ve w.r.t. cathode, and, therefore no current flow between cathode and grid. When terminal 1 is negative w.r.t. terminal 2, the potential at *a* is positive, grid becomes positive w.r.t. cathode and anode and electronic current flows through R_2, R_1 from grid to cathode, through R_3 and to terminal 1. In a few cycles the capacitor C_2 will get charged to the maximum voltage between *a* and 1. This is due to the large value of time constant. The capacitor does not discharge much during the negative half cycle of the grid voltage. Resistance R_1 limits the grid-cathode circuit current to a safe value and also determines the number of cycles in which C_2 will be fully charged. So long switch S remains open, capacitor C_2 remains charged by the grid rectification action. As soon as switch S is closed, the grid becomes very much negative w.r.t. cathode and there is no capacitor charging current through grid rectification. Capacitor will, therefore, start discharging through R_2 and –ve bias* of the grid will gradually decrease depending upon the time constant R_2C_2 of the discharge circuit. Conduction in thyratron tube will start during positive half cycle of anode voltage when the grid voltage instantaneously rises to critical grid voltage. Current through relay coil is rectified half waves. As such to avoid relay terminals chatter, a capacitor C_1 is connected across relay coil.

6.12.4. Energy Storage Welding Processes

To meet the demand of heavy current of very high conductivity metals such as aluminium and magnesium *energy storage welding circuits* are used. There are basically two such circuits namely electrostatically stored energy circuits and electromagnetically stored energy circuits.

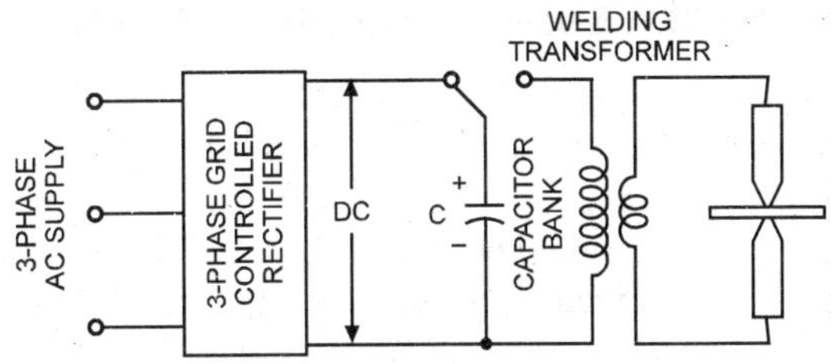

Fig. 6.23. *Capacitor Discharge Welding Circuit*

1. ***Capacitor Discharge Welding Circuit.*** As shown in Fig. 6.23, condenser C (capacitor bank of capacity of 2,000 to 3,000 μF) is charged to about 3,000 volts from grid controlled rectifier. When the condenser is connected to the primary of welding transformer by ignitron contactor, it will discharge and thus high transient current will be produced in the secondary to weld the material. The noteworthy points in connection with this circuit are :

(*i*) As the voltage of condenser approaches the voltage of the source of supply, charging rate becomes lower, therefore to charge condenser to about 3,000 V at high charging rate voltage of about 5,000 V to 6,000 V will be required. A voltage regulating circuit cuts off the rectifier from the bank when the voltage of the bank becomes 3,000 V.

(*ii*) If there is residual magnetism near saturation, it will result in low rate of change of flux linkages in the secondary and, therefore, in production of low heat. Hence in the welding transformer core flux should not be present.

2. ***Magnetic Energy Storage Welding Circuit.*** In this type of welding, energy stored in magnetic circuit is used in the welding operation. The dc voltage of the rectifier is suitably controlled so that the current in the primary of the transformer rises gradually without inducing large current in the secondary. This is necessary to avoid preheating of metals at the weld joint. Preheating in aluminium, magnesium etc. is undesirable as it causes deformation.

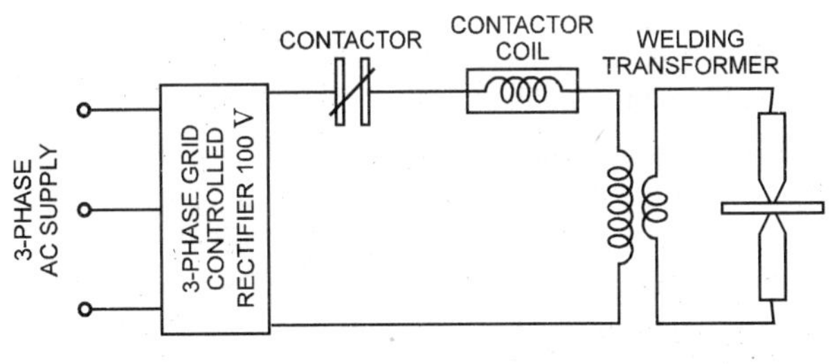

Fig. 6.24. *Magnetic Energy Storage Welding Circuit*

When sufficient energy has been stored up in the transformer core, the contactor opens, dc flow ceases and there is a rapid collapse of magnetic field. The decay of flux induces heavy currents in the secondary of the transformer for welding.

The kVA demand on the line in magnetic energy storage welding is higher as compared to that in capacitor discharge welding but a high voltage rectifier and costly capacitor bank are not required.

* Because of the combination of R_2 and C_2, the grid has two voltages impressed on it : one is the ac component (ac voltage between *a* and 2) and the other is the dc component due to the voltage to which the capacitor C_2 is charged. The dc component dies out at exponential rate.

6.13. POWER SUPPLY FOR ARC WELDING

Both dc and ac are used for electric arc welding, each having its particular applications; in some cases either is suitable.

The electric properties of the arc determines the requirements that should be met by source of power supply for arc welding. To initiate an arc, a higher voltage is required than to maintain it under normal conditions. For this reason the open-circuit voltage of the power supply source (when there is no arc drawn between the electrode and the work) must be higher than the arc, or closed-circuit voltage (when there is an arc drawn between the electrode and the work).

With dc the open-circuit voltage must be at least 30 or 35 volts, while with ac it should be not lower than 50 or 55 V. An open arc will be sustained at 18 to 25 volts. Open-circuit voltage usually ranges from 50 to 90 volts.

To initiate an arc, the electrode is touched to the work and is then withdrawn a short distance. On contact, the electrode shorts the welding circuit and an inadvertent short circuit may give rise to an excessive current in the circuit, which may dangerously heat and even burn the insulation. A power supply source for welding should, therefore, be able to limit the surge of current, when the arc is initiated, to a small percentage above the desired value.

In metal arc welding the globules of molten electrode metal are carried across the arc, thus changing the resistance of the arc and consequently the arc voltage and current. In fact the arc resistance is continually varying within broad limits in very short periods of time. The point is that the molten globules of metal pass from the electrode to the plate at the rate of 30 or more per second, continually causing approximate short circuits from the electrode to the work. If the arc is not to go out, the power supply source must be able to quickly vary its voltage with the resistance of the arc.

Of special importance is the ability of a power supply source to restore the voltage quickly after a globule of molten metal has broken away from the electrode. The arc will only restrike at not less than 25 volts. The voltage should, therefore, recover to 25 volts within 0.05 second.

For manual welding with coated electrodes the static characteristic (curve giving relation between the source voltage and the welding current) is represented by curve I in Fig. 6.25. This is a drooping or negative characteristic. OA represents the open-circuit voltage, OB represents the arc or closed circuit voltage, OD represents the arc current and OE represents the short-circuit current. The point C on the curve I shows the instant of striking the arc. With this type of volt-ampere characteristic the short-circuit current in the welding circuit is reduced to a limit safe enough for the windings of the welding generator or transformer.

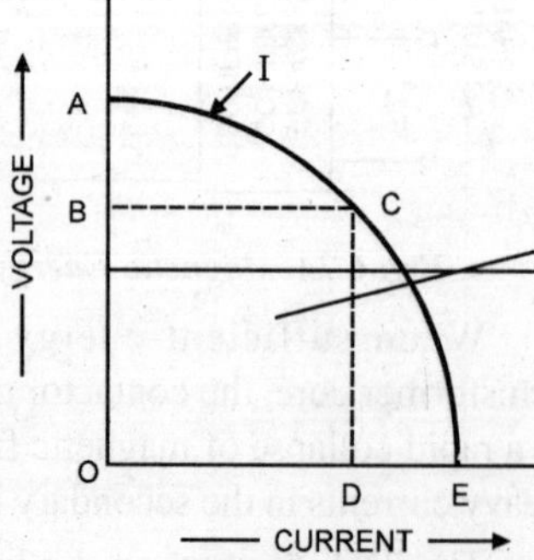

Fig. 6.25

If the arc has a flat (curve III) or a rising volt-ampere characteristic (curve II) such as in shielded arc welding, automatic welding with large currents the source of power supply should also have a flat or a rising characteristic respectively.

To sum up, the power supply for manual arc welding should meet the following requirements :

1. The open-circuit voltage must be sufficient for an arc to strike without difficulty, but safe for the operator (below 80 volts).
2. The short-circuit current must be within limits of safety for generator or transformer windings.
3. The voltage of the source of power supply should vary rapidly with changes in arc length.
4. The wattage of the source of power supply should be sufficient to give the desired arc current.

6.14. ELECTRIC WELDING EQUIPMENT

The electric welding sets may be either dc or ac type.

1. *DC Welding Sets.* Such sets are of two types namely (*i*) generator type and (*ii*) rectifier type.

Generator type welding set consists of a differential compound wound dc generator, giving drooping volt-ampere characteristic, driven by any type of prime mover (a squirrel cage induction motor or a petrol or diesel engine). In a differential compound wound dc generator the terminal voltage falls automatically with the increase in load current. The control may be obtained by tapping the series field or by providing a suitable shunt across the series field winding. The open-circuit voltage is adjusted from shunt field.

If supply from existing dc distribution system is to be used for welding then a ballast (resistance) is put in series with the equipment and control is obtained by varying this external resistance. This method is also suitable when a number of operators are working on the same supply system. In such cases each operator is provided with a separate ballast. The special field of this method is where the service of each arc welding circuit is infrequently in use (say 25% of the time or less). In such cases the loss in the series resistor is less than the no-load loss of a generator type unit.

Another type of dc welding set is a *dry type rectifier* used in conjunction with a multi-phase, high leakage reactance transformer. Many of these rectifier type welders use selenium rectifiers which are forced air cooled. Rectifier type welders are said to combine some of the desirable arcing characteristics of dc welding, such as easy arc starting, with those of welding transformers, such as reduced no-load losses. DC voltage is controlled by regulating the transformer output in this case.

2. *AC Welding Sets.* Single phase or 3-phase step-down transformers which provide low voltage (80-100 voltage on open-circuit) power for welding with some means of output control. One or two taps are provided on the primary side to take care of the voltage variations. The secondary winding of an air-cooled set is generally made of bare wound-on-edge copper strip. The welding transformers may be air or oil-cooled type. The air-cooled sets are lighter in weight and permit the use of insulating materials of the highest thermal classification. They are also cheaper, less hazardous and easier to maintain. On the other hand, the oil-cooled sets are more compact. Being heavy, they are more suitable for stationary applications in workshops. Synthetic liquid filled sets may be employed in hazardous locations. Some machines are provided with an *arc booster* that provides a momentary surge of current to give an arc a good start when it is struck.

In transformer type welding machines the current control is achieved by using (*a*) magnetic shunt or (*b*) a choke coil or reactor placed in series with primary or secondary winding or (*c*) tap changing switch in the primary winding.

In the magnetic shunt type, an adjustable gate of iron laminations is placed between the primary and the secondary windings. Part of the flux is diverted through this gate without linking with the secondary winding, thus providing a simple and reliable means of stepless voltage control.

The choke coil type regulator is a variable reactance with a movable iron plunger. The reactor is such that it operates well below the saturation point. Though it improves the arc stability but reduces the power factor.

The tap-changer type regulator is prone to give trouble on account of arcing on the contacts and the moving finger.

The use of series resistance can also be made for current control but efficiency is reduced.

6.15. ARC WELDING WITH DC AND AC

AC and dc both can be used in arc welding, each having its distinctive merits and applications.

The advantages of arc welding with dc lie in the higher arc stability and the degree to which the work is heated. It is best suited for thinner sheet metal (below 6 mm) and also for welding of non-ferrous metals.

AC welding, due to the absence of 'arc blow', is considered superior for production welding involving large size electrodes. AC welding has a number of economical advantages such as given below :

1. As power supply in factories is usually ac so a rotating dc generator or a rectifier is required for dc welding whilst for ac welding only transformer is required which is relatively cheaper in initial cost.
2. Maintenance of a transformer is less difficult and expensive than that of a dc generator.
3. Operating costs of ac equipment are also lower than those of dc equipment.
4. The electric energy consumption per kg of deposited metal in ac welding ranges from 3 to 4 kWh while for dc welding it is as high as 6 to 10 kWh.

One disadvantage of ac welding is the comparatively low power factor (about 0.4 lagging) which can be corrected by the use of capacitors.

Comparison between dc welding and ac welding is given in tabular form :

S.No.	Particulars	DC Welding	AC Welding
1.	Equipment	Motor–generator set or rectifier is required in case of availability of ac supply, otherwise oil engine-generator set is required.	In case of availability of ac supply only a transformer is required Transformer costs less and its maintenance cost is also low.
2.	Prime cost	Two to three times as compared to that of a transformer.	Comparatively low.
3.	Operating efficiency	Low; consequently high cost of electrical energy.	High (about 85%).
4.	No-load voltage	Low (safer).	Frequently too high (dangerous).
5.	No-load requirements	Very high.	Low (advantageous).
6.	Power factor	Comparatively higher.	Low, requiring capacitors for correction.
7.	Arc blow	Pronounced.	Not so pronounced with ac (advantageous).
8.	Electrodes	Both bare (non-coated) and thus cheap electrodes can be used.	Only coated electrodes—expensive ones.
9.	Heating	Uniform	Not so uniform as in case of dc.
10.	Connected load (cross-sectional area of conductors and fuses)	Normal.	Considerably higher because of low power factor.
11.	Welding of non-ferrous metals	Suitable.	—
12.	Arc stability	Higher.	—

6.16. COMPARISON BETWEEN RESISTANCE AND ARC WELDING

S.No.	Particulars	Resistance Welding	Arc Welding
1.	Supply.	Usually ac only.	AC or dc.
2.	Voltage.	Very low.	The striking voltage is high so requires voltage control.
3.	Power factor.	Very low.	Poor.
4.	Additional material requirement.	No material is added in any form to get the two pieces joined.	Suitable filler metal electrodes are necessary to get proper strength.
5.	External pressure.	External pressure is required.	No external pressure is required hence the equipment is more simple and easy to control.

S.No.	Particulars	Resistance Welding	Arc Welding
6.	Development of heat.	Heat is developed due to flow of current through the contact resistance mainly.	Heat is developed due to arc between electrode and the work piece.
7.	Temperature.	The temperature attained is not very high as in case of arc welding.	The temperature of the arc is very high and so likely to damage the work if not properly handled.
8.	Applications.	It cannot be used for repair work. It is most suitable for mass production.	It is not suitable for mass production. It is most suitable for repair work and where more metal is to be deposited.

Exercises

1. Describe with neat sketches the various methods of electric resistance welding. Give its merits or demerits with respect to arc welding.
[Pb. Univ. Electric Drives and Utilization of Electric Energy August 1984]
2. Name and describe various types of resistance welding processes.
[M.D. Univ. Utilization of Electric Power & Traction May 2009]
3. Why projection welding is considered superior to spot welding ?
[Pb. Technical Univ. Electric Drives and Utilization May 2009]
4. Explain with neat sketches, the construction, principle of operation and applications of three different types of resistance welding. Give its demerits with respect to arc welding.
[U.P. Technical Univ. Utilization of Electrical Energy and Electric Traction 2005-06]
5. Explain the principle of electric spot welding and seam welding.
Discuss the merits and demerits of electric resistance and arc welding. Indicate when each find its use.
[Gorakhpur Univ. Utilization of Elect. Power & Traction 1982]
6. Explain various types of electric welding. Explain seam welding in detail.
[Rajasthan Technical Univ. Utilization of Electrical Power 2011]
7. Explain the role of ignition contactor in welding process. What are its advantages over mechanical switches ?
[U.P. Technical Univ. Utilization of Electrical Energy & Traction 2009-10]
8. What is the fundamental difference between electric arc welding and resistance welding ? Explain with neat sketch how the spot welding is carried out by spot welding machine.
[Gorakhpur Univ. Utilization of Electrical Power & Traction Sept. 1983]
9. Explain the various types of resistance and arc welding. How the mechanical pressure, amount of current and its duration is controlled in case of resistance welding ? What is atomic hydrogen welding ? Explain. Give its relative merits. **[Pb. Univ. Electric Drives & Utilization of Electric Energy Nov. 1983]**
10. Describe briefly various types of arc welding processes used in industry.
[Pb. Technical Univ. Electric Drives & Utilization 2003]
11. Why a drooping characteristics of supply voltage is essential for maintaining steady arc in electric welding and how this characteristic is obtained in case of a dc source.
[Pb. Technical Univ. Electric Drives & Utilization May 2009]
12. Differentiate between carbon arc and metal arc welding.
[Rajasthan Technical Univ. Utilization of Electrical Power 2011]
13. What are the advantages of using coated electrodes ? Mention types of coating used.
[U.P. Technical Univ. Utilization of Electrical Energy and Electric Traction 2005-06]
14. Compare the performance of various electrodes used in electric arc furnaces.
[M.D. Univ. Utilization of Electric Power & Traction Dec. 2009]
15. Describe various welding schemes specifying the equipment for controlling temperature, pressure etc.
Draw a neat sketch of welding transformer showing primary and secondary circuits.
[Pb. Univ. Electric Drives & Utilization of Electric Energy April 1982]
16. Discuss laser welding and electron beam welding.
[Pb. Technical Univ. Electric Drives & Utilization Dec. 2007]
17. Explain electron beam welding and give its disadvantages.
[U.P. Technical Univ. Utilization of Electrical Energy & Traction 2009-10]
18. Explain different types of electric welding. What are qualities of good weld.
[M.D. Univ. Utilization of Electric Power & Traction May 2010]
19. With suitable example explain the requirements of good welding. **[U.P. Technical Univ. Utilization of Electrical Energy & Traction 2009-10]**
20. Compare dc and ac welding methods.
[Pb. Technical Univ. Electric Drives & Utilization May 2008]
21. What are the basic components of dc and ac arc welding sets and explain their working.
[Pb. Technical Univ. Electric Drives & Utilization May 2008]
22. Discuss the need of dc supply for electric welding. Why dc supply is preferred over ac supply for welding ?
[U.P. Technical Univ. Utilization of Electrical Energy 2009-10]
23. What is welding ? What are the different types of electric welding ? What type of electric supply is suitable for electric welding ? Explain.
[Rajasthan Univ. Utilization of Electrical Power 2003, 2005, 2006]
24. What is the fundamental difference between electric arc welding and resistance welding ? Give their merits and demerits.
[M.D. Univ. Utilization of Electric Power & Traction Dec. 2009]
25. Compare in details, the electric arc welding with resistance welding. Explain with the help of a neat sketch the process of spot welding.
[Rajasthan Technical Univ. Utilization of Electrical Power 2011]
26. Explain the basic difference between electric arc welding and resistance welding.
[Pb. Technical Univ. Electric Drives and Utilization Dec. 2009]
27. Write short notes on the following :
(*i*) Resistance welding. **[Rajasthan Univ. 2007]**
(*ii*) Arc welding.
(*iii*) Resistance and arc welding. **[Agra Univ. 1983]**
(*iv*) Atomic hydrogen welding.
(*v*) Ultrasonic welding. **[G.B.T.U. 2010-11]**
(*vi*) Laser welding. **[W.B. Univ. of Technology 2009]**

(vii) Advantages of coated electrodes. **[Pb. Univ. Nov. 1980]**

(viii) Power control in arc welding. **[G.B.T.U. 2010-11]**

(ix) Electric supply for arc welding.

(x) Welding transformer. **[Rajasthan Univ. 2004, 2007]**

(xi) Welding generator and welding transformer. **[M.D. Univ. Dec. 2008]**

Short Answer Type Questions With Answers

Q. 1. What is welding ?

Ans. Welding is a materials-joining process that produces coalescence of materials by heating them to the welding temperature, with or without application of pressure or by the application of pressure alone, and with or without the use of filler metal.

Q. 2. "All metals can be joined by one welding process or another". Comment on this statement.

Ans. The above statement needs slight modification and may be stated as follows.

All metals commercially employed for structural or strength parts are weldable. In general, metal that have a low melting point or low strength would not be welded.

Q. 3. What are the different types of electric welding ? **[P.T.U. May 2007]**

Ans. Elecuric welding may be resistance welding or arc welding. Resistance welding includes butt welding, spot welding, projection welding seam welding and percussion welding.

Arc welding is subdivided into five groups namely carbon arc welding, metal arc welding, atomic hydrogen arc welding, inert gas metal arc welding and submerged arc welding.

Q. 4. What is the difference between plastic welding and fusion welding ? **[Gorakhpur Univ. 1983]**

Ans. In the plastic welding, also called the pressure welding, the pieces of metal to be joined are heated to a plastic state and then forced together by external pressure whereas in fusion welding, also called the non-pressure welding, the material at the joint is heated to a molten state and allowed to solidify. Former one includes forge welding, resistance welding, thermit welding and gas welding while the latter includes gas welding, arc welding and thermit welding without pressure.

Q. 5. What is resistance welding ?

Ans. Resistance welding is a group of welding processes that produce coalescence of the faying surfaces with the heat obtained from resistance of the work to the flow of welding current in a circuit of which the work is a part and by the application of pressure.

Q. 6. What is the fundamental difference between electric arc welding and resistance welding ? **[Gorakhpur Univ. Sept., 1983]**

Ans. The resistance welding processes differ from arc welding in that pressure is used but filler metal or fluxes are not.

Q. 7. Why alternating current is found most suitable for resistance welding ? **[Gorakhpur Univ. 1980]**

Ans. Alternating current is found most suitable for resistance welding as it can provide any desired combination of current and voltage by means of a transformer.

Q. 8. Explain the theory of arc production. **[P.T.U. Dec. 2006]**

Ans. An electric arc is produced when two conductors (electrode and metal piece) are connected to a suitable source of electric current momentarily in contact and then separated by a small distance. Both dc and ac can be used for producing an electric arc, however, dc provides more steady arc.

Basically an arc is nothing but an electric discharge through a path of ionised particles of air. The electrical energy is converted into heat at a temperature of about 5,000°C.

Q. 9. What is meant by negative resistance characteristic of an electric arc and what precautions are taken for ill effects due to this ?

Ans. Negative resistance characteristic of an arc means decrease in arc resistance with the increase in temperature or current. The reduced resistance causes further increase in current and, therefore, steady arc cannot be obtained. For obtaining steady arc, supply source should have a drooping characteristic (*i.e.,* voltage decreasing with the increase in current).

Steady arc for dc welding can be obtained by using differential compound wound generator. The control may be obtained by tapping the series field or by providing a suitable shunt across the series field winding. The open-circuit voltage is adjusted from shunt field.

In an ac welding sets, welding transformer is used and drooping characteristic is obtained by using external reactor.

Q. 10. What are the fluxes used in arc welding ? **[P.T.U. Dec. 2009]**

Ans. In arc welding, flux coated electrodes are used. The fluxes used are silicon oxiders, graphite, sodium silicate, calcium oxide and aluminium in right proportions. Sometimes, the electrodes are washed (lightly coated) of several tenths of millimetres thickness, lime mixed with soluble glass.

The heavy coated electrodes have coating composed of ionising (chalk) deoxidising (aluminium, ferro-manganese etc.), gas generating (starch), sag forming (kaoline), alloying and binding materials.

Q. 11. What are the qualities of good weld ? **[P.T.U. Dec. 2007; Dec. 2008]**

Ans. The weld is said to be good weld if it is free from external defects such as irregular width and heights of beads, deviation of the weld from the prescribed dimensions, unfilled craters on the surface of welds, undercuts, slag on the surface of welds, porosity in the outer layer of the welds, visible cracks in and around the weld and internal defects such as poor fusion, hidden porosity in the deposited metal, hidden cracks in the weld and parent metal, poor fusion between layers in multiple layer welding, and entrapment of slag and oxides.

Q. 12. What is main drawback of electron beam process ?

Ans. The main drawback of electron beam process is its high capital cost. The price of the equipment required is very high, and it is expensive to operate, due to the requirement of vacuum pumps. In addition, fitup must be precise and locating the parts w.r.t. the beam must be perfect.

Q. 13. The location of the focal point of the beam w.r.t. the surface of the work piece is very important in laser beam welding. Why ?

Ans. Maximum penetration occurs when the beam is focused slightly below the surface. Penetration is less when the beam is focused on the surface or deep within the surface.

Q. 14. What is meant by welding electrode ?

Ans. An electrode is a piece of wire or rod (of a metal or alloy), with or without flux covering, which carries current for welding.

Q. 15. What type of dc generator is used in electric arc welding ?

Ans. Differential compound wound dc generator.

CHAPTER 7

Illumination Engineering

INSIDE THIS CHAPTER

7.1. Introduction **7.2.** Terms Used in Illumination **7.3.** Laws of Illumination **7.4.** Polar Curves **7.5.** Photometry **7.6.** Integrating Sphere **7.7.** Measurement of Illumination **7.8.** Sources of Light **7.9.** Arc Lamps **7.10.** Incandescent Lamps **7.11.** Gaseous Discharge Lamps **7.12.** Comparison Between Tungsten Filament Lamps and Fluorescent Tubes **7.13.** Basic Principles of Light Control **7.14.** Types of Lighting Schemes **7.15.** Design of Lighting Schemes **7.16.** Factory Lighting **7.17.** Methods of Lighting Calculations **7.18.** Street Lighting **7.19.** Flood Lighting.

7.1. INTRODUCTION

Light is the prime factor in the human life as all activities of human beings ultimately depend upon the light. Where there is no natural light, use of artificial light is made. Artificial lighting produced electrically, on account of its cleanliness, ease of control, reliability, steady output, as well as its low cost, is playing an increasingly important part in modern everyday life. Apart from its aesthetic and decorative aspects, good lighting has a strictly utilitarian value in increasing production, reducing workers fatigue, protecting their health, eyes and nervous system, and reducing accidents. The science of illumination engineering is, therefore, becoming of major importance.

Nature of Light. Light is a form of radiant energy. Various forms of incandescent bodies are the sources of light and the light emitted by such bodies depend upon the temperature of bodies. Heat energy is radiated into the medium by a body which is hotter than the medium surrounding it. The heat of the body, as seen, can be classified as red-hot or white-hot. While the body is red-hot the wavelength of radiated energy will be sufficiently large and the energy available is in the form of heat. When the temperature increases the body changes from red-hot to white-hot state, the wavelength of the energy radiated becomes smaller and enters into the range of the wavelength of light. Actually the hot bodies emit heat as well as light energy, the velocity of these being equal to 3×10^8 m/s. The light waves have wavelengths varying from 0.00075 mm to 0.0004 mm. The wavelength of light is usually expressed in Angstrom ; 1 Angstrom being equal to 10^{-10} metre. The ratio of the energy emitted by the body in the form of light to the total energy emitted by the body is known as the '*radiant efficiency*' of the body, which depends upon the temperature. Higher the temperature of the body ; lower the wavelength of the radiant energy and higher the radiant efficiency. The radiant efficiency will be maximum when the temperature of the body will be such that the wavelength of the shortest wave radiated by the body is 0.0004 mm (4,000 Å). It is found that maximum radiant efficiency would occur at about 6,250°C and even then it would be 20%. Since this temperature is far above the highest that has yet been obtained in practice, it is obvious that the actual efficiency of all artificial light sources, *i.e.*, those depending upon the temperature of incandescence is quite low.

Colour. The sensation of colour is due to the difference in the wavelengths of the light radiations. Visible light can have a wavelength between 4,000 Å and 7,500 Å and the colour varies in the way, as shown in Figs. 7.1 and 7.2.

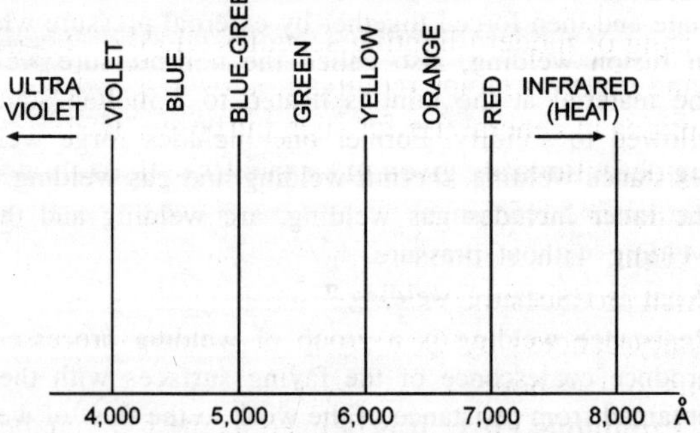

Fig. 7.1. *Wavelength and Colour*

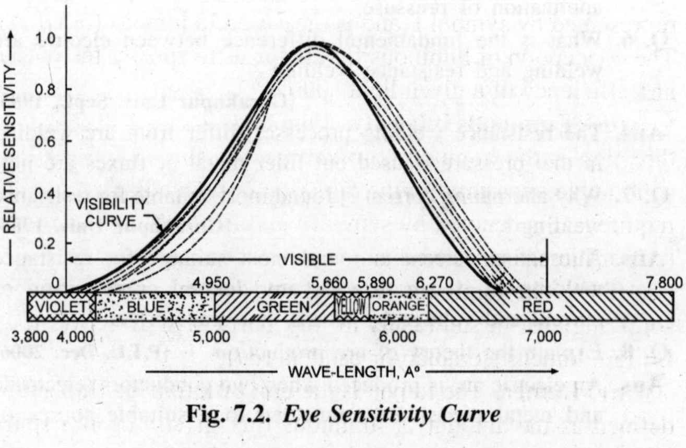

Fig. 7.2. *Eye Sensitivity Curve*

Relative Sensitivity. The sensitivity of the eye to the lights of different wavelengths varies from person to person and according to the age. The average relative sensitivity is shown in Fig. 7.2. The eye has greatest sensitivity for wavelengths of about 5,500 Å ; that is, yellow-green can be seen under poor conditions of illumination when blue or red cannot be seen under dim illumination, the sensitive curve tends to shift, as shown by the shaded region in Fig. 7.2. Therefore, violet disappears first and red remains visible. Yellow disappears last as the illumination becomes very dim. As each colour disappears, it becomes a grey shade, and finally black. The sensitivity of the eye to yellow-green radiation is taken as unity or 100% and the sensitivity to other wavelengths is expressed as a fraction or percentage of it. The relative sensitivity at any particular wavelength λ is written $K\lambda$ and is known as *relative luminosity factor.*

Illumination differs from light very much, though generally these terms are used more or less synonymously. Strictly speaking light is the cause and illumination is the result of that light on surfaces on which it falls. Thus the illumination makes the surface look more or less bright with a certain colour and it is this brightness and colour which the eye sees and interprets as something useful, or pleasant or otherwise.

Light may be produced by passing electric current through filaments as in the incandescent lamps, through arcs between carbon or metal rods, or through suitable gases as in neon and other gas tubes. In some forms of lamps the light is due to fluorescence excited by radiation arising from the passage of electric current through mercury vapour.

Some bodies reflect light in some measure, and when illuminated from an original source they become secondary sources of light. A good example is the Moon, which illuminates the earth by means of the reflected light originating in the Sun.

7.2. TERMS USED IN ILLUMINATION

The devising of modern lighting schemes and the selection of fittings and types of lamps requires a knowledge of the terms and quantities in general use for such purposes. Therefore, the following definitions are given in simple form to facilitate easy identification and reference.

(*i*) Light. It is defined as the radiant energy from a hot body which produces the visual sensation upon the human eye. It is usually denoted by Q, expressed in lumen-hours and is analogous to watt-hours.

(*ii*) Luminous Flux. It is defined as the total quantity of light energy emitted per second from a luminous body. It is represented by symbol F and is measured in lumens (or *cd-sr*). The conception of luminous flux helps us to specify the output and efficiency of a given light source.

(*iii*) Luminous Intensity. Luminous intensity in any given direction is the luminous flux emitted by the source per unit solid angle, measured in the direction in which the intensity is required. It is denoted by symbol I and is measured in candela (*cd*) or lumens per steradian.

If F is the luminous flux radiated out by source within a solid angle of ω steradians in any particular direction then $I = F/\omega$ lumens/steradian or candela (*cd*).

(*iv*) Lumen. The lumen is the unit of luminous flux and is defined as the amount of luminous flux given out in a space represented by one unit of solid angle by a source having an intensity of one candle power in all directions.

i.e., $\text{Lumens} = \text{Candle power} \times \text{solid angle} = \text{CP} \times \omega$

or total lumens given out by source of one candela is 4π lumens.

(*v*) Candle Power. Candle power is the light radiating capacity of a source in a given direction and is defined as the number of lumens given out by the source in a unit solid angle in a given direction. It is denoted by symbol CP.

$$i.e., \ \text{CP} = \frac{\text{Lumens}}{\omega}$$

(*vi*) Illumination. When the light falls upon any surface, the phenomenon is called the *illumination.* It is defined as the number of lumens, falling on the surface, per unit area. It is denoted by symbol E and is measured in lumens per square metre or lux or metre-candle.

If a flux of F lumens falls on a surface of area A, then the illumination of that surface is

$$E = \frac{F}{A} \text{ lumens/m}^2 \text{ or lux } (lx)$$

(*vii*) Lux or Metre-Candle. It is the unit of illumination and is defined as the luminous flux falling per square metre on the surface which is everywhere perpendicular to the rays of light from a source of one candle power and one metre away from it.

(*viii*) Foot-Candle. It is also the unit of illumination and is defined as the luminous flux falling per square foot on the surface which is everywhere perpendicular to the rays of light from a source of one candle power and one foot away from it.

$$i.e., \ 1 \text{ Foot-candle} = 1 \text{ lumen/ft}^2$$

$$= \frac{1 \text{ lumen}}{\left(\frac{1 \text{ m}}{3.28}\right)^2} = 10.76 \text{ metre-candle or lux.}$$

(*ix*) Candela. It is the unit of luminous intensity. It is defined as 1/60th of the luminous intensity per cm^2 of a black body radiator at the temperature of solidification of platinum (2,043 K).

(*x*) Mean Horizontal Candle Power (MHCP). It is defined as the mean of candle powers in all directions in the horizontal plane containing the source of light.

(*xi*) Mean Spherical Candle Power (MSCP). It is defined as the mean of candle powers in all directions and in all planes from the source of light.

(*xii*) Mean Hemispherical Candle Power (MHSCP). It is defined as the mean of candle powers in all directions above or below the horizontal plane passing through the source of light.

(*xiii*) Reduction Factor. Reduction factor of a source of light is the ratio of its mean spherical candle power to its mean horizontal candle power.

$$i.e., \ \text{Reduction factor} = \frac{\text{MSCP}}{\text{MHCP}}$$

(*xiv*) Lamp Efficiency. It is defined as the ratio of the luminous flux to the power input. It is expressed in lumens per watt.

(*xv*) Specific Consumption. It is defined as the ratio of the power input to the average candle power. It is expressed in watts per candela.

(*xvi*) Brightness or Luminance. When the eye receives a great deal of light from an object we say it is bright, and 'brightness' is an important quantity in illumination. It is all the same whether the light is produced by the object or merely reflected from it. The object sends out light as though each small piece of its surface were of a certain luminous intensity. Generally, the brightness of an object is not the same from all points of view. When brightness is considered in terms of measurable quantity, the term *luminance* is employed.

Brightness or *luminance* is defined as the luminous intensity per unit projected area of either a surface source of light or a reflecting surface and is denoted by L.

If a surface of area A has an effective luminous intensity of I candelas in a direction θ to the normal, then the luminance (brightness) of that surface is

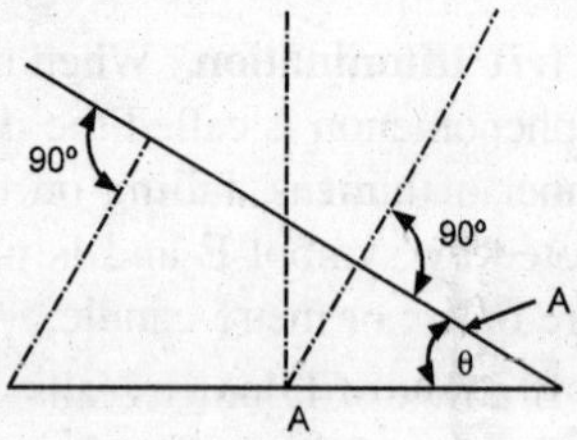

$$L = \frac{I}{A \cos\theta} \text{ candela/m}^2 \text{ or nits}$$

Fig. 7.3

i.e., nit is defined as candela per square metre. Bigger unit of luminance is *stilb* which is defined as candelas per square cm. Lambert is also the unit of brightness which is lumens/cm^2. Foot-lambert is lumens/ft^2.

***(xvii)* Glare.** The size of the opening of the pupil in the human eye is controlled by its iris. If the eye is exposed to a very bright source of light, the iris automatically contracts in order to reduce to amount of light admitted and prevent damage to retina ; this reduces the sensitivity, so that other objects within the field of vision can be only imperfectly seen. This effect is referred to 'glare' and is familiar in connection with motor-car headlights. *In other words glare may be defined as the brightness within the field of vision of such a character as to cause annoyance, discomfort, interference with vision or eye fatigue.*

***(xviii)* Space-height Ratio.** It is defined as the ratio of horizontal distance between adjacent lamps and height of their mountings.

$$\textit{i.e.}, \text{ Space-height ratio} = \frac{\text{Horizontal distance between two adjacent lamps}}{\text{Mounting height of lamps above working plane}}$$

***(xix)* Utilization Factor or Coefficient of Utilization.** It is defined as the ratio of total lumens reaching the working plane to total lumens given out by the lamp.

$$\textit{i.e.}, \text{ Utilization factor} = \frac{\text{Total lumens reaching the working plane}}{\text{Total lumens given out by the lamp}}.$$

***(xx)* Maintenance Factor.** Due to accumulation of dust, dirt and smoke on the lamps, they emit less light than that they emit when they are new ones and similarly the walls and ceilings, etc., after being covered with dust, dirt and smoke do not reflect the same output of light, which is reflected when they are new.

The ratio of illumination under normal working conditions to the illumination when the things are perfectly clean is known as maintenance factor.

i.e., Maintenance factor

$$= \frac{\text{Illumination under normal working conditions}}{\text{Illumination when everything is perfectly clean}}.$$

***(xxi)* Depreciation Factor.** This is merely the inverse of the maintenance factor and is defined as the ratio of initial metre-candles to the ultimate maintained metre-candles on the working plane. Its value is more than unity.

***(xxii)* Waste Light Factor.** Whenever a surface is illuminated by a number of sources of light, there is always a certain amount of waste of light on account of overlapping and falling of light outside the edges of the surface. The effect is taken into account by multiplying the theoretical value of lumens required by 1.2 for rectangular areas and 1.5 for irregular areas and objects such as statues, monuments etc.

***(xxiii)* Absorption Factor.** In the places where atmosphere is full of smoke fumes, such as in foundries, there is a possibility of absorption of light. The ratio of total lumens available after absorption to the total lumens emitted by the source of light is called the absorption factor. Its values varies from unity for clean atmosphere to 0.5 for foundries.

***(xxiv)* Beam Factor.** The ratio of lumens in the beam of a projector to the lumens given out by lamps is called the *beam factor.* This factor takes into account the absorption of light by reflector and front glass of the projector lamp. Its value varies from 0.3 to 0.6.

***(xxv)* Reflection Factor.** When a ray of light impinges on a surface it is reflected from the surface at an angle of incidence, as shown in Fig. 7.4. A certain portion of incident light is absorbed by the surface. The ratio of reflected light to the incident light is called the '*reflection factor*'. It is always less than unity.

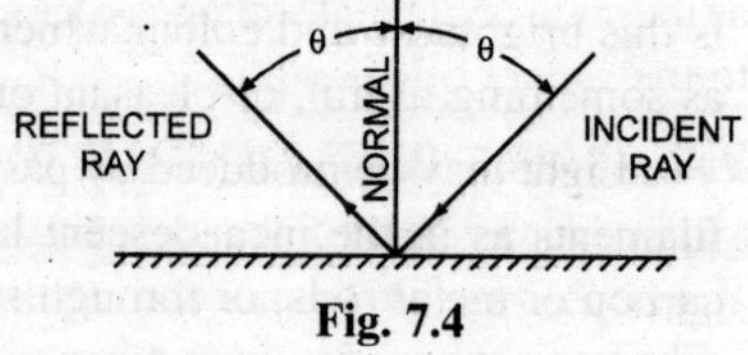

Fig. 7.4

***(xxvi)* Solid Angle.** Plane angle is subtended at a point in a plane by two converging straight lines and its magnitude is given by

$$\theta = \frac{\text{Arc}}{\text{Radius}} \text{ radians}$$

The largest angle subtended at a point is 2π radians.

Solid angle is the angle generated by the surface passing through the point in space and the periphery of the area. Solid angle is denoted by ω, expressed in steradians and is given by the ratio of the area of the surface to the square of the distance between the area and the point.

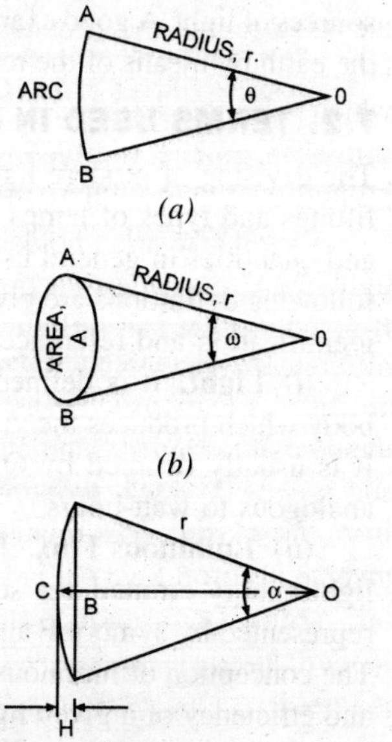

(a)

(b)

(c) Sectional View for Solid Angle

Fig. 7.5

$$\textit{i.e.}, \omega = \frac{\text{Area}}{(\text{Radius})^2} = \frac{A}{r^2}$$

The largest solid angle subtended at a point is that due to a sphere at its centre. If *r* is the radius of any sphere, its surface area is $4\pi r^2$ and the distance of its surface area from the centre is *r*, therefore, solid angle subtended at its centre by its surface,

$$\omega = \frac{4\pi r^2}{r^2} = 4\pi \text{ steradians.}$$

Relationship between plane and solid angle can be obtained as follows.

Consider a curved surface of a spherical segment of height H and radius r [Fig. 7.5. (c)].

Surface area of such a segment = $2\pi r H$

From Fig. 7.5 (c)

$$H = OC - OB = r - r\cos\frac{\alpha}{2} = r\left(1-\cos\frac{\alpha}{2}\right)$$

Surface area of the surface under consideration

$$= 2\pi r H = 2\pi r \times r\left(1-\cos\frac{\alpha}{2}\right) = 2\pi r^2\left(1-\cos\frac{\alpha}{2}\right)$$

$$\text{Solid angle, } \omega = \frac{\text{Surface area}}{(\text{Radius})^2} = \frac{2\pi r^2\left(1-\cos\frac{\alpha}{2}\right)}{r^2} = 2\pi\left(1-\cos\frac{\alpha}{2}\right)$$

Thus solid angle subtended at a point can be determined directly from the above relation knowing the plane angle subtended by its section.

(*xxvii*) Steradian. It is the unit of solid angle and is defined as the solid angle that subtends a surface on the sphere equivalent to the square of the radius.

7.3. LAWS OF ILLUMINATION

There are two laws of illumination (*i*) Law of inverse squares and (*ii*) Lambert's cosine law.

1. *Law of Inverse Squares*. If a source of light which emits light equally in all directions be placed at the centre of a hollow sphere, the light will fall uniformly on the inner surface of the sphere, that is to say, each square mm of the surface will receive the same amount of light. If the sphere be replaced by one of the larger radius, the same total amount of light is spread over a larger area proportional to the square of the radius. The amount which falls upon any square mm of such a surface will, therefore, diminish as the radius increases, and will be inversely proportional to the square of the distance.

A similar relation holds if we have to deal with a beam of light in the form of a cone or pyramid, as shown in Fig. 7.6 (a). If we consider parallel surfaces which cut the pyramid at different distances from the source, the areas of these surfaces are proportional to the square of these distances, and, therefore, the amount of light which falls on one unit of the area of these surfaces, is inversely proportional to the square of the distance from the source. This relationship is referred to as the *law of inverse squares*.

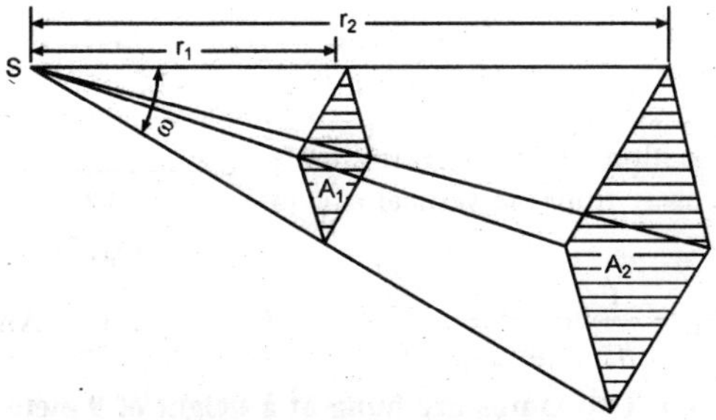

Fig. 7.6 (*a*). *Inverse Square Law*

Mathematically it can be proved as follows :

Let us consider surface area A_1 and surface area A_2 at distances r_1 and r_2 respectively from the point source S of luminous intensity I and normal to the rays, as shown in Fig. 7.6 (a).

Let the solid angle subtended be ω steradians

Luminous flux radiated per steradians = I

Total luminous flux radiated = Iω lumens

Illumination on the surface of area A_1

$$= \frac{I\omega}{A_1} \text{ lumens/unit area}$$

and area $A_1 = \omega r_1^2$

∴ Illumination on the surface of area A_1,

$$E_1 = \frac{I\omega}{\omega r_1^2} = \frac{I}{r_1^2} \text{ lumens per unit area}$$

Similarly illumination on the surface of area A_2,

$$E_2 = \frac{I\omega}{A_2} = \frac{I\omega}{\omega r_2^2} = \frac{I}{r_2^2} \text{ lumens per unit area.}$$

Hence the illumination of a surface is inversely proportional to the square of the distance between the surface and the light source provided that the distance between the surface and the source is sufficiently large so that the source can be regarded as a point source.

2. Lambert's Cosine Law. Very often the illuminated surface is not normal to the direction of light as AC in Fig. 7.6 (b) but is inclined as AB. The area over which the light is spread is then increased in the ratio

$$\frac{AB}{AC} = \frac{1}{\cos\theta}$$

and the illumination decreases in the ratio $\frac{\cos\theta}{1}$.

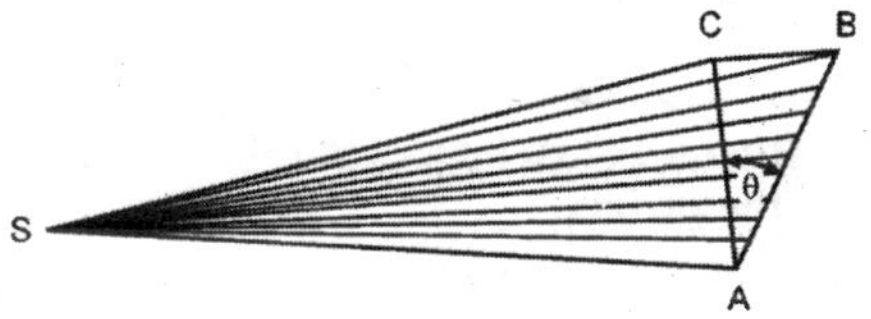

Fig. 7.6 (*b*). *Lambert's Cosine Law*

The expression for the illumination then becomes

$$E = \frac{I\cos\theta}{r^2}$$

According to this law the illumination at any point on a surface is proportional to the cosine of the angle between the normal at that point and the direction of luminous flux.

Example 7.1. The flux emitted by a lamp in all directions is 1,000 lumens. Calculate its MSCP.

[*A.M.I.E. Sec. B. Utilization of Elec. Power Winter 1984*]

Solution : Total luminous flux emitted, F = 1,000 lumens

$$\text{MSCP of the lamp} = \frac{F}{4\pi} = \frac{1{,}000}{4\pi} = 80 \textbf{ Ans.}$$

Example 7.2. State laws of illumination. A 250 V lamp has a total flux of 1,500 lumens and takes a current of 0.4 A. Calculate (*i*) Lumen per watt (*ii*) MSCP per watt.

[*M.D. Univ. Utilization of Electric Power and Traction Dec. 2009*]

Solution : Wattage of lamp = V × I = 250 × 0.4 = 100 watts

$$\text{Mean spherical candle power of lamp, MSCP} = \frac{F}{4\pi} = \frac{1{,}500}{4\pi} = 120$$

$$(i) \text{ Lumens per watt} = \frac{\text{Output of lamp in lumens}}{\text{Wattage of lamp in watts}} = \frac{1{,}500}{100} = 15 \textbf{ Ans.}$$

$$(ii) \text{ MSCP per watt} = \frac{\text{MSCP of lamp}}{\text{Wattage of lamp}} = \frac{120}{100} = 1.2 \textbf{ Ans.}$$

Example 7.3. A 0.4 metre diameter diffusing sphere of opal glass (20 per cent absorption) encloses an incandescent lamp with a luminous flux of 4,850 lumens. Calculate the average luminance of the sphere.

Solution : Flux emitted by the globe, F = 0.8 × 4,850 = 3,880 lumens

Surface area of the globe $= 4\pi r^2 = 4\pi (0.2)^2 = 0.16\,\pi$

$$\text{Average luminance of sphere} = \frac{\text{Flux emitted}}{\text{Surface area}}$$

$$= \frac{3{,}880}{0.16\pi} = 7{,}720 \text{ lumens/m}^2 \quad \textbf{Ans.}$$

Example 7.4. Filament lamp of 200 W is suspended at a height of 5 metres above working plane and gives uniform illumination over an area of 4 metre diameter. Assume efficiency of reflector as 50%. Determine the illumination on the working plane. Efficiency of lamp is 0.89 watt/CP.

[*West Bengal Univ. of Technology Utilization of Electric Power 2009*]

Solution : Candle power of lamp,

$$CP = \frac{\text{Wattage of lamp}}{\text{Efficiency of lamp in watt / CP}} = \frac{200}{0.89} = 224.72$$

Lumens output of lamp $= 4\pi CP = 4\pi \times 224.72 = 2{,}823.9$ lumens

Flux emitted by reflector

= Reflector efficiency × total output lumens

0.5 × 2,823.9 = 1,411.95 lumens

$$\text{Area of working plane, A} = \frac{\pi}{4}d^2 = \frac{\pi}{4}\times 4^2 = 4\pi \text{ m}^2$$

Illumination on the working plane

$$= \frac{\text{Lumens output}}{\text{Working plane area, A}}$$

$$= \frac{1{,}411.95}{4\pi^2} = 112.36 \text{ lumens/m}^2 \quad \textbf{Ans.}$$

Example 7.5. Determine the MSCP of a lamp emitting 1,000 lumens. A surface inclined at an angle of 60° to the rays is kept 5 metres away from a 100 cp lamp. Find the average intensity of illumination on the surface.

Solution : MSCP of a lamp emitting 1,000 lumens

$$= \frac{1{,}000}{4\pi} = 80 \quad \textbf{Ans.}$$

Average intensity of illumination on the surface inclined at an angle 60° to the rays and kept 5 metres away from a 100 CP lamp.

$$= \frac{CP}{d^2}\cos\theta = \frac{CP}{d^2}\cos(90° - 60°) = \frac{100}{(5)^2} \times 0.866$$

$$= 3.464 \text{ lux} \quad \textbf{Ans.}$$

Example 7.6. Deduce the relation to find illumination at any point on the surface due to light source suspended at a height *h* from the plane surface.

Solution : Let us consider a point P on the plane surface AB where illumination due to light source S of candle power CP at a height *h* from the plane AB is to be determined.

Let the distance between point P and light source S be *r*.

By laws of illumination,

Illumination at point P

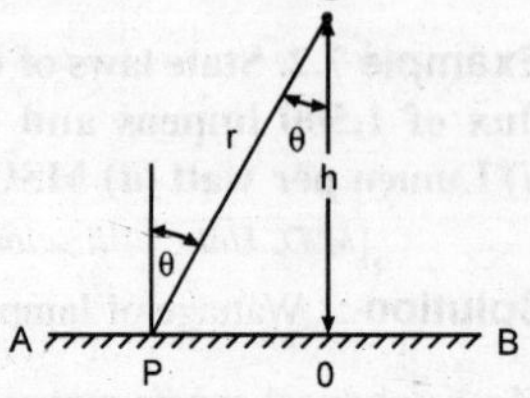

Fig. 7.7

$$= \frac{CP}{r^2}\cos\theta \quad \text{where } \cos\theta = \frac{h}{r}$$

$$= \frac{CP}{(h/\cos\theta)^2}\cos\theta = \frac{CP}{h^2}\cos^3\theta$$

and illumination at any point O, vertically below the source of light

$$= \frac{CP}{h^2}.$$

Hence illumination at any point on a plane is $\cos^3\theta$ *times of illumination at a point just vertically below the light source, where* θ *is the angle between the normal to the surface at the point and rays of light.*

Example 7.7. Define and explain 'illumination' and 'luminous flux intensity'.

A lamp of 500 W having a mscp of 1,250 is suspended 2.7 metres above the working plane. Calculate :

(*i*) Illumintion directly below the lamp at the working plane

(*ii*) Lamp efficiency

(*iii*) Illumination at a point 3 m away on the horizontal plane from vertically below the lamp.

[*U.P. Technical Univ. Utilization of Electrical Energy and Electric Traction 2005-06*]

Solution : Luminous flux intensity of lamp, I = 1,250

Height of lamp, *h* = 2.7 m

(*i*) Illumination directly below the lamp at the working plane,

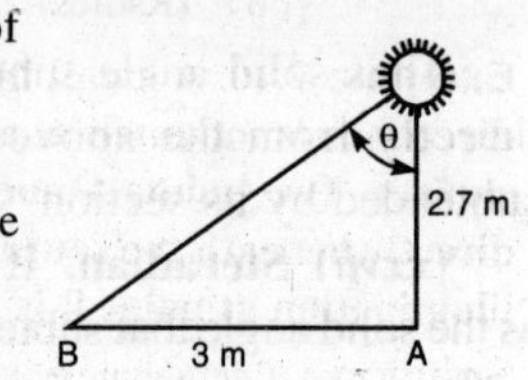

Fig. 7.8

$$E_A = \frac{1{,}250}{(2.7)^2} = 171.47 \text{ lux } \textbf{Ans.}$$

$$(ii) \text{ Lamp efficiency} = \frac{\text{Luminous flux}}{\text{Wattage of lamp}}$$

$$= \frac{4\pi \times \text{mscp}}{W} = \frac{4\pi \times 1{,}250}{500} = 31.42 \text{ lumens/watt } \textbf{Ans.}$$

(*iii*) Illumination at a point 3 metres away on the horizontal plane from vertically below the lamp,

$$E_B = \frac{I}{h^2}\cos^3\theta = \frac{1{,}250}{(2.7)^2} \times \frac{2.7^3}{[3^2 + 2.7^2]^{3/2}}$$

$$= 51.33 \text{ lux } \textbf{Ans.}$$

Example 7.8. The illumination at a point on a working plane directly below the lamp is to be 100 lumens/m². The lamp gives 256 CP, uniformly below the horizontal plane. Determine the height at which the lamp is suspended.

Also find illumination at a point on the working table 1.2 metres away from the vertical axis of the lamp.

Solution : Luminous intensity of the lamp, I = 256 cd

Illumination at a point on a working plane directly below the lamp,

$$E = 100 \text{ lumens/m}^2$$

Let the height at which the lamp is suspended be *h* metres

$$\because E = I/h^2$$

$$\therefore 100 = 256/h^2$$

$$\text{or } h = \sqrt{\frac{256}{100}} = 1.6 \text{ metres} \quad \textbf{Ans.}$$

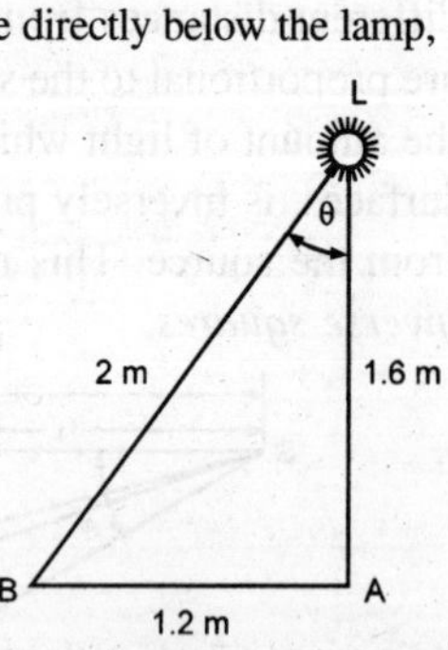

Fig. 7.9

Illumination on the working table 1.2 metres away from the vertical axis of the lamp,

$$E_B = \frac{I}{h^2}\cos^3\theta = \frac{256}{(1.6)^2} \times (0.8)^3 = 51.2 \text{ lux} \quad \textbf{Ans.}$$

Example 7.9. Two lamps are hung at a height of 9 metres from the floor level. The distance between the lamps is 2 metres. Lamp one is of 500 CP. If the illumination on the floor vertically below this lamp is 20 lux, find the candle power of the lamp two.

Solution : Candle power of lamp L_1 = 500 CP

Let the candle power of lamp L_2 be CP_x

Illumination at the point A on the floor level vertically below lamp L_1

= Illumination due to lamp L_1

+ Illumination due to lamp L_2

$$20 = \frac{500}{(9)^2} + \frac{CP_x}{(9)^2}\cos^3\theta$$

$$\text{or} \quad 20 = \frac{500}{81} + \frac{CP_x}{81}\left(\frac{9}{\sqrt{85}}\right)^3$$

$$\text{or } CP_x = \frac{1{,}620 - 500}{\left(\frac{9}{\sqrt{85}}\right)^3}$$

$$= \frac{1{,}120}{0.93} = 1{,}204 \text{ CP} \quad \textbf{Ans.}$$

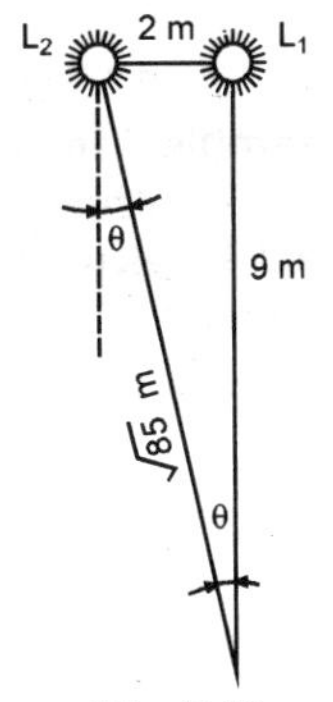

Fig. 7.10

Example 7.10. A small light source with intensity uniform in all directions is mounted at a height of 10 metres above a horizontal surface. Two points A and B both lie on the surface with point A directly beneath the source. How far is point B from point A if the illumination at point B is only 1/10 as great as at point A ?

Solution : Let the intensity of the lamp L be I candela and the distance between points A and B be x metres.

$$\text{Illumination at point A,} \quad E_a = \frac{I}{(10)^2} = \frac{I}{100}$$

$$\text{Illumination at point B,} \quad E_B = \frac{I}{(10)^2} \times \left(\frac{10}{\sqrt{(10)^2 + x^2}}\right)^3$$

$$= \frac{10I}{(100 + x^2)^{3/2}}$$

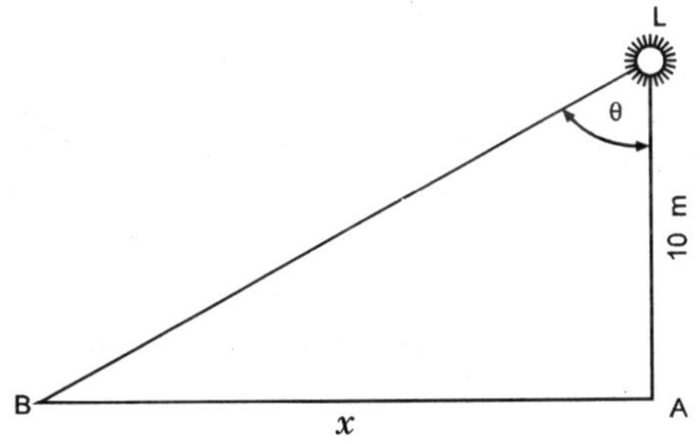

Fig. 7.11

$$\because \quad E_B = \frac{1}{10}E_A$$

$$\therefore \quad \frac{10I}{(100 + x^2)^{3/2}} = \frac{1}{10} \times \frac{I}{100}$$

$$\text{or} \quad (100 + x^2)^{3/2} = 10{,}000$$

$$\text{or} \quad (100 + x^2) = (10{,}000)^{2/3} = 464$$

$$\text{or} \quad x = \sqrt{464 - 100}$$

= 19.1 metres **Ans.**

Example 7.11. A light source having an intensity of 400 CP in all directions is fitted with a reflector so that it directs 80% of its light along a beam having a divergence of 15°. Determine the total light flux emitted along the beam. Also determine the average illumination produced on a surface normal to the beam direction at a distance of 8 metres.

[Gorakhpur Univ. Utilization of Electrical Power and Traction Sept. 1984]

Solution : Intensity of light source, I = 400 cd

Total light flux emitted along the beam $= 0.8 \times 4\pi \times 400$

= 4,020 lumens

Beam angle, $\theta = 15°$

Distance of sphere from light source, $l = 8$ m

Radius of circle of the surface to be illuminated,

$$r = l \tan\frac{\theta}{2} = 8 \tan\frac{15°}{2} = 1.05322 \text{ m}$$

Area of surface to be illuminated,

$$A = \pi r^2 = \pi \times (1.05322)^2 = 3.485 \text{ m}^2$$

$$\text{Average illumination} = \frac{4{,}020}{3.485} = 1{,}153 \text{ lux} \quad \textbf{Ans.}$$

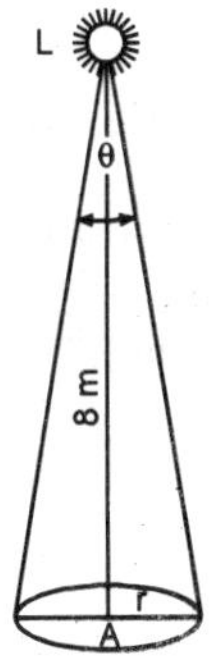

Fig. 7.12

Example 7.12. A lamp with a reflector is mounted 12 m above the centre of a circular area of 24 metres diameter. If the combination of the lamp and reflector gives a uniform CP of 1,000 over the circular area, determine the maximum and minimum illumination produced on the area.

[Pb. Univ. Elec. Drives and Utilization of Elec. Energy Nov. 1983]

Solution : Candle power of the lamp, CP = 1,000

The maximum illumination will occur directly below the lamp, *i.e.*, at point A and

$$= \frac{CP}{h^2} = \frac{1{,}000}{(12)^2} = 6.94 \text{ lux} \quad \textbf{Ans.}$$

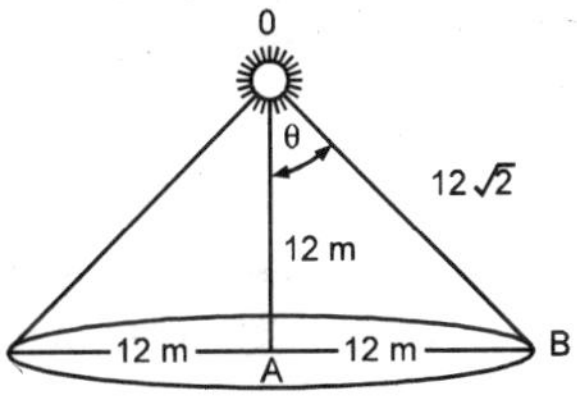

Fig. 7.13

Minimum illumination will occur at the periphery of the circular area, *i.e.*, at point B and

$$= \frac{CP}{h^2}\cos^3\theta = \frac{1{,}000}{(12)^2} \times \left(\frac{12}{\sqrt{12^2 + 12^2}}\right)^3$$

$$= \frac{1{,}000}{(12)^2} \times \frac{1}{(\sqrt{2})^3} = 2.46 \text{ lux} \quad \textbf{Ans.}$$

Example 7.13. A lamp having a uniform CP of 200 in all directions is provided with a reflector which directs 60 per cent of the total light uniformly on to a circular area of 10 m diameter. The lamp is hung 6 m above the area. Calculate the illumination

(*i*) at the centre and (*ii*) at the edge of the surface with and without the reflector.

Determine also the average illumination over the area without the reflector.

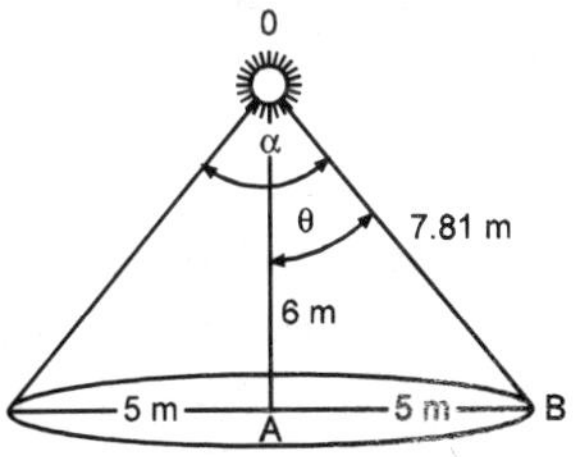

Fig. 7.14

Solution : Candle power of the lamp, CP = 200

(*i*) Illumination at the centre of the area without reflector

$$= \frac{CP}{h^2} = \frac{200}{(6)^2} = 5.56 \text{ lux} \quad \textbf{Ans.}$$

(*ii*) Illumination at the edge of the surface without reflector

$$= \frac{CP}{h^2}\cos^3\theta \text{ where } \cos\theta = \frac{6}{\sqrt{5^2 + 6^2}} = 0.768$$

$$= \frac{200}{(6)^2} \times (0.768)^3 = 2.52 \text{ lux} \quad \textbf{Ans.}$$

With the reflector the illumination at the edge and at the centre will be the same since the reflector directs the light uniformly on the surface.

Total lumens given out, $\phi = 4\pi \text{ CP} = 4\pi \times 200 = 800\,\pi$

Total lumens reaching the surface $= 0.6 \times \phi = 0.6 \times 800\,\pi$

Total surface area, $A = \pi r^2 = \pi \times (5)^2 = 25\,\pi\ m^2$

$$\text{Average illumination} = \frac{0.6 \times 800\pi}{25\pi} = 19.2 \text{ lux} \quad \textbf{Ans.}$$

Average illumination of the surface without reflector will be found by first determining the solid angle subtended by the surface at the lamp and then finding out the luminous flux emitted in that solid angle.

Solid angle subtended by the area at the lamp,

$$\omega = 2\pi\left(1 - \cos\frac{\alpha}{2}\right) = 2\pi\,(1 - \cos\theta) = 2\pi\,(1 - 0.768)$$

$$= 0.464\,\pi \text{ steradians}$$

$$\text{Total flux reaching the surface} = I\omega = 200 \times 0.464\,\pi$$

$$= 92.8\,\pi \text{ lumens}$$

$$\text{Average illumination} = \frac{92.8\pi}{\pi(5)^2} = 3.7 \text{ lux} \quad \textbf{Ans.}$$

Example 7.14. To lamp posts are 16 m apart and are fitted with a 100 CP lamp each at a height of 6 m above ground. Calculate the illumination on the ground (*a*) under each lamp (*b*) midway between the lamps.

[*Pb. Univ. Electric Drives and Utilization of Elec. Energy December 1982; Pb. Technical Univ. Electric Drives & Utilization December 2008; Rajasthan Technical Univ. Utilization of Electrical Power 2011*]

Solution : Candle power of each lamp = 100 CP

Height of lamp from the ground, h = 6 metres

Distance between the two lamp posts = 16 metres

Illumination midway between the lamps,

E_C = Illumination due to lamp L_1 + Illumination due to lamp L_2

$$= \frac{100}{(6)^2} \times (0.6)^3 + \frac{100}{(6)^2} \times (0.6)^3 = 1.2 \text{ lux} \quad \textbf{Ans.}$$

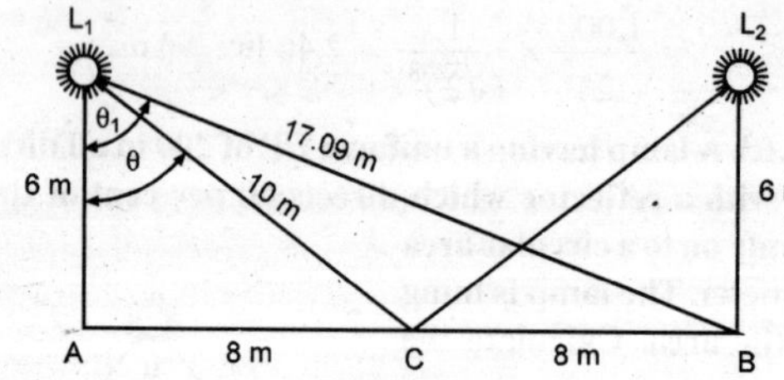

Fig. 7.15

Illumination under either of the lamps, say under lamp L_2,

E_B = Illumination due to lamp L_1 + Illumination due to lamp L_2

$$= \frac{100}{(6)^2} \times \left(\frac{6}{17.09}\right)^3 + \frac{100}{(6)^2} = 2.9 \text{ lux} \quad \textbf{Ans.}$$

Example 7.15. Explain laws of illumination:

In a street lighting scheme, lamps having uniform CP of 800 are hung at a height of 10 m. Distance between consecutive lamp posts is 12 metres. Find illumination under lamp and at the centre in between the lamp posts.

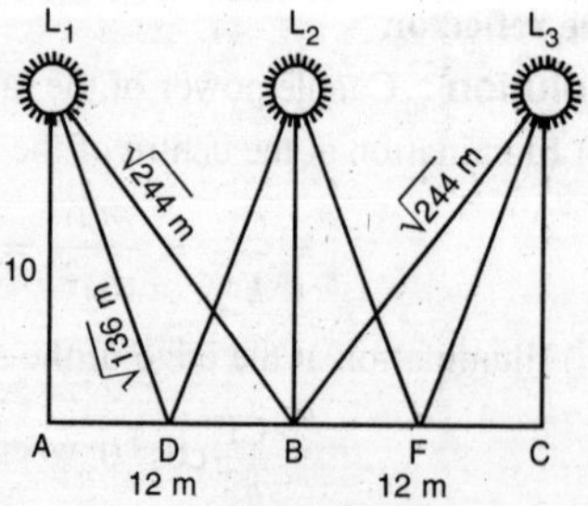

Fig. 7.16

[*U.P. Technical Univ. Utilization of Electrical Energy & Traction 2009-10*]

Solution : Illumination under lamp L_2

E_B = Illumination due to lamp L_2 + Illumination due to lamp L_1 + Illumination due to lamp L_3

$$= \frac{800}{10^2} + \frac{800}{10^2} \times \frac{10^3}{(244)^{3/2}} + \frac{800}{10^2} \times \frac{10^3}{(244)^{3/2}}$$

$$= 8 + 2.1 + 2.1 = 12.2 \text{ lux} \quad \textbf{Ans.}$$

Illumination at the centre in between the lamp posts

$$E_D \text{ or } E_F = \frac{800}{10^2} \times \frac{10^3}{(136)^{3/2}} + \frac{800}{10^2} \times \frac{10^3}{(136)^{3/2}}$$

$$= 5.044 + 5.044 = 10.088 \text{ lux} \quad \textbf{Ans.}$$

Example 7.16. Four lamps are hung at a height of 10 m from the floor in corner of a square of 20 m × 20 m. If each lamp is of 400 CP, calculate the illumination on the floor at the centre of the square.

[*U.P. Technical Univ. Utilization of Electrical Energy & Traction 2006-07*]

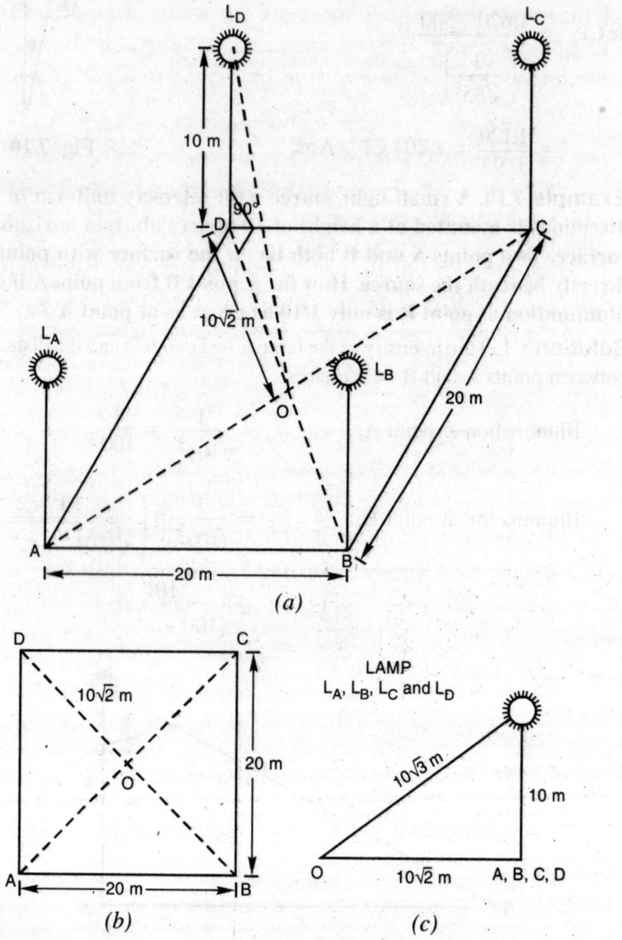

Fig. 7.17

Solution : Candle power of lamp = 400 CP

Height of lamp from ground, h = 10 m

Distance between lamp posts = 20 m

Distance of the centre of square from either corner or lamp post

$$= \frac{1}{2}\sqrt{20^2 + 20^2} = 14.142 \text{ m}$$

Illumination on the floor at the centre of the square,

E_D = Illumination due to lamp L_A + illumination due to lamp L_B + illumination due to lamp L_C + illumination due to lamp L_D

$$= \frac{400}{10^2} \times \left(\frac{10}{\sqrt{300}}\right)^3 + \frac{400}{10^2} \times \left(\frac{10}{\sqrt{300}}\right)^3 + \frac{400}{10^2} \times \left(\frac{10}{\sqrt{300}}\right)^3 + \frac{400}{10^2} \times \left(\frac{10}{\sqrt{300}}\right)^3$$

∵ Distance of each lamp from the centre of the square is same

$$= 0.7698 + 0.7698 + 0.7698 + 0.7698 = 3.08 \text{ lux} \quad \textbf{Ans.}$$

Example 7.17. Two similar lamps having uniform intensity of 500 candle power in all directions below the horizontal are mounted at a height of 4 metres. What must be the maximum spacing between the lamps so that the illumination on the ground midway between the lamps shall be at least one half the illumination directly under the lamps.

[*Gorakhpur Univ. Utilization of Elec. Power and Traction 1982*]

Solution : Candle power of lamp = 500 CP

Height of lamps from the ground, h = 4 metres

Let the maximum spacing between the lamps be of d metres

Illumination midway between the lamps,

$$E_C = 2 \times \text{Illumination due to either lamp}$$

$$= 2 \times \frac{500}{(4)^2} \times \frac{(4)^3}{\left(4^2 + \frac{d^2}{4}\right)^{3/2}} = \frac{4{,}000}{\left(4^2 + \frac{d^2}{4}\right)^{3/2}}$$

Illumination under either of the lamps, say under lamp L_2,

$$E_B = \text{Illumination due to lamp } L_1 + \text{Illumination due to lamp } L_2$$

$$= \frac{500}{(4)^2} \times \frac{(4)^3}{(4^2 + d^2)^{3/2}} + \frac{500}{(4)^2} = \frac{2{,}000}{(4^2 + d^2)^{3/2}} + 31.25$$

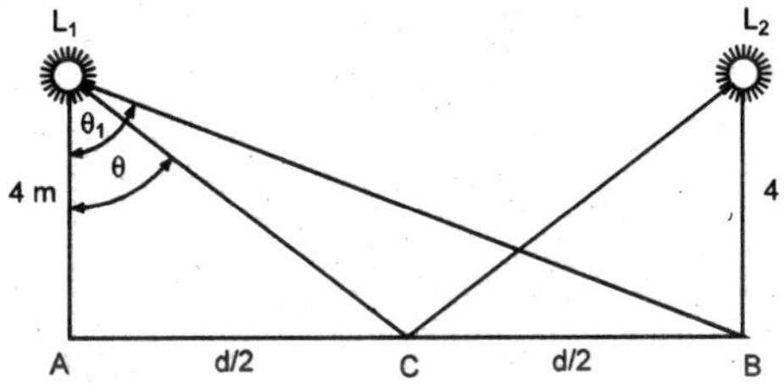

Fig. 7.18

Since $E_C = \frac{1}{2} E_B$

$$\therefore \quad \frac{4{,}000}{\left(4^2 + \frac{d^2}{4}\right)^{3/2}} = \frac{1}{2}\left[31.25 + \frac{2{,}000}{(4^2 + d^2)^{3/2}}\right]$$

or $$\frac{4{,}000}{\left(4^2 + \frac{d^2}{4}\right)^{3/2}} = 15.625 + \frac{1{,}000}{(4^2 + d^2)^{3/2}}$$

or d = 9.56 m **Ans.**

Example 7.18. A small area of 5 metres in diameter is to be illuminated by a lamp at a height of 4.5 m above the centre of the area. A lamp having an efficiency of 24 lumens/watt is fitted with a reflector which directs the light output only over the surface to be illuminated. If the utilization coefficient is 0.65 and illumination 800 lux, determine the wattage of the lamp.

[*M.D. Univ. Utilization of Electric Power and Traction May 2009*]

Solution : Area of working plane,

$$A = \frac{\pi}{4} d^2 = \frac{\pi}{4} \times (5)^2 = 6.25\pi \text{ m}^2$$

Illumination on working plane, E = 800 lux

Luminous flux reaching the working plane,

$$F = EA = 800 \times 6.25\pi \text{ lumens}$$

Total luminous flux emitted by lamp

$$= \frac{F}{\text{Utilization coefficient}} = \frac{800 \times 6.25\pi}{0.65}$$

$$= 24{,}166 \text{ lumens}$$

Wattage of the lamp $= \frac{24{,}166}{24} = 1{,}006.92 \text{ W} \simeq 1 \text{ kW}$ **Ans.**

7.4. POLAR CURVES

All discussions made so far were based on the assumption that the luminous intensity or the candle power from a source is uniformly distributed over the surrounding surface. But none of the practical type of lamp provides light uniformly distributed in all directions because of its un-symmetrical shape. It is often necessary to know the distribution of light in various directions to ascertain how the candle power of a light source varies in different directions. The luminous intensity in all the directions can be represented by *polar curves*. If the luminous intensity in a horizontal plane passing through the lamp is plotted against angular position, a curve known as *horizontal polar curve is obtained*. If the luminous intensity in a vertical plane is plotted against the angular position, a curve known as *vertical polar curve* is obtained. The typical polar curves for an ordinary filament lamp are shown in Fig. 7.19.

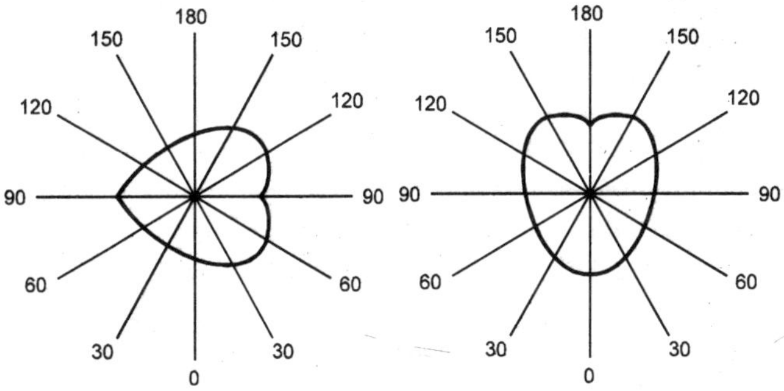

(a) Polar Curve For Horizontal Plane *(b) Polar Curve For Vertical Plane*

Fig. 7.19

The polar curves are used to determine the mean horizontal candle power (mhcp) and mean spherical candle power (mscp). These are also used to determine the actual illumination of a surface by employing the candle power in that particular direction as read from the vertical polar curve in illumination calculations.

The mean horizontal candle power of a lamp can be determined from the horizontal polar curve by taking the mean value of the candle power in a horizontal direction.

Mean spherical candle power can be determined from the vertical polar curve by Rousseau's construction.

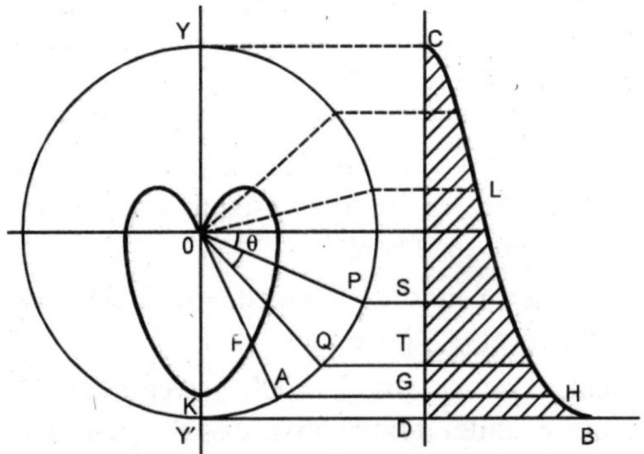

Fig. 7.20. *Rousseau's Construction*

Rousseau's Construction. The construction is illustrated in Fig. 7.20. A semicircle of any convenient radius is drawn with the pole of the polar diagram as centre. The line CD is drawn equal and parallel to the vertical diameter YY. Now from this line CD ordinate equal to corresponding radius on the polar curve are set up such as DB = OK, GH = OF and so on. The curve obtained by joining the ends of these ordinates is known as *Rousseau's curve*. The mean ordinate of this curve gives the mscp of the lamp having polar curve given in the Fig. 7.20.

$$\text{The mean ordinate of the curve} = \frac{\text{Area CSTGDBHLC}}{\text{Length of CD}}$$

The area under the curve can either be determined on a graph paper or found by Simpson's rule.

7.5. PHOTOMETRY

We shall now discuss the comparison and measurement of candle powers.

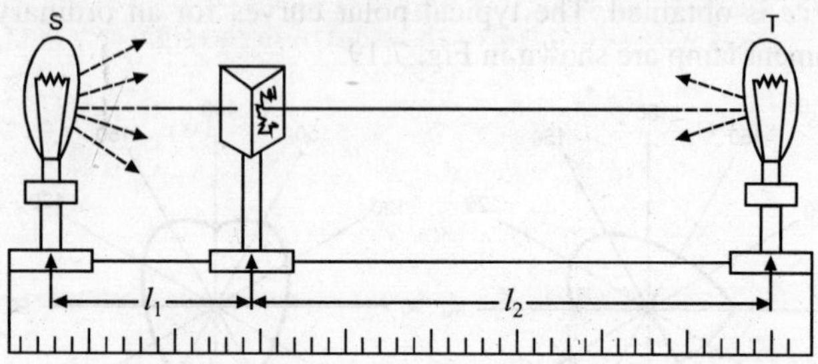

Fig. 7.21. ***Measurement of Candle Power***

The candle power of a source in any given direction is measured by comparison with a standard or substandard source employing photometer bench and some form of photometer. The experiment is performed in a dark room with dead black walls and ceiling in order to eliminate errors due to reflected light.

The photometer bench consists essentially of two steel rods which carry the stands or saddles for holding the two sources, the carriage for the photometer head and for any other apparatus employed in making measurements. One of the bar carries a bar strip, graduated in centimetres and millimetres. The carriages which slides upon the bench have except that carrying the photometer head, a circular table which can be rotated in a horizontal plane and clamped in any position. The circular table is provided with a scale graduated in degrees round its edge so that the angle of rotation of lamp from the direction of the axis of bench can be measured. The bench should be rigid so that the source being compared may be free from vibrations and the carriage holding the photometer head should be capable of moving smoothly and with very little effort. The photometer head acts as a screen for comparison of illumination of the standard source and the source under test. There are different types of photometers, which can be used for the purpose. Some of them will be described here.

The principle of most of the methods of measurement is based upon the Inverse Square Law.

The standard source, whose candle power is known (say S) and the source under test whose candle power is to be determined, are set on the bench at a distance apart with some type of screen in line with, and between, them, as shown in Fig. 7.21. The photometer head or screen is moved in between the two fixed sources until the illumination on both sides of the screen is same. If the distances of the standard source S and source under test T from photometer head are l_1 and l_2 respectively then according to inverse square law

$$\frac{\text{Candle power of source under test}}{\text{Candle power of standard source}} = \frac{l_2^2}{l_1^2}$$

$$\text{or} \quad \text{Candle power of source under test} = S \times \frac{l_2^2}{l_1^2}$$

Since the square of distances are involved in measurement, distances should be measured accurately. In order to obtain distance exactly two points are determined at which there is a perceptible difference in illumination from the two sides and the point half way between them is then taken as the position of equal illumination.

Most of the photometer heads consist of some device by means of which the illumination of two surfaces, side by side– one illuminated by standard source and other by the source whose candle power is to be determined may be compared under exactly similar conditions without movement of the eye. The photometers which are most common in use are the Bunsen and the Lummer Brodhun photometers. These photometers are best for use if the sources to be compared give light of same or approximately similar colours. In case the light from the two sources to be compared differ in colour, a Flicker photometer is best suited.

Bunsen Grease Spot Photometer. It consists of a piece of tissue paper with grease spot in the centre, held vertically in a carrier between the two light sources to be compared. The position of the carrier is then adjusted until the semi-transparent spot and the opaque parts of the paper are equally bright, *i.e.*, the grease spot is invisible. The distances of the tissue paper from both light sources are measured. The candle power of the source under test is then determined from the relation.

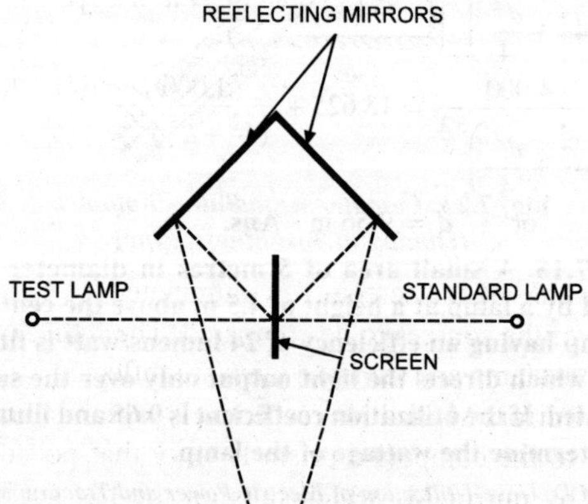

Fig. 7.22. ***Bunsen Grease Spot Photometer Head***

$$\frac{\text{Candle power of source under test}}{\text{Candle power of standard source}} = \left(\frac{\text{distance of source under test from photometer}}{\text{distance of standard source from photometer}}\right)^2$$

The Bunsen Grease spot photometer can be used in other way also. In this method of use, the photometer is fixed at one position and one side of the paper is illuminated first by test lamp and then by a standard lamp and distance of the lamp from the photometer head is measured in each case on disappearing of grease spot.

If l_1 and l_2 are the distances then the candle power of source under test

$$= \text{Candle power of standard lamp} \times \frac{l_1^2}{l_2^2}$$

The simplest and best method of use of Bunsen Grease spot photometer for comparison of illumination on either side of the screen is to use two mirrors placed behind the paper and view the illumination on both the sides through the mirrors simultaneously by standing in a plane passing through the photometer head and perpendicular to the lines joining the lamps. The position of the photometer head for equal contrast in illumination between the opaque and transparent portions of the paper on the two sides is determined. Now the candle power of the lamp under test can be calculated as before.

There are two types of Lummer Brodhun heads—one is the equality of brightness type and second is the contrast type. Contrast type is more accurate and, therefore, more used in photometric measurements.

Equality of Brightness Type Photometer Head. The photometer head consists of a plaster of paris screen, two mirrors M_1 and M_2, a compound prism P and a telescope. The compound prism is made up of two right angled glass prisms, one of which has principal surface as spherical one with a small flat portion at the centre. This small flat portion at the centre makes optical contact with the flat surface of the other prism, as shown in Fig. 7.23.

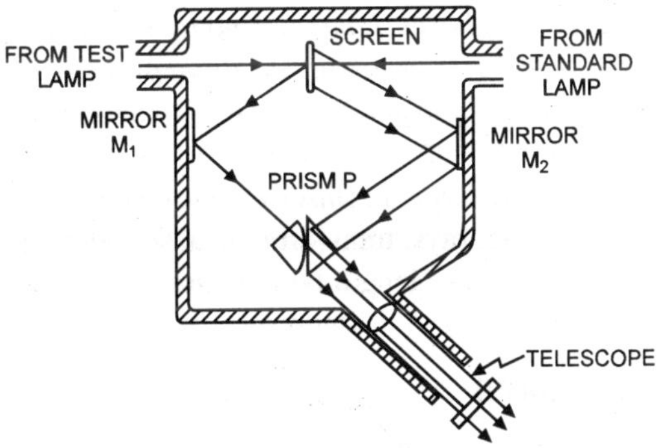

Fig. 7.23. ***Lummer Broadhun Photometer (Quality of Brightness Type)***

Each side of the screen S is illuminated by light entering from one of the sources to be compared. The light rays coming from the test lamp and standard lamp are reflected by the screen on to the mirrors M_1 and M_2 respectively which in turn reflects the same to the compound prism P. Only that portion of light reflected by mirror M_1 which falls on the flat surface of the spherical prism, passes straight into the telescope while the rest of it is reflected back. The portion of light reflected by mirror M_2, which falls on the surface of contact between the two prism passes straight through the compound prism while the rest of it is reflected through the telescope. Thus the observer sees the centre portion of the circular area illuminated by the test lamp and outer ring illuminated by the standard lamp.

While carrying out measurements, the photometer head is moved until the dividing line between the centre portion and surrounding disappears. The disappearance of this dividing line indicates the equality of brightness. Now the distance of the photometer head from the two lamps is noted and the candle power of lamp under test can be calculated by using *inverse square law principle*.

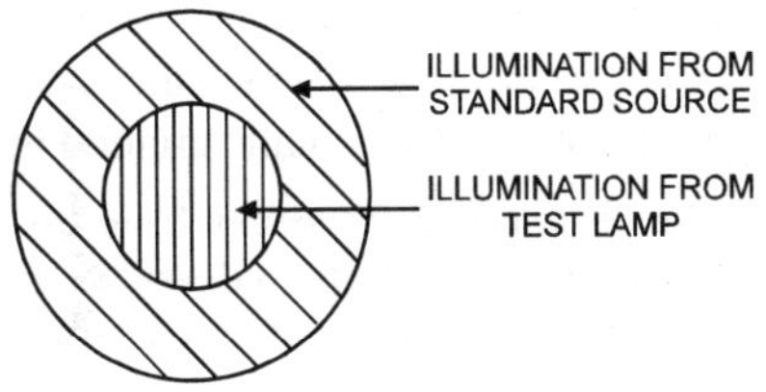

Fig. 7.24

Contrast Type of Lummer Brodhun Photometer Head. The compound prism, in the contrast type photometer head, is made up of right angled prisms with their principal surfaces face to face. One of the prism [the left hand one in Fig. 7.25 (*a*)] has its hypotenuses surface etched away at X, Y and Z so that it forms a pattern similar to that shown in the figure.

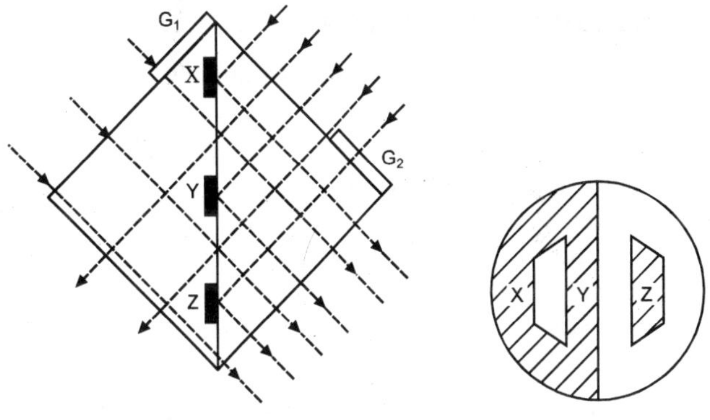

Fig. 7.25. ***Lummer Brodhun Photometer Head (Contrast Type)***

The light falling on the compound prism from both sides of the screen will pass through the un-etched portions of the prism at the junction but will be reflected by the etched surfaces. The etched portion will have difference in illumination as compared with the un-etched portion in case the illumination of the surfaces of reflections are different, as shown in Fig. 7.25 (*b*). G_1 and G_2 are two glass plates which give a little reflected light in order to maintain some difference between illumination of the etched and un-etched areas for all the positions of the photometer head. The difference is about 8% in balance position of photometer head. The photometer head will be in balance position when the difference in illumination of etched and the un-etched portions will be same on each half of the circular area. In case the balance position is disturbed, the difference in illumination of area Z and its surrounding area will decrease while the difference in illumination of area XY and the inner trapezium will increase. The photometer head is moved until the equal contrast is obtained. This type of head gives accuracy within 1% when lights of similar colours are compared.

Flicker Photometers. These photometers are employed when the two sources giving light of different colours are to be compared. The working of these photometers depend upon the fact that if two illuminated surfaces are presented to the eye with rapid alternations, the flicker disappears when the surfaces are of equal brightness. These photometers are not affected by the difference in colour of two lights to be compared as much as photometers of the steady comparison type are affected since the colour difference between two alternating fields of light disappears at a lower speed of alternation than the speed at which difference in brightness disappears. The speed of alternation should be kept as low as possible at which the disappearance of the flicker can be obtained for the small variation in brightness.

Example 7.19. The distances of the screen from the lamp under test and standard lamp when adjusted are 1.6 m and 0.9 m respectively. If the standard lamp be of 80 CP what is the CP of the lamp under test ?

Solution : Candle power of standards lamp, $CP_s = 80$

Distance of test lamp from screen, $l_1 = 1.6$ m

Distance of standard lamp from screen, $l_2 = 0.9$ m

Candle power of lamp under test, CP_t

$$= \frac{CP_s}{l_2^2} \times l_1^2 = \frac{80}{(0.9)^2} \times (1.6)^2 = 252.84 \text{ CP } \textbf{Ans.}$$

7.6. INTEGRATING SPHERE

It is a piece of apparatus which is now commonly employed for measurement of mean spherical candle power. In this method the sphere is used to measure the total flux radiated by the lamp, which when divided by 4π gives the mscp. Since in this method the flux radiated in all directions is taken into account, this method is better than that described in which it was assumed that the candle power distribution is same in all vertical planes, an assumption which may not always be justifiable.

The integrating sphere consists of a hollow sphere whose diameter is large compared to the lamp to be tested having a smooth inner surface with a uniform coating of white paint. If the lamp is hung inside the sphere, the light is so diffused that an uniform illumination is produced over the whole surface. A small window of translucent glass provided at one side of the sphere is illuminated by reflected light from the inner surface of the sphere. The small screen is inserted in between the lamp and window in order to prevent the light from the lamp reaching the window directly.

The measurement of mscp is done as follows :

The lamp, whose mscp is to be determined, is placed at the centre of sphere and brightness of window is measured with the help of some form of illuminometer. The test lamp is replaced by a standard lamp whose mscp is known and brightness of the window is again measured. Since the mscp of the two lamps are proportional to the respective brightness of the window, therefore, the mscp of the lamp under test can be determined by knowing the mscp of standard lamp and brightness of window in the two cases.

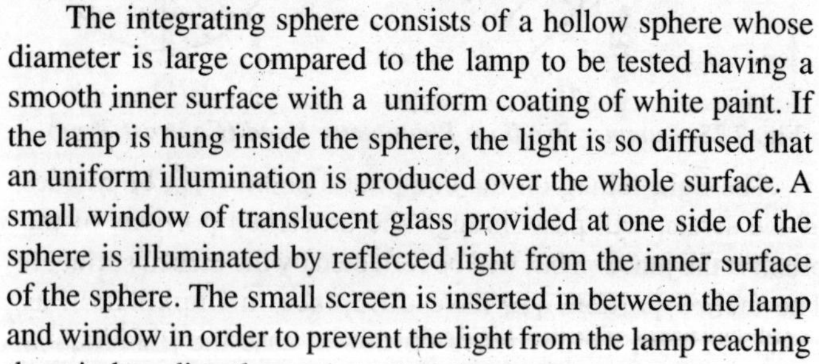

Fig. 7.26

A Lummer Brodhun photometer head can also be employed for the measurement of brightness of window with some modifications. A small mirror is placed along with the screen, as shown in Fig. 7.26 in order to reflect the light from the window of the sphere into compound prism of the photometer and balance is obtained using a lamp on the other side of the photometer. Balance is obtained by moving the comparison lamp first with the standard or substandard lamp in the sphere, and then with the lamp test inside the sphere. The squares of the corresponding distance of the comparison lamp from the photometer head is inversely proportional to the mscp of lamps.

7.7. MEASUREMENT OF ILLUMINATION

The best and simplest method of measuring illumination is by a portable type illumination photometer calibrated to read directly in lux. No shadow should be allowed to fall on the meter during measurements, and the meter should be held perfectly horizontal and stationary for accurate results. The meter should be colour-corrected type so that it can be used for all kinds of light sources.

7.8. SOURCES OF LIGHT

According to principle of operation the light sources may be grouped as follows :

1. ***Arc Lamps.*** Electric discharge through air provides intense light. This principle is utilised in arc lamps.

2. ***High Temperature Lamps.*** Oil and gas lamps and incandescent filament type lamps, which emit light when heated to high temperature.

3. ***Gaseous Discharge Lamps.*** Under certain conditions, it is possible to pass electric current through a gas or metal vapour, which is accompanied by visible radiations. Sodium and mercury vapour lamps operate on this principle.

4. ***Fluorescent Type Lamps.*** Certain materials, when exposed to ultraviolet rays, transform the absorbed energy into radiations of longer wave length lying within the visible range. This principle is employed in fluorescent lamps.

7.9. ARC LAMPS

In an arc lamp electric current is made to flow through two electrodes in contact with each other which are drawn apart. The result is an arc being struck. The arc maintains the current, and is very efficient source of light. There are various forms of arc lamps such as carbon-arc, flame-arc or magnetic-arc lamps.

7.9.1. Carbon-Arc Lamp

This is earliest type and is still used in cinema projectors and search lights. When two hard carbon rods are placed end to end and connected to the terminals of dc supply mains of not less than 45 volts, the current flows through them and the ends of carbon rods soon become incandescence due to high resistance. If they are slightly pulled (say about 2 or 3 mm) an arc will be formed between two carbon rods and white light will be produced. The arc is maintained by transfer of carbon particles from one rod to another one. It is found that these particles travel from the positive carbon rod to negative one.

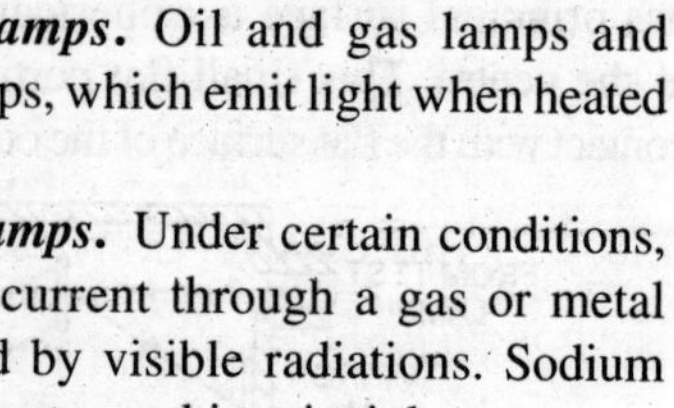

Fig. 7.27. Carbon-Arc Lamp

That is why, the positive rod after sometime of use becomes hollow and negative becomes a pointed pencil.

Due to transfer of particles and oxidation both rods burn away by and by. The rate of burning of positive rod is twice of that of negative rod, that is why, the x-section of the positive rod is made twice of that of negative one. In case of ac supply the rate of burning of both rods is same and, therefore, they are made of equal x-section. Since the carbon is consumed during operation, therefore, requires adjustment of distance between two rods now and then. Except in small lamps, an automatic device is required for feeding the carbon into the arc as they burn away. The voltage required for maintaining the arc is given approximately by $V = (39 + 2.8\, l)$ where l is the length of the arc in mm.

The carbon arc is unstable, since as the current increases the amount of vaporised carbon increases, and the resistance decreases so much that the product of current and resistance, *i.e.*, the voltage drop across the arc falls. A series resistance is, therefore, used for stabilising the arc, and this leads to a low overall luminous efficiency. The voltage drop across the arc is about 45 to 60 volts and the supply voltage is about 70 to 100 volts. In case of ac supply an inductor is used in place of series resistance for stabilising the arc.

In a carbon-arc lamp about 85% of light is given out by the positive carbon, 10% by the negative carbon and 5% by the air. The temperature of positive carbon is between 3,500°C and 4,000°C and that of negative one is about 2,500°C. The luminous efficiency of such a lamp is 12 lumens per watt.

7.9.2. Flame-Arc Lamp

The principle of operation is similar to that of carbon-arc lamp. The electrodes of such a lamp has 5 to 15% fluoride (called the flame material) and 85 to 95% carbon. Generally core type carbon electrodes are used and the cavities are filled with fluoride. The fluoride has a characteristic which radiates light energy efficiently from a very high heated arc stream. Fluoride turns into vapour along with the carbon and these fluoride

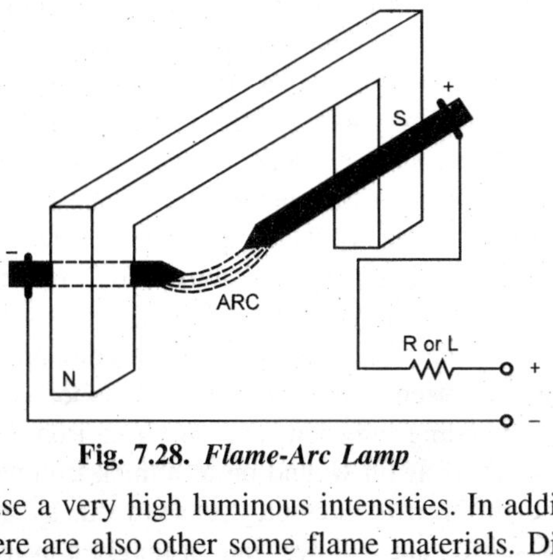

Fig. 7.28. *Flame-Arc Lamp*

vapours cause a very high luminous intensities. In addition to fluoride, there are also other some flame materials. Different flame materials will produce different colours. Most of these colours do not appeal to eyes and at the same time they produce strain on them. The arc can be drawn out to one side with the help of a magnet. Though the arc is very efficient but owing to objection to its colours it has now been superseded by the electric discharge lamps. The luminous efficiency of such a lamp is 8 lumens per watt.

7.9.3. Magnetic-Arc Lamp

In such a lamp positive electrode made of copper and negative electrode made of magnetic oxide of iron are used. The arc is struck in the similar way as in case of carbon-arc lamp. Such lamps are rarely used.

7.10. INCANDESCENT LAMPS

For sometimes the arc lamps were used for general lighting purpose but as the carbon-arc lamps were complicated so these have been superseded by filament lamps for general lighting.

When an electric current is passed through a fine metallic wire, heat is produced and the temperature of the wire increases. At low temperatures the wire radiates heat energy, as the temperature of the wire increases due to heating it radiates heat as well as light energy. The higher the temperature of the wire, higher is the amount of light energy radiated. A black body when heated to 6,250°C emits the maximum energy in the visible spectrum range.

The incandescent or filament type lamp consists of a glass globe completely evacuated and a fine wire, known as *filament*, within it. The glass globe is evacuated to prevent the oxidization and convection currents of the filament and also to prevent the temperature being lowered by radiation.

The material, which can be used for the filaments of incandescent lamps, must possess the properties of high melting point, low vapour pressure, high resistivity, low temperature coefficient, ductility and sufficient mechanical strength to withstand vibrations during use.

The materials which can be used for the filaments in incandescent lamps are carbon, osmium, tantalum and tungsten. Though the carbon has a melting point of 3,500°C but its main drawback is that it starts vaporizing at very fast rate if it is operated beyond 1,800°C, moreover its temperature coefficient of resistance is negative, *i.e.*, its resistance decreases with the rise in temperature and, therefore, it takes more current from mains. Because of its low operating temperature its efficiency is quite low and is of the order of 3.5 lumens per watt. Osmium is very rare and expensive metal. The melting point of osmium is 2,600°C. Average efficiency of osmium lamp is of the order of 5 lumens per watt. The melting point of tantalum is only 2,800°C and, therefore, it cannot be operated at higher temperature to give more efficiency. The efficiency of tantalum filament lamps is about 5 lumens per watt. Nowadays tungsten is the most commonly used metal for filament due to its high melting point (3,400°C), high resistivity, low temperature coefficient (0.0051), low vapour pressure, being ductile and mechanically strong to withstand vibration during use. The hot resistance of tungsten filament is about 15 times the cold resistance and, therefore, there is a inrush of current at the switching instant. However, the inrush current is not beyond 15 times the normal current. Since the filament heats up immediately and resistance increases almost instantaneously, so the inrush current attains a maximum value in 0.003 second and falls to normal operating value in 0.2 second. The average efficiency of tungsten filament lamp is about 10 lumens per watt. The light spectrum of an incandescent lamp is continuous, *i.e.*, it contains all the colours, but contains relatively excess of red and yellow radiations and less of blue and violet radiations.

Aging Effects. The light output of an incandescent lamp decreases gradually. As a tungsten lamp ages, its light output decreases for two reasons. Evaporation of the filament tends to cause the bulb to blacken. Also, evaporation makes the filament slowly decrease in diameter, which means that the resistance of the filament increases. Therefore, an old filament draws less current and operates at a lower temperature, which reduces its light output. In turn the efficiency of the lamp (lumens output/watt input) also decreases with the passage of time. The current drawn and the power consumed by the filament decrease at the same rate as the lamp ages. However, the efficiency decreases about four times as fast, and the light output decreases approximately five times as fast. The total depreciation of light output is roughly 15% over the useful life range.

Effects of Voltage Variations. The operating characteristics of an incandescent lamp are materially affected by departure from its normal operating voltage. An increase of 5% in operating voltage increases the lumens output by 20% but shortens the life of the lamp by 50%. On the other hand, reduction of 5% in operating voltage causes a reduction of approximately 20% in lumens output but doubles the life of the lamp. The efficiency of a lamp (lumens/watt) increases with the increase in voltage owing to increase in temperature and is proportional to the square of the voltage. 1% change in applied voltage modifies lamp wattage by 1.5%, efficiency by 2%, lumens output by 3.5%. The normal life of a filament lamp is about 1,000 working hours. The variations in power consumption, lumens output, efficiency and life of incandescent lamps with the variation in voltage are shown in Fig. 7.29.

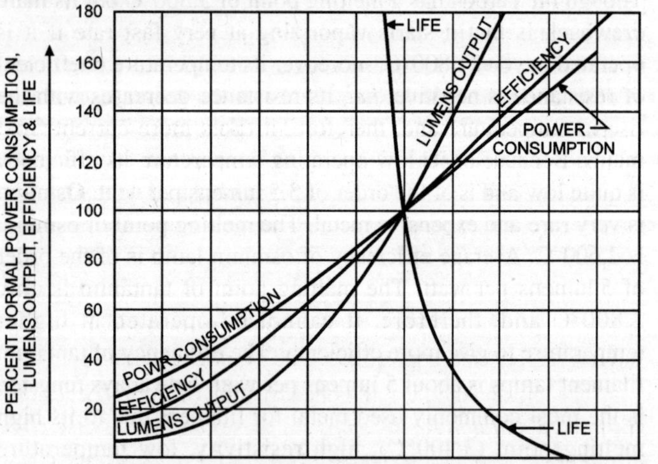

Fig. 7.29

Analytically these relationships are given as below :

Lumens output $\propto (V)^{3.55}$

Power consumption $\propto (V)^{1.55}$

Luminous efficiency $\propto (V)^2$

Life $\propto (V)^{-13}$ for vacuum lamps;

$\propto (V)^{-14}$ for gas-filled lamps

Advantages. 1. Operating power factor unity.

2. Direct operation on standard distribution voltage.

3. Availability in various shapes and shades.

4. Good radiation characteristic in the luminous range.

5. No effect of surrounding air temperature.

Filament Dimensions. Depending upon the voltage and wattage the diameter of a tungsten lamp filament may be as small as 10 microns (about one-sixth of diameter of human hair).

The influence of the diameter of the filament on the heating is found easily if it is assumed that the heat lost by convection is negligible compared with that lost by radiation.

The power absorbed by the filament

$$= I^2R = I^2\rho\frac{l}{a} = I^2\rho\frac{l}{\frac{\pi}{4}d^2} = \frac{4\rho l}{\pi d^2}I^2$$

where I is filament current in amperes, ρ is the resistivity of filament material at the operating temperature, l is length of the filament and d is diameter of the filament.

The power radiated by the filament

$$= eK \times \text{surface area} \times (T_1^4 - T_0^4) = eK \times \pi dl \times (T_1^4 - T_0^4)$$

where e is the emissivity of the surface, K is a constant, T_1 is the temperature of filament and T_0 is the temperature of the surrounding medium.

Since the power absorbed = Power radiated

$$\frac{4\rho l}{\pi d^2}I^2 = eK \times \pi\, dl \times (T_1^4 - T_0^4)$$

$$\text{or} \quad I = d^{3/2}\sqrt{\frac{eK\pi^2}{4\rho}(T_1^4 - T_0^4)}$$

For two filaments of same material operating at the same temperature the diameter d is proportional to $I^{2/3}$.

Gas Filled Lamps. A metal filament can work in an evacuated bulb up to 2,000°C without oxidation and if it is worked beyond this temperature it vaporizes quickly and blackens the lamp. For higher efficiency it is necessary to use working temperature more than 2,000°C keeping down the evaporation, which is possible by filling the bulb with an inert gas–argon with a small percentage of nitrogen. Nitrogen is added to reduce the possibility of arcing. Krypton is the best gas for this purpose but it is so expensive that it is used only in special purpose lamps, such as miners lamps. The tungsten filament can safely be burn at temperature of 2,400°C to 2,750°C according to the size of the lamp. However, due to presence of gas there is heat loss due to convection currents. This loss depends upon the surface area of filament. As such coiled coil filaments, which take much less space compared with coiled filaments are used with such lamps. A coiled coil filament is made by winding tungsten wire on a fine iron wire to form a spiral which is again wound on to a thick iron wire to form a coiled coil. (iron is later on dissolved out by acid). Helical filaments also have a slower rate of tungsten evaporation. This evaporation eventually causes bulb blackening because the tungsten vapour condenses as a black film on the inner surface of the bulb. In a gas filled lamp, the hot gas carries the tungsten vapour upward. Therefore, a black spot forms at the top of the bulb instead of spreading over the entire inner surface, as in a high-vacuum bulb. Chemicals called *'getters'* are often placed inside the bulb to capture tungsten vapour and thereby reduce the rate of blackening. A piece of wire mesh called a *collector grid* may also be attached to each lead-in-wire to attract the particles of tungsten vapour. Efficiency of coiled-coil lamp is about 12 lumens/watt.

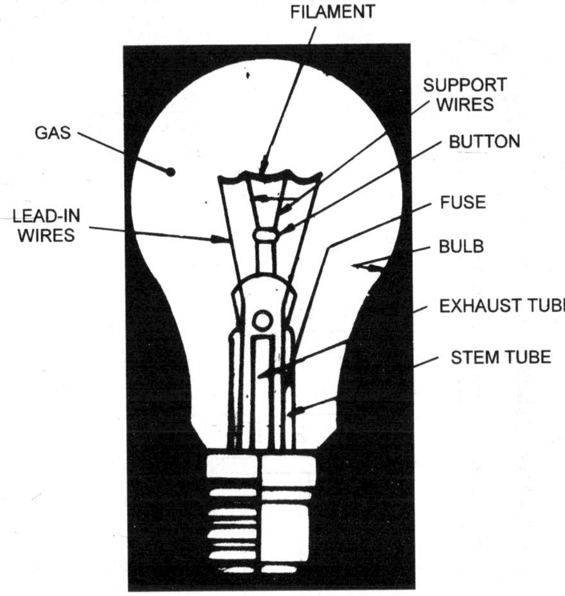

Fig. 7.30. ***Gas Filled Lamp***

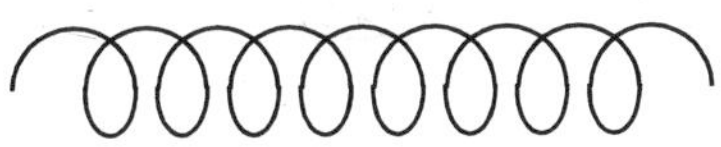

(a) Coiled Filament

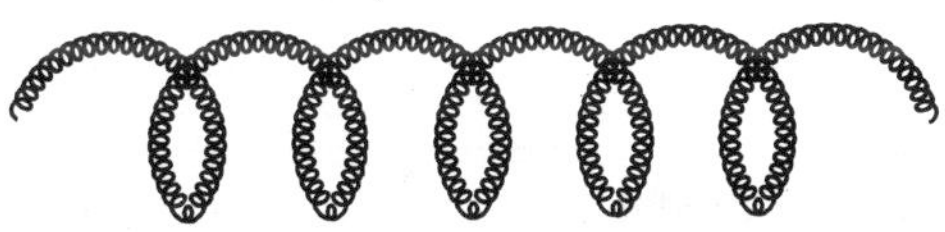

(b) Coiled Coil Filament

Fig. 7.31

For low wattage lamps, however, the heat loss due to introduction of gas is more than in medium wattage lamps, so for low wattage (up to 40 watts) vacuum type lamps are used.

Halogen Lamp. The halogen lamp is the latest member in the family of incandescent lamps. It possesses numerous advantages over the ordinary incandescent lamp. As already stated, the life and efficiency of an incandescent lamp fall off with use—partly due to slow evaporation of the filament and partly due to black deposit formed on the inside of the bulb. The addition of a small amount of halogen vapour to the filling gas restores part of the evaporated tungsten vapour back to the filament by means of a chemical reaction, *i.e.*, there is a sort of '*regenerative cycle*'.

Halogen lamps possess the following advantages :

(*i*) No blackening of lamp, hence no depreciation of lumens output.

(*ii*) High operating temperature with increased luminous efficiency varying from 22 lumens per watt to 33 lumens per watt.

(*iii*) Reduced dimensions of lamps–miniature size.

(*iv*) Long life—2,000 hours.

(*v*) Better colour rendition.

Halogen lamps, which are being manufactured in sizes up to 5 kW, are suitable for outdoor illumination of buildings, playing fields, large gardens, fountains, car parks, airport runways etc. and for lighting of public halls, factories, sport-halls, photo film and TV studios etc.

7.11. GASEOUS DISCHARGE LAMPS

Incandescent lamp suffers from two disadvantages—low efficiency and coloured light. The gaseous discharge lamps have been developed to overcome these drawbacks.

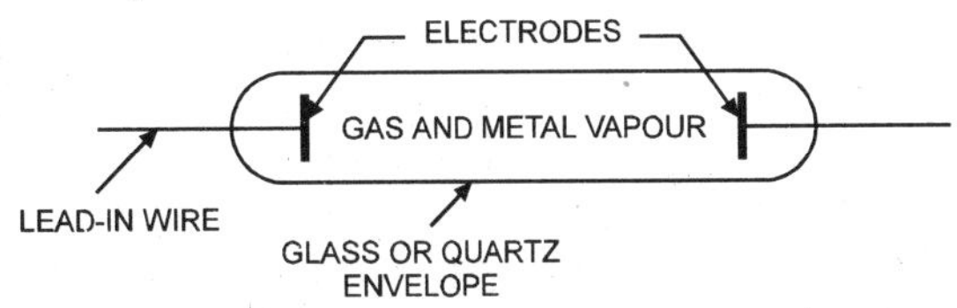

Fig. 7.32. ***Gaseous Discharge Lamp***

The basic principle of a gaseous discharge lamp is illustrated in Fig. 7.32. Gases are normally poor conductors, specially at atmospheric and higher pressures, but application of suitable voltage, *called the ignition voltage*, across the two electrodes can result in a discharge through the gas, which is accompanied by electromagnetic radiation. The wavelength of this radiation depends upon the gas, its pressure, and the metal vapour used in lamp. Argon gas and sodium and mercury vapours are commonly employed in the manufacture of gaseous discharge lamps.

Once the ionisation has commenced in the gas, it has a tendency to increase continuously accompanied by a fall in the circuit resistance, *i.e.*, gaseous discharge lamp possesses a *negative resistance* characteristic. In order to limit the current to a safe value use of a choke or a ballast is made. The choke performs the dual functions of providing the ignition voltage initially, and limiting the current subsequently. Since due to use of choke the power factor becomes poor (0.3—0.4), therefore, in order to improve the power factor of the gaseous discharge lamp use of a condenser is made. The light spectrum obtained is, however, discontinuous (*i.e.*, it consists of one or more coloured lines). The colour of the light obtained depends upon the nature of the gas or vapour used. Discharge lamps are of two types:

(*i*) Those which give the light of the same colour as produced by the discharge through the gas or vapour such as sodium vapour, mercury vapour and neon gas lamps.

(*ii*) Those which use the phenomenon of fluorescence and are known as *fluorescent lamps*. In these lamps, the discharge through the vapour produces ultraviolet waves which cause fluorescence in certain materials called as *phosphors*. The inside of a fluorescent lamp is coated with a phosphor which absorbs invisible ultraviolet rays and radiates visible rays. Example is fluorescent mercury vapour tube.

The gaseous discharge lamps are, in general, considered superior to metal filament lamps. However, they suffer from the following drawbacks:

(*i*) High initial cost and poor power factor.

(*ii*) Starting is somewhat complicated requiring starters in some cases and transformer in others.

(*iii*) These take time to attain full brilliancy.

(*iv*) Ballasts are necessary for stabilising the current since such lamps have negative resistance characteristic.

(*v*) Light output fluctuates at twice the supply frequency. The flicker causes *stroboscopic* effect.

(*vi*) These lamps can be used only in particular position.

7.11.1. Sodium Vapour Discharge Lamp

Principally the sodium vapour discharge lamp consists of a bulb containing a small amount of metallic sodium, neon gas, and two sets of electrodes connected to a pin type base. The presence of neon gas serves to start the discharge and to develop enough heat to vaporize the sodium. Since long discharge paths are necessary, therefore, the discharge envelope is usually bent into U shape. The lamp operates at a temperature-like 300°C and in order to conserve the heat generated and assure the lamp operating at normal air temperatures the discharge envelope is enclosed in a special vacuum envelope designed for this purpose. The lamp must be operated horizontally, or nearly so, to keep the sodium well spread out along the tube, although some small lamps may be operated vertically, lamp cap up. Care should be taken in handling these lamps, particularly when replacing inner U-tube, for if it is broken and sodium comes in contact with moisture fire will result.

The sodium vapour lamp is only suitable for alternating current, and therefore, requires choke control. This requirement is met by operating the lamp from a stray field, step-up, tapped auto-transformer with an open-circuit secondary voltage of 470 to 480 volts. The uncorrected power factor is very low, about 0.3, and a capacitor must be used to improve the power factor to about 0.8.

When the lamp is not in operation, the sodium is usually in the form of solid deposited on the side walls of the tube, therefore, at first when it is connected across the supply mains the discharge takes place in the neon gas and gives red-orange glow. The metallic sodium gradually vaporizes and then ionizes, thereby producing the characteristic monochromatic yellow light, which makes objects appear as grey. The lamp will come up to its rated light output in approximately 15 minutes. It will restart immediately should the power supply be momentarily interrupted since the presence of vapour is quite low and the voltage is sufficient to restrike the arc.

The wiring diagram of sodium vapour lamp(s) is given in Fig. 7.33 (*c*).

The efficiency of a sodium vapour lamp under practical conditions is about 40—50 lumens/watt. The major application of this type of lamp is for highway and general outdoor lighting where colour discrimination is not required, such as street lighting, parks, rail yards, storage yards etc. Such lamps are manufactured in 45, 60, 85 and 140 watt ratings. The average life is about 3,000 hours and is not affected by voltage variations. At the end of this period the light output will be reduced by 15% due to aging.

The lamp fails to operate when

(*i*) the filament breaks or burns out,

(*ii*) the cathode stops to emit electrons,

(*iii*) the sodium particles may concentrate on one side of the tube,

(*iv*) the lamp tube is blackened owing to sodium vapour action on the glass, in which case the output will be reduced.

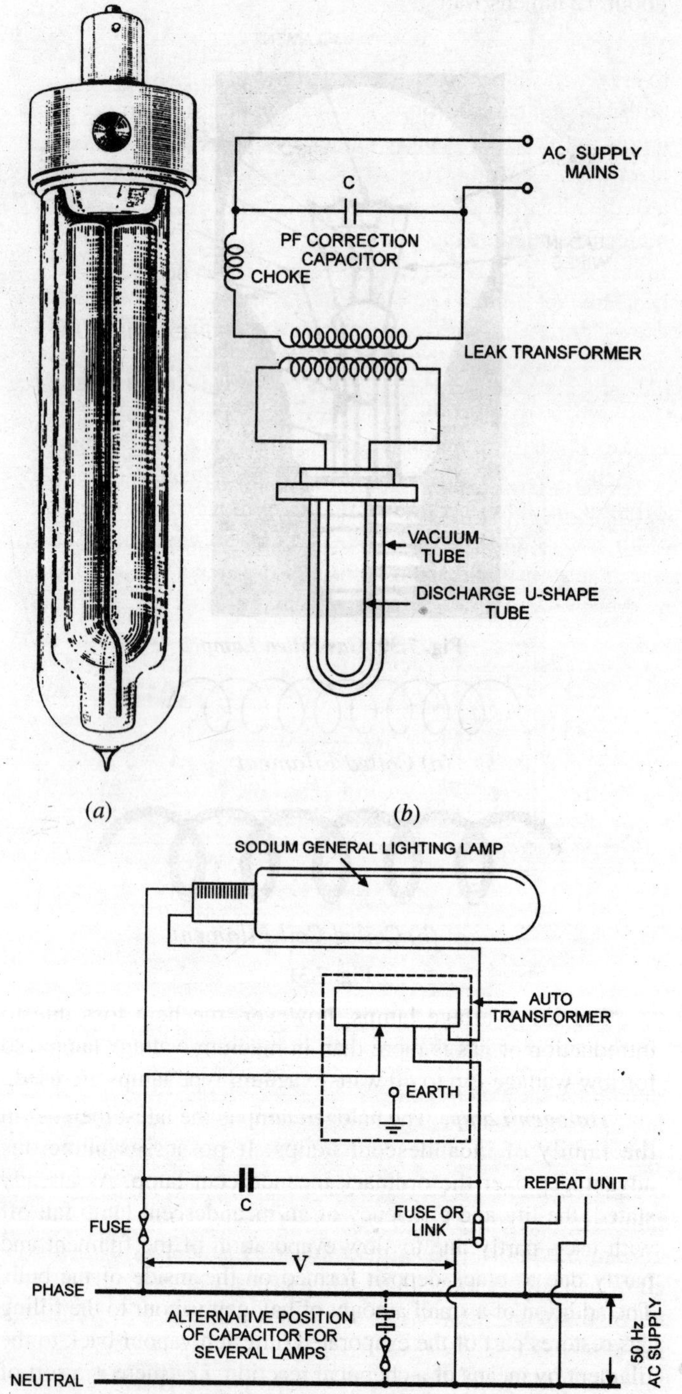

Fig. 7.33. ***Sodium Vapour Discharge Lamp***

7.11.2. High Pressure Mercury Vapour Discharge Lamp

The extensive use of the mercury vapour discharge lamp depends entirely upon the versatility of the mercury vapour as regards pressure, temperature, voltage and other characteristics, each change resulting in a lamp of different spectral quality and efficiency.

The mercury vapour discharge lamp is similar in construction to the sodium vapour discharge lamp. It consists of a discharge envelope enclosed in an outer bulb of ordinary glass. The discharge envelope may be of hard glass or quartz. The space between the bulb is partially or completely evacuated to prevent heat loss by convection from the inner bulb. The outer bulb absorbs harmful ultraviolet rays. The inner bulb contains argon and a certain quantity of mercury. In addition to two main electrodes a starting (auxiliary) electrode is connected through a high resistance (as shown in Fig. 7.34) is also provided. The main electrodes are made of tungsten wire in the shape of helix. In this case no separate heater is required for the cathode which is heated by the constant bombardment of the heavy mercury ions.

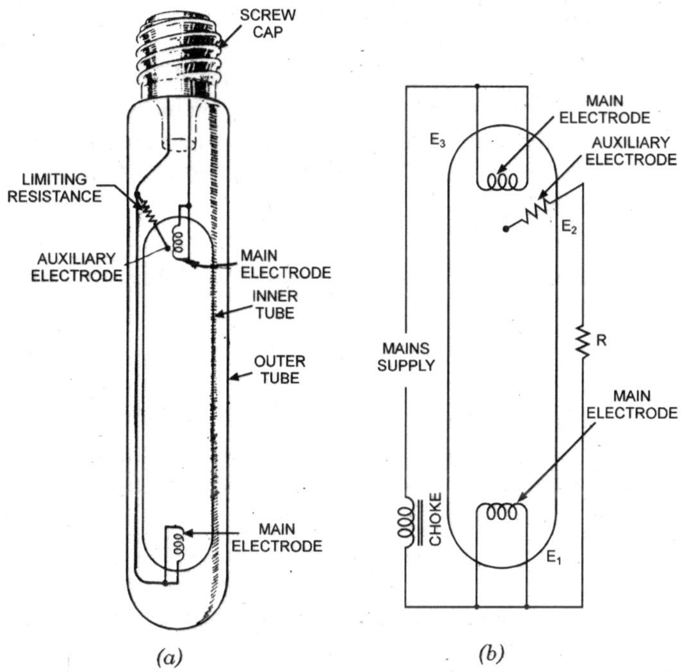

Fig. 7.34. ***High Pressure Mercury Vapour Discharge Lamp***

The lamp has to have auxiliary equipment for use with standard mains voltage, and the necessary connections are shown in Fig. 7.35. The choke is provided to limit the current to a safe value. This choke lowers the power factor, so a capacitor is connected across the circuit to improve the power factor. These lamps must be operated vertically, since if they are used horizontally convection will cause the discharge to touch the glass bulb, which will fail. Lamps which are intended to operate horizontally are fitted with a magnetic device which will hold the luminous column central.

Complete wiring diagram for high pressure mercury vapour discharge lamp is shown in Fig. 7.35.

When the supply is switched on, full mains voltage is applied between the auxiliary electrode and neighbouring main electrode; this breaks down the gap and a discharge through the argon gas takes place. This enables the main discharge to commence. As the lamp warms up, mercury is vaporised, increasing the vapour pressure and the luminous column becomes brighter and narrower. The lamp requires 4 or 5 minutes to attain full brilliancy. If the supply is interrupted, the lamp must cool down and the vapour pressure be reduced before it will start. This takes 3 or 4 minutes. The temperature of operation inside the inner bulb is about 600°C. It gives greenish blue colour light, which causes colour distortion. The efficiency is about 30—40 lumens/watt. These lamps are manufactured in 250 and 400 watt ratings for use on 200—250 volts ac supply mains. The pressure of vapour in these lamps is 2-3 atmosphere. Lamps of this type are used for general industrial lighting, railway yards, ports, work area, shopping centres etc., where greenish blue colour light is not objectionable.

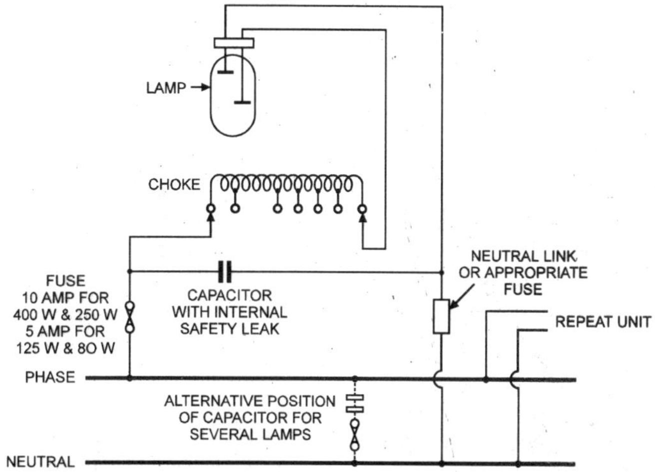

Fig. 7.35. ***Complete Wiring Diagram of HPMV Lamp***

The lamp described above is MA type. Another type, which is manufactured in 300 and 500 watt rating for use on ac as well as dc supply mains, is MAT type. This is similar to MA type except that choke is not used as a ballast. Space between two tubes instead of being evacuated consists of a tungsten filament in series with a discharge tube which acts as a ballast. When the supply is switched on, it operates as a filament lamp, its full output being given by the outer tube. At the same time the discharge or inner tube begins warming up and at a particular temperature a thermal switch operates cutting a part of the filament and thereby increasing the voltage across the discharge tube. The filament contributes a considerable portion of red rays. The combination of the rays from the filament and the blue radiations from the discharge tube produce a useful colour. As the filament acts as a resistance, the overall power factor of the lamp is about 0.95 and therefore, capacitor is not required.

Lower wattage lamps, such as 80 and 125 watts, are manufactured in a different design and using high vapour pressure of about 5-10 atmosphere. These are known as MB types. These operate in a manner similar to MA type except that resistance in series with starting electrode is large and outer bulb is of quartz, (not of ordinary glass) in order to withstand high temperature so that these lamps can be used in any position.

7.11.3. Mercury Iodide Lamps

These lamps are similar in construction to high pressure mercury vapour lamps but in addition to mercury, a number of iodides are added which fill the gaps in the light spectrum, and thus, improve the colour characteristic of the light. Their efficiency is also higher (75—90 lumens/watt). A separate ignition device, in addition to the choke, is required for the mercury iodide lamp. Such lamps are suitable for application in the fields of floodlighting, industrial lighting and public lighting.

7.11.4. Neon Lamp

It is a cold cathode lamp and consists of a glass bulb filled with neon gas with a small percentage of helium. These lamps give orange pink coloured light. Electrodes are of pure iron and are spaced only few mm apart so that lamps can be made for voltages as low as 110 volts ac or 150 dc. For use on ac the electrodes are of equal size. On dc the gas glows near the negative electrode, therefore, negative electrode is made larger in size. The efficiency of neon lamp lies between 15-40 lumens/watt. Owing to discharge of the gas between the electrodes in the form of an arc, it may cause the current drawn by the lamp to increase indefinitely. This is prevented by connecting a high resistance of few thousand ohms in series and mounting it in the cap. The lamp of this type is of the size of an ordinary incandescent lamp. The power consumption is of the order of 5 watts.

Neon lamps are used as indicator lamps, night lamps, for determination of polarity of dc mains and in larger sizes as neon tubes for the purpose of advertising.

7.11.5. Neon Tubes

The popularity of high voltage neon lighting arose almost entirely from its use in advertising, for signs, or in the decorative treatment of buildings, but later the lighting field became important. The neon tube, which is used in varying lengths up to about 9 metres, may be bent into almost any desired shape during manufacture. It consists of a length of glass tubing containing two electrodes, normally cylindrical in shape, of iron, steel, or copper.

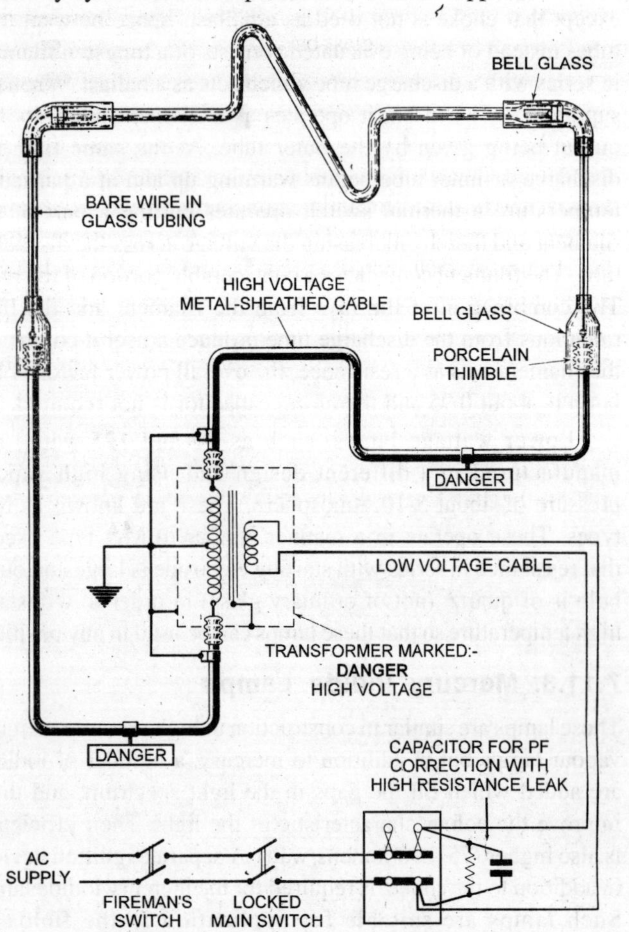

Fig. 7.36. *Neon Tube Installation*

The true neon tube contains neon, but the term is now used also for tubes with fillings of other rare gases. By varying the composition of glass and adding different substances to neon gas different colours such as orange, red, yellow, green etc. are obtained. The diameters of the tube vary, and common sizes of 10, 15, 20 and 30 mm carry currents of 25, 35, 60 and 150 mA respectively. Voltage required may vary from 300 V to 1,000 V per metre of tube length and for starting the discharge a striking voltage, about 1½ times this value is required. Such voltage is obtained by making use of a step-up transformer having a high leakage reactance so that it gives a drooping characteristic. The usual operating voltage is 6,000 volts.

The tubes are mounted either on a wooden frame or a metal base. These are matched with step-up transformers by connecting suitable tappings for the rated current. Connections between letters are made by nickel wires, the glass tubings being slipped over them. The power factor of neon tubes is quite low and is improved by using capacitors. The capacitors can, however, be placed only on the low voltage side of the transformer.

A number of precautions are prescribed for supply to high voltage luminous tube sign installations under IER 1956 (Rule No. 71).

Faults. Flickering may be due to;

(*i*) Transformer secondary voltage too low; this can be rectified by adjustment of the transformer tappings.

(*ii*) Reduction of gas pressure in tube due to absorption of the gas into the electrodes, the tube may be removed and refilled by the manufacturers.

Neon tubes installed in the open requires frequent cleaning, say 4 times per year. Care must be taken that the key of the opened lock switch is removed and held by the operator while work is being done.

7.11.6. Fluorescent Tubes

Fluorescent lighting has a great advantage over other light sources in many applications. The tubes can be obtained in a variety of lengths, with illumination in a variety of colours. It is possible to achieve quite high lighting intensities without excessive temperature rise and, owing to the nature of light sources, the danger of glare is minimised. It must not, however, be thought that fluorescent lamps can be used indiscriminately without giving careful consideration to the type of reflector to be employed or to the correct positioning of the light source. An exposed tube in the line of vision is usually as bad as a tungsten lamp would be. The efficiency of fluorescent lamp is about 40 lumens per watt, about three times the efficiency of an equivalent tungsten filament lamp. The fluorescent tube consists of a glass tube 25 mm in diameter and 0.38 m–1.52 m in length. The inside surface of the tube is coated with a thin layer of fluorescent material in the form of a powder.

The coating materials used depend upon the colour effect desired and may consist of zinc silicate, cadmium, silicate or calcium tungstate. These organic chemicals are known as *phosphors* which transforms short-wave invisible radiations into visible light. By mixing the various powders light of any desired colour including day light can be obtained. The tube contains small quantity of argon gas at a pressure of 2.5 mm of mercury and one or two drops of mercury. It is provided with two electrodes coated with electron emissive material. A starting switch is provided in the circuit, which puts the electrodes directly across the supply mains at the time of starting, so that electrodes may get heated and emit sufficient electrons. A stabilising choke is connected in series with it, which acts as a ballast in running condition and provides a voltage impulse for starting. A capacitor is connected across the circuit to improve the power factor.

Fluorescent tubes are available in the following sizes :

Length	Wattage
38 cm	14 W
46 cm	15 W
61 cm	20 W
100 cm	25 W
122 cm	40 W
152 cm	65 W
152 cm	80 W

The fluorescent lamp in common with most electrical devices may cause a certain amount of radio interference. This interference may be caused due to one of the following factors :

1. Direct radiation from the bulb to the antenna. This effect diminishes rapidly as the radio is separated from the lamp. Thus, for example, at a radius of 2.8 metres interference from this cause is negligible.
2. Line radiation from the electric supply line to the antenna.
3. Line feedback from the lamp through the line to the radio.

The radio interference effect may be reduced to a minimum by connecting a small capacitor (0.05 μF) across starter terminals, as shown in Figs. 7.37 (*a*) and 7.37 (*b*).

The starting switches are of two types, namely the thermal type and the glow type.

The connections of a fluorescent tube incorporating a thermal type starter are shown in Fig. 7.37 (*a*). The thermal starter is a current operated device and consists of two metallic strips and a heater coil. The bimetallic strips are in contact with each other when the lamp is not in operation. When the supply is switched on, the two electrodes get connected in series through the thermal switch, the relatively large current rising them to incandescence. The current also flows through the heater element as a result of which bimetallic strips break contact. This causes interruption in the current flowing through the circuit, which further results in a high voltage surge across the electrodes of the tube, which is enough to strike the arc between the electrodes. This arc is then maintained by the normal lamp voltage. Bombardment of the electrode's surface by the positive mercury ions maintains their temperature so that they continue to emit electrons. Thus the tube is put in operation. The thermal switch is now generally obsolete because of its more complicated construction, greater cost and greater power loss.

The connections of a fluorescent tube incorporating a glow type starter are shown in Fig. 7.37 (*b*). The glow type starter is a voltage operated device and consists of two bimetallic electrodes enclosed in a glass bulb filled with a mixture of helium and hydrogen. Normally the contacts are open. When the supply is switched on, the potential across bimetallic electrodes causes a small glow discharge at a small current not enough to heat up the tube electrodes (filament). This discharge is enough, however, to heat the bimetallic strips of the switch causing them to bend and make contact. The result is a large current through the electrodes, their temperature being raised to incandescence and the gas in the immediate neighbourhood is ionised. After one or two seconds the bimetallic strips cool down and the contacts open. This opening of contacts in series with the choke causes a momentary high voltage, which is sufficient to start the discharge in the main tube. The starter ceases to glow as the voltage is now too low. A small capacitor is placed inside the starter to suppress arcing and radio interference.

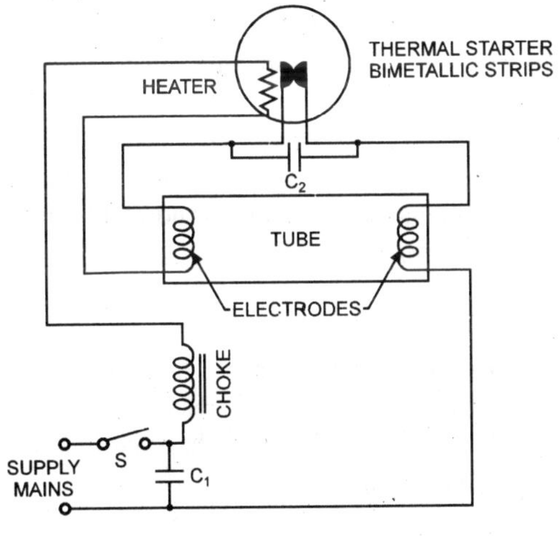

(*a*) *Incorporating a Thermal Type Starter*

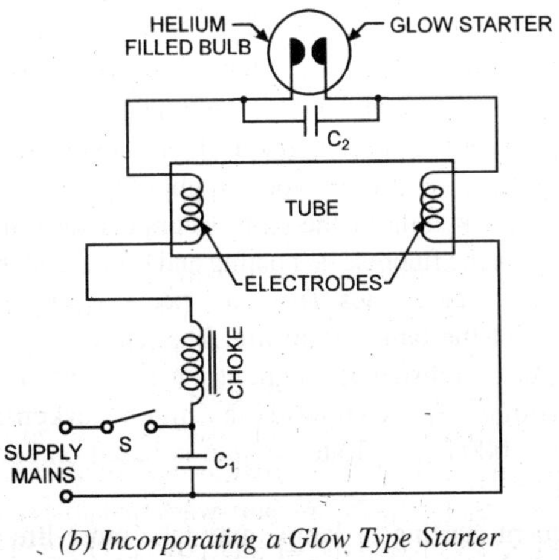

(*b*) *Incorporating a Glow Type Starter*

Fig. 7.37. ***Connection Diagram For a Fluorescent Tube***

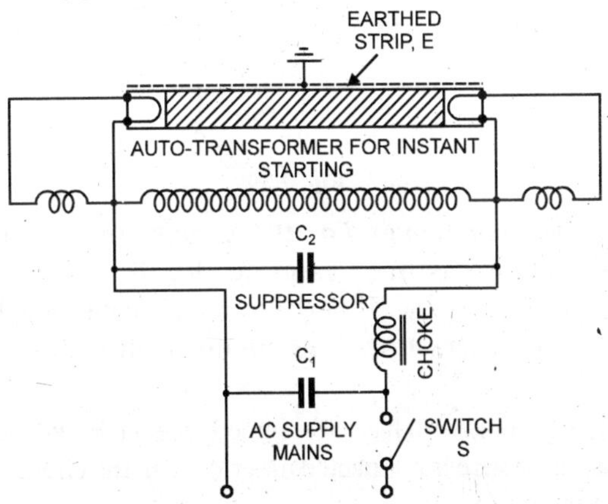

Fig. 7.38. ***Circuit For Instant Starting of Fluorescent Tube***

Nowadays starterless circuits, as instant start, rapid start etc. for such tubes are also employed (Fig. 7.38). The primary winding of an auto-transformer is connected in parallel with the tube and receives practically full mains voltage. The filaments are provided with current from secondary tappings. When the lamp has started the transformer receives the normal tube voltage and the filament currents are correspondingly reduced.

Stroboscopic Effect. At the usual alternating current supply frequency of 50 Hz, a discharge lamp will be extinguished 100 times per second. Although this effect is seldom noticeable in normal conditions. It is possible that this may, in some instances, give rise to a stroboscopic effect, that is moving objects such as rotating parts of machinery, illuminated by this light, may appear to be moving in disunity or jerks, or rotating more slowly than their actual speed.

In actual applications, however, where this effect might cause annoyance, it can be practically eliminated in a three lamp unit by connecting each lamp on a separate phase of a 3-phase system and it can be greatly reduced in a two lamp unit by the use of a two lamp control unit, which employs a condenser in the ballast of one of the lamps. (Fig. 7.39). The current through the lamps is almost 90° out of phase and under these conditions light output of one of the lamps is at a maximum. This method has an additional advantage of giving an overall power factor of nearly unity for the unity of two lamps. In this arrangement one circuit remains at a low power factor at about 0.5 lagging, while the other circuit incorporating a series capacitor C remains at a power factor of about 0.5 leading.

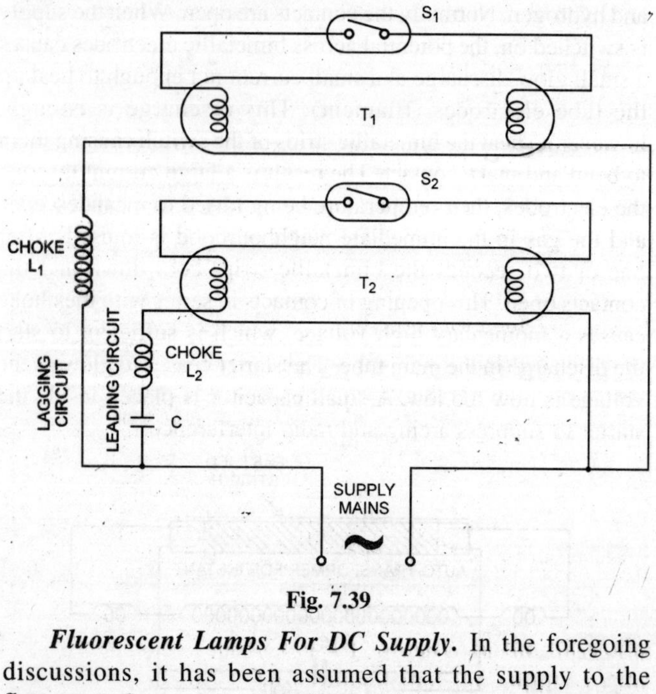

Fig. 7.39

Fluorescent Lamps For DC Supply. In the foregoing discussions, it has been assumed that the supply to the fluorescent lamps is ac. If however, the available supply is dc some special accessories and circuit modification will be required.

(*i*) The choke coil has a low impedance on dc and therefore, a ballast resistance is connected in series with the choke in order to limit the current.

(*ii*) On systems below 220 V, starting becomes less certain on dc. Only thermal type starters should be used.

(*iii*) The positive end becomes relatively dark on account of the tendency of the mercury vapour to migrate towards the negative end of the tube. In order to overcome this defect a reversing switch is included in the circuit between the supply and the fitting, as shown in Fig. 7.40.

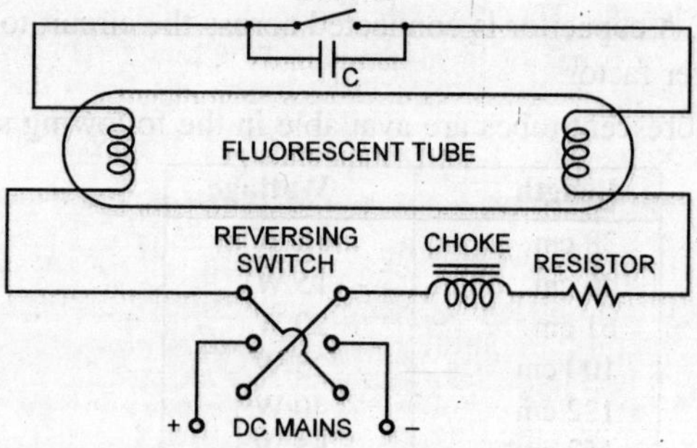

Fig. 7.40. ***Operation of Fluorescent Tube on DC***

In dc operation of fluorescent tube there is no problem of power factor correction and no stroboscopic effect. Its disadvantages are—low efficiency due to power loss in ballast series resistance, increased cost of the ballast resistance and reversing switch and less life of the tube (about 80% of that with ac operation).

Colour of Fluorescent Light. As already mentioned the colour of fluorescent light depends upon the fluorescent powder used and the vapour pressure in the lamp. A few important variations of white colour with their applications are given below.

(*i*) ***Daylight.*** This is suitable for industrial applications where checking of material is carried out at high illumination levels (say 1,000 lux or more) and for normal lighting service.

(*ii*) ***Warm-White.*** This is mostly used for street lighting where colour rendition, due to low level of lighting (say 10 lux or less), is of little importance.

(*iii*) ***Warm-White Deluxe.*** It is used for illuminating rooms of large gathering such as shops, restaurants, and dwelling houses. It combines well with the light of incandescent lamps, as in show-windows etc.

(*iv*) ***White.*** It is useful for use in offices, drawing offices, schools and factories. It harmonizes exceptionally well with the daylight, and therefore, light is a good supplement to daylight.

White-Deluxe. It is very suitable for use in offices, schools and shops where colour rendition is important.

Useful Lamp Life. In general, fluorescent lamps lose their usefulness because of reduction in light output before they fail to operate. Darkening of the bulb occurs because of the effect of mercury on the fluorescent coating and because of the material given off by the electrodes. The latter specially causes darkening at the ends of the bulb late in life. The rate of depreciation in light output diminishes throughout the life ; the first hundred hours produce approximately as much darkening as the following 1,000 hours. Rated output is based on conditions at 100 hours.

Frequent starting of lamps may take more life out of the electrodes than long hours of burning because momentarily there is a higher than normal voltage drop at the electrodes which causes the active material to sputter or evaporate off. If a lamp be started once a minute, for example, the hours of burning will be shorter than normal, but if it be switched on and burned continuously, its life will be longer than normal. When the active material on the electrodes is nearly exhausted, the voltage required for starting will rise and may equal or exceed the available supply. This may occur after the lamp has been started thousands of times or burned beyond its rated life. Sometimes the end of life is indicated by the lamp flashing momentarily and then going out.

The normal life of a fluorescent lamp is 7,500 hours. The average life is for three burning hours per switching operation. The actual life may vary from 5,000 to 10,000 hours depending upon the operating conditions. Light output is reduced by 15-20% after 4,000 hours of operation and it is, therefore, a good practice to replace the fluorescent lamp after 4,000-5,000 hours burning on economical grounds.

Performance Curves. Performance curves are shown in Fig. 7.41.

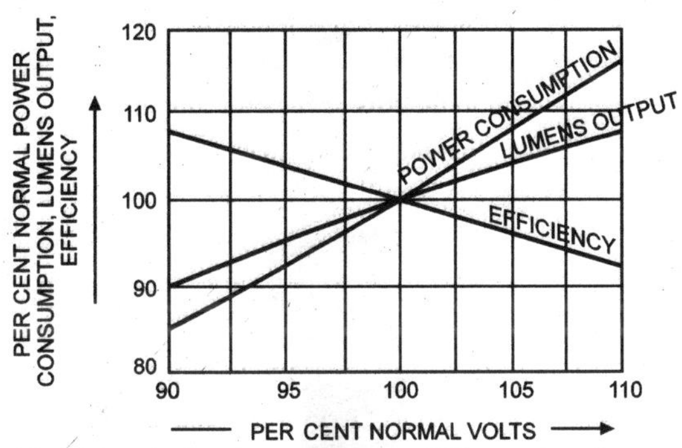

Fig. 7.41. ***Performance Curves of Fluorescent Lamps***

The effect of voltage variation in the case of fluorescent lamps is less marked as compared to the incandescent lamps. However, their life and performance are adversely affected both by low and high voltage. With increased voltage there is a greater heating of electrodes and they lose emissive material by evaporation. With reduced voltage, the current reduces causing sputtering at the electrodes shortening their life.

The best performance of fluorescent lamps is obtained at 20-25°C operating temperature. It decreases rapidly when a lamp is operated at a lower temperature or is exposed to cold wind drafts. The fluorescent lamps may be enclosed in outer envelops for their protection from cold weather in exposed locations. For operation at high temperature, the fittings with provision for air circulation should be employed.

Operating Instructions. (*i*) A fluorescent lamp should start up with little blinking. Blinking indicates a defective starter, low voltage or a defective tube. Both the choke and the starter will get damaged due to overheating, if this condition is allowed to persist. The tube or the starter must be replaced immediately.

(*ii*) Correct voltage must be ensured.

(*iii*) The starter should be replaced every time a tube is changed. A bad starter shortens the life of the tube.

(*iv*) Frequent switching operations should be avoided.

Merits and Demerits. As already stated the fluorescent lamps have got merits of high luminous efficiency, long life, low running cost, low glare level and less heat output. The demerits are stroboscopic effect, small wattage requiring large number of fittings and magnetic hum associated with the choke causing disturbance. The problem of noise can be solved to some extent by mounting chokes on resilient pads.

Trouble-Shooting. Various faults occurring in fluorescent tubes, their possible causes and remedies are given below :

Symptoms	Possible Causes	Remedies
1. Blinking on and off.	1. Low circuit voltage. 2. Loose contact. 3. Defective starter. 4. Defective tube. 5. Wrong connections.	Change ballast tapping if required. Check all connections. Replace starter. Replace tube. Check all connections.
2. End of a tube remain lighted.	1. Short-circuited starter.	Replace starter.
3. Failure of lamp to light up.	1. Defective tube. 2. Defective starter. 3. Defective choke. 4. Defective and loose holders.	Replace tube. Replace starter. Replace choke. Turn tube through 90°. Try another starter and check connections.
4. Blackening of the end (early life) (normal life)	1. Too low or too high voltage. 2. Mercury deposits at the ends.	Adjust ballast tapping. It will evaporate as the lamp operates.
5. Burn out electrodes	1. Control unit in circuit or choke short circuited.	Test with a new choke.
6. Dark streaks along lamp	1. Mercury globules.	Rotate tube through 180°.
7. Snaking spiralling (with glow type starter)	1. Insufficient heating due to either starter opening quickly or wrong tappings made on the ballast.	Replace starter or ballast. In case of new lamp, it cures after few days.

7.11.7. Compact Fluorescent Lamps (CFLs)

A compact fluorescent lamp (CFL), also known as a *compact fluorescent light* or energy saving light, is a type of fluorescent lamp which combines the energy efficiency of fluorescent lighting with the convenience and popularity of incandescent fixtures. CFLs can replace incandescents that are roughly 3-4 times their wattage, saving up to 75% of the initial light energy. Although CFLs costs 3-10 times more than comparable incandescent lamps, they last about 10 times as long (10,000 hours).

CFLs are most cost effective in areas where lights are ON *for long periods of time.*

Because of the potential of reducing electric consumption and pollution, some organizations have encouraged the adoption of CFLs. Some electric utilities and local governments have subsidized CFLs as a means of reducing electric demand. In India also almost all the states are encouraging the adoption of CFLs in place of incandescent lamps and are providing incentives or even distributing CFLs to people free of cost to save electricity.

CFLs radiate a light spectrum different from that of incandescent lamps. Improved phosphor formulations have improved the perceived colour of the light emitted by CFLs such that some sources rate the best "soft white" CFLs as subjectively similar in colour to standard incandescent lamps.

Like all fluorescent lamps, CFLs contain mercury, which makes their disposal complicated.

Operation. CFLs work much like standard fluorescent lamps. They consist of two components : a gas-filled tube, and a magnetic or electronic ballast. The gas in the tube glows with ultraviolet light when switched ON and electric current from the ballast flows through it. This in turn excites a white phosphor coating on the inside of the tube, which emits visible light throughout the tube surface.

CFLs with magnetic ballasts flicker slightly at start. They are also heavier than those with electronic ballasts. This may make them too heavy for some light fixtures. Electronic ballasts are more expensive, but light immediately (especially at low temperature). They are also more efficient than magnetic ballasts. The tubes will last about 10,000 hours and the ballast about 50,000 hours. Most currently available CFLs have electronic ballasts.

Electronic ballasts contain a small circuit board with rectifiers, a filter capacitor and usually two switching transistors connected as a high-frequency resonant series dc to ac inverter. The resulting high frequency (about 40 kHz or higher) is applied to the lamp tube. Since the resonant converter tends to stabilize lamp current and so the light produced over a range of input voltages, standard CFLs do not respond well in dimming applications and special lamps are required for dimming services.

CFLs are designed to operate within a specified temperature range. Temepratures below the range cause reduced output. Most of the CFLs are for indoor use, but there are models available for outdoor use. Outdoor CFLs need installation in enclosed fixtures so as to minimize the adverse effects of low temperatures.

Types of CFLs. CFLs are available in a variety of styles or shapes. Some have two, four, or six tubes while others have circular or spiral-shaped tubes. The size or total surface area of the tube(s) determines how much light it produces.

Some CFLs have the tubes and ballast permanently connected. Other CFLs have separate tubes and ballasts which facilitates in replacement of tubes without changing the ballast. They are also of types enclosed in a glass tube. These look somewhat similar to conventional incandescent lamps, except they are larger.

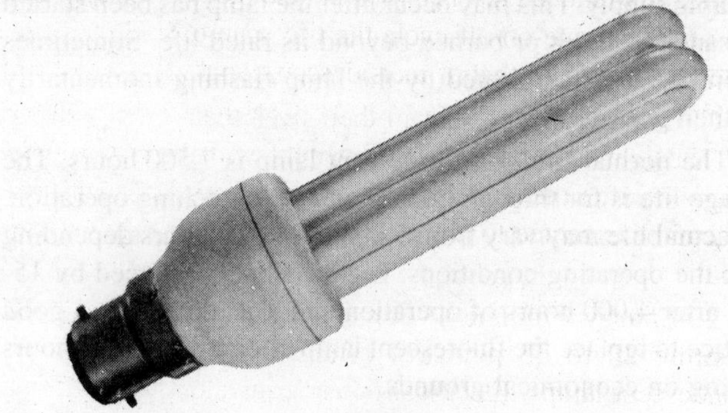

(a) The Tubular Type Compact Fluorescent Lamp

(b) Helical Integrated CFL

Fig. 7.42

CFLs are of two types : integrated and non-integrated lamps.

Integrated lamps combine a tube, an electronic ballast and either an Edison screw or a bayonet in a single unit. The incandescent lamps can be easily replaced with CFLs. Integrated CFLs work well in many standard incandescent light fixtures, reducing the cost of conversion.

Non-integrated CFLs have the ballast permanently installed in the luminaire, and only lamp is usually replaced at the end of life. Since the ballasts are placed in the light fixture, they are larger and last longer in comparison to the integrated ones. Non-integrated CFL housings can be both more expensive and sophisticated. They have two types of tubes : bi-pin tubes and quad-pin tubes. Bi-pin tubes are designed for conventional ballasts, and quad-pin tubes are designed for electronic ballasts or conventional ballasts with external starters. A bi-pin tube contains an integrated starter which obviates the need for external heating pins but causes incompatibility with electronic ballasts.

CFLs are produced for operation on both dc and ac supply. DC CFLs are popular for use in recreational vehicles and off-the-grid housing. CFLs can also be operated with solar powered street lights, using solar panels located on the top or sides of a pole and light fixtures that are specially wired to use the lamps.

Advantages of CFLs Over Incandescent Lamps

1. *Life Span.* As already mentioned, the average rated life of a CFL is about 10 times that of incandescent lamps.

The lifetime of any lamp depends on various factors, such as operating voltage, exposure to voltage spikes, mechanical shock, frequency of cycling on and off, lamp orientation, ambient temperature, manufacturing quality etc. The life of a CFL is significantly reduced if it is turned on and off frequently. In the case of a 5 minute on/off cycle the life span of a CFL can be reduced to "close to that of incandescent lamp".

CFLs provide less light in their lives than when they are new. The light output decay is exponential, with the fastest losses being soon after the lamp is first used. By the end of their lives, they can be expected to produce 70-80% of their original light output.

2. *Energy Efficiency.* For a given light output, CFLs consume 20 to 33 per cent of the power of equivalent incandescent lamps. Electrical power equivalents for different lamps are indicated below.

Electrical Power Consumption in Watts		Minimum Light Output in Lumens
CFLs	Incandescent Lamps	
8-9	40	450
9-15	60	800
15-20	75	1,100
20-25	100	1,600
25-45	150	2,600

3. *Heating and Cooling.* If in a building, indoor incandescent lamps are replaced by CFLs, the heat produced due to lighting will be significantly reduced. In warm climates office/ industrial buildings where air-conditioning is usually requried, CFLs would reduce the load on the cooling system resulting in saving in electrical consumption, in addition to the energy efficiency savings of using CFLs instead of incandescent lamps. However, in cooler climates in which buildings need heating, the heating system would need more electrical energy. While the CFLs are still saving electricity, total greenhouse gas emissions may increase in certain scenarios, such as the operation of natural gas furnace to replace unintended heating from CFLs running on low-GHG electricity. Thus overall energy cost saving depends on the climate, increased heating demand offsets some of the lighting energy saved.

4. *Efficiency.* The typical luminous efficiency of CFLs is 60 to 72 lumens/watt, and that of normal domestic incandescent lamps is 13 to 18 lumens/watt. Compared to a theoretical 100% efficient lamp, these figures are equivalent to lighting efficiency ranges of 9 to 11% for CFLs (60/680 to 72/680), and 1.9 to 2.6% for incandescent (13/680 to 18/680).

5. *Embodied Energy.* No doubt, CFLs need more energy in manufacturing than the incandescent lamp but this embodied energy is offset by their longer life and small energy consumption than equivalent incandescent lamps.

6. *Cost.* While the purchase price of an integrated CFL is typically 3 to 10 times higher than that of an equivalent incandescent lamp, the extended lifetime and lower energy consumption will more than compensate for the higher initial cost. CFLs are extremely cost-effective in commercial buildings when used to replace incandescent lamps.

7. *Starting Time.* Incandescent lamps attain full brilliancy a fraction of second after being switched on, CFLs turn on within a second, but many still take time to warm up to full brightness. The light colour may be slightly different immediately after being switched on. Some CFLs are marketed as "instant on" and have no noticeable warm-up period, but others can take up to a minute to attain full brightness, or longer in very cold temperatures. Some that make use of a mercury amalgam can take up to three minutes to attain full brightness. This and shorter life of CFLs when turned on and off for short periods may make CFLs less suitable for applications like motion-activated lighting.

Disadvantages of CFLs

1. *Health Issues.* If individuals are exposed to the light produced by some single-envelope CFLs for long periods of time at distances of less than 20 cm, it could lead to ultraviolet exposures approaching the current workplace limit set to protect workers from skin and retinal damage. The ultraviolet radiation received from CFLs is too small to contribute to skin cancer and the use of double-envelope CFL lamps "largely or entirely" mitigates any other risks.

2. *Environmental Issues.* CFLs, like all fluorescent lamps, contain small amounts of mercury as vapour inside the glass tubing. Because mercury is poisonous, even these small amounts are a concern for landfills and waste incinerators where the mercury from lamps may be released and contribute to air and water pollution. Health and environmental concerns about mercury have promoted many jursdictions to require used lamps to be properly disposed or recycled rather than being included in the general waste stream sent to land fills. So disposal of used CFLs required proper care.

7.12. COMPARISON BETWEEN TUNGSTEN FILAMENT LAMPS AND FLUORESCENT TUBES

	Tungsten Filament Lamps	Fluorescent Tubes
1.	Voltage fluctuation has comparatively more effect on the light output.	Voltage fluctuation has comparatively low effect on light output as the variations in voltage are absorbed in the choke.
2.	Luminous efficiency increases with the increase in voltage of the lamp.	Luminous efficiency increases with the increase in wattage and increase in length of tube.
3.	It gives light close to natural light, therefore, objects are properly seen.	It does not give light close to natural light, therefore, colour rendering is defective.
4.	Luminous efficiency of coloured filament lamps is poor because coloured glasses are used for this purpose.	Different colour lights can be obtained by using different composition of fluorescent powder. Hence efficiency is high and better colours are obtained.

	Tungsten Filament Lamps	Fluorescent Tubes
5.	Due to comparatively high working temperature heat radiations are also present.	Due to low working temperature heat radiation is low.
6.	Its brightness is more.	Its brightness is less.
7.	With the time light output is reduced.	With the time light output is gradually reduced.
8.	No stroboscopic effect.	It has objectionable stroboscopic effect.
9.	Though the life of filament lamps varies with the working voltage, however, its normal life is 1,000 working hours.	Life of fluorescent tubes is not affected so much by variations in voltage but it depends on the frequency of starting. The life of the tube is about 7,500 working hours.
10.	The initial cost per lamp is quite low.	The initial cost per tube is more.
11.	For same lumens output more lamps are required and wiring cost is more. Life of lamp is also low. Hence overall cost of maintenance is more.	For same lumens output, lesser number of tubes is required, the wiring cost is low. Since its life is comparatively more, replacement cost is low. Hence overall maintenance cost is low.

Example 7.20. Two 220 V lamps, one of 60 W and the other of 75 W are connected in series across a 440 V supply. Calculate the pd across each lamp, neglecting any variation in resistance. Assuming the candle power to be proportional to the fourth power of the voltage, calculate the candle power of each lamp under this condition as a percentage of its value under normal operation at 220 V.

Solution : Resistance of 60 watt, 220 V lamp,

$$R_1 = \frac{V^2}{W_1} = \frac{(220)^2}{60} = 806.67\ \Omega$$

Resistance of 75 watt, 220 V lamp,

$$R_2 = \frac{V^2}{W_2} = \frac{(220)^2}{75} = 645.33\ \Omega$$

Current flowing through the lamps when connected in series across 440 V,

$$I = \frac{440}{806.67 + 645.33} = 0.303\ \text{A}$$

PD across 60 watt lamp, $V_1 = IR_1 = 0.303 \times 806.67 = 245$ V **Ans.**

PD across 75 watt lamp, $V_2 = IR_2 = 0.303 \times 645.33 = 195$ V **Ans.**

Normal operating voltage, V = 220 V

Candle power of 60 watt lamp under this condition

$$= \left(\frac{V_1}{V}\right)^4 \times \text{normal candle power}$$

$$= \left(\frac{245}{220}\right)^4 \times 100 \text{ per cent of candle power under normal operation at 220 V}$$

= 154% of candle power under normal operation at 220 V **Ans.**

Similarly candle power of 75 watt lamp under this condition

$$= \left(\frac{195}{220}\right)^4 \times 100\% \text{ of CP under normal operation at 220 V}$$

= 62% of CP under normal operation at 220 V **Ans.**

Example 7.21. A lamp of 84 watts operates at 240 volts and power factor 0.7. Its power factor is to be corrected to unity. Determine the capacitance required for the condenser.

Solution : Wattage of lamp = 84 watts

Uncorrected pf, $\cos\phi = 0.7$

$$\text{Current draw by lamp, } I = \frac{\text{Wattage of lamp}}{V \times \cos\phi} = \frac{84}{0.7 \times 240} = 0.5\ \text{A}$$

Relative volt-amperes drawn by the lamp

$$= VI \sin\phi = 240 \times 0.5 \times 0.714 = 85.7$$

Corrected power factor, $\cos\phi' = 1$

Reactive volt-amperes drawn by the lamp after pf correction

$$= VI \sin\phi' = 0$$

Reactive volt-amperes to be compensated by the capacitor, $2\pi f CV^2 = 85.7$

$$\text{or} \quad C = \frac{85.7}{2\pi f V^2} = \frac{85.7}{2\pi \times 50 \times (240)^2} = 4.74\ \mu\text{F} \quad \textbf{Ans.}$$

Note : Supply frequency be assumed as 50 Hz.

Example 7.22. A 110-V lamp develops 16 CP and lamp of the same material and working at the same efficiency develops 25 CP on 220 V. Compare the diameter and length of the filaments.

[Pb. Technical Univ. Electric Drives and Utilization May 2009]

Solution : As the efficiency are the same, their operating temperatures must be the same.

Let d_1 and d_2 be the diameters of the filaments, l_1 and l_2 their respective lengths and I_1 and I_2 their respective currents.

Assuming that the power intakes of the two lamps are proportional to their output, we have

$$16 \propto 110\, I_1 \quad \text{and} \quad 25 \propto 220\, I_2$$

$$\text{or} \quad \frac{I_1}{I_2} = \frac{16}{25} \times \frac{220}{110} = \frac{32}{25}$$

Since the material and operating temperatures are same

$$\therefore \quad \frac{d_1}{d_2} = \left(\frac{I_1}{I_2}\right)^{2/3} = \left(\frac{32}{25}\right)^{2/3} = 1.18 \quad \textbf{Ans.}$$

Since output $\propto l_1 d_1$

$$\therefore \quad \frac{16}{25} = \frac{l_1 d_1}{l_2 d_2}$$

$$\text{or} \quad \frac{l_1}{l_2} = \frac{16}{25} \times \frac{d_2}{d_1} = \frac{16}{25} \times \frac{1}{1.18} = 0.543 \quad \textbf{Ans.}$$

Example 7.23. A 60 CP, 250 volt metal filament lamp has a measured candle power of 71.5 CP at 260 volts and 50 CP at 240 volts.

(a) Find the constant for the lamp in the expression $C = aV^b$ where C = candle power and V = voltage.

(b) Calculate the change of candle power per volt at 250 V. Determine the percentage variation of candle power due to a voltage variation of ± 4% from the normal value.

Solution : (*a*) Substituting C = 71.5 CP when V = 260 volts and C = 50 CP when V = 240 volts in expression $C = aV^b$ we get

$$71.5 = a\,(260)^b \quad \text{...(i)}$$

$$\text{and} \quad 50 = a\,(240)^b \quad \text{...(ii)}$$

Dividing Eq. (*i*) by Eq. (*ii*), we get

$$\left(\frac{260}{240}\right)^b = 1.43$$

$$\text{or} \quad b = 4.5 \quad \textbf{Ans.}$$

Substituting $b = 4.5$ in Eq. (*ii*), we get

$$a = \frac{50}{(240)^{4.5}} = 0.98 \times 10^{-9} \quad \textbf{Ans.}$$

Thus expression becomes $C = 0.98 \times 10^{-9}\, V^{4.5}$ ...(*iv*)

(*b*) Differentiating Eq. (*iii*) w.r.t. V for V = 250 volts, we get

$$\frac{dC}{dV} = 0.98 \times 10^{-9} \times 4.5\, V^{3.5} = 0.98 \times 10^{-9} \times 4.5 \times (250)^{3.5}$$

$$= 4.4\ V \quad \textbf{Ans.}$$

When voltage increases by 4%, $\frac{C_2}{C_1} = (1.04)^{4.5}$

and percentage change in candle power

$$= \frac{C_2 - C_1}{C_1} \times 100 = [(1.04)^{4.5} - 1] \times 100$$

$$= (1.193 - 1) \times 100 = 19.3\% \quad \textbf{Ans.}$$

When voltage falls by 4 percentage, $\frac{C_2}{C_1} = (0.96)^{4.5}$

and percentage change in candle power

$$= \frac{C_2 - C_1}{C_1} \times 100 = [(0.96)^{4.5} - 1] \times 100$$

$$= (0.832 - 1) \times 100 = -16.8\% \quad \textbf{Ans.}$$

7.13. BASIC PRINCIPLES OF LIGHT CONTROL

When light falls on a surface, depending upon the nature of the surface, some portion of light energy is reflected, some portion is transmitted through the medium of the surface and the rest is absorbed. *The ratio of reflected light energy to the incident light energy is known as reflection factor.*

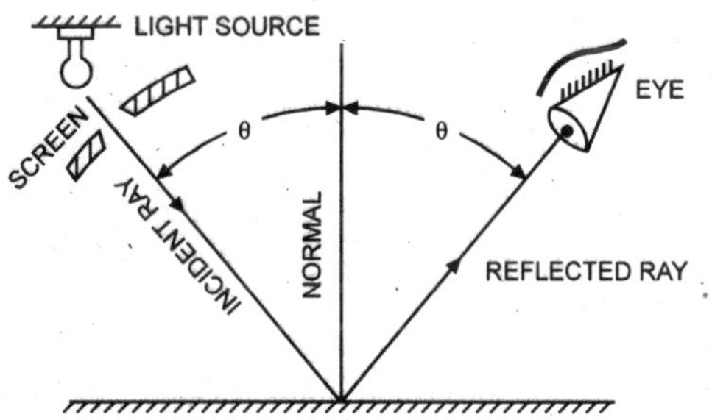

(a) Specular Reflection

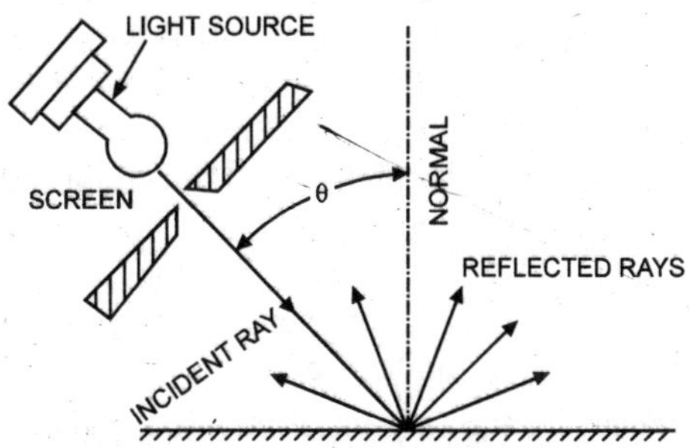

(b) Diffuse Reflection

Fig. 7.43. ***Basic Types of Reflections***

There are two basic types of reflection (*i*) mirror or specular reflection and (*ii*) diffuse reflection. In case of specular reflection a beam of light is reflected but not scattered, and unless the eye is placed in the path of the reflected beam, the viewer is unaware of the existence of light. Moreover, if his eye is placed in the path of the reflected beam, he sees, not the illuminated surface, but the light source. Surfaces causing specular reflection are silvered mirrors, highly polished metals etc. With diffuse reflection the reflected light is scattered in all directions, and the viewer sees the illuminated surface, not the light source. Surfaces causing diffuse reflection are paper, frosted glass, chalk, dry earth, plaster etc. If a surface that is uniformly illuminated by a beam of light appears to be equally bright when viewed from all possible angles the reflection is said to be perfectly diffused. Perfect mirror surfaces and perfect diffusing surfaces are ideals that do not exist in nature. The reflection from any actual surface is partly specular and partly diffuse, the proportional varying widely. A surface that is almost free from mirror reflection is called a *mat* surface.

7.14. TYPES OF LIGHTING SCHEMES

The distribution of the light emitted by lamps is usually controlled to some extent by means of reflectors and translucent diffusing screens, or even lenses.

The interior lighting schemes may be classified as (*i*) direct lighting (*ii*) semi-direct lighting (*iii*) semi-indirect lighting (*iv*) indirect lighting and (*v*) general lighting.

1. Direct Lighting. It is the most commonly used type of lighting scheme. In this lighting scheme more than 90 per cent of total light flux is made to fall directly on the working plane with the help of deep reflectors. Though it is most efficient but causes hard shadows and glare. It is mainly used for industrial and general outdoor lighting.

2. Semi-Direct Lighting. In this lighting scheme 60 to 90 per cent of the total light flux is made to fall downwards directly with the help of semi-direct reflectors, remaining light is used to illuminate the ceiling and walls. Such a lighting scheme is best suited to rooms with high ceilings where a high level of uniformly distributed illumination is desirable. Glare in such units is avoided by employing diffusing globes which not only improve the brightness towards the eye level but also improve the efficiency of the system with reference to the working plane.

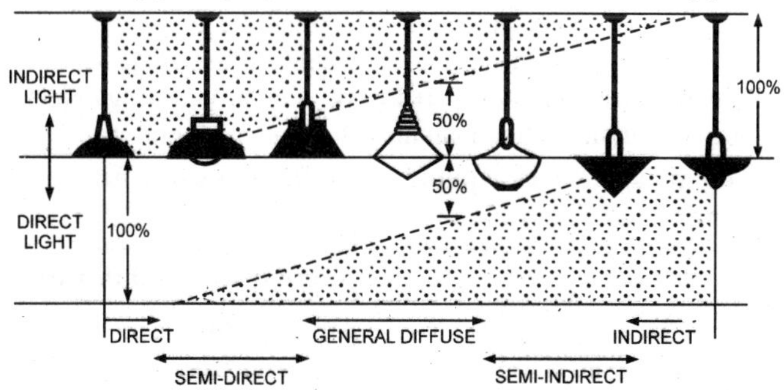

Fig. 7.44. ***General Classification of Luminaires***

3. Semi-Indirect Lighting. In this lighting scheme 60 to 90 per cent of total light flux is thrown upwards to the ceiling for diffuse reflection and the rest reaches the working plane directly except for some absorption by the bowl. This lighting scheme is with soft shadows and glare free. It is mainly used for indoor light decoration purposes.

4. Indirect Lighting. In this lighting scheme more than 90 per cent of total light flux is thrown upwards to the ceiling for diffuse reflection by using inverted or bowl reflectors. In such a system the ceiling acts as the light source, and the glare is reduced to minimum. The resulting illumination is softer and more diffused, the shadows are less prominent and the appearance of the room is much improved over that which results from direct lighting. It is used for decoration purposes in cinemas, theatres and hotels etc. and in workshops where large machines and other obstructions would cause troublesome shadows if direct lighting is employed.

5. General Lighting. In this scheme lamps made of diffusing glass are used which give nearly equal illumination in all directions.

7.15. DESIGN OF LIGHTING SCHEMES

The lighting scheme should be such that it may,

(*i*) provide adequate illumination,

(*ii*) provide light distribution all over the working plane as uniform as possible,

(*iii*) provide light of suitable colour, and

(*iv*) avoid glare and hard shadows as far as possible.

The following factors are required to be considered while designing the lighting scheme.

1. *Illumination Level.* This is the most vital factor because a sufficient illumination is the basic means where by we are able to see our surroundings, unless they are themselves light sources, since only when illuminated do the objects take on the necessary brightness. It is the task of illumination to give objects a distributed brightness. Body colours have property of reflecting light in different degrees. It is this differential brightness which gives essential perception of details. For each type of work there is a range of brightness most favourable to output, *i.e.*, which causes minimum fatigue and gives maximum output in terms of quality and quantity. Degree of illumination, to give necessary brightness to the objects depends upon (*i*) the size of the object to be seen and its distance from the observer–greater the distance of the object from observer and smaller the size of the object, greater will be the illumination required for its proper perception and (*ii*) contrast between the object and back ground—greater the contrast between the colour of the object and its back ground, greater will be the illumination required to distinguish the object properly. Objects which are seen for longer duration of time require more illumination than those for casual work. Similarly moving objects require more illumination than those for stationary object.

Illumination level required, as per ISI, in various parts of a building is given below:

Location	Illumination Level in Lux
Entrances, hallways	100
Living room	300
Dining room	150
Bed room General	300
Dressing tables, bed heads	200
Tables games	300
Games or recreation room	100
Kitchen	200
Kitchen sink	300
Laundry	200
Bathroom	100
Bathroom mirror	300
Sewing	700
Workshop	200
Stairs	100
Garage	70
Study	300

Illumination level required, as per ISI, for various types of traffic routes is given below :

Classification of Lighting Installation	Type of Road	Average Level of Illumination on Road Surface
Group A_1	Important traffic routes carrying fast traffic.	30
Group A_2	Other main roads carrying mixed traffic like main city streets, arterial roads, through ways etc.	15
Group B_1	Secondary roads with considerable traffic like principal local traffic routes, shopping streets etc.	8
Group B_2	Secondary roads with light traffic.	4

2. *Uniformity of Illumination.* The human eye adjusts itself automatically to the brightness within the field of vision. If there is a lack of uniformity, pupil or iris of the eye has to adjust more frequently and thus fatigue is caused to the eye and productivity is reduced. It has been found that visual performance is best if the range of brightness within the field of vision is not greater than 3 : 1, which can be achieved by employing general lighting in addition to localised lighting. Apart from the consideration of causing fatigue, local lighting without using matching general lighting creates psychological feeling of loneliness, gloom and unfriendliness. The modern trend is thus towards 'localised lighting plus general lighting' and towards the adoption of "general lighting oriented towards the working surface" especially in mass production factories, offices, drawing offices, shops etc.

3. *Colour of Light.* The appearance of the body colour entirely depends upon the colour of the incident light. In general the composition of the light should be such that the colour appears natural, *i.e.*, its appearance by artificial light is not appreciably different from that by daylight. Daylight fluorescent tubes nowadays make it possible to illuminate economically even large spaces with artificial daylight giving good colour rendering and at sufficiently high level. For certain applications such as street lighting, colour of light does not matter much if different components have not to be distinguished from each other by their colours, highly efficient discharge lamps, which cause colour distortion, can be used.

4. *Shadows.* In lighting installations, formation of long and hard shadows causes fatigue to eyes and, therefore, is considered to be a shortcoming. Complete absence of shadows altogether again does not necessarily mean an ideal condition of lighting installations. Contrary, perhaps to popular opinion, a certain amount of shadow is desirable in artificial lighting as it helps to give shape to the solid objects and makes them easily recognised. Objects illuminated by shadowless light appear flat and uninteresting, contours are lost, and it is difficult for the eye to form a correct judgement of the shape of an object. However, there is one exception to this, *i.e.*, in drawing offices, where we are to see flat surfaces, shadowless light is essential otherwise shadows will hinder the work. Hard and long shadows can be avoided by (*i*) using large number of small luminaries mounted at a height not less than 2.5 metres and (*ii*) by using wide surface sources of light using globes over filament lamps or by using indirect lighting system.

5. *Glare.* It may be direct or reflected, *i.e.*, it may come directly from the light source or it may be reflected brightness such as from a desk top, nickelled machine parts, or calendered paper.

Direct glare from a source of light is the more common, and is more often a hindrance to vision. A glance at the Sun proves that an extremely bright light source causes acute eye discomfort. Light sources of far less brilliancy than the Sun, such as the filament of an incandescent lamp, or the incandescent mantle of a gas lamp, also cause discomfort by a direct glare. Reflected glare is a glare which comes to the eyes as glint or reflection of the light source in some polished surface.

Toleration of bright light sources in the intermediate vicinity is made possible by locating them at such a height as to place them above the ordinary range of vision.

Metal reflectors for industrial lighting are ordinarily provided with a skirt around the rim of the reflector.

6. *Mounting Height.* The mounting height will largely be governed by the type of the building and type of lighting scheme employed. In the case of direct lighting, in rooms of large floor area, the luminaries should be mounted as close to the ceiling as possible. Lowering them not only will make the illumination less uniform, but will also bring them more into the field of vision, thus increasing the glare, without causing an appreciable increase in the coefficient of utilization. In the unusual case of small rooms with high ceilings, there is something to be gained by lowering the luminaries, but even here a better solution might be to use filament lamps with focusing reflectors and to mount them high.

In the case of indirect and semi-indirect lighting, it would of course be desirable to suspend the luminaries far enough down from the ceiling in order to give reasonably uniform illumination on the ceiling. In practice this is usually taken to mean that the length of the suspension tubes should be one-quarter to one-third the horizontal spacing between the rows of luminaries. Unfortunately it is often impossible to suspend them as far down as this, because it is quite generally recognized that there should be a minimum clearance of 2.5 metres between the luminaries and the floor. It follows that in rooms having low ceilings, the illumination on the ceiling is unavoidably non-uniform. This is not objectionable, because the ceiling is not a working plane.

7. *Spacing of Luminaries.* Correct spacing is of great importance to provide uniform illumination over the whole area and thus do away with comparatively dark areas which are so often found when the fittings are badly spaced.

In the case of direct and semi-direct luminaries the ratio of the horizontal spacing between rows to the height of the luminaries above the working plane depends to quite an extent on the candle power-distribution curve of the luminaire. With fluorescent luminaries it is good practice to aim at a value of unity for this ratio, and to set an upper limit of 3/4. In the case of tungsten lamps combined with focussing reflectors, the ratio of spacing to height should be about 0.6.

In the case of indirect and semi-indirect luminaries, it is good practice to aim at a horizontal spacing between rows approximately equal to a height of the ceiling above the working plane, and in no case should the horizontal spacing exceed 1½ times this height.

In the case of fluorescent luminaries, it is a common practice to join two or more luminaries end to end so that they can share a common outlet. In fact it often works out well to use continuous rows of luminaries, especially when the specified illumination is fairly high.

8. *Colour of Surrounding Walls.* The illumination in any room depends upon the light reflected from the walls and ceilings. White walls and ceilings reflect more light as compared to coloured ones.

7.16. FACTORY LIGHTING

Adequate lighting of factories is of vital importance, as it provides improved amenities for the employees, increased production and has a definite economic value in reducing accidents with their consequent loss of time and compensation payments.

General Requirements and Types of Installations. A factory lighting installation, in common with other indoor equipments should provide an adequate illumination on the working plane and give a good distribution of light, employ simple and easily cleaned fittings and avoid glare. It is essential not only to avoid glare from the lamp itself but also reflected glare from any polished surface, which may be within the line of vision.

General Lighting. The usual scheme in factories and workshops is to mount a number of lamps at a sufficient height so that uniform distribution of light over the working plane is obtained. In large machine shops the height is governed by the necessity of keeping the lamps above the travelling crane. In such cases it is often desirable to supplement the main lighting by side lighting in order to give additional illumination on a vertical plane. Since light coloured walls and ceiling add to the effectiveness of an installation, therefore; it is necessary to get whitewashing or painting done.

Local Lighting. On some points fairly intense illumination is required. For this purpose local lighting can be provided by means of adjustable fittings attached to the machine or bench in question or mounted on portable floor standards. Such lamps should be mounted in deep reflectors so that glare is avoided.

Portable hand lamps attached to wall plugs by means of trailing leads are used for maintenance work and emergency lighting. Low voltage lamps of not more than 50 volts are recommended for use as portable hand lamps because such lamps have thicker filament and are, therefore, more robust than those for normal voltage and danger of shock is also avoided in these few volt lamps. The supply for such lamps can be obtained from a special low voltage distribution system running through out the factory or by means of a small transformer for each individual lamp.

Local lighting should never be employed alone, good general lighting is essential so that the dark places between the local lighting units are avoided. Dark places between the local lighting units cause fatigue to the eyes on account of its continually having to adjust itself to new conditions.

Emergency Lighting. Some lights, such as for (*i*) internal pilot lighting required for safe and speedy evacuation of personnel after main lighting circuit is off (*ii*) external pilot lighting, provided with careful shades leading to shelters required for evacuation of personnel (*iii*) for control posts, first aid centres etc. (*iv*) dials and gauges in important plants required to be watched regularly are required during an air raid when all the factory lights are off as a matter of air-raid precaution. The circuit supplying the above emergency lights should be independently controlled. It is very desirable to provide auxiliary lighting from the source other than the main electric supply preferably from batteries or from small petrol driven generator set. If, however, emergency lighting circuits are operated from main electric supply, these should be completely separated from main lighting circuit.

7.16.1. Industrial Lighting Fittings

Reflectors for industrial purposes must be simple in design and easily cleaned. The requirements of most of the installations can be met by one of the following types of fittings.

Standard Reflectors. These reflectors are made to accommodate lamps of ratings from 40 to 1,500 watts and designed so that they give adequate and uniform illumination when they are mounted at a spacing equal to about 1.5 times their mounting height above the working plane.

Diffusing Fittings. When more diffused light is required than that given by the standard reflector a diffusing glass screen may be fixed across a standard type of reflector. Such fittings are used where highly polished articles are dealt with.

Concentrating Reflectors. A reflector with a concentrated beam is employed in large machine shops and foundaries, where the fittings are to be mounted on a considerable height above the working plane. In such places an ordinary reflector would have too wide angle of divergence and would waste a great deal of light on the walls.

Enclosed Diffusing Fittings. An opal globe completely enclosing the lamp giving a very even and well diffused light is used when light coloured walls and ceiling are there.

Angle Reflectors. Angle reflectors are used to provide illumination in a vertical plane where concentrating type reflectors are used. These can be mounted on suitable stanchions or the walls.

7.16.2. Maintenance

In order to maintain the fittings in a condition of reasonable efficiency it is necessary to clean the light fittings periodically. The frequency of cleaning depends on the conditions in the particular factory under consideration and varies from once or twice a week for very dirty surroundings to every four or six weeks under the best conditions.

7.16.3. Types of Lamps

The discharge lamps have been used in where colour rendering is not important. The fluorescent lamps are widely employed on account of its natural day—light colour, its even illumination and absence of glare and in some cases, the fact that it gives rise to considerably less heat than filament lamps of the same light output.

7.17. METHODS OF LIGHTING CALCULATIONS

A number of methods have been employed for lighting calculations, among which may be mentioned :

1. Watts per square metre method.
2. Lumen or light flux method.
3. Point to point or inverse-square law method.

1. *Watts Per Square Metre Method.* This is principally a "rule of thumb" method, very handy for rough calculation or checking. It consists in making an allowance of watts per square metre of area to be illuminated according to the illumination desired on the assumption of an average figure of overall efficiency of the system.

2. *Lumen or Light Flux Method.* This method is applicable to those cases where the sources of light are such as to produce an approximate uniform illumination over the working plane or where an average value is required.

From the size of lamp or lamps employed and from their efficiency total lumens output are determined. Multiplying the total lumens output from the source by coefficient of utilization, the lumens received on the working plane are determined. If the lamps and surroundings are not perfectly clean, then in determination of lumens received on working plane, the depreciation factor or maintenance factor should be included, *i.e.*,

Lumens received on working plane

$$= \text{Number of lamps} \times \text{wattage of each lamp} \times \text{efficiency of each lamp in terms of lumens/watt} \times \frac{\text{coefficient of utilization}}{\text{depreciation factor}}$$

Also Lumens received on working plane

= Number of lamps × wattage of each lamp × efficiency of each lamp in terms of lumens per watt × coefficient of utilization × maintenance factor.

Coefficient of Utilization or Utilization Factor. The whole light radiated by the lamps does not reach the working plane. The ratio of lumens reaching the working plane to the total lumens given out by the lamp or lamps, is known as *utilization factor* or *coefficient of utilization*. Higher the value of utilization factor, more lumens will reach the working plane for the given lumens output of the lamps. The value of utilization factor depends upon (*i*) the mounting height of lamps—utilization factor decreases with the increase in mounting height of lamps (*ii*) area to be illuminated—for given height, proportion of direct light becomes more and more if floor area increases, *i.e.*, utilization factor increases with the increase in area to be illuminated (*iii*) type of lighting—more for direct lighting and low for indirect lighting and (*iv*) colours of surroundings etc.—more for light colours and less for dark colours. Its value varies from 0.25 to 0.5 and from 0.1 to 0.25 for direct and indirect lighting schemes respectively.

Maintenance Factor. The illumination produced by a lighting installation is considerably less after a year or two of use than it was initially. The loss is due partly to the aging of the lamps and partly to the accumulation of dust on the lamps, on the reflecting and transmitting surfaces of the fixtures and on the ceiling and walls. This fact is taken into account by including the maintenance factor, which is defined as the ratio of the ultimate maintained metre-candles on the working plane to the initial metre-candles. Its value is more if the lamp fittings are cleaned regularly, say 0.8, and less if there is much dust etc. say 0.6.

Depreciation Factor. This is merely the inverse of the maintenance factor and is defined as the ratio of the initial metre-candles to the ultimate maintained metre-candles on the working plane. Its value is more than unity.

3. ***Point to Point or Inverse-Square Law Method.*** This method is applicable where the illumination at a point due to one or more sources of light is required, the candle powers of the sources in the particular direction under consideration being known.

If a polar curve of lamp and its reflector giving candle powers of the lamp in different directions is known, the illumination at any point within the range of the lamp can be calculated from the inverse square law. If two and more than two lamps are illuminating the same working plane, the illumination due to each can be calculated and added. This method is not much used because of its complicated and cumbersome applications. It is employed only in some special problems, such as flood lighting, yard lighting etc.

Example 7.24. A small assembly shop 16 m long, 10 m wide and 3 m up to trusses is to be illuminated to a level of 200 lux. The utilization and maintenance factors are 0.74 and 0.8 respectively. Calculate the number of lamps required to illuminate the whole area if the lumen output of the lamp selected is 3,000 lumens.

[M.D. Univ. Utilization of Electric Power & Traction Dec. 2009]

Solution : Working area, $A = 16 \times 10 = 160 \text{ m}^2$

Required illumination level, $E = 200$ lux

Total lumens required $= E \times A = 200 \times 160 = 32{,}000$ lumens

Total lumens to be given out by the lamp

$$= \frac{\text{Total lumens required on working area}}{\text{Utilization factor} \times \text{Maintenance factor}} = \frac{32{,}000}{0.74 \times 0.8} = 54{,}000 \text{ lumens}$$

$$\text{Number of lamps required} = \frac{\text{Total lumens given out by lamps}}{\text{Lumen output of each lamp}} = \frac{54{,}000}{3{,}000} = 18 \textbf{ Ans.}$$

Example 7.25. A minimum illumination of 80 lumens/m² is required in a factory shed of 100 m × 10 m. Calculate the number, location and the wattage of the units to be used. Assume that the depreciation factor is 0.8, coefficient of utilization is 0.4 and efficiency of the lamp is 40 lumens/watt.

[Pb. Technical Univ. Electric Drives & Utilization May 2008]

Solution : Area to be illuminated, $A = 100 \times 10 = 1{,}000 \text{ m}^2$

Illumination required, $E = 80 \text{ lumens/m}^2$

Total lumens required $= A \times E = 1{,}000 \times 80 = 80{,}000$

Utilization factor = 0.4

Depreciation factor = 0.8

$$\text{Gross lumens required, } \phi = \frac{A \times E}{UF \times DF} = \frac{80{,}000}{0.4 \times 0.8} = 250{,}000$$

$$\text{Total wattage required} = \frac{\phi}{\text{Lamp efficiency}} = \frac{250{,}000}{40} = 6{,}250$$

42 lamps of 150 W rating in 2 rows, each row having 21 lamps, can be used giving spacing of 4.76 m in length and 5 metres in width.

Disposition of lamps is shown in Fig. 7.45.

Example 7.26. An illumination on the working plane of 75 lux is required in a room 72 m × 15 m in size. The lamps are required to be hung 4 m above the work bench. Assuming a suitable space-height ratio, a utilization factor of 0.5, a lamp efficiency of 14 lumens per watt and a candle power depreciation of 20%, estimate the number, rating and disposition of lamps.

[Pb. Univ. Elec. Drives and Utilization of Elec. Energy August 1984; Pb. Technical Univ. Electric Drives & Utilization May 2009]

Solution : Area to be illuminated, $A = 72 \times 15 = 1{,}080 \text{ m}^2$

Illumination required, $E = 75$ lux

Total lumens required $= A \times E = 1{,}080 \times 75 = 81{,}000$

Utilization factor = 0.5

Maintenance factor = 1 – candle power depreciation = 1 – 0.2 = 0.8

$$\text{Gross lumens required, } \phi = \frac{A \times E}{UF \times MF} = \frac{81{,}000}{0.5 \times 0.8} = 202{,}500$$

$$\text{Total wattage required} = \frac{\phi}{\text{Lamp efficiency}} = \frac{202{,}500}{14} = 14{,}464$$

80 lamps in 4 rows, each row having 20 lamps, can be used giving spacings of 3.6 metres in length and 3.75 metres in width and space-height ratios of 0.9 and 0.9375 respectively.

$$\text{Wattage of each lamp} = \frac{14{,}464}{80} = 180.8 \simeq 200 \text{ W (say)} \quad \textbf{Ans.}$$

Disposition of lamps is shown in Fig. 7.46.

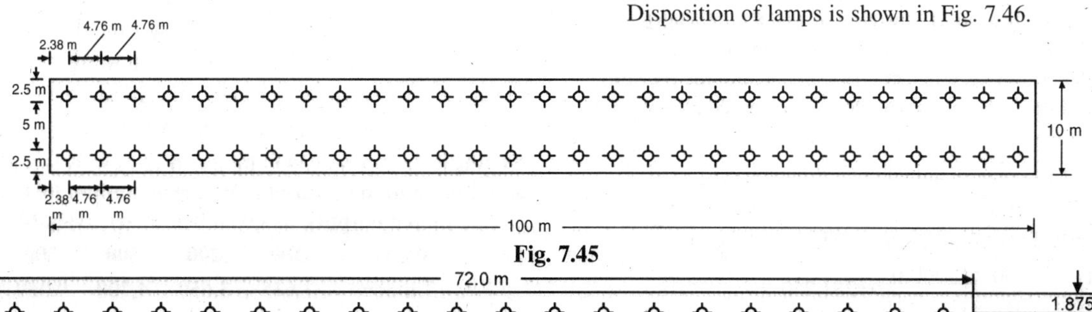

Fig. 7.45

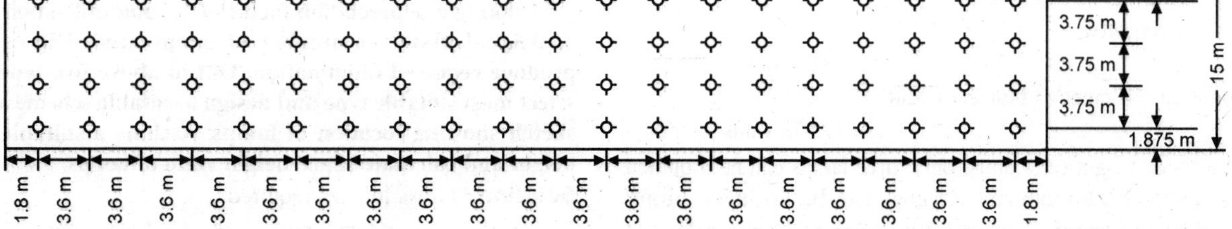

Fig. 7.46

Example 7.27. A class room of size 30 m × 30 m is to be illuminated with 75 lux. The lamps are required to be hung 5 m above the workbench. Assume a space-height ratio of 0.9-1.0, a utilization factor of 0.5, lamp efficiency of 15 lumens/watt, candle power depreciation of 20%. Estimate number, rating and disposition of lamps.

[*U.P. Technical Utilization of Electrical Energy and Electric Traction 2006-07*]

Solution : Gross lumens required,

$$\phi = \frac{A \times E}{UF \times DF} = \frac{30 \times 30 \times 75}{0.5 \times (1 - 0.2)} = 168{,}750$$

Total wattage required,

$$W = \frac{\phi}{\text{Lamp efficiency}} = \frac{168{,}750}{15} = 11{,}250$$

Total number of lamps required, assuming wattage rating of each lamp to be 300 W

$$N = \frac{11{,}250}{300} = 37.5 \text{ (say 42)}$$

6 lamps can be mounted along one side of class room (giving space-height ratio unity) and 7 lamps can be mounted along the other side of the room (giving space-height ratio of $\frac{30/7}{5}$ i.e. 0.857).

Disposition of lamps is shown in Fig. 7.47.

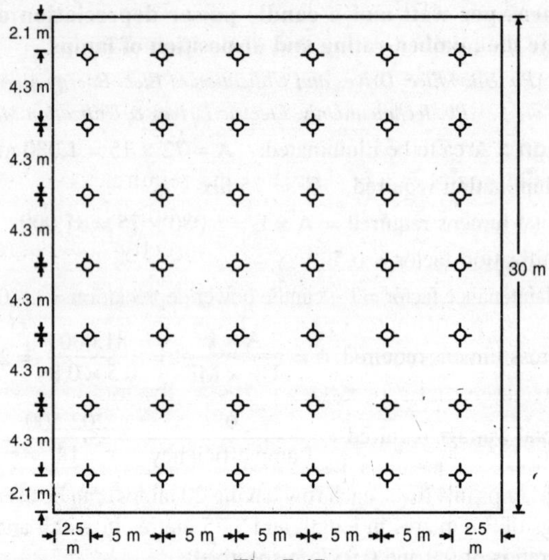

Fig. 7.47

Example 7.28. Estimate the number and wattage of lamps which would be required to illuminate a workshop space 60 × 15 metres by means of lamps mounted 5 metres above the working plane. The average illumination required is about 100 lux, coefficient of utilization = 0.4, luminous efficiency 16 lumens per watt. Assume a space-height ratio of unity and a candle power depreciation of 20%.

[*Gorakhpur Univ. Utilization of Elec. Power and Traction 1983*]

Solution : Gross lumens required,

$$\phi = \frac{\text{Area to be illuminated} \times \text{average illumination required}}{\text{Coefficient of utilization} \times \text{maintenance factor}}$$

$$= \frac{A \times E}{UF \times MF} = \frac{60 \times 15 \times 100}{0.4 \times (1 - 0.2)} = 281{,}250$$

Total wattage required,

$$W = \frac{\phi}{\text{Luminous efficiency in lumens / watt}} = \frac{281{,}250}{16}$$

$$= 17{,}578 \text{ watts}$$

For a space-height ratio unity, only three lamps can be mounted along the width of the room and 12 lamps can be arranged along the length of the room. Thus total number of lamps required = 12 × 3 = 36 **Ans.**

$$\text{Wattage of each lamp} = \frac{\text{Total wattage required}}{\text{Number of lamps}} = \frac{17{,}578}{36} = 488.3$$

We will take the nearest standard lamp of 500 watts. Location of the lamps is shown in Fig. 7.48.

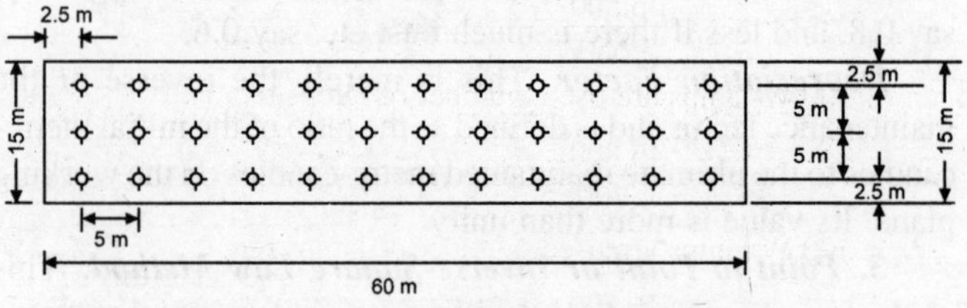

Fig. 7.48

Example 7.29. A hall 30 metres by 15 metres with a ceiling height of 5 metres is to be provided with a general illumination of 120 lumens/m². Taking a coefficient of utilization of 0.5 and depreciation factor of 1.4, determine the number of fluorescent tubes required, their spacing, mounting height and total wattage. Take luminous efficiency of fluorescent tube as 40 lumens per watt for 80 watt tubes.

Solution : Gross lumens required, $\phi = \frac{A \times E \times DF}{UF}$

$$= \frac{30 \times 15 \times 120 \times 1.4}{0.5}$$

$$= 151{,}200$$

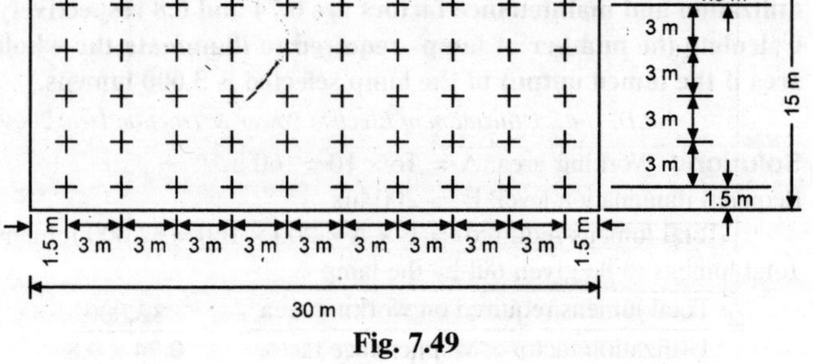

Fig. 7.49

$$\text{Total wattages required} = \frac{151{,}200}{40} = 3{,}780 \text{ watts} \quad \textbf{Ans.}$$

$$\text{Number of tubes required} = \frac{\text{Total wattage required}}{\text{Wattage of each tube}}$$

$$= \frac{3{,}780}{80} = 47.25 \cong 48 \text{ (say)}$$

50 tubes in 5 rows, each row having 10 tubes, can be used giving spacings of 3 metres in length as well as in width. These tubes may be mounted at an height of 3 metres giving space-height ratio of unity. Arrangement is shown in Fig. 7.49. **Ans.**

Example 7.30. A hall 30 m long and 12 m wide is to be illuminated and illumination required is 50 metre candles. Five types of lamps having lumen outputs, as given below, are available :

Watts	**100**	**200**	**300**	**500**	**1,000**
Lumens	**1,615**	**3,650**	**4,700**	**9,950**	**21,500**

Taking a depreciation factor of 1.3 and utilization coefficient of 0.5, calculate the number of lamps needed in each case to produce required illumination. Out of above five types of lamps select most suitable type and design a suitable scheme and make a sketch showing location of lamps. Assume a suitable mounting height and calculate space-height ratio of lamps.

Solution : Gross lumens required,

$$\phi = \frac{A \times E \times DF}{UF} = \frac{30 \times 12 \times 50 \times 1.3}{0.5} = 46{,}800$$

If 100 W lamps are used, number of lamps required

$$= \frac{46,800}{1,615} \cong 29 \quad \textbf{Ans.}$$

If 200 W lamps are used, number of lamps required

$$= \frac{46,800}{3,650} \cong 13 \quad \textbf{Ans.}$$

If 300 W lamps are used, number of lamps required

$$= \frac{46,800}{4,700} \cong 10 \quad \textbf{Ans.}$$

If 500 W lamps are used, number of lamps required

$$= \frac{46,800}{9,950} \cong 5 \quad \textbf{Ans.}$$

If 1,000 W lamps are used, number of lamps required

$$= \frac{46,800}{21,500} \cong 2 \quad \textbf{Ans.}$$

Let the mounting height be of 5 metres.

Most suitable type of lamps will be 300 W lamps requiring 10 lamps in two rows, each row having 5 lamps giving spacing of 6 metres in length as well as in width and space-height ratio of 6/5, *i.e.*, 1.2.

Lamps of lower wattage will require more number of fittings and, therefore, will increase the cost where as lamps of higher wattage will be few in requirement giving space-height ratio much more than desirable. Arrangement is shown in Fig. 7.50.

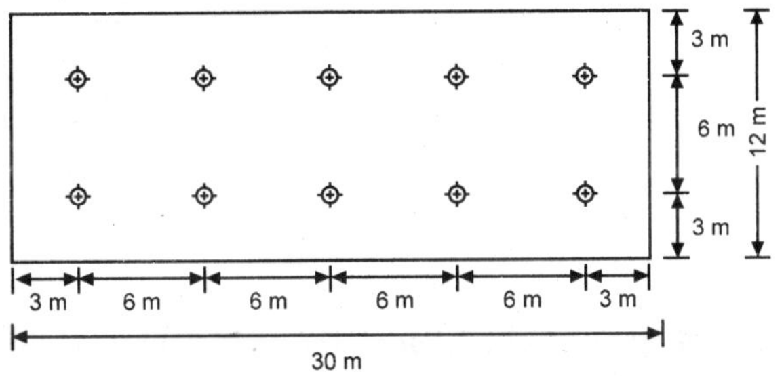

Fig. 7.50

7.18. STREET LIGHTING

The main objects of street lighting are :

(*i*) To make the traffic and obstructions on the road clearly visible in order to promote safety and convenience.

(*ii*) To make the street more attractive.

(*iii*) To increase the community value of the street.

The principle employed for street lighting is different from that of interior lighting. There are no walls and ceiling which reflect or diffuse light, hence only direct lighting scheme can be employed and hard shadows and high contrast cannot be avoided. Because of areas to be illuminated being large, the value of illumination for economic reasons, is very low compared to that for indoor lighting and the question of colour rendering is also of minor importance. High level of illumination is not necessary, because firstly one has seldom to look continuously at nearby objects, secondly it is the impression of whole scene that is required. In fact in case of interior lighting the objects are seen by the light reflected by them but in case of street lighting the objects are seen in relatively bright back ground *i.e.*, silhouetted against a relatively bright back ground. Further more, owing to the low illumination, the eye is in its most sensitive state and, therefore, glare must be avoided.

Two general principles are usually employed in the design of street lighting installations, namely (*i*) the diffusion principle and (*ii*) the specular reflection principle.

1. *Diffusion Principle*. In this case the lamps fitted with suitable reflectors are used. The reflectors are so designed that they may direct the light downwards and spread as uniformly as possible over the road surface. In order to avoid glare the reflectors are made to have a cut-off of between 30° to 45° so that the filament is not visible except from underneath it. The diffusing nature of the road surface causes the reflection of a certain proportion of the incident light in the direction of the observer, and therefore, the road surface appears bright to the observer. The illumination at any point on the road surface is calculated by applying point to point or inverse-square law method. Over certain proportions of the road the surface is illuminated from two lamps and the resultant illumination is the sum of the illuminations due to each lamp.

2. *Specular Reflection Principle*. In this case the reflectors are curved upwards so that the light is thrown on the road at a very large angle of incidence. It is observed that a motorist requires to see objects about 30 metres away. Thus in Fig. 7.51 the observer is shown about 30 metres from the object. Much of the light from the lamp L_3 is not reflected towards the observer, where as most of the light from lamps L_1 and L_2 is reflected towards him. Thus the object will appear silhouetted against the bright road surface due to lamps at a longer distance. The requirement of a pedestrian, who requires to see objects in his immediate neighbourhood, is also fulfilled in this method as some light from the lamps falls directly downwards. This method of street lighting is only suitable for straight sections of the road. This method is more economical also as compared to the diffusion method of lighting but it suffers from the disadvantage that it produces glare for the motorists.

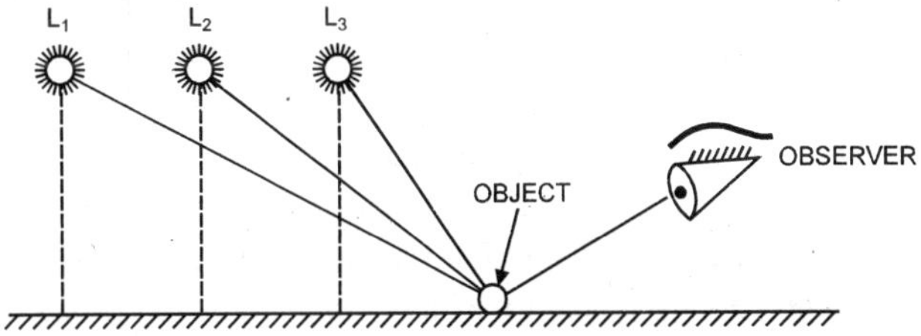

Fig. 7.51

***Illumination Level For Street Lighting and Mounting Height of Lamps*.** The illumination required depends upon the class of street lighting installations. In class A installations, *i.e.*, in important shopping centres and road junctions, illumination level of 30 lumens/m^2 is required where as a in poorly lighted suburban streets, illumination level of 4 lumen/m^2 is sufficient. An average well-lighted street, will require illumination level between 8 to 15 lumens per square metre. Excellent illumination is considered when the distance apart is not more than 8 times the height of the luminaires ; that is, with the luminaires 8 metres above street level, they should be spaced not more than 64 metres (normal spacing for standard lamps is 50 metres with a mounting height of 8 metres). Lamp posts should always be fixed at the junction of roads and as far as possible lamps near large trees should be avoided.

Types of Lamps For Street Lighting. Mercury vapour and sodium discharge lamps have been found to have certain particular advantages for street lighting purposes ; the most important of these is the lower power consumption for a given amount of light, which, inspite of the higher cost of the lamps, makes the overall cost of an installation with discharge lamps less than that employing filament lamps. The colour and monochromatic nature of the light produced by discharge lamps do not matter much in street lighting installations.

7.19. FLOODLIGHTING

Floodlighting means flooding of large surfaces with light from powerful projectors. It is employed to serve one or more of the following purposes.

1. *Aesthetic Floodlighting.* For enhancing beauty of building at night such as public places, ancient buildings and monuments, religious buildings on important festive occasions etc.

2. *Industrial and Commercial Floodlighting.* For illuminating railway yards, sports stadiums, car parks, construction sites, quarries etc.

3. *Advertising.* For illuminating advertisement boards and showcases.

For floodlighting it is necessary to concentrate the light from the light source into a relatively narrow beam. The particular type of reflector and its housing used for concentrating the light into narrow beam is known as *floodlight projector.* Since it is usually installed in remote positions, therefore, it must be robust and weatherproof in construction. The reflecting surface is the most important part in a projector. This may be made of silvered glass or chromium plate or stainless steel ; the efficiency of silvered glass is about 90% while that of polished metal is only about 70%. Metal reflectors being more robust are usually preferred.

The casing and its mounting are arranged so that the inclination of the beam can be varied in both a vertical and a horizontal direction on site. For permanent installations use of cast-metal cases is made to achieve robustness and protection against weather and for temporary installations or those in sheltered situations, use of sheet metal casing is made. When 500 or 1,000 watt lamps are used in projectors, ventilation may have to be provided for adequate cooling. This may be achieved by providing sufficient radiating surface. The front of the projector is usually of clear glass, often bowed outwards to protect it from the heat of the lamp ; use of diffusing glass is made when a diffuse beam is required. As far as possible the projectors should not be visible to the passers-by. In some cases the projectors may be housed in ornamental stand standards.

Projectors are classified according to the beam spread.

(*i*) *Narrow beam projectors* with beam spread between 12-25°. These are used for distance beyond 70 metres.

(*ii*) *Medium angle projectors*—Projectors with beam spread between 25-40°. These are used for distance between 30 to 70 metres.

(*iii*) *Wide angle projectors*—projectors with beam spread between 40-90°. These are used for distance below 30 metres.

For economic reasons use of wide angle projector with high wattage lamp is encouraged over narrow beam projector with low wattage lamp. High wattage lamp is more efficient than low wattge lamp used in narrow beam projectors. Standard gas filled tungsten filament lamps of 250, 500 or 1,000 watts are used in medium and wide angle projectors, although for narrow angle beams which require accurate location of the light source at the focal point of the reflector, special lamps having bunched filaments and known as *projector lamps* are generally employed.

Location and Mounting of Projectors. One of the most important factor which affects the choice of projector is the location of the projector. There are two possible locations of projectors in practice. Fig. 7.52 (*a*) shows symmetric projector kept 20 to 35 metres away from the surface to be floodlighted and providing approximately parallel beam having beam spread of 25° to 30°. Fig. 7.52 (*b*) shows the case when the projector cannot be located away from the building. In such a case, an unsymmetric reflector mounted in a basement area or on a bracket attached to the building is used which directs more intense light towards the top of the building.

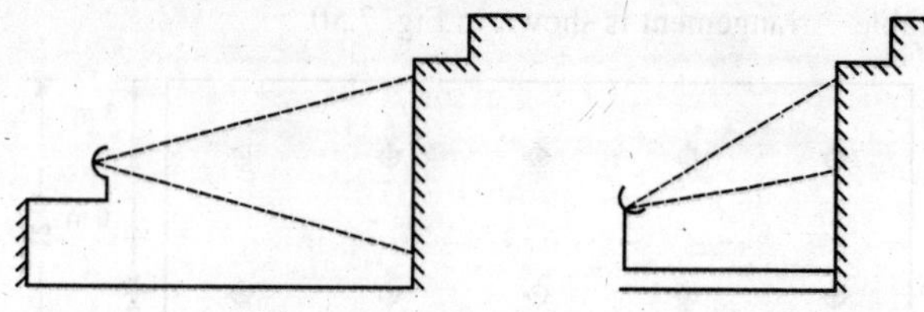

(*a*) (*b*)

Fig. 7.52

Floodlighting Calculations. The problem of flood-lighting calculations may be roughly separated into three steps.

First Step : Illumination Level Required. The illumination level (lumens/m^2) required depends upon the type of building, the purpose of the floodlighting, the amount of conflicting light in the vicinity etc. Typical figures are given below :

Building Surfaces	Reflection Factor	Lux Recommended For Brightness of Locality		
		Low	Medium	High
White terra cotta Cream terra cotta Light marble	60—80	50	100	150
Light gray lime stone Bed Ford lime stone Buff lime stone Smooth buff face brick	40—60	70	120	200
Briar hill sand stone Smooth gray brick Medium gray lime stone Common tan brick	20—40	100	150	250
Dark field gray brick Common red brick Brown stone	10—20	120	180	300

Second Step : Type of Projector. Two considerations enter into the choice of a projector, viz., beam size and light output. The former determines the area covered by the beam and the latter the illumination provided. Beam angle of the projector is decided keeping in view the distance of projector from the surface.

Third Step : Number of Projectors. For any desired intensity over a definite surface the number of projectors required is obtained from the following relation

$$N = \frac{A \times E \times \text{depreciation factor} \times \text{waste light factor}}{\text{Utilization factor} \times \text{wattage of lamp} \times \text{lumious efficiency of lamp}}$$

where N = Number of projectors
A = Area of surface to be illuminated in square metres
E = Illumination level required in lumens/m^2

The terms used in flood lighting calculations are described below :

1. ***Waste Light Factor.*** Whenever a surface is illuminated by a number of light sources, there is always a certain amount of waste light on account of overlapping and falling of light beyond the edges of the area to be illuminated. This effect is taken into account by multiplying the theoretical value of lumens required by 1.2 for rectangular areas and 1.5 for irregular areas and objects such as statues and monuments etc.

2. ***Depreciation Factor.*** Dirt and dust depositing on the reflectors surface and on the source of light reduces the effectiveness of the projector and 50 to 100 per cent more light must be provided so that the illumination shall be adequate at the end of the interval between the cleaning periods. Depreciation factor is defined as the ratio of illumination under ideal conditions to the illumination under normal conditions.

3. ***Coefficient of Utilization.*** It is also known as beam factor and is defined as the ratio of beam lumens to the lamp lumens. Its value lies between 0.3 and 0.5. This factor is due to the fact that all the light emitted by the projector is not along the direction of the beam but some of it is absorbed by the reflector and by the front glass. When the above losses are considerable the value of utilization is low. The coefficient of utilization factor goes up with deep reflectors having a good reflecting surface, such as silvered glass.

Example 7.31. The front of a building 50 m × 16 m is illuminated by sixteen 1000-W lamps arranged so that uniform illumination on the surface is obtained. Assuming a luminous efficiency of 17.4 lumens/watt and a coefficient of utilization of 0.4, determine the illumination on the surface. [*Madras Univ., Utilization of Elec. Energy 1970*]

Solution : Area to be floodlighted A = 50 × 16 = 800 m^2

Total lumens given out by 16 lamps

ϕ = Number of lamps × wattage of each lamp × luminous efficiency of each lamp

= 16 × 1,000 × 17.4 = 278,400 lumens

Total lumens reaching the surface to be floodlighted

$$= \frac{\phi \times \text{utilization factor}}{\text{Depreciation factor} \times \text{waste light factor}}$$

$$= \frac{278{,}400 \times 0.4}{1.3 \times 1.2} = 71{,}400 \text{ lumens}$$

Assuming depreciation factor = 1.3 and waste light factor = 1.2

$$\text{Illumination on the surface, } E = \frac{71{,}400}{800} = 89.25 \text{ lumens/m}^2$$ **Ans.**

Example 7.32. The following table gives the watts and lumens

Watts	300	500	1,000	1,500
Lumens	5,000	9,000	18,000	27,000

The front of the building measuring 60 m and 15 m is to be floodlighted by means of projectors placed at a distance of 8 m from the wall. The average illumination required is 50 lux. Determine the number and size of projectors required. Assume waste light factor of 1.2, depreciation factor 0.8 and coefficient of utilization 0.5.

[*G.B. Technical Univ. Utilization of Electrical Energy and Traction 2010-11*]

Solution : Area to be illuminated, A = 60 × 15 = 900 m^2

Total lumens required on the surface = A × E = 900 × 50

= 45,000 lumens

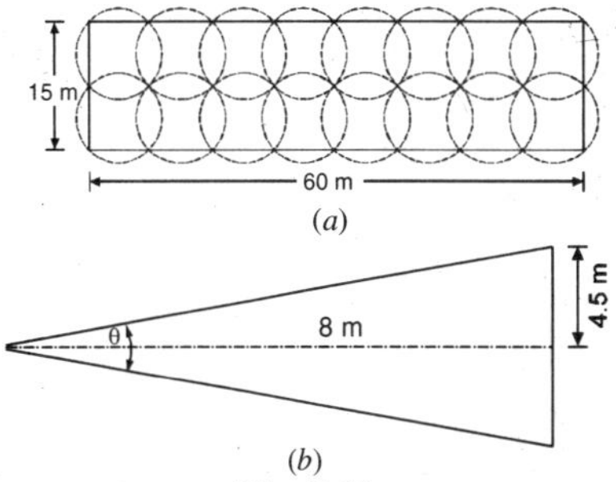

Fig. 7.53

Total lumens required to be given out by lamps

$$\phi = \frac{\text{Total lumens required on the surface} \times \text{waste light factor}}{\text{Utilization factor} \times \text{depreciation factor}}$$

$$= \frac{45{,}000 \times 1.2}{0.5 \times 0.8} = 135{,}000 \text{ lumens}$$

Assuming that 500 watt lamps having a lumens output of 9,000 lumens are used.

Number of projectors required

$$= \frac{\phi}{\text{Lumens output of each 500 W lamp}} = \frac{135{,}000}{9{,}000} = 15; \text{ say (16)}$$

$$\because \text{Tan } \frac{\theta}{2} = \frac{r}{l}$$

where r is the radius of illuminated circle at distance l from light source whose beam angle is θ

So, angle of spread, $\theta = 2 \tan^{-1} \frac{4.5}{8} = 60°$ **Ans.** [Fig. 7.53 (*b*)]

Hence 16 projectors of 500 watts each with beam angle of 60° will be required.

Example 7.33. A 50 m × 15 m of a concrete building in an industrial town is to be illuminated by floodlighting projectors 25 m away. If the required illumination is 100 lux, coefficient of utilization 0.5, depreciation factor 1.5, waste light factor 1.2, estimate the number and size of projectors.

[*Rajasthan Technical Univ. Utilization of Electric Power 2011*]

Solution : Area to be illuminated, A = 50 × 15 = 750 m^2

Total lumens required on the surface = A × E = 750 × 100

= 75,000 lumens

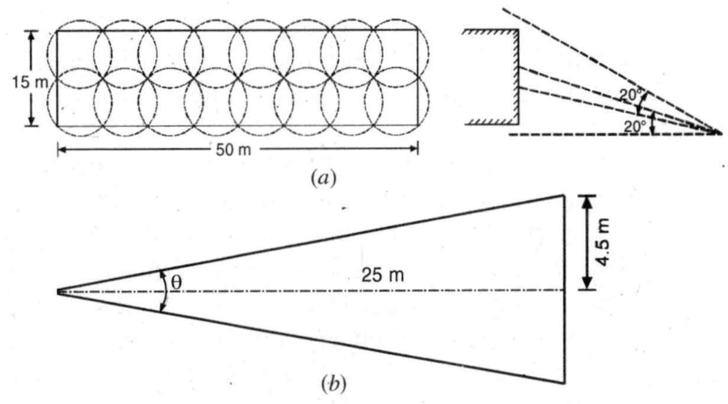

Fig. 7.54

Total lumens required to be given out by lamps

$$\phi = \frac{\text{Total lumens required on the surface} \times \text{depreciation factor} \times \text{waste light factor}}{\text{Utilization factor}}$$

$$= \frac{75{,}000 \times 1.5 \times 1.2}{0.5} = 270{,}000$$

Assuming that 1,000 watt lamps having a luminous efficiency of 17 lumens/watt are used.

Number of projectors required

$$= \frac{\phi}{\text{Wattage of lamp} \times \text{luminous efficiency of lamp}} = \frac{270{,}000}{1{,}000 \times 17}$$

$$= 16 \textbf{ Ans.}$$

Angle of spread $\theta = 2 \tan^{-1} \frac{4.5}{25} = 20°$ **Ans.** [Fig. 7.54 (*b*)]

Hence 16 projectors of 1,000 watts each with beam angle of 20° will be required.

Exercises

1. Explain the basic nature of light.
2. Define and explain the following terms : (*i*) Mean horizontal candle power (*ii*) Mean spherical candle power (*iii*) Mean hemi-spherical candle power (*iv*) Flux (*v*) Solid angle (*vi*) Specific consumption (*vii*) Glare (*viii*) Absorption factor (*ix*) Beam factor and (*x*) Waste light factor.
3. Explain laws of illumination. **[U.P. Technical Univ. Utilization of Electrical Energy & Traction 2009-10; M.D. Univ. Utilization of Electric Power and Traction May 2010]**
4. Derive relation between E, I and L for a uniform diffuse source. **[M.D. Univ. Utilization of Electric Power & Traction Dec. 2008]**
5. Derive expressions for the illumination on a surface (*i*) when it is normal and (*ii*) when it is inclined to the axis of a beam of incident light. **[Gorakhpur Univ. Utilization of Elec. Power & Traction 1981]**
6. Define : (*i*) plane angle and solid angle, (*ii*) Depreciation factor, (*iii*) Cosine inverse law, (*iv*) Polar curve. **[Rajasthan Technical Univ. Utilization of Electrical Power 2011]**
7. What are polar curves ? Mention the uses of polar curves. **[W.B. Univ. of Technology Utilization of Electric Power 2009]**
8. What are polar curves ? Explain Rousseau diagram and its importance in illumination engineering ? **[Rajasthan Technical Univ. Utilization of Electrical Power 2009]**
9. Explain Rousseau diagram and its importance in illumination engineering. **[M.D. Univ. Utilization of Electric Power & Traction Dec. 2008]**
10. Why is tungsten selected as the filament material and on what factors does its life depend ? **[Agra Univ. Traction & Utilization of Elec. Power 1983]**
11. Prove that in a filament lamp the diameter of filament is directly proportional to $I^{2/3}$, where I is the current flowing in the filament. **[Gorakhpur Univ. Utilization of Elec. Power and Traction 1983]**
12. Explain the advantages of using inert gas in filament lamps and the purpose to get the filament as coiled coil.
13. Describe the construction and working of a filament lamp. Compare it with fluorescent lamp.
14. Discuss the effects of voltage variation on the life and illumination as regards tungsten filament lamp and fluorescent lamp.
15. Explain the working of halogen lamp. **[U.P. Technical Univ. Utilization of Electrical Energy & Electric Traction 2006-07]**
16. Describe with a neat sketch the principle of electric discharge lamp. State the advantages and disadvantages of discharge lamps over the filament lamp and give their applications. **[B.H.U Elec. Traction & Utilization 1978]**
17. Explain with a neat diagram the principle of operation of a sodium vapour lamp. Mention its use. **[M.D. Univ. Utilization of Electric Power & Traction Dec. 2009]**
18. Describe the construction and principle of operation and application of mercury vapour lamp. **[M.D. Univ. Utilization of Electric Power & Traction Dec. 2008]**
19. Explain the operation of the following with circuit diagrams: (*a*) LP sodium vapour lamp, (*b*) HP mercury vapour lamp. **[W.B. Univ. of Technology Utilization of Electric Power 2009]**
20. Write construction and working of Neon lamp. **[Pb. Technical Univ. Electric Drives & Utilization Dec. 2006]**
21. Explain the working of a fluoescent tube with the help of the circuit giving the function of various parts. How stroboscopic effect is eliminated in fluorescent tube lighting. **[Pb. Technical Univ. Electric Drives and Utilization May 2009]**
22. Describe the working of fluorescent tube. What are the functions of starter and choke. Explain street lighting. Mention its principles and explain in detail. **[Pb. Technical Univ. Electric Drives and Utilization May 2008]**
23. With the help of circuit diagrams explain the working of the following light sources : (*a*) High pressure mercury vapour lamp (*b*) Fluorescent tube (*c*) Carbon-arc lamp. What are the usual values of power factors for the above lamps ? Indicate the systems of illumination where above lamps are used. **[Bombay Univ. 1970]**
24. Describe the construction and principle of operation and application of fluorescent lamp. **[M.D. Univ. Utilization of Electric Power May 2009]**
25. Discuss the construction and principle of operation and application of (*a*) filament lamp (*b*) sodium vapour lamp, (*c*) mercury vapour lamp (*d*) fluorescent lamp. **[Rajas. Technical Univ. Utilization of Electric Power 2010, 2011]**
26. Compare metal filament lamp with discharge lamp. **[M.D. Univ. Utilization of Electric Power & Traction May 2010]**
27. Compare the merits of alternative lighting schemes using incandescent lamps, sodium vapour lamps or fluorescent lamps for industrial lighting. **[Gorakhpur Univ. Utilization of Elec. Power & Traction 1982]**
28. Why CFL lamps are becoming more popular, nowadays ? Discuss technically. **[U.P. Technical Univ. Utilization of Electrical Energy and Electric Traction 2006-07]**
29. What do you understand by direct, indirect, and semi-indirect lighting ? **[Agra Univ. Traction and Utilization of Elec. Power, 1982]**
30. State and describe various types of lighting schemes. **[Pb. Univ. Elec. Drives and Utilization of Elec. Energy April, 1981]**
31. Describe in brief the requirements of good lighting. Enumerate the factors to be considered while designing an indoor lighting scheme. **[Gorakhpur Univ. Utilization of Elec. Power & Traction Sept., 1984; Pb. Technical Univ. Electric Drives & Utilization Dec. 2008]**
32. How does designing of indoor and outdoor system differ ? Explain method for designing an indoor system. **[U.P. Technical Univ. Utilization of Electrical Energy and Electric Traction 2005-06]**

33. What are the requirements of good street lighting ?
[Pb. Univ. Elec. Drives and Utilization of Elec. Power Nov. 1980; Pb. Technical Univ. Electric Drives & Utilization May 2009]

34. What are the general principles that are usually employed in the design of street lighting explain.
[Pb. Univ. Elec. Drives and Utilization of Elec. Energy April, 1983; Pb. Technical Univ. Electric Drives & Utilization Dec. 2008]

35. Explain the specular reflection principle of street lighting.
[Allahabad Univ. Utilization of Elec. Energy and Traction, 1975]

36. Enumerate the various factors to be considered while designing street lighting.
[Gorakhpur Univ. Utilization of Elec. Power and Traction 1983; Pb. technical Univ. Electric Drives & Utilization May 2009]

37. What are the aims of street lighting and how are they achieved ? List types of lamps used for street lighting.
[U.P. Technical Univ. Utilisation of Electrical Energy and Traction 2006-07]

38. Explain the various factors to be taken into account for designing schemes for (*i*) street lighting (*ii*) floodlighting.
[M.D. Univ. Utilization of Electric Power & Traction May 2009]

39. Explain the factors to be taken into account for designing schemes for (*i*) street lights, (*ii*) floodlighting and (*iii*) highway lighting.
[W.B. Univ. of Technology Utilization of Electric Power 2009]

40. Write short notes on the following:
(*i*) Laws of illumination.
[P.T.U. December 2009; W.B. Univ. of Technology 2009]
(*ii*) Polar curves. **[Pb. Univ. Nov. 1980]**
(*iii*) Rousseau's diagram. **[Agra Univ. 1976]**
(*iv*) Sodium vapour lamp.
[M.D. Univ. Dec. 2008; May 2009; G.B.T.U. 2010-11]
(*v*) Mercury vapour lamp. **[M.D. Univ. Dec. 2009]**
(*vi*) Halogen lamp.
(*vii*) Fluorescent lamps. **[G.B.T.U. 2010-11]**
(*viii*) CFL
(*ix*) Street lighting. **[Pb. Univ. April, 1978]**
(*x*) Essential requirement of good lighting.
(*xi*) Floodlighting. **[P.T.U. Dec. 2009]**

Short Answer Type Questions With Answers

Q. 1. What are the advantages of artificial lighting produced electrically ?

Ans. Artificial lighting produced electrically have the advantages of cleanliness, ease of control, reliability, steady output as well as low cost.

Q. 2. What is Angstrom unit (AU) ?

Ans. Angstrom unit (AU) is unit of wavelength of light and equals to 10^{-10} m.

Q. 3. What is difference between illumination and light ?

Ans. Light is the cause and illumination is the result of that light on surface on which it falls.

Q. 4. Explain candle power and mean horizontal candle power. **[U.P.T.U. 2005-06]**

Ans. **Candle power.** Candle power is the light radiating capacity of a source in a given direction and is defined as the number of lumens given out by the source in a unit solid angle in a given direction. It is denoted by symbol CP.

$$i.e., \text{ CP} = \frac{\text{Lumens}}{\omega}$$

Mean horizontal candle power. It is defined as the mean of candle powers in all directions in the horizontal plane containing the source of light.

Q. 5. Define and explain "illumination" and "luminous intensity". **[U.P.T.U: 2005-06]**

Ans. **Illumination.** When the light falls upon any surface, the phenomenon is called the *illumination.* It is defined as the number of lumens, falling on the surface, per unit area. It is denoted by symbol E and is measured in lumens per square metre or lux or metre-candle.

If a flux of F lumens falls on a surface of area A, then the illumination of that surface is

$$E = F/A \text{ lumens/m}^2 \text{ or lux } (lx)$$

Luminous Intensity. Luminous intensity in any given direction is the luminous flux emitted by the source per unit solid angle, measured in the direction in which the intensity is required. It is denoted by symbol I and is measured in candela (*cd*) or lumens per steradian.

If F is the luminous flux radiated out by source within a solid angle of ω steradians in any particular direction then $I = F/\omega$ lumens/steradian or candela (*cd*).

Q. 6. Differentiate between candela and lux. **[U.P.T.U. 2006-07]**

Ans. Candela or lumens per steradian is the unit of luminous intensity while lux is the unit of illumination.

Q. 7. Define luminous intensity. **[P.T.U. Dec. 2006]**

Ans. Please refer to Answer of Q. 5.

Q. 8. Define illumination. Mention properties of good illumination. **[P.T.U. May 2008]**

Ans. When the light falls upon any surface, the phenomenon is called the *illumination.* It is defined as the number of lumens, falling on the surface, per unit area. It is denoted by symbol E and is measured in lumens per square metre or lux or metre-candle.

If a flux of F lumens falls on a surface of area A, then the illumination of that surface is

$$E = F/A \text{ lumens/m}^2 \text{ or lux } (lx)$$

Good illumination means adequate illumination level, uniform distribution of light all over the working plane, suitable light colour and absence of glare and hard shadows.

Q. 9. What do you understand by foot candle and luminance ? **[P.T.U. Dec. 2008]**

Ans. **Foot-Candle.** It is also the unit of illumination and is defined as the luminous flux falling per square foot on the surface which is every where perpendicular to the rays of light from a source of one candle power and one foot away from it.

$$i.e., \text{ 1 Foot-candle} = 1 \text{ lumen/ft}^2 = \frac{1 \text{ lumen}}{\left(\dfrac{1 \text{ m}}{3.28}\right)^2}$$

$$= 10.76 \text{ metre candle or lux.}$$

Luminance. When the eye receives a great deal of light from an object we say it is bright, and 'brightness' is an important quantity in illumination. It is all the same whether the light is produced by the object or merely reflected from it. The object sends out light as though each small piece of its surface were of a certain luminous intensity. Generally, the brightness of an object is not the same from all points of view. When brightness is considered in terms of measurable quantity, the term *luminance* is employed.

Brightness or *luminance* is defined as the luminous intensity per unit projected area of either a surface source of light or a reflecting surface and is denoted by L.

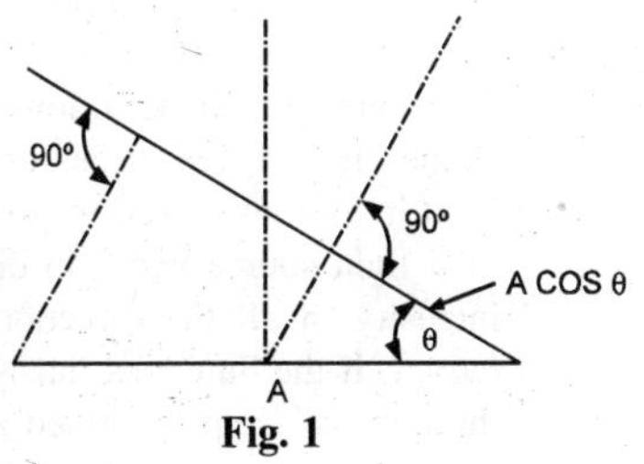

Fig. 1

If a surface of area A has an effective luminous intensity of I candelas in a direction θ to the normal, then the luminance (brightness) of that surface is

$$L = \frac{I}{A \cos \theta} \text{ candela/m}^2 \text{ or nits}$$

i.e., nit is defined as candela per square metre. Bigger unit of luminance is *stilb* which is defined as candelas per square cm. Lambert is also the unit of brightness which is lumens/cm^2. Foot-lambert is lumens/ft^2.

Q. 10. What is luminance ? **[P.T.U. Dec. 2009]**

Ans. Please refer to Answer of Q. 9.

Q. 11. Differentiate between illumination and luminous intensity. **[P.T.U. December 2009]**

Ans. Please refer to Answer of Q. 5.

Q. 12. What is relation between candela (cd) and lumens per steradian ?

Ans. Candela (cd) and lumens per steradian, both, are units of luminous intensity. 1 Candela = 1 lumen per steradian.

Q. 13. What is relation between lux and lumens/m^2 ?

Ans. Lux and lumens/m^2, both, are units of illumination. 1 lux = 1 lumen/m^2.

Q. 14. Differentiate between lamp efficiency and specific consumption.

Ans. Lamp efficiency is the ratio of the luminous flux in lumens to the power input in watts while the specific consumption is the ratio of the power input in watts to the average candle power in candela.

Q. 15. What is difference between maintenance factor and depreciation factor ?

Ans. Maintenance factor is defined as the ratio of illumination under normal working conditions to the illumination when every thing is perfectly clean. Depreciation factor is merely inverse of the maintenance factor.

Q. 16. What are the two laws of illumination ?

Ans. (*i*) Law of inverse squares (*ii*) Lambert's cosine law.

Q. 17. What are laws of illumination ? **[P.T.U. May 2007]**

Ans. There are two laws of illumination (*i*) Law of inverse squares and (*ii*) Lambert's cosine law.

According to law of inverse square the illumination of a surface is inversely proportional to the square of the distance between the surface and the light source provided that the distance between the surface and the source is sufficiently large so that the source can be regarded as a point source.

According to Lambert's cosine law, the illumination at any point on a surface is proportional to the cosine of the angle between the normal at that point and the direction of luminous flux.

Q. 18. Why none of the practical type of lamp provides uniformly distributed light in all directions ?

Ans. Because of unsymmetrical shape of the practical types of lamps.

Q. 19. What do you understand by polar curves ? **[P.T.U. May 2009; Pb. Univ. April 1980]**

Ans. None of the practical type of lamp provides light uniformly distributed in all directions because of its un-symmetrical shape. It is often necessary to know the distribution of light in various directions to ascertain how the candle power of a light source varies in different directions. The luminous intensity in all the directions can be represented by *polar curves*. If the luminous intensity in a horizontal plane passing through the lamp is plotted against angular position, a curve known as *horizontal polar curve is obtained.* If the luminous intensity in a vertical plane is plotted against the angular position, a curve known as *vertical polar curve* is obtained.

Q. 20. How is polar curve useful to an illumination engineer ?

Ans. The polar curve are used to determine the mhcp and mscp of a lamp. They are also used to determine the actual illumination of a surface by using the cp in that particular direction as read from the vertical polar curve in illumination calculations.

Q. 21. What for an integrating sphere is used ?

Ans. Integrating sphere is a piece of apparatus which is commonly used for measurement of mscp.

Q. 22. What are various methods of producing light ? **[P.T.U. Dec. 2007]**

Ans. Artificial light can be produced by any one of the following methods :

1. By passing electric current through a wire raising its temperature to incandescence.
2. By establishing an arc between two electrodes.
3. By maintaining an electric discharge through a vapour or gas.

Q. 23. List various light sources ? **[P.T.U. May 2008]**

Ans. Refer to Art. 7.8.

Q. 24. Why the glass globe of a filament lamp is completely evacuated ?

Ans. The glass globe of a filament of incandescent lamp is completely evacuated to prevent the oxidization and convection currents of the filament and also to prevent the temperature being lowered by radiation.

Q. 25. Why is tungsten selected as the filament material ? **[Agra Univ. 1983]**

Ans. Nowadays tungsten is the most commonly used metal for filament due to its high melting point (3,400°C), high resistivity, low temperature coefficient (0.0051), low vapour pressure, being ductile and mechanical strong to withstand vibration during use.

Q. 26. Why does the efficiency of a filament lamp increase with the increase in operating voltage ?

Ans. The efficiency (lumens/watt) of a lamp increases with the increase in operation voltage owing to increase in temperature and is proportional to the square of the operating voltage.

Q. 27. Why sodium vapour discharge lamps are not used for general lighting ? **[Gorakhpur Univ. 1981]**

Ans. Sodium vapour discharge lamps are not used for general lighting because they have got drawback of colour discrimination.

Q. 28. Why thermal type starter for use in fluorescent tubes have become obsolete nowadays ?

Ans. Because of more complicated construction, greater cost and greater power loss.

Q. 29. What is the role of condenser and choke used in fluorescent lamp ? **[P.T.U. May 2009]**

Ans. When the supply is switched on to a fluorescent lamp, the starter provides a path through the electrodes and choke for providing preheating. The starter switch then opens automatically thereby interrupting the current. Since this heating current reduces to zero instantaneously in the choke circuit also, the choke field collapses releasing stored energy and so developing a high voltage between the electrodes and thus causes the preionized tube to strike.

Thus choke performs two functions : (*i*) it provides inductive voltage surge to start the discharge of electrons and (*ii*) it limits the current in the lamp circuit during normal working. Power factor of fluorescent lamp is about 0.5 (lagging) due to presence of choke, so condenser is connected across the supply to improve the power factor. Overall pf of the fluorescent tube circuit is about 0.95 (lagging).

Q. 30. What is meant by stroboscopic effect ? How this effect is eliminated in fluorescent tube lightings ?
[Gorakhpur Univ. Sept. 1983]

Ans. At the usual alternating current supply frequency of 50 Hz, a discharge lamp will be extinguished 100 times per second. Although this effect is seldom noticeable in normal conditions. It is possible that this may, in some instances, give rise to a stroboscopic effect, that is moving objects such as rotating parts of machinery, illuminated by this light, may appear to be moving in disunity or jerks, or rotating more slowly than their actual speed.

In actual applications, however, where this effect might cause annoyance, it can be practically eliminated in a three lamp unit by connecting each lamp on a separate phase of a 3-phase system and it can be greatly reduced in a two lamp unit by the use of a two lamp control unit, which employs a condenser in the ballast of one of the lamps. The current through the lamps is almost 90° out of phase and under these conditions light output of one of the lamps is at a maximum. This method has an additional advantage of giving an overall power factor of nearly unity for the unity of two lamps. In this arrangement one circuit remains at a low power factor at about 0.5 lagging, while the other circuit incorporating a series capacitor C remains at a power factor of about 0.5 leading.

Q. 31. What are the drawbacks of direct lighting system ?
[P.T.U. May 2007]

Ans. The drawbacks of direct lighting scheme are that hard shadows and glare are caused.

Q. 32. What system of lighting should be recommended for drawing office ? **[P.T.U. Dec. 2007]**

Ans. Direct lighting scheme may be adopted on the working table in drawing office. Light should be concentrated by reflector and shadow and glare should be avoided.

Q. 33. What is floodlighting ?

Ans. Floodlighting means flooding of large surfaces with light from powerful projector(s).

Q. 34. How projectors are classified according to the beam spread ?

Ans. Projectors, according to beam spread, are classified as narrow beam angle projectors (beam spread 12-25°); medium beam angle projectors (beam spread 25-40°) and wide beam angle projectors (beam spread 40-90°)

Problems

1. The efficiency of a certain filament lamp is stated to be 15 lumens per watt. What is the mean spherical power of the lamp if its intake is 100 watts? **[Ans.** 119.4]
2. Find the candle power of a lamp to produce an illumination of 8 lux on a surface 5 m from the lamp when the light rays are (*i*) normal to the surface; (*ii*) inclined to normal at 60° (*iii*) parallel to the surface.
[Pb. Univ. Elec. Drives & Utilization of Elec. Energy Nov. 1977]
[Ans. (*i*) 200 (*ii*) 400 (*iii*) ∞]
3. A filament lamp of 20 W is suspended at a height of 5 metres above working plane and gives uniform illumination over an area of 4 m diameter. Assume efficiency of reflector as 50%. Determine the illumination on the working plane. Efficiency of lamp is 0.89 W per candle power.
[Pb. Technical Univ. Electric Drives & Utilization May 2007; Dec. 2007; Dec. 2008] [Ans. 11.236 lumens/m^2]
4. A corridor is lighted by four lamps spaced 10 metres apart and suspended at a height of 5 metres above the centre line of the floor. If each lamp gives 200 cp in all directions below the horizontal, find the illumination at the point on the floor midway between the second and third lamp.
[Bombay Univ. October 1971] [Ans. 6.16 lumens/m^2]
5. Two lamps A and B of 240 candela and 400 candela respectively are situated 100 metres apart. The height of A above the ground level is 10 metres and that of B is 20 metres. If a photometer is placed at the centre of the line joining the two lamp posts, calculate its reading.
[Gujarat Univ. 1970] [Ans. 0.066 lumens/m^2]
6. It is desired to illuminate a corridor. 100 CP lamps, three number, 8 metres apart are suspended 5 metres above floor level. Calculate the intensity of illumination exactly vertically below the lamp position on the surface of the floor. **[Ans.** 4.7 lux; 5.19 lux; 4.7 lux]
7. Two lamps, each emitting a luminous flux of 3,000 lumens below the horizontal plane, are suspended 6 metres above the centre line of a horizontal roadway and are 30 metres apart. If the road is 12 metres wide between kerbs, calculate the illumination at the middle of the kerb between the lamps. Assume that the luminous intensity of each lamp is uniform in all directions below the horizontal plane. **[Ans.** 7 lux]
8. A 100 cp lamp is hung 2 metres above the centre of a circular area of 3 metre diameter. Determine the illumination at (*i*) the centre of the area, (*ii*) circumference of the area, and (*iii*) average illumination. Also calculate the average illumination, if a reflector of 50% efficiency is used.
[Ans. (*i*) 25 lux (*ii*) 12.8 lux (*iii*) 17.8 lux (*iv*) 88.0 lux]
9. A lamp having a uniform candle power of 200 in all directions is provided with a reflector which directs 60% of the total light uniformly on to a circular area of 10 metre diameter. The lamp is hung 7 metres above the area. Calculate the illumination at the edge of the surface with and without the reflector. Determine also the average illumination over the area without the reflector.
[Gorakhpur Univ. Utilization of Elec. Power With Traction 1980]
[Ans. 19.2 lux; 2.2 lux; 2.98 lux]
10. A light source having an intensity of 700 CP in all directions is fitted with a reflector so that it directs 70% of its light along a beam having a divergence of 20°. What would be the average illumination produced on a surface normal to the beam direction at a distance of 25 metres ? Calculate also the CP of the lamp and the reflector as viewed from the illuminated surface.
[U.P.S.C. Engineering Service Examination 1974]
[Ans. 100.8 lux; 64,480 CP]
11. In a street lighting scheme lamps with candle power of 500 are hung at a height of 5 m. The distance between the posts is 10 m. Determine the illumination under lamps and at the mid point between the posts.
[M.D. Univ. Utilization of Electric Power & Traction Dec. 2008]
[Ans. 23.58 lux; 14.14 lux]
12. Two lamp posts are 14 m apart and are fitted with 200 CP lamp each at a height of 5 m above the ground. Calculate :
(*i*) illumination mid way between them.
(*ii*) illumination under each lamp.
[M.D. Univ. Utilization of Electric Power & Traction May 2010]
[Ans. 3.142 lux; 8.304 lux]

13. In a street lighting scheme, lamps having uniform candle power of 520 candela are hung at a height of 6 m. The distance between consecutive lamp posts is 8 m. Find the illumination under lamp and a distance midway between the lamp posts.
[Rajasthan Technical Univ. Utilization of Electrical Power 2010]
[**Ans.** 20.684 lux; 16.64 lux]

14. The relation between the voltage (V) and candle power (C) of a metal filament lamp is given by $C = aV^b$. Calculate the percentage change in candle power for a change of voltage of ± 5% from the rated voltage if $b = 4.5$.
[**Ans.** 24.6%; –20.6%]

15. Find the illumination due to 20 bulbs of 200 W in a room of 15 m × 8 m assuming following data.
Coefficient of utilization 0.6, Depreciation factor 1.2, MSCP of 200 W lamp is 250. [**Ans.** 262 lux]

16. It is required to provide an illumination of 100 lux in a factory hall 30 m × 12 m. Assume that the depreciation factor is 0.8, the coefficient of utilization 0.4, and the efficiency of proposed lamps 11 lumens per watt. Calculate the number of lamps and their disposition.
[Pb. Univ. Elec. Drives & Utilization of Elec. Energy April 1982]
[**Ans.** 40 lamps of 200 W each in 4 rows, each row having 10 lamps]

17. A minimum illumination of 80 lumens/m^2 is required in the factory shed of 50 m × 12 m. Calculate the number, the location and wattage of the units to be used. Assume that the depreciation factor is 0.8, coefficient of utilization is 0.4 and the efficiency of lamp units is 14 lumens/watt.
[M.S. Univ. of Baroda, Power System I, 1971]
[**Ans.** 36 lamps of 300 watts in 3 rows of 12 lamps in each row]

18. A uniform illumination of 50 lux is to be obtained on the floor of a room measuring 10 m × 10 m by arranging electric lights suitably. The system to be employed should be shadowless and glare free. Calculate the number of lamps and watt rating of each lamp if the efficiency of lamp is 15 lumens/watt. Assume suitable values of coefficient of utilization and depreciation. Indicate the layout.
[Bombay Univ. Elec. Power System I October 1972]
[**Ans.** 4 lamps of 250 W rating]

19. A building measuring 30 m × 20 m is to be floodlit on the front side with brightness of 25 lumens/m^2. Coefficient of reflection of building surface is 0.25. Lamps of 500 W having lumens output of 8,000 each are used. Assume beam factor as 0.6, waste light factor 1.2 and maintenance factor 0.75, determine the number of lamps required.
[Rajasthan Univ. Utilization of Electrical Power, 2006]
[**Ans.** 20 lamps]

20. Estimate the number of 1,000 W floodlight projectors required to illuminate the upper 75 metres of one face of a 96 metre tower of width 13 metres and approximate initial average luminance is to be 6.85 cd/m^2. The projectors are mounted at ground level 51 metres from base of tower. Utilization factor is 0.2, reflection factor of wall is 25% and efficiency of lamp is 18 lumens per watt. [**Ans.** 24]

21. A building frontage 50 m × 15 m is to be illuminated by floodlighting projectors situated 25 metres away. If the illumination is 100 lux, coefficient of utilization 0.5, depreciation factor 1.5, waste light factor 1.2, estimate the number and size of projectors. Sketch the projectors recommended indicating the usual adjustments provided.
[A.M.I.E. Sec B. Utilisation of Elec. Power. May, 1968]
[**Ans.** 16 projectors of 1,000 W each with beam angle of 20°]

22. A factory requires 3,20,000 lumens for proper illumination and the total lighting period in a year is 3,000 hours. Compare the annual cost of energy and the cost of lamp renewals for the following schemes.
(*a*) 100 watt filament lamps having a life of 1,000 hours, a luminous efficiency of 13.4 lumens/watt and renewal cost of Rs. 20 per lamp.
(*b*) Twin 40 watt fluorescent lamps having a life of 3,000 hours, a luminous efficiency of 30 lumens/watt and renewal cost of Rs. 200 per lamp.
The factory gets its power supply at the rate of Rs. 2,500 per kW per year plus Rs 1.20 per unit.
[**Ans.** Scheme A : Annual cost of energy Rs. 1,46,400
and cost of lamp renewals Rs. 14,400.00
Scheme B : Annual cost of energy Rs. 65,400
and cost of lamp renewals Rs. 53,600.00]

❑❑❑

CHAPTER 8

Electrolytic Processes

INSIDE THIS CHAPTER

8.1. Introduction 8.2. Electrolysis—Basic Principle 8.3. Faraday's Laws of Electrodeposition 8.4. Calculation of Current Required for Depositing Given Amount of Metal 8.5. Current Efficiency 8.6. Voltage 8.7. Energy Efficiency 8.8. Extraction of Metals 8.9. Refining of Metals 8.10. Production of Chemicals 8.11. Electrodeposition 8.12. Electroplating 8.13. Deposition of Alloys 8.14. Electroforming 8.15. Electrotyping 8.16. Electrofacing 8.17. Electrometallisation 8.18. Electrodeposition of Rubber 8.19. Anodizing 8.20. Electropolishing 8.21. Power Supply For Electrolytic Processes.

8.1. INTRODUCTION

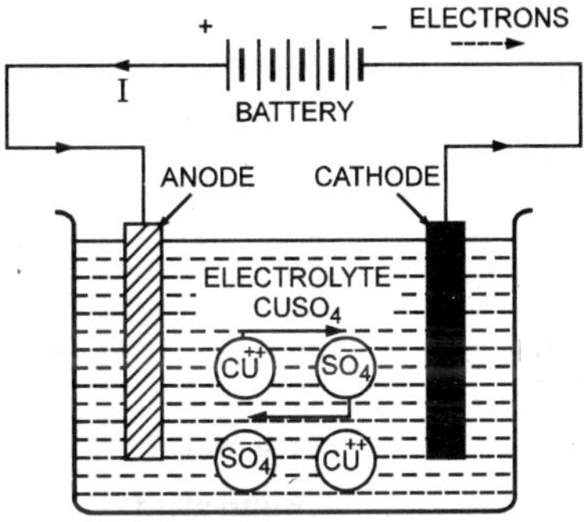

Fig. 8.1

The fact that electrical energy can produce chemical changes and the processes based on it, called the '*electrolytic processes*', are widely used for the extraction of pure metals from their ores (such as aluminium, zinc, copper, magnesium, sodium etc.), manufacturing of various chemicals (such as caustic soda, potassium permanganate, hydrogen, oxygen, chlorine etc.), electrodeposition of metals including electroplating, electrotyping, electroforming, building up of worn-out parts in metallurgical, chemical and other industries. Though the various processes mentioned are different in apparent detail but fundamentally they are alike, being based on the principle of electrolysis.

8.2. ELECTROLYSIS—BASIC PRINCIPLE

In nature atoms of electrolyte (chemical compound) are closely bound together but bond becomes weaker, when dissolved and the molecules of the electrolyte split up into two types of ions carrying electric charges, called the *cations* and *anions*, and moving freely in the solution. Now if two electrodes are dipped into the electrolyte and connected to the dc supply, ions associated with positive charge (cations) and moving freely in the solution are attracted by the cathode (electrode connected to the negative terminal of the supply) and the ions associated with negative charge (anions) and moving freely in the solution are attracted by the *anode* (electrode connected to the positive terminal of the supply).

For example, when copper sulphate ($CuSO_4$) is dissolved in water, immediately it gets dissociated into +vely charged copper ions (Cu^{++}) and negatively charged sulphions (SO_4^{--}) moving freely in the solution and if a potential difference is applied between the two electrodes immersed in the solution the +vely charged copper ions will move towards the cathode and the –vely charged sulphions will move towards the anode. Each of the positively charged copper ions reaching the cathode will take two electrons from it and become a metallic atom of copper, and similarly each of the negatively charged sulphions reaching the anode will give up two electrons to it and cease to be anion. Thus the electrons will move from anode to cathode in the external circuit and constitute flow of current from anode to cathode in the electrolyte. Thus the function of source of supply seems only to serve as an electron pump pumping electrons from +ve side of the supply and supply to the –ve side of the supply.

As mentioned above each +vely charged copper ion on reaching the cathode takes two electrons from it, becomes atom of metallic copper and deposite there. Similarly each –vely charged sulphion on reaching the anode gives up two electrons to it and becomes SO_4 radical but since SO_4 radical cannot exist in the electrical neutral state so it will attack the anode and will form the corresponding sulphate of the material of the anode; for example if the anode is of copper then copper sulphate ($CuSO_4$) will be formed, but if the anode is of such a material which cannot be attacked by SO_4, for example if the anode is of carbon the SO_4 will react with water and form sulphuric acid and liberate oxygen according to chemical reaction

$$2SO_4 + 2H_2O \rightarrow 2H_2SO_4 + O_2$$

The whole process described above is called the *electrolysis* and the effect is that the copper gets dissolved from the anode and deposited on the cathode. During the process there is no accumulation of charge at any point in the circuit and the mass of copper deposited at the cathode is exactly equal to that removed from the anode.

8.3. FARADAY'S LAWS OF ELECTRODEPOSITION

The laws governing the electrolytic processes were formulated by Michael Faraday, an English Scientist, and are known after his name. These may be stated as below :

Faraday's First Law. According to this law the chemical deposition due to flow of current through an electrolyte is directly proportional to the quantity of electricity (coulombs) passed through it.

i.e., mass of chemical deposition,

$$m \propto \text{Quantity of electricity, Q}$$

$$\text{or} \quad m \propto It \qquad \because \quad Q = It$$

$$\text{or} \quad m = ZIt \qquad ...(8.1)$$

where I is the steady current in amperes flowing through the electrolyte for t seconds and Z is a constant of proportionality and is known as the *electrochemical equivalent* of the substance.

If I = 1 A and t = 1 second, then Z = m

Thus electrochemical equivalent, Z, of a substance is defined as the amount of the substance deposited on passing a steady electric current of 1A for one second through its solution. It is usually expressed in terms of milligrams per coulomb. The SI unit of electrochemical equivalent Z is the kilogram per coulomb (kg C^{-1}).

Faraday's Second Law. This law states that when the same quantity of electricity is passed through several electrolytes, the mass of the substances deposited are proportional to their respective chemical equivalents or equivalent weights.*

From this law it follows that the constant of proportionality Z in Eq, (8.1) is proportional to the chemical equivalent.

Table 8.1 gives the values of atomic weight, valency, chemical equivalent and electrochemical equivalent of various elements.

8.4. CALCULATION OF CURRENT REQUIRED FOR DEPOSITING GIVEN AMOUNT OF METAL

The theoretical value of current required for depositing a given quantity of metal and the time for which this current should be passed through the electrolyte, can be calculated from the Faraday's laws, if the electrochemical equivalent of the metal is known. The following example will make the method of calculation clear.

Example 8.1. Calculate the ampere-hours required to deposit a coating of silver 0.05 mm thick on a sphere of 5 cm radius. Assume electrochemical equivalent of silver = 0.001118 and density of silver to be 10.5.

Solution : Surface area of the sphere, $S = 4\pi r^2 = 4\pi \times (5)^2 = 314 \text{ cm}^2$

Thickness of coating, t = 0.05 mm = 0.005 cm

Mass of silver to be deposited, $m = S \times t \times \text{density of metal}$

$$= 314 \times 0.005 \times 10.5$$

$$= 16.493 \text{ gm}$$

$$= 0.016493 \text{ kg}$$

ECE of silver, Z = 0.001118 gm/coulomb

$$= \frac{0.001118 \times 3{,}600}{1{,}000}$$

$$= 0.0040248 \text{ kg/A-h}$$

$$\text{Ampere-hours required} = \frac{m}{Z} = \frac{0.016493}{0.0040248}$$

$$= 4.12 \quad \textbf{Ans.}$$

Table 8.1

Elements	Symbol	Atomic Weight	Valency	Chemical Equivalent	Electrochemical-Equivalent Kg/Coulomb
Aluminium	Al	27.00	3	9.00	9.4×10^{-8}
Cadmium	Cd	112.4	2	56.2	58.2×10^{-8}
Calcium	Ca	40.08	2	20.04	20.75×10^{-8}
Chlorine	Cl	35.45	1	35.37	36.74×10^{-8}
Chromium	Cr	52.01	3,6	17.3, 8.6	17.9×10^{-8}, 9.1×10^{-8}
Copper (Cupric)	Cu	63.54	2	31.8	32.9×10^{-8}
Gold	Au	197	3	65.4	67.68×10^{-8}
Hydrogen	H	1.008	1	1.0	1.045×10^{-8}
Ferrous	Fe	55.9	2	27.95	29.02×10^{-8}
Ferric	Fe	55.5	3	18.63	19.35×10^{-8}
Lead	Pb	207.21	2	103.2	107.16×10^{-8}
Magnesium	Mg	24.32	2	11.97	12.43×10^{-8}
Nickel	Ni	58.7	2	29.3	30.43×10^{-8}
Nitrogen	N	14	3	4.67	4.85×10^{-8}
Oxygen	O	16	2	7.98	8.296×10^{-8}
Potassium	K	39.1	1	39.04	40.54×10^{-8}
Silver	Ag	107.88	1	107.67	111.81×10^{-8}
Sodium	Na	22.99	1	22.99	23.87×10^{-8}
Tin	Sn	118.7	2	59.3	61.4×10^{-8}
Zinc	Zn	65.38	2	32.45	33.88×10^{-8}

* The chemical equivalent or equivalent weight of a substance is defined as the weight of that substance which will combine with or displace unit weight of hydrogen. The chemical equivalent of hydrogen is, thus, unity. Since valency of a substance is equal to the number of hydrogen atoms, which it can replace or with which it can combine, the chemical equivalent of a substance, therefore, may be defined as the ratio of its atomic weight to its valency.

$$\text{Thus chemical equivalent} = \frac{\text{Atomic weight}}{\text{Valency}}$$

Example 8.2. If 18.258 gm of nickel is deposited by 100 A flowing for 10 minutes, how much copper would be deposited by 50 A for 6 minutes? Atomic weight of nickel = 58.6 and that of copper 63.18. Valency of both is 2.

[Rajasthan Technical Univ. Utilization of Electrical Power 2010]

Solution : ECE of nickel, $Z = \dfrac{m}{I \times t} = \dfrac{18.258 \times 10^{-3}}{100 \times 10 \times 60}$

$= 3.043 \times 10^{-7}$ kg / C

ECE of copper, $Z' = Z \times \dfrac{\text{Chemical equivalent of copper}}{\text{Chemical equivalent of nickel}}$

$= 3.043 \times 10^{-7} \times \dfrac{63.18/2}{58.6/2}$

$= 3.2808 \times 10^{-7}$ kg/C

Mass of copper deposited, $m' = Z' \times I' \times t'$

$= 3.2808 \times 10^{-7} \times 50 \times 6 \times 60$

$= 0.0059$ kg or 5.9 gm **Ans.**

Example 8.3. A steady current of 40 A is passed through a solution of copper sulphate for 1 hour. Find the weight of copper deposited, if ECE of copper is 0.01044 mgm/C.

[Pb. Technical Univ. Electric Drives and Utilization Dec. 2006]

Solution : ECE of copper, Z = 0.01044 mg/C = 1.044×10^{-8} kg/C

Current strength, I = 40 A

Time for which current is passed,

$t = 1 \times 60 \times 60 = 3{,}600$ seconds

Mass of copper deposited, $m = Z \times I \times t$

$= 1.044 \times 10^{-8} \times 40 \times 3{,}600$

$= 0.0015$ kg or 1.5 gm **Ans.**

Example 8.4. Find the thickness of copper deposited on a plate area of 0.00025 m² during electrolysis if a current of one ampere is passed for 100 minutes. Density of copper is 8,900 kg/m³ and ECE of copper = 32.95 × 10⁻⁸ kg/coulomb.

[Gorakhpur Univ. Utilization of Elec. Power and Traction, Sept., 1983; Pb. Technical Univ. Electric Drives & Utilization May 2009]

Solution : ECE of copper, Z = 32.95×10^{-8} kg C^{-1}

Current strength, I = 1 A

Copper density, D = 8,900 kg/m³

Time for which current is passed, $t = 100 \times 60 = 6{,}000$ seconds

Mass of copper deposited, $m = Z I t$

$= 32.95 \times 10^{-8} \times 1 \times 6{,}000$

$= 0.001977$ kg

Volume of copper deposited, $v = \dfrac{m}{D} = \dfrac{0.001977}{8{,}900}$

$= 0.222 \times 10^{-6}$ m³

Thickness of copper deposited, $t = \dfrac{v}{A} = \dfrac{0.222 \times 10^{-6}}{0.00025}$

$= 0.888 \times 10^{-3}$ m

$= 0.888$ mm **Ans.**

Example 8.5. Find the thickness of copper deposited on a plate area of 0.00025 m² during electrolysis, if a current of 1.5 A is passed for one hour. Density of copper is 8,900 kg/m³ and ECE of copper = 32.95 × 10⁻⁸ kg/C.

[U.P. Technical Univ. Utilization of Electrical Energy and Traction 2006-07]

Solution : Mass of copper deposited,

$m = Z \times I \times t = 32.95 \times 10^{-8} \times 1.5 \times 60 \times 60 = 0.0017793$ kg

Volume of copper depsotied,

$v = \dfrac{m}{D} = \dfrac{0.0017793}{8{,}900} = 0.199 \times 10^{-6}$ m³

Thickness of copper deposited,

$t = \dfrac{v}{A} = \dfrac{0.199 \times 10^{-6}}{0.00025} = 0.0008$ m or 0.8 mm **Ans.**

Example 8.6. In a copper-sulphate voltameter, the copper cathode is increased in weight by 50 gm in two hours, when the current maintained was constant. Calculate the value of this current. Given that

Atomic weight of copper = 63.5

Atomic weight of hydrogen = 1

Atomic weight of silver = 108

Electrochemical equivalent of silver = 0.001118 gm

Solution : Mass of copper deposited, m = 50 gm

Time of flow of current, t = 2 hours

$= 2 \times 60 \times 60$

$= 7{,}200$ seconds

ECE of silver = 0.001118 gm

$= 111.8 \times 10^{-8}$ kg C^{-1}

Chemical equivalent of silver $= \dfrac{\text{Atomic weight}}{\text{Valency}}$

$= \dfrac{108}{1} = 108$

Chemical equivalent of copper $= \dfrac{63.5}{2} = 31.75$

ECE of copper, Z = ECE of silver $\times \dfrac{\text{Chemical equivalent of copper}}{\text{Chemical equivalent of silver}}$

$= 111.8 \times 10^{-8} \times \dfrac{31.75}{108} = 32.867 \times 10^{-8}$ kg C^{-1}

Strength of current, $I = \dfrac{m}{Z \times t} = \dfrac{50 \times 10^{-3}}{32.867 \times 10^{-8} \times 7{,}200}$

$= 21.13$ A **Ans.**

Example 8.7. A copper refining plant, using 500 electrolytic cells, carries a current of 6,000 A, voltage per cell being 0.25 volt. If the plant were to work 40 hours/week, calculate the energy consumption per tonne, assuming ECE of copper as 0.3281 mg/coulomb of electricity.

Solution : ECE of copper, Z = 0.3281 mg/C

$= 32.81 \times 10^{-8}$ kg C^{-1}

$= 32.81 \times 10^{-8} \times 3{,}600$ kg/A-h

$= 0.00118116$ kg/A-h

Total number of ampere-hours per annum

$= 500 \times 6{,}000 \times 40 \times 52 = 624 \times 10^{7}$ Ah

Annual output of plant $= 0.00118116 \times 624 \times 10^{7}$ kg

$= 7{,}370{,}438$ kg $= 7{,}370.438$ tonnes

Energy consumed per annum = A-h per annum × voltage per cell

$= 624 \times 10^{7} \times 0.25$

$= 156 \times 10^{7}$ watt-hours

$= 156 \times 10^{4}$ kWh

Energy consumption per tonne $= \dfrac{\text{Annual consumption in kWh}}{\text{Annual output in tonnes}}$

$= \dfrac{156 \times 10^{4}}{7{,}370.438}$

$= 211.66$ kWh/tonne **Ans.**

Example 8.8. Determine the maximum voltage required for electrolysis of water if one kg of hydrogen on oxidation to water liberates 14.212×10^7 joules and electrochemical equivalent of hydrogen is 1.0384×10^{-8} kg C^{-1}.

[*G.B. Technical Univ. Utilization of Electrical Energy & Traction 2010-11*]

Solution : During electrolysis of water the energy is required to decompose water into hydrogen and oxygen and this is equal to the energy expanded in the circuit in forcing the quantity of electricity through the electrolyte.

Energy expanded during electrolysis $= \frac{1}{Z} \times V$ watt-second per kg

$= \frac{V}{Z}$ joules per kg

Energy liberated by 1 kg of hydrogen when it combines with oxygen

$= 14.212 \times 10^7$ joules

$\therefore \quad \frac{V}{Z} = 14.212 \times 10^7$

or $\quad V = 14.212 \times 10^7 \times Z = 14.212 \times 10^7 \times 1.0384 \times 10^{-8}$

$= 1.476$ volts **Ans.**

8.5. CURRENT EFFICIENCY

Owing to impurities, which causes secondary reactions, the quantity of substance or substances liberated is slightly less than that calculated from Faraday's laws. This is taken into account by employing a factor, called the *current efficiency*.

The current efficiency is defined as the ratio of the actual quantity of substance liberated or deposited to the theoretical quantity, as calculated from Faraday's laws.

i.e., Current efficiency

$$= \frac{\text{Actual quantity of substance liberated or deposited}}{\text{Theoretical quantity of substance liberated or deposited}}$$

Its value usually lies between 90 and 98%.

In certain cases this efficiency is very low. For example in chromium plating it is roughly 12 to 15 per cent. It is because only 15 per cent of the total current passed through the electrolyte consisting of some chromium acid solution, is used in depositing chromium and the rest is wasted in producing oxygen and hydrogen gases, which for the purpose in hand, are useless.

Example 8.9. How much aluminium will be produced from aluminium oxide in 24 hours if the average current is 3,000 A and current efficiency is 92%. Aluminium is trivalent and atomic weight is 27. The chemical equivalent weight of silver is 107.98 and 0.00111 gm of silver is deposited by one coulomb.

[*Gorakhpur Univ. Utilization of Elec. Power and Traction 1981; M.D. Univ. Utilization of Electric Power and Traction May 2010*]

Solution : ECE of silver = 0.00111 gm/C or 111×10^{-8} kg C^{-1}

Chemical equivalent weight of silver = 107.98

Chemical equivalent weight of aluminium $= \frac{27}{3} = 9$

$\therefore$ ECE of aluminium,

$$Z = \frac{\text{ECE of silver} \times \text{chemical equivalent weight of aluminium}}{\text{Chemical equivalent weight of silver}}$$

$$= 111 \times 10^{-8} \times \frac{9}{107.98} = 9.252 \times 10^{-8} \text{ kg C}^{-1}$$

Current efficiency = 92% = 0.92

Average current, I = 3,000 A

Duration of flow of current, $t = 24 \times 60 \times 60 = 86{,}400$ seconds

Mass of aluminium produced, $m = ZIt \times$ current efficiency

$= 9.252 \times 10^{-8} \times 3{,}000 \times 86{,}400 \times 0.92$

$= 22.062$ kg **Ans.**

Example 8.10. A worn-out shaft of 10 cm diameter and 25 cm long is to be coated with 2 mm thick layer of nickel. Determine the quantity of electricity required and time taken if current density of 160 A/sq metre is assumed. Assume current efficiency of 90%. Density of nickel may be taken as 8.9 gm/CC.

[*U.P. Technical Univ. Utilization of Electrical Energy & Traction 2009-10*]

Solution : Surface area of circular shaft portion to be coated,

$S = \pi \times d \times l = \pi \times 0.1 \times 0.25 = 0.07854 \text{ m}^2$

Mass of nickel to be deposited,

$m = S \times$ thickness of coating $\times$ density of metal

$= 0.07854 \times 2 \times 10^{-3} \times 8.9 \times 10^3 = 1.398$ kg

Electrochemical equivalent of nickel,

$Z = 30.43 \times 10^{-8}$ kg/C ...Refer to Table 8.1

Theoretical value of quantity of electricity required,

$$Q = \frac{m}{Z} = \frac{1.398}{30.43 \times 10^{-8}} = 0{,}045942 \times 10^8 \text{ C or A-s}$$

$$= \frac{0.045942 \times 10^8}{3{,}600}$$

$= 1{,}276$ A-h

Actual value of quantity of electricity required

$$Q' = \frac{Q}{\text{Current efficiency}} = \frac{1{,}276}{0.9} = 1{,}418 \text{ A-h } \textbf{Ans.}$$

Since the current density is 160 A/sq meter and surface area of sheet is 0.07854 m^2, therefore current used

$I = 160 \times 0.07854 = 12.5664$ A

Time required,

$$t = \frac{1{,}418}{12.5664} = 112.84 \text{ hours } \textbf{Ans.}$$

8.6. VOLTAGE

The voltage necessitated to pass the current through any electrolytic cell is equal to the sum of voltage drop in the resistance of the electrolyte and the voltage drops at electrodes (anode and cathode). From the resistivity of the electrolyte and the x-sectional area and the length of the electric current path, the resistance of electrolyte can be determined. For economy the electrolyte resistance should be reduced to minimum and to achieve it in many cases special conducting agents are added to the electrolyte. The addition of sulphuric acid to copper sulphate solution in copper plating is an example of this. There is some potential difference between the cathode and the electrolyte and between the electrolyte and anode. This potential difference is a measure of the tendency of the metal to go into the solution and is known as *electrode potential*. The electrode potential depends upon the exact conditions (*i.e.*, temperature and concentration), and also upon the nature of the metal and the electrolyte. Under ideal conditions the value of electrode potential for most of the substances lie between 0.5 to 1.0 volt. The total potential difference required to pass the necessary current through electrolytic cell is about 1 or 2 volts.

8.7. ENERGY EFFICIENCY

On account of various secondary effects and reactions the substance deposited by a given quantity of electricity is less than that determined theoretically from Faraday's laws. Voltage required is also higher than that determined theoretically. Hence actual energy consumption will be higher than that determined theoretically for depositing a given quantity of the substance.

The ratio of theoretical energy required to the actual energy required for depositing a given quantity of metal is known as *energy efficiency*.

$$i.e., \quad \text{Energy efficiency} = \frac{\text{Theoretical energy required}}{\text{Actual energy required}}$$

8.8. EXTRACTION OF METALS

Extraction of metal is an electrochemical process used for the production of metal with commercially acceptable purity. There are two methods of extraction of metals depending upon the physical state of the ore. In one of the processes the ore is treated with a strong acid to obtain a salt and the solution of such a salt is electrolysed to liberate the metal. The second process is employed when the ore is available in molten state or can be fused and in this process the ore, which is in a molten/fused state, is electrolysed in a furnace.

The details of the processes adopted for some of the metals are given in Table 8.2.

Table 8.2. Extraction of Metals

Metal to be Extracted	Treatment of Ore	Solution	Energy Consumption in kWh/Tonne
Aluminium	—	Fused cryolite	20,000—25,000
Copper	Roasted and treated with sulphuric acid	Copper sulphate	2,000—2,500
Magnesium	—	Fused magnesium chloride or carnalite	17,000—20,000
Sodium	—	Fused sodium hydrate or sodium chloride and sodium nitrate	10,000—20,000
Zinc	Treated with concentrated sulphuric acid	Zinc chloride and zinc sulphate	3,000—5,000

The methods adopted for extracting zinc and aluminium are explained below :

1. *Extraction of Zinc.* This is an example in which an aqueous solution of the salt is used. The ore, consisting largely of zinc oxide, is treated with concentrated sulphuric acid, roasted, and passed through various chemical processes in order to remove impurities (such as cadmium, copper etc.) by precipitation. The zinc sulphate solution so obtained is then electrolysed. The electrolysis of zinc sulphate is accomplished in large lead-lined wooden boxes having a number of aluminium cathodes and lead anodes. Zinc gets deposited on the cathodes and is removed periodically (once or twice a day). The current density on the cathodes is about 1,000 A/m^2. The voltage per cell is about 3.5 V and usually 100 to 150 cells are used in series requiring a pressure of roughly 500 V. Energy consumption is 3,000 to 5,000 kWh/tonne.

2. *Extraction of Aluminium.* This is an example of fused electrolyte process. Aluminium is produced from bauxite containing aluminium oxide or alumina (70% in case of high grade bauxite), silica (silicon oxide) and iron oxide. The bauxite ore is first reduced to aluminium oxide by chemical treatment and then it is dissolved in fused cryolite. Cryolite is a solution of aluminium fluoride and fluoride of either of sodium, potassium or calcium. The mixture thus obtained is electrolysed. The fusion and electrolysis are accomplished in a large shallow rectangular steel bath lined with carbon; carbon anodes projecting downwards into the bath and the bottom of the bath forms the cathode. The charge is melted by the arc struck between the carbon anodes and cathode and is then maintained in a molten state by the heating action of the electric current flowing through the charge. The liquid metal deposits at the cathode and settles at the bath bottom and is periodically siphoned out into large ladles from which it is poured into pig or ingot moulds. Fresh alumina is fed into the bath at short intervals to replace that which has been decomposed by the current and the process is, therefore, a continuous one. The aluminium obtained by this process is 99.5% pure. A furnace having an area of about 15 square metres will need a voltage of about 6 volts and a current of about 40,000 amperes. Energy consumption is 20,000 to 25,000 units/tonne. Almost the whole of the aluminium required in the present day's industry is produced in this way. As the electrolytic process requires large amount of electric power and process is continuous, so such plants are installed near hydroelectric power stations. The high temperature (1,000°C) necessary to keep the ores in a fused state is maintained by the ohmic losses due to the current flowing through the electrodes and electrolyte.

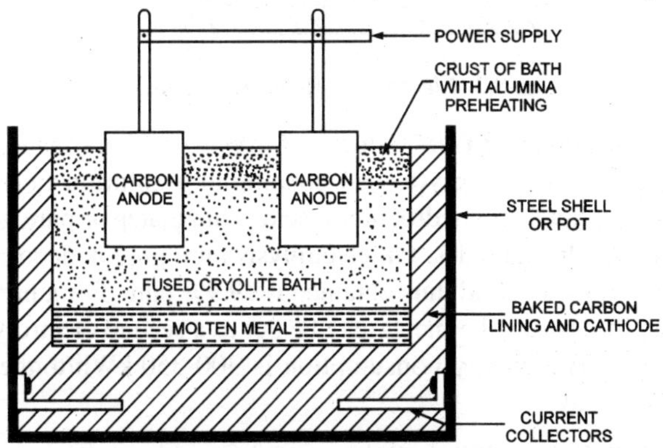

Fig. 8.2. *Extraction of Aluminium*

8.9. REFINING OF METALS

Refining is the process whereby a highly concentrated mixture of metals is subjected to electrochemical treatment for recovering not only the principal metal in pure form, but also the precious metals like gold, silver, bismuth etc., which may be present in the form of minute traces. Some of the processes are explained below.

1. ***Refining of Copper.*** The impure copper is cast into plates which are made the anodes in a bath of copper sulphate acidified with sulphuric acid. The cathodes consist of sheets of pure copper which are coated with graphite so that the electrolyte deposit of copper may be easily stripped off. When the current is passed, copper ions are discharged at the cathode and are deposited as pure copper while hydroxyl ions discharged at the anode bring an equivalent amount of cupric ions into the solution. Thus there is a transfer of pure copper from the anode to the cathode while the impurities like nickel, zinc and iron, pass into the solution as sulphates and impurities like silver and gold, which are not affected by sulphuric acid—copper sulphate solution settles down as the *anode sludge or mud.* At regular intervals the deposit is stripped from the cathode and the anode is replaced. Copper extracted from its ore, known as blister copper, is 98 to 99 per cent pure. Copper of purity as high as 99.95% is obtained from blister copper by electrolytic process. Energy consumption in copper refining by electrolytic process is 150—300 kWh/ tonne of refined copper.

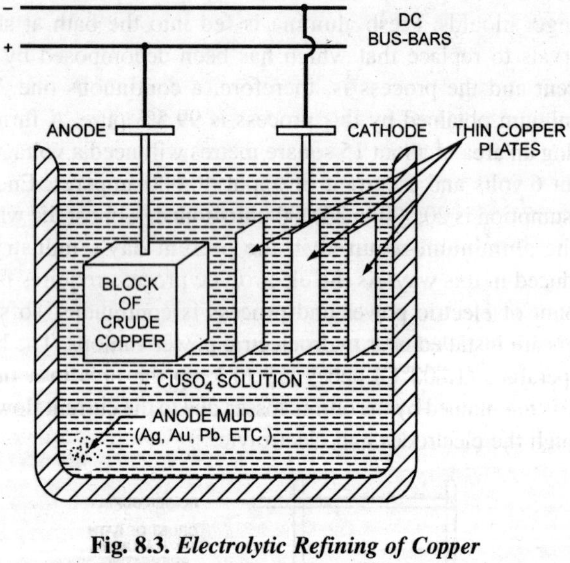

Fig. 8.3. ***Electrolytic Refining of Copper***

2. ***Refining of Gold.*** Blocks of impure gold are made the anodes in an electrolytic cell containing a solution of gold chloride ($AuCl_3$) in hydrochloric acid. Thin plates of pure gold constitute the cathodes. On electrolysis pure gold is deposited on the cathode. Gold thus obtained is 99.9% pure. For a higher purity, if desired, the metal may be put to a second electrolytic refining. The energy consumption is 300-350 kWh/tonne of refined gold.

3. ***Refining of Silver.*** Solution of nitric acid and silver nitrate is required for refining of silver with electric energy consumption of 400-425 kWh/tonne of refined silver. The process is similar to that for refining of copper.

4. ***Refining of Nickel.*** The *matter* is first roasted and treated with dilute sulphuric acid at 80°C to remove copper. The residue consisting mostly of nickel compounds is made the anode in a solution of nickel ammonium sulphate. Iron plates are made the cathode. On electrolysis nickel deposits on these plates. The energy consumption is 2,500-4,000 kWh/tonne of refined nickel.

5. ***Refining of Lead.*** The desilverised lead can be purified further by electrolysing a solution of lead fluorosilicate ($PbSiF_6$) and hydrofluorosilicic acid (H_2SiF_6) containing a little gelatin. The desilverised lead is cast into anodes while the cathodes are sheets of pure lead. Gelatin facilitates the formation of a coherent deposit of lead at the cathode. The metallic impurities which are more electropositive than lead, such as iron and tin, go into solution while the rest of the impurities are thrown down as anode mud. The energy consumption is 100-120 kWh per tonne of refined lead.

6. ***Refining of Zinc.*** The crude ore is roasted and leached with dilute sulphuric acid. The resulting solution is purified from elements present such as copper, cadmium, silver, cobalt, antimony, arsenic by treatment with the powdered zinc and with ferric hydroxide. The solution of purified zinc sulphate is then electrolysed in large concrete tank lined with lead or with rubber, using chemically pure lead as anode and aluminium cathode. The electrolysis is carried at 35-40°C using a current density of 300-400 A/m^2 and a cell voltage of 3.25-3.5 volts. The zinc is stripped from the aluminium cathodes and then melted down to give a material of high purity (99.99 per cent zinc).

7. ***Refining of Iron.*** In case of purification of iron, the iron-ammonium sulphate is used as electrolyte and energy consumption 1,000—1,600 kWh per tonne of refined iron.

The summary of the some of the important refining processes is given in Table 8.3.

Table 8.3. Refining of Metals

Metal to be Refined	Solution	Energy Consumption in kWh/Tonne
Copper	Copper sulphate	150—300
Gold	Gold chloride	300—350
Iron	Iron-ammonium sulphate	1,000—1,600
Lead	Lead fluosilicate	100—120
Nickel	Nickel-ammonium sulphate	2,500—4,000
Silver	Nitric acid and silver nitrate	400—425

8.10. PRODUCTION OF CHEMICALS

There are various industrial applications of electrolysis in the production of chemicals such as production of caustic soda, chlorine, potassium permanganate, ammonium per sulphate, hydrogen and oxygen by electrolysis.

The most important processes of production caustic soda by electrolysis of brine and production of hydrogen and oxygen gases by electrolysis of water are explained below.

1. ***Production of Caustic Soda.*** The oldest process is *Diaphragm process.* There are a number of variations, but all essentially consist of an anode compartment separated from a cathode compartment by a porous diaphragm which prevents the mechanical mixing of the two solutions. Chlorine is formed at the anode, and most of it is evolved as a gas, a small part going into solution. Sodium is discharged at the cathode and reacts with the hydroxyl ions to form sodium hydroxide and

hydrogen gas is liberated at the cathode. Usually, the brine is fed into the anode compartment to resist the flow of hydroxyl ions towards the anode.

Another method of production of caustic soda by electrolytic process is *mercury-cathode* process. Mercury cells may be built in small (1,000 A) to very large sizes (3,000 A) or even as high as 50,000 A or more per unit. Purified saturated brine is fed to the cells and commonly circulated through them with a reduction in salt concentration. Brine is resaturated for another cycle. Each unit consists of two sections ; one which is closed and has a chlorine outlet, graphite anodes, and a mercury cathode; and a second, the denuder section, containing a mercury amalgam anode and iron cathodes. Circulation of mercury between the two sections constitutes the electrical connection. The current efficiencies depend upon the concentration of the amalgam formed in the first section and the effectiveness of the circulation of the amalgam. Current densities in mercury cells are considerably higher than in diaphragm units, being of the order of 1.5 mA or more per square mm of anode surface.

Energy efficiencies are of the order of 50-60%. Anode consumption, when graphite anodes are employed, is of the order of 3-3.5 kg per tone of chlorine. The voltage requirement is about 4 volts per cell.

Other productions, that can be obtained from electrolysis of brine by interaction of the cathodic and the anodic products are sodium hydrochloride, bleaching powder and hydrogen when the solution is cold and sodium chlorate ($NaClO_3$) and hydrogen when the solution is hot.

2. *Production of Hydrogen and Oxygen Gas By Electrolysis of Water.* Gas obtained by this process are of high purity and at a cheap cost because of the low energy consumption. The electrolyte consists of 15-20% solution of caustic soda or its equivalent caustic potash, and electrodes are of iron. Sulphuric acid is no longer used. With caustic soda as electrolyte the chemical reactions are :

Cathode reaction $2Na + 2H_2O = 2NaOH + H_2$

Anode reaction $2OH = H_2O + \frac{1}{2}O_2$

Table 8.4. Production of Chemicals

Solution	Product	Energy Consumption in kWh/Tonne
Brine	Caustic soda	3,000—4,000
Brine	Chlorine, compounds of sodium (chlorates, per-chlorates, hypochlorates)	3,000—7,000
Potassium-manganate	Potassium-permanganate	70—80
Water	Hydrogen and oxygen	About 6 kWh per cubic metre of hydrogen and $\frac{1}{2}$ cubic metre of oxygen
Ammonium-sulphate	Ammonium persulphate	2,000—2,500

Thus hydrogen and oxygen gases are liberated at cathode and anode respectively and water disappears while the quantity of caustic soda remains constant. It is, therefore, necessary to add water to the solution periodically.

The voltage requirement is 2-2.2 V per cell during operation and 2.3-2.5 V per cell during starting period. Energy consumption is about 6 kWh per cubic metre of hydrogen and 1/2 cubic metre of oxygen.

Various industrial applications of electrolysis in the production of chemicals are summarized is Table 8.4.

8.11. ELECTRODEPOSITION

The process of depositing a coating of one metal over an other metal or non-metal electrically is called the *electrodeposition.* It is used for protective, decorative and functional purposes and includes such processes as electroplating, electroforming, electrotyping, electrofacing, electrometallisation, electrodeposition of rubber and building up of worn-out parts for repairs.

The factors, on which the quality of electrodeposition depends, are given below.

1. *Nature of Electrolyte.* The formation of smooth deposit largely depends upon the nature of electrolyte employed. The electrolyte from which complex ions can be obtained, such as cyanides, provides a smooth deposit.

2. *Current Density.* Electrodeposition depends upon the rate at which crystals grow and the rate at which fresh nuclei are formed, therefore, at low current densities the deposits are coarse and crystalline in nature. The deposit of metal will be uniform and fine-grained if the current density is used at rate higher than that at which the nuclei are formed. In case the rate of formation of nuclei is very high due to very high current density, there is a chance that the limiting value of the electrolyte is exceeded. At such instances, the deposit will be spongy and porous.

3. *Temperature.* A low temperature of the solution favours formation of small crystals of metal; and a high temperature, large crystals. In some cases this is very marked, a difference of only 15°C resulting in a 50% decrease in strength of the metal deposited. On the other hand, high temperature may give beneficial results due to (*a*) increased solubility of the salts, permitting greater metal concentration and higher current densities ; (*b*) increased conductivity, which also permits higher current densities and reduces the tendency to form trees ; (*c*) decreased occlusion of hydrogen in the deposited metal, which in many cases is the case of bad deposits. Since both (*a*) and (*b*) tend to decrease crystal size, they may in some cases counteract the tendency of temperature alone to increase the crystal size.

4. *Conductivity.* The use of a solution of good conductivity is important from the standpoint of view of economy in power consumption and also because it reduces the tendency to form trees and rough deposits.

5. *Electrolytic Concentration.* Higher current density, which is necessary to obtain uniform and fine-grain deposit, can be achieved by increasing the concentration of the electrolyte.

6. *Additional Agents.* The addition of acids or other substances to the electrolyte reduces its resistance, as already mentioned. There is another class of additional agents which takes little or no direct part in the chemical reactions but influences the nature of deposit, sometimes even making an otherwise unworkable process into one of practical importance. Such additional agents are glue, gums, dextrose, dextrin, gelatine, agar, alkaloids, albumen, phenol, glycrin, sugar, glucose, rubber etc. The crystal nuclei absorbs the additional agent added in the electrolyte. This prevents it to have large growth and thus deposition will be fine-grained. For obtaining satisfactory deposit of zinc from zinc sulphate solution addition of glucose or certain types of sugar is necessary.

7. *Throwing Power.* This is the ability of electrolyte to produce uniform deposit on an article of irregular shape and is one of the most important characteristics of plating or deposition bath. The distance between the various portions of cathode and anode will be different due to irregular shape of the cathode. Due to unequal distance, the resistance of the current path through the electrolyte for various portions of the cathode will be different but the potential difference between the anode and any point on the article to be plated (cathode) will, of course be the same and the result will be that the current density will be more on the portion nearer to anode and it will cause uneven deposit of the metal. Throwing power can be improved in two ways—firstly by increasing the distance between the anode and cathode and secondly by reducing the voltage drop at the cathode surface. In some cases decrease of current density causes a decrease in voltage drop at cathode, leaving more voltage available for overcoming the resistance of the electrolyte, thus tending to counteract any change in current concentration. This is the reason that solutions of the cyanides of metals usually have a better throwing power than solutions of the sulphates.

8. *Polarization.* The rate of deposition of metal increases with the increase in electroplating current density up to a certain limit after which electrolyte surrounding the base metal becomes so much deplated of metal ions that the increase in current density does not cause increase in rate of deposition. Use of current density beyond this limit causes electrolysis of water and hydrogen liberation on the cathode. This hydrogen evolved on the cathode blankets the base metal which reduces the rate of metal deposition. This phenomenon is called the *polarization.* Blanketing effect can be reduced by agitating the electrolyte. With reverse current electroplating, in which at regular intervals plating current is reversed for a second or so, sufficient electron concentration is established around the base metal and the polarization effect becomes negligible even with very high overall speed of plating.

The other advantages of *reverse current plating* are (*i*) unsound and inferior metal is deplated during reverse current period and flat level surfaces are produced. (*ii*) Metal surface is brightened causing elimination of buffing or polishing operation.

8.12. ELECTROPLATING

Electroplating is an art of depositing a superior or a more noble metal on an inferior or a base metal by means of electrolysis of an aqueous solution of a suitable electrolyte. For example, metals like iron which are easily corroded by atmospheric air, moisture and carbon dioxide, are coated electrolytically with deposits of nickel or chromium which are more resistant to chemical attack. Picture frames and machinery parts are often chromium-plated to protect them from corrosion and at the same time to give them a good polish.

Sometimes, electroplating is done, with a view to repair worn-out parts of a machinery. In such cases the suitable material is deposited electrolytically on the effected parts of the machinery.

Electroplating is also done occasionally for ornamentation and decoration purposes. For example, several articles made of copper or its alloys, such as tablewares, decoration pieces, are coated with silver or gold.

The electrolytic deposits are crystalline in nature. The crystals must be very fine in order to get firm, coherent and uniform deposits. For this purpose, suitable electrolytes should be used in the electrolytic bath and current density used should have an appropriate value. The temperature should also be maintained at a proper level. By experiments, a certain optimum value of current density and temperature have been worked out for each electrolyte. For example, for $Na_3\,[Cu(CN)_4]$ bath, the optimum current density lies between 32-65 A/m^2 and optimum temperature is 45°C. The optimum conditions for $[Ag(CN)_2]$ bath, are 42 A/m^2 of current density and 20°C as the temperature. The articles to be coated with the noble metals should be in as high a state of purity as possible. These conditions are briefly discussed below.

Preparation For Plating. The preparation of an object for plating may involve any or all of the following operations :

1. Removal of oil, grease, or other organic material.
2. Removal of rust, scale, oxides, or other inorganic coatings adhering to the metal.
3. Mechanical preparation of the surface of the metal to receive the deposited metal, by polishing, buffing etc.

For the first, soaps, hot alkali solutions, or organic solvents such as gasoline or carbon tetrachloride are used, for the second, various acids, alkali and salt solutions, mechanical abrasion, and electrolytic cleaning, and for the third mechanical abrasion and polishing are used.

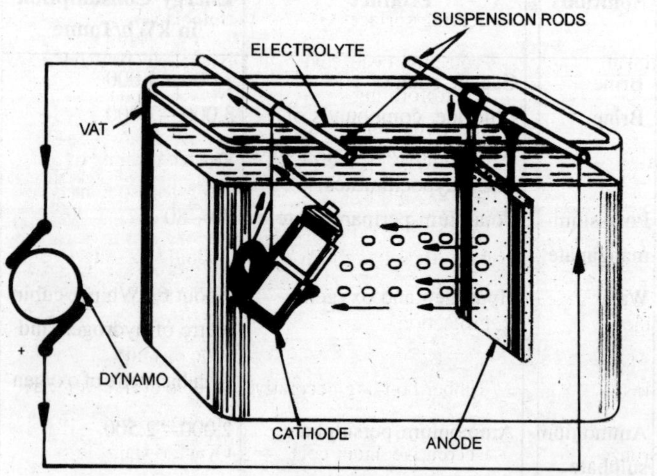

Fig. 8.4

Cleaning Methods. In case the object to be electroplated is not cleaned, polished and degreased, the deposit formed may not be well adherent to the base metal and is likely to peel off. For smooth, bright and strong deposit, the surface upon which a layer of a noble metal is required, should be thoroughly cleaned first mechanically by grinding or scratching (against a rough surface) or sandblasting and then chemically by treatment with hot alkalies (if the surface is greasy) or with dilute acids (if the surface has oxide films) or with organic solvent.

Electrolytic Bath. The electrolyte used in the electrolytic bath depends upon the nature of the metal to be deposited.

For *copper plating* two types of electrolytic baths are used. In acid type bath, solution is made of 150-200 gm of copper sulphate and 25-35 gm of sulphuric acid per 1,000 cc of solution. Current density used is 150-400 A/m^2 and temperature of 25 to 50°C. Deposit obtained is thick and rough requiring polishing. In cyanide bath solution is made of 25 gm of copper cyanide, 28 gm of sodium cyanide, 6 gm of sodium carbonate and 6 gm of sodium bisulphate per 1,000 cc of solution. Current density used is 50-150 A/m^2 and the temperature required is 25-40°C. It provides thin and smooth deposits. Copper anodes are used in both of the baths.

For *silver plating* solution consisting of 24 gm of silver cyanide, 24 gm of potassium carbonate and 36 gm of potassium cyanide per 1,000 cc is used. The required current density and temperature are 50-150 A/m^2 and temperature of 20-35°C respectively.

For *gold plating* solution used consists of 18 gm of potassium gold cyanide, 12 gm of potassium cyanide, 6 gm of potassium sulphate and 12 gm of caustic potash per 1,000 cc Anode employed is of stainless steel. Current density of 50—150 A/m^2 and temperature of 50–70°C are used.

For *chromium plating* solution most commonly used consists of 180-300 gm of chromic acid and 2-3 gm of sulphuric acid per 1,000 cc. Current density employed is 1,500-2,500 A/m^2 and the working temperature is 35-50°C. Current density used is higher for hard chromium plating than for decorative plating. Anodes are of antimonial lead. Vats used for chromium plating are of steel with lead lining. Chromic acid is added in the solution when required. Arrangements for removal of the fumes are also to be provided.

For *nickel plating* the solution consists of 180-240 gm of nickel sulphate, 36 gm of nickel chloride and 24 gm of boric acid per 1,000 cc. The current density used is 100-200 A/m^2. The working temperature is 25-40°C. Anode is of pure nickel.

Summary of the various processes is given in Table 8.5.

Table 8.5. Summary of Electroplating Practice

Metal Deposited	Principal Uses	Solution	Current Density in A/m^2	Voltage	Temperature of Solution	Throwing Power
Cadmium	Protection	Cadmium oxide and sodium cyanide	100—400	1—4	Room temperature	Good
Chromium	Decorative Engineering (hard) cylinder liners (porous)	Chromic acid	1,500—2,500	6—8	35—50°C	Poor
Copper	Electroforming under coat for other metals	Copper sulphate	150—400	1—2	25—50°C	Fair
		Cuprous cyanide and sodium cyanide	50—150	1.5—3	25—40°C	Good
Gold	Decorative	Double cyanide of potassium and gold	50—150	5—15	50—70°C	Good
Iron	Electroforming	Iron chloride	600	—	80–90°C	—
	Repair	Iron sulphate	200	—	Room temperature	—
Lead	Protection	Fluoborate	100—800	0.5	Room temperature	Good
	Bearing surfaces	Sulphate	50—400	3—8	25—50°C	Poor
Nickel	Protection decorative Electroforming under coat for Cr etc.	Nickel chloride and Nickel sulphate	100—200	0.5—3	25—40°C	Fair
Silver	Decorative Protective Bearing surfaces	Double cyanide of silver and sodium	50—150	1	20—35°C	Good
Tin	Protection Food and dairy Bearings Electrical industry	Sodium stannate	400	4—8	65—90°C	Excellent
		or Stannous Stannate	400	1—3	Room	Fair
Zinc	Protection	Zinc sulphate	150—4,000	—	25—40°C	Fair
		Zinc cyanide	100—500	—	40°C	Good
Brass	Rubber bonding decorative	Double cyanide of zinc and potassium	30—100	2—3	25—40°C	Good
Bronze	Decorative under coat for chromium	Cyanide-stannate	200—1,000	3—6	70°C	Excellent

8.13. DEPOSITION OF ALLOYS

It is possible to deposite alloys also provided the electrode potentials of the constituent metals are not much different. Electroplating of brass and bronze are example of it. For deposition of alloys the anode is made of an alloy to be deposited and the electrolyte consists of a mixture of electrolytes which would have been employed for separate deposition of the metals consumed. For brass plating solution of double cyanides of zinc and potassium and of copper and potassium is used as electrolyte and current density of 25-40 amperes/m^2 is used.

8.14. ELECTROFORMING

Reproduction of objects by electrodeposition on some sort of a mould or form is known as *electroforming.* This is another application of electrodeposition.

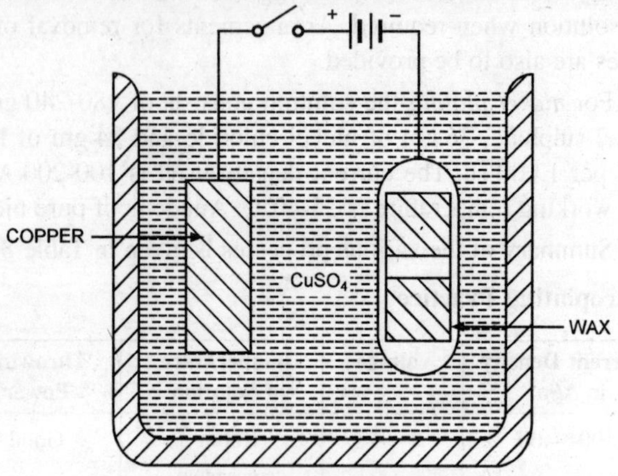

Fig. 8.5. *Electroforming*

In the reproduction of coins, medals, engravings and the like, a mould is first made by impressing the object, say, in wax. The surface of the wax, which bears exact impressions of the object, is coated by powdered graphite in order to make it conducting. The mould is then dipped in an electroforming cell as a cathode. After obtaining coating of desired thickness, the article is removed and the wax core is melted out of the metal shell.

This process is employed in the manufacture of gramophone records, in which the original recording is done on a record made of wax composition. This mould is then coated with powdered gold to make it conducting and dipped into a blue vitriol electrolyte as the cathode. The solution is kept saturated by using a copper anode. The plating of copper on to the wax record produces a negative master plate which is used to stamp a large number of shellac discs.

Production of seamless tubes is another application of electroforming. Ferric chloride solution is used as an electrolyte, cast iron as anodes and revolving mandrel as cathode for producing iron tubes. For producing copper tubes copper sulphate acid solution is used as an electrolyte and cast copper is made as anode. A thin coating of material, which renders easy withdrawl of tube of material deposited, is given to the mandrel.

8.15. ELECTROTYPING

This is a special application of electroforming and it is used to reproduce printers type, engraving and medals etc. The process is same as for electroforming.

8.16. ELECTROFACING

It is a process of coating of metallic surface with a harder metal by electrodeposition in order to increase its durability.

8.17. ELECTROMETALLISATION

It is a process of depositing metal on conducting base for protection and decoration. Non-conductive base is made conductive by a coating of graphite employed as cathode.

8.18. ELECTRODEPOSITION OF RUBBER

Rubber latex, as obtained from the tree, consists of very fine colloidal particles of rubber suspended in water. Like other colloidal solutions, particles of rubber are negatively charged. On electrolysis of the solution the rubber particles migrate towards the anode and deposit on it. Current density of roughly 100 A/m^2 is used.

8.19. ANODIZING

Anodizing is a process of anodic oxidation in which a thin uniform passive (tough thin, uniform and impermeable) film is produced artificially by the passage of electric current. Passive film is formed to protect the base metal from further corrosion. In the process of anodizing electric current of 10 to 30 amperes per square metre is passed through an electrolyte of 3% solution of chromic acid for a period of half an hour to one hour, the article to be anodized is thoroughly cleaned and made anode and carbon rod is made cathode. Aluminium and magnesium have capacity of producing such passive films and, therefore, anodizing is applicable to them. In case of aluminium great feature of these coatings is their submicroscopic porosity. This makes the passive films to take dyes which are absorbed and affect the entire firm length.

8.20. ELECTROPOLISHING

Closely allied to bright dipping is electropolishing, which utilises anodic treatment in specially formulated electrolytes to bring up a polished surface on such metals as stainless steel. Electropolishing is also useful as a tool in preparing metallic surfaces for microscopic examination. Both bright dipping and electropolishing depend on the more rapid eating away by the solution of micro-projections on the metal, so that a smoother surface results.

Electropolishing has one advantage over buffing. Unlike buffing electropolished surfaces are free from cold worked surface cemented layer embedding abrasive particles of mop in the metal.

The process of polishing, in principle, consists of making the work as anode in a suitable electrolytic solution. This develops insoluble compounds that are broken down by more anodic action on the hills than on valleys of surface. Thus smooth surface is produced.

8.21. POWER SUPPLY FOR ELECTROLYTIC PROCESSES

Power supply required for electrolytic processes is direct current and at very low voltage. The power required for electrodeposition is usually very small (between 100 and 200 amperes at 10 or

12 volts) and can be obtained either by employing a motor-generator set consisting of a standard induction motor driving a heavy-current low-voltage dc generator (preferably separately excited) or by employing the copper oxide rectifier. The latter is preferred because of low maintenance cost, occupying less space and higher operating efficiency. Mercury-arc rectifier cannot be used because it has low efficiency at low output dc voltage on account of constant voltage drops at cathode and anodes. The plate rectifier unit is usually placed along with its transformer in the oil so that it may be protected from the corrosive fumes of the electrolyte. Recently the solid state rectifying devices employing germanium and silicon diodes have been developed for use. These solid state devices occupy very small space even as compared to metal rectifiers. Output dc voltage can be controlled by controlling the excitation of the generator in case of motor-generator set and by means of continuously variable auto-transformer in case of rectifier supply. This method of control is suitable where only one bath is being supplied. In case, more than one bath are supplied, a variable resistance is connected in series with each bath so that the supply to each bath can be controlled independently. With the development of SCRs or thyristors, which can control output voltages of power supplies, their use in power supplies for electrochemical processes has increased. They are also compact and light in weight, even cooling attachments are included. Voltage control is by using output transformers.

Power supply required for extraction and refining of metals and large scale manufacture of chemicals is in very large amounts. Since most of the processes are continuous, therefore, have a load factor of 100 per cent. Because of power requirements in huge amount and at 100 per cent load factor, such plants are located near the hydroelectric power stations or atomic power stations even if extra transportation of raw material is necessitated. The advantage of a high load factor is greater with such stations than with steam stations and also transmission costs are eliminated. Nangal fertilizer factory producing calcium ammonium nitrate and heavy water and utilizing power of 180 MW from left bank Bhakra power house and Sri Ram Fertilizer factory located at Kota (Rajasthan) are instances in the support of the above statement.

The voltage of each cell is about 10 volts, but if many cells are connected in series, current of the order of several thousand amperes will be required at the voltage of the order of 500 to 800 volts. Thus by employing heavy current motor-generators, rotary convertors or even mercury-arc rectifiers, the required supply may be obtained from the modern grid network.

Exercises

1. State Faraday's laws of electrolysis and explain them clearly.
[U.P. Technical Univ. Utilization of Electrical Energy & Electric Traction 2005-06; M.D. Univ. Utilization of Electric Power & Electric Traction Dec. 2009; Pb. Technical Univ. Electric Drives & Utilization Dec. 2009]
2. Discuss various laws of electrolysis. What are different applications of electrolysis.
[Pb. Technical Univ. Electric Drives & Utilization May 2007]
3. What is electrolysis? State Faraday's laws of electrolysis and explain them clearly.
[M.D. Univ. Utilization of Electric Power & Traction Dec. 2008]
4. What are the various applications of electrolysis ? Explain extraction of metals. (Take example of any metal).
[Gorakhpur Univ. Utilization of Electrical Power & Traction September 1984]
5. Briefly describe the process of extraction of aluminium from the ore.
[U.P. Technical Univ. Utilization of Electrical Energy and Electric Traction 2006-07]
6. Explain with the help of a neat sketch the extraction process of aluminium.
[Rajasthan Technical Univ. Utilization of Electrical Power 2011]
7. Explain the terms 'polarisation'; 'throwing power' and 'electrodeposition'. How are zinc and copper refined from their base metals electrically ?
[Gorakhpur Univ. Utilization of Electrical Power and Traction September, 1983]
8. What is electrolysis? Explain the following terms used in electrolytic process :
(*i*) Current efficiency (*ii*) Throwing power (*iii*) Electro-chemical equivalent ?
[M.D. Univ. Utilization of Electric Power & Traction May 2009]
9. Define and explain 'current efficiency', 'voltage required' and 'energy efficiency' when referred to electrolytic processes. Also state and explain Faraday's laws of electrolysis.
[G.B. Technical Univ. Utilization of Electrical Energy & Traction 2010-11]
10. Describe principle of electrode position in electrolytic process.
[U.P. Technical Univ. Utilization of Electrical Energy & Traction 2009-10]
11. What is electrodeposition ? Explain the various factors on which the quality of deposition depends.
[Pb. Technical Univ. Electric Drives & Utilization Dec. 2008]
12. Explain the basic principle of electrodeposition. Give its applications.
[U.P. Technical Univ. Utilization of Electrical Energy & Electric Traction 2005-06]
13. State and explain Faraday's laws of electrolysis. What factors govern the rate of electrodeposition process?
[Rajasthan Technical Univ. Utilization of Electrical Power 2010]
14. What is electroplating and what for is it done? Describe various operations involved in electroplating.
[M.D. Univ. Utilization of Electric Power & Traction Dec. 2008]
15. Why is it necessary to clean a job before electroplating and how it is performed?
[Rajasthan Technical Univ. Utilization of Electrical Power 2010]
16. Describe in detail the process of nickel electroplating in industry.
[M.D. Univ. Utilization of Electric Power & Traction May 2009]
17. What do you understand by 'reverse current process' in electroplating? Discuss its advantages.
[M.D. Univ. Utilization of Electric Power and Traction Dec. 2009, May 2010]
18. Give the detailed description of equipment and process used for chromium plating. What is the composition of electrolyte used ?
[U.P. Technical Univ. Utilization of Electrical Energy & Electric Traction 2005-06]
19. List the major applications of electrolysis. What do you understand by anodising ?
[U.P. Technical Univ. Utilization of Electrical Energy and Electric Traction 2005-06]

20. Explain briefly : (*i*) Electrodeposition of rubber (*ii*) Anodic oxidation of aluminium.

[G.B. Technical Univ. Utilization of Electrial Energy & Traction 2010-11]

21. What is meant by anodising ? Explain the process of anodising and describe the equipment used for it.

22. List the major applications of electrolysis. Explain the basic principle of electrodeposition. Discuss in detail the power supply requirements for different electrolytic processes.

23. State various types of power supplies used for electrolytic process. Discuss any one in detail.

[Rajasthan Technical Univ. Utilization of Electrical Power 2010]

24. Why it is preferred to supply large chemical works from hydroelectric or atomic power stations?

[Rajasthan Technical Univ. Utilization of Electrical Power 2011]

25. Write short notes on the following :

(*i*) Electrolysis and its laws.

(*ii*) Applications of electrolysis. **[UPTU 2006-07]**

(*iii*) Nickel plating. **[UPTU 2006-07]**

(*iv*) Extraction of metals using electrolytic processes.

(*v*) Anodizing and its applications.

[W.B. Univ. of Technology 2009]

(*vi*) Reverse current process of electroplating.

[UPTU 2006-07]

(*vii*) Electroforming. **[UPTU 2006-07]**

(*viii*) Manufacture of chemicals by electrolytic processes.

(*ix*) Power supply requirements for electrolytic processes.

(*x*) Extraction and refining of metals by electrolysis.

Short Answer Type Questions With Answers

Q. 1. What is electrolysis ? **[P.T.U. December 2009]**

Ans. The breaking of a substance (known as electrolyte) into ions by flow of electric current through it is known as electrolysis.

Q. 2. What happens when an electric current is passed through solution of copper sulphate ? **[P.T.U. December 2009]**

Ans. When an electric current is passed through solution of copper sulphate, the electrolyte ($CuSO_4$) is decomposed into cations (Cu^{++}) and anions (SO_4^{--}) according to chemical reaction

$$CuSO_4 \longrightarrow Cu^{++} + SO_4^{--}$$

The cations move, toward the cathode (or negative electrode) and each of the positively charged copper ions reaching the cathode takes two electrons from it and becomes a metallic atom of copper. The negatively charged sulphions (SO_4^{--}) move towards the anode (or positive electrode) and each of the negatively charged sulphions reaching the anode gives up two electrons to it and ceases to be anion. Thus, metallic copper is deposited on the cathode, while the sulphions are collected at the anode and combine with water forming H_2SO_4 and O_2 according to chemical reaction

$$2SO_4 + 2H_2O \longrightarrow 2H_2SO_4 + O_2$$

The oxygen gas so formed is liberated at the anode.

Q. 3. Write the applications of electrolysis. **[P.T.U. Dec. 2006]**

Ans. Electrolytic processes are widely used for the extraction of pure metals from their ores (such as aluminium, zinc, copper, magnesium, sodium etc.), manufacturing of various chemicals (such as caustic soda, potassium permanganate, hydrogen, oxygen, chlorine etc.), electrodeposition of metals including electroplating, electrotyping, electroforming, building up of worn-out parts in metallurgical, chemical and other industries.

Q. 4. What is meant by electrochemical equivalent of a substance ?

Ans. Electrochemical equivalent (ECE) of a substance is the amount of substance deposited on passing a steady electric current of 1 A for one second through its solution.

Q. 5. What is relation between chemical equivalent, atomic weight and valency of any substance ?

Ans. Chemical equivalent of any substance may be defined as the ratio of its atomic weight to its valency

i.e., $\text{Chemical equivalent} = \dfrac{\text{Atomic weight}}{\text{Valency}}$

Q. 6. Define current efficiency in electrolytic process.

[P.T.U. Dec. 2006]

Ans. The current efficiency is defined as the ratio of the actual quantity of substance liberated or deposited to the theoretical quantity, as calculated from Faraday's laws.

i.e., Current efficiency

$$= \frac{\text{Actual quantity of substance liberated or deposited}}{\text{Theoretical quantity of substance liberated or deposited}}$$

Its value usually lies between 90 and 98%.

Q. 7. What is electrodeposition ?

Ans. The process of depositing a coating of one metal over another metal or non-metal electrically is called the electro-deposition.

Q. 8. What for electrodeposition process is used ?

Ans. The electrodeposition process is used for protective, decorative and functional purposes and includes electroplating, electroforming, electrotyping, electrofacing, electrometallisation, electrodeposition of rubber, building up of worn-out parts for repairs etc.

Q. 9. Enumerate the factors on which the quality of electro-deposition depends.

Ans. The factors on which the quality of electrodeposition depends are (*i*) nature of electrolyte (*ii*) current density (*iii*) temperature (*iv*) conductivity and concentration of electrolyte (*v*) additional agents (*vi*) throwing power and (*vii*) polarisation.

Q. 10. What is meant by 'throwing power' and how it can be improved ? **[Gorakhpur Univ. Sept. 1981]**

Ans. Throwing power is the ability of electrolyte to produce uniform deposit on an article of irregular shape and is one of the important characteristics of plating or depositing bath. It can be improved in two ways—firstly, by increasing the distance between the anode and cathode, and secondly, by reducing the voltage drop at the cathode surface.

Q. 11. What is polarization and how can it be made negligible ?

Ans. Use of current density, in electrodeposition process, beyond a certain limit causes electrolysis of water and hydrogen liberation on the cathode. This hydrogen evolved on the cathode blankets the base metal which reduces the rate of metal deposition. This phenomenon is called polarization.

Polarization can be reduced by agitating the electrolyte and can be made negligible by reverse current process.

Q. 12. What do you understand by "reverse current process" in electroplating ?

Ans. In reverse current process in electroplating, at regular intervals plating current is reversed for a second or so.

Q. 13. What is electroplating ?

Ans. Electroplating is an art of depositing a superior or more noble metal on an inferior or a base metal by means of electrolysis of an aqueous solution of a suitable electrolyte.

Q. 14. What is use of electroplating ? **[P.T.U. May 2007; Dec. 2007]**

Ans. Electroplating is done occasionally for ornamentation and decoration purposes. Sometimes, it is done, with a view to repair worn-out parts of a machinery.

Q. 15. Why the article to be electroplated is thoroughly cleaned, polished and degreased before putting it for electroplating ? **[Gorakhpur Univ. 1980]**

Ans. In case the object to be electroplated is not cleaned, polished, and degreased, the deposit formed may not be well adherent to the base metal and is likely to peel off.

Q. 16. What is electroforming ? **[P.T.U. Dec. 2008]**

Ans. Reproduction of objects by electrodeposition on some sort of a mould or form is known as *electroforming*. This is another application of electrodeposition.

Q. 17. What is meant by anodising ?

Ans. Anodizing is a process of anodic oxidation in which a thin uniform passive (tough, thin, uniform and impermeable) film is produced artificially by the passage of electric current.

Q. 18. What is anodizing and Faraday constant ? **[P.T.U. May 2009]**

Ans. For anodizing refer to Answer of Q. 17.

According to Faraday's second law of electrolysis; the chemical equivalent weight is proportional to electrochemical equivalent

i.e. $e \propto Z$ or $e = FZ$

where F is called the Faraday constant. The quantity of electricity which liberates one *gm* equivalent of an element is known as Faraday constant *i.e.* 1 F = 96,540 coulombs.

Q. 19. What type of power supply is required for electrolytic processes ?

Ans. Power supply required for electrolytic processes is dc and at very low voltage ; usually very small in magnitude (between 100 and 200 A at 10 or 12 V)

Q. 20. Why it is preferred to instal large chemical plants near hydroelectric or atomic power plants ?

Ans. Because of huge power requirement and 100 per cent load factor, large chemical plants are located near the hydroelectric or atomic power plants. The advantage of a high load factor is greater with such plants than with steam power plants.

Problems

1. A coating of silver 0.005 mm thick is to be deposited on a vessel whose surface area is 0.8 m^2. The current is maintained at 5 A. The weight of the silver per cc is 10.5 gm. The ECE of silver is 0.001118 gm per coulomb. Calculate the time taken for finishing the job.
[**Ans.** 2 hours 5 minutes and 15 seconds]

2. State Faradays laws of electrolysis and explain them clearly. If 20 gm of nickel is deposited by 120 A current flowing for 10 minutes, how much copper would be deposited by 100 A current in 5 minutes ? Atomic weight of nickel and copper are 58.6 and 63.18 respectively and valency of both is two. **[U.P. Technical Univ. Utilization of Electrical Energy and Electric Traction 2005-06]**
[**Ans.** 9 gm]

3. Find the thickness of copper deposited on a plate of 2.25 cm^2 during electrolysis if a current of one ampere is passed for 150 minutes. Density of copper is 8.9 gm per cc and electrochemical equivalent of copper is 0.0003295.
[Pb. Technical Univ. Electric Drives & Utilization May 2008]
[**Ans.** 1.48 mm]

4. Find the thickness of copper deposited on a plate of 2.25 cm^2 during electrolysis if a current of one ampere is passed for 100 minutes. Density of copper is 8.9 gm per cc and electrochemical equivalent of copper is 0.0003295.
[**Ans.** 0.9872 mm]

5. A group of 400 cells in a copper refining plant carries a current of 5,000 A. The voltage per cell is 0.25 V. If the plant working time per week is 100 hours, determine the annual output of refined copper from the cells and the energy consumption per tonne. ECE of copper = 0.327 mg per coulomb. [**Ans.** 12,243 tonnes; 212.4 kWh per tonne]

6. A copper refining plant employs 5,000 cells carrying a current of 5,000 A. The voltage per cell is 0.28 volt. If the plant is worked for 48 hours per week, determine the annual output of refined copper and energy consumption per tonne. ECE of copper is 32.95 × 10^{-8} kg C^{-1}.
[Gorakhpur Univ. Utilization of Elec. Power & Traction 1982]
[**Ans.** 74,019 tonnes; 236 kWh]

7. A copper refining plant, employing 500 cells carries a current of 4,000 A. The voltage per cell is 0.25 V. Determine the annual output of the plant and energy consumption per tonne, if the plant works for 80 hours per week and ECE of copper is 32.81 × 10^{-8} kg C^{-1}.
[Gorakhpur Univ. Utilization of Elec. Power & Traction 1982]
[**Ans.** 9,827 tonnes; 212 kWh/tonne]

8. A 20 cm long portion of a circular shaft having 10 cm diameter is to be coated with a layer of 1.5 mm nickel. Determine the quantity of electricity in Ah and the time taken for the process. Assume a current density of 195 A/m^2 and a current efficiency of 92%. Specific gravity of nickel is 8.9 and its ECE is 1.0954 kg per 1,000 Ah.
[**Ans.** 832.33 Ah; 68 hours]

❑❑❑

CHAPTER 9

Refrigeration and Air Conditioning

INSIDE THIS CHAPTER

9.1. Introduction **9.2.** Terminology **9.3.** Refrigeration Cycle **9.4.** Refrigeration Systems **9.5.** Compound or Multi-Stage Vapour Compression Refrigeration System **9.6.** Refrigerants **9.7.** Domestic Refrigerator **9.8.** Water Cooler **9.9.** Desert Cooler **9.10.** Air Conditioning **9.11.** Comfort Air Conditioning **9.12.** Industrial Air Conditioning **9.13.** Effective Temperature **9.14.** Summer Air Conditioning System **9.15.** Winter Air Conditioning System **9.16.** Year-Round Air Conditioning System **9.17.** Types of Air Conditioning Systems **9.18.** Room Air Conditioners **9.19.** Central Air Conditioning Systems **9.20.** Heating of Buildings **9.21.** Calculation of Rating of Electrical Equipment.

9.1. INTRODUCTION

Refrigeration and air conditioning has become a necessity in every sphere of human utility and industrial set-ups whether it is in houses, schools, colleges, departmental stores, cinema houses, dancing halls, night clubs, restaurants, cars, buses, railways, manufacturing and measurement or inspection departments/shops of the industry, power houses. In the early years of the twentieth century it used to be considered a luxury but with the change in the atmospheric set-ups, changing environments and standards of living it has become necessity. It is employed for the comforts of human beings both as domestic appliances and industrial equipment for producing low air temperature and for cooling of storage chambers in which perishable foods, fruits, medicines, chemicals etc. are preserved. In industry it is employed to maintain the temperature of the factories, tool rooms, and working place below that of surroundings so that the workers, various types of materials and equipments can perform their duties efficiently. It is also used for production of ice, for liquifying of gases and vapours in chemical and pharmaceutical industries and in air conditioning of buildings, coaches, milk vans etc.

9.2. TERMINOLOGY

The important terms used in refrigeration and air conditioning are defined below:

1. ***Refrigeration.*** The term '*refrigeration*' may be defined as the process of removing heat from a substance under controlled conditions. It also includes the process of reducing and maintaining the temperature of a body below the general temperature of its surroundings. In other words, the refrigeration means a continued extraction of heat from a body whose temperature is already below the temperature of its surroundings.

2. ***Refrigeration System.*** The mechanism used for lowering or producing low temperature in a body or a space, whose temperature is already below the temperature of its surroundings, is called the refrigeration system. Here the heat is being generally pumped from the low level to the higher one and is rejected at high temperature.

3. ***Refrigerator.*** Theoretically, a refrigerator is a reversed heat engine or a heat pump which takes out heat from a cold body and delivers it to a hot body.

4. ***Refrigerant.*** The refrigerant is a heat carrying medium which during their cycle (*i.e.*, compression, condensation, expansion, and evaporation) in the refrigeration system absorbs heat from a low temperature system (cold body) and delivers it to a higher temperature system (hot body).

5. ***Air Conditioning.*** 'Air conditioning' involves the control of the temperature, humidity (moisture content), and motion of air in an enclosure. Air conditioners usually incorporate filters for removing lint and dust and may include equipment for services such as removing odours or killing bacteria.

6. ***Tonne of Refrigeration (TR).*** It is a practical unit of refrigeration and is defined as the amount of refrigeration effect produced by the uniform melting of one tonne (1,000 kg) of ice from and at 0ºC in 24 hours. It is equivalent to heat extraction rate of 50 kcal/minute. While large commercial plants are designated in tonnes, domestic refrigerators are designated by the capacity (volume) of the cabinet. One tonne capacity of a refrigeration plant needs approximately 3/4 kW output motor.

9.3. REFRIGERATION CYCLE

For lowering the temperature of a body below that of its surroundings and then maintaining its temperature at this lower level, a continuous extraction of the heat from the body is required. The lowering or producing of low temperature in a body or a space is accomplished by a mechanism called the *refrigeration system.* Here the heat is being generally pumped from the low level to the higher one and is rejected at that high temperature. This rejection of heat from the low level to the high level of temperature can only be performed with the help of external work according to second law of thermodynamics. The total amount of heat being rejected to the outside body consists of two parts—heat extracted from the body to be cooled and the heat equivalent to the mechanical work required for extracting it.

A refrigerator in the theoretical sense is nothing but a reverse heat engine run in the reverse direction by means of external aid. Every type of refrigeration system used for producing cold must have the following four basic units.

(*i*) Low temperature thermal sink to which the heat is rejected for cooling the space.

(*ii*) Means of extracting the heat energy from the sink, raising its level of temperature before delivering it to heat receiver.

(*iii*) A receiver is a storage to which the heat is transferred from the high temperature, high pressure refrigerant.

(*iv*) Means of reducing the pressure and temperature of the refrigerant before it returns to the sink.

The refrigeration cycle (reversed heat engine cycle) is shown in Fig. 9.1 in which the four basic units or processes of the cycle are opposite to those of the heat engine cycle's evaporation, expansion, condensation and compression *i.e.*, evaporation, compression, condensation and expansion. Thus we see that by reversing a heat engine cycle completely and by changing the working agent also to the refrigerant, a refrigeration cycle is obtained.

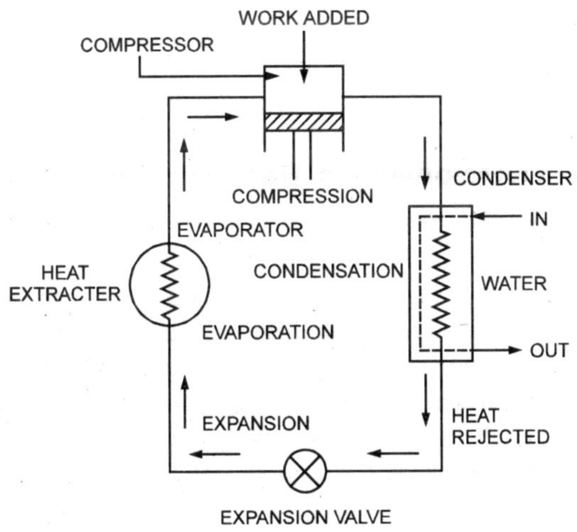

Fig. 9.1. *Referigeration Cycle*

9.4. REFRIGERATION SYSTEMS

There are three refrigeration systems :

(*a*) Vapour compression refrigeration system.

(*b*) Vapour absorption refrigeration system.

(*c*) Thermoelectric refrigeration system.

9.4.1. Vapour Compression Refrigeration System

This is the most important system from the standpoint of commercial and domestic utility and most practical form of refrigeration. The working fluid, known as a *refrigerant*, used in this refrigeration system readily evaporates and condenses or changes alternately between the vapour and liquid phases without leaving the refrigerating plant. During evaporation it absorbs heat from the cold body and in condensing or cooling or liquifying it rejects heat to the external hot body. The heat absorbed from cold body during evaporation is used as its latent heat for converting it from liquid to vapour. Thus a cooling effect is created in the working fluid. This system of refrigeration, thus act as a *latent heat pump* since it pumps its latent heat from the cold body or brine and rejects it or deliver it to the external hot body or the cooling medium. According to the laws of thermodynamics, this can be done only on the expenditure of energy which is supplied to the system in the form of electrical energy driving the compressor.

The main advantages of this system of refrigeration over the other systems are the smaller size for a given refrigerating capacity (except for some recent development in air refrigeration system in the aircraft cooling), higher coefficient of performance or lower power requirements for a given capacity and less complexity in both design and operation. It can be used over a large range of temperatures.

The main disadvantages like toxity, inflammability and leakage of vapours have largely been eliminated by improvements in design resulting in greater safety and prevention of leakages and by development of non-toxic, nonflammable vapours to be used as refrigerants.

The vapour compression cycle is used in most of the modern refrigeration systems in large industrial plants requiring the rapid removal of large quantities of heat and in most of the domestic refrigerators. The vapour in this cycle is circulated through the various components of the system, where it undergoes a number of changes in its state or condition. Each cycle of operation consists of the four fundamental changes of state or processes (*i*) expansion (*ii*) vaporisation (*iii*) compression and (*iv*) condensation.

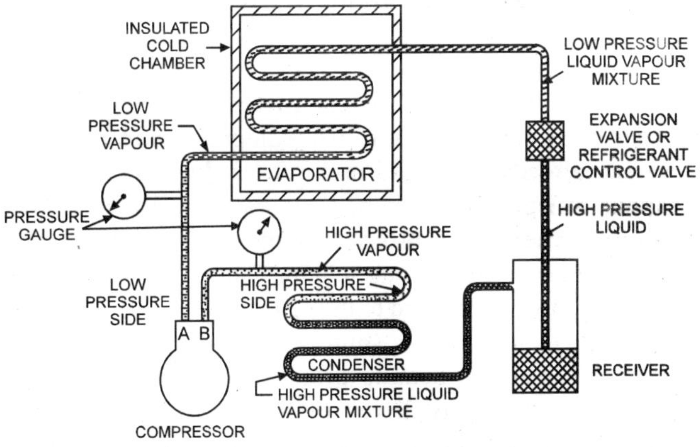

Fig. 9.2. *Simple Vapour Compression Refrigeration System*

The schematic diagram of a vapour compression refrigeration system is shown in Fig. 9.2. It consists of the following five essential parts.

1. *Compressor.* The low pressure and temperature vapour refrigerant from evaporator is drawn into the compressor through the inlet or suction valve A, where it is compressed to a high pressure and temperature. The high pressure and temperature vapour refrigerant is discharged into the condenser through the delivery or discharge valve B.

2. *Condenser.* The condenser or cooler consists of coils of pipe in which the high pressure and temperature vapour refrigerant is cooled and condensed. The refrigerant, while passing through the condenser, rejects its latent heat to the surrounding condensing medium which is normally air or water. Thus hot refrigerant vapour received from compressor is converted into liquid form in condenser.

3. *Receiver.* The condensed liquid refrigerant from the condenser is stored in a vessel, known as *receiver,* from where it is supplied to the expansion valve or refrigerant control valve.

4. *Expansion Valve or Throttle Valve.* The function of this valve is to allow the liquid refrigerant under high pressure and temperature to pass at a controlled rate after reducing its pressure and temperature. Some of the liquid refrigerant evaporates as it passes through the expansion valve, but the greater portion is vaporised in the evaporator at the low pressure and temperature.

5. *Evaporator.* An evaporator consists of coils of pipe in which the liquid vapour refrigerant at low pressure and temperature is evaporated and changed into vapour refrigerant at low pressure and temperature.

During evaporation process, the liquid vapour refrigerant absorbs its latent heat of vaporization from the medium (air, water or brine) which is to be cooled.

9.4.2. Vapour Absorption Refrigeration System

The vapour absorption refrigeration system is one of the oldest method of producing refrigerating effect. The refrigerant, commonly used in this system, is ammonia. This system uses heat energy, instead of mechanical energy as in vapour compression systems, in order to change the conditions of the refrigerant required for the operation of the refrigeration cycle.

In the vapour absorption refrigeration system, the compressor is replaced by an absorber, a pump, a generator and a pressure reducing valve. These components in vapour absorption refrigeration system, perform the same function as that of a compressor in vapour compression refrigeration system, already discussed in Art 9.4.1. In this system, the vapour refrigerant from the evaporator is drawn into an absorber where it is absorbed by the weak solution of the refrigerant forming a strong solution. This strong solution is pumped to the generator where it is heated by some external source. During the heating process, the vapour refrigerant is driven off by the solution and enters into the condenser where it is liquified. The liquified refrigerant then flows into the evaporator and thus the cycle is completed.

The simple vapour absorption refrigeration system is shown in Fig. 9.3. It essentially consists of an absorber, a pump, a generator and a pressure reducing valve, in addition to condenser, receiver, expansion valve and evaporator, as in the vapour compression refrigeration system.

In this system, the low pressure ammonia vapour leaving the evaporator, enters the absorber where it is absorbed by the cold water. The water has the ability to absorb very large quantity of ammonia vapour and the solution so formed, is known as *aqua-ammonia.* The absorption of ammonia vapour in water lowers the pressure in the absorber which in turn draws more ammonia vapour from the evaporator and thus the temperature of the solution is raised. Some form of cooling arrangement (usually water cooling) is used in the absorber for removal of heat from the solution evolved there. This is necessary so that the absorption capacity of the water is increased. The strong solution thus formed in the absorber is pumped to the generator by the liquid pump. The pump increases the pressure of the solution up to 10 kg/cm^2.

The strong solution of ammonia in the generator is heated by some external source such as gas or steam. During the heating process, the ammonia vapour is driven off the solution at high pressure leaving behind the hot weak ammonia solution in the generator. This weak ammonia solution flows back to the absorber at low pressure after passing through the pressure reducing valve. The high pressure ammonia vapour from the generator is condensed in the condenser to a high pressure liquid ammonia. The liquid ammonia is passed to the expansion valve through the receiver and then to the evaporator. Thus the simple absorption cycle is completed.

This system has a least number of moving parts. They are quite in operation and are used both in commercial as well as in domestic installations. In recent years, they have been used extensively in recreation vehicles, camper-trailers and boats as portable units due to their operation from a small gas cylinder.

9.4.3. Thermoelectric Refrigeration System

This system of refrigeration works on the principle of *Peltier effect* according to which heat energy is evolved at one junction and absorbed at the other one when direct current is passed through a junction of two dissimilar metals like antimony and bismuth. It consists of a number of thermoelectric module assemblies in series joined by copper strips, as shown in Fig. 9.4. Each thermoelectric module is built up of a large number of thermoelectric couples. When the direct current

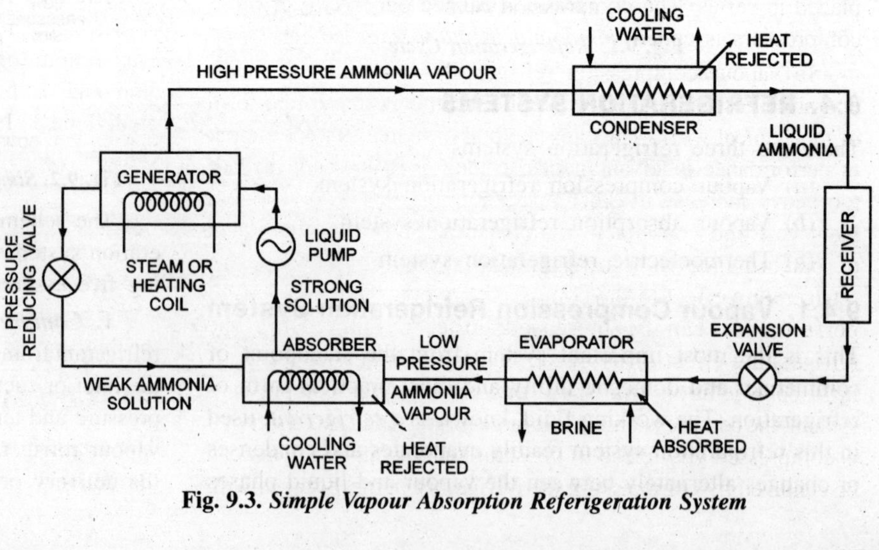

Fig. 9.3. ***Simple Vapour Absorption Referigeration System***

is passed through the thermoelectric module assemblies in the direction shown in Fig. 9.4, junctions at the top of the assembly are cooled and those at the bottom are heated up. Thus top junctions abstract heat from the surroundings and produce refrigerating effect and bottom junctions require cooling by water. Module assembly, therefore, abstracts heat from the medium at top and rejects the same to the medium at bottom. If the direction of flow of current is reversed, heat production at the junctions is reversed, *i.e.*, now bottom of the modules will be cooled and top ones get heated up. If it is an air conditioner, air will get heated up instead of getting cooled. Thus summer and winter operating of air conditioner can easily be obtained just by reversing the direction of flow of current. The main advantages of this system of refrigeration are absence of moving parts and ease of automatic control by controlling the magnitude of current. The only drawback of this system of refrigeration is its very high initial cost.

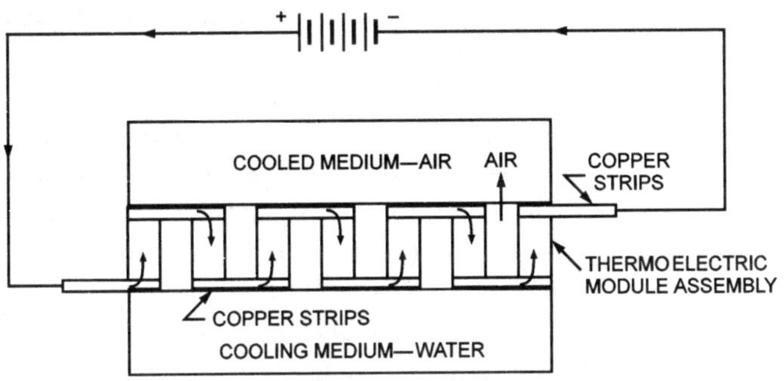

Fig. 9.4. *Thermoelectric Refrigeration System*

9.5. COMPOUND OR MULTI-STAGE VAPOUR COMPRESSION REFRIGERATION SYSTEM

In Art. 9.4.1, we have discussed the simple vapour compression refrigeration system in which the low pressure vapour refrigerant from the evaporator is compressed in single stage (or in a single compressor) and then delivered to a condenser at a high pressure. But sometimes, the vapour refrigerant is required to be delivered at a very high pressure as in the case of low temperature refrigerating systems. In such cases either the vapour refrigerant should be compressed by employing a single stage compressor with a very high pressure ratio between the condenser and evaporator or it should be compressed in two or more compressors placed in series. The compression carried out in two or more compressors is called *compound* or *multi-stage compression.*

In vapour compression refrigeration systems, the major operating cost is the energy supplied to the system in the form of mechanical work. Thus any method of increasing coefficient of performance is advantageous so long as it does not involve too heavy increase in other operating costs, as well as initial plant cost and consequent maintenance, Since the coefficient of performance of a refrigeration system is the ratio of refrigerating effect to the compression work, the coefficient of performance can be increased either by increasing the refrigerating effect or by decreasing the compression work. A little consideration will show that in a vapour compression system, the compression work is largely reduced if the refrigerant is compressed very close to the saturated vapour line. This can be achieved by compressing the refrigerant in more stages with intermediate intercooling. But it is economical only where the pressure ratio is considerable as would be the case when very low evaporator temperatures are desired or when high condenser temperature may be required. The compound compression is generally economical in large plants.

The refrigerating effect can be increased by maintaining the condition of the refrigerant in more liquid state at the entrance to the evaporator. This can be achieved by expanding the refrigerant very close to the saturated liquid line. It may be noted that by subcooling the refrigerant and by removing the flashed vapour, as they are during multi-stage expansion, the expansion can be brought close to the liquid line.

9.6. REFRIGERANTS

The suitability of a refrigerant for a certain application is determined by its physical, thermodynamic, chemical properties and by various practical factors. There is no one refrigerant which can be used for all types of applications, *i.e.,* there is no ideal refrigerant. If one refrigerant has certain advantages, it will have some drawbacks also. Hence, a refrigerant is chosen which has more advantages and less disadvantages.

For safe, economical and thermodynamically efficient production of cold, refrigerant should possess as may as practicable of the following desirable properties.

1. Low boiling point.
2. High critical temperature.
3. High latent heat of vaporisation.
4. Low specific volume of vapour.
5. Low specific heat of liquid.
6. Low cost.
7. Easy to liquify at moderate pressure and temperature.
8. Easy of locating leaks by odour or suitable indicator.
9. It should be noncorrosive.
10. It should be noninflammable.
11. It should be non-toxic.
12. It should be free from objectionable odour.

Various types of refrigerants used for domestic purposes and various industrial utilities in modern refrigerating equipment and plants are classified as primary refrigerants and secondary refrigerants. Primary refrigerants are those working mediums or heat carriers which actually traverse the whole cycle of evaporation, compression, condensation and liquidification *e.g.*, ammonia, carbon dioxide, sulphur dioxide, methyl chloride, methylene chloride, ethyl chloride and freon (mixture of chlorine, fluorine and carbon). Secondary refrigerants are those circulating cold substances which only transfer heat from a remote point to the evaporator of the refrigeration system, *e.g.*, ice, solid carbon dioxide etc. Ammonia is used for commercial and industrial refrigeration for ice manufacture, food processing, cold storage etc., as it is cheap and readily available and has the highest efficiency.

9.7. DOMESTIC REFRIGERATOR

The applications of refrigeration for domestic purposes are mainly in the form of domestic refrigerators and home freezers.

The main purpose of this type of refrigeration is to provide low temperature for storage and distribution of foods and drinks. It represents a significant portion of the refrigeration industry due to the use of these units in large number.

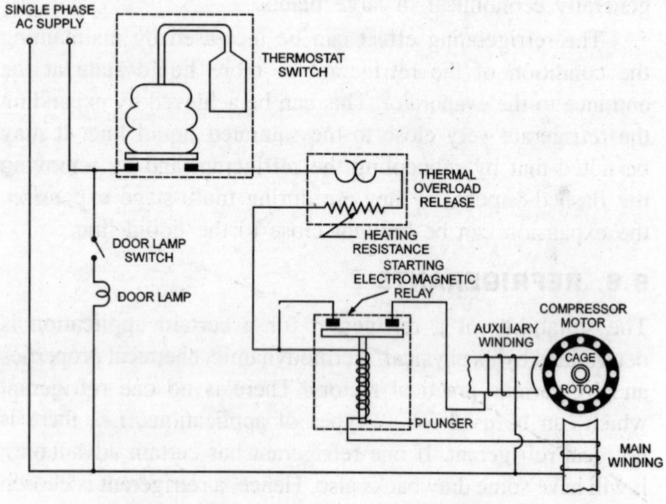

Fig. 9.5. ***Electrical Circuit of a Refrigerator***

The growth of food destroying micro-organisms is quite fast (about 100 times) at 10ºC than at a temperature lower than 4ºC. In order to preserve the fruits, vegetables, fish, meat, milk, medicines etc. against spoilage at higher temperature (generally during summer season), the use of refrigerators is very much emphasized. The storage of perishable products at low temperature greatly reduces the activity of both the enzymes and micro-organisms.

For domestic preservation, the storage is generally short-term or temporary storage. The domestic refrigerators and freezers used for the purpose are usually small in sizes with ratings ranging from 1/20 to 1/2 tonne. The units are usually self-contained and hermetically sealed. Due to short-term storage the domestic refrigerator load is intermittent. The requirements of domestic refrigerator is that it should be simple in construction, automatic in action, nominal in initial cost, dependable, and without any necessity of expert inspection and repair. Non-irritant and non-toxic refrigerant should be used. Generally methylene chloride, Freon-12 and Freon-11 are used as the refrigerants.

The common type of domestic refrigerator have a cabinet shaped with compressor motor-fan assembly, the condenser and receiver fitted in their basement. The expansion valve and evaporator coils are exposed in the storage cabinet with the pipings, carrying liquid refrigerant passing through the body. The heat of the bodies to be cooled is carried to the evaporator coils by means of air trapped in the cabinet. The refrigerator used may be the open compressor type; hermetically sealed compressor type or absorption type. Usually hermetically sealed type refrigerators are used. The main advantages are that the problem of gas leakage is minimised as no moving part extends through the sealed housing, it has relatively silent operation and it does not require any maintenance. In open type compressor, the motor drives the compressor using pulley and belt arrangement. This is employed in large commercial refrigerators.

Refrigeration is not only provided with double walled cabinet packed with materials having high thermal insulation such as fibreglass, cork or expanded rubber but also all round the inside of door flap soft rubber seal is used which makes the cabinet airtight. Also the door is provided with automatic closing mechanism–door hinges are provided in such a way that door flap when left in open position automatically comes to closing position due to gravity and as it approaches closing position it is attracted by the magnetic strip fitted behind the sealing rubber ring and thus the door is closed with snap action. All this is done to prevent leakage of atmospheric heat inside the refrigerator.

Main precaution to be taken is that very hot things should not be put in refrigerator, if it is done it will quickly evaporate the refrigerant in evaporator coils producing a large vapour pressure resulting in increase in the duty of the compressor and as such motor has to work for longer duration in order to bring down the temperature in the freezer compartment. It may damage the motor, which is short-time rated. Also moisture released from hot things will freeze around the freezer compartment which has to be defrosted.

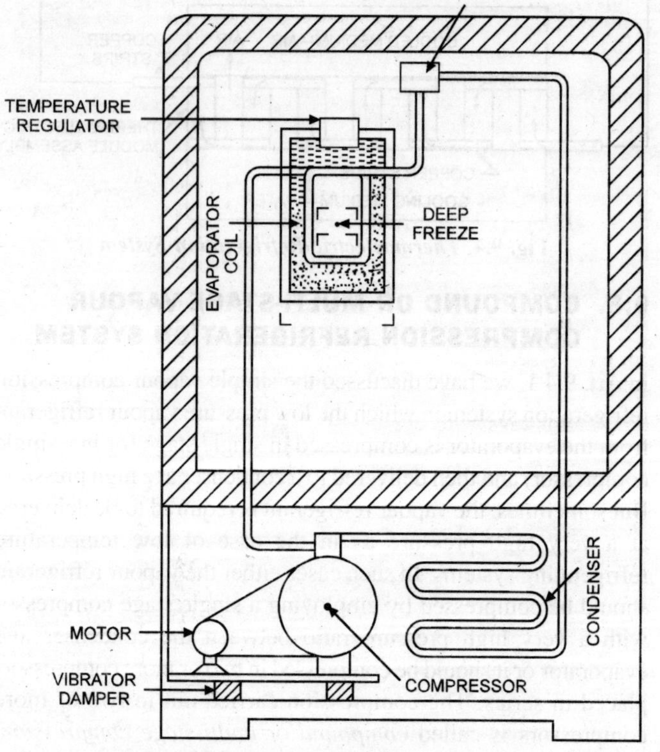

Fig. 9.6. ***Domestic Refrigerator***

Electrical circuit of a refrigerator is shown in Fig. 9.5. Refrigerator is provided with a door push switch, which closes on opening of refrigerator and puts the lamp on. Capacitor-start single phase induction motor is used in open type refrigerators and split-phase single phase induction motor is used in sealed unit refrigerators. Electromagnetic relay is provided to connect auxiliary winding on the start and disconnect it when the motor picks up the speed. Thermal overload release is provided to protect the motor from damage against flow of overcurrent. Thermostat switch is provided to control the temperature inside the refrigerator. Temperature inside the refrigerator can be adjusted by means of temperature control screw. Greater the distance between contacts, greater will be the temperature inside the refrigerator and vice versa. To protect the motor against undervoltage use of automatic voltage regulator is essential since in case of fall in applied voltage, motor will draw heavy current to develop the required torque and will become hot, thermal overload relay will, therefore, repeatedly disconnect and connect the motor to supply, eventually burning it out.

Domestic refrigerator employing vapour compression refrigeration system is shown in Fig. 9.6.

9.7.1. Trouble Shooting of Refrigerator

Various troubles often found in a refrigerator are listed below.

Symptoms	Causes
1. Motor fails to start on giving supply.	(*i*) Blown fuses in the main switch. (*ii*) Thermostat contacts open. (*iii*) Overload release open. (*iv*) Open circuit in the main winding. (*v*) Open circuit in the auxiliary winding. (*vi*) Grounded winding. (*vii*) Burnt or shorted winding (*viii*) Excessive overload (*ix*) Worn-out or tight bearings. (*x*) Bent rotor shaft.
2. Motor runs slower than normal speed.	(*i*) Low voltage. (*ii*) Overload. (*iii*) Shorted main winding. (*iv*) Defective electromagnetic relay. (*v*) Worn-out bearings.
3. Motor runs hot.	(*i*) Low voltage. (*ii*) Overload. (*iii*) Shorted or grounding winding. (*iv*) Worn-out bearings.
4. Motors does not start and gives humming noise.	(*i*) Low voltage. (*ii*) Overload. (*iii*) Open circuit in the auxiliary winding. (*iv*) Defective electromagnetic (starting) relay-contacts being not closed.
5. Motor runs with noise	(*i*) Shorted winding. (*ii*) Improperly connected poles. (*iii*) Loose rotor bar. (*iv*) Worn-out bearings. (*v*) Too much end play. (*vi*) Foreign material in the rotor.
6. Motor keeps on running but (*a*) cooling is nil	(*a*) (*i*) No refrigerant gas in the system (indicated by the unit, tubing and condenser being at ambient temperature). (*ii*) Complete choking of capillary.
(*b*) cooling is insufficient	(*b*) (*i*) Less refrigerant gas in the system. (*ii*) Partial choking of capillary.
7. Motor keeps on running even though it is very cold inside the refrigerator.	(*i*) Wrong setting of thermostat. (*ii*) Defective thermostat—shorted wiring or sticking contacts.
8. Motor starts very frequently.	(*i*) Bad door seal. (*ii*) Wrong setting of thermostat.
9. Motor operates normal with normal cooling in freezer but cooling in the rest portion is unsatisfactory.	(*i*) Bad door seal. (*ii*) Opening of door too frequent.
10. Motor operates normal with good cooling but defrosting starts all of a sudden. Cooling again starts after some time	(*i*) Presence of moisture in the refrigerant cycle.
11. Too much frosting around the freezer.	(*i*) High atmospheric humidity or steaming hot liquids stored.

9.8. WATER COOLER

Water is one of the most needed thing for a person. In summer season cold water gives life to a thirsty person. At 10ºC water is most refreshing. Thus cooling of water in summer season becomes necessary. Water coolers are used to produce cold water at about 7 to 13ºC. The temperature of water is controlled with the help of a thermostatic switch.

Water coolers may be classified as follows:

(*i*) Bottle type water cooler,

(*ii*) Pressure type water cooler,

(*iii*) Self-contained or remote type water cooler.

9.8.1. Instantaneous Type Water Coolers

In this type of coolers the cooling coil is wrapped round the pipeline such that by the time water reaches the tank it is cooled to desired temperature. Various types of instantaneous type water coolers are described below.

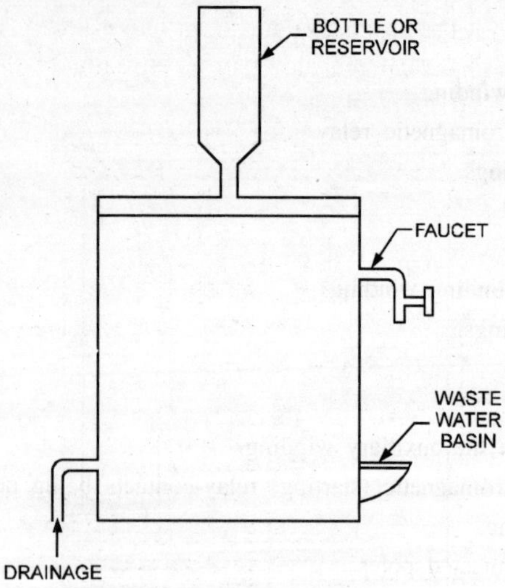

Fig. 9.7 (*a*). *Bottle Type Water Cooler*

1. *Bottle Type Water Cooler.* In this type of water cooler, water to be cooled is stored in a bottle or reservoir. For filling glass tumblers or containers faucet or similar means are provided. The dripping water from the faucet is collected in the waste water basin or water drip, as illustrated in Fig. 9.7 (*a*). The usual size is 25 litres and is suitable for places where plumbing installation is expensive and drains are available.

2. *Pressure Type Water Cooler.* In this type of cooler shown in Fig. 9.7 (*b*), water is supplied under pressure. For filling glass tumblers or containers faucets or similar means are provided. A valve is employed to control an appropriate flow of water or projected stream of water from a bubbler. An arrangement should be made to collect water and allow complete collection of water spreading from the bubbler. The temperature of waste water is low, it is used for cooling the supply water by passing through a pipe coil wrapped round the drainage line. By doing so, the cooling load for cooler is reduced. Since the water is supplied under pressure the cold water can be obtained from the top mounted at any height of the cooler. In case of bottle type, faucet has to be at a height up to which syphoned water can be obtained from the tank of the cooler. The refrigeration system is usually mounted at the bottom structure of the cooler body and a cooling coil is wrapped round the water tank. To ensure good surface contact between the evaporative tube and tank, either the tank surface is corrugated to accommodate pipe or pipes are secured using soft solder to provide metal contact. Sometimes a helical or U-type coil is immersed in the water tank. Although such as arrangement provides high heat transfer from water to the coil but formation of undesirable salt due to chemical reaction between water contaminant and the copper surface proves to be a great drawback in this system.

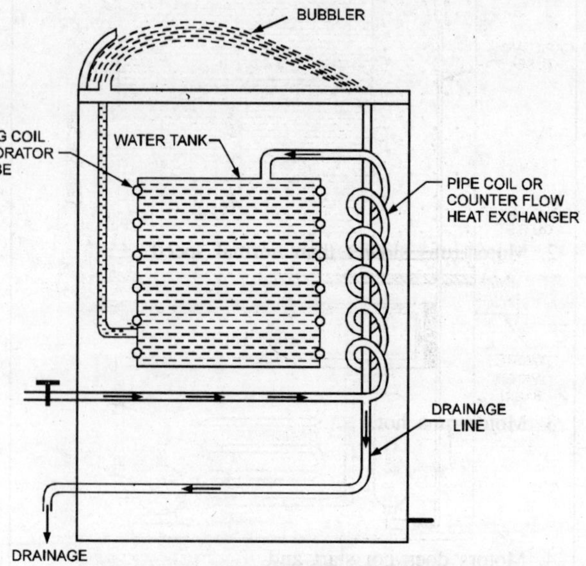

Fig. 9.7 (*b*). *Pressure Type Water Cooler*

3. *Self-Contained or Remote Type Water Cooler.* This type of cooler [Fig. 9.7 (*c*)] employs mechanical refrigeration system and is a factory assembled unit. A remote type water cooler cools the water which is supplied to the desired drinking place (away from the system). It is quite a useful unit as it does not need extra space near the place of work.

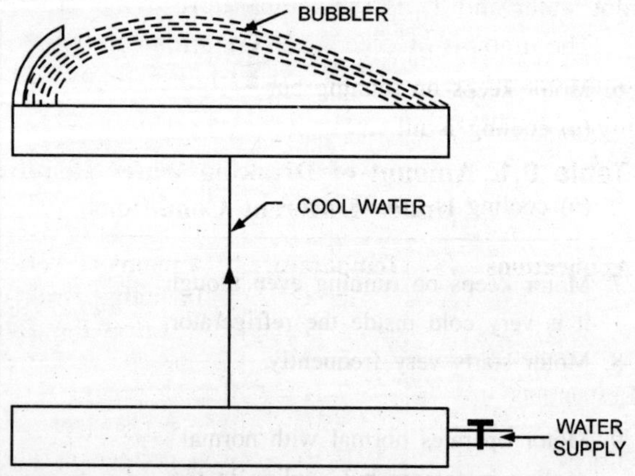

Fig. 9.7(*c*). *Self-Contained or Remote Type Water Cooler*

9.8.2. Storage Type Water Cooler

Such type of water coolers are used where continuous supply of water is not available. Fig. 9.8 shows a schematic storage type water cooler which is a self-explanatory. Here water is filled in the storage tank and level of water is kept the same by the use of a float valve. The storage tank is surrounded by an evaporator coil through which flows a low pressure liquid refrigerant which takes away the heat of water and thus makes it cold. When the water attains the desired temperature the thermostat operates and disconnects the power supply to the motor. The motor employed is capacitor-start capacitor-run single phase induction motor.

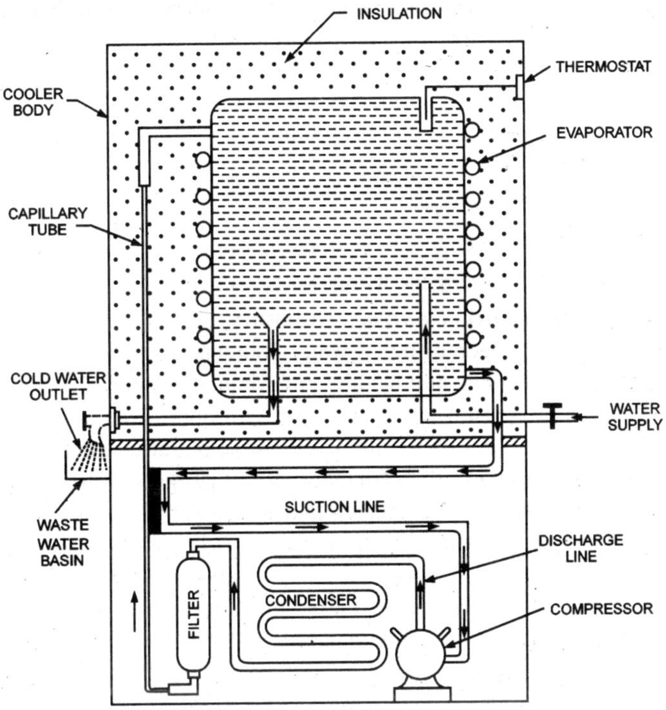

Fig. 9.8. ***Storage Type Water Cooler***

9.8.3. Capacity of Water Coolers

For determination of cooling load for the water cooler (Q) the following relation may be used

$$Q = m_w \times C_p (T_i - T_0)$$

where m_w is rate of water consumption in kg, C_p is the specific heat of water in kJ/kg/K, T_i is the temperature of inlet water and T_0 is the temperature of outlet water.

The amount of cold water requirements under different conditions is given in Table 9.1 (figures based on extensive statistical survey).

Table 9.1. Amount of Drinking Water Requirements Under Different Conditions

Applications	Temperature in °C	Amount of Water Used including Drainage in Litres Per Hour
Restaurants	7-10	0.5 per person
Schools	7-10	0.56 per student
Offices	7-10	0.56 per person
Stores	7-10	4.5 per 100 customers
Hotels	7-10	0.4 per room
Hospitals	7-10	0.4 per bed and per attendent
Theatres	7-10	4.5 per 100 seats
Light Duty Manufacturing Unit	10-13	0.9 per person
Heavy Duty Manufacturing Unit	10-14	1.1 per person

* R-12 is the most common refrigerant up to 1 tonne refrigeration capacity and R-22 for 2 tonnes refrigeration capacity, and appropriate combinations for units of larger size.

** While estimating the amount of water consumption, the amount of wastage of cold water should be taken into consideration.

*** For ensuring significant heat transfer through insulation, usually heavy insulation around 4 cm to 6 cm thick glass wool or thermocole is provided.

9.9. DESERT COOLER

As the name implies, it is suitable for places where the humidity is quite low and temperature quite high. Such conditions are in conformity with desert areas. Hence the coolers are called the desert coolers. The principle on which a desert cooler operates is "*evaporative cooling*". Evaporative cooling is a process is which sensible heat is removed and moisture added to the air. When the air passes through a spray of water it gives up heat to water, some of water evaporates and picks up heat from the air equivalent to its latent heat. The vapours thus formed are carried along in air stream. In this way air is cooled and humidified.

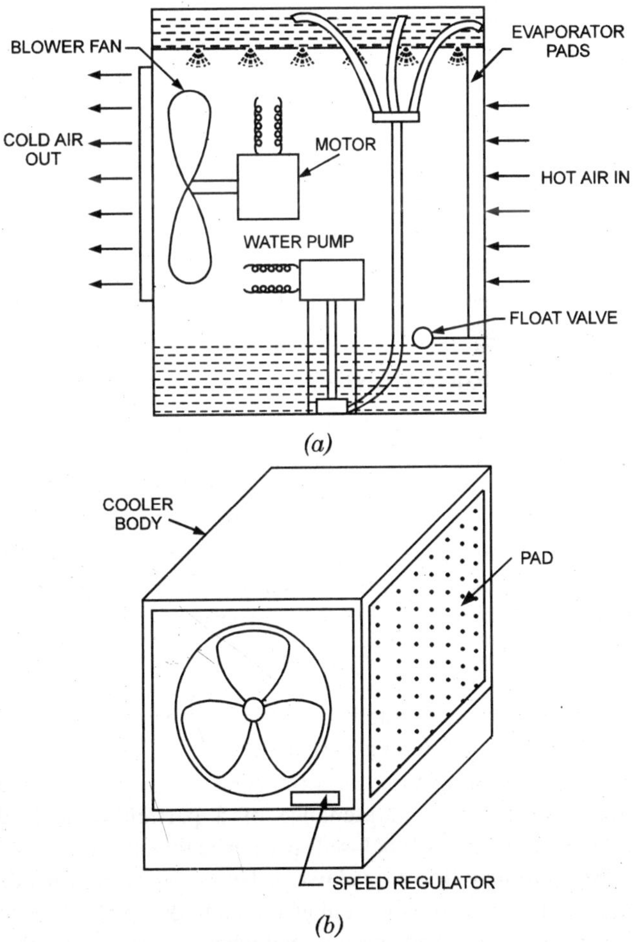

Fig. 9.9. ***Desert Cooler***

Construction and Working. A desert cooler consists of the following main parts:

1. Blower/Fan 2. Water circulating pump 3. Water wetted pads. 4. Water tank 5. Float valve 6. Water traps.

The water is filled in the sump of the cooler from water supply mains, the level of which is controlled by a float valve. A water pump lifts the water and supplies it at the top of the cooler to the water distribution system which consists of small branches of copper pipe or so equipped with orifices which deliver equal amount of water to the troughs which in turn supply water to the wetted pads. The water which drops back from the pads is recirculated. The pump may be made of brass, stainless steel or even plastic.

The blower pulls the air through the wetted pads and delivers it to the space to be cooled through and opening in the fourth side of the desert cooler cabinet. The air sucked through the pads is cooled by the *principle of evaporative* cooling. The blower provides adequate velocity to the air before it is delivered to the space to be cooled.

To have long life of the desert cooler and better performance, pads should be changed every year and holes for water distribution system should be cleaned. The tank should be cleaned just after the season and coated with the corrosion resisting paint like red oxide and some other paints.

The efficiency of desert cooler is given as

$$\eta_{desert} = \frac{t_{db} - t_{dbg}}{t_{db} - t_{wb}}$$

where t_{db} is dry bulb temperature of ambient air, t_{dbg} is the temperature of air leaving the grille, and t_{wb} is the wet bulb temperature of ambient air.

According to ISS, the velocity across the pads should be around 40 m/minute. The desert cooler efficiency should be about 75%. Though the high efficiency of desert cooler corresponds to cooled air, the large humidity causes discomfort. The desert cooler efficiency above 80% is not desirable. For good performance the efficiency of a desert cooler should not fall below 70%.

9.10. AIR CONDITIONING

In this modern age everybody is trying to improve his surroudings in order to make living conditions more comfortable. Air conditioning is one of the important contribution in this respect. The technique of maintaining of air condition inside a building in such a way that produces comfort to human beings, ensures trouble free operation of delicate equipment or helps in industrial manufacturing operations is called *air conditioning*. It involves control of temperature, cleanliness, humidity and circulation of air.

The important factors involved in a proper, efficient and complete air conditioning systems are :

1. *Temperature Control.* In air conditioning, temperature control means the maintenance of a particular desired temperature within an enclosed space even though the outside temperature may be either above or below the desired inside temperature. This requires either the addition or removal of heat from the enclosed space as and when necessary. For this purpose the outside air to be supplied in the room is either cooled or heated.

In summer, air is cooled by one of the following methods :

(*i*) Circulation of cold water.

(*ii*) Evaporative system using water.

(*iii*) Ice activated system.

(*iv*) System using absorption principle.

(*v*) Vapour compression system employing a compressor, evaporator and a suitable refrigerant. This is commonly used method. Three arrangements are possible under it namely (*a*) individual window type or floor mounted package type air conditioners (*b*) a central plant employing chilled water in combination with unit conditioners located at various parts of the building and (*c*) a central plant with a ducting system to convey cooled air to the different parts of the building.

In winter, air is warmed by one of the following methods :

(*i*) Electric heaters

(*ii*) Electromechanical heat pumps and

(*iii*) Heat exchangers using waste system.

2. *Humidity Control.* Humidity control means the increasing or decreasing of water contents of air in order to create comfortable and healthy environments depending upon a particular season–summer or winter. Control of humidity is not only necessary for human comfort but also it increases the production and the efficiency of the workers. It has been found to improve the quality and to reduce the cost of the production of products like automobiles, clay products, drugs, chemicals, electric goods, foods, textiles etc. It is seen in factories during winter season or summer season, that though there is a normal environmental temperature yet the workers feel uneasy. That is due to lack of proper water contents or humidity of air. The humidity control is achieved by the process of humidification (addition of humidity or moisture) or dehumidification (reduction of humidity or water vapour). The humidification is accomplished either by supplying or spraying steam or hot water or cold water into the air. Dehumidification may be accomplished by using an air washer or by using absorbents. In the air washer system, the outside air is cooled below dew point temperature so that it loses moisture by condensation. The moisture removal is also accomplished when the spray water is chilled water and its temperature is lower than the dew point of the outside air. The absorbent materials are activated alumina, silicagel and calcium chloride. These are generally used for window type small room coolers. In general, for summer air conditioning, the relative humidity should not be less than 60% while for winter air conditioning it should not exceed 40%.

3. *Air-movement and Circulation.* In order to keep constant temperature throughout the conditioned rooms, it is necessary that there should be equi-distribution of air throughout the space to be air conditioned. For this purpose the air movement and its circulation should be controlled.

4. *Air-filtration, Cleaning and Purification.* Proper filtration, cleaning and purification of air is necessary to keep it free from dust, dirt and other impurities which are very injurious to health. The dust particles are liable to clog the nose, throat and lungs of the worker seriously affecting their health. The dirt present in air acts as a carrier of germs and is responsible for disease spreading. In case of certain industries like chemicals the dirt produced as a byproduct is very harmful for health if inhaled. The presence of dirt in the air used for air cooling of electrical machines and diesel sets in power plants is liable to clog their air passage resulting in reduced efficiency and life. Various types of air impurities other than gas or vapours called the Aerosols are also present. These are dust (earthly wet soil dust, dust of coal, cement, wheat, lint—a string like dust from cotton or wool); fumes (solid particles due to condensation and solidification of gaseous substances, zinc oxide fumes from zinc vapours); smoke (due to combustion of organic materials like tobacco, wood, oil, coal etc.); pollen (from weed, tree, flower and other vegetation—may be harmless or harmful); bacterial (living micro-organisms) etc.

There are various methods of cleaning, filtering and purifying the air from the above mentioned impurities. Mists or fogs in the air are usually removed by raising the temperature of air. This results in the formation of small suspended water particles into vapours making them invisible. The impurities like gases and vapours are removed by the process of condensation. Pollen grains responsible for hay fever, rose fever and other respiratory difficulties are removed by filters or by using a wet space. The particle of pollen grains will adhere to these wetted surfaces.

Bacteria present in the air is removed either by filter chemicals or by using a sterilising light of ultraviolet rays. For proper air sterilisation special ultraviolet ray lamps of radiation 2,600 Angstrom units are used. These are generally placed near the outlet grille on its inside end. They are shielded properly to avoid exposure of skin or eyes. Ultraviolet lamps are largely used in meat packing houses and storage rooms to avoid the growth of bacteria. Fine powder of propylene or triethylene glycol is also sprayed in the air to kill bacteria.

To remove dust particles from the air four methods using equipments like centrifugal devices, water filters or air washers, adhesive impregnated filters, electrostatic precipitators are used.

9.11. COMFORT AIR CONDITIONING

This is the type of air conditioning whose objective is to produce comfortable and healthy conditions for occupants. In other words, it is the process of treating air so as to control simultaneously the prevailing conditions of temperature, humidity, cleanliness and distribution in the conditioned space to meet the comfort and health requirements of the occupants. It includes the air conditioning of spaces, places and surroundings which are occupied by human beings.

Some of the important places intended for human occupancy and requiring comfort air conditioning are the residences, departmental stores, public buildings including theatres, cinema halls, dancing halls, skating rings, restaurants and night clubs, hospitals etc. In order to understand the comfort air conditioning the human requirements should be known.

Human Requirements. Our body is thermostat with inner temperature of 98.4°F (37°C) and outer temperature of about 26.5°C. Under normal conditions our body emits 2,500 k cals per 24 hours; 45% of which is dissipated by radiation, 30% by convection and remainder 25% by evaporation. Heat dissipated by radiation, convection and evaporation is governed by surrounding temperature, air temperature and humidity of air. It is necessary for our body to dissipate the above proportion of heat in order that we may feel convenient and comfortable. For producing comfort it is necessary to maintain surrounding atmosphere at a temperature between 22°C to 25°C during summer and 17°C to 20°C during winter and humidity between 40 to 60 per cent. It is not desirable however, to maintain more than a 8 to 11°C difference between indoor and outdoor temperatures since a higher temperature difference will cause discomfort when a person goes out of a room and comes back.

In *Indian Railway air-conditioned coaches the temperature is maintained at 24°C during summer and 20°C during winter, as these are most accepted temperature by passengers.*

9.12. INDUSTRIAL AIR CONDITIONING

A large number of manufacturing, inspection, measuring processes require a close control of temperature and relative humidity. Industrial air conditioning is generally defined as the process of simultaneously controlling the temperature, relative humidity, ventilation, cleaning and movement of air within an enclosure intended for the manufacture, inspection and measurement of a product which is sensitive to atmospheric conditions.

It covers a large number of industries requiring controlled air conditions for their efficient and accurate manufacture as well as healthy and comfortable environments for the workers. Air conditioning is very essential for the manufacture of rayon and various plastics, paper, colour printing, in tobacco industry, in the precision manufacturing and measurement works for year-round production, textiles, felt goods, leather, electrical equipment, photographic material manufacturing, candy etc.

9.13. EFFECTIVE TEMPERATURE

Comfort requirements for human beings excluding radiation, air odours and cleanliness generally depend upon the dry- and wet-bulb temperatures of the air and its movement rate. An arbitrary index which combines in a single value the degree of sensation warmth or coolness felt by the human body due to the effect of air, temperature, humidity and air movement is called the *effective temperature.* Its numerical value cannot be measured directly but is taken as that temperature of the saturated still air (at a velocity of 4.5 to 7.5 metres per minute) which induces the same sensation of warmth or coolness as those created by the air surrounding a person. An atmospheric condition will have an effective temperature of 20°C if it induces the same sensation of warmth or coolness as created by saturated still air at 20°C.

From the investigations and results of various studies it has now been proved that the effect of humidity shown by effective temperature scale are not great if the air temperature and the human activities are such that very little heat is lost by evaporation of the perspiration. For an individual doing an activity in sitting position with air temperature between 23°C and 25°C a relative humidity in the range of 30 to 65% will have no effect on his comfort. At the same temperature, a higher rate of activity may result in a sensation of warmth but a relative humidity of 40% would be more comfortable than 60% relative humidity under similar conditions. In case of a large number of persons doing different activities like sitting, playing, dancing etc. in the same space, a 23.5ºC temperature with 40% or less relative humidity generally provides most comfortable condition for all.

9.14. SUMMER AIR CONDITIONING SYSTEM

It is the most important type of air conditioning, in which air is cooled and generally dehumidified. Figure 9.10 shows a schematic arrangement of a typical summer air conditioning system.

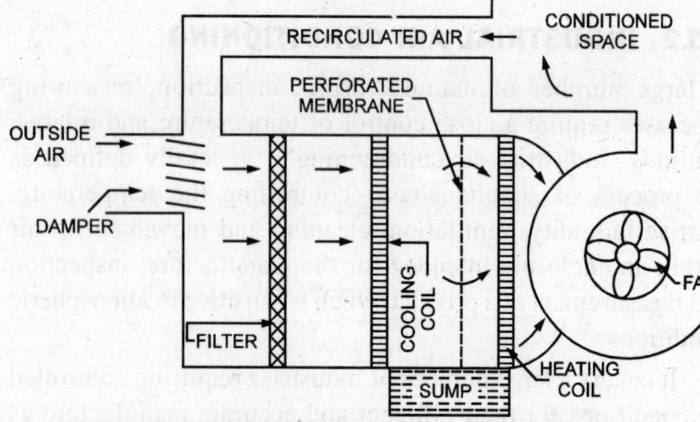

Fig. 9.10. *Summer Air Conditioning System*

The outside air flows through the damper, and mixes up with recirculated air (which is had from the conditioned space). The mixed air passes through a filter for removing dirt, dust and other impurities. The air now passes through a cooling coil, which has a temperature much below the required dry bulb temperature of air in the conditioned space. The cooled air passes through a perforated membrane and loses its moisture in the condensed form which is collected in a sump. After that, air is made to pass through a heating coil to heat up the air slightly. This is accomplished for bringing the air to the designed dry bulb temperature and relative humidity.

Now the conditioned air is supplied to the conditioned space by a fan. From the conditioned space, a part of the used air is exhaused to atmosphere by the exhaust fans or ventilators. The remaining part of the used air, known as *recirculated* air, is again conditioned as illustrated in Fig. 9.10. The outside air is sucked and made to mix with the recirculated air in order to make up for the loss of conditioned (or used) air through exhaust fans or ventilators from the conditioned space.

9.15. WINTER AIR CONDITIONING SYSTEM

In winter air conditioning, the air is heated, which is usually accompanied by humidification. The schematic arrangement of winter air conditioning system is shown in Fig. 9.11.

In this system, the outside air flows through a damper and mixes up with the recirculated air coming from the conditioned space. The mixed air is passed through a filter for removing dirt, dust and other impurities. The air is now passed through a preheater so as to avoid the possible freezing of water and to control the evaporation of water in the humidifier. After that, the air is made to pass through a reheat coil to bring the air to the designed dry bulb temperature. Now, the conditioned air is supplied to the conditioned space by fan. From the conditioned space, a part of the used air is exhausted to atmosphere by the exhaust fans or ventilators. The remaining used air is again conditioned as shown in Fig. 9.11. The outside air is sucked and made to mix with recirculated air, in order to make up for the loss of conditioned (or used) air through exhaust fans or ventilators from the conditioned space.

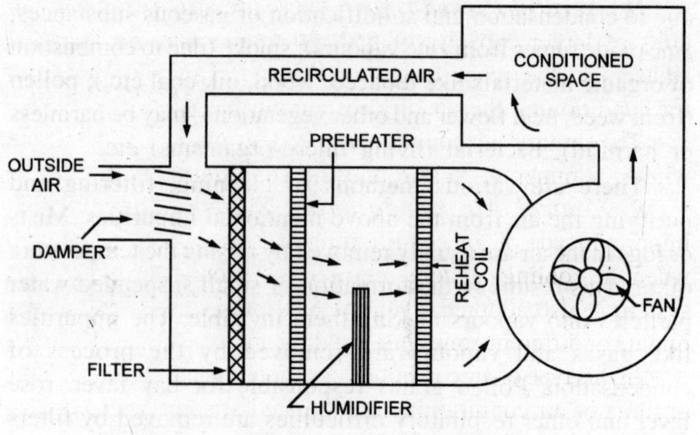

Fig. 9.11. *Winter Air Conditioning System*

9.16. YEAR-ROUND AIR CONDITIONING SYSTEM

The year-round air conditioning system has equipment for both the summer and winter air conditioning. The schematic arrangement of a modern year-round air conditioning system is shown in Fig. 9.12.

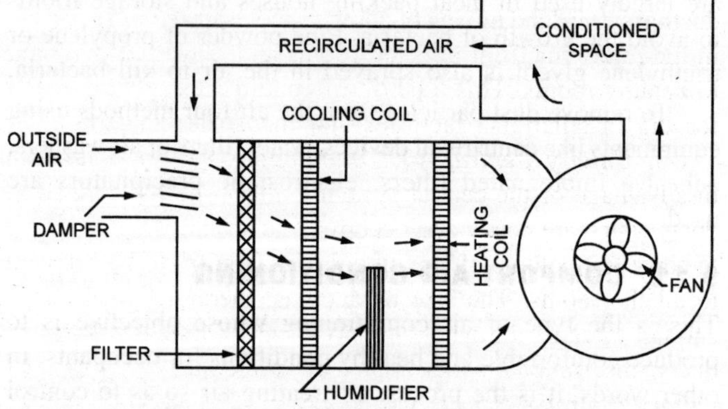

Fig. 9.12. *Year-around Air Conditioning System*

The outside air flows through the damper and mixes up with the recirculated air coming from the conditioned space. The mixed air is passed through a filter so as to remove dirt, dust and other impurities. In summer, the cooling coil is operated to cool the air to the desired value. The dehumidification is obtained by operating the cooling coil at a temperature below than the dew point temperature (apparatus dew point). In winter, the cooling coil is made inoperative and the heating coil is operated to heat the air. The spray type humidifier is also used in the dry season to humidify the air.

9.17. TYPES OF AIR CONDITIONING SYSTEMS

There are three types of air conditioning systems.

1. ***Unitary Air Conditioning System.*** In this system, factory assembled air conditioners are installed in or adjacent to the space to be conditioned. The unitary air conditioning systems are of the following two types.

(a) Window Units. These are self-contained units of small capacity of 1 tonne to 3 tonnes, and are mounted in a window or through the wall. They are employed to condition the air of one room only. If the room is of bigger size, two or more units are installed. Now smaller units of 1 tonne, even 3/4 tonne are also available.

(b) Vertical Packed Units. These are also self-contained units of larger capacity of 5 to 20 tonnes and are installed adjacent to the space to be conditioned. This is very useful for conditioning the air of a restaurant, bank or small office.

The unitary air conditioning system may be adopted for winter, summer or year-round air conditioning.

2. *Central Air Conditioning System.* This is most important type of air conditioning system, which is adopted, when the cooling capacity required is 25 tonnes or more. This system is also adopted when the air flow exceeds 300 m^3 per minute or different zones in a building are to be air conditioned.

In this type of air conditioning various essential components, or units of air conditioning are situated at some central place usually in basement from where conditioned air is led through sheet metal ducts to the various rooms or space of building. There are return ducts for carrying air from these rooms back to the central place where it is dehumidified, cooled and recharged with fresh air. The circulation of air is achieved either naturally or by a fan. Advantages of this type of air conditioning system are high efficiency of operation and robustness of the equipment. The main disadvantages are absence of any adjustment of individual room temperature, necessity of costly ducting system and mixing of odours, smoke and bacteria present in the return air from infected rooms and redistributing them to healthy rooms. This type of air conditioning is best suited for big installations, factories and industrial concerns.

3. *Unitary Central Air Conditioning System.* The main disadvantage of the central system is that it needs large sized ducts which are costly and occupy large space. The ducts have to carry full volume of air (outdoor plus return) to be supplied to all the rooms. The size of ducts is, therefore, made large enough to carry the whole supply of air, Similarly for return ducts, the same remarks apply. To overcome this problem, *unitary central system* is used.

In this type of air conditioning the features of both the central type and unitary type air conditioning systems have been combined in order to use it for all purposes and to make greater life expectancy. In this type of air conditioning heating and cooling equipments are installed in the central room. The various unit air conditioners in different rooms are supplied with heat energy (in the form of steam or hot water) or chilled water through the pipes. The air conditioning unit in this case comprises of heating and cooling coils, filters, blowers, dampers, valves, control and other associated equipment. All the disadvantages of central type are done away with. There will be no return air duct, no mixing of odours and bacterial of one room and redistribution of the same among other rooms. Temperature of each space unit can be controlled by room unit. Fresh air duct required to supply only 15% fresh air, rest of 85% air being recirculated, will be of very much small size. Fresh air introduced will displace corresponding quantity of air which will escape through openings etc.

9.18. ROOM AIR CONDITIONERS

It is a common type of air conditioner used to condition the air of a particular space occupied by human beings *i.e.*, office room, residential room etc. It has automatic operation to cool and humidify the air.

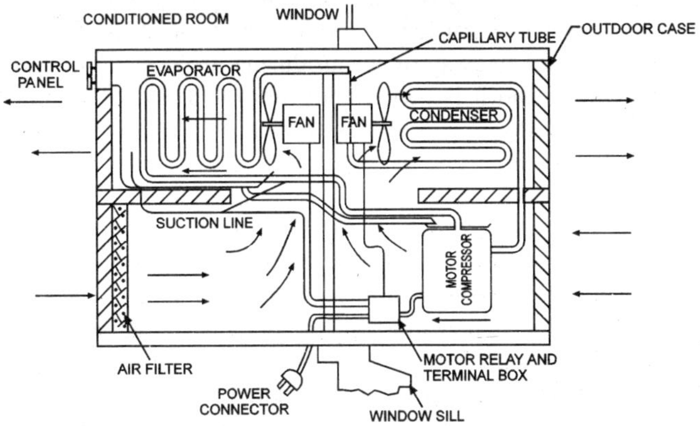

Fig. 9.13. ***Room Air Conditioner (Window Type)***

A window type air conditioner is shown in Fig. 9.13. It consists of a case divided into two parts by a partition with a small opening at the top as the outdoor part and indoor part. The outdoor portion consists of a hermetically sealed motor compressor unit, condenser, motor driven fan and a tray. This portion partitioned by a portion L into two parts is provided with a left hand side opening.

The indoor portion consists of evaporator, motor driven fan, remote bulb, refrigerant, control, a control panel, an air filter, power connector and a tray. This portion is further subdivided into two parts by opening on the right hand side. A pipeline connects the two trays in the inner and outer parts. A capillary line control through a refrigerant filter connects the condenser to the evaporator. Evaporator is connected to the compressor by a suction pipeline. These units are usually fitted flush with window still on inside having attractive frontage. On outside it is supported on brackets and projects beyond the wall. The front and back of the inner and outer portions of the cooler is fitted with shutters. These shutters are adjustable at different inclinations according to the requirements.

The working substance is a vapour which readily evaporates and condenses. The most popular substances are ammonia, freon, carbon dioxide etc.

As the unit is put into operation, the low pressure vapour through the suction pipeline is drawn from the evaporator and passed to the compressor. The compressor delivers it at high pressure to the condenser. In the condenser, the vapour gets condensed and the heat is removed from the refrigerant vapour. The liquid refrigerant collected at its lower coils is passed through the filter into the capillary tube control before flowing back at low pressure into the evaporator coils.

The low pressure liquid refrigerant inside the evaporator coils quickly picks up from its surface. The air is drawn from inside the room through a filter from the lower portion of the unit by a motor driven fan. This air is passed over the evaporator coils. The cooled air from the evaporator coils surface is supplied back into the room. Due to dehumidification of moisture from the circulating air, the moisture after flowing downward from the surface of the evaporator coils is collected at the bottom of the evaporator into a pan. This moisture in the pan flows due to gravity into another pan placed in the outdoor portion at the bottom of the condenser. The moisture helps to cool the compressor and condenser due to evaporation.

In the compressor and condenser compartment, the fan draws in the outside air from the lower portion of the unit, circulates it over the condenser and discharges from the upper portion of unit to the outside due to mounting of the compressor and condenser in a particular design. The heat of the vapour refrigerant is taken away by the air during its passage through the condenser and it is cooled into the liquid form. The unit automatically stops as soon as the desired temperature is reached inside the room. The automatic operation of the unit is accomplished by a valve at the control panel.

Main advantages of room air conditioners are given below :

1. Saving in installation and field assembly labour.
2. Exact requirement of each separate room is met whereas a central system cannot meet the individual needs of separate rooms.
3. Zoning and duct work is eliminated.
4. Only those rooms which need cooling will have their units running, whereas the central plant will have to operate all the time for the sake of only a few rooms.
5. Failure of unit affects a single room where as all the rooms are affected when failure occur in the central system.
6. Low initial cost.
7. Flexibility of operation.
8. Starting with only a few rooms, the air conditioning may be progressively extended to other rooms without spending a large sum from the very beginning.

9.18.1. Comparison Between Desert Cooler and Window Type Air Conditioner

Particulars	Desert Cooler	Window Type Air Conditioner
Initial Cost	Low	Quite high
Main Components	Motor and fan	Complete refrigeration unit: compressor, condenser, evaporator, and controls etc.
Humidity Control	Not possible	Possible
Power Required Per Unit Cooling	Less	More
Location	Inside or in the window of the room.	Fitted either in the window or wall.
Utility	Can be used to cool the air and that too in the dry season.	Can be used for both cooling and heating purposes.

9.19. CENTRAL AIR CONDITIONING SYSTEMS

The central air conditioning systems serving the needs of large building or spaces are generally called the year-round air conditioning systems. They have a cooling capacity of 25 tonnes and circulate about 300 m^3/minute of conditioned air. Depending upon its type such a system may contain some of the following equipments :

(*a*) Heating coils supplied with steam or hot water.

(*b*) Cooling and dehumidifying coils—both for chilled water and for direct expansion of the refrigerants.

(*c*) Blower and driving motor.

(*d*) Sprays for cooling and dehumidifying or for washing the air.

(*e*) Air-cleaning equipment containing filters, electrostatic precipitators, odour removing equipment and germicidal lamps.

(*f*) Various types of control devices.

The devices assembled as an air-conditioning unit are housed and enclosed in sheet metal housings or plenums. The plenum is usually covered with insulating board over its entire outer surface and is stoutly braced. The refrigeration condensing units and steam or hot water boilers for air conditioning in this system are either located in the same room or in the basement or in an adjoining room.

The various factors affecting the choice of the equipment for design and installation of the central air-conditioning systems are capacity of the plant, filters and mixing plenums, refrigeration and heating equipment, insulation, noise and vibration.

1. *Capacity of the plant.* For each 30 m^3/minute of air handled, heating and cooling coils handling air must provide 0.185 m^2 of face area (four row coil) at face velocities of about 150 m/minute. One of the limiting factors influencing the design of this system is the physical size of the coil and filter banks.

The horizontal space between flows of a building is usually kept between 3.65 and 4.87 metres because the operating expenses of the system depends upon the floor area utilised. Thus for a single air conditioning unit a practical limit of 1,200 m^3/minute or 100 tonnes of cooling capacity is set. For the spaces or buildings requiring more than this capacity two or more zones may be planned with an air conditioning system installed to handle each unit.

The extent of the zoning to be specified is generally determined by the nature of the building. A theatre or departmental store installation requiring spaces of large cubical structure are most economically conditioned by only a few (preferably one) central station air conditioning units. Office buildings, schools, colleges and similar buildings requiring many rooms necessitate greater zoning. Air conditioning units of 300—900 m^3/minute are usually employed. Considering all the factors discussed above, it seems more economical to supply chilled water, hot water, steam or refrigerant through pipes to various smaller air conditioning units than erect massive supply and return ducts for serving an extremely large unit.

2. *Filters and mixing plenums.* Before filtering, the return air and fresh outside air should be brought into a plenum for proper mixing. The amount of outside air to be introduced is usually controlled by motorised dampers. Depending upon the nature of occupancy, the dampers are controlled by a thermostat to admit less outside air during extreme outdoor temperature or by smoke and odour sensing devices to increase the amount of ventilation air according to requirement in the conditioned space.

Vertically assembled and cleanable types of filter banks are generally used. For central systems of less than 300 m^3/minutes capacity disposable filters are used. The various types of filters and air cleaning devices commercially available accomplish all the following cleaning operations.

(*a*) Removal of solid particles of visible size (dust, dirt, lint, pollen, cinders etc.)

(*b*) Removal of particles of microscopic size (smoke, haze, etc). Electrostatic precipitators are used for this purpose.

(*c*) Proper control of odour— for this purpose activated charcoal is usually used.

(*d*) Killing of bacteria—for this purpose germicidal lamps and certain chemicals are used.

3. *Refrigeration and Heating Equipment.* The design of central station system is greatly affected by these equipments. The direct-expansion (DX) refrigerant coils and hot water coils are unusually employed in systems requiring only one or two zones. The air conditioning units in these systems are generally located close to the cooling and heating equipments. The chilled water is used and circulated as a cooling medium in the air conditioning coils for many zones at remote locations which are served by a single cooling and heating plant. The heating medium is generally the hot water or steam. For serving the various zones water is generally pumped to the mechanical rooms.

4. *Insulation.* If the supply ducts are not running through the conditioned spaces, they should be properly insulated. Generally 25 mm thick insulating board is used to insulate the outer surface of the entire housing for the central station air conditioning unit. In order to prevent the heat losses and gains, all the steam, hot water, chilled water and suction refrigerant piping should be properly insulated.

5. *Noise and Vibration.* The noise and vibrations produced in the air conditioning units are very important factors for considerations while designing and selecting equipments. The undue noise and vibrations are considered undesirable.

9.20. HEATING OF BUILDINGS

The purpose of most forms of space-heating installation is to provide comfortable conditions in which people can work or relax. Space heating is far from being an exact science because opinions vary considerably as to what constitutes a comfortable environment. It is, however, reasonable to say that in order to be comfortable the feet and head should be at about the same temperature. The head should not be subjected to intense radiation and there should be as little draught as possible. Under these circumstances an ambient temperature between 20°C and 22°C will be acceptable to most people.

9.20.1. Factors Affecting The Choice of Heating Systems

The type of heating equipment used is influenced by several considerations. Probably the most significant of these is whether heat is to be required for long or short periods. *Short-term heating* generally relies on radiation which can be very effective in no time. Convected heat is best suited to *long-term* heating in which the time taken to 'warm up' is not so important.

The construction of building and the conditions existing in it are important. Where ceilings are low and changes of air are limited it may be practical to heat all the air in the room. If on the other hand the air is changing continually or there is a very high ceiling some sort of directional heating may be required.

Some forms of heating can only be installed effectively and economically during the construction of a new building whilst other may be used to improve an existing building.

Complete or Partial Heating Systems. In the past when coal or wood fires were the normal heat source it was customary to treat each room as a separate room. This persisted for many years after electricity was first used for space heating and in many cases a portable electric heater simply took the place of an open fire. This approach to heating produced result that tended to be both uncomfortable and uneconomical. Heat being continually lost from warm parts of the building to the cold and the wide differences of temperature experienced preventing all parts being used to equal advantage. Better results are obtained by designing a heating installation to suit an entire building. With this is done, a comfortable temperature can be achieved throughout the building and the points at which heat losses occur can be readily identified and the necessary steps taken. These steps would normally include double glazing of windows, draught-sealing doors and insulating the roof space with glass wool or some other thermal insulant.

One- and Two-Level Heating. Having decided to design a comprehensive heating system for a building another choice must be made. This is whether to settle upon a single value of temperature and heat the entire building to it or not. In office blocks and factories where every effort is made to get the maximum use out of the available floor space there is little waste involved in heating to the same temperature throughout.

In domestic installations the amount of use given to different parts varies largely and also there are considerable differences in levels of activity from one time of the day to another. To allow for such variations a two-level heating system is often employed. In such a case general heating is provided to a fairly low comfort level in the region of 15°C. This is called the *background heating* and keeps all the house or flat comfortable for housework or similar activities. Additional appliances are available, controlled by the consumer, to raise the temperature of specific areas a further 5° or 6° when he or she is sitting still for any length of time. This is called *topping up.* Background heating is usually provided by convection heaters and radiant heaters are used for topping up.

9.20.2. Types of Heating Equipments Used For Space Heating

The different types of equipments used for heating of buildings are given below :

1. *High Temperature Radiators.* The high temperature radiator essentially consists of a high resistance element wound on porcelain or refractory former and mounted into decorative frames. Resistance elements get heated to high temperature (1,400°C to 1,600°C) due to flow of electric current through them and 50% to 70% of heat so produced is dissipated by radiation and remaining by convection. Advantages of this type of heater are that it is portable, cheap in initial cost, simple in construction and attains its highest temperature quickly. The main drawback of this type of heater is that heat dissipated is localised. These are used for domestic purposes or where intermittent heating is required. The usual sizes are from 0.5 kW to 3 kW.

2. *Tubular Heaters or Low Temperature Convectors.* The most common type of low temperature heater is 5 cm diameter steel tubes of length ranging from 0.6 to 5 metres and containing resistance elements arranged on mica, fire clay, or porcelain former. Such tubes are mounted around the skirting board of room to be heated. The normal loading is 200 watts per metre length of the convector. The temperature of the external surface is maintained at about 93° C. About 90 per cent of the heat is transferred by convection. These are spaced to produce convection currents in the room so as to heat whole of it. These are specially suitable for thermostatic control, by which means they are switched on and off automatically as the room temperature falls below or rises above the desired value. When fixed at the base of a cold wall or underneath a window they prevent cold down draughts, which cause cold feet to the occupants of the room.

3. *Convector Heaters.* In these heaters, heating elements are mounted within sheet-metal cases which admits cool air from bottom. Air after getting heated up, comes out from grills at the top. The front and top may be of moulded plastic material for the sake of pleasing appearance. These heaters may be either portable or fixed, and may be handled without fear of burns. In these type of heaters, heating element attains low temperature of 66° to 93°C and proportion of heat transmitted by convection is large as compared to heat transmitted by radiation.

A very satisfactory method of heating a room is to use convector heaters to warm the room to about 13°C, together with a small radiant heater for local heating.

4. *Panel Heaters.* In this system of heating large panels of fire clay or some other suitable heat resisting materials are used along the walls or ceiling of a room. The resistance elements are embedded in these panels. The heat produced is utilised in the form of convection or radiations. This system of heating of buildings is advantageous from the decoration point of view and gives a uniform distribution of heat. Loadings are of the order of 400 watts/m^2 for the low temperature panels and 6,000 watts/m^2 for the high temperature panels. Low temperature panels operating at temperature from 27°C to 65°C may be fixed in walls or ceiling of a room and plastered over. High temperature panels operating at temperature up to 288°C are used on walls and ceiling but are not fixed in the wall surface. They are sometimes suspended from the ceiling.

5. *Underfloor Heating.* This is one of the systems that may be installed in a building whilst it is being built.

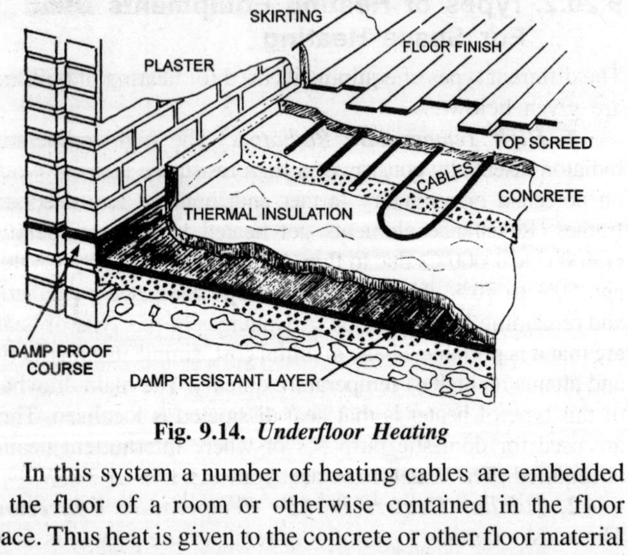

Fig. 9.14. *Underfloor Heating*

In this system a number of heating cables are embedded in the floor of a room or otherwise contained in the floor space. Thus heat is given to the concrete or other floor material slowly and heat from the floor is used in warming the air in the room, and in warming the walls and the furniture. The electrical power intake is controlled by thermostat fixed on the room wall in a suitable position. The cables used for heating are made from alloys of nickel, chromium, iron or other resistance materials and are insulated with asbestos mineral insulation, butyl rubber, or other heat-resisting insulating materials. For mechanical protection lead, copper, aluminium, or PVC sheathing can be used. Loading of from 100 to 150 watts per metre^2 of floor area are generally satisfactory.

In this system also the advantage of cheaper power available during night hours can be taken but in such a case it will be desirable that one or more convectors or radiant heaters for use towards the end of the floor-warming output cycle are installed.

6. *Thermal Storage System.* Roughly this system is similar to well-known central-heating system employing a coal- or coke-fired boiler except that the heat is supplied to the storage tank electrically. Hot water is circulated through the heating pipes and radiators situated throughout the building. It must be noted that in electrical heating system the tank and piping must be lagged to prevent loss of heat, a feature which is usually omitted with fuel-fired systems, since with such systems the efficiency is so low that the additional loss due to lack of lagging is relatively insignificant. There are two types of heaters used in this system : the immersion heater for a load up to 100 to 200 kW and the electrode boiler type for loads exceeding 200 kW.

The *immersion heater* used for heating the water in a storage tank consists of a spiral of resistance wire embedded in refractory insulating medium like magnesium oxide inside a copper or steel tube, which is bent to the desired shape. The tube is provided with a flange at one end, so that it can be bolted to the vertical side of the tank with the heating element projecting horizontally into the water. The loading of such a tube is usually between 16 and 24 kW/m^2 of face, so that a tube of length 305 mm and diameter 51.0 mm will be rated at about 1 kW. Single tubes in ratings of 4-5 kW are available. For higher ratings group of tubes are used.

The *electrode boiler* is only suitable for use on ac, usually three phase, since dc would electrolyse the water. In electrode boilers water is heated by placing cast iron electrodes directly in the water. For 3-ϕ supply three electrodes are placed in the water and the tank is earthed so as to form a star-connected system. The resistance of average cold tap water varies between about 20 and 50 Ω-m. Sodium carbonate (soda) is, therefore, added for increasing the conductivity of water. Another noteworthy point is that resistance of water decreases with the increase of its temperature, the value at 100°C being one-third of that at 20°C. Automatic control of the power supply is, therefore, required if a constant input is required. Control of power input may be carried out by varying the height of the electrodes in the water, by means of a movable shield surrounding the electrodes, by moving one electrode relative to the other, or in large units by varying the water level by means of a pump. A low-voltage boiler operating at 400 V is usually employed for a requirement up to 500-700 kW while for larger requirement a high-voltage boiler operating at voltages up to 11,000 V or even higher is used.

The advantage of this system is that the water can be heated during night (off-peak periods) and is supplied throughout the day, thus use of cheap power available during off-peak period can be made. This system of heating also provides hot water service for the building.

9.21. CALCULATION OF RATING OF ELECTRICAL EQUIPMENT

The amount of heat required for personal comfort in an enclosed space such as a room or office depends upon the following factors :

1. Number of changes of air per hour.
2. Area of windows.
3. Situation of walls, external or internal (most rooms have at least one external wall).
4. The exposure of the ceiling (open to the roof or with heated bedroom over).

5. Material of which walls, floor and ceilings are composed.
6. The temperature of outside air and
7. Whether the building is to be heated continuously or intermittently.

The method of calculations will be clear from the following examples.

Example 9.1. It is proposed to condition 3,000 m^3 of air per hour from a temperature of 5° C to 20° C. It is further necessary to evaporate 5 kg of moisture per 1,000 m^3 of air per hour to control humidity. Estimate the power required if heat required to raise the temperature of 1 m^3 of air through 1° C is 1,220 J and latent heat of evaporation is 2,450 × 10^3 J/kg.

Solution : Heat required to raise the temperature of air,

H_1 = Volume of air to be conditioned × heat required to raise the temperature of 1 m^3 of air through 1°C × difference of temperatures

$= 3{,}000 \times 1{,}220 \times (20 - 5)$ Joules $= 54.9 \times 10^6$ J

The moisture present in the air $= \dfrac{5 \times 3{,}000}{1{,}000} = 15$ kg

Latent heat required to evaporate the moisture

$$H_2 = 15 \times 2{,}450 \times 10^3 = 36.75 \times 10^6 \text{ J}$$

Total heat required, $H = H_1 + H_2$

$= (54.9 + 36.75) \times 10^6 = 91.65 \times 10^6$ J/hour

$$\text{Power required} = \frac{91.65 \times 10^6}{3{,}600 \times 1{,}000} \simeq 25 \text{ kW } \textbf{Ans.}$$

Example 9.2. A drawing office 6 m × 5 m × 4 m size is to be heated. The temperature of the room is to be kept at 25°C when the outside temperature is 5°C. The air is to be renewed every 15 minutes. Find the necessary rating of the electric heater. The heat loss from the walls is 836 × 10^3 J/°C/hour. Assume heat required to raise temperature of 1 m^3 air is 1,220 J.

[*Pb. Technical Univ. Electric Drives & Utilization Dec. 2008*]

Solution : Volume of air to be conditioned

$$= 6 \times 5 \times 4 \times \frac{60}{15} = 480 \text{ m}^3\text{/hour}$$

Heat required to raise the temperature of 1 m^3 of air through 1°C = 1,220 J

Heat required to raise the temperature of 480 m^3 air through (25 – 5), *i.e.*, 20°C

$H_1 = 480 \times 1{,}220 \times 20 = 11.712 \times 10^6$ J/hour

Heat losses from the walls $= 836 \times 10^3 \times 20 \times 1$

$= 16.72 \times 10^6$ J/hour

Total heat required, $H = H_1 + H_2$

$= 11.712 \times 10^6 + 16.72 \times 10^6$

$= 28.432 \times 10^6$ J/hour.

$$\text{Rating of heater} = \frac{28.432 \times 10^6}{3{,}600 \times 1{,}000} \simeq 8 \text{ kW } \textbf{Ans.}$$

Exercises

1. Some people have an idea that refrigeration is luxury but others feel it is a necessity. What is your opinion in respect of this particular controversial point. Give some concrete examples in support of your opinion.
2. Give a diagram describing a basic refrigeration system. **[Pb. Technical Univ. Electric Drives & Utilization May 2008]**
3. Write about vapour compression cycle in refrigerator. **[Pb. Technical Univ. Electric Drives & Utilization Dec. 2006]**
4. Enumerate the successive operations to be performed on a refrigerant in (*i*) a compression system and (*ii*) an absorption system. Why is ammonia the only commonly refrigerant used in the absorption system ?
5. Discuss in details
   (*a*) Vapour compression refrigeration system.
   (*b*) Vapour absorption refrigeration system.
   (*c*) Thermoelectric refrigeration system.
   [Pb. Technical Univ. Electric Drives & Utilization Dec. 2009]
6. Give the merits and demerits of vapour absorption system over vapour compression system.
7. Enlist the main requirements of a good refrigerant. What are the primary and secondary refrigerants ? Name the refrigerants generally used.
8. Explain the working principle of thermoelectric refrigeration system. List out the merits and demerits of the thermoelectric refrigeration system over other refrigeration systems. What are the field of its application ? **[Pb. Univ. Electric Drives and Utilization of Electric Energy August 1984]**
9. Draw the electric circuit of a refrigerator and explain its working. How can temperature inside the refrigerator be adjusted ? **[Pb. Technical Univ. Electric Drives & Utilization May 2009]**
10. Explain vapour compression refrigeration cycle and show how cooling effect is produced ? Describe electric circuit of a refrigerator. **[Pb. Technical Univ. Electric Drives & Utilization Dec. 2007]**
11. With the help of a circuit diagram, explain the working of a water cooler.
12. What is air conditioning ? On what factors it depends ? **[Pb. Technical Univ. Electric Drives & Utilization Dec. 2008]**
13. Define air conditioning. On what factors does the air conditioning depend ? Explain in detail. **[Pb. Univ. Elec. Drives & Utilization of Elec. Energy April 1981]**
14. Discuss the role of air conditioning in our day-to-day life.
15. What are the different conditions which constitute human comfort ? **[Pb. Univ. Elec. Drives & Utilization of Elec. Energy April 1981]**
16. Differentiate between comfort air conditioning and industrial air conditioning. **[Pb. Univ. Elec. Drives & Utilization of Elec. Energy April 1982]**
17. What is air conditioning ? How is air from micro-organism, gaseous contaminants and odours purified ? **[Pb. Univ. Elec. Drives & Utilization of Elec. Energy Nov. 1990]**
18. Describe in detail functioning of window air conditioner. **[U.P. Technical Univ. Utilization of Electrical Energy & Traction 2009-10]**
19. Explain with the help of diagram the working of window type room air conditioner. **[Pb. Technical Univ. Electric Drives & Utilization Dec. 2009]**
20. Explain the working of central air conditioning system. **[Pb. Technical Univ. Electric Drives & Utilization May 2009]**
21. Describe complete arrangement of centrally air conditioning plant. **[Pb. Univ. Elec. Drives & Utilization of Elec. Energy April 1982]**
22. Discuss the effects of temperature and humidity in the atmosphere on the human body from the view point of comfort. Define the effective temperature in air conditioning. Specify suitable values of effective temperature and relative humidity for summer as well as winter air conditioning of temperature and tropical climate assuming approximately still air.
23. Discuss different methods of heating of buildings. **[Agra Univ. Traction & Utilization Elec. Power 1982]**
24. What are different space heating systems ? Discuss them briefly. **[Pb. Univ. Elec. Drives & Utilization of Elec. Energy Nov. 1981; April 1981]**
25. What are different space heating systems ? Discuss them. **[Pb. Technical Univ. Electric Drives & Utilization May 2008]**
26. Write short notes on the following :
    (*i*) Thermoelectric refrigeration. **[Pb. Univ. Nov. 1983, April 1983]**
    (*ii*) Refrigerator **[G.B.T.U. 2010-11]**
    (*iii*) Water cooler **[G.B.T.U. 2010-11]**
    (*iv*) Air conditioning

(*v*) Effective temperature **[G.B.T.U. 2010-11]**
(*vi*) Window air conditioner **[G.B.T.U. 2010-11]**
(*vii*) Room air conditioner **[Pb. Univ. Nov. 1983, April 1983]**
(*viii*) Heating of buildings. **[Pb. Univ. Nov. 1983, April 1983]**
(*ix*) Heating of rooms and buildings **[Agra Univ. 1983]**
(*x*) Space heating systems. **[Pb. Univ. April 1980]**

Short Answer Type Question With Answers

Q. 1. What is practical unit of refrigeration ? Define it.

Ans. Tonne of refrigeration (TR) is the practical unit of refrigeration and is defined as the amount of refrigeration effect produced by the uniform melting of one tonne (1,000 kg) of ice from and at 0ºC in 24 hours.

Q. 2. Draw the neat sketch of refrigeration cycle. **[P.T.U. Dec. 2009]**

Ans. Refer to Fig. 9.1.

Q. 3. What are the applications of vapour compression refrigeration cycle ?

Ans. The vapour compression refrigeration cycle is used in most of the modern refrigeration systems in large industrial plants requiring rapid removal of large amount of heat and in most of the domestic refrigerators.

Q. 4. On what principle does thermoelectric refrigeration system operate ?

Ans. Thermoelectric refrigeration system operates on the principle of 'Peltier effect'.

Q. 5. What are the desirable properties of refrigerants ? **[P.T.U. Dec. 2009]**

Ans. For safe, economical and thermodynamically efficient production of cold, refrigerant should possess as many as practicable of the following desirable properties.

1. Low boiling point.
2. High critical temperature.
3. High latent heat of vaporisation.
4. Low specific volume of vapour.
5. Low specific heat of liquid.
6. Low cost.
7. Easy to liquify at moderate pressure and temperature.
8. Easy of locating leaks by odour or suitable indicator.
9. It should be noncorrosive.
10. It should be noninflammable.
11. It should be non-toxic.
12. It should be free from objectionable odour.

Q. 6. Why is ammonia the only common refrigerant used in the absorption system ?

Ans. Because ammonia is cheap, readily available and has the highest efficiency.

Q. 7. Why thermostat is used in a domestic refrigerator ?

Ans. Thermostat switch is provided in a domestic refrigerator to control the temperature inside a refrigerator.

Q. 8. With the aid of a sketch of a thermostat explain how the temperature inside a refrigerator is adjusted. **[P.T.U. Dec. 2006]**

Ans. Thermostat is a thermal switch that is provided to control the temperature inside the refrigerator. Temperature inside the refrigerator can be adjusted by means of temperature control screw. Greater the distance between contacts, greater will be the temperature inside the refrigerator and vice versa.

Q. 9. Why it is not advisable to put very hot things into the refrigerator ?

Ans. Main precaution to be taken is that very hot things should not be put in refrigerator, if it is done it will quickly evaporate the refrigerant in evaporator coils producing a large vapour pressure resulting in increase in the duty of the compressor and as such motor has to work for longer duration in order to bring down the temperature in the freezer compartment. It may damage the motor, which is short-time rated. Also moisture released from hot things will freeze around the freezer compartment which has to be defrosted.

Q. 10. Why is an automatic door closing mechanism provided in a modern refrigerator ? **[P.T.U. Dec. 2006]**

Ans. The door is provided with automatic closing mechanism– door hinges are provided in such a way that door flap when left in open position automatically comes to closing position due to gravity and as it approaches closing position it is attracted by the magnetic strip fitted behind the sealing rubber ring and thus the door is closed with snap action. All this is done to prevent leakage of atmospheric heat inside the refrigerator.

Q. 11. What will happen if automatic starter relay stops to operate in refrigerator ? **[P.T.U. Dec. 2007]**

Ans. Capacitor-start single phase induction motor is used in open type refrigerators and split-phase single phase induction motor is used in sealed unit refrigerator. Electromagnetic relay is provided to connect auxiliary winding on the start and disconnect it when the motor picks up the speed. If the automatic starter relay keeps the auxiliary winding open after the motor has been operating, little harm should occur since the motor is provided with overcurrent protection. The motor simply won't start on the next attempt. On the other hand, should the switch remain closed, the motor will start and run. The auxiliary winding will get quickly overheated, and if not detected, or cut out by a protective relay, this winding may cause damage to the main winding due to their proximity.

Q. 12. Mention the requirements of air conditioning. **[P.T.U. May 2008]**

Ans. In this modern age every body is trying to improve his surroundings in order to make living conditions more comfortable. Air conditioning is one of the important contribution in this respect. The technique of maintaining of air condition inside a building in such a way that produces comfort to human beings, ensures trouble free operation of delicate equipment or helps in industrial manufacturing operations is called *air conditioning.* It involves control of temperature, cleanliness, humidity and circulation of air.

Q. 13. What is comfort air conditioning ? **[P.T.U. May 2009]**

Ans. This is the type of air conditioning whose objective is to produce comfortable and healthy conditions for occupants. In other words, it is the process of treating air so as to control simultaneously the prevailing conditions of temperature, humidity, cleanliness and distribution in the conditioned space to meet the comfort and health requirements of the occupants. It includes the air conditioning of spaces, places and surroundings which are occupied by human beings.

Q. 14. Define the effective temperature in air conditioning ?

Ans. An arbitrary index which combines into a single value the degree of sensation warmth or coolness felt by a human body due to the effect of air, temperature, humidity and air movement is called the effective temperature.

Problems

1. In an air-conditioning plant it is required to raise the temperature of 2,000 m^3 of air per hour from 0ºC. It is further necessary to evaporate 0.5 kg of moisture per 100 m^3 of air per hour to control the humidity. The density of air may be taken as 1.3 kg/m^3, its specific heat 0.25 and latent heat of evaporation as 500 calories per gram. Estimate the power required.
[Pb. Univ. Elec. Drives & Utilization of Elec. Energy Nov. 1977]
[Ans. 13.4 kW]

2. A room measures 4 m × 4 m × 9 m. The air in it is to be renewed every 30 minutes and maintained at a temperature of 10ºC above that of incoming air. Find the necessary rating of an electric heater ignoring losses. Find the rating if the heat loss from the walls etc. is 3,00,000 calories per ºC per hour. Take for air : density 1.28 g per litre, specific heat 0.24.
[Pb. Univ. Elec. Drives & Utilization of Elec. Energy Nov. 1979]
[Ans. 4.5 kW]

❑❑❑

SECTION - C

ELECTRIC TRACTION

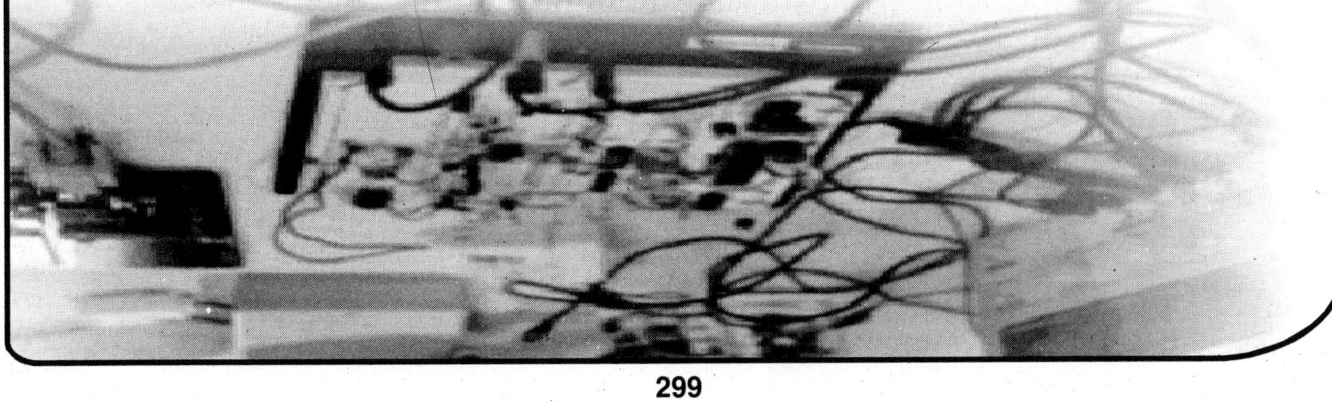

CHAPTER 10

Traction Systems

INSIDE THIS CHAPTER

10.1. Introduction **10.2.** Different Systems of Traction **10.3.** Systems of Electric Traction **10.4.** Systems of Track Electrification **10.5.** Comparison Between DC and AC Systems of Railway Electrification From The Point of View of Main Line and Suburban Line Railway Services.

10.1. INTRODUCTION

Act of drawing or state of being drawn; propulsion of vehicle is called the *traction* and the system of traction involving the use of electricity is called the *electric traction system.* There are various systems of traction prevailing in our country such as steam engine drive, internal combustion engine drive, diesel electric drive, battery electric drive, straight electric drive. These systems of traction may be classified broadly into two groups namely (*i*) the traction systems which do not involve the use of electricity at any stage and called the non-electric traction systems such as steam engine drive, internal combustion engine drive etc. (*ii*) the traction systems which involve the use of electricity at some stage or the other, and called the electric traction systems such as diesel electric drive, straight electric drive etc. Systems of electric traction may further be subdivided into two main groups: (*i*) the group consisting of vehicles which receive electric power from a distributing network fed at suitable points from either a central power station or substations suitably spaced such as tramways, trolley buses, electric railways etc. (*ii*) the group consisting of self-contained locomotives such as diesel electric trains and ships, petrol electric trucks and lorries, battery-driven road vehicles etc.

Each system of traction has got its own merits, demerits and fields of applications and no single system fulfills the requirements of an ideal traction system which are given below:

1. High starting tractive effort in order to have rapid acceleration.
2. Self-contained and compact locomotive or train unit so that it may be able to run on any route.
3. Equipment capable of withstanding large temporary overloads.
4. Minimum wear on the track.
5. Braking should be such that minimum wear is caused on the brake shoes, and if possible the energy should be regenerated and returned to the supply during braking period.
6. Equipment required should be minimum, of high efficiency, and low initial and maintenance cost.
7. No interference to the communication lines (telephone and telegraph lines) running along the track.
8. Easy speed control.
9. It should be pollution free.

10.2. DIFFERENT SYSTEMS OF TRACTION

The various traction systems commonly used are described below.

1. ***Direct Steam Engine Drive.*** The steam engine drive used to be widely employed for railway work. In this type of drive, the reciprocating steam engine is invariably used for getting the necessary motive power because of its inherent simplicity, operational dependability, simplified maintenance, the simplicity of connections between the cylinders and the driving wheels and easy speed control. The locomotive and train unit is self-contained, therefore, it is not tied to a route. It causes no interference to the communication lines running along the track. The capacity is very high in comparison to direct internal combustion engine drive, and battery electric drive etc. Initial investment required is low in comparison to that of electric drive as no overhead structure, distribution system and power generation is required. It is cheap for low density traffic areas and in initial stages of communication by rail.

Steam engine drive is losing ground because of the following drawbacks:

1. Because of difficulty of installing a condenser on a locomotive, the steam engines employed are of non-condensing type, therefore, efficiency is very poor (6 to 8 per cent). Further fires have to be banked even when locomotive is idle.
2. Adequate supplies of feed water at regular intervals is required.

3. The overload capacity of steam locomotive is limited.
4. Steam locomotive has to carry sufficient quantity of coal and water which correspondingly reduces the payload that can be hauled.
5. Limited availability and cruising range which are influenced not only due to quantity of water and coal to be carried in the locomotive but also due to periodic stops necessary to clean the ash pan. For standard passenger locomotives this cruising range is about 160 km.
6. Performance of steam locomotive is governed by the rate of firing of the coal. Single fireman can maintain a firing rate of about 2 tonne/hour for short periods and with two firemen it may be about 2.7 tonne/hour.
7. More number of crew (one driver and two firemen) are required, therefore, wage bill is increased.
8. Extensive and costly ancilliary equipment, such as coaling cranes, ash and fire cleaning pits, turntable, water supply plants, is required. The maintenance of these facilities also costs money.
9. Steam locomotive requires more repair and maintenance. The repair costs of steam locomotive are heavy specially for the boiler. Steam locomotive requires large number of unskilled and semi-skilled staff for maintenance.
10. Larger size of running sheds and workshops are required.
11. Owing to coal dust, cinder and ash the steam engine drive is not clean drive.
12. Steam locomotive cannot be put into service at any moment as time is required for steam raising.
13. Since driving wheels are very close, more concentrated adhesive weight is required.
14. Steam locomotive causes considerable wear on the track owing to unbalanced reciprocating parts and corrosion of steel structures by the smoke emitted.
15. Because of unbalanced reciprocating masses of steam engine riding qualities are not good.
16. Because of unbalanced forces produced by reciprocating masses, coefficient of adhesion of steam locomotive is low (0.25 as compared with that of electric ranging from 0.3 to 0.45).
17. Due to low coefficient of adhesion power-weight ratio of steam locomotive is low.
18. Owing to high centre of gravity of steam locomotive, speed is limited.
19. Owing to smoke, steam engine drive cannot be employed for underground railways.
20. Steam locomotive is only available for hauling for about 60 per cent of its working day, the remainder of the time is spent in coaling, fire cleaning, watering and servicing.
21. Because of low coefficient of adhesion it is not possible to have acceleration and retardation with steam engine drive as high as with electric drive. Steam engine drive is, therefore, not suitable for urban or suburban services where distance between stations is small. It is not also suitable for hauling heavy loads on steep gradients.

A few steam turbine locomotives have been built having a condenser and reduction gearing between the turbine and driving wheels for experimental purposes. Though the efficiency of such locomotives is between 10 and 15 per cent but due to their high initial and maintenance costs they could not be put into general use.

Gas-turbine locomotives have also been built on an experimental basis.

2. *Direct Internal Combustion Engine Drive.* Direct internal combustion engine drive is widely employed for road transport (buses, trucks, cars etc.). The efficiency of internal combustion engine at its normal speed is about 25 per cent. It is self-contained unit and therefore, it is not tied to any route. Initial investment required, being only the cost of vehicle and garage, is very low. Speed control and braking system employed is very simple. It is cheap drive for the outer suburbs and country districts.

The disadvantages of this type of drive are:

1. Its overload capacity is limited on account of the speed-torque characteristics of the engine and even 10% of overloading may result in stalling of the engine.
2. Its operation at any but the normal speed is very uneconomical.
3. As torque exerted by an internal combustion engine is approximately constant, hence an engine capable of giving high starting torque required is not used to its full capacity when running at full speed unless some form of gear is employed.
4. Speed control is possible only by employing a gearbox.
5. It is not suitable for heavy railway work as the arrangement possible for speed control and utilizing the torque of engine to full capacity when running at full speed is not altogether satisfactory for it.
6. The life of propulsive equipment is much shorter than that of electrical equipment of a tramcar or a trolley bus.
7. The maintenance and running costs are higher.
8. Since the fuel oil is to be imported in our country, therefore, there is drain on foreign exchange which is vital for expanding economy of the country. In wartime this system may get completely paralysed.
9. The oil engined bus accommodate slightly less passengers than the trolley bus.
10. It has no starting torque and must be started by some auxiliary means such as compressed air or electrically.

3. *Steam Electric Drive.* A few locomotives employing steam turbine for driving a generator used for supplying current to electric motors have been built for experimental purposes. Such locomotives have not been put into general use because of some mechanical difficulties and complications.

4. *Internal Combustion Engine Electric Drive.* In this drive the reduction gear and gearbox are eliminated as the diesel engine is to drive the dc generator coupled to it at a constant speed. This type of drive has found considerable favour for railway work and locomotives of this type are becoming widely used. The advantages and disadvantages of this drive are given below.

Advantages. 1. No modification of existing track is required for conversion from steam traction to diesel engine electric traction.

2. Initial investment required is low as no overhead structure distribution system and equipment is required.
3. The diesel engine electric vehicle can accommodate more passengers than steam engine vehicle and because of its higher acceleration and braking retardation the schedule speed over a given route will be higher than that of the latter vehicle, so that fewer diesel-electric locos will be required for operating a given service.
4. Power loss in speed control is very low because it can be carried out by the field control of generator.
5. The locomotive and train is a self-contained unit and, therefore, it is not tied to any route.
6. Though the maintenance cost is fairly heavy but the time spent in maintenance and overhaul is very less (only 5 to 10 per cent of its working day). Hence such a locomotive is available for hauling for almost $6\frac{1}{2}$ days in a week for 24 hours duty.
7. It can be put into service at any moment.
8. Overall efficiency is greater than that of steam locomotives (about 25 per cent in comparison of 8 per cent or so in case of steam locomotives).

Disadvantages. 1. Its overload capacity is limited as diesel engine cannot be overloaded.

2. Life of the diesel engine is comparatively shorter.
3. Special cooling system is required for cooling the diesel engine in addition to motor generator set.
4. Running and maintenance costs are high.
5. The dead weight of such locomotives is more. It requires more axles (six for diesel electric locomotive in comparison with four for electric locomotive).
6. Since the fuel oil is to be imported in our country, therefore, there is drain on foreign exchange which is vital for expanding economy of the country. In wartime this system may be completely paralysed.
7. For the same power output diesel electric locomotive is costlier than steam or electric locomotive because it carries diesel engine, main generator and traction motors, all capable of handling full locomotive power. This also makes diesel electric locomotive considerably heavier than plain electric locomotive.
8. Regenerative braking cannot be used with such drives but there is no bar in making use of rheostatic braking.

5. *Petrol Electric Traction.* This system has been used in heavy lorries and buses. Due to electric conversion it provides a very fine and continuous control which makes the vehicle capable of moving slowly at an imperceptible speed and creeping up the steepest slope without throttling the engine.

6. *Battery Electric Drive.* In this drive the locomotive carries the secondary batteries which supply power to dc motors employed for driving the vehicle. Such a drive is well suited for frequently operated service such as for local delivery of goods in large towns with maximum daily run of 50 to 60 km, shunting and traction in industrial works and mines. Battery driven vehicle is easy to control and very convenient to use. Other advantages are low maintenance cost and absence of fumes. The major limitation of this type of drive is the small capacity of the batteries and the necessity for frequent charging, speed range is also limited. Battery vehicles are started by series-parallel grouping of batteries—in parallel for starting and running at the speed up to half maximum speed and in series for running at full maximum speed.

7. *Electric Drive.* The drive of this type is most widely used. In this system of traction the vehicle draws electrical energy from the distribution system (from a contact wire suspended above the track or from an additional rail laid along side it) fed at suitable points from either a central power station or substations. The advantages and disadvantages of electric drive are as follows:

Advantages. 1. It is the cleanest of all other types of systems of traction. Due to this only it is ideally suitable for the underground and tube railways.

2. No water and coaling depots are required in electric drive.
3. Ancilliary equipment, such as coaling cranes, ash and fire cleaning pits, water supply plants, is not required. Size of running sheds and workshops is comparatively smaller.
4. The electrical energy required for lights and fans of the train can be drawn from the lines directly and, therefore, there is no need of providing Rosenberg generators.
5. Owing to better speed control in electric drive locomotion on steep gradients at slow speed is possible.
6. It has got another advantage of rapid acceleration and braking retardation. An electric locomotive has an acceleration of 1.5 to 3.5 kmphps for urban service. Trolley buses have acceleration of 4 to 6 kmphps. The steam locomotive has an acceleration of only 0.8 kmphps. Hence the schedule speed may be from 50 to 100 per cent higher than that corresponding to steam operation over short runs and with frequent stops.
7. Because of *larger* passenger carrying capacity and higher schedule speed the electric railway can handle the traffic of any amount—up to double the amount possible with steam railway.
8. The electric train can be divided and run in sections during the periods of light traffic, thereby enabling a frequent service to be maintained.
9. Less terminal space is required on account of speedy movement and lesser number of units required for operating a given service.
10. An electric locomotive requires much less time for maintenance and repairs than a steam locomotive and such vehicles can, if desired, be kept in service for 95 per cent or more of the working day.
11. Maintenance and repair cost is about 50 per cent of that of steam locomotive.
12. The electric locomotive can be put into service immediately whereas steam locomotive takes about two hours to get up steam.

13. There is no damage to the plant, equipment and building due to corrosive smoke fumes because of absence of unwanted gases and fire hazards.
14. This system is healthier from the hygienic point of view owing to complete absence of smoke and fumes.
15. Electric traction is most economical in areas of high traffic density particularly if electrical energy is cheap.
16. Centre of gravity of electric locomotive is lower than that of steam locomotive and, therefore, electric locomotive is able to negotiate curves at comparatively higher speeds.
17. The vibrations in electrically operated vehicles are less as the torque exerted by the electric motors is continuous.
18. In electric traction it is possible to instal power units in two or more power cars in the train and to control them from one point whereas in steam traction each locomotive is to be manned by its own crew.
19. The coefficient of adhesion is better in electric traction than that in steam traction. This is due to continuous torque of electric motors and distribution of driving wheels along the length of the train. Higher coefficients of adhesion not only reduces the ratio of weight to output kW of the locomotive but also results in improvement in riding qualities and reduction in wear and tear of the track.
20. Electric equipment can withstand large temporary overloads and can draw relatively large power from the distributing system.
21. Electric traction helps in saving of high grade coal which is limited in quantity in our country.
22. Power requirement for railway electrification has been of the order of 50 kW/track km. This is a sizeable load which tends to increase power development schemes. Traction load has high load factor of the order of 60 to 70%. Electric traction, therefore, provides a most important base load. This, therefore, enables the use of large generator units having high thermal efficiency possible in thermal stations. High base load also affords economic development of hydroelectric potential.
23. Railway electrification encourages rural electrification as no special transmission lines have to be run for this purpose.
24. Electric braking is superior to mechanical braking because of less wear on brake shoes, wheels and track. In case of regenerative braking a part of energy is returned to the supply instead of being wasted as heat in the brake shoes, thus saving in overal consumption is affected. Goods traffic on gradients is more safe and speedy.
25. In case of steam traction brake blocks and wheels are likely to overheat on continuous grades and sometimes trains have to be stopped for allowing these brake blocks and wheels to cool down but in case of electric traction it is not so. In electric traction dynamic braking is applied to hold the train under control and brake blocks are applied for stopping the train and thus wheels and brake blocks are always cool whenever required for stopping the train.

Disadvantages. 1. The greatest disadvantage of electric traction is that it involves a heavy initial expenditure for the power supply system and, in the case of tramways, for the track rails and electrical return system, in addition to the cost of vehicles, depots etc.

2. Number of somewhat severe statutory regulations, concerning leakage currents from overhead conductors and the voltage drop in the track rails, have to be complied with.
3. Failure of power supply for a few minutes may paralyse the whole system. Such an occurrence is, however, rare nowadays due to increased reliability of power supply.
4. Steam locomotives can use their steam for heating the compartments in cold weather, very cheaply whereas the electric locomotive have to do it at extra cost.
5. Electric traction system is tied up to only electrified routes.
6. In case of ac traction the communication lines running along the track experience considerable interference from the power lines. The communication lines therefore must, either be removed away from the track, or replaced by special expensive cables. The cost incurred for doing this may be as high as 15% of the capital outlay.
7. Additional equipment is required for regeneration. In case of dc series motors regeneration is not simple process.
8. In cold countries a service locomotive is required to run up and down the line in order to prevent the formation of a layer of ice on the conductor rails.

The main consideration in the electrification of railways or other traction services is one of the economics. If operation by electric traction proves to be cheaper than any other method (*i.e.*, the electrical energy is available at cheaper rates and the density of the traffic is high) and necessary capital can be raised then no other factor or disadvantage would prevent the conversion of steam railway to electric railway. There is no technical difficulty in adoption of electric traction.

So far as Indian Railways are concerned the route kilometrage on which electrification is justified is quite considerable. The reasons are:

1. About 72% of route kilometrage comprises of single line section and average traffic density even at sixties level of traffic is very much higher than that necessary for justification of electrification of single line sections.
2. The traffic densities for justification of electrification of railways in case of graded sections are much less due to relative increase in fuel energy charges and savings in the energy affected by employing regenerative braking and in our country nearly 30% of railway track is having gradients steeper than 0.33%.
3. In our country electric energy is available at cheaper rates comparatively because of availability of substantial sources of hydropower generation and nuclear power generation. By electrification of railways high grade coal, which is limited in quantity in our country, will be saved which can be used for some other useful purposes.

Efficiences of different types of locomotives are given below in tabular form.

Table 10.1

S.No.	Traction System	Efficiency
1.	Steam locomotive	6–8%
2.	Gas turbine electric locomoive	10%
3.	Diesel electric locomotive	25–30%
4.	Electric locomotive with thermal power plant	About 35%
5.	Electric locomotive with hydroelectric power plant	About 40%

10.3. SYSTEMS OF ELECTRIC TRACTION

As already mentioned, systems of electric traction may be divided into two main groups.

(1) The group consisting of vehicles which receive electric power from a distributing network fed at suitable points from either a central power station or substations suitably spaced. This group is further subdivided into:

(*a*) Systems operating with dc such as trolley buses, tramways and railways.

(*b*) Systems operating with ac, such as railways.

(2) The group consisting of self-contained locomotives. This group is further subdivided into:

(*a*) Diesel electric trains and ships.

(*b*) Petrol electric trucks and lorries.

(*c*) Battery driven road vehicles..

Out of these self-contained electric vehicles, only the diesel electric traction system is widely employed and hence keeping in view the limitations discussion will be limited to it.

10.3.1. Tramways

The power is supplied to the tramcar, usually at 600 dc volts, from a single overhead conductor (of positive polarity) fed at suitable points from a central power station or substations, the track rail forms the return conductor. The tramcar is provided with at least two driving axles so that necessary adhesion is secured and control is possible from either end. The use of two or more driving axles lends itself to the use of two motors with series-parallel control. Field weakening method may be employed to obtain higher running speeds for suburban service. Two drum type controllers, one at each end, usually employed for control of vehicle, are connected in parallel but are provided with some suitable interlocking arrangements so that operation is possible from one end at a time. With the increasing outputs required by present-day tramcars the controller is tending to become so large that its operation causes fatigue to the driver, and master controller equipment similar to that used on railways is employed in recent designs. The great disadvantage of the tramway is that it needs laying of overhead supply system and track for its use which is costly to maintain and constitutes a source of danger to other road users. For very dense traffic in large cities the tramway is most economical means of transportation. Nowadays the trolley bus or internal combustion engine omnibus is preferred on account of its greater manoeuvrability. The equipment required for tramcars is similar to that used in railways but of considerably smaller output (not exceeding 45 or 55 kW). Life of tramcar equipment is much more than that of internal combustion engine omnibus. The conditions for regenerative braking are not very favourable but it may be used in hilly areas and on level track up to a speed of 20 kmph. Rheostatic and mechanical brakings are employed for normal service, Mechanical braking is employed when the speed becomes too low that electrical braking becomes ineffective. For mechanical braking, electromechanical drum brakes are employed. Magnetic brakes are also used for providing better retardation. The magnetic brakes are in fact electromagnets suspended on springs. These are attracted to the rail track, where they exert sufficient braking force. Thus dual braking effect is obtained. Tramcars, if provided with magnetic brakes, can be excited by the rheostatic braking current.

10.3.2. Trolley Buses

Serious drawback of tramway is lack of manoeuvrability in congested areas and noise. These are overcome by trolley bus drive. The trolley bus is an electrically operated pneumatic tyred vehicle and is fed usually at 600 dc volts from two overhead contact wires by means of two collectors. Although the overhead equipment required by the trolley buses is more expensive and more unsightly than that for the tram, this drawback is usually more than offset by the elimination of the necessity of the track in the roadway, which is costly to maintain and also constitutes a source of danger to the other road users.

Another advantage of trolley bus is that it is able to manoeuvre in a traffic over a width of several metres on each side of the centre line of the trolley wires. As the adhesion between a rubber-tyred wheel and the roadways is sufficiently high, therefore, only a single driving axle and so a single motor is normally used. A dc compound motor of output of 50 to 100 kW is usually employed.

Speed control is obtained by field weakening (by providing a resistance in the shunt field or the tappings on series field or divertors with the series field) method. Foot operated master controllers are used so that driver may have his hands free to steer the vehicle and apply handbrake. One pedal controls the starting, speed control and regenerative braking if any, and second pedal controls rheostatic and compressed air brakes. Regenerative braking is usually not employed in trolley bus drive because of difficulty of ensuring that supply system is always in a position to absorb the energy regenerated. In order to avoid damage to the power transmission gear, which may occur, in case plain rheostatic braking is used, due to high adhesion between road and rubber tyres, stabilised rheostatic braking is used. The lighting system in the car is low-voltage dc supplied from a motor-generator set connected in parallel with a battery. The vehicles are usually provided with secondary batteries so that the vehicles can be manoeuvered in case of emergency. As the body of the vehicle is insulated from earth because of tyred wheels, care must be taken to ensure adequate insulation resistance. The insulation resistance is checked at the end of the day when it is hot and damp and, therefore, in its worst condition. The trolley bus can accommodate slightly more passengers than an oil engined bus, and because of its high acceleration and braking retardation the schedule speed over a given route will be higher than that of the latter vehicle, so that fewer trolley buses will be required for operating a given service. Trolley buses are, therefore, used for the inner suburbs with medium traffic density.

Although oil-engined buses are more expensive to maintain and operate than a trolley bus, urban and rural services operating with a service interval of more than 15 or 20 minutes are most economically dealt with by these vehicles, as the cost of overhead equipment for a trolley bus service would be too large relative to the number of passengers carried and the revenue earned.

10.3.3. Diesel Electric Traction

The high initial cost of any system of track electrification renders it uneconomical except where the traffic density is high or where heavy grades and loads preclude any other type of traction. Under other conditions self-contained motive power units are required, and these may employ either steam or internal combustion engines as the prime movers. No doubt for road transport the IC engine is supreme but for railway the choice lies between steam and heavy-oil (diesel) engines. The factors governing the choice of traction system are, therefore, largely economic and depend on the natural resources of the country/region concerned. For instance countries (like Denmark, South America and Siam) where fuel has to be imported and water is sometimes scarce, will like to adopt diesel electric traction. The diesel electric traction system is quite economical where the cost of diesel oil is low as in USA. On the other hand countries rich in coal, like Great Britain, may like to have electric traction for urban areas and steam traction for suburban areas. This system is also adopted in the areas where coal for steam engines is not available in plenty and water is also scarce as in Rajasthan. This avoids transportation of large quantities of coal over long distances. Diesel locomotive consumes very small amount of water as compared to steam locomotives.

Diesel locomotives were introduced for the first time in our country in 1945 for shunting purposes on BG sections and in 1956 for main line services on MG sections. Indian Railway started using diesel locomotives extensively for main line services since 1958.

The diesel-electric locomotives employed in practice are of following four types:

(*i*) The main-line diesel electric locomotive having engines of output not exceeding, 1,500 kW and speeds 160 kmph.

(*ii*) The shunting diesel electric locomotive having an engine of 225-375 kW output and speed between 25 to 50 kmph.

(*iii*) The diesel-electric multiple unit stock of which each motor has an engine of 135-150 kW output and the train is capable of having speeds between 80 to 110 kmph.

(*iv*) The diesel-electric rail car having an engine of 75-450 kW output which may operate as a single car or with one or more trailer coaches.

In diesel electric traction system the electric motors employed for driving the locomotive are fed from a dc generator driven by the diesel engine put on the same locomotive. Thus diesel electric locomotive is self-contained motive power unit and, therefore, heavy expenditure on track electrification and substations etc. is eliminated.

The diesel engine employed for diesel electric locomotives is usually high speed type with three or four different running speeds with different output power. The generator may be self-excited type or separately excited. In case of a separately excited generator excitation is obtained from an auxiliary generator connected in parallel with the battery, so that standby supply is available. Similarly diesel engine is made self starting employing the supply from the battery. The voltage control of main generator becomes easier.

The characteristic of diesel engine is that the torque obtainable is approximately constant and independent of speed and the output power increases fairly uniformly with the speed, as illustrated in Fig. 10.1. It has no starting torque and must be started by some auxiliary means such as compressed air or electrically, as in the ordinary motor car. The overload capacity of the diesel engine is also very small (not exceeding 10%).

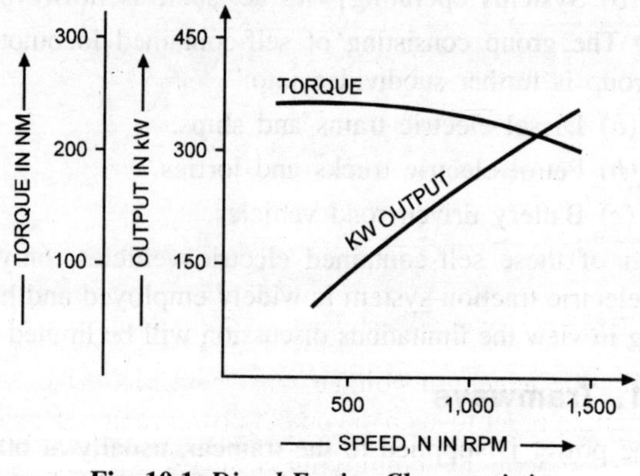

Fig. 10.1. *Diesel Engine Characteristics*

The torque required in traction service is roughly inversely proportional to the speed being a maximum at start and at low speeds so that the constant torque characteristic of a diesel engine is not entirely suitable; if such an engine is directly coupled to the driving wheels and made of such a size that it provides the required torque at very low speeds, then at full load it would be loaded up to 20 to 25 per cent of its full load.

Some form of torque converter must, therefore, be included in the transmission between the engine shaft and the driving axles so as to convert the constant torque at full engine speed into a torque which decreases with the increase in speed, maintaining the kilowatts output (proportional to the product of torque and speed) approximately constant and, therefore, enabling the available output of the diesel engine to be used to the full at all speeds.

There are two types of transmission systems which are usually employed.

1. ***Mechanical transmission system*** employs a variable gear between the engine shaft and driving wheels as in the ordinary motor car. This method is almost universally employed for road and light railway vehicles but it is not suitable for railway work where output required, exceeds 300 kW due to the mechanical difficulties with the gears and gear changing equipment. This system of transmission of power is light and more efficient than electrical transmission system but it is not so flexible as the latter one.

2. ***Electrical transmission system*** employs a diesel engine coupled to a dc generator which supplies current to ordinary dc series motors on the driving axles. This system makes use of the simple speed control properties and desirable speed-torque characteristic of the dc series motors.

Assuming that the diesel engine is running at a constant speed and the generator is giving a constant voltage, any increase in the tractive effort required (say, due to its meeting an up gradient) will reduce the speed of the traction motors owing to their characteristics. The reduction in speed will not, however, be sufficient to compensate for the increased tractive effort, and, therefore, output power required will increase and the engine will be overloaded.

In order that the power delivered by the diesel remains the same, the product of tractive effort and speed *i.e.*, $T_E \times N$ must remain constant.

$$\text{Hence speed} \quad N \propto \frac{1}{T_E}$$

$$\text{But also} \quad N \propto \frac{V}{I} \qquad \text{If } \phi \propto I$$

and $$T_E \propto I^2$$

$\therefore$ For constant output

$$\frac{V}{I} \propto N \propto \frac{1}{T_E} \propto \frac{1}{I^2}$$

$$\text{or} \quad V \propto \frac{1}{I}$$

Thus the generator voltage must be varied inversely as the current in order to keep the output power required constant.

This is done by employing a series winding on the generator or its exciter, which is so connected in the circuit of traction motor that the increase in traction motor current demagnetises the field. This method does not, however fulfil the above condition very closely and a large number of modifications and additions to it have been developed and employed by different manufacturers to obtain the constant power output curve.

In diesel electric traction system the series motors are supplied power directly from the generator instead of through dropping resistance in series as employed in straight electric traction system. The starting and speed control is obtained by varying the excitation of the generator.

Diesel engine driven Ward Leonard drive is depicted in Fig. 10.2. Such drive employs a non-electrical prime mover (diesel engine or a gas turbine) for driving a dc generator that feeds a dc motor. The generator-motor combination operates as a torque converter, like a stepless gear, to provide to the motor speed-torque characteristics required by the load. The motor runs at variable speed, the prime mover and the generator coupled to it runs at a fixed higher speed resulting in reduced their size and cost and optimized efficiency. Regenerative breaking is not possible bcause the flow of energy in reverse direction is not possible but the rheostatic braking can be employed if required.

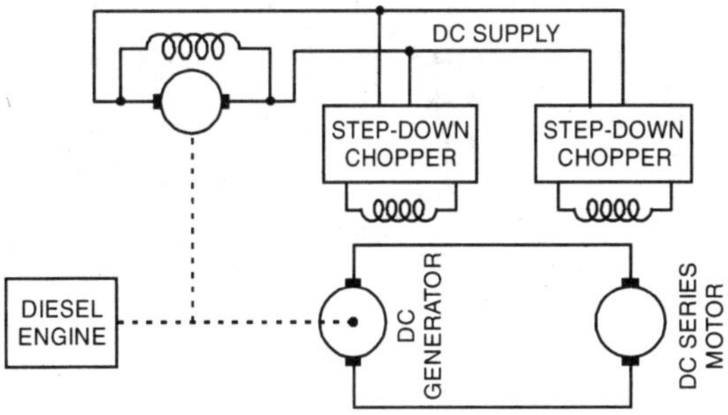

Fig. 10.2. ***Diesel Engine Driven Ward Leonard Drive***

The dc generator speed is restricted to a value because of commutation difficulties. This may not permit the prime mover to be driven at an optimum speed. Further, commutator also imposes restrictions on the maximum power rating of a dc generator. In some large power applications, the generator is to feed a number of motors. The generator, therefore, should be of a size larger than what a dc generator can be. Furthermore, a dc generator needs frequent maintenance because of commutator. All these limitations and problems can be avoided if a synchronous generator and an uncontrolled rectifier bridge are used instead of a dc generator. The voltage across the dc series motor is controlled by varying the excitation of the synchronous generator.

Several electric transmission have been employed, the most common one are :

(*a*) Diesel engine driven dc generator feeding traction motors (dc series motors).

(*b*) Diesel engine driven three phase synchronous generator feeding traction motors through semiconductor diode rectifier.

(*c*) Diesel engine driven three phase synchronous generator feeding three phase squirrel-cage induction motors through semiconductor diode rectifier followed by a 3-phase VSI (voltage source inverter).

(*a*) Diesel Engine Driven DC Generator Feeding DC Series Motors. Block diagram of a diesel electric traction using a dc generator and dc series motors is shown in Fig. 10.3. Diesel engine does not have starting torque. It is made self starting employing the supply from the battery to feed the generator.

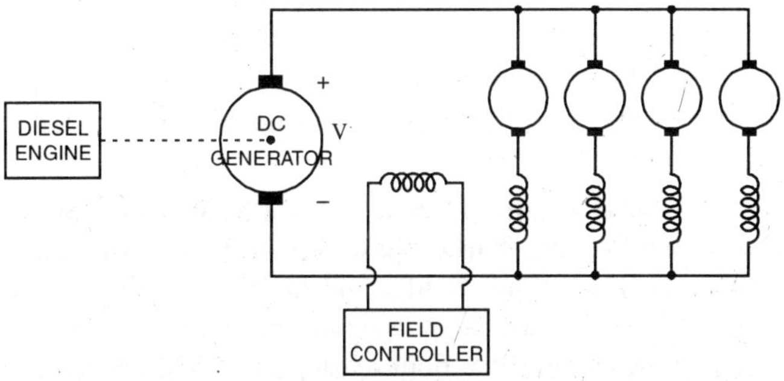

Fig. 10.3. ***Diesel Engine Driven DC Generator Feeding Traction Motors (DC Series Motors)***

(*b*) Diesel Engine Driven Three Phase Synchronous Generator Feeding DC Series Motors Through Semiconductor Diode Rectifier. Block diagram of diesel electric traction

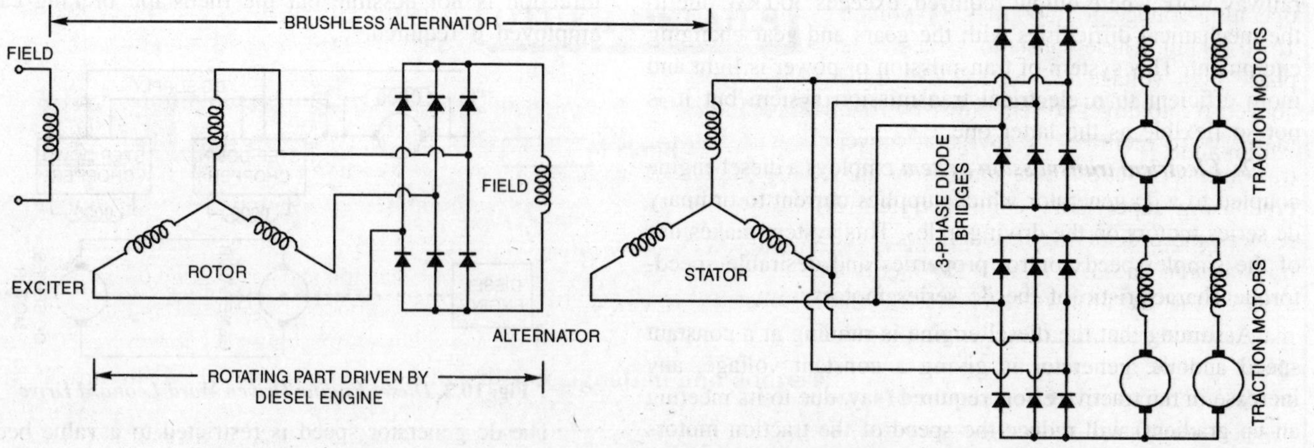

Fig. 10.4. ***Diesel Engine Driven Brushless Alternator Feeding DC Series Traction Motors***

using brushless alternator and dc series traction motors is shown in Fig. 10.4.

3-Phase synchronous generator is excited by supplying dc into the rotor by means of slip-rings and brushes which need frequent maintenance. The dc series traction motors are fed through two diode bridges. In the scheme shown in Fig. 10.4, slip rings and brushes are eliminated by making use of brushless excitation system. It employs an exciter which is an alternator with inverted rotor and stator functions *i.e.,* field on the stator and three phase winding on the rotor. Exciter is mounted on the alternator shaft. Three phase voltages induced in the exciter rotor are rectified by a diode bridge which is mounted on the rotor shaft, and therefore, revolves with it. The rectified voltage is supplied to the field of synchronous generator. The excitation to alternator is adjusted by controlling exciter field current. The advantages of this scheme over dc generator scheme shown in Fig. 10.3 are :

(*i*) Because of restricted maximum speed, the maximum power rating of a dc generator is low. But with this scheme there is no restriction.

(*ii*) Because of brushless alternator, maintenance is reduced.

(*iii*) Transmission efficiency is improved by 2-4%.

(*iv*) Advantage of good adhesion is retained due to parallel connection of all motors.

(*v*) Reduced weight of generating unit (around 50%) brings down the volume and cost of generating unit.

(*vi*) The scheme is more economical because of reduced maintenance and reduced cost of alternator rectifier.

The only drawback of this scheme is that it needs a separate motor for its starting because the engine cannot be started by this arrangement.

(*c*) Diesel Engine Driven Three Phase Synchronous Generator Feeding Three Phase Squirrel-Cage Induction Motors Through Diode Bridge and Inverter. The block diagram of diesel electric drive using inverter fed induction motors is shown in Fig. 10.5. Both six-step and PWM inverter can be employed.

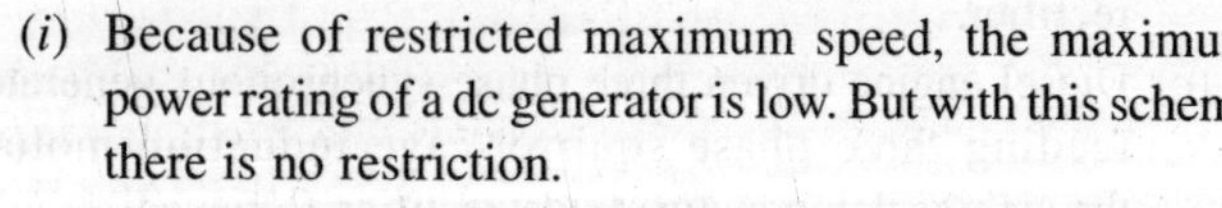

Fig. 10.5. ***Diesel Electric Drive Using Inverter Fed Induction Motors***

With the use of 6-step inverter, voltage is controlled by adjusting field current of alternator and with the use of PWM inverter, voltage is controlled by the inverter itself and the rectifier output voltage can be maintained constant at the rated value. This allows reduction in the current rating of alternator and diode rectifier; further resulting in reduction in the cost, weight and size of the generator unit. A cycloconverter can also be employed instead of diode rectifier and inverter. But the alternator must be driven at a higher speed. This scheme has the following advantages :

(*i*) High power rating, high efficiency due to small power losses and reduced maintenance.

(*ii*) High ratio of power output to locomotive weight. So heavier trains on faster schedules can be pulled by the locomotive.

(*iii*) All motors can be connected in parallel providing best condition from the point of view of adhesion.

10.4. SYSTEMS OF TRACK ELECTRIFICATION

These comprise (*i*) the dc system (*ii*) the single phase ac system (*ii*) the three phase ac system and (*iv*) the composite system (*a*) single phase-three phase system and (*b*) single phase-direct current system.

1. *DC System.* In this system the electric motors employed for getting the necessary motive power are usually dc series motors, although the compound motors are coming into favour for tramways and trolley buses where regenerative braking is desired. The operating voltage is from 600 volts to 750 volts for tramways and for many suburban railways and from 1,500 volts to 3,000 volts for main line railways. The driving motors receive power from distribution system, (in dc as well as single phase ac system). The distribution system consists of one contact wire in case of tramways and two contact wires in case of trolley buses (the running rail acts as current returning conductor in case of tramways) which is fed from substations (except in case of very short railways where a direct supply from a dc generating station may be possible) suitably spaced. The spacing of the substations depends upon the operating voltage and the traffic density of the route. For suburban railways where the operating voltage is 600 volts the substations are spaced 3 to 5 km apart and for main line railways having operating voltage of 1,500 to 3,000 volts they may be spaced 15 to 40 km. These substations receive ac power from a 3-phase high voltage (33 to 132 kV) transmission lines and convert it into direct current either by rotary convertors or by mercury-arc rectifiers or thyristor converters with suitable transformation ratio. These substations are usually unattended type using supervisory control system. This arrangement reduces the cost of transmission lines and copper losses in transmission lines. As regards the merits and demerits of dc system the dc motors have better characteristics, low maintenance cost and a smaller weight per kW output, better speed control, and efficient regenerative braking but because of low operating voltage conductor of large x-section for the distribution system and more number of substations spaced at shorter distances for feeding the distribution system are required for a given service. The cost of substations is also more due to heavy cost of converting machinery required for converting ac into dc and of additional equipment such as negative boosters etc. required to keep the voltage of return rail within limits.

The dc system is preferred for suburban services and road transport where stops are frequent and distances of run are small.

2. *Single Phase AC System.* In single phase ac system ac series motors are used for getting the necessary motive power. The voltage employed for distribution network is 15,000 to 25,000 volts at $16\frac{2}{3}$ or 25 Hz, which is stepped down on the locomotive to a low voltage (300 to 400 V) suitable for supplying to single phase ac series motors by means of a step-down transformer carried on the locomotive. The distribution network may be fed directly, at high voltage, from a generating station when the extent of electrification is within a radius of about 30 km from the generating station. For longer distances the economic voltage for the power transmission system is higher than that desirable for traction system, and therefore, transformer substations are employed for feeding the distribution network. The spacing of substations is 50 to 80 km because of low current requirement at high voltage. The substations receive power at high voltage either from the existing extra high voltage transmission lines or from a separate system employing single phase generating stations and steps down the voltage to the required value and/or change the supply frequency through frequency convertors. The change of supply frequency becomes necessary because of better performance of the ac series motors at low frequency. Low frequency operation of ac series motor improves its commutation properties, power factor and efficiency. Low frequency operation of overhead transmission line reduces the line reactance and hence the voltage drop directly.

In case the supply at low frequency is obtained directly from a generating station exclusively meant for traction purposes, there is no problem as power corresponding to this frequency can always be generated. If, however, electric supply is taken from industrial frequency network, which is usually the case, substations in addition to transformers have necessarily the frequency converters to convert 3-phase, 50 Hz into single phase $16\frac{2}{3}$ or 25 Hz as required.

AC single phase system is invariably adopted for main line service where the cost of overhead structure is of more importance and rapid acceleration and retardation is not so important as for suburban railways.

3. *Three Phase AC System.* In this system 3-phase induction motors operating at 3,000 to 3,600 volts and normal frequency of $16\frac{2}{3}$ Hz are employed for getting the required motive power. The three phase induction motors are simple and robust in construction and have a high operating efficiency and property of automatic regenerative braking without requiring any additional equipment. However, the induction motors have few drawbacks such as low starting torque, high starting current and absence of speed control. The distribution system consists of two overhead wires and track rail for the third phase and receives power either directly from the generating station or through the transformer substations. The substations receive power from high voltage transmission lines at power frequency and step down the voltage and change the frequency. Since two trolley wires per track are necessary, the line voltage of the distribution systems has to be limited to values below the highest voltage used for single phase railways, and as the 3-phase induction motors can be built economically for operation at moderately high voltage, the line voltage is usually chosen so that the motors may be supplied directly from the trolley wires. This system is employed in some hilly areas, where output power required is high and regeneration on large scale is possible. On account of disadvantages of two overhead contact wires and constant speed characteristic of induction motors this system is no longer likely to be adopted.

4. *Composite Systems.* Efforts have been made to combine the advantages of the dc/ac and 3-phase/single-phase systems and composite systems have been evolved. The single phase ac sysem is preferable from the viewpoint of the distribution and contact wire system. It can be converted into either three-phase ac or dc at low voltage in the locomotive. At present there are two composite systems in vogue.

(*i*) Kando system or single phase-to-three phase system.

(*ii*) Single phase ac-to-dc system.

(a) Single Phase-To-Three Phase System or Kando System. In this system single phase high voltage ac system is employed for distribution purposes and three phase induction motors for getting the necessary motive power in order to have the advantage of low cost of single phase overhead distribution system together with desirable characteristics of three phase induction motors. The voltage used for distribution network is 15,000 volts at 50 Hz. The locomotive carries a phase convertor for converting single phase supply into 3-phase supply at relatively low frequency. The development of silicon controlled rectifier (SCR) used as invertor has made possible to get variable low frequency supply ($\frac{1}{2}$ to 10 Hz) at which induction motor develops high starting torque without drawing excessive current from the system and speed control is also conveniently obtained by varying the supply frequency. There exists good possibility of this traction system to develop for further adoption.

(b) Single Phase-To-Direct Current System. In this system the advantages of the two systems are combined together by employing high voltage ac single phase system for overhead distribution and dc series motors for the necessary drive so that high efficiency of the ac distribution system is obtained together with the desirable characteristics of dc series motors. The locomotive carries transformer and converting machinery to step down the voltage and convert into dc. The voltage employed for overhead distribution system is 25 kV at normal supply frequency of 50 Hz. This system is adopted in all the areas of major traffic density on main lines. It has been decided to adopt this system of traction employing 25 kV, industrial frequency (50 Hz) ac supply for track electrification in our country and this is being installed on sections which are being electrified.

This system has got numerous advantages over dc system as enumerated below:

1. The line current for a given demand of power is reduced on account of high system voltage. It causes reduction in the cross section of overhead conductor which leads to a saving in the cost of supporting structures and their foundations being lighter in weight and simpler. This fact will be obvious from the following datas:

   The weight of copper required per track km for the 1,500 V dc system is 3,700 kg and for the 3,000 V dc system is 3,300 kg whereas for ac system it is 1,600 kg. The weight of steel required per track km in ac system is only about half of that required for dc system.
2. On account of high voltage the substations can be spaced at longer distances (50 to 80 km) whereas the substations are spaced at 12 to 30 km in case of 3,000 V dc system and at 5 to 12 km in case of 1,500 V dc system. It affects substantial saving in the capital expenditure on traction substations and transmission lines on account of reduced number of substations to be erected and flexibility offered in the location of site for substations. AC substations are simpler and cheaper than dc substations as they only have step-down transformers and associated switchgear.
3. A very special feature of the ac locomotive in comparison with the dc locomotive is the greater amount of adhesion obtained with ac locomotive. In the case of ac rectifier locomotive employing dc traction motors, coefficient of adhesion is obtained as much as 45% against the value of 27% in case of dc locomotives. The greater adhesion is obtained because of possibility of connecting of all dc series motors in parallel permanently and controlling their speeds by varying the voltage across each motor by tap changing auto-transformer on the ac side. In this system slipping of a particular motor does not affect the voltage across the remaining motors and so adhesion is maintained but in case of dc traction employing resistance and series—parallel control slipping of a particular motor throws its load on the other motor which may reduce the voltage across the other motor and may, therefore, cause slipping of this motor also. The increase in coefficient of adhesion increases the hauling capacity of the locomotives (75 tonne ac locomotive can haul the same load as the 110 tonne dc locomotive).
4. The starting efficiency of an ac locomotive is higher than that of a dc locomotive as in case of ac locomotive supply voltage for starting is reduced by means of 'on-load tap changer' installed on either the primary or the secondary side of the transformer where as in case of dc locomotive it is obtained by using starting resistors. Elimination of starting bulky resistors not only increases the starting efficiency but also makes the torque-speed curve more steep and thus improves adhesion. This in turn increases the power-weight ratio in the case of ac locomotives.
5. Since the dc series motors having ideal traction characteristics are employed in this system for getting the required propelling power, therefore, this system have got the advantages of the dc system.
6. AC locomotive has less kW demand at starting than that of a dc locomotive owing to elimination of starting resistances in ac locomotives. This is of particular importance at peak traffic hours specially when a number of locomotives are to be started simultaneously after an interruption of supply.

There are, however, some drawbacks with the single phase ac system such as unbalance imposed on the power supply system and interference with neighbouring communication lines. Fortunately both of the shortcomings can be minimised by taking suitable precautions such as

(*i*) Taking of supply for the traction substations at very high voltage (say 132 or 220 kV) and from a supply system having high capacity.

(*ii*) Balancing of traction loads equally on all the three phases which is possible by connecting different traction substations across different phases in rotation.

(*iii*) Employing the Scott or T-connected three phase/two phase transformers in the traction substations.

(*iv*) Replacing of open aerial communication lines by lead or aluminium sheathed underground cables and earthing the sheathing of the cables at regular intervals.

(*v*) Providing of isolating transformers which reduce the induced voltages and keep them below 60 V under operating conditions.

(*vi*) Making use of booster transformer and return conductor for suppressing interference at the source in case of thickly populated areas.

10.5. COMPARISON BETWEEN DC AND AC SYSTEMS OF RAILWAY ELECTRIFICATION FROM THE POINT OF VIEW OF MAIN LINE AND SUBURBAN LINE RAILWAY SERVICES

1. *Main Line Railway Service.* The most important requirements of main line railway service are:

1. Higher maximum speed.
2. Minimum cost of overhead structure.
3. Acceleration and retardation are not so important for main line service as in suburban service.

The cost of overhead structure can be reduced by employing single phase ac system.

Single phase ac system is preferred for main line railway service on account of special advantages enumerated below:

(*a*) Because of possibility of employing high voltage the cost of distribution system employing conductor of smaller section to carry the small current required is reduced.

(*b*) Also because of possibility of employing high voltage for distribution purposes, the spacing of the substations is higher and, therefore, number of substations required is less.

(*c*) The initial, maintenance and operating costs of ac substations are less as the substation equipment required in single phase ac system is less, cheap and efficient.

Single phase-to-three phase composite system can also be employed for approximately same results.

2. *Suburban Line Railway Service.* The most important requirements of suburban line railway service are:

1. Rapid acceleration and retardation as frequent stopping and starting is required.
2. The working of motors should not be affected by voltage fluctuations.
3. There should be no interference to the communication lines (telephone and telegraph lines) running along the track on account of power system adopted.

The dc system fulfills all of the requirements mentioned above for suburban service, and therefore, invariably adopted. The advantages and disadvantages of dc system are as follows.

Advantages. 1. In similar conditions the energy consumption in dc system is less as compared with that in ac system.

2. For exerting the same torque, current required in dc system is less than that in ac system.
3. The dc locomotive and motor coach equipment is lighter in weight, cheaper in initial as well as in maintenance cost and more efficient.

Disadvantages. 1. Due to low voltage employed for distribution purpose, the current required is high, therefore, the section of conductor required is large, the spacing of substations required is less and the number of substations required is more.

2. Substations are more costly as converting machinery is required.
3. Efficiency of substations is less.
4. Additional equipment such as negative booster etc. is required for keeping the voltage of return rail within limits.

The above disadvantages are partly eliminated by adopting composite system single phase ac-to-dc system.

Exercises

1. Explain the requirements for ideal traction and show which drive satisfy almost all the requirements.
[M.D. Univ. Utilization of Electric Power & Traction May 2009]
2. State the main requirements of an ideal traction system. Explain different traction systems.
[M.D. Univ. Utilization of Electric Power & Traction May 2010]
3. Discuss various types of traction systems.
[Pb. Technical Univ. Elecric Drives & Utilization Dec. 2009]
4. State merits and demerits of electric traction with reference to electric power crisis in India.
[Pb. Technical Univ. Electric Drives and Utilization Dec. 2006]
5. State the main requirements for an ideal traction system. What factors govern the operation of dieselisation or electrification of a steam operated railway section ? Which system of electric traction is now being preferred for electrification of main line sections and why ?
[Gorakhpur Univ. Utilization of Electrical Power & Traction Sept. 1981]
6. What are the advantages and disadvantages of electric traction over other types of traction system.
[Rajasthan Univ. Utilization of Electrical Power 2005, 2006, 2007, 2008]
7. What are the different traction systems ? Discuss them in brief.
Indicate the general trend in the adoption of traction systems for city road and rail transport, and discuss their merits and demerits. **[Gorakhpur Univ. Utilization of Electrical Power & Traction 1983]**
8. Describe briefly a battery powered vehicle.
[A.M.I.E. Sec B. Electrical Drives and Their Control Winter 1998]
9. Briefly discuss the general considerations for a tramway.
[Pb. Univ. Electric Traction December 1992]
10. What are the disadvantages of a tramcar (tramway) system of electric traction ? How are these overcome in the trolley bus system ? **[Pb. Univ. Electric Traction December 1991]**
11. Explain the trolley bus system of electric traction.
[Pb. Technical Univ. Electric Drives & Utilization Dec. 2008]
12. The diesel engine has an approximately constant torque at all speeds. Explain how it is electrically transmitted to motors so that their tractive effort increases as the speed decreases.
13. What are the main characteristics of diesel engine with special reference to its application for traction purposes ?
[Pb. Univ. Electric Traction December 1991]
14. Discuss the problems associated with diesel electric traction. Suggest ways to overcome these in actual system.
[Pb. Univ. Electric Drives and Utilization May 1989]
15. Discuss diesel electric traction system. Write the situation, when diesel electric traction is preferred over electric traction.
[U.P. Technical Univ. Utilization of Electrical Energy & Electric Traction 2005-06]
16. Explain working of a diesel engine driven dc generating feeding dc series motors systems.
[U.P. Technical Univ. Utilization of Electrical Energy & Traction 2009-10]
17. What are the different systems of railway electrification? Which one is being commonly adopted in our country and why ?
Under what conditions is diesel electric traction preferred to track electrification ?
[Gorakhpur Univ. Utilization of Electrical Power & Traction Sept. 1983]
18. Compare the following systems of track electrification as regards operating voltages, economy in operation and suitability to the topography of regions, where they are best suited.
(*i*) DC system (*ii*) AC system (single phase) and (*iii*) Composite system. **[Pb. Univ. Electric Traction Nov. 1973]**
19. Discuss the merits and demerits of the dc and single phase ac systems for the main and suburban line electrification of the railways.

Which system you consider to be the best for the suburban railways in the vicinity of large cities ? Give reasons for your answer. **[Bombay Univ. Elec. Power 1971]**

20. Compare dc and ac systems of railway electrification from the point of main line and suburban line railway service. **[Rajasthan Univ. Utilization of Electrical Power 2008]**

21. Explain the factors that determine the choice of supply system for main line traction and give voltage and frequency of the supply as adopted by Indian Railways. **[A.M.I.E. Sec. B. Electric Drives & Their Control Winter 1996]**

22. What are the systems of electric traction ? Explain why single phase ac electrification is superseding the other types of system. **[Pb. Technical Univ. Electric Drives and Utilization 2003]**

23. Give the nature and specifications of supply sysetms for electric traction used in India. **[U.P. Technical Univ. Utilization of Electrical Energy & Electric Traction 2005-06]**

24. Discuss the advantages and disadvantages of the system of traction using 25 kV, 50 Hz supply and dc series motor. **[Pb. Technical Univ. Electric Drives & Utilization May 2008]**

25. State and explain the track electrification system generally employed for the following services.

(*i*) Tramway service in a busy town area.
(*ii*) Electric suburban railway service.
(*iii*) Main line services.

[Rajasthan Technical Univ. Utilization of Electrical Power 2011]

26. Write advantages of 25 kV ac system over dc system. **[Pb. Technical Univ. Electric Drives & Utilization Dec. 2006]**

27. Write disadvantages of 25 kV ac traction systems. **[Pb. Technical Univ. Electric Drives & Utilization Dec. 2007]**

28. Compare ac and dc systems of traction. **[Pb. Technical Univ. Electric Drives & Utilization May 2007; Dec. 2007]**

29. Compare dc and ac systems of railway electrification from the view of main line and suburban line railway services. **[Pb. Technical Univ. Electric Drives & Utilization May 2009]**

30. Describe the following systems of railway electrification and write down the relative merits of each:
(*i*) DC system (*ii*) Single phase ac system, (*iii*) 3-phase ac system, (*iv*) Composite system.
[Rajasthan Univ. Utilization of Electrical Power 2010]

31. Write short notes on the following:

(*i*) Traction systems in India.
(*ii*) Battery driven vehicles and their applications.
(*iii*) Trolley bus system of traction. **[Pb. Univ. July 1992]**
(*iv*) Diesel electric traction system. **[Pb. Univ. Nov. 1988]**
(*v*) Comparison of various systems of track electrification.
(*vi*) Relative merits of ac single phase and dc systems of supply in electric traction.

Short Answer Type Questions With Answers

Q. 1. Why steam engine drive is not suitable for urban and suburban services where distance between stations is small ? **[Gorakhpur Univ. 1980]**

Ans. Because of low coefficient of adhesion it is not possible to have acceleration and retardation with steam engine drive as high as with electric drive. Steam engine drive is, therefore, not suitable for urban or suburban services where distance between stations is small.

Q. 2. Why steam engine drive is not suitable for underground railway?

Ans. Owing to smoke, steam engine drive cannot be employed for underground railways.

Q. 3. How is speed controlled in direct IC engine drive ?

Ans. By using a gearbox.

Q. 4. What is the operational efficiency of IC engine electric drive ?

Ans. About 25 per cent.

Q. 5. What are the limitations of battery electric drive ?

Ans. The major limitations of the battery electric drive are the small capacity of the batteries and the necessity for frequent charging and limited speed range.

Q. 6. What is the major drawback of electric traction ?

Ans. The greatest drawback of electric traction is that it involves a heavy initial expenditure for the power supply system and, in the case of tramways, for the track rails and electrical return system, in addition to the cost of vehicles, depots etc.

Q. 7. What are the major disadvantages of the tramway ?

Ans. The great disadvantage of tramway is that it needs laying of overhead supply system and track for its use which is costly to maintain and constitutes a source of danger to other road users. The other main drawback of tramway is lack of manoeuvrability in congested areas and noise.

Q. 8. Why electric traction system is considered superior over other systems of traction in India, even when there is a power crisis in India.

Ans. In our country electric energy is available at cheaper rates comparatively because of availability of substantial sources of hydropower generation and nuclear power generation. By electrification of railways high grade coal, which is limited in quantity in our country, will be saved which can be used for some other useful purposes.

Q. 9. Which of the system (dc or ac) is preferred for suburban services and road transport where stops are frequent and distances of run are small ?

Ans. DC system.

Q. 10. What is the value of voltage used in overhead supply for trains in India ? **[A.M.I.E. Sec. B. Winter 1999]**

Ans. 25 kV.

Q. 11. Why single phase system is preferred for main line railway service ? **[P.T.U. Dec. 2008; May 2009]**

Ans. AC single phase system is invariably adopted for main line service where the cost of overhead structure is of more importance and rapid acceleration and retardation is not so important as for suburban railways.

Q. 12. What are the most important requirements of the main line railway service ?

Ans. High maximum (crest) speed and minimum cost of overhead structure.

Q. 13. Which of the system is preferred for main line railway service ?

Ans. Single phase ac system.

Q. 14. What are the most important requirements of suburban line railway service ?

Ans. Rapid acceleration and retardation, no interference to nearby communication lines and operation of motors free from adverse effects of voltage fluctuations.

CHAPTER 11

Train Movement and Energy Consumption

INSIDE THIS CHAPTER

11.1. Introduction **11.2.** Typical Speed-Time Curves **11.3.** Crest Speed, Average Speed and Schedule Speed **11.4.** Factors Affecting Schedule Speed **11.5.** Simplified Speed-Time Curves **11.6.** Mechanics of Train Movement **11.7.** Tractive Effort For Propulsion of Train **11.8.** Power Output From The Driving Axles **11.9.** Energy Output From Driving Axles **11.10.** Determination of Specific Energy Output Using Simplified Speed-Time Curve **11.11.** Factors Affecting Specific Energy Consumption of an Electric Train Operating on a Given Schedule Speed **11.12.** Dead Weight, Accelerating Weight and Adhesive Weight.

11.1. INTRODUCTION

The movement of trains and their energy consumption can be most conveniently studied by means of speed-time and speed-distance curves, which show respectively the speed at different time instants after the start of run and the speed at different distances from the starting point. Of the two, the speed-time curve is generally the more useful.

The curve drawn between speed and time, taking speed (in km/hour) on the Y-axis and time (in seconds or minutes) on X-axis, is known as *speed-time curve*. The speed-time curve provides complete information of the motion of the train. This curve gives the speed at various time instants after the start of run directly. Slope of the curve at any point gives the acceleration at the corresponding instant or speed. The area covered by the curve, the time axis and the ordinates through the instants between which the time is taken, represents the distance covered in the corresponding time.

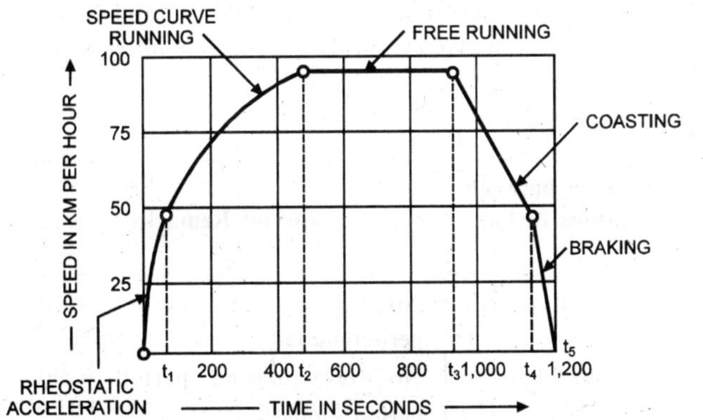

Fig. 11.1. ***Typical Speed-Time Curve For Main Line Service***

Speed-time curve (Fig. 11.1) mainly consists of (*i*) initial acceleration (*ii*) constant speed run or free run (*iii*) coasting and (*iv*) retardation. Typical speed-time curve for main line service is shown in Fig. 11.1.

(*i*) ***Acceleration.*** It consists of two parts known as (*a*) constant acceleration or acceleration while notching up and (*b*) speed curve running or acceleration on the speed curve.

(*a*) ***Constant Acceleration or Acceleration During Notching Up.*** During notching up period (0 to t_1) the current is maintained approximately constant and the voltage across the motor is gradually increased by cutting out the starting resistance. Thus tractive effort is constant and, therefore, acceleration remains constant during this period.

(*b*) ***Speed Curve Running or Acceleration on Speed Curve.*** During speed curve running (t_1 to t_2) the voltage acting across the motor remains constant and current starts decreasing with the increase in speed according to the characteristics of the motor and finally the current taken by the motor becomes constant. During this period, though the train accelerates but acceleration decreases with the increase in speed and finally becomes zero at the speed at which the tractive effort developed by the motor becomes exactly equal to the resistance to motion of the train.

(*ii*) ***Free Run or Constant Speed Run.*** At the end of speed curve running *i.e.*, at t_2 the train attains the maximum speed. During this period the train runs with constant speed attained at t_2 and constant power is drawn.

(*iii*) ***Coasting.*** At the end of free running period (*i.e.*, at t_3) power supply is cut off and the train is allowed to run under its own momentum. The speed of train starts decreasing on account of resistance to the motion of train. The rate of decrease of speed during coasting period is known as *coasting retardation*.

(iv) Retardation or Braking Period. At the end of coasting period (*i.e.,* at t_4) the brakes are applied to bring the train to rest. During this period speed decreases rapidly and finally reduces to zero.

11.2. TYPICAL SPEED-TIME CURVES

1. *Urban or City Service.* In urban or city service the distance between the two stops is comparatively very short (say 1 km or so). The time required for this run is very small (few minutes). The acceleration as well as retardation is required to be high so that high average speed and short time of run is obtained. The acceleration and retardation for urban service is between 1.5 and 4 kmphps and between 3 and 4 kmphps respectively. Free run is not present in this run. The coasting retardation is about 0.15 kmphps. The coasting period is also small. Typical speed time curve for urban service is shown in Fig. 11.2.

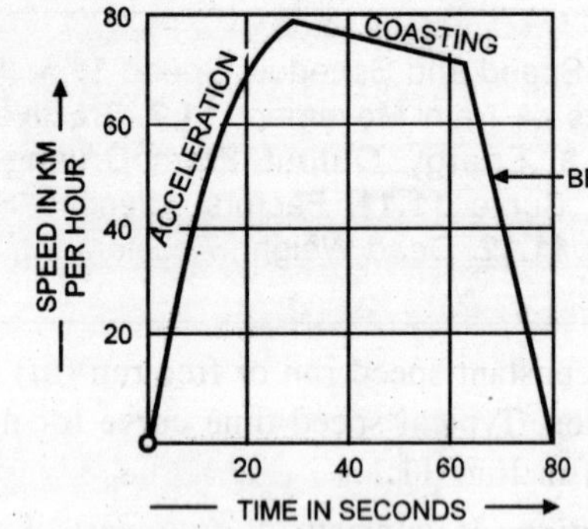

Fig. 11.2. *Typical Speed-Time Curve For Urban Service*

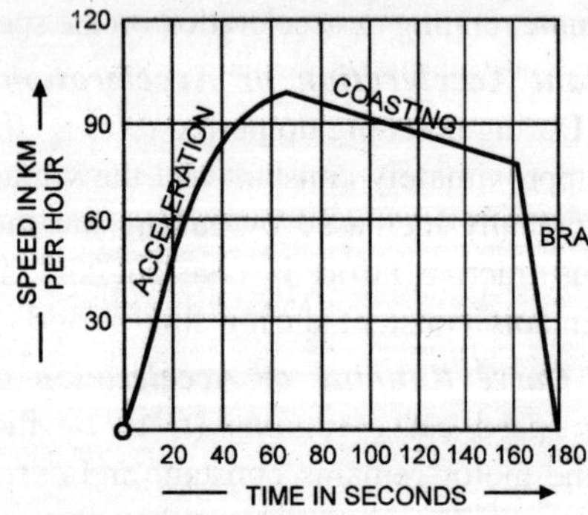

Fig. 11.3. *Typical Speed-Time Curve For Suburban Service*

2. *Suburban Service.* In this service the distance between the stops is little longer than urban service but smaller than main line service (say between 2 and 5 km). Free run is still not possible. Coasting is for a comparatively longer period. Acceleration and retardation required are as high as for urban service.

3. *Main Line Service.* The distance between two stops in main line service is considerably more (usually more than 10 km). In this service free run is of longer duration. The duration of acceleration and retardation is a small fraction of total running time. Since rate of acceleration and retardation does not affect the average and schedule speeds, therefore, these are of little importance in main line service.

Characteristics of various types of services are given in Table 11.1.

The acceleration and free-running periods are governed by the speed-torque characteristics of the electric traction motors, and these must be available before this part of the curve can be determined. Coasting and braking are governed by the train resistance and the allowable braking retardation respectively. The actual shape of any particular speed-time curve may be somewhat different from the above due to the existence of gradients and with different lengths of run.

11.3. CREST SPEED, AVERAGE SPEED AND SCHEDULE SPEED

Crest Speed. The maximum speed attained by the vehicle during the run is known as *crest speed.*

Average Speed. The mean of the speeds from start to stop *i.e.,* the distance covered between two stops divided by the actual time of run is known as *average speed.*

$$\text{Mathematically average speed} = \frac{\text{Distance between stops}}{\text{Actual time of run, T}}$$

Schedule Speed. The ratio of distance covered between two stops and total time of run including time of stop is known as *schedule speed.*

Mathematically schedule speed

$$= \frac{\text{Distance between stops}}{\text{Actual time of run + stop time}}$$

From above it is obvious that the schedule speed is always smaller than the average speed. The difference is large in case of urban and suburban services and is negligibly small in case of main line service.

11.4. FACTORS AFFECTING SCHEDULE SPEED

The schedule speed of a given train when running on a given service (*i.e.,* with a given distance between stations) is affected by the following factors:

Table 11.1

S.No.	Type of Service	Acceleration in kmphps	Retardation in kmphps	Maximum Speed in kmph	Distance between Stations in km	Special Remarks
1.	Urban	1.5 to 4.00	3 to 4	120	1	No free running period, coasting period—small
2.	Suburban	1.5 to 4.00	3 to 4	120	2 to 5 km	No free running period, coasting period—long
3.	Main Line	0.6 to 0.8	1.5	160	More than 10 km	Free running and coasting periods—long. Acceleration and braking periods—comparatively small.

1. Acceleration and braking retardation.
2. Maximum or crest speed.
3. Duration of stop.

1. *Effect of Acceleration and Braking Retardation.* For a given run and with fixed crest speed the increase in acceleration will result in decrease in actual time of run and, therefore, increase in schedule speed. Similarly increase in braking retardation will affect the schedule speed. Variation in acceleration and retardation will have more effect on schedule speed in case of shorter distance run in comparison to longer distance run.

2. *Effect of Maximum Speed.* For a constant distance run and with fixed acceleration and retardation the actual time of run will decrease, and therefore, schedule speed will increase with the increase in crest speed. The effect of variation in crest speed on schedule speed is considerable in case of long distance run.

3. *Effect of Duration of Stop.* For a given average speed the schedule speed will increase by reducing the duration of stop. The variation in duration of stop will affect the schedule speed more in case of shorter distance run as compared to longer distance run. It is because stops in case of urban and suburban services are kept very small (say 15 to 20 seconds). When the distance between stations is much larger and duration of actual run is long as in case of main line service, station stop has less conspicuous effect on the schedule speed.

11.5. SIMPLIFIED SPEED-TIME CURVES

In order to study the performance of a service at different schedule speeds the speed-time curves are replaced by simple geometric shaped curves. From these simplified curves the relationships between acceleration, retardation, average speed and distance can be easily worked out. These can have either quadrilateral or trapezoidal shape.

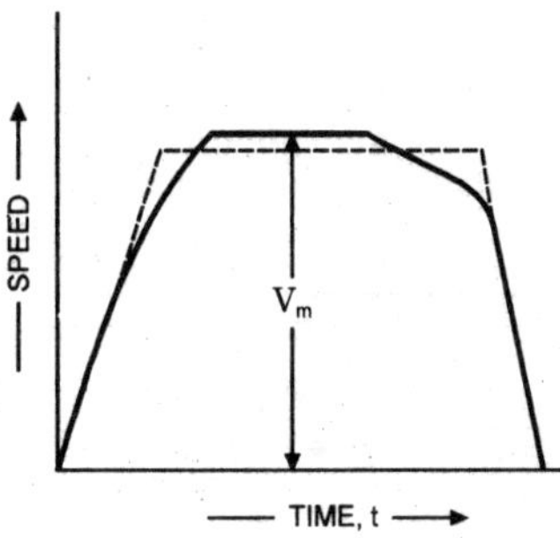

(a)

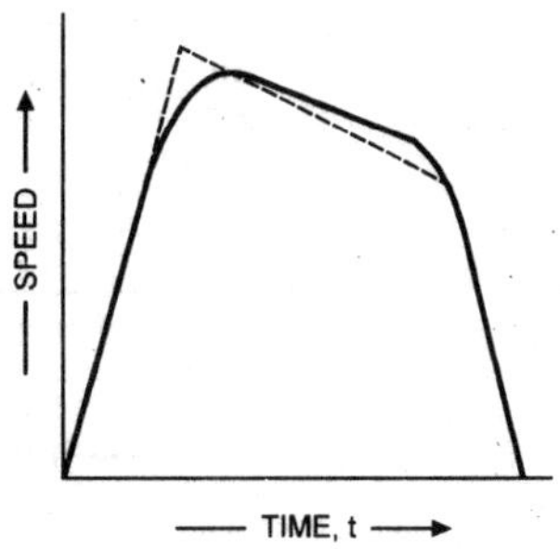

(b)

Fig. 11.4. ***Approximate Speed-Time Curves***

The speed-time curve of an urban service can be replaced by an equivalent speed-time curve of simple quadrilateral shape. The speed-time curve of a main line service is best and most easily replaced by a trapezoid. Since the area of the speed-time curve represents the total distance travelled hence the areas of the two curves should be same. The values of acceleration and retardation are also kept the same as those of the original speed-time curve. In case of simplified trapezoidal speed-time curve, speed curve running and coasting periods are replaced by constant speed period, as illustrated in Fig. 11.4 (*a*). While in case of quadrilateral speed-time curve, initial acceleration and coasting periods are extended, as illustrated in Fig. 11.4 (*b*).

The following examples illustrate the methods of calculations.

1. *Calculations By Trapezoidal Speed-Time Curve*

Let α = Acceleration in kmphps

β = Retardation in kmphps

V_m = Crest speed in kmph

T = Total time of run in seconds.

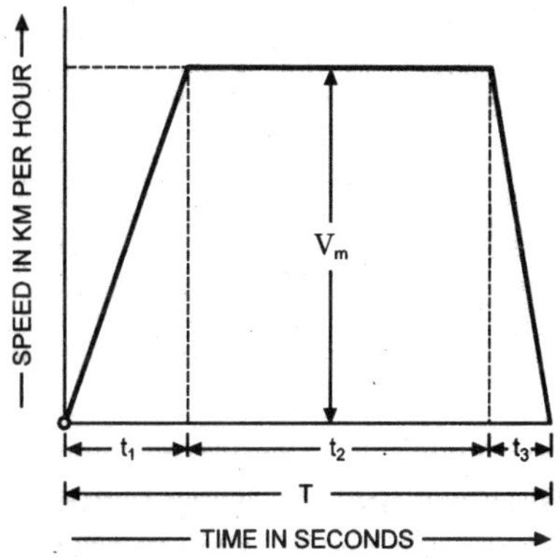

Fig. 11.5. ***Trapezoidal Speed-Time Curve***

Time for acceleration in seconds, $t_1 = \dfrac{V_m}{\alpha}$

Time for retardation in seconds, $t_3 = \dfrac{V_m}{\beta}$

Time for free running in seconds, $t_2 = T - (t_1 + t_3)$

$$= T - \left(\frac{V_m}{\alpha} + \frac{V_m}{\beta}\right)$$

Total distance of run in km,

S = Distance travelled during acceleration
+ distance travelled during free run
+ distance travelled during braking

$$= \frac{1}{2} V_m \frac{t_1}{3{,}600} + V_m \frac{t_2}{3{,}600} + \frac{1}{2} V_m \frac{t_3}{3{,}600}$$

Substituting $t_1 = \dfrac{V_m}{\alpha}$, $t_3 = \dfrac{V_m}{\beta}$ and $t_2 = T - \left(\dfrac{V_m}{\alpha} + \dfrac{V_m}{\beta}\right)$ we have

$$S = \frac{V_m^2}{7{,}200\alpha} + \frac{V_m}{3{,}600}\left[T - \left(\frac{V_m}{\alpha} + \frac{V_m}{\beta}\right)\right] + \frac{V_m^2}{7{,}200\beta}$$

or
$$S = \frac{V_m^2}{7{,}200\alpha} + \frac{V_m}{3{,}600}T - \frac{V_m^2}{3{,}600\alpha} - \frac{V_m^2}{3{,}600\beta} + \frac{V_m^2}{7{,}200\beta}$$

$$= \frac{V_m T}{3{,}600} - \frac{V_m^2}{7{,}200\alpha} - \frac{V_m^2}{7{,}200\beta} \qquad \text{...(11.1)}$$

or
$$\frac{V_m^2}{3{,}600}\left(\frac{1}{2\alpha} + \frac{1}{2\beta}\right) - \frac{V_m T}{3{,}600} + S = 0$$

or
$$V_m^2\left(\frac{1}{2\alpha} + \frac{1}{2\beta}\right) - V_m T + 3{,}600 S = 0$$

This is a quadratic equation for V_m. Substituting $\dfrac{1}{2\alpha} + \dfrac{1}{2\beta} = K$, we get

$$KV_m^2 - V_m T + 3{,}600\,S = 0$$

$$\text{or } V_m = \frac{T \pm \sqrt{T^2 - 4K \times 3{,}600S}}{2K}$$

$$= \frac{T}{2K} \pm \sqrt{\frac{T^2}{4K^2} - \frac{3{,}600\,S}{K}}$$

The +ve sign cannot be adopted, as value of V_m obtained by using +ve sign will be much higher than that is possible in practice. Hence –ve sign will be used and, therefore, we have

$$V_m = \frac{T}{2K} - \sqrt{\frac{T^2}{4K^2} - \frac{3{,}600\,S}{K}} \qquad ...(11.2)$$

From the above equation unknown quantity can be determined by substituting the value of known quantities.

2. *Calculation by Quadrilateral Speed-Time Curve.* Let

α = Acceleration in kmphps

β_c = Coasting retardation in kmphps

β = Braking retardation in kmphps

V_1 = Maximum speed at the end of acceleration in kmph

V_2 = Speed at the end of coasting in kmph

T = Total time of run in seconds

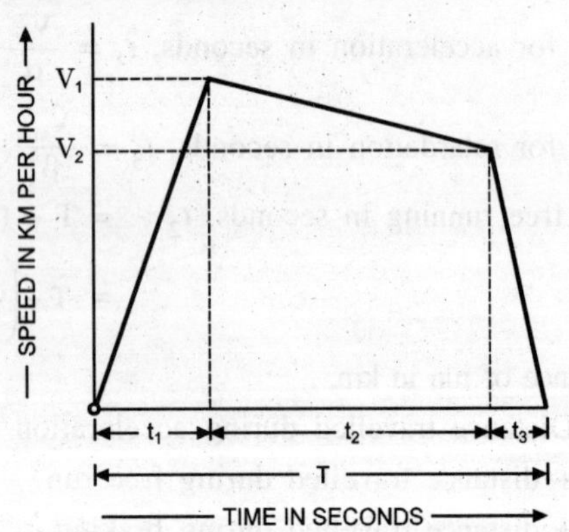

Fig. 11.6. ***Quadrilateral Speed-Time Curve***

Time of acceleration in seconds, $t_1 = \frac{V_1}{\alpha}$

Time of coasting in seconds, $t_2 = \frac{V_1 - V_2}{\beta_c}$

Time of braking in seconds $t_3 = \frac{V_2}{\beta}$

Total distance travelled in km,

S = Distance travelled during acceleration + distance travelled during coasting + distance travelled during retardation

$$= \frac{1}{2} V_1 \times \frac{t_1}{3{,}600} + \frac{V_1 + V_2}{2} \times \frac{t_2}{3{,}600} + \frac{1}{2} V_2 \times \frac{t_3}{3{,}600}$$

$$= \frac{V_1 t_1}{7{,}200} + \frac{V_1 t_2}{7{,}200} + \frac{V_2 t_2}{7{,}200} + \frac{V_2 t_3}{7{,}200}$$

$$= \frac{V_1}{7{,}200}(t_1 + t_2) + \frac{V_2}{7{,}200}(t_2 + t_3)$$

$$\text{or } S = \frac{V_1}{7{,}200}(T - t_3) + \frac{V_2}{7{,}200}(T - t_1)$$

Since $t_1 + t_2 + t_3 = T$

$$\text{or } S = \frac{T}{7{,}200}(V_1 + V_2) - \frac{V_1 t_3}{7{,}200} - \frac{V_2 t_1}{7{,}200}$$

$$\text{or } S = \frac{T}{7{,}200}(V_1 + V_2) - \frac{V_1}{7{,}200} \times \frac{V_2}{\beta} - \frac{V_2}{7{,}200} \times \frac{V_1}{\alpha}$$

$$= \frac{T}{7{,}200}(V_1 + V_2) - \frac{V_1 V_2}{7{,}200\,\beta} - \frac{V_1 V_2}{7{,}200\,\alpha}$$

$$\text{or } 7{,}200S = T(V_1 + V_2) - V_1 V_2\left(\frac{1}{\alpha} + \frac{1}{\beta}\right) \qquad ...(11.3)$$

We have $V_2 = V_1 - \beta_c t_2 = V_1 - \beta_c(T - t_1 - t_3)$

$$= V_1 - \beta_c\left(T - \frac{V_1}{\alpha} - \frac{V_2}{\beta}\right)$$

$$\text{or } \left(V_2 - \frac{\beta_c}{\beta} V_2\right) = V_1 - \beta_c\left(T - \frac{V_1}{\alpha}\right)$$

$$\text{or } V_2 = \frac{V_1 - \beta_c T + \frac{\beta_c}{\alpha} V_1}{1 - \frac{\beta_c}{\beta}} \qquad ...(11.4)$$

Solving Eqs. (11.3) and (11.4) values of S, V_1, V_2 etc. can be obtained.

Example 11.1. A train service consists of following :

Uniform acceleration of 5 km/hr/sec for 30 seconds followed by free running for 10 minutes, then uniform braking at 5 km/hr/sec to stop followed by a stop of 5 minutes.

Draw the speed Vs time curve and calculate

(*i*) Distance between the stations

(*ii*) Average speed and (*iii*) Scheduled speed.

[*U.P. Technical Univ. Utilization of Electrical Energy & Electric Traction 2005-06*]

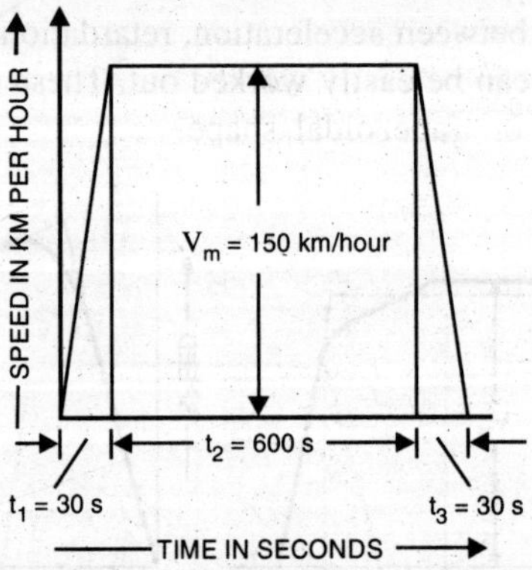

Fig. 11.7

Solution : Acceleration, α = 5 kmphps

Accelerating period, t_1 = 30 seconds

Maximum speed, $V_m = \alpha t_1 = 5 \times 30 = 150$ kmph

Time for free running, $t_2 = 10 \times 60 = 600$ seconds

Retardation, β = 5 kmphps

Time for retardation, $t_s = \frac{V_m}{\beta} = \frac{150}{5} = 30$ seconds

Distance travelled during acceleration period,

$$S_1 = \frac{1}{2}\frac{V_m t_1}{3{,}600}$$

$$= \frac{1}{2} \times \frac{150 \times 30}{3{,}600} = 0.625 \text{ km}$$

Distance travelled during free running,

$$S_2 = \frac{V_m \times t_2}{3{,}600} = \frac{150 \times 600}{3{,}600} = 25 \text{ km}$$

Distance travelled during braking period,

$$S_3 = \frac{1}{2} \times \frac{V_m \times t_3}{3{,}600} = \frac{1}{2} \times \frac{150 \times 30}{3{,}600} = 0.625 \text{ km}$$

(*i*) Total distance between stations, S

$$= S_1 + S_2 + S_3 = 0.625 + 25 + 0.625$$
$$= 26.25 \text{ km} \quad \textbf{Ans.}$$

$$\text{Average speed, } V_a = \frac{S \times 3{,}600}{T} = \frac{26.25 \times 3{,}600}{30 + 600 + 30}$$
$$= 143.18 \text{ kmph} \quad \textbf{Ans.}$$

$$\text{Scheduled speed, } V_s = \frac{S \times 3{,}600}{T + \text{stop time}} = \frac{26.25 \times 3{,}600}{660 + 5 \times 60}$$
$$= 98.44 \text{ kmph} \quad \textbf{Ans.}$$

Example 11.2. An electric train has an average speed of 42 kmph on a level track between stops 1,400 m apart. It is accelerated at 1.7 kmphps and is braked at 3.3 kmphps. Draw the speed-time curve for the run.

[*Pb. Technical Univ. Electric Drives & Utilization Dec. 2008*]

Solution : Acceleration, α = 1.7 kmphps

Retardation, β = 3.3 kmphps

Distance of run, S = 1,400 m = 1.4 km

Average speed, V_a = 42 kmph

$$\text{Time of run, } T = \frac{S}{V_a} \times 3{,}600 = \frac{1.4}{42} \times 3600$$
$$= 120 \text{ seconds}$$

$$\text{Using equation } V_m = \frac{T}{2K} - \sqrt{\frac{T^2}{4K^2} - \frac{3{,}600\, S}{K}}$$

$$\text{where } K = \frac{1}{2\alpha} + \frac{1}{2\beta} = \frac{1}{3.4} + \frac{1}{6.6} = 0.4456$$

Maximum speed,

$$V_m = \frac{120}{2 \times 0.4456} - \sqrt{\frac{(120)^2}{(2 \times 0.4456)^2} - \frac{3600 \times 1.4}{0.4456}}$$
$$= 52 \text{ kmph}$$

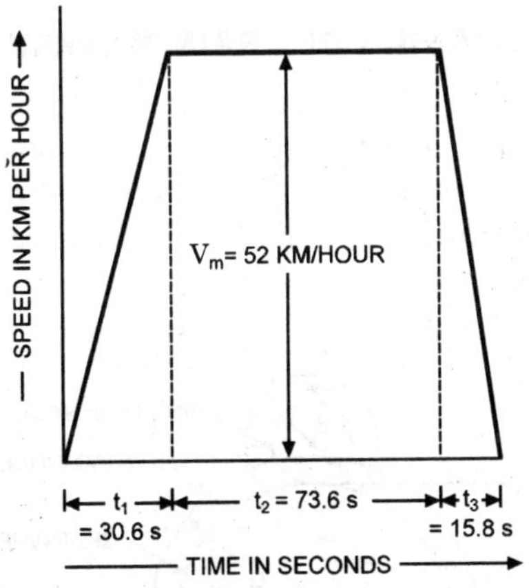

Fig. 11.8

$$\text{Acceleration period, } t_1 = \frac{V_m}{\alpha} = \frac{52}{1.7} = 30.6 \text{ seconds}$$

$$\text{Braking period, } t_3 = \frac{V_m}{\beta} = \frac{52}{3.3} = 15.8 \text{ seconds}$$

$$\text{Free running period, } t_2 = T - (t_1 + t_2) = 120 - (30.6 + 15.8)$$
$$= 73.6 \text{ seconds.}$$

The desired speed-time curve is shown in Fig. 11.8.

Example 11.3. A schedule speed of 45 km per hour is required between two stops 1.5 km apart. Find the maximum speed over the run if the stop is of 20 second duration. The values of acceleration and retardation are 2.4 kmphps and 3.2 kmphps respectively. Assume a simplified trapezoidal speed-time curve.

[*Pb. Technical Univ. Electric Drives and Traction 2003*]

Solution : Acceleration, α = 2.4 kmphps

Retardation, β = 3.2 kmphps

Distance of run, S = 1.5 km

Schedule speed, V_s = 45 kmph

$$\text{Schedule time, } T_s = \frac{S \times 3{,}600}{V_s} = \frac{1.5 \times 3{,}600}{45}$$
$$= 120 \text{ seconds}$$

$$\text{Actual time for run, } T = T_s - \text{stop duration}$$
$$= 120 - 20 = 100 \text{ seconds}$$

Using Eq. (11.2) we have

$$\text{Maximum speed, } V_m = \frac{T}{2K} - \sqrt{\frac{T^2}{4K^2} - \frac{3{,}600\, S}{K}}$$

$$\text{where } K = \frac{1}{2\alpha} + \frac{1}{2\beta} = \frac{1}{2 \times 2.4} + \frac{1}{2 \times 3.2} = 0.3646$$

$$\text{So, } V_m = \frac{100}{2 \times 0.3646} - \sqrt{\frac{100^2}{4 \times (0.3646)^2} - \frac{3{,}600 \times 1.5}{0.3646}}$$
$$= 74 \text{ kmph} \quad \textbf{Ans.}$$

Example 11.4. An electric train is to have acceleration and braking retardation of 0.8 km/h/s and 3.2 km/h/s respectively. If the ratio of maximum to average speed is 1.3 and time for stops 26 seconds, find schedule speed for a run of 1.5 km. Assume simplified trapezoidal speed-time curve.

[*Gorakhpur Univ. Utilization of Electrical Power & Traction Sept. 1984; M.D. Univ. Utilization of Electric Power & Traction May 2010; Rajasthan Technical Univ. Utilization of Electrical Power 2008, 2010*]

Solution : Let the actual time of run be T seconds

$$\text{Average speed, } V_a = \frac{3{,}600S}{T} = \frac{3{,}600 \times 1.5}{T} = \frac{5{,}400}{T} \text{ km per hour}$$

$$\text{Maximum speed, } V_m = 1.3V_a = 1.3 \times \frac{5{,}400}{T} = \frac{7{,}020}{T} \text{ km per hour}$$

$$\text{Since } V_m^2\left(\frac{1}{2\alpha} + \frac{1}{2\beta}\right) - V_m T + 3{,}600\, S = 0$$

$$\therefore V_m^2 = \frac{V_m T - 3{,}600\, S}{\frac{1}{2\alpha} + \frac{1}{2\beta}}$$

$$= \frac{\frac{7{,}020}{T} \times T - 3{,}600 \times 1.5}{\frac{1}{2 \times 0.8} + \frac{1}{2 \times 3.2}} = \frac{7{,}020 - 5{,}400}{5/6.4}$$

$$\text{or } V_m = \sqrt{1{,}620 \times \frac{6.4}{5}} = 45.53 \text{ kmph}$$

$$\text{and } V_a = \frac{V_m}{1.3} = \frac{45.53}{1.3} = 35.028 \text{ kmph}$$

$$\text{Actual time of run } T = \frac{3{,}600\, S}{V_a} = \frac{3{,}600 \times 1.5}{35.028} = 154 \text{ seconds}$$

Schedule time, T_s = Actual time of run + time of stop

$= 154 + 26 = 180$ seconds

$$\text{Schedule speed, } V_s = \frac{S \times 3{,}600}{T_s} = \frac{1.5 \times 3{,}600}{180} = 30 \text{ kmph} \quad \textbf{Ans.}$$

Example 11.5. A train has schedule speed of 30 kmph over a level track, distance between stations being 1 km. Station stopping time is 20 seconds. Assuming braking retardation of 3 kmphps and maximum speed 25 per cent greater than average speed, calculate acceleration required to run the service if the speed-time curve is approximated by a trapezoidal curve.

[Rajasthan Univ. Utilization of Electrical Power 2005]

Solution : Schedule time of run, $T_s = \frac{S \times 3{,}600}{V_s} = \frac{1 \times 3{,}600}{30}$

$= 120$ seconds

Actual time of run, $T = T_s$ – duration of stop

$= 120 - 20 = 100$ seconds

$$\text{Average speed, } V_a = \frac{S \times 3{,}600}{T} = \frac{1 \times 3{,}600}{100} = 36 \text{ kmph}$$

Maximum speed, $V_m = 1.25\ V_a = 1.25 \times 36$

$= 45$ kmph

Using Eq. (11.1) we have

$$V_m^2\left(\frac{1}{2\alpha} + \frac{1}{2\beta}\right) - V_m T + 3{,}600\ S = 0$$

$$\text{or} \quad \frac{1}{2\alpha} + \frac{1}{2\beta} = \frac{V_m T - 3{,}600\ S}{V_m^2}$$

$$= \frac{45 \times 100 - 3{,}600 \times 1}{(45)^2} = \frac{4}{9}$$

$$\text{or} \quad \frac{1}{2\alpha} = \frac{4}{9} - \frac{1}{2\beta} = \frac{4}{9} - \frac{1}{2 \times 3} = 0.2777$$

$$\text{or } \alpha = \frac{1}{2 \times 0.2777} = 1.8 \text{ kmphps} \quad \textbf{Ans.}$$

Example 11.6. A suburban electric train has a maximum speed of 70 kmph. The schedule speed including a station stop of 30 seconds is 45 kmph. If the acceleration is 1.5 kmphps, find the value of retardation when the average distance between stops is 4 km. *[A.M.I.E. Sec B. Utilization of Elec. Power Winter 1984; Pb. Technical Univ. Electric Drives & Utilization May 2009]*

Solution : Schedule time of run, $T_s = \frac{3{,}600 \times S}{V_s} = \frac{3{,}600 \times 4}{45}$

$= 320$ seconds

Actual time of run, $T = T_s$ – duration of stop

$= 320 - 30 = 290$ seconds

$$\text{Since } V_m^2\left(\frac{1}{2\alpha} + \frac{1}{2\beta}\right) - V_m T + 3{,}600\ S = 0$$

$$\text{or} \quad \frac{1}{2\alpha} + \frac{1}{2\beta} = \frac{V_m T - 3{,}600\ S}{V_m^2}$$

$$= \frac{70 \times 290 - 3{,}600 \times 4}{(70)^2} = 1.204$$

$$\text{or} \quad \frac{1}{2\beta} = 1.204 - \frac{1}{2\alpha} = 1.204 - \frac{1}{2 \times 1.5} = 0.87$$

$$\text{or } \beta = \frac{1}{2 \times 0.87} = 0.575 \text{ kmphps} \quad \textbf{Ans.}$$

Example 11.7. A train is required to run between two stations 1.6 km apart at an average speed of 40 kmph. The run is to be made to a simplified quadrilateral speed-time curve. If the maximum speed is to be limited to 64 kmph, acceleration to 2.0 kmphps and coasting and braking retardation to 0.16 kmphps and 3.2 kmphps respectively, determine the duration of acceleration, coasting and braking periods.

[Gorakhpur Univ. Utilization of Electrical Power with Traction 1980; Pb. Univ. Electric and Utilization January 1991; Rajasthan Technical Univ. Utilization of Electrical Power 2011]

Solution : Distance of run, S = 1.6 km

Average speed, V_a = 40 km per hour

Maximum speed, V_1 = 64 km per hour

Acceleration, α = 2.0 kmphps

Coasting retardation, β_c = 0.16 kmphps

Braking retardation, β = 3.2 kmphps

$$\text{Duration of acceleration, } t_1 = \frac{V_1}{\alpha} = \frac{64}{2.0} = 32 \text{ seconds} \quad \textbf{Ans.}$$

$$\text{Actual time of run, } T = \frac{3{,}600\ S}{V_a} = \frac{3{,}600 \times 1.6}{40} = 144 \text{ seconds}$$

Let the speed before applying brakes be V_2

$$\text{then duration of coasting, } t_2 = \frac{V_1 - V_2}{\beta_c} = \frac{64 - V_2}{0.16} \text{ seconds}$$

$$\text{Duration of braking, } t_3 = \frac{V_2}{\beta} = \frac{V_2}{3.2} \text{ seconds}$$

Since actual time of run, $T = t_1 + t_2 + t_3$

$$\therefore\ 144 = 32 + \frac{64 - V_2}{0.16} + \frac{V_2}{3.2}$$

$$\text{or } V_2\left(\frac{1}{0.16} - \frac{1}{3.2}\right) = 32 + 400 - 144$$

$$\text{or } V_2 = \frac{288}{6.25 - 0.3125} = 48.5 \text{ km per hour}$$

$$\text{Duration of coasting, } t_2 = \frac{V_1 - V_2}{\beta_c} = \frac{64 - 48.5}{0.16} = 96.85 \text{ seconds} \quad \textbf{Ans.}$$

$$\text{Duration of braking, } t_3 = \frac{V_2}{\beta} = \frac{48.5}{3.2} = 15.15 \text{ seconds} \quad \textbf{Ans.}$$

11.6. MECHANICS OF TRAIN MOVEMENT

Essential driving mechanism of an electric locomotive is shown in Fig. 11.9. The armature of the driving motor has a pinion of diameter *d′* attached to it. The tractive effort at the edge of the pinion is transferred to the driving wheel by means of a gearwheel.

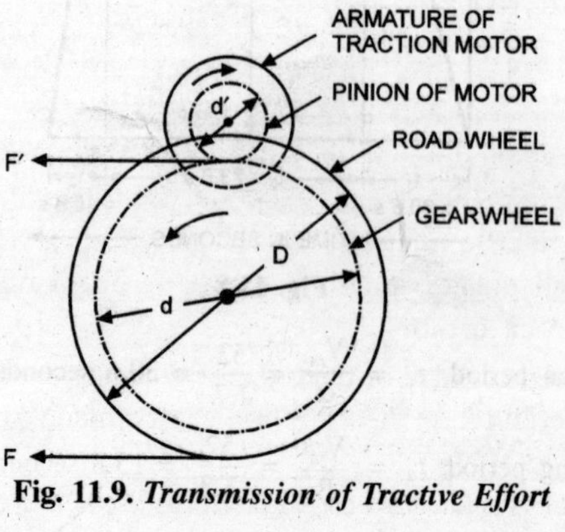

Fig. 11.9. *Transmission of Tractive Effort*

Let the driving motor exert a torque T in Nm.

Tractive effort at the edge of pinion is given by the equation

$$T = F'\frac{d'}{2} \quad \text{or} \quad F' = \frac{2T}{d'}$$

Tractive effort transferred to the driving wheel

$$F = \eta F'\left(\frac{d}{D}\right) = \eta\frac{2T}{d'}\left(\frac{d}{D}\right) = \eta T\left(\frac{2}{D}\right)\left(\frac{d}{d'}\right)$$

$$= \eta T\frac{2\gamma}{D} \qquad ...(11.5)$$

where d is diameter of gearwheel in metres, D is diameter of driving wheel in metres and η is the efficiency of transmission, γ is the gear ratio and is equal to $\frac{d}{d'}$.

The maximum frictional force between the driving wheel and the track = μW where μ is the *coefficient of adhesion* between the driving wheel and the track and W is the weight of the train on the driving axles (called *adhesive weight*). Slipping will not take place unless tractive effort $F > \mu W$. For motion of trains without slipping tractive effort F should be less than or at the most equal to μW but in no case greater than μW.

The magnitude of the tractive effort that can be employed for propulsion, therefore, depends upon the weight coming over the driving wheels and the coefficient of adhesion between the driving wheel and the track. The coefficient of adhesion is defined as

i.e., Coefficient of adhesion,

$$\mu = \frac{\text{Maximum tractive effort that can be applied without slipping of wheels}}{\text{Adhesive weight (Weight on the driving wheels)}} \qquad ...(11.6)$$

The coefficient of adhesion reduces with the increase in speed as shown below in Table 11.2.

Table 11.2. Variation of coefficient of adhesion with speed on dry rails.

Speed in kmph	0	15	30	45	60	75
Coefficient of Adhesion	0.25	0.18	0.14	0.12	0.10	0.09

The normal value of coefficient of adhesion with clean dry rails is 0.25 and with wet or greasy rails the value may be as low as 0.08.

Electric traction has a very important advantage over steam traction and that is on account of greater adhesive weight - in a motor coach 100 per cent of the weight is on the driving wheels, in an electric locomotive 70 per cent whereas in a steam locomotive less than 50 per cent. The coefficient of adhesion in electric traction is also greater than that in steam traction due to the following two reasons.

(*i*) In electric traction the torque exerted is continuous whereas in steam traction the torque is pulsating which causes jolting and skidding.

(*ii*) In electric traction the driving wheels are distributed over a much greater length whereas in steam traction they are close to each other.

Since higher value of tractive effort can be used in electric traction, therefore, an electric train can be made to accelerate at a faster rate. This results in saving of time, especially when the distance between stops is small *e.g.* in urban and suburban areas.

11.7. TRACTIVE EFFORT FOR PROPULSION OF TRAIN

The effective force, necessary to propel the train, at the wheels of locomotive is called the *tractive effort*. It is tangential to the driving wheels and measured in newtons.

Total tractive effort required to run a train on track

= Tractive effort required for linear and angular acceleration + tractive effort to overcome the effect of gravity + tractive effort to overcome the train resistance.

or $F_t = F_a \pm F_g + F_r$ ...(11.7)

1. *Tractive Effort For Acceleration.* According to laws of dynamics force is required to accelerate the motion of the body and is given by the expression

Force = Mass × acceleration

Consider a train of weight W tonnes being accelerated at α kmphps

The weight of train = 1,000 W kgf

Mass of train, m = 1,000 W kg

Acceleration = α kmphps

$$= \alpha \times \frac{1{,}000}{3{,}600} \text{ m/s}^2$$

$$= 0.2778\ \alpha \text{ m/s}^2$$

Tractive effort required for linear acceleration,

$$F_a = m\alpha$$

$$= 1{,}000\ W \times 0.2778\ \alpha$$

$$= 277.8\ W\alpha \text{ newtons} \qquad ...(11.8)$$

With the linear acceleration of the train, the rotating parts of the train such as wheels and motors also accelerate in an angular direction, and therefore, the tractive effort required is equal to the arithmetic sum of tractive effort required to have the angular acceleration of rotating parts and tractive effort required to have the linear acceleration. The tractive effort required to have the angular acceleration depends upon the individual weight, radius of gyration etc. of the rotating parts requiring angular acceleration. Hence the equivalent or accelerating weight of the train is taken as W_e which is higher than the dead weight W requiring linear acceleration to consider the tractive effort for the angular acceleration. In practice W_e is higher than W by 8 to 15%. The normal value lies between 10 and 12 per cent.

Hence tractive effort required for acceleration

$$F_a = 277.8\ W_e\ \alpha \text{ newtons} \qquad ...(11.9)$$

2. *Tractive Effort For Overcoming The Effect of Gravity.* When a train is on a slope, a force of gravity equal to the component of the dead weight along the slope acts on the train and tends to cause its motion down the gradient or slope. Hence force due to gradient, F_g = 1,000 W sin θ kg ...(11.10)

But in railway work gradient is expressed as rise in metres in a track distance of 100 metres and is denoted as '*percentage gradient*' (G%).*

* Some authors define percentage gradient as rise in metres in a track distance of 1,000 metres.

i.e., $G = \sin\theta \times 100$

or $\sin\theta = \dfrac{G}{100}$

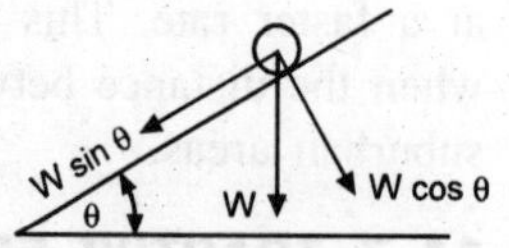

Fig. 11.10

Substituting $\sin\theta = \dfrac{G}{100}$ in Eq. (11.10) we have,

$$F_g = 1{,}000\ W \times \frac{G}{100}$$
$$= 10\ WG\ kg = 10\ WG \times 9.81$$
$$= 98.1\ WG\ newtons \quad ...(11.11)$$

When the train is going up a gradient, the tractive effort will be required to balance this force due to gradient but while going down the gradient, the force will add to the tractive effort.

3. ***Tractive Effort For Overcoming Train Resistance.*** Train resistance consists of all the forces resisting the motion of a train when it is running at uniform speed on a straight and level track. Under these circumstances the whole of the energy output from the driving axles is expended against train resistance. Train resistance is due to (*i*) the friction at the various parts of the rolling stock (*ii*) friction at the track and (*iii*) air resistance. The first two components constitute the *mechanical resistance* component of train resistance. The train resistance depends upon various factors, such as shape, size and condition of track etc., and is expressed in newtons per tonne of the dead weight. For a normal train the value of specific resistance has been 40 to 70 newtons/tonne.

The general equation for train resistance is given as

$$R = k_1 + k_2 V + k_3 V^2 \quad ...(11.12)$$

where k_1, k_2 and k_3 are constants depending upon the train and the track, R is the resistance in newtons and V is the speed in kmph. The first two terms represent the mechanical resistance and the last term represents air resistance.

Tractive effort required to overcome the train resistance,

$$F_r = W \times r \text{ newtons} \quad ...(11.13)$$

where r is the specific resistance in newtons per tonne of the dead weight.

Total tractive effort required,

$$F_t = F_a \pm F_g + F_r = 277.8\ W_e \alpha \pm 98.1\ WG + Wr \quad ...(11.14)$$

+ ve sign for the motion up the gradient and

– ve sign for the motion down the gradient.

11.8. POWER OUTPUT FROM THE DRIVING AXLES

Power, P = Rate of doing work

$$= \text{Tractive effort} \times \frac{\text{distance}}{\text{time}}$$
$$= \text{Tractive effort} \times \text{speed}$$
$$= F_t \times v \quad \text{where } F_t \text{ is in newtons and } v \text{ is in m/s}$$

$$\text{or } P = F_t \times V \times \frac{1{,}000}{3{,}600} \text{ watts} = \frac{F_t \times V}{3{,}600} \text{ kW} \quad ...(11.15)$$

where V is in kmph.

Example 11.8. An electric train weighing 200 tonne has eight motors geared to driving wheels, each wheel is 90 cm diameter. Determine the torque developed by each motor to accelerate the train to a speed of 48 kmph in 30 seconds up a gradient of 1 in 200. The tractive resistance is of 50 newtons per tonne, the effect of rotational inertia is 10% of the train weight, the gear ratio is 4 to 1 and gearing efficiency is 80 per cent.

[Agra Univ. Traction & Utilization of Elec. Power 1983]

Solution : Weight of train, W = 200 tonnes

Diameter of driving wheels, D = 90 cm = 0.9 m

Percentage gradient, $G = \dfrac{1}{200} \times 100 = 0.5\%$

Tractive resistance, r = 50 newtons/tonne

Gear ratio, $\gamma = 4$

Gearing efficiency, $\eta = 0.80$

Equivalent accelerating weight of the train

$$W_e = 1.10\ W = 1.1 \times 200 = 220 \text{ tonnes}$$

Maximum speed, V_m = 48 kmph

Time taken in accelerating the train to a speed of 48 kmph,

t_1 = 30 seconds

$$\text{Acceleration, } \alpha = \frac{V_m}{t_1} = \frac{48}{30} = 1.6 \text{ kmphps}$$

Tractive effort required,

$$F_t = 277.8\ W_e \alpha + 98.1\ WG + Wr$$
$$= 277.8 \times 220 \times 1.6 + 98.1 \times 200 \times 0.5 + 200 \times 50$$
$$= 117{,}596 \text{ N}$$

Total torque developed,

$$T = \frac{F_t \times D}{\eta \times 2\gamma} = \frac{117{,}596 \times 0.9}{0.8 \times 2 \times 4} = 16{,}537 \text{ N-m}$$

Torque developed by each motor $= \dfrac{16{,}537}{8} = 2{,}067$ N-m **Ans.**

Example 11.9. A 200 tonne motor coach having 4 motors each developing 6,000 N-m torque during acceleration, starts from rest. If the gradient is 30 in 1,000; gear ratio 4, gear transmission efficiency 90%, wheel radius 45 cm, train resistance 50 N/tonne; addition of rotational inertia 10%, calculate the time taken to attain a speed of 50 kmph. If the line voltage is 3,000 V dc and efficiency of motors 85%, find the current during notching period.

[Pb. Univ. Electric Drives and Utilization Dec. 1989; Rajasthan Univ. Utilization of Electrical Power 2006]

Solution : Weight of train, W = 200 tonnes

Diameter of driving wheels, D = 2 × 45 = 90 cm = 0.9 m

Percentage gradient, $G = \dfrac{30}{1{,}000} \times 100 = 3\%$

Tractive resistance, r = 50 N/tonne

Gear ratio, $\gamma = 4$

Gear transmission efficiency, η = 90% or 0.9

Equivalent accelerating weight of the train,

$$W_e = 1.10 \times 200 = 220 \text{ tonnes}$$

Total torque developed, T = 4 × 6,000 = 24,000 N-m

Tractive effort,

$$F_t = \frac{\eta T 2\gamma}{D} = \frac{0.9 \times 24{,}000 \times 2 \times 4}{0.9} = 192{,}000 \text{ N}$$

Let the acceleration of the train be α kmphps

$$\text{Then } F_t = 277.8\ W_e \alpha + 98.1\ WG + Wr$$
$$\text{or } 192{,}000 = 277.8 \times 220\alpha + 98.1 \times 200 \times 3 + 200 \times 50$$

$$\alpha = \frac{123{,}140}{277.8 \times 220} = 2.015 \text{ kmphps}$$

Time taken for the train to attain a speed of 50 kmph

$$t = \frac{V_m}{\alpha} = \frac{50}{2.015} = 24.82 \text{ seconds} \quad \textbf{Ans.}$$

Power output from the driving axles

$$= \frac{F_t \times V}{3,600} = \frac{192,000 \times 50}{3,600} = 2,666.7 \text{ kW}$$

$$\text{Power input} = \frac{\text{Power output}}{\eta_m} = \frac{2,666.7}{0.85} = 3,137 \text{ kW}$$

$$\text{Total current drawn} = \frac{\text{Power input in kW} \times 1,000}{V}$$

$$= \frac{3,137 \times 1,000}{3,000} = 1,046 \text{ A}$$

$$\text{Current drawn per motor} = \frac{1,046}{4} = 261.5 \text{ A} \quad \textbf{Ans.}$$

Example 11.10. A train weighing 120 tonnes is to be driven up an incline of 2 per cent at a speed of 36 kmph. If the train resistance at this speed is 2 kg per tonne, find the current required at 1,500 V dc if the efficiency of the motors and gearing is 88 per cent. If the current were cut off, how long would the train take to come to rest

Solution : Tractive effort,

$$F_t = 98.1 \text{ WG} + \text{W}r$$
$$= 98.1 \times 120 \times 2 + 120 \times 2 \times 9.81 = 25,900 \text{ N}$$

Speed, V = 36 kmph

$$\text{Power output of driving axles} = \frac{F_t \times V}{3,600} = \frac{25,900 \times 36}{3,600} = 259 \text{ kW}$$

$$\text{Power input} = \frac{\text{Power output}}{\text{Efficiency of motors and gears}} = \frac{259}{0.88} = 294.3 \text{ kW}$$

$$\text{Current required} = \frac{\text{Power input in kW} \times 1,000}{\text{Voltage of the line}}$$

$$= \frac{294.3 \times 1,000}{1,500} = 196.2 \text{ amperes } \textbf{Ans.}$$

Let the coasting retardation be β_c

then $F_t = 277.8\, W_e(-\beta_c) + 98.1 \text{ WG} + \text{W}r$

Assuming W_e = 110 per cent of dead weight, we have

$$0 = -277.8 \times 1.1 \text{ W} \beta_c + 98.1 \times 2\text{W} + 2 \times 9.81 \text{ W}$$

$$\text{or } \beta_c = \frac{98.1 \times 2.2}{277.8 \times 1.1} = 0.7 \text{ kmphps}$$

$$\text{Time taken to come to rest} = \frac{V}{\beta_c} = \frac{36}{0.7} = 51 \text{ seconds} \quad \textbf{Ans.}$$

Example 11.11. An electric train accelerates uniformly from rest to a speed of 48 km/hour in 24 seconds. It then coasts for 69 seconds against a constant resistance of 58 N/tonne and is braked to rest at 3.3 km/hour/second in 11 seconds. Calculate *(i)* the acceleration *(ii)* coasting retardation and *(iii)* the schedule speed, if the station stops are of 20 second duration. What would be the effect on schedule speed of reducing the station stops to 15 second duration, other conditions remaining same? Allow 10% for rotational inertia.

Solution : Time of acceleration, t_1 = 24 seconds

Time of coasting, t_2 = 69 seconds

Time of braking, t_3 = 11 seconds

Maximum speed, V_1 = 48 kmph

$$(i) \text{ Acceleration, } \alpha = \frac{V_1}{t_1} = \frac{48}{24} = 2 \text{ kmphps} \quad \textbf{Ans.}$$

Let the coasting retardation be β_c

$$F_t = 277.8\, W_e(-\beta_c) + 98.1 \text{ WG} + \text{W}r$$
$$= -277.8 \times 1.1 \text{ W} \beta_c + 98.1 \text{ W} \times 0 + 58 \text{ W}$$

$$(ii) \quad \text{or } \beta_c = \frac{58}{277.8 \times 1.1} = 0.19 \text{ kmphps} \quad \textbf{Ans.}$$

Speed at the end of coasting period V_2

$$= V_1 - \beta_c \times t_2 = 48 - 0.19 \times 69 = 34.9 \text{ kmph}$$

Distance travelled,

$$S = \frac{V_1 t_1}{7,200} + \frac{(V_1 + V_2)t_2}{7,200} + \frac{V_2 t_3}{7,200} \text{ km}$$

$$= \frac{48 \times 24}{7,200} + \frac{(48 + 34.9) \times 69}{7,200} + \frac{34.9 \times 11}{7,200} = 1.0 \text{ km}$$

Schedule time, $T_s = t_1 + t_2 + t_3$ + duration of stop = 24 + 69 + 11 + 20 = 124 seconds

$$(iii) \text{ Schedule speed, } V_s = \frac{S \times 3,600}{T_s} = \frac{1.0 \times 3,600}{124}$$
$$= 29 \text{ kmph} \quad \textbf{Ans.}$$

When the duration of stop is 15 seconds

Schedule time, T_s' = 24 + 69 + 11 + 15 = 119 seconds

$$\text{and schedule speed, } V_s' = \frac{1.0 \times 3,600}{119} = 30.25 \text{ kmph} \quad \textbf{Ans.}$$

Example 11.12. An electric train has quadrilateral speed-time curve as follows:

***(i)* Uniform acceleration from rest at 2 kmphps for 30 seconds;**

***(ii)* Coasting for 50 seconds; *(iii)* Uniform braking to rest for 20 seconds.**

If the train is moving a uniform upgradient of 10/1,000, train resistance is 40 N/tonne, rotational inertia effect 10% of dead weight and duration of stop 30 seconds, find the schedule speed. *[Pb. Technical Univ. Electric Drives & Utilization May 2009]*

Solution : Time of acceleration, t_1 = 30 seconds

Time of coasting, t_2 = 50 seconds

Time of braking, t_3 = 20 seconds

Acceleration, α = 2 kmphps

Maximum speed, $V_1 = \alpha t_1 = 2 \times 30 = 60$ km/hour

$$\text{Gradient, G} = \frac{10}{1,000} \times 100 = 1\%$$

Let the coasting retardation be β_c

$$F_t = 277.8\, W_e(-\beta_c) + 98.1 \text{ WG} + \text{W}r$$
$$\text{or } 0 = -277.8 \times 1.1 \text{ W} \beta_c + 98.1 \times 1 \times \text{W} + 40 \text{ W}$$

$$\text{or } \beta_c = \frac{98.1 + 40}{277.8 \times 1.1} = \frac{138.1}{277.8 \times 1.1} = 0.452 \text{ kmphps}$$

$$V_2 = V_1 - \beta_c t_2 = 60 - 0.452 \times 50 = 37.4 \text{ km per hour}$$

$$\text{Braking retardation, } \beta = \frac{V_2}{t_3} = \frac{37.4}{20} = 1.87 \text{ kmphps}$$

$$\text{Distance travelled, S} = \frac{V_1 t_1}{7,200} + \frac{(V_1 + V_2)t_2}{7,200} + \frac{V_2 t_3}{7,200} \text{ km}$$

$$= \frac{60 \times 30}{7,200} + \frac{60 + 37.4}{7,200} \times 50 + \frac{37.4 \times 20}{7,200}$$
$$= 1.03 \text{ km}$$

Schedule time, $T_s = t_1 + t_2 + t_3$ + duration of stop

= 30 + 50 + 20 + 30 = 130 seconds

$$\text{Schedule speed, } V_s = \frac{3,600 \times S}{T_s} = \frac{3,600 \times 1.03}{130}$$
$$= 28.52 \text{ km per hour} \quad \textbf{Ans.}$$

Example 11.13. The speed-time curve of an electric train on a uniform rising gradient of 1 in 100 comprises

***(i)* Uniform acceleration from rest at 2 kmphps for 30 seconds**

***(ii)* Coasting with power off for 70 seconds**

***(iii)* Braking at 3 kmphps to standstill.**

The weight of the train is 250 tonnes, the train resistance on level track being 5 kg/tonne, and allowance for rotary inertia 10%.

Calculate the maximum power developed by traction motors and total distance travelled by the train. Assume transmission efficiency as 97%.

[*Agra Univ. Traction & Utilization of Elec. Power 1979*]

Solution : Weight of train, W = 250 tonnes

Effective weight of train, $W_e = \left(1+\frac{10}{100}\right)W = 1.1\ W$

Train resistance, r = 5 kg/tonne = 5 × 9.81 = 49.05 N/tonne

Gradient, G = 1 per cent

Maximum velocity, $V_m = \alpha t_1 = 2 \times 30 = 60$ km/hour

Tractive effort required,

$$F_t = 277.8\ W_e\ \alpha + 98.1\ WG + Wr$$

$$= 277.8 \times 1.1 \times 250 \times 2 + 98.1 \times 250 \times 1 + 250 \times 49.05$$

$$= 189{,}577 \text{ newtons}$$

Maximum power output from the driving axles

$$= \frac{F_t V_m}{3{,}600} = \frac{189{,}577}{3{,}600} \times 60 = 3{,}160 \text{ kW}$$

Maximum power developed by traction motors

$$= \frac{3{,}160}{0.97} = 3{,}258 \text{ kW} \quad \textbf{Ans.}$$

Let the coasting retardation be β_c then

$$F_t = 277.8\ W_e(-\beta_c) + 98.1\ WG + Wr$$

or $0 = -277.8 \times 1.1\ W \times \beta_c + 98.1\ W \times 1 + W \times 49.05$

or $\beta_c = \frac{98.1 + 49.05}{277.8 \times 1.1} = 0.4815$ kmphps

$$V_2 = V_1 - \beta_c t_2 = 60 - 70 \times 0.4815 = 26.3 \text{ km/hour}$$

Braking period, $t_3 = \frac{V_2}{\beta} = \frac{26.3}{3} = 8.77$ seconds

Total distance travelled by the train,

$$S = \frac{V_1 t_1}{7{,}200} + \frac{(V_1 + V_2)t_2}{7{,}200} + \frac{V_2 t_3}{7{,}200}$$

$$= \frac{60 \times 30}{7{,}200} + \frac{60 + 26.3}{7{,}200} \times 70 + \frac{26.3 \times 8.77}{7{,}200} = 1.12 \text{ km} \quad \textbf{Ans.}$$

11.9. ENERGY OUTPUT FROM DRIVING AXLES

Assuming the run according to trapezoidal speed-time curve

Total energy required for the run

= Energy required during acceleration + energy required during free run

= Average power during acceleration × acceleration period + average power during free run × duration of free run

$$= \frac{1}{2}\frac{F_t V_m}{3{,}600} \times \frac{t_1}{3{,}600} + \frac{F_t' V_m}{3{,}600} \times \frac{t_2}{3{,}600} \text{ kWh} \quad ...(11.16)$$

where V_m is the maximum speed in kmph, t_1 is the time of acceleration in seconds, t_2 is the time of free run in seconds, F_t is the tractive effort required during acceleration in newtons and F_t' is the tractive effort required during free run in newtons. Instead of expressing the energy in kWh, it is more convenient for the purpose of comparison to introduce the weight of the train and the distance of run and to express the energy in watt-hours per tonne-km.

i.e.,
$$\frac{\text{Energy output in watt - hours}}{\text{Weight of the train in tonnes} \times \text{distance of run in km}} \quad ...(11.17)$$

This quantity is called *specific energy output*, and is used for comparing the dynamical performances of trains operating to different schedules.

The energy input to the motors is called the *energy consumption* of the train, since it is the energy consumed for propelling the train. The total energy drawn from the distribution system will be greater than this by the quantity required for lighting, heating, control and braking.

The energy consumption can be expressed in watt-hours per tonne-km.

i.e.,
$$\frac{\text{Energy consumption of train in watt-hours}}{\text{Weight of train in tonnes} \times \text{distance of run in km}} \quad ...(11.18)$$

The above quantity is called the *specific energy consumption*.

11.10. DETERMINATION OF SPECIFIC ENERGY OUTPUT USING SIMPLIFIED SPEED-TIME CURVE

Let the track have a gradient of G per cent through out its run.

Energy output to accelerate the train from rest to a speed V_m

$$= \frac{1}{2}\frac{F_t V_m}{3{,}600} \times \frac{t_1}{3{,}600} \text{ kWh} = \frac{1}{2}\frac{F_t V_m}{3{,}600} \times \frac{V_m}{3{,}600\alpha} \text{ kWh}$$

since $t_1 = \frac{V_m}{\alpha}$

$$= \frac{1}{2}\frac{V_m^2}{(3{,}600)^2\alpha}[277.8\ W_e\alpha + 98.1\ WG + Wr] \text{ kWh} \quad ...(11.19)$$

since $F_t = 277.8W_e\alpha + 98.1WG + Wr$

Energy output to run the train at the speed V_m against the gradient and resistance to motion

$$= \frac{F_t' V_m}{3{,}600} \times \frac{t_2}{3{,}600} \text{ kWh} = \frac{F_t' \times S'}{3{,}600} \text{ kWh}$$

$$= [Wr + 98.1WG] \times \frac{S'}{3{,}600} \text{ kWh} \quad ...(11.20)$$

since $\frac{V_m t_2}{3{,}600} = S'$, the distance travelled during free run

Total energy output for the run

$$= \frac{V_m^2}{2\alpha(3{,}600)^2}[277.8W_e\alpha + 98.1WG + Wr] + [Wr + 98.1WG] \times \frac{S'}{3{,}600} \text{ kWh}$$

$$= \frac{V_m^2 \times 1{,}000}{2\alpha \times (3{,}600)^2}[277.8W_e\alpha + 98.1WG + Wr] + [Wr + 98.1WG] \times \frac{S' \times 1{,}000}{3{,}600} \text{ Wh}$$

$$= \frac{V_m^2}{2\alpha} \times \frac{1{,}000}{(3{,}600)^2} \times 277.8W_e\alpha + \frac{V_m^2}{2\alpha} \times \frac{1{,}000}{(3{,}600)^2}[98.1WG + Wr] + \frac{1{,}000S'}{3{,}600}[Wr + 98.1WG]$$

$$= 0.01072V_m^2W_e + [98.1WG + Wr]\left[\frac{V_m^2}{2\alpha \times 3{,}600} + S'\right] \times \frac{1{,}000}{3{,}600}$$

$$= 0.01072V_m^2W_e + \frac{1{,}000}{3{,}600}[98.1WG + Wr][S'' + S']$$

where $\frac{V_m^2}{2\alpha \times 3{,}600} = S''$, distance travelled during accelerating period

$$= 0.01072\, V_m^2 W_e + 0.2778\,(98.1\,WG + Wr)S_1 \quad ...(11.21)$$

where S_1 is the distance travelled during acceleration and free run in km

Specific energy output

$$= \frac{\text{Energy output for the run in watt-hours}}{\text{Weight of train in tonnes} \times \text{distance of run in km}}$$

$$= \frac{0.01072 V_m^2 W_e + 0.2778(98.1WG + Wr)S_1}{W \times S}$$

$$= \frac{0.01072\, V_m^2}{S} \times \frac{W_e}{W} + 0.2778 \frac{S_1}{S}(98.1\,G + r) \text{ watt-hours per tonne-km} \quad ...(11.22)$$

If the track is level one, G = 0

then specific energy output

$$= \frac{0.01072\, V_m^2}{S} \times \frac{W_e}{W} + 0.2778 \frac{S_1}{S} r \text{ watt-hours/tonne-km} \quad ...(11.23)$$

Specific energy consumption

$$= \frac{\text{Specific energy output at driving wheels}}{\text{Overall efficiency of the motors and gearing}}$$

$$= \frac{0.01072 V_m^2}{S\eta} \times \frac{W_e}{W} + 0.2778 \frac{S_1}{S} \times \frac{r}{\eta} \text{ watt-hours/tonne-km} \quad ...(11.24)$$

11.10.1. Factors Affecting Energy Consumption

Energy consumption in propelling the train is required for

1. having the linear and angular acceleration;
2. working against gravity while moving up the gradient;
3. working against the resistance to motion;
4. overcoming the losses due to gearing system and
5. supplying losses in motors and other electrical system.

Energy required being the product of power and time depends upon the power required and the duration for which the power is taken.

1. The power (in watts) required in accelerating the train on level track depends upon the rate of acceleration α, the effective weight of the train W_e and the speed at the end of acceleration V_m, being equal to $77.17\, V_m W_e \alpha$ whereas the energy (in watts-hours) depends upon the rate of acceleration, effective weight of train, maximum speed and the time for which the train accelerates, being equal to $0.01072\, V_m W_e \alpha\, t$.
2. The power (in watts) required in overcoming the force of gravity while going up the gradient being equal to $27.334\, WG\, V_m$ depends upon the gradient. While going down the gradient the energy is returned back. The energy required or returned (in watt-seconds) is given by the expression $27.334\, WG\, V_m t$ and, therefore, depends upon the gradient, dead weight of train, maximum speed and duration of run.
3. The power (in watts) required in overcoming the resistance to motion, is equal to $0.2778\, Wr\, V_m$ and energy required $= 0.2778\, Wr\, V_m t$ watt-seconds. The expression for energy required indicates that it depends upon the resistance to motion, maximum speed and time of run.

The energy output of train for a given speed-time curve is independent of the type of drive employed, but energy consumption depends upon the type of drive employed.

11.11. FACTORS AFFECTING SPECIFIC ENERGY CONSUMPTION OF AN ELECTRIC TRAIN OPERATING ON A GIVEN SCHEDULE SPEED

The specific energy consumption of a train operating at a given schedule speed depends upon the following factors

(*a*) distance between stops (*b*) acceleration (*c*) retardation (*d*) maximum speed (*e*) nature of route and (*f*) the type of train equipment.

Specific energy output is given by the expression

Specific energy output

$$= \frac{0.01072\, V_m^2 W_e}{SW} + \frac{0.2778\,(98.1\,G + r)S_1}{S} \text{ watt-hours/tonne-km} \quad ...(11.25)$$

Specific energy output is independent of locomotive overall efficiency but the specific energy consumption being equal to specific energy output divided by locomotive overall efficiency depends upon the overall efficiency of the locomotive. Greater the overall efficiency lesser will be the specific energy consumption for a given specific energy output at axles.

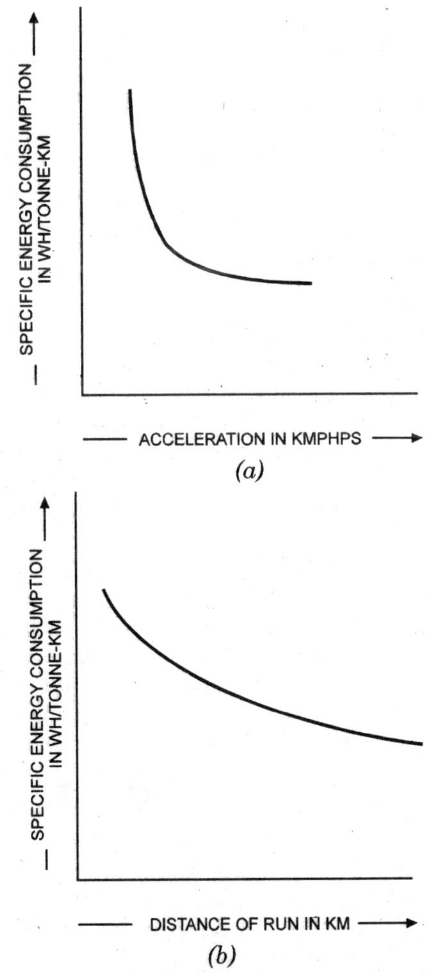

Fig. 11.11

The Eq. (11.25) clearly shows that specific energy consumption depends upon the maximum speed V_m, the distance travelled by the train while power is on, S_1, the specific resistance *r*, gradient G and distance between stops. Greater the distance between stops lesser will be the specific energy consumption. For a given run at a given schedule speed, greater the value of acceleration and retardation, more will be the period of coasting and, therefore, lesser the period during which power is on *i.e.*, S_1 will be small and, therefore, specific energy consumption will accordingly be less. Steep gradient will involve more energy consumption even if regenerative braking is used. Similarly more the train resistance, greater will be the specific energy consumption.

How the specific energy consumption falls with the increase in value of acceleration (retardation) and distance of run is illustrated in Figs. 11.11 (*a*) and 11.11 (*b*) respectively.

Typical values of specific energy consumption are (*i*) 50 – 75 watt-hours per tonne-km for suburban services and (*ii*) 20 – 30 watt-hours per tonne-km for main line service.

Example 11.14. An electric train has an average speed of 42 kmph on a level track between stops 1,400 m apart. It is to be accelerated at 1.7 kmphps and is braked at 3.3 kmphps. Draw the speed-time curve of the run. Estimate the energy consumption at the axles of the train per tonne-km. Take tractive resistance constant at 50 N per tonne and allow 10 per cent for rotational inertia. [*W.B. Univ. of Technology Utilization of Electric Power 2009*]

Solution : From solution of Example 11.2 we have

Maximum speed, V_m = 52 kmph

Duration of braking, t_3 = 15.8 seconds

$$\text{Distance travelled during braking} = \frac{1}{2} \times V_m \times \frac{t_3}{3{,}600}$$

$$= \frac{1}{2} \times 52 \times \frac{15.8}{3{,}600} = 0.114 \text{ km}$$

$$S_1 = S - 0.114 = 1.4 - 0.114 = 1.286 \text{ km}$$

Resistance to motion, r = 50 N/tonne

$$\frac{W_e}{W} = 1.1$$

So energy consumption at the axle of the train per tonne-km

$$= \frac{0.01072 V_m^2}{S} \times \frac{W_e}{W} + 0.2778 r \frac{S_1}{S}$$

$$= \frac{0.01072 \times 52^2}{1.4} \times 1.1 + 0.2778 \times 50 \times \frac{1.286}{1.4} = 35.534 \text{ Wh} \quad \textbf{Ans.}$$

Example 11.15. Calculate the specific energy consumption if a maximum speed of 12.2 m/s and for a given run of 1,525 metres an acceleration of 0.366 m/s² are desired. Train resistance during acceleration is 52.6 newtons/1,000 kg and during coasting is 6.12 newtons/1,000 kg. Effective rotating mass is 10% of the dead weight. The efficiency of the equipment during the acceleration period is 65%. Assume a quadrilateral speed-time curve.

[*U.P. Technical Univ. Utilization of Electrical Energy and Electric Traction 2005-06*]

Solution : Effective weight of train W_e = 1.1 W

where W is the weight of train in kg

Acceleration α = 0.366 m/s²

Maximum speed, V_m = 12.2 m/s

Distance of run, S = 1,525 metres

$$\text{Acceleration period, } t_1 = \frac{V_m}{\alpha} = \frac{12.2}{0.366} = 33.3 \text{ seconds}$$

Train resistance during acceleration,

r = 52.6 N/1,000 kg = 0.0526 N/kg

Tractive effort required during acceleration,

$$F_t = W_e \alpha + Wr = 1.1 \text{ W} \times 0.366 + 0.0526 \text{ W} = 0.4552 \text{ W newtons}$$

Total energy required for the run

= Energy required during acceleration as there is no free run

= Average power during acceleration × acceleration period

$$= \frac{1}{2} F_t V_m \times t_1 = \frac{1}{2} \times 0.4552 \text{ W} \times 12.2 \times 33.3 \text{ joules}$$

= 92.56 W joules or watt-seconds

$$= \frac{92.56 \text{ W}}{3{,}600} = 0.0257 \text{ W watt-hours}$$

$$\text{Specific energy output} = \frac{\text{Energy required for the run}}{\text{W} \times \text{S}}$$

$$= \frac{0.0257 \text{ W}}{\text{W} \times 1{,}525}$$

$$= 1.685 \times 10^{-5} \text{ Wh/kg-m}$$

$$\text{Specific energy consumption} = \frac{\text{Specific energy output}}{\eta}$$

$$= \frac{1.685 \times 10^{-5}}{0.65}$$

$$= 2.59 \times 10^{-5} \text{ Wh/kg-m} \quad \textbf{Ans.}$$

Example 11.16. An electric train has quadrilateral speed-time curve as follows:

(*i*) Uniform acceleration from rest at 2 kmphps for 30 seconds

(*ii*) Coasting for 50 seconds

(*iii*) Braking period of 20 seconds.

The train is moving a uniform upgradient of 1%, tractive resistance is 40 newtons per tonne, rotational inertia effect 10% of dead weight, duration of station stop 15 seconds and overall efficiency of transmission gear and motor as 75%. Calculate the value of its schedule speed and specific energy consumption of run. [*Pb. Univ. Electric Drives and Utilization Nov. 1988; Rajasthan Univ. Utilization of Electric Power 2003*]

Solution : Maximum speed, $V_1 = \alpha t_1 = 2 \times 30 = 60$ kmph

Refer to Fig. 11.7 and also solution to Example 11.12.

Coasting retardation, β_c = 0.452 kmphps

V_2 = 37.4 kmph

Braking retardation, β = 1.87 kmphps

Total distance travelled, S = 1.03 km

Schedule time, $T_s = t_1 + t_2 + t_3$ + stop duration

= 30 + 50 + 20 + 15 = 115 seconds

$$\text{Schedule speed, } V_s = \frac{3{,}600 \times S}{T_s} = \frac{3{,}600 \times 1.03}{115}$$

$$= 32.24 \text{ kmph} \quad \textbf{Ans.}$$

Distance travelled when power is on,

S_1 = Distance travelled during acceleration period

$$= \frac{V_1 t_1}{7{,}200} = \frac{60 \times 30}{7{,}200} = 0.25 \text{ km}$$

Specific energy output

$$= \frac{0.01072 V_m^2}{S} \cdot \frac{W_e}{W} + 0.2778\,(98.1\,G + r)\frac{S_1}{S}$$

$$= \frac{0.01072 \times 60^2}{1.03} \times 1.1 + 0.2778\,(98.1 \times 1 + 40) \times \frac{0.25}{1.03}$$

= 50.5 Wh/tonne-km

$$\text{Specific energy consumption} = \frac{50.5}{0.75} = 67.34 \text{ Wh/tonne-km} \quad \textbf{Ans.}$$

Example 11.17. An electric train has quadrilateral speed-time curve as follows :

(*i*) Uniform acceleration from rest at 2 kmphps for 30 seconds.

(*ii*) Coasting for 50 seconds.

(*iii*) Braking period of 20 seconds.

The train is moving a uniform down gradient of 1%, tractive resistance 40 newtons per tonne, rotational inertia effect 10% of dead weight, duration of stop 15 seconds and overall efficiency of transmission gear and motor as 75%. Calculate its schedule speed and specific energy consumption of run.

[Pb. Univ. Electric Drives and Utilization May 1989]

Solution : Maximum speed, $V_1 = \alpha t_1 = 2 \times 30 = 60$ kmph

Let the coasting retardation be β_c

Tractive effort, $F_t = 277.8\ W_e\ (-\beta_c) - 98.1\ WG + Wr$

or $0 = -277.8 \times 1.1\ W\beta_c - 98.1 \times W \times 1 + 40\ W$

$$\text{or } \beta_c = \frac{-58.1}{277.8 \times 1.1} = -0.19 \text{ kmphps}$$

$$V_2 = V_1 - \beta_c t_2 = 60 - (-0.19) \times 50 = 69.5 \text{ kmph}$$

$$\text{Distance travelled, } S = \frac{V_1 t_1}{7{,}200} + \frac{V_1 + V_2}{7{,}200} \times t_2 + \frac{V_2 t_3}{7{,}200} \text{ km}$$

$$= \frac{60 \times 30}{7{,}200} + \frac{60 + 69.5}{7{,}200} \times 50 + \frac{69.5 \times 20}{7{,}200}$$

$$= 0.25 + 0.899 + 0.193 = 1.342 \text{ km}$$

Schedule time, $T_s = t_1 + t_2 + t_3$ + stop duration

$= 30 + 50 + 20 + 15 = 115$ seconds

$$\text{Schedule speed, } V_s = \frac{3{,}600 \times S}{T_s} = \frac{3{,}600 \times 1.342}{115}$$

$= 42$ kmph **Ans.**

Specific energy output

$$= \frac{0.01072 V_1^2}{S} \times \frac{W_e}{W} + 0.2778\,(98.1G + r)\frac{S_1}{S}$$

$$= \frac{0.01072 \times 60^2}{1.342} \times 1.1 + 0.2778\,[98.1(-1) + 40] \times \frac{0.25}{1.342}$$

$= 28.6$ Wh/tonne-km

$$\text{Specific energy consumption} = \frac{28.6}{0.75} = 38.14 \text{ Wh/tonne-km}$$ **Ans.**

Example 11.18. An electric train weighing 100 tonnes has a rotational inertia of 10%. This train while running between two stations which are 2.5 km apart has an average speed of 50 km/hour. The acceleration and retardation during braking are respectively 1 km/hour/second and 2 km per hour per second. The percentage gradient between these two stations is 1% and the train is to move up the incline. The track resistance is 40 N/tonne. If the combined efficiency of the electric train is 60%, determine (*i*) maximum power at the driving axle (*ii*) total energy consumption and (*iii*) specific energy consumption. Assume that journey estimation is being made on simplified trapezoidal speed-time curve.

Solution : Effective weight of train,

$$W_e = \left(1 + \frac{10}{100}\right) W = 1.1\ W = 1.1 \times 100 = 110 \text{ tonnes}$$

Distance of run, $S = 2.5$ km

Average speed, $V_a = 50$ kmph

$$\text{Duration of run} = \frac{3{,}600 \times S}{V_a} = \frac{3{,}600 \times 2.5}{50} = 180 \text{ seconds}$$

$$K = \frac{1}{2\alpha} + \frac{1}{2\beta} = \frac{1}{2 \times 1} + \frac{1}{2 \times 2} = 0.75$$

$$\text{Maximum speed, } V_m = \frac{T}{2K} - \sqrt{\left(\frac{T}{2K}\right)^2 - \frac{3{,}600\,S}{K}}$$

$$= \frac{180}{2 \times 0.75} - \sqrt{\left(\frac{180}{2 \times 0.75}\right)^2 - \frac{3{,}600 \times 2.5}{0.75}}$$

$= 71$ kmph

$$\text{Acceleration period, } t_1 = \frac{V_m}{\alpha} = \frac{71}{1} = 71 \text{ seconds}$$

$$\text{Braking period, } t_3 = \frac{V_m}{\beta} = \frac{71}{2} = 35.5 \text{ seconds}$$

Time of free run, $t_2 = 180 - (71 + 35.5) = 73.5$ seconds

Tractive effort required,

$F_t = 277.8\ W_e\alpha + 98.1\ WG + Wr$

$= 277.8 \times 110 \times 1 + 98.1 \times 100 \times 1 + 100 \times 40 = 44{,}368$ N

(*i*) Maximum power at the driving axle

$$= \frac{F_t \times V_m}{3{,}600} = \frac{44{,}368 \times 71}{3{,}600} = 875 \text{ kW}$$ **Ans.**

Total energy output

$$= \frac{1}{2} F_t \times \frac{V_m}{3{,}600} \times \frac{t_1}{3{,}600} + \frac{F_t' V_m}{3{,}600} \times \frac{t_2}{3{,}600}$$

where F_t' is effort required during free run and is equal to $(98.1WG + Wr)$

i.e., $98.1 \times 100 \times 1 + 100 \times 40 = 13{,}810$ N

$\therefore$ Total energy output

$$= \frac{1}{2} \times 44{,}368 \times \frac{71}{3{,}600} \times \frac{71}{3{,}600} + 13{,}810 \times \frac{71}{3{,}600} \times \frac{73.5}{3{,}600}$$

$= 14.19$ kWh

$$(ii) \text{ Total energy consumption} = \frac{14.19}{0.6} = 23.65 \text{ kWh}$$ **Ans.**

(*iii*) Specific energy consumption

$$= \frac{\text{Total energy consumption in watt-hours}}{\text{Weight of train in tonnes} \times \text{distance of run in km}}$$

$$= \frac{23.65 \times 1{,}000}{100 \times 2.5} = 94.6 \text{ Wh/tonne-km}$$ **Ans.**

Example 11.19. A train weighing 203 tonnes accelerates uniformly from rest to a speed of 45 kmph up a gradient of 1 in 500, the time taken being 30 seconds. The power is then cut off and train coasts down a uniform gradient of 1 in 1,000 for a period of 40 seconds when brakes are applied for a period of 15 seconds so as to bring the train uniformly to rest on this gradient. Calculate (*i*) the maximum power output from the driving axles (*ii*) the energy taken from the conductor rails in kWh, assuming an efficiency of 60%.

Assume train resistance to be 44 newtons per tonne at all speeds, and allow 10% for rotational inertia.

[Agra Univ. Traction & Utilization of Elec. Power 1978]

Solution : Weight of train, $W = 203$ tonnes

Effective weight of train, $W_e = 1.1\ W = 1.1 \times 203 = 223.3$ tonnes

The train resistance, $r = 44$ newtons/tonne

$$\text{Gradient, } G = \frac{1}{500} \times 100 = 0.2\%$$

Maximum speed, $V_m = 45$ kmph

Acceleration period, $t_1 = 30$ seconds

$$\text{Acceleration, } \alpha = \frac{V_m}{t_1} = \frac{45}{30} = 1.5 \text{ kmphps}$$

Tractive effort required,

$F_t = 277.8\ W_e\alpha + 98.1\ WG + Wr$

$= 277.8 \times 223.3 \times 1.5 + 98.1 \times 203 \times 0.2 + 203 \times 44$

$= 105{,}964$ newtons

(*i*) The maximum power output from driving axles

$$= \frac{F_t \times V_m}{3,600} = \frac{105,964}{3,600} \times 45 = 1,324.55 \text{ kW} \quad \textbf{Ans.}$$

Total energy required for the run

= Energy required during acceleration as there is no free run

$$= \frac{1}{2}\frac{F_t V_m}{3,600} \times \frac{t_1}{3,600} \text{ kWh} \quad = \frac{1}{2} \times \frac{105,964 \times 45}{3,600} \times \frac{30}{3,600}$$

$$= 5.52 \text{ kWh}$$

(*ii*) The energy taken from conductor rails $= \frac{5.52}{0.6} = 9.2$ kWh **Ans.**

Example 11.20. It is proposed to put an electric trolley service in a city. The schedule speed is to be 45 kmph. The distance between stops is 2.8 km. The track is assumed to be level. Each stop is of 30 seconds duration. Using simplified speed-time curve, calculate the maximum speed assuming the acceleration to be 2 kmphps and retardation to be 3.2 kmphps. The dead weight of the car is 16 tonnes, rotational inertia is 10 per cent of the dead weight and track resistance is 40 newtons per tonne. If the overall efficiency is 80%, calculate (*a*) maximum power output from driving axles (*b*) the specific energy consumption in watt-hours per tonne-km.

Solution : Schedule speed, $V_s = 45$ kmph

$$\text{Schedule time of run, } T_s = \frac{S \times 3,600}{V_s} = \frac{2.8 \times 3,600}{45}$$

$$= 224 \text{ seconds}$$

Actual time of run, T = 224 − 30 = 194 seconds

$$K = \frac{1}{2\alpha} + \frac{1}{2\beta} = \frac{1}{4} + \frac{1}{6.4} = \frac{13}{32} = 0.40625$$

Maximum speed,

$$V_m = \frac{T}{2K} - \sqrt{\left(\frac{T}{2K}\right)^2 - \frac{3,600\, S}{K}}$$

$$= \frac{194}{2 \times 0.40625} - \sqrt{\left(\frac{194}{2 \times 0.40625}\right)^2 - \frac{3,600 \times 2.8}{0.40625}}$$

= 58 kmph **Ans.**

$$\text{Acceleration time, } t_1 = \frac{V_m}{\alpha} = \frac{58}{2} = 29 \text{ seconds}$$

$$\text{Duration of braking, } t_3 = \frac{V_m}{\beta} = \frac{58}{3.2} = 18.125 \text{ seconds}$$

Time of free run, $t_2 = T - (t_1 + t_3)$

$= 194 - (29 + 18.125) = 146.875$ seconds

Tractive effort during acceleration,

$F_t = 277.8\, W_e \alpha + Wr$

$= 277.8 \times 16 \times 1.1 \times 2 + 16 \times 40 = 10,418.56$ Newtons

(*a*) Maximum power output

$$= \frac{F_t V_m}{3,600} \text{ kW} = \frac{10,418.56 \times 58}{3,600} = 167.855 \text{ kW} \quad \textbf{Ans.}$$

(*b*) Distance travelled during braking period

$$= \frac{1}{2}\frac{V_m \times t_3}{3,600} = \frac{1}{2} \times \frac{58 \times 18.125}{3,600} = 0.146 \text{ km}$$

Distance travelled with power on $S_1 = 2.8 - 0.146 = 2.654$ km

Specific energy output

$$= \frac{0.01072\, V_m^2}{S} \times \frac{W_e}{W} + 0.2778 r \frac{S_1}{S} \text{ Wh/tonne-km}$$

$$= \frac{0.01072 \times 58^2}{2.8} \times 1.1 + 0.2778 \times 40 \times \frac{2.654}{2.8}$$

= 24.7 Wh/tonne-km

Specific energy consumption $= \frac{24.7}{0.8} = 30.9$ Wh/tonne-km **Ans.**

Example 11.21. An electric train weighing 300 tonnes runs 1% upgradient with following speed-time curve:

(*i*) Uniform acceleration of 1.5 kmphps for 30 seconds

(*ii*) Constant speed for 40 seconds

(*iii*) Coasting for 30 seconds

(*iv*) Braking at 2.5 kmphps

Calculate the specific energy consumption if the tractive effort is 45 N/tonne, rotational inertia effect 10% and overall efficiency of transmission and motor 75%.

[*Utilization of Electrical Power Rajasthan Univ. 2009; Rajasthan Technical Univ. 2010*]

Solution : Speed at the end of accelerating period,

$V_1 = \alpha t_1 = 1.5 \times 30 = 45$ kmph

Let the coasting retardation be β_c, then

$F_t = 277.8\, W_e (-\beta_c) + 98.1\, WG + Wr$

or $0 = -277.8 \times 1.1\, W \times \beta_c + 98.1 \times 1\, W + 45\, W$

$$\text{or } \beta_c = \frac{143.1}{277.8 \times 1.1} = 0.4683 \text{ kmphps}$$

Speed at the end of coasting period,

$V_2 = V_1 - \beta_c t_3 = 45 - 0.4683 \times 30 = 30.95$ kmph

$$\text{Braking duration, } t_4 = \frac{V_2}{\beta} = \frac{30.95}{2.5} = 12.38 \text{ seconds}$$

Total distance travelled,

$$S = \frac{V_1 t_1}{7,200} + \frac{V_1 t_2}{3,600} + \frac{V_1 + V_2}{7,200} \times t_3 + \frac{V_2 t_4}{7,200}$$

$$= \frac{45 \times 30}{7,200} + \frac{45 \times 40}{3,600} + \frac{45 + 30.95}{7,200} \times 30 + \frac{30.95 \times 12.38}{7,200}$$

$= 0.1875 + 0.5 + 0.3165 + 0.0532 = 1.0572$ km

Distance travelled when power is on,

$$S_1 = \frac{45 \times 30}{7,200} + \frac{45 \times 40}{3,600} = 0.1875 + 0.5 = 0.6875 \text{ km}$$

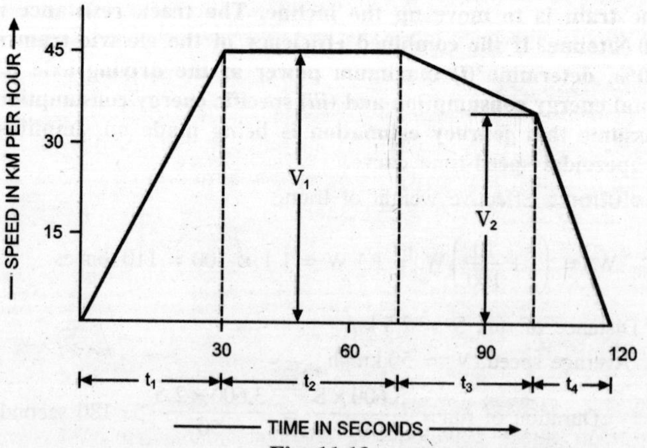

Fig. 11.12

Specific energy output

$$= \frac{0.01072\, V_m^2}{S} \frac{W_e}{W} + 0.2778\,(98.1 G + r) \frac{S_1}{S}$$

$$= \frac{0.01072 \times 45^2 \times 1.1}{1.0572} + 0.2778\,(98.1 \times 1 + 45) \times \frac{0.6875}{1.0572}$$

$= 22.587 + 25.852 = 48.44$ Wh/tonne-km

Specific energy consumption $= \dfrac{48.44}{0.75} = 64.587$ Wh/tonne-km **Ans.**

Example 11.22. An electric train weighing 500 tonnes climbs up gradient with G = 8 and following speed-time curve.

(*i*) uniform acceleration of 2.5 km/hr/sec for 60 seconds.

(*ii*) Constant speed for 5 minutes.

(*iii*) Coasting speed for 3 minutes.

(*iv*) Dynamic braking at 3 kmphps to rest.

The train resistance is 25 N/tonne, rotational inertia effect 10% and combined efficiency of transmission and motor is 80%. Calculate the specific energy consumption.

[*U.P. Technical Univ. Utilization of Electrical Energy & Traction 2009-10*]

Solution : Speed at the end of accelerating period,

$V_1 = \alpha t_1 = 2.5 \times 60 = 150$ kmph

Let the coasting retardation be β_c, then

$F_t = 277.8\, W_e\,(-\beta_c) + 98.1 WG + Wr$

or $0 = -277.8 \times 1.1\, W \times \beta_c + 98.1\, W \times \dfrac{8}{1{,}000} \times 100 + 25\, W$

or $\beta_c = \dfrac{103.48}{277.8 \times 1.1} = 0.3386$ kmphps

Speed at the end of coasting period,

$V_2 = V_1 - \beta_c \times t_3 = 150 - 0.3386 \times 3 \times 60 = 89.052$ kmph

Braking retardation,

$t_4 = \dfrac{V_2}{\beta} = \dfrac{89.052}{3} = 29.684$ seconds

Total distance travelled,

$$S = \frac{V_1 t_1}{7{,}200} + \frac{V_1 t_2}{3{,}600} + \frac{V_1 + V_2}{7{,}200} \times t_3 + \frac{V_2 t_4}{7{,}200}$$

$$= \frac{150 \times 60}{7{,}200} + \frac{150 \times 5 \times 60}{3{,}600} + \frac{150 + 89.052}{7{,}200} \times 3 \times 60 + \frac{89.052 \times 29.684}{7{,}200}$$

$= 1.25 + 12.5 + 5.9763 + 0.3671 = 20.0934$ km

Distance travelled when power is on,

$$S_1 = \frac{150 \times 60}{7{,}200} + \frac{150 \times 5 \times 60}{3{,}600} = 1.25 + 12.5 = 13.75 \text{ km}$$

Specific energy output

$$= \frac{0.01072\, V_m^2}{S} \times \frac{W_e}{W} + 0.2778(98.1G + r)\frac{S_1}{S}$$

$$= \frac{0.01072 \times 150^2}{20.0934} \times 1.1 + 0.2778\left(98.1 \times \frac{8}{1{,}000} \times 100 + 25\right) \times \frac{13.75}{20.0934}$$

$= 13.204 + 19.67 = 32.874$ Wh per tonne-km

Specific energy consumption

$= \dfrac{32.874}{0.8} = 41.09$ Wh/tonne-km **Ans.**

11.12. DEAD WEIGHT, ACCELERATING WEIGHT AND ADHESIVE WEIGHT

Dead Weight. The total weight of locomotive and train to be pulled by the locomotive is known as *dead weight.*

Accelerating Weight. The dead weight of the train *i.e.*, the weight of locomotive and train can be considered to be divided into two parts.

1. The weight, which requires angular acceleration such as weight of wheels, axles, gears etc. and
2. The weight, which requires linear acceleration.

Hence the effective weight, which is greater than dead weight is called the *accelerating weight.* Accelerating weight is taken 5 to 10 per cent more than dead weight.

Adhesive Weight. The total weight to be carried on the driving wheels is known as the *adhesive weight.*

Example 11.23. 400 tonne goods train is to be hauled by a locomotive up a gradient of 2% with acceleration of 1 kmphps. Coefficient of adhesion is 20%, track resistance 40 N/tonne and effective rotating masses 10% of the dead weight. Find the weight of the locomotive and number of axles if the axle load is not to increase beyond 22 tonnes.

[*Agra Univ. Traction & Utilization of Elec. Power 1979*]

Solution : Let the weight of the locomotive be W_L tonnes

Dead weight of train and locomotive, $W = (400 + W_L)$ tonnes

Gradient, $G = 2$ per cent

Coefficient of adhesion, $\mu = 0.2$

Effective weight of train and locomotive, $W_e = 1.1\,(400 + W_L)$

Tractive effort,

$F_t = 277.8\, W_e\, \alpha + 98.1\, WG + Wr$

$= \left[277.8 \dfrac{W_e}{W} \alpha + 98.1\, G + r\right] W$

$= [277.8 \times 1.1 \times 1 + 98.1 \times 2 + 40]\, W$

$= 541.78\, W$ newtons

$= \dfrac{541.78\, W}{9.81 \times 1{,}000} = 0.05523\, W$ tonnes

$= 0.05523\,(400 + W_L)$ tonnes ...(*i*)

Maximum value of tractive effort that can be possible with W_L adhesive weight of locomotive

$F_t = \mu\, W_L = 0.2\, W_L$ ...(*ii*)

Comparing Eqs. (*i*) and (*ii*) we get

$0.2\, W_L = 0.05523\,(400 + W_L)$

or $W_L = \dfrac{0.05523 \times 400}{0.2 - 0.05523} = \dfrac{22.092}{0.14477} = 152.6$ tonnes **Ans.**

Number of axles required $= \dfrac{W_L}{\text{Allowable load}} = \dfrac{152.6}{22} = 7$ **Ans.**

Example 11.24. An electric locomotive is required to haul a train of 12 coaches each weighing 30 tonnes on the main line service requiring an initial acceleration of 0.8 kmphps up a gradient of 1 in 100. Estimate the adhesive weight and hence the number of driving axles the locomotive must have if the permissible axle loading is 20 tonnes per axle assuming rotational inertia to be 4% for the coaches and 15% for the locomotive. Maximum coefficient of adhesion is 0.2 and the tractive resistance 5 kg/tonne. [*B.H.U. Elec. Traction & Utilization 1978*]

Solution : Dead weight of train,

$W = 12 \times 30 = 360$ tonnes

Accelerating weight of train,

$W_e = 1.04 \times 360 = 374.4$ tonnes

Let adhesive weight of locomotive $= W_L$

Accelerating weight of locomotive $= 1.15\, W_L$

Tractive resistance, $r = 5$ kg per tonne $= 5 \times 9.81 = 49.05$ N/tonne

Coefficient of adhesion, $\mu = 0.2$

Tractive effort,

$F_t = (277.8\,\alpha)(W_e + 1.15\,W_L) + (98.1\,G + r)(W + W_L)$

$= (277.8 \times 0.8)(374.4 + 1.15\,W_L) + (98.1 + 49.05)(360 + W_L)$ newtons

$= \dfrac{222.24\,(374.4 + 1.15\,W_L) + 147.15(360 + W_L)}{1{,}000 \times 9.81}$ tonnes

$= (13.882 + 0.041\,W_L)$ tonnes ...(*i*)

Maximum value of tractive effort that can be possible with W_L adhesive weight of locomotive

$= \mu W_L = 0.2\,W_L$ ...(*ii*)

Equating Eqs. (*i*) and (*ii*), we get

$0.2\,W_L = 13.882 + 0.041\,W_L$

or $W_L = \dfrac{13.882}{0.159} = 87.3$ tonnes

Number of driving axles $= \dfrac{87.3}{20} = 4.37 \simeq 5$ **Ans.**

Example 11.25. A locomotive of 100 tonnes can just accelerate train of 500 tonnes with an acceleration of 1 kmphps up gradient of 1 %. Adhesive weight of locomotive is 70% of total dead weight, tractive resistance 45 newtons/tonne and inertia 10%. If this locomotive is helped by another locomotive of 130 tonnes with 100% adhesive weight, find out :

(*i*) Trailing weight that can be hauled up the same gradient under the same conditions.

(*ii*) The maximum gradient, trailing hauled load remaining unchanged.

[*U.P. Technical Univ. Utilization of Electrical Energy & Traction 2009-10*]

Solution : Dead weight of train and locomotive,

$W = 100 + 500 = 600$ tonnes

Effective weight,

$W_e = 1.1\,W = 1.1 \times 600 = 660$ tonnes

Tractive effort required,

$F_t = 277.8 W_e \alpha + 98.1 WG + Wr$

$= 277.8 \times 660 \times 1 + 98.1 \times 600 \times 1 + 600 \times 45$

$= 183{,}348 + 58{,}860 + 27{,}000 = 269{,}208$ newtons ...(*i*)

Maximum tractive effort for the first locomotive

$= \mu W_L \times \dfrac{\text{Adhesive weight}}{\text{Dead weight}}$

$= \mu \times 100 \times 1{,}000 \times 9.81 \times 0.7 = 686{,}700\,\mu$ newtons ...(*ii*)

Equating Eqs. (*i*) and (*ii*), we have

$686{,}700\,\mu = 269{,}208$

or $\mu = \dfrac{269{,}208}{686{,}700} = 0.392$

Adhesive weight of second locomotive

$= 130 \times 1 = 130$ tonnes

Total adhesive weight with both locomotives

$= 100 \times 0.7 + 130 = 200$ tonnes

Total tractive effort developed

$= \mu \times$ adhesive weight of both locomotive

$= 0.392 \times 200$

$= 78.4$ tonnes or $78.4 \times 9.81 \times 1{,}000$ *i.e.*, 769,104 newtons ...(*iii*)

Let the trailing weight that can be hauled, up the same gradient and under the same condition be W_1 tonnes

Now the total dead weight,

$W' = (100 + 130 + W_1) = (230 + W_1)$ tonnes

Tractive effort required,

$F'_t = (277.8 \times 1.1 \times 1 + 98.1 \times 1 + 45)W'$

$= 448.68\,W'$ ...(*iv*)

Equating Eqs. (*iii*) and (*iv*), we have

$448.68W' = 769{,}104$

or $W' = \dfrac{769{,}104}{448.68} = 1{,}714$ tonnes

or Trailing weight,

$W_1 = W' - 230 = 1{,}714 - 230 = 1{,}484$ tonnes **Ans.**

(*ii*) Total hauled load, $W'' = 100 + 500 + 130 = 730$ tonnes

Tractive effort required,

$F''_t = \left[277.8 \times \dfrac{W_e}{W''} \times \alpha + 98.1G + g\right] W''$

$= (277.8 \times 1.1 \times 1 + 98.1G + 45) \times 730$

$= (350.8 + 98.1G) \times 730$ newtons ...(*v*)

Equating Eqs. (*iii*) and (*v*), we have

$(350.8 + 98.1G) \times 730 = 769{,}104$

or $G = \left(\dfrac{769{,}104}{730} - 350.8\right) \times \dfrac{1}{98.1} = 7.164\%$ **Ans.**

Example 11.26. Determine acceleration in Example 11.25 if gradient and hauled load remains the same.

Solution : Tractive effort required,

$F''_t = \left[277.8 \times \dfrac{W_e}{W''} \times \alpha' + 98.1G + g\right] W''$

$= [277.8 \times 1.1 \times \alpha' + 98.1 \times 1 + 45] \times 730$

$= [305.58\alpha' + 143.1] \times 730$ newtons ...(*vi*)

Equating Eqs. (*iii*) and (*vi*), we have

$(305.58\alpha' + 143.1) \times 730 = 769{,}104$

or $\alpha' = \left(\dfrac{769{,}104}{730} - 143.1\right) \times \dfrac{1}{305.58} = 2.98$ kmphps **Ans.**

Example 11.27. A train with a locomotive having 4 motors has a total mass of 250 tonnes. Starting from rest the train attains a speed of 40 kmph in 20 seconds on a 1% up gradient. The gear ratio is 3, the gear efficiency 95%, the wheel diameter 95 cm, train resistance (average) is 40 N per tonne and rotational inertia is 10%. Find the torque developed by each of the motors and the minimum weight of the locomotive, given the adhesive coefficient is 0.25. [*Pb. Univ. Electric Drives and Utilization June 1991*]

Solution : Total mass of train and locomotive, W = 250 tonnes

Diameter of driving wheels, D = 95 cm = 0.95 m

Percentage gradient, G = 1%

Train resistance, r = 40 N/tonne

Gear ratio, $\gamma = 3$

Gear efficiency, $\eta = 95\%$ or 0.95

Equivalent accelerating weight of train,

$W_e = \dfrac{100 + 10}{100} \times 250 = 275$ tonnes

Time taken in accelerating the train to a speed of 40 kmph,

$t_1 = 20$ seconds

Acceleration, $\alpha = \dfrac{V_m}{t_1} = \dfrac{40}{20} = 2$ kmphps

Tractive effort required,

$F_t = 277.8\,W_e\,\alpha + (98.1\,G + r)W$

$= 277.8 \times 275 \times 2 + (98.1 \times 1 + 40) \times 250$

$= 1{,}87{,}315\,\text{N} = \dfrac{1{,}87{,}315}{9.81 \times 1{,}000} = 19.0943$ tonnes

Total torque developed,

$$T = \frac{F_t \times D}{\eta \times 2\gamma} = \frac{1{,}87{,}315 \times 0.95}{0.95 \times 2 \times 3} = 31{,}220 \text{ Nm}$$

Torque developed by each motor

$$= \frac{31{,}220}{4} = 7{,}805 \text{ Nm} \quad \textbf{Ans.}$$

Coefficient of adhesion, $\mu = 0.25$

$$\text{Adhesive weight, } W_L = \frac{F_t}{\mu} = \frac{19.0943}{0.25} = 76.4 \text{ tonnes}$$

Assuming an adhesive weight of locomotive to be 75% of its dead weight, we have

$$\text{Dead weight of locomotive} = \frac{W_L}{0.75} = \frac{76.4}{0.75} = 102 \text{ tonnes} \quad \textbf{Ans.}$$

Exercises

1. What do you understand by speed-time curves ? What is its use in practice ? **[Gorakhpur Univ. Utilization of Electrical Power & Traction 1983]**
2. Draw the speed-time curves for each of the following traction systems:
(*i*) Tramcar (*ii*) Diesel electric (*iii*) Trolley buses (*iv*) Suburban electric train and (*v*) Urban services.
3. What is a speed-time curve ? Discuss briefly the distinct periods in a typical speed-time curve of a train running on main line. **[Pb. Univ. Electric Traction Dec. 1991]**
4. Draw a typical speed-time curve for train movement, and explain (*a*) acceleration (*b*) free running (*c*) coasting (*d*) retardation.
Draw the speed-time curve for urban service and suburban service train movement and simplify these curves by approximation.
5. Draw and explain general speed-time curve of a train running between two stations. How can this curve be approximated for (*a*) main line service (*b*) suburban service ? **[U.P. Technical Univ. Utilization of Electrical Energy & Traction 2009-10]**
6. Sketch the speed-time curves for urban, suburban and main line services. Give the approximate values of acceleration, braking and coasting periods in each case. **[Rajasthan Univ. Utilization of Electrical Power 2004, 2005]**
7. Discuss the main features of the various train services. What type of train services corresponds to trapezoidal and quadrilateral speed-time curves ? **[M.D. Univ. Utilization of Electric Power & Traction Dec. 2008; May 2009; G.B. Technical Univ. Utilization of Electrical Energy & Traction 2010-11]**
8. Draw and explain a typical speed-time curve for an electric train and explain what do you understand by crest speed, average speed and schedule speed. **[Agra Univ. Traction & Utilization of Electric Power 1982; M.D. Univ. Utilization of Electric Power & Traction Dec. 2009]**
9. Define (*i*) crest speed, (*ii*) average speed, (*iii*) schedule speed. **[Rajasthan Technical Univ. Utilization of Electrical Power 2010]**
10. Draw a main line speed-time curve and label its various parts. Derive an expression for maximum speed of a suburban train in terms of total distance travelled, acceleration, retardation and average speed. Assume the speed-time curve to be trapezoidal. **[Utilization of Electrical Power Rajasthan Univ. 2009; Rajasthan Technical Univ. 2010]**
11. Define 'crest speed' and 'schedule speed' and discuss the factors which affect the schedule speed of a train. **[B.H.U. Elec. Traction & Utilization 1978]**
12. The speed-time curve of a locomotive going down gradient is represented by a quadrilateral speed-time curve. The locomotive is accelerated to V_1 speed in time t_1 seconds coasted for V_2 speed in t_2 seconds and braked to standstill in t_3 seconds. Obtain expressions for total distance covered and speed V_2. Choose acceleration during starting and coasting as α and β_c and retardation during braking as β. **[U.P. Technical Univ. Utilization of Electrical Energy and Electric Traction 2006-07]**
13. Define the terms crest speed, average speed and schedule speed. What are different methods of approximation of speed-time curves ? Derive expressions for distance travelled using quadrilateral approximation method of $V(t)$ curve. **[Gorakhpur Univ. Utilization of Electrical Power & Traction 1984]**
14. Explain speed-time curve for a main line service and derive the expression
$\frac{1}{2}\left[\frac{1}{\alpha}+\frac{1}{\beta}\right] = \frac{3{,}600\, D}{V_m^2}\left[\frac{V_m}{V_a}-1\right]$ where the symbols have their usual meanings. **[Pb. Univ. Electric Traction December 1992]**
15. Explain mechanics of train movement. **[Rajasthan Univ. Utilization of Electric Power 2003]**
16. Derive an expression for tractive effort required for propulsion of a train during (*a*) acceleration (*b*) free running (*c*) coasting (*d*) moving down a gradient. **[W.B. Univ. of Technology Utilization of Electric Power 2009]**
17. Derive an expression for :
(*i*) The tractive effort for propulsion of a train on level track.
(*ii*) The tractive effort for propulsion of a train up and down a gradient.
[G.B. Technical Univ. Utilization of Electrical Energy & Traction 2010-11]
18. Explain various functions that the tractive effort exerted by traction unit is supposed to perform and derive an expression for total tractive effort. **[U.P. Technical Univ. Utilization of Electrical Energy & Traction 2009-10]**
19. Deduce the expressions for:
(*i*) The tractive effort exerted by road wheel in terms of wheel diameter, motor torque, gear ratio and efficiency of transmission of power through gears.
(*ii*) The tractive effort for the propulsion of train up and down the gradient.
[Gorakhpur Univ. Utilization of Electrical Power & Traction 1982]
20. What is coefficient of adhesion ? How does it affect slipping of the driving wheels of the traction unit ? **[M.D. Univ. Utilization of Electric Power & Traction May 2010]**
21. Comment on the following statements:
(*i*) The adhesion coefficient of a locomotive employing dc shunt motor will be higher than the using dc series motor.
(*ii*) A higher ratio of adhesive weight to total weight is desirable for suburban trains, where the distance between the consecutive stops is very small.
22. Explain the terms (*i*) coefficient of adhesion (*ii*) tractive effort and (*iii*) gradient. **[Pb. Univ. Electric Traction December 1992]**

23. What do you mean by specific energy consumption ? What are the factors that affect specific energy consumption of a train ?
[W.B. Univ. of Technology Utilization of Electric Power 2009]

24. What are the factors which affects the specific energy consumption in electric trains ?
[Utilization of Electric Power Rajasthan Univ. 2000, 2004, 2006; Rajasthan Technical Univ. 2010]

25. Discuss various factors affecting specific energy consumption in the traction.
[G.B. Technical Univ. Utilization of Electrical Energy & Traction 2010-11]

26. Derive equation of motion of a vehicle. How does the effective mass of a vehicle differ with its physical (actual) mass ? Explain significance of equation of motion in energy consumption calculations.
[Gorakhpur Univ. Utilization of Electrical Power & Traction September 1984]

27. Derive a simple expression for the maximum power output of traction motors in terms of tractive effort, maximum speed and efficiency of transmission gear.

28. Derive an expression for specific energy consumption of a suburban train for a typical inter-station run.
[B.H.U. Elec. Traction & Utilization 1978]

29. Differentiate between adhesive weight and total weight of an electric locomotive.
[Gorakhpur Univ. Utilization of Electrical Power & Traction 1983]

30. Explain the terms 'dead weight', 'effective weight' and 'adhesive weight' in a locomotive.
[B.H.U. Elec. Traction & Utilization 1978]

31. Define (*i*) adhesive weight (*ii*) coefficient of adhesion (*iii*) Tractive effort. **[Rajasthan Technical Univ. Utilization of Electrical Power 2008, 2010]**

32. Write short notes on the following
(*i*) Speed-time curve.
(*ii*) Mechanics of train movement.
[Rajasthan Univ. 2003; M.D. Univ. May 2009; Dec. 2009]
(*iii*) Tractive effort for propulsion of train.
(*iv*) Factors affecting energy consumption in propelling a train.
(*v*) Specific energy consumption. **[M.D. Univ. May 2010]**
(*vi*) Dead weight, accelerating weight and adhesive weight of a train.
[Agra Univ. Traction & Utilization of Elec. Power 1982]

Short Answer Type Questions With Answers

Q. 1. What is meant by speed-time curve ?
Ans. The curve drawn between speed and time taking speed on the Y-axis and time on the X-axis is known as speed-time curve.

Q. 2. Define terms 'coasting' and 'braking'. **[U.P.T.U. 2006-07]**
Ans. ***Coasting.*** At the end of free running period (*i.e.,* at t_3) power supply is cut off and the train is allowed to run under its own momentum. The speed of train starts decreasing on account of resistance to the motion of train. The rate of decrease of speed during coasting period is known as *coasting retardation.*

Braking. At the end of coasting period (*i.e.,* at t_4) the brakes are applied to bring the train to rest. During this period speed decreases rapidly and finally reduces to zero.

Q. 3. What is meant by crest speed ?
Ans. The maximum speed attained by the vehicle during the run is known as crest speed.

Q. 4. What is meant by schedule speed of a train ?
[Gorakhpur Univ. 1975]
Ans. The ratio of distance covered between two stops and total time of run including time of stop is known as schedule speed.

Q. 5. What are those factors which affect the average speed of a train ?
Ans. The average speed of a given train when running on a given service (*i.e.,* distance between two stops) is affected by the (*i*) acceleration and retardation and (*ii*) crest speed.

Q. 6. What are the factors which affect the schedule speed of a train ?
Ans. The schedule speed of a given train when running on a given service is affected by (*i*) acceleration and retardation (*ii*) maximum or crest speed and (*iii*) time duration of stop.

Q. 7. How does increase in acceleration and braking retardation affect the schedule speed of a train for a given run with fixed crest speed ?
Ans. For a given run and with fixed crest speed the increase in acceleration will result in decrease in actual time of run, and therefore, increase in schedule speed. Similarly increase in braking retardation will affect the schedule speed.

Q. 8. How does increase in crest speed affect the schedule speed for a given run with fixed acceleration and retardation ?
Ans. For a given run with fixed acceleration and retardation the actual time of run will decrease, and therefore, schedule speed will increase with the increase in crest speed.

Q. 9. How does increase in duration of stop affect the schedule speed ?
Ans. For a given average speed, the schedule speed will increase with the decrease in duration of stop.

Q. 10. What is meant by coefficient of adhesion ?
Ans. The coefficient of adhesion is defined as the ratio of the tractive effort just necessary to slip the wheels on the track and adhesive weight.

Q. 11. How does coefficient of adhesion vary with the increase in speed ?
Ans. The coefficient of adhesion decreases with the increase in speed.

Q. 12. What is coefficient of adhesion ? Why a train driven by separately excited motors has better adhesion than a train driven by series dc motors ? **[U.P.T.U. 2005-06]**
Ans. Coefficient of adhesion (μ) provides a quantitative measure of the tendency of wheels to slip and is defined as the ratio of the tractive effort just necessary to slip the wheels on the track to the adhesive weight (weight on the driving axles).

The coefficient of adhesion depends on the nature of speed-torque characteristics of driving motor and has a higher value for motor with low regulation of speed *i.e.,* when for a given increase of speed drop in torque is large. When a wheel slips, the speed of the driving motor increases. The torque drops by a large amount in motors with low regulation of speed and the wheels regain their grip on rails immediately avoiding slip.

Since separately excited dc motors have low speed regulation than dc series motors, a train driven by separately excited dc motors has better adhesion than a train driven by dc series motors.

Q. 13. What do you understand by train resistance and on what factors does this depend ? **[Rajasthan Univ. 2006]**

Ans. ***Tractive Effort For Overcoming Train Resistance.*** Train resistance consists of all the forces resisting the motion of a train when it is running at uniform speed on a straight and level track. Under these circumstances the whole of the energy output from the driving axles is expended against train resistance. Train resistance is due to (*i*) the friction at the various parts of the rolling stock (*ii*) friction at the track and (*iii*) air resistance. The first two components constitute the *mechanical resistance* component of train resistance. The train resistance depends upon various factors, such as shape, size and condition of track etc., and is expressed in newtons per tonne of the dead weight. For a normal train the value of specific resistance has been 40 to 70 newtons/tonne. The general equation for train resistance is given as

$$R = k_1 + k_2V + k_3V^2$$

where k_1, k_2 and k_3 are constants depending upon the train and the track, R is the resistance in newtons and V is the speed in kmph. The first two terms represent the mechanical resistance and the last term represents air resistance.

Q. 14. What is meant by dead weight ?

Ans. The total weight of locomotive and train to be pulled by the locomotive is known as dead weight.

Q. 15. Why the accelerating weight of an electric train is greater than its dead weight ? **[Gorakhpur Univ. 1981]**

Ans. With the linear acceleration of the train, the rotating parts of the train such as wheels and motors also accelerate in an angular direction, and therefore, the tractive effort required is equal to the arithmetic sum of tractive effort required to have the angular acceleration of rotating parts and tractive effort required to have the linear acceleration. Hence the equivalent or accelerating weight of the train is taken higher than the dead weight.

Q. 16. What is meant by the term 'specific energy consumption' ?

Ans. Specific energy consumption is the total energy consumption of the train divided by the product of the weight of the train and distance of run.

Q. 17. What are those factors which affect the specific energy consumption of a train ?

Ans. The specific energy consumption of a train operating at a given schedule speed depends upon (*i*) distance between stops (*ii*) acceleration (*iii*) retardation (*iv*) crest speed (*v*) nature of route and (*vi*) the type of train equipment.

Q. 18. How is specific energy consumption affected by variation in distance between stops ?

Ans. Greater the distance between stops lesser will be the specific energy consumption.

Q. 19. How does the value of acceleration and retardation affect the specific energy consumption for a given run at a given schedule speed ?

Ans. For a given run at a given schedule speed, greater the value of acceleration and retardation more will be the period of coasting and, therefore, lesser the period during which power is on and, therefore specific energy consumption will accordingly be less.

Q. 20. What is meant by the term 'adhesive weight' ?

Ans. The total weight to be carried on the driving wheels is known as adhesive weight.

Problems

1. A train runs at an average speed of 50 kmph between stations situated 2.5 km apart. It accelerates at 2 kmphps and retards at 3 kmphps. Draw the speed-time curve for the run and find its maximum speed assuming a trapezoidal speed-time curve. **[Ans.** 57.75 kmph**]**

2. A train runs at an average speed of 45 kmph between stations 2.5 km apart. The train accelerates at 2 kmphps and retards at 3 kmphps. Find its maximum speed assuming a trapezoidal speed-time curve. Calculate also the distance travelled by it before the brakes are applied. Derive the formula used for calculation of maximum speed. **[B.H.U. Elec. Traction & Utilization 1978]** **[Ans.** 50.263 kmph; 2.383 km**]**

3. The speed-time curve of a train consists of :
   (*i*) Uniform acceleration of 6 km/h/s for 25 s;
   (*ii*) Free running for 10 minutes;
   (*iii*) Uniform deceleration of 6 km/h/s to stop the train;
   (*iv*) A stop of 5 minutes.
   Find the distance between the stations, the average and schedule speeds. **[A.M.I.E. Sec. B. Electric Drives and Their Control Winter 1996]** **[Ans.** 26.042 km; 144.23 kmph and 98.68 kmph**]**

4. A train runs with an average speed of 72 kmph. Distance between stops is 6 km. The train accelerates at 2 kmphps and retards at 3 kmphps. Determine the crest speed and schedule speed of the train. Duration of stops is 60 seconds. Assume trapezoidal speed-time curve. **[Ans.** 81.2 kmph; 60 kmph**]**

5. An electric train is to have a braking retardation of 3.2 kmphps. If the ratio of maximum speed to average speed is 1.3, the time for stops is 26 seconds, and acceleration is 0.8 kmphps, find its schedule speed for a run of 1.5 km. Assume trapezoidal speed-time curve. **[Agra Univ. Traction & Utilization of Elec. Power 1982]** **[Ans.** 30 kmph**]**

6. The schedule speed with a 200 tonne train on an electric railway with stations 777 metres apart is 27.3 kmph and the maximum speed is 20 per cent higher than the average running speed. The braking rate is 3.22 kmphps and the duration of stop is 20 seconds. Find the acceleration required. Assume a simplified speed-time curve with free running at the maximum speed. **[Agra Univ. Traction & Utilization of Elec. Power 1976 Supp.]** **[Ans.** 2.73 kmphps**]**

7. An electric train operating on a suburban service has a maximum speed of 60 kmph. The average distance between stops is 4 km and the schedule speed including a station stop of 30 seconds is 40 kmph. Assuming a simplified speed-time curve, calculate the necesary acceleration if the retardation is 4 kmphps. **[Ans.** 0.364 kmphps**]**

8. A train is required to run between two stations 2 km apart at an average speed of 40 km per hour. The run is to be made according to a simplified quadrilateral speed-time curve. If the maximum speed is to be limited to 60 km per hour, acceleration to 2 kmphps, coasting retardation to 0.15 kmphps and braking retardations 3 kmphps, determine the duration of accelerating, coasting and braking periods. **[Rajasthan Univ. Utilization of Electrical Power 2004]** **[Ans.** 30 seconds; 137 seconds, 13 seconds**]**

9. A 200 tonne motor coach with train has four motors each delivering a shaft torque of 300 N-m during acceleration from rest. Find out the time taken for the train to attain a speed of 40 kmph starting from rest on 1/200 gradient with a wheel diameter of 1,140 mm and gear ratio 4.5. Assume train resistance 0.5 kg/tonne, rotary inertia as 10% and gear efficiency as 95%. **[Gorakhpur Univ. Utilization of Elec. Power & Traction 1981]** **[Ans.** 23.7 seconds**]**

10. An electric train is accelerated uniformly from rest to a speed of 40 kmph, the period of acceleration being 20 seconds. If it coasts for 60 seconds against a constant resistance

of 50 N/tonne and is brought to rest in a further period of a 10 seconds by braking, determine (*i*) the acceleration (*ii*) the coasting retardation (*iii*) the braking retardation (*iv*) distance travelled and (*v*) the schedule speed with station stops of 10 second duration. Allow 10% for rotational inertia.

[Pb. Univ. Elec. Traction 1973]

[**Ans.** (*i*) 2 kmphps (*ii*) 0.1636 kmphps (*iii*) 3 kmphps (*iv*) 0.736 km and (*v*) 27.5 kmph]

11. An electric train weighing 300 tonnes is accelerated from rest up a gradient 1 in 200 at a uniform rate of 0.8 kmphps until a speed of 64 kmph is attained. Power is then cut off and the train coasts for 1 minute along level track. The brakes are now applied and brought to rest with retardation of 3.2 kmphps. Find specific energy consumption. Allow 10% rotational inertia and train resistance 5 kg/tonne.

[Gorakhpur Univ. Utilization of Elec. Power & Traction 1981]

[**Ans.** 46.5 Wh per tonne-km]

12. A 250 tonne train with 10% rotational inertia effect is started with uniform acceleration and reaches a speed of 50 kmph in 25 seconds on level road. Find the specific energy consumption if the journey is to be made according to a simplified trapezoidal speed-time curve, the acceleration is 2 kmphps, braking retardation 3 kmphps and the distance between the two stations is 2.4 km, efficiency of motors is 0.9, track resistance 5 kg/tonne.

[Pb. Univ. Elec. Traction April 1975]

[**Ans.** 28.3 Wh/tonne-km]

13. An electric train weighing 200 tonnes runs a uniform upgradient of 1% with the following speed-time curve:

(*i*) Uniform acceleration 2 kmphps for 30 seconds

(*ii*) Constant speed for 40 seconds

(*iii*) Coasting for 30 seconds

(*iv*) Braking at 2.5 kmphps to rest

(*v*) Stop at station 15 seconds.

If the tractive resistance is 40 N/tonne, rotational inertia effect 10% of dead weight and overall efficiency of transmission and motor is 75% determine:

(*i*) schedule speed (*ii*) specific energy consumption (*iii*) total energy consumption (*iv*) distance between two stations.

[Pb. Univ. Electric Traction June 1993]

[**Ans.** (*i*) 39.88 kmph (*ii*) 69.93 Wh/tonne-km (*iii*) 20.7 kWh (*iv*) 1.48 km]

14. An electric train weighing 200 tonnes accelerates uniformly on the level track to a speed of 43 kmph in 20 seconds. Power is then cut off and the train coasts until the speed is 34 kmph when the brakes are applied and the train is brought to rest in 10 seconds. Allowing a train resistance of 5 kg per tonne during acceleration and 6.5 kg per tonne during coasting, and 12% for rotational inertia, calculate (*i*) the maximum horse power developed at the driving axles, and (*ii*) the specific energy output.

[**Ans.** (*i*) 2,340 hp (metric); 37.4 Wh/tonne-km]

15. A 40 tonne train starts a gradient of 1 in 75 at the rate of 1.6 kmphps. The tractive resistance is 66.75 N per tonne and allowance for rotational inertia is 10%. Calculate:

(*a*) the energy in kWh usually employed in attaining a speed of 48 kmph from rest,

(*b*) the specific energy consumption in Wh per tonne-km when running at a steady speed of 56 kmph. up this gradient if the overall efficiency of equipment is 70%.

[Gorakhpur Univ. Utilization of Elec. Power & Traction Sept. 1982]

[**Ans.** (*a*) 2.18 kWh; 78.4 Wh/tonne-km]

16. An electric train weighing 400 tonnes has to maintain an average speed of 40 km per hour between two stations 3 km apart on an incline of 1 in 200. The train accelerates at 2 kmphps and retards at 4 kmphps. The tractive resistance is 5 kg per tonne, and allowance for rotational inertia 10%. Assuming a trapezoidal speed-time curve, find the energy consumption for the run while going up the incline. The overall efficiency may be taken as 65%. [**Ans.** 64.58 kWh]

17. What is meant by 'adhesive weight' of a locomotive? Calculate the weight of a locomotive to haul a train weighing 500 tonnes up an incline of 1 in 75 with an acceleration of 1 kmphps. The tractive resistance is 5 kg per tonne and allowance for rotational inertia 10%. Coefficient of adhesion 0.25. [**Ans.** 120 tonnes]

18. A 1,500 V dc locomotive is required to start up a train weighing 1,000 tonnes (including loco) on an up gradient of 1/200 and accelerates it at the rate of 0.5 kmphps, to a speed of 60 kmph. The train resistance is 7 kg/tonne and rotational inertia 8%. Making your own assumptions for any missing data, determine the minimum weight of the locomotive and its horse power.

[**Ans.** 213 tonnes; 47,333 hp (metric)]

19. Differentiate between adhesive and total weight of an electric locomotive. Determine the minimum adhesive weight of a loco required to start a 2,000 tonnes (inclusive of loco) on 1 : 150 gradient and accelerate it at 0.1 kmphps. Assume coefficient of adhesion as 0.25, train resistance 4 kg/tonne and rotary inertia as 8%.

[Gorakhpur Univ. Utilization of Elec. Power & Traction 1983]

[**Ans.** 110 tonnes]

20. A 500 tonne goods train is to be hauled by a locomotive up a gradient of 2% with an acceleration of 1.2 kmphps. Coefficient of adhesion is 25%, track resistance 40 N/tonne and effecting rotating masses 10% of dead weight. Find the weight of the locomotive and number of axles if the axle load is not to exceed 20 tonnes. [**Ans.** 160 tonnes; 8]

21. A locomotive of 100 tonnes can just accelerate a train of 400 tonnes with an acceleration of 1.5 kmphps up a gradient of 1 in 100. Assume an adhesive weight of locomotive 70% of total dead weight. The tractive resistance is 40 N/tonne and rotational inertia increases the dead weight by 10%. If the above locomotive is aided by another locomotive of 150 tonnes with 80% adhesive weight, determine:

(*i*) The additional trailing weight that can be hauled up the same gradient under the same conditions;

(*ii*) The maximum gradient, trailing hauled load remaining unchanged;

(*iii*) The acceleration, if both gradient and hauled load remain unchanged.

[**Ans.** (*i*) 1,106 tonnes, (*ii*) 7.6%, (*iii*) 3.62 kmphps]

22. An electric locomotive is required to haul a train of 12 coaches, each weighing 30 tonnes on the main line service requiring an initial acceleration of 0.8 kmphps up a gradient of 1 in 100. Estimate the adhesive weight needed and hence the number of driving axles the locomotive must have if the permissible axles load is 20 tonnes. Allowance for rotational inertia to be 4% for the coaches and 15% for the locomotive. Maximum coefficient of adhesion is 0.2 and the tractive resistance is 3.5 kg per tonne.

[Gorakhpur Univ. Utilization of Elec. Power & Traction 1980, 1982]

[**Ans.** 80 tonnes; 4]

❑❑❑

CHAPTER 12

Electric Traction Motors

INSIDE THIS CHAPTER

12.1. General Features of Traction Motors 12.2. Operating Characteristics of DC Motors 12.3. DC Series Motor 12.4. DC Shunt Motor 12.5. AC Series Motor 12.6. Three Phase Induction Motor 12.7. Linear Induction Motor 12.8. Rating and Ventilation.

12.1. GENERAL FEATURES OF TRACTION MOTORS

The chief requirements of electric motors used for traction work are:

1. *Mechanical Features.* (*i*) A traction motor must be robust and capable to withstand continuous vibrations since service conditions are extremely severe.

(*ii*) The weight of the traction motor should be minimum in order to increase the payload capacity of the vehicle. This is achieved by using high speed motors, upper limit being fixed by excessive centrifugal stresses.

(*iii*) The traction motor is located underneath a motor coach (except in few electric locomotives). The space underneath a motor coach is limited by the size of driving wheels and the track gauge. The traction motor, therefore, must be small in overall dimensions specially in its overall diameter.

(*iv*) The traction motor must be totally enclosed type, particularly when mounted beneath the locomotive or motor coach, to provide protection against ingress of dirt, dust, water, mud etc.

Thus, for magnetic circuit of a traction motor cast iron, which cannot withstand continuous vibrations, is not suitable. Use of cast steel or fabricated steel, which gives more mechanical strength, is made in place of cast iron. Those parts of the motor, which are not highly stressed; must be made of pressed or fabricated steel plates and light alloys.

2. Electrical Characteristics. (i) High starting torque. A traction motor must be capable of developing high starting torque, specially when the train is to be accelerated at a reasonably high rate such as in case of urban or suburban services.

(ii) Simple speed control. The traction motor should be amenable to simple speed control methods as an electric train has to be started and stopped very often.

(iii) Self-relieving property. The speed-torque characteristics of the motor should be such that the speed may fall with the increase in load. The motors having such speed-torque characteristics are self protective against excessive overloading as power output of a motor is proportional to the product of torque and speed.

(iv) Possibility of dynamic or regenerative braking. The traction motor should be amenable to easy and simple methods of rheostatic and/or regenerative braking.

(v) Capability of withstanding voltage fluctuations. Traction motor should be capable of withstanding rapid fluctuations in supply voltage without undue effect on its performance since in traction work rapid voltage fluctuation owing to heavy current inrush at start, is common feature.

(vi) Capability of withstanding temporary interruption of supply. Traction motor should be capable of withstanding temporary interruptions of supply without undue rush of current, since it occurs at the instant of crossing over the cross-overs and section insulators.

(vii) Overload capacity. Traction motor should be capable of taking excessive load as it is subjected to very arduous duty.

(viii) Parallel running. In traction work, usually more than one motor (two or four motors per motor car) are required. Traction motors, therefore, should be of such speed-torque and current-torque characteristics that when they are operated in parallel and mechanically coupled, they share the load almost equally.

No motor meets all the requirements mentioned above. Most suitable motors for dc systems are the series and compound motors whereas for ac systems the single phase series motors and 3-phase induction motors are employed.

12.2. OPERATING CHARACTERISTICS OF DC MOTORS

The performance and, therefore, suitability of a dc motor is determined from its characteristics, known as *performance characteristics*. The important characteristics of dc motors are:

1. ***Torque-Armature Current Characteristic.*** This characteristic curve gives relation between torque developed in the armature, T and armature current I_a. This is also known as *electrical characteristic*.

2. ***Speed-Armature Current Characteristic.*** This characteristic curve gives relation between speed N and armature current I_a.

3. ***Speed-Torque Characteristic.*** This characteristic gives relation between speed N and torque developed in armature, T. This is also known as *mechanical characteristic*. This curve may be derived from the two characteristic curves mentioned above.

The important relations to be kept in mind while discussing motor characteristics are:

(*i*) $I_a = \dfrac{V - E_b}{R_a}$ (*ii*) $N \propto \dfrac{E_b}{\phi}$ and (*iii*) $T \propto \Phi I_a$.

1. Torque developed by a dc motor is found as follows:

Let T_e be the electromagnetic torque developed in newton-metres by the motor running at *n* rps.

$$\text{Power developed} = \text{Work done per second} = T_e\omega = T_e \times 2\pi n \text{ watts} \quad ...(12.1)$$

Electrical equivalent of mechanical power developed by the armature, also

$$= E_b I_a \text{ watts} \quad ...(12.2)$$

Comparing Eqs. (12.1) and (12.2), we have

$$T_e \times 2\pi n = E_b I_a$$

$$\text{or } T_e = \frac{E_b I_a}{2\pi n} = 0.159 \frac{E_b I_a}{n} \text{ N-m} \quad ...(12.3)$$

$$\text{Also } T_e = \frac{E_b I_a}{2\pi \frac{N}{60}} = 9.55 \frac{E_b I_a}{N} \text{ N-m} \quad ...(12.4)$$

where N is speed in rpm

Substituting $E_b = \Phi Z \dfrac{N}{60} \times \dfrac{P}{A}$ in Eq. (12.4), we have

$$T_e = 9.55 \times \Phi \times Z \times \frac{N}{60} \times \frac{P}{A} \times \frac{I_a}{N}$$

$$= 0.159 \Phi Z P \frac{I_a}{A} \text{ newton-metres} \quad ...(12.5)$$

2. The relation between speed and armature current is found as follows :

As the back emf developed in a dc motor is nothing but a generated emf, therefore, it is given by the expression

$$E_b = \frac{\phi Z N}{60} \times \frac{P}{A} \text{ volts} \quad ...(12.6)$$

where ϕ is the flux per pole in webers, Z is the number of armature conductors, N is the speed of armature in rpm, P is the number of poles and A is the number of parallel paths.

$$\text{Back emf also} = V - I_a R_a \quad ...(12.7)$$

Comparing Eqs. (12.6) and (12.7) we get

$$\frac{\phi Z N}{60} \times \frac{P}{A} = V - I_a R_a$$

$$\text{or } N = \frac{V - I_a R_a}{\phi Z} \times \frac{60 A}{P} \quad ...(12.8)$$

or $N \propto \dfrac{V - I_a R_a}{\phi}$ since Z, P and A, for a particular machine, are constant.

or $N \propto \dfrac{V}{\phi}$ since $I_a R_a$ is very small as compared to applied voltage.

or $N \propto \dfrac{1}{\phi}$ if applied voltage V is constant.

3. The relation between speed and torque is determined as follows.

Torque developed, $T = 0.159 \phi Z I_a \dfrac{P}{A}$ [Refer to Eq. (12.5)]

$$\text{or } \frac{A}{\phi Z P} = \frac{0.159 I_a}{T}$$

Substituting $\dfrac{A}{\phi Z P} = \dfrac{0.159 I_a}{T}$ in Eq. (12.8), we get

$$N = \frac{(V - I_a R_a) \times 60 \times 0.159 I_a}{T}$$

$$= \frac{9.55 (V - I_a R_a) I_a}{T} \quad ...(12.9)$$

The operating characteristics of dc motors have already been dealt in chapter 1 (Art. 1.16.1)

Example 12.1. The torque-armature current characteristics of a series traction motor are given as:

Armature Current (Amp)	5	10	15	20	25	30	35	40
Torque (N-m)	20	50	100	155	215	290	360	430

The motor resistance is 0.3 Ω. If this motor is connected across 230-V dc supply deduce the speed-armature current characteristic.

Solution : Supply voltage, V= 230 V

Resistance of motor, $R_m = R_a + R_{se} = 0.3\ \Omega$

Armature current, I_a in amperes	5	10	15	20	25	30	35	40
Torque, T in N-m	20	50	100	155	215	290	360	430
Back emf, $E_b = (V - I_a R_m)$ in volts	228.5	227.0	225.5	224.0	222.5	221.0	219.5	218.0
Speed, N $= \dfrac{9.55(V - I_a R_m)}{T} . I_a$ in RPM	545	434	323	276	247	218	204	194

The deduced speed-armature current characteristic is shown in Fig. 12.1.

Example 12.2. The following figures give the magnetisation curve of a dc series motor when working as a separately excited generator at 600 rpm:

Field Current (amperes)	**20**	**40**	**60**	**80**
EMF (volts)	**215**	**381**	**485**	**550**

The total resistance of the motor is 0.8 ohm. Deduce the speed-torque curve for this motor when operating at a constant voltage of 600 volts.

Solution : Voltage applied across the motor, V = 600 volts

Resistance of the motor, $R_m = (R_a + R_{se}) = 0.8\ \Omega$

Speed, $N_1 = 600$ rpm

Field current (amperes)	20	40	60	80
Back emf, E_1 (volts) at speed N_1 (600 rpm) = EMF generated by the armature (given)	215	381	485	550
Back emf, E (volts) at speed N (to be determined) = $V - I_a R_m$	584	568	552	536
Speed, N (to be determined in RPM) = $\frac{N_1}{E_1} \times (V - I_a R_m)$	1,630	895	683	585
Torque, T = $\frac{9.55(V - I_a R_m) I_a}{N}$ in N-m	68.4	240	462	700

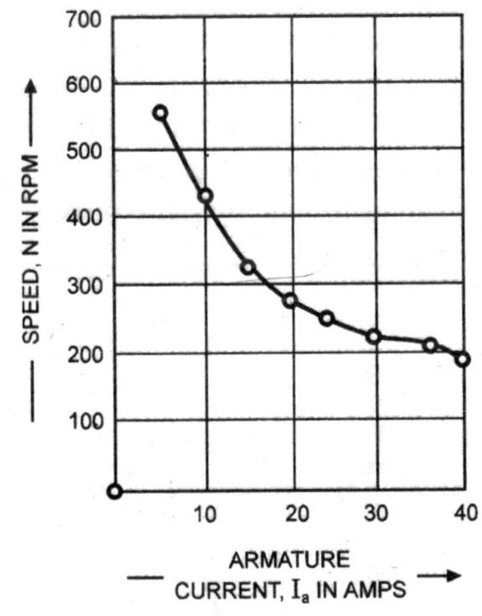

Fig. 12.1

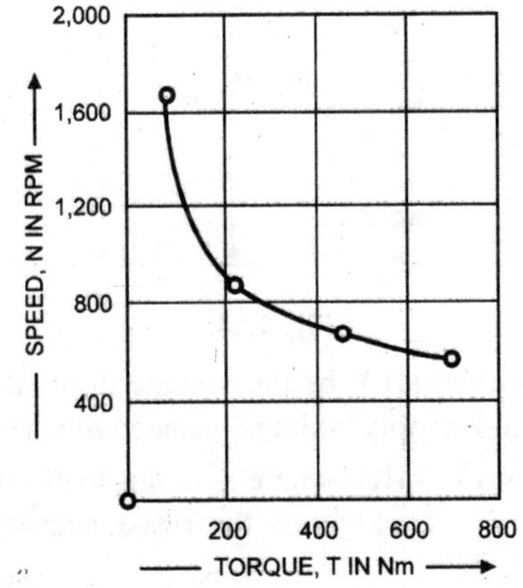

Fig. 12.2

The deduced speed-torque curve for the motor is shown in Fig. 12.2.

Example 12.3. A dc series motor drives a load. The torque of the motor varies as the square of the speed. The motor takes a current of 15 A when the speed is 640 rpm. Calculate the motor speed and current drawn when the field winding is shunted by a diverter of the same resistance as that of the field winding. Neglect all motor losses and assume that the magnetic circuit is unsaturated.

Solution : Speed, $N_1 = 640$ rpm

Series field current, $I_{se_1} = I_{a1} = I_1 = 15$ A

Let the speed be N_2 rpm, line current I_2 when the field winding is shunted by a diverter of the same resistance as that of field winding.

Series field current, $I_{se_2} = \frac{1}{2} I_2$

$\because$ because line current I_2 divides equally between series field and diverter, being of equal values.

Since torque developed, $T \propto \phi I_a \propto \phi I$ $\qquad \because I_a = I$

$$\therefore \quad \frac{T_2}{T_1} = \frac{\phi_2 I_2}{\phi_1 I_1} = \frac{\frac{1}{2} I_2^2}{I_1^2}$$

$\because \phi \propto I_{se}$; magnetic circuit is unsaturated

$$\text{or} \quad \frac{T_2}{T_1} = \frac{I_2^2}{2 I_1^2} \qquad ...(i)$$

$$\text{Also} \quad \frac{T_2}{T_1} = \frac{N_2^2}{N_1^2} \qquad ...(ii) \qquad \because \text{Given that } T \propto N^2$$

Comparing Eqs. (*i*) and (*ii*) we have

$$\frac{N_2^2}{N_1^2} = \frac{I_2^2}{2 I_1^2}$$

$$\text{or} \quad \frac{N_2}{N_1} = \frac{1}{\sqrt{2}} \frac{I_2}{I_1} \qquad ...(iii)$$

Neglecting voltage drop in armature and series field and assuming applied voltage constant

$$N \propto \frac{1}{\phi}$$

$$\frac{N_2}{N_1} = \frac{\phi_1}{\phi_2} = \frac{I_1}{\frac{1}{2} I_2} = \frac{2 I_1}{I_2} \qquad ...(iv)$$

Comparing Eqs. (*iii*) and (*iv*) we have

$$\frac{1}{\sqrt{2}} \frac{I_2}{I_1} = \frac{2 I_1}{I_2}$$

Current drawn, $I_2 = \sqrt{2\sqrt{2} I_1^2} = \sqrt{2\sqrt{2} \times 15^2} = 25.227$ A **Ans.**

Speed, $N_2 = \frac{2 I_1}{I_2} N_1 = \frac{2 \times 15}{25.227} \times 640 = 761$ rpm **Ans.**

12.3. DC SERIES MOTOR

The characteristics of dc series motors have already been discussed in chapter 1 [Art. 1.16.1 (*b*)]. In this chapter dc series motor will be analysed from the point of view of requirements of traction service.

1. The dc series motor develops high torque at start which is an essential requirement of traction service.

2. The series motor is amenable to simple speed control methods. Various methods of speed control of dc series motors will be discussed in forthcoming chapter *i.e.,* chapter 13.
3. In case of dc series motors

$$\text{Speed N} \propto \frac{V}{\phi} \propto \frac{1}{I} \quad ...(i)$$

assuming flux proportional to exciting current and applied voltage constant.

$$\text{Also } T \propto \phi I \propto I^2 \quad ...(ii)$$

Comparing Eqs. (*i*) and (*ii*) we have

$$N \propto \frac{1}{I} \propto \frac{1}{\sqrt{T}}$$

But since power output of a motor is proportional to the product of torque and speed

$$\text{Power output} \propto TN \propto T \times \frac{1}{\sqrt{T}} \propto \sqrt{T}$$

Thus power drawn from supply mains varies as the square root of the load torque in case of dc series motors *i.e.,* the series motors to some extent are self-relieving ones.
4. Series motors, as discussed in chapter 1 (Art. 1.20.3) are not suitable for regenerative braking as these are not electrically stable unless certain measures are taken.
5. In case of dc series motors commutation is excellent up to twice full load, therefore, replacement of brushes etc. is not required frequently.
6. In case of dc series motors the flux varies as, the armature current, torque corresponding to a given armature current, therefore, is independent of line voltage and thus unaffected by variations in the supply voltage.

With the sudden increase in supply voltage, the armature current tends to increase, but with the increase in armature current the flux also increases which causes increase in back emf and thus armature current restores to its initial value. Thus series motors are less susceptible to sudden change in supply voltage.
7. In case of dc series motors, up to the point of magnetic saturation, torque developed is proportional to the square of the armature current. Therefore, dc series motor requires comparatively less increased power input with the increase in load torque. Thus the series motors are capable of withstanding excessive loads.
8. Because of low time constant of field in case of series motor, the field flux dies away in very short time with the result that back emf of the motor ceases with temporary interruption of supply. In case of dc series motors, therefore, there is undue rush of initial current on temporary interruption of supply.
9. The dc series motors, when operated in parallel to drive a vehicle by means of different driving axles, share load almost equally even there is unequal wear of different driving wheels. This is explained as below.

Consider two identical motors of series type (say of speed-current characteristic shown in Fig. 12.3) connected to different driving axles of the vehicle. For a given speed of vehicle, the peripheral speed of all the driving wheels will be same, but should there be a slight difference in the diameters of the driving wheels (which is possible on account of difference in wear) there will be a corresponding difference in the speeds of the driving axles to which they are connected and also, therefore, a difference in speed of the motors. The speed of the motors will be given by the relation $\frac{N_2}{N_1} = \frac{D_1}{D_2}$.

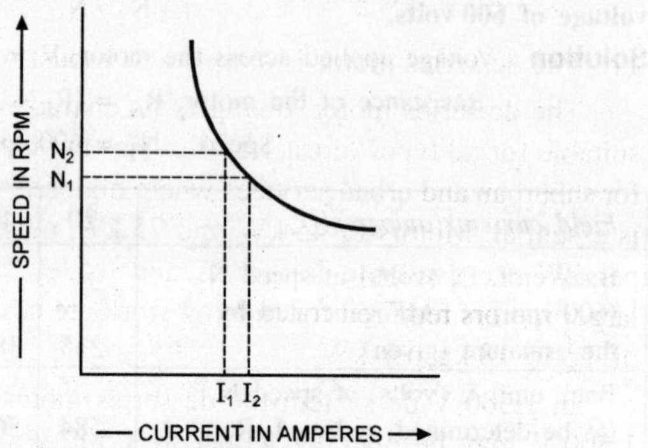

Fig. 12.3

When the motors are working in parallel the current drawn by each motor (determined from speed-current characteristic) due to slight difference in the speed does not differ much. As voltage across each motor is same, therefore motors share the load almost equally.
10. When the series motors are operating in series (Fig. 12.4), the current drawn by each motor will be same but due to slight difference in speeds, the back emfs will differ slightly. This will result in slight difference in voltage acting across each motor and also, therefore, slight difference in load shared by each motor.

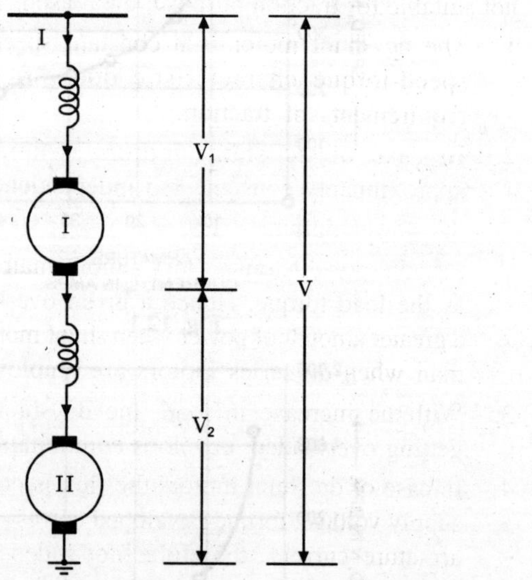

Fig. 12.4.

Let V_1 and V_2 and V be the voltage drops across motors I and II and supply voltage respectively (Fig. 12.4).

Since $N \propto (V - IR)$ where I is the load current and R is the resistance of each motor

$$\therefore \frac{N_2}{N_1} = \frac{V_2 - IR}{V_1 - IR} \quad ...(12.10)$$

Also $V_1 + V_2 = V$ ...(12.11)

Solving above Eqs. (12.10) and (12.11) we have

$$V_1 = \frac{\frac{N_1}{N_2}(V - IR) + IR}{1 + \frac{N_1}{N_2}} \quad ...(12.12)$$

and $$V_2 = \frac{\frac{N_2}{N_1}(V - IR) + IR}{1 + N_2 / N_1} \quad ...(12.13)$$

11. The dc series motor is simple and robust in construction.

The dc series motor, owing to its characteristics, is most suitable for all types of traction services but more particularly for suburban and urban services where high rate of acceleration is essential. Motors of 45 kW and 75 kW, mass 5.4 to 7.2 kg per kW are available for trams and trolley buses although larger motors required for railway work are usually of masses of 12 to 18 kg per kW.

In 1,500 V dc system the dc series motors insulated for 1,500 V may be operated either at 1,500 V or at 750 V by connecting two of them permanently in series. In 3,000 V dc system the dc motors are insulated for 3,000 V but operated at 1,500 V by connecting two in series, as shown in Fig. 12.4.

The main drawbacks of dc series motors are owing to commutator such as restricted speed, voltage and current. There is a risk of flash-over and the brush gear requires considerable maintenance. The speed is also limited due to centrifugal forces acting on the rotor winding.

12.4. DC SHUNT MOTOR

The characteristics of dc shunt motors have already been discussed in chapter 1 [Art 1.16.1 (*a*)]. DC shunt motors are not suitable for traction purpose, the reasons are given below:

1. The dc shunt motor is a constant speed motor and its speed-torque characteristic does not meet with the requirements of traction.
2. In case of dc shunt motors, where the speed is approximately constant and independent of load torque, the power output $\propto TN \propto T$

   i.e., the power drawn from supply mains varies directly as the load torque. Hence a given overload will require a greater amount of power when shunt motors are employed than when dc series motors are employed.
3. With the increase in load, the dc shunt motor besides getting overloaded, develops commutation troubles also.
4. In case of dc shunt motors the flux is not independent of supply voltage, torque developed corresponding to a given armature current, therefore, not independent of supply voltage. Thus the torque developed is considerably affected by variation in supply voltage.

Due to large inductance of field winding, the rate of change of current in this winding is low. Thus the current increases with the increase in supply voltage. With the build up of current in the field winding due to increase in supply voltage, the back emf increases and hence the value of increased current starts decreasing. The current inrush in case of shunt motors will, therefore, be of long duration and high initial value in comparison to that in series motors.

5. In case of dc shunt motors, the field flux remains constant and torque developed is directly proportional to the armature current. Hence for a given increase in torque developed a greater amount of current will be drawn from supply mains when dc shunt motor is used than when dc series motor is used.
6. Because of the large time constant of field winding of dc shunt motor and because it is connected across the armature, the back emf during interruption of supply is maintained even though it is decreasing. Now when the supply voltage is restored, the inrush current in the shunt motor will be determined by the resultant voltage in the armature circuit and impedance of armature circuit. Thus, inrush current in the shunt motor due to interruption of supply will be less in comparison to dc series motor.
7. In case of two identical shunt motors connected in parallel, the difference in currents drawn from supply mains due to small difference in speeds will be considerable, as obvious from Fig. 12.5, and therefore, motors will be loaded very unequally.

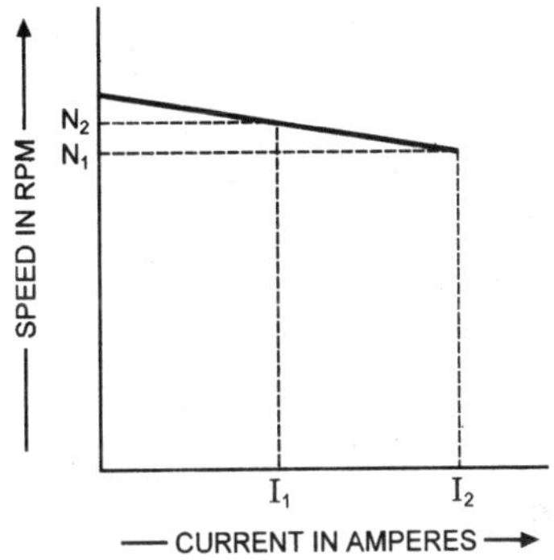

Fig. 12.5

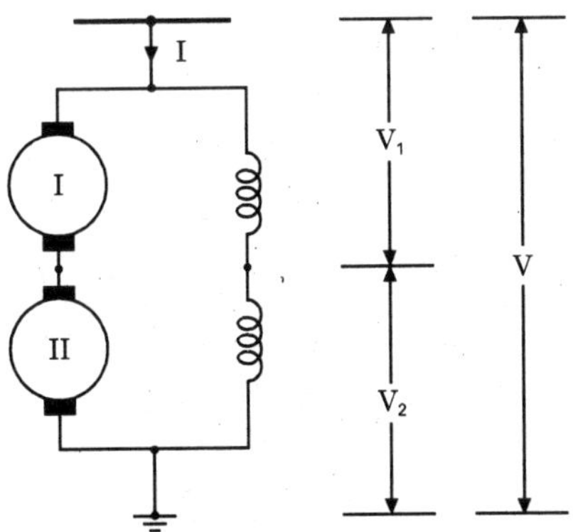

Fig. 12.6. ***Series Operation of DC Shunt Motors***

In case of two identical shunt motors connected in series (Fig. 12.6) there will be a slight difference in the voltage acting across each motor due to slight difference in the speeds and thus the motors will share the load almost equally.

From the above discussions it is obvious that shunt motor is not suitable for traction purposes.

Example 12.4. Two dc traction motors run at speed 700 rpm and 750 rpm respectively when each takes a current of 50 A from 500 V mains. Each motor has an effective resistance of

0.3 Ω. Calculate the speed and voltage across each machine when mechanically coupled and electrically connected in series and taking a current of 50 A from 500 V mains, the resistance of each motor being unchanged.

Solution : Let the two motors be 1 and 2 of speed N_1 = 700 rpm and N_2 = 750 rpm, respectively.

Resistance of each motor R_m = 0.3 Ω

Applied voltage across each motor, V = 500 V

Back emf of motor A when taking a current of 50 A

$$E_{b_1} = V - IR_m = 500 - 50 \times 0.3 = 485 \text{ V}$$

Back emf of motor B when taking a current of 50 A

$$E_{b_2} = V - IR_m = 500 - 50 \times 0.3 = 485 \text{ V}$$

When the machines are mechanically coupled and connected in series, the speed of each motor will be same, say N, current will be same and equal to 50 A (given) and the sum of voltage across the two motors will be equal to 500 V.

Let the voltage across motors 1 and 2 be V_1 and V_2 respectively

$$\text{Now } V_1 + V_2 = 500 \qquad ...(i)$$

$$\text{Back emf of motor 1, } E'_{b_1} = E_{b_1} \times \frac{N}{N_1} = \frac{485}{700} \times N$$

$$\text{Voltage across motor 1, } V_1 = E'_{b_1} + IR_m = \frac{485}{700} N + 50 \times 0.3 = \frac{485}{700} N + 15$$

$$\text{Back emf of motor 2, } E'_{b_2} = E_{b_2} \times \frac{N}{N_2} = \frac{485}{750} N$$

$$\text{Voltage across motor 2, } V_2 = E'_{b_2} + IR_m = \frac{485}{750} N + 15$$

Substituting $V_1 = \frac{485}{700} N + 15$ and $V_2 = \frac{485}{750} N + 15$ in Eq. (i) we get

$$\frac{485}{700} N + 15 + \frac{485}{750} N + 15 = 500$$

$$\text{or } \left(\frac{485}{700} + \frac{485}{750}\right) N = 470$$

$$\text{or } N\left(\frac{1}{700} + \frac{1}{750}\right) = \frac{470}{485}$$

$$\text{or } N = \frac{470}{485} \times \frac{10{,}500}{15 + 14} = 351 \text{ rpm } \textbf{Ans.}$$

$$\text{PD across machine 1, } V_1 = \frac{485\text{ N}}{700} + 15 = \frac{485 \times 351}{700} + 15 = 258.0 \text{ V } \textbf{Ans.}$$

$$\text{PD across machine 2, } V_2 = \frac{485\text{ N}}{750} + 15 = \frac{485 \times 351}{750} + 15 = 242.0 \text{ V } \textbf{Ans.}$$

Example 12.5. A tramcar is equipped with two motors which are operating in parallel. Calculate the current drawn from the supply mains at 500 volts when the car is running at a steady speed of 40 kmph and each motor is developing a tractive effort of 1,800 newtons. The resistance of each motor is 0.4 ohm. The friction, windage and other losses may be assumed, as 3,200 watts per motor.

Solution : Voltage across each motor, V = 500 volts

Maximum speed, V_m = 40 kmph

Tractive effort, F_t = 1,800 newtons

Motor resistance, R_m = 0.4 Ω

Losses per motor = 3,200 watts

Power output of each motor

$$= \frac{F_t \times V_m \times 1{,}000}{3{,}600} \text{ watts} = \frac{1{,}800 \times 40 \times 1{,}000}{3{,}600} = 20{,}000 \text{ watts}$$

Constant losses = 3,200 watts

Copper losses = $I^2 R_m = 0.4 I^2$

where I is the current drawn from supply mains.

Input to motor = Motor output + constant losses + copper losses

$$VI = 20{,}000 + 3{,}200 + 0.4 I^2$$

$$\text{or } 0.4 I^2 - 500 I + 23{,}200 = 0 \qquad \because V = 500 \text{ volts}$$

$$\text{or } I = \frac{500 \pm \sqrt{(500)^2 - 4 \times 0.4 \times 23{,}200}}{2 \times 0.4}$$

or I = 48.25 A or 1,201.75 A

or Current drawn by each motor = 48.25 A

∵ 1,201.75 A , being unreasonably high, cannot be accepted

Total current drawn from supply mains

= 48.25 × 2 = 96.5 A **Ans.**

Example 12.6. A locomotive exerts a tractive effort of 35,300 N in hauling a train at 48 kmph on the level track. If the motor is to haul the same train on a gradient of 1 in 50 and the tractive effort required is 55,180 N, determine the power delivered by the locomotive if it is driven by (*i*) dc series motors (*ii*) induction motors. [*Pb. Technical Univ. Electric Drives & Utilization, Dec. 2008*]

Solution : Tractive effort, F_t = 35,300 N

Maximum speed, V = 48 kmph

Power output required on the level track

$$= \frac{F_t \times V \times 1{,}000}{3{,}600} = \frac{35{,}300 \times 48 \times 1{,}000}{3{,}600} = 470{,}667 \text{ W or } 470.667 \text{ kW}$$

Power delivered by locomotive on upgradient track with dc series motors

$$= 470.667 \sqrt{\frac{55{,}180}{35{,}300}} = 588.5 \text{ kW } \textbf{Ans.}$$

$$\because \text{Power output} \propto \sqrt{T} \propto \sqrt{F_t}$$

Power delivered by locomotive on upgradient track with induction motors

$$= 470.667 \times \frac{55{,}180}{35{,}300} = 735.7 \text{ kW } \textbf{Ans.}$$

$$\because \text{Power output} \propto T \propto F_t$$

Example 12.7. A dc traction motor has speed-tractive effort and armature current characteristics as given below with driving wheels having 85 cm wheel diameter and gear ratio as 71 : 21.

Armature current (amperes)	60	120	180	240	300	360
Speed (kmph)	80	50	45	42	38	35
Tractive Effort in thousand newtons	1.7	5	10	14	16	20

Determine new characteristics if driving wheels are replaced by 90 cm diameter wheels and gear ratio changed to 74 : 19.

[*Pb. Univ. Electric Traction July 1992*]

Solution : Diameter of driving wheel, $D_1 = 85$ cm = 0.85 m

Diameter of new driving wheel, $D_2 = 90$ cm = 0.9 m

Gear ratio, $\gamma_1 = 71 : 21$

Gear ratio, $\gamma_2 = 74 : 19$

$$\text{New speed, } V_2 = \frac{D_2}{D_1} \times \frac{\gamma_1}{\gamma_2} \times V_1 = \frac{0.9}{0.85} \times \frac{71/21}{74/19} = 0.919\, V_1$$

$$\text{New tractive effort, } F_{t_2} = \frac{D_1}{D_2} \times \frac{\gamma_2}{\gamma_1} F_{t_1} = \frac{0.85}{9} \times \frac{74/19}{71/21} = 1.088\, F_{t_1}$$

New characteristics are:

Armature current (amperes)	60	120	180	240	300	360
Speed (kmph), $V_2 = 0.919\, V_1$	73.52	45.95	41.36	38.6	34.92	32.17
Tractive effort in thousand Newtons $F_{t_2} = 1.088\, F_{t_1}$	1.85	5.44	10.88	15.23	17.41	21.76

Example 12.8. A motor coach is being driven by two identical dc series motors. First motor is geared to driving wheel having diameter of 90 cm and other motor to driving wheel having diameter of 86 cm. The speed of the first motor is 500 rpm when connected in parallel with the other across 600 V supply. Find the motor speeds when connected in series across the same supply. Assume armature current to remain same and armature voltage drop of 10% at this current.

Solution : Speed of first motor, $N_1 = 500$ rpm

$$\text{Back emf, } E_{b_1} = 600 - \frac{10}{100} \times 600 = 540 \text{ volts}$$

When the motors are connected in series across 600 V supply, as shown in Fig. 12.7.

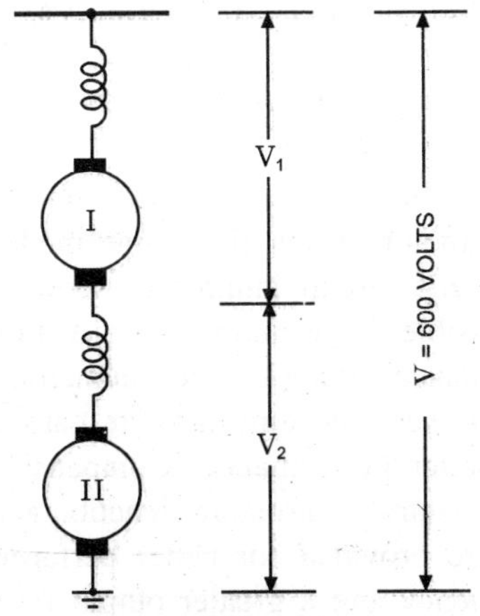

Fig. 12.7

Let the supply voltage across motors I and II be V_1 and V_2 volts and speed N'_1 and N'_2 respectively.

$$\text{Since speed, } N \propto \frac{V - IR}{\phi}$$

Current through the motors remains the same, therefore, flux produced by it also remains the same and $N \propto (V - IR)$

$$\therefore \quad \frac{N'_1}{N'_2} = \frac{V_1 - IR}{V_2 - IR} = \frac{V_1 - \frac{10}{100} \times 600}{V_2 - \frac{10}{100} \times 600} = \frac{V_1 - 60}{V_2 - 60} \quad ...(i)$$

And also $N'_1 D_1 = N'_2 D_2$

$$\text{or} \quad \frac{N'_1}{N'_2} = \frac{D_2}{D_1} = \frac{86}{90} \quad ...(ii) \text{ Since peripheral speed is equal}$$

$$\therefore \quad \frac{V_1 - 60}{V_2 - 60} = \frac{86}{90}$$

or $90V_1 - 5{,}400 = 86V_2 - 5{,}160$

or $90V_1 - 86V_2 = 5{,}400 - 5{,}160 = 240$ ...(*iii*)

and also $V_1 + V_2 = V = 600$ V ...(*iv*)

Solving Eqs. (*iii*) and (*iv*) we get

$V_1 = 294.55$ V and $V_2 = 305.45$ V

Now the speeds of the motors can be calculated as follows:

$$\frac{N'_1}{N_1} = \frac{E'_{b_1}}{E_{b_1}}$$

$$\text{or } N'_1 = N_1 \times \frac{E'_{b_1}}{E_{b_1}} = 500 \times \frac{294.55 - 60}{600 - 60} = 217 \text{ rpm } \textbf{Ans.}$$

$$\text{and } N'_2 = N'_1 \times \frac{D_1}{D_2} = 217 \times \frac{90}{86} = 277 \text{ rpm } \textbf{Ans.}$$

Example 12.9. The characteristics of a dc series motor when operating at 600 volts and driving 107 cm wheels are as follows :

Current (amperes)	500	400	300	200
Train speed (km/hour)	35.6	38.5	42.5	50.5
Tractive effort (kg)	2,580	1,960	1,380	725

Two of such motors are mounted on a motor coach. The wheels driven by one motor A are 107 cm in dia and other motor B are 102 cm in dia.

Calculate the power input and tractive efforts of each motor when

(*i*) motors are connected in parallel and train speed is 40 kmph.

(*ii*) motors are in series and current is 400 A.

Assume resistance of each motor as 0.08 Ω.

[*Pb. Univ. Elec. Traction July, 1974*]

Solution : The characteristics of a dc series motor when operated at 600 V and driving 107 cm diameter wheels are drawn from the given datas (refer to Fig. 12.8(*a*)).

The characteristic of a dc series motor when operated at 600 V and driving 102 cm diameter wheel are derived as follows and drawn (Refer to Fig. 12.8 (*b*)).

Amperes	500	400	300	200
Train speed km/hr, $V_2 = \frac{V_1 \times 1.02}{1.07}$	33.9	36.7	40.5	48.2
Tractive effort kg $F_{t_2} = F_{t_1} \times \frac{1.07}{1.02}$	2,710	2,060	1,450	760

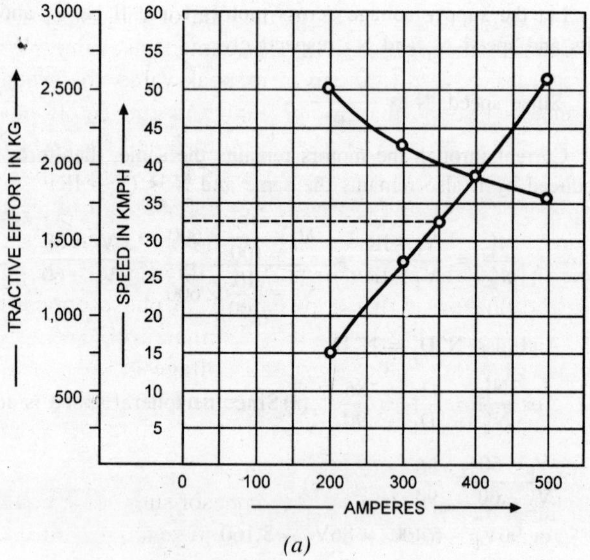

(a)

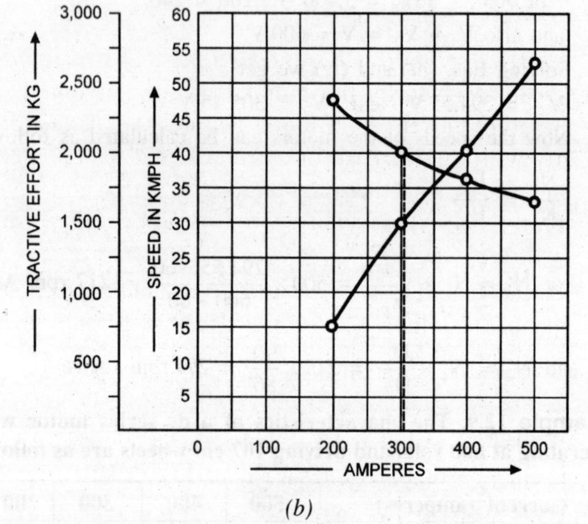

(b)

Fig. 12.8

Let the two motors be A and B of wheel diameters 107 and 102 cm respectively.

(i) When motors are connected in parallel and train speed is 40 kmph

(*a*) From speed-current characteristic corresponding to speed of 40 kmph

Current drawn by motor A, $I_A = 350$ A

Current drawn by motor B, $I_B = 305$ A

$$\text{Power input to motor A, } P_A = \frac{VI_A}{1,000} = \frac{600 \times 350}{1,000} = 210 \text{ kW}$$

$$\text{Power input to motor B, } P_B = \frac{600 \times 305}{1,000} = 183 \text{ kW}$$

(*b*) From tractive effort-current characteristics

Tractive effort of motor A, $F_{t_A} = 1,625$ kg **Ans.**

Tractive effort of motor B, $F_{t_B} = 1,480$ kg **Ans.**

(ii) When the motors are connected in series and current is 400 A

Resistance of each motor, $R_m = 0.08\ \Omega$

Applied voltage across each motor, V = 600 volts

Back emf of motor A when taking current of 400 A and at a speed of 38.5 kmph

$$E_{b_A} = 600 - 400 \times 0.08 = 568 \text{ V}$$

Back emf of motor B when taking current of 400 A and at a speed of 36.7 kmph

$$E_{b_B} = 600 - 400 \times 0.08 = 568 \text{ V}$$

After motors being connected in series, let the peripheral speed be *v* kmph, which will be same for both motors.

$$\text{Now back emf of motor A, } E'_{b_A} = E_{b_A} \times \frac{v}{v_A} = \frac{568}{38.5} \times v$$

$$\text{Voltage across motor A, } V_A = E'_{b_A} + IR_m = \frac{568}{38.5} \times v + 400 \times 0.08 = \frac{568v}{38.5} + 32$$

$$\text{Back emf of motor B, } E'_{b_B} = \frac{568}{36.7} \times v$$

$$\text{Voltage across motor B, } V_B = E'_{b_B} + IR_m = \frac{568}{36.7}v + 32$$

$$\text{Since } V_A + V_B = 600$$

$$\therefore \quad \frac{568v}{38.5} + 32 + \frac{568v}{36.7} + 32 = 600$$

$$\text{or} \quad 14.8v + 15.5v = 536$$

$$\text{or } v = \frac{536}{30.3} = 17.7 \text{ kmph}$$

$$\text{PD across motor A, } V_A = \frac{568}{38.5} \times 17.7 + 32 = 294 \text{ volts}$$

$$\text{PD across motor B, } V_B = \frac{568}{36.7} \times 17.7 + 32 = 306 \text{ volts}$$

$$(a) \text{ Power input to motor A, } P_A = \frac{V_A \times I}{1,000} = \frac{294 \times 400}{1,000} = 117.6 \text{ kW}$$

$$\text{Power input to motor B, } P_B = \frac{V_B \times I}{1,000} = \frac{306 \times 400}{1,000} = 122.4 \text{ kW}$$

(*b*) From tractive effort-current characteristics corresponding to 400 A of current,

Tractive effort of motor, A, $F_{t_A} = 1,960$ kg **Ans.**

Tractive effort of motor B, $F_{t_B} = 2,060$ kg **Ans.**

12.5. AC SERIES MOTOR

Many single phase ac motors have been developed for traction purposes but only compensated series type commutator motor is found to be best suited for traction. Single phase induction motors have been abandoned as they are not capable of developing high starting torque.

The construction of an ac series motor is very similar to a dc series motor except that some modifications (such as whole magnetic circuit laminated, series field with as few turns as possible, large number of armature conductors, use of high resistance carbon brushes, numerous poles with lesser flux per pole, very short air gaps etc.) are incorporated so as to obtain better performance. Compensating windings are provided to neutralize armature reaction and commutating or interpoles are provided for better performance in terms of higher efficiency and a greater output from a given size of armature core.

A schematic diagram of a single phase series motor with interpole and compensating windings is illustrated in Fig. 12.9.

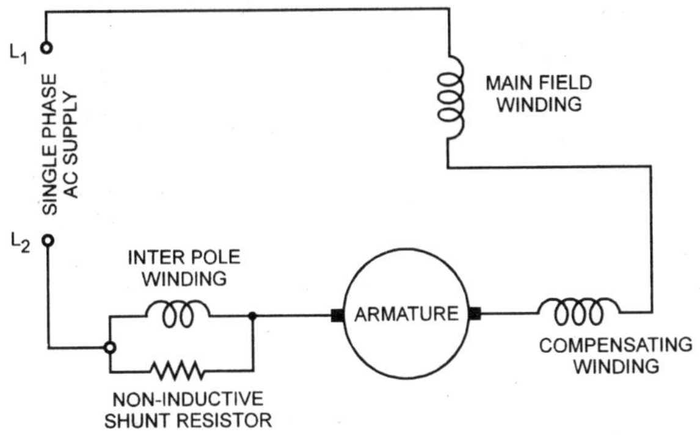

Fig. 12.9. ***AC Series Motor With Interpoles and Compensating Windings***

The average value of the torque on the motor shaft is given as

$$T = \frac{1}{2\sqrt{2}\pi} Z \frac{P}{A} \phi_{max} I \cos\theta \quad ...(12.14)$$

where I is the effective value of current, ϕ_{max} is the peak value of the flux per pole and θ is the phase angle between phasors φ and I.

The phasor diagram of a single phase ac series motor is given in Fig. 12.10. Flux φ is produced by current I. Flux φ lags behind the current I by an amount too small to be important, so it has been shown in phase with current I. The resistive drops IR_{se}, IR_p, IR_c and IR_a due to resistances of series field, interpole circuit, compensating winding and of armature respectively are in phase with current I. The reactive drops IX_{se}, IX_p, IX_c and IX_a due to reactances of series field, interpole circuit, compensating winding and of armature respectively are in quadrature with current I and leading. Here it is noteworthy that reactive drop due to series field is much greater than that of either armature or the compensating field even all the measures to reduce series field inductance, are taken. Unlike the armature cross-flux φ′, the flux that causes this reactive drop cannot be neutralized because it is the necessity for the development of torque. The interpole circuit consists of the interpole winding in parallel with the non-inductive shunt, as illustrated in Fig. 12.9. R_p and X_p are the respective equivalent resistance and reactance of the interpole parallel circuit.

The phase angle of E, in the phasor diagram is most easily determined by considering instantaneous values. When the alternating field flux φ is at its peak value, the armature conductors cut the maximum flux, and the rotational or speed emf is, therefore, a maximum. When the field flux is zero, the rotational or speed emf, E is zero. Since the graphs of flux φ and generated emf E pass through zero at the same instant, and also through maximum at the same instant, the phasor E must be either exactly in phase with the phasor of flux φ or exactly in phase opposition. The first possibility is readily eliminated by noting that in case of a motor the generated voltage is in opposition to the direction of flow of current, and, therefore, E is in phase opposition with current and is called the counter emf.

The terminal voltage V is the phasor sum of the counter-emf reversed and all the resistive and reactive volt drops.

Series Motor Characteristics

1. *Power-Output Characteristic.* Mechanical power developed is given by the product of counter emf E and the armature current I . Though counter emf E decreases slightly with the increase in current, but if it is neglected in comparison with the magnitude of counter emf E, mechanical power developed increases almost in proportion with current I.

Power available at shaft or power output is equal to the mechanical power developed less rotational losses. Power output characteristic is shown in Fig. 12.14.

2. *Power Factor Characteristic.* The cosine of the phase angle φ is the power factor of the motor.

From phasor diagram shown in Fig. 12.10

$$\sin\phi = \frac{I(X_{se} + X_p + X_c + X_a)}{V} = \frac{IX}{V}$$

where $X = X_{se} + X_p + X_c + X_a$ and = effective reactance of the motor

$$\text{or Power factor, } \cos\phi = \sqrt{1 - \left(\frac{IX}{V}\right)^2} \quad ...(12.15)$$

In order to have high power factor the reactance drops must be low and counter emf high. The reactance drops are lowest and the counter emf is highest at light loads, and therefore, the power factor of the single phase series motor is highest (nearly unity) at light loads and falls with the increase in current. This is the reason that power factor of a series motor is low at overloads and at starting. This is the reverse of the power factor relation that exists in case of the induction motor and transformer.

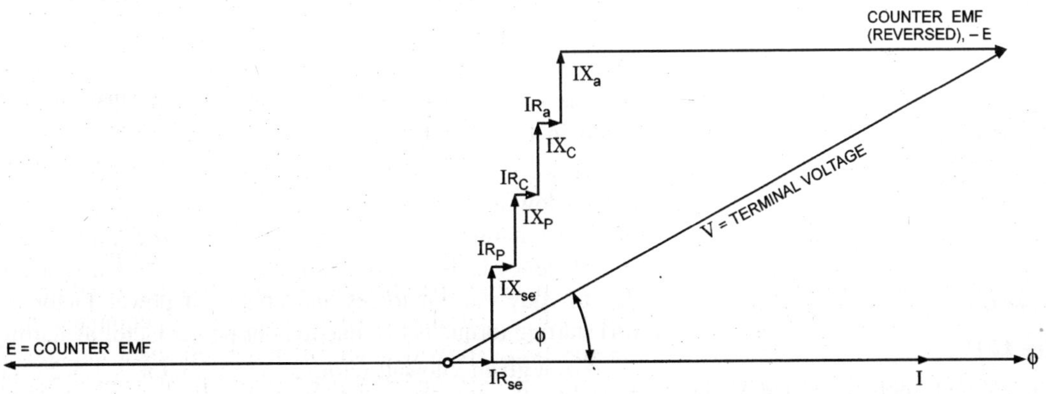

Fig. 12.10. ***Phasor Diagram For Single Phase Series Motor***

3. *Speed-Current Characteristic.* Speed of a commutator motor is proportional to counter emf E (E_r or E_b) or proportional to supply voltage less voltage drops because

$$E \propto \phi N \text{ and } E = V - \text{voltage drops}$$

The current causes resistive drops in series field and armature in case of dc operation but in case of ac operation it also causes large reactance drops in series field, armature, compensating winding and interpole circuit in addition to resistive drops in these windings. Hence with ac operation, counter emf E developed is much less than that with dc operation. It means speed-current characteristic for ac series motors are more drooping as compared to those for dc series motors, as illustrated in Fig. 12.11.

Let the same series commutator motor be operated on dc as well as on ac and with same current I.

The back emf developed in case of dc operation is given as $E_b = \dfrac{\phi Z N_{dc}}{60} \times \dfrac{P}{A}$ volts

and $E_r = \dfrac{\phi_{max}}{\sqrt{2}} \dfrac{Z N'_{ac}}{60} \times \dfrac{P}{A}$ volts

Neglecting saturation, $\phi_{max} = \sqrt{2}\,\phi$ as the peak value of ac flux ϕ_{max} is set up by the peak value of current $\sqrt{2}$ while dc flux ϕ is set up by current I. So from above two equations

$$\frac{N_{ac}}{N_{dc}} = \frac{E_r}{E_b} = \frac{V\cos\phi - IR}{V - IR} = \text{Cos }\phi \qquad \text{...(12.16)}$$

if the supply voltage is same in both of the cases and resistive drops IR are neglected in comparison with supply voltage V

where ϕ is the phase angle between supply voltage and current.

4. *Torque-Current Characteristic.* From Eq. (12.14) $T \propto \phi I$ if phase angle between flux ϕ and current I is neglected or $T \propto I^2$ neglecting saturation. Under saturated condition, flux ϕ becomes constant and, therefore, torque becomes directly proportional to current I. Torque-current characteristic for an ac series motor is shown in Fig. 12.12.

5. *Torque-Speed Characteristic.* The torque-speed characteristic for an ac series motor can be derived from speed-current and torque-current characteristics given in Figs. 12.11 and 12.12 respectively.

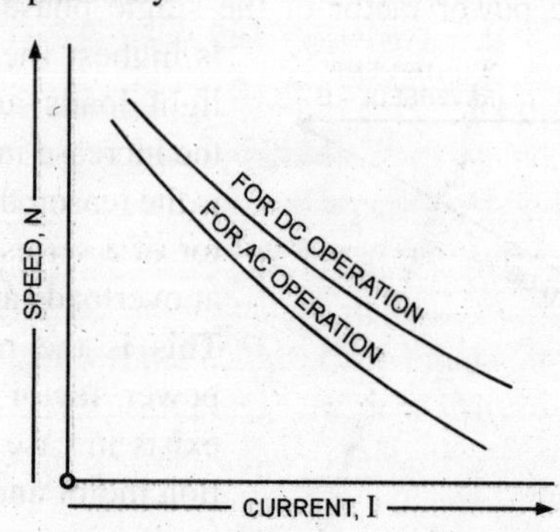

Fig. 12.11

For given value of torque T and applied voltage V, the armature current is same but voltage drop in case of ac series motor is much more than that in case of a dc series motor so speed of an ac series motor for a given developed torque is less than that of a dc series motor, as illustrated in Fig. 12.13.

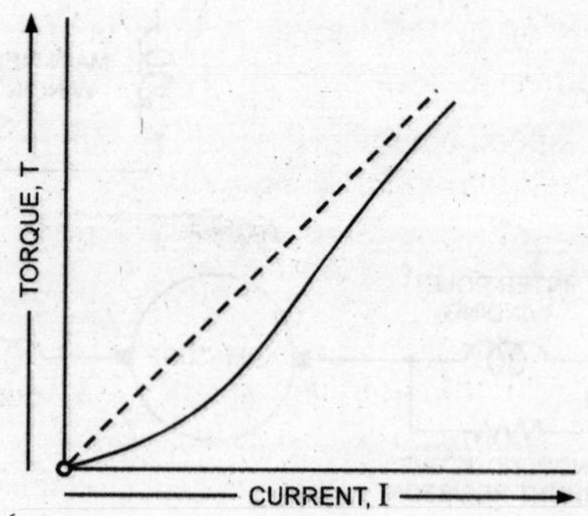

Fig. 12.12

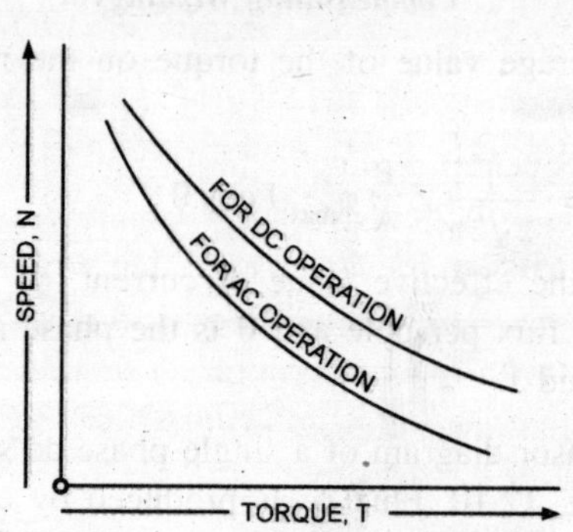

Fig. 12.13

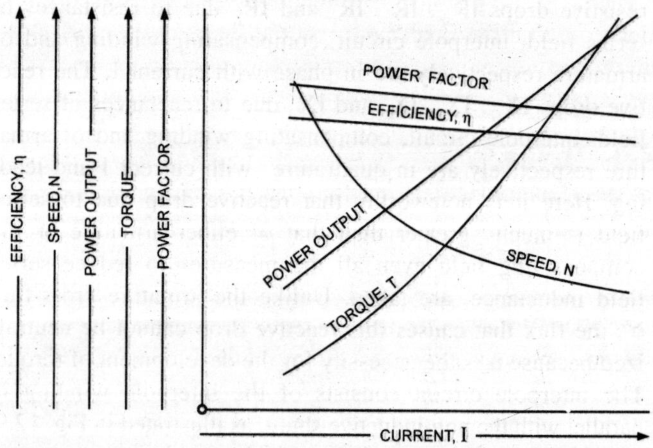

Fig. 12.14

The single phase ac series motor has practically the same operating characteristics as the dc series motor. The torque, or tractive effort, varies nearly as the square of the current and the speed varies inversely as the current. This is illustrated in Fig. 12.14.

However, in case of an ac series motor (*i*) power factor is very low at starting and on overloads on account of high inductive nature of the series field and armature circuits (*ii*) efficiency is not as good as in a corresponding dc machine due to eddy current losses and effects of power factor and (*iii*) starting torque is low due to poor power factor at starting.

For a given kilowatt rating ac series motor is 1.5-2 times in size and weight of the corresponding dc series motor. The construction cost of an ac series motor is much more than that of a dc series motor.

The speed of an ac series motor may be controlled efficiently by taps on a transformer, which is not possible in case of a dc series motor.

The torque-speed characteristic of the single phase series motor is similar to that of the dc series motor *i.e.*, high starting torque and decrease in speed with increase in load making it to have self-relieving property from heavy excessive load, so such a machine is particularly useful for traction services.

These motors are used for main line services. These are not suitable for urban and suburban services because of low starting torque and poor power factor at start.

However, single phase series motors have better performance (improved power factor, higher efficiency, improved commutation, better starting and fewer poles for a given output) on reduced supply frequency (say $16\frac{2}{3}$ or 25 Hz). The higher frequency results in higher leakage reactances and hence a relatively poor power factor. The weight per kW output is also greater for the higher frequency because of larger dimensions of the motor.

The 50 Hz motors have steeper speed-torque characteristics as they operate at lower flux densities than a $16\frac{2}{3}$ or 25 Hz motors. This results in fewer voltage steps at starting and the division of load between parallel connected motors will be less affected by wear of the driving wheels.

The operating voltage is kept low (300 to 400 V) in order to reduce the inductance. Because high voltage motor with proportionately low current would require a large number of turns to produce the given flux.

12.6. THREE PHASE INDUCTION MOTOR

The three phase induction motors have the advantages of simple and robust construction, trouble free operation, less maintenance, high voltage operation consequently requiring reduced amount of current and automatic regeneration. But due to their flat speed-torque characteristics, constant speed operation, developing low starting torque, drawing high starting current, complicated speed control system and complicated overhead feeding systems they are not suitable for electric traction work. It is also impossible to employ 3-phase induction motors for a multiple unit system in which two or more locomotives are used for propelling a heavy train. In traction system employing 3-ϕ induction motors it becomes necessary to couple all the driving axles through a connecting rod so that there is no possibility of a difference in speeds.

With the development of thyristorised inverter circuits, it has now been possible to invert the supply and obtain a variable frequency supply which could be used for the 3-phase induction motor and a very smooth speed control can be obtained.

Induction motor gives good power factor and good efficiency at the speed near to the synchronous speed of the motor. Now in conventional methods of speed control, change of slip method is normally used. But in this method at low speeds of motor or at high value of slip, rotor losses will be more so good efficiency cannot be obtained. To get good efficiency and good power factor, synchronous speed of the motor itself can be brought to lower speed near to the desired actual speed of the motor. This can be done by supplying the motor with the help of variable frequency supply.

Starting current of motor also decreases when motor is started at low frequency.

When 3-phase induction motor is used, distribution system consists of two overhead wires and track rail for the third phase to feed power to locomotive. This gives a complicated overhead structure and also a person who comes in contact of third live rail might be in danger. This drawback is removed by employing Kando system. In this system single phase HV supply is given to locomotive with the help of single overhead wire. In locomotive, this single phase supply is converted into three phase supply with the help of phase converters and fed to the motor. Kando system is in use in Hungary and in some sections of Italian State Railway.

12.7. LINEAR INDUCTION MOTOR

It is a special type of induction motor which gives linear motion instead of rotational motion as in the case of a conventional induction motor. It operates on the same principle on which a conventional induction motor operates *i.e.*, "whenever there occurs a relative motion between the field and the short-circuited conductors, currents are induced in them which results in electromagnetic forces and under the influence of these forces, according to Lenz's law, the conductors try to move in such a way as to eliminate the induced currents". In case of a conventional induction motor, movement of field is rotary about an axis so the movement of the conductors is also rotary. But is case of a linear induction motor, the movement of the field is rectilinear and so the movement of conductors.

In its simplest form, a linear induction motor consists of field system having a 3-phase distributed winding placed in slots as shown in Fig. 12.15. The field system may be a single primary system or double primary system (Fig. 12.15). The secondary of the linear induction motor is normally a conducting plate made of either copper or aluminium in which interaction currents are induced. Either member can be the *stator*, the other being the *runner* in accordance with the particular requirements imposed by the duty for which motor is intended.

In a single primary system a ferromagnetic plate is usually placed on the other side of the conducting plate to provide a path of low reluctance to the main flux. However, the ferromagnetic plate gets attracted towards the primary on energisation of the field and this causes unequal gap length on two sides of the conducting plate. This problem can be overcome by employing double primary system. [Fig. 12.15 (*b*)].

Which of the two primary and secondary will be shorter in length compared to the other depends upon the use of the motor. When the operating distance is large, the primary is made shorter than the secondary because it is uneconomical to wind a very long 3-phase primary. The short secondary form [Fig. 12.15 (*c*)] is useful with limited operating distance.

When the three phase primary winding of a linear induction motor is energized from a balanced three phase source, a magnetic field moving in a straight line from one end to the other at a linear synchronous speed v_s is developed. The linear synchronous speed is given as

$$v_s = 2\tau f \text{ metres/second} \quad ...(12.17)$$

where τ is the pole pitch in metres and f is the supply frequency in Hz. *It is to be noted here that the synchronous speed does not depend on the number of poles, but only on the pole pitch and the stator supply frequency.*

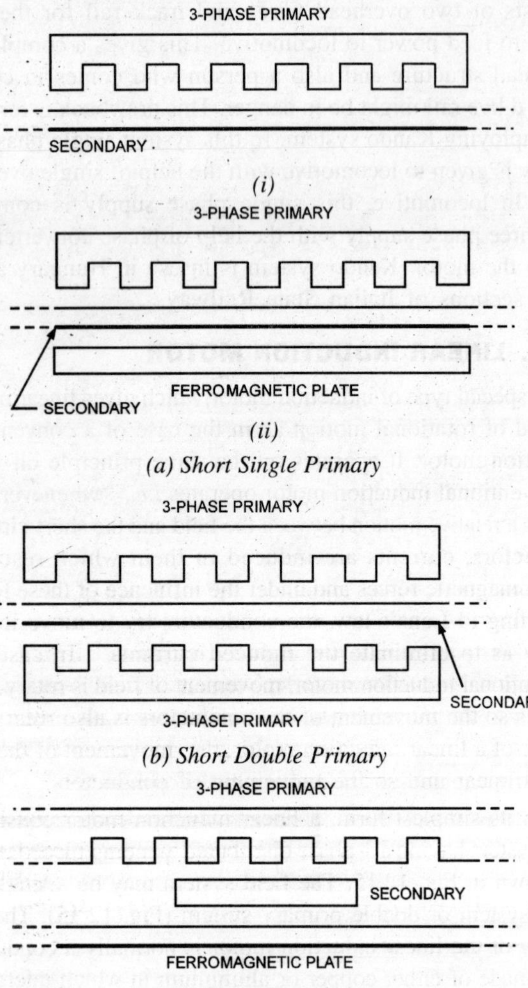

Fig. 12.15

As the flux moves linearly, it drags rotor plate alongwith it in the same direction. This reduces the relative speed of travel of the flux with respect to rotor plate. If the speed of the rotor plate is equal to that of the magnetic field, latter would be stationary when viewed from rotor plate. This is corresponding to the synchronous speed of induction motor. If the rotor plate is moved faster than this speed, the direction of the force would be reversed and a form of regenerative braking based on the principle of induction generator will come into being.

Slip of a linear induction motor is given as

$$s = \frac{v_s - v}{v_s} \quad ...(12.18)$$

where v is the actual speed of rotor plate.

Thrust or force or tractive effort is given as

$$F = \frac{P_2}{v_s} \quad ...(12.19)$$

where P_2 is the actual power supplied to the rotor

Active power flow is similar to that in a conventional rotary induction motor *i.e.*,

$$\text{Copper losses in rotor} = sP_2 \quad ...(12.20)$$

$$\text{and mechanical power developed, } P_{mech} = (1 - s)\, P_2 \quad ...(12.21)$$

Tractive effort will be a function of slip *s i.e.*, $(v_s - v)$. The thrust (or tractive effort)-speed characteristics of a linear induction motor, as shown in Fig. 12.16, are similar to the torque-speed characteristics of a conventional rotary induction motor.

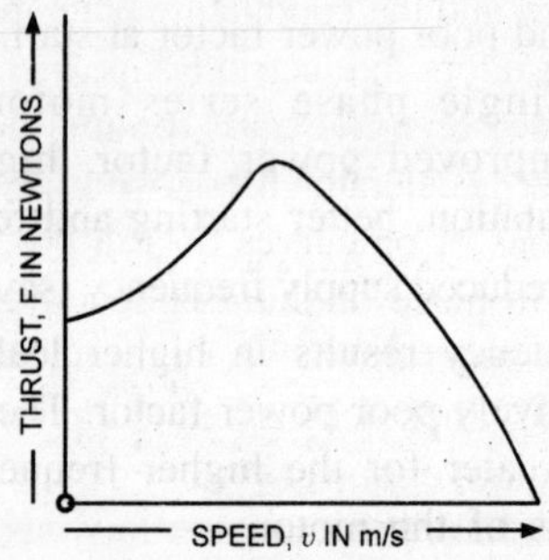

Fig. 12.16

Tractive effort, F can be controlled by varying both frequency and voltage simultaneously so that induction density remains constant.

There are two peculiar effects, which are encountered in a linear induction motor but not in a conventional rotary induction motor. These effects are *transverse edge effect* and *end effect*.

The paths of the induced currents in the secondary are not well defined because the secondary of a linear induction motor is a solid conducting plate. The portion of the current paths parallel to the direction of motion of secondary does not make any contribution towards the production of useful thrust but only contributes towards losses. This effect reduces the effective thrust and increases the losses and is known as the *transverse edge effect* because the current paths parallel to the direction of motion are more towards the edges of the conducting plate.

In the case of linear induction motors with short primary, the current paths towards the end of field structure on the conducting plate go beyond the field structure and such portions of current paths do not contribute to useful thrust but only towards motor losses. This is called the *end effect*. The end effect can be effectively reduced by increasing the number of poles on the motor.

The **advantages** of linear induction motors are (*i*) low initial cost (*ii*) low maintenance cost because of absence of rotating parts (*iii*) simplicity (*iv*) no limitation of tractive effort due to adhesion between the wheel and the rail (*v*) no limitation of maximum speed due to centrifugal forces (*vi*) no overheating of rotor because the motor moves continuously over cool rotor plate leaving behind heated rotor portion and (*vii*) better power to weight ratio.

The **disadvantages** of linear induction motors are (*i*) poor utilisation of motor due to transverse edge effect and end effect (*ii*) larger air gap and non-magnetic reaction rail (rotor plate) need more magnetizing current resulting in poor efficiency and low pf (*iii*) very high capital cost of reaction rail fixed along the centre line of the track (*iv*) complications and high cost involved in providing three phase collector system along the track (*v*) difficulties encountered in maintaining adequate clearances at points and crossings.

Applications. Linear induction motors can be used in conveyors, travelling cranes, haulers, electromagnetic pumps, high-speed rail traction etc. Such motors may be employed in applications in which the field system moves and the conducting plate remains stationary such as travelling crane motors. These motors can also be used in applications where the field system remains stationary and the conducting plate moves such as in automatic sliding doors in electric trains, metallic belt conveyors etc. It can be used on trolley cars for internal transport in workshop, as booster accelerator for moving heavy trains from rest or up the inclines or on curves or as a propulsion unit in marshalling yards in place of shunting locomotives. Linear induction motor has superiority over conventional rotary motor for speeds over 200 kmph. Linear induction motor provides excellent source of motive power for magnetically suspended trains where conventional rotary motor fails because the latter depends for torque conversion to linear tractive force upon the adhesive weight on driving wheels. The use of linear induction motor is limited to only a few applications till now owing to design difficulties and economic considerations.

Example 12.10. A single-sided linear induction motor has 98 poles, and its pole pitch is 50 cm. The motor is used for propelling an electric vehicle. Determine the linear synchronous velocity and the vehicle speed in km per hour if the frequency is 50 Hz and the slip is 0.25.

Solution : Supply frequency, $f = 50$ Hz

Pole pitch, $\tau = 50 \text{ cm} = 0.5 \text{ m}$

Slip, $s = 0.25$

Linear synchronous velocity $v_s = 2\tau f = 2 \times 0.5 \times 50$

$$= 50 \text{ m/s} = \frac{50 \times 60 \times 60}{1{,}000}$$

$= 180$ kmph **Ans.**

Vehicle speed, $v = v_s(1 - s) = 180\,(1 - 0.25)$

$= 135$ kmph **Ans.**

12.8. RATING AND VENTILATION

The motors are designed for intermittent loading as well as continuous loading. The motors employed in traction work are subject to intermittent loading particularly in case of tramway and suburban services, therefore, traction motors are given one-hour rating as well as continuous rating. The ratings have been fully defined in Indian Standard Specifications (ISS). The continuous rating of an ordinary self ventilated motor is approximately one-third to one-fourth of its one hour rating while the continuous rating of a forced ventilated motor is about four-fifth of its one hour rating.

The motors employed in traction work are usually totally enclosed so that protection against the ingress of dust, dirt, mud and water may be provided. Since the weight and overall dimensions of the motor are to be kept to the minimum, efficient cooling is, therefore, essential. For cooling purpose either self ventilation (by using a fan mounted on the motor shaft which blows air through suitable ducts in the motor) or forced ventilation (by employing an external fan) is used. In either case care must be taken in designing the ducts to prevent their becoming choked with dirt and axial rather than radial ducts are always used.

Forced ventilation is invariably employed in case of locomotives, as it is operative even when the vehicle is stationary but for motor coaches and road vehicles, self ventilation is usually employed as it does not cause noise due to fan employed for this purpose.

Exercises

1. Briefly discuss the special design features of traction motors. **[Pb. Univ. Electric Traction Dec. 1992]**
2. Discuss briefly the desirable properties of traction motors. **[M.D. Univ. Utilization of Electric Power & Traction Dec. 2009]**
3. Explain various characteristics that an ideal traction motor should possess. Discuss suitability of electric motors for main line, suburban and urban traction.
4. Discuss the suitability of dc series motor for its application in electric locomotive for traction duty. **[Rajasthan Technical Univ. Utilization of Electrical Power 2010]**
5. Discuss the advantages of series motors for electric traction on suburban railways. **[Rajasthan Univ. Utilization of Electrical Power 2006]**
6. Explain the merits of dc series motor that make it a more suitable drive for electric traction than other drives. **[Pb. Univ. Electric Traction Dec. 1992]**
7. Explain load sharing with respect to traction motors (*i*) separately excited dc motors (*ii*) dc series motors. **[A.M.I.E. Sec. B. Electric Drives and Their Control Winter 1999]**
8. What is the effect of changing wheel diameter and gear ratio on the characteristics of a motor ? **[Pb. Univ. Electric Traction April 1975]**
9. Explain how the difference in driving wheel diameters due to unequal wear affects the sharing of load by two similar series motors, working in parallel, driving an electric train. **[Agra Univ. Traction & Utilisation of Elec. Power 1976]**
10. What type of ac motor is usually employed for single phase electric traction ? Discuss briefly the principal features in the construction of the motor and analytically how good commutation and high power factor are obtained. For what frequency and voltage are such motors usually built and why ? **[Pb. Univ. Electric Traction Nov. 1973]**
11. Explain briefly the construction and characteristics of ac series motor, pointing out how they differ from the dc type. In what way is the good commutation and high power factor assured ? **[Agra Univ. Traction & Utilisation of Elec. Power 1976 Supp.; 1978]**
12. How a good power factor and commutation can be obtained in ac series motors ? **[Agra Univ. Traction & Utilisation of Elec. Power 1982]**

13. Compare the use of dc series motor and ac series motor in electric traction. **[Pb. Univ. Electric Traction June 1993]**

14. Compare the suitability of dc series motor for traction duty versus dc shunt motor and single phase ac series motor.

15. Describe with neat sketches the construction and working principle of high acceleration linear induction motor. Discuss its advantages and disadvantages.

[Pb. Univ. Electric Traction April 1975]

16. Explain the theory, working and characteristics of linear induction motor for traction purposes.

[Pb. Univ. Electric Traction June 1993]

17. Write short notes on the following:

(*i*) Design features of traction motors. **[Pb. Univ. Dec. 1991]**

(*ii*) Traction motors.

[M.D. Univ. May 2010; Rajasthan Univ. 2003, 2008]

(*iii*) Series motor for electric traction. **[Pb. Univ. May 1989]**

(*iv*) Suitability of ac and dc series motors for traction service.

(*v*) Suitability of squirrel-cage induction motor for traction drives. **[U.P.T.U. 2005-06]**

(*vi*) Linear induction motor. **[Pb. Univ. Dec. 1991; Dec. 1992]**

(*vii*) Selection of traction motors.

Short Answer Type Questions With Answers

Q. 1. Why the magnetic circuit of a traction motor is not made of cast iron ?

Ans. Because cast iron cannot withstand continuous vibrations but traction motors operate in extremely severe service conditions.

Q. 2. Why the traction motor must be small in overall dimensions specially in its overall diameter ?

Ans. The traction motor is located underneath a motor coach (except in few electric locomotives) and the space underneath a motor coach is limited by the size of the driving wheels.

Q. 3. Why the traction motor must be of totally enclosed construction ?

Ans. The traction motor must be totally enclosed type, particularly when mounted beneath the locomotive or motor coach, to provide protection against ingress of dirt, dust, water, mud etc.

Q. 4. Why the traction motor must be light in weight, robust and capable to withstand continuous operation ?

Ans. The traction motor must be light in weight in order to increase the payload capacity of the vehicle.

A traction motor must be robust and capable to withstand continuous vibrations since service conditions are extremely severe.

Q. 5. Why a dc series motor is ideally suited for traction purposes ? **[A.M.I.E. Sec. B Winter 1995]**

Or

Why series motors are preferred for electric traction ?

[Pb. Univ. November 1988]

Ans. The series motors are ideally suited for traction purposes because of the following reasons :

(*i*) The dc series motor develops high torque at start which is an essential requirement of traction service.

(*ii*) Simple speed control.

(*iii*) Power drawn from supply mains varies as the square of the load torque and so series motors are self relieving ones.

(*iv*) Commutation is excellent up to twice full load.

(*v*) The series motors are less susceptible to sudden change in supply voltage.

(*vi*) The series motors are capable of withstanding excessive overloads.

(*vii*) The dc series motors, when operated in parallel to drive a vehicle by means of different driving axles, share load almost equally even there is unequal wear of different driving wheels. This is explained as below.

Consider two identical motors of series type (say of speed-current characteristic shown in Fig. 12.3) connected to different driving axles of the vehicle. For a given speed of vehicle, the peripheral speed of all the driving wheels will be same, but should there be a slight difference in the diameters of the driving wheels (which is possible on account of difference in wear) there will be a corresponding difference in the speeds of the driving axles to which they are connected and also, therefore, a difference in speed of the motors. The speed of the motors will be given by the relation $\frac{N_2}{N_1} = \frac{D_1}{D_2}$

When the motors are working in parallel the current drawn by each motor (determined from speed-current characteristic) due to slight difference in the speed does not differ much. As voltage across each motor is same, therefore motors share the load almost equally.

(*viii*) The dc series motor is simple and robust in construction.

Q. 6. Why are the ac single phase series motors not suitable for urban and suburban services ?

Ans. Single phase ac series motors develop low starting torque and poor power factor at start and for urban and suburban services large starting torque is an essential feature to provide high acceleration and frequent starting is involved.

Q. 7. Why ac series motors used for traction services are designed to operate at low frequency *i.e.,* $16\frac{2}{3}$ or 25 Hz ?

Ans. Single phase ac series motors have better performance (improved pf, higher efficiency, improved commutation, better starting etc.) if operated on low frequency ($16\frac{2}{3}$ or 25 Hz).

Q. 8. Why ac series motors used in traction services are designed to operate at low voltage (300 to 400 V) ?

Ans. The operating voltage is kept low (300 to 400 V) in order to reduce inductance. Because high voltage motor with proportionately low current would require a large number of turns to produce the required flux.

Q. 9. Why 3-phase induction motors are not used in electric traction ?

Ans. Because of flat speed-torque characteristics, constant speed operation, developing low starting torque, drawing higher current at start, complicated speed control system and complicated overhead feeding system, 3-phase induction motors are not suitable for electric traction.

Q. 10. Traction motors are given one-hour rating as well as continuous rating ?

Ans. The motors are designed for intermittent loading as well as continuous loading. The motors employed in traction work are subject to intermittent loading particularly in case of tramway and suburban services, therefore, traction motors are given one-hour rating as well as continuous rating.

Problems

1. The relation between the current and torque of a series motor determined by a static test is as follows:

Current (amps.)	20	40	60	80	100	120	140	160
Torque (N-m)	67.5	195	346	525	705	918	1,152	1,366

Deduce the speed-current curve of the motor when supplied at a constant voltage of 600 volts, having given the resistance of the main circuit of motor 0.3 ohm.

2. The following figures give the magnetisation curve of a dc series motor when working as a separately excited generator at 500 rpm.

Field current (amps)	10	20	30	40
EMF (volts)	144	254	324	366

The total resistance of the motor is 1.2 Ω. Deduce the speed-torque curve for this motor when operating at a constant voltage of 400 volts.

3. The following figures refer to the speed-current and tractive effort-current characteristic of a dc series traction motor:

Current (amps)	40	80	120	160	200
Speed (kmph)	51.5	33.6	28.8	26.0	24.6
Tractive effort (N)	1,200	4,200	7,400	10,700	14,000

The diameter of a car wheel is 96 cm, the gear ratio is 4 : 1. The car wheel is changed to 92 cm, the gear ratio to 3.5 : 1. Give the new characteristics.

4. Two dc series wound traction motors run at 800 and 850 rpm when each is connected to 700 volt dc mains and takes 100 A. The effective resistance of each motor is 0.5 Ω. If the two motors are connected in series to a 1,500 volt supply and draw 100 A and are also mechanically coupled, at what speed will they run and what will be the voltage across each motor ? **[Ans.** 887 rpm; 771 V; 729 V]

5. A locomotive exerts a tractive effort of 28,200 N in hauling a train at 48 kmph on the level track. If the motor is to haul the same train on a gradient and the tractive effort required is 44,000 N. Determine the power delivered by the locomotive in case it is driven by (*i*) dc series motors (*ii*) induction motors. **[Ans.** 375 kW; 470 kW]

6. A motor coach is equipped with two dc series motors, the characteristics of each at 675 V with 107 cm wheels are as follows:

Speed (kmph)	55.5	46.3	41	37.4
Tractive effort (kg)	610	1,120	1,680	2,240

If the wheels driven by one motor (A) are 107 cm in diameter and those driven by the other motor (B) are 101.5 cm in diameter, determine the total tractive effort at a train speed of 43 kmph when the motors are operating in parallel at normal voltage.

[Agra Univ. Traction & Utilisation of Elec. Power 1976]

[Ans. 2,625 kg]

7. A coach is driven by two similar series motors A and B. A drives 105 cm wheels and B drives 100 cm wheels. Characteristics of motor A are as follows:

Current (amperes)	200	300	400	500
Speed (kmph)	50.5	42.5	38.3	35.3
Tractive effort (kg)	730	1,340	1,970	2,580

Calculate the input to and tractive effort of each motor when (*a*) motors are in parallel and speed is 40 km per hour (*b*) motors are in series taking 400 A. For each motor, total resistance in the armature circuit is 0.8 W. Assume the motors are connected to 600 V dc supply system.

[Ans. (*a*) 213 kW; 183 kW; 1,675 kg, 1450 kg.
(*b*) 117.4 kW, 122.6 kW; 1,970 kg; 2,068 kg]

❑❑❑

CHAPTER 13

Control of Traction Motors

INSIDE THIS CHAPTER

13.1. Introduction **13.2.** Starting and Speed Control of DC Traction Motors **13.3.** Plain Rheostatic Starting (Notching) **13.4.** Series-Parallel Starting **13.5.** Transition Methods **13.6.** Drum Controller **13.7.** Contactor Type Controller **13.8.** Buck and Boost Method of Speed Control **13.9.** Metadyne Control **13.10.** Thyristor Control of Traction Motors **13.11.** Speed Control and Starting of Single Phase Series Motors **13.12.** Starting and Speed Control of 3-Phase Induction Motors.

13.1. INTRODUCTION

If the traction motors are connected across the supply when the vehicle is at standstill, an enormous current would flow through them and they would develop an immense tractive effort. The result would be disastrous both electrically and mechanically; in fact, satisfactory operation would be impossible. Some means, therefore, is necessary to limit the starting current. It is also desirable to provide more than one economical running speed and electrical braking—rheostatic or regenerative. Thus control equipment is required for (*i*) connecting the motor or motors to supply without taking excessive current at start and provide smooth acceleration without causing sudden shock so as to avoid damage to the couplings and inconvenience to the passengers, (*ii*) adjusting the speed as per requirement, and (*iii*) providing rheostatic or regenerative braking.

13.2. STARTING AND SPEED CONTROL OF DC TRACTION MOTORS

As already discussed, only series and compound dc motors are suitable for traction work. With a dc series motor, the current and torque produced at standstill may be reduced by strengthening the field or lowering the terminal voltage or both. Motors may be placed in series, reducing the terminal voltage of each without loss in external resistance. External resistance may be placed in series with the motors to limit the starting current to any desired value, and by varying the resistance the current may be kept constant during notching up period as desired, as the back emf is being built up. Since maximum torque while starting demands full field strength, any shunt or reduced field connections are usually thrown out of action in starting.

With the dc compound motor the start may be made with full armature current in the series field coils and maximum current in the shunt field coils. A starting resistance inserted in the armature circuit is reduced in steps until the armature and series field are connected across the line. Further speed increase is affected by reducing the shunt field current in steps to the point where the shunt field winding is disconnected, and the action is then identical with that of a plain series motor.

Various methods used for starting and controlling of speed of dc series motors are discussed below.

1. *Rheostatic Control.* A series motor can be started by connecting an external resistance (starter) in series with the main circuit of the motor. At the starting instant, since the back emf developed by the motor is zero, therefore, the resistance connected in series with the motor is maximum and is of such a value that the voltage drop across it with full load rated current is equal to line voltage. As the motor speeds up, the back emf developed by the motor increases, therefore, the external resistance is gradually reduced in order to maintain the current constant throughout the starting or accelerating period. Basic traction motor circuit with rheostatic starting is shown in Fig. 13.1. In this method there is a considerable loss of energy in the external circuit.

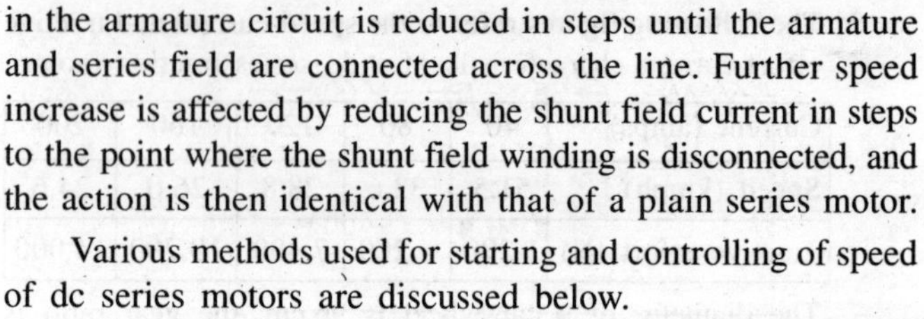

Fig. 13.1

The resistors employed are designed for short-time rating and not for continuous rating as they are required to carry current only during starting of motors. The motor can, therefore, have only one speed characteristic.

2. *Series-Parallel Control.* The main disadvantage of wastage of electrical energy in rheostatic control is partly overcome in this method when there are two or more motors.

In case of two motors, the motors are first connected in series with each other and a starting or control resistance as illustrated in Fig. 13.2 (*a*). The additional resistance is gradually cut out by the controller as the motors attain the speed and finally the control resistance is totally removed, then each motor has one half of the line voltage across it, as shown in Fig. 13.2 (*b*). This is the first running position. In this position for any given value of armature current, each motor will run at half of its normal speed.

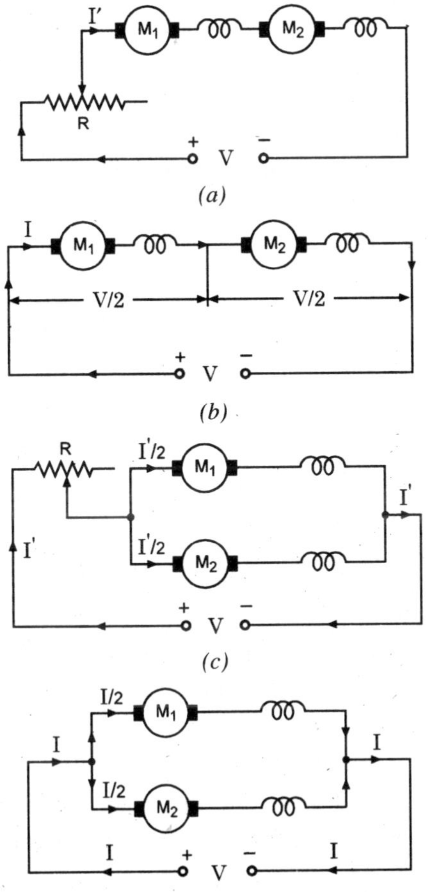

Fig. 13.2

Since there is no external resistance in the circuit, there is no wastage of energy and so motors operate at an efficiency nearly equal to that obtainable with full line voltage across the terminal of each motor.

In the next step the two motors are connected in parallel and in series with a variable resistance R, as shown in Fig. 13.2 (*c*). This resistance is gradually cut out as the motors attain the speed and finally when this resistance is totally removed from the circuit, as illustrated in Fig. 13.2 (*d*), the second running position is obtained. In this position each motor is connected across the full line voltage. A diagram of connections illustrating the switching sequence is given in Fig. 13.3.

3. *Field Control.* As the speed of the motor is inversely proportional to the flux (assuming line voltage constant), therefore, the speed can be varied by varying the flux. In case of series motors the flux can be varied either (*i*) by connecting a variable resistance known as *diverter* in parallel with the series field winding (refer to Fig. 13.4) or (*ii*) by cutting out some of the series field turns (refer to Fig. 13.5). Since in both the cases the flux can be only reduced, this method is known as *field weakening method* and speeds above normal can be obtained. By this method speed can be raised to the extent of 15 to 30 per cent of normal speed owing to design difficulties arising with traction motors.

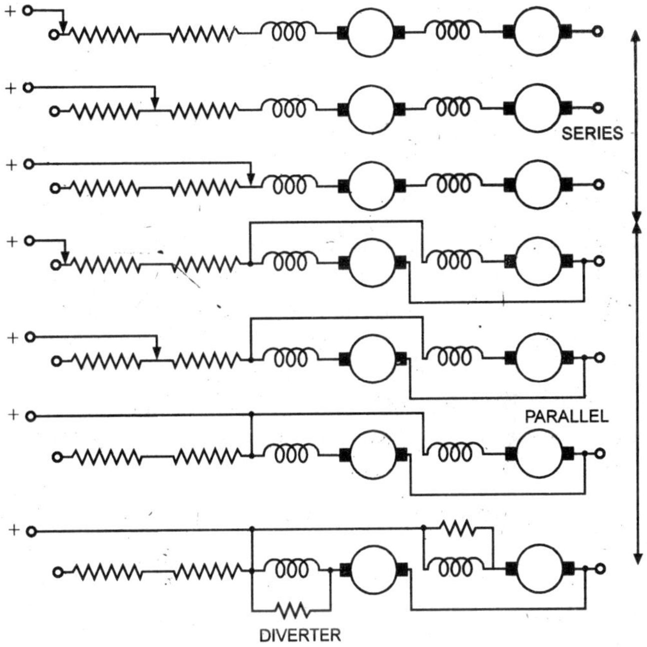

Fig. 13.3. *Combination of Series-Parallel and Resistance Control of DC Series Motors*

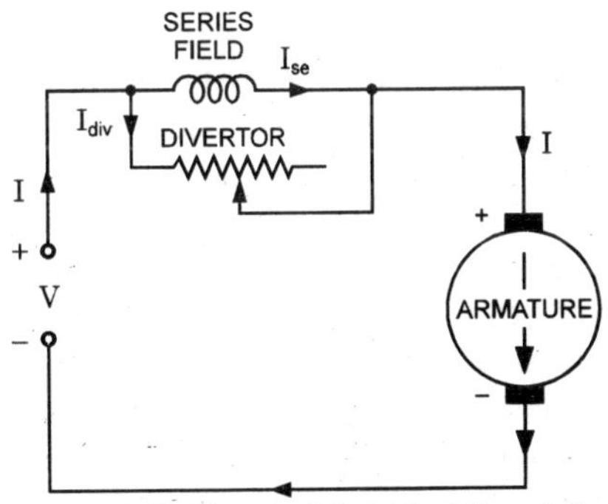

Fig. 13.4. *Field Divertor Method of Speed Control For DC Series Motors*

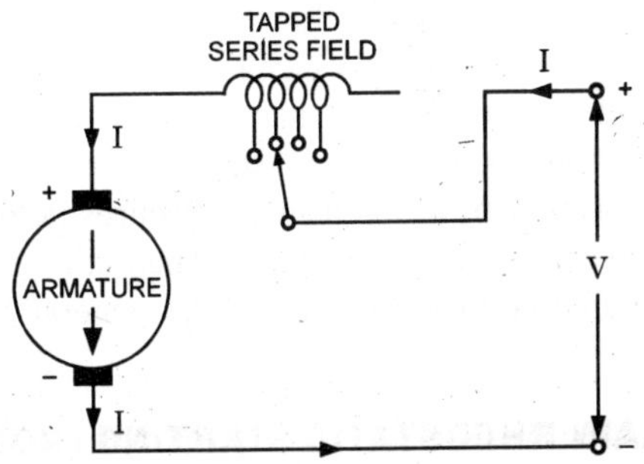

Fig. 13.5. *Tapped Field Control For DC Series Motors*

The field weakening method is of no use for starting purpose. This method is used for increasing the speed of traction motors up to the extent of 10 or 15 per cent when they have attained maximum possible speed by series-parallel control system. The advantage of this system is that it increases the flexibility of the train utility.

For instance for city service, the speed required is low and frequent starting and stopping are required, the equipment can operate with the full field. When the same vehicle has to be employed for suburban and inter-urban lines where higher speeds are required, it would need a change in the gear ratio. By using field weakening method of speed control, the necessity of changing the gear ratio can be eliminated. Thus one type of equipment can be used to operate various types of services with a reasonable energy consumption.

4. *Motor-Generator Locomotive Control.* The motor-generator set usually consists of a single phase synchronous motor driving one or two dc generators and an exciter on the same shaft. The set is usually started and brought up to speed by a starting motor and automatically synchronised when it attains speed. The generator voltage is regulated by means of field control from exciter, therefore, neither resistance is required in traction motor supply circuit nor series-parallel control is required to obtain economical running speeds or save rheostat losses in the traction motor control. A series-parallel arrangement is sometimes used to permit economy in the size and weight of the generator.

5. *Diesel Electric Locomotive Control.* As already discussed in chapter 10, diesel electric locomotive have come into prominence in the past few decades and have largely replaced steam locomotive, in many parts of the world. Primarily a diesel-electric drive, as used on locomotive, is a self-contained system of power conversion in which a diesel engine supplies power to the traction motors through a dc generator which it operates. Thus there are three fundamental parts in the diesel locomotive namely the diesel engine, the dc generator and the dc series traction motors. The diesel engine drives the generator, and the generator supplies dc power to the traction motors, which in turn supply motive power to drive the vehicle. The field of the generator is indirectly excited by a battery the voltage of which may be varied by a rheostat. The battery voltage in turn varies the voltage of an auxiliary exciter or an amplidyne and this excites the main generator. The voltage of the generator is thus under control and this form of control makes unnecessary resistance grids in the main motor circuits to protect the traction motors in starting and accelerating. The engine speed is normally controlled by a load regulator which loads the engine according to the controller setting. This regulator automatically maintains a constant output, because if the engine demands more fuel than called for the setting, the regulator reduces field excitation and thus prevents stalling the engine. There are several types of diesel engine control, but they all in general follow this principle.

13.3. PLAIN RHEOSTATIC STARTING (NOTCHING)

As already mentioned in this method the voltage across the motor is gradually increased from zero to full (normal) voltage and the current is kept constant at normal rated value by reducing the external resistance gradually. Let the starting period be T seconds, constant current I amperes and line voltage V volts.

At the instant of switching on the supply, the back emf is zero and it starts building up immediately the motor starts rotating. At any instant the supply voltage is equal to emf developed by motor plus voltage drop in the series field and armature of motor plus voltage drop in starting rheostat.

$$i.e., \; V = E_b + IR_m + IR \qquad ...(13.1)$$

At the end of starting period, the back emf becomes approximately equal to line voltage and is given as.

$$E_b = V - IR_m \qquad ...(13.2)$$

$\because$ external resistance is reduced to zero, and therefore, voltage drop in the external resistance is zero.

Energy drawn from the supply during starting

= VIT watt-seconds

Energy lost in external resistance

= Average voltage drop across the external resistance × current × time

$$= \frac{V - IR_m + 0}{2} \times I \times T = \frac{V}{2} \times I \times T$$

$$= \frac{1}{2} VIT \text{ watt-seconds}$$

if voltage drop in series field and armature of motor is neglected.

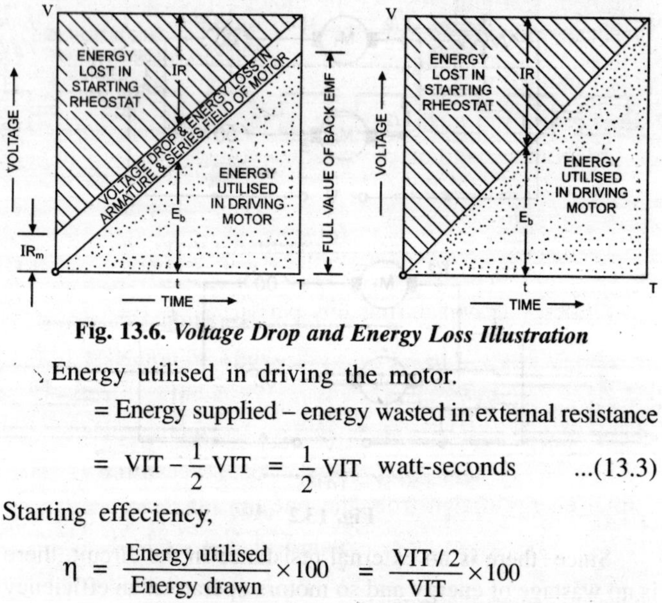

Fig. 13.6. ***Voltage Drop and Energy Loss Illustration***

Energy utilised in driving the motor.

= Energy supplied – energy wasted in external resistance

$$= VIT - \frac{1}{2} VIT = \frac{1}{2} VIT \text{ watt-seconds} \qquad ...(13.3)$$

Starting effeciency,

$$\eta = \frac{\text{Energy utilised}}{\text{Energy drawn}} \times 100 = \frac{VIT/2}{VIT} \times 100$$

$$= 50\% \textbf{ Ans.} \qquad ...(13.4)$$

The energy lost and the voltage drop can be illustrated with the help of illustrations shown in Fig. 13.6.

13.4. SERIES-PARALLEL STARTING

This method is usually employed in electric traction as motors employed for traction work are usually two, four, six or more than this even.

First consider two motors. Let the current be constant and equal to I amperes throughout, the line voltage be V volts and accelerating period be T seconds.

At the starting instant, both the motors are connected in series along with the starting resistance, the voltage across the motors (neglecting voltage drop in armature and series field resistance) is approximately equal to zero and voltage drop across external resistance is maximum and equal to V volts. As the motors speed up, the external resistance is gradually reduced to zero. During this period the voltage acting across each motor gradually increases from 0 to $\frac{V}{2}$ volts. Let the time required be half of total accelerating period *i.e.*, $\frac{T}{2}$ seconds.

Energy utilised by each motor

= Average voltage acting across each motor × current × time.

$$= \frac{0+\frac{V}{2}}{2} \times I \times \frac{T}{2} = \frac{VIT}{8} \text{ watt-seconds}$$

Energy utilised by two motors

$$= 2 \times \frac{VIT}{8} = \frac{VIT}{4} \text{ watt-seconds} \quad ...(13.5)$$

Energy drawn during this period

$$= VI \times \frac{T}{2} = \frac{VIT}{2} \text{ watt-seconds} \quad ...(13.6)$$

When the motors are changed from series grouping to parallel grouping external resistance is again inserted in the motors circuit. The voltage across each motor is equal to $\frac{V}{2}$ volts, the current per motor is I amperes and current drawn from the line is 2I amperes. As the motors speed up, the external resistance is gradually reduced to zero and voltage across each motor increases gradually from $\frac{V}{2}$ to V volts.

Energy utilised in driving the motors

= Number of motors × average voltage across each motor × current × time

$$= 2 \times \frac{\frac{V}{2}+V}{2} \times I \times \frac{T}{2} = \frac{3VIT}{4} \text{ watt-seconds} \quad ...(13.7)$$

Energy drawn during this period

$$= V \times 2I \times \frac{T}{2} = VIT \text{ watt-seconds} \quad ...(13.8)$$

Total energy utilised in driving the motors

$$= \frac{VIT}{4} + \frac{3VIT}{4} = VIT \text{ watt-seconds} \quad ...(13.9)$$

Total energy drawn from the line

$$= \frac{VIT}{2} + VIT = \frac{3}{2}VIT \text{ watt-seconds} \quad ...(13.10)$$

Overall starting efficiency,

$$\eta = \frac{\text{Energy utilised}}{\text{Energy input}} \times 100 = \frac{VIT}{\frac{3}{2}VIT} \times 100 = 66.67\% \quad ...(13.11)$$

In the case of 4 motors all the four motors are connected in series with starting resistance first. The voltage across each motor is gradually increased from 0 to $\frac{V}{4}$ volts by reducing the external resistance gradually to zero. The current per motor is equal to I amperes. The current drawn from the line is also equal to I amperes. The time required for increasing the voltage across each motor from 0 to $\frac{V}{4}$ volts is one-fourth of the total accelerating period *i.e.*, $\frac{T}{4}$ seconds.

In second step two groups, each consisting of two motors connected in series are put in parallel. The voltage across each motor increases from $\frac{V}{4}$ to $\frac{V}{2}$ volts. The current per motor is equal to I amperes and current drawn from the line is 2I amperes. The period is $\frac{T}{4}$ seconds.

In third step all the four motors are placed in parallel. The voltage across each motor gradually increases from $\frac{V}{2}$ to V volts. The current per motor is I amperes. Line current is 4I amperes. The period is $\frac{T}{2}$ seconds.

Energy utilised in driving the motors

$$= 4 \times \frac{\left(0+\frac{V}{4}\right)}{2} \times I \times \frac{T}{4} + 4\frac{\left(\frac{V}{4}+\frac{V}{2}\right)}{2} \times I \times \frac{T}{4} + 4\frac{\left(\frac{V}{2}+V\right)}{2} \times I \times \frac{T}{2}$$

$$= \frac{VIT}{8} + \frac{3VIT}{8} + \frac{3VIT}{2} = 2VIT \text{ watt-seconds} \quad ...(13.12)$$

$$\text{Total energy input} = V \times I \times \frac{T}{4} + V \times 2I \times \frac{T}{4} + V \times 4I \times \frac{T}{2}$$

$$= 2.75 \text{ VIT watt-seconds} \quad ...(13.13)$$

Overall starting efficiency

$$= \frac{\text{Energy utilised}}{\text{Energy input}} \times 100 = \frac{2VIT}{2.75VIT} \times 100 = 72.72\% \quad ...(13.14)$$

In case of 6 motors all the six motors are connected in series with starting resistance. The voltage across each motor is gradually increased from 0 to $\frac{V}{6}$ volts by reducing the external resistance gradually to zero. The current drawn per motor is equal to I amperes. The time required for increasing the voltage across each motor from 0 to $\frac{V}{6}$ volts is one-third of the total accelerating period *i.e.*, $\frac{T}{3}$ seconds.

In second step two groups, each consisting of three motors connected in series are put in parallel. The voltage across each motor increases from $\frac{V}{6}$ to $\frac{V}{3}$ volts. The current per motor is equal to I amperes and current drawn from the line is 2I amperes. The time required will be $\frac{T}{3}$ seconds.

In third step three groups each consisting of two motors connected in series are put in parallel. The voltage across each motor increases from $\frac{V}{3}$ to $\frac{V}{2}$ volts. The current per motor is equal to I amperes and current drawn from the line is 3I amperes. The time required will be again $\frac{T}{3}$ seconds.

Energy utilised in driving the motors

$$= 6 \times \frac{\left(0+\frac{V}{6}\right)}{2} \times I \times \frac{T}{3} + 6\frac{\left(\frac{V}{6}+\frac{V}{3}\right)}{2} \times I \times \frac{T}{3} + 6\left(\frac{\frac{V}{3}+\frac{V}{2}}{2}\right) \times I \times \frac{T}{3}$$

$$= \frac{VIT}{6} + \frac{VIT}{2} + \frac{5VIT}{6} = 1.5VIT \text{ watt-seconds} \quad ...(13.15)$$

Total energy input

$$= \frac{VIT}{3} + V \times 2I \times \frac{T}{3} + V \times 3I \times \frac{T}{3} = 2VIT \text{ watt-seconds} \quad ...(13.16)$$

$\therefore$ Starting efficiency

$$= \frac{\text{Energy utilised}}{\text{Energy input}} \times 100 = \frac{1.5VIT}{2VIT} \times 100 = 75\% \quad ...(13.17)$$

The system of double series-parallel control is used on heavy locomotives requiring a large number of rheostatic steps so as to provide smooth starting. With suburban trains, however, the slight saving in energy of double series-parallel control compared with series-parallel control is entirely offset by the additional cost, weight and maintenance of the double series parallel control equipment.

Advantages of Sereis-Parallel Starting. (i) Higher starting efficiency. 66.67% with two motors, 72.72% with four motors and 75% with six motors whereas only 50% efficiency is obtained with plain rheostat starting.

(ii) Different economical speeds. Two with two motors in the ratio of 1 : 2 : three with four motors in the ratio of 1 : 2 : 4 and three with six motors in the ratio of 1 : 2 : 3.

(iii) For the same power input the torque of different magnitude is obtained. Torque in parallel combination is 1/4th of that in series combination but speed is four times of that with series combination. For the same motor current, the torque remains the same but speed in parallel combination becomes double the speed of that of series combination.

(iv) Owing to low energy loss in the starting rheostats, they are not of cumbersome size.

The main difficulty associated with series-parallel control is in obtaining a satisfactory method of transition from series to parallel without interrupting the torque or permitting any heavy rush of current. Leaving smaller units it is quite inadmissible to open the circuit to affect the change-over.

13.4.1. Practical Requirements

As already stated above, starting resistance is inserted in series with the motors connected in series or series-parallel control at the start and then gradually reduced to zero with the increase in speed of motors. In practice it is not possible to cut off the resistance from the circuit gradually but it is affected with a limited number of steps in the starting rheostats so as to avoid excessive cost and complication of the controller. Hence the motor current during starting cannot be maintained absolutely constant as assumed but it varies between two definite upper and lower limits as the sections of the rheostat are cut out (Fig. 13.7). The number of steps to be provided

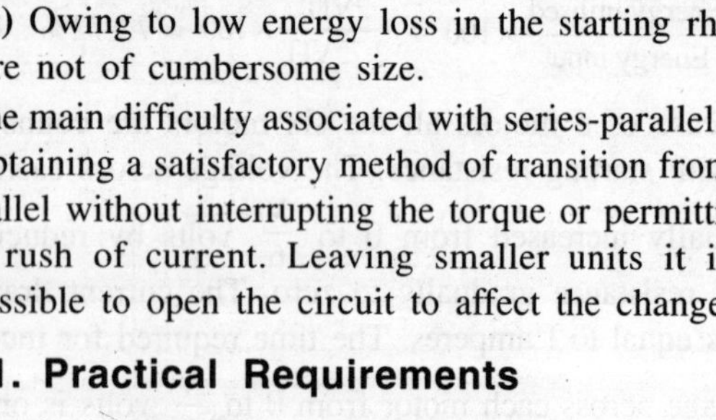

Fig. 13.7. *Representation of Current Variation During Starting Period*

and the value of resistance cut out depends upon the ratio of maximum to minimum current values which also decides the proportional increase in the tractive effort taking place when a section of resistance is cut out. Because of heavy loads involved, shock tolerance limit is lower in the cases of locomotive trains than in the case of multiple unit trains. This, therefore, requires more rheostat steps in the case of controllers for locomotive trains than for multiple unit trains.

In order to maintain the average value of starting current constant it becomes necessary to change-over from series grouping to parallel grouping when all the starting rheostat has been cut out and the starting current is at its minimum value or at the lower limit. It is also essential to reinsert the starting resistance in the circuit while changing-over from series grouping to parallel grouping so that the motor current is limited to the prescribed value. Transition can be affected by any of the following methods.

13.5. TRANSITION METHODS

The methods of changing-over the connections from one grouping to another are known as *transition methods.* The change-over from *full series* (when the two motors are connected in series and voltage across each motor is equal to half of the supply voltage) to first parallel (when the two motors are connected in parallel and part of the starting resistance is reinserted in the circuit to cause the voltage drop equal to half of the supply voltage) is possible by three methods.

1. Open-circuit transition 2. Shunt transition and 3. Bridge transition.

1. *Open-Circuit Transition.* The various steps involved in open-circuit transition are shown in Fig. 13.8. In open-circuit transition first the series connection between the motors is disconnected and power is switched off with full load current. Then some of the starting resistance is reinserted in the circuit, the two motors are connected in parallel and supply is restored. It causes interruption of torque, and therefore, its use is limited to very small units.

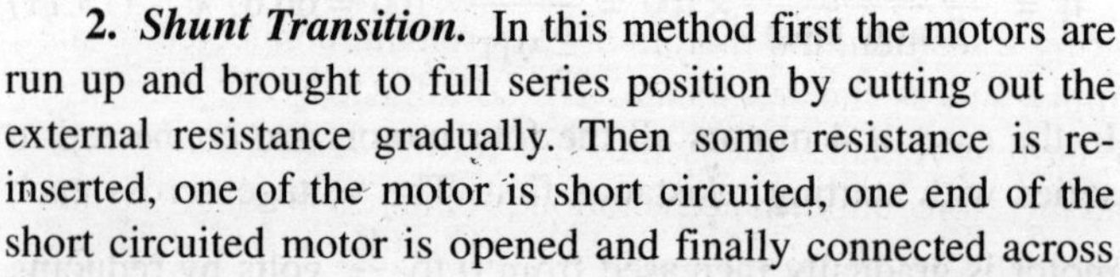

Fig. 13.8. *Open-Circuit Transition*

2. *Shunt Transition.* In this method first the motors are run up and brought to full series position by cutting out the external resistance gradually. Then some resistance is re-inserted, one of the motor is short circuited, one end of the short circuited motor is opened and finally connected across in such a way that motors are placed in first parallel. The series resistance is now gradually reduced to zero and the motors are placed in full parallel. There is a jerk in this system as one motor is shorted and ceases to act and then other jerk when it is reinserted. This method is employed in tramways, industrial locomotives and main line locomotives. It is preferred for voltages above 600 volts.

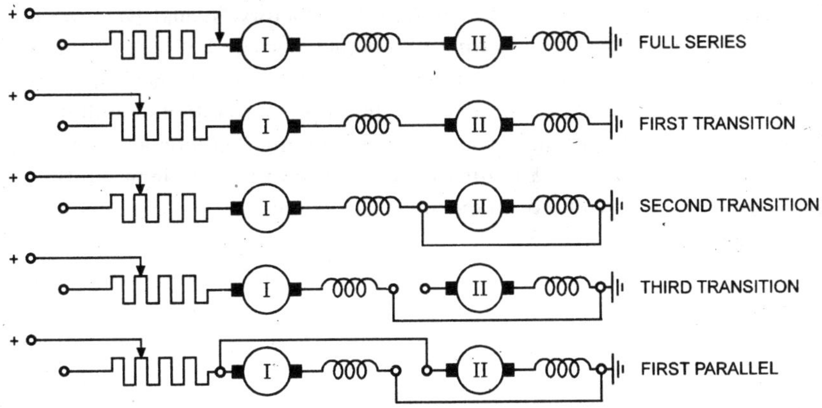

Fig. 13.9. *Shunt Transition*

3. ***Bridge Transition.*** In this method the starting resistance is split up into two equal parts. The motors and starting resistances are connected in series, as shown in Fig. 13.10 (first series). The starting resistances are gradually cut out and the motors come in full series, as shown in the figure. In the transition each motor have a section of the starting resistance in parallel with it and two such combinations are put in series with each other, then the bridge link is removed and the motors are placed in first parallel. Now the external resistance is gradually reduced to zero and the motors come in full parallel.

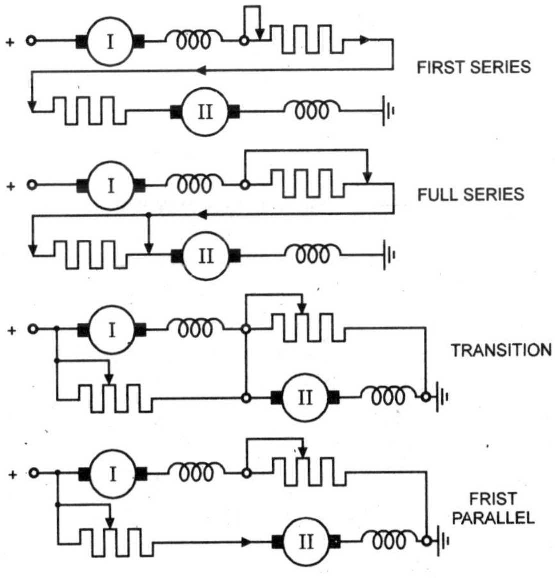

Fig. 13.10. *Bridge Transition*

The advantage of the bridge transition method is that during transition, the motors always remain connected to the supply and as the resistances are so adjusted that the value of current remains the same, the torque remains constant, and therefore, uniform acceleration is obtained without causing inconvenience to the passengers. This method is used for railway traction. In case this method is employed for voltage above 600 V, double contacts at series links are required in order to avoid heavy arcing due to heavy current.

13.6. DRUM CONTROLLER

A series-parallel controller (with shunt transition) for two series motors is shown in Fig. 13.11.

In addition to its primary function of providing series-parallel operation of dc series traction motors it also controls (*i*) speed and direction of motion of the vehicle (*ii*) magnitude of retardation during braking period and (*iii*) provides means to remove faulty motor if any fault occurs during the operation and operating the vehicle with the remaining motor and (*iv*) if possible it must also provide means to prevent its moving backward when stopped on a steep gradient.

The controller is in the form of a rotating drum having insulated and interconnected segments in the form of strips, which makes contact with the fixed points (*known as fingers*). The controller has 8 positions, 4 positions for series and four for parallel running. The working positions of the controller (*known as notches*) are shown by vertical dotted lines. The segments on the rotating drum are shown by black rectangles. Fixed contacts, *known as fingers*, are shown by vertical row of large circular dots on the left. Across these fingers, starting resistances and reversing contacts are connected. The reversing arrangement is provided by providing an additional rotating drum known as *reversing drum*.

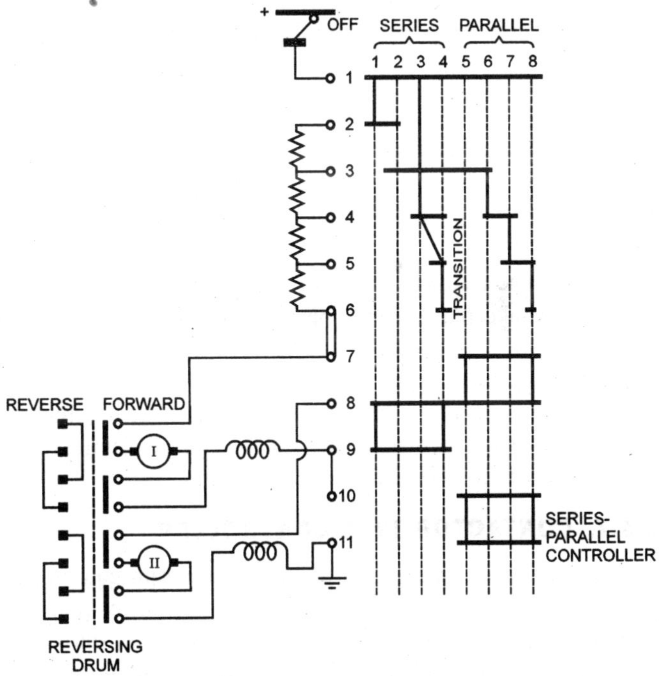

Fig. 13.11. *Series-Parallel Controller With Reversing Drum*

Now when the reversing drum is put in forward position and controlling drum is set in position 1, the connections with fixed points are along vertical line 1. In this position segments make contact with fixed points 1 and 2, 8 and 9. In this position current flows from + ve fixed point 1 through the drum segment, fixed point 2, all the starting resistances 2 – 6, fixed point 7, reversing drum to armature of motor No. I, series field of motor No. I, fixed points 9 and 8, armature and series field of motor No. II and returns back to the –ve terminal.

In the second position the segments of the controller drum come in contact with fixed points 1 and 3, 8 and 9 and thereby bringing some of the starting resistance out of circuit, therefore, speed of the motors increases. In the subsequent 3rd position additional step of resistance is brought out of the circuit and finally in fourth position starting resistance is reduced to zero and half of the supply voltage acts across each motor.

Further movement of the drum reinsert some of the starting resistance in the circuit and short circuit the motor No. II as the current from +ve fixed point 1 passes to fixed point 3, starting resistance, fixed points 6 and 7, reversing drum, armature and series field of motor No. 1, fixed points 9 and 10 and returns to the –ve terminal through lower two segments on the controller drum (on lines 5 to 8).

As the controller drum reaches vertical line No. 5 the motor No. II is again inserted in the circuit and operates in parallel with motor No. I. As the controller drum moves further the starting resistance is reduced and finally in vertical position 8 of controlling drum, both the motors are directly connected across the supply mains and are in full parallel.

For reversing the direction of rotation of motor, it is essential that the direction of current of either of the two (armature or field) is changed. Generally it is achieved by reversing the armature current. In the forward position of reversing drum, the current enters the armatures and series field of motors No. I and II through reversing drum from the left but when the reversing drum is placed in reverse position, the current enters the armatures of motors No. I and II from the right while the direction of field currents remains the same. Thus the direction of rotation of motors are reversed by changing the position of the reversing drum which has two fixed positions.

When the position of controller drum is changed, the arcing will take place. For suppression of the arc magnetic blow-outs are employed. The blow-out coil system consists of flat iron inserted in the movable part of the rotating drum. The reversing drum is mechanically interlocked with the controlling drum so that it may not be operated unless the controlling drum is in the 'OFF' position.

13.7. CONTACTOR TYPE CONTROLLER

Schematic power diagram of a 1,500 V dc train equipment consisting of 4 dc series motors wired for series-parallel operation during starting, in its simplified form is shown in Fig. 13.12. Motors M_1 and M_2 are connected in series and have three breaks L_1 and L_2 and D. Motors M_3 and M_4 are also connected in series and have three breaks L_3, L_4 and C. Bridging contactors B_1 and B_2 have full line voltage of 1,500 V across them when the motors are in parallel and for this reason there are two breaks. Resistance X is inserted in the circuit to protect the system at switch on in case of fault in any motor.

In the starting position line switches L_3, L_4 and bridging contractor B close and then line switches L_1 and L_2. Now motors M_1, M_2, M_3 and M_4 are in series with full starting resistance in the circuit. Then contactors R operate on the following notches and takes the starting resistances out of circuit in steps, the motors are running in series on the full line voltage. Then bridging contactors B_1 and B_2 close and B opens, the motors still operate in series but with no external resistance in the circuit. The contactors R then open, B_1 and B_2 open, C and D close, placing the motors in parallel with full starting resistance in the circuit. The contactors R then take the starting resistances out of circuit in steps and finally the required arrangement is obtained.

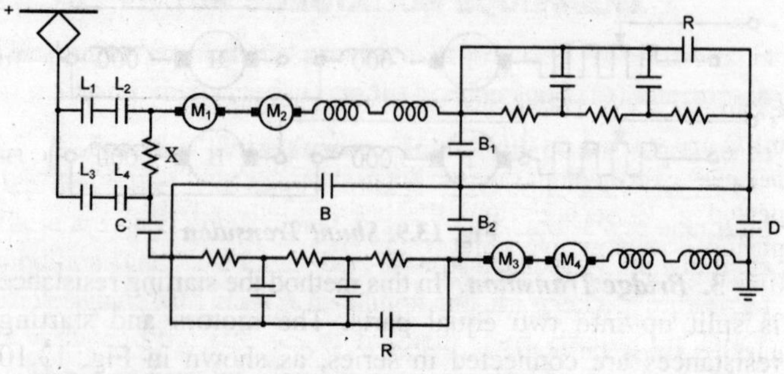

Fig. 13.12. ***Power Diagram of 1,500 V DC Train Equipment (Bridge Transition)***

Example 13.1. Two 600 V dc series motors of coach are started by series-parallel method with the current per motor being 350 A during starting time of 20 seconds and each motor has 0.15 Ω resistance. Calculate (*i*) energy lost in starting rheostat (*ii*) energy lost in motor (*iii*) total energy input from line.

[U.P. Technical Univ. Utilization of Electrical Energy and Electric Traction 2006-07]

Solution : Line voltage, V = 600 V

Current per motor, I = 350 A

Starting period, T_s = 20 second

Motor resistance, R = 0.15 Ω

Back emf of each motor in full series position,

$$E_{bse} = \frac{V}{2} - IR = \frac{600}{2} - 350 \times 0.15 = 247.5 \text{ V}$$

Back emf of each motor in full parallel position

$$E_{bp} = V - IR = 600 - 350 \times 0.15 = 547.5 \text{ V}$$

Assuming smooth acceleration, back emf will be built up at constant rate. Since motors take 20 seconds to build up 547.5 V, the time taken to build up 247.5 V emf,

$$i.e., \; T_{series} = \frac{247.5}{547.5} \times 20 = 9.04 \text{ seconds}$$

$$T_{parallel} = 20 - 9.041 = 10.96 \text{ seconds}$$

(*i*) Voltage drop in the starting rheostat in series combination at the starting instant

$$= V - 2IR = 600 - 2 \times 350 \times 0.15 = 495 \text{ V}$$

which reduces to zero in full series position

Energy dissipated in starting rheostat during series combination

$$= \frac{(V - 2IR) + 0}{2} \times I \times \frac{T_{series}}{3{,}600} = \frac{495 + 0}{2} \times 350 \times \frac{9.04}{3{,}600}$$

$$= 217.525 \text{ Wh}$$

Voltage drop across starting resistance in first parallel position is equal to $\frac{600}{2}$ *i.e.*, 300 V which is gradually reduced to zero.

Energy dissipated in starting resistance during parallel combination

$$= \frac{\frac{V}{2} + 0}{2} \times 2I \times \frac{T_{parallel}}{3600} = \frac{\frac{600}{2} + 0}{2} \times 2 \times 350 \times \frac{10.96}{3{,}600}$$

$$= 319.667 \text{ Wh}$$

Energy lost in starting rheostat

= 217.525 + 319.667 = 537.2 Wh or 0.5372 kWh **Ans.**

(*ii*) Energy lost in motors

$$= \frac{2I^2RT_s}{3{,}600 \times 1{,}000} = \frac{2 \times 350^2 \times 0.15 \times 20}{3{,}600 \times 1{,}000} = 0.102 \text{ kWh } \textbf{Ans.}$$

(*iii*) Total energy input from line

$$= \frac{VIT_{se} + 2VIT_P}{3{,}600 \times 1{,}000} = \frac{600 \times 350 \times 9.04 + 2 \times 600 \times 350 \times 10.96}{3{,}600 \times 1{,}000}$$

= 1.806 kWh **Ans.**

Example 13.2. Two 600 V motors each having a resistance of 0.1 Ω are started on the series-parallel system, the mean current per motor throughout the starting period being 300 A. The starting period is 15 seconds and the train speed at the end of this period is 29 km per hour. Calculate (*i*) the rheostatic losses (in kWh) during (*a*) the series and (*b*) the parallel combinations of motors (*ii*) the train speed at which transition from series to parallel must be made.

[*Agra Univ. Traction & Utilization of Elec. Power 1976 Supp.*]

Solution : Line voltage, V = 600 volts

Current per motor, I = 300 amperes

Starting period, T_s = 15 seconds

Motor resistance, R = 0.1 Ω

Maximum speed, V_m = 29 kmph

Back emf of each motor in full series position,

$$E_{bse} = \frac{V}{2} - IR = \frac{600}{2} - 300 \times 0.1 = 270 \text{ volts}$$

Back emf of each motor in full parallel position,

$$E_{bp} = V - IR = 600 - 300 \times 0.1 = 570 \text{ volts}$$

Assuming acceleration smooth, back emf will be built up at constant rate.

Since motors take 15 seconds to build up 570 volts, therefore, time taken to build up 270 volts emf

i.e., $T_{series} = 15 \times \frac{270}{570} = 7.1$ seconds

$T_{parallel}$ = 15 − 7.1 = 7.9 seconds

(*i*) (*a*) Voltage drop in the starting rheostat in series combination at the starting instant

= V − 2IR = 600 − 2 × 300 × 0.1 = 540 volts,

which reduces to zero in full series position.

Energy dissipated in starting resistance during series combination

$$= \frac{(V - 2IR) + 0}{2} \times I \times \frac{T_{series}}{3{,}600}$$

$$= \frac{540 + 0}{2} \times 300 \times \frac{7.1}{3{,}600} = 160 \text{ watt-hours}$$

= 0.16 kWh **Ans.**

(*b*) Voltage drop across the starting resistance in first parallel position is equal to $\frac{V}{2}$ *i.e.*, 300 volts which gradually reduces to zero.

Energy dissipated in starting resistance during parallel combination

$$= \frac{\frac{V}{2} + 0}{2} \times 2I \times \frac{T_{parallel}}{3{,}600} = \frac{\frac{600}{2} + 0}{2} \times 2 \times 300 \times \frac{7.9}{3{,}600}$$

= 198 watt-hours

= 0.198 kWh **Ans.**

(*ii*) Acceleration, $\alpha = \frac{\text{Maximum speed}}{\text{Starting period}} = \frac{V_m}{T_s} = \frac{29}{15}$ kmphps

Speed at the end of series period

$$= \alpha T_{se} = \frac{29}{15} \times 7.1 = 13.7 \text{ km/hour } \textbf{Ans.}$$

Example 13.3. Of the four dc series motors of an electric train each has the resistance of 0.1 Ω and takes 200 A from 600 V supply during the starting period of 20 seconds. Calculate (*a*) the starting efficiency and (*b*) the speed at which the series connections are to be changed. Assume that the acceleration during the starting period is uniform and the speed just after the starting period is 80 kmph.

[*Allahabad Univ. Utilization of Elec. Energy & Traction 1975*]

Solution : Line voltage, V = 600 V

Current per motor, I = 200 A

Motor resistance, R = 0.1 Ω

Starting period, T_s = 20 seconds

Back emf of each motor during series combination, (assuming two motors in series and two such motor circuits in parallel)

$$E_{bse} = \frac{V}{2} - IR = \frac{600}{2} - 200 \times 0.1 = 280 \text{ volts}$$

Back emf of each motor during parallel combination when all the four motors are in parallel giving 4 circuits

$$E_{bp} = V - IR = 600 - 200 \times 0.1 = 580 \text{ volts}$$

Time of operation in series grouping,

$$T_{series} = T_s \times \frac{\frac{V}{2} - IR}{V - IR} = 20 \times \frac{280}{580} = 9.66 \text{ seconds}$$

Time of operation in parallel grouping,

$$T_{parallel} = T_s - T_{series} = 20 - 9.66 = 10.34 \text{ seconds.}$$

Motor input per motor

$$= \frac{VIT_{se}}{2 \times 3{,}600} + \frac{VIT_p}{3{,}600} = \frac{600 \times 200 \times 9.66}{2 \times 3{,}600} + \frac{600 \times 200 \times 10.34}{3{,}600}$$

= 505.7 watt-hours

Energy lost in rheostat per motor

$$= \frac{1}{2} \times \frac{(V - 2IR) + 0}{2} \times I \times \frac{T_{series}}{3{,}600} + \frac{\frac{V}{2} + 0}{2} \times I \times \frac{T_{parallel}}{3{,}600}$$

$$= \frac{1}{2} \times \frac{600 - 2 \times 200 \times 0.1}{2} \times 200 \times \frac{9.66}{3{,}600} + \frac{\frac{600}{2} + 0}{2} \times 200 \times \frac{10.34}{3{,}600}$$

= 75.1 + 86.2 = 161.3 watt-hours

Energy lost per motor

$$= \frac{I^2RT_s}{3{,}600} = \frac{(200)^2 \times 0.1 \times 20}{3{,}600} = 22.2 \text{ watt-hours}$$

Output per motor

= Input to each motor – energy lost in rheostat per motor – energy lost per motor

= 505.7 − 161.3 − 22.2 = 322.2 watt-hours

Starting efficiency,

$$\eta = \frac{\text{Output}}{\text{Input}} \times 100 = \frac{322.2}{505.7} \times 100 = 63.7\% \textbf{ Ans.}$$

(*b*) Acceleration, $\alpha = \frac{V_m}{T_s} = \frac{80}{20} = 4$ kmphps

Speed at which series connections are to be changed

$$= \alpha T_{series} = 4 \times 9.66 = 38.64 \text{ km per hour } \textbf{Ans.}$$

Example 13.4. A motor coach weighing 150 tonnes is equipped with four 600-V motors for series-parallel control. If during series-parallel starting the current per motor is maintained at 300 A.

Calculate:

(a) The durations of the starting period.

(b) The speed of the train at transition.

(c) The rheostatic losses during (i) series (ii) parallel steps and start.

At 300 A 600 V, the tractive effort is 15,000 N per motor and the train speed is 30 kmph. Assume the train is started up a gradient of 1 per cent and train resistance is 10 N per tonne. Allow 10 per cent for the effect of rotational inertia. Each motor has a resistance of 0.1 Ω.

[Gorakhpur Univ. Utilization of Elec. Power and Traction 1975]

Solution : Dead weight of train, W = 150 tonnes

Accelerating weight of train, $W_e = 1.1\,W = 1.1 \times 150 = 165$ tonnes

Average current per motor, I = 300 A

Maximum speed, V_m = 30 kmph

Supply voltage, V = 600 volts

Train resistance, r = 10 N per tonne

Motor resistance, R = 0.1 Ω

Tractive effort, $F_t = 4 \times 15{,}000 = 60{,}000$ N

$$\because F_t = 277.8 \times W_e \times \alpha + Wr + 98.1\,WG$$

$$\therefore 60{,}000 = 277.8 \times 165 \times \alpha + 150 \times 10 + 98.1 \times 150 \times 1$$

$$\text{or } \alpha = \frac{43{,}785}{277.8 \times 165} = 0.955 \text{ kmphps}$$

(a) Duration of starting period,

$$T_s = \frac{V_m}{\alpha} = \frac{30}{0.955} = 31.4 \text{ seconds. } \textbf{Ans.}$$

Back emf of each motor in series (2 circuits, each consisting of two motors in series)

$$E_{bse} = \frac{V}{2} - IR = \frac{600}{2} - 300 \times 0.1 = 270 \text{ V}$$

Back emf of each motor in full parallel combination,

$$E_{bp} = V - IR = 600 - 300 \times 0.1 = 570 \text{ V}$$

Duration for series running,

$$T_{series} = T_s \times \frac{E_{bse}}{E_{bp}} = 31.4 \times \frac{270}{570} = 14.9 \text{ seconds} \quad \textbf{Ans.}$$

Duration for parallel running,

$$T_{parallel} = T_s - T_{series} = 31.4 - 14.9 = 16.5 \text{ seconds}$$

(b) Speed of the train at transition

$$= \alpha \times T_{series} = 0.955 \times 14.9 = 14.23 \text{ kmph} \quad \textbf{Ans.}$$

(c) (i) Rheostat losses during series operation

$$= \frac{(V - 2IR) + 0}{2} \times 2I \times \frac{T_{series}}{3{,}600}$$

$$= \frac{(600 - 2 \times 300 \times 0.1) + 0}{2} \times 2 \times 300 \times \frac{14.9}{3{,}600}$$

$$= 270 \times 600 \times \frac{14.9}{3{,}600} = 670.5 \text{ watt-hours} \quad \textbf{Ans.}$$

(ii) Rheostat losses during parallel operation

$$= \frac{\frac{V}{2} + 0}{2} \times 4I \times \frac{T_{parallel}}{3{,}600} = \frac{300 + 0}{2} \times 4 \times 300 \times \frac{16.5}{3{,}600}$$

$$= 825 \text{ watt-hours} \quad \textbf{Ans.}$$

Rheostat losses during start = 670.5 + 825 = 1,495.5 watt-hours **Ans.**

Example 13.5. A series motor is required to produce constant full-load torque whilst its speed is raised 25% above the full-load value by means of a diverter. Calculate the diverter resistance required as a percentage of the field resistance. Neglect losses in the motor. The relation between field current and flux is as follows:

Current	**50**	**75**	**100%**	**of full-load value**
Flux	**60**	**86**	**100%**	**of full-load value**

Solution : Let the full-load speed be N_f, flux ϕ_f and armature current I_{af}

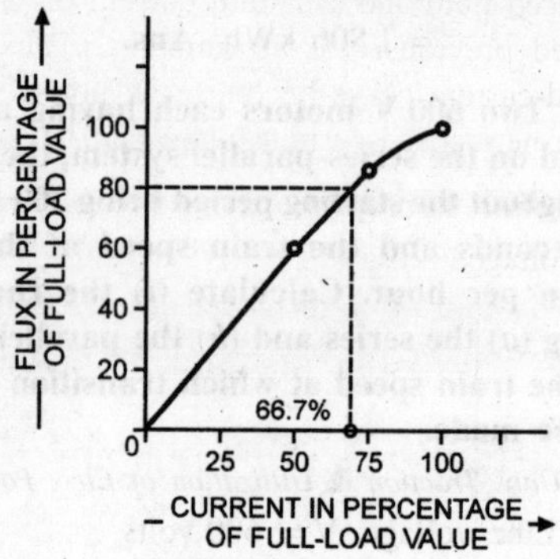

Fig. 13.13

After connecting diverter speed, $N_2 = 1.25\,N_f$

Since speed is inversely proportional to flux, therefore flux required,

$$\phi_2 = \frac{\phi_f N_f}{N_2} = \frac{\phi_f \times N_f}{1.25\,N_f} = 0.8\,\phi_f$$

From magnetic characteristics when $\phi_2 = 0.8\,\phi_f$

$$I_{se_2} = 0.667\,I_{sef}$$

Since torque is constant,

$$\therefore I_{a_2} I_{se_2} = I_{af} I_{sef}$$

$$\text{or } I_{a_2} = I_{af}\frac{I_{sef}}{I_{se_2}} = \frac{I_{af}}{0.667} = 1.5\,I_{af}$$

or $I_{a_2} = 1.5\,I_{sef}$ Since without diverter $I_a = I_{se}$

and $I_{se_2} = 0.667\,I_{sef}$

While using diverter, the armature current is equal to the sum of field current plus diverter current.

$$\therefore I_{a_2} = I_{se_2} + I_{div}$$

$$\text{or } 1.5\,I_{sef} = 0.667\,I_{sef} + I_{div}$$

$$\text{or } I_{div} = 0.883\,I_{sef}$$

Since current is inversely proportional to the resistance

$$\therefore I_{div} R_{div} = I_{se_2} R_{se}$$

$$0.883\,I_{sef} \times R_{div} = 0.667\,I_{sef} \times R_{se}$$

$$\text{or } R_{div} = \frac{0.667}{0.883} R_{se} = 75\% \text{ of series field resistance.} \quad \textbf{Ans.}$$

Example 13.6. A motor has characteristics as follows:

Current in amperes	**40**	**60**	**80**	**100**	**120**	**160**	**200**
Speed in kmph	**58.1**	**47.4**	**40.3**	**35.8**	**33.9**	**29.8**	**28.5**
Drawbar pull in kg	**185**	**440**	**700**	**970**	**1,245**	**1,800**	**2,360**

Deduce the corresponding approximate characteristics for the motor working with two-thirds of the normal field turns in the circuit.

Solution : Speed-armature current and draw bar pull-current curves are shown in Fig. 13.14.

Now when the field turns in the circuit are made two-third, the flux produced by armature current I_a will be equal to that, created by armature current $\frac{2}{3}I_a$ with all turns in series. Thus when the armature current is 120 amperes, the flux is that produced by a normal field coil with $\frac{2}{2}\times 120 = 80$ amperes.

Assuming supply voltage as constant and ignoring voltage drop in armature, speed is inversely proportional to the flux. The speed with the reduced field and armature current of 120 amperes is that which occured previously at 80 amperes. Thus point P_1 on the original speed-current curve becomes the point P_2. Similarly other points on the new speed-current curve can be located.

Tractive effort is proportional to ϕI_a *i.e.*, proportional to $\frac{I_a}{N}$ at constant voltage.

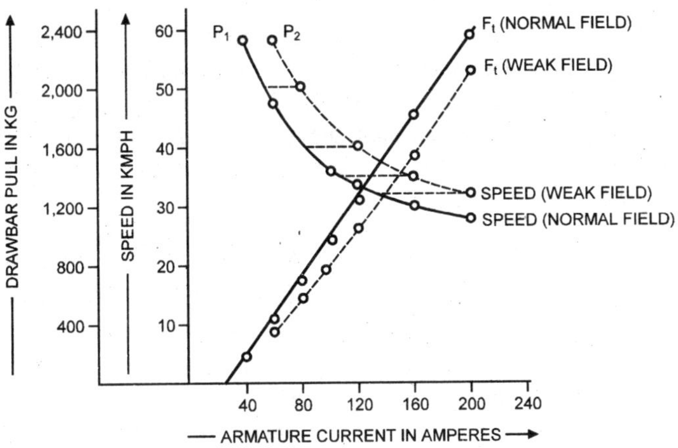

Fig. 13.14

The tractive effort with weak field

$$= \frac{\text{Tractive effort with normal field} \times \text{speed with normal field}}{\text{Speed with weak field}}$$

$$= F_{t_1} \times \frac{N_1}{N_2} \text{ for any armature current}$$

When armature current $I_a = 200$ A

$F_{t_1} = 2{,}360$ kg; $N_1 = 28.5$ kmph; $N_2 = 32$ kmph

$$\therefore F_{t_1} = F_{t_2} \times \frac{N_1}{N_2} = 2{,}360 \times \frac{28.5}{32} = 2{,}100 \text{ kg}$$

Similarly,

$$F_{t_2} \text{ for armature current of 160 A} = \frac{1{,}800 \times 29.8}{35} = 1{,}530 \text{ kg}$$

$$F_{t_2} \text{ for armature current of 120 A} = \frac{1{,}245 \times 33.9}{40.3} = 1{,}050 \text{ kg}$$

$$F_{t_2} \text{ for armature current of 100 A} = \frac{970 \times 35.8}{45} = 770 \text{ kg}$$

$$F_{t_2} \text{ for armature current of 80 A} = \frac{700 \times 40.3}{50} = 564 \text{ kg}$$

$$F_{t_2} \text{ for armature current of 60 A} = \frac{440 \times 47.4}{58.1} = 359 \text{ kg}$$

The corresponding approximate characteristics for the motor working with two-thirds of the normal field turns in the circuit are as follows:

Current in amperes	60	80	100	120	160	200
Speed in kmph	58.1	50	45	40.3	35	32
Drawbar pull in kg	359	564	770	1,050	1,530	2,100

13.8. BUCK AND BOOST METHOD OF SPEED CONTROL

The circuit diagram is shown in Fig. 13.15. The armatures of both the traction motors I and II and M-G set are connected in series. The whole series combination is connected across the supply. When the generator terminal voltage is equal to supply voltage in magnitude but of opposite polarity and the main contactor MC is closed, the voltage across the traction motors is zero and thus their speed is also zero. On reducing the generator voltage, voltage across the traction motors will start increasing and so their speed too. When the generator voltage becomes equal to zero, full line voltage will appear across both the motors *i.e.*, one-half of the supply voltage across each motor. In case polarities of both supply source and generator are same and generator voltage is equal to supply voltage, voltage across each traction motor will be equal to supply voltage. Thus the voltage across traction motors can be adjusted (bucked or boosted up) by adjusting the excitation of the generator.

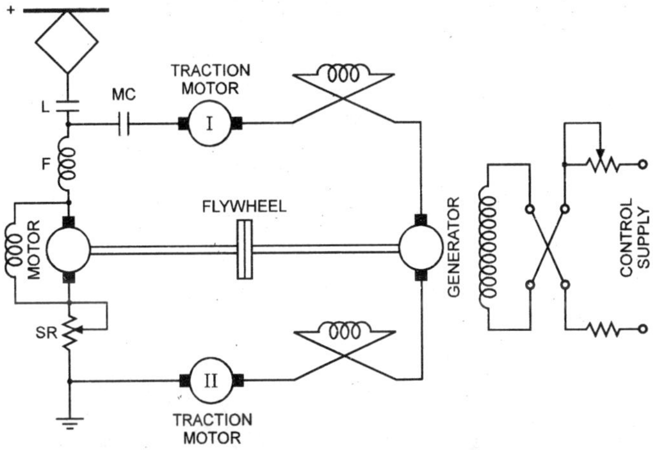

Fig. 13.15. ***Buck and Boost Method of Speed Control***

The above method of speed control have the following advantages:

(*i*) Traction motors can be operated on any speed, while in case of resistance control only a few speeds are possible.

(*ii*) There is no energy loss in the starting resistance of the traction motors. However, there is a loss of energy in the starting resistance of the M-G set.

(*iii*) In case of temporary interruption in the supply, the kinetic energy of the flywheel can be utilized in generating energy from the M-G set and supplied to the traction motors.

The main drawback of this method of speed control is requirement of M-G set. There is a loss of energy in M-G set.

13.9. METADYNE CONTROL

The metadyne control system is based on constant current system of speed control. In resistance control or series- parallel control a great deal of energy is dissipated in the starting resistance and jerks are experienced when the controller of the starter moves from one position to another position. In metadyne control, current throughout the accelerating period remains constant, therefore, uniform tractive effort is developed and very smooth control, without causing any wastage of energy in the starting resistance, is achieved.

The essential part of the metadyne control is metadyne converter. The metadyne converter is a cross-field machine which behaves like a transformer on direct current. The transformation ratio of a metadyne can be varied continuously. It takes power at constant voltage and variable current and delivers the same at constant current and varying voltage.

The metadyne converter essentially consists, in its simplest form, of a 2 pole dc armature with two pairs of brushes and a four pole field magnet, as shown in Fig. 13.16. One pair of the brushes (say A and C) are connected across constant voltage dc supply while the other pair (B and D) are connected to the load (normally a dc series motor).

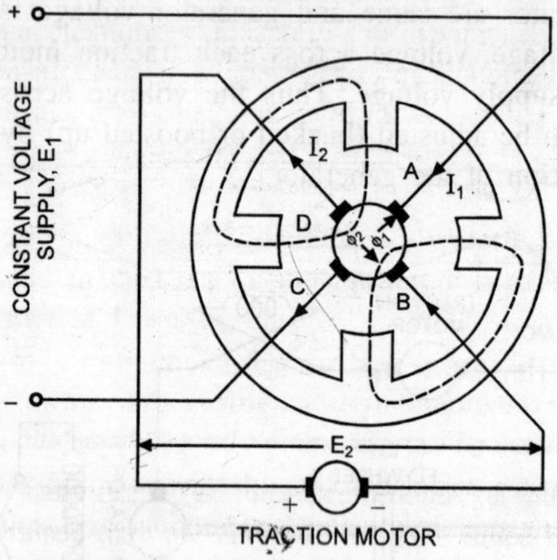

Fig. 13.16. ***Simple Metadyne Converter***

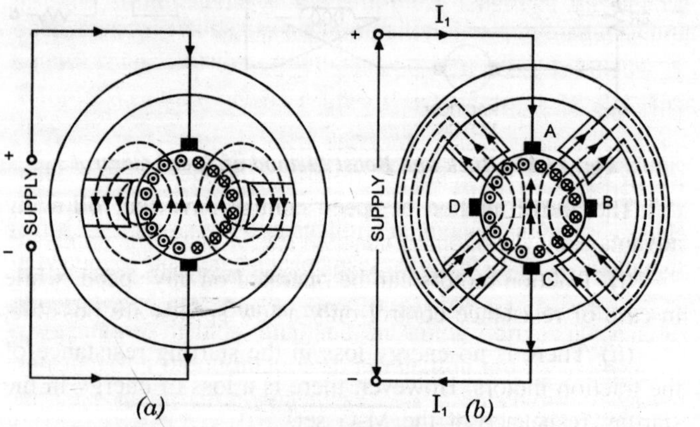

Fig. 13.17. ***Illustration of Metadyne Principle***

For understanding the working of a metadyne converter consider first an ordinary dc machine with two poles and two brushes supplied with a current flowing in the direction shown in Fig. 13.17 (*a*). It will cause armature current distribution, as illustrated in the figure with corresponding cross flux, mainly confined to the poles.

Now consider that metadyne converter (a dc machine with two pairs of brushes and two pairs of poles) is running at constant speed and drawing a current I_1 from the dc supply main, which flows through the armature conductors via the brushes A and C, as shown in Fig. 13.17 (*b*). An armature reaction flux ϕ_1 set up in usual way is provided with a fairly low reluctance path through the yoke by the four poles, as shown in the figure. Due to rotation of armature conductors in this primary flux, and emf $E_2 = KI_1$ is set up between the brushes B and D. When these brushes B and D are connected to a load, a current I_2 flows through the load. The load current I_2 sets up another flux ϕ_2 known as *secondary flux*, at the right angles to the first, the distribution is shown in Fig. 13.17 (*c*). This secondary flux ϕ_2 causes an emf, $E_1 = KI_2$ between brushes A and C opposing the applied voltage. As the applied voltage is constant, the resistance drop is negligible so the back emf E_1 opposing applied voltage and the current I_2 producing E_1 are also constant.

$$\text{Since input} = E_1I_1 = KI_2I_1 = KI_2 \times \frac{E_2}{K} = E_2I_2 = \text{output},$$

therefore, power required to drive the metadyne is very small being equal to the running losses of the machine.

This simple metadyne converter transforms the constant voltage supply into a constant current variable voltage supply to feed the load. The arrangement, therefore, is quite suitable for starting dc traction motors. With this arrangement the load current I_2 and supply voltage V remain constant and as the load increases on account of building of back emf in dc traction motor E_2 and I_1 increases to meet with the increased load.

The metadyne described above has no winding on the poles and is capable of delivering only a single value of constant current but for supplying dc traction motors, after the motor has gained speed, the load current I_2 has to be reduced to the running value. For this purpose the field magnet poles are provided with variator and regulator windings, as shown in Fig. 13.18.

The variator winding sets up a flux in the same direction as that set up by the load current I_2. Total flux ϕ_2 required in this axis being constant in order to produce a back emf E_1 equal to the constant supply voltage, therefore if some of this flux is set up by a separate winding, known as *variator winding*, the load current I_2 will decrease and can be of a smaller constant value. Similarly the load current can be increased by causing the current to flow in the variator winding in the opposite direction.

If the output current I_2 is, say, reduced in this way, the voltage remaining the constant, the total output will be reduced but input will remain the same and therefore, the set will speed up. In order to keep the speed of the metadyne converter constant, an additional winding known as *regulator winding* is provided. By adjusting the current in the regulator winding, the input current can be varied and therefore, input power can be adjusted equal to output power, the speed of the converter remaining the same.

The regulator winding is supplied from a small dc shunt generator mounted on the shaft of the metadyne. Any tendency towards a change in speed of the metadyne will cause corresponding change in the emf set up by the shunt machine and as it acts in the opposite direction of the supply voltage, so corresponding change in the regulator winding current will result in.

The variator winding is supplied excitation from an exciter mounted on the same shaft, as shown in Fig. 13.18.

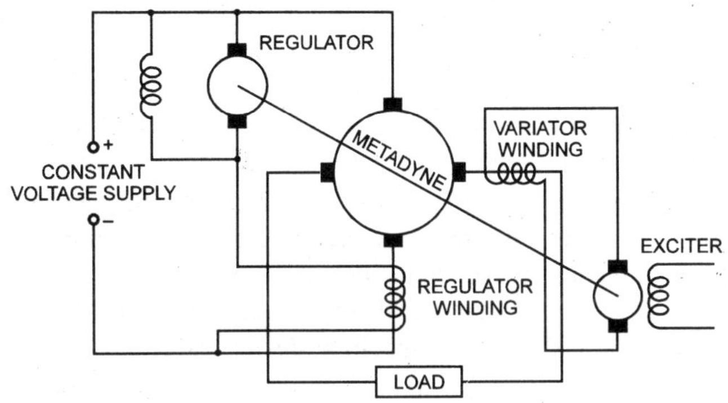

Fig. 13.18. *Metadyne Control System*

With metadyne converter, regenerative braking can be accomplished very easily by reversing the field of the traction motor. This causes the reversal of direction of induced emf E_2 which in turn will change the direction of current I_1. Thus current I_1 can be supplied back to the supply source. By controlling the magnitude of reversed excitation of traction motors supplied by metadyne, the magnitude of regenerative braking can be regulated.

The metadyne is employed whenever control of dc motors is required. The control provided by the metadyne is smooth and does not require any switching. Thus switchgear and arcing are avoided. In some cases it is cheaper than the Ward Leonard system in initial cost. In traction it provides smooth acceleration without skill on the part of driver and regenerative braking down to very slow speeds. The savings due to these items may easily counterbalance the additional cost of the more complicated equipment required and its additional maintenance cost. It is already being employed in the underground railway.

13.10. THYRISTOR CONTROL OF TRACTION MOTORS

The modern trend is towards the use of dc motors (both separately excited and dc series motors) equipped with thyristor chopper control. The operating voltages are 600 V or 1,000 V. Brakings employed are mechanical, rheostatic and regenerative. Thyristorised converters provide accurate control and fast response. Main advantages of thyristor control are the absence of bulky on-load tap changer and electromagnetic devices, saving of energy, notchless control, incease in pulling ability of the motive power, and minimum wear and tear because of absence of conventional moving parts in the motor control circuits.

Various methods of speed control and electric braking employing thyristors have already been discussed in chapter 3.

In addition to ordinary phase control methods, *cycle selection methods* of control of SCR for varying the voltage applied to the traction motors are also employed. In this method the required average voltage is obtained by accepting or rejecting a certain number of complete half cycles. In practice, at the start only one half cycle out of eight is accepted and as the speed builds up, it is gradually raised to $\frac{2}{8}, \frac{3}{8}$ and finally $\frac{8}{8}$ for full-power operation. This method is advantageous due to low frequency harmonics, low rate of rise of current, better power factor etc.

In *chopper control* of traction motors, at start, the '*on*' period of pulse is kept very short which lengthens during the period of controlled acceleration. Thus the average voltage applied across the traction motors is gradually increased keeping the mean value of the input current close to the desired value.

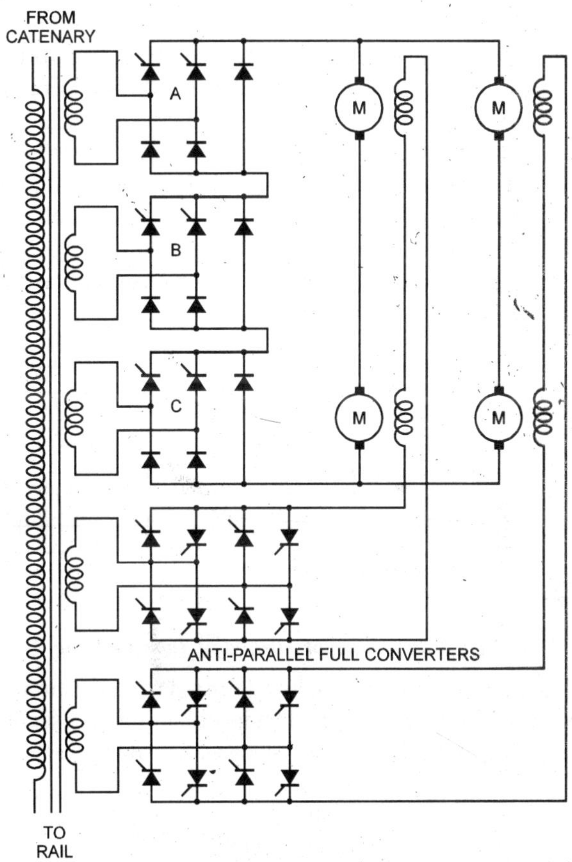

Fig. 13.19. *Thyristorised DC Electric Traction*

Figure 13.19 shows a typical thyristorised dc traction system supplying a group of four separately excited motors. The armatures are supplied from half controlled bridge converters. However, it is desirable to feed the field windings through fully controlled bridge converters so as to reduce the ripple in the field current. Low ripple in the field current ensures low iron losses in the machines. However, if regenerative braking is required then the armatures should be supplied from fully controlled bridges. Freewheeling diodes are connected as illustrated to ensure good waveform of armature current. The armatures are connected in series-parallel arrangement to ensure good starting and running characteristics. It is seen that armatures are supplied by three bridges connected in series. For starting first only bridge A is triggered and firing angle is advanced as speed builds up. When bridge A is fully conducting (*i.e.,* when $\alpha = 0$), bridge B is triggered and then bridge C is triggered. During starting field currents are set to maximum to provide high starting torque. The use of three bridges ensures better power factor than would be possible with a single bridge.

13.11. SPEED CONTROL AND STARTING OF SINGLE PHASE SERIES MOTORS

The control of speed and torque in case of single phase ac series motor is carried out by varying the applied voltage. As the operating voltage for a single phase series motor is very low (about 300 volts), thus, transformer forms an essential part of the locomotive equipment and, therefore, variable voltage for both starting and speed control can be obtained by means of tappings on the transformer without requiring any additional equipment for this purpose. The transformer tappings are usually provided on *lt* side of the transformer in order to avoid insulation difficulties but 11,000 volt locomotives have been built with hv tappings, which enables a lighter construction to be used on account of much lower currents to be dealt with. As no external resistance is required for starting purpose, so there is no loss of power and each control point becomes a running point. The regulation of voltage is affected by the contactor method of tap changing, as shown in Fig. 13.20.

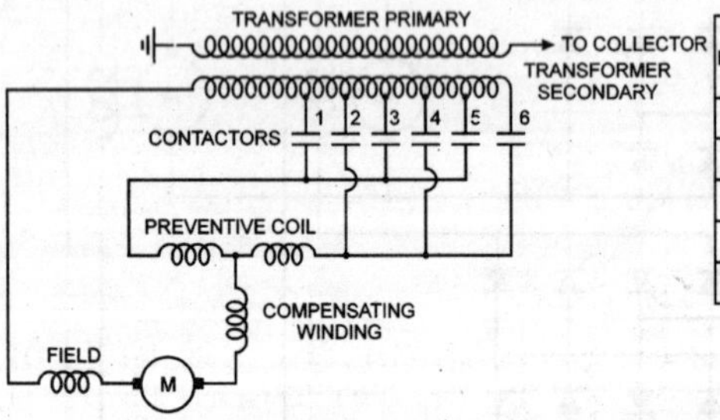

SEQUENCE OF CONTACTORS

NOTCH	CONTACTORS 1	2	3	4	5	6
1	•	•				
2		•	•			
3			•	•		
4				•	•	
5					•	•

Fig. 13.20. *Contactor Method of Tap Changing*

A preventive coil is employed for reducing the circulating current flowing on account of short circuiting of a section of the winding of the transformer between the tappings. Preventive coil (choking coil having high resistance) is inserted between the terminals of two groups consisting of tappings 1, 3, 5 and 2, 4, 6 as shown in the Fig. 13.20. The motor is connected to the centre point of the preventive coil. Two contactors, connected to the adjacent tappings (1 and 2 or 2 and 3 or 3 and 4 or 4 and 5 or 5 and 6) are closed on each notch of the controller. Each contactor, therefore, carries approximately one half of the total current required. For changing-over voltage from one value to another one contactor is opened and another contactor belonging to the same group is closed.

An important advantage of this method is that each notch provides a running position, so that many running speeds are available. Between 6 and 12 tappings are usually provided, giving a corresponding number of economical speeds.

In case larger currents are to be handled by the contactors (for large capacity traction motors), a larger number of contactors are to be employed simultaneously so that the current is divided between them and each contactor has not to deal more than a pre-specified design value.

Contactor method of tap changing for larger currents (over 1,000 A) is shown in Fig. 13.21. In this method four adjacent contactors are connected to the same notch of the controller. There are three preventive coils, also called *'bridging coils'* arranged in such a fashion as to facilitate the correct division of current among the contactors.

The ac contactors are quite different from dc contactors. In general, dc contactor is much lighter than the ac contactor and needs less power for its operation. Other special features of ac contactors are (*i*) laminated magnetic circuit (*ii*) provision of shading coil in order to avoid chattering of contactors.

13.12. STARTING AND SPEED CONTROL OF 3-PHASE INDUCTION MOTORS

The method employed for speed control of 3-phase induction motors are (*i*) rheostatic control (*ii*) pole changing control (*iii*) cascade control and (*iv*) combination of cascading and pole changing control.

Different methods of speed control of 3-phase induction motors have already been discussed in chapter 1 (Art. 1.18.2).

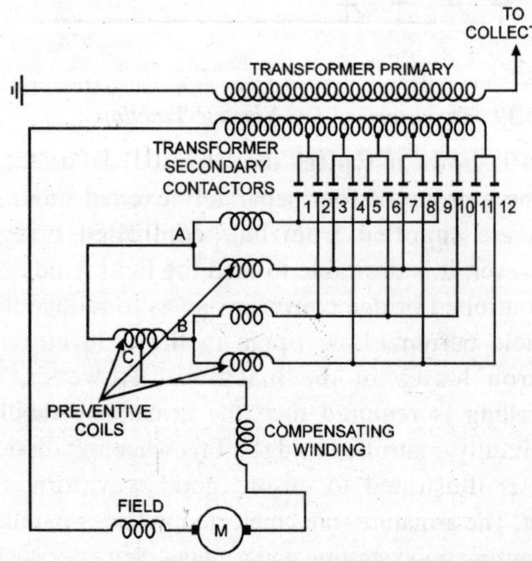

SEQUENCE OF CONTACTORS

STEP	CONTACTORS 1	2	3	4	5	6	7	8	9	10	11	12
1	•	•	•	•								
2		•	•	•	•							
3			•	•	•	•						
4				•	•	•	•					
5					•	•	•	•				
6						•	•	•	•			
7							•	•	•	•		
8								•	•	•	•	
9									•	•	•	•

Fig. 13.21. *Contactor Method of Tap Changing For Larger Currents*

Thyristor control of 3-phase induction motors have been discussed in chapter 3 (Art. 3.13).

Rheostatic control is the simplest but the least efficient method of speed control of 3-phase induction motors. This method of speed control is employed in light locomotives and motor coaches where a single economical speed is sufficient and energy consumption is of no importance.

Pole-changing control is the simplest of the multi-speed control methods. This method has the advantage of simplicity, good speed regulation for each setting, high operation efficiency, and moderate first cost and maintenance. The choice of the number of poles on a pole-changing winding is in the ratio of 2 : 1, 3 : 2, 4 : 3.

Cascade control is employed for the two-speed goods locomotives (24 kmph and 50 kmph).

In *combination of cascading and pole-changing control method*, both of the motors employed are wound for the same number of poles and windings so arranged that the number of poles can be changed. With equal number of poles on both the machines one speed is obtained and then more speeds are made available by changing the number of poles. This method has been employed for the four-speed passenger locomotives (37, 51, 74 and 102 kmph) in some types of locomotives.

Exercises

1. Describe with the aid of diagrams following methods of speed control for dc traction motors, pointing out the merits of each:
   (*i*) series-parallel control. (*ii*) control by weakening field.
   Explain also the conditions under which the speed control by weakening of field is advantageous.
2. Explain with suitable connection diagrams, the resistance control method employed in dc traction drives.
   [A.M.I.E. Sec. B. Electric Drives & Their Control Winter 1996]
3. Under what circumstances is the field weakening adopted for controlling speed of traction motors and what are the merits and demerits of this method of control ? Illustrate your answer with the help of tractive effort-speed characteristics of the locomotive.
4. Explain the method of series-parallel starting of traction motors. Mention its advantages.
   [Pb. Univ. Electric Traction Dec. 1992]
5. Discuss the advantages of series-parallel control of starting as compared to the rheostatic starting for a pair of dc traction motors.
   [W.B. Univ. of Technology Utilization of Electric Power 2009]
6. Explain the starting of dc series motor type traction motor and prove that with the series-parallel method of starting the efficiency of starter is increased from 50% in case of normal rheostatic starting to 66.67%.
7. Distinguish between shunt and bridge transition methods of starting a pair of dc traction motors.
   [Rajasthan Univ. Utilization of Electrical Power 2005]
8. Draw and explain the complete view of drum controller giving shunt transition between two series traction motors.
   [Pb. Univ. Electric Traction 1975]
9. Explain the short-circuit transition method of control of a pair of tram motors and give the developed view of the drum of a controller arranged for this method of control, indicating also the connections of the reversing barrel.
10. Give a schematic power diagram of 1,500 V dc train equipment of 2 dc series motors wired for series-parallel operation during starting and explain it.
    [B.H.U. Electric Traction & Utilization 1978]
11. Explain with relevant diagrams the method of metadyne control of electric traction motors.
    [B.H.U. Elec. Traction & Utilization 1978, Agra Univ. Traction & Utilization of Elec. Power 1983]
12. Describe with a neat connection diagram the principle of operation of a metadyne set to be used for starting the series traction motor. What are the advantages of variator and regulator windings ?
    [Agra Univ. Traction & Utilization of Elec. Power 1979]
13. Describe with a neat diagram the principle of operation of a metadyne set to be used for starting of the series traction motor. **[Rajasthan Univ. Utilization of Electrical Power 2004, 2005, 2006, 2007]**
14. Discuss the advantages and disadvantages of thyristor control of electric traction.
    [M.D. Univ. Utilization Electric Power and Traction Dec. 2008]
15. Discuss the advantages and disadvantages of thyristor control of traction motors.
    Explain the chopper method of control of dc motors. What are the merits and demerits of the above method of control ?
    [Pb. Univ. Electric Traction June 1993]
16. Describe the contactor method of tap changing for the control of speed and torque of a single phase series motor.
17. Compare the use of dc series motor and ac series motor in electric traction. Discuss the controllers used for such motors.
    [Pb. Univ. Electric Traction June 1993]
18. Explain any one method of speed control of induction motor. What speed range is obtainable by this method ?
    [Pb. Univ. Electric Traction Dec. 1991]
19. What are the requirements of electric drives for traction ? Explain a static control scheme used in ac traction employing ac motors. **[A.M.I.E. Sec. B. Electric Drives & Their Control Winter 1998]**
20. Write short notes on the following:
    (*i*) Speed control of dc series motors.
    (*ii*) Series-parallel control of starting for pair of dc traction motors. **[G.B.T.U. 2010-11]**
    (*iii*) Series-parallel control of dc series motors.
    [U.P.T.U. 2009-10]
    (*iv*) Energy saving with series-parallel starting.
    [Rajasthan Univ. 2000, 2004, 2006; Rajasthan Technical Univ. 2010]
    (*v*) Tramway controller.
    (*vi*) Metadyne control. **[D.E.I. Agra 1989]**
    (*vii*) Thyristor control of electric traction.
    [M.D. Univ. December 2009]
    (*viii*) Speed control and starting of single phase ac series motors.
    (*ix*) Drum controller. **[Pb. Univ. July 1974]**

Short Answer Type Questions With Answers

Q. 1. The resistors employed for control of traction motors are designed for short-time rating. Why ?

Ans. Because control resistors are required only during starting of traction motors.

Q. 2. Under what conditions is field weakening adopted for controlling the speed of a dc series traction motor ?

Ans. The field weakening method of speed control is used for increasing the speed of traction motors when they have attained maximum possible speed by series-parallel control system.

Q. 3. Under what circumstances series-parallel method of speed control of dc series traction motors adopted ?

Ans. The sereis-parallel method of speed control of dc series motors is adopted when motors employed are two, four, six or more than this even.

Q. 4. What is the main difficulty associated with series-parallel control ?

Ans. The main difficulty associated with series-parallel control is in obtaining a satisfactory method of transition from series to parallel without interrupting the torque or permitting any heavy rush of current.

Q. 5. What are the advantages of thyristor control of traction motors ?

Ans. The advantages of thyristor control of traction motors are accurate control and fast response, elimination of bulky on-load tap changer and electromagnetic devices, saving of energy, notchless control, increase in pulling ability of the motive power, and minimum wear and tear.

Q. 6. Enumerate the methods employed for speed control of 3-phase induction motors used in electric traction.

Ans. The methods employed for speed control of 3-phase induction motors used in electric traction are (*i*) rheostatic control (*ii*) pole-changing control, (*iii*) cascade control (*iv*) combination of cascading and pole-changing control method and (*v*) thyristorised control.

Q. 7. In ac locomotive tap-changer is provided on high voltage side of transformer. Give suitable reason. **[U.P.T.U. 2006-07]**

Ans. The transformer tappings are usually provided on low voltage side of the transformer to avoid insulation difficulties but 11,000 V ac locomotives have been built with high voltage tappings, which enables a lighter construction to be used on acount of much lower currents to be dealt with.

Problems

1. Two motors of a motor coach are started in series-parallel, the current per motor being 300 A (considered as being maintained constant during the starting period of 24 seconds). If the acceleration is uniform, the line voltage is 750 volts and the resistance of each motor is 0.1 Ω, determine
   (*a*) the time during which the motors are operating in series.
   (*b*) the time during which the motors are operating in parallel.
   (*c*) the energy drawn from the supply mains.
   (*d*) the energy dissipated in starting resistances.
   (*e*) the starting efficiency and
   (*f*) the speed at which series connections should be changed to parallel if the speed at the end is to be 48 kmph.
   [**Ans.** (*a*) 11.5 seconds (*b*) 12.5 seconds (*c*) 2.28125 kWh (*d*) 0.72125 kWh (*e*) 68.4% (*f*) 23 kmph]

2. Two motors of a motor coach are started on series-parallel system, the current per motor being 350 A (considered as being maintained constant during the starting period which is 18 seconds). If the acceleration is uniform, the line voltage is 600 volts and the resistance of each motor is 0.1 Ω, calculate
   (*a*) the time during which motors are operating in series
   (*b*) loss in the rheostats during starting period.
   [DEI Agra Utilization of Electric Power 1989; Rajasthan Univ. 2007]
   [**Ans.** (*a*) 8.44 seconds (*b*) 496 watt-hours]

3. A train equipped with two 600 V dc series motors is started on series-parallel system, the main accelerating current being 300 amperes. The train speed at the end of starting period of 15 seconds is 30 kmph. Determine the train speed at the time of transition and rheostatic losses in kWh during (*a*) series period (*b*) parallel period. Take resistance of each motor as 0.1 Ω. **[P.U. B.Sc. (Engg.) Nov. 1974]**
   [**Ans.** (*i*) 14.2 kmph (*ii*) (*a*) 0.16 kWh (*b*) 0.18 kWh]

4. Two 600 V dc series motors are started by series-parallel control. Each motor takes 500 A during starting time for 20 seconds and has 0.1 Ω resistance. Calculate (*i*) energy lost in starting rheostat (*ii*) energy lost in motor (*iii*) motor output and (*iv*) efficiency of starting.
   [**Ans.** (*i*) 0.77 kWh (*ii*) 0.139 kWh (*iii*) 0.7635 kWh (*iv*) 59.3%]

5. Two 600 volt dc traction motors each having a resistance of 0.15 Ω, are started with series-parallel control. The starting time is 20 seconds and the current in each motor during starting is maintained constant at 300 A. Calculate the energy in kWh dissipated in the starting rheostats.
   [**Ans.** 0.466 kWh]

6. A car driven by two dc traction motors takes current of 200 A from a 600 V line with motors in full series. Each motor has a resistance of 0.2 Ω. The motors are to be switched into parallel through a resistance. If the transition is to be affected without shock, calculate the value of this resistance.
   [**Ans.** 1.5 Ω]

7. A series traction motor has following characteristics:

Current	40	80	120	160	200 amperes
Speed	51.5	33.6	28.8	26	24.6 kmph
Tractive effort	120	420	740	1,070	1,400 kg

   Draw the new characteristics when 25% of the field turns are cut off. **[Pb. Univ. Electric Traction April, 1975]**

8. The tractive effort and speed characteristics of a motor are given below. Find the new characteristics if one third of the field turns are cut off.

Current (amperes)	50	75	100	150	200	250
Speed (kmph)	73.4	59.6	50.7	42.7	38.7	36
Tractive effort (kg)	207	496	794	1,400	2,030	2,660

   [Pb. Univ. Electric Traction July 1974]

CHAPTER 14

Braking; Mechanical Considerations and Control Equipment

INSIDE THIS CHAPTER

14.1. Braking **14.2.** Mechanical Regenerative Braking **14.3.** Mechanical Braking **14.4.** Hydraulic Braking **14.5.** Magnetic Track Brakes **14.6.** Electromechanical Drum Brakes **14.7.** Eddy Current Brakes **14.8.** Mechanical Considerations **14.9.** Control Equipment **14.10.** Multiple Unit Control **14.11.** Auxiliary Equipment.

14.1. BRAKING

Electrical and mechanical, both types of braking are used in electric traction. In electric braking the braking energy is converted into electrical energy instead of converting it into heat energy at the break shoes and either dissipated in the resistances mounted on the vehicle or returned to the supply system. Electric braking reduces the wear of the brake shoes and wheel tyres considerably and gives higher rate of braking retardation, thus brings the vehicle quickly to rest and shortens the running time to a considerable extent. In case regenerative braking is employed, the braking energy is returned to the supply system thereby a considerable saving is affected in the running cost, higher speeds are possible while going down the gradients as more braking power is available and heavier trains can be propelled down the gradients with safety and speed without dividing it into sections. Electric braking cannot replace the ordinary mechanical brakes, as the vehicle cannot be held stationary by it.

The desirable requirements of a braking system are given below:

1. The braking system should be simple, robust, quick, and reliable in action.
2. Maintenance needs should be minimum and braking system must be easy for driver to control and operate.
3. The system should apply brakes simultaneously over all the vehicles.
4. Normal service application of brakes should be very gradual and smooth so as to avoid damage to the goods and discomfort to the passengers.
5. The braking force applied to each axle should be proportional to the axle load so as to obtain uniform deceleration.
6. In case of emergency braking, safety consideration is the main consideration. As such retardation rate would be maximum consistent with the safety, so as to make unfailing halt in the minimum possible distance.
7. Kinetic energy of the train should as far as possible be storable during braking which could subsequently be used during acceleration of the train.
8. The braking system should be inexhaustible.
9. There should be automatic slack adjustment for constant piston stroke as a result of wear on the rim and the brake blocks in the case of mechanical braking.

The requirements of a braking system on a main line locomotive differ from those of motor coaches—the former usually needs braking to hold the train at the steady speed on a long down grade, whereas with the latter braking is primarily required for stopping the train. Rheostatic braking is employed in both the cases but regenerative braking is mainly confined to main line locomotives due to the complications involved in providing regenerative braking with dc series motors.

As already discussed in Art. 1.19, there are three methods of applying electric braking namely (*i*) plugging (*ii*) rheostatic braking and (*iii*) regenerative braking.

14.1.1. Plugging

This is the simplest type of braking. In this method of braking the torque of the motor is reversed, which brings the motor and its driven machine to standstill. This method of applying electric braking with dc motors, 3-phase induction motors and synchronous motors has already been discussed in chapter 1. [Arts. 1.20.1, 1.21.1 and 1.22].

14.1.2. Rheostatic or Dynamic Braking

In this method of braking, the motor is disconnected from the supply and operated as a generator driven by the kinetic energy of the rotating parts of the motor and its driven machines. Thus the kinetic energy of rotation is converted into electrical energy, which is dissipated in the external resistance connected across the motor at the braking instant. This method of applying electric braking with dc motors, 3-phase synchronous motors and induction motors (with a separate source of dc excitation during braking) has already been discussed in chapter 1. [Arts. 1.20.2, 1.21.2 and 1.22].

In traction work where two or more motors are employed, they are put in parallel across a resistance for braking, as shown in Fig. 14.1, as the series connection would produce too high voltage. During the braking period the motors are driven as generators owing to the kinetic energy of the train and the electrical energy so generated is dissipated in the form of heat in the resistances connected across them. An equaliser connection, as shown in Fig. 14.1 (*a*), is used in order to ensure that the two machines share the load equally. If the equaliser connection were not done, the machine that would build up first would send a large current through the other in the opposite direction, causing it to excite with reversed voltage and so the two machines would be short circuited on themselves. A large braking torque would be developed in this way, but it would be quite uncontrollable and the currents would be dangerously large. An alternative method of avoiding this situation is cross-connection of the two machines, as illustrated in Fig. 14.1 (*b*).

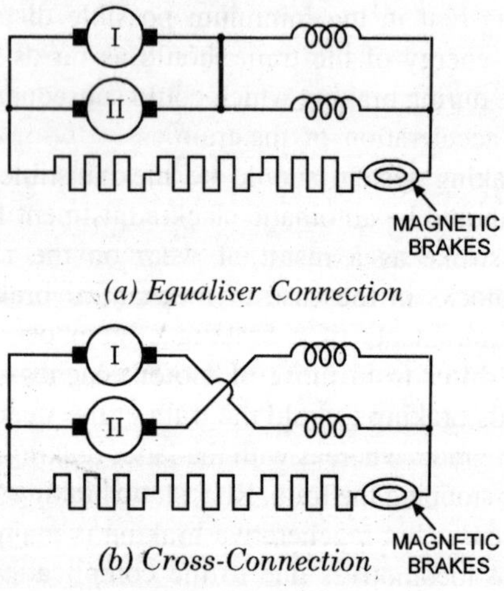

(*a*) *Equaliser Connection*

(*b*) *Cross-Connection*

Fig. 14.1. ***Rheostatic Braking***

In case the voltage of one of the machines becomes larger than that of the other, the first machine will cause flow of a higher current through the field winding of the second machine. As a result the second machine will have higher voltage while the field current in the first machine due to second one is smaller and hence it will cause lower voltage across first machine. Thus, an automatic compensation of the unbalancing is achieved.

The second method is advantageous to the first one as if the direction of rotation of the machine armatures reverses (may be on account of run back due to upgradient), the machines will fail to excite with equaliser connection and, therefore, no braking effect will be produced which may prove fatal for the passengers whereas with cross-connection the machines will build up in series and being short circuited on themselves will provide an emergency braking and the train will not be allowed to run back on the gradient.

Rheostatic braking cannot be employed with 3-phase induction motors.

In case of ac series motors the rheostatic braking is obtained by operating the machines as generators excited from the supply or as self-excited dc generators supplying power to resistance load. In the former case the fields are energised at low voltage from a suitable tapping on the main transformer while in the latter case the fields of the motors are excited from one of the motors acting as a series generator and in this case dc will be generated in rotors of the motors and the kinetic energy of the rotors will be dissipated as dc power in the loading resistors.

14.1.3. Regenerative Braking

In the previous two methods of electric braking namely plugging and rheostatic braking, stored energy of the rotating parts of the motor and its driven machines is wasted whilst in plugging extra energy is drawn during the braking period and wasted. In the regenerative braking the motors remain connected to the supply and return the braking energy to the supply.

For regenerative braking it is necessary that traction motors must generate power at a voltage higher than the supply voltage and at a reasonable constant voltage.

Regenerative braking is an inherent characteristic of *dc shunt wound motors* and does not require any change of connections. DC shunt motors are electrically stable also *i.e.,* braking torque is independent of line voltage fluctuations. Regenerative braking with dc shunt motors has already been discussed in Art. 1.20.3(1).

The dc series motors cannot be used readily for regenerative braking, as already discussed in Art. 1.20.3(2), and some modification of shunt winding or separate excitation of the series field at low voltage is necessary to be employed to enable the motor to act as a generator. One method of obtaining regenerative braking with series motors is *French method,* which has already been discussed in Art. 1.20.3(2).

Inherently a dc series generator has no stability of voltage, and therefore, an external means is required. In Fig. 14.2 (*a*) a separate motor driven exciter has a separately excited field winding F_1 and also another field winding F_2 connected in the main motor circuit in such a fashion that the field created by it opposes the field created by separately excited field winding F_1 during regeneration. The stabilising resistance R is employed to prevent current surges when the tram crosses from one section of the supply to another, and to compensate for variable line voltage. In case the line voltage falls, regenerated current will tend to increase resulting in strengthening of field F_2 which will weaken the field F_1 and, therefore, reduction in emf generated by the exciter. Thus the field of the traction motors will be weakened resulting in reduction of emfs generated by the traction motors operating as dc series generators. Thus compensation for a decrease in the line voltage is automatically provided.

The arrangement shown in Fig. 14.2 (*b*) has the exciter armature connected in the circuit of the field windings of

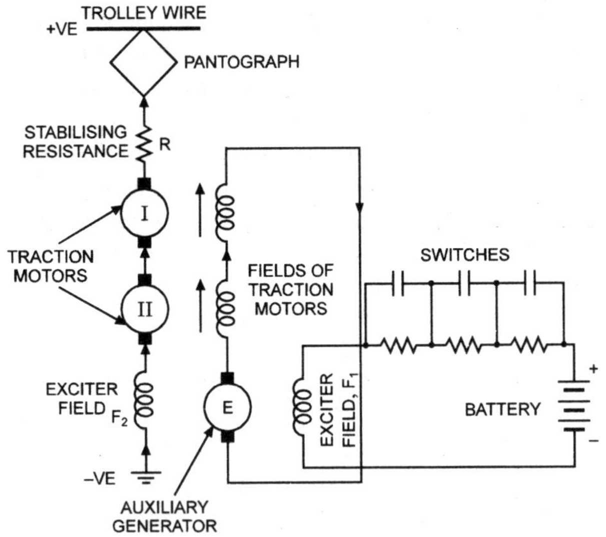

(a) Differentially Compounded Exciter

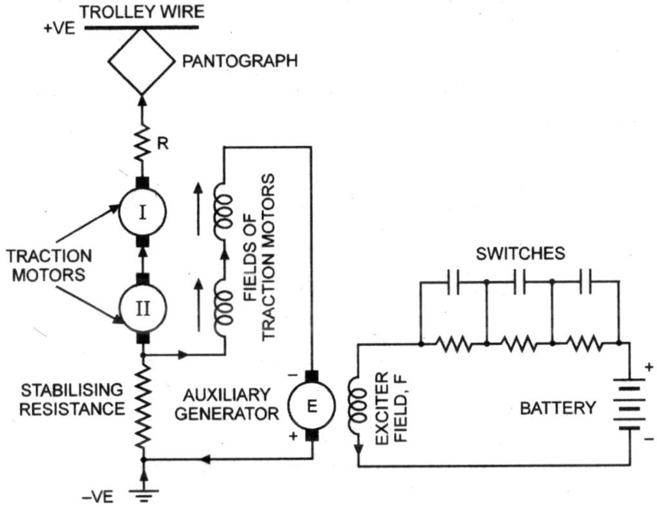

(b) Separately Excited Exciter

Fig. 14.2. *Regenerative Braking With Separate Exciter*

traction motors and the stabilising resistance. The current through the stabilising resistance is the sum of the exciter current and the regenerated current. The voltage of the exciter circuit can be regulated by either varying its field strength or manipulating resistances in series with the armature. In case the line voltage falls the regenerated current will tend to increase resulting in increase in voltage drop across the stabilising resistance and, therefore, reduction in the voltage available in the exciter armature circuit causing a reduction in the excitation current of the traction motors operating as dc series generators. This reduces the emf's generated and thus compensation is provided automatically.

Regenerative braking is an inherent characteristic of an *induction motor* since an induction motor operates as an induction (non-synchronous) generator when run at speeds above synchronous speed and it feeds power back to the supply line. The machine, however, is not self-exciting as a generator and is required to be connected to a system supplied from synchronous generators. This system supplies the excitation and determines the frequency at which the induction generator operates. Torque-speed characteristics of an induction machine are shown in Fig. 14.3. From the torque-speed characteristics of an induction machine it is obvious that without any extra resistance in the rotor circuit, there is only a slight variation of speed with the torque whereas with the external resistance inserted in the rotor circuit speed increases for a particular braking torque. Thus we see that with no extra resistance in the rotor circuit the speed during braking remains almost constant and independent of the gradient and the weight of the train. This is a great advantage with the induction motor when used for traction. But if increased speeds are necessary with light loads, these can be obtained by inserting external resistance in the rotor circuit. It is advantageous on mountain railways. It returns about 20% of the total energy on certain railway run and saves a great deal of brake shoe wear.

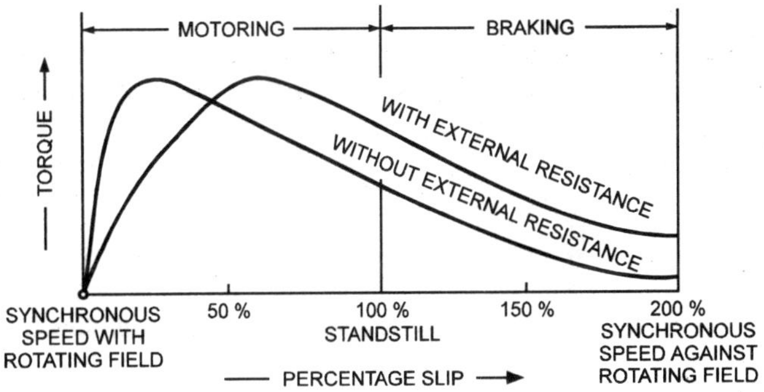

Fig. 14.3

Regenerative braking with ac series motors is more difficult than that with the dc series motor. The difficulties encountered are as follows.

During the regeneration period, the machine should not operate as a self-excited generator. Generally the high power factor is the main consideration in order to achieve a reasonable braking torque. Circuit for regeneration at high power factor usually suffers from self excitation, and those which are inherently stable and free from self excitation usually operate at a poor power factor.

For regenerative braking the regenerated power should have the same frequency as that of the main supply. This necessitates the energizing of the motor field winding from the ac mains. The regenerated current must be in phase opposition to the applied voltage and also the flux ϕ, so that power may be supplied back into the supply system.

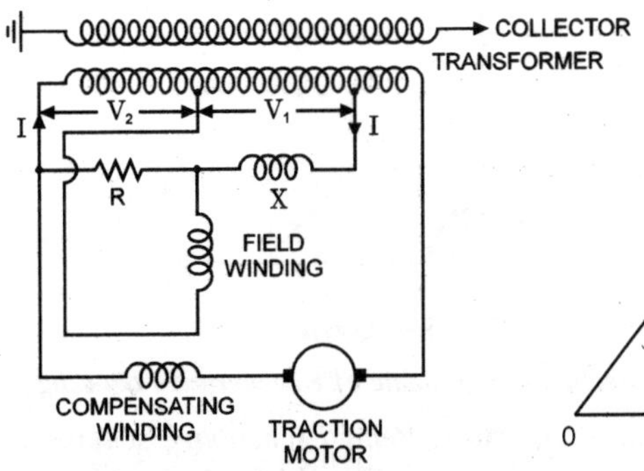

(a) Connection Diagram

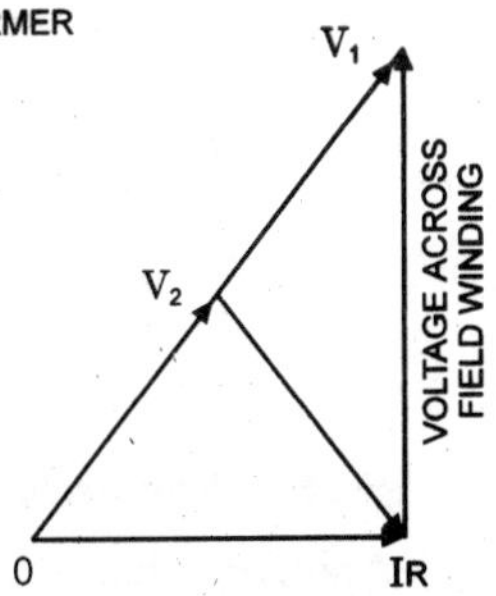

(b) Phasor Diagram

Fig. 14.4. *Regenerative Braking With Single Phase AC Series Motor*

The arrangement to provide the above conditions is illustrated in Fig. 14.4.

In another satisfactory arrangement one of the traction motors is used as a generator to supply current to excite the fields of the remaining motors. As the speed of the locomotive falls, the voltage of the motors can be controlled by increasing the excitation current. This is accomplished by increasing the voltage on the exciting motor by employing another transformer tap. Further flexibility can be obtained by changing the tap from the generating motor (Fig. 14.5).

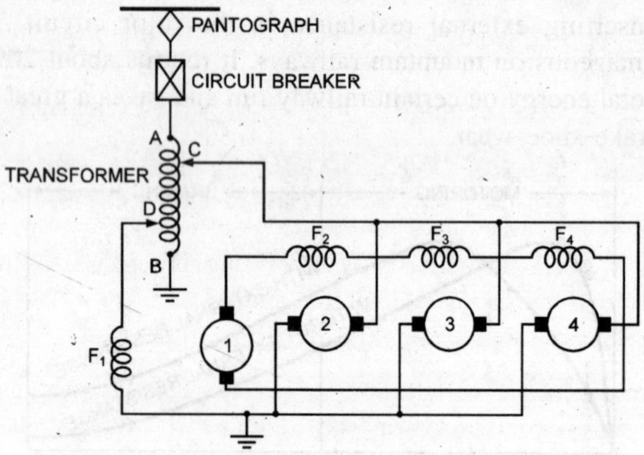

Fig. 14.5. ***Connection Diagram For Regeneration With Single Phase AC Series Motor***

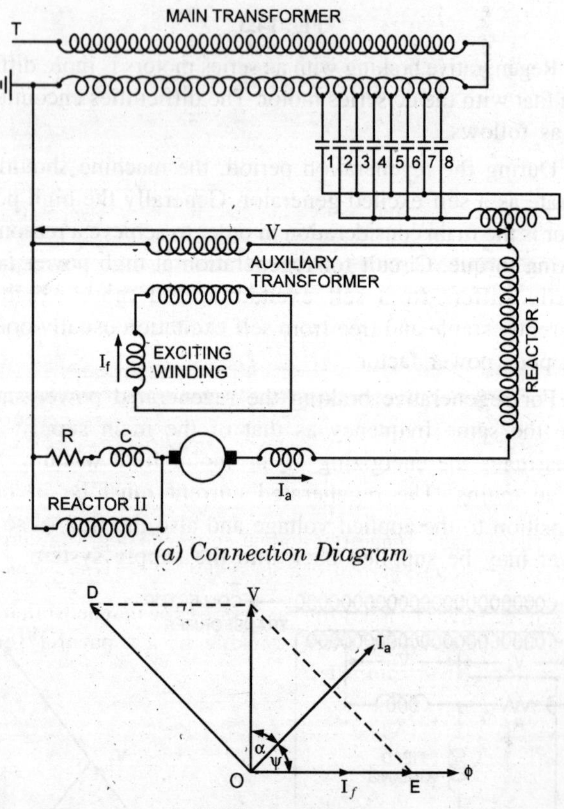

(a) Connection Diagram

(b) Phasor Diagram

Fig. 14.6. ***Behn Eschenburg Scheme of Regenerative Braking***

Another scheme known as *Behn Eschenburg scheme* of regenerative braking is shown in Fig. 14.6 (*a*). In this case an auxiliary transformer is used to excite the exciting winding of the traction motor. The armature of the traction motor is connected to the main transformer through tap changer. In between the motor armature and tap changer is inserted a choking coil or iron-cored reactor in series, as shown in Fig. 14.6 (*a*). Commutating pole winding C in series with a resistor R is shunted with another iron-cored reactor in order to obtain the correct (leading) phase of the commutating flux.

The current flowing in the exciting winding, I_f will lag behind the voltage V at the tap changer approximately by 90° and the emf generated in the armature of the motor, E will be in phase with the exciting current I_f. The voltage equal to phasor difference of voltage V and emf E will act across the reactor I. The armature current I_a lags behind the voltage acting across the reactor I (represented by phasor OD) by roughly 90°.

For a given voltage across the tap changer the braking torque produced will be proportional to $I_a \cos \psi$ and the power returned to the supply mains will be proportional to $I_a \cos \alpha$ if losses are neglected.

From the phasor diagram we conclude that, for constant excitation,

(*i*) The braking torque is approximately constant at all speeds.

(*ii*) Volt-amperes absorbed by reactor I are more than those generated by the armature.

(*iii*) The power factor of the generated power is about 0.7, which is considered as low power factor. But this is not considered a serious drawback as the arrangement provides simplicity of operation, reliability, stability and free from self excitation.

Calculations of Energy Returned During Regeneration. When the train is accelerated up to a certain speed, it acquires energy, known as *kinetic energy,* corresponding to that speed (KE = $\frac{1}{2}mv^2$). While coasting a part of this stored energy is utilised in propelling the train against frictional and other resistances to motion and, therefore, the speed falls. Under ideal conditions (no resistance to motion of the train) the speed of the train would have not decreased.

Similarly while the train going down the gradient or moving on level track, the speed remaining the same or reduced, this stored energy can be converted into electrical energy and returned back to the lines.

The amount of energy returned to the line depends upon (*i*) the initial and final speeds during regenerative braking (*ii*) the train resistance and gradient of the track also in case the train is moving down the gradient and (*iii*) efficiency of the system.

Let the initial and final speeds of the train be V_1 and V_2 kmph respectively. The kinetic energy stored in the train at a speed of V_1 kmph

$$= \frac{1}{2} \times \frac{1{,}000\,W_e}{9.81}\left(\frac{1{,}000\,V_1}{3{,}600}\right)^2 \text{ kg-m}$$

$$= \frac{1}{2} \times 1{,}000\,W_e \left(\frac{1{,}000\,V_1}{3{,}600}\right)^2 \text{ N-m or watt-seconds}$$

$$= \frac{1}{2} \times 1{,}000\,W_e \left(\frac{1{,}000\,V_1}{3{,}600}\right)^2 \times \frac{1}{3{,}600} \text{ watt-hours}$$

$$= 0.01072\,W_e\,V_1^2 \text{ watt-hours}$$

Similarly kinetic energy at speed

$V_2 = 0.01072\ W_e V_2^2$ watt-hours

Energy available during regeneration

$= 0.01072\ W_e (V_1^2 - V_2^2)$ watt-hours ...(14.1)

Some of the energy is lost to overcome the resistance to motion and the losses in the traction system including traction motors.

Energy lost to overcome the resistance to motion

$= \frac{W \times r \times S \times 1{,}000}{3{,}600}$ watt-hours $= 0.2778\ WrS$ Wh ...(14.2)

where r is the specific resistance in newton/tonne.

While going down the gradient in the hilly track service, energy is supplied as tractive effort due to the gradient and energy is added up to the energy available during regeneration. Energy available due to motion down the gradient.

$= \frac{98.1GW \times S \times 1{,}000}{3{,}600} = 27.25\ GSW$ ...(14.3)

Hence total energy available during regeneration

$= [0.01072W_e(V_1^2 - V_2^2) + 27.25\ GSW - 0.2778\ WrS]$ watt-hours ...(14.4)

Taking η as the efficiency of the system, Energy returned to the line

$= [0.01072W_e(V_1^2 - V_2^2) + 27.25\ GSW - 0.2778WrS] \times \eta$ watt-hours ...(14.5)

Advantages and Disadvantages of Regenerative Braking

Advantages. 1. A part of energy is returned to the supply system, so energy consumption for the run is considerably (about 20 to 30 per cent) reduced thereby affecting a considerable saving in the operating cost.

2. The wear of the brake shoes and wheel tyres is reduced to considerable extent, therefore, their life is increased and replacement cost is reduced.
3. Higher value of braking retardation is obtained so that the vehicle can be brought to rest quickly and running time is considerably reduced.
4. Small amount of brake dust is produced when the mechanical brakes are applied.
5. Higher speeds are possible while going down the gradients because the high braking retardation can be obtained with regenerative braking.
6. Propulsion of heavier trains on gradients is possible without dividing them into sections with speed and safety.

Disadvantages. 1. Additional equipment is required for control of regeneration and for protection of equipment and machines, hence initial as well as maintenance cost is increased.

2. The dc machines required in case of regenerative braking are of large size and cost more than those ordinarily employed, therefore, the weight of the locomotive and thus the required mechanical strength and cost increase.
3. Owing to recuperated energy the operation of the substations becomes complicated and difficult.
4. In case of substations employing mercury-arc rectifiers for conversion purpose, additional equipment is required either to deal with regenerated energy separately or to change one or more of the ordinary rectifiers over to inverted operation. No such difficulty is experienced in case of substations employing rotary convertors or motor-generator sets for converting purpose.
5. Regenerative braking is employed down to a speed of 16 kmph, then rheostatic braking to about 6.5 kmph and then mechanical braking is required to bring the locomotive to rest.

In most of the cases, however, and especially with motor-coach trains the increased cost of the train equipment and the additional features required in order to obtain regenerative braking, combined with the increase in the maintenance cost of the electrical equipment, may entirely offset the economics in the energy consumption and the other items.

For tramways, trolley buses, the regenerative braking is not recommended as it will unnecessarily increase the initial cost as well as increase the operating problems. Generally regenerative braking is desirable and necessary for the service lines having long gradients exceeding 0.6%.

Example 14.1. The following figures refer to the speed-current and torque-current characteristics of a 600 V dc series traction motor.

Current, amperes	**50**	**100**	**150**	**200**	**250**
Speed, kmph	**73.6**	**48**	**41.1**	**37.3**	**35.2**
Torque, N-m	**150**	**525**	**930**	**1,335**	**1,750**

Determine the braking torque at a speed of 48 kmph when operating as self-excited dc generator. Assume resistance of motor and braking rheostat to be 0.6 Ω and 3.0 Ω respectively.

Solution : **As Motor.** Terminal voltage, V = 600 volts

The motor current at a speed of 48 kmph (from speed-current characteristic curve), I = 100 A

Back emf developed by the motor,

$E_b = V - I R_m = 600 - 100 \times 0.6 = 540$ V

As Generator**.** At the instant of applying rheostatic braking at a speed of 48 kmph the terminal voltage of the machine will be equal to the emf developed by the machine *i.e.,* 540 V.

Total resistance in the circuit $= R_m + R_{rheostat} = 0.6 + 3 = 3.6\ \Omega$

Current delivered by the machine, $I = \frac{540}{3.6} = 150$ A

The braking torque (the torque corresponding to a current of 150 A from torque-current characteristic) is 930 N-m **Ans.**

Example 14.2. The following figures give the magnetisation curve of a dc series motor when working as a separately excited generator at 640 rpm :

Field current, amperes	**20**	**40**	**60**	**80**
EMF volts	**215**	**381**	**485**	**550**

The total resistance of the motor is 0.8 Ω. When connected for rheostatic braking against a load giving a torque of 40 kg-m it is desired to limit the speed to 600 rpm. What resistance must be connected across the motor during braking ?

Solution : Magnetisation characteristic corresponding to 600 rpm will be as follows :

Field current in amperes	20	40	60	80
EMF in volts	202	357	455	516

Load torque, T = 40 kg-m = 40 × 9.81 = 392.4 N-m

$$\text{Input to load} = T \times \frac{2\pi N}{60} = 392.4 \times \frac{2\pi \times 600}{60} = 24{,}650 \text{ W}$$

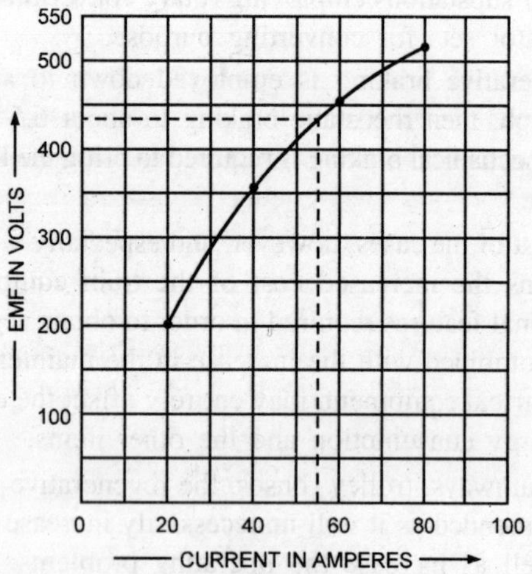

Fig. 14.7

During rheostatic braking the machine operates as a series generator. Assuming no mechanical losses, the generated power will be equal to the input to the load *i.e.*, 24,650 watts.

From curve shown in Fig. 14.7 we find that for the excitation current of 56 A, the induced voltage is about 440 V, which provides generated power of 24,650 watts.

$$\text{Total resistance of the circuit} = \frac{440}{56} = 7.857\ \Omega$$

External resistance to be connected across the motor during braking = 7.857 – 0.8 = 7.057 Ω **Ans.**

Example 14.3. A train weighing 400 tonnes has speed reduced by regenerative braking from 40 to 20 kmph over a distance of 2 km on a down gradient of 2 %. Calculate the electrical energy and average power returned to the line. Tractive resistance is 40 newtons/tonne and allow rotational inertia of 10% and efficiency of conversion 75%.

[*Agra Univ. Traction and Utilization of Elec. Power 1976*]

Solution : Accelerating weight of train,

W_e = 1.1 W = 1.1 × 400 = 440 tonnes

Distance travelled during period, S = 2 km

Gradient, G = 2%

Efficiency of conversion, η = 75% = 0.75

Energy available due to reduction in speed

$= 0.01072\, W_e (V_1^2 - V_2^2)$

$= 0.01072 \times 440 \times (40^2 - 20^2)$

= 5,660 watt-hours = 5.66 kWh

Tractive effort required while going down the gradient

$F_t = Wr - 98.1\, WG$

= 400 × 40 – 98.1 × 400 × 2 = – 62,480 newtons

i.e., tractive effort of 62,480 newtons is available.

Hence energy available on account of moving down the gradient a distance of 2 km

$$= \frac{F_t \times S \times 1{,}000}{1{,}000 \times 3{,}600} \text{ kWh} = \frac{62{,}480 \times 2 \times 1{,}000}{1{,}000 \times 3{,}600} = 34.7 \text{ kWh}$$

Total energy available = 5.66 + 34.7 = 40.37 kWh

Energy returned to the line = 0.75 × 40.37 = 30.28 kWh **Ans.**

$$\text{Average speed} = \frac{40 + 20}{2} = 30 \text{ kmph}$$

$$\text{Average time taken to cover 2 km} = \frac{2}{30} = \frac{1}{15} \text{ hours}$$

$$\text{Average power, } P = \frac{\text{Energy returned to the line}}{\text{Time}} = \frac{30.28}{1/15}$$

= 454 kW **Ans.**

Example 14.4. A 2,340 tonne train (including loco) proceeds down a gradient of 1 in 80 for 5 minutes, during which period, its speed gets reduced form 60 kmph to 36 kmph by application of regenerative braking. Find the energy returned to the lines, if the tractive resistance is 5 kg/tonne, rotational inertia 10% and the overall efficiency of the motors during regeneration as 70%.

Solution : Accelerating weight of train,

W_e = 1.1 × W = 1.1 × 2,340 = 2,574 tonnes

Track resistance, *r* = 5 × 9.81 N per tonne

$$\text{Gradient, } G = \frac{100}{80} = 1.25\%$$

Regenerative period, *t* = 5 minutes = 5 × 60 = 300 seconds

Overall efficiency, η = 0.7

Energy available due to reduction in speed

$= 0.01072\, W_e (V_1^2 - V_2^2)$

$= 0.01072 \times 2{,}574 \times (60^2 - 36^2)$ = 63,575 Wh or 63.575 kWh

Tractive effort required while going down the gradient

= W*r* – 98.1 WG

= 2,340 × 5 × 9.81 – 98.1 × 2,340 × 1.25 = – 172,165.5

i.e., tractive effort of 172,165.5 newtons is available.

Distance moved during regeneration

$$= \frac{V_1 + V_2}{2} \times \frac{1{,}000}{3{,}600} \times t = \frac{60 + 36}{2} \times \frac{1{,}000}{3{,}600} \times 300$$

= 4,000 metres

Energy available on account of moving down the gradient over distance of 4,000 metres.

$$= \frac{172{,}165.5 \times 4{,}000}{1{,}000 \times 3{,}600} = 191.3 \text{ kWh}$$

Total energy available = 63.575 + 191.3 = 254.875 kWh

Energy returned to the line = 0.7 × 254.875 = 178.4 kWh **Ans.**

Example 14.5. A train weighing 500 tonnes is going down a gradient of 20 in 1,000. It is desired to maintain train speed at 40 kmph by regenerative braking. Calculate the power fed into the line. Tractive resistance is 40 N/tonne and allow rotational inertia of 10% and efficiency of conversion of 75%.

Solution : Dead weight of train, W = 500 tonnes

Train speed, V = 40 kmph

Tractive resistance, *r* = 40 N/tonne

$$\text{Gradient, } G = \frac{20 \times 100}{1{,}000} = 2 \text{ per cent}$$

Efficiency of conversion, η = 75% = 0.75

Tractive effort required while going down the gradient

= W*r* – 98.1 WG = 500 × 40 – 98.1 × 500 × 2

= –78,100 newtons

i.e., tractive effort of 78,100 newtons is available

$$\text{Power available, } P = \frac{F_t \times V}{3{,}600} \text{ kW} = \frac{78{,}100 \times 40}{3{,}600} = 867.8 \text{ kW}$$

Power fed into the line = Power available × efficiency of conversion

= 867.8 × 0.75 = 651 kW **Ans.**

Example 14.6. The emf generated between brushes of a single phase series traction motor is 500 V and brushes are connected through a 0.5 Ω reactor to a 400 V tapping on the supply transformer. Neglecting losses, calculate the power generated.

[*Pb. Univ. Electric Traction June 1993*]

Solution : Voltage across the reactor

= Voltage difference of V and E

= Voltage represented by phasor OD = 640 V

$$\text{Armature current, } I_a = \frac{\text{Voltage across reactor}}{\text{Ohmic value of reactor}} = \frac{640}{0.5} = 1{,}280 \text{ A}$$

Phase angle between V and I_a,

$\alpha = 38.66°$

Power generated, $P = VI_a \cos \alpha$

$= 400 \times 1{,}280 \cos 38.66°$

$= 399{,}804$ watts $= 399.804$ kW **Ans.**

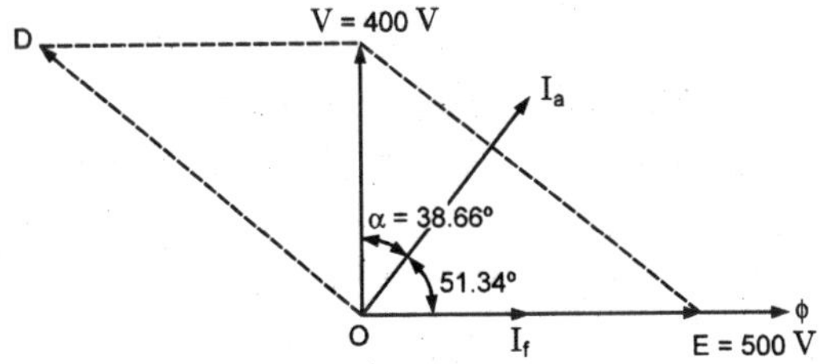

Fig. 14.8

14.2. MECHANICAL REGENERATIVE BRAKING

When the train is accelerated up to a certain speed it acquires energy, known as *kinetic energy,* corresponding to that speed ($KE = \frac{1}{2}mv^2$). While coasting a part of this stored energy is utilised in propelling the train against frictional and other resistances to motion and, therefore, the speed falls. Thus coasting may be considered as a form of *mechanical regenerative braking*. The saving in energy consumption can be considerably affected by increasing the coasting but it will reduce the schedule speed. If, however, the original schedule speed is to be maintained, an increase in the acceleration is required which is possible with large motors. Thus the saving in energy consumption may be neutralised by the increased train weight and the additional cost of the equipment.

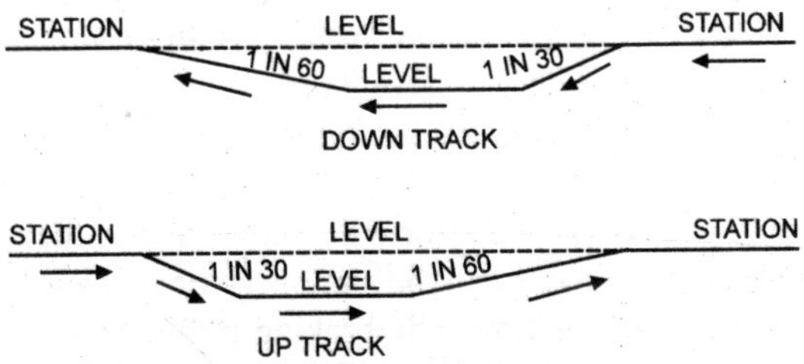

Fig. 14.9. ***Graded Track Construction***

With modern city and suburban traffic conditions, the coasting period is about 20 to 50 per cent of the total running period and consequently a large percentage of kinetic energy has to be wasted in brakes. One of the way of tackling this problem based on the principle of mechanical regenerative braking is adoption of graded track systems, as shown in Fig. 14.9. It facilitates the saving of energy consumption to about 25%.

In the graded track system the stations are situated at a certain height above level track, slope of track while leaving station is 1 in 30 and while approaching station is 1 in 60. In this system the kinetic energy of the train instead of being wasted in brakes is utilised and stored as potential energy as the train has to climb up a slope of 1 in 60 while approaching the station. This potential energy is utilised in moving the train down gradient and imparting acceleration to it. By using this system small motors can be used for driving the train and specific energy consumption can be reduced to 75 per cent of that when the level track was employed. This system can be employed only in underground railways because on surface railway difficulties will be experienced in the construction of the required track.

14.3. MECHANICAL BRAKING

For stopping the train and holding it stationary, in addition to electrical brakes, mechanical brakes are also required. The modern tendency is to employ regenerative braking down to a speed of 16 km per hour, then rheostatic braking down to about 6.5 km per hour, and finally mechanical braking to a standstill. Mechanical brakes are usually applied by brake block shoes pressed with force against the tread of the wheels. Braking force is caused by the movement of the piston which is transmitted to the brake blocks through a system of levers.

Mechanical brakes are of two types namely the *compressed air brakes* and *vacuum brakes*, the former is extensively used in electric railways whereas the latter is used in steam railways. The compressed air brakes are little advantageous to the vacuum brakes as compressed air can conveniently be stored up and released for quick action whereas in the vacuum brakes the necessary vacuum is to be created by the pump. However, this drawback is overcome by use of vacuum reservoir.

1. *Compressed Air Brakes.* The compressed air brake system consists of a reservoir of compressed air, a brake cylinder, compressed air pipe, a valve, springs, piston and a piston rod, as shown in Fig. 14.10. The piston is connected

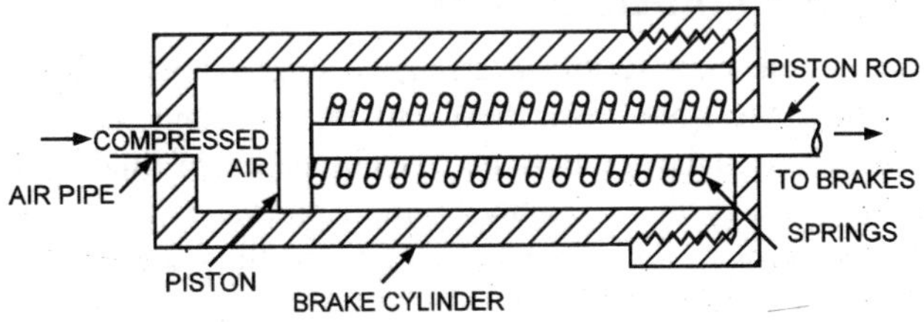

Fig. 14.10. ***Action of Compressed Air Brakes***

to brake shoe through piston rod and levers. The brakes are kept in the 'off' position by means of springs provided for this purpose. While applying brakes, compressed air is allowed to enter into the brake cylinder through the air pipe and valve, which presses the piston against the force of springs. The force with which the brakes are applied depends upon the quantity of compressed air allowed to enter into the cylinder. A valve placed at the inlet of brake cylinder controls the flow of air into it. The brakes are released by exhausting the air. The piston alongwith the brakes return to its original position under the influence of spring force on exhausting the air. The compressed air is used at about 5.5 kg/cm^2 pressure. The compressed air brakes are extensively employed on electric railways.

2. ***Vacuum Brakes.*** The vacuum brake system consists of a vertical cyclinder with a piston inside, the piston is an easy fit in the cylinder and is provided with rolling rubber ring. The piston is connected to the piston rod which operates the braking arrangement through a system of levers. Vacuum is always maintained on the top and the under side of the piston so that in normal conditions the piston rests at the bottom of the cylinder.

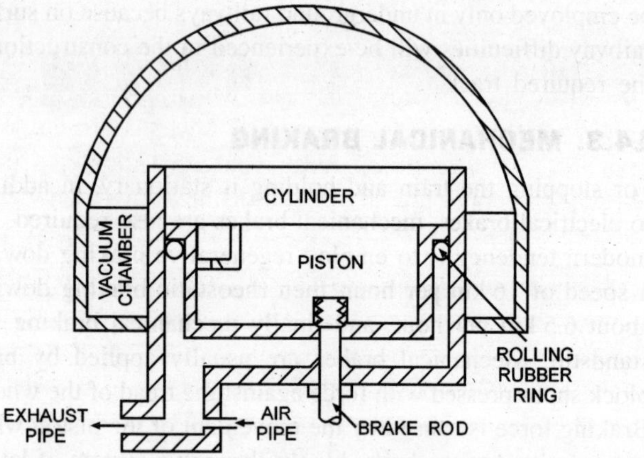

Fig. 14.11. ***Vacuum Brakes***

While applying brakes air at atmospheric pressure is gradually admitted beneath the piston through air pipe and valve the piston moves up and applies the brakes. The force with which the brakes are applied depends upon the rapidity with which the vacuum is destroyed.

The brakes are released by recreating the vacuum. The vacuum is recreated by drawing air from lower side of the piston through the pipe. The piston falls to bottom of the cylinder on account of its own weight.

Vacuum brakes are applied with steam locomotives.

14.4. HYDRAULIC BRAKING

This system consists of an assembly similar to a hydraulic coupling, the rotor being keyed to the axle and the stator being keyed to the bogie frame. The circulation of fluid (water or oil) through the assembly by the rotor vanes produces a braking torque. The brake is applied by filling the coupling by means of a pump.

14.5. MAGNETIC TRACK BRAKES

It comprises a bipolar electromagnet with elongated pole faces a short distance apart (about 6 mm) and along with track. The body of the electromagnet is made of cast steel, the pole faces are made of soft steel and can be easily replaced and the winding is enclosed in a water tight metal case.When the line voltage is impressed on these coils, the poles are strongly attracted to the rails and produce a retarding effect due partly to the magnetic pull (which increases the effective weight of the coils) and partly to mechanical drag.

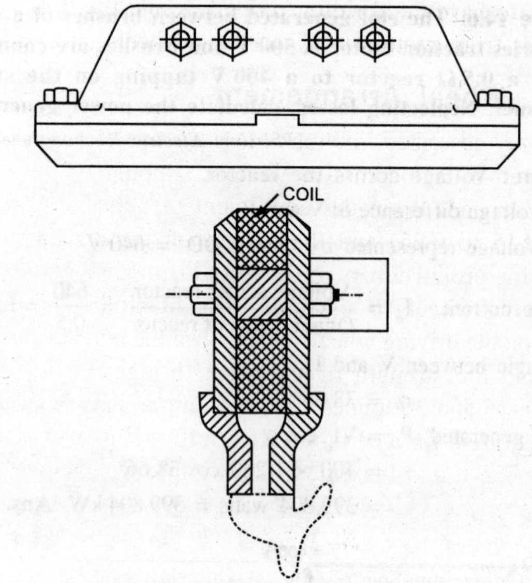

Fig. 14.12. ***Magnetic Track Brake***

The flux created due to flow of current is perpendicular to the pole faces and track and force of attraction between the track and magnet is given as

$$F = \frac{B^2 A}{2 \times 4\pi \times 10^{-7}} \text{ newtons} \quad ...(14.6)$$

where B is flux density in Wb/m^2 and A is the area of the pole face in square metres. The braking force or the drag = μF where μ is the coefficient of friction. Magnetic brakes are usually employed on trolley buses and cars.

14.6. ELECTROMECHANICAL DRUM BRAKES

In this type the brake drum is mounted on the motor shaft. The brake shoe is pressed against it by springs. The brakes are released by means of a solenoid energised from a battery. This type of brake is usually employed on trolley buses and cars.

14.7. EDDY CURRENT BRAKES

Eddy current brakes are of two types viz linear type and rotary type. In the *linear type brakes*, shoe carrying the excitation current acts as a primary and the rail itself forms the secondary element of the system. During braking, traction motors are made to operate as generators and supply power to the shoe of the eddy current brakes. In the *rotary system*, the rotor, which acts as a secondary member is attached to the motor shaft and the dc excitation is supplied to the primary member which is fixed in space. The torque developed due to induced currents in the rotor will tend to reduce the relative speed between the rotor and the stationary field. Thus braking effect will come in action. The braking torque can be varied by the varying the dc excitation of the primary member (stator).

The main advantage of eddy current braking is the same as that of the disc brake and is used satisfactorily in conjunction with the friction brakes for bringing the vehicle to rest.

14.8. MECHANICAL CONSIDERATIONS

The main considerations in connection with the locomotive are the wheel arrangement and general disposition of equipment in order to have good riding properties and the transmission of power from motors to the driving wheels.

14.8.1. Wheel Arrangement

As already described, the maximum tractive effort which can be exerted by a locomotive without slipping of the driving wheels depends upon the coefficient of adhesion (which is about 0.15 to 0.25) and the weight on the driving wheels. Hence for propulsion of train, certain maximum weight on the driving wheels is essential. The maximum permissible weight on the driving wheels depends on the number of driving axles and the strength of the track, bridges etc. (which is between 15 and 30 tonnes). The minimum number of driving axles required, for a certain tractive effort F, given coefficient of adhesion μ and maximum permissible weight per driving axle.

$$= \frac{F}{\mu \times \text{maximum permissible weight per driving axle}} \quad ...(14.7)$$

For example, if the maximum tractive effort required is 30,000 kg, coefficient of adhesion is 0.2 and maximum allowable weight per driving axle is 19 tonnes, then number of driving axles required $= \frac{30{,}000}{0.2 \times 1{,}000 \times 19} = 8.$

Such a train will be hauld by two locomotives each having 4 driving axles making 8 in all.

In case of dc locomotives since the electrical equipment is lighter in weight, so whole of the locomotive weight is placed on the driving axles so that required tractive effort may be obtained.

In case of ac locomotives, since the electrical equipment is comparatively heavy, so the required tractive effort can be obtained by placing only a portion of the weight on the driving wheels and supporting the rest on trailing axles. The riding qualities of high speed locomotives are improved by the trailing axles, which are often used on dc locomotives for this purpose.

14.8.2. Drive

There are two methods of driving the electric train. These are individual drive and collective drive. In individual drive separate motor of smaller size is employed for each driving axle whereas in collective drive a single large sized motor is employed for driving all the axles through connecting rods as in steam locomotives. The collective drive has the following advantages and disadvantages over individual drive.

Advantages. 1. The wheels when coupled have got less tendency to slip.

2. The motor can be mounted well up in main body of the locomotive, therefore, the centre of gravity is raised thereby reducing the wear on track.
3. The motor for collective drive need not be paid so much attention in designing with regard to dimensions as much that for individual drive.

Disadvantages. Collective drive is not suitable for high speed locomotives because of stresses and vibrations in the connecting rods at high speeds.

From the above it can be concluded that individual drive is advantageous only for high-speed locomotives whereas for low-speed locomotives, which are usually employed for passenger and goods trains, collective drive is preferred.

14.8.3. Transmission of Drive

The electric locomotives are specifically designed to have springs between the driving axles and the main body and it is desirable to have as much of locomotive weight spring-borne as possible. This reduces not only vibrations in the locomotive but also reduces substantially the damage to the track owing to hammer blows. In case the motors are fixed on the locomotive or bogie frame in order to have maximum locomotive weight spring-borne, there will be a relative motion between the motor armature and the driving axle owing to compression and extension of the springs. Hence flexible drive between the two is required, which is one of the most difficult problems of the electric locomotive*. There are several methods of transmitting power from the armatures of the electric motors to the driving wheels and only a few will be described here.

1. *Gearless Drive.* (*a*) *Direct Drive.* The simplest way is to employ bipolar motors, whose armatures are mounted directly on the driving axles with the field attached to the frame of the locomotive, as shown in Fig. 14.13. The pole pieces of the motors are approximately flat so that the armature can move relatively to them without affecting the operation. The size of motor armature is limited by the diameter of driving wheels. In case driving wheels of large diameters are employed, the speed of the motor will have to be limited to lower one. Because of these facts and relatively large unsprung weight, poor utilization of material and the low centre of gravity of the locomotives the applications of this method are limited.

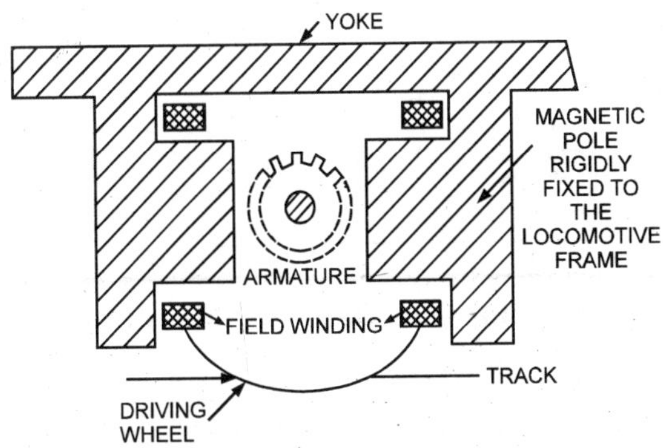

Fig. 14.13. *Direct Drive*

(b) Direct Quill Drive. The quill is a hollow shaft which surrounds the driving axle and is connected to it by springs or some other flexible device, as shown in Fig. 14.14. The armature of the motor is the directly mounted on the quill and, therefore, it is entirely spring-borne. Thus the unsprung weight is reduced to a minimum and motor of normal design can be employed. The speed and size of armature are still limited by the diameter of the driving wheels.

* The problem does not arise in the steam locomotive, as the piston (attached to the driving axle) is free to move in the cylinder (attached to the frame).

One class of devices for obtaining a flexible connection between the quill flange and the driving wheel uses springs; other manufacturers avoid the use of springs and possible troubles owing to breakages by employing a link motion.

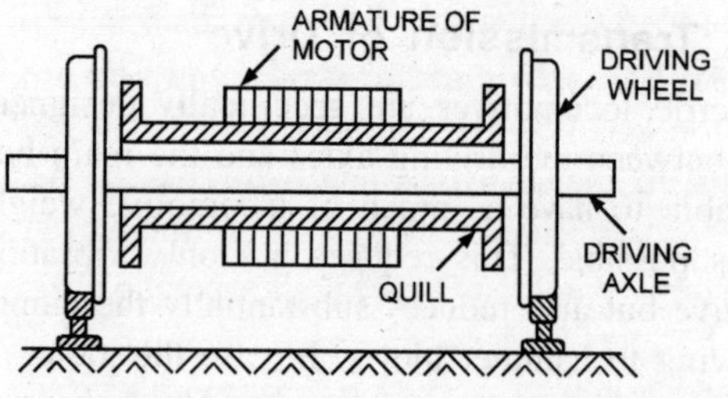

Fig. 14.14. *Direct Quill Drive*

2. *Geared Drive.* The size of the motor is reduced with the increase in speed for a given output, so the gear drive is necessary in order to reduce the size of the motor for given output by running it at a higher speed. The gearing ratio usually employed is 3 to 5 : 1.

(a) Nose-Suspension Geared Drive. The most common type of geared drive is nose suspended motor shown in Fig. 14.15. In nose suspension the motor is partly supported from the driving axles and partly between springs which are put on the frame of the vehicle. More than 50 per cent of the total weight of the motor and gear is thus spring-borne. The motor is coupled to the driving axle by means of gear, one being on the same axle and the other on the shaft of the motor armature. Because of axle support the space available is short and limits the size of the motor to about 337.5 kW for normal gauge vehicles. The method is usually employed for motor coaches and tramways. Because of the relatively large unsprung weight this system is not used on large locomotives. This limitation is also imposed by the available space which will not accommodate motor of larger size.

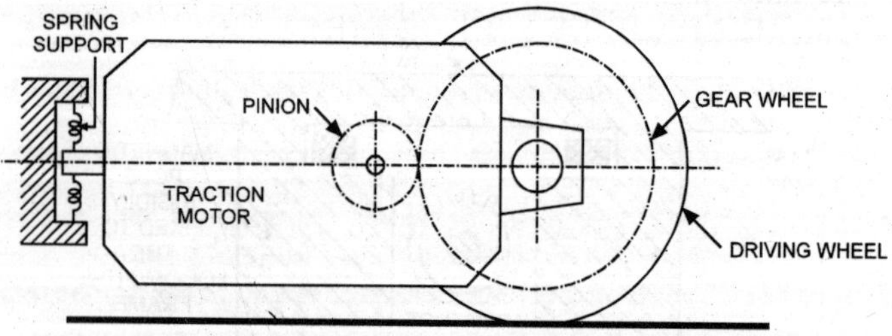

Fig. 14.15. *Nose-Suspension Geared Drive*

(b) Geared Quill Drive. Geared quill drive is employed when the motor is to be placed higher in the frame than is possible with the nose suspension so that a motor of larger size (output) can be employed, centre of gravity can be raised and when unsprung weight is to be reduced. In this case quill surrounds the driving axle, as in direct quill drive and gear is mounted on the quill instead of motor armature.

One of the methods already discussed can be used for providing a flexible connection between the quill and the axle.

3. *Brown Boveri Individual Drive.* In this drive, flexibility is achieved by providing a special link arrangement between the gearwheel and the driving wheel, as illustrated in Fig. 14.16. This arrangement can be employed with considerable success on both high- and low-speed locomotives.

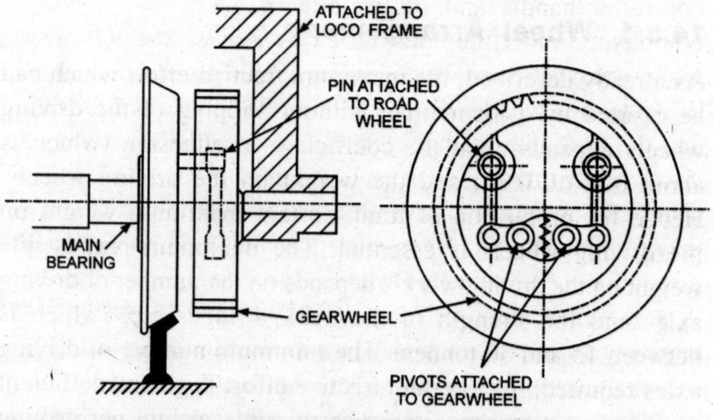

Fig. 14.16. *Brown Boveri Individual Drive*

4. *Transmission Arrangement in Collective Drive.*

(a) Collective Drive With Connecting rod. Common method of collective drive is to drive with coupled wheels by means of a connecting rod attached to a crank mounted on the motor shaft or on a "jack" shaft geared to the motor shaft. The connecting rod is fixed in such a way as to remain nearly horizontal as in the steam locomotive. Such an arrangement eliminates the chances of any relative motion between the motor shaft and the driving wheels. Such an arrangement is illustrated in Fig. 14.17 (*a*).

(a) Collective Drive With Connecting Rod

(b) Collective Drive With Scotch Yoke

Fig. 14.17. *Transmission Arrangement in Collective Drive*

(b) Collective Drive With Scotch Yoke. In this drive also, known as Scotch Yoke, or Kando type of drive, a triangular frame work having two sides projecting from the apex is used. The apex of the triangular framework is attached to the coupling rod and the other two corners are attached to the cranks mounted on two motor shafts. Vertical play is allowed in the bearings to secure the required flexibility between motor shaft and driving wheels. Such an arrangement is shown in Fig. 14.17 (*b*).

14.9. CONTROL EQUIPMENT

As already mentioned in Art. 13.1, the conrol equipment is required for (*i*) connecting the motor or motors to the supply mains without taking excessive current at start and provide smooth acceleration without causing sudden shock so as to avoid damage to the coupling (*ii*) adjusting the speed as per requirement, and (*iii*) providing rheostatic or regenerative braking.

The operation may be *by hand* in which the closing of the accelerating contactors is governed by the position of the controller handle and so the rate of acceleration remains completely under the control of driver or *semi-automatic* in which the position of the controller handle governs the number of contactors which will close, but the sequence of operation and the time interval between their individual operations are governed by time-lag devices, as in fully automatic equipment. The hand-operation is preferred in locomotives, where there is a probability of divergences in the hauling loads and possibility of slipping of wheels. Semi-automatic operation is sometimes employed for suburban services in order to ensure uniform acceleration with a minimum energy consumption.

Plain drum controllers of the type shown in Fig. 13.11 are employed by the trams and small industrial locomotives but in railway work, where the currents to be dealt are too high for this type of equipment, *master controller system* is employed. In master controller system the contact instead of being made or broken by hand are closed or opened by the switches known as *contactors* operated by the solenoids.

The master conroller is usually equipped with '*dead man's handle*' the function of which is to stop the train in case the driver faints or becomes incapacitated. This consists of a contact attached to the knob of the controller handle. The driver is to hold it always while the train is running. As soon as the dead man's handle is released, circuit of the operating solenoid will get opened. This will lead to the opening of the motor circuit and bringing the train to a stop by automatic application of brakes.

There are various methods for closing and opening of the contactors by the master controller and the methods, usually employed are:

(*i*) All electric method (*ii*) Electropneumatic method and (*iii*) Cam shaft method.

1. *All Electric Method.* In this method the contactors are operated (closed or opened) by the solenoids energised from the master controller. The main disadvantages of this method is that the wide fluctuations in the supply voltage, which are liable to occur, may cause uncertainty of action of the contactors. The difficulty however, can be overcome by employing a special low-voltage generator to provide a supply at 100 V or lower for the operating circuits.

2. *Electropneumatic Method.* In this method, contactors are operated by the compressed air. The valves, which admit the air into the operating cylinders, are operated electrically by the solenoids energised from the master controller. This system is preferred to all electric operation system because

1. The contactor groups are more compact and robust.
2. Each valve magnet requires only a few watts for operation whereas the solenoids in all electric operation system may require from 100 to 200 watts.
3. A low voltage control supply can be employed, which simplifies the interlocking and auxiliary contacts on the contactors and their circuits, reduces the size of the master controller and gives reliability to the whole of the control apparatus and circuits.

As electric trains are almost universally equipped with compressed air brakes, the supply of air for electropneumatic operation is, therefore, always available.

3. *Cam Shaft Drive.* In this method the contactors are closed or opened directly by cams, which are mounted on a cam shaft driven by an electric or pneumatic motor. The electrical connections are very simple, therefore, this method is widely used. The position of the master controller handle governs the angle through which the cam shaft is turned and therefore, fixes the number of contactors, which are closed.

This method has also the advantages of (*i*) less wear and tear (*ii*) light in weight (*iii*) easy maintenance and (*iv*) occupying less space.

14.10. MULTIPLE UNIT CONTROL

For city and suburban services motor coach units are usually employed. Each motor coach unit consists of one motor coach and two or three trailing coaches for passengers. Motor coach carries driving electric motors and other controlling equipment. The number of motor coaches employed at different hours of the day are different, which depends upon the traffic to be handled at the various hours. In peak traffic hours number of motor coach units are combined to clear of the traffic. The units are so combined that they can easily be split

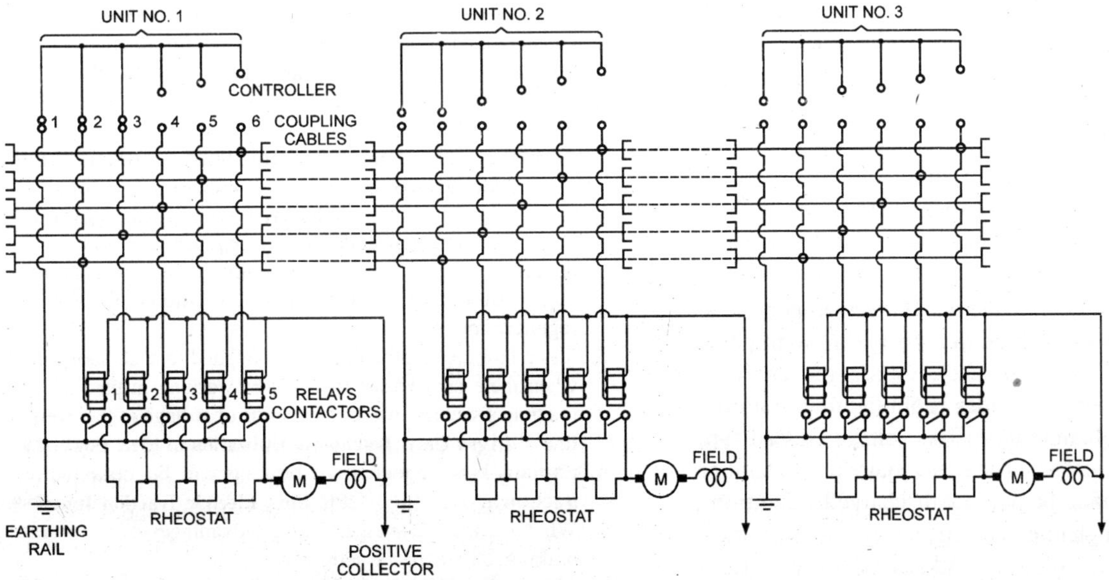

Fig. 14.18. ***Multiple Unit Control***

up and run individually during the light traffic hours. The major coach unit (having number of motor coach units combined) thus formed must be capable of control from a single point. This is achieved in the *multiple unit control.*

In multiple unit control method each motor coach unit consists of two or more traction motors provided with a master controller, series-parallel controller (with rheostat and reverser), control-circuit multi-core cables (control bus line) with coupler sockets, acclerating relay; also a motor-generator set, battery or tapped line connected resistor (potentiometer) if the control circuit is to be supplied at low voltage. The trailer coach comprises master controller (when required), control-circuit multi-core cables with coupler sockets. Each motor coach is also equipped with overcurrent relay with driver's control switch, isolating switch for master controller, cut-out switch and fuses for isolating the control circuit of the motor controller from the control bus line.

All these individual coach units are put in parallel and are controlled from a single point by means of a master controller. Control circuits of various coach units are put in parallel by means of multi-core cables known as coupling cables. There being more than one master controller, it becomes necessary to incorporate interlocks so that only one master controller can be operated at a time.

Multiple unit control for three motor coach units is shown in Fig. 14.18. The controller of any one unit can be employed for master control over the whole train. If control of unit no. 1 is operated to start the motor, it is seen from Fig. 14.18 that relay 1 of all units are energised, whole of the starting resistance is placed in series with the motors and all the motors start simultaneously. In next step some of the starting resistance will be taken out of the circuits of all the motors by controller of unit no. 1. In the same way other operations will be carried out by controller of unit no. 1 which will be same and simultaneous for each unit.

14.11. AUXILIARY EQUIPMENT

Auxiliary equipment, in addition to main traction motors, required to be installed on the electric locomotives are:

1. *Motor-Generator Set.* The motor-generator set is of the two bearing type and consists of a series motor and a shunt generator. The speed at light load is limited by the ventilating fan and the voltage of the generator is controlled by an automatic regulator. It is to supply lighting, control system and other power circuits at low voltage from 30 V to 100 volts. The set must be of special design so that fluctuation of line voltage do not have any effect on the low voltage supply.

2. *Battery.* Provision of batteries is very important as some source of energy is required to raise the pantograph, to operate the air blast circuit breaker, to run the auxiliary compressor. Arno-converter provides cabin lighting. Batteries may be lead type or alkaline type. The voltage of lead type battery is 2 volts/cell. Lead acid batteries require more attention and maintenance, cannot be left in discharged conditions, and cannot withstand high rate of charge and discharge without reducing their life. The voltage of alkaline bettery is 1.2 V per cell. Alkaline cells can be left in discharged conditions and can withstand the short-circuit conditions even without any effect on their life. The battery is either charged by a separate rectifier or from a dc generator. The capacity of battery depends on the vehicle and may be between 175 and 375 AH at the 5 hours. The batteries are usually connected in parallel with the motor-generator set.

3. *Air Compressor.* The compressor is usually of two cylinder type and is directly connected to the motor. With compressors for motor coaches, a moderate speed motor is employed so that the weight is reduced and the compressor is driven through double-helical spur gearing.

Automatic starting and stopping of the motor to maintain the air pressure within desired limits is achieved by employing a pressure operated switch, the moving contact of which is actuated, through a system of levers and a toggle mechanism by the air pressure on a spring loaded diaphram. It is for supplying compressed air for operating brakes, pantograph operating gear and electropneumatic control gear. At 1,500 V or below the compressor motor can be supplied directly from the line, otherwise, it is usually fed from the auxiliary generator.

4. *Heating Equipment.* It is required for heating the locomotive and train if desired. This is supplied directly from the line up to 3,000 V. Where an electric locomotive has to haul steam-heated rolling stock an electrically heated boiler is often provided.

5. *Pantograph Operating Equipment.* Raising and lowering of pantograph is usually carried out by compressed air and suitable switches for controlling the solenoid-operated valves of the air supply are required to be provided.

6. *Traction Motor Blowers.* These are required on locomotives when forced ventilated motors are used. In some cases a single blower is used which delivers air at a low pressure into a central duct built into the under frame of the locomotive body from which it is distributed to the motors. In other cases with large frame mounted motors, each motor is provided with a separate blower. The blowers are then mounted on the motor frames and two or more blowers are coupled together and are driven by a single motor.

Exercises

1. Discuss a method of electric braking for traction motors. **[Rajasthan Technical Univ. Utilization of Electrical Power 2011]**
2. Describe how plugging and rheostatic braking are employed with dc motors.
3. With the help of neat diagram, briefly explain 'French method of regenerative braking' with series motors. **[Pb. Univ. Elec. Traction April 1975]**
4. State the condition under which regenerative braking with dc series motors is possible. With the aid of diagram of connection, explain the various methods of providing regeneration.
5. Show how regenrative braking is applied to a dc series motor when there are at least two motors working in parallel. How to compensate the effect of variation of voltage in the overhead line ? **[Agra Univ. Traction & Utilization of Elec. Power 1978]**
6. Explain how regenerative braking can be obtained with traction motors. **[Pb. Univ. Electric Traction June 1993]**
7. What are the advantages and disadvantages of regenerative braking of electric traction motor ? **[M.D. Univ. Utilization of Electric Power & Traction May 2010]**

8. Explain briefly how regenerative braking differs from other methods of braking traction motors and also write the usefulness of this method of braking with (*a*) dc series motors and (*b*) ac series motors in traction. **[M.D. Univ. Utilization of Electric Power & Traction Dec. 2009]**
9. What are the conditions necessary to achieve electric regenerative braking ?
   Explain the suitability of (*a*) dc series motor and (*b*) three-phase induction motor for the purpose of electric regenerative braking. **[Pb. Univ. Electric Traction July 1992]**
10. What are the advantages of regenerative braking ? Explain how regenerative braking can be obtained in dc locomotives. **[B.H.U. Elec. Traction and Utilization 1978]**
11. Explain how hydraulic and eddy current brake works. What are their advantages ?
    What is mechanical regenerative braking and what are its advantages ?
12. Explain with proper diagram the following braking systems used in traction:
    (*a*) Vacuum brake (*b*) Compressed air brake (*c*) Magnetic track brake (*d*) Electromechanical drum brake (*e*) Regenerative braking with 3-ϕ induction motors.
13. Describe the various methods of drives which have been applied for transmitting motive power from motor shaft to driving wheels. **[Pb. Univ. Elec. Traction 1973]**
14. Explain the method of Quill drive of transmitting power from the motor to the driving wheels. **[B.H.U. Electric Traction and Utilization 1978]**
15. State with sketches, the three important methods of transmission of mechanical power from traction motor to the driving wheels. **[Agra Univ. Traction & Utilization 1976 Supp.]**
16. Explain multiple unit control of dc motors. **[M.D. Univ. Utilization of Electric Power & Traction May 2009]**
17. Describe the auxiliary electrical equipment for an electric locomotive. **[Pb. Univ. Elec. Traction Dec. 1992]**
18. Write short notes on the following:
    (*i*) Electric braking in traction motors. **[U.P.T.U. 2005-06]**
    (*ii*) Plugging.
    (*iii*) Reverse current braking. **[Pb. Univ. July 1992]**
    (*iv*) Regenerative braking.
    (*v*) Vacuum brakes.
    (*vi*) Quill drive.
    (*vii*) Nose suspended motor.
    (*viii*) Methods of transmission of power from motors to driving wheels.
    (*ix*) Multiple unit control. **[M.D. Univ. Dec. 2008; Dec. 2009]**
    (*x*) Auxiliary equipment for electric locomotives.

Short Answer Type Questions With Answers

Q. 1. Why ordinary mechanical braking cannot be totally replaced by electrical braking ?

Ans. Because electric braking cannot hold the vehicle stationary.

Q. 2. On what factors does the amount of energy returned to the line depend ?

Ans. The amount of energy returned to the line depends upon (*i*) the initial and final speeds during regenerative braking (*ii*) the train resistance and gradient of the track also in case the train is moving down the gradient and (*iii*) efficiency of the system.

Q. 3. Why is regenerative braking not recommended for tramways and trolley buses ?

Ans. Because it will unnecessarily increase the capital cost as well as enhance the operating problems.

Q. 4. What is meant by mechanical regenerative braking ?

Ans. During coasting a part of the stored energy (kinetic energy) is utilized in propelling the train against frictional and other resistances to motion and, therefore, speed falls. Thus coasting may be considered as a form of mechanical regenerative braking.

Q. 5. Why gear drive is necessary ?

Ans. The size of the motor is reduced with the increase in speed for a given output, so gear drive is necessary to reduce the size of the motor for given output by operating it at higher speed.

Q. 6. Why is master controller usually equipped with 'dead man's handle' ?

Ans. So as to stop the train in case the driver faints or becomes incapacitated.

Problems

1. A train weighing 300 tonnes has its speed reduced from 80 kmph to 30 kmph while going down an incline of 1 in 100 through a distance of 3 km by employing regenerative braking. Taking tractive resistance of 4 kg per tonne and allowance for rotational inertia as 8%, calculate the kWh returned to the line assuming an overall efficiency of 75%. **[Ans.** 25.34 kWh**]**
2. A 300 tonne train on a down grade of 1 in 75 has its speed reduced from 60 kmph to its steady value 40 kmph. Calculate the energy returned to the line, in a distance of 1,000 metres by adopting regenerative braking. Assume overall efficiency of 85 per cent, track resistance of 5 kg per tonne and allowance for rotary inertia as 10 per cent. **[Ans.** 13.28 kWh**]**
3. A 400 tonne train has its speed reduced by regenerative braking from 90 to 50 kmph in 2,000 metres on a down gradient of 1 in 100. Assume overall efficiency 80 per cent and allow 10 per cent for rotary inertia and 5 kg per tonne for tractive resistance. Find the energy in kWh returned to the line. **[Pb. Univ. Elec. Traction 1973] [Ans.** 29.81 kWh**]**
4. Rheostatic braking is used to limit the speed of a dc crane motor (series motor) to 500 rpm when the supply is cut off. The motor has a resistance of 1.2 Ω and the descending load exerts a torque of 45 kg-m. Calculate the value of the resistance to be connected across the motor terminals. State assumption made.
   The magnetisation characteristics of the motor at 500 rpm is given by

Current amperes	20	30	40	50
EMF volts	360	440	490	520

   [Ans. 10.155 Ω**]**
5. A dc series motor is to be used for rheostatic braking on a tram car. With the field respectively excited on test, the open-circuit voltage across armature at a speed corresponding to 24 kmph was:

Field amperes	10	20	40	60	80
Armature volts	145	280	500	680	700

   When the machine is being used for braking a tramcar, what resistance would have to be put in the motor circuit to obtain a current of 70 A at a speed of 16 kmph ? The resistance of the motor armature and field windings may be neglected. **[Ans.** 6.4 Ω**]**
6. A 525 V series traction motor has the following data:

Current (amperes)	50	70	80	100
Speed (kmph)	33.8	26.9	25.1	23.8
Torque (N-m)	216	344	422	500

   What will be the braking torque at a speed of 26 kmph when operating as a self-excited series generator, the resistance of the braking rheostat being 7.5 ohms, and that of motor being 0.5 ohm ? **[Ans.** 300 N-m**]**

CHAPTER 15

Power Supply For Electric Traction

INSIDE THIS CHAPTER

15.1. Introduction **15.2.** Current Collection Systems **15.3.** Current Collectors For Overhead System **15.4.** Overhead Construction For Tramways and Trolley Buses and Railways **15.5.** Sag and Tension Calculation For a Trolley Wire **15.6.** DC Track Electrification **15.7.** Low Frequency Single Phase System **15.8.** AC Track Electrification **15.9.** Power Supply Arrangement For AC Track Electrification **15.10.** Block Diagram of an AC Electric Locomotive **15.11.** Major Substation Equipment **15.12.** Remote Control Centre **15.13.** Signalling **15.14.** Overhead Equipment **15.15.** Interference in Telecommunication Circuits **15.16.** Booster Transformer **15.17.** Negative Booster **15.18.** Protective Devices.

15.1. INTRODUCTION

As already discussed in chapter 10, the various systems of railway electrification are (*i*) dc system (*ii*) single phase ac system operating at low frequency (*iii*) three phase ac system operating at low frequency and (*iv*) composite system (*a*) Kando system and (*b*) single phase—dc system.

Electric traction was first introduced in Britian in the year 1890 at 600 V dc using third rail for urban and suburban services. For its application to long distance main-line service the developments were towards increasing the operating voltage and as a result the line rail system utilising 1,200 V dc was tried in Manchester-Bury line. This system of supply was not satisfactory due to inherent drawbacks, such as insufficient protection provided by the wooden planks and choking up of grooves by snow during winter season, and therefore, overhead contact system was adopted employing 1,500 V dc later on. This system is still used for suburban and main lines in many countries including our country.

In India electric traction was first introduced for suburban service in Bombay in 1925 and in Madras in 1931. It was later extended to Poona and Igatpuri. The system adopted was 1,500 V dc. The power used to be supplied to traction motors at 1,500 V dc from rotary converter substations. For 1,500 V supply two 750 V rotary converters, each of 2,500 kW capacity, were employed in series. Mercury-arc rectifiers were introduced for the first time for supplying dc power when Madras–Tambaram route was electrified. The rectifiers employed were multianode continuously evacuated steel tank type. Later on grid controlled mercury-arc rectifiers were used. Multianode air-cooled sealed type steel tank rectifiers having on-load tap changing arrangement for variation of voltage have also been employed. Nowadays solid state devices are finding increasing use in these dc substations.

After about 20 years, electric traction was introduced on the Howrah—Burdwan and Sheoraphuli-Tarakeshwar branch lines on 3,000 V dc, as at that time there was a general trend towards adopting of 3,000 V dc for electric traction.

The early applications of the three phase system were to Swiss mountain railways, and the first application to main-line railways (operating at 3,000 V) was in 1902. Subsequent development of the 3-ϕ system have been entirely for main-line electrification, and the principal applications are in Northern Italy.

The single phase ac system operating at low frequency ($16\frac{2}{3}$ Hz) found extensive applications in Switzerland, Germany, Austria and Scandinavia for suburban as well as main-line services. In America about 800 km (on the Pennsylvania and New Haven lines) are electrified on the single phase system, but at 25 Hz frequency and 1,100 V voltage. The single phase 50 Hz system has also been adopted for the electrification of the Katanga lines in the Belgian Congo.

Kando system (single phase-three phase system) was developed in Hungary in 1932.

From the reports obtained from French National Railways on relative merits and demerits of ac and dc systems it was found that ac system employing commercial frequency of 50 Hz was advantageous both financially and operationally. In 1957 Railway Board decided to adopt 25 kV, 50 Hz single phase system of electrification for all future schemes. This system gives considerable saving in the cost of overhead equipment, substations and locomotives. The size of the overhead system is reduced because of adoption of high voltage leading to saving in the cost of supporting structures and their foundations. On account of the high voltage, the traction substations can be spaced at longer distances. AC substations are simpler than dc substations because they have only step-down transformers and necessary switchgear. However, ac traction systems have some drawbacks also like induced voltage in aerial signalling equipment and interference with neighbouring communication lines, and therefore, it will need some remedial devices to overcome such drawbacks.

In this system 25 kV, 50 Hz single phase system is employed for supplying power to the locomotives. The locomotive draws power from the overhead equipment (OHE) through sliding contact by means of pantograph. The locomotive essentially consists of (*i*) air circuit breaker, provided on the roof of the locomotive, mainly to disconnect the locomotive from hv supply in the event of fault in the equipment and to isolate the locomotive equipment at a voltage change point or phase change point in the overhead equipment such as before entry into the neutral section (*ii*) step-down transformer to step down the supply voltage to utilization level (*iii*) on-load tap changer for providing variable voltage (*iv*) converting machinery for converting ac into dc, (*v*) smoothing chokes for reducing the magnitude of ac component of rectified undulating dc and (*vi*) dc traction motors.

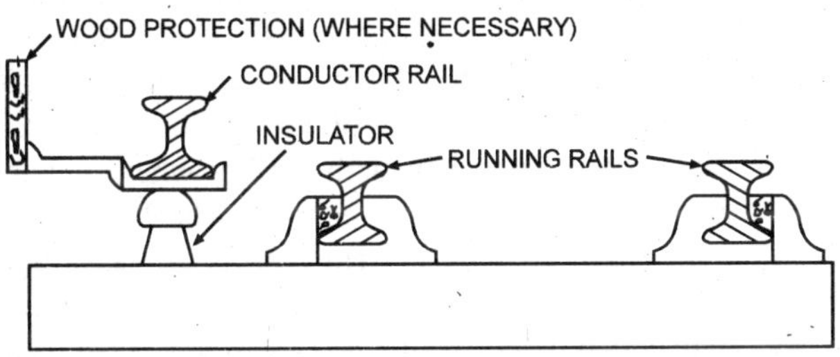

Fig. 15.1. ***Conductor Rail System***

15.2. CURRENT COLLECTION SYSTEMS

There are mainly two systems of current collection for locomotives, tramways or trolley buses viz (*i*) conductor rail system and (*ii*) overhead system. Current collection from overhead system is far superior to that from the conductor rail system. This is because both theoretically and experimentally current collection is more difficult from a rigid body than from an elastic one. Further the insulation of the third rail at high voltages used on single phase ac traction would also be impracticable and endanger the safety of the personnel.

15.2.1. Conductor Rail System

The conductor rail system is employed at 600 V for suburban services on account of its relative cheapness and easier inspection and maintenance. The supporting structure does not interfere in the visibility of the signals. Also, it is not necessary to protect the rail from accidental contact by plate players etc., except at stations and in sidings. In this system, current is supplied to the electrical operated vehicle either through one rail conductor or through two rail conductors. In case of one rail conductor the track rail is employed as the return conductor. The advantage of using insulated return rail is elimination of electrolytic action due to return currents on other public services buried in the vicinity of the railway tunnels. The rails are mounted on insulators parallel with the track rails at a distance of 0.3 to 0.4 m from the running rail with their upper most surface acting as contact surface and are fed at suitable points from substations. Conductor rail with a side running contact, as shown in Fig. 15.2 (*a*), is also employed. Some railways employ an under-running contact. Latter systems are supposed to be more protective against accidental contact. The wearing of the rail conductor is due only to the friction of the collector shoes and pitting which usually occurs at starting when the current drawn is very large. The rails are designed from electrical considerations rather than mechanical. The main considerations are (*i*) electrical conductivity (*ii*) cost, (*iii*) wearing qualities (*iv*) contact surface available for the collector shoes and (*v*) shape and size of the conductor rail, keeping in view the type of insulators used. A special steel alloy is used for the rails for economy reasons. The composition is such that it provides high conductivity and at the same time satisfies all other requirements. A typical composition is iron 99.63%, carbon 0.05%, manganese 0.2%, phosphorous 0.05%, silicon 0.02% and sulphur 0.05%. It has a resistance of about 12.0×10^{-8} Ω-m at 18°C (about seven times of that of copper and about one quarter or one-fifth of that of the running rails). To reduce the voltage drop at joints, conductor rails are bonded together by short length flexible copper conductors riveted or welded to the rails. The conductor rail is not fixed rigidly to the insulators in order to take care of the contraction and expansion of rails. To prevent creepage due to friction of the collector shoes, however, it is anchored at intervals of 100 to150 metres.

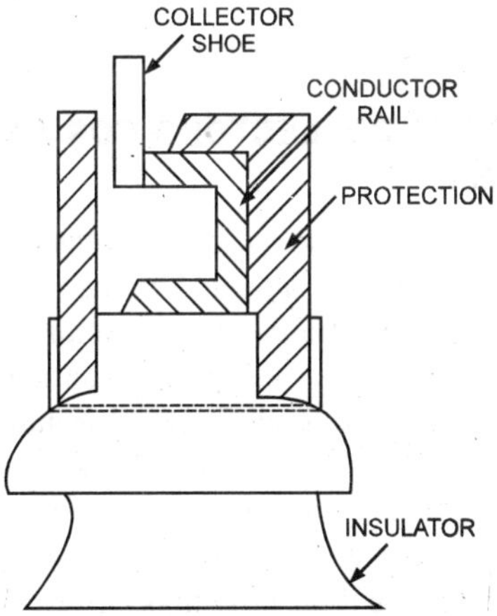

(*a*) *Conductor Rail With Side Running Contact*

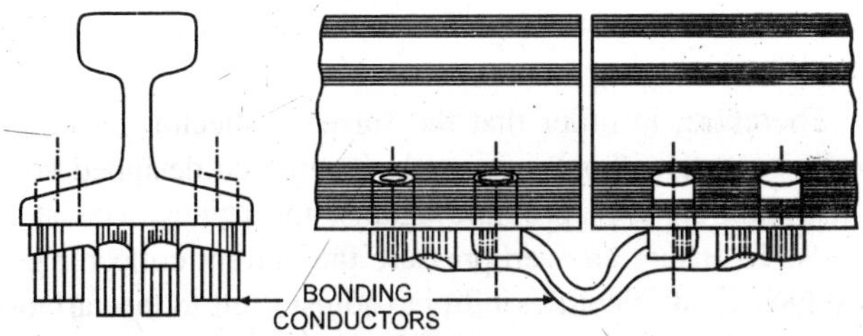

(*b*) *Bonding of Conductor Rails*

Fig. 15.2

The current is conveyed from the conductor rail to the train equipment by means of collector shoe (flat steel shoe about 20 cm in length and 7.6 cm in width) which presses on to the rail with a force of about 15 kg. In case of top contact the necessary contact force is obtained by gravity and in case of side and under-running contacts springs are employed to obtain necessary contact force. The current which can be collected is about 300 to 500 A. Sometimes it may not be convenient to have the conductor rail on the same side of the track, therefore, it is desirable to provide shoes on both sides of the locomotive or train. Also, there are going to be gaps in the rails at points and crossings, at least two shoes must be provided on each side so as to avoid discontinuity in the current flow. This system is suitable for heavy current collection, top contact system for voltage up to 750 V and side contact system up to 1,200 V.

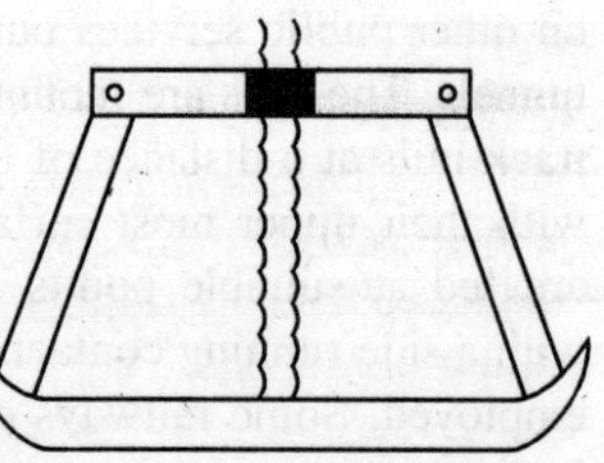

Fig. 15.3. ***Current Collecting Shoe***

15.2.2. Overhead System

This system is adopted when the trains are to be supplied at high voltage (1,500 volts or above). In this system with high operating voltage the trains requiring high power may be supplied through conductors of relatively small cross section and the collection of the current required by a train can be done by a collector with a sliding contact. Overhead construction is universal for all ac railways and is also used with dc tramways, trolley buses and locomotives operating at voltages of 1,500 volts and above. In all these cases the running rails are utilized as the return conductor, therefore, with dc and single phase system only one overhead wire is required for each track.

15.3. CURRENT COLLECTORS FOR OVERHEAD SYSTEM

The current in the overhead system is collected with the help of sliding contact collector mounted on the roof of the vehicle. The main requirement of a current collector is that it should, under no circumstances, leave the contact of overhead equipment. Contact wire in all practical installations is never perfectly horizontal, it rises and falls depending upon the weight of the contact wire, and distance between droppers. Also contact wire comes very low under bridges and tunnels and rises high over public crossings. The current collector has to rise and come down according to the speed of the electric vehicle in order to maintain the contact with overhead equipment.

Therefore, in order that the current collector gear picks up the current without any spark, it must be designed so as to maintain an even pressure against the trolley wire at all speeds. To ensure an even pressure the current collector gear must have a sufficient flexibility to adjust itself to the variation of the level of the conductor at high speeds. Either a spring or compressed air is used to maintain the pressure. When the contact wire rises the current collector has to move upwards so as to maintain contact, and the pressure exerted on the contact wire is the pressure of the spring or compressed-air piston less the force required to accelerate the current collector to its new position and any force necessary to overcome the frictional resistance. To design a satisfactory current collector therefore, particularly for high speed work, the frictional resistance must be reduced to a minimum and the moment of inertia must be kept low. Furthermore, means must be provided for lowering when idle and reasonable precautions taken to prevent the collector leaving the wire at points and crossings.

As the tramcars and trolley buses draw their power from overhead conductors suspended about 6.7 m (from 4.7 m to 7.5 m) above the road surface, the current collector must, therefore, be mounted on the vehicle. Three types of current collectors are in common use, viz, the trolley collector, bow collector and the pantograph collector.

1. *Trolley Collector.* The trolley collector is universally employed with tramways and trolley buses. This consists of a grooved gun metal wheel or grooved slider shoe with carbon insert carried at the end of a long pole. The other end of this pole is hinged to a swivelling base fixed to the roof of the vehicle. Necessary upward pressure for the pole and current collector is achieved by means of springs. As two trolley wires are required for a trolley bus a separate trolley collector is provided for each wire, the bases being mounted side by side. The pressure for wheel is approximately 10 kg and for a carbon insert slider is approximately 17 kg. The main drawback of trolley collector is that it has to be rotated through 180° for reversing the direction of motion of the vehicle. Another drawback particularly with trolley wheel collector is that there is poor contact between the wheel and trolley wire. This gives rise to high current density in gassy mines where trolley collector with grooved slider shoe is preferred.

The trolley collector is suitable for comparative low speeds (say 24 to 32 kmph) beyond this speed there is every possibility of its jumping off the trolley wire.

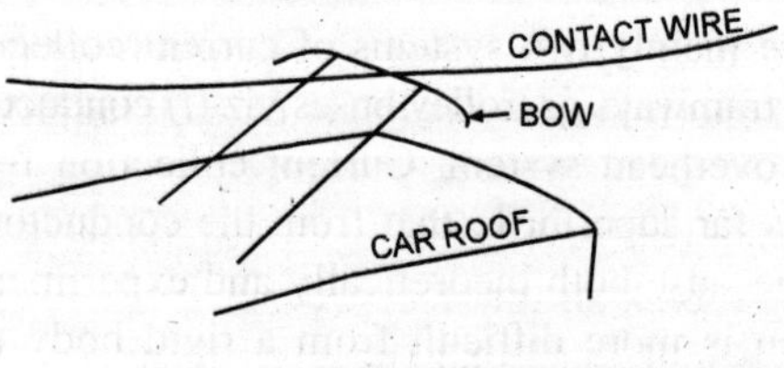

Fig. 15.4. *Bow Collector*

2. *Bow Collector.* The bow collector is also employed for collecting the current with tramways. The bow collector consists of light metal strip or bow 0.6 or 0.9 metre wide pressing against the trolley wire and attached to a framework mounted on the roof of vehicle. Although provision of metal strip enables vehicle to run at higher speeds without running the risk of leaving the contact of trolley wire, its manoeuvrability is lost. It is as such not suitable for trolley buses. In order to avoid jumping off of the trolley wire at high speeds, it is desired that the wire be accurately located above the track and staggered about 15 cm to each side of centre line to avoid the wearing of the groove in the contact strip. Collection strip is made of soft material such as copper, aluminium or carbon so that it should wear instead of trolley wire as it is easy to replace worn out collection strip than trolley wire. Necessary upward pressure is obtained by springs. The length of the poles is kept sufficient to have lateral movement along the mean path of trolley bus. This arrangement enables vehicle to run at higher speed without danger of the jumping off wire. Bow collector has always to run trailing just like trolley collector. It, therefore, requires either provision of duplicate bows or an arrangement of reversing the bows for motion in the reverse direction. The bow collector has smaller inertia but is not so readily adoptable for collection of large currents as is a pantograph collector.

3. ***Pantograph Collector.*** The pantograph is employed in railways for collection of current where the operating speed is as high as 100 or 130 kmph and the currents to be collected are as large as 2,000 or 3,000 amperes. Pantographs are mounted on the roof of the vehicles and usually carry a sliding shoe for contact with the overhead trolley wire. These consist of a jointed frame usually of steel tubing. The contact shoes are usually about 1.2 metres long. There may be a single shoe or two shoes on each pantograph. The shoes may be straight throughout their lengths or cambered slightly or may be in the form of an auxiliary bow. Material used for pantograph is often steel with sometimes, wearing plates of copper or bronze inserted. The pressure varies from 5 to 15 kg.

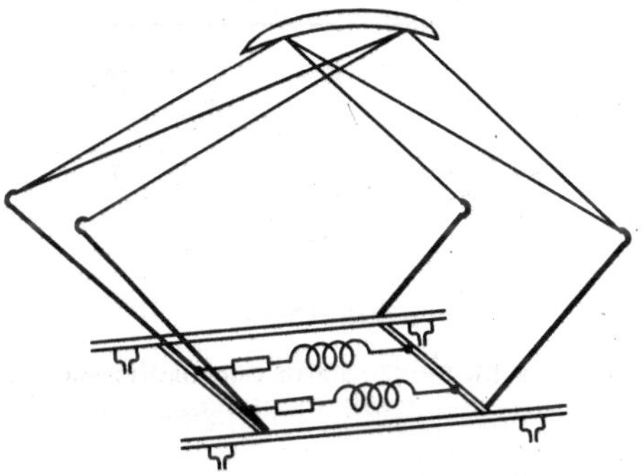

Fig. 15.5. ***Pantograph Collector***

The pantograph is raised or lowered from the driver cab by one of the following methods or with some modification of it.

(*i*) air raised, gravity lowered. (*ii*) air raised, spring lowered. (*iii*) spring raised, air lowered.

The pantographs should be of such construction that they maintain continuous contact with the overhead wire at reasonably constant pressure at any height. They must be light enough in order to follow the overhead wire as it changes height above the rails to pass under overhead crossing etc., with minimum inertia effects even at high speeds they must be sufficiently strong and rigid to resist air pressures both head on, due to speed of train, and transverse, due to wind; also they must resist stresses due to the sway of the locomotive at high speeds and to some extent, stresses due to blows from the overhead system at deflectors or turnouts etc. and from striking birds. If, however, there is a serious trouble on the overhead system, the pantograph must be sufficiently pliable so that it will be damaged rather than pulling down the overhead structure.

The above arrangement has the following advantages over other types of collectors.

(*i*) It can operate in either direction of motion.

(*ii*) There is no risk of leaving wire junction etc.

(*iii*) The erection of the overhead network is very simple due to absence of points and grooved crossings required for bows.

(*iv*) Its height can be varied from the drivers' cabin by carrying out simple operations.

15.4. OVERHEAD CONSTRUCTION FOR TRAMWAYS AND TROLLEY BUSES AND RAILWAYS

As already mentioned in Art. 15.2, for trams, trolley buses and for railways operating at 1,500 V and above overhead contact wire is used. The design of overhead system is more complex in comparison to conductor rail system because of greater difficulty of ensuring that the collector and contact wire remain in contact at a fairly even pressure.

15.4.1. Tramways and Trolley Buses

The overhead construction for tramways has to conform the following regulations:

1. The trolley wire is to be erected at a minimum height of 5.185 metres above the street surface (except under bridges).
2. The trolley wire is to be supported at intervals not exceeding 36.6 metres.
3. Each trolley wire must be divided into sections not exceeding 0.8 km in length with an emergency switch between every two sections.
4. The trolley wire should be capable of withstanding stresses due to variation in temperature.

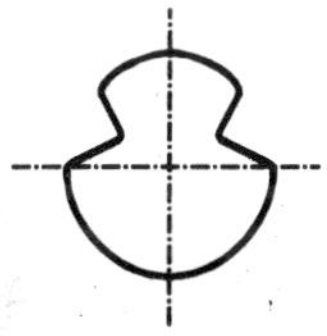

Fig. 15.6

The materials employed for trolley wire are hard drawn copper or alloys of copper (cadmium copper or silicon bronze). The alloy is preferred because it has uniform hardness throughout the x-section whereas in case of hard drawn copper the hardness is confined to the outer surface *i.e.,* the skin of the trolley wire only. The uniform hardness gives uniform wearing of conductor, which is highly desirable in electric traction. The trolley wire is usually of circular grooved cross section, as illustrated in Fig. 15.6. The cross-sectional area of a standard trolley wire is 80 mm^2.

The trolley wire is supported and insulated from a transverse span wire by means of a steel bolt screwed into a gunmetal ear. The gunmetal ear is attached to the trolley wire, and the bolt is insulated from and held in a hanger attached to the span wire. The span wire is supported from poles erected facing each other on both sides of the road and insulated there from.

On straight runs the span wire is above the trolley wire but at curves the span wire is in level with trolley wire. Hangers employed for curved paths are known as '*pull-offs*'. The span wire is of stranded galvanised steel. The span wires may be extended over the whole width of the road (in which case these are attached to poles on each side of the road) or may be of a shorter length (in which case these are attached to brackets carried from horizontal arms attached to the poles at one side of the road). The former arrangement is known as *span wire construction* and the latter *bracket arm construction.*

15.4.2. Railways

Since in railways a sliding current collector (bow or pantograph) is employed, therefore, the contact wire is required to be suspended with very small sag so that the contact between the wire and collector may be maintained at high speeds. This is achieved without excessive tension in the wire by employing spans of comparatively shorter length (3 to 4.5 metres). For such short spans an indirect method of suspension is adopted, in which the trolley wire is supported by another wire known as the "*Catenary*" or "*Messenger*" and is fixed with considerable sag (2 to 3 per cent of the span) between supports at suitable distance. The wire is usually made of more than seven strands of steel. To avoid the use of insulated droppers the whole of the messenger wire is usually insulated, and it may therefore be used if required for carrying current. The position of contact wire remains practically constant at all temperatures which may cause large amount of sag in the catenary wire. Nowadays the catenary construction employed is of either the single catenary or compound catenary.

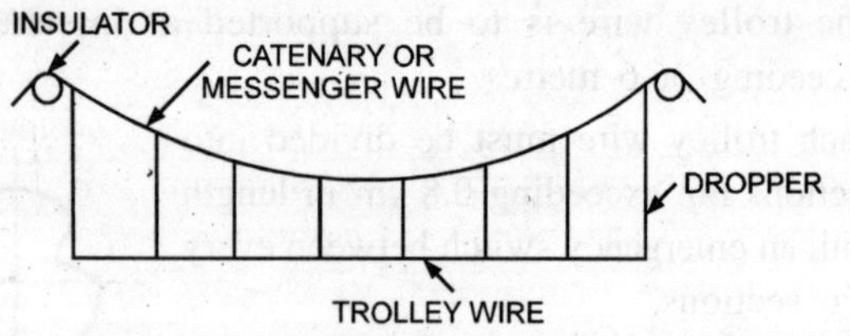

Fig. 15.7. ***Single Catenary Construction***

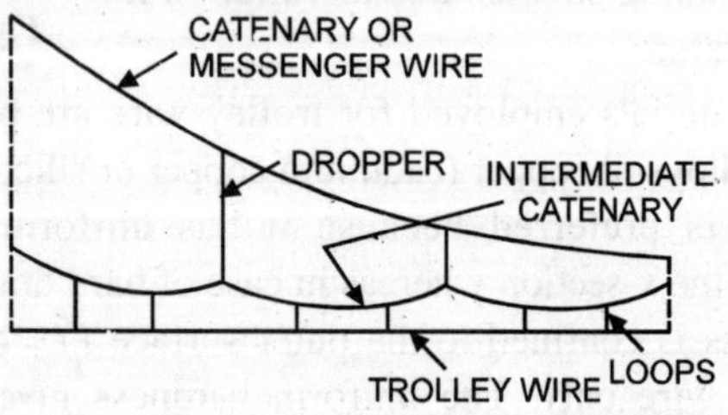

Fig. 15.8. ***Compound Catenary Construction***

In *single catenary construction,* a steel wire with 7 or more strands supports the trolley wire by means of droppers clipped to catenary and trolley wire at equidistant horizontal intervals. The span of catenary wire may be from 45 to 90 metres with a sag of from 1 to 2 metres respectively for straight track. The distance of droppers is between 3 and 5 metres. On curved track the spans of shorter lengths are employed and the trolley wire is kept in position by pull-off wires. Such construction provides sufficient flexibility for speeds up to 120 kmph without any problem of current collection. Besides this it is relatively cheaper, easy to erect and does not require heavy maintenance. Single catenary construction is considered suitable for most light weight ac systems and for branch and loop lines and secondary routes where traffic density is low and operating speeds are low.

The compound catenary construction is employed for heavy current *i.e.,* where traffic density is high. It consists of three wires, all in the same vertical plane. The upper wire is catenary wire, which is insulated from the supporting structures. From this wire another wire, known as *intermediate catenary,* is supported by droppers clipped to both wires. The third wire *i.e.,* trolley wire is maintained in definite and constant tension by means of automatic tightening gear. Both contact and catenary wires are divided in sections of 150 metres and each section is anchored at the centre. The purpose of intermediate catenary besides increasing the current carrying capacity is to provide more uniformity in elasticity.

15.5. SAG AND TENSION CALCULATION FOR A TROLLEY WIRE

A flexible wire when suspended horizontally between two supports it would assume the form of catenary. But in case of tramways the sag in the trolley wire rarely exceeds 1 or $1\frac{1}{2}$ per cent of the span length and, therefore, the wire may be considered to hang in the form of a parabola.

If $2l$ is the span length in metres, w is the weight in kg per metre length of wire and T is the tension of wire in kg.

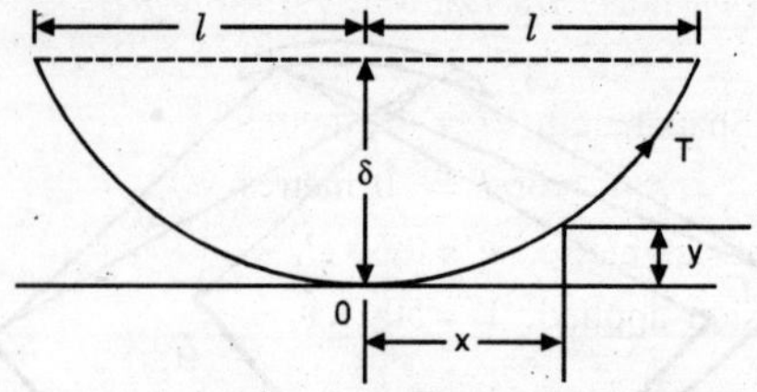

Fig. 15.9

Then by taking moments about one of the supports, we get

$$T\delta = wl \times \frac{l}{2}$$

$$\text{or} \quad \delta = \frac{wl^2}{2T} \qquad \text{...(15.1)}$$

The equation of parabola is given by the expression

$$y = \frac{\delta x^2}{l^2} = \frac{w}{2T}x^2 \qquad \text{...(15.2)}$$

We know that

$$\frac{ds}{dx} = \sqrt{1+\left(\frac{dy}{dx}\right)^2}$$

$$\text{or} \quad \frac{ds}{dx} = \sqrt{1+\left(\frac{wx}{T}\right)^2} \qquad \text{Since } y = \frac{wx^2}{2T}$$

$$\text{or} \int ds = \int\left(1+\frac{w^2x^2}{T^2}\right)^{1/2} dx \qquad \therefore \frac{dy}{dx} = \frac{wx}{T}$$

$$\text{or } s = \int\left(1+\frac{w^2x^2}{T^2}\right)^{1/2} dx$$

$$= \int\left(1+\frac{w^2x^2}{2T^2}+\frac{3w^4x^4}{8T^4}+...\right)dx$$

$$= x+\frac{w^2x^3}{6T^2}+...$$

When $x = l$ *i.e.*, half of the span, the length of conductor in half of the span,

$$s = l + \frac{w^2 l^3}{6T^2} = l + \left(\frac{2T\delta}{l^2}\right)^2 \times \frac{l^3}{6T^2}$$

$$= l + \frac{2}{3} \cdot \frac{\delta^2}{l} \quad ...(15.3)$$

$$\text{Since } \delta = \frac{wl^2}{2T}$$

The change in temperature causes a change in length which will cause the change in sag and tension in a span. It should be noted that decrease in temperature reduces the sag and increases the tension. The reduction in the sag causes stretching which on the contrary tends to decrease the sag. In view of this combined effect the elasticity of the conductor material play an important role in determining the exact values of the sag at different temperatures.

Example 15.1. A trolley wire of a tramway is suspended from two poles 40 metres apart. If the diameter of the wire is 1 cm and weight per metre is 0.5 kg, find its sag if the tension applied is 500 kg. Find also the total length of wire required. Derive the necessary formula used.

[*Pb. Univ. Elec. Traction Nov. 1973; Pb. Technical Univ. Electric Drives & Utilization, May 2009*]

Solution : Span length, $2l = 40$ metres

or $l = 20$ metres

Weight per metre, $w = 0.5$ kg

Tension applied, $T = 500$ kg

$$\text{Sag, } \delta = \frac{wl^2}{2T} = \frac{0.5 \times 20^2}{2 \times 500} = 0.2 \text{ metre} \quad \textbf{Ans.}$$

Total length of wire required

$$2s = 2\left[l + \frac{2}{3}\frac{\delta^2}{l}\right] = 2\left[20 + \frac{2}{3} \times \frac{(0.2)^2}{20}\right]$$

$$= 40.00267 \text{ m} \quad \textbf{Ans.}$$

Example 15.2. Calculate the sag to be given to a trolley wire 1 cm in diameter and weighing 0.72 kg per metre when erected with a span of 60 metres and if the stress in the wire erected is 640 kg/cm².

Solution : Span length, $2l = 60$ metres

or $l = 30$ metres

Weight per metre, $w = 0.72$ kg

$$\text{Tension of wire, } T = Ea = 640 \times \frac{\pi}{4} \times (1)^2 = 503 \text{ kg}$$

$$\text{Sag, } \delta = \frac{wl^2}{2T} = \frac{0.72 \times (30)^2}{2 \times 503}$$

$$= 0.644 \text{ m} = 64.4 \text{ cm} \quad \textbf{Ans.}$$

Example 15.3. In a catenary system the catenary wire is of steel with a permissible tension of 1,000 kg. If the span is 60 metres, what shall be the minimum sag of the catenary ? The weight of the catenary wire is 0.9 kg per metre run, the weight of the trolley wire is 1.2 kg per metre run, and the allowance for droppers and fittings may be taken as 20% of the trolley wire weight.

Solution : Span length, $2l = 60$ metres

or $l = 30$ metres

Average weight of catenary wire, $w_1 = 0.9$ kg/metre run

Average weight of trolley wire, $w_2 = 1.2$ kg/metre run

Average weight of droppers and fittings,

$$w_3 = \frac{20}{100} \times 1.2 = 0.24 \text{ kg/metre run,}$$

Average weight of catenary wire, trolley wire and dropper per metre run,

$$w = w_1 + w_2 + w_3 = 0.9 + 1.2 + 0.24 = 2.34 \text{ kg}$$

Sag of the catenary,

$$\delta = \frac{wl^2}{2T} = \frac{2.34 \times (30)^2}{2 \times 1{,}000} = 1.053 \text{ metres} \quad \textbf{Ans.}$$

15.6. DC TRACK ELECTRIFICATION

The substations are necessary for control of electric supply to a considerably long network for electric traction.

15.6.1. DC Substations

These substations are required for tramways, trolley buses and railways. Their functions are to (*i*) step down the transmission voltage to the utilization level (*ii*) convert it into dc and (*iii*) sectionalizing the overhead equipment. Such substations, therefore, consist of step-down transformers, converting machinery, ac circuit breaker, dc circuit breaker and smoothing and metering equipment.

Mercury-arc rectifiers used to be employed as converting machinery because of their much higher all-day efficiency, 100% overload capacity for few minutes, low maintenance cost, noiseless and reliable operation at all voltages and adaptability to remote control with simplicity of operations. Nowadays solid state devices are finding increasing use in these substations.

Since the rectifier used to be non-reversible, hence suitable arrangements were required to be made to receive the energy returned during regenerative braking. In cases where regenerated energy was considerable, inverters (grid - controlled rectifiers arranged to operate with reversed voltage) were employed. In other cases the energy used to be dissipated in a load resistor connected to the dc bus bars.

15.6.2. Transmission System For Traction Substations

The substations lying along the railway track are fed by means of double transmission line running along the track. Double transmission line is employed to ensure continuity of supply. Normally, the current carried by each line is half of the total current but in case of fault on any line, the total current is carried by the healthy line.

Line diagram of duplicate transmission system is shown in Fig. 15.10. In case of fault on any section, the faulty section is opened out by the line circuit breaker and the voltage drop in the other line of this section is doubled because it has to carry double the current. For each single unit substation there are four line circuit breakers and one transformer circuit breaker and since the cost of high voltage equipment increases very rapidly with higher voltages, it is clear that transmission lines operating at 220 or 132 kV must be tapped as seldom as possible.

In case the ht supply is available from the grid, the substations are fed from the grid supply through step-down transformers. An arrangement for feeding the substations from the grid is shown in Fig. 15.11. The EHT supply is tapped at suitable points A, B and C, its voltage is stepped down by the transformers at substations at A, B and C. Now either a single ht transmission line may be taken to the substations from the transformer substations, as shown in Fig. 15.11(*a*) or duplicate transmission lines are provided between substations 1-2-3, 4-5-6, 7-8-9, as shown in Fig. 15.11(*b*) in order to avoid the risk of discontinuity of supply due to one of the faulty line.

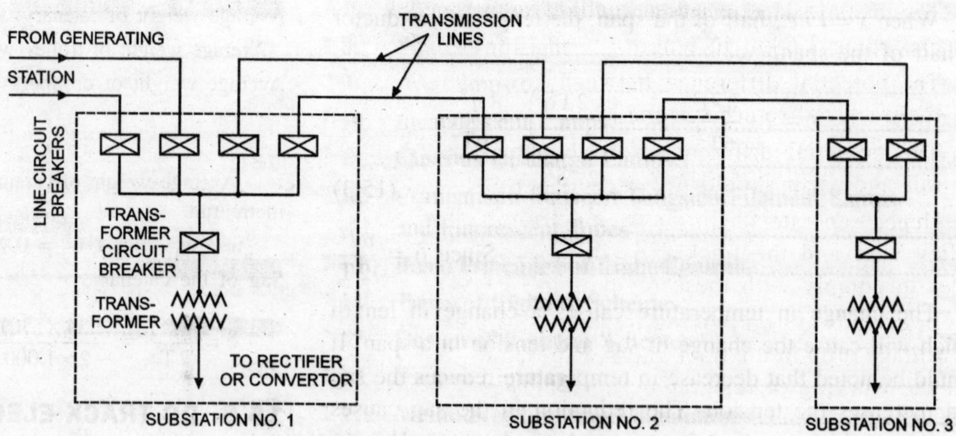

Fig. 15.10. ***Duplicate Transmission System***

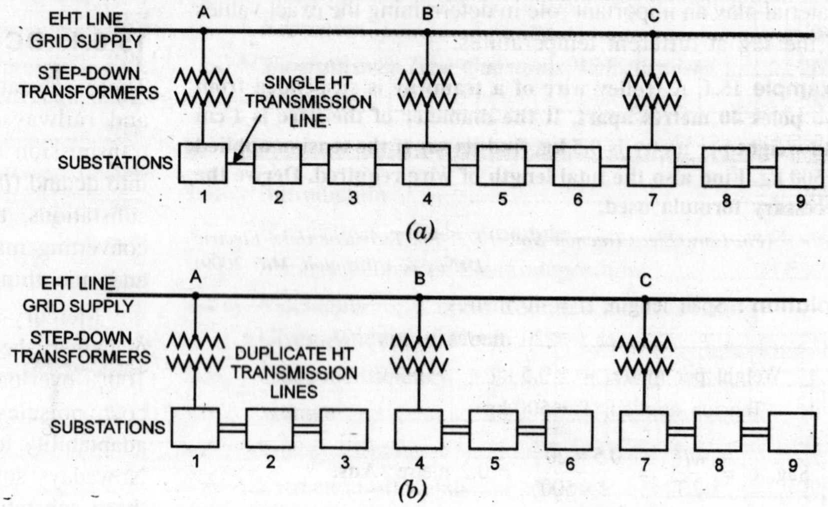

15.6.3. Location of DC Substations

Since in case of trolley bus both the conductors are trolley wires, therefore, more voltage drop takes place in it and the spacing between substations is reduced in comparison to tramway substations for similar capacity. Hence the substations required are more in number and of lesser capacity. Spacing of substations is kept 4.8 km for single track and 3.2 km for double track (up and down). Each section of 0.8 km length is individually fed, as shown in Fig. 15.12.

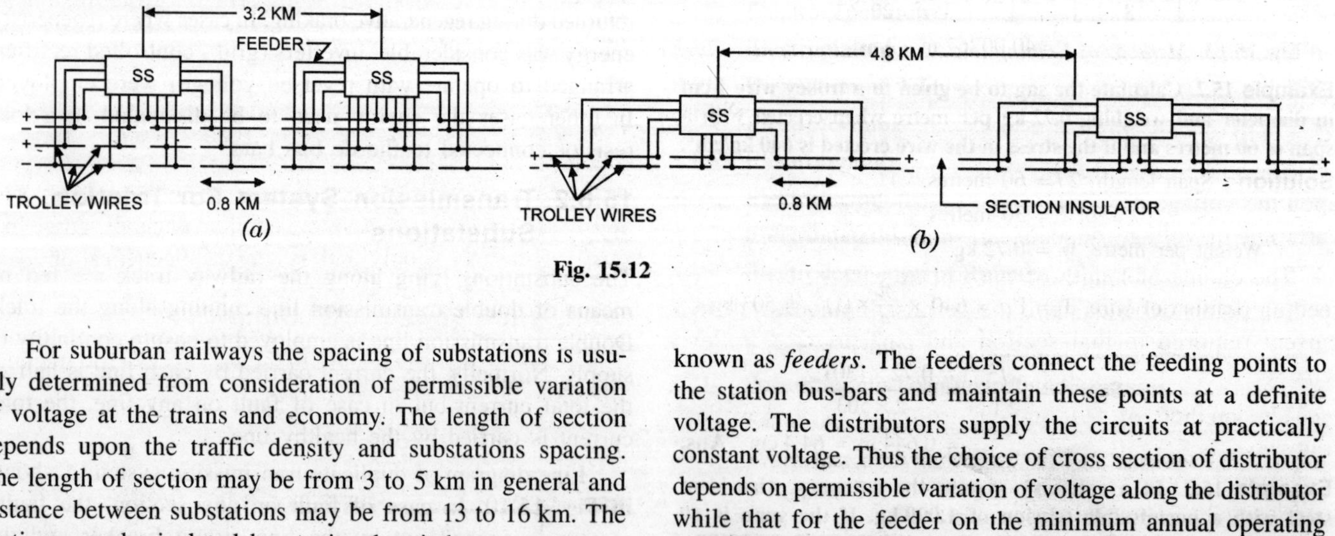

Fig. 15.12

For suburban railways the spacing of substations is usually determined from consideration of permissible variation of voltage at the trains and economy. The length of section depends upon the traffic density and substations spacing. The length of section may be from 3 to 5 km in general and distance between substations may be from 13 to 16 km. The sections can be isolated by sectional switches.

15.6.4. Feeding and Distributing Systems

The electrical energy is supplied from a power station (or a substation) to a number of circuits at a constant voltage through distributor cables (known as *distributors*), which are fed at suitable points (known as feeding points) by other cables known as *feeders*. The feeders connect the feeding points to the station bus-bars and maintain these points at a definite voltage. The distributors supply the circuits at practically constant voltage. Thus the choice of cross section of distributor depends on permissible variation of voltage along the distributor while that for the feeder on the minimum annual operating expenses. In electric traction systems the trolley wires and track rails (or two sets of conductor rails) form the distributors.

The tramways and trolley bus systems must conform to certain statutory regulations, chief among which are:

1. The voltage at the trolley wire is not to exceed 550 volts and that at generating stations (or substations) 650 volts.

2. The trolley wire shall be divided into sections not exceeding 1.6 km in length.
3. The potential difference between any two points of the track rail return systems shall not exceed 7 volts.
4. The potential difference between any point on the rail and earth should not exceed 4 volts.

Hence separate feeding systems are required for the trolley wire and the track rails in case of tramways in order to comply with these regulations.

In dc railways employing track rails as a return path, the voltage drop in the track rails is limited to such values that the working of telegraphic apparatus is not affected by earth currents. The positive portion of the distributing system need not to conform to statutory regulations with respect to voltage drop and sectionalising at every 0.8 km.

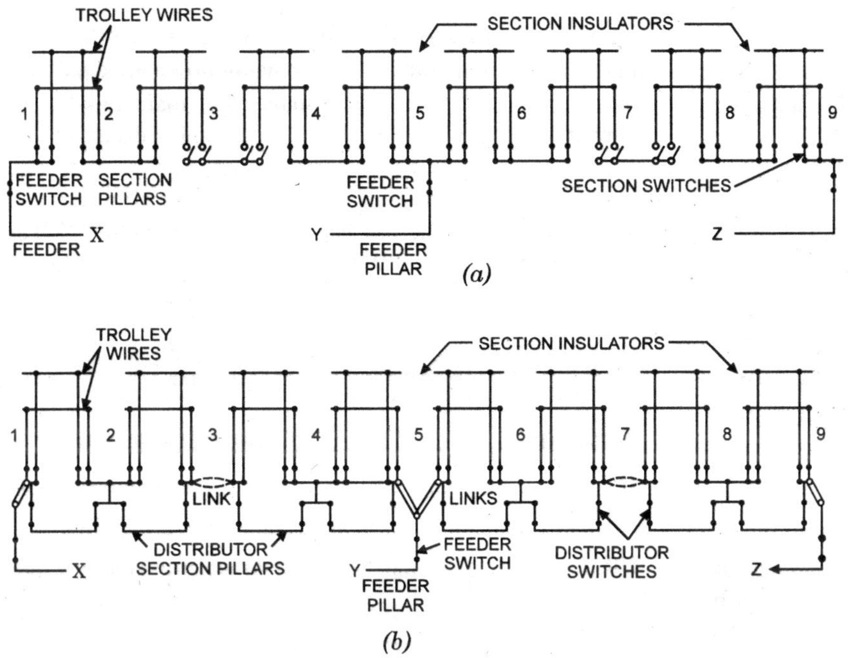

Fig. 15.13. ***Methods of Feeding and Sectionalizing Trolley Wires For Tramways***

The variation of voltage at the tramcars depends mainly upon the voltage drop in the trolley wire. The permissible variation in voltage depends upon service conditions.

The choice of length of the trolley wire between two feeding points depends upon the permissible voltage drop, current required in that section and particular of trolley wire.

The method of feeding the trolley wire in case of tramways is largely influenced by traffic density and economical considerations. For heavy traffic a separate feeder supplied from substation is required for each section of 0.8 km length while with light traffic, several 0.8 km sections of the trolley wire are fed from a single feeder because of economical considerations. The two methods are shown diagrammatically in Fig. 15.13.

In one case [Fig. 15.13 (*a*)] the trolley wire is used as a distributor and each distribution section comprises two or more sections of trolley wires. In other case [Fig. 15.13 (*b*)] feeders of large x-section are run to a few points in the system from which current is distributed to the various sections of trolley wire by means of graded distributing cables. The second method is more economical and the variation of voltage in the adjacent sections is less.

The voltage along the track is kept within the limits of 7 volts above earth by employing *negative booster* (Refer to Art. 15.17). Sometimes more number of negative feeders of comparatively low resistance are used.

15.7. LOW FREQUENCY SINGLE PHASE SYSTEM

If the supply available is from a National grid the transformers and frequency changers are required to convert the supply available (high voltage, commercial frequency supply) into low voltage (15,000 volts) and low frequency $\left(16\frac{2}{3}\text{ Hz}\right)$ supply in order to suit the traction requirements in addition to switchgear and protective and metering equipment.

If the primary supply system is designed solely for the purpose of feeding a traction system on single phase as is with most of the European railways, the generators employed at the generating stations are single phase, low frequency $\left(16\frac{2}{3}\text{ Hz}\right)$ and 15,000 volt machines. The substations are fed from generating stations through high voltage (220 kV/132 kV) transmission lines and step down the voltage to 15,000 volts by means of step-down transformers. Modern substations are of outdoor type because of low cost.

15.8. AC TRACK ELECTRIFICATION

As even in single phase electrification, the substations are necessarily required to keep the voltage drop within permissible limits. The type of substations required depends upon the nature of primary supply and system of electrification.

15.8.1. High Voltage AC Supply

Arrangements of power supply to the various electric traction substations in India are as follows:

1. Power is purchased from supply authorities who are responsible for the operation and maintenance of 220/132 kV transmission lines and grid substations up to 25 kV outgoing terminals of substations. However, 25 kV feeder circuit breaker is controlled from remote control centre by traction power controller (TPC).
2. All the 220/132 kV and 25 kV equipment of substation is owned, installed and maintained by the supply authorities whereas 25 kV circuit breakers are owned, installed, operated and maintained by the railway authorities. This

arrangement ends the dual control of 25 kV circuit breakers (operation by railways and installation and maintenance by supply authorities.)

3. For better control and expeditious execution of electrification programme Railways considered necessary to own, install, operate and maintain 220 kV/132 kV equipment at the substation themselves. The Railway authorities take high voltage supply of 220 kV/132 kV at the substations from the supply authorities. This arrangement exists on Madras-Villupuram, Rourkela-Durg and Virar-Sabarmati sections.
4. In another arrangement 220/132 kV 3-ϕ supply is taken at one point from the supply authorities and then it is distributed by the railway authorities to different traction substations through their own transmission lines as in case of Kanpur-Tundla section.

15.9. POWER SUPPLY ARRANGEMENT FOR AC TRACK ELECTRIFICATION

As mentioned in Art. 15.8.1, electric power is taken from grid supply at 220/132 kV and then stepped down to 25 kV single phase at traction substations through step-down transformers and suitable switchgear. The electric power at 25 kV ac is transmitted to feeding posts through feeders. Feeding posts are located opposite the track. At feeding posts, this power is fed to OHE above track for traction work. The power supply arrangement is described in details in the forthcoming sub-articles.

15.9.1. Sectioning Arrangement For Power Feeding

Power is generated and transmitted, by supply authorities, by 3-phase system and it is distributed in such a way that the loading of all the three phases remain almost same. Unbalanced

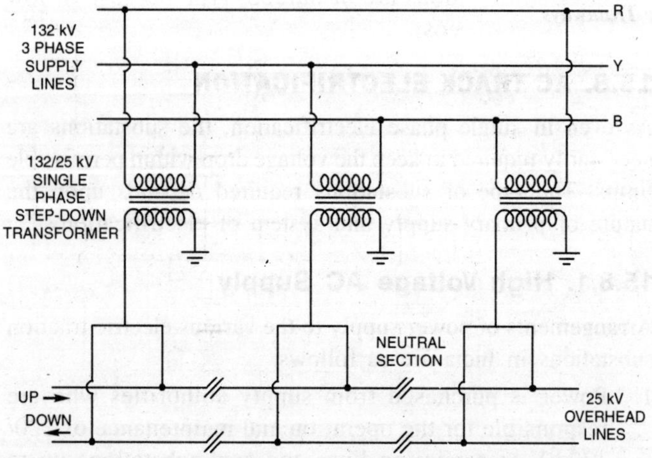

Fig. 15.14. ***Sectioning Arrangement For Power Feeding***

loading of the three phase system results in phase shift and also causes rotor heating of alternators. Electric traction is a single phase load so in order to avoid unbalanced loading consecutive substations are supplied by different phases in rotation as depicted in Fig. 15.14. In dc track electrification, all the substations feed OHE in parallel but in ac track electrification, section supplied by one substation works as an independent unit and never work in parallel. As different sections fed by different substations work on different phases, a neutral section is provided approximately in the middle of two feeding posts. The purpose of providing neutral section is to isolate the two sections supplied by different phases. Thus, the pantograph of the locomotive does not bridge two different live phases of 25 kV supply while passing from one section to another one. Neutral section will be discussed in detail later on (refer to Art. 15.14.4)

15.9.2. Substations

Figure 15.15 depicts the layout diagram of a typical traction substation. Traction substation is supplied by 220/132 kV ac supply either from two sources of supply or by double circuit 220/132 kV transmission line for the sake of reliability of supply. For rapid clearance of fault(s) occurring on transmission line or substation suitable protective equipment is provided. Two sets of 220/132 kV bus-bars (one main bus-bar and other transfer bus-bar) are provided in each substation. Both of these bus-bars are connected through bus coupler circuit breaker. The incoming and outgoing transmission lines are connected to both the bus-bars through the circuit breakers, as shown in Fig. 15.15. With such an arrangement of the two

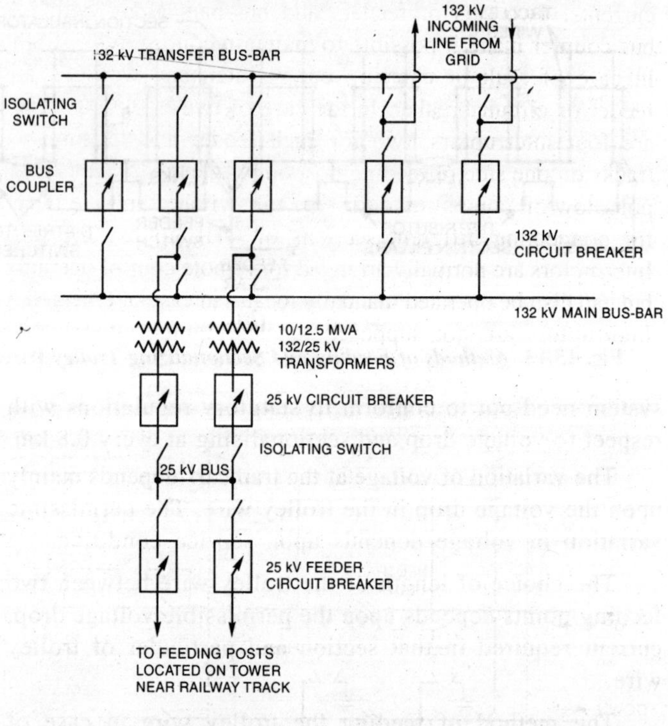

Fig. 15.15. ***Layout Diagram of a Typical Traction Substation***

bus-bars and isolators on both sides of circuit breakers continuity of supply can be maintained while doing maintenance work on any bus-bar or circuit breaker on primary side of transformer. Normally two single phase transformers of rating 10 or 12.5 MVA, 132/25 kV are installed in a traction substation. Of the two transformers only one will be in operation normally, the other one acting as standby. Output voltage at 25 kV is maintained within –5% to +10% with the help of tappings provided on the transformer. One 25 kV transformer circuit breaker and one 25 kV feeder circuit breaker equipped with an overload relay and an impedance relay are provided on the output side of transformer for clearance of all faults on OHE. One terminal of 25 kV winding of transformer is connected to earth solidly at substation and return feeders from all the rails of electrified track opposite to feeding post are also connected to this earthed end solidly. Provision of 25 kV bus-bar on the output side facilitates the maintenance of any transformer or its associated circuit breaker without any interruption of power supply. These substations are unattended type and are controlled by remote control operation. Spacing of these substations are 50 to 90 km and they are located closed to feeding posts, maximum distance being 2 km.

15.9.3. Feeding Post

Figure 15.16 depicts the layout diagram of feeding post. It is located opposite the electrified track on towers. Two 25 kV incoming feeders from traction substation are connected to two sets of bus-bars in feeding post through circuit breakers. The circuit breakers are controlled from a remote control centre of the railway. Each feeder consists of two conductors, one insulated for 25 kV connected to the bus-bar and the other insulated for 3 kV connected to the track for return currents. Two sets of feeders and bus-bars along with one bus coupler make it possible to maintain continuity of supply in case of fault or carrying out maintenance work on any feeder or circuit breaker. In the case of two track lines there are four interruptors, two for each feeder and feeding two tracks on one side of feeding post only. Interruptors are single pole, low oil content outdoor type oil switches and are meant for connecting different sections of OHE to feeding post. Interruptors are normally arranged for remote control operation but can also be operated manually locally in case of emergency. Interruptors are not supposed to clear faults automatically like circuit breakers as these are not provided with any protective relays. These are supposed to operate (open or close) in normal load conditions only.

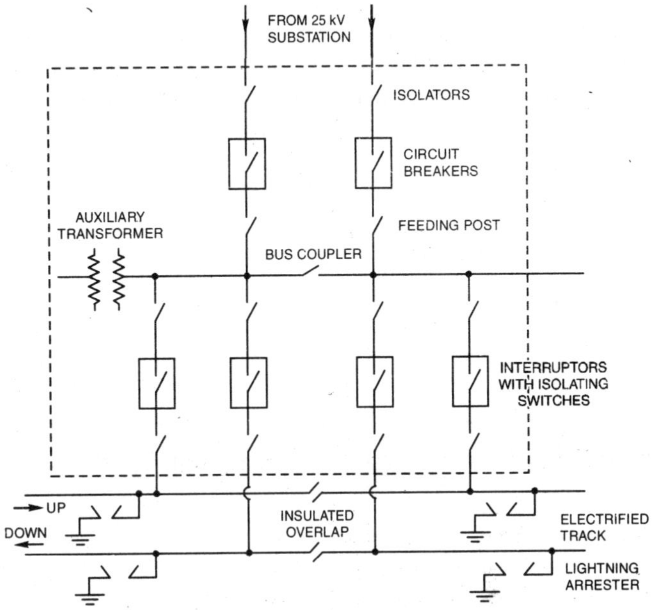

Fig. 15.16. *Feeding Post*

One auxiliary transformer of rating 25 kV/230 V, 10 kVA is also installed for 230 V supply at feeding post for charging batteries. Batteries are installed for providing uninterrupted power supply for remote control equipment, signalling and lighting circuits.

15.9.4. Sectioning and Paralleling Post

Zone covering up and down tracks between feeding post and neutral section is called one section and so each feeding post supplies two sections. The supply to each section is controlled by a circuit breaker. Figure 15.17 illustrates the power supply

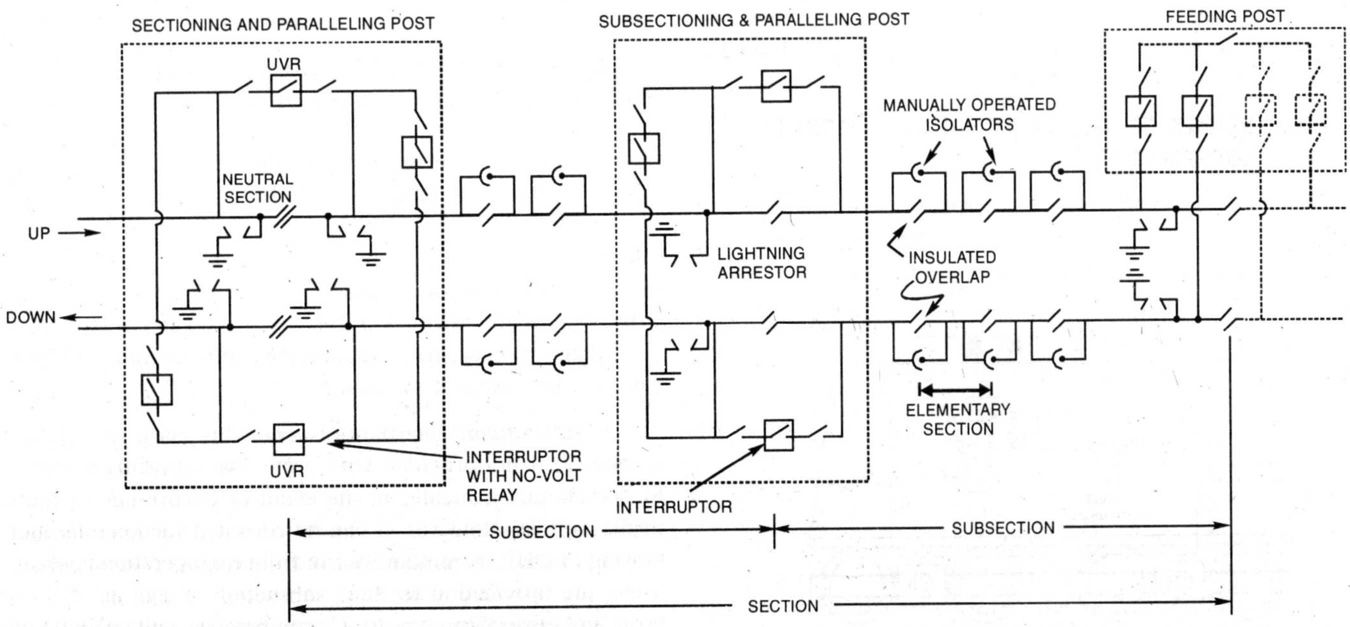

Fig. 15.17. *Power Supply Arrangement For a Two Track Line in Between a Feeding Post and a Neutral Section*

arrangement for a double track line in between a feeding post and a neutral section. Two adjacent sections supplied by different substations are connected through neutral section to avoid bridging of different phases of supply by pantograph. In case of failure of supply from one substation, supply is extended to it by adjacent section through two bridging interruptors–one for each track provided in sectioning and paralleling post. These posts also consists of two paralleling interruptors to parallel OHE of up and down tracks, one on each side of neutral section. The bridging interruptors remain normally open and are closed in case of emergency only. Bridging interruptors are provided with no-volt relay to avoid short circuiting of two different live phases. These are closed after giving warning to driver to drop the pantograph while passing in front of feeding post to which supply is extended. This precaution is necessary because no neutral section is provided in front of a feeding post except only an insulated overlap.

15.9.5. Subsectioning and Paralleling Post

Each section is subdivided into two or more sections by sectioning and paralleling posts, normally at a distance of every 10 to 15 km, so as to isolate the faulty section. Each subsectioning and paralleling post consists of two interruptors for bridging adjacent subsections and one interruptor for paralleling up and down tracks.

Sometimes only subsectioning post is provided and so there is no interruptor for paralleling up and down tracks. This is provided very rarely.

At some stations with large yards, change-over switches are sometimes provided so that the supply may be derived either from OHE of one section or the other. Normally, it is locked in one position but in an emergency it can be changed over after taking necessary precautions.

15.9.6. Elementary Section

Each subsection is further divided into elementary sections of length approximately 1 km so as to ensure rapid isolation of fault on OHE and facilitate maintenance work. Two adjacent elementary sections are separated by insulated overlap bridged by isolators, which are manually operated on no load.

15.10. BLOCK DIAGRAM OF AN AC ELECTRIC LOCOMOTIVE

Block diagram of an ac electric locomotive is depicted in Fig. 15.18. It collects current from overhead equipment (OHE)

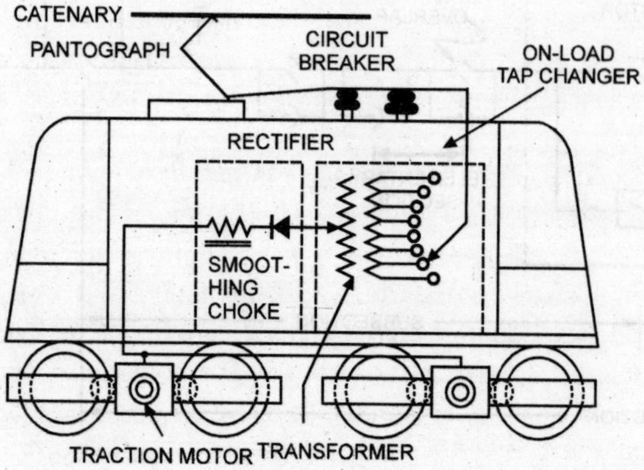

Fig. 15.18. ***Block Diagram of AC Electric Locomotive***

by means of a pantograph collector which usually carries a sliding shoe for contact with the overhead trolley wire. The locomotive essentially consists of (*i*) air circuit breaker (*ii*) on-load tap changer (*iii*) step-down transformer (*iv*) converting machinery (*v*) smoothing chokes and (*vi*) dc traction motors, as already mentioned in Art. 15.1.

15.11. MAJOR SUBSTATION EQUIPMENT

The major equipment provided at traction substations is (*i*) traction transformer (*ii*) circuit breakers and (*iii*) interruptor.

1. *Traction Transformer.* Transformers are required to step down the grid supply voltage (220 or 132 kV) to 25 kV. These are specially designed so as to withstand severe operating conditions and have large short-time peak capacity. Its windings are designed with class A insulation and are specially reinforced to withstand the stresses of frequent short circuits on 25 kV overhead equipment. Transformers are designed to have large per cent impedance (not less than 8.5% in case of 10 MVA and 10% for 12.5 MVA transformers) so as to limit the effect of short circuit. Transformers need not to be provided with on-load tap changer as voltage at substations is maintained within permissible limits by the supply authorities. However, off-load tappings are provided on the hv side at – 10%, – 5%, 0% and +10% of rated voltage so as to provide voltage between 27,500 V and 23,750 V. This has the advantage of less copper in the taps and smaller switches, which also may be all immersed in the oil used for cooling the transformers. These transformers are provided with standard accessories such as conservator tank, Buchholz relay, breather, oil level indicator, drain valve, dial gauge thermometer, relief pipe and two thermostats to indicate the oil temperature rise. One is meant to give an alarm indication while the other is for tripping of circuit breaker in case of excessive temperature rise. The cooling arrangement is obviously oil cooling, which is provided by radiators. The radiators can be easily removed for maintenance and repair simply by closing valves in the inlet and outlet pipes connecting radiators and transformer.

2. *Circuit Breakers.* As already mentioned in Art. 15.6.2, for each single unit substation there are four line circuit breakers and one transformer circuit breaker. Line circuit breaker trips in case of fault in the overhead equipment. Transformer circuit breaker on 25 kV side usually remains closed and is intended to trip on the occurrence of internal fault on the transformer. The circuit breakers provided are of low oil content type and of rating 750 A at 25 kV and rupturing capacity of 500 MVA. Circuit breakers can be operated electrically by local push button or by remote control. These can also be operated manually.

3. *Interruptor.* Interruptor is a non-automatic type circuit breaker. It is not provided with protective relays to enable it to operate automatically in the event of occurrence of fault in the system. However, it can be operated for opening and closing of OHE by remote control from remote control centre. These are provided at feeding, subsectioning and paralleling posts and subsectioning posts. Circuit breakers and interruptors are provided with the following interlocks.

1. Double pole switches cannot be operated unless the interruptor or circuit breaker is open so that sparking at the switches can be avoided.
2. The door, if any, into the bay of circuit breaker or interruptor, cannot be opened unless the double pole switch is locked in open position.
3. The circuit breaker/interruptor cannot be operated from remote control when its key is placed for operation locally.

15.12. REMOTE CONTROL CENTRE

It is provided to have a comprehensive picture of supply conditions and arrangement to operate switchgear with minimum delay so that continuity of electric supply to the track, which is of vital importance, can be maintained. Incharge of remote control centre, known as *traction power controller* (TPC), looks after it all the 24 hours and is directly responsible for carrying out all the switching operations on the electric supply system. Each remote control centre is connected to grid substations, feeding posts, sectioning and subsectioning posts, station master offices, important signalling cabins, divisional office and traffic controller office. It controls all the interruptors provided at different control posts and circuit breakers provided at different substations by remote control system that operates on transmission of voice frequency signals. This signalling system is accomplished by pilot wires with one pair for taking the operating signal and another for incoming indicating signalling. This is to avoid magnetically induced disturbance due to transient currents of switching operations. A mimic diagram depicting diagrammatically the entire OHE system as well as different substations and control posts is provided at each remote control centre to provide actual visual position of supply system to traction power controller. The mimic diagram is generally not lit up but whenever there is any change in the interruptor position at the switching post, corresponding point on the diagram gets illuminated and there is an alarm also to draw the TPC attention. The availability of the electrical supply to each and every section of OHE is monitored by the PTs at switching stations. Thus entire electrical supply system remains under the control of TPC who is fully equipped with facilities for instantaneous control of any switching operation for quick isolation of the faulty sections and restoration of power supply to the adjoining healthy sections and also for the isolation of sections for maintenance work anywhere in the system that may extend up to 250 km in every direction of the centre.

15.13. SIGNALLING

The purpose of railroad signals is to keep trains apart but permit spacing as close as safety permits. For optimum utilization of track capacity with utmost safety, automatic power-operated signalling systems are employed. Power operation also facilitates effective interlocking which reduces reliance upon human factors to the minimum. The signalling may be with semaphores arms, coloured lights or a combination of both. With semaphores or movable arms a lamp is usually so placed that, as the arm moves to one or two more positions, coloured lenses are shifted in front of the lamp to provide colour indications to the passing engine crew to correspond to the indication of the arms. Colour light signals use lamps that are brilliant enough to be seen in the brightest daylight.

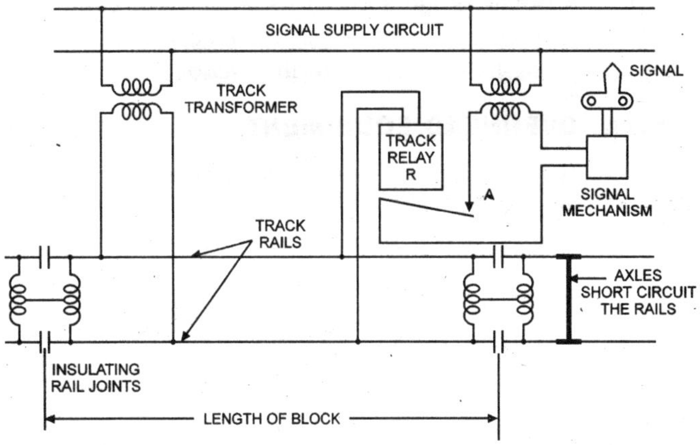

Fig. 15.19. *Simplified Diagram of Automatic Signal Track Circuits*

The rail track is divided in blocks. A block is a length of track which is controlled or protected by a signal at the entering end. A block system is a series of consecutive blocks and each block is insulated from the adjacent block.

Automatic signals are operated by the presence of a train in the block. Automatic signals may employ either battery current or ac supply. The track circuits are energized by power between 2 and 8 volts either alternating or direct current. The rails at each end of the block are cut and connected by insulated joints. There is sufficient insulation in the wood ties so that the rails constitute conductors in a closed circuit with signal power supply at one end and a relay at the other end to actuate the signal (Fig. 15.19). When a train passes the insulating joints, the axles short circuit the rails and causes the relay to drop. An advantage of this arrangement is that a broken rail or a failure of power will also cause the relay to drop and provide a stop-and-proceed indication to the signal. Track relay R holds contact A closed when the relay is energised and allows it to open by gravity if for any reason the relay is de-energized. If ac power is used, the relay is a frequency type, usually 50 or 100 Hz, so if it should be energized at a lower frequency than that for which it is designed to operate, it will open its contacts, causing the signal to go to stop position. The source of energy is of as low voltage as practicable, to prevent operation in the event of short circuit by the train is not complete— and if there should be stray currents present.

DC is used where tracks are free from stray currents from adjacent traction systems. AC operated system is employed when there is a possibility of stray currents or where the track is employed for return current. In the latter case, the blocks are coupled across the insulated joints by impedance bonds. Their function is to permit the flow of traction current around the insulated joints by offering low resistance to it and offer high impedance to the signalling current. An *impedance bond* consists of a massive laminated iron core carrying a few turns of heavy copper. Two end terminals are connected to the two rails and the mid-point or neutral is brought out which is connected to a similar bond beyond the insulated joint.

When the track is unoccupied, the relay circuit is completed, via the two rails and a clear signal is hoisted. When a train enters the section, its axles short circuit the rails and causes the relay to drop. "Track occupied" signal is hoisted as long as the train is in the section.

15.14. OVERHEAD EQUIPMENT

Simplest type of overhead equipment consists of a single contact wire supported either by bracket or by overhead span. The maximum distance between two consecutive supports with such a system is restricted to 30 m and speed is limited to 30 kmph. The single contact wire system, therefore, finds application for tramcars or in complicated yards and terminal stations where simplicity of layout is desired and speeds are low.

Catenary construction (either single catenary or compound catenary), as described in Art. 15.4.2, is used for railways.

15.14.1. Automatic Weight Tension and Temperature Compensation

For sparkless current collection by pantograph collector under high speeds it is necessary that the contact wire should not only remain horizontal at the time of erection but should remain so under all conditions of temperatures and wind pressures likely to be encountered in the service. In an unregulated OHE, in which there is no provision for temperature compensation, sag and tension in catenary and contact wire are so adjusted that contact wire remains approximately horizontal at the time of erection. This type of unregulated OHE is not suitable for current collection at high speeds as encountered on main lines as tension varies inversely as temperature. It affects the stiffness of the line and its dynamic behaviour.

In order that the contact wire remains at the same level under all weather conditions, it is necessary to keep the tension in the catenary and contact wire constant under all temperatures likely to be encountered in service and allow movements of wires along the track which result due to changes in temperature. This is achieved by using regulated OHE at either end of the tension length (Fig. 15.20). The tensioning device consists of a pulley block or a winch with suitable reduction ratio (about 5). Increase in tension reduces the static elasticity of the OHE and makes it more uniform and thus improves the dynamic behaviour and current collection. Contact and catenary wires are given tension of about 1,000 kg each so as to limit the lateral stresses exerted on the masts and foundations due to increase in tension. Cast iron weights of 400 kg are required by the automatic tensioning device (Fig. 15.20) in order to develop tension of 1,000 kg in each of contact and catenary wire. In order to prevent longitudinal to and fro oscillations of OHE in the running condition of train, it is fixed in the centre, which is known as *anticreep* (Fig. 15.21). For the temperature range experienced in electrified sections in India, maximum distance between the anticreep and the end of tension length *i.e.,* half tension length is restricted to 750 metres.

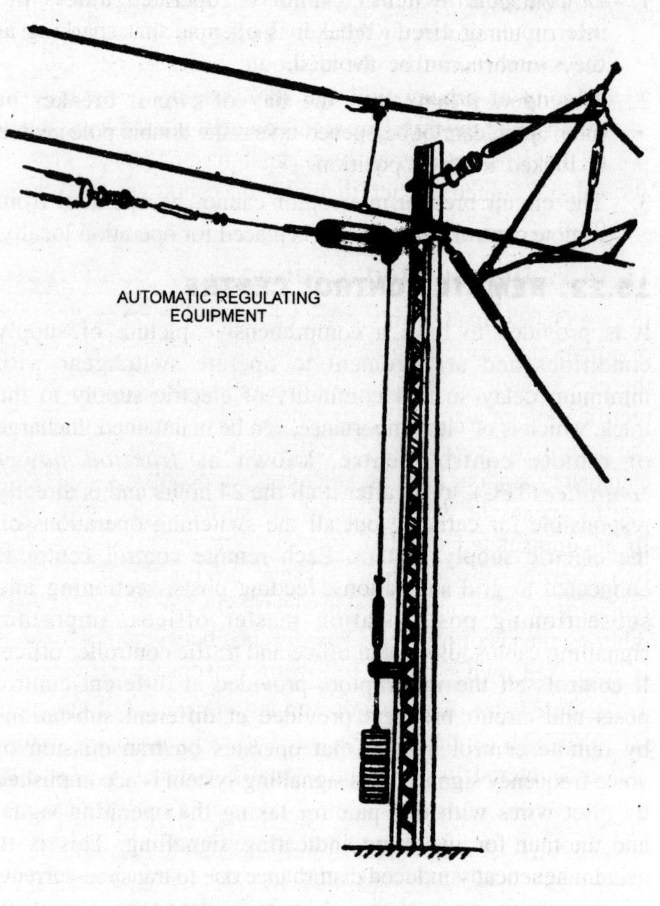

Fig. 15.20. ***Automatic Tensioning Device***

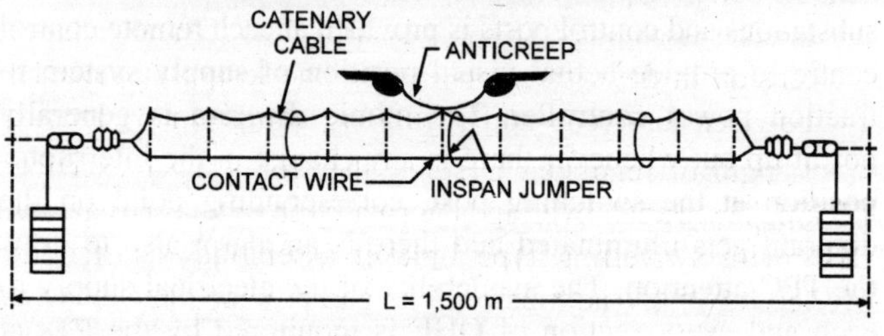

Fig. 15.21

Weight tensioning has the following advantages:

1. Sparkless current collection under all atmospheric conditions with consequent reduction of wear and tear of both contact wire and collecting strips of pantograph.
2. Automatically taking up of creep of the conductors by the tensioning device with the passage of time.
3. Some economy in the design of the supports and foundations because below 40°C temperature, the highest tensions attained are considerably less than with unregulated OHE.
4. Alleviation of problem of providing adequate clearances under overline structures and in tunnels as with fixed tension in the OHE, fluctuations in its height are eliminated.
5. Owing to contact tension in the conductors, there is only one critical velocity of propagation of waves in

the contact wire and this velocity is usually greater than that of unregulated OHE. In high speed traction, it is very important that there should be only one critical velocity of propagation and this should be higher than the speed of the train. From this point of view, the regulated OHE is advantageous because it has only one critical velocity much higher than the maximum speed of the train.

15.14.2. Uninsulated Overlap

In case OHE is made continuous, which is possible by splicing of conductors, long lengths of it would be affected whenever there is a fault or damage. It will also need long time for rectification of damage. It is, therefore, essential to divide the length of elementary section into a number of sub-elementary sections each of which is separated from its adjacent one by means of what is known as an *uninsulated overlap*. In an uninsulated overlap, two contact wires belonging to two adjacent sub-elementary sections are run parallel to each other for at least one span and electrically connected to each other by means of jumpers. The distance between the overlaps is restricted owing to the following reasons:

1. Limited size of the conductor and the weight of the drum on which conductor is rolled. This is to avoid difficulties encountered in handling in transportation.
2. Necessity of isolation of portions of OHE affected by the fault for effective operation and easy maintenance. It makes necessary that overlaps are provided at lesser intervals.
3. The friction of components used on swivelling type of bracket assemblies. This makes difficult to maintain uniform tension in the conductors in case the length of the OHE between anticreep and regulating equipment is long.

Length of regulated OHE is, therefore, never exceeded 1.5 km. Three span overlap is normally provided on regulated OHE with swivelling type bracket assemblies as standard overlap for 25 kV ac traction. The gap provided between two contact wires, in parallel running portion is about 20 cm which is maintained by mechanical linkage of the conductors.

15.14.3. Insulated Overlap

Insulated overlaps are installed chiefly for providing isolating facilities required for the operation and maintenance of OHE. In case two lengths of OHE belong to two adjacent elementary sections, there will be insulated overlap span. Insulated overlaps are also provided at feeding posts, subsectioning posts and booster stations. Insulated overlaps are similar to uninsulated overlaps except for the following:

1. In order to avoid accidental contact between the contact wires, gap between them is kept more in case of insulated overlap.
2. Electrical continuity between the two sections is maintained by means of jumper connections in series with isolator or interruptor. With such an arrangement, a section can be made dead and work can be carried out on the other equipment away from the overlap.
3. The anchoring portion of the catenary and contact wire is separated from the rest of the portion by means of insulators provided at a distance of 2.5 m from the supports and within overlap span. The tail ends of the anchoring OHEs are connected to the other nearby equipment so as to prevent it from attaining any other potential.

In dc traction, single span insulated overlap is employed. Lateral separation between the two equipments is kept as 0.3 m for 1,500 V and 0.4 m for 3,000 V. This separation is maintained by means of insulated rods clamped on to the wires. In 25 kV ac traction system, this lateral separation is increased to 0.5 m. Three span overlap is the normal type of insulated overlap provided on 25 kV ac OHE with swivelling type of bracket assemblies.

15.14.4. Neutral Section

In dc traction, supply to OHE can be provided from both ends so that voltage drop can be reduced and in such a case substations operate in parallel. In case it is felt necessary, say for sake of convenience in maintenance, to have some boundary of supply between two substations, it can be had simply by means of insulated overlap. There will be no damage in this case even if pantograph bridges the OHE. But in single phase ac traction above arrangement is not permissible because adjacent substations tap different phases of the three-phase system so as to equalise load on the three phases. So momentary passing of pantograph under insulated overlap will result in short circuit between two phases of the supply system and damage the pantograph and OHE. This situation is avoided by interposing a small length of OHE, called the *neutral section*. It is insulated from both sides and is not connected to any source of power supply. It's main function is to allow physically smooth and electrically sparkless passage of pantograph from one section to the other. As per Indian Railway practice, neutral section is located in the middle* of two feeding posts and in front of sectioning and paralleling post, as illustrated in Fig. 15.17.

Important precautions in connection with neutral section are given below:

1. Neutral section should be located far away from stopping signals so that the train can pick up enough momentum to enable it to coast through the neutral section without difficulty because there is no power supply in the neutral section.
2. Neutral section should neither be located on upgradient nor on curves because neutral section has to be negotiated with power off.

   The best position for its location would be at mid sections between two feeding posts, away from stopping signals and on down gradient in case of single track or straight level section in case of multi-track.

* In some countries neutral section is provided in front of feeding post also so as to avoid short circuit between the phases in the case of emergency when there is shutdown on the feeding post and power supply is extended from adjacent feeding posts. But Indian Railways did not consider it necessary as it is required only during emergency and additional expenditure will have to be incurred.

3. It is essential that the length of the neutral section should be as small as possible but not of length less than the spacing between extreme pantographs. This is because the train is to negotiate the neutral section with power off and if it gets stalled in this section, there will be dislocation in the traffic and arrangements will have to be made to feed the neutral section. In case its length is shorter than the spacing between extreme pantographs, they will short circuit the OHEs supplied from different phases.

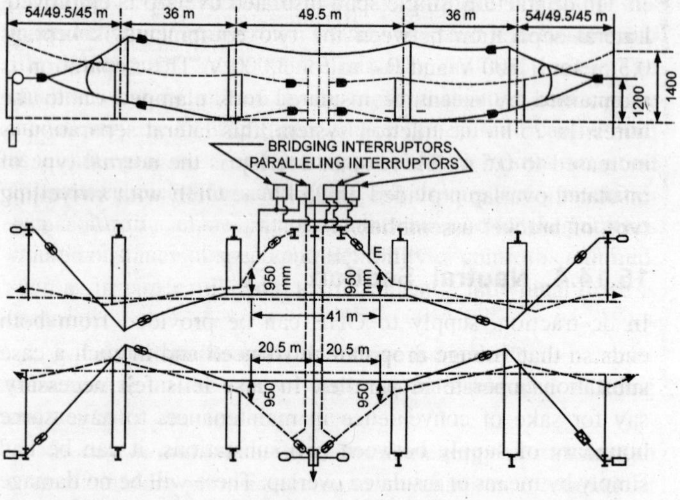

Fig. 15.22. *Standard Neutral Section Adopted By Indian Railways*

When pantographs at the extreme ends of multiple head trains are raised they are about 40 m apart. Since both the air gaps at the extreme ends of the neutral section have not to be bridged simultaneously, the minimum length of the neutral section adopted by Indian Railways is 41 metres. Standard neutral section as adopted by Indian Railways is illustrated in Fig. 15.22.

Long neutral sections are usually provided in suburban sections and steep graded sections. These neutral sections make use of section insulators in place of air gaps at the extreme ends of standard neutral section. These section insulators are fixed within the span immediately after supports in the same direction in case of double tracks and in opposite direction to each other in case of single tracks, as illustrated in Fig. 15.23.

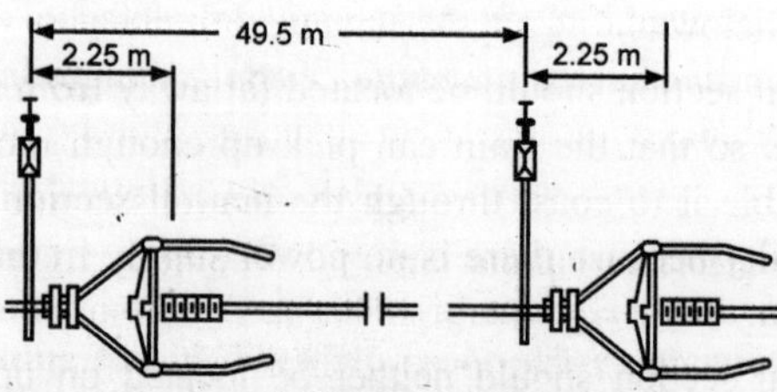

Fig. 15.23

15.14.5. Section Insulators

Section insulators are provided for insulating OHE of one elementary section from the OHE of another adjacent elementary section such that at cross-overs from one track to the other, from main line to siding etc. as shown in Fig. 15.17. Section insulator, as illustrated in Fig. 15.24, consists of straining insulator with two runners connected to one of the contact wires. Runners are at the same height as the contact wire on the other side and so shaped as to permit smoother passage of pantograph as the train passes underneath. Runners extend and overlap the contact wire on the other side so that the locomotive draws interruption free current.

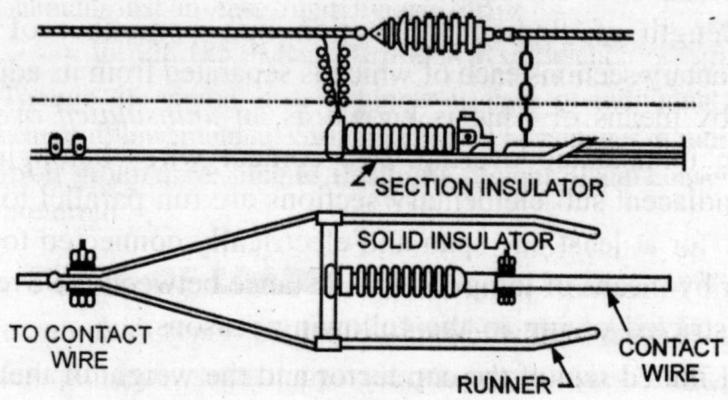

Fig. 15.24. *Section Insulator*

15.15. INTERFERENCE IN TELECOMMUNICATION CIRCUITS

In all traction systems track is used as a return path for the current. Some part of the current flows through the track and the remaining portion of the current flows through earth, as shown in Fig. 15.25. Current flowing through earth causes interference with communication lines. Now to constrain the return current to flow through restricted path *i.e.*, along the track some steps are required to be taken. In ac traction system use of booster transformer, described in Art. 15.16, is made while in dc traction system negative booster, described in Art. 15.17, is used to reduce the potential drop across the track.

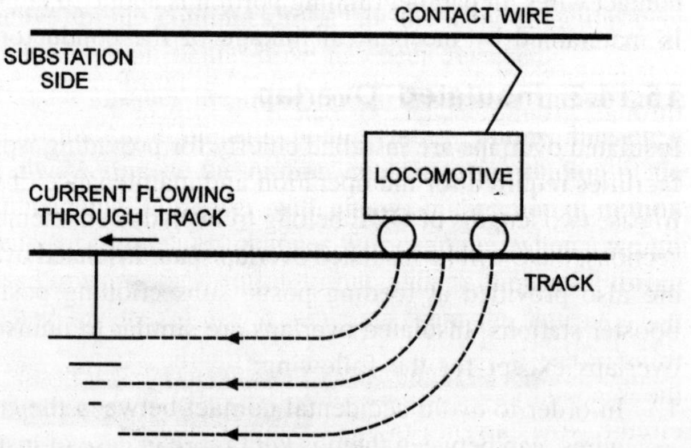

Fig. 15.25

Path of return current is also constrained by reducing the track resistance. The resistance per km length of continuous track rail is of order of 0.05 Ω but the resistance of fish-plate joints employed in jointing the commercial length rail mechanically is more than that of a rail. Hence in order to reduce the resistance of the track as a whole the joints are supplemented with a good conductor or "bond". Modern bonds are of the flexible type and are designed to be as short as the conditions of locations permit. The bonds are usually fixed under the fish plates as they are protected from damage while carrying out the repairs to the paving.

In single phase ac traction system presence of harmonics in supply system also causes interference in telecommunication systems.

Interference in telecommunication and signalling lines is also caused due to electromagnetic induction. Current flowing through contact wire sets up a magnetic field around itself. This magnetic field links with nearby circuits and induces emf in circuits.

Electrostatic induction in signalling and telecommunication circuits also causes interference in these circuits. There occurs a voltage gradient between hv contact wire and earth. Now statically induced emf in communication and singnalling lines is produced because of its positioning voltage gradient.

The interference effects on telecommunication circuits owing to reasons mentioned above (last three paras) can be reduced by taking following steps.

(*i*) All telecommunication circuits are installed underground in cables instead of in overhead lines.

(*ii*) Telecommunication or signalling lines are provided with aluminium sheath instead of lead.

(*iii*) Use of PVC or rubber covering is made to prevent sheath corrosion.

(*iv*) Telecommunication circuits may be removed to the distant place, to the extent practically possible, from the railway track.

(*v*) Long telecommunication circuits are broken by providing isolating transformers.

(*vi*) Booster transformer employing return feeder is also used to avoid magnetic linking of telecommunication lines with power lines.

15.16. BOOSTER TRANSFORMER

In ac traction system return current mainly flows to the substation through track leaving a small portion, which flows through ground because of leakage of current from track to the ground. It has already been described that this leakage current causes a heavy interference with communication lines which are running near the track. So it is very necessary to avoid flow of leakage current through ground. For this purpose booster transformer is used. Booster transformer consists of the windings of unity turn-ratio. Primary winding of booster transformer is inserted in series with the contact wire and the secondary is inserted in series with the return current circuit. Now current flowing through contact wire also passes through primary winding which induces current in the secondary winding in equal amount. This induced current in secondary winding forces the return current to flow through only return circuit, and so avoids leakage current.

There are two methods of connecting booster transformers.

1. *Rail-Connected Booster Transformer.* Circuit diagram for such connections is shown in Fig. 15.26. Primary winding of the booster transformer is connected in series with the contact wire and the secondary is connected in series with the rails. Now induced voltage in the secondary winding due to contact wire current flowing through primary winding causes flow of return current only through the rails. The main drawback of this system is that rail joints are insulated at the location of booster transformer and there may occur an insulation puncture of insulated rail joint which makes the booster transformer ineffective because of short circuiting the secondary winding. There may be also a voltage rise between rails and between rail and earth which may be dangerous for human life.

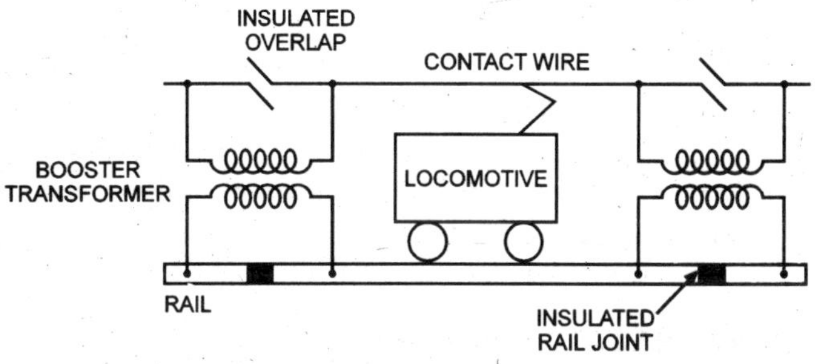

Fig. 15.26. ***Rail-Connected Booster Transformer***

2. *Booster Transformer Using Return Feeder.* Connection diagram for booster transformer using return feeder is shown in Fig. 15.27. In this method track is not used for return current flow to substations, but the return current flows through a return feeder which is provided in addition to contact wire. This system is more effective than rail-connected booster transformer system and, therefore, used more extensively. Turn ratio of booster transformer employed in this system also is unity. Primary winding of booster transformer is connected in series of contact wire and the secondary winding is connected in series of return feeder.

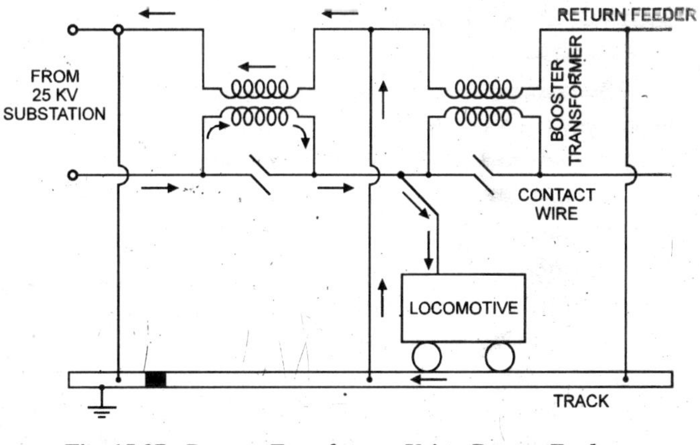

Fig. 15.27. ***Booster Transformer Using Return Feeder***

Return feeder is connected to rail at mid points between two booster transformers. Now current flowing through primary winding induces a voltage in the secondary winding which forces the current to flow only through return feeder.

Drawback of booster transformer is that it is connected in series of traction circuit causing increase in impedance of the circuit. Thus regulation of voltage in locomotive is increased. To improve the voltage regulation it is necessary to locate feeding posts closer, otherwise a capacitor of proper value may also be introduced in series of traction circuit. Former is costlier method and, therefore, the later is mostly used.

Precautions in Connecting Booster Transformers.

1. Booster transformers should be able to withstand severe mechanical stresses which will develop under short-circuit conditions when current of thousands of amperes flows.
2. Primary of booster transformer is connected at the two sections of insulated overlap. When the train passes under the insulated overlap, pantograph short circuits the primary winding for a moment and makes booster transformer ineffective. So precautions should be taken that booster transformer should not be located near the signals or the places where the train is likely to stop.
3. Insulation of primary winding of booster transformer is kept at the level of 25 kV. Secondary windings insulation level is kept at 3 kV because a high potential is created in secondary winding under fault conditions.

15.17. NEGATIVE BOOSTER

In case of dc traction system, return current flows through the rails, to the substations. So due to this current flow through rails some voltage drop exists along the track. Due to potential difference along the track some of the return currents leak out to the earth and follow low resistance paths provided by water pipes, gas mains, cable sheaths etc. as shown in Fig. 15.28. These leakage currents cause electrolytic corrosion to occur at places where the current leaves the pipes as they carry the metal of the pipe (acting as anode) to the electrolytic solution (moisture and salts in the surrounding soil).

The statutory regulations, according to which the voltage drop in the rails is limited to 7 volts and the voltage between any pipe and the rails is limited to 1 volt positive and 3 volts negative, are to minimize the corrosion of pipes and cable sheaths adjacent to track.

The voltage drop in the track rails is maintained within the limit of 7 volts by adopting any one of the following methods.

1. By making return path of very low resistance, which may be achieved by providing good bonds or by employing insulated negative feeders.
2. By discouraging the entry of current into the pipes by inserting occasional insulating joints.
3. By using negative booster.

As shown in Fig. 15.29, the negative bus-bar is connected to the track rail through negative feeder and the positive bus-bar is connected to the trolley wire near the generating stations. There are two boosters, one is positive booster which adds voltage to the line and the other is the negative booster which lowers the potential of the point to which it is connected. Both the boosters are mechanically coupled together and driven by a dc motor. Negative booster is separately excited dc generator whose armature is connected with negative busbar at the far end of the track through negative feeder. Field of negative booster is connected in series with positive feeder supplying the corresponding sections of the trolley wire. The boost on the feeder is adjusted by means of a diverter rheostat connected in parallel with the field winding.

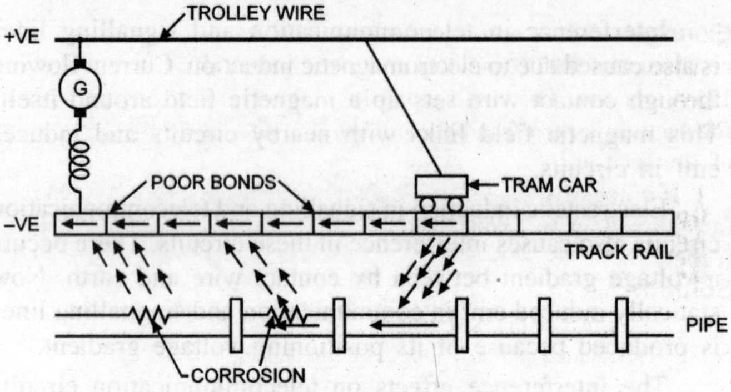

Fig. 15.28. ***Electrolysis by Earth Currents***

Sometimes a positive booster is also connected to the + ve side in addition to – ve booster on – ve side in order to compensate the voltage drop in trolley wire. Positive booster is a self excited series generator which is connected across the + ve busbar and at the far end of trolley wire in series with the field of the negative booster.

Example 15.4. In a section of tramway system the track resistance is 0.08 Ω and the total current which may be considered to be taken uniformly along the length of the section is 300 amperes. Calculate the current to be carried by a negative feeder connected to the distant end of the section to reduce the voltage drop between the ends of the section to 6 volts.

Solution : Let the current to be carried by a negative feeder connected to the distant end of the section to reduce the voltage drop between the ends of the section to 6 volts be of I′ amperes.

Total current, I = 300 amperes

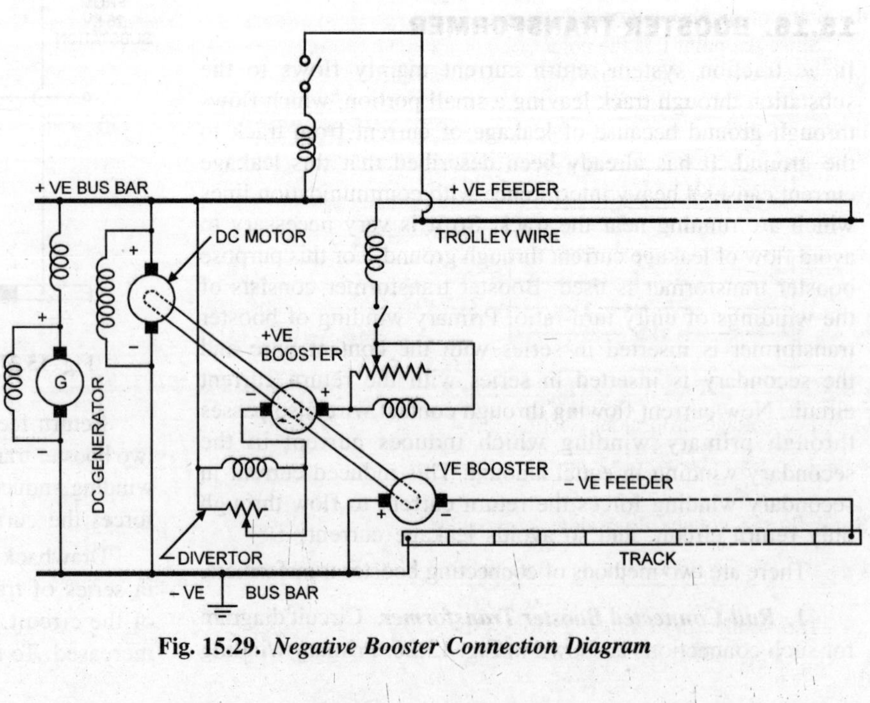

Fig. 15.29. ***Negative Booster Connection Diagram***

Total track resistance, R = 0.08 Ω

Since voltage drop between the ends of the section $= \frac{IR}{2} - I'R$

$$\therefore R\left(\frac{I}{2} - I'\right) = 6$$

$$\text{or} \quad 0.08\left(\frac{300}{2} - I'\right) = 6$$

$$\text{or } I' = 150 - \frac{6}{0.08} = 75 \text{ A} \quad \textbf{Ans.}$$

Example 15.5. A section of a tramway 3 km long has a track resistance of 0.02 Ω per km and a uniformly distributed load of 125 A per km. The earth potential of the track at the far end of the section is zero. Find the potential of the track at the far end of the section.

If a negative feeder and booster are employed to reduce the far end voltage to 7 volts, find the current carried by the negative feeder.

Solution :

Current per km length of the track, i = 125 A

Resistance per km length of the track, r = 0.02 Ω

Length of track, l = 3 km

Potential of the track at the far end

= Potential drop along the track

$$= \frac{irl^2}{2} = \frac{125 \times 0.02 \times (3)^2}{2} = 11.25 \text{ volts} \quad \textbf{Ans.}$$

Let the current carried by the negative feeder be of I amperes

Since the potential drop along the track $= \frac{irl^2}{2} - Irl$

$$\therefore \frac{125 \times 0.02 \times (3)^2}{2} - I \times 0.02 \times 3 = 7$$

$$\text{or } I = 70.833 \text{ amperes} \quad \textbf{Ans.}$$

Example 15.6. A section MN of a tramway 5 km long has a resistance of 0.02 Ω/km and a loading of 200 A/km which may be assumed to be uniformly distributed. A negative feeder is connected to the track at a point X, 3 km from M. If the end M is earthed and the potential of X is 1 volt above M, calculate the current carried by the negative feeder.

Solution : Let the point P at a distance of x km from point M, be the point of highest potential.

Since the point P is the point of highest potential on the section MX, therefore, current at this point is zero and increases towards M and X. Direction of currents in different sections of track are shown in Fig. 15.30.

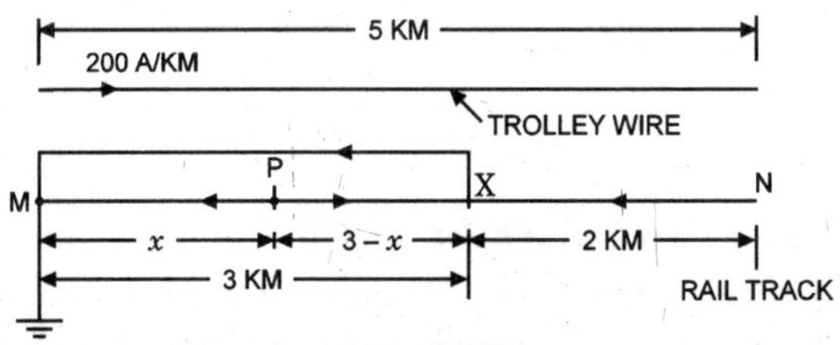

Fig. 15.30

The voltage drop along section PM,

$$V_{PM} = \frac{ix \times rx}{2} = \frac{irx^2}{2} = \frac{200 \times 0.02 x^2}{2} = 2x^2$$

Similarly the voltage drop along section PX,

$$V_{PX} = \frac{ir(3-x)^2}{2} = \frac{200 \times 0.02\,(3-x)^2}{2} = 2(3-x)^2$$

The voltage drop along section XM,

$$V_{XM} = V_{PM} - V_{PX}$$

$$\text{or } 1 = 2x^2 - 2(3-x)^2$$

$$\text{or } x = \frac{19}{12} \text{ km}$$

$$\therefore I_{PX} = 200 \times \left(3 - \frac{19}{12}\right) = 283.33 \text{ A}$$

and $I_{NX} = 200 \times 2 = 400$ A

∴ Current through negative booster

$= I_{PX} + I_{NX} = 283.43 + 400 = 683.33$ A **Ans.**

Example 15.7. A section ABC of a tramway rail system is 4.8 km long. A is earthed and B is 3.2 km from A. The resistance of the rails is 0.02 Ω/km. The loading is 250 A per km and may be assumed uniformly distributed.

A negative feeder (connected to a booster) is tapped to the rails at B and maintains B at 2 volts above earth potential. Calculate the maximum pd across any two points on the rails. Also determine the voltage and kW capacity of the booster used. Resistance of negative feeder is 0.02 Ω.

Solution : The rail track is earthed at point A *i.e.*, the potential of point A, $V_A = 0$

Potential of point B, $V_B' = 2$ volts

Current per km length of track, i = 250 A

Resistance per km length of track, r = 0.02 Ω

Let the point D at a distance of x km from point A be point of highest potential on the section AB.

Since point D is the point of highest potential on the section AB, current at this point is zero and increases towards A and B. Directions of currents in different sections are shown in Fig. 15.31.

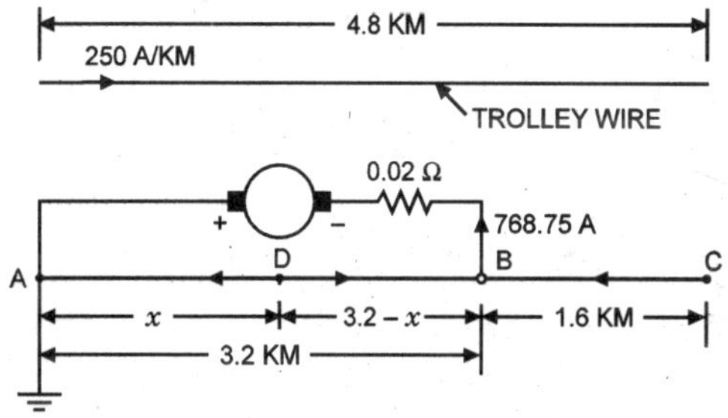

Fig. 15.31

The voltage drop along section DA;

$$V_{DA} = \frac{irx^2}{2} = \frac{250 \times 0.02 x^2}{2} = 2.5x^2$$

The voltage drop along section DB,

$$V_{DB} = \frac{ir(3.2-x)^2}{2} = \frac{250 \times 0.02\,(3.2-x)^2}{2}$$

$$= 2.5\,(3.2 - x)^2$$

The voltage drop along section BA = $V_{DA} - V_{DB}$

$\therefore 2 = 2.5x^2 - 2.5\,(3.2 - x)^2$

$\therefore x = 1.725$ km

The current distribution will be as follows :

At point C the current is zero and gradually increases till point B is reached, where it is 1.6 × 250 *i.e.*, 400 amperes. At point D current is zero and gradually increases towards points A and B. At point A the current is 250 × 1.725 = 431.25 amperes and at point B the current is 250 (3.2 – 1.725) *i.e.*, 368.75 amperes.

The points D and C, where the current is zero, are at maximum potential.

Potential of point C

= Potential of point B + voltage drop is section BC

$$= 2 + \frac{250 \times 0.02 \times (1.6)^2}{2} = 8.4 \text{ volts}$$

Potential of point D

$$= \frac{irx^2}{2} = \frac{250 \times 0.02 \times (1.725)^2}{2} = 7.439 \text{ volts}$$

Thus the maximum potential difference on any two points on the rails = 8.4 volts **Ans.**

The current carried by the negative booster

= 400 + 368.75 = 768.75 A

Voltage drop in negative feeder = 768.75 × 0.02 = 15.375 V

Boost to be given by negative booster = 15.375 – 2 = 13.375 V

$$\text{Capacity of the booster} = \frac{13.375 \times 768.75}{1{,}000} = 10.28 \text{ kW} \quad \textbf{Ans.}$$

Example 15.8. A section of tramway ABC is 6 km long and earthed at A. Its resistance is 0.01 Ω per km and loading of 200 A/km uniformly distributed. Negative feeder with booster is taken off from point B, 4 km from A. If potential of point B is 4 V above earth, calculate the rating of the booster, negative feeder resistance being 0.02 Ω. [*Electric Traction Pb. Univ. December 1991; Rajasthan Univ. Utilization of Electrical Power 2001*]

Solution : Let the point D, at a distance of *x* km from point A, be the point of highest potential on section AB.

Since point D is the point of highest potential on the section AB, current at this point is zero and increases towards A and B. Directions of currents in different sections are shown in Fig. 15.32.

Voltage drop along section DA,

$$V_{DA} = \frac{irx^2}{2} = \frac{200 \times 0.01 \times x^2}{2} = x^2$$

$\because i = 200$ A/km and $r = 0.01$ Ω/km

Voltage drop along section DB,

$$V_{DB} = \frac{ir(4-x)^2}{2} = \frac{200 \times 0.01 (4-x)^2}{2} = (4-x)^2$$

Voltage drop along section

$BA = V_{DA} - V_{DB}$

or $4 = x^2 - (4-x)^2$

or $x = 2.5$ km $\qquad \because$ potential of point B is 4 V

At point C the current is zero and gradually increases till point B is reached where it is 2 × 200 = 400 A. At point D current is zero and gradually increases towards points A and B. At point A current is 2.5 × 200 = 500 A and at point B current is (4 – 2.5) × 200 = 300 A.

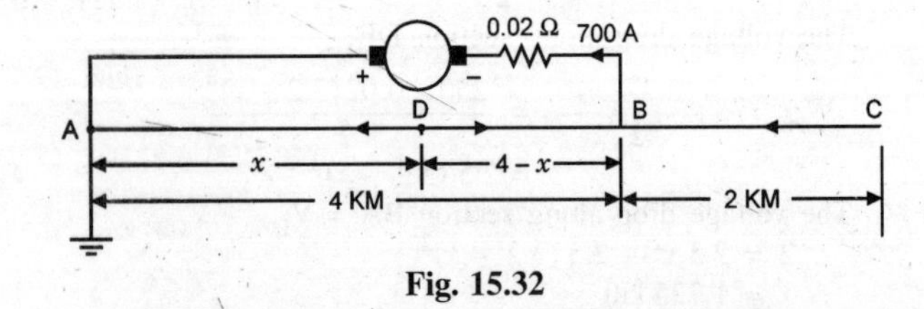

Fig. 15.32

Current carried by negative booster = 400 + 300 = 700 A

Voltage drop in negative feeder = 700 × 0.02 = 14 V

Boost to be given by negative booster = 14 – 4 = 10 V

$$\text{So rating of the booster} = \frac{700 \times 10}{1{,}000} = 7 \text{ kW} \quad \textbf{Ans.}$$

Example 15.9. A section of tramway by using positive and negative booster arrangements is supplied by the dc generator of voltage rating 600 V output. When trolley is at the far end of the section it should get voltage difference between trolley wire and track of at least 578 V. Voltage drop across the trolley wire is 60 V and that across the track is 12 V when the trolley is at far end.

(*a*) Determine the voltage available to the trolley when it is at the far end, and boosters are not used.

(*b*) How much would positive booster boost to keep the voltage at the far end within limit if negative booster provides boost of 10 volts.

Solution : Voltage drop along the trolley wire up to the far end,

$V_W = 60$ V

Voltage drop along the track up to the far end, $V_T = 12$ V

Total voltage drop up to the far end = 60 + 12 = 72 V

(*a*) So voltage available to trolley when it is at the far end, without using boosters.

$V_A = V_S - (V_W + V_T) = 600 - (60 + 12) = 528$ V **Ans.**

(*b*) Boost given by negative booster, $V_N = 10$ V

Let the boost given by the positive booster be V_P volts

Recommended voltage to the trolley when it is at the far end, $V_R = 578$ V

Supply voltage at the generator side, $V_S = 600$ V

Voltage drop along the track with – ve booster in use

$V_{TN} = V_T - V_N = 12 - 10 = 2$ V

Voltage drop along the trolley with + ve booster in use

$V_{WP} = V_W - V_P = (60 - V_P)$ volts

Total voltage drop at the far end with boosters in use

$= V_{TN} + V_{WP} = 2 + (60 - V_P) = (62 - V_P)$ volts

Admissible maximum voltage drop = 600 – 578 = 22 V

$\therefore 62 - V_P = 22$ V

or $V_P = 62 - 22 = 40$ V

So positive booster should provide a boost of 40 V. **Ans.**

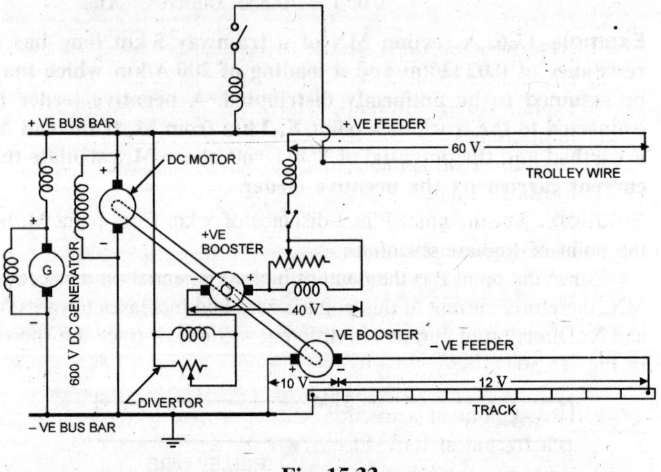

Fig. 15.33

15.18. PROTECTIVE DEVICES

1. *Circuit Breakers.* The circuit breaker is required to protect the electrical equipment against excessive overloads automatically. It is connected between the current collector and the main wiring. One circuit breaker is required in case of tramcar and two circuit breakers are required in case of trolley bus, one in each trolley circuit. The main contacts of the circuit breaker are located in an arc chute and are provided with a magnetic blow-out. The blow-out coil actuates also the tripping mechanism. It is also provided with a handle to trip the circuit breaker when it is moved to the "off" position by hand.

2. *Radio Interference Suppressor.* The emfs of radio frequency may be induced in the wiring due to arcing of sparking at the collector, controller, motors etc. and radiated out from the trolley wires. The most serious interference occurs at the contacts of the master controller in systems with electromagnetic contactors operated at line voltage. Radio-interference suppressor consists of two radio-frequency chokes, one connected in each supply lead to the master controller, a capacitor filter comprising three star-connected capacitors connected between the trolley wires and chassis. The capacity of each capacitor is about 0.1 μF.

3. *Lightning Arrestor.* Only the vehicles operating in the area liable to lightning storms are provided with lightning arrester or surge diverter. One type consists of a small cylindrical block of carborundum held between two electrodes. One electrode with a small air gap in series is connected to the positive trolley wire and the other one to the – ve trolley wire or earth.

Exercises

1. What are the requirements that a current collector gear of electric vehicle should fulfil ?
   Describe with a neat diagram the pantograph collector.
2. What are different types of current collectors? Give their merits and demerits. **[Rajasthan Univ. Utilization of Electrical Power 2005, Rajasthan Technical Univ. 2011]**
3. Discuss and compare various arrangements of current collection used in electric traction.
   [M.D. Univ. Utilization of Electric Power & Traction Dec. 2008]
4. Discuss different types of catenary construction for traction system.
   [Pb. Technical Univ. Electric Drives and Utilization May 2007]
5. Discuss the limitations of single catenary construction and explain some alternative construction to remove these limitations. **[Pb. Univ. Electric Traction June 1993]**
6. Describe with suitable diagram the compound catenary in overhead lines.
7. Describe a typical traction substation with the help of a sketch of substation equipment layout.
   [Rajasthan Univ. Utilization of Electrical Power 2002, 2006, 2007, 2008]
8. Discuss the location of substations with regard to ac and dc systems of electric traction.
   [Pb. Univ. Electric Traction Dec. 1991]
9. Give layout of a traction substation and describe it.
   [Pb. Technical Univ. Electric Drives & Utilization May 2008]
10. What are the functions of dc traction substation ? What is the major requirement for dc substation ?
    [Pb. Technical Univ. Electric Drives & Utilization Dec. 2007]
11. Discuss the various feeding and distributing systems used in electric traction.
    [Gorakhpur Univ. Utilization of Electric Power & Traction 1983]
12. State the various 'statutory regulations' that should be observed in cases of feeding and distributing systems for tramways. Why are such regulations imposed ?
13. Explain the working of a remote control centre. What are its advantages ? What is mimic diagram ?
14. Briefly discuss the salient design features of OHE (overhead equipment) giving suitable diagrams.
    Differentiate between insulated and uninsulated overlaps. Also explain the importance of neutral section in OHE.
    [Pb. Univ. Electric Traction Dec. 1992]
15. Explain the importance of neutral section in traction system.
    [Rajasthan Technical Univ. Utilization of Electrical Power 2008, 2010]
16. Explain why neutral section is provided in the OHE of ac traction system but not in dc traction system. What factors determine the span length of the OHE ?
17. What is critical velocity, discuss its significance.
    [Pb. Univ. Electric Traction Dec. 1991]
18. With the help of neat diagrams, briefly explain the following :
    (*a*) Pantograph (*b*) Negative booster
    (*c*) Interference effect of railway electrification on communication circuits. **[Pb. Univ. Electric Traction April 1975]**
19. What steps can be taken to reduce the interference in the telecommunication circuits ? Explain the use of booster transformer. **[Pb. Univ. Electric Traction July 1992]**
20. In traction system employing the running rails as return conductors, describe measures and precautions that may be necessary to keep the voltage within reasonable limits.
21. Describe the negative booster for use on a traction system.
    [D.E.I. Agra 1989]
22. Explain with relevant diagram the necessity of using positive and negative boosters in dc electric traction.
    [B.H.U. Elec. Traction & Utilization 1978]
23. Explain the functions of (*i*) feeder booster (*ii*) negative booster in an electric tramway system.
    [Agra Univ. Traction & Utilization of Elec. Power 1982]
24. Draw and describe the complete connection diagram of a negative booster along with the motor starter set in a 650 V dc tramway line. If the potential at the point of connection is zero, show the distribution of current and voltage in the rail at least in two successive sections.
    [Agra Univ. Traction & Utilization of Elec. Power 1978]
25. Discuss the protection of underground installation against corrosion due to railway electrification. Discuss the application of negative booster. **[Pb. Univ. Electric Traction June 1993]**
26. Write short notes on the following:
    (*i*) Comparison of various arrangements of current collection used in electric traction. **[U.P.T.U. 2005-06]**
    (*ii*) Conductor rail system. **[W.B. Univ. of Technology 2009]**
    (*iii*) Collector gear for overhead equipment. **[Agra Univ. 1976]**
    (*iv*) Overhead collector gear employed on trams and electric trains. **[Agra Univ. 1978]**
    (*v*) Pantograph. **[Allahabad Univ. 1975]**
    (*vi*) Methods of supporting overhead control contact wire equipment for electric tramways and railways. **[Agra Univ. 1978]**
    (*vii*) Compound catenary. **[Agra Univ. 1976; Pb. Univ. Dec. 1991]**
    (*viii*) Effect of temperature on level of contact wires. **[Pb. Univ. Nov. 1974]**
    (*ix*) Traction substations.
    (*x*) System of remote control.
    (*xi*) Booster transformer.
    (*xii*) Protection of underground installations against corrosion. **[Pb. Univ. July 1992]**
    (*xiii*) Phenomenon of electrolysis by earth currents. **[G.B.T.U. 2010-11]**
    (*xiv*) Negative booster.
    [Pb. Univ. December 1992; Rajasthan Univ. 2003, 2005]

Short Answer Type Questions With Answers

Q. 1. What is the arrangement of power supply on Kanpur-Tundla section ?

Ans. 132 kV supply is taken at one point from the supply authorities and then it is distributed by the railway authorities.

Q. 2. Name the current collection systems employed for electric locomotives, tramways or trolleys.

Ans. There are mainly two systems of current collection for locomotives, tramways or trolleys viz. (*i*) conductor rail system and (*ii*) overhead system.

Q. 3. The design of overhead system is more complex as compared to conductor rail system. Why ?

Ans. Because great difficulty is experienced in ensuring that the current collector and contact wire remain in contact at a fairly even pressure.

Q. 4. Why current collection from overhead system is considered superior over conductor rail system ?

Ans. Current collection from overhead system is considered superior over conductor rail system because both theoretically and experimentally current collection is more difficult from a rigid body than from an elastic one. Further the insulation of the third rail at high voltages used on single phase ac traction would also be impracticable and endanger the safety of the personnel.

Q. 5. What is intermediate category ? **[P.T.U. December 2009]**

Ans. Compound catenary construction consists of three wires, all in the same vertical plane. The upper wire is catenary wire, which is insulated from the supporting structures. From this wire another wire, known as *intermediate catenary,* is supported by droppers clipped to both wires. The third wire *i.e.,* trolley wire is maintained in definite and constant tension by means of automatic tightening gear.

Q. 6. What are the functions of dc traction substations ? **[P.T.U. December 2007]**

Ans. The function of dc traction substations are to (*i*) step down the transmission voltage to the utilization level (*ii*) convert it into dc and (*iii*) sectionalizing the overhead equipment. Such substations, therefore, consist of step-down transformers, converting machinery, ac circuit breaker, dc circuit breaker and smoothing and metering equipment.

Q. 7. Why separate feeding systems are required for the trolley wire and the track rails ?

Ans. Separate feeding systems are required for the trolley wire and track rails in case of tramways in order to comply with the following regulations.

1. The voltage at the trolley wire is not to exceed 550 volts and that at generating stations (or substations) 650 volts.
2. The trolley wire shall be divided into sections not exceeding 1.6 km in length.
3. The potential difference between any two points of the track rail return systems shall not exceed 7 volts.
4. The potential difference between any point on the rail and earth should not exceed 4 volts.

Q. 8. What for TPC stands ?

Ans. TPC stands for traction power controller.

Q. 9. How mimic diagram enables the TPC to visualize the whole of the power supply system ?

Ans. A mimic diagram depicting diagrammatically the entire OHE system as well as different substations and control posts is provided at each remote control centre to provide actual visual position of supply system to traction power controller. The mimic diagram is generally not lit up but whenever there is any change in the interruptor position at the switching post, corresponding point on the diagram gets illuminated and there is an alarm also to draw the TPC attention. The availability of the electrical supply to each and every section of OHE is monitored by the PTs at switching stations. Thus entire electrical supply system remains under the control of TPC who is fully equipped with facilities for instantaneous control of any switching operation for quick isolation of the faulty sections and restoration of power supply to the adjoining healthy sections and also for the isolation of sections for maintenance work anywhere in the system that may extend upto 250 km in every direction of the centre.

Q. 10. Why are insulated overlaps installed ?

Ans. Insulated overlaps are installed mainly to provide isolating facilities required for the operation and maintenance of OHE.

Q. 11. How insulated overlap is different from uninsulated overlap ?

Ans. **Uninsulated overlap.** In case OHE is made continuous, which is possible by splicing of conductors, long lengths of it would be affected whenever there is a fault or damage. It will also need long time for rectification of damage. It is, therefore, essential to divide the length of elementary section into a number of sub-elementary sections each of which is separated from its adjacent one by means of what is known as an *uninsulated overlap.* In an uninsulated overlap, two contact wires belonging to two adjacent sub-elementary sections are run parallel to each other for at least one span and electrically connected to each other by means of jumpers.

Insulated Overlap. Insulated overlaps are installed chiefly for providing isolating facilities required for the operation and maintenance of OHE. In case two lengths of OHE belong to two adjacent elementary sections, there will be insulated overlap span. Insulated overlaps are also provided at feeding posts, sub-sectioning posts and booster stations. Insulated overlaps are similar to uninsulated overlaps except for the following:

1. In order to avoid accidental contact between the contact wires, gap between them is kept more in case of insulated overlap.
2. Electrical continuity between the two sections is maintained by means of jumper connections in series with isolator or interruptor. With such an arrangement, a section can be made dead and work can be carried out on the other equipment away from the overlap.
3. The anchoring portion of the catenary and contact wire is separated from the rest of the portion by means of insulators provided at a distance of 2.5 m from the supports and within overlap span. The tail ends of the anchoring OHEs are connected to the other nearby equipment so as to prevent it from attaining any other potential.

Q. 12. Why neutral section is provided in the OHE of ac traction system but not in dc traction system ?

Ans. In dc traction, supply to OHE can be provided from both ends so that voltage drop can be reduced and in such a case substations operate in parallel. In case it is felt necessary, say for sake of convenience in maintenance, to have some boundary of supply between two substations, it can be had simply by means of insulated overlap. There will be no damage in this case even if pantograph bridges the OHE. But in single phase ac traction above arrangement is not permissible because adjacent substations tap different phases of the three-phase system so as to equalise load on the three phases. So momentary passing of pantograph under insulated overlap will result in short circuit between two phases of the supply system and damage the pantograph and OHE. This situation is avoided by interposing a small length of OHE, called the *neutral section.* It is insulated from both sides and is not connected to any source of power supply. It's main function is to allow physically smooth and electrically sparkless passage of pantograph from one section to the other. As per Indian Railway practice, neutral section is located in the middle of two feeding posts and in front of sectioning and paralleling post, as illustrated in Fig. 15.17.

□□□

SECTION - D

ECONOMIC CONSIDERATIONS

CHAPTER 16

Economic Considerations

INSIDE THIS CHAPTER

16.1. Introduction 16.2. Classification of Costs 16.3. Cost Analysis of Power Plants 16.4. Interest and Depreciation 16.5. Methods of Determination of Depreciation 16.6. Economics of Power Generation 16.7. Significance of Load Factor and Diversity Factor 16.8. Capacity Scheduling 16.9. Load Sharing Between Base Load and Peak Load Plants 16.10. Choice of Size and Number of Generating Units 16.11. Load Forecasting.

16.1. INTRODUCTION

In all engineering works, the question of cost is of first importance. In most cases the cost decides whether a certain project will be carried out or not, although political and other considerations may intervene sometimes. The electrical power supplier is required to supply power to a large number of consumers (whenever apply for it) to meet their requirements. While designing electrical power generating stations and other systems efforts are made to achieve overall economy so that the per unit cost of generation is the lowest possible. This will enable the supplier to supply electrical energy to its consumers at reasonable rates. The problem of determining the cost of any scheme is often difficult, because the cost varies considerably with time, tariff conditions and even with convention. The schemes involving the use of aluminium, which used to be considered very expensive a century ago, are now considered to be cheap due to lower cost. Economic problem occurs in the fields of generation, transmission, distribution and utilization of electrical power. The electrical engineer has to adopt the cheapest and most convenient scheme after having study of the relative costs of the possible schemes.

16.2. CLASSIFICATION OF COSTS

In general the cost of generating electrical energy can be divided into the following three elements :

1. ***Fixed Cost.*** This cost is independent of maximum demand and energy output. It is due to the annual cost of central organisation, interest on the capital cost of land (especially if some land is held for future development) and salaries of high officials.

The annual expenditure on central organisation and salaries of high officials is fixed since it has to be met whether the plant has high or low maximum demand or it generates less or more energy (units). Further, the capital investment on the land is fixed and hence the amount of interest is also fixed.

2. ***Semi-fixed Cost.*** This cost depends upon the maximum demand but is independent of energy output. The semi-fixed cost is due to annual interest and depreciation on the capital cost of the generating plant, transmission and distribution network, buildings and other civil works, all types of taxes and insurance charges and salaries of management and clerical staff. Also yearly compensation given to the workers is included under this head. The semi-fixed cost is approximately proportional to maximum demand.

The size and cost of the installation is governed by the maximum demand on the power plant. The greater the maximum demand on the power plant, the greater is its size and cost of installation. Further, the taxes, insurance charges and the strength of clerical staff depend upon the size of the plant and hence upon the maximum demand.

3. ***Running or Operating Cost.*** This cost depends upon the number of hours the plant is in operation or upon the number of units of electrical energy generated. The running or operating cost is due to annual cost of fuel, lubricating oil, water, maintenance and repair cost of equipment and wages and salaries of operational and maintenance staff and salaries of supervisory staff engaged in the running of the plant. The operating cost is approximately proportional to units generated.

Total annual cost incurred in the power generation is represented by the expression

$$E = a + b \text{ kW} + c \text{ kWh}$$

where a, b and c are constants. Fixed and semi-fixed cost being independent of the amount of energy generated is also called "*standing cost*".

In deciding any scheme for any given service the choice must be such that the total operating cost (sum of annual fixed cost, semi-fixed cost and operating cost) be minimum.

16.3. COST ANALYSIS OF POWER PLANTS

The generating cost per unit (kWh) of energy depends upon the cost covering the purchase, installation and erection of equipment, cost of fuel, labour, repair etc. The generation cost can, broadly be divided into *fixed cost* and the operating or *running cost*.

1. Fixed Cost. The *annual fixed cost* consists of the interest on the total investment, all types of taxes and insurance charges, salaries of high officials, management and clerical staff. The annual contribution to the fund to pay off the capital expended after useful life of the plant and yearly compensation paid to the workers is also included in the annual fixed cost. This cost varies approximately in direct proportion with the installed capacity of the plant and it does not depend upon the fact whether the plant supplies any energy or not.

The total investment or capital cost of a plant includes the preliminary cost, cost of land and other real estate, cost of design and planning, cost of building and equipment, cost of transportation, erection and installation of equipment and overheads etc. In case of hydroelectric power plants, the capital cost of the plant includes the costs of dam, earthwork, excavation, railhead, highways and other civil works and compensation to land/property owner (whose lands would be submerged in the reservoir) in addition to the costs mentioned above.

Cost of land and building will depend upon the location of the plant. If the plant is situated near a city, the land will be costlier than that in case the plant is located far from the city or residential areas.

The cost of steam plant located near a river, where adequate supply of cooling water is available, would be lesser than that of one in which cooling towers or ponds would have to be made available. In general a larger plant costs less per kW of installed capacity than a smaller one. A plant having a smaller number of large sized generating units costs less than a plant of the same total generating capacity but having a large number of small sized units. The capital costs of the plants vary widely, even in case of plants of similar types, and depend to a large extent on the market conditions.

2. Operating or Running Costs. The operating cost of a plant means the expenses which vary with the extent of operation or the amount of energy generated. This cost is due to annual cost of fuel, lubricating oil, water, maintenance and repair cost of equipment and wages and salaries of operational and maintenance staff and salaries of supervisory staff engaged on the running of the plant. The operating cost is approximately proportional to units generated. In steam, diesel and gas plants, the operating costs form a major portion of total annual cost whereas in hydro, nuclear and solar plants, the fixed costs overshadow the small operating costs.

3. Investor Profit. If the power plant is the public property, as is the case in India, then the customers will be taxpayers to share the burden of the Government. For this purpose, there is an item in the rates to cover taxes in place of the investors profit. These taxes will be paid by the consumers in the form of electric consumption bill. This amount is collected in twelve instalments per annum or six instalments per annum.

16.3.1. Fixed and Operating Costs of Hydropower Plants

The total capital outlay or the investment on a hydropower station includes the costs of preliminary surveys and investigations of the topography and geology of the proposed plant site, purchase of water rights and the land required for providing sufficient storage and pondage, compensation to the ousters, costs of preparation of detailed designs and specifications, costs of testing of materials of construction, costs of carrying out experimental work and model tests on designs for hydraulic structures, costs of construction and the purchase and installation of equipment, interest on the capital during construction, costs of new roads, railway lines, residential houses and other civil engineering works. The capital cost is very much affected by topographical and geological conditions.

The annual fixed costs include interest on the capital cost, depreciation, all types of taxes and insurance charges and salaries of high officials, management and clerical staff. The total cost of construction of a hydropower plant is invariably higher than that of a steam power plant of the same capacity. The fixed charges of a hydropower plant are about 60 to 70 per cent of the total cost of power and these are independent of the plant output.

The operating costs of hydropower plants include annual costs of lubricating oil, maintenance and repair cost of equipment and wages and salaries of operational and maintenance staff and salaries of supervisory staff engaged in the running of the plant. Because of the absence of fuel cost, the operating cost of a hydropower plant is very small.

16.3.2. Fixed and Operating Costs of Steam Power Plants

The capital cost of a steam power plant includes the costs of land, design and planning specifications, installation, powerhouse building, equipment, installation, testing, commissioning, overheads etc.

The range of cost for steam power plants is smaller than that for hydropower plants, the reason being that in case of steam plants, only about 15% of the cost is influenced by topographical and geological conditions, whereas in the case of hydropower plants, around 75% of the cost is affected by these conditions.

The annual fixed costs include interest on the capital, depreciation, all types of taxes and insurance charges and salaries of high officials, management and clerical staff.

The operating costs of a steam power plant include cost of fuel including its handling and ash handling, cost of water for boiler feed, condensers, cooling and house service, costs of oil, waste, stores and other supplies, repairs and maintenance cost of equipment, and salaries and wages of operational and maintenance staff and salaries of supervisory staff engaged in the running of the plant.

In the case of a steam power plant, the cost of coal can be split into two components—a small fixed component (required to keep the plant in readiness to meet the demand for power and independent of energy generated) and a large running component (cost of coal equivalent of the heat converted into electric power plus the heat rejected in the condenser). The latter component is directly proportional to the energy generated. The amount of fuel required per unit generated can be determined by dividing the heat rate by the calorific value of the coal. The total amount of coal consumed per unit (kWh) of energy generated and its cost depend upon the plant capacity, its efficiency, load factor, transportation charges and cost of storage etc. The fuel cost influences the total cost of steam power quite considerably—a change in fuel cost by 20% may cause about 8% variation in the total cost per kWh generated.

As already discussed, a number of environmental problems are created in case of steam power plants, and therefore, the flue gases will need cleaning *i.e.*, removal of particulates, SO_2 and NO_2 etc. so as to keep the pollutants within the safe limits. The fixed and operating costs of these air purification systems are to be included in the cost analysis of steam power plants.

16.3.3. Fixed and Operating Costs of Nuclear Power Plants

The components constituting the capital cost of a nuclear power plant are almost the same as those for steam power plants. However, the annual fixed cost in case of nuclear power plants is quite high in comparison to that of steam power plants, because of the following reasons.

1. The sophisticated technology is required for the manufacture of nuclear reactors and, therefore, the capital cost becomes pretty high.
2. The cooling water requirements of a nuclear power plant are very heavy and so the cooling tower cost becomes high as compared to that of a steam power plant of the same capacity.

In case of nuclear power plants though the quantity of fuel required is very small, problems of fuel transportation, fuel handling and ash handling are eliminated (as in case of steam power plants) but the cost of the fuel required is enormously high. Consequently the operating cost of a nuclear power plant is lesser than that of a steam power plant but higher than that of a hydropower plant. The operating efficiency of a nuclear power plant is high when it is operated at a high load factor. This is the reason that nuclear power plants are always operated as base load plants.

16.4. INTEREST AND DEPRECIATION

Usually money is borrowed from banks or insurance companies or other financial institutions for big projects such as generating plant and at the end of a year, the undertaking is required to pay the interest on the capital cost. Even if equipment is purchased out of reserve cash, the normal rate of interest must still be allowed for, since the money might otherwise be earning this interest. If a capital outlay required for a certain installation is P and the rate of interest per unit is r per annum then an amount of rP per annum must be provided as interest, and this is added to the annual operating cost. Rate of interest to be paid depends upon the money market position and the credit of the borrower and varies between 8 to 10 per cent (0.08 to 0.10 per unit) per annum.

If the installation were to last for ever, the interest is the only charge that would have to be made. In practice after a certain time (from ten to forty or fifty years) the plant and machinery have to be replaced, due to its getting old, obsolete and inadequate and it is, therefore, necessary to set aside certain amount annually to produce a sufficient sum at the end of the probable life for replacing the plant and machinery by new ones. The amount is known as *depreciation* and depends upon the type and make of machines. There are several methods of determining the yearly subscription towards this fund and some of these methods will be discussed in Art 16.5.

16.4.1. Life of Generating Plant

The depreciation charge represents the amount which is set aside from income every year and placed in depreciation reserve. For calculating this charge, it is necessary to estimate the useful life of the plant.

The expected physical life of *conventional steam power plant* is 40–45 years or even more. There is, however, the possibility that technological developments will make the plant obsolete before its physical life is over. Financial practice, therefore, is to write it off over 25–30 years. Conventional steam power plant of this age is quite serviceable, however, and it would not be economical or desirable under present conditions to scrap such plant. This, of course, was not known 30 years before nor do we know for certain what the conditions will warrant 30 years ahead. In current planning, economic appraisal is done on the basis of the capital being amortized over a life of 30 years. In the event, the plant would not necessarily be taken out of service at the end of 30 years, but in planning for the period 8 to 10 years ahead it is assumed that the old plant would be retained until it has reached 35 years of age. Even at this stage the plant would probably be still capable of further operational life and, before a decision was taken to scrap it, consideration would be given as to whether the old plant should be retained for peak load operation for a year or so longer or whether it would pay to scrap it (and save repairs, renewals, surveys, wages etc.), or to replace it by new capacity involving capital investment but with prospects of considerable saving in running costs, including fuel.

The current practice to assume for **nuclear plant** a financial life of 25 years.

Hydropower plant and pumped storage plant contain a large element of civil works (dams and tunnels, etc) having a very long physical life, say 80 years or even longer. The mechanical and electrical items (turbines, generators, etc) have, however, a life comparable to similar items in steam power stations and may be regarded for financial purposes a life of 30 years. Once a hydro-scheme has been built its operating costs are so low as to make its unlikely to be displaced by technological developments.

16.5. METHODS OF DETERMINATION OF DEPRECIATION

Depreciation is the most important item in the fixed costs and it represents the reduction in the value of the equipment and other property of the plant every year due to continuous wear and tear and also due to obsolescence. The first factor can be reduced by proper maintenance of the equipment and the buildings, but the second factor is quite unpredictable. To account for the depreciation, a certain fixed amount is set aside every year so that by the time the life span of the plant is over, the total amount accumulated equals the replacement cost (original cost of the equipment less the expected salvage value). The methods commonly used for determination of annual depreciation charges are (*i*) straight line method, (*ii*) diminishing value method and (*iii*) sinking fund method.

1. *Straight Line Method.* This method assumes that certain depreciation occurs according to the straight line law and, therefore, in this method a constant depreciation charge is made every year on the basis of total depreciation (initial cost – scrap or salvage value) and useful life of the equipment/property.

i.e., Annual depreciation charge

$$= \frac{\text{Initial cost} - \text{salvage value}}{\text{Number of years of useful life}} = \frac{P - S}{n} \quad \text{...(16.1)}$$

Though this method is very popular because of its simplicity but it does not take into account the amount of interest earned by the amount set aside yearly. The assumption of constant depreciation every year is also not correct.

Figure 16.1 illustrates this method graphically.

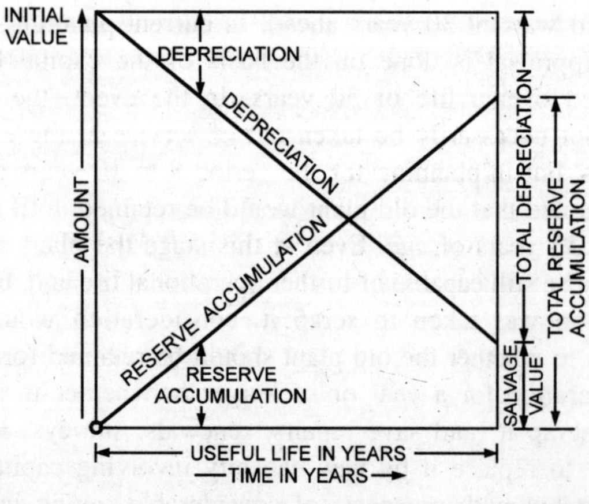

Fig. 16.1. ***Straight Line Method of Determination of Depreciation***

2. *Diminishing Value or Declining Balance Method.* In this method provision is made for setting each year a fixed rate, first applied to the original cost and then to the diminishing value; such rate depending upon the useful life of the plant. This method results in distribution of the total expenses on the plant (depreciation plus maintenance cost) over its total useful life because in this method depreciation charges are heavy in early years when maintenance charges are low and depreciation charges are low in the late years when maintenance charges are heavy. The serious drawback of this method is that an heavy burden towards depreciation charges is imposed in the early years when the plant is to develop and build up its income. In this method also the amount of interest earned by the reserve accumulation is ignored.

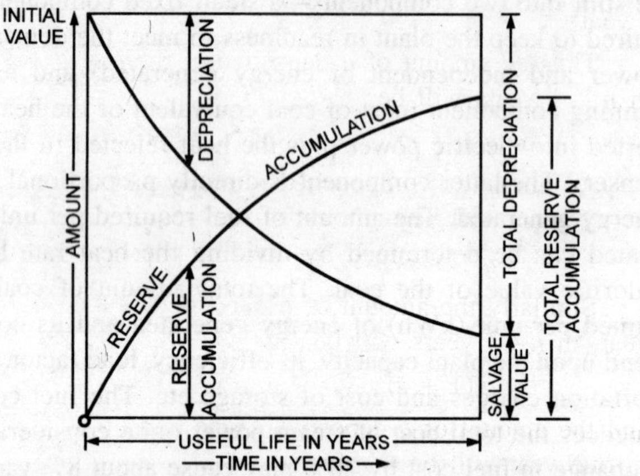

Fig. 16.2. ***Diminishing Value Method of Determination of Depreciation***

Let the capital cost of the plant be P, salvage value after useful life of equipment of *n* years be S and annual unit depreciation be *x*.

The value of plant after one years $= P(1 - x)$

The value of plant after two years $= P(1 - x)(1 - x)$

$= P(1 - x)^2$

The value of plant after *n* years, $S = P(1 - x)^n$

$$\text{or } (1 - x) = \left(\frac{S}{P}\right)^{\frac{1}{n}}$$

$$\text{or } x = 1 - \left(\frac{S}{P}\right)^{\frac{1}{n}} \quad \text{...(16.2)}$$

Deposit to be made at the end of first year of completion of the plant

$$= xP = P\left[1 - \left(\frac{S}{P}\right)^{\frac{1}{n}}\right] \quad \text{...(16.3)}$$

Deposit to be made at the end of second year of completion of the plant

$$= x(1 - x)P = P\left(\frac{S}{P}\right)^{\frac{1}{n}} \; P\left[1 - \left(\frac{S}{P}\right)^{\frac{1}{n}}\right] \text{ and so on}$$

Figure 16.2 illustrates this method graphically.

3. *Sinking Fund Method.* In this method provision is made for setting aside each year such a sum as, invested at certain interest rate compounded annually, will give the amount equal to replacement cost of the installation at the end of its useful life. As compared to straight line method, it requires smaller annual amounts and also the amounts for annuity are uniform.

Let P be the initial value of the equipment, n be the useful life of equipment in years, S be the scrap or salvage value after useful life of equipment of n years, and r is the rate of interest per annum expressed as a fraction.

If the annual deposit be q, it will earn interest rq in one year, so that it will worth $q + rq = q(1 + r)$ at the end of one year. Thus its value will be multiplied by the ratio $(1 + r)$ every year and it will worth $q(1 + r)^2$ at the end of 2 years, $q(1 + r)^3$ at the end of 3 years and $q(1 + r)^n$ at the end of n years.

Thus an amount of q deposited at the end of first year, in $n - 1$ years will be

$$= q(1 + r)^{n-1}$$

An amount of q deposited at the end of second year, in $n - 2$ years will be

$$= q(1 + r)^{n-2}$$

Similarly the amount of q deposited at the end of third year, in $n - 3$ years will be

$$= q(1 + r)^{n-3}$$

and the amount of q deposited at the end of $(n - 1)$th year, in one year will be $= q(1 + r)$

Thus after n years total sinking fund will be

$$= q[(1 + r)^{n-1} + (1 + r)^{n-2} + (1 + r)^{n-3} + \ldots (1 + r)]$$

The above terms are in geometrical progression and

$$\text{Total sum} = \frac{\frac{1}{1+r}q - q(1+r)^{n-1}}{\frac{1}{1+r} - 1}$$

$$= q\frac{1-(1+r)(1+r)^{n-1}}{1-(1+r)} = q\frac{(1+r)^n - 1}{r}$$

This must be equal to the cost of replacement Q which is equal to (P – S).

$$\text{So,} \quad Q = q\frac{(1+r)^n - 1}{r}$$

$$\text{or annual deposit, } q = \frac{Qr}{(1+r)^n - 1} \qquad \ldots(16.4)$$

This method is illustrated graphically in Fig. 16.3.

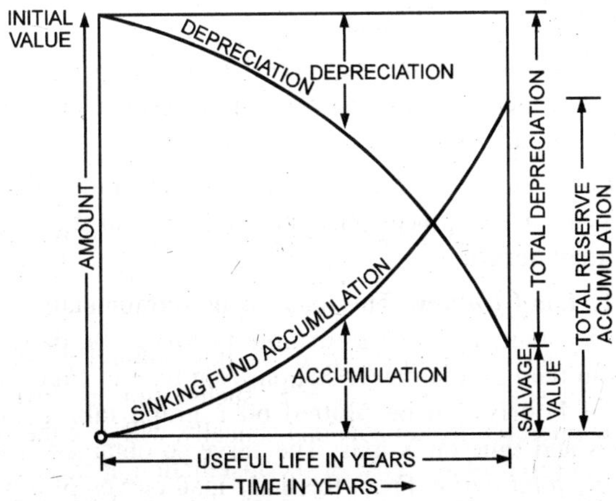

Fig. 16.3. ***Sinking Fund Method of Determination of Depreciation***

Example 16.1. A 300 kVA distribution transformer costs Rs 20,000 and has a salvage value of Rs 1,000 at the end of 20 years. Determine the depreciated value of the plant at the end of 10 years on the following methods of assessment :

(*i*) Straight line depreciation (*ii*) Sinking fund depreciation of 8% compounded annually.

[Pb. Univ. Electrical Power II (Generation) Dec. 1995]

Solution : Initial cost of the plant, P = Rs 20,000.00

Salvage value of the plant, S = Rs 1,000.00

Useful life of the plant, n = 20 years

(*i*) By Straight Line Method. Annual deposit to be made,

$$q = \frac{P-S}{n} = \text{Rs } \frac{20{,}000 - 1{,}000}{20} = \text{Rs } 950.00$$

Depreciated value of the plant at the end of 10 years

$$= p - q \times 10 = \text{Rs } 20{,}000 - \text{Rs } 950 \times 10$$

$$= \text{Rs } 10{,}500.00 \textbf{ Ans.}$$

(*ii*) By Sinking Fund Method. Rate of interest, r = 8% = 0.08

Annual deposit in the Sinking Fund,

$$q = \text{Rs } (P - S)\left[\frac{r}{(1+r)^n - 1}\right]$$

$$= \text{Rs } \frac{(20{,}000 - 1{,}000) \times 0.08}{(1+0.08)^{20} - 1} = \text{Rs } 415.20$$

Total amount collected at the end of 10 years

$$= q\frac{(1+r)^{10} - 1}{r} = \text{Rs } 415.20 \times \frac{(1+0.08)^{10} - 1}{0.08}$$

$$= \text{Rs } 6{,}015.00$$

Depreciated value of the plant at the end of 10 years

$$= \text{Rs } 20{,}000 - \text{Rs } 6{,}015 = \text{Rs } 13{,}985 \textbf{ Ans.}$$

Example 16.2. A plant costs Rs. 7.56 × 10^5 and it is estimated that after 25 years it will have to be replaced by a new one. At that instant its salvage value will be Rs 1.56 × 10^5. Calculate (*i*) annual deposit to be made in order to replace the plant after 25 years, (*ii*) the value of plant after 10 years on the following basis :

(*i*) reducing balance depreciation method (*ii*) sinking fund depreciation method at 8% annual compound interest.

[Pb. Univ. Electrical Power II (Generation) Dec. 1994]

Solution : Initial cost of the plant, P = Rs 7.56 × 10^5

Salvage value of the plant, S = Rs 1.56 × 10^5

Useful life of the plant, n = 25 years

(*i*) By Reducing Balance Depreciation Method.

Annual unit depreciation,

$$x = 1 - \left(\frac{S}{P}\right)^{\frac{1}{n}} = 1 - \left(\frac{1.56 \times 10^5}{7.56 \times 10^5}\right)^{\frac{1}{25}} = 0.0612 \textbf{ Ans.}$$

Value of plant after 10 years

$$= \text{Rs P } (1 - x)^{10} = \text{Rs } 7.56 \times 10^5 \, (1 - 0.0612)^{10}$$

$$= \text{Rs } 4.02 \times 10^5 \textbf{ Ans.}$$

(*ii*) By Sinking Fund Depreciation Method.

Rate of interest, r = 8% = 0.08

Cost of replacement, Q = P – S

$$= \text{Rs } (7.56 \times 10^5 - 1.56 \times 10^5)$$

$$= \text{Rs } 6 \times 10^5$$

Annual deposit in the sinking fund,

$$q = \text{Rs}\ \frac{Qr}{(1+r)^n - 1} = \text{Rs}\ \frac{6 \times 10^5 \times 0.08}{(1+0.08)^{25} - 1} = \text{Rs } 8{,}207$$

Total amount collected at the end of 10 years

$$= q\frac{(1+r)^{10} - 1}{r} = \text{Rs } 8{,}207 \times \frac{(1+0.08)^{10} - 1}{0.08}$$

$$= \text{Rs. } 1.19 \times 10^5$$

Depreciated value of the plant at the end of 10 years

$$= \text{Rs } (7.56 \times 10^5 - 1.19 \times 10^5) = \text{Rs } 6.37 \times 10^5 \textbf{ Ans.}$$

Example 16.3. The cost of a small preheater is Rs 12,000 and its expected life is 16 years. The net annual instalment to recover the cost of equipment is Rs 425. The interest is 5%. Using sinking fund method find out the salvage value of the preheater after 16 years of service. *[A.M.I.E. Sec. B. Power Plant Engineering Winter 1996]*

Solution : Initial cost of preheater, P = Rs 12,000

Useful life of the preheater, n = 16 years

Rate of interest, r = 5% = 0.05

Annual deposit in the sinking fund, q = Rs 425

Total amount collected at the end of 16 years

$$= q\frac{(1+r)^{16} - 1}{r} = \text{Rs } 425\ \frac{(1+0.05)^{16} - 1}{0.05} = \text{Rs } 10{,}054$$

Salvage value,

S = Depreciated value of the plant at the end of 16 years

= Rs 12,000 – Rs. 10,054 = Rs 1,946 **Ans.**

Example 16.4. The equipment in a power station cost Rs 1,560,000 and has a salvage value of Rs 60,000 at the end of 25 years. Determine the depreciated value of the equipment at the end of 20 years by sinking fund method at 5% compound interest annually. *[Rajasthan Univ. Generation of Electrical Power 2004, 2005]*

Solution : Initial cost of equipment, P = 1,560,000

Useful life of equipment, n = 25 years

Rate of interest, r = 5% = 0.05

Salvage value of equipment S = Rs 60,000

Replacement cost, Q = P – S = Rs 1,560,000 – Rs 60,000 = 1,500,000

$$\text{Annual deposit, } q = \frac{Qr}{(1+r)^n - 1} = \frac{1{,}500{,}000 \times 0.05}{(1+0.05)^{25} - 1} = \text{Rs } 31{,}428.69$$

The amount collected at the end of 20 years

$$= \frac{q(1+r)^{20} - 1}{r} = 31{,}428.69 \times \frac{(1+0.05)^{20} - 1}{0.05} = \text{Rs } 1{,}039{,}219$$

Depreciated value of the equipment at the end of 20 years

= Rs (1,560,000 – 1,039,219) = Rs 520,781 **Ans.**

16.6. ECONOMICS OF POWER GENERATION

The generation of electrical energy economically is not an ordinary matter, rather it requires a long experience to decide about the type, location and the rating of generating stations.

The generating stations may be steam, hydro, nuclear, diesel or any other type. This factor mainly depends upon the natural sources available in the areas. Steam power stations are best suited near the coal fields and also adopted where coal supply is available in plenty at reasonable rates, large amount of power is required to be generated and financial, climatic and geographical conditions are not favourable to hydro and nuclear power stations. Hydropower stations are best suited in case water is available at certain height and nuclear power stations are best suited in area far away from collieries and where fuel costs are high and alternative cheap hydropower is not available as in Rajasthan. Diesel power stations are installed where supply of coal and water is not available in sufficient quantity or where power is to be generated in small quantity or where standby sets are required for continuity of supply such as in hospitals, telephone exchanges, radio-stations and cinemas.

The power station should be as near as possible to the centre of the load so that the transmission cost and losses are minimum. This factor is important when dc supply system is adopted. However, in case of ac supply system where transformation of energy from lower to higher voltage and vice versa is possible, power station can be located at places other than that of centre of load provided other conditions are favourable.

The other considerations for the design of the power station are reliability, minimum capital and operating costs. For this it is necessary that the layout should be such that the maintenance and repairs can be carried out easily; the design must be compact and well planned; equipment used must be standard one so that capital cost is reduced and replacement of worn-out parts becomes easy and equipment used should be simple so that it can be operated by semi-skilled workers. The design of power station should be such that the station could be divided into a number of sections to avoid complete shutdown of the station when the fault occurs. Another advantage of sectionalizing is that the erection can be completed and put into service in parts. The equipment must be automatic whenever possible in order to save the labour cost provided the reliability of the station is not affected. The automatic equipment should be such that it could be operated manually in case it is required. The scheme employed should be such that extension could be made to meet with the increase in demand in future without incurring further heavy expenditure.

Before a power project is taken into hand, the project engineer should have the following information with him.

(*i*) Estimate of probable load.

(*ii*) Future load conditions.

(*iii*) Location of the loads, specially in case of hydroelectric generating stations because the cost of transmission is also required to be considered.

For deciding the type and rating of generating plant it is necessary that engineer may be familiar with the following important terms.

1. *Load Curves*. The load on the power station is seldom constant; it varies from time to time. The daily variations in load on the power station from time to time—hourly or half hourly can be plotted on a graph taking load on Y-axis and time on X-axis. The curve so obtained is known as *daily load curve*. From the daily load curves of a particular month, the monthly load curve can be plotted by calculating the average value of power at a particular time of the day. Similarly if we consider such monthly load curves of a particular year, and find the average value of the power at a particular time of the day, the annual load curve can be obtained. The load curve bears a great importance in generation. The load curves supply the following informations :

(*i*) The variation of the load during different hours of the day.

(*ii*) The area under the curve represents the total number of units generated in a day.

(*iii*) The peak of the curve represents the *maximum demand* on the station on the particular day.

(*iv*) The area under the load curve divided by the number of hours represents the average load on the power station.

(*v*) The ratio of the area under the load curve to the total area of the rectangle in which it is contained gives the *load factor*.

The type of variations in load or demand on a plant depends on the domestic and industrial users (their type and habits). The load requirements of an individual consumer are different at different timings in the day. The load curve for Sunday will be different from those for other days of the week. The load curves for winter and summer will also differ. The curves for rural, suburban and urban areas will also differ.

The load curves for Sunday, other weekdays, summer and winter for a typical metropolitan area are shown in Fig. 16.4.

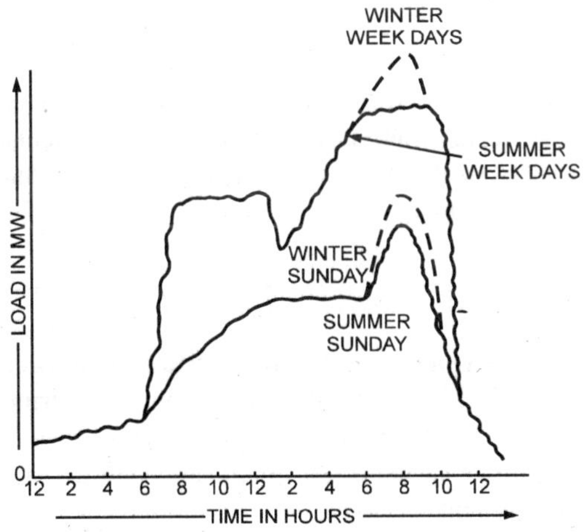

Fig. 16.4. ***Chronological Load Curves***

During the early morning hours the demand is always low. Around 5 AM the load starts increasing because of increasing industrial and traction load. Around 9 AM the load reaches a high value and remains almost constant till evening except for some dip during lunch hours. The load again starts increasing in the evening hours as the residential and traction load start coming up. The peak occurs around 7–9 PM and then the load starts decreasing. On Sundays the load demand is only due to residential consumers and so low as compared to those of other weekdays.

Thus, on the basis of above informations the load curves help in deciding the size of the units to be installed and also in preparing the schedule of operation of the generating units.

2. ***Load Duration Curve.*** This is another type of curve which indicates the variation of load, but with the loads arranged in descending order of magnitude *i.e.,* the greatest load on the left, lesser loads towards the right and the least load at the extreme right. This curve gives the number of hours for which a particular load lasts during the day. The area under this curve like *load curve* or *chronological load curve* gives the total number of units generated for the period considered. From this curve the load factor of the station can also be determined. From these curves the distribution of load between various generating units can also be predicted. (For illustration refer to Example 16.6)

3. ***Integrated Load Duration Curve.*** This curve gives the total number of units generated for the given demand. The ordinate represents the demand in kW or MW and the abscissa represents the units (kWh) generated at or below a given demand (in kW or MW). Such a curve can be obtained from the load duration curve keeping the abscissa corresponding to each ordinate equal to the area of the duration curve up to the value of that ordinate.

Such curves are useful in that with a given number of units (kWh) available, say from a river flow, the load (kW) that could be carried at the base or peak may be easily determined. (For illustration refer to Example 16.6)

4. ***Mass Curve.*** This curve is plotted with units (kWh) as ordinate and time as abscissa. Thus a mass curve gives the total energy consumed by the load up to a particular time in a day. This curve can be easily plotted from the chronological load curve by summing up the energy consumed up to different times starting at the zero time.

This curve is used in the study of variations between the rate of water flow and the electrical load and for the determination of the necessary storage.

5. ***Connected Load.*** The sum of the continuous ratings of all electrical equipment connected to the supply system is known as *connected load.*

6. ***Maximum Demand.*** It is not necessary that all the connected load be switched to a system at a time. The greatest of all "short time interval averaged" (15 minutes or $\frac{1}{2}$ hour or 1 hour) during a given period (a day, a month or a year), on the power station is called the *maximum demand.* It is sometimes also called as *system peak.* It is to be clarified here that the maximum demand never means the greatest instantaneous maximum demand but the greatest short time average demand occurring during a long period of time under consideration. Hence whenever the maximum demand is to be mentioned it is essential to indicate the short interval of time over which averaging has been carried out for the measurement of demand and the period of load under consideration. It is the maximum demand which determines the size and the cost of the installation.

7. ***Demand Factor.*** The ratio of actual maximum demand on the system to the total rated load connected to the system is called the *demand factor.* It is always less than unity.

$$\text{Mathematically demand factor} = \frac{\text{Maximum demand}}{\text{Connected load}} \quad ...(16.5)$$

The idea of a demand factor was introduced due to the fact that all the equipment connected to the system will not be worked at a time in practice and the kW or kVA maximum demand of a group of electricity consuming devices will be always less than the sum of the kW or kVA ratings or capacities of these devices.

The minimum capacity of the generating plant must be such as to meet the maximum demand, the value of demand factor, therefore, determines the capacity and hence cost of the power equipment required to serve a given load.

8. *Average Load or Demand.* The average load or demand on the power station is the average of loads occurring at the various events. It can also be stated as energy delivered in a given period divided by the number of hours in that period. Depending upon the duration of time period such as a day, a month or a year, we get daily, monthly or annual average load.

$$\text{Daily average load} = \frac{\text{kWh supplied in a day}}{24} \quad ...(16.6)$$

$$\text{Monthly average load} = \frac{\text{kWh supplied in a month}}{24 \times 30} \quad ...(16.7)$$

$$\text{Annual average load} = \frac{\text{kWh supplied in a year}}{24 \times 365} \quad ...(16.8)$$

9. *Load Factor.* The ratio of average load to the maximum demand during a certain period of time such as a day or a month or a year is called the *load factor*. Since average load is always less than the maximum demand, load factor is therefore, always less than unity.

$$\text{Load factor} = \frac{\text{Average demand}}{\text{Maximum demand}} \quad ...(16.9)$$

Load factor may also defined as the ratio of the number of units actually generated in a given period to the number of units which could have been generated had the load or demand remained at the maximum value throughout this period.

i.e., Load Factor

$$= \frac{\text{Units generated in a given period}}{\text{Maximum demand} \times (\text{number of hours of operation in the given period})} \quad ...(16.10)$$

10. *Diversity Factor.* The maximum demands of all the consumers supplied from an installation do not occur usually at the same time, maximum demand on the installation is, therefore, always less than the sum of individual maximum demands of all consumers connected to it.

The ratio of sum of the individual maximum demands of all the consumers supplied by it to the maximum demand of the power station is called the *diversity factor.*

i.e., Diversity factor

$$= \frac{\text{Sum of individual maximum demands}}{\text{Maximum demand of power station}} \quad ...(16.11)$$

It is always greater than unity.

11. *Coincidence Factor.* The reciprocal of the diversity factor is called the coincidence factor.

i.e., Coincidence factor

$$= \frac{\text{Maximum demand of power station}}{\text{Sum of individual maximum demands}} \quad ...(16.12)$$

It is always less than unity.

12. *Capacity Factor or Plant Factor.* Every plant has to have a reserve capacity so as to take care of the future expansion and increase in load and, therefore, total installed capacity of the plant is usually greater than that actually required (maximum demand).

Capacity (or plant) factor is defined as the ratio of the average load to the rated capacity of the power plant.

i.e., Capacity (or plant) factor

$$= \frac{\text{Average demand}}{\text{Rated capacity of power plant}} \quad ...(16.13)$$

Plant capacity factor may also be defined as the actual energy generated divided by the maximum possible energy that the plant might have generated during a given period.

i.e., Plant capacity factor

$$= \frac{\text{Units generated in a given period}}{\text{Rated capacity of the power plant in kW} \times (\text{number of hours in the given period})} \quad ...(16.14)$$

The capacity factor depicts the extent of the use of the power plant. It is different from load factor and the difference is an indication of the reserve capacity.

Capacity factor

$$= \frac{\text{Maximum demand}}{\text{Rated capacity}} \times \text{load factor} \quad ...(16.15)$$

13. *Utilization Factor.* It is a measure of the utility of the power plant capacity and is the ratio of maximum demand to the rated capacity of the power plant. It is always less than unity.

i.e., Utilization factor

$$= \frac{\text{Maximum demand on the power station}}{\text{Rated capacity of the power station}} \quad ...(16.16)$$

14. *Plant Operating Factor or Plant Use Factor.* It is defined as the ratio of actual energy generated during a given period (say a year) to the product of capacity of the plant and the number of hours the plant has been actually in operation during the period.

i.e., Plant use factor

$$= \frac{\text{Total kWh generated}}{\text{Rated capacity of the plant} \times \text{number of operating hours}} \quad ...(16.17)$$

15. *Installed Capacity.* The total of station capacities available to supply the system load is called the *installed capacity.*

16. *Firm Power.* It is the power intended to be always available (even under emergency conditions).

17. *Prime Power.* It is the power (mechanical, hydraulic or thermal) that is always available for conversion into electrical power.

18. ***Dump Power.*** It is power in excess of the load requirements and it is made available by surplus water. It refers to hydropower plants.

19. ***Cold Reserve.*** It is that reserve generating capacity which is available for service but not normally ready for immediate loading.

20. ***Hot Reserve.*** It usually refers to boiler excess capacity, which is kept hot and with steam pressure, ready for use.

21. ***Operating Reserve.*** It refers to capacity in service in excess of peak load.

22. ***Spinning Reserve.*** It is the generating capacity connected to the bus and ready to take load.

In order to take instantaneous peak loads, the system must have capacity in *readiness to serve.* For this purpose certain machines are kept running on no load *i.e., floating* on the system. This is called the *spinning reserve* of the *system.*

16.7. SIGNIFICANCE OF LOAD FACTOR AND DIVERSITY FACTOR

Load factor and diversity factor play an important part in the cost of the supply of electrical energy. *Higher the values of load factor and diversity factor, lower will be the overall cost per unit generated.*

Higher load factor means greater average load, resulting in greater number of units generated for a given maximum demand. Thus, the standing charges, which are proportional to maximum demand and independent of number of units generated, can be distributed over a larger number of units supplied and therefore overall cost per unit of electrical energy generated will be reduced.

The capital cost of the power station depends upon the capacity of the power station. Lower the maximum demand of the power station, the lower is the capacity required and therefore lower is the capital cost of the plant. With a given number of consumers the higher the diversity factor of their loads, the smaller will be the capacity of the plant required and consequently the fixed charges due to capital investment will be much reduced.

The suppliers should always try to improve the load factor as well as diversity factor by inducing the consumers to use the electrical energy during off-peak hours and they may be charged at lower rates for such schemes.

Typical values of demand factors and diversity factors for different types of consumers are given below :

Type of Consumers	Demand Factors	Diversity Factors
Domestic lighting up to 1 kW	0.5–0.7	3–5
Domestic power	0.5	1.5–2
Commercial loads mostly of lighting	0.5–0.7	1.5–2
Industrial power	0.5–0.8	1.5–2

16.8. CAPACITY SCHEDULING

Capacity scheduling means the order in which the various generating units should be placed in operation in order to meet the total plant or system load. Certain units will be put in operation only during peak load demand hours. To ensure continuity of power supply, sufficient generating capacity should be in operation at all times so that if one unit fails, there is no interruption in power supply. If the load on the system/plant is equal to or less than the capacity of one generating unit, then it is always advisable to run two units to supply the load so that noninterrupted power supply can be maintained. The order of putting the machines in operation depends upon their efficiencies. The machines are loaded in the descending order of the efficiency. The load is divided among the various units on the basis of incremental rates.

The capacity of the generating unit in active operation is known as the 'spinning capacity'. Spinning capacity should always be more than the load in order to avoid interruption. "Spinning reserve" is the difference between the spinning capacity and the load on the system. The minimum magnitude of the spinning reserve is usually equal to the capacity of the largest unit.

16.9. LOAD SHARING BETWEEN BASE LOAD AND PEAK LOAD PLANTS

Base load plants run throughout the year and have high load factors. The economic characteristics of base load plants should be such that they supply power at high capital costs but low cost of operation. Hydro and nuclear power plants are usually classified as base load plants. Peak load plants run for a few hours in the year and work at low load factors. The economic characteristics of peak load plants should be such that they supply power at low capital costs, although at high cost of operation. Peak load power plants should be capable of starting quickly and should be inexpensive in starting and shutting down operations.

Let the operating costs of base load and peak load plants be Rs (a_1kW + b_1kWh) and Rs (a_2kW + b_2kWh) where $a_1 > a_2$ and $b_2 > b_1$.

Let P be the maximum demand on the power plant and x be the total number of units generated.

If P_1 is the maximum demand on base load plant and x_1 is the number of units generated by the base load plant.

Peak load on peak-load plant, $P_2 = P - P_1$

Units generated by peak-load plant, $x_2 = x - x_1$

Total annual cost of operation of the system,

$$C = a_1P_1 + a_2(P - P_1) + b_1x_1 + b_2(x - x_1)$$
$$= P_1(a_1 - a_2) + a_2P + x_1(b_1 - b_2) + b_2x$$

or $$\frac{dC}{dP_1} = (a_1 - a_2) + \frac{dx_1}{dP_1}(b_1 - b_2)$$

For minimum cost, $$\frac{dC}{dP_1} = 0$$

$$\therefore \ (a_1 - a_2) + \frac{dx_1}{dP_1}(b_1 - b_2) = 0$$

or $$\frac{dx_1}{dP_1} = \frac{a_1 - a_2}{b_2 - b_1} \text{ hours} \qquad ...(16.18)$$

i.e., for economic load sharing, peak-load power plant will operate for $\frac{a_1 - a_2}{b_2 - b_1}$ hours per year.

16.10. CHOICE OF SIZE AND NUMBER OF GENERATING UNITS

The selection of units and their operation play an important role in the working of a power station and on the economics of power generation. The number of units and the size of each unit is decided from the load curve. The following factors should be considered while deciding the number of units and preparing the operating schedule.

1. The total capacity of the generating units must be capable of meeting the peak demand of the power station.
2. Since the machines operate with maximum efficiency at the three-fourth of the rated capacity, hence the number and size of units must be so selected that they operate at maximum efficiency and better overall efficiency and load factor of the power station is had.
3. Reliability of service is a very important factor. Cheaper power without reliability of supply is of no use. There should be a spare set of capacity of that largest unit in the power station so that maintenance and repairs or overhaul of the working units may be carried out without any disturbance in power supply.
4. The growth of the demand in near future should be kept in view.
5. The capacity of the power station should be 15 or 20% more than expected maximum demand.

The minimum number of generating units chosen could be one having a capacity equal the maximum demand on the power station. Large size units have lower capital cost per kW, need less floor area, require less operating labour and have better efficiency. Thus large sized units are economical both from the point of view of initial investment and operating cost.

The size of thermal units, in India, has been progressively increasing for the past many years. Around 1955 only 30 MW units were popular while now 500 MW units are being installed. The number and size of units in hydropower stations are governed by the availability of head and water. The earliest hydro units in India installed before independence were of 5 MW capacity while the largest hydro units in operation now are 165 MW at Dehar power plant of Beas-Sutlej project in Himachal Pradesh.

However, selection of a single generating unit having a capacity equal to the maximum demand on the system or slightly more has got following drawbacks.

1. As the load on the system varies at different times of the day, there is a considerable period during the day when the load on the system would be much less than the peak load. During this light load period the generating unit may be running at 50 per cent of the rated output or even practically on no load. Hence the unit would not be running at all times under conditions best suited for its operation to give maximum efficiency. So generating cost per unit would be high.
2. There will be complete failure of supply due to breakdown or maintenance and repairs to be carried out on the generating unit for weeks together.

If the plant is an isolated one and another unit of equal capacity is installed, as standby unit, so as to ensure continuity of supply, there will be substantial increase in the initial investment. In general the system reserve requirements increase with increase in unit size.

Alternatively the size and number of generating units may be so chosen as to fit the load curve as closely as possible. Then each unit can be made to operate in such a way that it operates almost at full load or at a load to give maximum efficiency. The reserve required in this case would be only one unit of the largest size chosen and would be much smaller than the maximum reserve capacity needed in the former case. As the reserve capacity required would be lesser, it would result in better plant capacity factor and plant use factor.

If all the machines of same ratings and identical in operation are chosen, the initial cost would be less, space required would be less, requirements of spare parts would be less, the appearance would be symmetrical and good looking.

The drawbacks and difficulties involved in this alternative of selecting large number of smaller units are given below :

1. The floor area required and initial cost per kW would be increased.
2. There would be substantial increase in maintenance cost and in personnels for handling and operating the equipment.

From the above discussion it is concluded that the number of units in a power station should neither be very large nor very small. As such the size of units for a power station has to be compatible with the total capacity of the power station.

Example 16.5. A residential consumer has 10 lamps of 40 W each connected at his premises. His demand is as follows :

From 12 midnight to 5 AM – 40 W
From 5 AM to 6 PM – No load
From 6 PM to 7 PM – 320 W
From 7 PM to 9 PM – 360 W
From 9 PM to 12 midnight – 160 W

Plot the load curve. Find the average load, maximum load, load factor and electric energy consumption during the day.

[Pb. Univ. Electrical Power II (Generation) May 1995]

Solution : Load curve is drawn as shown in Fig. 16.5.

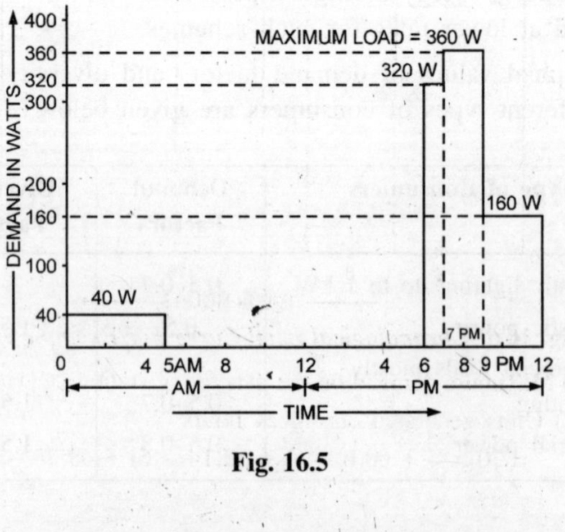

Fig. 16.5

Electrical energy consumption during the day

= Area of load curve

= 40 × 5 + 320 × 1 + 360 × 2 + 160 × 3

= 1,720 Wh or 1.72 kWh **Ans.**

$$\text{Average load} = \frac{\text{Energy consumed during 24 hours}}{24}$$

$$= \frac{1{,}720}{24} = 71.67 \text{ W} \textbf{ Ans.}$$

Maximum load = 360 W

$$\text{Load factor} = \frac{\text{Average load}}{\text{Maximum demand}} = \frac{71.67}{360}$$

$$= 0.1991 \text{ or } 19.91\% \textbf{ Ans.}$$

Example 16.6. Define the terms (*i*) load factor, (*ii*) diversity factor and discuss their influence on the cost of generation.

A generating station has daily load cycle as

Time (Hrs)	0–6	6–10	10–12	12–16	16–20	20–24
Load (MW)	40	50	60	50	70	40

Draw the load curve and load duration curve and find (*i*) maximum demand (*ii*) units generated and (*iii*) load factor.

[U.P. Technical Univ. Power Station Practice 2007]

Solution: From the given data, the load curve is drawn as shown in Fig. 16.6.

For drawing load duration curve, the load oridinates are rearranged in descending order *i.e.,* greatest to the left, lesser loads towards the right and the least load at the extreme right, as shown in Fig. 16.7 (data given below)

Load in MW	Duration of Hours
70	4
60 and above	6
50 and above	14
40 and above	24

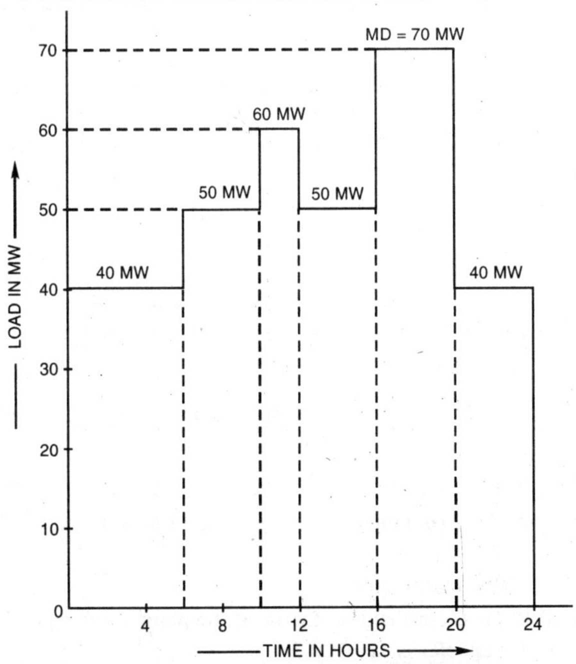

Fig. 16.6. ***Chronological Load Curve For Exampie 16.6.***

(*i*) Maximum demand on the generating station, MD = 70 MW

(*ii*) Units generated during 24 hours

= 70 × 4 + 60(6 − 4) + 50(14 − 6) + 40(24 − 14)

= 280 + 120 + 400 + 400

= 1,200 MWh or 12 × 10^5 kWh **Ans.**

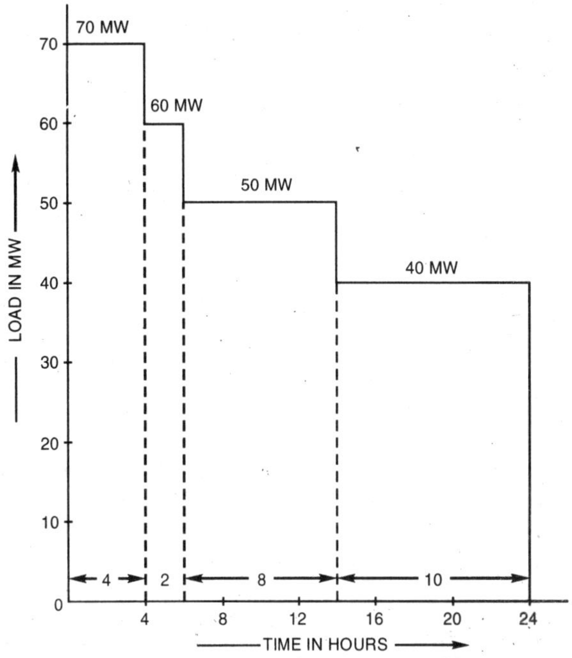

Fig. 16.7. ***Load Duration Curve For Example 16.6.***

(*iii*) Average load

$$= \frac{\text{Energy generated in MWh during the day}}{24}$$

$$= \frac{1{,}200}{24} = 50 \text{ MW} \textbf{ Ans.}$$

$$\text{Load factor} = \frac{\text{Average load}}{\text{Maximum demand}} = \frac{50}{70} = 0.714 \text{ or } 71.4\% \textbf{ Ans.}$$

Example 16.7. The load on a power plant on a typical day is as under :

Time	12-5 AM	5-9 AM	9-6 PM	6-10 PM	10-12 Midnight
Load (MW)	20	40	80	100	20

Plot the chronological load curve and load duration curve. Find the load factor of the plant and the energy supplied by the plant in 24 hours.

[Pb. Univ. Electrical Power II (Generation) January 1993]

Solution : From the given data, the load curve is drawn, as shown in Fig. 16.8.

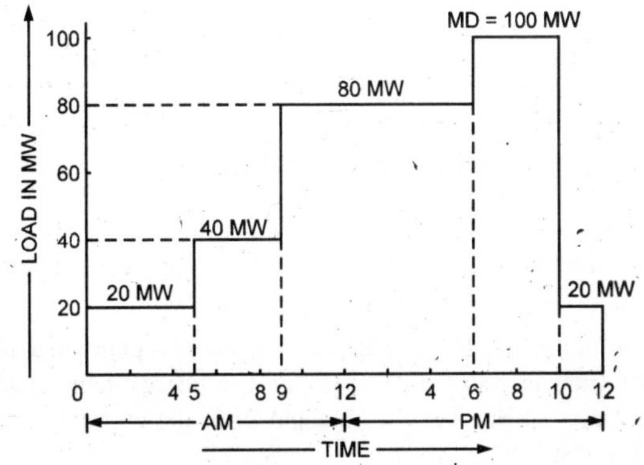

Fig. 16.8. ***Chronological Load Curve For Example 16.7***

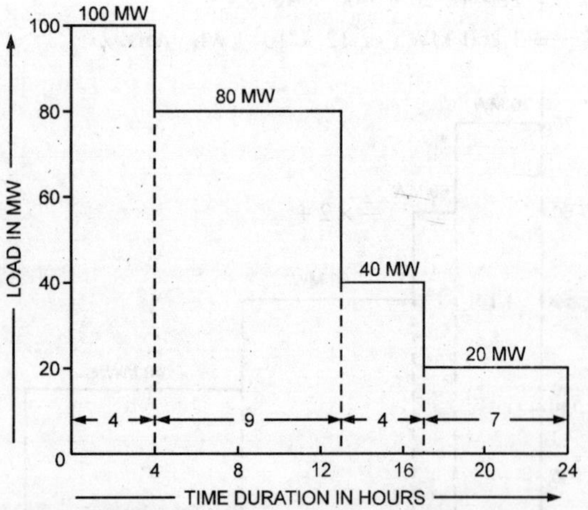

Fig. 16.9. ***Load Duration Curve For Example 16.7***

For drawing the load duration curve, the load ordinates are re-arranged in descending order *i.e.,* the greatest on the left, lesser loads towards the right and the least load at the extreme right, as illustrated in Fig. 16.9. (data given below).

Load in MW	Duration of Hours
100	4
80 and above	13
40 and above	17
20 and above	24

Maximum demand on the power plant, MD = 100 MW

Energy supplied during 24 hours

= 100 × 4 + 80 × (13 – 4) + 40 (17 – 13) + 20 (24 – 17)

= 1,420 MWh **Ans.**

Average load

$$= \frac{\text{Energy generated in MWh during the day}}{24}$$

$$= \frac{1{,}420}{24} = 59.167 \text{ MW}$$

Load factor

$$= \frac{\text{Average load}}{\text{Maximum demand}} = \frac{59.167}{100} = 0.59167 \text{ or } 59.167\% \textbf{ Ans.}$$

Example 16.8. Draw the integrated load duration curve and mass curve for the chronological load curve given in Example 16.7.

Solution : Integrated load duration curve is drawn between the MW demand and the total energy generated up to the MW demand and is obtained from the load duration curve. The following table gives the energy at different load levels.

Load in MW	20	40	80	100
Energy Generated in MWh	20 × 24 = 480	480 + 20 × 17 = 820	820 + 40 × 13 = 1,340	1,340 + 20 × 4 = 1,420

Integrated load duration curve (energy load curve) is drawn as illustrated in Fig. 16.10.

Mass curve is plotted between the energy generated up to a particular time in a day and the time. The energy generated up to different times of the day are obtained from the chronological load curve as given.

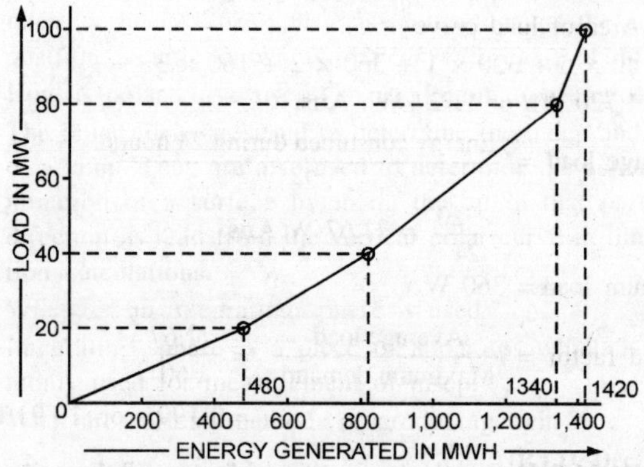

Fig. 16.10. ***Integrated Load Duration Curve***

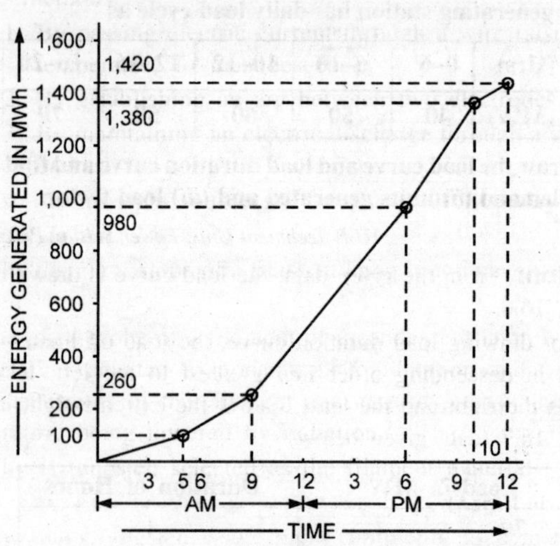

Fig. 16.11. ***Mass Curve***

Time	0	5 AM	9 AM	6 PM	10 PM	12 Mid-night
Energy Generated in MWh up to That Time	0	20 × 5 = 100	100 + 40 × 4 = 260	260 + 80 × 9 = 980	980 + 100 × 4 =1,380	1,380 + 20 × 2 = 1,420

Mass curve is drawn as illustrated in Fig. 16.11.

Example 16.9. An electrical system experiences linear changes in load such that its daily load curve is defined as follows :

Time	12 PM	2 AM	6 AM	8 AM	12 NOON	12.30	1 PM	5 PM	6 PM	12 Midnight
Load (MW)	24	12	12	60	60	48	60	60	84	24

(*i*) **Plot the chronological curve and load duration curve of the system.**

(*ii*) **Find the load factor.**

(*iii*) **What is the utilization factor of the plant serving this load if its capacity is 120 MW?**

[M.D. Univ. Power Plant Engineering May 2007]

Solution : (*i*) From the given data, the chronological curve is drawn as shown in Fig 16.12.

From Fig. 16.12, it is observed that

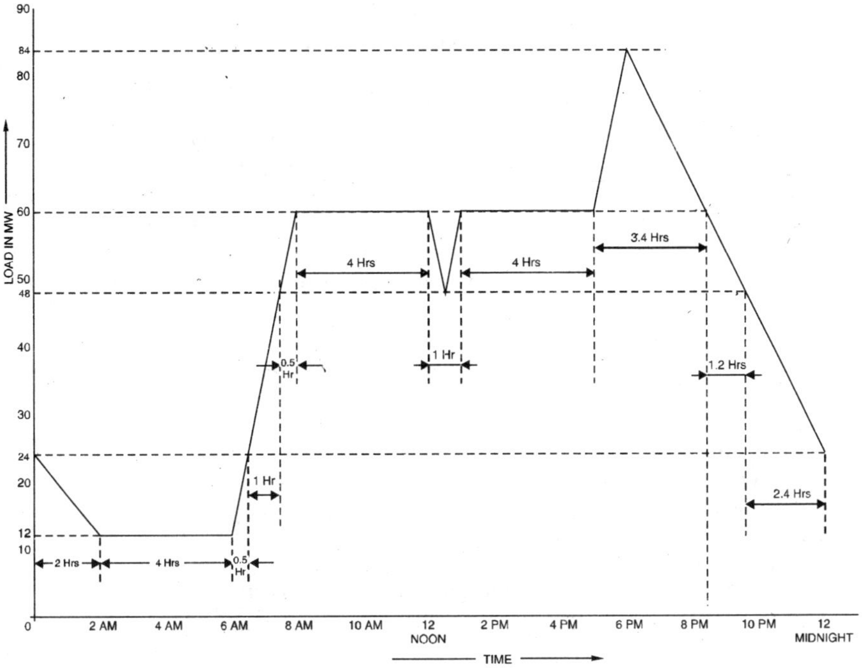

Fig. 16.12

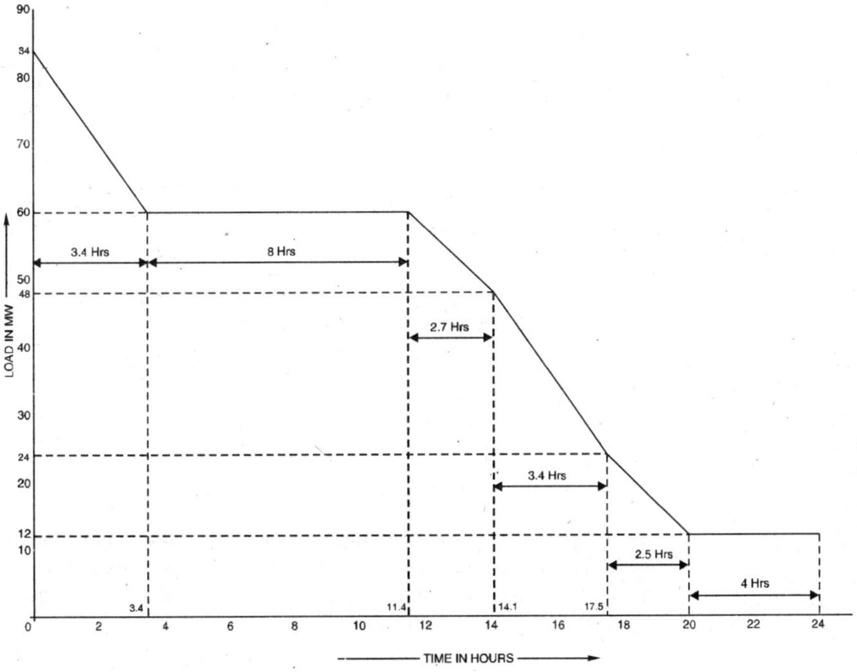

Fig. 16.13

Load falls from 84 MW to 60 MW during the time duration of 3.4 hours (5 PM to 8.24 PM)

Load remains constant at 60 MW for time duration of 8 hours 8 AM to 12 AM and 1 PM to 5 PM)

Load varies from 60 MW to 48 MW during the time duration of 0.5 + 1 + 1.2 *i.e.* 2.7 hours (7.30 to 8.00 hours, 12 to 13.00 hours and 20.24 hrs to 21.36 hrs)

Load varies from 48 MW to 24 MW during the time duration of 1 + 2.4 *i.e.* 3.4 hours (6.30 AM to 7.30 AM and 9.36 to 12.00 midnight)

Load varies from 24 MW to 12 MW during the time duration of 2 + 0.5 *i.e.* 2.5 hours (0 to 2 AM and 6 AM to 6.30 AM)

Load remains constant at 12 MW for time duration of 4 hours (2 AM to 6 AM)

From the above datas the load duration curve is drawn as shown in Fig. 16.13.

(*ii*) Energy required by the system in 24 hours

$$= \frac{24+12}{2} \times 2 + 12 \times 4 + \frac{12+60}{2} \times 2 + 60 \times 4 + \frac{60+48}{2} \times \frac{1}{2} + \frac{48+60}{2} \times \frac{1}{2} + 60 \times 4 + \frac{60+84}{2} \times 1 + \frac{84+24}{2} \times 6$$

$$= 36 + 48 + 72 + 240 + 27 + 27 + 240 + 72 + 324$$

$$= 1{,}086 \text{ MWh}$$

Average load

$$= \frac{\text{Energy required in MWh in 24 hours}}{24}$$

$$= \frac{1{,}086}{24} = 45.25 \text{ MW}$$

Load factor

$$= \frac{\text{Average load}}{\text{Maximum demand}} = \frac{45.25}{84}$$

$$= 0.5387 \text{ or } 53.87\% \textbf{ Ans.}$$

(*iii*) Utilization factor

$$= \frac{\text{Maximum demand in MW}}{\text{Plant capacity in MW}} = \frac{84}{120} = 0.7 \textbf{ Ans.}$$

Example 16.10. The following data were collected from the daily load curves of a power system during a year :

Load kW	Duration hours
15,000	**87**
12,000	**and over 876**
10,000	**and over 1,752**
8,000	**and over 2,628**
6,000	**and over 4,380**
4,000	**and over 7,000**
2,000	**and over 8,760**

Construct the annual load duration curve and find the load factor of the system.

[*Pb. Univ. Electrical Power II (Generation) Dec. 1995*]

Solution : Annual load duration curve is plotted as shown in Fig. 16.14.

Units generated in 8,760 hours

$$= 15{,}000 \times 87 + 12{,}000\,(876 - 87) + 10{,}000\,(1{,}752 - 876) + 8{,}000\,(2{,}628 - 1{,}752) + 6{,}000\,(4{,}380 - 2{,}628) + 4{,}000\,(7{,}000 - 4{,}380) + 2{,}000\,(8{,}760 - 7{,}000)$$

$$= 51{,}053{,}000$$

$$\text{Average load} = \frac{\text{Units generated}}{\text{Time in hours}} = \frac{51{,}053{,}000}{8{,}760} = 5{,}828 \text{ kW}$$

$$\text{Load factor} = \frac{\text{Average load}}{\text{Maximum demand}} = \frac{5{,}828}{15{,}000}$$

$$= 0.3885 \text{ or } 38.85\% \textbf{ Ans.}$$

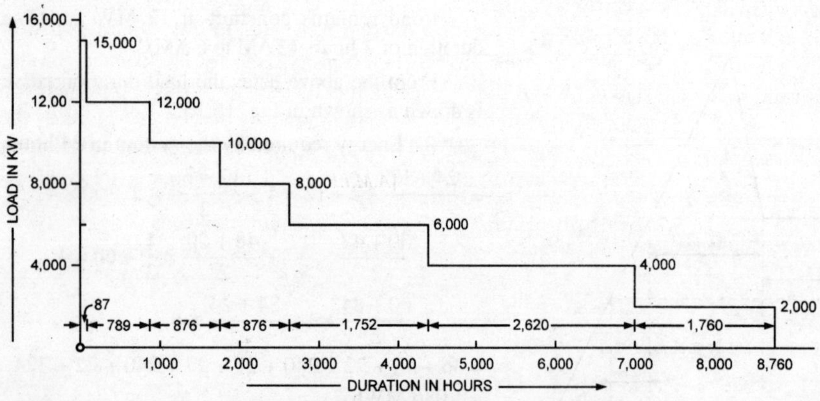

Fig. 16.14

Example 16.11. Make two points of consideration for favouring the choice of each of the following alternatives of units for a thermal power plant with a maximum demand of 1,000 MW. (*i*) One unit of 1,000 MW (*ii*) Two units of 500 MW (*iii*) One unit of 400 MW and three units of 200 MW. [*A.M.I.E. Sec. B. Power Systems Summer 1993*]

Solution : (*i*) One unit of 1,000 MW

Capital cost per kW would be less, requirement of floor area and operating staff will be less.

(*ii*) Two units of 500 MW Each

Reserve required in this case would be much smaller than that required in the former case (500 MW capacity against 1,000 MW capacity) and high operating efficiency during light load hours.

(*iii*) One unit of 400 MW and Three units of 200 MW Each

There is possibility of operating each machine in such a way that it operates at full load or at a load to give maximum efficiency and reserve required would be much smaller (400 MW capacity only).

Example 16.12. A generating station a has maximum demand of 10,000 kW and the daily load on the station is as follows :

Time	kW
11 PM to 6 AM	2,000
6 AM to 8 AM	3,500
8 AM to 12 Noon	8,000
12 Noon to 1 PM	3,000
1 PM to 5 PM	7,500
5 PM to 7 PM	8,500
7 PM to 9 PM	10,000
9 PM to 11 PM	4,500

Choose the size and number of generating units. What reserve plant would be necessary ? Find the load factor and plant capacity factor. [*Pb. Univ. Electrical Power II (Generation) January 1993*]

Solution : The chronological load curve is shown in Fig. 16.15.

Maximum demand = 10,000 kW

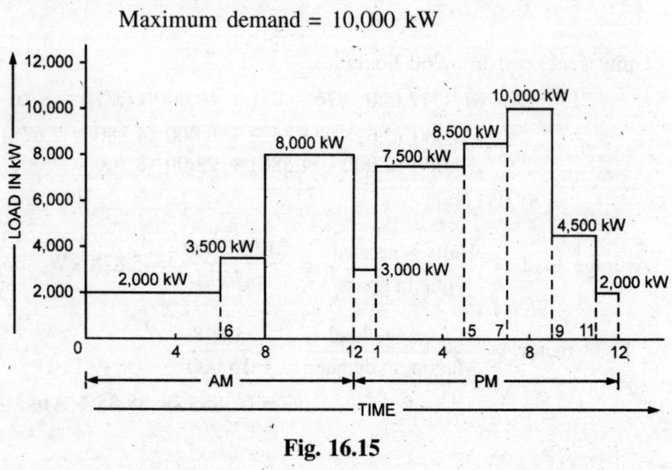

Fig. 16.15

Energy generated in 24 hours

= 2,000 × 7 + 3,500 × 2 + 8,000 × 4 + 3,000 × 1 + 7,500 × 4 + 8,500 × 2 + 10,000 × 2 + 4,500 × 2

= 132,000 kWh

$$\text{Load factor} = \frac{\text{Energy generated in 24 hours}}{\text{Maximum demand} \times 24}$$

$$= \frac{132{,}000}{10{,}000 \times 24} = 0.55 \text{ or } 55\% \textbf{ Ans.}$$

From the shape of the load curve it is obvious that three generating units of capacities 5,000, 3,000 and 2,000 kW would be the best choice to meet the given daily load on the station (see operating schedule).

Reserve plant capacity

= Capacity of the unit of the largest size = 5,000 kW **Ans.**

Installed capacity

= 5,000 × 2 + 3,000 + 2,000 = 15,000 kW

Plant capacity factor

$$= \frac{\text{Energy generated in 24 hours}}{\text{Rated capacity of power station in kW} \times 24}$$

$$= \frac{132{,}000}{15{,}000 \times 24} = 0.3667 \text{ or } 36.67\% \quad \textbf{Ans.}$$

Operating schedule will be as follows :

One unit of 2,000 kW from 11 PM to 6 AM.

One unit of 2,000 kW in parallel with one unit of 3,000 kW from 6 AM to 8 AM.

One unit of 5,000 kW in parallel with one unit of 3,000 kW from 8 AM to 12 noon. (2,000 kW unit stopped)

One unit of 3,000 kW with 5,000 kW unit floating on the system from 12 noon to 1 PM.

One unit of 5,000 kW in parallel with one unit of 3,000 kW from 1 PM to 5 PM.

All three units of 5,000 kW, 3,000 kW and 2,000 kW in parallel from 5 PM to 9 PM.

One unit of 3,000 kW and one unit of 2,000 kW in parallel from 9 PM to 11 PM (5,000 kW unit stopped)

At 11 PM unit of 3,000 kW will also be stopped.

Example 16.13. A generating station has a connected load of 23 MW and a maximum demand of 20 MW, the units generated being 61.5 × 10^6 per annum, calculate (*a*) demand factor (*b*) Average demand (*c*) Load factor.

[*Generation of Electrical Power Rajasthan Univ. 2007; Rajasthan Technical Univ. 2011*]

Solution : Maximum demand = 20 MW

Connected load = 23 MW

Units generated = 61.5 × 10^6 kWh

$$\text{Average load} = \frac{\text{Units generated per annum in kWh}}{24 \times 365}$$

$$= \frac{61.5 \times 10^6}{24 \times 365} = 7{,}020 \text{ kW}$$

$$(a) \text{ Demand factor} = \frac{\text{Maximum demand}}{\text{Connected load}} = \frac{20}{23} = 0.869 \textbf{ Ans.}$$

(*b*) Average demand = 7,020 kW **Ans.**

$$(c) \text{ Load factor} = \frac{\text{Average load}}{\text{Maximum demand}} = \frac{7{,}020}{20 \times 1{,}000}$$

= 35.1% **Ans.**

Example 16.14. A consumer has the following connected loads: 10 lamps of 60 W each and two heaters of 1,000 W each. His maximum demand is 1,500 W. On the average he uses 8 lamps 5 hours a day and each heater for 3 hours a day. Find his average load, monthly energy consumption and load factor.

[*A.M.I.E. Sec. B. Power Systems Winter 1993*]

Solution : Maximum demand, MD = 1,500 W = 1.5 kW

Connected load = 10 × 60 + 2 × 1,000 = 2,600 W

Daily energy consumption

= Number of lamps used × wattage of each lamp × working hours per day + number of heaters × wattage of each heater × working hours per day

= 8 × 60 × 5 + 2 × 1,000 × 3 = 8,400 Wh or 8.4 kWh

Average load

$$= \frac{\text{Daily energy consumption in Wh}}{24} = \frac{8{,}400}{24} = 350 \text{ W}$$ **Ans.**

Monthly energy consumption

= Daily energy consumption × 30 = 8.4 × 30 = 252 kWh **Ans.**

Monthly load factor

$$= \frac{\text{Monthly energy consumption}}{\text{MD} \times 24 \times 30} = \frac{252}{1.5 \times 24 \times 30}$$

= 0.2333 or 23.33% **Ans.**

Example 16.15. A residential consumer has the following connected load : 8 bulbs of 100 W each; 2 fans of 60 W each, 2 light plug points of 100 W each. His use of electricity during a day is as under.

12 midnight–5 AM = One fan

5 AM–7 AM = 2 Fan + 1 light point

7 AM–9 AM – Nil

9 AM–6 PM = 2 Fans

6 PM–Midnight = 2 Fans + 4 bulbs

Find (*a*) Maximum demand (*b*) Connected load (*c*) Demand factor (*d*) Energy consumed during 24 hours.

[*Rajasthan Univ. Generation of Electrical Power 2007*]

Solution : Daily load cycle of the consumer is given below:

Time	*Watts*
12 midnight – 5 AM	1 × 60 = 60 watts
5 AM to 7 AM	2 × 60 + 1 × 100 = 220 watts
7 AM to 9 AM	Nil
9 AM to 6 PM	2 × 60 = 120 watts
6 PM to Midnight	2 × 60 + 4 × 100 = 520 watts

(*a*) Maximum demand, MD = 520 watts **Ans.**

(*b*) Connected load = Number of bulbs × wattage of each bulb + Number of fans × wattage of each fan + Number of plug points × wattage of each plug point

= 8 × 100 + 2 × 60 + 2 × 100

= 800 + 120 + 200 = 1,120 W **Ans.**

(*c*) Demand factor $= \frac{\text{Maximum demand}}{\text{Connected load}} = \frac{520}{1{,}120} = 0.4643$ **Ans.**

(*d*) Energy consumed during 24 hours

= 60 × 5 + 220 × 2 + 120 × 9 + 520 × 6

= 300 + 440 + 1,080 + 3,120 = 4,940 Wh or 4.94 kWh **Ans.**

Example 16.16. A power station is to supply four regions of loads whose peak loads are 15,000 kW, 10,000 kW, 9,000 kW and 11,000 kW respectively. The diversity factor of the loads at the station is 1.5 and the average annual load factor is 50%. Calculate the maximum demand on the station and the annual energy from the station. Suggest the installed capacity and the number of units taking all aspects into account. [*A.M.I.E. Sec. B. Power Systems Summer 1995*]

Solution : Sum of individual maximum demands

= 15,000 + 10,000 + 9,000 + 11,000 = 45,000 kW

Diversity factor = 1.5

Maximum demand on the power station,

$$\text{MD} = \frac{\text{Sum of individual maximum demands}}{\text{Diversity factor}} = \frac{45{,}000}{1.5}$$

= 30,000 kW **Ans.**

Annual average load

= MD × average annual load factor = 30,000 × 0.5 = 15,000 kW

Annual energy supplied

= Annual average load × 365 × 24 = 15,000 × 8,760

= 131.4×10^6 kWh **Ans.**

Consider that there is an increase of maximum demand by 50% in the next 5 year period. Thus maximum demand will become 30,000 × 1.5 = 45,000 kW or 45 MW. Hence three similar units of 15 MW each can be chosen to meet the present demand with one unit standby to meet emergency breakdown and carry out routine maintenance and repairs. It would necessitate the minimum number of spare parts and appearance would be symmetrical and good looking. One additional unit of 15 MW will be installed with the growth of demand, for which provision will be made now.

Example 16.17. A power station has to meet the following demand:

Group A : 200 kW between 8 AM and 6 PM

Group B : 100 kW between 6 AM and 10 AM

Group C : 50 kW between 6 AM and 10 AM

Group D : 100 kW between 10 AM and 6 PM and then 6 PM and 6 AM.

Plot daily load curve and determine (*i*) Diversity factor (*ii*) Units generated per day (*iii*) Load factor.

[*U.P. Technical Univ. Power Station Practice 2008-09*]

Solution : The given load cycle can be tabulated as under :

Time in Hours	0–6	6–8	8–10	10–18	18–24
Group A	—	—	200 kW	200 kW	—
Group B	—	100 kW	100 kW	—	—
Group C	—	50 kW	50 kW	—	—
Group D	100 kW	—	—	100 kW	100 kW
Total load on Power Station	100 kW	150 kW	350 kW	300 kW	100 kW

From the above table, it can be seen that load on the power station is as follows :

0–6 hours 100 kW

6–8 hours 150 kW

8–10 hours 350 kW

10–18 hours 300 kW

18–24 hours 100 kW

The load curve is drawn, from the given data, as shown in Fig. 16.16.

Maximum demand, MD = 350 kW

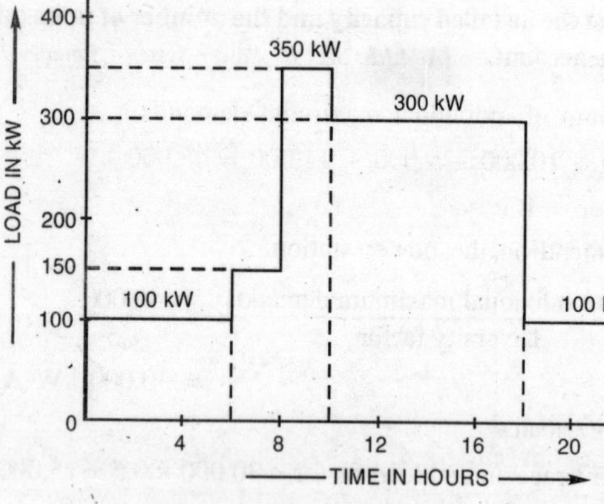

Fig. 16.16. ***Daily Load Curve For Example 16.17***

Sum of individual maximum demands of all the four groups

$= 200 + 100 + 50 + 100 = 450$ kW

(*i*) Diversity factor

$$= \frac{\text{Sum of individual maximum demands}}{\text{Maximum demand on power station}} = \frac{450}{350} = 1.286 \textbf{ Ans.}$$

(*ii*) Units generated per day

= Area (in kWh) under load curve

$= 100 \times 6 + 150 \times 2 + 350 \times 2 + 300 \times 8 + 100 \times 6$

$= 4{,}600$ kWh **Ans.**

(*iii*) Average load

$$= \frac{\text{Units generated per day}}{24} = \frac{4{,}600}{24} = 191.7 \text{ kW}$$

Load factor

$$= \frac{\text{Average load}}{\text{Maximum demand}} = \frac{191.7}{350} = 0.548 \text{ or } 54.8\% \textbf{ Ans.}$$

Example 16.18. The daily demands of three consumers are given below :

Time	Consumer 1	Consumer 2	Consumer 3
12 Mid night to 8 AM	No Load	200 W	No load
8 AM to 2 PM	600 W	No load	200 W
2 PM to 4 PM	200 W	1,000 W	1,200 W
4 PM to 10 PM	800 W	No load	No Load
10 PM to Midnight	No Load	200 W	200 W

Plot the load curve and find : (*i*) maximum demand of individual consumer (*ii*) diversity factor (*iii*) load factor of individual consumer (*iv*) load factor of the station.

[U.P. Technical Univ. Power Station Practice 2007-08]

Solution : Daily demand of all the three consumers is as under :

0–8 hours	200 W
8–14 hours	600 + 200 = 800 W
14–16 hours	200 + 1,000 + 1,200 = 2,400 W
16–22 hours	800 W
22–24 hours	200 + 200 = 400 W

The load curve is drawn as illustrated in Fig. 16.17.

(*i*) Maximum demands of

Consumer 1 = 800 W **Ans.**

Consumer 2 = 1,000 W **Ans.**

Consumer 3 = 1,200 W **Ans.**

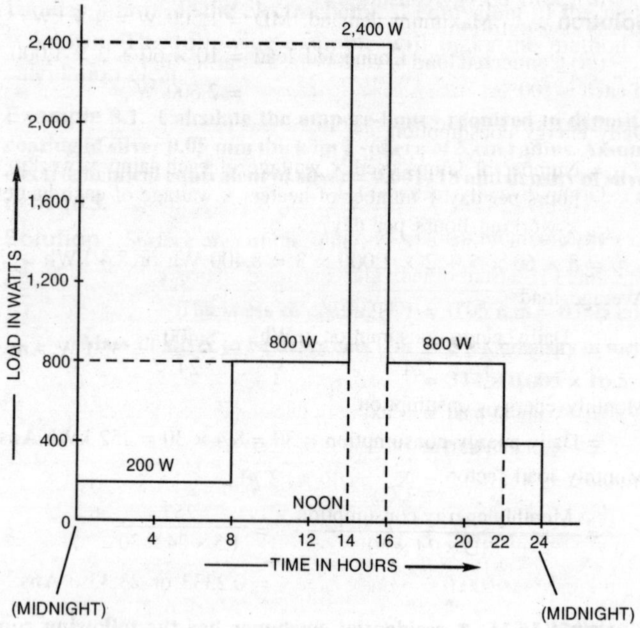

Fig. 16.17

(*ii*) Sum of individual

Maximum demands of all the three consumers

$= 800 + 1{,}000 + 1{,}200 = 3{,}000$ W

Simultaneous maximum demand on the power stations is 2,400 W that occurs from 2 PM to 4 PM

$$\text{Diversity factor} = \frac{\text{Sum of individual maximum demands}}{\text{Maximum demand on power station}}$$

$$= \frac{3{,}000}{2{,}400} = 1.25 \textbf{ Ans.}$$

(*iii*) Load factor of consumer 1

$$= \frac{\text{Energy consumed per day}}{\text{Maximum demand} \times 24} = \frac{600 \times 6 + 200 \times 2 + 800 \times 6}{800 \times 24}$$

$= 0.4583$ or 45.83% **Ans.**

$$\text{Load factor of consumer 2} = \frac{200 \times 8 + 1{,}000 \times 2 + 200 \times 2}{1{,}000 \times 24}$$

$= 0.1667$ or 16.67% **Ans.**

$$\text{Load factor of consumer 3} = \frac{200 \times 6 + 1{,}200 \times 2 + 200 \times 2}{1{,}200 \times 24}$$

$= 0.1389$ or 13.89% **Ans.**

(*iv*) Load factor of the power station

$$= \frac{200 \times 8 + 800 \times 6 + 2{,}400 \times 2 + 800 \times 6 + 400 \times 2}{2{,}400 \times 24}$$

$= 0.2916$ or 29.16% **Ans.**

Example 16.19. A power station has the following load :

Residential lighting load :	**Maximum load = 1,000 kW**
	Load factor = 20%
	Diversity between consumers = 1.3
Commercial load :	**Maximum load = 2,000 kW**
	Load factor = 30%
	Diversity between consumers = 1.1

Industrial load : Maximum demand = 5,000 kW

Load factor = 80%

Diversity between consumers = 1.2

Overall diversity factor may be taken as 1.4.

Find (*i*) maximum demand on system.

(*ii*) Daily energy consumption of each type of load and total energy consumption.

(*iii*) Connected load of each type, assuming the demand factor of each is 100%. [*Rajasthan Univ. Generation of Electrical Power 2000*]

Solution : The sum of maximum demands of three types of loads

= 1,000 + 2,000 + 5,000 = 8,000 kW

Group diversity factor = 1.4

(*i*) Maximum demand on the system

$$= \frac{\text{Sum of individual maximum demands}}{\text{Group diversity factor}}$$

$$= \frac{8,000}{1.4} = 5,714.28 \text{ kW} \quad \textbf{Ans.}$$

(*ii*) Average load = Maximum demand × load factor

So Average demand of domestic load

$$= 1,000 \times \frac{20}{100} = 200 \text{ kW}$$

Average demand of commercial load

$$= 2,000 \times \frac{30}{100} = 600 \text{ kW}$$

Average demand of industrial load

$$= 5,000 \times \frac{80}{100} = 4,000 \text{ kW}$$

∵ Daily consumption = Average load in kW × 24

Daily consumption for lighting load

= 200 × 24 = 4,800 kWh **Ans.**

Daily consumption for commercial load

= 600 × 24 = 14,400 kWh **Ans.**

Daily consumption for industrial load

= 4,000 × 24 = 96,000 kWh **Ans.**

Total energy consumption = 4,800 + 14,400 + 96,000

= 115,200 kWh **Ans.**

(*iii*) Maximum demand of domestic consumers

= Maximum domestic demands × diversity factor

= 1,000 × 1.3 = 1,300 kW

Maximum demand of commercial consumers

= 2,000 × 1.1 = 2,200 kW

Maximum demand of industrial consumers

= 5,000 × 1.2 = 6,000 kW

$$\because \text{ Connected load } = \frac{\text{Maximum demand}}{\text{Demand factor}}$$

So connected load of domestic consumers

$$= \frac{1,300}{1} = 1,300 \text{ kW} \quad \textbf{Ans.}$$

Connected load of commercial consumers

$$= \frac{2,200}{1} = 2,200 \text{ kW} \quad \textbf{Ans.}$$

Connected load of industrial consumer

$$= \frac{6,000}{1} = 6,000 \text{ kW} \quad \textbf{Ans.}$$

Example 16.20. A new housing development is to be added to the lines of public utility. There are 1,000 apartments each having a connected load of 4 kW, also stores and services are included of the characteristics shown in the following table :

Stores or Services	Connected Load (kW)	Demand Factor, Per cent
1 Laundry	20	68
2 Churches	10 each	56
1 Restaurant	60	52
1 Book store	5	66
2 Medical stores	10 each	79
2 Grocery stores	5 each	73
1 Theatre	100	48

The demand factor for the apartments is 45 per cent. The group diversity factor of the residential load for this system is 3.6 and peak diversity factor is 1.4. The commercial load group diversity factor is 1.5 and the peak diversity factor is 1.1.

Find the increase in peak demand on the total system delivery from the station bus resulting from the addition of this development on the distribution system. Assume line losses at 5% of the delivered energy.

[*A.M.I.E. Sec. B. Power Plant Engineering Summer 1994*]

Solution : Maximum demand of each apartment

= Connected load of each apartment × demand factor

= 4 × 0.45 = 1.8 kW

Maximum demand of 1,000 apartments = 1,000 × 1.8 = 1,800 kW

Demand of 1,000 apartments at the time of the system peak

$$= \frac{\text{Maximum demand of 1,000 apartments}}{\text{Group diversity factor} \times \text{peak diversity factor}} = \frac{1,800}{3.6 \times 1.4}$$

= 357 kW

Maximum demand

$$= \frac{\text{Connected load} \times \text{demand factor}}{\text{Group diversity factor}}$$

So maximum demand of all stores/services

$$= \left(\frac{20 \times 0.68 + 2 \times 10 \times 0.56 + 60 \times 0.52 + 5 \times 0.66 + 2 \times 10 \times 0.79 + 2 \times 5 \times 0.73 + 100 \times 0.48}{1.5} \right)$$

= 86.933 kW

Demand of all stores/services at the time of system peak

$$= \frac{86.933}{1.1} = 79 \text{ kW}$$

Increase in demand on the distribution system

= 357 + 79 = 436 kW

$$\text{Increase in demand on the station bus} = \frac{436}{1 - 0.05} = 459 \text{ kW} \quad \textbf{Ans.}$$

Example 16.21. The maximum demand of a power plant is 40 MW. The capacity factor is 0.5 and utilization factor is 0.8. Find

(*i*) Load factor (*ii*) Plant capacity (*iii*) Reserve capacity (*iv*) Annual energy production.

[*Rajasthan Technical Univ. Generation of Electrical Power 2008, 2009*]

Solution : Maximum demand, MD = 40 MW

Capacity factor = 0.5

Utilization factor = 0.8

Dividing Eq (16.15) by Eq (16.16), we have

$$(i)\text{ Load factor} = \frac{\text{Capacity factor}}{\text{Utilization factor}} = \frac{0.5}{0.8} = 0.625 \textbf{ Ans.}$$

$$(ii)\text{ Plant capacity} = \frac{\text{Maximum demand}}{\text{Utilization factor}} = \frac{40}{0.8} = 50 \text{ MW } \textbf{Ans.}$$

(*iii*) Reserve capacity = Plant capacity – maximum demand
= 50 – 40 = 10 MW **Ans.**

(*iv*) Annual energy production
= Maximum demand × load factor × 365 × 24
= 40 × 0.625 × 365 × 24 kWh
= 219,000 kWh **Ans.**

Example 16.22. A generating station has a maximum demand of 20 MW, a load factor of 60%, plant capacity factor of 48% and a plant use factor of 80%. Find (*a*) the daily energy consumed (*b*) The reserve capacity of the plant (*c*) the maximum energy that could be produced daily, if the plant were running all the time and (*d*) the maximum energy that could be produced daily, if the plant were running fully loaded. [*Pb. Technical Univ. Generation of Electric Power May 2010*]

Solution : Maximum demand, MD = 20 MW
Load factor, LF = 60% = 0.6
Average load = MD × LF = 20 × 0.6 = 12 MW
(*a*) Daily energy consumed = Average load × 24
= 12 × 24 = 288 MWh **Ans.**
(*b*) Generating station installed capacity

$$= \frac{\text{Average load}}{\text{Plant capacity factor}} = \frac{12}{0.48} = 25 \text{ MW}$$

Reserve capacity of the plant
= Installed capacity – maximum demand
= 25 – 20 = 5 MW **Ans.**

(*c*) The maximum energy that could be produced daily if the plant were running all the time
= Rated capacity × 24 = 25 × 24 = 600 MWh **Ans.**

(*d*) The maximum energy that could be produced daily if the plant are fully loaded
= Plant use factor × rated capacity of the plant × number of operating hours
= 0.8 × 25 × 24 = 480 MWh **Ans.**

Example 16.23. The annual load duration curve of a small hydro plant shows 438 × 10⁴ kWh of energy during the year. It is a peak load plant with 20% annual load factor. Find the station capacity if plant capacity factor is 15%. Find the reserve capacity of the plant. [*Pb. Technical Univ. Generation of Electric Power May 2008*]

Solution : Average load

$$= \frac{\text{Energy generated per annum}}{24 \times 365} = \frac{438 \times 10^4}{24 \times 365} = 500 \text{ kW}$$

Maximum demand, MD

$$= \frac{\text{Average load}}{\text{load factor}} = \frac{500}{0.2} = 2{,}500 \text{ kW or } 2.5 \text{ MW}$$

Plant capacity

$$= \frac{\text{Average load}}{\text{Capacity factor}} = \frac{500}{0.15} = 3{,}333 \text{ kW or } 3.333 \text{ MW } \textbf{Ans.}$$

Reserve capacity
= Installed capacity – maximum demand
= 3.333 – 2.5 = 0.833 MW **Ans.**

Example 16.24. The load connected to a station is 100 MW and the demand factor of the consumers is 1.25. Estimate the capacity of the generating station when load factor is 60%, capacity factor is 50%, diversity factor of distribution is 1.33 and that of transmission is 1.5. Take the efficiency of transmission and distribution as 90%. [*Pb. Univ. Electrical Power II (Generation) Dec. 1993*]

Solution : Maximum demand on the distribution network

$$= \frac{\text{Connected load to distribution network}}{\text{Demand factor of consumers}} = \frac{100}{1.25} = 80 \text{ MW}$$

Maximum demand of the generating station

$$= \frac{\text{Maximum demand on the distribution network}}{\text{Diversity factor of distribution} \times \text{diversity factor of transmission} \times \text{efficiency of transmission and distribution}}$$

$$= \frac{80}{1.33 \times 1.5 \times 0.9} = 44.556 \text{ MW}$$

Capacity of the generating station

$$= \frac{\text{Maximum demand on the generating station}}{\text{Plant capacity factor}} \times \text{load factor}$$

$$= \frac{44.556 \times 0.6}{0.5} = 53.5 \text{ MW } \textbf{Ans.}$$

Example 16.25. The peak load on a power station is 30 MW. The loads having maximum demands of 25 MW, 10 MW, 5 MW and 7 MW are connected to power station. The capacity of the plant is 40 MW and the annual load factor is 50%. Find (*i*) average load on the power station (*ii*) energy supplied per annum (*iii*) demand factor and (*iv*) diversity factor. [*M.D. Univ. Power Plant Engineering May 2009*]

Solution : (*i*) Average load = Maximum demand × load factor
= 30 × 0.5 = 15 MW

(*ii*) Energy supplied per annum
= Average load × 24 × 365
= 15 × 24 × 365 = 131, 400 MWh or 131.4 × 10⁶ kWh **Ans.**

$$(iii)\text{ Demand factor} = \frac{\text{Maximum demand}}{\text{Connected load}} = \frac{30}{25+10+5+7} = 0.6383 \textbf{ Ans.}$$

$$(iv)\text{ Diversity factor} = \frac{\text{Sum of individual maximum demands}}{\text{Maximum demand of power station}}$$

$$= \frac{25+10+5+7}{30} = 1.567 \textbf{ Ans.}$$

Example 16.26. The peak load on a 50 MW power station is 39 MW. It supplies power through four transformers whose connected loads are 17, 12, 9 and 10 MW. The maximum demands on these transformers are 15, 10, 8 and 9 MW respectively. If the annual load factor is 50% and the plant is operating for 65% of the period in the year, find out

(*i*) average load on the station (*ii*) energy supplied per year (*iii*) demand factor (*iv*) diversity factor (*v*) use factor for the power station. [*U.P. Technical Univ. Power Station Practice 2009-10; M.D. Univ. Power Plant Engineering May 2007*]

Solution : Power station rated capacity = 50 MW
Maximum demand on the power station = 39 MW
Sum of connected loads = 17 + 12 + 9 + 10 = 48 MW
Sum of maximum demands on transformers
= 15 + 10 + 8 + 9 = 42 MW
Annual load factor = 50% or 0.5
Plant operating period = 0.65 × 8,760 = 5,694 hours
(*i*) Average load on station = MD × LF = 39 × 0.5
= 19.5 MW **Ans.**
(*ii*) Energy supplied per year = Average load × 8,760
= 19.5 × 8,760
= 170,820 MWh **Ans.**
(*iii*) Demand factor

$$= \frac{\text{Maximum demand}}{\text{Sum of connected loads}} = \frac{39}{48} = 0.8125 \text{ or } 81.25\% \textbf{ Ans.}$$

(*iv*) Diversity factor

$$= \frac{\text{Sum of maximum demands}}{\text{Maximum deamand}} = \frac{42}{39} = 1.077 \textbf{ Ans.}$$

(v) Use factor $= \dfrac{\text{Energy generated per year}}{\text{Rated capacity} \times \text{number of operating hours}}$

$$= \frac{170{,}820 \times 10^3}{50 \times 10^3 \times 5{,}694} = 0.6 \text{ or } 60\% \textbf{ Ans.}$$

Example 16.27. From a load duration curve, the following data are available :

The maximum demand on the system is 25 MW. The load supplied by two units is 15 MW and 12.5 MW. Unit no 1 acts as a base load unit and no 2 as a peak load unit. The base load unit works for 100% of the time and peak load unit for only 40% of the time. The energy generated by unit no 1 is 1 × 10^8 units and that by no 2 is 1 × 10^7 units. Determine the load factor, plant capacity factor, and plant use factor of each unit and load factor of the total plant.

[*Pb. Univ. Electrical Power II (Generation) May 1994*]

Solution : For Standby Unit. Assuming maximum demand on base load unit equal to its rated capacity *i.e.,* 15 MW.

Peak load on the standby unit

= Maximum demand of the system – maximum demand on base load unit

= 25 – 15 = 10 MW.

Capacity of standby unit = 12.5 MW

Annual standby unit output = 1 × 10^7 kWh

Hours of operation of standby unit during the year

= 0.4 × 8760 = 3,504 hours

Annual average load of standby unit

$$= \frac{\text{Output in kWh}}{24 \times 365} = \frac{1 \times 10^7}{8{,}760} = 1{,}142 \text{ kW}$$

Annual load factor

$$= \frac{\text{Annual average load}}{\text{Maximum demand}} = \frac{1{,}142}{10 \times 10^3} = 0.114 \text{ or } 11.4\% \textbf{ Ans.}$$

Plant capacity factor

$$= \frac{\text{Average Load}}{\text{Rated Capacity}} = \frac{1{,}142}{12.5 \times 10^3} = 0.0913 \text{ or } 9.13\% \textbf{ Ans.}$$

Plant use factor

$$= \frac{\text{Total units generated}}{\text{Rated capacity of unit} \times \text{number of operating hours}}$$

$$= \frac{1 \times 10^7}{12.5 \times 10^3 \times 3{,}504} = 0.2283 \text{ or } 22.83\% \textbf{ Ans.}$$

For Base Load Unit

Annual output = 1 × 10^8 kWh

$$\text{Annual average load} = \frac{\text{Output in kWh}}{8{,}760} = \frac{1 \times 10^8}{8{,}760} = 11{,}416 \text{ kW}$$

$$\text{Annual load factor} = \frac{\text{Annual average load}}{\text{Maximum demand}}$$

$$= \frac{11{,}416}{15 \times 10^3} = 0.761 \text{ or } 76.1\% \textbf{ Ans.}$$

$$\text{Capacity factor} = \frac{\text{Annual average load}}{\text{Rated capacity}}$$

$$= \frac{11{,}416}{15{,}000} = 0.761 \text{ or } 76.1\% \textbf{ Ans.}$$

$$\text{Use factor} = \frac{\text{Total kWh generated}}{\text{Rated capacity of unit} \times \text{number of operating hours}}$$

$$= \frac{1 \times 10^8}{15{,}000 \times 8{,}760} = 0.761 \text{ or } 76.1\% \textbf{ Ans.}$$

Load factor of the total plant

$$= \frac{\text{Total kWh generated by both units}}{\text{Rated capacity of the plant} \times 8{,}760}$$

$$= \frac{1 \times 10^8 + 1 \times 10^7}{(15 + 12.5) \times 10^3 \times 8{,}760} = 0.4566 \text{ or } 45.66\% \textbf{ Ans.}$$

Example 16.28. The yearly duration curve of a certain plant can be considered as a straight line from 140 MW to 30 MW. Power is supplied with one generating unit of 95 MW capacity and two units of 45 MW capacity each. Determine :

(*i*) Installed capacity (*ii*) Load factor (*iii*) Plant capacity factor (*iv*) Maximum Demand (*v*) Utilization factor.

[*Pb. Univ. Electrical Power II (Generation) Dec. 1993*]

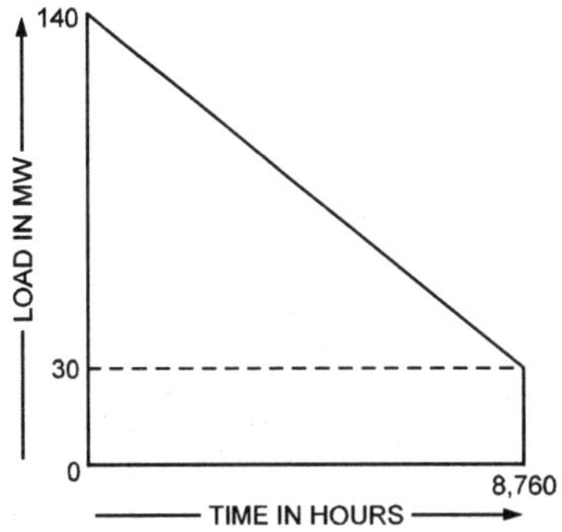

Fig. 16.18. *Load Duration Curve For Example 16.28*

Solution : (*i*) Installed capacity = 95 + 2 × 45 = 185 MW **Ans.**

(*iv*) Maximum demand = 140 MW

$$\text{Average load} = \frac{140 + 30}{2} = 85 \text{ MW}$$

$$(ii) \text{ Load factor} = \frac{\text{Average load}}{\text{Maximum demand}}$$

$$= \frac{85}{140} = 0.607 \text{ or } 60.7\% \textbf{ Ans.}$$

$$(iii) \text{ Plant capacity factor} = \frac{\text{Average load}}{\text{Rated capacity}}$$

$$= \frac{85}{185} = 0.46 \text{ or } 46\% \textbf{ Ans.}$$

$$(v) \text{ Utilization factor} = \frac{\text{Maximum demand}}{\text{Rated capacity}}$$

$$= \frac{140}{185} = 0.757 \text{ or } 75.7\% \textbf{ Ans.}$$

Example 16.29. For a certain industrial organisation the maximum and minimum power demands are 50 MW and 10 MW respectively. Its load duration curve can be approximated as shown in Fig. 16.19. The hydro power available at the time of minimum

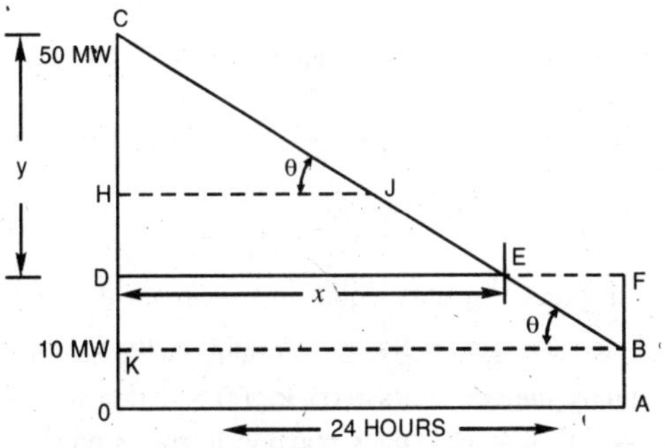

Fig. 16.19

regulated flow is just sufficient to take a peak load of 60 MWh per day. Further, it is observed that it will be economical to pump water from tail race to the reservoir tank by utilizing the steam power plant during off peak periods and thus running the station at 100 per cent load factor. If the efficiency of the steam electric conversion is 60%, calculate the maximum capacity of each type of plant.

[*U.P.S.C.I.E.S. Electrical Engineering-II 2006*]

Solution : Let the load of 60 MWh per day be represented by ΔCJH.

So $\frac{1}{2}$HJ × CH = 60 MWh

or HJ × CH = 120 ...(*i*)

From Fig. 16.19

$$HJ = \frac{CH}{\text{Tan}\,\theta} = \frac{CH}{CK/BK} = \frac{CH}{(50-10)/24} = \frac{24}{40}CH = 0.6\,CH$$

Substituting HJ = 0.6 CH in Eq (*i*), we have

$$0.6\,CH \times CH = 120$$

$$\text{or } CH = \sqrt{\frac{120}{0.6}} = 14.142 \text{ MW}$$

So Maximum capacity of hydropower plant = 14.142 MW

Maximum capacity of steam power plant

$$= \frac{50-14.142}{0.6} = 59.763 \text{ MW} \quad \textbf{Ans.}$$

Example 16.30. A generating station has an installed capacity of 50,000 kW and delivers 220 × 10⁶ units per annum. If the annual fixed charges are Rs 160 per kW installed capacity and running charges are 4 paise per kWh, determine the cost per unit generated.

[*Pb. Technical Univ. Generation of Electric Power December 2010*]

Solution : Installed capacity = 50,000 kW

Fixed charges = Rs 160 per kW installed capacity

Total fixed charges = Rs 160 × 50,000 = Rs 8,000,000

Total number of units generated per annum

$$= 220 \times 10^6$$

Running cost = 4 paise per kWh

$$\text{Total running cost} = \text{Rs } \frac{4}{100} \times 220 \times 10^6 = \text{Rs } 8,800,000$$

Total annual charges = Fixed charges + running charges

= Rs 8,000,000 + Rs 8,800,000

= 16,800,000

$$\text{Cost per unit generated} = \frac{\text{Total annual charges per annum}}{\text{Total delivered units per annum}}$$

$$= \text{Re } \frac{16,800,000}{220 \times 10^6}$$

= Re 0.0764 or paise 7.64 **Ans.**

Example 16.31. The output of generating station is 500 × 10⁶ kWh per annum. The average load factor is 60%. If the annual fixed charges are Rs 60 per kW of installed capacity and the annual running charges are 5 paise per kWh, what is the cost of energy per kWh at the bus-bars ? [*Agra Univ. Generation and Switchgear Practice 1982*]

Solution : Output energy per annum = 500 × 10⁶ kWh

$$\text{Average load} = \frac{500 \times 10^6}{24 \times 365} = 57,000 \text{ kW}$$

$$\text{Maximum demand} = \frac{\text{Average load}}{\text{Load factor}} = \frac{57,000}{0.6} = 95,000 \text{ kW}$$

Assuming installed capacity equal to maximum demand

Fixed charges = Rs 60 × 95,000 = Rs 5,700,000

Annual running charges = Rs 0.05 × 500 × 10⁶ = Rs 25,000,000

Total annual charges = Rs 5,700,000 + Rs 25,000,000

= Rs 30,700,000

Cost of energy at the bus-bar

$$= \frac{30,700,000}{500 \times 10^6} = \text{Re } 0.0614 \text{ per kWh or } 6.14 \text{ P per kWh } \textbf{Ans.}$$

Example 16.32. A generating plant has a maximum capacity of 100 kW and costs Rs 160,000. The fixed charges are 12% consisting of 5% interest, 5% depreciation and 2% taxes. Find the fixed charges per kWh if the load factor is (i) 100% and (ii) 50%.

[*U.P. Technical Univ. Power Station Practice 2006-07*]

Solution : Assuming maximum demand equal to generating plant capacity

Maximum demand = 100 kW

Annual fixed charges = Rs 0.12 × 160,000 = Rs. 19,200

(*i*) *When load factor is 100% i.e. unity*

Units generated per annum

= Maximum demand × LF × hours in a year

= 100 × 1 × 8,760 = 876,000 kWh

$$\text{Fixed charges per kWh} = \frac{\text{Annual fixed charges}}{\text{Units generated / annum}}$$

$$= \frac{19,200}{876,000} = \text{Re } 0.0219 \text{ or } 2.19 \text{ paise } \textbf{Ans.}$$

(*ii*) *When load factor is 50% i.e., 0.5*

Units generated per annum = 100 × 0.5 × 8,760 = 438,000 kWh

$$\text{Fixed charges/kWh} = \frac{19,200}{438,000} = \text{Re. } 0.0438 \text{ or } 4.38 \text{ paise } \textbf{Ans.}$$

Example 16.33. The annual fixed and operating cost of a 500 MW station are Rs 3,000 per kW of the installed capacity per annum and Rs 0.60 per kWh respectively. Plot the variation of unit cost of energy with load factor. [*A.M.I.E. Sec. B. Power Systems Summer 2000*]

Solution : Installed capacity = 500 MW or 5 × 10⁵ kW

Assuming maximum demand on the power station to be equal to installed capacity *i.e.*, 5 × 10⁵ kW

Fixed cost = Rs. 3,000 × 5 × 10⁵ = Rs. 15 × 10⁸

Units generated per annum for unity load factor

= 5 × 10⁵ × 8,760 = 438 × 10⁷ kWh

Units generated per annum for load factors 0.2, 0.4, 0.6 and 0.8 will be 0.2, 0.4, 0.6 and 0.8 times of units generated per annum for unity load factor *i.e.*, 87.6 × 10⁷ kWh, 175.2 × 10⁷ kWh, 262.8 × 10⁷ kWh 350.4 × 10⁷ kWh respectively.

Total cost of electrical energy generated per unit for station load factor of unity

= Fixed cost per kWh + operating cost per kWh

$$= \frac{\text{Total fixed cost}}{\text{Total units generated}} + \text{operating cost per kWh}$$

$$= \text{Re } \frac{15 \times 10^8}{438 \times 10^7} + \text{Re } 0.60 = \text{Re } 0.34 + \text{Re } 0.60 = \text{Re } 0.94/\text{unit}$$

Similarly total cost per unit of electrical energy generated for station load factor of 0.8

$$= \text{Re } \frac{15 \times 10^8}{350.4 \times 10^7} + \text{Re } 0.60 = \text{Re } 0.43 + \text{Re } 0.60 = \text{Rs } 1.03/\text{unit}$$

Total cost per unit of electrical energy generated for station load factor of 0.6

$$= \text{Re } \frac{15 \times 10^8}{262.8 \times 10^7} + \text{Re } 0.60$$

= Re 0.57 + Re 0.60 = Rs 1.17/unit

Total cost per unit of electrical energy generated for station load factor of 0.4

$$= \text{Re}\ \frac{15 \times 10^8}{175.2 \times 10^7} + \text{Re } 0.60 = \text{Re } 0.86 + \text{Re } 0.60 = \text{Rs } 1.46/\text{unit}$$

Total cost per unit of electrical energy generated for station load factor of 0.2

$$= \text{Re}\ \frac{15 \times 10^8}{87.6 \times 10^7} + \text{Re } 0.60 = \text{Rs } 1.71 + \text{Re } 0.60 = \text{Rs } 2.31/\text{unit}$$

Variation of unit cost of energy with load factor is shown in Fig. 16.20.

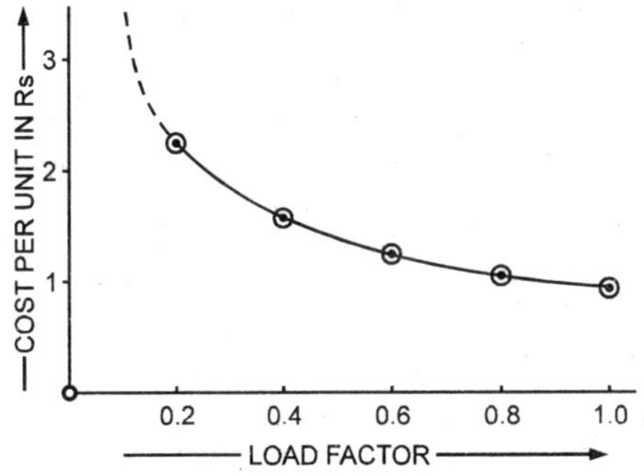

Fig. 16.20

Example 16.34. A peak load on the thermal plant of 12 MW capacity is 10 MW. The plant annual load factor is 0.7. Design a two part tariff from the following data :

Cost of plant Rs 7,000/kW installed capacity

Interest and depreciation = 10% of the capital cost

The capital cost of transmission and distribution = Rs 3 × 10^5

Interest and depreciation on it = 5%

Operating cost = Rs 3 × 10^5/year

Cost of fuel = Rs 50/ton

Plant maintenance cost = Rs 25,000/year (fixed)

Plant maintenance cost = Rs 35,000/year (running)

Coal used = 3 × 10^4 ton per year

[*M.D. Univ. Power Plant Engineering May 2009*]

Solution : Installed capacity = 12 MW

Maximum demand, MD = 10 MW

Plant annual load factor = 0.7

Total units generated/annum = MD × load factor × 8,760

= 10 × 0.7 × 8,760

= 61,320 MWh = 61.32 × 10^6 kWh

Fixed costs

Capital cost of plant = Rs 7,000 × 12 × 10^3 = Rs 84 × 10^6

Interest and depreciation (fixed cost)

$$= \text{Rs } \frac{10}{100} \times 84 \times 10^6 = \text{Rs } 8.4 \times 10^6$$

Capital cost of transmission and distribution = Rs 3 × 10^5

Interest and depreciation on transmission and distribution

$$= \text{Rs } \frac{5}{100} \times 3 \times 10^5 = \text{Rs } 0.15 \times 10^5$$

Plant maintenance cost (fixed) = Rs 25,000

Total fixed cost = Rs (8.4 × 10^6 + 15,000 + 25,000)

= Rs 8,440,000

So fixed cost per kW per annum

$$= \text{Rs } \frac{8{,}440{,}000}{\text{MD}} = \text{Rs } \frac{8{,}440{,}000}{10{,}000} = \text{Rs } 844 \text{ per kW/annum}$$

Running costs

Operating cost = Rs 300,000 per annum

Maintenance cost = Rs 35,000 per annum

Fuel cost = Rs 50 × 3 × 10^4 = Rs 1,500,000

Total running cost = Rs 1,500,000 + 300,000 + 35,000

= Rs 1,835,000

$$\text{Running cost per kWh} = \frac{\text{Total running cost/annum}}{\text{Units generated/annum}}$$

$$= \text{Re } \frac{1{,}835{,}000}{61.32 \times 10^6}$$

= Re 0.03/kWh or 3 paise per kWh

So two part tariff is Rs 844 per kW of maximum demand per annum + 3 paise per kWh **Ans.**

Example 16.35. What are the different factors related to plants and consumers for power plant economics ?

Estimate the generating cost per kWh delivered from a generating station from the following data :

Plant capacity = 50 MW

Annual load factor = 40%

Capital cost = Rs 1.2 crores, annual cost of wages, taxation etc. Rs. 4 lacs, cost of fuel, lubrication, maintenance etc. = 1.0 paise/kWh generated, interest 5% per annum, depreciation 6% per annum of initial value.

[*U.P. Technical Univ. Power Station Practice 2006-07*]

Solution : Assuming maximum demand equal to the capacity of generating plant

Maximum demand = 50 MW

Capital cost = Rs 1.2 crores = Rs 1.2 × 10^7

Interest and depreciation per annum

$$= \text{Rs } \frac{5+6}{100} \times 1.2 \times 10^7 = \text{Rs } 13.2 \times 10^5$$

Wages and taxation = Rs 4 × 10^5

Total annual fixed charges = Rs (13.2 × 10^5 + 4 × 10^5)

= Rs 17.2 × 10^5

Units generated per annum = MD × LF × 8,760 kWh

$$= 50 \times 10^3 \times \frac{40}{100} \times 8{,}760$$

= 1.752 × 10^5 kWh

Annual cost of fuel, lubrication and maintenance etc.

= Rs 0.01 × 1,752 × 10^5 = Rs 17.52 × 10^5

Total annual charges

= Total annual fixed charges + Total running charges

= Rs (17.20 × 10^5 + 17.52 × 10^5) = Rs 34.72 × 10^5

Generating cost per kWh

$$= \frac{\text{Total annual charges}}{\text{Total units generated / annum}}$$

$$= \text{Rs } \frac{34.72 \times 10^5}{1{,}752 \times 10^5} = \text{Re } 0.0198 \text{ or } 1.98 \text{ paise}$$ **Ans.**

Example 16.36. A power plant of 180 MW installed capacity has the following data :

Capital cost = Rs 2,160/kW installed

Interest and depreciation = 12%

Annual load factor = 60%

Annual capacity factor = 50%
Annual running charges = Rs 36 × 10^6
Energy consumed by power auxiliaries = 60%
Calculate (*i*) reserve capacity (*ii*) Cost of generation per kWh

[*M.D. Univ. Power Plant Engineering December 2009*]

Solution : Total installed capacity of plant = 180 MW

Annual capacity factor = 50% = 0.5

Average load = Plant installed capacity × capacity factor
= 180 × 0.5 = 90 MW

Annual load factor = 60% = 0.6

$$\text{Maximum demand, MD} = \frac{\text{Average load}}{\text{Load factor}} = \frac{90}{0.6} = 150 \text{ MW}$$

(*i*) Reserve capacity = Installed capacity – MD = 180 – 150 = 30 MW **Ans.**

Total units generated per annum
= Average load × 8,760 = 90 × 10^3 × 8,760 = 788.4 × 10^6 kWh

Net energy delivered per annum
= (Total units generated – energy consumed by power auxiliaries) per annum

$$= 788.4 \times 10^6 - \frac{6}{100} \times 788.4 \times 10^6 = 741.096 \times 10^6 \text{ kWh}$$

Capital cost of plant = Rs 2,160 × 180 × 10^3 = Rs 388.8 × 10^6

Interest and depreciation on plant capital cost

$$= \text{Rs } 388.8 \times 10^6 \times \frac{12}{100} = \text{Rs } 46.656 \times 10^6$$

Annual running charges = Rs 36 × 10^6

Total annual cost = Rs 46.656 × 10^6 + Rs 36 × 10^6 = Rs 82.656 × 10^6

Cost of generation per kWh

$$= \frac{\text{Total annual cost}}{\text{Energy delivered per annum}}$$

$$= \text{Re } \frac{82.656 \times 10^6}{741.096 \times 10^6}$$

= Re 0.1115 per unit or 11.15 paise per unit **Ans.**

Example 16.37. Determine the generation cost per unit of energy from the following plant data.

Installed capacity = 120 MW
Capital cost of plant = Rs 10,000 per kW
Interest and depreciation = 15%
Fuel consumption = 0.64 kg/kWh
Fuel cost = Rs 500 per 1,000 kg
Salaries, wages, repairs and other operating costs per annum = Rs 10,000,000/-
Peak load = 100 MW
Load factor = 60%.

[*Rajasthan Technical Univ. Generation of Electrical Power 2011*]

Solution : Energy generated per annum
= Maximum demand × load factor × 8,760

$$= 100 \times 10^3 \times \frac{60}{100} \times 8{,}760 = 525.6 \times 10^6 \text{ kWh}$$

Fixed cost (annual interest and depreciation on capital cost)

$$= \text{Rs } 10{,}000 \times 120 \times 10^3 \times \frac{15}{100} = \text{Rs } 180 \times 10^6/\text{annum}$$

Fuel consumption = 0.64 × 525.6 × 10^6 = 336.384 × 10^6 kg

$$\text{Fuel cost} = \text{Rs } \frac{500}{1{,}000} \times 336.384 \times 10^6 = \text{Rs } 168.192 \times 10^6$$

Annual other cost = Rs 10 × 10^6

Total annual cost = Rs (180 × 10^6 + 168.192 × 10^6 + 10 × 10^6)
= Rs 358.192 × 10^6

$$\text{Generation cost} = \frac{\text{Total annual cost}}{\text{Units generated per annum}}$$

$$= \text{Re } \frac{358.192 \times 10^6}{525.6 \times 10^6} = \text{Re } 0.68/\text{kWh}$$

or 68 paise per kWh **Ans.**

Example 16.38. A 200 MW steam power station at Panki (Kanpur) is estimated to cost Rs 1,000 millions. Other costs are as follows :

(*a*) Insurance and taxes Rs 20 lacs/annum.
(*b*) Fuel and lubricants Rs 200 lacs/annum.
(*c*) Transport and storage Rs 30 lacs/annum.
(*d*) Salaries and wages Rs 36 lacs/annum.
(*e*) Miscellaneous Rs 6 lacs/annum.

Reckoning interest and depreciation at 15% per annum of capital cost, determine the cost of energy generated per unit if the power station works at an average load factor of 60%. What would it be, if the load factor be increased to 75% with a consequent increase in fuel costs by 10%, other costs remaining the same.

Solution : Power station capacity = 200 MW or 200 × 10^3 kW

Power station cost = Rs 1,000 millions = Rs 1,000 × 10^6

Annual cost of insurance and taxes = Rs 2 × 10^6

Annual cost of fuel and lubricants = Rs 20 × 10^6

Annual cost of transport and storage = Rs 3 × 10^6

Annual cost of salaries and wages = Rs 3.6 × 10^6

Annual miscellaneous costs = Rs 0.6 × 10^6

$$\text{Annual interest and depreciation} = \text{Rs } 1{,}000 \times 10^6 \times \frac{15}{100}$$

= Rs 150 × 10^6

Total annual cost
= Rs (2 + 20 + 3 + 3.6 + 0.6 + 150) = Rs 179.2 × 10^6

Total energy generated per annum
= Maximum demand × average load factor × 8,760
= 200 × 10^3 × 0.6 × 8,760 = 1,051.2 × 10^6 kWh

Generation cost

$$= \frac{\text{Total annual cost}}{\text{Total generation}} = \text{Re } \frac{179.2 \times 10^6}{1{,}051.2 \times 10^6}$$

= Re 0.17 or Paise 17/kWh **Ans.**

If the load factor is improved to 75%

Total energy generated per annum
= 200,000 × 0.75 × 8,760 = 1,314 × 10^6 kWh

Increase in operating cost

$$= 10\% \text{ of fuel cost} = \text{Rs } 20 \times 10^6 \times \frac{10}{100} = \text{Rs } 2 \times 10^6$$

Hence total operating cost
= Rs 179.2 × 10^6 + Rs 2 × 10^6 = Rs 181.2 × 10^6

Generation cost

$$= \text{Re } \frac{181.2 \times 10^6}{1{,}314 \times 10^6} = \text{Re } 0.138 \text{ or paise } 13.8/\text{kWh } \textbf{Ans.}$$

Note : Maximum demand has been assumed equal to power station capacity.

Example 16.39. The following data refer to a 100 MW thermal power plant :

Capital cost = Rs 3 × 10^9; Annual interest and depreciation rate = 12%; Annual taxation = Rs 10^7 ; Annual salary and wages = Rs 2 × 10^7; Cost of coal of Rs 700/tonne; Heating value of coal = 14,000 kJ/kg; Thermal efficiency = 38% at 100% load and 30% at 50% load. Calculate unit cost in Rs/kWh at 100% and 50% load. Comment on your answer.

[*A.M.I.E. Sec. B. Power Plant Engineering Winter 1994*]

Solution : Capital cost = Rs 3×10^9

Annual interest and depreciation = 12% of Rs 3×10^9

= Rs 360×10^6

Annual taxation = Rs 10×10^6

Annual salary and wages = Rs 20×10^6

Units generated at 100% capacity factor

$= 100 \times 10^3 \times 8,760 = 876 \times 10^6$ kWh

Coal required per annum

$$= \frac{876 \times 10^6 \times 1,000 \times 3,600}{0.38 \times 14,000 \times 1,000} = 592.782 \times 10^6 \text{ kg}$$

$\because$ 1 kWh = 1,000 × 3,600 watt-seconds

$$\text{Annual fuel cost} = \text{Rs } \frac{700 \times 592.78 \times 10^6}{1,000} = \text{Rs } 414.95 \times 10^6$$

Total annual costs = Rs $(360 + 10 + 20 + 414.95) \times 10^6$

= Rs 804.95×10^6

$$\text{Generating cost} = \text{Re } \frac{804.95 \times 10^6}{876 \times 10^6} = \text{Re } 0.92\text{/kWh}$$ **Ans.**

Units generated at 50% capacity factor

$= 100 \times 10^3 \times 0.5 \times 8,760 = 438 \times 10^6$ kWh

Coal required per annum

$$= \frac{438 \times 10^6 \times 1,000 \times 3,600}{0.3 \times 14,000 \times 1,000} = 375.4286 \times 10^6 \text{ kg}$$

Annual fuel cost

$$= \text{Rs } \frac{700 \times 375.4286 \times 10^6}{1,000} = \text{Rs } 262.80 \times 10^6$$

Total annual costs

= Rs $(360 + 10 + 20 + 262.80) \times 10^6$ = Rs 652.80×10^6

$$\text{Generating cost} = \text{Rs } \frac{652.80 \times 10^6}{438 \times 10^6} = \text{Rs } 1.49\text{/kWh}$$ **Ans.**

As the capacity factor or load factor falls, generating cost per unit increases due to fixed annual costs of interest and depreciation, taxation, salary and wages and poor operating efficiency of the plant at low capacity or load factor.

Example 16.40. An industrial consumer is offered electrical energy from *(a)* private diesel generating station *(b)* public supply undertaking. Calculate which scheme will be economical for the consumer having a maximum demand of 2,000 kW at 0.8 pf lagging at 40% load factor in each case. Given that

Private Plant : Capital cost Rs 1,200,000, cost of fuel oil Rs 70.00 per tonne, consumption 0.3 kg/kWh, cost of lubricating oil 0.15 paise/kWh. Taxation and insurance 0.25 paise per kWh generated, cost of repairs 0.2 paise/kWh. Wages Rs 30,000 per annum. Rate of interest and depreciation is 12%.

Public Supply Undertaking : The tariff is Rs 50.00 per kVA of maximum demand plus 2.0 paise per kWh consumed.

[Pb. Univ. Elec. Power II (Generation) December 1994]

Solution : Maximum demand, MD = 2,000 kW

Load factor, LF = 40% or 0.4

Average load = MD × LF = 2,000 × 0.4 = 800 kW

Units generated per annum

$= 800 \times 8,760 = 7.008 \times 10^6$ kWh

Maximum demand in kVA

$$= \frac{\text{kW}}{\text{Power factor}} = \frac{2,000}{0.8} = 2,500$$

***(a)* Private Generating Plant :**

Capital cost = Rs 1.2×10^6

Annual interest and depreciation

= 12% of Rs 1.2×10^6 = Rs 14.4×10^4

$$\text{Annual fuel cost per kWh} = \text{Re } \frac{70}{1,000} \times 0.3 = \text{Re } 0.021 \text{ or paise } 2.1$$

Annual wages = Rs 30,000

Total annual charges

= Annual charges of interest and depreciation + annual wages + annual fuel cost + annual lubricating oil cost + annual taxation and insurance + annual repair cost

$$= \text{Rs } 144,000 + \text{Rs } 30,000 + \text{Rs } \frac{7.008 \times 10^6}{100}(2.1 + 0.15 + 0.25 + 0.2)$$

= Rs 144,000 + Rs 30,000 + $7.008 \times 10^4 \times 2.7$ = Rs 363,216.00

***(b)* Public Supply Undertaking :**

Maximum demand charges = Rs 50 × 2,500 = Rs 125,000.00

$$\text{Electrical energy charges} = \text{Rs } \frac{2}{100} \times 7.008 \times 10^6 = \text{Rs } 140,160.00$$

Total annual charges = Rs (125,000 + 140,160) = Rs 265,160.00

Evidently alternative of supply *(b)* is quite economical.

Example 16.41. Compare the cost of generation per kWh between *(i)* a nuclear station and *(ii)* hydroelectric station when the system requires a load of 400 MW with a load factor of 80%. The capital cost in the case of nuclear station is Rs 3,000.00 per kW of installed capacity and in case of hydroelectric station it is Rs 2,000 per kW of installed capacity. The operating cost for the nuclear unit and the hydro unit are 2 P and 1.25 P respectively. Transmission cost per kWh is negligible in case of nuclear plant, while it is 0.25 P per kWh in hydro plant. The rate of interest and depreciation may be taken as 12% in both the cases.

[U.P.S.C. I.E.S. Generation, Transmission & Distribution of Power 1972]

Solution : Maximum load = 400 MW = 400,000 kW

Load factor = 80% = 0.8

Unit generated per annum = Maximum load × load factor × 365 × 24

= 400,000 × 0.8 × 8,760 = 28×10^8 kWh

***(i)* Nuclear Power Station :**

Capital cost = Rs 3,000 × 400,000 = Rs 12×10^8

Annual cost towards interest and depreciation

$$= \text{Rs } 12 \times 10^8 \times \frac{12}{100} = \text{Rs } 1.44 \times 10^8$$

Annual operating cost = Rs $0.02 \times 28 \times 10^8$ = Rs 0.56×10^8

Total annual cost = Rs 1.44×10^8 + Rs 0.56×10^8 = Rs 2.0×10^8

$$\text{Cost of generation per kWh} = \frac{2.0 \times 10^8}{28 \times 10^8}$$

= Re 0.0714/kWh

= Paise 7.14/kWh **Ans.**

***(ii)* Hydro Power Station :**

Capital cost = Rs 2,000.00 × 400,000 = Rs 8×10^8

Annual interest and depreciation

$$= \text{Rs } 8 \times 10^8 \times \frac{12}{100} = \text{Rs } 0.96 \times 10^8$$

Annual operating cost = Rs $0.0125 \times 28 \times 10^8$ = Rs 0.35×10^8

Annual transmission cost = $0.0025 \times 28 \times 10^8$ = Rs 0.07×10^8

Total annual cost = Rs 0.96×10^8 + Rs 0.35×10^8 + Rs 0.07×10^8

= Rs 1.38×10^8

Cost of generation per kWh

$$= \text{Rs } \frac{1.38 \times 10^8}{28 \times 10^8} = \text{Re } 0.0493\text{/kWh} = \text{Paise } 4.93\text{/kWh}$$ **Ans.**

Example 16.42. A power system, having a maximum demand of 100 MW, has a load factor of 30%. It can be supplied by one of the following schemes :

***(i)* A steam station capable of supplying the whole load.**
***(ii)* A nuclear station capable of supplying the whole load.**
***(iii)* A hydro station capable of supplying the whole load.**
***(iv)* A steam power station in conjunction with a hydro station, the latter supplying 100×10^6 units per annum with a maximum output of 40 MW.**

	Steam	Hydro	Nuclear
Capital cost per kW installed capacity	Rs 600	Rs 1,500	Rs 2,000
Interest and depreciation on capital	12%	10%	10%
Operating cost per unit	5 P	1 P	2 P
Transmission cost per unit	negligible	0.5 P	negligible

Workout total cost of the scheme per annum in each case.

[*U.P.S.C. I.E.S. Generation, Transmission and Distribution of Power 1971*]

Solution :

Maximum demand on the power system = 100 MW = 100,000 kW

Load factor = 30% = 0.3

Average load = 100,000 × 0.3 = 30,000 kW

Total number of units generated per annum

$= 30{,}000 \times 8{,}760 = 26.28 \times 10^7$ kWh

***(i)* Steam Power Station :**

Capital cost $= 600 \times 100{,}000 = \text{Rs } 6 \times 10^7$

Annual interest and depreciation

$= \text{Rs } 6 \times 10^7 \times 0.12 = \text{Rs } 0.72 \times 10^7 = \text{Rs } 72$ lacs

Annual operating cost $= \text{Rs } 0.05 \times 26.28 \times 10^7 = \text{Rs. } 131.4$ lacs

Annual transmission cost = negligible

Total annual cost = Rs 72 lacs + Rs 131.4 lacs

= Rs 203.4 lacs **Ans.**

***(ii)* Nuclear Power Station :**

Capital cost $= \text{Rs } 2{,}000 \times 100{,}000 = \text{Rs } 20 \times 10^7$

Annual interest and depreciation $= \text{Rs } 20 \times 10^7 \times 0.1 = \text{Rs } 200$ lacs

Annual operating cost $= \text{Rs } 0.02 \times 26.28 \times 10^7$

= Rs 52.56 lacs

Annual transmission cost = negligible

Total annual cost = Rs 200 lacs + Rs 52.56 lacs

= Rs 252.56 lacs **Ans.**

***(iii)* Hydropower Station :**

Capital cost $= \text{Rs } 1{,}500 \times 100{,}000 = \text{Rs } 15 \times 10^7$

Annual interest and depreciation

$= \text{Rs } 15 \times 10^7 \times 0.1 = \text{Rs } 150$ lacs

Annual operating cost $= \text{Rs } 0.01 \times 26.28 \times 10^7 = \text{Rs } 26.28$ lacs

Annual transmission cost $= \text{Rs } 0.005 \times 26.28 \times 10^7 = \text{Rs } 13.14$ lacs

Total annual cost = Rs (150 + 26.28 + 13.14) lacs

= Rs 189.42 lacs **Ans.**

***(iv)* Steam Power Station in conjunction with a Hydropower Station :**

Maximum demand on hydropower station = 40,000 kW

Annual units generated by hydropower station $= 10 \times 10^7$ kWh

Capital cost of hydropower station = Rs 1,500 × 40,000

= Rs 600 lacs

Annual interest and depreciation = Rs 600 × 0.1 = Rs 60 lacs

Annual operating cost $= \text{Rs } 0.01 \times 10 \times 10^7 = \text{Rs } 10$ lacs

Annual transmission cost $= \text{Rs } 0.005 \times 10 \times 10^7 = \text{Rs } 5$ lacs

Maximum demand on steam power station

= 100,000 – 40,000 = 60,000 kW

Annual units generated by steam power station

$= 26.28 \times 10^7 - 10 \times 10^7 = 16.28 \times 10^7$ kWh

Capital cost of steam power station

= Rs 600 × 60,000 = Rs 360 lacs

Annual interest and depreciation

= Rs 360 × 0.12 = Rs 43.2 lacs

Annual operating cost

$= \text{Rs } 0.05 \times 16.28 \times 10^7 = \text{Rs } 81.40$ lacs

Annual transmission cost = negligible

Total annual cost = Rs (60.00 + 10.00 + 5.00 + 43.20 + 81.40) lacs

= Rs 199.60 lacs **Ans.**

Example 16.43. A load having a maximum demand of 100 MW at 30% load factor may be supplied by one of the following schemes :

***(i)* A steam plant capable of supplying the whole load.**

***(ii)* A steam plant in conjunction with a pumped storage plant capable of supplying 10^6 kWh energy per year with a maximum load of 40 MW. Using the following data, find the most economic scheme among the two.**

Capital cost of steam station is Rs 1,000/kW of installed capacity. Capital cost of pumped storage plant is Rs 700/kW of installed capacity. Operating cost of steam station = 2.5 P/kWh. Interest and depreciation together on capital cost is 15%. Assume no reserve capacity is required for both the schemes.

[*A.M.I.E. Sec. B. Power Plant Engineering Summer 1993*]

Solution : Maximum demand, MD = 100 MW or 100,000 kW

Load factor, LF = 30% or 0.3

Average load = MD × LF = 100,000 × 0.3 = 30,000 kW

Total number of units generated/annum

$= 30{,}000 \times 8{,}760 = 26.28 \times 10^7$ kWh

(i) When supplied by steam power station alone

Capital cost $= \text{Rs } 1{,}000 \times 100{,}000 = \text{Rs } 10 \times 10^7$

Annual interest and depreciation

$$= \text{Rs } 10 \times 10^7 \times \frac{15}{100} = \text{Rs } 1.5 \times 10^7$$

$$\text{Annual operating cost} = \text{Rs } \frac{2.5}{100} \times 26.28 \times 10^7 = \text{Rs } 0.657 \times 10^7$$

Total annual cost $= \text{Rs } (1.5 + 0.657) \times 10^7 = 2.157 \times 10^7$

$$\text{Cost per kWh} = \frac{\text{Total annual cost}}{\text{Total units generated}}$$

$$= \text{Re } \frac{2.157 \times 10^7}{26.28 \times 10^7} = \text{Re } 0.082 \text{ or Paise } 8.2$$ **Ans.**

When supplied by steam power station in conjunction with a pumped storage scheme.

Maximum demand on pumped storage station

= 40 MW or 40,000 kW

Maximum demand on steam power station

= 100 – 40 = 60 MW or 60,000 kW

Energy supplied by pumped storage plant

$= 10^6$ kWh $= 0.1 \times 10^7$ kWh

Energy supplied by steam power station

$= 26.28 \times 10^7 - 0.1 \times 10^7 = 26.18 \times 10^7$ kWh

Total capital cost $=$ Rs $1{,}000 \times 60{,}000$ + Rs $700 \times 40{,}000$

$=$ Rs 8.8×10^7

Annual interest and depreciation

$$= \text{Rs } 8.8 \times 10^7 \times \frac{15}{100} = \text{Rs. } 1.32 \times 10^7$$

Total annual operating cost

$$= \text{Rs } \frac{2.5}{100} \times 26.18 \times 10^7 = \text{Rs } 0.6545 \times 10^7$$

neglecting operating cost of pumped storage plant

Total annual cost $=$ Rs 1.32×10^7 + Rs 0.6545×10^7

$=$ Rs 1.9745×10^7

$$\text{Cost per kWh} = \text{Re } \frac{1.9745 \times 10^7}{26.28 \times 10^7} = \text{Re } 0.075 \text{ or Paise } 7.5$$ **Ans.**

Obviously scheme (*ii*) is economical.

Example 16.44. The annual electricity requirements of an industry are 20,000 MWh with a maximum demand of 50 MW.

The requirements can be met from a utility charging Rs 800 per kW of maximum demand plus 30 paise per kWh. Alternatively the industry can set up a private steam plant, which will have a capacity of 60 MW. The following different plants for setting up the steam plants are available

	Plant A	Plant B	Plant C
Total capital cost of power plant	**9,500/kW**	**10,000/kW**	**11,000/kW**
Heat rate kcal/kWh	**3,500**	**3,000**	**2,600**

Taxes 4%, insurance 0.5%, interest rate 8%, Depreciation rate may be calculated by straight line method taking life as 20 years. Heat value of coal = 5,000 kcal/kg, fuel cost = Rs 400 per 1,000 kg ; Annual salaries, supply and maintenance = Rs 620,000 per year. Compare the different plants and find most economical scheme. [*Rajasthan Univ. Generation of Electrical Power 2005*]

Solution : Depreciation rate $= \frac{P - S}{n} = \frac{P}{n}$ assuming zero salvage value

$= \frac{P}{n \times P}$ where n is the plant life in years

$= \frac{1}{n} = \frac{1}{20} = 0.05$

Total fixed charge rate

= Interest + depreciation + taxes + insurance

$= 0.08 + 0.05 + 0.04 + 0.005 = 0.175$

Capital investment of

Plant A = Rs $9{,}500 \times 60 \times 10^3$ = Rs 57×10^7

Plant B = Rs $10{,}000 \times 60 \times 10^3$ = Rs 60×10^7

Plant C = Rs $11{,}000 \times 60 \times 10^3$ = Rs 66×10^7

Annual fixed cost of

Plant A = Rs $0.175 \times 57 \times 10^7$ = Rs 99.75×10^6

Plant B = Rs $0.175 \times 60 \times 10^7$ = Rs 105×10^6

Plant C = Rs $0.175 \times 66 \times 10^7$ =Rs 115.5×10^6

Annual electricity consumption = 20,000 MWh = 20×10^6 kWh

Annual fuel cost

$$= \text{Annual energy consumption} \times \frac{\text{Heat rate in k cal/kWh}}{\text{Heat value of coal in kcal/kg}} \times \text{fuel cost in Rs per kg}$$

Annual fuel cost of

$$\text{Plant A} = \text{Rs } 20 \times 10^6 \times \frac{3{,}500}{5{,}000} \times \frac{400}{1{,}000} = \text{Rs } 5.6 \times 10^6$$

$$\text{Plant B} = \text{Rs } 20 \times 10^6 \times \frac{3{,}000}{5{,}000} \times \frac{400}{1{,}000} = \text{Rs } 4.8 \times 10^6$$

$$\text{Plant C} = \text{Rs } 20 \times 10^6 \times \frac{2{,}600}{5{,}000} \times \frac{400}{1{,}000} = \text{Rs } 4.16 \times 10^6$$

Annual salaries, supply and maintenance charges

= Rs 620,000 = Rs 0.62×10^6

Comparison of plant annual costs :

Cost \ Plant	A	B	C
Annual fixed cost	Rs 99.75×10^6	Rs 105×10^6	Rs 115.5×10^6
Fuel cost	Rs 5.6×10^6	Rs 4.8×10^6	Rs 4.16×10^6
Salaries, supply and maintenance charges	Rs 0.62×10^6	Rs 0.62×10^6	Rs 0.62×10^6
Total	Rs 105.97×10^6	Rs 110.42×10^6	Rs 120.28×10^6

Annual cost of purchasing electricity from utility

= Demand charge in Rs/kW × MD + Energy charges in Re/kWh × energy consumption

$$= \text{Rs } 800 \times 50 \times 10^3 + \text{Rs } \frac{30}{100} \times 20 \times 10^6$$

= Rs 40×10^6 + Rs 6×10^6 = Rs 46×10^6

Thus it would be cheaper to purchase electricity from the utility as compared to set up own private plant.

In case supply from utility is not available, plant A with the lowest annual cost should be preferred.

Example 16.45. A region has a maximum demand of 500 MW at a load factor of 50%. The load-duration curve can be assumed to be a triangle. The utility has to meet this load by setting up a generating system which is partly hydro and partly thermal. The costs are as under:

Hydro plant Rs 6,000 per kW per annum and operating expenses of 30 paise per kWh.

Thermal plant Rs 3,000 per kW per annum and operating expenses of Rs 1.30 per kWh.

Determine the capacity of hydro plant, energy generated annually by each and overall generation cost per kWh.

Solution : Maximum demand of the region

= 500 MW or 500,000 kW

Load factor, LF = 50% or 0.5

Total energy generated/annum

$= \text{MD} \times \text{LF} \times 8{,}760 = 5{,}00{,}000 \times 0.5 \times 8{,}760$

$= 219 \times 10^7$ kWh

The load duration curve is shown in Fig. 16.21. Hydro plant, being cheaper in operating cost, would be used as base-load power station.

Let the capacity of hydro plant be P kW and the energy generated by it be E kWh/annum

Capacity of thermal plant $= (500{,}000 - \text{P})$ kW

Energy generated by thermal plant $= (219 \times 10^7 - \text{E})$ kWh

Annual generation cost of hydro plant = Rs (6,000 P + 0.30 E)

Annual generation cost of thermal plant

= Rs $[3{,}000\,(500{,}000 - \text{P}) + 1.3\,(219 \times 10^7 - \text{E})]$

Total cost C

$$= 6{,}000\,P + 0.30\,E + 3{,}000\,(500{,}000 - P) + 1.3\,(219 \times 10^7 - E)$$

or $C = 434.7 \times 10^7 + 3{,}000\,P - E$

For minimum cost, $\dfrac{dC}{dP} = 0$

$$\text{or } 3{,}000 - \frac{dE}{dP} = 0$$

$$\text{or } \frac{dE}{dP} = 3{,}000$$

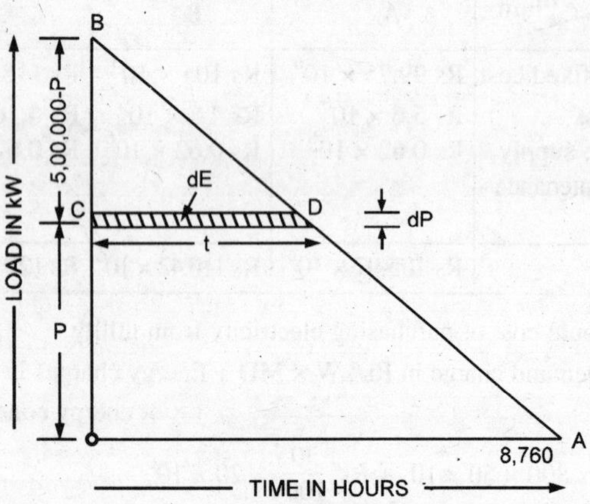

Fig. 16.21

From Fig. 16.21, $dE = dP \times t$

$$\therefore\ t = \frac{dE}{dP} = 3{,}000 \text{ hours}$$

From triangles BCD and BOA, we get

$$\frac{500{,}000 - P}{500{,}000} = \frac{3{,}000}{8{,}760}$$

$$\text{or } P = 500{,}000 - \frac{3{,}000}{8{,}760} \times 500{,}000$$

$$= 328{,}767 \text{ kW} = 330{,}000 \text{ kW (say) or } 330 \text{ MW}$$

Capacity of thermal plant = 500 – 330 = 170 MW

Energy generated by thermal plant

$$= \frac{170{,}000 \times 3{,}000}{2} = 255 \times 10^6 \text{ kWh } \textbf{Ans.}$$

Energy generated by hydro plant,

$$E = (219 \times 10^7 - 255 \times 10^6) = 1{,}935 \times 10^6 \text{ kWh } \textbf{Ans.}$$

Total annual generation cost

$$= \text{Rs } (434.7 \times 10^7 + 3{,}000 \times 330{,}000 - 1{,}935 \times 10^6)$$

$$= \text{Rs } 340.2 \times 10^7$$

Generation cost per kWh

$$= \frac{\text{Total cost}}{\text{Total generation}} = \text{Rs } \frac{340.2 \times 10^7}{219 \times 10^7} = \text{Rs } 1.553 \textbf{ Ans.}$$

Example 16.46. A generating station has a maximum demand of 100 MW and a yearly load factor of 40%. Generating costs inclusive of station capital costs are Rs 900 per annum per kW demand plus 15 paise per kWh transmitted. The annual capital charge for the transmission system is Rs 4 × 10^7 and for distribution system is Rs 3 × 10^7, the respective diversity factors being 1.2 and 1.25. The efficiency of transmission system is 90% and that of distribution system inclusive of substation losses 85%. Find the yearly cost per kW and cost per kWh supplied (*i*) at the substation and (*ii*) on the consumer premises.

Solution : Maximum demand on generating station

$$= 100 \text{ MW} = 10^5 \text{ kW}$$

Annual capital generating cost = Rs 900 × 10^5 = Rs 9 × 10^7

Annual capital charges for transmission system = Rs 4 × 10^7

Hence total fixed charges associated with the supply of energy on the substations

$$= \text{Rs } 9 \times 10^7 + \text{Rs } 4 \times 10^7 = \text{Rs } 13 \times 10^7$$

Aggregate of all maximum demands on the substations

$$= \text{MD on generating station} \times \text{diversity factor}$$

$$= 100{,}000 \times 1.2 = 1.2 \times 10^5 \text{ kW}$$

Annual cost per kW demand on the substations

$$= \text{Rs } \frac{13 \times 10^7}{1.2 \times 10^5} = \text{Rs } 1{,}083.33 \textbf{ Ans.}$$

Annual capital charges for distribution system

$$= \text{Rs } 3 \times 10^7$$

Total fixed charges associated with the supply of energy on the consumer premises

$$= \text{Rs } 13 \times 10^7 + \text{Rs } 3 \times 10^7 = \text{Rs } 16 \times 10^7$$

Aggregate of maximum demands of all the consumers

= Aggregate of all maximum demands on substations × diversity factor

$$= 1.2 \times 10^5 \times 1.25 = 1.5 \times 10^5 \text{ kW}$$

Annual cost per kW demand on the consumer premises

$$= \text{Rs } \frac{16 \times 10^7}{1.5 \times 10^5} = \text{Rs } 1{,}066.67$$

Since transmission and distribution efficiencies are 90% and 85% respectively so for every kWh transmitted only 0.9 kWh reaches the substation and 0.9 × 0.85 *i.e.,* 0.765 kWh reaches the consumer premises and costs 15 paise.

∴ Cost per kWh delivered on the substations

$$= \text{Paise } \frac{15}{0.9} = \text{Paise } 16.67 \textbf{ Ans.}$$

Cost per kWh delivered on the consumer premises

$$= \text{Paise } \frac{15}{0.9 \times 0.85} = \text{Paise } 19.6 \textbf{ Ans.}$$

16.11. LOAD FORECASTING

When planning exploitation of the natural energy sources of a country, progress takes time and needs a lot of capital investment. So advance decisions are required to be taken for judicious and profitable investment in various projects to make them effective, economical and useful. For this purpose, the anticipated electrical energy demand should be known. The resources available in the country for electrical power generation *e.g.,* for steam power stations, hydropower stations, and nuclear power stations, can then be developed considering the electrical power and energy needs, and the locations or regions where the demand is expected.

Forecasts may be long term forecasts, medium term forecasts or short term forecasts.

Long term forecasts covering a 20 year period provide a basis for examining the energy supply problems of the country in a broader perspective in relation to various other modes of development, including conversion and substitution of different forms of primary and secondary energy.

Construction and installation of equipment in power stations takes 4 to 6 years. Medium term forecasts of 5 or 6 years are, therefore, very significant for planning the size of power stations, Forecasts of demand and energy are required to estimate the additional installed capacity required to facilitate the plant maintenance programme and to estimate the firm capacity of the restricted hydroplants.

Short term forecasts of one or two years are mainly required for deciding operating procedures, and for preparing budget estimates.

Forecasting of load requirements is an essential part of power project design. The forecast is based upon (*i*) past development of power projects in the region (*ii*) applications received from different types of consumers and (*iii*) trends towards the development of industries and other potentialities for load development. The methods, that are generally used for forecasts or estimates of future demand of electrical energy are :

1. Load survey methods
2. Methods of extrapolation
3. Mathematical methods
4. Mathematical methods using economic parameters

1. Load Survey Methods : In load survey methods, the area of interest is visited and the existing and future load requirements are estimated, taking into considerations of the communities in the region and the factors which are likely to increase the load demand of the region in the near future. Loads are grouped under different categories such as residential/ domestic, commercial, industrial etc. Detailed information, categorywise and areawise is collected to estimate the demand (and its growth rate), duration and time of occurrence of load and the energy requirements. The existing data on loads in the various categories and the expected increase in each category of load yearly for the coming 5 or 6 years is then used in the preparation of load demand and energy forecasts.

2. Methods of Extrapolation : These methods are essentially statistical. These methods involve the study of data collected from different records. These methods are used in comparing the trends towards increase in demand and in energy consumption during past few years comparable with the period of the forecasts. These methods are used in determination of the average rate of increase of energy consumption in the past, and also in the countries with similar economic structure. Then, these datas are used in forecasting and estimating the future needs. These methods, however, have a drawback that these do not consider unforseen developments, such as the establishment of large and major industries which may alter the future load requirements altogether. For example, in our country since independence there has been considerable emphasis on power projects under various five year plans, which could not have been predicted from pre-independence conditions.

3. Mathematical Methods : These methods use linear or exponential curve fitting techniques for extrapolation of the curves indicating energy consumption for past several years. Thee methods can be applied only when adequate statistical data over a sufficiently long period on the past consumption of electrical energy is available.

4. Mathematical Methods Using Economic Parameters : These methods are based on the assumption that per capital consumption of electrical energy in a country depends upon economic factors, such as specific gross investments, industrial production, specific GNP, steel consumption, housing transport, etc. Indices of these parameters are found and suitably used for extrapolation by linear or exponential approximations.

Exercises

1. Differentiate between fixed and operating costs of power plants. List the items which constitute the fixed and operating cost.
[Generation of Electrical Power, Pb. Technical Univ, May 2008; May 2009; Rajasthan Univ. 2000, 2007; Rajasthan Technical Univ. 2008;2009]
2. Discuss and explain the various components of (*i*) fixed cost and (*ii*) operating cost and make a comparative assessment of these in thermal, hydro and nuclear power plants. **[A.M.I.E. Sec. B. Power Systems Summer 1996]**
3. What is depreciation reserve? Discuss the methods to calculate the depreciation charges.
[Pb. Technical Univ. Generation of Electric Power Dec. 2007]
4. What is depreciation reserve ? Why is it necessary to maintain it ? Discuss the methods to calculate the depreciation charges.
[Pb. Univ. Electrical Power II (Generation) October 1991; U.P. Technical Univ. Power Station Practice 2007-08; Rajasthan Technical Univ. Generation of Electrical Power 2008, 2010]
5. What is depreciation reserve?
Briefly explain the sinking fund method to find it. Prove that the depreciation charge by the sinking methods is given by :
$$q = (C - S)\left[\frac{r}{(1+r)^n - 1}\right]$$
where C = Initial cost of plant, S = Salvage cost, r = Annual rate of interest, n = Useful life of plant.
[Rajasthan Univ. Generation of Electrical Power 2000, 2004, 2005]
6. Explain the difference between chronological load curve and load duration curve.
[A.M.I.E. Sec. B. Power Systems Summer 1998]
7. Explain the load duration curves and write the various factors that may be calculated.
[Pb. Technical Univ. Generation of Electric Power May 2010]
8. Convert a chronological load curve into load duration curve.
[Rajasthan Univ. Generation of Electrical Power 2006]
9. Explain the load duration curve and mass curve with their complete information.
[Rajasthan Univ. Generation of Electrical Power 2007]
10. Explain (*i*) chronological load curve (*ii*) load duration curve and (*iii*) energy load curve.
[Pb. Univ. Electrical Power II (Generation) January 1993]
11. What are the different factors related to plants and consumer for power plant economics.
[U.P. Technical Univ. Power Station Practice 2006-07]
12. Define the following:
(*i*) Diversity factor (*ii*) Load factor (*iii*) Plant capacity factor (*iv*) Plant use factor.
[U.P.S.C. I.E.S. Electrical Engineering II, 2004]

13. Define load factor, utility factor, capacity factor, demand factor, and diversity factor. **[Pb. Univ. Electrical Power II (Generation) Dec. 1995]**
14. Define installed capacity, installed reserve, cold reserve, operating reserve, and spinning reserve. **[Pb. Univ. Electrical Power II (Generation) Dec. 1995]**
15. Explain in brief about plant capacitor factor. **[M.D. Univ. Power Systems-I, 2010-2011]**
16. Differentiate between fixed and operating costs of power plants. List the items which constitute the fixed and operating costs. Also discuss the role of load factor on the cost of electrical energy. **[G.B. Technical Univ. Power Station Practice 2010-11]**
17. Discuss the effect of load factor on the cost of generation in a power system. Explain the following terms : Connected load, Maximum demand and Demand factor. **[U.P. Technical Univ. Power Station Practice 2006-07]**
18. Explain the terms load factor and diversity factor. How do these factors influence the cost of generation? **[Rajasthan Technical Univ. Generation of Electrical Power 2008]**
19. What is the effect of load factor and load diversity factor on power system economics? Describe fixed and operating costs of hydroelectric power systems. **[Rajasthan Technical Univ. Generation of Electrical Power 2011]**
20. Define the terms (*i*) load factor, (*ii*) diversity factor and discuss their influence on the cost of generation. **[U.P. Technical Univ. Power Station Practice 2007]**
21. Discuss the role of load factor on the cost of electrical energy. Briefly discuss the factors affecting economics of generation and distribution of power and how to reduce the power generation cost. **[U.P. Technical Univ. Power Station Practice 2009-10]**
22. Discuss the factors affecting economics of generation and distribution of power and how to reduce the power generation cost. **[G.B. Technical Univ. Power Station Practice 2010-11]**
23. How the number and size of units in an isolated power plant are selected ? **[Pb. Univ. June 1991]**
24. Write short notes on the following:
    (*i*) Generation cost and annual fixed cost. **[Raj Univ. 2001]**
    (*ii*) Load curves. **[Pb. Technical Univ. December 2010]**
    (*iii*) Load division on hydro-steam system. **[Pb. Technical Univ. December 2010]**
    (*iv*) Load factor. **[Pb. Technical Univ. December 2010]**
    (*v*) Plant capacity factor. **[Pb. Technical Univ. Dec. 2010]**
    (*vi*) Utilization factor. **[Pb. Technical Univ. Dec. 2010]**
25. Write short notes on demand factor, capacity factor, utilization factor and mass curve. **[Raj Univ. 2002]**

Short Answer Type Questions With Answers

Q. 1. State factors on which generating cost depends. **[P.T.U. May 2008]**

Ans. The generating cost per unit (kWh) of energy depends upon the cost covering the purchase, installation and erection of equipment, cost of fuel, labour, repair etc.

Q. 2. What items constitute the fixed and running charges of generation of electrical energy ?

Ans. The fixed charges include the annual interest and depreciation on the capital cost of generating plant, transmission and distribution network, buildings and other civil works, all types of taxes and insurance charges and salaries of high officials, management and clerical staff and compensation given to the workers.

The running or operating cost includes annual costs of fuel, lubricating oil, water, maintenance and repairs of equipment and salaries of operational and maintenance staff and salaries of supervisory staff engaged in the running of the plant.

Q. 3. Why is fixed cost independent of maximum demand and units generated? **[P.T.U. December 2010]**

Ans. Fixed cost is due to the annual cost of central organisation, interest on the capital cost of land (especially if some land is held for future development) and salaries of high officials. The annual expenditure on central organisation and salaries of high officials is fixed since it has to be met whether the plant has high or low maximum demand or it generates less or more energy (units). Further, the capital investment on the land is fixed and hence the amount of interest is also fixed. As such fixed cost is independent of maximum demand and units generated.

Q. 4. What is depreciation reserve? **[Rajasthan Univ. 2000, 2004, 2005]**

Ans. The depreciation charge represents the amount which is set aside from income every year and the amount so accumulated is called the depreciation reserve.

Q. 5. What do you mean by off peak energy utilization? **[Rajasthan Univ. 2001]**

Ans. Off peak energy utilization means encouraging the consumers to use electricity during off peak hours by giving a special discount. Its benefits are : 1. Improvement of plant capacity factor resulting in reduction of capital cost and better utilization of investment. 2. Reduced requirement for peaking units resulting in lower plant capacity and lower production and maintenance costs of fossil peaking units.

Q. 6. How does the load duration curve differ from chronological load curve ?

Ans. The daily variations in load on the power station from time to time—hourly or half hourly can be plotted on a graph taking load on Y-axis and time on X-axis. The curve so obtained is known as chronological load curve. The load duration curve indicates the variation of load, but with the load arranged in descending order of magnitude *i.e.*, the greatest load on the left, lesser loads towards the right hand and the least load at the extreme light.

Q. 7. What informations can be supplied by load curves ?

Ans. The informations that can be had from load curves are (*i*) variations of load during different hours of the day (*ii*) total number of units generated in a day (*iii*) maximum demand on the power station (*iv*) average load and (*v*) load factor.

Q. 8. Define energy load curve. **[Rajasthan Univ. 2001]**

Ans. Energy load curve gives the total number of units generated for the given demand. The ordinate represents the demand in kW or MW and the abscissa represents the units (kWh) generated at or below a given demand (in kW or MW). Such a curve can be obtained from the load duration curve keeping the abscissa corresponding to each ordinate equal to the area of the duration curve up to the value of that ordinate.

Such curves are useful in that with a given number of units (kWh) available, say from a river flow, the load (kW) that could be carried at the base or peak may be easily determined.

Q. 9. How does energy load curve differ from mass curve ?

Ans. The curve drawn between demand in kW or MW on Y-axis and units generated (kWh) on X-axis is known as integrated load duration curve or energy load curve while mass curve is plotted with units (kWh) generated as ordinate and time as abscissa.

Q.10. What is load factor ? **[M.D. Univ. 2010-11]**

OR

Define load factor. **[P.T.U. December 2008]**

OR

Define load factor of electric installation. **[P.T.U. May 2010]**

Ans. The ratio of average load to the maximum demand during a certain period of time such as a day or a month or a year is called the *load factor*. Since average load is always less than the maximum demand, load factor is therefore, always less than unity.

$$\text{Load factor} = \frac{\text{Average demand}}{\text{Maximum demand}}$$

Load factor may also defined as the ratio of the number of units actually generated in a given period to the number of units which could have been generated had the load or demand remained at the maximum value throughout this period.

i.e., Load Factor

$$= \frac{\text{Units generated in a given period}}{\text{Maximum demand} \times (\text{number of hours of operation in the given period})}$$

Q. 11. Can the load factor of the sysem be 100% ? **[P.T.U. May 2010]**

Ans. Since average load is always less than the maximum demand, load factor being the ratio of average load to maximum demand is always less than unity. As such load factor of the system cannot be 100%.

Q. 12. Why even if the maximum demand and load factor of the two systems are equal, their load duration curves may not be similar? **[P.T.U. December 2008]**

Ans. Load factor is the ratio of average load to maximum load. Even if load factor and maximum demands are the same, the load duration curve, which is plotted between the load in MW and time (load arranged in descending order of magnitude) and indicates the total number of units generated for the period considered, may not be same. This is because the maximum demand may not occur simultaneously. Thereby, the area under load duration curves might be same but they might differ in shape.

Q. 13. Define demand factor. **[P.T.U. December 2008]**

Ans. The ratio of actual maximum demand on the system to the total rated load connected to the system is called the *demand factor*. It is always less than unity.

$$\text{Mathematically demand factor} = \frac{\text{Maximum demand}}{\text{Connected load}}$$

The idea of a demand factor was introduced due to the fact that all the equipment connected to the system will not be worked at a time in practice and the kW or kVA maximum demand of a group of electricity consuming devices will be always less than the sum of the kW or kVA ratings or capacities of these devices.

The minimum capacity of the generating plant must be such as to meet the maximum demand, the value of demand factor, therefore, determines the capacity and hence cost of the power equipment required to serve a given load.

Q. 14. Can demand factor be more than 1? **[P.T.U. May 2009]**

Ans. Demand factor, defined as the ratio of maximum demand on the system to the total rated load connected to the system, cannot be more than unity because maximum demand on the system can never be more than rated load connected to the system.

Q. 15. What is diversity factor? **[M.D. Univ. 2010-11]**

OR

Define diversity factor. **[P.T.U. December 2009]**

Ans. The maximum demands of all the consumers supplied from an installation do not occur usually at the same time, maximum demand on the installation is, therefore, always less than the sum of individual maximum demands of all consumers connected to it.

The ratio of sum of the individual maximum demands of all the consumers supplied by it to the maximum demand of the power station is called the *diversity factor.*

$$i.e., \text{ Diversity factor} = \frac{\text{Sum of individual maximum demands}}{\text{Maximum demand of power station}}$$

It is always greater than unity.

Q. 16. Differentiate between diversity factor and load factor. **[P.T.U. May 2008]**

Ans. Diversity factor is defined as the ratio of the sum of the individual maximum demands to maximum demand of the group whereas load factor is defined as the ratio of average load to the maximum demand.

Q. 17. Discuss why, maximum demand of a group of consumers is always less than the sum of their individual maximum demands? **[P.T.U. May 2009]**

Ans. The maximum demands of all the consumers supplied from an installation do not occur usually at the same time, maximum demand on the installation is, therefore, always less than the sum of individual maximum demands of all consumers connected to it.

Q. 18. What is diversity factor and how does it influence the cost of generation ?

Ans. The ratio of sum of individual maximum demands of all the consumers supplied by it to the maximum demand of the power station is called the diversity factor. It is always greater than unity. Higher the diversity factor better it is. With a given number of consumers the higher the diversity factor of their loads, the smaller will be capacity of the power plant required and consequently the fixed charges due to capital investment will be much reduced.

Q. 19. How is diversity helpful in reducing the costs? **[P.T.U. Dec. 2007; Dec. 2008]**

Ans. Diversity helps in reducing the costs as the diversity would mean that the maximum demands of all the consumers supplied from an installation would not occur at the same time, this would reduce total maximum demand on the installation. With a given number of consumers the higher the diversity factor of their loads, the smaller will be capacity of the power plant required and consequently the fixed charges due to capital investment will be much reduced.

Q. 20. What is capacity factor? **[P.T.U. December 2007]**

Ans. Every plant has to have a reserve capacity so as to take care of the future expansion and increase in load and, therefore, total installed capacity of the plant is usually greater than that actually required (maximum demand).

Capacity (or plant) factor is defined as the ratio of the average load to the rated capacity of the power plant.

i.e., Capacity (or plant) factor $= \dfrac{\text{Average demand}}{\text{Rated capacity of power plant}}$

Plant capacity factor may also be defined as the actual energy generated divided by the maximum possible energy that the plant might have generated during a given period.

i.e., Plant capacity factor

$$= \frac{\text{Units generated in a given period}}{\text{Rated capacity of the power plant in kW} \times \text{(number of hours in the given period)}}$$

The capacity factor depicts the extent of the use of the power plant. It is different from load factor and the difference is an indication of the reserve capacity.

$$\text{Capacity factor} = \frac{\text{Maximum demand}}{\text{Rated capacity}} \times \text{load factor}$$

Q. 21. Discuss why utilization factor of a plant may even be greater than unity? **[Rajasthan Univ. 2001]**

Ans. Utilization factor is a measure of the utility of the power plant capacity and is defined as the ratio of maximum demand to the rated capacity of the power plant. It is always less than unity because power plant is never operated beyond its rated capacity.

Q. 22. Explain various types of reserves in power system. **[Rajasthan Univ. 2006]**

Ans. Every system must have certain amount of reserve generating capacity to provide for overhaul of generating equipment, forced outage of equipment etc. Various types of reserves are given below.

Operating reserve refers to capacity in service in excess of peak load. It is provided for regulation within the hour to cover minute to minute variations, load forecasting errors, loss of equipment and maintenance of equipment. It consists of spinning and/or non spinning reserve.

(*a*) **Spinning Reserve :** It is the generating capacity connected to the bus and ready to take load.

In order to take instantaneous peak loads, the system must have capacity in *readiness to serve.* For this purpose certain machines are kept running on no load *i.e., floating* on the system. This is called the *spinning reserve* of the *system.*

(*b*) Non Spinning Reserve is that operating reserve which is capable of being connected to the bus and loaded within a specified time.

Q. 23. Distinguish between operating reserve and spinning reserve. **[P.T.U. May 2008]**

Ans. Refer to Answer of Q.22.

Q. 24. What is no spinning reserve? **[P.T.U. 2009]**

Ans. Non Spinning Reserve is that operating reserve which is capable of being connected to the bus and loaded within a specified time.

Q. 25. What do you understand by (*i*) base load and (*ii*) peak load of a power plant? **[P.T.U. December 2010]**

Ans. The unvarying load, which occurs almost the whole day on the power plant is called the base load whereas the various peak demands of the load over and above the base load of the power plant is called the peak load.

Q. 26. Differentiate between base load plants and peak load plants. **[P.T.U. May 2010]**

Ans. Base load plants run throughout the year and have high load factors. The economic characteristics of base load plants should be such that they supply power at high capital costs but low cost of operation. Hydro and nuclear power plants are usually classified as base load plants. Peak load plants run for a few hours in the year and work at low load factors. The economic characteristics of peak load plants should be such that they supply power at low capital costs, although at high cost of operation. Peak load power plants should be capable of starting quickly and should be inexpensive in starting and shutting down operations.

Q. 27. Generally which type of generation is used for base load demand? **[P.T.U. May 2009]**

Ans. The economic characteristics of base load plants should be such they supply power at high capital cost but low operating cost. Hydro and nuclear power plants are usually used for supplying base load demand.

Q. 28. Define load forecasting.

Ans. Prediction of load on the system for future is termed as load forecasting.

Problems

1. A plant costs Rs. 80,000/- and has useful life of 15 years. If the salvage value of the equipment is Rs 5,000/-, determine the amount which should be saved annually to replace the equipment at the end of that time by (*i*) straight line method (*ii*) the sinking fund method. Assuming that the annual cost of compound interest is 5 per cent.

[Rajasthan Univ. Generation of Electric Power 2001]

[Ans, (*i*) Rs 5,000 per annum (*ii*) Rs. 3,476 per annum]

2. The equipment in power station costs Rs 1,560,000 and has salvage value of Rs 60,000 at the end of 25 years. Determine the depreciation value of the equipment at the end of 20 years on the following methods (*i*) Diminishing value method (*ii*) sinking fund method at 5% compound interest annually.

[Rajasthan Univ. Generation of Electric Power 2003]

[Ans. (*i*) Rs. 115,615 (*ii*) Rs 520,781]

3. A plant costing Rs 65,000 has a useful life of 15 years. Find the amount which should be saved annually to replace the equipment at the end of that time (*i*) by the straight line method and (*ii*) by the sinking fund method if the annual rate of compound interest is 15%. Assume that the salvage value of the equipment is Rs 5,000.

[Pb. Univ. Generation December 1982]

[Ans. (*i*) Rs 4,000.00 (*ii*) Rs 1,261.00]

4. A plant costs Rs 85,000 and has a useful life of 20 years. What amount should be saved annually to replace the equipment at the end of its useful life by sinking fund method, the annual rate of compound interest being 6%? Assume the salvage value of the equipment to be Rs 5,000.

[Pb. Univ. Electrical Power II (Generation) May 1994]

[Ans. Rs 2,175**]**

5. The load curve of an electrical system is linear with the following values at different times of the day :

Time	12	5 AM	9	6	8 PM	10	12
Load(MW)	50	50	100	100	150	80	50

Plot chronological load curve. Find the energy required by the system in one day and the system daily load factor.

[Pb. Univ. Electrical Power II (Generation) October 1991; Pb. Technical Univ. Generating Electric Power Dec. 2007]

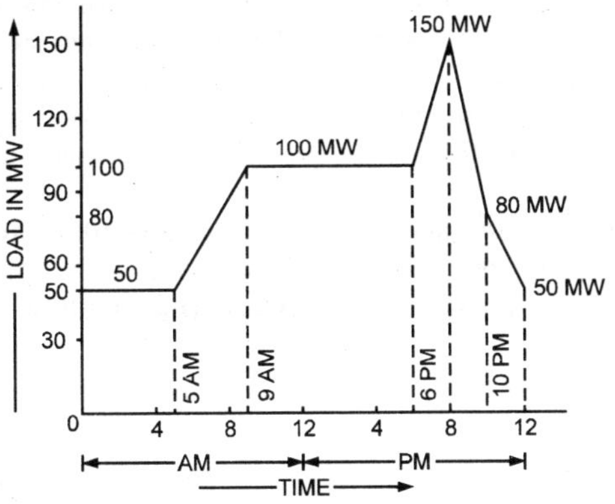

Fig. 1

Chronological load curve drawn is shown in Fig. 1.

[Ans. 2,060 MWh, 57.22%]

6. A generating station has the following daily loads :

0 to 6 hr	4,000 kW
6 to 8 hr	3,000 kW
8 to 12 hr	8,000 kW
12 to 14 hr	2,000 kW
14 to 18 hr	7,500 kW
18 to 20 hr	3,000 kW
20 to 24 hr	4,500 kW

Sketch load duration curve and determine the load factor and plant capacity factor assuming capacity of the plant 11,000 kW. **[Ans.** 62.5%; 45.45%]

7. A proposed station has to supply load as follows:

Time in hours	Load in kW
01 – 08	10,000
08 – 12	20,000
12 – 17	25,000
17 – 20	18,000
20 – 23	35,000
23 – 01	20,000

After drawing the load curve, find out the load factor. Also choose suitable generator units to supply this load, maintaining reliability of supply. Prepare operation schedule for the machines and calculate plant use factor.

[Ans. 56.43% ; Plant use factor with 4 units of 10 MW each and one unit of 5 MW – 98.75% ; Plant use factor with 2 units of 15 MW each and two units of 10 MW each – 92%]

8. The daily load curve of a power plant is given by the table below :

Time	12	2	4	6	8	10	12	2	4	6	8	10	12
Load kW × 10^3	2	2.5	3	4	6	6.5	6.5	5	6	8	9	5	2

(*a*) Find the daily load factor.

(*b*) All loads in excess of 400 kW are carried by unit no 2 rated at 600 kW. Find its use factor.

[Ans. (*a*) 0.814 (*b*) 41.7%]

9. Define load duration curve. Draw the daily load duration curve for a power station which has the following load cycle.

Load	100 kW	5,000 kW	500 kW	3,000 kW	2,000 kW
Time	4 hrs	3 hrs	6 hrs	6 hrs	5 hrs

[Rajasthan Univ. Generation of Electrical Power 2005]

Ans.

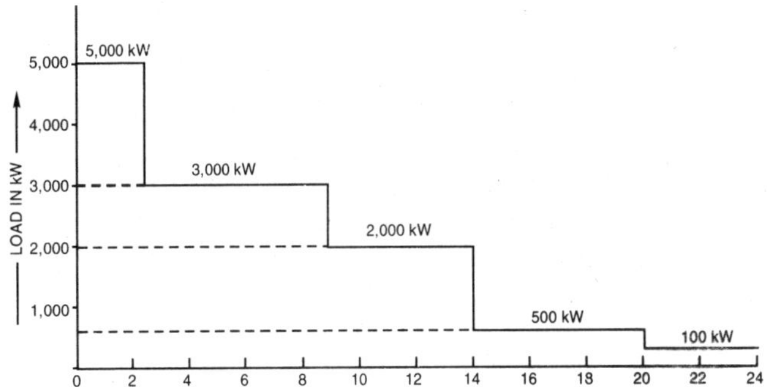

Fig. 2

10. A diesel station supplies the following loads to various consumers :

Industrial consumers = 1,500 kW
Commercial load = 750 kW
Domestic load = 100 kW
Other load = 450 kW

If the maximum demand on the station is 2,500 kW and the number of kWh generated per year is 45 × 10^5, determine: (*i*) Diversity factor (*ii*) Annual load factor.

[Rajasthan Univ. Generation of Electrical Power 2003]

[Ans. (*i*) 1.12 (*ii*) 0.2055]

11. A power station has a maximum demand of 15,000 kW. The annual load factor is 50% and plant capacity factor is 40%. Determine the reserve capacity of the plant.

[Pb. Technical Univ. Generation of Electric Power Dec. 2010]

[Ans. 3,750 kW]

12. A generating station has a maximum demand of 450 MW. The annual load factor is 45% and capacity factor is 40%. Find the reserve capacity of the plant.

[Pb. Technical Univ. Generation of Electric Power Dec. 2009]

[Ans. 56.25 MW]

13. The maximum demand of a power plant is 80 MW. The capacity factor is 0.5 and the utilization factor is 0.8. Find (*i*) load factor (*ii*) plant capacity (*iii*) reserve capacity (*iv*) annual energy production.

[Pb. Univ. Electrical Power II (Generation) April 1993]

[Ans. (*i*) 62.5% (*ii*) 100 MW (*iii*) 20 MW (*iv*) 438,000 MWh]

14. The maximum demand of a power station is 80 MW, the capacity factor is 0.6 and the utilization factor is 0.9. Find (*i*) the load factor (*ii*) plant capacity (*iii*) reserve capacity (*iv*) annual energy production.

[Rajasthan Univ. Generation of Electrical Power 2001]

[Ans. (*i*) 0.6667 (*ii*) 88.889 MW (*iii*) 8.889 MW (*iv*) 467,200 MWh]

15. The yearly load duration of a power plant is a straight line. The maximum load is 30 MW and the minimum load is 20 MW. The capacity of the plant is 35 MW. Find (*i*) plant capacity factor (*ii*) load factor (*iii*) utilization factor.

[Rajasthan Univ. Generation of Electrical Power 2006]

[Ans. (*i*) 0.71428 (*ii*) 0.833 (*iii*) 0.857]

16. A power station is to supply four regions of loads whose peak loads are 10,000 kW, 5,000 kW, 8,000 kW and 7,000 kW. The diversity factor of the load at the station is 1.5 and average annual load factor is 60%. Calculate the maximum demand on the station and annual energy supplied from the station. Suggest the installed capacity and number of units taking all aspects into account.

[Pb. Univ. Electrical Power II (Generation) Dec. 1993, Dec. 1994]

[Ans. 20 MW; 105.12 × 10^6 kWh ; 3 units, each of 10 MW capacity]

17. A substation supplies power by four feeders to its consumers. Feeder no. 1 supplies six consumers whose individual demands are 70 kW, 90 kW, 20 kW, 50 kW, 10 kW and 20 kW, while maximum demand on the feeder is 200 kW. Feeder no. 2 supplies four consumers whose daily maximum demands are 60 kW, 40 kW, 70 kW and 30 kW while the maximum demand on feeder no. 2 is 160 kW. Feeders nos. 3 and 4 have daily maximum demands of 150 kW and 200 kW respectively, while the maximum demand on the station is 600 kW.

Determine the diversity factors for the consumers of feeder no. 1 and feeder no. 2 and for the four feeders.

[**Ans.** 1.3; 1.25 and 1.183]

18. Given the following data of a power station:

(*i*) Annual maximum demand on station – 100 MW

(*ii*) Maximum demands of different types of load supplied – 40 MW, 30 MW, 25 MW and 20 MW

(*iii*) Average load factor of the station – 50% and

(*iv*) Capacity of station – 2 units of 50 MW each and one unit of 25 MW.

Determine (*a*) the number of units (kWh) supplied annually (*b*) diversity factor and (*c*) utilization factor.

[**Ans.** (*a*) 438×10^6 kWh (*b*) 1.15 (*c*) 0.83]

19. A generating station supplies the following loads: 15 MW, 12 MW, 8.5 MW, 6 MW and 0.45 MW. The annual load factor of the power station is 48% Calculate (*i*) the number of units supplied annually (*ii*) diversity factor, and (*iii*) demand factor, if the station has a maximum demand of 22 MW. [**Ans.** (*i*) 92.5×10^6 kWh (*ii*) 1.907 (*iii*) 52.5%]

20. A power station has a maximum demand of 200 MW, a plant load factor of 0.6, a plant capacity factor of 0.5 and a plant use factor of 0.8. Find (*i*) the daily energy consumed (*ii*) the reserve capacity of the plant and (*iii*) maximum energy that can be produced daily.

[A.M.I.E. Sec. B. Power Plant Engineering Summer 1996]

[**Ans.** (*i*) 2,880 MWh (*ii*) 40 MW (*iii*) 5,706 MWh]

21. The maximum demand of a power supply system is 30 MW. The annual factors are as follows.

Load factor = 60% ; Capacity factor = 50% ; Use factor = 75%.

Determine (*i*) the annual energy production (*ii*) reserve capacity and (*iii*) the hours during which the plant is not in service per year.

[A.M.I.E. Sec. B. Power Plant Engineering Winter 1993]

[**Ans.** (*i*) 157.68×10^6 kWh (*ii*) 6 MW (*iii*) 2,920 hours]

22. A generating station has following data :

Installed capacity = 300 MW, capacity factor = 50%. Annual load factor = 60%. Annual fuel cost = Rs 9×10^7. Capital cost = Rs 10^7, annual interest and depreciation = 10%. Calculate (*i*) minimum reserve capacity of the station and (*ii*) the cost per kWhr generated.

[Rajasthan Univ. Generation of Electrical Power 2006]

[**Ans.** (*i*) 50 MW (*ii*) $1,314 \times 10^6$ kWh]

23. A base load station having a capacity of 18 MW and a standby station having a capacity of 20 MW share a common load. Find the annual load factor, use factor, and capacity factor of the two power stations from the following data :

Annual standby station output = 7.35×10^6 kWh

Annual base load station output = 101.35×10^6 kWh

Peak load on the standby station = 12 MW

Hours of use of standby station during the year = 2,190

[**Ans.** Standby station : 7% ; 16.78% ; 4.2%.
Base load Station : 64.28% ; 64.28% ; 64.28%]

24. A generating station has got maximum demand of 50 MW. Calculate the cost per kWh delivered from the following data:

(*i*) Capital cost of Rs 95×10^6 ; (*ii*) Annual cost of fuel and oil Rs 9×10^6 ; (*iii*) Taxes, wages and salaries Rs 6×10^6.

The rate of interest and depreciation is 10% ; Annual load factor is 50%.

[Pb. Univ. Electrical Power II (Generation) Dec. 1993]

[**Ans.** Paise 11.2/kWh]

25. The output of a generating station is 500×10^6 kWh per year and average load factor is 70%. If the annual fixed charges are Rs 500 per kW of installed plant capacity and annual running charges are 50 paise per kWh, what is the cost per kWh of energy at the bus bar ?

[**Ans.** paise 58.2 per kWh]

26. A small generating unit of 5,000 kW capacity supplies the following loads :

(*i*) Domestic consumers with a maximum demand of 3,000 kW at a load factor of 20 per cent :

(*ii*) Small industrial load with a maximum demand of 2,000 kW at a load factor of 50 per cent.

Find the overall energy rate for both type of consumers. Use the following data :

Capital cost of plant = Rs 2,000 per kW of installed capacity.

Total running cost = Rs 600,000 per year.

Annual rate of depreciation and interest on capital cost is 10 per cent. [**Ans.** (*i*) Paise 15.7 per unit ; Paise 8.85 per unit]

27. The fixed costs of a thermal station are Rs 262.80 per kW of installed capacity per year. The fuel and operating costs are 3 paise per kWh generated. Find the cost of the electrical energy generated per kWh. at station load factors of 100%, 75%, 50% and 25%.

Plot the curves showing the variations of the cost of energy per kWh generated.

[Agra Univ. Generation & Switchgear Practice 1976]

[**Ans.** Paise 6, Paise 7, Paise 9, Paise 15 per kWh respectively]

28. A 100 MW, steam power station is estimated to cost Rs 100 crores. The operating expenses are estimated as follows : Cost of fuel and oil = Rs 700 lacs/annum ; Transportation and storage = Rs 100 lacs/annum.

Salaries and wages = Rs 100 lacs/annum ; Miscellaneous = Rs 100 lacs/annum.

Reckoning interest and depreciation 10% of capital cost, calculate the cost of generation per unit if the average load factor of the power station is 60%.

What economics would be affected if the load factor was improved to 80% and the operating expenses increasing by only 10% thereby ?

[**Ans.** 38 paise/kWh ; Reduction in cost/kWh by 21.25%]

29. A system with a maximum demand of 100,000 kW and a load factor of 30% is to be supplied by either (*a*) steam station alone or (*b*) a steam station in conjunction with a water storage scheme, the latter supplying 100 million units with a maximum output of 40,000 kW. The capital cost of the steam and storage stations are Rs 600 per kW and Rs 1,200 per kW respectively. The corresponding operating costs are 15 paise and 3 paise per kWh respectively. The interest on capital cost is 15% per annum. Calculate the overall generating cost per kWh and state which of the two projects will be economical.

[**Ans.** Paise 18.425 per kWh ; Paise 15.23 per kWh]

30. A steam power station having a capital cost of Rs 1,000/-per kW and maintenance cost of 2 P per kWh, using coal of calorific value of 6,110 k cal/kg. The overall thermal efficiency of the station is estimated at 25% or (*i*) a hydro station having a capital cost including transmission lines etc. of Rs 3,000/- per kW and a running cost of 0.5 P/kWh. Allowing 12% for the steam station and 9% for the hydro station for interest and depreciation per year calculate the price of coal above which steam generation is uneconomical, given 1 kWh = 860 kcals. [**Ans.** Rs 34.2 per tonne]

31. An electric supply undertaking having maximum load of 100 MW generates 375 million kWh per year, and supplies consumers having an aggregate maximum demand of 165 MW. The annual expenses including capital charges are Rs 32×10^5, fixed expenses connected with generation Rs 48×10^5, transmission and distribution expenses Rs 56×10^5. Assuming that 90 per cent of the fuel cost is assigned to the running charges and that the losses in transmission and distribution are 15% of the kWh generated, deduce a two part tariff to represent the actual cost of supply to the consumer. [**Ans.** Rs 65 per kW of maximum demand plus paise 0.9 per kWh]

32. A power system, having a maximum demand of 100 MW, has a load factor 30%. It can be supplied by one of the following schemes :

(*i*) A steam station capable of supplying the whole load (*ii*) A nuclear station capable of supplying the whole load (*iii*) A hydro station capable of supplying the whole load.

	Steam	Hydro	Nuclear
Capital cost/kW installed capacity	Rs 600	Rs 1,500	Rs 2,000
Interest and depreciation on capital	12%	10%	10%
Operating cost per unit	5 P	1 P	2 P
Transmission cost per unit	Negligible	0.5 P	Negligible

Workout total cost of the scheme per annum in each case.

[Pb. Univ. Electrical Power II (Generation) May 1995]

[**Ans.** Steam – Rs 203.4 lacs; Hydro – Rs 189.42 lacs; Nuclear – Rs 252.56 lacs]

33. A generating station has a maximum demand of 80,000 kW and yearly load factor of 50%. Generating costs inclusive of station capital cost are Rs 70 per annum per kW demand plus 1.5 P per kWh transmitted. The annual capital charge for the transmission system is Rs 30 lakhs and for distribution system Rs 25 lakhs, respective diversity factors being 1.2 and 1.3. The efficiency of the transmission system is 90% and that of distribution system inclusive of substation losses is 83%. Find the yearly cost per kW demand and the cost per kWh supplied on the consumer premises.

[**Ans.** Rs 88.94 per annum per kW demand ; Paise 2.0 per kWh delivered at the consumer premises]

34. Compute the generation cost per kWh from the following data :

Installed capacity = 200 MW; Capital cost = Rs 3,000 per kW; Interest and depreciation = 12%; Fuel consumption = 0.9 kg/kWh; Fuel cost = Rs 70 per 1,000 kg; Other operating costs = 30% of fuel costs; Load factor = 80%; Peak Load = 170 MW.

[Pb. Univ. Electrical Power II (Generation) October 1991, January 1993]

[**Ans.** 14.23 paise per kWh]

❑❑❑

CHAPTER 17

Tariffs and Power Factor Improvement

INSIDE THIS CHAPTER

17.1. Tariffs **17.2.** Types of Tariffs **17.3.** Spot Pricing **17.4.** Types of Consumers and Their Tariffs **17.5.** Power Factor **17.6.** Disadvantages of Low Power Factor **17.7.** Causes of Low Power Factor **17.8.** Methods of Power Factor Improvement **17.9.** Advantages of Power Factor Improvement **17.10.** Economics of Power Factor Improvement **17.11.** Economical Comparison of The Two Methods of Increasing The Power Supplied **17.12.** Choice of Equipment.

17.1. TARIFFS

Tariff means the schedule of rates or charges. Tariff, in case of electric supply, means the schedule or rates framed for supply of electrical energy to different classes of consumers.

17.1.1. Objective of Tariff

The main objective of the tariff is to distribute equitably the cost of supplying energy among the various classification of use. Therefore, a tariff must cover the following items :

(*i*) Recovery of cost of capital investment in generating, transmitting and distributing equipment.

(*ii*) Recovery of cost of operation, supplies and maintenance of equipment.

(*iii*) Recovery of cost of metering equipment, billing, collection costs and miscellaneous services and

(*iv*) A satisfactory return on the total capital investment.

17.1.2. Principal Factors Affecting Framing of Tariff

The principal factors involved in fixing of a tariff are enumerated below :

1. Proper return is secured from each consumer. While fixing the tariff for different classes of consumers, it has to be taken into account whether the tariff will result in a revenue meeting all the expenditure of the supply authority. In addition, the tariff should bring forth sufficient money to enable future expansion to meet an anticipated load requirement.
2. The consumers are encouraged to make more extended use of electricity.
3. The tariff should be simple and capable of easy explanation to the public. A complicated tariff may cause an opposition from the public which is generally distrustful of the supply authorities.
4. The consumers are charged according to what the energy costs. The tariff should be such as to satisfy the consumers of all categories. A big consumer should be charged at a lower rate than a small consumer. This is because increased energy consumption spreads the fixed charges over a greater number of units, thus reducing the overall cost of producing electrical energy.
5. The consumers are encouraged to use power during off-peak hours and penalised for high loads demanded at system peak by making a provision for higher demand charges.
6. The consumers are penalised for poor power factor.

The type of service rendered by the supply authority also determines the tariff that must be charged to a consumer. This is explained below by giving some examples. Electric supply to a domestic consumer, having usually light and fan points, makes the life of an ordinary citizen convenient and pleasant. But such a class of consumers cannot be charged at higher rates. This is because in a welfare state like India, the state has the responsibility to provide some basic necessities to its citizens. Similarly, the farmer for supply of electrical energy to agricultural loads cannot be charged at a higher rate because of the national concern for agricultural output. However, the consumers using electricity for extra comforts such as for air-conditioning and for commercial purposes can be charged at higher rates.

17.2. TYPES OF TARIFFS

Various types of tariffs, which are commonly used, are described below.

1. *Flat Demand Tariff.* This is one of the earliest form of tariffs used for charging the consumers for electrical energy consumption. In this case, the total demand and the energy consumption were fixed.

In early days, when the use of electricity was mainly restricted to very few applications such as for lamps etc. and the number of hours of use of the lamps were fixed, they were charged on the basis of number of lamps installed or total load connected in kW and fixed number of hours of use in a month or in a quarter of a year or in a year. Since the total demand and energy consumption were fixed and all the consumers were of alike class, so the rate could be expressed directly as certain amount per lamp or per unit of load installed. If x is the number of lamps or load connected in kW and a is the rate per lamp or per kW of connected load then

$$\text{Energy charges} = \text{Rs } ax$$

In this type of tariff the metering equipment, meter reading, billing and accounting costs are eliminated resulting in overall decrease of total cost. The main disadvantage of this tariff is that it does not differentiate between the consumers who use same appliance or equipment but for different duration of time. It encourages the consumers to keep their appliances connected to the supply mains even when not required. Nowadays the tariff of this type is restricted to use such as in street lighting, signal systems, sign lighting and other loads where the energy consumption in quantity could be readily predicted.

2. *Simple Tariff.* This is the simplest type of tariff according to which the cost of energy is charged on the basis of units consumed. The rate can be derived as

$$\text{Cost/kWh} = \frac{\text{Annual fixed cost + annual operating cost}}{\text{Total number of units supplied to the consumers per annum}}$$

Although this method of deriving the rate is simple and also the drawback of encouraging the consumers to keep their appliances connected to the supply mains even when not required is removed as they have to pay for all the energy consumed but suffers from the following drawbacks :

(*i*) There is no discrimination among the different types of consumers (domestic, industrial, bulk) having different load factor, diversity factor and power factor.

(*ii*) The cost per kWh delivered is higher.

The simple tariff can be made suitable by having the following modifications.

(*i*) A discount be allowed to the consumer who consumes more electricity than an average consumer.

(*ii*) Special tariffs be framed for the various types of consumers just as the domestic consumers using electricity for light and fans be charged at higher rates as compared to power consumers. The reason for it is that the load factor of light and fan load is low and the demand is during peak hours of the day.

(*iii*) The consumers be encouraged to use electricity during off-peak hours by giving a suitable discount. Such a tariff is called the *off-peak tariff.*

(*iv*) The consumers be encouraged to use electricity at high load factor and power factor by framing special types of tariffs such as maximum demand tariff and power factor tariff.

3. *Flat Rate Tariff.* This type of tariff differs from the former one in the sense that the different types of consumers are charged at different rates *i.e.,* the flat rate for light and fan loads is slightly higher than that for power load. The rate for each category of consumers is arrived at by taking into account its load factor and diversity factor.

The method is usually most popular with the public since it can easily be understood by the consumers and the calculations at the supplier's end are simple. If the energy consumed during the billing period is x units and the flat rate is Rs a per unit then charges are Rs ax.

The disadvantages of this tariff are :

(*i*) Separate meters are required for different types of supply.

(*ii*) Difficulty is experienced to derive the load factor and diversity factor for various types of loads to be employed in deciding the tariff.

(*iii*) The consumer is charged for the total quantity of energy consumed at the same rate irrespective of the magnitude of energy consumed while increased generation or consumption spread the fixed charges over a greater number of units and so the overall cost per unit decreases as the consumption increases.

4. *Step Rate Tariff.* The step rate tariff is a group of flat rate tariffs of decreasing unit charges for higher range of consumption, say for example

Rs 4.0 per unit if the consumption does not exceed 50 kWh

Rs 3.50 per unit if the consumption exceeds 50 kWh but does no exceed 200 kWh

Rs 3.0 per unit if the consumption exceed 200 kWh.

This type of tariff takes into account the fact of lower generation cost owing to higher energy consumption consequent to improvement of load factor and therefore promotes the use of electricity. The drawback of this tariff is however that by increasing the energy consumption so as just to enter the next range from the final stage of previous range, the total energy cost is reduced. Thus there is tendency with the consumer, just approaching the limit of the step to, anyhow cross the step and enter the next one in order to reduce the total energy cost. This drawback is removed in *block rate tariff* explained below.

5. *Block Rate Tariff.* In this type of tariff a given block of energy is charged at higher rate and succeeding blocks of energy are charged at progressively reduced rates, say for example

The first 25 units may be charged at the rate of Rs 4.0 paise/unit.

The next 40 units may be charged at the rate of Rs 3.50 per unit.

The consumption exceeding 65 units may be charged at the rate of Rs 3.0 per unit.

The advantage of such a tariff is that the consumer gets an incentive for consuming more electrical energy. This increases the load factor of the system and hence the generation cost is reduced. Its main drawback is that it lacks a measure of the consumer's demand. This tariff is most popular nowadays among domestic, commercial and small industrial consumers.

6. *Hopkinson Demand Rate or Two Part Tariff.* The total charge to be made to the consumer is split into two components namely fixed charge and running charge. Since fixed charge is independent of energy consumed and proportional to the maximum demand so the fixed charge is made at a certain amount per kW of maximum demand, which can be assessed on the basis of the rateable value of the premises or on the number of rooms excluding bath rooms, attics etc. or on the connected load or on the total combined kW capacity of all the consuming devices owned by a particular consumer. Running charge is made at a certain amount per kWh for the total energy consumed.

This type of tariff is expressed by the expression

Total energy charges, e = Rs $a \times$ kW + Rs b kWh where Rs a is the charge per kW of maximum demand assessed and Rs b is the charge per kWh of energy consumed.

This tariff is mostly applicable to medium industrial consumers.

In this tariff charge made on maximum demand recovers the fixed charges such as interest and depreciation on the capital cost of building and equipment, taxes and insurance charges and operating cost which is independent of energy supplied by it and varies with variation of maximum demand. Charge made on total energy consumption recovers the operating cost, which varies with variation in energy supplied.

The drawback of this tariff from the point of view of consumer is that he is to pay his fixed charge irrespective of the consumption. For example, if during any month any industry remains closed, the owner will be required to pay the fixed charges unnecessarily.

7. *Maximum Demand Tariff or Wright Demand Rate.* This tariff is similar to that of two part tariff except that in this case maximum demand is actually measured by a maximum demand indicator instead of merely assessing it on the basis of rateable value. In this tariff the drawback of the two part tariff is removed. This tariff is almost applicable to all bulk supplies and large industrial consumers, who have a control over their maximum demand.

Such a tariff induces the consumer to keep his maximum demand at a low value.

8. *Power Factor Tariffs.* Since the efficiency of plant and equipment depends upon the power factor, therefore, in order to increase the utility of plant and equipment to the maximum, the plant must be operated at the most economical power factor. That is why, sometimes consumers are penalised for poor power factor by applying the following types of power factor tariffs :

(*a*) kVA Maximum Demand Tariff. It is a modified form of two part tariff. In this case maximum demand is measured in kVA instead of in kW. This type of tariff encourages the consumers to operate their machines and other equipment at improved power factor, because low power factor will cause more demand charges.

(*b*) kWh and kVARh Tariff. In this type of tariff kWh and kVARh are measured and charged separately. Since kVARh decreases with the increase in power factor, therefore, the consumer tries to improve the power factor of his installation in order to decrease the charges on account of kVARh recorded.

(*c*) Sliding Scale or Average Power Factor Tariff. In this type of tariff, an average power factor, say 0.8 lagging may be taken as reference and a surcharge for each 0.01 by which the average power factor falls below this figure may be made. Similarly a discount may be allowed for each 0.01 by which the average power factor rises above this figure. Such tariffs are rarely used.

9. *Three Part Tariff or Doherty Rate.* In this tariff total charge is split into three elements namely fixed charge, semi-fixed charge and variable charge.

Constant or fixed charge recovers the cost on account of expenses in giving a supply (of any sort or of any magnitude) service connection (incurred once only but charged for either by a capital or an annum sum or charged each billing period), office and service expenses including accounting, metering and establishment charges.

Semi-fixed charge, which are levied on the basis of maximum demand in kW or kVA, recovers the plant initial cost as well as that operating cost which is independent of total energy supplied by it and varies with the variation in maximum demand.

Variable charge, which are levied on the basis of energy consumed, recovers the cost varying with the variation in energy supplied such as on account of fuels etc.

So the general expression for the recovery of the cost split into three sections mentioned above can be written as

Total charges = Rs $a + b$ kW + c kWh

where a is a constant charge made each billing period.

b = Unit charge in Rs per kW of metered maximum demand in kW during billing period. In some cases it is also charged in Rs per kVA so that the consumers are penalised for poor power factor.

and c = Unit charge for energy in Rs per kWh of energy consumed.

This type of tariff is usually applicable to bulk supplies.

10. *Off-Peak Tariff.* The load on the power station usually has pronounced peak loads in the morning and early evening and a very low load during the night (from 10 PM to 6 AM). During the night, therefore, and other off-peak period which may occur, a large proportion of the generating and distribution equipment will be lying idle. In case the consumers are encouraged to use electricity during off peak hours by giving a special discount, the energy can be supplied without incurring an additional capital cost and should, therefore, prove very profitable.

This type of tariff is very advantageous for certain processes such as water heating by thermal storage, pumping, refrigeration etc.

17.3. SPOT PRICING

The major concern of electric utilities all over the world is very large demand during peak load hours and surplus energy during off peak hours. This put the utilities in awkward situation of capacity shortages during peak load hours and large surplus during off peak hours. In spot princing the consume is charged based on the time at which consumption occur. This is resorted to offset the imbalance caused because of large demand during peak load conditions and off peak conditions. The price is varied for reflecting the utility cost of providing the electrical energy at a given time. The spot pricing encourages the consumer to shift the maximum demand to off peak hours when the supply is abundant and prices are low. The spot pricing rate can be expressed as

Instantaneous spot rate
= Incremental operating cost + quality of supply component.

The quality of supply component is choosen so that the resultant price reflects the marginal value of unserved energy. Spot pricing rates are economically rational as they embody the principles of marginal cost theory and result in efficient location of resources. Spot pricing is also sometimes termed as 'load adaptive pricing', 'flexible pricing', 'homeostatic control pricing', 'real time pricing' or 'response pricing'. It is also known as 'Time of Day Tariff'.

Because of recent advancement in microcomputers and telecommunications and a drastic reduction in metering and communication costs, spot pricing has become feasible and many utilities in developed countries have adopted at least partial spot pricing in their industrial and commercial rate structures.

17.4. TYPES OF CONSUMERS AND THEIR TARIFFS

The main types of load on a system are domestic, commercial agricultural, industrial, municipal, traction etc. Accordingly the consumers may be categorised as domestic consumers, commercial consumers, agricultural consumers, industrial consumers (small, medium and large), bulk consumers etc.

17.4.1. Domestic Consumers

Residential load consists of lights, fans, and appliances such as radios, TVs, heaters, electric irons, refrigerators, electric water heaters, washing machines, coolers, air-conditioners, domestic pump sets etc. Domestic consumers are given single phase supply up to a load of 5 kW and a 3-phase supply for loads exceeding 5 kW.

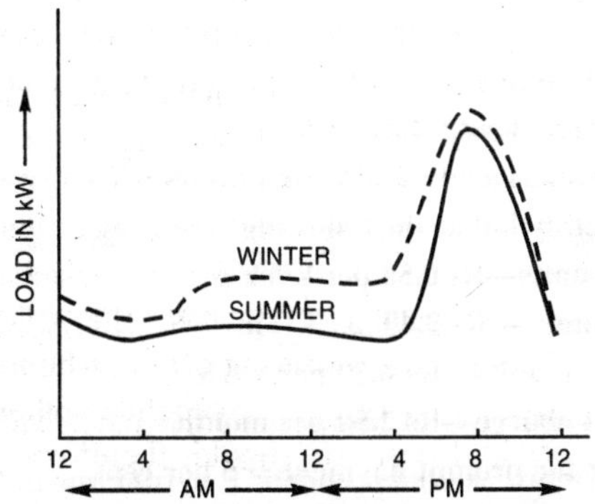

Fig. 17.1. ***Typical Chronological Load Curve For Domestic Consumers***

The small consumers generally use all the light points simultaneously so their demand factor is high (almost unity). Big domestic consumers may have demand factor low (around 0.5). During summer the major part of domestic load may consist of fan load during day and light and fan load during evenings and early mornings and some air-conditioner load. During winter, the major load is light load during evenings and early mornings and some heater load. Typical chronological load curve for domestic consumers is given in Fig. 17.1. The tariffs applicable to domestic consumers are simple tariff, flat rate tariff or block rate tariff. In addition meter rent and electricity duty are also charged from the consumers.

17.4.2. Commercial Consumers

Non-residential premises, such as shops, business-houses, cinemas, hotels, public offices, clubs etc. fall under this category. The load mainly consists of lights, fans and small electric appliances. The load remains fairly constant from around morning 10 to evening 9 hours. During night the load may consist of some lighting load. The demand factor is fairly high. Such consumers are given single phase supply for loads up to 5 kW and three phase supply for loads exceeding 5 kW. Typical chronlogical load curve for commercial consumers is shown in Fig. 17.2. The tariffs applicable to commercial consumers are also simple tariff, flat rate tariff or block rate tariff but charges per unit are higher in comparison to those in case of domestic consumers. In addition, meter rent and electricity duty are also charged from the consumers. Sometimes there is also a provision for minimum charges in case the energy consumption remains below a certain prescribed limit.

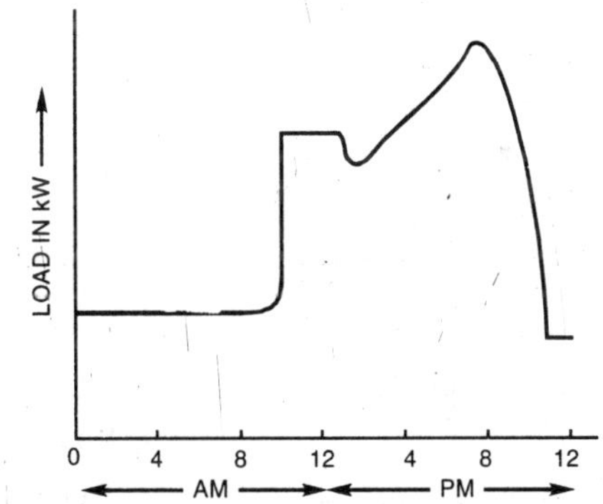

Fig. 17.2. ***Typical Chronological Load Curve For Commercial Consumers***

17.4.3. Agricultural Consumers

Consumers drawing power up to 20 kW for irrigation pumping units are categorized as agricultural consumers. Such

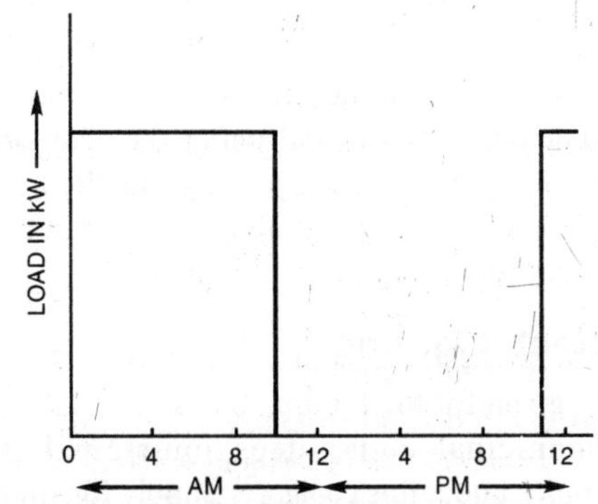

Fig. 17.3. ***Typical Chronological Load Curve For Agricultural or Irrigation***

consumers are given a three phase supply. The loads of the tubewells used for irrigation constitute a substantial portion of the system load. The demand factor and diversity factor are both almost unity. Figure 17.3 depicts the typical chronological load curve for agricultural or irrigation load.

Agricultural consumers are charged at a flat rate tariff which may be either on the basis of a fixed charge per unit consumed or a fixed charge per kW or hp of connected load.

17.4.4. Industrial Consumers

Industrial consumers may further be categorised as small industrial consumers, medium industrial consumers and large industrial consumers according to the rating of loads.

Small industrial consumers are owners of small workshops, atta chakkis, wheat threshers, saw machines and other small manufacturing and repair shops with load not exceeding 20 kW. The demand factor depends on the nature of load but usually it is high (around 0.8). Such consumers are given 3-phase supply at 415 V. A block tariff is usually offered to such consumers.

Industrial consumers with loads exceeding 20 kW but not exceeding 100 kW fall under the category of *medium industrial consumers*. They are given three phase supply at 415 V and are usually charged on two part tariff.

Industrial consumers with loads exceeding 100 kW are categorized as *large industrial consumers*. They are supplied power at 11 or 33 kV, and in rare cases, at 415 V three phase depending on the requirement of the consumers. The demand factor may be around 0.5. Such consumers are usually charged kVA maximum demand tariff. Typical chronological load curve for industrial load is depicted in Fig. 17.4.

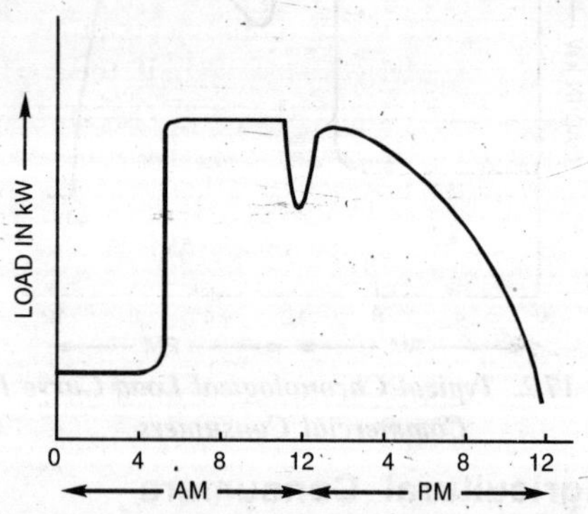

Fig. 17.4. *Typical Chronological Load Curve For Industrial Load*

17.4.5. Bulk Consumers

Power consumers such as railways, public work departments, educational institutions, military establishments, hospitals having loads exceeding 100 kW fall under the category of bulk consumers. Bulk consumers are usually supplied by 3-phase supply at 415 V or 11 kV depending on their requirements. Such consumers are charged at flat rate.

17.4.6. Street Lighting

Power supply given for the lighting of parks, roads and streets under the municipal committees, municipal boards or panchayats comes under this category. Supply for street lighting is given at 415 V three phase or 240 V single phase. Such a load has demand factor and diversity factor of unity. The switching on the lights and their switching off is synchronised with dusk and dawn respectively. Separate distributors are run for street lighting to enable their switching simultaneously. The tariff charged for street lighting is such that it recovers the cost of the energy consumed as well as the cost of replacement of lamps. Chronological load curve for street lighting is shown in Fig. 17.5.

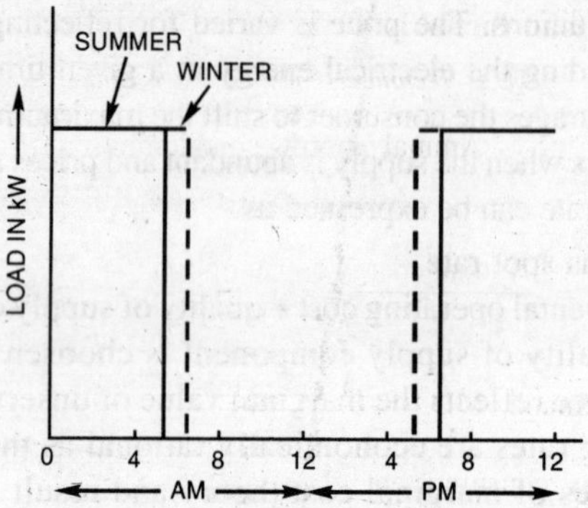

Fig. 17.5. *Typical Chronological Load Curve For Street Lighting*

17.4.7. Water Supply

The load for water supply is for pumping water to the overhead tanks. It is generally possible to fit this load during system off peak hours, usually during night hours.

Example 17.1. The daily load of an industrial consumer is as follows :

100 kW for 9 hours, 125 kW for 6 hours, 50 kW for 7 hours and 5 kW for 2 hours.

The tariff rate is Rs 800 per kW of maximum demand per year plus Rs 2.50 per kWh. Determine the energy consumption per year (365 days) and yearly bill. [*West Bengal Univ. of Technology 2009-10*]

Solution : Daily energy consumption

$= 100 \times 9 + 12.5 \times 6 + 50 \times 7 + 5 \times 2$

$= 900 + 750 + 350 + 10 = 2{,}010$ kWh

Annual energy consumption $= 2{,}010 \times 365 = 733{,}650$ kWh **Ans.**

Maximum demand = 125 kW

Demand charges per annum = Rs 800×125 = Rs 100,000

Energy charges per annum = Rs $2.50 \times 733{,}650$ = Rs 1,834,125

Yearly bill = Demand charges per annum + energy charges per annum

= Rs 100,000 + Rs 1,834,125 = Rs 2,834,125 **Ans.**

Example 17.2. Light load : 6 tube lights 40 watts each working for 4 hours daily

Fan load : 6 fans 100 watts each working for 6 hours daily

Refrigerator load : 2 kWh daily

Miscellaneous load : 2 kW for 2 hours daily

Find the monthly bill at the following rate :

First 20 units—Rs 0.50 per kWh

Next 30 units—Rs 0.40 per kWh

Remaining units—Rs 0.30 per kWh

Constant charge—Rs 2.50 per month

Discount for prompt payment = 5 per cent.

[*Rajasthan Technical Univ. Generation of Electrical Power 2008, 2009*]

Solution : Daily energy consumption

$$= \frac{(6 \times 40 \times 4) + (6 \times 100 \times 6)}{1{,}000} + 2 + 2 \times 2 = 10.56 \text{ kWh}$$

Monthly (30 days) consumption

$= 10.56 \times 30 = 316.8$ kWh or units

Monthly bill = (Energy charges + constant charge) (1 – discount)

$= \text{Rs } [20 \times 0.5 + 30 \times 0.4 + (316.8 - 50) \times 0.3 + 2.5]\,(1 - 0.05)$

$= \text{Rs } (10 + 12 + 80.04 + 2.5)\,(0.95) = \text{Rs } 99.31$ **Ans.**

Example 17.3. A consumer has an annual consumption of 70,080 kWh. The charge is Rs 100 per kW of maximum demand plus 5 paise per kWh. Find the annual bill and the overall costs per kWh, if the load factor is 40%.

[*U.P. Technical Univ. Power Station Practice 2003-04*]

Solution : Annual energy consumption = 70,080 kWh

Annual load factor, LF = 40% or 0.4

$$\text{Average load} = \frac{\text{Annual energy consumption}}{8{,}760} = \frac{70{,}080}{8{,}760} = 8 \text{ kW}$$

$$\text{Maximum demand, MD} = \frac{\text{Average load}}{\text{LF}} = \frac{8}{0.4} = 20 \text{ kW}$$

Annual energy consumption bill

= Demand charges/annum + energy charges/annum

$= \text{Rs } 100 \times 20 + \text{Rs } 0.05 \times 70{,}080$

= Rs 2,000 + Rs 3,504 = Rs 5,504 **Ans.**

Example 17.4. An industrial undertaking has connected load of 200 kW. The maximum demand is 150 kW. On average each machine works for 70% of time. Find yearly expenditure on electricity if the tariff is Rs 3,000 + Rs 700 per kW of maximum demand per year + Rs 0.60 per kWh.

[*Rajasthan Univ. Generation of Electrical Power 2003*]

Solution : Yearly energy consumption

$= 200 \times 0.7 \times 365 \times 24 = 1{,}226{,}400$ kWh

Fixed charges/annum = Rs 3,000

Demand charges/annum $= \text{Rs } 700 \times 150 = \text{Rs } 105{,}000$

Energy charges/annum $= \text{Rs } 0.60 \times 1{,}226{,}400 = \text{Rs } 735{,}840$

Yearly expenditure on electricity

= Rs 3,000 + Rs 105,000 + Rs 735,840 = Rs 843,840 **Ans.**

Example 17.5. The annual consumption of domestic customer is $1{,}500 \times 10^3$ units and the annual load factor is 40%. What will be the saving in the average unit cost if the load factor improves to 100%? The two part tariff is Rs 500 per kW of maximum demand per year + 40 paise per kWh.

[*Rajasthan Univ. Generation of Electrical Power 2000*]

Solution : Annual energy consumption $= 1{,}500 \times 10^3$ kWh

$$\text{Average load} = \frac{\text{Annual energy consumption}}{365}$$

$$= \frac{1{,}500 \times 10^3}{365 \times 24} = 171.233 \text{ kW}$$

Load factor = 40% = 0.4

$$\text{Maximum demand} = \frac{\text{Average load}}{\text{Load factor}} = \frac{171.233}{0.4} = 428 \text{ kW}$$

In case the consumer improves the load factor to 100 per cent, his maximum demand will be reduced to $\frac{171.233}{1}$ *i.e.* 171.233 kW

Annual energy bill with 40% load factor

$= \text{Rs } 500 \times 428 + \text{Rs } 0.40 \times 1{,}500 \times 10^3 = \text{Rs } 814{,}000$

$$\text{Unit cost} = \text{Rs } \frac{814{,}000}{1{,}500 \times 10^3} = \text{Re } 0.5427 \text{ or paise } 54.27$$

Annual energy bill with 100% load factor

$= \text{Rs } 500 \times 171.233 + \text{Rs } 0.40 \times 1{,}500 \times 10^3 = \text{Rs } 685{,}616.50$

$$\text{Unit cost} = \text{Rs } \frac{685{,}616.50}{1{,}500 \times 10^3} = \text{Re } 0.4571 \text{ or paise } 45.71$$

Saving in the average unit cost

= Paise 54.27 – paise 45.71 = Paise 8.56/kWh **Ans.**

Example 17.6. An industrial consumer has single phase 230V supply. His monthly energy consumption is 2,020 kWh. A maximum demand indicator installed at his premises indicates 40 A which is charged at unity power factor for 2 hours daily at Rs 3.50 per kWh. The remaining units are charged at Rs 1.80 per kWh. Find the monthly bill for 30 days and average tariff per kWh.

[*Generation of Electrical Power, Pb. Technical Univ. May 2008; Rajasthan Univ. 2001, 2004, 2005, Rajasthan Technical Univ. 2008*]

Solution : Monthly energy consumption corresponding to maximum demand

$$= \frac{\text{Voltage} \times \text{maximum demand of current} \times \text{pf} \times \text{working hour} \times 30}{1{,}000}$$

$$= \frac{230 \times 40 \times 1 \times 2 \times 30}{1{,}000} = 552 \text{ kWh}$$

Cost of energy corresponding to maximum demand

$= \text{Rs } 3.50 \times 552 = \text{Rs } 1{,}932$

Energy cost for remaining energy consumption

$= \text{Rs } 1.80 \times (2{,}020 - 552) = \text{Rs } 2{,}642.40$

Total monthly bill = Rs 1,932 + Rs 2,642.40 = Rs 4,514.40 **Ans.**

$$\text{Average tariff} = \frac{\text{Monthly energy cost}}{\text{Monthly energy consumption}}$$

$$= \text{Rs } \frac{4{,}514.40}{2{,}020} = \text{Rs } 2.265 \text{ per kWh}$$ **Ans.**

Example 17.7. An industrial consumer having a maximum demand of 100 kW, maintains a load factor of 60%. The tariff rates are Rs 900 per kVA of maximum demand per annum plus Rs 1.80 per kWh of energy consumed. If the average power factor is 0.8 lagging, calculate the total energy consumed per annum and the annual electricity bill. Also workout the overall cost per kWh consumed.

Solution : Maximum demand = 100 kW

Average power factor = 0.8

$$\text{Maximum demand in kVA} = \frac{\text{Maximum demand in kW}}{\text{Average power factor}} = \frac{100}{0.8} = 125$$

Load factor = 60% or 0.6

Total energy consumed per annum

= Maximum demand in kW × load factor × 8,760

$= 100 \times 0.6 \times 8{,}760 = 525{,}600$ kWh **Ans.**

Annual demand charges $= \text{Rs } 900 \times 125 = \text{Rs } 112{,}500$

Annual energy charges $= \text{Rs } 1.80 \times 525{,}600 = \text{Rs } 946{,}080$

Annual electricity charges = Rs (112,500 + 946,080)

= Rs 1,058,580 **Ans.**

$$\text{Overall cost per kWh supplied} = \text{Rs } \frac{1{,}058{,}580}{525{,}600} = \text{Rs } 2.02$$ **Ans.**

Example 17.8. A consumer takes a steady load of 250 kW at a power factor of 0.8 lagging for 10 hours per day and 300 days per annum. Estimate the annual payment under each of the following tariffs

(*i*) Rs 1.20 per kWh + Rs 1,200 per kVA per annum.

(*ii*) **Rs 1.20 per kWh + Rs 1,200 per kW per annum + 25 paise per kVARh.**

Solution : Maximum demand = 250 kW

Annual energy consumption

= Steady load in kW × working hours per day × working days per annum

= 250 × 10 × 300 = 750,000 kWh

$$\text{Maximum demand in kVA} = \frac{\text{kW}}{\text{PF}} = \frac{250}{0.8} = 312.5$$

kVARh consumed per annum

= kWh × tan ($\cos^{-1}$ pf)

= 750,000 × tan ($\cos^{-1}$ 0.8) = 750,000 × 0.75 = 562,500

(*i*) Annual payment under tariff (*i*)

= Rs 1.20 × 750,000 + Rs 1,200 × 312.5

= Rs 1,275,000.00 **Ans.**

(*ii*) Annual payment under tariff (*ii*)

= Rs 1.20 × 750,000 + Rs 1,200 × 250 + Rs 0.25 × 562,500

= Rs (900,000 + 300,000 + 140,625) = Rs 1,340,625.00 **Ans.**

Example 17.9. Calculate the number of units to be consumed so that the annual bill on the basis of two part tariff is same from the following data :

Maximum demand = 10 kW

Two part tariff—Rs 1,200 per annum per kW of maximum demand plus Rs 1.80 per unit consumed.

Flat rate tariff—Rs 2.40 per unit.

Solution : Maximum demand = 10 kW

Let the consumption be x units so that the annual bill on the basis of two part tariff and flat rate tariff is the same.

Annual bill under two part tariff

= Rs 1,200 × 10 + Rs 1.80 × x = Rs (1.8x + 12,000)

Annual bill under flat rate tariff

= Rs 2.40 × x = Rs 2.4x

Annual bill will be same if (1.8x + 12,000) = 2.4x

$$\text{or } x = \frac{12{,}000}{(2.4 - 1.8)} = 20{,}000 \text{ units}$$ **Ans.**

Example 17.10. The monthly reading of a consumer's meter are as follows :

Maximum demand = 50 kW

Energy consumed = 36,000 kWh

Reactive energy = 23,400 kVARh

If the tariff is Rs 80 per kW of maximum demand plus 8 paise per unit plus 0.5 paise per unit for each 1% of power factor below 86%, calculate the monthly bill of the consumer.

[*Generation of Electrical Power Rajasthan Univ. 2003, 2004, 2005, 2006; Rajasthan Technical Univ. 2011*]

Solution : Average load

$$= \frac{\text{Monthly energy consumption}}{24 \times 30} = \frac{36{,}000}{24 \times 30} = 50 \text{ kW}$$

$$\text{Power factor angle } \phi = \text{Tan}^{-1}\frac{\text{kVA hr}}{\text{kWh}} = \text{Tan}^{-1}\frac{23{,}400}{36{,}000} = 33.024°$$

Power factor = Cos ϕ = Cos 33.024° = 0.838444 or 83.444%.

Power factor surcharge = Rs 0.005 × 36,000 (86 – 83.8444) = Rs 388

Monthly bill = Rs (80 × 50 + 0.08 × 36,000 + 388) = Rs 7,268 **Ans.**

Example 17.11. A factory works for 16 hours a day for 300 days in a year. The following two systems of tariff are available :

High voltage supply at Re 1 per unit plus Rs 50 per month per kVA of maximum demand. Low voltage supply at Rs 60 per month per kVA of maximum demand plus Rs 1.10 per unit.

The factory has an average load of 250 kW at 0.8 power factor and a maximum demand of 300 kW at the same power factor.

The high voltage equipment costs Rs 500 per kVA and losses can be taken as 5%. Interest and depreciation charges are 12%. Calculate the difference in the annual cost between the two systems. [*A.M.I.E. Sec. B. Power Plant Engineering Summer 1998*]

Solution : Average load = 250 kW

Working hours per annum

= Number of working hours/day × number of working days in a year

= 16 × 300 = 4,800

Number of units consumed on low voltage supply

= 250 × 4,800 = 1,200,000

$$\text{Maximum demand on low voltage supply} = \frac{300}{0.8} = 375 \text{ kVA}$$

Annual cost under low voltage supply tariff

= Rs 60 × 12 ×·375 + Rs 1.10 × 1,200,000 = Rs 1,590,000.00

High voltage equipment losses = 5% or 0.05

Rating of high voltage equipment

$$= \frac{\text{Maximum demand in kW}}{\text{Power factor} \times (1 - \text{losses})} = \frac{300}{0.8 \times 0.95} = 395 \text{ kVA}$$

Number of units consumed on high voltage supply

$$= \frac{1{,}200{,}000}{0.95} = 1{,}263{,}158$$

Annual cost of high voltage equipment

= Rs 500 × 395 = Rs 197,500

Annual interest and depreciation charges

$$= \text{Rs } 197{,}500 \times \frac{12}{100} = \text{Rs } 23{,}700.00$$

Annual cost under high voltage supply tariff

= Annual interest and depreciation charges on hv equipment + annual maximum demand charges + annual energy charges

= Rs 23,700.00 + Rs 50 × 12 × 395 + Rs 1.0 × 1,263,158

= Rs 1,523,858.00 **Ans.**

Thus high voltage supply will be cheaper and saving per annum

= Rs 1,590,000.00 – Rs 1,523,858 = Rs 66,142.00

17.5. POWER FACTOR

The cosine of the angle between voltage and current in an ac circuit is known as *power factor.*

In an ac circuit, there is generally a phase difference between voltage and current. In an inductive circuit, the current lags behind the applied voltage and the power factor of the circuit is referred to as lagging. In a capacitive circuit the current leads the applied voltage and therefore, the power factor of the circuit is said to be leading.

Consider an inductive circuit, which draws a current I from the supply mains lagging behind the supply voltage V by an angle ϕ, known as phase angle, the phasor diagram is shown in Fig. 17.6.

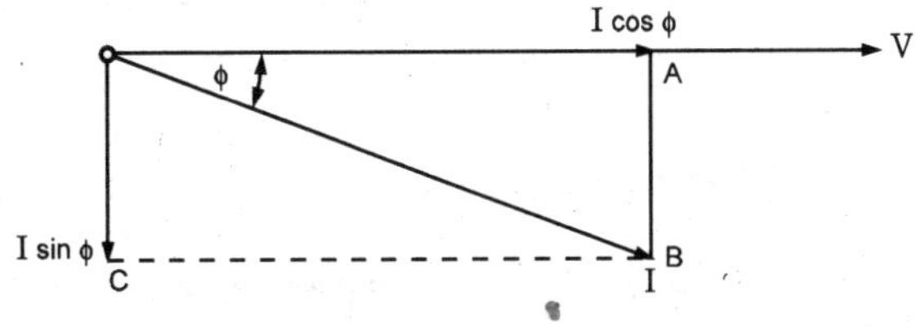

Fig. 17.6

The current I can be resolved into two components, one along the voltage phasor and the other perpendicular to it. The component along the voltage phasor, I cos ϕ is called the *in- phase* or *active* component of current, and the one perpendicular to the voltage phasor, I sin ϕ is called the *out of phase* or *wattless* or *reactive* component of current.

If all these components are multiplied by voltage V, the product of voltage V and in-phase component of current I cos ϕ *i.e.,* VI cos ϕ will represent the true power of the circuit in watts or kW, whereas the product of voltage V and the quadrature component of current I sin ϕ *i.e.,* VI sin ϕ will represent the reactive power in VARs or kVARs and the product of voltage V and current I *i.e.,* VI will represent the apparent power in volt-amperes or kVA. Thus we get a power triangle, as shown in Fig. 17.7.

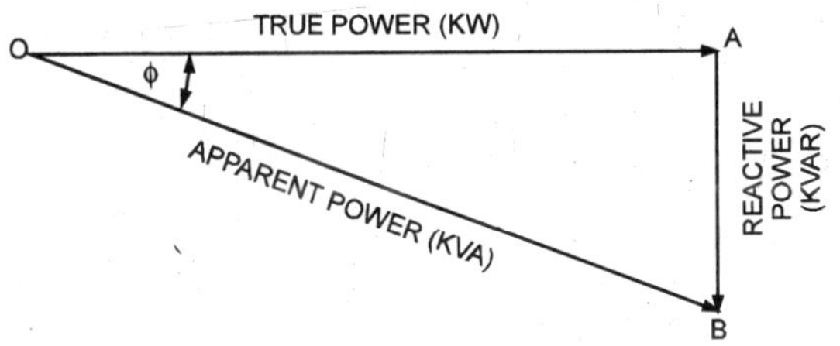

Fig. 17.7

From power triangle OAB shown in Fig. 17.7.

OA = kW component of power

AB = kVAR component of power

$$\text{Cos}\ \phi = \frac{OA}{OB} = \frac{kW}{kVA}, \text{ the ratio of true power and apparent power} \quad ...(17.1)$$

$$\text{Reactive power in kVAR} = \text{Apparent power} \times \sin\phi$$
$$= \text{kVA} \sin\phi$$
$$= (\text{kVA} \cos\phi) \times \frac{\sin\phi}{\cos\phi} = \text{kW} \tan\phi$$

i.e., smaller the reactive component of power, the smaller is the phase angle and the higher is the power factor.

For leading currents the triangle becomes reversed. This fact provides a key to the power factor improvement. If a device drawing leading reactive power is connected in parallel with the inductive load, then the lagging reactive power of the load will be partly neutralised, resulting in improvement of the power factor of the system.

17.6. DISADVANTAGES OF LOW POWER FACTOR

The current for a given load supplied at constant voltage will be higher at a lower power factor and lower at higher power factor. For example if load P is to be supplied at terminal voltage V and at power factor cos ϕ by a 3-phase balanced system then load current is given by

$$I_L = \frac{P}{\sqrt{3}\, V \cos\phi}$$

If P and V are constant, the load current, I_L is inversely proportional to power factor, cos ϕ *i.e.,* lower the power factor, higher the current and vice versa. The higher current due to poor power factor affects the system and results in following disadvantages.

(*i*) Rating of generators and transformers are proportional to their output current hence inversely proportional to power factor, therefore, large generators and transformers are required to deliver same load but at low power factor.

(*ii*) The cross-sectional area of the bus-bar, and the contact surface of the switchgear is required to be enlarged for the same power to be delivered but at low power factor.

(*iii*) For the same power to be transmitted but at low power factor, the transmission line or distributor or cable have to carry more current. The size of the conductor will have to be increased if current density in the line is to be kept constant. Thus more conductor material is required for transmission lines, distributors and cables to deliver the same load but at low power factor.

(*iv*) Energy losses are proportional to the square of the current hence inversely proportional to the square of the power factor *i.e.,* more energy losses incur at low power factor, which results in poor efficiency.

(*v*) Low lagging power factor results in large voltage drop in generators, transformers, transmission lines and distributors which results in poor regulation. Hence extra regulating equipment is required to keep the voltage drop within permissible limits.

(*vi*) Low lagging power factor reduces the handling capacity of all the elements of the system.

Thus we see that the low power factor leads to a high capital cost for the alternators, switchgears, transformers, transmission lines, distributors and cables etc.

Keeping in view the various drawbacks associated with the low power factor, the power suppliers insist on a power factor of 0.8 or above for industrial establishments. The power tariffs are devised to penalize the consumers with low lagging power factor and to encourage them to instal power factor correction devices or equipment.

17.7. CAUSES OF LOW POWER FACTOR

(*i*) All ac motors (except overexcited synchronous motors and certain type of commutator motors) and transformers operate at lagging power factor. The power factor falls with the decrease in load. For example an induction motor has a reasonable higher power of 0.85 at full load, 0.8 at 75% of full load, 0.7 at half-full load, 0.5 at 25% of full load and as low as 0.1 on no load.

(*ii*) Arc lamps and electric discharge lamps operate at low lagging power factor.

(*iii*) Due to increased supply mains voltage, which usually occurs during low-load periods such as lunch hours, night hours etc, the magnetizing current of inductive reactances increase and power factor of the electrical plant as a whole comes down.

(*iv*) The power factor at which motors operate falls due to improper maintenance and repairs of motors. In repaired motors, less wire is sometimes used than originally wound motors, therefore, in such motors leakage of magnetic flux increases and power factor of the motor decreases.

In case of heavily worn-out bearings, the rotor may catch at the stator.

Some metal is sometimes removed from the rotor by turning instead of replacing the defective bearings. In doing so, the length of air gap between stator and rotor increase, due to which greater magnetising current is required and, therefore, power factor drops.

(*v*) Industrial heating furnaces such as arc and induction furnaces operate on very lagging power factor.

The average power factors of some of the common appliances are given below :

Type of Load	Power Factor
Incandescent lamps	0.98 – 1.0
Fluorescent lamps	0.6 – 0.8
Neon lamps used for advertisements	0.4 – 0.5
Arc lamps used in cinemas	0.3 – 0.7
Fans	0.5 – 0.8
Induction motors	0.5 – 0.85
Fractional kW motors	0.4 – 0.75
Induction heaters	0.85
Resistance furnaces	0.6 – 0.9
Arc furnaces	0.85
Induction furnaces	0.6
Arc welders	0.3 – 0.4
Resistance welders	0.4 – 0.75

17.8. METHODS OF POWER FACTOR IMPROVEMENT

As we have already indicated, the low power factor is almost invariably due to inductive nature of load and, therefore, the logical corrective is to connect such devices across the load, which take leading reactive power such as static capacitors, synchronous machines or synchronous condensers. The leading reactive component of current drawn by power factor correcting device neutralises the lagging reactive component of current drawn by the load partly or completely. Power factor of the system will become unity when lagging reactive component of load current is completely neutralised by the leading reactive component of current drawn by power factor correcting device.

Let the current drawn by an inductive circuit be I lagging behind the applied voltage by an angle ϕ. The leading current required to neutralize the lagging reactive component of current drawn by the inductive circuit (equipment) to give unity power factor.

$$= I \sin \phi = I\sqrt{(1-\cos^2 \phi)} = I\sqrt{1-(pf)^2} \qquad ...(17.2)$$

17.8.1. By Use of Static Capacitors

Power factor can be improved by connecting the capacitors* in parallel with the equipment operating at lagging power factor such as induction motors, fluorescent tubes. Static capacitors have the advantages of small losses (less than $\frac{1}{2}$ per cent) or higher efficiency (say 99.6%), low initial cost, little maintenance owing to absence of rotating parts, easy installation being lighter in weight and capability to operate under ordinary atmospheric conditions. However, they have drawbacks of short service life (8 to 10 years), getting damaged on overvoltages and uneconomical repair. The current drawn by induction motors or fluorescent tubes can be resolved into two components ; the active component, which is in phase with the supply voltage and the quadrature or wattless component of constant magnitude. The capacitors draw current leading the supply voltage by 90º approximately and neutralise the quadrature or wattless component of current drawn by the equipment across which these are connected. In case of 3-phase loads, capacitors can be connected either in star or delta, as shown in Figs. 17.8 and 17.9. These capacitors remain connected permanently across the equipment and are across the supply mains, whenever the equipment is switched on.

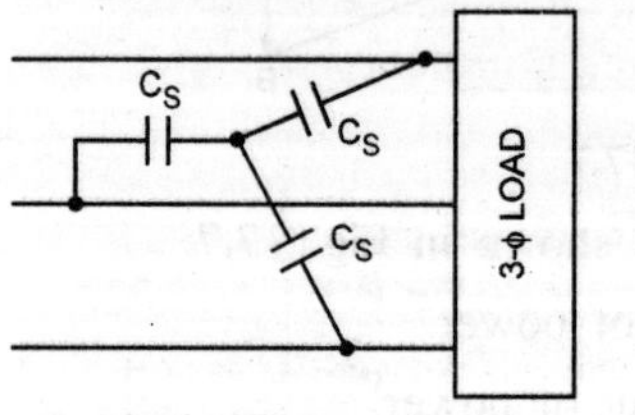

Fig. 17.8. ***Star-Connected Capacitors***

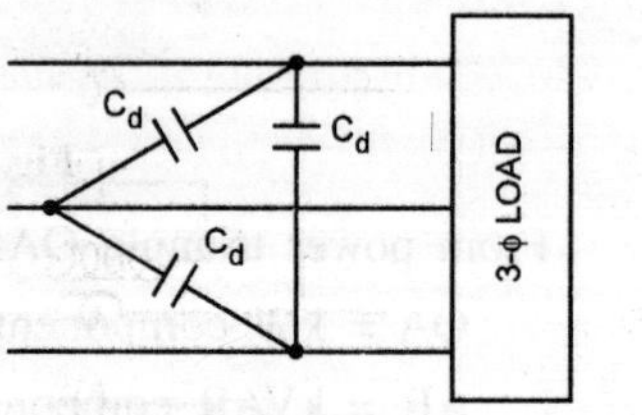

Fig. 17.9. ***Delta-Connected Capacitors***

The value of the static capacitors for the improvement of the power factor can be determined as follows :

The leading current required to neutralise the lagging reactive component of the current drawn by the equipment to give unity power factor is expressed as

$$I_C = I_L = I \sin \phi = I\sqrt{(1-\cos^2 \phi)} = I\sqrt{1-(pf)^2}$$

The value of capacitance in star bank is given by

$$C_S = \frac{I_C}{2\pi f V} = \frac{I}{2\pi f V}\sqrt{1-(pf)^2} \qquad ...(17.3)$$

where V is the phase voltage, I is the phase current and f is supply frequency.

For given kVAR and line voltage the delta value will be one-third of star value.

* The capacitors are generally designated as static capacitors to distinguish from the so called synchronous condensers which are synchronous motors operating under no-load condition and taking leading current.

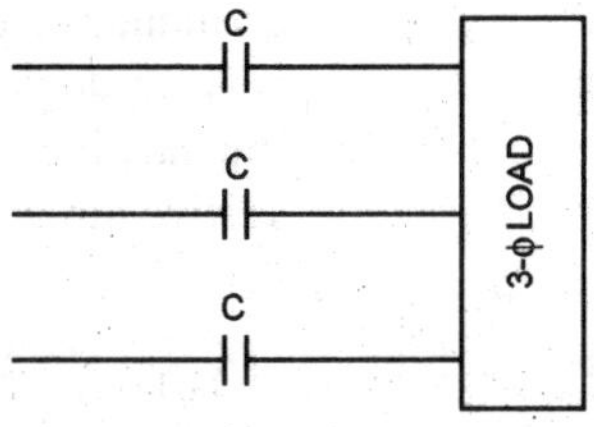

Fig. 17.10

Power factor can also be improved by connecting static capacitors in series with the line, as shown in Fig. 17.10. Capacitors connected in series with the line neutralize the line reactance. The capacitors, when connected in series with the line, are called the *series capacitors,* and when connected in parallel with the equipment, are called the *shunt capacitors.*

Shunt capacitors are used in factories, plants and also on transmission lines.

Series capacitors are used on long transmission lines as they provide automatic compensation with the variations in load. The capacity of the capacitors to neutralize the line reactance is given by

$$C = \frac{1}{(2\pi f)^2 L}$$

where f is the supply frequency and L is the inductance of the line per phase.

The value of reactance required is usually very large but reduced to reasonable value by use of a transformer, as shown in Fig. 17.11.

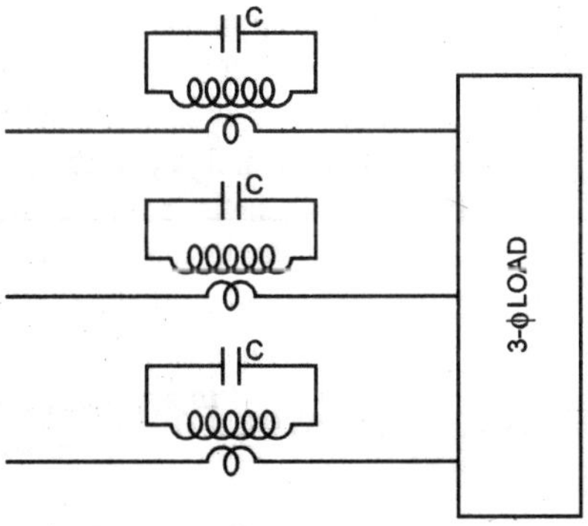

Fig. 17.11

Shunt capacitors are used in ratings from 15 kVAR to 10,000 kVAR. Small capacitors, up to a few hundred rating are used on individual distribution circuits of customers. Capacitors banks of 500 – 3,000 kVAR ratings are employed in small distribution substations and those with larger rating at big substations.

Three phase capacitor banks can be connected in star earthed, star unearthed or in delta arrangements. Ungrounded star connection is preferred because of easier protection. In this method, the fault current in case of a fault in any unit in one of the phases is restricted by the capacitors in the sound phases. This results in the use of smaller fuses and less protection materials. The capacitor must be provided with a suitable discharge device to dissipate the stored energy, and to reduce the residual voltage to a safer value within a short period (50 V or less within one minute in case of medium voltage capacitors and within five minutes in case of high voltage capacitors as per ISS 2834–1963). The discharge resistance is usually incorporated within the 'unit' itself in the case of medium voltage capacitors and in case of high voltage capacitors, potential transformers of the circuit breakers are generally utilised as a discharge device.

The reactive output of the capacitors in kVAR is given by

$\frac{\text{Voltage} \times \text{line current}}{1{,}000}$ kVAR or $2\pi f CV^2 \times 10^{-9}$ kVAR in case of single phase circuit ...(17.4)

where V is the line voltage, f the supply frequency and C is the capacitance in microfarads.

and $\frac{\sqrt{3}\ \text{voltage} \times \text{line current}}{1{,}000}$ kVAR or $2\pi f CV_L^2 \times 10^{-9}$ kVAR in case of 3-phase star-connected circuits. ...(17.5)

and $= 6\pi f CV_L^2 \times 10^{-9}$ kVAR in case of 3-phase delta-connected circuits ...(17.6)

where V_L is the line voltage, f is the supply frequency in Hz and C is the capacitance in μF between the line terminals.

Thus we see that the corrective capacity of the capacitors is a function of the line voltage and supply frequency, varying in accordance with the square of the voltage and directly with the supply frequency. The units as manufactured are designed for a variation of voltage of ±10% of normal voltage. It is, therefore, impossible to overload these units so long as normal voltage and frequency are maintained.

The characteristics of capacitors, in general, are similar to those of synchronous condensers except that the corrective kVAR of the synchronous condenser is adjustable and may be controlled automatically whereas that of the capacitor is fixed unless there is a possibility of change in the number of units connected.

17.8.2. By Use of Synchronous or High Power Factor Machines

Synchronous machines are excited by dc, and the power factor may be controlled by controlling the field excitation. The various synchronous machines available for power factor correction comprise synchronous motors, synchronous condensers, synchronous converters, synchronous phase modifiers, phase advancers, and synchronous-induction motors.

1. ***By Use of Synchronous Motors.*** These motors have characteristics that make them adaptable for a wide range of applications. The speed is constant, the efficiency is high and uniform from light loads up to considerable overloads, and the starting characteristics compare favourably with those of induction motors. Another desirable characteristic of the synchronous motor is its tendency to maintain a constant load voltage even if there are variations in the supply voltage. When the line voltage increases, the leading reactive kVA falls and when the line voltage falls, the leading reactive kVA increases. The usual practice is to keep the field excitation constant at a value corresponding to normal full-load rating as regards output and power factor. Synchronous motors are designed for 1.0 – 0.8 leading power factors at full load. The unity power factor motor costs less and has a higher efficiency, but if fully loaded, it cannot furnish leading reactive kVA to compensate for lagging reactive kVA in the system.

2. *By Use of Synchronous Condensers.* An overexcited synchronous motor running on no load is called the *synchronous condenser* or *synchronous phase advancer* and behaves like a capacitor, the capacitive reactance of which depends upon the motor excitation. Power factor can be improved by using synchronous condensers like shunt capacitors connected across the supply.

In phasor diagram (Fig. 17.12), phasor I_L represents the current drawn by the industrial load, lagging behind the applied voltage V by a large angle ϕ_L and phasor I_M represents the current drawn by the synchronous condenser leading the applied voltage V by the angle ϕ_M. The resultant current I is the phasor sum of I_L and I_M and now angle of lag ϕ is much smaller than ϕ_L. Thus overall power factor is improved from cos ϕ_L to cos ϕ by the use of the synchronous condenser. In this way the power factor can be made unity even.

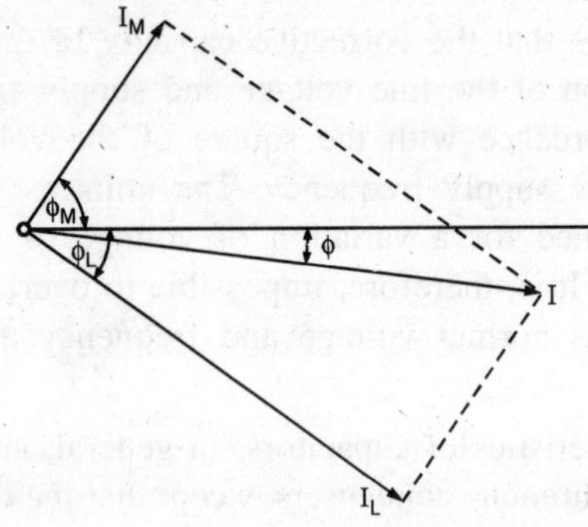

Fig. 17.12

Synchronous condensers are usually built in large units and are employed where a large quantity of corrective kVAR (say 5,000 kVAR or more) is required.

The advantages of synchronous condensers over static capacitors as a power factor correction devices are (*i*) a finer control can be obtained by variation of field excitation; (*ii*) inherent characteristic of synchronous condensers of stabilizing variations in the line voltage and thereby automatically aid in regulation, (*iii*) possibility of overloading a synchronous condenser for short periods, and (*iv*) improvement in the system stability and reduction of the effect of sudden changes in load owing to inertia of synchronous condenser. By use of synchronous condensers at intermediate stations, the voltage of the line can be kept constant at various points along its length, thereby, increasing the current carrying capacity of the line and improvement of power factor.

The disadvantages of synchronous condensers over static capacitors as power factor correcting devices are : (*i*) except in size above about 5,000 kVAR, the cost is higher than that of static capacitors of the same rating ; (*ii*) comparatively higher maintenance and operating costs ; (*iii*) comparatively lower efficiency (say 97%) due to losses in rotating parts and heat losses, (*iv*) noise is produced in operation (*v*) an auxiliary equipment is required for starting synchronous condensers ; (*vi*) possibility of synchronous condensers falling out of synchronism causing in interruption of supply ; and (*vii*) increase of short-circuit currents when the fault occurs near the synchronous condenser.

Synchronous condensers are largely employed by utilities at large substations for improving the power factor and voltage regulation. Machines up to 100 MVAR rating or even higher have been used. The excitation current is regulated automatically to give a desired voltage level.

3. *By Use of Phase Advancers.* The power factor of an induction motor falls mainly due to its exciting current drawn from the ac supply mains, because exciting current lags behind the voltage by $\pi/2$. It may be improved by equipping the set with an ac exciter or phase advancer which supplies this exciting current to the rotor circuit at slip frequency. Such an exciter may be mounted on the same shaft as the main motor or may be suitably driven from it. Use of phase advancer is not generally economical in connection with motors below 150 kW output but above this size, phase advancers are frequently employed. Shunt and series type of phase advancers are available according to whether the exciting winding of the advancer is connected in parallel or series with the rotor winding of the induction motor.

There are two main advantages of phase advancers (*i*) lagging kVAR drawn by the motor are considerably reduced due to supply of exciting ampere-turns at slip frequency and (*ii*) the phase advancers can be conveniently employed where the use of synchronous motor is inadmissible.

4. *By Use of Synchronous-Induction Motors.* These are special types of motors which operate at certain loads as synchronous motors and at other loads as induction motors.

5. *By Use of High Power Factor Motors.* Besides synchronous motors or synchronous-induction motors there are other several types of motors which operate at a power factor of approximately unity such as compensated induction motors, and Schrage motors. These motors are more expensive and have higher maintenance cost than ordinary induction motors.

17.8.3. Location of Power Factor Correction Equipment

The best location for the power factor correction equipment to be installed is where the apparatus or equipment responsible for low power factor is operating. Synchronous condensers are used at load centres where considerable corrective kVAR is required whereas static capacitors are justifiably used in smaller units and may be placed closer to the point where the load of inductive nature is installed and thereby relieving the distributors and feeders from carrying excessive currents owing to low power factor.

In case of transmission system, if synchronous condensers are to be employed for power factor improvement then these should be installed at the receiving end so that not only the generators but also the transmission lines are relieved of carrying excessive current due to poor power factor. However, if synchronous condensers are installed near the generators then only generators will be relieved from the excessive current component and the transmission lines will have to carry it.

17.9. ADVANTAGES OF POWER FACTOR IMPROVEMENT

Modern alternators are designed to have high reactance in order that alternators may not get damaged at the time of short circuit and normally reactance is 20 times of resistance. Any change in current and power factor causes the change in terminal voltage, so voltage can be kept fairly constant by power factor control.

If the power factor of the supply or power station is raised to unity, the current for the same amount of power to be supplied is reduced to minimum. This results in reduction of transmission line copper losses, and reduction of voltage drop in transmission line and in alternator windings, as copper losses are directly proportional to the square of supply current and voltage drop is directly proportional to the current.

The terminal voltage of an alternator is given by the phasor difference of induced emf and voltage drop in synchronous impedance. Neglecting the resistance as compared to synchronous reactance, the synchronous impedance can be considered purely inductive, so synchronous impedance voltage drop leads the load current by $\pi/2$.

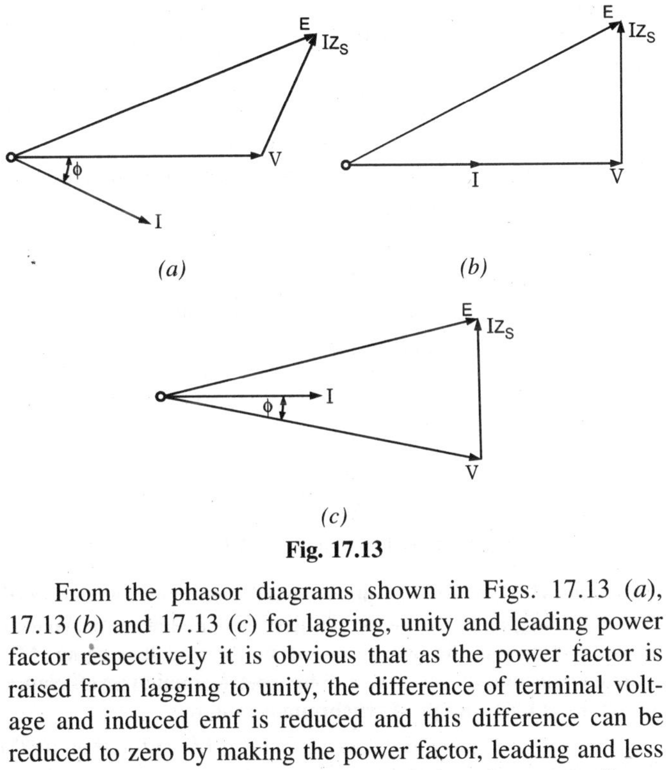

Fig. 17.13

From the phasor diagrams shown in Figs. 17.13 (*a*), 17.13 (*b*) and 17.13 (*c*) for lagging, unity and leading power factor respectively it is obvious that as the power factor is raised from lagging to unity, the difference of terminal voltage and induced emf is reduced and this difference can be reduced to zero by making the power factor, leading and less than unity. Hence to have zero regulation the power factor should be made leading one, so that no other regulating equipment is required.

It is not economical to raise the power to unity or leading one due to the following reasons :

1. In case the power factor is improved to unity for full-load conditions, it would become leading for loads less than full load (unless some capacitors are switched off which is usually difficult). At power factors lower than unity (lagging or leading) the current for the same amount of power to be supplied is increased and thereby energy losses in generators, transformers, transmission lines and distribution lines will be increased.
2. As the power factor approaches unity, the capacity of power factor correction device increases more rapidly *i.e.*, the power factor of an installation can be improved from 0.7 or 0.8 to 0.8 or 0.9 by a much smaller capacitive kVAR than that required for raising the power factor from 0.9 to unity.

The advantages of good (or improved) power factor are : (*i*) reduction in load current ; (*ii*) increase in voltage level across the load ; (*iii*) reduction in energy losses in the system (generators, transformers, transmission lines and distributors) due to reduction in load current ; (*iv*) reduction in kVA loading of the generators and transformers which may relieve an overloaded system or release capacity for additional growth of load and (*v*) reduction in kVA demand charge for large consumers.

Example 17.12. Find the power factor at an installation, supplying the following loads :

(*a*) 300 kW at unity power factor (*b*) 1,000 kW at 0.9 lagging power factor and (*c*) 1,500 kW at 0.8 lagging power factor. Also find the maximum load at unity power factor which can be supplied by the substation.

[*Rajasthan Univ. Generation of Electrical Power 1997, 2003, 2004*]

Solution : Total load on the substation

$= 300 + 1,000 + 1,500 = 2,800$ kW

Total reactive load on the substation

$= 300 \tan(\cos^{-1} 1.0) + 1,000 \tan(\cos^{-1} 0.9) + 1,500 (\cos^{-1} 0.8)$

$= 300 \tan 0° + 1,000 \tan 25.842° + 1,500 \tan 36.87°$

$\because$ kVAR = kW $\tan\phi$ and $\phi = \cos^{-1}$pf

$= 0 + 484.3 + 1,125 = 1609.3$ kVAR

Total kVA load on the substation

$= \sqrt{(\text{kW})^2 + (\text{kVAR})^2} = \sqrt{(2,800)^2 + (1,609.3)^2}$

$= 3,229.53$ kVA

Power factor at the substation $= \dfrac{\text{kW}}{\text{kVA}} = \dfrac{2,800}{3,229.53} = 0.867$ (lagging) **Ans.**

The maximum unity power factor load which can be supplied by the substation = Total kVA × pf = 3,229.53 × 1.0 = 3,229.53 kW **Ans.**

Example 17.13. A single phase motor connected to a 240V, 50 Hz supply takes a current of 20A at a pf of 0.75 lagging. A capacitor is shunted across the motor terminals to improve the pf to 0.9. Determine the capacitance of the capacitor to be used.

[*Rajasthan Univ. Generation of Electrical Power 2003*]

Solution : Initial motor current, $I_1 = 20$A

Initial power factor, $\cos\phi_1 = 0.75$

Lagging reactive component of initial motor current

$= I_1 \sin\phi_1 = I_1\sqrt{1-\cos^2\phi} = 20\sqrt{1-0.75^2} = 13.229$A

In-phase component of initial motor current

$= I_1 \cos\phi = 20 \times 0.75 = 15$A

Motor current after improvement of power factor

$$I_2 = \frac{\text{Active component of initial current}}{\text{Improved power factor}} = \frac{15}{0.9} = 16.667\text{A}$$

Lagging reactive component of motor current after pf improvement,

$= I_2 \sin\phi_2 = I_2\sqrt{1-\cos\phi_2{}^2} = 16.667\sqrt{1-0.9^2}$

$= 7.625$ A

Leading reactive component drawn by the capacitor

$I_C = I_1 \sin\phi_1 - I_2 \sin\phi_2 = 13.229 - 7.265 = 5.964$A

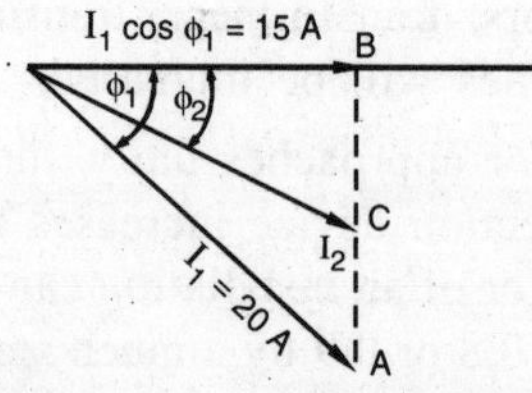

Fig. 17.14

Capacitance of the capacitor, $C = \dfrac{I_C}{2\pi f V}$ $\quad \because I_C = 2\pi f CV$

$$= \frac{5.964}{2\pi \times 50 \times 240} = 791\ \mu F \textbf{ Ans.}$$

Example 17.14. A 400 V, 50 Hz, 3-phase line delivers 200 kW at 0.8 pf lagging. It is desired to raise the line power factor to unity by installing shunt capacitors. Calculate the capacitance of each unit if they are connected in (*i*) star and (*ii*) delta.

[Rajasthan Technical Univ. Generation of Electrical Power 2008]

Solution : Load P = 200 kW

Power factor, $\cos \phi_1 = 0.8$ (lag) ; $\phi_1 = \text{Cos}^{-1} 0.8 = 36.87°$ and Tan $36.87° = 0.75$

Power factor, $\cos \phi_2 = 1.0$; $\phi_2 = \text{Cos}^{-1} 1.0 = 0°$ and Tan $0° = 0$

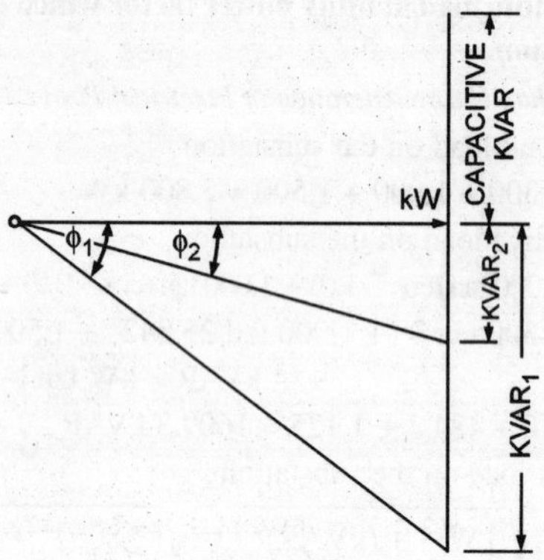

Fig. 17.15

Leading kVAR supplied by static capacitors,

Capacitive kVAR = $\text{kVAR}_1 - \text{kVAR}_2$

$= P (\tan \phi_1 - \tan \phi_2) = 200 (0.75 - 0) = 150$

Since reactive kVAR output of capacitors when connected in star (Eq. 17.5) is given by $2\pi f C V_L^2 \times 10^{-9}$, where C is in μF

$\therefore\ 2\pi f C V_L^2 \times 10^{-9} = 150$

or $C = \dfrac{150 \times 10^9}{2\pi f V_L^2} = \dfrac{150 \times 10^9}{2\pi \times 50 \times (400)^2} = 2{,}984\ \mu F$ **Ans.**

Since reactive kVAR output of capacitors when connected in delta (Eq. 17.6) is given as $6\pi f C V^2 \times 10^{-9}$ where C is in μF

$\therefore\ C = \dfrac{150 \times 10^9}{6\pi f V_L^2} = \dfrac{150 \times 10^9}{6\pi \times 50 \times (400)^2} = 995\ \mu F$ **Ans.**

Example 17.15. The load on the mains of a supply system is 1,000 kW at a power factor of 0.707 lagging. What must be the kVA rating of the power factor improvement equipment which takes leading current at a power factor of 0.1 in order to raise the power factor of the entire system to 0.95?

[Pb.Technical Univ. Generation of Electric Power May 2010]

Solution : Load, P = 1,000 kW

Phase angle of load, $\phi_1 = \cos^{-1} 0.707 = 45°$ (lagging)

Improved phase angle of load, $\phi_2 = \cos^{-1} 0.95 = 18.195°$ (lagging)

Required rating of pf improvement equipment

$= P (\tan \phi_1 - \tan \phi_2)$

$= 1{,}000 (\tan 45° - \tan 18.195°)$

$= 1{,}000 (1.0 - 0.32868) = 671.32$ kVAR

Phase angle of pf improvement equipment,

$\phi = \cos^{-1} 0.1 = 84.26°$ (leading)

Rating of pf improvement equipment

$= \dfrac{\text{Reactive kVAR supplied by the equipment}}{\sin \phi}$ kVA

$= \dfrac{671.32}{\sin 84.26°} = 674.7$ kVA **Ans.**

Example 17.16. A synchronous motor improves the power factor of a load of 500 kW from 0.707 lagging to 0.95 lagging. Simultaneously the motor carries a load of 100 kW. Find (*i*) the leading kVAR supplied by the motor (*ii*) kVA rating of the motor and (*iii*) power factor at which the motor operates.

[Rajasthan Univ. Generation of Electrical Power 1999]

Solution : Load $P_1 = 500$ kW

Initial power factor of load, $\cos \phi_1 = 0.707$ (lagging)

Phase angle $\phi_1 = \cos^{-1} 0.707 = 45$ (lagging)

Motor load, $P_2 = 100$ kW

Motor load including its own load, P

$= P_1 + P_2 = 500 + 100 = 600$ kW

Improved power factor, $\cos \phi_2 = 0.95$ (lagging)

Phase angle $\phi_2 = \cos^{-1} 0.95 = 18.195°$ (lagging)

(*i*) Reactive kVA supplied by the synchronous motor

= Reactive kVA drawn by the load of 500 kW – reactive kVA drawn by the combined load of 600 kW

$= P_1 \tan \phi_1 - P \tan \phi_2$

$= 500 \tan 45° - 600 \tan 18.195°$

$= 500 - 197.21 = 302.79$ kVAR (leading) **Ans.**

(*ii*) kVA rating of synchronous motor

$= \sqrt{P_2^2 + (\text{kVAR supplied by the motor})^2}$

$= \sqrt{(100^2 + 302.79)^2} = 318.875$ kVA (leading) **Ans.**

(*iii*) Power factor of synchronous motor

$= \dfrac{100}{318.875} = 0.3136$ (leading) **Ans.**

Example 17.17. A 400 V, 3-phase installation draws a current of 50A at 0.8 lagging power factor. It is desired to install a synchronous motor to improve the overall power factor to 0.95 lagging. The synchronous motor will drive a 25 hp (metric) load at an efficiency of 0.9. Find kVA and pf of synchronous motor.

[Rajasthan Univ. Generation of Electrical Power 1998]

Solution : Load of installation,

$P_1 = \sqrt{3}\, V_L I_L \cos \phi_1 = \sqrt{3} \times 400 \times 50 \times 0.8$

$= 27{,}713$ W or 27.713 kW

kVAR rating of installation,

$Q_1 = P_1 \tan \phi_1 = 27.713 \tan (\cos^{-1} 0.8) = 20.785$ kVAR

kW input to synchronous motor, $P_2 = \dfrac{25 \times 735.5}{0.9 \times 1{,}000} = 20.43$ kW

Total load, $P = P_1 + P_2 = 27.713 + 20.43 = 48.143$ kW

Overall power factor angle, $\phi = \text{Cos}^{-1} 0.95 = 18.195°$ (lagging)

Total reactive kVA, $Q = P \tan \phi = 48.143 \tan 18.195°$

$= 48.143 \times 0.3287 = 15.82$ (lagging)

kVAR supplied by synchronous motor

$$Q_2 = Q_1 - Q = 20.785 - 15.82 = 4.965 \text{ kVAR (leading)}$$

$$\text{kVA of synchronous motor} = \sqrt{P_2^2 + Q_2^2} = \sqrt{(20.43)^2 + (4.965)^2}$$

$$= 21.025 \text{ kVA } \textbf{Ans.}$$

$$\text{Power factor of synchronous} = \frac{20.43}{21.025} = 0.9717 \text{ (leading) } \textbf{Ans.}$$

Example 17.18. A generator station is connected with the following loads :

(*i*) A lighting load of 200 kW (*ii*) an induction motor of 200 hp (metric) pf 0.7 lagging and efficiency 80% (*iii*) A synchronous motor of 100 hp (metric) efficiency 90%. Calculate the power factor of the synchronous motor so that the combined power factor becomes unity.

[*P.b. Univ. Electrical Power II (Generation) December 1995*]

Solution : Input of induction motor

$$P_M = \frac{\text{Output in hp} \times 735.5}{\text{Motor efficiency}} = \frac{200 \times 735.5}{0.8}$$

$$= 183{,}875 \text{ W or } 183.875 \text{ kW}$$

Total load

= Lighting load + induction motor load + synchronous motor load

$$= 200 + 183.875 + \frac{100 \times 735.5}{0.9 \times 1{,}000} = 200 + 183.875 + 81.722$$

$$= 465.6 \text{ kW}$$

Leading kVAR drawn by synchronous motor

= Lagging kVAR drawn by induction motor

$$= 183.875 \tan(\cos^{-1} 0.7) = 183.875 \times 1.02 = 187.6$$

∵ pf of lighting load is considered unity and pf after improvement is also unity

Power factor of synchronous motor

$$= \cos \tan^{-1} \frac{\text{Leading kVAR drawn by synchronous motor}}{\text{Power drawn by synchronous motor}}$$

$$= \cos \tan^{-1} \frac{187.6}{81.722} = 0.4 \text{ (leading) } \textbf{Ans.}$$

Example 17.19. A 3-phase synchronous motor having a mechanical load (including losses) of 122 kW is connected in parallel with a load of 510 kW at 0.8 pf lagging. The excitation of the motor is adjusted so that the kVA input to the motor becomes 140 kVA. Determine the new power factor of the whole system.

Solution : Load, $P_1 = 510$ kW

Load power factor, $\cos \phi_1 = 0.8$ lagging

kVAR (lagging) drawn by the load, kVAR_1

$$= P_1 \tan \phi_1 = 510 \tan(\cos^{-1} 0.8) = 382.5$$

kVA input to the synchronous motor = 140

kW input to the synchronous motor, $P_2 = 122$ kW

kVAR (leading) drawn by the synchronous motor,

$$\text{kVAR}_2 = \sqrt{(\text{kVA})^2 - (P_2)^2} = \sqrt{(140)^2 - (122)^2} = 68.673$$

Total kVAR drawn by the whole system

$$= \text{kVAR}_1 - \text{kVAR}_2 = 382.5 - 68.673 = 313.827$$

Total load on the system, $P = P_1 + P_2 = 510 + 122 = 632$ kW

$$\text{New power factor of the system, } \cos \phi = \cos \tan^{-1} \frac{313.827}{632}$$

$$= 0.8956 \text{ (lagging) } \textbf{Ans.}$$

Example 17.20. A 3 φ, 50 Hz, 400 V motor develops 74.6 kW, the power factor being 0.75 lagging and efficiency 93%. A bank of capacitors is connected in delta across the supply terminals and power factor is improved to 0.95 lagging. Each of the capacitance units is built of 4 similar 100V capacitors. Determine the capacitance of each capacitor.

[*Rajasthan Univ. Generation of Electrical Power 2006*]

Solution : $\text{Input to motor} = \dfrac{\text{Motor input}}{\eta} = \dfrac{74.6}{0.93} = 80.2 \text{ kW}$

Initial power factor, $\cos \phi_1 = 0.75$ (lagging)

Power factor after improvement, $\cos \phi_2 = 0.95$ (lagging)

kVAR rating of capacitor bank

$$= P(\tan \phi_1 - \tan \phi_2)$$

$$= 80.2 [\tan(\cos^{-1} 0.75) - \tan(\cos^{-1} 0.95)]$$

$$= 80.2 (0.881917 - 0.328684) = 44.37 \text{ kVAR}$$

Since reactive kVA output of the capacitors when connected in delta is given by $6\pi f C V_L^2 \times 10^{-9}$; where C is the capacitance in μF, V_i is the line voltage in volts and f is supply frequency in Hz.

$$\therefore C = \frac{44.37}{6\pi \times 50 \times 400^2 \times 10^{-9}} = 294 \ \mu\text{F}$$

The bank of capacitors used to improve the pf is connected in delta so phase voltage is 400 V, and therefore, each phase of bank will have 4 capacitors connected in series.

So capacitance of each unit = 4 × 294 = 1,176 μF **Ans.**

Example 17.21. A 35 kW induction motor has pf 0.9 and efficiency 0.9 at full load, power factor 0.6 and efficiency 0.7 at half full-load. At no load the current is 25% of full-load current and power factor 0.1. Capacitors are supplied to make the line power factor 0.8 at half full-load. With these capacitors in circuit, find the line power factor at (*i*) full load, and (*ii*) no load.

Solution : Input to motor at full load,

$$P_f = \frac{\text{Rated output}}{\eta} = \frac{35}{0.9} = 38.88 \text{ kW}$$

kVAR (lagging) drawn by the motor at full-load,

$$\text{kVAR}_{f_1} = P_{f_1} \tan \phi_{f_1} = 38.88 \tan(\cos^{-1} 0.9) = 18.835$$

Input to motor at half full-load,

$$P_{h_1} = \frac{\frac{1}{2} \times 35}{0.7} = 25 \text{ kW}$$

kVAR (lagging) drawn by the motor at half full-load

$$\text{kVAR}_{h_1} = P_{h_1} \tan(\cos^{-1} \phi_{h_1}) = 25 \tan(\cos^{-1} 0.6) = 33.33$$

Full-load current,

$$I_f = \frac{35 \times 1{,}000}{\sqrt{3} V_L \times 0.9 \times 0.9} = \frac{24{,}947}{V_L} \text{ amperes}$$

$$\text{No-load current, } I_0 = 0.25 I_f = \frac{24{,}947 \times 0.25}{V_L} = \frac{6{,}237}{V_L} \text{ amperes}$$

$$\text{No-load input, } P_0 = \frac{\sqrt{3} V_L I_0 \cos \phi_0}{1{,}000} = \frac{\sqrt{3} \times V_L \times 6{,}237}{1{,}000 \times V_L} \times 0.1$$

$$= 1.08 \text{ kW}$$

kVAR (lagging) drawn by the motor at no load,

$$\text{kVAR}_{01} = 1.08 \tan(\cos^{-1} 0.1) = 10.75$$

kVAR (lagging) drawn from mains at half load with capacitors,

$$kVAR_{h2} = P_h \tan\phi_{h2} = 25 \tan(\cos^{-1} 0.8) = 18.75$$

kVAR supplied by capacitors,

$$kVAR_C = kVAR_{h1} - kVAR_{h2} = 33.33 - 18.75 = 14.58$$

kVAR drawn from mains at full load with capacitors,

$$kVAR_{f2} = kVAR_{f1} - kVAR_C = 18.835 - 14.58 = 4.255$$

(*i*) Line power factor at full load

$$= \cos\left(\tan^{-1}\frac{kVAR_{f2}}{P_f}\right) = \cos\left(\tan^{-1}\frac{4.265}{38.88}\right)$$

$$= 0.994 \text{ lagging} \quad \textbf{Ans.}$$

kVAR drawn from mains at no load with capacitors,

$$kVAR_{0_2} = kVAR_{0_1} - kVAR_C = 10.75 - 14.58 = -3.83$$

(*ii*) Line power factor at no load

$$= \cos\left(\tan^{-1}\frac{kVAR_{0_2}}{P_0}\right) = \cos\left(\tan^{-1}\frac{-3.83}{1.08}\right)$$

$$= 0.272 \text{ leading} \quad \textbf{Ans.}$$

17.10. ECONOMICS OF POWER FACTOR IMPROVEMENT

When the power factor is improved it involves an expenditure on account of the power factor correcting equipment. Improvement of power factor will result in reduction of maximum demand and thus affect an annual saving over the maximum demand charge but on the other hand an expenditure is to be incurred every year in the shape of interest and depreciation on account of the investment made over the power factor correcting equipment. The limit of the power factor at which the net saving (saving in annual maximum demand charges less annual expenditure incurred on power factor correcting equipment) is maximum is known as *economical limit of power factor correcting.* It will be seen that the economical limit of power factor correction is governed by the relative costs of the supply and power factor correcting equipment.

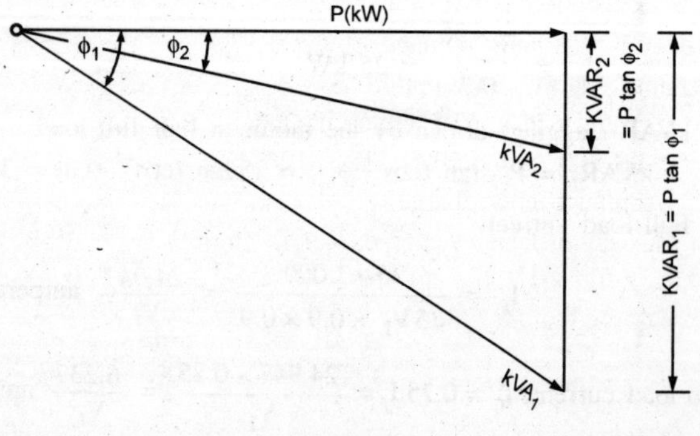

Fig. 17.16

Consider a consumer taking a peak load of P kW at a power factor of $\cos\phi_1$ and charged at the rate of Rs x per kVA of maximum demand per annum. Let the expenditure per kVAR per annum of the power factor correction equipment be Rs y.

Corresponding to maximum demand of P kW at power factor of $\cos\phi_1$

$$\text{Maximum demand in kVA, } kVA_1 = \frac{P}{\cos\phi_1} = P\sec\phi_1$$

$$\text{Reactive power, } kVAR_1 = P\tan\phi_1$$

Let by installing the power factor correction equipment (such as phase advancer) power factor become $\cos\phi_2$.

New maximum demand in kVA,

$$kVA_2 = \frac{P}{\cos\phi_2} = P\sec\phi_2$$

New reactive power, $kVAR_2 = P\tan\phi_2$

Annual saving in maximum demand charges

$$= \text{Rs } x(kVA_1 - kVA_2) = \text{Rs } x(P\sec\phi_1 - P\sec\phi_2)$$
$$= \text{Rs } P\,x(\sec\phi_1 - \sec\phi_2).$$

Leading kVAR supplied by power factor correction equipment

= Difference of the reactive power drawn before and after power factor correction

$$= P(\tan\phi_1 - \tan\phi_2)$$

Annual cost of pf correction equipment

$$= \text{Rs } y \times P(\tan\phi_1 - \tan\phi_2)$$

Net annual saving,

$$S = xP(\sec\phi_1 - \sec\phi_2) - yP(\tan\phi_1 - \tan\phi_2)$$

This saving will be maximum if the differentiation of it with respect to ϕ_2 is zero.

i.e., $\frac{dS}{d\phi_2} = 0$

i.e., $\frac{dS}{d\phi_2}[xP(\sec\phi_1 - \sec\phi_2) - yP(\tan\phi_1 - \tan\phi_2)] = 0$

or $-xP\sec\phi_2\tan\phi_2 + yP\sec^2\phi_2 = 0$

or $\tan\phi_2 = \frac{y}{x}\sec\phi_2$

or $\sin\phi_2 = \frac{y}{x}$

or $\cos\phi_2 = \sqrt{1-\sin^2\phi_2} = \sqrt{1-\left(\frac{y}{x}\right)^2}$ ...(17.7)

From, the above expression, value of most economical power factor $\cos\phi_2$ can be determined which is independent of original factor $\cos\phi_1$ and is governed by the relative costs of supply and power factor correction equipment.

Example 17.22. Calculate the value of the new power factor when the tariff is Rs 1,350 per kVA of maximum demand plus a flat rate paise 80 per kWh. Assume additional cost of condensers etc. at Rs 1,050 per kVA of such plant. Rate of interest and depreciation together is taken as 10%.

Solution : Maximum demand charges,

$$x = \text{Rs } 1{,}350 \text{ per kVA/annum}$$

Cost of phase advancing plant = Rs 1,050 per kVA

Expenditure on phase advancing plant,

y = Annual interest and depreciation

$$= \text{Rs } 1{,}050 \times \frac{10}{100} = \text{Rs } 105/\text{kVAR/annum}$$

Most economocial power factor,

$$\cos \phi_2 = \sqrt{1-\left(\frac{y}{x}\right)^2} = \sqrt{1-\left(\frac{105}{1,350}\right)^2} = 0.997 \text{ (lagging)} \quad \textbf{Ans.}$$

Example 17.23. A factory operates at 0.8 pf lagging and has a monthly demand of 750 kVA. The monthly power rate is Rs 8.50 per kVA. To improve the power factor, 250 kVA capacitors are installed in which there is negligible loss. The installed cost of equipment is Rs 20,000 and fixed charges are estimated as 10% per year. Calculate the annual savings affected by the use of capacitors.

[Pb. Technical Univ. Generation of Electric Power December 2010]

Solution : Maximum demand = 750 kVA

Load power factor, $\cos \phi$ = 0.8

Power factor angle $\phi = \cos^{-1} 0.8 = 36.87°$

kW component of load, P = kVA $\cos \phi$ = 750 × 0.8 = 600 kW

kVAR component of load, Q_1 = kVA $\sin \phi$ = 750 sin 36.87° = 450 (lagging)

Leading kVAR supplied after pf improvement

$Q = Q_1$ – kVAR supplied by capacitors = 450 – 250 = 200

kVA demand after pf improvement

$$= \sqrt{P^2 + Q^2} = \sqrt{600^2 + 200^2} = 632.45$$

Reduction in kVA demand = 750 – 632.45 = 117.55

Annual savings on kVA charge = Rs 8.50 × 12 × 117.55 = Rs 11,990

Fixed charges per annum on the capacitors

$$= \text{Rs } 20,000 \times \frac{10}{100} = \text{Rs } 2,000$$

Annual savings = Rs 11,990 – Rs 2,000 = Rs 9,900 **Ans.**

Example 17.24. An industry consumes 80,000 units of electricity in a year at an average pf 0.707 lag. The recorded maximum demand is 500 kVA. The tariff is Rs 120/kVA maximum demand + 2.5 paise per kWhr. Calculate the annual cost of supply and find out annual saving in cost by installing phase advancing plant costing Rs 50 per kVAR which raises the power factor from 0.707 to 0.9 lagging. Allow 10% per year on the cost of phase advancing plant to cover all additional costs. *[Rajasthan Univ. Generation of Electrical Power 2006]*

Solution : Maximum demand = 500 kVA

Annual energy consumption = 80,000 units

Annual cost of supply

= Annual maximum demand charges + annual energy charges

= Rs 120 × 500 + Rs 0.025 × 80,000 = Rs 62,000 **Ans.**

Initial power factor, $\cos \phi_1 = 0.707$

Phase angle $\phi_1 = \text{Cos}^{-1}\, 0.707 = 45°$ (lagging)

Power factor after improvement, $\cos \phi_2 = 0.9$

Phase angle $\phi_2 = \text{Cos}^{-1}\, 0.9 = 25.842$ (lagging)

Load, P = 500 × 0.707 = 353.5 kW

Leading kVAR supplied by phase advancing plant

= P ($\tan \phi_1 - \tan \phi_2$) = 353.5 (tan 45° – tan 25.842°) = 182.3 kVAR

KVA demand after improvement of power factor

$$= \frac{P}{\text{Cos}\,\phi_2} = \frac{353.5}{0.9} = 392.8 \text{ kVA}$$

Reduction in kVA demand = 500 – 392.8 = 107.2 kVA

Annual saving in kVA demand charges

= Rs 120 × 107.2 = Rs 12,864

Annual cost on phase advancing plant

$$= \text{Rs } 50 \times 182.3 \times \frac{10}{100} = \text{Rs } 911.50$$

Annual saving = Rs 12,864 – Rs 911.50 = Rs 11,952.50 **Ans.**

Example 17.25. A consumer takes a steady load of 300 kW at a lagging power factor of 0.7 for 3,000 hours a year. The tariff is Rs 1,300 per kVA of maximum demand per annum plus Re 0.80 per kWh. The annual cost of the phase advancing plant is Rs 130 per kVAR. Determine the annual saving if the pf of the load is improved.

Solution : Maximum demand charges,

x = Rs 1,300 per kVA/annum

Expenditure on phase advancing plant,

y = Rs 130 per kVAR/annum

Let the power factor be improved to most economical limit

Most economical new power factor,

$$\cos \phi_2 = \sqrt{1-\left(\frac{y}{x}\right)^2} = \sqrt{1-\left(\frac{130}{1,300}\right)^2} = 0.995 \text{ (lagging)}$$

Capacity of phase advancing plant

= Load in kW ($\tan \phi_1 - \tan \phi_2$)

= 300 [tan ($\cos^{-1}$ 0.7) – tan ($\cos^{-1}$ 0.995)] = 276 kVAR

Annual saving in demand charges

$$= \text{Rs } x\left(\frac{P}{\cos\phi_1} - \frac{P}{\cos\phi_2}\right) = \text{Rs } 1,300\left(\frac{300}{0.7} - \frac{300}{0.995}\right)$$

= Rs 165,180

Annual expenditure on phase advancing plant

= Rs y ($\text{kVAR}_1 - \text{kVAR}_2$) = Rs 130 × 276 = Rs 35,880

Net saving in annual cost = Rs (165,180 – 35,880)

= Rs 129,300 **Ans.**

kWh charges will be same as there will be no change in energy consumption.

Example 17.26. The load on a certain installation may be considered constant at 1,200 kVA, 0.75 lagging power factor for 3,000 hours per annum. The tariff is Rs 1,300 per kVA MD plus 80 paise per kWh.

(*i*) Determine the annual charge of energy.

(*ii*) Power factor improving apparatus is installed to improve the pf to 0.95 lagging. Determine the kVAR required and the new annual charge of energy if the power factor improving apparatus costs Rs 1,200 per kVAR, annual interest and depreciation charges are 10% of the capital cost.

Solution : Maximum demand = 1,200 kVA

Annual energy consumption

= Average load in kW × working hours/annum

= 1,200 × 0.75 × 3,000 = 27 × 10^5 kWh

(*i*) Annual energy charges

= Maximum demand charges + unit charges

= Rs 1,300 × 1,200 + Rs 0.80 × 27 × 10^5 = Rs 3,720,000 **Ans.**

(*ii*) kVAR required = P ($\tan \phi_1 - \tan \phi_2$)

= 1,200 × 0.75 [tan $\cos^{-1}$ (0.75) – tan $\cos^{-1}$ (0.95)]

= 900 (0.8819 – 0.3287) = 498 **Ans.**

Capital cost of pf improving apparatus

= Rs 1,200 × 498 = Rs 597,600

Annual interest and depreciation charges

$$= \text{Rs } 597,600 \times \frac{10}{100} = \text{Rs } 59,760$$

Maximum demand

$$= \frac{P}{\text{Cos}\,\phi_2} = \frac{1,200 \times 0.75}{0.95} = 947 \text{ kVA}$$

New annual charges

$= \text{Rs } 1{,}300 \times 947 + \text{Rs } 27 \times 10^5 \times 0.80 + \text{Rs } 59{,}760$

$= \text{Rs } 1{,}231{,}100 + \text{Rs } 2{,}160{,}000 + \text{Rs } 59{,}760$

$= \text{Rs } 3{,}450{,}860$ **Ans.**

17.11. ECONOMICAL COMPARISON OF THE TWO METHODS OF INCREASING THE POWER SUPPLIED

The increase in power demand on the generating station can be met either by increasing the capacity of the generating plant working at the same pf or by raising the power factor of the system by installation of phase advancers. Owing to improvement of power factor in the beginning the saving in the generating and distributing plant outweighs the extra cost of the pf correction equipment in most of the cases but as the power factor is raised further its cost begins to approximate to the saving and finally any saving over the plant is obtained by incurring a greater expenditure on the pf correcting equipment. Thus there is a limit beyond which it is not economical still further to improve the power factor. The maximum value to which the power factor can be economically raised entirely depends upon the relative costs of the generating plant and phase advancing plant.

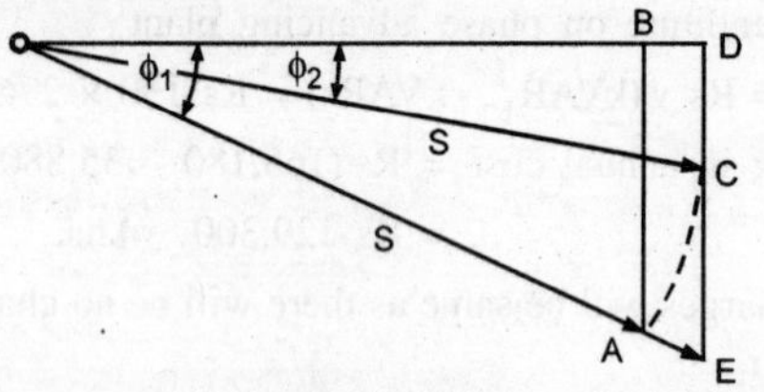

Fig. 17.17

Let the rating of generating station be S in kVA, supplying load at power factor $\cos \phi_2$.

Assume new demand can be met by improving the power factor from $\cos \phi_1$ to $\cos \phi_2$ with the same capacity of the plant.

Existing load in kW, $P_1 = S \cos \phi_1$

New load in kW, $P_2 = S \cos \phi_2$

(i) Cost by increasing the capacity of the generating station

Increase in load $= P_2 - P_1 = S (\cos \phi_2 - \cos \phi_1)$

Increase in the capacity of the generating plant operating at pf $\cos \phi_1$

$$= \frac{\text{Increase in load in kW}}{\text{Power factor of plant}} \text{ kVA}$$

$$= \frac{P_2 - P_1}{\text{Cos } \phi_1} = \frac{S(\cos \phi_2 - \cos \phi_1)}{\text{Cos } \phi_1} \text{ kVA}$$

Increase in annual cost due to increase in capacity of the plant

$$= \text{Rs } xS \frac{\cos \phi_2 - \cos \phi_1}{\cos \phi_1} \quad ...(17.8)$$

where x is the annual cost per kVA of generating plant

(ii) Cost on Power Factor Correction Equipment

Reactive power drawn by load operating at the old power factor, $\cos \phi_1$

$= P_2 \tan \phi_1 = S \cos \phi_2 \tan \phi_1$

Reactive power supplied by the plant

$= \text{Capacity of the plant in kVA} \sin \phi_2 = S \sin \phi_2$

since kVA rating of the plant remains the same

kVAR rating of the pf correction equipment

= Reactive power drawn by load – reactive power supplied by the plant

$= S \cos \phi_2 \tan \phi_1 - S \sin \phi_2 = S (\cos \phi_2 \tan \phi_1 - \sin \phi_2)$

Annual cost on pf correction equipment

$$= \text{Rs } y \, S (\cos \phi_2 \tan \phi_1 - \sin \phi_2) \quad ...(17.9)$$

where y is the annual cost per kVAR rating of power factor correction equipment.

Power factor correction equipment will be cheaper if annual cost on it is less than annual cost on account of increasing the generating capacity.

$$\text{or } Sy (\tan \phi_1 \cos \phi_2 - \sin \phi_2) < Sx \frac{\cos \phi_2 - \cos \phi_1}{\cos \phi_1}$$

$$\text{or } y (\tan \phi_1 \cos \phi_2 - \sin \phi_2) < x \frac{\cos \phi_2 - \cos \phi_1}{\cos \phi_1}$$

In the limiting case, the maximum cost of pf correction equipment, which would justify its installation, should be equal to cost on account of increasing the generating plant capacity.

$$\text{i.e., } y (\tan \phi_1 \cos \phi_2 - \sin \phi_2) = \frac{x(\cos \phi_2 - \cos \phi_1)}{\cos \phi_1}$$

$$\text{or } y \left(\frac{\sin \phi_1 \cos \phi_2 - \sin \phi_2 \cos \phi_1}{\cos \phi_1} \right) = x \frac{\cos \phi_2 - \cos \phi_1}{\cos \phi_1}$$

$$\text{or } y \sin (\phi_1 - \phi_2) = x (\cos \phi_2 - \cos \phi_1)$$

$$\text{or } y = \frac{x (\cos \phi_2 - \cos \phi_1)}{\sin (\phi_1 - \phi_2)} \quad ...(17.10)$$

Example 17.27. A power plant is working at its maximum kVA capacity with a lagging pf of 0.7. It is now required to increase its kW capacity to meet the demand of additional load. This can be done

***(i)* By increasing the pf to 0.85 (lagging) by pf correction equipment.**

or

***(ii)* By installing additional generation plant costing Rs 800 per kVA.**

What is the maximum cost per kVA of pf correction equipment to make its use more economical than the additional plant ?

[U.P. Technical Univ. Power Station Practice 2008-09]

Solution : Let the rating of the power plant be S kVA

Increase in load

$= P_2 - P_1 = S(\cos \phi_2 - \cos \phi_1) = S(0.85 - 0.7) = 0.15 \text{ S}$

Increase in capacity of the generating plant

$$= \frac{P_2 - P_1}{\text{Plant } pf} = \frac{0.15 \text{ S}}{0.7} = 0.2143 \text{ S}$$

Cost of additional plant = Rs 800 × 0.2143 S = Rs 171.43 S

Let the rate of interest and depreciation be m per cent per annum

Increase in annual cost due to increase in capacity of power plant

$$= \text{Rs } \frac{m}{100} \times 171.43 \text{ S} = \text{Rs } 1.7143 \text{ mS} \quad ...(i)$$

Reactive power drawn by load

$= P_2 \tan \phi_1 = 0.85 \text{ S} \tan (\cos^{-1} 0.7) = 0.867 \text{ S}$

$\because P_2 = S \cos \phi_2$ and load pf is the same

Reactive power supplied by the plant

$= S \sin \phi_2 = S \sin (\cos^{-1} 0.85) = 0.5268 \text{ S}$

$\because$ new pf of the plant is 0.85

kVAR rating of power factor correction equipment

= kVAR drawn by load – kVAR supplied by the plant

= 0.867 S – 0.5268 S = 0.3402 S

If y is the cost per kVAR of pf correction equipment then total cost on pf correction equipment

$= 0.3402y \text{ S}$

Assume rate of interest and depreciation the same as in case of plant *i.e.*, m per cent per annum

The annual cost on pf correction equipment

$$= \text{Rs } \frac{m}{100} \times 0.3402 \ y \text{ S} = \text{Rs } 0.003402 my \text{ S} \quad ...(ii)$$

In limiting case both the cost must be equal

i.e., $0.003402 \ m \ y \text{ S} = 1.7143 \text{ mS}$

$$\text{or } y = \frac{1.7143}{0.003402} = \text{Rs } 503.60 \text{ per kVAR} \ \textbf{Ans.}$$

Alternative Method. Annual cost per kVA of generating plant,

$$x = \text{Rs } 800 \times \frac{m}{100} = \text{Rs } 8 \ m$$

where m is the rate of interest and depreciation per annum

Original power factor,

$\cos \phi_1 = 0.707$ and therefore $\phi_1 = \text{Cos}^{-1} 0.7 = 45.573°$

Improved power factor,

$\cos \phi_2 = 0.87$ and therefore $\phi_2 = \text{Cos}^{-1} 0.85 = 31.788°$

Annual cost per kVAR of pf correction equipment,

$$y = \frac{x(\cos \phi_2 - \cos \phi_1)}{\sin(\phi_1 - \phi_2)} = \frac{8m(0.85 - 0.7)}{\sin(45.573° - 31.788°)} = 5.036 \ m$$

Cost per kVAR of power factor correction equipment

$$= \text{Rs } 5.036 \ m \times \frac{100}{m} = \text{Rs } 503.60 \ \textbf{Ans.}$$

17.12. CHOICE OF EQUIPMENT

At the time of purchasing equipment usually several alternatives are made available, each possessing one or more special features. The choice between alternative types of electrical equipment usually depends upon the relative costs of the alternative equipment. In selecting equipment for a particular service, the choice must fall upon that which makes the sum of the fixed charges (annual interest and depreciation on the capital cost of equipment) and the operating charges minimum. It is usually found that a saving can be affected only by incurring increased fixed costs and vice versa, for instance, an electric motor of a particular type and size of lower operating efficiency will usually cost less than a similar motor of higher efficiency, although the latter will have a lower operating cost on account of its lower loss and may, therefore, have a lower operating cost.

Typical problems on choice of a transformer and motors are worked out below.

Example 17.28. It is necessary to choose a transformer to supply a load which varies over 24 hour period in the manner given below :

500 kVA for 4 hours, 1,000 kVA for 6 hours, 1,500 kVA for 12 hours and 2,000 kVA for the rest of the period. Two transformers each rated at 1,500 kVA have been quoted. Transformer I has iron loss of 2.7 kW and full-load copper loss of 8.1 kW, while transformer II has an iron loss and full-load copper loss of 5.4 kW each.

(*i*) Calculate the annual cost of supplying losses for each transformer, if electrical energy costs Rs 2.5 per kWh.

(*ii*) Determine which transformer should be chosen if the capital cost of the transformer I is Rs 25,000 more than that of transformer II and annual charges of interest and depreciation are 10%.

(*iii*) What difference in capital cost will reverse the decision made in (*ii*) above ?

Solution : Transformer I

Energy loss per day due to iron loss = 24 × 2.7 = 64.8 kWh

Energy loss per day due to copper loss

$$= \left(\frac{500}{1,500}\right)^2 \times 8.1 \times 4 + \left(\frac{1,000}{1,500}\right)^2 \times 8.1 \times 6 + \left(\frac{1,500}{1,500}\right)^2 \times 8.1 \times 12 + \left(\frac{2,000}{1,500}\right)^2 \times 8.1 \times (24 - 4 - 6 - 12)$$

= 3.6 + 21.6 + 97.2 + 28.8 = 151.2 kWh

Total energy loss per annum

= (64.8 + 151.2) × 365 = 78,840 kWh

Annual cost of supplying losses

= Rs 2.50 × 78,840 = Rs 197,100 **Ans.**

Transformer II

Energy loss per day due to iron loss = 24 × 5.4 = 129.6 kWh

Energy loss per day due to copper loss

$$= \left(\frac{500}{1,500}\right)^2 \times 5.4 \times 4 + \left(\frac{1,000}{1,500}\right)^2 \times 5.4 \times 6 + \left(\frac{1,500}{1,500}\right)^2 \times 5.4 \times 12 + \left(\frac{2,000}{1,500}\right)^2 \times 5.4 \times 2$$

= 2.4 + 14.4 + 64.8 + 19.2 = 100.8 kWh

Total energy loss per annum = (129.6 + 100.8) × 365 = 84,096 kWh

Annual cost of supplying energy losses

= Rs 2.5 × 84,096 = Rs 210,240 **Ans.**

(*ii*) The capital cost of transformer I

= Rs 25,000 more than that of transformer II

Extra annual charges in case transformer I is chosen

$$= \text{Rs } 25,000 \times \frac{10}{100} = \text{Rs } 2,500$$

Annual saving in energy cost due to losses

= Rs 210,240 – Rs 197,100 = Rs 13,140

Since annual saving in cost of supplying energy losses is much more than extra annual charges due to extra cost of transformer I, therefore, transformer I will be chosen **Ans.**

(*iii*) Decision taken in (*ii*) above will be reversed *i.e.*, transformer II will be chosen in case annual charges due to extra capital cost of transformer I exceeds the annual saving in the energy cost due to losses.

i.e., $x \times \frac{10}{100} >$ Rs 13,140

where x is the capital cost of transformer I in excess

or $x >$ Rs 13,140 $\times \frac{100}{10}$ or x exceeds Rs 131,400 **Ans.**

Example 17.29. The motor of a 30 hp condensate pump has got burnt beyond economical repairs. Two alternatives have been proposed to replace it by :

Motor A-Cost = Rs 150,000 ; η at full load = 90% ; η at half load = 86%

Motor B-Cost = Rs 100,000 ; η at full load = 85% ; η at half load = 82%

The life of each is 20 years and its salvage value is 10% of the initial cost. The rate of interest is 5% annually. The motor operates at full load for 25% of the time and at half load for the remaining period. The annual maintenance cost of motor A is Rs 10,500 and that of motor B is Rs 6,000. The energy rate is Rs 2.50 per kWh. Which motor will you recommend and why ?

Solution : Motor A

Capital cost = Rs 150,000

Annual interest on capital cost

$$= \text{Rs } 150{,}000 \times \frac{5}{100} = \text{Rs } 7{,}500.00$$

Annual depreciation charges

$$= \frac{\text{Original cost} - \text{salvage value}}{\text{Life of motor in years}}$$

$$= \text{Rs } \frac{150{,}000 - \frac{10}{100} \times 150{,}000}{20}$$

= Rs 6,750.00

Annual maintenance cost = Rs 10,500.00

Energy consumed per annum

$$= \frac{30 \times 735.5}{0.9 \times 1{,}000} \times 0.25 \times 8{,}760 + \frac{\frac{1}{2} \times 30 \times 735.5}{0.86 \times 1{,}000} \times 0.75 \times 8{,}760$$

= 53,691 + 84,283 = 137,974 kWh

Annual energy cost = Rs 2.50 × 137,974 = Rs 344,935

Total annual cost = Rs (7,500 + 6,750 + 10,500 + 344,935)
= Rs 369,685.00

Motor B Capital cost = Rs 100,000

Annual interest on capital cost

$$= \text{Rs } 100{,}000 \times \frac{5}{100} = \text{Rs } 5{,}000.00$$

Annual depreciation charges

$$= \text{Rs } \frac{100{,}000 - \frac{10}{100} \times 100{,}000}{20} = \text{Rs } 4{,}500$$

Annual maintenance charges = Rs 6,000

$$\text{Energy input per annum} = \frac{30 \times 0.7355}{0.85} \times 0.25 \times 8{,}760 + \frac{\frac{1}{2} \times 30 \times 0.7355}{0.82} \times 0.75 \times 8{,}760$$

= 56,850 + 88,395 = 145,245 kWh

Annual energy cost = Rs 2.50 × 145,245 = Rs 363,112

Total annual cost = Rs (5,000 + 4,500 + 6,000 + 363,112)
= Rs 378,612

Since total annual cost of motor B is more by Rs (378,612 – 369,685) *i.e.*, Rs 8,927 than that of motor A, so motor A is recommended.

Example 17.30. A consumer requires an induction motor of 50 hp (metric). He is offered two motors of the following specifications :

Motor (A)-efficiency 88% and power factor 0.9 ;

Motor (B)-efficiency 90% and power factor 0.81 ;

The consumer is being charged on a two part tariff of Rs 1,400 per kVA of the maximum demand plus Re 1.0 per unit. The power factor of the motor B is to be raised to 0.89 by installing condensers. The motor B costs Rs 3,000.00 less than A.

The cost of condensers is Rs 1,200 per kVAR. Determine which motor is more economical and by how much. Assume rate of interest and depreciation 10% and working hours of motors 2,400 hours in a year.

Solution : Motor input $= \frac{\text{Motor output}}{\eta} = \frac{50 \times 0.7355}{0.88} = 41.8$ kW

Motor input in kVA $= \frac{41.8}{pf} = \frac{41.8}{0.9} = 46.43$

Annual energy consumption
= Motor input in kW × number of working hours/annum
= 41.8 × 2,400 = 100,320 kWh

Annual energy charges
= Rs 1.0 × 100,320 + Rs 1,400 × 46.43 = Rs 165,320

Fixed annual charges of motor A in excess over those of motor B

$$= \text{Rs } 3{,}000 \times \frac{10}{100} = \text{Rs } 300.00$$

Motor B : Motor input $= \frac{50 \times 0.7355}{0.9} = 40.86$ kW

Motor input in kVA with condensers $= \frac{40.86}{0.89} = 45.91$

Energy consumed per annum = 40.86 × 2,400 = 98,064 kWh

Annual energy charges
= Rs 1,400 × 45.91 + Rs 1.0 × 98,064 = Rs 162,338

Leading kVAR supplied by the condensers
$= 40.86 [\tan (\cos^{-1} 0.81) - \tan (\cos^{-1} 0.89)]$
= 40.86 (0.724 – 0.512) = 8.66

Capital cost of condensers = Rs 1,200 × 8.66 = Rs 10,392

Annual interest and depreciation charges

$$= \text{Rs } 10{,}392 \times \frac{10}{100} = \text{Rs } 1{,}039$$

Total annual charges
= 162,338 + 1,039 = 163,377.00

Motor B will be economical and annual saving will be Rs (165,320 + 300 – 163,377) = Rs 2,243.00 **Ans.**

Exercises

1. **What are the main points to be considered while fixing the tariff ?**
2. **What are the main objectives in framing a tariff ? Discuss various types of tariffs.**

[U.P. Technical Univ. Power Station Practice 2002-03]

3. What are the different types of tariff used by the electricity authority ? Discuss any two of them and indicate the type of consumers where such tariffs are used.
[U.P. Technical Univ. Power Station Practice 2007-08]
4. Discuss different types of tariffs used for charging the consumers of electrical energy.
[G.B. Technical Univ. Power Station Practice 2010-11]
5. Explain giving examples :
(*i*) Flat rate tariff (*ii*) Block rate tariff (*iii*) Two part tariff (*iv*) Power factor tariff.
[Pb. Univ. Electrical Power II (Generation) Dec. 1995]
6. Explain the straight meter rate tariff and block meter rate tariff.
[Pb. Technical Univ. Generation of Electric Power May 2008]
7. What do you understand by tariff? Explain briefly various types of tariffs.
[Pb. Technical Univ. Generation of Electric Power Dec. 2010]
8. Discuss the different types of tariffs used for charging the consumers electrical energy.
[Pb. Technical Univ. Generation of Electric Power May 2009]
9. What is the special feature of two part tariff ? For which category of consumers is it used ?
Discuss the importance of encouraging customers to use electricity during off-peak hours.
[Pb. Univ. Electrical Power II (Generation) Dec. 1993]
10. What is the objective of tariff ? What type of tariff is employed for domestic consumers ? Why this tariff is not employed for bulk consumers ? Suggest and explain the tariff which encourages the consumers to keep load factor and power factor high.
11. In modern LT distribution stations the VAR meters are installed. What is the mechanism of charging the tariff related to reactive power and how the above can be effectively installed ?
[U.P. Technical Univ. Power Station Practice 2005-06]
12. A distribution company wants to charge and penalize the consumers in case of excessive loads of poor power factor. What charges should be made in present existing substation so as to meet the above ?
[U.P. Technical Univ. Power Station Practice 2005-06]
13. Give the basis for expressing the service cost of an electric supply as A + B kW + C kWh and explain the factors on which A, B and C depend.
14. What are objective of tariffs? What is the general form of tariff? Describe the straight meter rate and block meter rate tariffs. What do you mean by spot pricing?
[Generation of Elecrical Power Rajasthan Univ. 1997, Rajasthan Technical Univ. 2011]
15. What is spot pricing? What is its importance?
[Pb. Technical Univ. Generation of Electric Power Dec. 2008]
16. What are different types of loads and draw their chronological load curves.
[Pb. Technical Univ. Generation of Electric Power Dec. 2007]
17. What are the disadvantages of low power factor? Why do some agencies put a penalty for low power factor?
[Generation of Electrical Power Rajasthan Univ. 2004 ; Rajasthan Technical Univ. 2007]
18. What is the effect of low power factor ? Explain the different methods to improve the power factor of the system.
[U.P. Technical Univ. Power Station Practice 2003-04]
19. What are the causes and effects of poor power factor ? Discuss any one method to improve the power factor.
[U.P. Technical Univ. Power Station Practice 2007]
20. What are the advantages of power factor improvement?
[Pb. Technical Univ. Generation of Electric Power May 2009]
21. What are the advantages of power factor improvement. Explain how a synchronous capacitor can improve the power factor.
[U.P. Technical Univ. Power Station Practice 2007-08]
22. Explain the necessity of improving power factor.
[G.B. Technical Univ. Power Station Practice 2010-11]
23. Compare the advantages and disadvantages of using a synchronous condenser and a static capacitor for power factor improvement.
[Rajasthan Technical Univ. Generation of Electrical Power 2011]
24. What do you understand by power factor ? Explain the necessity of improving power factor. Also discuss the factor which determines the economical limit of power factor correction.
[U.P. Technical Univ. Power Station Practice 2009-10]
25. What factors determine the economical limit of power factor correction ? Show that the economical limit to which the power factor of a lagging power factor load can be raised is independent of the original value of the power factor if the tariff consists of a fixed charge per kVA of maximum demand plus a flat rate per kWh.
26. Calculate the most economical power factor when kW demand is constant. **[Rajasthan Technical Univ. Generation of Electrical Power 2008, 2009, 2011]**
27. Explain how the installation of a pf improvement device can augment the kW capacity of a generating station of a given kVA capacity.
Derive the conditions under which the installation of a pf improvement plant for increasing the kW capacity of a generating station is preferable to installing additional generating equipment.
28. List and explain the factors that influence the economic choice of equipment.
29. Write short notes on the following :
(*i*) Tariffs **[Pb. Univ. December 1993]**
(*ii*) Tariffs methods of electrical energy **[M.D. Univ. May 2007]**
(*iii*) Hopkinson demand rate **[Rajasthan Univ. 1995, 2001]**
(*iv*) Two part tariff. **[West Bengal Univ. of Technology 2010]**
(*v*) Power factor tariff.
(*vi*) Power factor improvement. **[Pb. Univ. Janvary 1993]**
(*vii*) Power factor improvement using synchronous condensers. **[Rajasthan Univ. 2002; 2005]**
(*viii*) Necessity of power factor improvement.
(*ix*) Economic limits of power factor improvement.
(*x*) Choice of equipment.

Short Answer Type Questions With Answers

Q. 1. What is meant by tariff? **[Rajasthan Univ. 2003]**

Ans. Tariff means the schedule of rates or charges. Tariff, in case of electric supply, means the schedule or rates framed for supply of electrical energy to different classes of consumers.

Q. 2. Name the different types of tariffs. **[Pb. Technical Univ. December 2008, May 2009]**

Ans. Different types of tariffs are :

(*i*) Flat demand tariff (*ii*) Simple tariff (*iii*) Flat rate tariff (*iv*) Step rate tariff (*v*) Block rate tariff (*vi*) Hopkinson demand or two part tariff (*vii*) Maximum demand tariff or Wright demand rate (*viii*) Power factor tariffs (*ix*) Three part tariff or Doherty rate tariff (*x*) Off peak tariff.

Q. 3. What is tariff? What is general form of tariff? **[Rajasthan Univ. 2005]**

Ans. Tariff means the schedule of rates or charges. Tariff, in case of electric supply, means the schedule or rates framed for supply of electrical energy to different classes of consumers.

General form of tariff is

Total energy charges, e = Rs $a + b$ kW + c kWh

where a is a constant charge made each billing period.

b = Unit charge in Rs per kW of metered maximum demand in kW during billing period. In some cases it is also charged in Rs per kVA so that the consumers are penalised for poor power factor.

and c = Unit charge for energy in Rs per kWh of energy consumed.

Q. 4. Why is tariff less for power load than lighting load? **[Rajasthan Univ. 2003]**

Ans. Power load spreads the fixed charges over greater number of units, thus reduces the overall cost of generation of electrical energy. Lighting load has low load factor and therefore cost of generation of electrical energy is higher. These are the reasons that tariff is less for power load than lighting load.

Q. 5. What type of tariff is employed for domestic consumers ?

Ans. Block rate tariff.

Q. 6. What is two part tariff ?

Ans. In the two part tariff, the total charge to be made to the consumer is divided into two components namely fixed charge (proportional to connected load or maximum demand) and the running or operating charges (proportional to the units consumed).

Q. 7. For which category of consumers two part tariff is used ?

Ans. Two part tariff is mostly applicable to medium industrial consumers.

Q. 8. How do demand factor, load factor and diversity factor in a power system affect the fixation of tariffs? **[Rajasthan Univ. 2004]**

Ans. The tariff must cover the following items :

(*i*) Recovery of cost of capital investment in generating, transmitting and distributing equipment.

(*ii*) Recovery of cost of operation, supplies and maintenance of equipment.

(*iii*) Recovery of cost of metering equipment, billing, collection costs and miscellaneous services and

(*iv*) A satisfactory return on the total capital investment.

The unit cost depends quite substantially on the load factor. The fixed cost remains constant irrespective of the load factor. At low load factor the fixed cost is shared by a smaller number of units of energy resulting in a relatively higher energy cost. At high load factor, the same fixed cost is shared by a large number of units of energy thereby reducing the unit energy cost.

The demand factor, being the ratio of maximum demand to connected load, determines the capacity of the system and hence the cost of power equipment required to serve a given load. Thus demand factor affects the fixtation of tariff.

Diversity between the loads of different consumers and different areas leads to reduction in peak demand resulting in reduction in generation, transmission and distribution costs. So the fixation of tariff is affected.

Q. 9. Explain reasons, why power factor tariff is imposed ? **[Rajasthan Univ. 2005]**

Ans. Since the efficiency of plant and equipment depends upon the power factor, therefore, in order to increase the utility of plant and equipment to the maximum, the plant must be operated at the most economical power factor. That is why, sometimes consumers are penalised for poor power factor by imposing the power factor tariff.

Q. 10. Which type of tariff encourages the consumers to keep the load factor and power factor high ?

Ans. KVA maximum demand tariff encourages the consumers to keep the load factor and power factor high.

Q. 11. Can a power factor be included in a tariff ?

Ans. Yes, power factor is included in power factor tariffs such as kVA maximum demand tariff, kWh and kVARh tariff and sliding scale or average power factor tariff.

Q. 12. What type of tariff is usually applied to bulk consumers.

Ans. Three part tariff.

Q. 13. What are the advantages of using industrial/agricultural loads during night hours ?

Ans. During night hours a large proportion of the generating and distribution equipment remains idle due to extremely low demand. In case industrial and agricultural consumers are encouraged to use electricity during night and other off-peak hours by giving a special discount, the energy can be supplied without incurring any additional capital cost.

Q. 14. What are the different types of loads ?

Ans. The main types of loads on a power system are domestic, commercial, agricultural, industrial, municipal, traction etc.

Q. 15. Draw an approximate load curve for domestic load during 24 hours. **[Pb. Technical Univ. May 2008]**

Ans. Refer to Fig. 17.1

Q. 16. Define power factor. **[Pb. Technical Univ. May 2009]**

Ans. The cosine of the angle between voltage and current in an ac circuit is known as power factor.

Q. 17. Define active, reactive and apparent power. **[Pb. Technical Univ. December 2008]**

Ans. The product of voltage V, current I and power factor cos ϕ *i.e.* VI cos ϕ is known as active power.

The product of voltage V, current I and sine of phase angle ϕ *i.e.*, VI sin ϕ is known as reactive power.

The product of voltage V and current I *i.e.*, VI is known as apparent power.

These three powers are related as follows :

$$\text{Apparent power} = \sqrt{(\text{Active power})^2 + (\text{reactive power})^2}$$

Q. 18. Why is the power factor not more than unity? **[Pb. Technical Univ. December 2010]**

Ans. Power factor being the cosine of the phase angle between voltage V and current I can never exceed unity.

Q. 19. What are the effects of low power factor. **[Pb. Technical Univ. December 2007]**

Ans. Refer to Art 17.6.

Q. 20. What is the effect of power factor on the cost of generation? **[Pb. Technical Univ. December 2010]**

Ans. Lower power factor means higher kVA for the given load, therefore, large sized generators, transformers, switchgear, bus-bars, more energy loss and poor voltage regulation thereby increasing the cost of generation.

Q. 21. What is the need of improving power factor? **[Pb. Technical Univ. December 2009]**

Ans. Low power factor causes the ratings of generators and transformers, cross-sectional area of the bus-bars and the contact surface of the switchgear, the size of the feeders and distributors, energy losses, voltage drops in generators, transformers, transmission lines and distributors to increase. So power factor needs to be improved to reduce the generation, transmission and distribution costs.

Q. 22. What are the advantages of power factor improvement? **[Pb. Technical Univ. December 2007]**

Ans. The advantages of good (or improved) power factor are : (*i*) reduction in load current ; (*ii*) increase in voltage level across the load ; (*iii*) reduction in energy losses in the system (generators, transformers, transmission lines and distributors) due to reduction in load current ; (*iv*) reduction in kVA loading of the generators and transformers which may relieve an overloaded system or release capacity for additional growth of load and (*v*) reduction in kVA demand charge for large consumers.

Q. 23. What is the importance of power factor? **[Pb. Technical Univ. Dec. 2010]**

Ans. Low power factor leads to high capital cost for generators, transformers, switchgears, transmission lines, distributors and cables. So the power factor is very important.

Q. 24. Where are shunt capacitors located? **[Pb. Technical Univ. December 2009]**

Ans. Refer to Art. 17.8.3.

Q. 25. Should the power factor to be raised to 1. **[Pb. Technical Univ. May 2009]**

OR

Why it is not economical to raise the power factor to unity? **[Rajasthan Univ. 2002, 2004]**

Ans. As the power factor approaches unity, the capacity of the power factor improving equipment increases more rapidly i.e. the power factor of an installation can be improved from 0.8 (lagging) to 0.9 (lagging) by a much small capacity kVAR than which will be required to raise the power factor from 0.9 to unity. So it is not economical to raise the power factor to unity.

Q. 26. What factors determine the economical limit of power factor correction ?

Ans. Economical limit of power factor correction is governed by the relative costs of the supply and power factor correcting equipment.

Problems

1. Calculate the annual energy cost for industrial consumer which takes a load of 20 kW for 1 hour per day, 150 kW for 7 hours per day and 50 kW for 8 hours per day. The tariff in force is Rs 300 per kW and Rs 1.50 per kWh. Assume 6 working days in a week. **[Ans.** Rs 732,960**]**
2. The yearly consumption of an industrial consumer is 5×10^6 kWh with a maximum demand of 1,500 kW. Compare the annual cost of electrical energy under the following two tariffs.
   (*a*) Rs 1,500 per kW plus 40 paise per kWh.
   (*b*) Flat rate of Rs 1.50 paise per kWh.
   [Ans. (*a*) Rs 4,250,000 (*b*) Rs 7,500,000**]**
3. Load factor of a consumer is 35% and monthly consumption is 504 kWh. If the rate of electricity is Rs 180 per kW of maximum demand plus Rs 2.00 per kWh, find the monthly bill and the average cost per kWh.
   [U.P. Technical Univ. Power Station Practice 2004-05]
   [Ans. Rs 1,037.60**]**
4. An industrial consumer having a maximum demand of 100 kW, maintains a load factor of 60%. The tariff rates are Rs 1,500 per kVA of maximum demand per annum plus paise 30 per kWh of energy consumed. If the average pf is 0.8 lagging, calculate, the total energy consumed per annum and annual bill. **[Ans.** 525,600 kWh ; Rs 345,180**]**
5. Compute the cost of electrical energy and average unit cost for consuming 450 kWh under block rate tariff quoted as below.
   First 50 kWh at Rs 3.80 per kWh
   Next 50 kWh at Rs 3.20 per kWh
   Next 50 kWh at Rs 2.80 per kWh
   Next 50 kWh at Rs 2.40 per kWh
   Excess over 200 kWh at Rs 2.10 per kWh.
   [Ans. Rs 1,135.00 ; Rs 2.52/kWh**]**
6. Calculate the number of units to be consumed which will justify the two part tariff over flat rate tariff from the following data :
   Maximum demand 5 kW, Two part tariff Rs 1,200 per annum per kW of maximum demand plus Rs 2.10 per unit consumed. Flat rate tariff is Rs 2.60 per unit.
   [Ans. 12,000 units**]**
7. A light industry has a maximum demand of 100 kW. Two alternative tariffs are as follows :
   (*i*) A fixed charge of Rs 900 per kW plus a running charge of Rs 1.50 per unit.
   (*ii*) A charge of Rs 1.80 per unit flat.
   If the factory runs for 3,000 hours with a load factor of 80% which tariff is economical ?
   [Ans. Flat rate is cheaper for Rs 18,000**]**
8. The maximum demand of a consumer is 1 MW and the units consumed per annum are 500,000 kWh. Calculate the reduction in cost if the power factor is raised from 0.5 to 0.8. The tariff is Rs 900 per annum per kW demand plus Rs 1.50 per kWh plus Re 0.30 per kVAh reactive. Calculate the flat rate tariff for a unity power factor load.
   [Ans. Rs 147,310.00; Rs 3.30 per kWh**]**
9. A supply is offered on the basis of fixed charges of Rs 900 per annum per kW plus 90 paise per unit or alternatively at the rate of Rs 1.80 per unit for the first 400 units per annum and Rs 1.50 per unit for all the additional units. The maximum demand is 1 kW. Find the number of units consumed per annum for which the cost under these two tariffs becomes the same. **[Ans.** 1,300 units**]**
10. A consumer, whose connected load is 400 kVA, consumes 1×10^6 kWh per annum and the maximum demand registered is 250 kVA. He is offered the following two alternative tariffs :
    Tariff A : Rs 1,000 per annum per kVA of connected load + first 150 kWh per kVA of connected load at Rs 3.0/kWh, next 50,000 kWh at Rs 2.00 per kWh and the rest at Re 1.0/kWh.
    Tariff B : Rs 1,200 per annum per kVA of maximum demand + first 150 kWh per kVA of maximum demand at Rs 3.0 per kWh, next 50,000 kWh at Rs 2.0/kWh and the rest at Re 1.0/kWh.
    Which tariff is economical and what is the annual gain ?
    [Ans. Tariff B is cheaper ; Annual gain = Rs 145,000.00**]**
11. A consumer has a yearly consumption of 200,000 units. The tariff is Rs 1,000.00 per annum per kW of maximum demand + Rs 2.0 per kWh. Find out the total annual cost and overall cost per unit (*i*) consumption were 200,000 units per annum with the load factor of 25% (*ii*) consumption were reduced by 25% with the same load factor.
    [Ans. (*i*) Rs 491,324.00 and Rs 2.46/kWh
    (*ii*) Rs 368,493.00 and Rs 2.46/kWh**]**
12. Calculate the minimum two part tariff to be charged from the following data :
    Generating cost per unit – Re 1.0

Annual generating cost per kW of maximum demand – Rs 1,250.00
Units supplied annually – 4,380 × 10^4
Annual load factor of the station – 50%
Annual charges for distribution – Rs 25 × 10^5
Diversity factor for distribution work – 1.25 and the distribution losses are 10%.

[**Ans.** Rs 1,250 per kW of maximum demand plus Rs 1.10/kWh]

13. Two systems of tariff are available for a factory, working 8 hours a day for 300 working days in a year.

(*i*) HV supply at Rs 1.25 per unit plus Rs 1,350 per annum per kVA of maximum demand.

(*ii*) LV supply at Rs 1,500 per annum per kVA of maximum demand plus Rs 1.375 per unit.

The factory has an average load of 200 kW at 0.8 pf and a maximum demand of 250 kW at the same power factor. The hv equipment costs Rs 1,250.00 per kVA and losses can be taken as 4%. Interest and depreciation charges are 12%. Calculate the difference in annual costs between the two systems. [**Ans.** HV supply is cheaper for Rs 15,470.00]

14. An industrial load requiring 500 kW at 0.8 pf can be supplied on the following tariff rates :

(*i*) HT supply at Rs 1,500 per kVA per annum + 60 paise per unit.

(*ii*) LT supply at Rs 1,650 per kVA per annum + 75 paise per unit.

Transformer and switchgear for ht supply costs Rs 1,125 per kVA, the full load losses being 2.5%. The fixed charges are 25 per cent per annum of the capital cost of ht plant. If the installation works on full load, find the number of working hours per week above which ht supply is cheaper. There are 48 weeks in a year. [**Ans.** 34.2 hours per week]

15. A synchronous motor having a power consumption of 50 kW is connected in parallel with a load of 200 kW having pf 0.8 lagging. The excitation of motor is adjusted until the combined pf becomes 0.9 lagging. Calculate the kVA input to the motor. [**Ans.** 57.76 kVA]

16. Describe the objectives and different forms of the tariff. The monthly readings of a consumer meter are as :
Maximum demand 50 kW. Energy consumed = 36,000 kWh and reactive energy = 23,400 kVAh.
If the tariff is Rs 80 per kW of maximum demand plus 8 paise per unit plus 0.5 paise per unit for each 1% of pf below 87%, calculate the monthly bill of the consumer.

[U.P. Technical Univ. Power Station Practice 2007]
[**Ans.** Rs 7,448]

17. It is desired to correct the power factor to 0.95 by means of static condensers connected across each phase of a 3-phase, 400 V, 50 Hz motor installation having a maximum load of 50 kVA at pf of 0.75 lagging. What must be the capacity of each delta-connected capacitor ? [**Ans.** 244 μF]

18. A 340 kW, 3,300 V, 50 Hz, 3-phase star-connected induction motor has full load efficiency of 85% and pf of 0.8 lagging. It is desired to improve the power factor to 0.96 lagging by using a bank of three capacitors. Calculate (*i*) The kVA rating of the condenser bank (*ii*) The capacitance of each limb of the condenser bank connected in delta (*iii*) The capacitance of each capacitor, if one each of the limbs of the delta-connected condenser bank is formed by using 6 similar 3,300 V capacitors.

[**Ans.** (*i*) 183.333 (*ii*) 17.86 μF (*iii*) 2.977 μF]

19. Derive the most economic angle of lag at which a consumer should take power when the charge is based on a fixed sum per kVA of maximum demand plus a flat rate per kWh.

Calculate this value when the tariff is Rs 1,000 per kVA of maximum demand plus a flat rate per kWh. The annual cost of condenser and additional equipment for power factor improvement is 10% of the capital cost of Rs 800 per kVA of such plant. [**Ans.** 0.9968 lagging]

20. A consumer is taking a load of 20 kW at 0.8 pf (lagging).

(*i*) Find the rating of the capacitor to raise the pf to 0.95 lagging.

(*ii*) If a phase advancing device is used which takes current at leading power factor of 0.1, find the rating of the device.

[U.P. Technical Univ. Power Station Practice 2002-03]
[**Ans.** (*i*) 8.426 kVA, (*ii*) 8.468 kVA]

21. A consumer has an annual consumption of 1 million kWh. The tariff is Rs 1,000 per year per kVA of maximum demand plus Re 1.0 per kWh. Determine the annual power cost if the power factor is 0.75 and the load factor is 50%. What would be most economical value of power factor, if the loss free phase advancing apparatus costs Rs 1,000 per kVAR, the rate of interest and depreciation 20% per annum.

[**Ans.** Rs 1,304,400; 0.98]

22. The power factor of an industrial three-phase load of 490 kW is to be improved from 0.7 lagging to 0.97 lagging by connecting loss free delta-connected capacitors across the 6.6 kV, 50 Hz supply. The cost of suitable capacitors and control gear is Rs 200 per kVAR and the annual tariff charge is Rs 120 per kVA maximum demand. The annual interest and depreciation charges are 15%. Calculate (*a*) the total kVAR rating of capacitors required. (*b*) the required value of capacitance per phase (*c*) the net annual saving.
[GATE 1992] [**Ans.** (*a*) 377 kVAR (*b*) 9.183 μF (*c*) Rs 12,072]

23. For increasing the kW capacity of a power plant working at 0.7 lagging power factor the necessary increase in power can be obtained by raising the power factor to 0.9 or by installing additional plant. What is the maximum cost per kVA of power factor correction apparatus to make its use more economical than addtional plant at Rs 2,400 per kVA?

[**Ans.** 1,422/kVA]

24. A system is working at its maximum kVA capacity with a lagging pf of 0.7. An anticipated increase of load can be met by one of the two following methods.

(*i*) By raising the pf of the system to 0.866 by installing phase advancing equipment.

(*ii*) By installing extra generating plant. If the total cost of generating plant is Rs 100 per kVA, estimate the limiting cost per kVA of phase advancing equipment to make its use more economical than the additional generating plant. Interest and depreciation charges may be assumed 10% in each case.

[U.P. Technical Univ. Power Station Practice 2007-08]
[**Ans.** Rs 61.88]

25. Tender A and B for a 1,000 kVA 0.8 pf transformer are :
A—Full-load efficiency = 98.3% and core loss = 7 kW at rated voltage.
B—Full-load efficiency = 98.7% and core loss = 4 kW but costs Rs 5,000.00 more than A.
The service needed is 1,800 hours per annum at 1,000 kVA, 600 hours at 600 kVA, 400 hours at 15 kVA and remaining period shutdown. Take the annual charges for interest and depreciation at 12½% of capital costs and energy costs at 8 paise per kWh. Which is better tender and what would be the annual savings? **[Allahabad Univ., 1967]**
[**Ans.** Tender B is cheaper; Rs 211.00]

❑❑❑

OBJECTIVE TYPE QUESTIONS

MULTIPLE CHOICE TYPE QUESTIONS

Note: Each of the following questions comprises multiple responses (two or more than two answers). Select the correct answers. In case you feel that there is more than one correct answer, select the answer which you consider the best. In any case choose ONLY ONE answer or response for each question.

1. The basic elements of an electric drive are
(a) electric motor and the transmission system.
(b) electric motor, the transmission and control system.
(c) the transmission and control system.
(d) electric motor and conversion equipment.

2. Electric drive is becoming more and more popular because
(a) it is simple, clean, compact and reliable.
(b) it provides easy and smooth control, flexibility in layout, easy starting and facility for remote control.
(c) it is cheaper in initial as well as in maintenance cost.
(d) all of the above.

3. The main drawback of electric drive is that
(a) it is cumbersome drive.
(b) it is costlier in initial as well as in maintenance cost.
(c) electrical power supply failure makes the drive standstill.
(d) all of the above.

4. An existing workshop is to be changed over from an engine drive to an electric drive. The type of drive likely to be adopted is
(a) individual drive. (b) group drive.
(c) multi-motor drive. (d) any of the above.

5. The type of drive used for a paper mill requiring constant speed operation and flexibility of control is
(a) group drive. (b) multi-motor drive.
(c) individual or multi-motor drive.
(d) individual drive.

6. Introduction of automation in production processes has become possible only because of use of
(a) an electric drive. (b) group drives.
(c) individual and multi-motor drives.
(d) individual drive.

7. The selection of an electric motor is governed by
(a) nature of load to be handled.
(b) environmental conditions.
(c) nature of electric supply available. (d) all of the above.

8. DC motors are still preferred for use in
(a) electric excavators, steel mills and cranes.
(b) lathes and machine tools.
(c) flour mills and jaw crushers.
(d) paper industry.

9. The least significant electrical characteristic in selection of electric motor for a flour mill is
(a) starting characteristics. (b) braking.
(c) running characteristics. (d) efficiency.

10. The least significant feature while selecting a motor for centrifugal pump is
(a) speed control. (b) power rating of motor.
(c) operating speed. (d) starting characteristics.

11. A typical active load is
(a) hoist. (b) blower. (c) pump. (d) lathe.
[A.M.I.E. Sec. B. Elec. Drives & Their Control Summer 1998]

12. An elevator drive is required to operate in
(a) one quadrant only. (b) two quadrants.
(c) three quadrants. (d) four quadrants.
[A.M.I.E. Sec. B. Electric Drives & Their Control Winter 1996; Summer 2000]

13. Load torque constant at all speeds is represented by a
(a) fan. (b) compressor.
(c) centrifugal pump load.
[A.M.I.E. Sec. B. Elec. Drives & Their Control Winter 1997]

14. An example of a motor having short-time duty is found in
(a) centrifugal pumps. (b) crane drives.
(c) drilling machines. (d) all of the above.
[A.M.I.E. Sec. B. Elec. Drives & Their Control Summer 2000]

15. The load, for which the motor always starts on load is
(a) fan motor. (b) conveyor motor.
(c) flour mill motor. (d) all of the above.

16. The load cycle for a motor driving a power press will be
(a) continuous. (b) variable.
(c) intermittent and variable.
(d) continuous but periodical.

17. The application(s) that need(s) frequent starting and stopping is/are
(a) paper mills. (b) lifts and hoists.
(c) blowers and fans. (d) grinding mills.

18. An electric motor, in jaw crushers, has to often start against
(a) heavy load. (b) normal load.
(c) medium load. (d) low load.

19. The application(s) in which the load on the motor changes in cyclic order is/are
(a) rolling mills. (b) cranes.
(c) electric shovels. (d) all of the above.

20. The machine having heavy fluctuation of load is
(a) lathe. (b) planer.
(c) punching machine. (d) printing machine.

21. Variable speed operation is preferred in
(a) water pump. (b) ceiling fan.
(c) exhaust fan. (d) refrigerator.

22. The application in which the motor is to start with high acceleration is
(a) lifts and hoists. (b) centrifugal pump.
(c) oil expeller. (d) flour mill.

23. While selecting a motor for air-conditioner the feature of utmost importance is
(a) type of enclosure. (b) type of bearing.
(c) noise.
(d) power transmission arrangement.

24. Belt conveyors offer...starting torque.
(a) high (b) medium
(c) low (d) zero

Answers: 1. (b), 2. (d), 3. (c), 4. (b), 5. (c), 6. (c), 7. (d), 8. (a), 9. (b), 10. (a), 11. (a), 12. (d), 13. (b), 14. (b), 15. (d), 16. (d), 17. (b), 18. (a), 19. (d), 20. (c), 21. (b), 22. (a), 23. (c), 24. (a),

25. The characteristic(s) of drive for crane hoisting and lowering is are
(a) fast speed control. (b) smooth movement.
(c) precise control. (d) all of the above.

26. The capacity of a crane is expressed in terms of
(a) span. (b) tonnes.
(c) type of drive. (d) any of the above.

27. The travelling speed of a crane varies from
(a) 20 to 25 m/s. (b) 10 to 20 m/s.
(c) 1.0 to 2.5 m/s. (d) 0.2 to 1.0 m/s.

28. Light duty cranes are generally used in
(a) power houses. (b) pumping stations.
(c) automobile workshops. (d) all of the above.

29. Heavy duty cranes are employed in
(a) power houses and pumping stations.
(b) automobile workshops.
(c) steel plants and ore handling plants.
(d) both (a) and (c).

30. In case of centrifugal pumps the starting torque is usually
(a) less than running torque. (b) same as running torque.
(c) slightly higher than running torque.
(d) double of running torque.

31. The size of an excavator is usually expressed in terms of
(a) crowd motion. (b) m^3.
(c) travel in metres. (d) angle of swing.

32. The range of rating of electric motor used for rolling mills is of the order of
(a) 10 to 25 kW. (b) 25 to 85 kW.
(c) 85 to 400 kW. (d) 400 to 1,000 kW.

33. If the liquid to be pumped has relative density of 1.25, the output power required in comparison to that required for pumping water in case of centrifugal pump will be
(a) 1.25 times. (b) the same.
(c) 0.8 times. (d) 15 times (roughly).

34. The load torque decreases with the increase in speed in case of
(a) lathes, boring machines, milling machines.
(b) hoist winches, machine tool feed mechanism.
(c) blowers, fans, centrifugal pumps.
(d) both (a) and (b).

35. In synthetic fibre mills motors with
(a) variable speeds are required.
(b) constant speeds are preferred.
(c) low starting torque are required.
(d) high starting torque are preferred.

36. In case of kiln drives, starting torque is
(a) more than double of the running torque.
(b) almost $\sqrt{2}$ times of the running torque.
(c) almost equal to running torque.
(d) almost zero.

37. The diameter of the rotor shaft for an electric motor depends upon
(a) speed only. (b) power output only.
(c) speed and power output.
(d) speed, power output and power factor.

38. Which one of the following is not necessarily the advantage of dc motors over ac motors ?
(a) Low cost.
(b) Excellent torque and speed operating characteristics.
(c) Flexible speed control. (d) None of the above.

39. An unsaturated dc shunt motor drives a constant torque load at a speed at 1,000 rpm taking an armature current of 20 A when 230 V is applied to the motor. If the applied voltage is reduced to 115 V, the motor runs at.
(a) 500 rpm taking an armature current of 20 A.
(b) 1,000 rpm taking an armature current of 20 A.
(c) 500 rpm taking an armature current of 40 A.
(d) 1,000 rpm taking an armature current of 40 A.
[A.M.I.E. Sec. B. Elec. Drives & Their Control Winter 1994]

40. An electric train employing a dc series motor is running at a fixed speed, when a sudden slight drop in the mains voltage occurs. This would result in
(a) drop in speed and rise in current.
(b) rise in speed and drop in current.
(c) rise in speed and rise in current.
(d) drop in speed with current unaltered. **[U.P.S.C.I.E.S. E.E.-I:, 1995]**

41. When an electric train is moving down a hill, the dc motor will operate as a dc
(a) series motor. (b) series generator.
(c) shunt motor. (d) shunt generator.

42. A dc series motor is used for an overhauling load. It can work stably if
(a) the armature is shunted by a resistor.
(b) the field winding is reversed.
(c) a resistor is put in series with the machine.
(d) a divertor is put across the field.
[A.M.I.E. Sec. B. Elec. Drives & Their Control Summer 1993]

43. DC series motors are very suitable for heavy duty applications such as electric railways, rolling mills because of
(a) low initial as well as maintenance cost.
(b) high starting torque. (c) possibility of speed control.
(d) nearly constant speed.

44. A flywheel is normally fitted to
(a) dc series motor driving a constant torque load.
(b) separately excited dc motor driving pulsed torque load.
(c) cumulatively compound motor driving pulsed torque load.
(d) differentially compound motor driving pulsed torque load.
[A.M.I.E. Sec. B. Elec. Drives & Their Control Winter 1994]

45. Which motor should not be used for centrifugal pumps ?
(a) Shunt. (b) Series.
(c) Cumulatively compound. (d) Differentially compound.
[U.P.S.C. I.E.S. E.E.-II, 1992]

46. For continuously running rolling mills with intermittent loading, the most suitable dc drive is
(a) dc series motor. (b) dc shunt motor.
(c) dc differentially compounded motor.
(d) dc cumulatively compounded motor. **[U.P.S.C. I.E.S. E.E.-II, 1992]**

47. For quick speed reversal the motor preferred is
(a) dc motor. (b) squirrel cage induction motor.
(c) slip-ring induction motor. (d) synchronous motor.

48. The motor, owing to its inherent characteristics, best suited for the rolling mills is
(a) dc shunt motor.
(b) dc cumulative compound wound motor.
(c) squirrel cage induction motor.
(d) synchronous motor.

Answers: 25. (d), 26. (b), 27. (c), 28. (d), 29. (c), 30. (a), 31. (b), 32. (c), 33. (a), 34. (a), 35. (b), 36. (a), 37. (c), 38. (a), 39. (d), 40. (a), 41. (b), 42. (a), 43. (b), 44. (c), 45. (b), 46. (d), 47. (a), 48. (b),

49. Squirrel cage induction motors with high slip and slip-ring induction motors develop maximum torque at standstill and are used for
(a) elevators. (b) machine tools.
(c) presses and punches. (d) all of the above.

50. Belted slip-ring induction motor is almost invariably employed for
(a) conveyors. (b) jaw-crushers.
(c) centrifugal blowers. (d) water pumps.

51. Belted wound induction motors are preferred for
(a) gyratory crushers. (b) screw pumps.
(c) machine tools. (d) water pumps.

52. The pair used for frequency converters is
(a) squirrel cage induction motor and synchronous motor.
(b) slip-ring induction motor and synchronous motor.
(c) slip-ring induction motor and squirrel cage induction motor.
(d) all of above.

53. For the same rating, the size of the single phase induction motor is about .... that of the corresponding three-phase induction motor.
(a) three lines (b) the same as
(c) 1.5 times (d) one-third

[A.M.I.E. Sec. B. Elec. Drives & Their Control Winter 1999]

54. A fluctuating voltage supply is detrimental to a refrigerator motor but not to a ceiling fan, although both are single phase induction motors because, the refrigerator motor
(a) is made more robust than the fan motor.
(b) is subjected to short duty cycle but the fan motor is subjected to continuous duty.
(c) is enclosed in a sealed unit while the fan motor is open to the environment.
(d) load is constant, but the fan motor load is voltage dependent.

[U.P.S.C. I.E.S. E.E.-II, 1993]

55. Ward-Leonard controlled dc drives are usually used for... duty excavators.
(a) light (b) medium (c) heavy

56. Speed control by variation of field flux results in
(a) constant power drive. (b) variable power drive.
(c) constant torque drive. (d) variable torque drive.

[A.M.I.E. Sec. B. Elec. Drives & Their Control Summer 1994]

57. In variable speed dc motor drive the armature voltage control leads to
(a) constant power operation.
(b) constant torque operation.
(c) variable torque operation.
(d) randomly varying power operation.

[A.M.I.E. Sec. B. Elec. Drives & Their Control Winter 1998]

58. Speed of an induction motor is controlled by injecting voltage into the rotor winding. Frequency of the injected voltage should be
(a) main frequency.
(b) frequency corresponding to required speed.
(c) frequency corresponding to rotor speed.
(d) slip frequency.

[A.M.I.E. Sec. B. Elec. Drives & Their Control Winter 1994]

59. The slip power recovery scheme is used in induction motor for speed control in the range
(a) above synchronous speed.
(b) below synchronous speed.
(c) both above and below synchronous speed.
(d) none of the above.

[A.M.I.E. Sec. B. Elec. Drives & Their Control Summer 1999]

60. The electric braking system commonly employed in rolling mills, elevators and printing presses is
(a) plugging. (b) rheostatic.
(c) dynamic. (d) regenerative.

61. For rheostatic braking of two series motors connected in parallel
(a) cross-connection is better.
(b) equaliser connection is better.
(c) both are equally good.
(d) none of the two is required.

62. Dynamic braking is employed to brake
(a) non-reversing drive. (b) reversing drive.
(c) in both (a) and (b).

[A.M.I.E. Sec. B. Elec. Drives & Their Control Summer 2000]

63. The condition of regenerative braking can be achieved by
(a) speed higher than no-load speed of overhauling load.
(b) increasing the excitation while supply voltage remains constant.
(c) increasing the armature current.
(d) either (a) or (b).

64. In a dc shunt motor, regenerative braking is limited by
(a) saturation of flux at high speed.
(b) saturation of flux at low speed.
(c) armature current at high speed.
(d) either (a) or (b).

65. For regenerative braking with dc series motors, its field windings are ______ excited.
(a) separately (b) series
(c) shunt (d) either (b) or (c)

66. The most economical method of electric braking is
(a) plugging
(b) dynamic braking with separate excitation.
(c) dynamic braking with self excitation.
(d) regenerative braking.

67. The plugging provides _______ braking torque in comparison to rheostatic and regenerative braking systems.
(a) negligible (b) small
(c) highest

68. Rheostatic braking may be applied to an induction motor provided
(a) it is a squirrel cage type.
(b) it is a wound type.
(c) separate dc source for field excitation is available.
(d) variable external resistance is available.

69. The motor enclosure used for collieries, chemical plants is
(a) flame proof type. (b) splash proof type.
(c) totally enclosed type. (d) pipe ventilated type.

70. The motor enclosure used in woodworking industry is
(a) protected type.
(b) totally enclosed fan cooled type.
(c) flame-proof type. (d) splash-proof type.

71. The motor enclosure used for industrial purposes is
(a) protected type. (b) drip proof type.
(c) totally enclosed type. (d) open type.

72. Rotor of a motor is usually supported on
(a) bush bearings. (b) thrust bearing.
(c) ball or roller bearings. (d) any of the above.

73. Ball bearings are
(a) used up to 75 kW motors.
(b) of long life and low friction loss.
(c) costlier and noisy particularly at high motor speed.
(d) all of the above.

Answers: 49. (c), 50. (b), 51. (a), 52. (a), 53. (c), 54. (d), 55. (c), 56. (a), 57. (b), 58. (d), 59. (c), 60. (a), 61. (a), 62. (c), 63. (d), 64. (b), 65. (a), 66. (d), 67. (c), 68. (c), 69. (a), 70. (b), 71. (a), 72. (c), 73. (d),

74. Sleeve bearings
(a) are normally of bronze.
(b) have self-lubricating properties.
(c) are used where noise is to be avoided.
(d) all of the above.

75. Hot bearings of a dc motor may be caused by
(a) poor ventilation. (b) incorrect voltage.
(c) loose coupling. (d) lack of or dirty lubricant.
[A.M.I.E. Sec. B. Electric Drives & Their Control Winter 1999]

76. Excessive motor vibration is caused by
(a) worn bearings. (b) open armature coil.
(c) bent shaft. (d) excessive brush tension.

77. Excessive sparking at the brushes may be caused due to
(a) loose coupling. (b) worn bearings.
(c) dirt on commutator. (d) open armature coil.

78. Intermittent sparking at the brushes of a dc motor may be caused due to
(a) intermittent load. (b) open armature coil.
(c) loose coupling. (d) incorrect voltage.

79. The least expensive drive is
(a) belt drive. (b) rope drive.
(c) chain drive. (d) gear drive.

80. For very high speed ratio the indispensable drive is
(a) direct drive. (b) rope drive.
(c) chain drive. (d) any of the above.

81. In rotating electrical machines, the insulation temperature limit for class B type is
(a) 105°C. (b) 130°C.
(c) 150°C. (d) 180°C.
[A.M.I.E. Sec. B. Electric Drives & Their Control Winter 1996]

82. For a particular motor, the cooling time constant is usually
(a) smaller than the heating time constant.
(b) greater than the heating time constant.
(c) equal to the heating time constant.
[A.M.I.E. Sec. B. Elec. Drives & Their Control Winter 1995]

83. For a particular type of motor, the heating time constant
(a) increases with the increase in size.
(b) decreases with the increase in size.
(c) same for all sizes.
[A.M.I.E. Sec. B. Elec. Drives & Their Control Winter 1995]

84. For estimation of power rating of an electric drive, its losses can be considered to be proportional to
(a) power. (b) $(\text{power})^2$.
(c) $(\text{power})^3$. (d) torque.
[A.M.I.E. Sec. B. Electric Drives & Their Control Winter 1996]

85. The duty cycle of a motor is

Load	Duration
20 kW	120 s
10 kW	120 s
stopped	280 s

The continuous rating of motor suitable for above application would be
(a) 20 kW. (b) 16.4 kW.
(c) 14.14 kW. (d) 9 kW.
[U.P.S.C. I.E.S. E.E.-II, 1992]

86. The heating time constant of a totally enclosed motor is relatively
(a) lower. (b) higher.
(c) independent of type of enclosure.

87. For a certain industrial application, an overrated ac motor was selected. It will lead to operation with
(a) higher efficiency and better pf.
(b) lower efficiency and poorer pf.
(c) higher efficiency and poorer pf.
(d) lower efficiency and better pf.
[A.M.I.E. Sec. B. Electric Drives & Their Control Winter 1998]

88. A flywheel is generally used in
(a) a cement mill drive. (b) a paper mill drive.
(c) a rolling mill drive. (d) a sugar centrifuge drive.
[A.M.I.E. Sec. B. Elec. Drives & Their Control Winter 1996; Summer 1998]

89. In overhead travelling cranes
(a) continuous rating motors are used.
(b) continuous minimum rating motors are used.
(c) short-time rating motors are used.
(d) any of the above.

90. 15 minute rated motors are suitable for
(a) light duty cranes. (b) medium duty cranes.
(c) heavy duty cranes. (d) all of the above.

91. For medium duty cranes the short-time rating motor used is
(a) 10 minutes. (b) 15 minutes.
(c) 30 minutes. (d) one hour.

92. Load equalisation is desirable in the case of
(a) very large refrigeration and airconditioning plants.
(b) rolling mills, electric hammers, presses, reciprocating pumps.
(c) lathes, woodworking machines, paper making machines, shapers and slotters.
(d) travelling cranes, lifts.

93. The phase controlled rectifiers used in speed control of dc motors converts fixed ac supply voltage into ________ output voltage.
(a) variable dc (b) variable ac
(c) variable frequency ac (d) full rectified ac

94. A four quadrant operation requires
(a) two full converters in series.
(b) two full converters connected back to back.
(c) two full converters connected in parallel.
(d) two semi converters connected back to back.
[A.M.I.E. Sec. B. Elec. Drives & Their Control Winter 1995]

95. The following converters can feed power in any of the four quadrants
(a) semi-converter. (b) full converter.
(c) dual converter.
(d) combination of a semi- and full-converters.
[A.M.I.E. Sec. B. Elec. Drives & Their Control Winter 1993]

96. For speed control of dc motors using controlled rectifiers, armature voltage control gives speed
(a) above rated speed. (b) below rated speed.
(c) above as well as below rated speed.
[A.M.I.E. Sec. B. Power Electronics Summer 1996]

97. Armature of a dc motor is fed from a phase controlled rectifier where as its field is supplied from a constant dc source. To reduce the speed of the motor regeneratively
(a) the firing angle of the converter should be increased.
(b) the polarity of the dc voltage should be reversed.
(c) the polarity of the armature induced voltage should be reversed.
(d) the firing angle should be varied in the range 90° to 180° simultaneously reversing the armature connections.
[A.M.I.E. Sec. B. Elec. Drives & Their Control Winter 1994]

Answers: 74. (d), 75. (d), 76. (a), 77. (c), 78. (b), 79. (a), 80. (b), 81. (b), 82. (b), 83. (a), 84. (b), 85. (c), 86. (b), 87. (b), 88. (c), 89. (c), 90. (a), 91. (c), 92. (b), 93. (a), 94. (b), 95. (c), 96. (b), 97. (a),

98. For the speed control of ac drive, the preferred method using thyristors is
(a) phase control. (b) integral cycle control.
(c) chopper control. (d) all are equally good.
[A.M.I.E. Sec. B. Elec. Drives and Their Control Winter 1995]

99. The semiconductor device used for speed control of single phase induction motors in every day domestic use is
(a) SCR. (b) power transistor.
(c) Triac. (d) power MOS.
[A.M.I.E. Sec. B. Elec. Drives & Their Control Winter 1999]

100. Heavy-duty steel-works cranes having wide load variations are equipped with
(a) dc series motors.
(b) plain squirrel cage induction motors.
(c) wound-rotor induction motors.
(d) synchronous motors.

101. The most suitable motor for a steel mill requiring a motor having high starting torque, wide speed range and precise speed control is
(a) plain squirrel cage induction motor.
(b) wound rotor induction motor.
(c) dc series motor. (d) dc shunt motor.

102. For automatic drives the preference is for
(a) synchronous motors.
(b) Ward-Leonard controlled dc shunt motors.
(c) plain squirrel cage induction motors.

103. The type of motor that can be used for hoisting machinery is
(a) dc compound motor.
(b) Ward-Leonard controlled dc shunt motor.
(c) wound rotor induction motor.
(d) any of the above.

104. The motor normally used for crane travel is
(a) wound rotor induction motor.
(b) synchronous motor.
(c) dc differentially compounded motor.
(d) plain squirrel cage induction motor.

105. The drive, that can be used for derricks and winches is
(a) pole-changing squirrel cage induction motor.
(b) ac slip-ring induction motor with variable resistance.
(c) dc shunt motor with Ward-Leonard control.
(d) any of the above.

106. Ward-Leonard controlled dc drives are generally used for ——— duty excavators
(a) light (b) medium (c) heavy (d) none of these.

107. The motor suitable for a reciprocating pump required to start under load is
(a) plain squirrel cage induction motor.
(b) double-squirrel cage induction motor.
(c) synchronous motor. (d) dc shunt motor.

108. A pole-changing squirrel cage induction motor employed in derricks has four, eight and twenty four poles. The lowest speed is used in
(a) hoisting. (b) landing the load.
(c) lifting. (d) lowering.

109. A pole-changing type squirrel cage induction motor employed in derricks has four, eight and twenty four poles. The medium speed is used in
(a) landing the load. (b) hoisting.
(c) lowering. (d) lifting.

110. Motor preferred for kiln drive is usually
(a) wound rotor induction motor.
(b) cascaded controlled ac motor.
(c) Ward-Leonard controlled dc shunt motor.
(d) any of the above.

111. The motor used in mines is
(a) flame proof squirrel cage induction or wound rotor motor.
(b) dc series motor.
(c) dc shunt motor. (d) any of the above.

112. The motor used in punches, presses and shears is
(a) dc series or shunt motors.
(b) dc cumulative compound motor.
(c) high slip squirrel cage or wound rotor induction motor.
(d) both (b) and (c).

113. A domestic mixer uses the following motor
(a) induction motor. (b) reluctance motor.
(c) universal motor.
(d) permanent magnet synchronous motor.
[A.M.I.E. Sec. B. Elec. Drives & Their Control Winter 1998]

114. A perfect black body is one which
(a) absorbs all incident radiations.
(b) reflects all incident radiations.
(c) transmits all incident radiations.
(d) all of the above.

115. For the transmission of heat from one body to another it is essential that
(a) both bodies are solids.
(b) the two bodies are at different temperatures.
(c) both bodies are in contact.
(d) at least one of the bodies has some source of heating.

116. Heat is transferred simultaneously by conduction, convection and radiation
(a) during melting of ice.
(b) from refrigerator coils to refrigerator freezer.
(c) inside boiler furnaces.
(d) through the surface of the insulated pipe carrying steam.

117. The highest value of thermal conductivity is for
(a) aluminium. (b) brass. (c) copper. (d) iron.

118. The highest value of thermal conductivity is for
(a) steam. (b) water. (c) melting ice. (d) solid ice.

119. The thermal conductivity will be highest in case of
(a) dry concrete.
(b) concrete having 10% moisture by volume.
(c) concrete having 0.4% reinforcement.
(d) concrete having 10% moisture by volume and 0.4% reinforcement.

120. Radiations from a black body are proportional to
(a) T^2. (b) T^3. (c) T^4. (d) $1/T^4$.

121. A body reflecting entire radiations incidenting on it is called the
(a) white body. (b) gray body.
(c) black body. (d) transparent body.

Answers: 98. (a), 99. (c), 100. (a), 101. (c), 102. (b), 103. (d), 104. (d), 105. (d), 106. (c), 107. (b), 108. (b), 109. (d), 110. (d), 111. (a), 112. (d), 113. (c), 114. (a), 115. (b), 116. (c), 117. (c), 118. (d), 119. (d), 120. (c), 121. (a),

122. The insulating material suitable for low temperature applications is
(a) cork. (b) diatomaceous earth.
(c) asbestos paper. (d) 75 percent magnesia.

123. The quantity of heat absorbed from the heater by convection depends upon
(a) the temperature of heating element above the surroundings.
(b) the surface area of the heater.
(c) the position of the heater.
(d) all of the above.

124. The material of the heating element should be
(a) such that it may withstand the required temperature without getting oxidised.
(b) of low resistivity. (c) of low melting point.
(d) of high temperature coefficient.

125. The material to be used for heating element should be of high resistivity so as to
(a) increase the life of the heating element.
(b) reduce the length of the heating element.
(c) reduce the effect of oxidation.
(d) produce large amount of heat.

126. The material to be used for heating element should be of low temperature coefficient so as to
(a) avoid initial rush of current.
(b) avoid change in kW rating with temperature.
(c) reduce the effect of oxidation.
(d) both (a) and (b) above.

127. Which of the following heating element will have the least temperature range ?
(a) Eureka. (b) Silicon carbon.
(c) Nichrome. (d) Kanthal.

128. Which of the following heating element can give highest temperature in resistance heating ?
(a) Nichrome. (b) Silicon carbide.
(c) Copper. (d) Nickel-Cr-Fe alloy.

129. Ni-Cr-Fe alloy wires can be safely used for temperatures upto
(a) 2,500°C. (b) 2,000°C.
(c) 1,150°C. (d) 850°C.

130. The heating element to be used in a furnace employed for heating around, 1,600°C should be of material
(a) nichrome. (b) eureka.
(c) molybdenum. (d) silicon-carbide.

131. For radiant heating around 2,000°C the heating elements used should be of material
(a) tungsten alloy. (b) copper alloy.
(c) carbon. (d) Stenleess.

132. The heating element to be used in vacuum furnaces should be of the material
(a) molybdenum because of high vapour pressure.
(b) tungsten because of low vapour pressure.
(c) molybdenum because of low vapour pressure.
(d) tungsten because of high vapour pressure.

133. In an electric press, mica is used
(a) for induction heating. (b) for dielectric heating.
(c) as an insulator.
(d) for improvement of power factor.

134. Maximum operating voltage is limited
(a) to 600 V. (b) from safety consideration.
(c) by electrical insulation at high temperatures.
(d) all of the above.

135. The heat produced in the heating element(s) is also to overcome the losses occurring because of
(a) heat used in raising the temperature of oven (or furnace) and container (or carriers).
(b) heat conducted through the walls.
(c) escapement of heat due to opening of door.
(d) all of the above.

136. In a resistance furnace, the temperature is controlled by
(a) variation of operating voltage.
(b) variation of resistance of heating circuit.
(c) switching on and off the supply periodically.
(d) all of the above.

137. For temperature control in resistance furnaces resistance variation can be affected by connecting resistance elements in
(a) series. (b) parallel.
(c) series-parallel. (d) star or delta.
(e) any of the above.

138. The simplest and most commonly used method for temperature control is
(a) external series resistance in the heating circuit.
(b) change of connections of heating circuit.
(c) use of variable number of heating elements.
(d) transfomer tappings.

139. The device necessarily used for automatic temperature control in a furnace is
(a) thermostat. (b) auto-transformer.
(c) thermocouple. (d) any of the above.

140. Control of power input to salt-bath furnace is affected by
(a) varying the depth of immersion of electrodes.
(b) varying the distance between the electrodes.
(c) both (a) and (b). (d) none of (a) and (b).

141. The temperature inside a furnace is usually measured by
(a) mercury thermometer. (b) optical pyrometer.
(c) alcohol thermometer. (d) any of the above.

142. In a resistance furnace the atmosphere is
(a) oxidising. (b) deoxidising.
(c) reducing. (d) neutral.

143. On increasing the thickness of refractory walls of the furnace
(a) energy consumption will increase.
(b) heat loss through furnace walls will increase.
(c) temperature on the outer surface of furnace walls will drop.
(d) temperature inside the furnace will drop.

144. Radiant heating is used for
(a) melting of ferrous metals.
(b) annealing of metals.
(c) drying of paints and varnishes.
(d) any of the above.

145. Direct resistance heating is used in
(a) electrode boiler. (b) salt-bath furnace.
(c) resistance welding. (d) all of the above.

146. Resistance ovens are used for
(a) domestic and commercial heating.
(b) vulcanizing and hardening of synthetic materials.
(c) drying of varnish coatings, drying and baking of potteries.
(d) all of the above.

147. The function(s) of an heating chamber is/are to
(a) store as much of the heat supplied as may be practicable and economical.

Answers: 122. (b), 123. (d), 124. (a), 125. (b), 126. (d), 127. (a), 128. (b), 129. (d), 130. (c), 131. (a), 132. (b), 133. (c), 134. (d), 135. (d), 136. (d), 137. (e), 138. (b), 139. (a), 140. (c), 141. (b), 142. (a), 143. (c), 144. (c), 145. (d), 146. (d), 147. (d),

(b) confine the atmosphere around the charge.
(c) control the cooling rate of the charge (if required) and control the distribution of heat within the chamber.
(d) all of the above.

148. For obtaining best results, the infrared lamps should be located from the object to be heated at distance of
(a) 10-20 cm. (b) 25-30 cm.
(c) 40-50 cm. (d) 50-60 cm.

149. In direct arc furnace which of the following is of high value ?
(a) Current. (b) Voltage.
(c) Power factor. (d) all of the above.

150. The power factor at which the direct arc furnace operates is
(a) low lagging. (b) low leading.
(c) unity. (d) high leading.

151. For arc heating, the electrodes used are made of
(a) copper. (b) graphite.
(c) tungsten. (d) aluminium.

152. In an arc furnace, the choke is provided to
(a) reduce the surge severity.
(b) stabilize the arc. (c) improve the power factor.
(d) all of the above.

153. It is desirable to operate the arc furnaces at a power factor of
(a) zero. (b) unity.
(c) 0.707 lagging. (d) 0.707 leading.

154. It is desirable to keep the arc length short in order to
(a) have better heating.
(b) have better stirring action and reduce oxidation problem.
(c) increase the life of roof refractory.
(d) all of the above.

155. Arc drawn between the two electrodes produces heat and has a temperature
(a) between 0 – 600°C. (b) between 500°C – 1,000°C.
(c) between 1,000°C – 3,500°C.
(d) of the order of 4,000°C.

156. The arc furnaces of conical shapes have the advantage(s) of
(a) large surface area per unit bath volume.
(b) low power consumption.
(c) reduced radiation losses.
(d) all of the above.

157. In arc furnace, energy losses that take place in the furnace are
(a) losses through walls.
(b) losses through escape gasses.
(c) losses through water cooling.
(d) all of the above.

158. In submerged arc furnaces the power is controlled by
(a) varying the spacing between the electrodes.
(b) varying the voltage applied to the electrodes.
(c) either (a) or (b).
(d) varying the arc length.

159. For arc furnaces the low voltage high current power supply is needed because
(a) heavy currents produce large amount of heat resulting in higher temperatures.
(b) maximum secondary voltage is also limited to 275 V (line-to-line on open circuit) owing to insulation and safety consideration.
(c) life of the roof refractory is increased.
(d) all of the above.

160. The main application of indirect arc furnace is to melt
(a) iron. (b) steel.
(c) non-ferrous metals. (d) none of the above.

161. For power transformers employed for arc furnaces, it is desirable to arrange the furnace and the transformer in such a way that leads are
(a) shorter in length and placed distant apart.
(b) shorter in length and placed close together.
(c) longer in length and placed close together.
(d) in any arrangement.

162. In induction heating
(a) heat is produced due to currents induced in the charge by electromagnetic action.
(b) the resistance of the charge must be low and voltage applied must be high in order to produce sufficient heat.
(c) magnetic materials can be easily treated in comparison to non-magnetic materials.
(d) all of the above.

163. In heating the ferromagnetic materials by induction heating, heat is produced owing to
(a) flow of induced current through the charge.
(b) hysteresis loss occuring below Curie temperature.
(c) hysteresis loss as well as eddy current loss occurring in the charge.
(d) any of the above factors.

164. Induction heating takes place in
(a) insulating materials.
(b) conducting and magnetic materials.
(c) conducting but non-magnetic materials.
(d) conducting materials may be magnetic or non-magnetic.

165. Induction hardening is possible in case of
(a) dc supply only. (b) ac supply only.
(c) ferrous materials only.
(d) non-conducting materials only.

166. Low frequency supply is necessary for direct core type induction furnaces because
(a) magnetic coupling between the primary and secondary circuit is poor.
(b) with normal frequency supply the electromagnetic forces cause severe stirring action in the molten metal.
(c) both (a) and (b). (d) none of (a) and (b).

167. In induction heating, which of the following is of high value?
(a) Frequency. (b) Current.
(c) Voltage. (d) Power factor.

168. Induction furnaces are used for
(a) heat treatment of castings.
(b) heating of insulators. (c) melting of aluminium.
(d) all of the above.

169. In induction heating, the depth up to which current will penetrate is proportional to
(a) $1/(\text{frequency})^{1/2}$. (b) 1/ frequency.
(c) frequency. (d) $(\text{frequency})^2$.

170. The advantage(s) of eddy current heating is/are
(a) easy temperature control, little wastage of heat and possibility of heating in vacuum or other special atmosphere.
(b) heat can be made to penetrate into metal surface to any desired depth.
(c) the area of surface over which heat is produced can be accurately controlled.
(d) all of the above.

Answers: 148. (b), 149. (a), 150. (a), 151. (b), 152. (b), 153. (c), 154. (d), 155. (c), 156. (d), 157. (d), 158. (c), 159. (d), 160. (c), 161. (b), 162. (d), 163. (c), 164. (d), 165. (b), 166. (c), 167. (a), 168. (a), 169. (a), 170. (d),

171. The supply frequency usually employed for high frequency eddy current heating is
(a) 10 MHz. (b) 10 kHz to 400 kHz.
(c) 5 kHz. (d) 1 kHz.

172. In case of core-less induction furnace, the charge should be in
(a) solid state. (b) molten state.
(c) either of (a) or (b).

173. High frequency induction heating is used for
(a) ferrous metals only. (b) non-ferrous metals only.
(c) both ferrous and non-ferrous metals.

174. In dielectric heating current flows through
(a) air. (b) dielectric.
(c) metallic conductor.
(d) ionic discharge between dielectric medium and metallic conductor.

175. Dielectric loss is proportional to
(a) frequency. (b) $(\text{frequency})^2$.
(c) $(\text{frequency})^3$. (d) $(\text{frequency})^{1/2}$.

176. The dielectric loss in a dielectric is proportional to
(a) voltage impressed on the dielectric.
(b) the square of the voltage impressed on the dielectric.
(c) the square root of the voltage impressed on the dielectric.
(d) none of the above.

177. For heating of plywood, the frequency should be
(a) 1–2 MHz. (b) 10–25 kHz. (c) 1 kHz. (d) 100 Hz.

178. The normal voltage used in dielectric heating is
(a) 1.5 kV. (b) 15 kV. (c) 33 kV. (d) 66 kV.

179. For dielectric heating, with the increase of mass of workpiece, the optimum frequency .... for maximum power transfer.
(a) increases (b) decreases
(c) independent of mass

180. For dielectric heating the range of frequency normally employed is
(a) 10 kHz to 100 kHz. (b) 100 kHz to 10 MHz.
(c) 1 MHz to 10 MHz. (d) 10 MHz to 40 MHz.

181. The power factor will be maximum in case of
(a) electric arc heating. (b) resistance heating.
(c) induction heating. (d) dielectric heating.

182. Furnaces used for cremation are
(a) electric resistance heating.
(b) electric arc heating. (c) dielectric heating.
(d) high frequency eddy current heating.

183. In an electric room heat convector the method of heating used is
(a) arc heating. (b) resistance heating.
(c) induction heating. (d) infrared heating.

184. When the composition of non-ferrous metal is to be varied frequently or when heating is intermittent .... furnace is desirably used.
(a) direct resistance (b) direct arc
(c) indirect resistance (d) Ajax Wyatt

185. Hysteresis loss and eddy current loss are used in
(a) resistance heating. (b) dielectric heating.
(c) induction heating of steel.
(d) induction heating of brass.

186. The power factor will be leading in case of
(a) dielectric heating. (b) induction heating.
(c) electric arc heating. (d) resistance heating.

187. .... is used for heating non-conducting materials
(a) eddy current heating. (b) arc heating.
(c) induction heating. (d) dielectric heating.
[West Bengal Univ. of Technology 2009-10]

188. The method appropriate for heating of non-ferrous metals is
(a) indirect resistance heating.
(b) radiant heating. (c) indirect arc heating.
(d) dielectric heating.

189. The method suitable for heating of conducting medium is
(a) induction heating. (b) indirect arc heating.
(c) eddy current heating. (d) radiant heating.

190. The most modern method for food processing is
(a) induction heating. (b) resistance heating.
(c) dielectric heating. (d) eddy current heating.

191. The main advantage of dielectric heating is that
(a) it can be used for drying the explosives.
(b) heating occurs in the material itself.
(c) heating occurs due to the high frequency.
(d) none of the above.

192. ________ is not an application of dielectric heating.
(a) Food processing. (b) Soldering.
(c) Gluing of wood. (d) Diathermy.

193. Which of the following falls under the category of plastic or non-fusion welding ?
(a) Resistance welding. (b) Electron beam welding.
(c) Electro-slag welding. (d) Arc welding.

194. Which of the following falls under the category of fusion or non-pressure welding?
(a) Resistance welding. (b) Metal-arc welding.
(c) Ultrasonic welding. (d) Explosive welding.

195. Proper selection of welding depends upon, in addition to cost involved,
(a) kinds of metals to be joined.
(b) nature of products to be fabricated.
(c) production technique used.
(d) all of the above.

196. The metal surfaces for electrical resistance welding must be
(a) cleaned. (b) lubricated.
(c) moistened. (d) rough.

197. Resistance welding cannot be used for
(a) ferrous materials. (b) non-ferrous materials.
(c) dielectrics. (d) any of the above.

198. In electrical resistance welding the greatest resistance is offered by
(a) metal surface.
(b) contact layer of metals to be welded.
(c) contact point of electrode with metal top.
(d) contact point of electrode with metal bottom.

199. In electric resistance welding
(a) the current required exceeds 100 A.
(b) the voltage required ranges from 4 to 12 V.
(c) the amount of power supplied to the weld usually ranges from 60 watts to 80 watts for each square mm of area.
(d) all of the above.

200. Resistance to the flow of current is made of
(a) resistance of current path in the work.
(b) resistance between the contact surfaces of the parts being welded.

Answers: 171. (b), 172. (c), 173. (c), 174. (b), 175. (a), 176. (b), 177. (a), 178. (a), 179. (b), 180. (d), 181. (b), 182. (a), 183. (b), 184. (c), 185. (c), 186. (a), 187. (d), 188. (c), 189. (a), 190. (c), 191. (b), 192. (b), 193. (a), 194. (b), 195. (d), 196. (a), 197. (c), 198. (b), 199. (d), 200. (d),

(c) resistance between the electrodes and the surface of the parts being welded.
(d) all of the above.

201. In resistance welding, the magnitude of current is controlled
(a) by varying the primary voltage of the welding transformer using an auto-transformer between supply and welding transformer.
(b) by changing the primary turns of the welding transformer.
(c) by varying the magnitude and waveform of the primary as well as secondary current by using thyratron or ignitron.
(d) any of the above.

202. The main drawback of resistance welding is
(a) high initial as well as maintenance cost.
(b) difficult shapes and sections cannot be welded.
(c) only similar metals can be welded.
(d) parent metal is affected.

203. Plain and butt welds may be used on materials up to thickness of about
(a) 5 mm. (b) 10 mm. (c) 25 mm. (d) 40 mm.

204. In upset butt welding
(a) the faces of the metal pieces to be joined are prepared for even contact.
(b) heating is obtained by the contact resistance of metal pieces to be welded.
(c) the voltage required is 2-8 V and current required ranges from 50 A to several hundred amperes depending upon material and the area to be welded at a time.
(d) all of the above.

205. In flash-butt welding
(a) no special preparation of the faces to be welded is necessary.
(b) clean and pure weld is obtained.
(c) power requirement is less.
(d) all of the above.

206. Spot welding process basically depends on
(a) generation of heat.
(b) application of forging pressure.
(c) both (a) and (b). (d) ohmic resistance.

207. In spot welding, composition and thickness of the base metal determines the
(a) holding time. (b) amount of weld current.
(c) amount of squeeze pressure.
(d) all of the above.

208. The tips of the electrodes, for spot welding are made of
(a) carbon.
(b) copper alloy or pure copper.
(c) mica. (d) porcelain.

209. The power factor of a spot welding machine is expected to be about
(a) 0.3 to 0.5 lagging. (b) 0.8 to 0.85 lagging.
(c) 0.75 to 0.85 leading. (d) unity.

210. During spot welding, the current flows for
(a) fraction of a minute.
(b) fraction of a second to several seconds.
(c) few milliseconds. (d) few microseconds.

211. Spot welding is employed for
(a) thin metal sheets (thickness being usually limited to 10-12 mm).
(b) castings only. (c) thick sections.
(d) rough and irregular surfaces.

212. Spot welding
(a) makes the weld air tight.
(b) makes the weld water tight.
(c) provides mechanical strength.
(d) all of the above.

213. In spot welding
(a) it is desirable to clean the sheets thoroughly before welding.
(b) the workpieces being welded are pressed together by mechanical pressure exerted through electrodes.
(c) current required is above 5,000 A and the voltage between the electrodes is usually less than 2 V (open-circuit voltage less than 12 V).
(d) all of the above.

214. Projection welding can be considered as a mass production form of
(a) seam welding. (b) spot welding.
(c) upset welding. (d) flash welding.

215. In comparison to spot welding, projection welding has the advantage(s) of
(a) simple welding process, more output and good finished appearance.
(b) electrode life is increased owing to use of low current density and low pressure.
(c) it is easy to weld certain parts which cannot be welded by spot welding .
(d) all of the above.

216. In seam welding
(a) the workpiece is fixed and disc electrodes move.
(b) the workpiece moves but rotating electodes are fixed.
(c) the electrodes used are of disc or roller shape.
(d) either (a) or (b) and (c).

217. In case of seam welding, the flow of current through the electrode should be
(a) intermittent. (b) continuous.
(c) either (a) or (b).

218. In .... welding it is essential that the surfaces to be welded are clean, dust and dirt free.
(a) butt (b) spot (c) seam
(d) both spot and seam

219. Seam welding is normally not recommended for
(a) aluminium alloys. (b) stainless and coated steels.
(c) copper and high copper alloys.
(d) alloys of nickel and magnesium.

220. In percussion welding
(a) the process used depends on the arc effect for heating and not on the resistance.
(b) the action of process is so rapid that there is little heating effect in the material adjacent to the weld.
(c) the equipment used is quite expensive.
(d) all of the above.

221. The basic electrical requirement in arc welding is that there should be
(a) high open-circuit voltage.
(b) no arc blow. (c) dc power supply.
(d) coated electrodes.

222. In arc welding the temperature of the arc is about
(a) 150°C. (b) 1,540°C.
(c) 3,500°C. (d) 10,000°C.
[West Bengal Univ. of Technology 2009-10]

Answers: 201. (d), 202. (a), 203. (c), 204. (d), 205. (d), 206. (c), 207. (d), 208. (b), 209. (a), 210. (b), 211. (a), 212. (c), 213. (d), 214. (b), 215. (d), 216. (d), 217. (a), 218. (c), 219. (c), 220. (d), 221. (a), 222. (c),

223. The electric arc has
(a) linear resistance characteristic.
(b) positive resistance characteristic.
(c) negative resistance characteristic.
(d) highly inductive characteristic.

224. In an electric arc welding, the voltage required to strike dc arc is about
(a) 50- 60 V. (b) 80-90 V. (c) 100-120 V. (d) 220 V.

225. In an electric arc welding, the voltage required to strike ac arc is about
(a) 50- 60 V. (b) 80-90 V. (c) 100-120 V. (d) 230 V.

226. In an electric arc welding, the voltage required to maintain the arc will be
(a) 250-500 V. (b) 150-250 V.
(c) 20-30 V. (d) below 10 V.

227. For an electric arc welding the current range is usually
(a) 50-1,000 A. (b) 30-50 A.
(c) 20-30 A. (d) below 20 A.

228. Increased heat due to shorter arc is harmful owing to
(a) burn through. (b) excessive porosity.
(c) undercutting of base material.
(d) all of the above.

229. Arc blow results in
(a) non-uniform weld beads.
(b) shallow weld puddle giving rise to weak weld.
(c) splashing out of metal from weld puddle.
(d) all of the above defects.

230. Arc blow effects occurring with dc welding machines can be reduced by
(a) shortening the arc column length.
(b) reducing the welding current or electrode size or rate of travel of the electrode.
(c) wrapping the welding electrode cable a few turns around the work.
(d) all of the above.

231. In an electric welding, major personal hazards are
(a) weld spatter. (b) flying sparks.
(c) harmful infrared and ultraviolet rays from the arc.
(d) all of the above.

232. During electric arc welding as the thickness of the metal to be welded increases
(a) voltage is increased keeping current the same.
(b) current is increased keeping voltage unchanged.
(c) current and voltage, both are increased.
(d) current and voltage, both are reduced.

233. In a dc arc welding
(a) both electrode as well as workpiece are made + ve.
(b) both electode as well as workpiece are made – ve.
(c) electrode is made –ve and workpiece + ve.
(d) electrode is made +ve and workpiece – ve.

234. The length of arc required depends on
(a) kind of electrodes used, its coating and its diameter.
(b) magnitude of current used.
(c) position of welding. (d) all of the above.

235. Overhead welding position is thought to be the most
(a) hazardous. (b) economical.
(c) useful. (d) difficult.

236. In arc welding best results are obtained when arc length is equal to
(a) 25 mm. (b) 19 mm.
(c) diameter of electrode.
(d) double the diameter of the electrode.

237. Flat position welding is cosidered to be the most
(a) easiest and economical. (b) hazardous.
(c) adaptable for welding of both ferrous and non-ferrous metals particularly for cast iron.
(d) both (a) and (c).

238. In direct current straight polarity (DCSP) or electrode negative welding
(a) heavily coated electrodes are used.
(b) base metal penetration is narrow and deep.
(c) electrode is the hottest.
(d) workpiece is relatively cooler.

239. In direct current reverse polarity or electrode positive welding
(a) workpiece is the hottest.
(b) bare and medium coated electrodes can be used.
(c) electrode is the hottest.
(d) base metal penetration is narrow and deep.

240. The electrode is coated in order to
(a) improve bead quality. (b) cleanse the base metal.
(c) provide shielding to weld pool.
(d) prevent atmospheric contamination.

241. In DCRP or electrode positive welding, the heat produced at the electrode is ______ of total heat produced.
(a) one-third. (b) two-third. (c) three-fourth. (d) one-fourth.

242. A 10 SWG electrode usually operates in the current range of
(a) 95-135 A. (b) 80-95 A. (c) 45-70 A. (d) 20-30 A.

243. Electrode of 8 SWG is to be employed for welding two 12.5 mm steel plates. The current required will of the order of
(a) 20 A. (b) 50 A. (c) 150 A. (d) 250 A.

244. The purpose of coating on arc welding electrodes is to
(a) provide a protective covering.
(b) provide slag for protection of the molten metal.
(c) stabilise the arc. (d) all of the above.

245. 20 per cent duty cycle of a welding machine means
(a) operation of machine for 2 minutes in a duration of 10 minutes.
(b) machine output is 20 per cent of rated output.
(c) machine operates at 20 per cent efficiency.
(d) operation of machine is for 1 minute in a duration of 5 minutes.

246. In carbon arc welding, if the electrode is connected to the + ve terminal of the dc supply
(a) arc will not strike. (b) arc will be dull.
(c) carbon arc will have tendency to go in to the weld joint.
(d) metal will not melt.

247. Carbon arc welding has the main drawback of
(a) occurrence of blow holes owing to magnetic arc blow especially while welding near edges of the workpiece.
(b) need of separate filler. (c) necessity of bare electrodes.
(d) fast consumption of electrodes.

248. Carbon arc welding
(a) uses carbon or graphite rod as a negative electrode and work to be welded as a positive.
(b) cannot be done with ac supply.
(c) is not suitable for vertical and overhead welding.
(d) all of the above.

Answers: 223. (c), 224. (a), 225. (b), 226. (c), 227. (a), 228. (d), 229. (d), 230. (d), 231. (d), 232. (b), 233. (c), 234. (d), 235. (a), 236. (c), 237. (d), 238. (b), 239. (c), 240. (a), 241. (b), 242. (a), 243. (c), 244. (d), 245. (a), 246. (c), 247. (a), 248. (d),

249. Carbon arc welding is suitable particularly for .... metals.
(a) ferrous. (b) nonferrous. (c) all.

250. For metal arc welding
(a) both dc and ac can be used but ac is preferred.
(b) bare electrodes are no longer used except for automatic welding having arrangement tc protect the weld area from the atmosphere.
(c) correct welding current, voltage and speed are very important.
(d) all of the above.

251. The atomic hydrogen welding is based on the principle(s) of
(a) obtaining atomic hydrogen by means of an electric arc between two tungsten electrodes, in an atmosphere of hydrogen at atmospheric pressure.
(b) development of very high temperature by recombination of atoms which occurs in the cooler regions immediately outside the arc.
(c) very large heat conductivity of hydrogen at high temperatures.
(d) all of the above.

252. In atomic hydrogen arc welding
(a) an arc is maintained between two tungsten electrodes.
(b) a stream of hydrogen gas under a pressure of 0.5 kg/cm^2 is passed through the arc and around the electrodes.
(c) ac supply is used.
(d) all of the above.

253. In atomic hydrogen welding, electrodes are of long life. It is because
(a) of being non-pressure process.
(b) of arc being in the shape of a fan.
(c) of ac supply.
(d) two arcs are used at a time.

254. Helium produces
(a) faster welding speeds. (b) deep penetration.
(c) narrower heat affected zone in base metal.
(d) all of the above.

255. In argon arc welding, argon is used as
(a) an agent for heat transfer.
(b) a shield to protect the work from oxidation.
(c) a source of heat. (d) a flux.

256. In argon arc welding, the argon is used to
(a) prevent oxidation of metal by coming in contact with air.
(b) create inert atmosphere around the work.
(c) obviate the necessity for using flux.
(d) all of the above.

257. Submerged arc process is characterized by
(a) high welding current.
(b) exceptionally smooth beads.
(c) deep penetration. (d) all of the above.

258. Electrode is not consumed in case of
(a) TIG welding.
(b) atomic hydrogen arc welding.
(c) MIG welding. (d) all of the above.

259. Unlike TIG welding, MIG welding
(a) needs no post-weld cleansing.
(b) needs no flux.
(c) uses consumable electrodes.
(d) provides complete protection from atmospheric contamination.

260. MIG welding is becoming more and more popular as it
(a) is easy in operation.
(b) has high metal deposite rate.
(c) both (a) and (b).
(d) can be used for both ferrous and nonferrous metals.

261. In MAG (metal active gas) welding the gas used is
(a) argon or helium.
(b) carbon dioxide or its mixture with other gas.
(c) hydrogen. (d) any of the above.

262. A weld bead of wineglass design is produced in
(a) plasma arc welding. (b) MAG welding.
(c) electron beam welding. (d) all of the above.

263. In electro-slag welding, theoretically there is no limit to the
(a) rate of metal deposit. (b) thickness of weld bead.
(c) temperature of salt bath.
(d) rate of slag consumption.

264. In manual arc welding of mild steel, the metal deposition rate will be about
(a) 10-15 kg per hour. (b) 5-10 kg per hour.
(c) 2-5 kg per hour. (d) below 1 kg per hour.

265. In ultrasonic welding, the frequency range is usually
(a) 20-60 kHz. (b) 50-100 kHz.
(c) 100-200 kHz. (d) above 250 kHz.

266. Welding is not done directly from the supply mains as
(a) its voltage is too high.
(b) it is impracticable to draw heavy currents directly from the supply mains.
(c) its voltage remains fluctuating.
(d) none of the above.

267. AC welding machine cannot be used for
(a) resistance welding. (b) submerged arc welding.
(c) MIG welding. (d) atomic hydrogen welding.

268. In electric welding, arc blow can be avoided by
(a) using ac machines. (b) increasing arc length.
(c) using bare electrodes.
(d) welding away from ground connections.

269. Welding leads have
(a) high current carrying capacity.
(b) high flexibility. (c) both (a) and (b).
(d) none of the above.

270. The welding load
(a) is always intermittent.
(b) is always continuous and constant.
(c) is always continuous but varying.
(d) may be any one of the above.

271. The load power factor using welding transformer depends on
(a) arc length. (b) material to be welded.
(c) type of electrode to be used.
(d) all of the above.

272. For power factor correction in a welding circuit, a capacitor is usually connected
(a) across the mains.
(b) across secondary side of welding transfomer.
(c) across primary side of welding transformer.
(d) across arcing electrodes.

273. Which of the following is different from the remaining ?
(a) Butt welding. (b) Electro-slag welding.
(c) TIG welding. (d) MIG welding.

Answers: 249. (b), 250. (d), 251. (d), 252. (d), 253. (c), 254. (d), 255. (b), 256. (d), 257. (d), 258. (a), 259. (c), 260. (c), 261. (b), 262. (a), 263. (b), 264. (c), 265. (a), 266. (b), 267. (c), 268. (a), 269. (c), 270. (a), 271. (b), 272. (c), 273. (a),

274. Chipping hammers are used
(a) for slag welding.
(b) for aligning the pieces to be welded.
(c) for removing slag from welding.
(d) for marking spots to be welded.

275. The danger of electric shock is maximum
(a) before welding. (b) during arcing.
(c) while inserting electrode into the holder.
(d) after welding.

276. The eyes of welding operator must be protected against
(a) infrared radiations. (b) ultraviolet radiations.
(c) both (a) and (b). (d) solar radiations.

277. Which of the following automatic welding processes is likely to give maximum rate of metal deposition ?
(a) Multiple power submerged arc.
(b) Gas shielded bare wire.
(c) Single wire submerged arc.
(d) Continuous flux covered electrode.

278. In a welded joint poor fusion is on account of
(a) improper current. (b) high welding speed.
(c) uncleaned metal surface.
(d) lack of flux.

279. Which of the following uses consumable electrodes?
(a) TIG. (b) MIG.
(c) Laser. (d) None of the above.

280. In which of the following welding methods the molten metal is poured for joining the metals?
(a) Gas welding. (b) Thermit welding.
(c) TIG welding. (d) Arc welding.

281. The porosity of weld joint may be caused by
(a) poor base metal. (b) incorrect size of electrode.
(c) low welding current. (d) any of the above.

282. In welding, weld spatter defect is generally because of too .... during welding.
(a) high voltage (b) high current
(c) low voltage (d) low current

283. High temperature metals, such as columbium can be easily welded by
(a) MIG welding. (b) TIG welding.
(c) electron beam welding. (d) flash butt welding.

284. .... welding is used for welding of car body.
(a) Gas (b) Spot
(c) Carbon-arc (d) Atomic hydrogen

285. Grey iron is usually welded by
(a) gas welding. (b) resistance welding.
(c) TIG welding. (d) arc welding.

286. The method recommended for the welding of aluminium alloy is ..... welding.
(a) dc arc (b) ac arc
(c) acetylene-oxygen gas (d) tungsten arc

287. Steel pipes are manufactured by
(a) arc welding. (b) argon arc welding.
(c) resistance welding. (d) thermit welding.

288. Steel rails are welded by
(a) argon arc welding. (b) thermit welding.
(c) gas welding. (d) resistance welding.
[West Bengal Univ. of Technology 2009-10]

289. Aluminium is difficult to weld as
(a) it has an oxide coating.
(b) it conducts away heat very rapidly.
(c) both (a) and (b). (d) none of the above.

290. Electronic components are joined by
(a) spot welding. (b) soldering.
(c) brazing. (d) none of the above.

291. Air craft body is
(a) rivetted. (b) seam welded.
(c) gas welded. (d) spot welded.

292. A seamless pipe has
(a) arc welded joint. (b) spot welded joint.
(c) no joint. (d) either (a) or (b).

293. The method usually not preferred for welding of chromium molybdenum steel is
(a) thermit welding. (b) resistance welding.
(c) submerged arc welding.
(d) oxyacetelyne gas welding.

294. Light
(a) is a form of heat energy.
(b) is a form of electrical energy.
(c) consists of electromagnetic waves.
(d) consists of shooting particles.

295. Radiant efficiency of the luminous source depends on
(a) temperature of the source.
(b) wavelength of light rays.
(c) shape of the source. (d) all of the above.

296. A substance which changes its electrical resistance when illuminated by light is called
(a) photoconductive. (b) photovoltaic.
(c) photoelectric. (d) none of the above.

297. Materials which reflect all wavelength in the spectrum of light appear to be
(a) opaque. (b) transparent.
(c) black to white. (d) green to red.

298. An object, which appears red to the eyes, absorbs
(a) blue radiations. (b) violet radiations.
(c) green radiations. (d) all of the above.

299. The sensors in the eye are known as
(a) retina and antenna. (b) wires and nerves.
(c) rodes and cones. (d) none of the above.

300. The colour temperature of daylight is about
(a) 100 K. (b) 200 K. (c) 500 K. (d) 600 K.

301. Sky appears blue due to
(a) radiation. (b) reflection.
(c) refraction.
(d) scattering of light over dust particles.

302. Colour of light depends upon
(a) frequency. (b) wavelength.
(c) speed of light.
(d) both frequency and wavelength.
[West Bengal Univ. of Technology 2009-10]

303. The wavelength of 5,500 Å will give light of
(a) green colour. (b) red colour.
(c) orange colour. (d) yellow-green colour.

304. Which of the following colour has wavelength between blue and yellow ?
(a) Violet. (b) Orange. (c) Red. (d) Green.

Answers: 274. (c), 275. (c), 276. (c), 277. (a), 278. (a), 279. (b), 280. (b), 281. (a), 282. (b), 283. (c), 284. (b), 285. (a), 286. (d), 287. (c), 288. (b), 289. (c), 290. (b), 291. (a), 292. (c), 293. (b), 294. (c), 295. (a), 296. (a), 297. (c), 298. (d), 299. (c), 300. (d), 301. (d), 302. (d), 303. (d), 304. (d),

305. Luminous flux is
(a) the rate of energy radiation in the form of light waves.
(b) the part of light energy, radiated by Sun that is received on earth.
(c) measured in lux.

306. Candle power is
(a) the luminous flux emitted by the source per unit solid angle.
(b) the light radiating capacity of a source in a given direction.
(c) the unit of illumination.

307. Glare is caused due to
(a) excessive luminance.
(b) excessive lighting contrast in the field of vision.
(c) either (a) or (b).

308. Unit of illumination is
(a) lumen. (b) lambert.
(c) lux. (d) steradians.
[West Bengal Univ. of Technology 2009-10]

309. The illumination at a surface due to a source of light placed at a distance 'd' from the surface varies as
(a) $1/d^2$. (b) $1/d$. (c) d. (d) d^2.

310. The illumination at various points on a horizontal surface illuminated by the same source varies as
(a) $\cos\theta$. (b) $\cos^2\theta$. (c) $\cos^3\theta$. (d) $1/\cos\theta$.

311. Carbon arc lamps are commonly used in
(a) cinema projectors. (b) domestic lighting.
(c) factory lighting. (d) street lighting.

312. For two filaments of same material operating at the same temperature, the diameter and current through filament are related by
(a) diameter directly proportional to current.
(b) diameter proportional to $(\text{current})^2$.
(c) diameter proportional to $(\text{current})^{2/3}$.
(d) none of these. **[West Bengal Univ. of Technology 2009-10]**

313. The melting point of tungsten is
(a) 3,400°C. (b) 2,800°C. (c) 2,600°C. (d) 2,400°C.

314. The vacuum inside an incandescent lamp is of the order of
(a) 10^{-2} mm Hg. (b) 10^{-4} mm Hg.
(c) 10^{-6} mm Hg. (d) 10^{-8} mm Hg.

315. The rate of evaporation of tungsten filament in a lamp depends upon
(a) exhaust tube diameter. (b) glass shell diameter.
(c) vapour pressure inside. (d) none of the above.

316. Heat from the lamp filament is transmitted to the surrounding mainly through
(a) circulation. (b) conduction.
(c) convection. (d) radiation.

317. Which of the following material is most commonly used for the filaments in incandscent lamps ?
(a) Tungsten. (b) Osmium.
(c) Tantalum. (d) Silver.

318. A zero watt lamp consumes power of
(a) zero watt. (b) 5–10 W.
(c) about 15 W. (d) about 25 W.

319. The safe operating temperature of a tungsten filament lamp is
(a) 1,000°C. (b) 2,000°C.
(c) 3,000°C. (d) 3,500°C.

320. The output of a tungsten filament depends on
(a) size of the shell. (b) size of the lamp.
(c) temperature of filament. (d) all of the above.

321. What percentage of the input energy is radiated by filament lamps ?
(a) 60 – 70 per cent. (b) 40 – 50 per cent.
(c) 25 – 30 per cent. (d) 10 – 15 per cent.

322. An electric bulb, when broken, produces bang. It is on account of
(a) vacuum inside the bulb.
(b) pressure inside is equal to that outside.
(c) pressure of air in the bulb. (d) none of the above.

323. In an incandescent lamp, bird cage filament is usually used in vacuum bulb so as to
(a) increase the life span of the filament.
(b) give uniform radiations.
(c) reduce the oxidation phenomenon.
(d) reduce the convection losses.

324. Filament lamp at starting will take current
(a) equal to its full running current.
(b) more than its full running current.
(c) less than its full running current.
(d) none of these. **[West Bengal Univ. of Technology 2009-10]**

325. Most affected parameter of a filament lamp due to variation in operating voltage is
(a) life. (b) light output.
(c) luminous efficiency. (d) wattage.

326. The lumens output of an incandescent lamp with the variation in operating voltage is given as lumens output α V^n where n is a constant. The value of n for tungsten is
(a) from 1.0 to 2.2 (b) from 3 to 4
(c) from 8 to 10 (d) from 10 to 12

327. Which gas is sometimes used in filament lamps ?
(a) Nitrogen. (b) Carbon dioxide.
(c) Argon. (d) Krypton.

328. Which of the following vapour/gas will give yellow colour in a filament lamp ?
(a) Helium. (b) Mercury. (c) Magnesium. (d) Sodium.

329. Magnesium vapour in a filament lamp gives
(a) green colour light. (b) pink colour light.
(c) blue colour light. (d) white colour light.

330. The gas used in a gas-filled filament lamp is
(a) helium. (b) oxygen. (c) nitrogen. (d) ozone.

331. In filament lamps, coiled coil filaments are used in
(a) coloured lamps. (b) gas-filled lamps.
(c) low wattage lamps. (d) higher wattage lamps.

332. In electric discharge lamps, light is produced by
(a) magnetic effect of current.
(b) heating effect of current. (c) cathode ray emission.
(d) ionisation in a gas or vapour.

333. In electric discharge lamps for stabilization of arc
(a) a choke is connected in series with the supply.
(b) a variable resistance is connected in series with the circuit.
(c) a condenser is connected across the supply.
(d) any of the above.

334. Glow lamps
(a) have hot cathode. (b) have cold cathode.
(c) draw more power. (d) cannot withstand shocks.

335. Halogen lamps have the advantage(s) of
(a) reduced dimensions of the lamp.
(b) better colour rendition and longer life (about 2,000 hours).
(c) high operating temperature with increased luminous efficiency.
(d) all of the above.

Answers: 305. (a), 306. (b), 307. (c), 308. (c), 309. (a), 310. (c), 311. (a), 312. (c), 313. (a), 314. (b), 315. (c), 316. (d), 317. (a), 318. (b), 319. (c), 320. (c), 321. (d), 322. (a), 323. (b), 324. (b), 325. (a), 326. (b), 327. (c), 328. (d), 329. (a), 330. (c), 331. (b), 332. (d), 333. (a), 334. (b), 335. (d),

336. The filament of a GLS lamp is made of
(a) aluminium. (b) tungsten.
(c) carbon. (d) copper.

337. Which of the following is not the GLS lamp of the standard rating ?
(a) 15 W. (b) 40 W. (c) 75 W. (d) 100 W.

338. The luminous efficiency of GLS lamps is normally in the range of......lumens/watt.
(a) 100 to 150 (b) 75 to 100
(c) 50 to 75 (d) 10 to 18

339. Nitrogen or argon is filled in GLS lamps so as to
(a) retard evaporation of tungsten filament.
(b) improve efficiency.
(c) change the colour of the light.
(d) reduce the glare.

340. In case of frosted GLS lamps, frosting of shell is done by
(a) acid etching. (b) ammonia.
(c) ozone. (d) salt water.

341. In case of GLS lamps the increase in supply voltage reduces
(a) power consumption. (b) lumens output.
(c) life. (d) efficiency.

342. Sodium vapour lamp needs an ionisation voltage of about
(a) 5 V. (b) 20 V. (c) 50 V. (d) 100 V.

343. The ignition voltage for a sodium lamp is about
(a) 100–125 V. (b) 200–240 V.
(c) 300–400 V. (d) 400–600 V.

344. When a sodium vapour lamp is switched on, initially the colour is
(a) red. (b) pink. (c) yellow. (d) blue.

345. The colour of light given out by a sodium vapour discharge lamp is
(a) pink. (b) bluish green. (c) yellow. (d) blue.

346. An auto transformer used with sodium vapour lamp should have
(a) high step-up ratio. (b) high leakage reactance.
(c) high step-down resistance.
(d) high efficiency. **[West Bengal Univ. of Technology 2009-10]**

347. Leak transformer in a sodium vapour lamp initially provides
(a) high current. (b) low voltage.
(c) high voltage.

348. In a sodium vapour discharge lamp the neon gas
(a) changes the colour of light.
(b) prevents vapourization of filament.
(c) acts as a shield around the filament.
(d) assists in developing enough heat to vapourize the sodium.

349. The capacitor is used in auto-transformer circuit of a sodium vapour lamp in order to
(a) regulate discharge voltage.
(b) improve the circuit power factor.
(c) control lamp illumination level.
(d) protect the lamp against overvoltage.

350. The sodium vapour lamp
(a) is only suitable for ac and so needs choke control.
(b) needs capacitor in its auto-transformer circuit to improve the power factor which is very low (about 0.3 lagging).
(c) comes up to its rated output in approximately 15 minutes.
(d) all of the above.

351. The luminous efficiency of a sodium vapour lamp is .... lumens per watt.
(a) 40-50 (b) 50-100 (c) 120-200 (d) 10-12

352. Phosphor zinc silicate produces light of
(a) pink colour. (b) green colour.
(c) red colour. (d) yellow colour.

353. A mercury vapour lamp gives light of
(a) pink colour. (b) yellow colour.
(c) greenish-blue colour. (d) red colour.

354. High pressure mercury vapour lamps
(a) are similar in construction to sodium vapour lamps.
(b) need 4 or 5 minutes to attain full brilliancy.
(c) are generally used for general industrial lighting, railway yards, work areas, shoping centres etc.
(d) all of the above.

355. In high pressure mercury vapour lamps
(a) the main electrodes are made of tungsten wire in the form of helix.
(b) the choke is provided to limit the current to a safe value.
(c) in addition to two main electrodes a starting (auxiliary) electrode is connected through a high resistance.
(d) all of the above.

356. The luminous efficiency of high pressure mercury vapour lamps ranges from ______ lumens per watt.
(a) 30 to 40 (b) 60 to 100 (c) 100 to 150 (d) 250 to 400

357. In mercury iodide lamps
(a) a number of iodides are added, in addition to mercury, to fill up the gaps in the light spectrum.
(b) a separate ignition device, in addition to the choke, is provided.
(c) both (a) and (b).
(d) none of the above.

358. Mercury iodide lamps
(a) are similar in construction to high pressure mercury vapour lamps.
(b) have luminous efficiency of the order of 80 lumens per watt.
(c) are suitable for applications in the field of floodlighting, industrial lighting and public lighting.
(d) all of the above.

359. Which of the following lamps is a cold cathode lamp ?
(a) Sodium vapour lamp. (b) Neon lamp.
(c) Mercury vapour lamp. (d) None of the above.

360. Neon lamps
(a) are of the size of ordinary incandescent lamps.
(b) have power consumption of the order of 5 watts.
(c) are used as indicator lamps, night lamps, for determination of polarity of dc mains.
(d) all of the above.

361. In neon signs argon gas is used to give
(a) red colour. (b) blue colour.
(c) pink colour. (d) green colour.

362. In neon signs, neon with a mixture of mercury is used to give
(a) blue colour. (b) green colour.
(c) pink colour. (d) yellow colour.

363. The electrodes of neon tube work at
(a) very low temperatures. (b) ordinary voltage.
(c) 2 kV to 6 kV. (d) 33 kV.

Answers: 336. (b), 337. (c), 338. (d), 339. (a), 340. (b), 341. (c), 342. (a), 343. (d), 344. (b), 345. (c), 346. (b), 347. (c), 348. (d), 349. (b), 350. (d), 351. (a), 352. (b), 353. (c), 354. (d), 355. (d), 356. (a), 357. (c), 358. (d), 359. (b), 360. (d), 361. (a), 362. (b), 363. (c),

364. Neon tubes are widely used for
(a) advertising. (b) indoor lighting.
(c) road signalling. (d) airport lighting.

365. The vapour discharge tube used for domestic light has
(a) one filament. (b) two filaments.
(c) four filaments. (d) no filament.

366. The fluorescent tube is coated from inside with a thin layer of fluorescent material in the form of powder in order to
(a) absorb invisible ultraviolet rays and radiate visible rays.
(b) improve the appearance.
(c) reduce glare. (d) improve life.

367. A stabilising choke is connected in the fluorescent tube circuit so as to
(a) reduce the flicker.
(b) act as a ballast in operating conditions and provide a voltage impulse for starting.
(c) act as a starter. (d) avoid radio-interference.

368. A capacitor is connected across the fluorescent tube circuit in order to
(a) eliminate the noise. (b) limit the current.
(c) improve the power factor of the tube circuit.
(d) all of the above.

369. Which of the following is present inside the fluorescent tube ?
(a) Helium and oxygen. (b) Argon and neon.
(c) Argon and carbon dioxide.
(d) Mercury vapour.

370. In a fluorescent tube, a ballast resistance is connected in series with the choke
(a) when the tube is operated on dc supply.
(b) when the tube is operated on ac supply.
(c) to reduce radio-interference.
(d) to reduce stroboscopic effects.

371. For operation of fluorescent tube on dc supply the additional device incorporated in the tube circuit is a
(a) transformer. (b) resistor. (c) inductor.

372. The light of fluorescent tube appears cooler than that from an incandescent lamp. This is due to the fact that
(a) the tube consumes less power.
(b) surface area of tube is larger than that of the incandescent lamp.
(c) tungsten is not used in the tube .
(d) none of the above.

373. Sometimes the wheels of the rotating machinery, in fluorescent tube lighting, appear to be stationary . This is due to
(a) luminescence effect. (b) light fluctuations.
(c) stroboscopic effect. (d) all of the above.

374. The flicker effect of fluorescent lamps is more pronounced at
(a) lower voltages. (b) higher voltages.
(c) higher frequencies. (d) lower frequencies.

375. The frequency of flickers in a fluorescent tube operating at 220 V, 50 Hz supply is
(a) 200 per second. (b) 100 per second.
(c) 50 per second. (d) 25 per second.

376. Standard wattage of a 1 m fluorescent tube is
(a) 25 W. (b) 65 W. (c) 80 W. (d) 100 W.

377. A fluorescent tube can be operated on
(a) both dc as well as ac.
(b) ac only. (c) dc only.
(d) satisfactorily only on dc.

378. Radio-interference from a fluorescent tube can be reduced by
(a) eliminating choke.
(b) connecting a small capacitor across starter terminals.
(c) putting two tubes in parallel.
(d) any of the above.

379. In a mercury vapour lamp light, red objects appear black. This is on account of
(a) colour mixing.
(b) high wavelengths of red colour.
(c) absence of red light from lamp radiations.
(d) absorption of red light by the lamp.

380. Stroboscopic effect is related to
(a) fluorescent lamp. (b) incandescent lamp.
(c) night lamp. (d) none of these.
[West Bengal Univ. of Technology 2009-10]

381. The cost of a fluorescent tube is more than that of an incandescent lamp on account of
(a) more components in the tube.
(b) more quantity of glass in the tube.
(c) more labour required. (d) all of the above.

382. The life of a fluorescent tube is affected by
(a) low voltage. (b) high voltage.
(c) frequency of switching and blinking.
(d) all of the above.

383. The disadvantage(s) of fluorescent tubes in comparison to filament lamps is/are
(a) high cost. (b) noise in choke.
(c) stroboscopic effect. (d) all of the above.

384. The ratio of radiant heat produced by a fluorescent tube and that by a filament lamp of the same rating is
(a) 0.2. (b) 0.35. (c) 0.6. (d) 0.75.

385. The lamp that cannot sustain much voltage fluctuations is
(a) sodium vapour lamp. (b) mercury vapour lamp.
(c) incandescent lamp. (d) mercury iodide lamp.

386. The lamp that gives nearly ultraviolet light is
(a) carbon arc lamp. (b) vacuum type filament lamp.
(c) argon gas filled filament lamp.
(d) all of the above.

387. For the same wattage the cheapest lamp is
(a) sodium vapour lamp. (b) GLS lamp.
(c) halogen lamp. (d) fluorescent tube.

388. Power factor is highest in case of
(a) sodium vapour lamp. (b) mercury vapour lamp.
(c) incandescent lamp. (d) neon lamp.

389. The lamp which gives nearly monochromatic light is
(a) sodium vapour lamp. (b) mercury vapour lamp.
(c) tungsten filament lamp. (d) fluorescent tube.

390. The lamp that does not have separate choke is
(a) mercury vapour lamp. (b) sodium vapour lamp.
(c) fluorescent tube. (d) all of the above.

391. For the same wattage the lamp that will produce minimum radiant heat is
(a) sodium vapour lamp. (b) fluorescent lamp.
(c) tungsten filament lamp. (d) mercury vapour lamp.

392. The lamp/tube preferred for air-conditioned spaces is
(a) sodium vapour lamp. (b) mercury vapour lamp.
(c) fluorescent tube. (d) GLS lamp.

Answers: 364. (a), 365. (b), 366. (a), 367. (b), 368. (c), 369. (d), 370. (a), 371. (b), 372. (c), 373. (c), 374. (d), 375. (b), 376. (a), 377. (a), 378. (b), 379. (c), 380. (a), 381. (d), 382. (d), 383. (d), 384. (a), 385. (c), 386. (a), 387. (b), 388. (c), 389. (a), 390. (b), 391. (b), 392. (c),

393. The lamp that causes radio-interference is
(a) fluorescent tube. (b) halogen lamp.
(c) mercury vapour lamp. (d) all of the above.

394. The lamp that cannot be used for dimming is
(a) cold cathode lamp. (b) fluorescent lamp.
(c) GLS lamp. (d) all of the above.

395. Which of the following application does not require ultraviolet lamps ?
(a) Aircraft cockpit dashboard lighting.
(b) Blue print machines. (c) Car lighting.
(d) None of the above.

396. Floodlighting is not used for
(a) industrial purposes. (b) advertising purposes.
(c) aesthetic purposes. (d) any of the above.

397. The lumens emitted from a 100 W incandescent lamp will not increase if
(a) supply voltage is increased.
(b) filament temperature is increased.
(c) the diameter of the glass shell is increased.
(d) any of the above.

398. Total flux required in any lighting scheme depends inversely on
(a) surface area. (b) space-height ratio.
(c) illumination. (d) coefficient of utilisation.

399. To avoid glare
(a) object should be viewed from a distance.
(b) object should be moved constantly.
(c) object should be viewed from a close vicinity.
(d) viewer should move constantly.

400. The illumination level on a surface is least affected by
(a) candle power of light source.
(b) ambient temperature.
(c) type of reflector used.
(d) distance of the light source.

401. Desired illumination level on the working plane depends upon
(a) age group of observers.
(b) whether the object is stationary or moving.
(c) size of the object to be seen and its distance from the observer.
(d) all of the above.

402. Which of the following will need the highest level of illumination ?
(a) Living room. (b) Kitchen.
(c) Proof reading. (d) Hospital wards.

403. Which of the following will need the lowest level of illumination ?
(a) Workshop. (b) Displays.
(c) Railway platform. (d) Garage.

404. The illumination level in houses is in the range of
(a) 20–50 lux. (b) 100–200 lux.
(c) 300–500 lux. (d) 700 lux.

405. The illumination level required for precision work is about
(a) 500–1,000 lux. (b) 250 lux.
(c) 100 lux. (d) 50 lux.

406. The illumination level required for important traffic routes carrying fast traffic is about
(a) 30 lux. (b) 100 lux.
(c) 200 lux. (d) 5 lux.

407. Illumination due to moon light is about
(a) 0.03 lumen/m^2. (b) 0.3 lumen/m^2.
(c) 30–50 lumens/m^2. (d) 300–500 lumens/m^2.

408. The luminous flux reaching the working plane least depends on
(a) proportion of the room.
(b) the lumen output of the lamps.
(c) colour of the working plane surface.
(d) reflectance of internal surfaces.

409. The depreciation factor depends upon
(a) ageing of the lamp(s). (b) lamp cleaning schedule.
(c) type of work carried out at the premises.
(d) all of the above.

410. Utilisation factor depends upon
(a) size of the room.
(b) space-height ratio of the lamps.
(c) colour of walls/ceiling. (d) all of the above.

411. Glare is reduced by
(a) increasing the mounting height of the lamp.
(b) using diffusers.
(c) using reflectors to cut-off the light at certain angle.
(d) all of the above.

412. The glass that transmits the maximum light is
(a) clear glass. (b) milk glass.
(c) serrated glass. (d) opalescent glass.

413. The glass that transmits the least amount of light is
(a) clear glass. (b) milk glass.
(c) serrated glass. (d) opalescent glass.

414. The surface that has the lowest reflection factor for white light is
(a) white oil paint. (b) white plaster work.
(c) blue curtains. (d) aluminium sheet.

415. The lamp is provided with a reflector in order to
(a) avoid glare. (b) provide better illumination.
(c) protect the lamp. (d) all of the above.

416. Reflectors are provided with slits at the top so as to
(a) reduce heating effect. (b) reduce colour effect.
(c) introduce chimney effect for cleaning.
(d) all of the above.

417. Heat from light source is particularly of importance in
(a) designing for air-conditioning.
(b) designing for illumination level.
(c) designing for floor space utilization.
(d) all of the above.

418. Dimming systems for lights are used in
(a) auditoriums. (b) theatres.
(c) ball room. (d) all of the above.

419. The voltage required to pass the necessary current through an electrolytic cell is of the order of
(a) 1–2 V. (b) 10–20 V.
(c) 100–120 V. (d) 150–200 V.

420. The energy consumption for production of ammonium sulphate is of the order of
(a) 2,000–2,500 kWh/tonne. (b) 3,000–4,000 kWh/tonne.
(c) 4,000–7,000 kWh/tonne. (d) 70–80 kWh/tonne.

421. Materials used for copper plating are
(a) copper sulphate and sulphuric acid.
(b) copper sulphate and nitric acid.
(c) copper carbonate and ammonium carbonate.
(d) copper nitrate and sulphuric acid.

Answers: 393. (a), 394. (b), 395. (c), 396. (a), 397. (c), 398. (b), 399. (a), 400. (b), 401. (d), 402. (c), 403. (c), 404. (b), 405. (a), 406. (a), 407. (b), 408. (c), 409. (d), 410. (d), 411. (d), 412. (a), 413. (b), 414. (c), 415. (d), 416. (d), 417. (a), 418. (d), 419. (a), 420. (a), 421. (a),

422. Materials used for chromium plating are
(a) chromium carbonate and sulphuric acid.
(b) chromic acid and sulphuric acid.
(c) chromium chloride and hydrochlroride acid.
(d) none of the above.

423. The gold plating is carried out
(a) with a current density of 150–250 A/m^2 at a voltage of 1–2 V.
(b) with a current density of 50–150 A/m^2 at a voltage of 5–15 volts.
(c) with a current density of 100–150 A/m^2 at a voltage of 1–4 volts.
(d) with a current density of 50–150 A/m^2 at a voltage of 1 volt.

424. Power supply required for electrolytic processes is
(a) alternating current (100–200 A) at very low voltage (10 or 12 V).
(b) direct current (100–200 A) at very high voltage.
(c) direct current (100–200 A) at very low voltage (10 or 12 V).
(d) alternating current at very high voltage.

425. The plants for extraction and refining of metals or large scale manufacturing are located near the
(a) atomic power station. (b) hydroelectric power station.
(c) steam power station.
(d) either atomic power station or hydroelectric power station.

426. The process of coating of a metallic surface with a harder metal by electrodepostion is called
(a) electrofacing. (b) electroforming.
(c) electrometallisation. (d) either (a) or (b).

427. Basically electroplating means
(a) formation of ions by two metallic plates in the acidic liquid.
(b) electrodeposition of metal on electrodes.
(c) electrodeposition of metal upon metallic surfaces.
(d) none of the above.

428. Electroplating is done for
(a) replacement of worn out material.
(b) protection of metals against corrosion.
(c) giving a shining appearance to articles.
(d) all of the above.

429. The preparation of an object for electroplating involves the
(a) removal of oil, grease or other organic material.
(b) removal of rust, scale, oxides, or other inorganic coatings adhering to the metal.
(c) mechanical preparation of the metal surface by polishing, buffing etc .
(d) any or all of the above operations.

430. A one tonne refrigerating machine means that
(a) the total weight of the machine is one tonne.
(b) the quantity of the refrigerant used is one tonne.
(c) one tonne of water can be converted into ice.
(d) one tonne of ice when melts from and at 0ºC in 24 hours, the refrigerating effect is equivalent to 14,000 kJ per hour.

431. The capacity of a refrigerating machine is expressed as
(a) inside volume of the cabinet.
(b) lowest temperature attained.
(c) gross weight of machine in tonnes.
(d) rate of abstraction of heat from the space being cooled.

432. Which of the following properties of a refrigerant is undesirable ?
(a) High critical temperature.
(b) Low specific heat of liquid.
(c) High boiling point.
(d) Low specific volume of vapour.
(e) High latent heat of vapourisation.

433. Various desirable properties which should be possessed by refrigerants are
(a) non-corrosive and non-inflammability.
(b) no-toxic and freedom from objectionable odour with low working pressure.
(c) both (a) and (b). (d) none of the above.

434. In a refrigeration cycle, the flow of refrigerant is controlled by
(a) compressor. (b) expansion valve.
(c) condenser. (d) evaporator.

435. Which part of the vapour compression refrigeration cycle, produces the refrigeration effect ?
(a) Evaporator. (b) Condenser.
(c) Expansion valve. (d) Compressor.

436. .... is used as refrigerant in a vapour absorption refrigerator.
(a) Acqua ammonia (b) Freon
(c) Water (d) Sulphur dioxide

437. The refrigerant Freon 12 is a compound consisting of
(a) carbon, fluorine and chlorine.
(b) carbon, chlorine and hydrogen.
(c) carbon, fluorine and hydrogen.
(d) chlorine, bromine and iodine.

438. Ammonia is preferred as a refrigerant in large commercial installations because
(a) it is non-tonic.
(b) it has a low working pressure.
(c) it is relatively cheap. (d) it has low latent heat.

439. In a vapour compression system, which of the following units is adversely affected by the presence of moisture ?
(a) Evaporator. (b) Expansion valve.
(c) Compressor. (d) Condenser.

440. In a vapour absorption system lithium bromide is used as
(a) refrigerant. (b) lubricant.
(c) absorbent. (d) cooling substance.

441. Vapour absorption refrigeration systems have the advantage(s) of
(a) absence of moving parts.
(b) universal power source.
(c) automatic defrostration. (d) both (a) and (b).

442. Thermoelectric refrigeration system has the advantage(s) of
(a) small initial cost. (b) absence of moving parts.
(c) ease of automatic control by variation of magnitude of current.
(d) both (b) and (c).

443. In a domestic refrigerator, compressor and motor are assembled in a single unit known as .... sealed unit.
(a) hermetically (b) homogeneously
(c) heterogeneously (d) none of these

444. Hermetically sealed units have the advantage(s) of
(a) relatively silent operation.
(b) minimum gas leakage problem because no moving part extends through the sealing housing.
(c) minimum maintenance requirement.
(d) all of the above.

445. Which of the following motors is used in household refrigerators ?
(a) DC shunt motor. (b) Reluctance motor.
(c) Single phase induction motor.
(d) Synchronous motor.

Answers: 422. (b), 423. (b), 424. (c), 425. (d), 426. (a), 427. (c), 428. (d), 429. (d), 430. (d), 431. (d), 432. (c), 433. (c), 434. (b), 435. (a), 436. (a), 437. (a), 438. (b), 439. (b), 440. (c), 441. (d), 442. (d), 443. (a), 444. (d), 445. (c),

446. The compressor of a domestic refrigerator has been noted to run for comparatively large durations, it indicates that
(a) thermostat has become defective.
(b) the machine is overloaded.
(c) capillary tube is choked.
(d) none of the above.

447. In a refrigerator, defrosting may be done by
(a) opening the door of the refrigerator.
(b) stopping the compressor for a while.
(c) placing trays of warm waters in the refrigerator.
(d) any of the above.

448. Types of water coolers are
(a) instantaneous or pressure type.
(b) storage type. (c) exhaust fan type.
(d) both (a) and (b).

449. Air-conditioning is the simultaneous control of .... in a confined space.
(a) temperature (b) humidity
(c) air movement
(d) temperature, humidity, purity and movement of air

450. .... is a process which is generally used for summer air-conditioning.
(a) Cooling with dehumidification
(b) Chemical dehumidification
(c) Adiabatic cooling
(d) Cooling with humidification

451. .... is a process that is generally used in winter air-conditioning.
(a) Humidification (b) Dehumidification
(c) Heating and humidification
(d) Heating and dehumidification

452. Air is dehumidified by
(a) cooling. (b) chemical absorption.
(c) heating. (d) both (a) and (b).

453. Compared to individual systems, the central air-conditioning system has .... overall efficiency.
(a) higher (b) lower (c) same (d) unpredictable

454. The main disadvantage(s) of central type air-conditioning is/are that
(a) dusting is very costly.
(b) absence of any individual room temperature adjustments.
(c) cigarette smoke and bacteria present in the return air from infected rooms redistribute to healthy rooms.
(d) all of the above.

455. Which of the following unitary-central systems are in common use ?
(a) Fan coil units. (b) Induction units.
(c) All air high velocity systems.
(d) All of the above.

456. The conditions conductive to comfort depends upon ....
(a) humidity (b) temperature
(c) air movement (d) air purity
(e) all of the above factors

457. Main traction systems used in India are those using .... locomotives.
(a) diesel engine (b) diesel electric
(c) electric (d) all of the above

458. The steam engine provided on steam locomotive is
(a) double acting condensing type.
(b) double acting non-condensing type.
(c) single acting condensing type.
(d) single acting non-condensing type.

459. In case of steam locomotives the tractive effort is provided by
(a) double cylinder, double acting steam engine.
(b) double cylinder, single acting steam engine.
(c) single cylinder, double acting steam engine.
(d) single cylinder, single acting steam engine.

460. In case of a steam locomotive an average coal consumption per km of run is around
(a) 5–10 kg. (b) 25–30 kg.
(c) 60–80 kg. (d) 100–150 kg.

461. The pressure of steam used in steam locomotives is about
(a) 1–5 kgf/cm^2. (b) 5–10 kgf/cm^2.
(c) 10–15 kgf/cm^2. (d) 20–30 kgf/cm^2.

462. The maximum horse power of steam locomotive is
(a) up to 1,500. (b) 1,500–2,000.
(c) 2,000–3,000. (d) 3,000–4,000.

463. The speed of a steam locomotive is controlled by
(a) applying brakes. (b) gear box.
(c) regulating steam flow to engine.
(d) flywheel.

464. The pulsating torque exerted by steam locomotives causes
(a) jolting and skidding. (b) hammer blow.
(c) pitching. (d) all of the above.

465. Steam engine drive
(a) has limited overload capacity.
(b) is not clean drive.
(c) is not suitable for urban or suburban services or for hauling heavy loads on steep gradients.
(d) all of the above..

466. Steam locomotives
(a) cannot be put into service at any moment.
(b) cause considerable wear on the track.
(c) need more repair and maintenance.
(d) all of the above.

467. Steam engine drive is not suitable for urban and suburban services as
(a) it is not clean drive.
(b) coefficient of adhesion of steam locomotives is quite low.
(c) steam locomotive causes considerable wear on the track.
(d) steam locomotive has limited speed.
(e) all of the above.

468. Direct internal combustion drive
(a) has higher maintenance and operating costs.
(b) has no starting torque and needs some auxiliary means for its starting.
(c) has shorter life of propulsive equipment.
(d) all of the above.

469. Diesel electric traction has comparatively limited overload capacity because
(a) diesel engine is a constant output prime mover.
(b) diesel engine has shorter life span.
(c) regenerative braking cannot be employed.
(d) diesel-electric locomotive is heavier than an ordinary electric locomotive.

Answers: 446. (b), 447. (d), 448. (d), 449. (d), 450. (a), 451. (c), 452. (d), 453. (a), 454. (c), 455. (d), 456. (e), 457. (d), 458. (b), 459. (a), 460. (b), 461. (c), 462. (a), 463. (c), 464. (a), 465. (d), 466. (d), 467. (e), 468. (d), 469. (a),

470. The range of horse power of diesel locomotives is
(a) 100–500. (b) 1,500–2,500.
(c) 3,000–4,500 (d) 4,500–5,000.

471. In India, diesel locomotives are manufactured at
(a) Varanasi. (b) Kolkata. (c) Bangalore. (d) Ajmer.

472. In diesel-electric drive
(a) initial investment required is low.
(b) locomotive and train is a self-contained unit.
(c) power loss in speed control is low.
(d) all of the above.

473. Battery driven vehicles
(a) are easy to control and very convenient to use.
(b) have low maintenance cost.
(c) cause no pollution. (d) all of the above.

474. Battery operated trucks are used
(a) in power stations.
(b) for local delivery of goods in large towns with maximum daily run up to 50–60 km.
(c) in narrow gauge traction. (d) for main line service.

475. Electric traction in comparison to other traction systems has the advantage(s) of
(a) higher acceleration and braking retardation.
(b) cleanest system and so ideally suitable for the underground and tube railways.
(c) better speed control. (d) all of the above.

476. Electric railway can handle the traffic up to double the amount possible with steam railway. It is because of
(a) better speed control.
(b) larger passenger carrying capacity.
(c) higher schedule speed. (d) both (b) and (c).

477. Electric traction in comparison to other traction systems has the drawback(s) of
(a) interference with communication lines running along the track.
(b) heavy initial expenditure in laying out overhead electric supply system.
(c) interruption of traffic for hours owing to short time power failure.
(d) all of the above.

478. The most vital factor against electric traction is
(a) its high maintenance cost.
(b) possibility of power failure.
(c) high initial cost in laying out overhead electric supply system.
(d) necessity of providing a negative booster.

479. The locomotive that has the highest operational availability is
(a) diesel-electric. (b) electric.
(c) steam. (d) steam-electric.

480. Unbalanced forces are maximum in case of
(a) diesel shunters. (b) diesel locomotives.
(c) steam locomotives. (d) electric locomotives.

481. Maintenance requirements are minimum in case of
(a) electric lcomotives. (b) steam locomotives.
(c) diesel electric locomotives.
(d) diesel engines.

482. A drive suitable for mines where explosive gas may exist, is
(a) steam engine. (b) battery locomotive.
(c) diesel engine. (d) any of the above.

483. A submarine is provided driving power through (while moving under water)
(a) diesel engines. (b) steam engines.
(c) batteries. (d) gas turbine.

484. Ordinary tramway is the most economical means of transport for
(a) very dense traffic in large cities.
(b) rural services. (c) suburban services.

485. Unlike a tramway, a trolley bus needs no
(a) driving axles. (b) running rail.
(c) hand brakes. (d) overhead contact wire.

486. The direct current system employed for tramways operates on
(a) 3,000 V. (b) 1,500 V. (c) 750 V. (d) 400 V.

487. For tramways the return circuit is
(a) track rails. (b) neutral wire.
(c) common earthing. (d) cables.

488. For 600 V dc line for tramcars, track is connected to the
(a) positive of the supply. (b) negative of the supply.
(c) midvoltage of 300 V. (d) none of the above.

489. In tramways
(a) the power is supplied at 600 V dc from a single overhead conductor of + ve polarity.
(b) the speed is controlled by field weakening or series-parallel control.
(c) rheostatic and mechanical brakings are employed for normal service.
(d) all of the above.

490. The main drawback of tramways is that
(a) the life of tramcar equipment is short.
(b) it has restricted manoeuvrability.
(c) it needs laying overhead supply system and track for its use, which is costly to maintain and constitutes a source of danger to other road users.
(d) all of the above.

491. The trolley bus
(a) is an electrically operated pneumatic tyred vehicle, usually fed at 600 V dc from two overhead contact wires.
(b) is able to manoeuvre in a traffic over a width of several metres on either side of the central line of the trolley wires.
(c) uses field weakening method for its speed control.
(d) all of the above.

492. The trolley buses
(a) do not require laying of track for its use as return conductor.
(b) make use of foot operated master controllers.
(c) make use of stabilised rheostatic braking.
(d) all of the above.

493. In diesel electric traction
(a) the driving motors are dc series type and fed from the dc generators driven by the diesel engine put on the same locomotive.
(b) the starting and speed control is affected by varying the excitation of the generator.
(c) regenerative braking is not possible.
(d) all of the above.

494. Suburban railways use
(a) 1,500 V dc. (b) 400 V, 3 phase ac.
(c) 3300 V, 3 phase ac. (d) 600 V, 3 phase ac.

495. Long distance railways operate on
(a) 600 V dc. (b) 25 kV single phase ac.
(c) 25 kV three phase ac. (d) 15 kV three phase ac.

Answers: 470. (b), 471. (a), 472. (d), 473. (d), 474. (b), 475. (d), 476. (d), 477. (d), 478. (c), 479. (b), 480. (c), 481. (a), 482. (b), 483. (c), 484. (a), 485. (b), 486. (c), 487. (a), 488. (b), 489. (d), 490. (c), 491. (d), 492. (d), 493. (d), 494. (a), 495. (b),

496. The electric locomotives run faster at curved routes in comparison to steam locomotives because
(a) its centre of gravity is lower than that of steam locomotive.
(b) its centre of gravity is higher than that of steam locomotive.
(c) the speed at curved routes is independent of locations of centre of gravity.
(d) none of the above.

497. Low frequency operation of overhead line in traction system
(a) increases the spacing between substations.
(b) reduces the spacing between substations.
(c) spacing is independent of frequency of supply.

498. In Kando system of track electrification
(a) single phase ac is converted into dc.
(b) single phase ac is converted into 3-phase ac.
(c) 3 phase ac is converted into dc.
(d) 3 phase ac is converted into single phase ac.

499. The composite system (single phase ac to dc system) has been chosen for all future track electrification in India as
(a) it needs light overhead catenary.
(b) it needs less number of substations.
(c) it combines the advantages of high voltage ac distribution at 50 Hz with dc series traction motors.
(d) it provides flexibility in the location of substations.

500. Single phase ac system is preferred for main line railway service because
(a) the initial, maintenance and operating costs of ac substations are less as the substation equipment required in single phase ac system is less, cheap and efficient.
(b) number of substations required is less.
(c) of reduced cost of distribution system.
(d) all of the above.

501. In underground traction, the supply system is
(a) 500 V to 1,000 V dc. (b) 25 kV, 50 Hz.
(c) 50 kV, 50 Hz. (d) 25 kV, 25 Hz.
[A.M.I.E. Sec. B. Elec. Drives & Their Control Winter 1996]

502. Problem(s) of single phase traction systems is/are
(a) current and voltage unbalance.
(b) electrostatic and electromagnetic induction.
(c) generation of harmonics. (d) all of the above.

503. The chief requirements of main line railway services are
(a) high maximum speed.
(b) minimum cost of overhead structure.
(c) high acceleration and braking retardation.
(d) both (a) and (b).

504. The main requirements of suburban railway services are
(a) rapid acceleration and braking retardation.
(b) no interference to the communication circuits running along the track.
(c) high maximum speed. (d) both (a) and (b).

505. The acceleration rate for urban or suburban services is
(a) 1.5–4 kmphps. (b) 3–4 kmphps.
(c) 5–10 kmphps. (d) 0.5–1.5 kmphps.

506. The braking retardation for urban or suburban services is
(a) 1.5–2.5 kmphps. (b) 3–4 kmphps.
(c) 5–10 kmphps. (d) 0.5–1.5 kmphps.

507. The coasting retardation for main line railway services is about
(a) 10 kmphps. (b) 3 kmphps.
(c) 0.16 kmphps. (d) 0.01 kmphps.

508. The maximum speed at which trains run on main line railway service is
(a) 160 kmph. (b) 120 kmph.
(c) 100 kmph. (d) 200 kmph.

509. In main line services as compared to urban and suburban services
(a) distance between the stops is more (exceeding 10 km).
(b) maximum speed attained is high.
(c) acceleration and braking retardation rates are low.
(d) all of the above.

510. The speed-time curve for urban service has no
(a) coasting period. (b) free running period.
(c) braking period. (d) acceleration period.

511. Free running and coasting periods are generally long in case of
(a) city service. (b) suburban serivce.
(c) main line service. (d) outer suburban service.

512. In suburban services as compared to urban services
(a) the coasting period is longer.
(b) the coasting period is smaller but free running period is longer.
(c) the coasting and free running periods are smaller.
(d) none of the above is true.
[A.M.I.E. Sec. B. Elec. Drives & Their Control Winter 1995]

513. Trapezoidal speed-time curve is a better approximation to the actual conditions for
(a) suburban services. (b) urban services.
(c) main line service. (d) urban and suburban services.
[West Bengal Univ. of Technology 2009-10]

514. Quadrilateral speed-time curve is the close approximation for
(a) urban service. (b) suburban service.
(c) urban/suburban service. (d) main line service.

515. The speed-time curve for a main-line service is illustrated in figure given below.
In figure given below AB represents
(a) coasting.
(b) free running.
(c) acceleration.
(d) average speed.

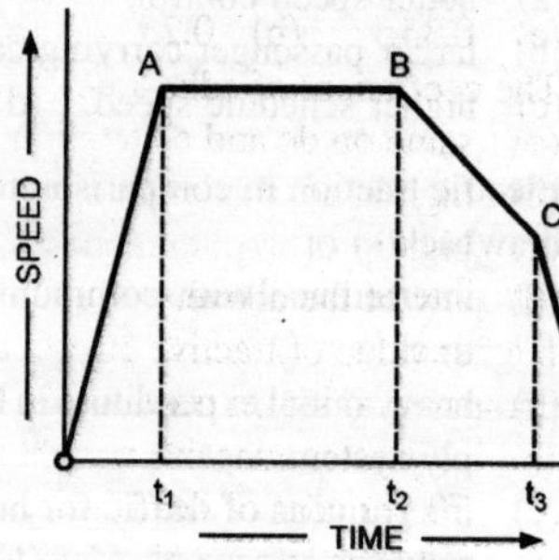

516. Area under the speed-time curve represents
(a) total distance travelled. (b) average speed.
(c) average acceleration. (d) none of the above.

517. The duration of coasting is represented by the duration
(a) $0 - t_1$ (b) $t_1 - t_2$ (c) $t_2 - t_3$ (d) $t_3 - t_4$

518. From the above fig it can be concluded that
(a) average acceleration is zero.
(b) during coasting and braking the acceleration is negative.
(c) acceleration during free running is zero.
(d) both (b) and (c).

519. The speed of train estimated taking into account the stoppage time at a station in addition to the actual running time between stops, is called the......speed.
(a) average (b) schedule (c) free running (d) notching

520. The average speed of a train is independent of
(a) duration of stops.
(b) acceleration and braking retardation.
(c) distance between stops. (d) running time.

521. The schedule speed of a given train when running on a given service (with given distance between stations) is affected by
(a) acceleration and braking retardation.

Answers: 496. (a), 497. (a), 498. (b), 499. (c), 500. (d), 501. (a), 502. (d), 503. (d), 504. (d), 505. (a), 506. (b), 507. (c), 508. (a), 509. (d), 510. (b), 511. (c), 512. (a), 513. (c), 514. (c), 515. (b), 516. (a), 517. (c), 518. (d), 519. (b), 520. (a), 521. (d),

(b) maximum or crest speed. (c) duration of stop.
(d) all of the above.

522. Skidding of a vehicle always occurs when
(a) braking effort exceeds its adhesive weight.
(b) brake is applied suddenly.
(c) it negotiates a curve.
(d) it passes over points and crossings.

523. The adhesive weight is the
(a) total weight of the locomotive and the train.
(b) weight coming over the driving wheels.
(c) same as the accelerating weight.
(d) none of the above.

524. Coefficient of adhesion is the ratio of tractive effort to slip the wheels and
(a) dead weight. (b) accelerating weight.
(c) adhesive weight. (d) none of the above.

525. The normal value of coefficient of adhesion is
(a) 0.25. (b) 0.35. (c) 0.50. (d) 0.65.

526. Coefficient of adhesion reduces due to the presence of
(a) dew on rails. (b) oil and grease on rails.
(c) dry sand on rails. (d) both (a) and (b).

527. Coefficient of adhesion improves due to presence of
(a) dry sand on rails. (b) rust on rails.
(c) dust on rails. (d) all of the above.

528. The value of coefficient of adhesion will be high when rails are
(a) wet. (b) cleaned with sand.
(c) greased. (d) sprayed with oil.

529. The coefficient of adhesion for wet or greasy rails is
(a) 0.35. (b) 0.25. (c) 0.08. (d) zero.

530. The coefficient of adhesion
(a) same on dc and ac traction.
(b) high in ac traction and low in dc traction.
(c) high in dc traction and low in ac traction.
(d) any of the above.

531. Higher value of tractive effort can be used in electric traction as compared to steam traction because
(a) of greater adhesive weight (the weight on driving wheels is 70 per cent of dead weight in case of electric locomotive whereas in case of steam locomotive it is 50 per cent).
(b) the torque exerted in electric traction is continuous whereas in steam traction it is pulsating one.
(c) the driving wheels are distributed over a much greater length in electric traction whereas in steam traction they are close to each other.
(d) all of the above.

532. When a bogie negotiates a curve, reduction in adhesion occurs resulting in sliding. This sliding is acute when
(a) degree of curvature is more.
(b) wheel base of axles is more.
(c) none of the above. (d) both (a) and (b).

533. For a given maximum axle load tractive effort with ac locomotive will be
(a) less than that of dc locomotive.
(b) more than that of dc locomotive.
(c) equal to that of dc locomotive. (d) none of the above.

534. The resistance encountered by a train in motion is on account of
(a) resistance offered by air.
(b) friction at the track.
(c) friction at various parts of the rolling stock.
(d) all of the above.

535. The air resistance to the movement of the train is proportional to
(a) 1/speed. (b) (speed). (c) $(\text{speed})^2$. (d) $(\text{speed})^3$

536. The friction at the track is proportional to
(a) speed. (b) $(\text{speed})^2$. (c) $(\text{speed})^3$. (d) $(1/\text{speed})^2$.

537. If the resistance to electric train movement is given by
$$F_r = a + bv + cv^2$$
In the above expression b is likely to cover
(a) air resistance. (b) track resistance.
(c) frictional resistance. (d) none of the above.

538. Tractive effort of an electric locomotive can be increased by
(a) using high output motors.
(b) increasing the supply voltage.
(c) increasing dead weight over the driving axles.
(d) both (a) and (c).

539. A locomotive with a mass of 50,000 kg on a track whose coefficient of adhesion is 20 per cent will produce a tractive effort of (approximately)
(a) 1 kN. (b) 100 kN. (c) 25 kN. (d) 250 kN.
[A.M.I.E. Sec. B. Drives & Their Control Winter 1996]

540. Energy consumption in propelling the train is required for
(a) accelerating of train mass.
(b) overcoming the gradient while moving up the gradient.
(c) overcoming the train resistance.
(d) all of the above.

541. Longer coasting period for a train results in
(a) higher schedule speed .
(b) lower specific energy consumption.
(c) higher retardation. (d) higher acceleration.

542. Specific energy consumption is affected by
(a) acceleration and retardation values.
(b) the crest speed and nature of route.
(c) distance between stops. (d) all of the above.

543. Specific energy consumption is minimum in..... services.
(a) main line (b) urban
(c) suburban (d) equal for all type of

544. For a given run and given schedule speed, the specific energy consumption is
(a) lower, the higher the acceleration and retardation.
(b) higher, the higher the acceleration and retardation.
(c) lower, the higher the acceleration and lower the retardation.
(d) lower, the lower the acceleration and higher the retardation.

545. Specific energy consumption
(a) increases with the increase in maximum speed.
(b) decreases with the increase in maximum speed.
(c) is independent of maximum speed.
(d) none of the above.

546. If the speed-time curves are similar (not identical), the specific energy consumption of the curve having higher maximum speed is .... than that of lower maximum speed.
(a) lower (b) equal (c) higher (d) none

547. Specific energy consumption becomes
(a) more when distance between stops is more.
(b) more with the higher values of acceleration (or retardation).
(c) more with high train resistance.
(d) less with the increase in crest speed.

548. The specific energy consumption for suburban services is usually ...... watt hours per tonne-km.
(a) 20-30 (b) 30-45 (c) 50-75 (d) 100-150

Answers: 522. (a), 523. (b), 524. (c), 525. (a), 526. (d), 527. (d), 528. (b), 529. (c), 530. (b), 531. (d), 532. (d), 533. (b), 534. (d), 535. (c), 536. (a), 537. (b), 538. (d), 539. (b), 540. (d), 541. (b), 542. (d), 543. (a), 544. (a), 545. (a), 546. (b), 547. (c), 548. (c),

549. The specific energy consumption for main line services is around .... watt hours per tonne-km.
(a) 20-30 (b) 30-45 (c) 50-75 (d) 100-150

550. Specific energy consumption is maximum in .... services.
(a) urban (b) suburban
(c) main line (d) equal for all types of

551. The electric motor used for traction work should be mechanically
(a) small in overall dimensions (specially in its overall diameter).
(b) light in weight and robust in construction.
(c) capable to withstand continuous vibrations.
(d) all of the above.

552. The electric motor used for traction work, electrically should be
(a) capable of developing high starting torque and withstanding voltage fluctuations, and temporary supply interruptions.
(b) amenable to simple speed control methods, self protective against excessive overloading and amenable to easy and simple methods of rheostatic and/or regenerative braking.
(c) of such characteristics that when they are operated in parallel and coupled mechanically, they share the load almost equal.
(d) all of the above.

553. In suburban trains, the traction motors are installed on
(a) locomotive only. (b) locomotive and coaches.
(c) coaches only.
[A.M.I.E. Sec. B. Elec. Drives & Their Control Winter 1996]

554. The type of dc motor used in electric traction is
(a) series. (b) shunt.
(c) separately excited. (d) none of the above.
[A.M.I.E. Sec. B. Electric Drives & Their Control Summer 1999]

555. The dc series motor is most suitable for traction services but more particularly for urban/suburban services because
(a) dc series motors are suitable for regenerative braking.
(b) dc series motors are capable of withstanding rapid fluctuations in supply voltage.
(c) dc series motors are capable of developing high torque at start.
(d) dc series motors are capable of withstanding temporary interruption of supply without undue rush of current.

556. The three phase induction motors have the advantages of simple and robust in costruction, trouble free operation, high voltage operation consequently requiring reduced amount of current and automatic regeneration but are not suitable for traction work. This is because of their
(a) flat speed-torque characteristics, constant speed operation and low starting torque.
(b) complicated speed control system and complicated overhead feeding systems.
(c) higher initial as well maintenance cost.
(d) both (a) and (b).

557. DC shunt motors are not suitable for traction services because of their
(a) hard characteristics.
(b) large time constant.
(c) power varying directly with developed torque.
(d) wide variation in torque and flux due to variation in voltage in electric traction.
(e) all of the above.

558. Two dc shunt motors having identical characteristics are used to drive a tramcar with unequal wheel diameters. The two motors share load equally if they are connected in
(a) series. (b) parallel.
(c) the loading will always be unequal.
[A.M.I.E. Sec. B. Elec. Drives & Their Control Winter 1995]

559. Two locomotives having identical tractive effort-slip characteristics haul a heavy train. The loco with larger diameter shares .... tractive effort.
(a) larger (b) smaller (c) equal

560. For single phase ac system of track electrification, low frequency is desirable as
(a) it improves commutation properties of ac motors.
(b) it increases efficiency of ac motors.
(c) it improves power factor of ac motors.
(d) all of the above.

561. At low frequency of the order of 1/2 Hz to 10 Hz, the induction motor develops
(a) high starting torque with excessive starting current.
(b) high starting torque without excessive starting current.
(c) low starting torque with excessive starting current.
(d) low starting torque without excessive starting current.

562. When two or more motors are used for traction service, the method of speed control used will be
(a) rheostatic control. (b) series-parallel control.
(c) field control. (d) motor generator control.

563. The advantage(s) of series-parallel starting is/are
(a) small energy loss and higher efficiency.
(b) economical speed control.
(c) higher reliability of operation.
(d) all of the above.

564. Parallel operation of traction motors is easier with
(a) dc shunt motors. (b) dc series motors.
(c) induction motors. (d) none of these.

565. Series traction motor operating on undulating dc supply has undulation
(a) $\frac{I_{max} - I_{min}}{I_{max} + I_{min}}$. (b) $\frac{I_{max} - I_{min}}{I_{max}}$.
(c) $\frac{I_{max}}{I_{max}}$. (d) $\frac{I_{max} - I_{min}}{2I_{min}}$.

566. In motor-generator locomotive control
(a) rheostatic control is used.
(b) series parallel control is used.
(c) the output voltage of generator is regulated by means of field control from exciter.
(d) any one of the above method is used.

567. In metadyne control,
(a) current throughout the accelerating period remains constant.
(b) tractive effort developed is uniform.
(c) there is no wastage of energy in starting rheostat.
(d) all of the above.

568. The method of speed control adopted in 25 kV, single phase, 50 Hz traction is
(a) tap changing control of transformer.
(b) reduced current method.
(c) series-parallel control.
(d) rheostatic control.

Answers: 549. (a), 550. (a), 551. (d), 552. (d), 553. (a), 554. (a), 555. (c), 556. (d), 557. (e), 558. (a), 559. (a), 560. (d), 561. (b), 562. (b), 563. (d), 564. (b), 565. (a), 566. (c), 567. (d), 568. (a),

569. The preferable method of speed control of linear induction motor is
(a) variable flux control. (b) PAM control.
(c) variable frequency and constant voltage control.
(d) variable frequency and variable voltage control.

570. The method that can bring the locomotive to standstill is
(a) rheostatic braking. (b) regenerative braking.
(c) plugging. (d) none of these.
[West Bengal Univ. of Technology 2009-10]

571. The braking system employed with steam locomotives is
(a) vacuum system. (b) pneumatic system.
(c) hydraulic system. (d) any of the above.

572. Braking system employed in diesel electric traction is
(a) hydraulic type. (b) vacuum type.
(c) regenerative type. (d) any of these.

573. The type of braking used in electric traction is
(a) mechanical braking. (b) vacuum brake system.
(c) electropneumatic braking.
(d) both (a) and (c).

574. Vacuum is created by
(a) vacuum pumps. (b) ejector.
(c) vacuum pump or ejector. (d) none of the above.

575. Bearings employed for supporting axles of rolling stock are
(a) bush bearings. (b) roller bearings.
(c) ball bearings. (d) journal bearings.

576. The wheels of a train, engine as well as bogies, are slightly tapered so as to
(a) reduce friction. (b) increase friction.
(c) facilitate in taking turns. (d) facilitate braking.

577. B_0B_0 locomotives have two bogies with
(a) two driving axles with individual drive motors.
(b) two driving axles with group drives.
(c) four driving axles with individual drive motors.
(d) four driving axles with group drives.

578. A locomotive for Indian Railways has been designated as WAM. In this W indicates that
(a) the locomotive is to run on metre gauge track.
(b) the locomotive is to operate on broad gauge track.
(c) the locomotive is for goods trains only.
(d) the locomotive is only for shunting duty.

579. Locomotives with manometer bogies
(a) are suited both for passengers as well as freight service.
(b) have better coefficient of adhesion.
(c) have better riding qualities owing to reduction of lateral forces.
(d) all of the above.

580. Factor(s) affecting the design of rolling stock is/are
(a) environmental condition.
(b) vibrations and shocks.
(c) unbalanced and fluctuating supply voltages for auxiliaries.
(d) all of the above.

581. In electric traction if contact voltage exceeds 1,500 V, current collection is invariably by
(a) a contact rail. (b) a conductor rail.
(c) overhead wire. (d) third rail.

582. Conductor rail system of supply is
(a) cheap and easy to repair and inspection.
(b) suitable for heavy current collection at voltages up to 1,200 V.
(c) universally used for all ac railways.
(d) both (a) and (b).

583. Overhead lines for power supply to tramcars are at a minimum height of a
(a) 10 m. (b) 8 m. (c) 5 m. (d) 15 m.

584. The current collector that can be employed with different speeds under all wind conditions and stiffness of OHE is known as the
(a) messenger collector. (b) pantograph collector.
(c) trolley collector. (d) bow collector.

585. DC track circuit consists of
(a) amplitude modulation equipment.
(b) a negative booster, feeding points and signals.
(c) insulated joint and track, track bonding, regulating resistance, track battery to track relay.
(d) all of the above.

586. In case of bow and pantograph current collectors
(a) the pantograph current collector is employed for large currents at higher speeds.
(b) the pantograph current collector is employed for large currents at lower speeds.
(c) the bow current collector is employed for large currents at higher speeds.
(d) the bow current collector is employed for large currents at lower speeds.

587. Contact with wire is maintained by
(a) air pressure in both bow and pantograph current collectors.
(b) springs in both bow and pantograph current collectors.
(c) spring in case of bow and air pressure in case of pantograph.
(d) spring in case of pantograph and air pressure in case of bow collectors.

588. Factors affecting quality of current collection in pantograph is/are
(a) contact pressure.
(b) effective mass of frame and span.
(c) adequate damping in the frame.
(d) all of the above.

589. Advantages of two-stage pantograph is/are
(a) inertia forces are reduced.
(b) takes care of variation in contact height and displacement of contact wire between supports.
(c) takes care of low amplitude displacement of contact wire between droppers.
(d) all of the above.

590. Span length between supports in electric traction is determined by
(a) weight of wire per unit length.
(b) maximum wind pressure.
(c) permissible tension in the wire and speed of train.
(d) all of the above.

591. Electrical circuit breaker is
(a) connected between the current collector and the main wiring.
(b) provided on the electric locomotive to protect the electrical equipment against excessive overloads automatically.
(c) provided with a handle to trip it when it is moved to the OFF position by hand.
(d) all of the above.

592. The characteristics of OHE depend on
(a) loss of contact and its duration.
(b) temperature and wear of contact wire.
(c) quality of current collection and pantograph pressure on contact wire.
(d) height, stagger, push up of an gradient of contact wire.
(e) all of the above

Answers: 569. (c), 570. (c), 571. (a), 572. (b), 573. (c), 574. (c), 575. (b), 576. (c), 577. (a), 578. (b), 579. (d), 580. (d), 581. (c), 582. (d), 583. (a), 584. (b), 585. (c), 586. (a), 587. (b), 588. (d), 589. (d), 590. (d), 591. (b), 592. (e),

593. Insulated overlaps are employed in electric traction to provide
(a) protection of transmission equipment in electric locomotives.
(b) isolating facilities required for the maintenance and operation of OHE.
(c) protection of signals in EMU trains.
(d) gap between adjacent terminal sections of railways.

594. Neutral sections are provided in single phase ac traction as
(a) to protect the system from lightning strokes.
(b) supply cannot be maintained from adjacent substations at reduced voltage drops.
(c) adjacent substations tap different phases of three phase system to achieve balancing.
(d) to provide return section for the communication purpose.

595. Use of booster transformers is restricted to urban areas because
(a) there are no earth and water currents.
(b) they are subjected to densely populated areas and so need protection.
(c) they complicate the OHE and cause more faults in the system and thus reduce the system reliability.
(d) they increase the impedance and need feeding posts to be located at close intervals.
(e) both (c) and (d). (f) all of the above.

596. Negative booster
(a) is a separately excited generator whose armature is connected to negative busbar and far end of the track through negative feeder.
(b) creates low voltage level in the track to allow all currents to flow through it.
(c) provides protection against earth fault currents.
(d) causes additional voltage drop necessary for smooth operation.
(e) both (a) and (b).

597. In long distance electric trains power for lighting in passenger coaches is provided
(a) through rails. (b) through locomotive.
(c) through individual generator of bogie and batteries.
(d) directly from overhead electric lines.

598. Power requirements of a train lighting and air-conditioning is met by
(a) servogenerators.
(b) chargeable batteries in each compartment.
(c) axle driven generators in conjunction with batteries.
(d) either of (b) or (c).

599. Coach air-conditioning equipment includes
(a) standard refrigeration equipment.
(b) mechanical power generating equipment.
(c) electrical equipment for control with accessories.
(d) an air-conditioner with its standard accessories.
(e) all of the above except (d).

600. Requirement(s) of a railway coach air-conditioning is/are
(a) to provide heating/cooling.
(b) to supply clean, fresh air at a controlled uniform temperature.
(c) sufficient power requirement.
(d) to account for charging passenger traffic.
(e) all of the above.

601. Annual operating cost of a generating plant consists of
(a) fixed charges. (b) semifixed charges.
(c) operating or running charges.
(d) all of the above.

602. For a power plant the expenditure on which of the following items is expected to be almost negligible ?
(a) Publicity. (b) Taxes.
(c) Insurance. (d) Maintenance.

603. In a steam power station which of the following is not a fixed cost ?
(a) Interest on capital cost of land and buildings.
(b) Salaries of high officials.
(c) Fuel and lubricating oil cost.
(d) Insurance charges.

604. In a power plant which of the following items does not fall in the category of operating cost ?
(a) Salaries of operational and maintenance staff.
(b) Maintenance and repair cost.
(c) Salaries of supervisory staff engaged on the running of the plant.
(d) Salaries of management and clerical staff.

605. In a power plant, which of the following items fall in the category of semifixed charges ?
(a) Annual interest and depreciation on capital cost.
(b) All types of taxes and insurance charges.
(c) Salaries of management and clerical staff.
(d) All of the above.

606. The interest on the capital cost is included in
(a) annual fixed cost. (b) annual operating cost.
(c) both (a) and (b). (d) neither (a) nor (b).

607. The capital cost of a power plant depends on
(a) total installed capacity only.
(b) total number of units only.
(c) total installed capacity and number of units as well.
(d) neither the installed capacity nor number of units.
[A.M.I.E. Sec. B. Power Systems Summer 1998; Summer 2000]

608. The long term load forecast is required for
(a) operation of plant. (b) economic operation of plant.
(c) planning the addition in generation capacity.
(d) both (b) and (c).

609. The annual depreciation reserve depends on
(a) capital cost only. (b) salvage value only.
(c) on the method of calculation depreciation reserve.
(d) all of the above.

610. Salvage value of a plant
(a) is always positive. (b) is always zero.
(c) is always negative. (d) may be any of the above.

611. An equipment purchased for Rs 25 lakhs, 3 years before, has now market value of 32 lakhs, It shows that
(a) the value has depreciated as per straight line method.
(b) the value has depreciated as per diminishing value method.
(c) the value has depreciated as per sinking fund method.
(d) the value has appreciated with the time.

612. Ideally depreciated value of the plant plus the accumulation in the depreciation fund should be equal to
(a) obsolescence rate. (b) sinking fund.
(c) original invested value. (d) salvage value.

613. In a straight line method, annual depreciation charges are calculated by
(a) the capital cost minus salvage value divided by the number of years of life.
(b) the capital cost divided by the number of years of life.

Answers: 593. (b), 594. (c), 595. (e), 596. (e), 597. (c), 598. (c), 599. (e), 600. (e), 601. (d), 602. (a), 603. (c), 604. (d), 605. (d), 606. (a), 607. (a), 608. (c), 609. (d), 610. (d), 611. (d), 612. (c), 613. (a),

(c) making a provision for setting each year a fixed rate, first applied to the original cost and then to the diminishing value, depending upon the useful life of the plant.
(d) none of the above.

614. Annual instalment towards depreciation increases with the decrease in interest rate in case of
(a) straight line depreciation. (b) sink fund depreciation.
(c) reducing balance depreciation.
(d) all of the above.

615. Annual depreciation of the plant is proportional to the earning capacity of the plant in case of
(a) straight line depreciation. (b) sinking fund depreciation.
(c) reducing balance depreciation.
(d) none of the above.

616. Annual estimated depreciation charges for a plant are heavy during early years in case of
(a) diminishing value method.
(b) sinking fund method. (c) straight line method.
(d) none of the above.

617. In a power plant insurance cover is provided for
(a) equipment only. (b) unskilled workers only.
(c) unskilled and skilled workers.
(d) equipment and skilled and unskilled workers.

618. A company can raise funds through
(a) fixed deposits. (b) shares.
(c) bonds. (d) any of the above.

619. All of the following are the sources of borrowing money except
(a) shares. (b) debentures.
(c) bonds. (d) fixes deposits.

620. Which of the following are not repayable after a stipulated period ?
(a) Fixed deposits. (b) Shares.
(c) Bonds. (d) Cash certificates.

621. During summer season load on the system increases, this is due to
(a) increase in demand of water supply.
(b) increased business activity.
(c) increased use of fans, air coolers and air conditioners.
(d) vacations in schools and institutions.

622. The connected load of a domestic consumer (medium income group) is usually about
(a) 2 – 5 kW. (b) 10 – 20 kW.
(c) 20 – 30 kW. (c) below 1 kW.

623. A load curve is a plot of
(a) load versus duration of time.
(b) load versus current. (c) load versus time.
(d) total number of units generated versus time.

624. The chronological load curve indicates
(a) variation in demand factor during 24 hours.
(b) variation in demand from instant to instant during 24 hours.
(c) the total energy consumed upto different times of the day.
(d) the total number of hours for which a particular load lasts during the day.

625. The area under the load curve represents
(a) system voltage. (b) current.
(c) energy consumed. (d) maximum demand.
(e) average demand. **[A.M.I.E. Sec. B. Summer 1994]**

626. The area under daily load curve divided by 24 gives
(a) average load for the day. (b) maximum demand.
(c) connected load. (d) demand factor.

627. Which of the following statements regarding load curve is true ?
(a) The area under the curve gives average demand.
(b) The ratio of area under the load curve to the total area of the rectangle in which it is contained gives the load factor for the day.
(c) The area of the load curve divided by the number of hours gives the load factor.
(d) The area of the load curve divided by the number of hours gives the demand factor.

628. Load curve of a power generating station is of always
(a) negative slope. (b) zero slope.
(c) positive slope.
(d) any combination of (a), (b) and (c).

629. Load curve helps in deciding the
(a) total installed capacity of the plant.
(b) size of the generating units.
(c) operating schedule of the generating units.
(d) all of the above.

630. Load duration curve gives
(a) the variations of load during different hours of the day.
(b) average load.
(c) the number of hours for which a particular load lasts during the day.
(d) none of the above.

631. The load duration curve for unity load factor will be of
(a) rectangular shape. (b) triangular shape.
(c) L-shape. (d) I-shape.

632. The mass curve represents
(a) average load.
(b) the total energy consumed by the load up to a particular time in a day.
(c) the number of hours for which a particular load lasts during the day.
(d) the variation of load during different hours of the day.

633. A mass curve can be plotted from
(a) load duration curve. (b) chronological load curve.
(c) energy load curve.
(d) both load duration curve and chronological load curve.
[A.M.I.E. Sec. B. Summer 1998, 2000]

634. Integrated load duration curve representing total number of units (kWh) generated for a given demand in kW can be plotted directly from
(a) load curve. (b) load duration curve.
(c) mass curve. (d) any of these.

635. In a load duration curve for an integrated power system, the uppermost crest represents the energy contributed by which one of the following?
(a) Base power stations. (b) Major thermal stations.
(c) Peaking hydro or gas turbine stations.
(d) Non-conventional power stations. **[U.P.S.C. I.E.S. E.E.-II, 2007]**

636. During which season the load on a power system is maximum ?
(a) Autumn. (b) Rainy. (c) Summer. (d) Winter.

637. Maximum demand on the power system is
(a) the greatest of all 'short time interval averaged' demand during a period.
(b) instantaneous maximum value of kW supplied during a period.
(c) instantaneous maximum value of kVA supplied during a period.
(d) maximum value of units (kWh) supplied during a period.

Answers: 614. (b), 615. (c), 616. (a), 617. (d), 618. (d), 619. (a), 620. (b), 621. (c), 622. (a), 623. (c), 624. (b), 625. (c), 626. (a), 627. (b), 628. (d), 629. (d), 630. (c), 631. (a), 632. (b), 633. (b), 634. (b), 635. (d), 636. (d), 637. (a),

638. The power system experiences peak demand from
(a) midnight to 8 A.M. (b) 8 A.M. to 2 P.M.
(c) 2 P.M. to 6 P.M. (d) 6 P.M. to 10 P.M.

639. Connected load means
(a) installed electrical load in the premises of a consumer.
(b) maximum load a consumer draws.
(c) load drawn by consumer at any instant.
(d) none of the above.

640. The maximum demand of a power station is
(a) sum of the maximum demands of all its consumers.
(b) greatest average load in a specified time.
(c) peak value of load in a specified time.
(d) all of the above.

641. Demand factor is defined as the ratio of
(a) average load to maximum demand.
(b) maximum demand to connected load.
(c) connected load to maximum demand.
(d) maximum demand to average load.

642. Demand factor on a power system is
(a) always greater than unity. (b) normally greater than unity.
(c) always lesser than unity. (d) normally lesser than unity.

643. The load factor is equal to
(a) average load/peak load. (b) peak load/average load.
(c) average load/connected load. (d) average load/base load.
[A.M.I.E Sec. B Design of Electrical Systems Summer 2005]

644. A power station's plant load factor is defined as the ratio of
(a) the energy generated to that of maximum energy that could have been generated.
(b) average load to peak load.
(c) minimum load to peak load.
(d) minimum load to average load. **[U.P.S.C. I.E.S. E.E.-II, 2006]**

645. Load factor of a power plant is
(a) generally equal to unity. (b) always less than unity.
(c) always more than unity. (d) normally more than unity.

646. The load factor for domestic loads may be taken as
(a) about 85%. (b) 50 – 60%.
(c) 25 – 50%. (d) 10 – 15%.

647. Load factor for heavy industries may be taken as
(a) 70 – 80%. (b) 40 – 50%.
(c) 25 – 40%. (d) 20 – 25%.

648. The annual load duration curve of a power supply system may be considered as a straight line from 40 MW to 8 MW. The load factor of the system is
(a) 20%. (b) 40%. (c) 60%. (d) 83.33%.
(e) none of the above. **[A.M.I.E. Sec. B. Winter 1993]**

649. An industrial consumer has a load pattern of 2,000 kW 0.8 lag for 12 hours and 1,000 kW unity power factor for 12 hours. The load factor is
(a) 0.5. (b) 0.75. (c) 0.6. (d) 2.0.
[GATE E.E. 1999]

650. The load curve of a system is shown in the figure. The load factor of the system is
(a) 1.66%.
(b) 6.013%.
(c) 16.6%.
(d) 60.13%.

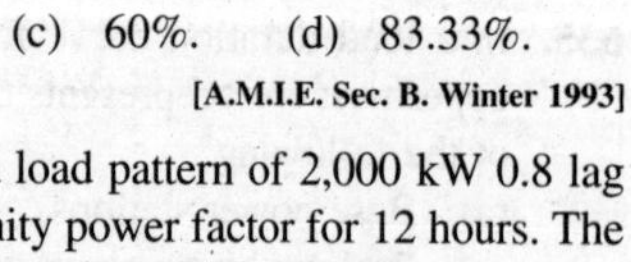

[U.P.S.C. I.E.S. E.E.-II, 1997]

651. Diversity factor is the ratio of
(a) sum of maximum demands of consumers/system maximum demand.
(b) maximum demand of consumers/average demand.
(c) demand of all consumers/average demand.
(d) none of the above. **[A.M.I.E. Sec. B. Power Systems, Winter 1998]**

652. The curve shown in the figure is
(a) chronological load curve.
(b) annual load duration curve.
(c) mass curve.
(d) energy load curve.

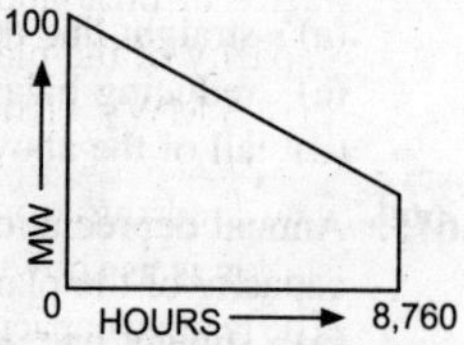

[A.M.I.E Sec. B Design of Electrical Systems Summer 2005]

653. The daily energy produced in thermal power station is 720 MWh at a load factor of 0.6. What is the maximum demand of the station?
(a) 50 MW (b) 30 MW (c) 72 MW (d) 720 MW
[U.P.S.C. I.E.S. E.E.-II, 2008]

654. Diversity factor in a power system is
(a) always less than unity. (b) normally less than unity.
(c) always more than unity. (d) normally more than unity.
[A.M.I.E. Sec. B. Summer 1993]

655. Diversity factor × maximum demand is
(a) average demand.
(b) sum of consumer's maximum demands.
(c) installed capacity. (d) generated power.
[A.M.I.E. Sec. B. Summer 1994]

656. Diversity factor has direct effect on
(a) fixed cost per unit generated.
(b) operating cost per unit generated.
(c) both (a) and (b). (d) neither (a) nor (b).

657. The knowledge of diversity factor helps in computing
(a) plant capacity. (b) average load.
(c) units (kWh) generated. (d) peak demand.

658. Plant capacity factor is
(a) actual energy produced/maximum possible energy that could have been produced (based on installed capacity).
(b) actual energy produced /plant capacity hours for which the plant has been in operation.
(c) energy generated in a given period/ (maximum demand x hours of operation in given period).
(d) none of the above. **[A.M.I.E. Sec. B. Winter 1993]**

659. Which of the following statement is correct ?
(a) Load factor = Capacity factor × utilisation factor.
(b) Utilisation factor = Capacity factor × load factor.
(c) Capacity factor = $\dfrac{\text{Load factor}}{\text{Utilisation factor}}$
(d) Capacity factor = Load factor × utilisation factor.
[A.M.I.E. Sec. B. Power Systems, Summer 2000]

660. Utilisation factor is defined as the ratio of
(a) average demand to rated capacity of the power plant.
(b) maximum demand on the power plant to the rated capacity of the power plant.
(c) rated capacity of the power plant to the maximum demand.
(d) none of the above.

661. Capacity factor will be very low when the power plant
(a) is operated as base load plant.
(b) is operated for supplying base loads as well as peak loads.
(c) is operated in emergency only.
(d) is under maintenance.

Answers: 638. (d), 639. (a), 640. (b), 641. (b), 642. (c), 643. (a), 644. (b), 645. (b), 646. (d), 647. (a), 648. (c), 649. (b), 650. (d), 651. (a), 652. (b), 653. (a), 654. (c), 655. (b), 656. (a), 657. (a), 658. (a), 659. (d), 660. (b), 661. (c),

662. A thermal generating station has an installed capacity of 15 MW and supplies a daily load of 10 MW for 12 hours and 5 MW of remaining 12 hours. The plant capacity factor for this station is
(a) 1. (b) 0.75. (c) 0.67. (d) 0.5.
[U.P.S.C. I.E.S. E.E.-II, 1999]

663. A generating station has a maximum demand of 30 MW, a load factor of 60% and a plant capacity factor of 50%. The reserve capacity of the plant is
(a) 5 MW (b) 4 MW (c) 6 MW (d) 10 MW
[U.P.S.C. I.E.S. E.E.-II, 2003]

664. The load duration curve for a power station is as shown in the given figure. The reserve capacity in the plant at 70% capacity factor is

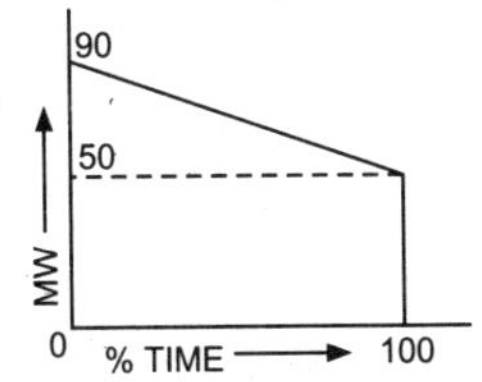

(a) zero.
(b) 10 MW.
(c) 30 MW.
(d) 50 MW. [U.P.S.C. I.E.S. E.E.-II, 2001]

665. If the rated plant capacity and maximum load of a power station are equal then
(a) load factor is unity. (b) capacity factor is unity .
(c) utilization factor is unity.
(d) load factor and capacity factor are equal.
(e) both (c) and (d).

666. Spinning reserve is
(a) the reserve generating capacity which is available for service but not in operation.
(b) the reserve generating capacity which is connected to the bus and is ready to take load.
(c) the reserve generating capacity which is in operation but is not in service.
(d) none of the above. [A.M.I.E. Sec. B. Winter 1993]

667. The power which must be available even under emergency conditions is known as
(a) spinning reserve. (b) cold reserve.
(c) firm reserve. (d) hot reserve.
[A.M.I.E. Sec. B. Power Systems, Summer 2000]

668. In a power plant, a reserve generating capacity which is in operation but not in service, in called the
(a) cold reserve. (b) hot reserve.
(c) spinning reserve. (d) firm reserve.

669. In a power plant, a reserve generating capacity, which is in operable condition and available for service, but not in operation is called the
(a) spinning reserve. (b) firm reserve.
(c) cold reserve. (d) hot reserve.

670. A power plant has a maximum demand of 15 MW. The load factor is 50% and the plant factor is 40%. The operating reserve is
(a) 3 MW. (b) 3.75 MW.
(c) 6 MW. (d) 7.5 MW.
[U.P.S.C. I.E.S. E.E.-II, 1996]

671. Which of the following relations holds good ?
(a) Maximum demand = Connected load × demand factor.
(b) Average load = Maximum load × load factor.
(c) Capacity factor = Utilization factor × load factor.
(d) All of the above.

672. Two areas A and B have equal connected loads. However, load diversity in area A is more than in area B. Then
(a) maximum demands of the two areas would be equal.
(b) maximum demand of A would be more than that of B.
(c) maximum demand of B would be more than that of A.
(d) maximum demand of A may be more or less than that of B.
[A.M.I.E. Sec. B. Power Systems, Summer 1998]

673. A large diversity factor of the load in a power system
(a) reduces the installation cost.
(b) increases the installation cost.
(c) does not affect the installation cost.
[A.M.I.E. Sec. B. Power Systems, Summer 1999]

674. As the load factor of a generating plant increases, the generation cost per kWh generated
(a) decreases. (b) increases.
(c) remains same. (d) none of these.
[West Bengal Univ. of Technology 2009-10]

675. In order to have a lower cost of electrical energy generation
(a) the load factor and diversity factor should be low.
(b) the load factor should be low but diversity factor should be high.
(c) the load factor should be high but the diversity factor should be low.
(d) the load factor and diversity factor should be high.
[GATE E.E. 1995]

676. In a power station, the cost of generation of power reduces most effectively when
(a) diversity factor alone increases.
(b) both diversity factor and load factor increase.
(c) load factor alone increases.
(d) both diversity factor and load factor decrease.
[U.P.S.C. I.E.S. E.E.-II, 1994)

677. Flat rate tariff can be charged on the basis of
(a) connected load. (b) units consumed.
(c) maximum demand. (d) both (a) and (b).

678. Domestic consumers are usally charged
(a) flat demand tariff. (b) block rate tariff.
(c) flat rate tariff. (d) off peak tariff.

679. If the tariff for electrical energy charges provides incentive by way of reduced charges for higher consumption, then it may be concluded that the
(a) load factor is high. (b) power factor is high.
(c) plant has sufficient reserve capacity.
(d) power is generated through hydropower plant.

680. Two part tariff is charged on the basis of
(a) connected load. (b) units consumed.
(c) maximum demand. (d) both (a) and (b).

681. Block rate tariff, where energy charge decreases with the increase in energy consumption,
(a) encourages the consumers for more consumption.
(b) discourages the consumers for more consumption.
(c) encourages the consumers to restrict their demands.
(d) encourages the consumers to improve the power factor.

682. The penalty for low power factor is imposed on
(a) residential and commercial consumers.
(b) industrial consumers. (c) agricultural consumers.
(d) all of the above.

683. Doherty rate tariff is applied to
(a) domestic consumers. (b) medium industrial consumers.
(c) bulk supplies. (d) municipal loads.

684. Maximum demand tariff is generally not applied to the domestic consumers owing to their
(a) low maximum demand. (b) low load factor.
(c) low power factor. (d) low energy consumption.

Answers: 662. (d), 663. (c), 664. (b), 665. (e), 666. (b), 667. (c), 668. (b), 669. (c), 670. (b), 671. (d), 672. (c), 673. (a), 674. (a), 675. (d), 676. (b), 677. (b), 678. (b), 679. (c), 680. (d), 681. (a), 682. (b), 683. (c), 684. (a),

685. Low power factor is usually not due to
(a) discharge lamps. (b) incandescent lamps.
(c) arc lamps. (d) induction furnaces.

686. The primary reason for low power factor is owing to installation of
(a) synchronous motors. (b) dc motors.
(c) induction motors. (d) commutator motors.

687. Low power factor has the drawback(s) of
(a) increased transmission and distribution losses.
(b) poor voltage regulation.
(c) high cost of equipment for a given load.
(d) all of the above.

688. The low power factor of an industrial plant is uneconomical for
(a) electric supply utility only.
(b) owner of the plant only.
(c) both the owner of the plant and the electric supply utility.
(d) either (a) or (b).

689. For a consumer the most economical power factor is usually
(a) 0.25 – 0.5 lagging. (b) 0.25 – 0.5 leading.
(c) 0.85 – 0.95 lagging. (d) 0.85 – 0.95 leading.

690. For power factor improvement static capacitors have the advantage(s) of
(a) small losses. (b) easy installation.
(c) low initial cost and little maintenance.
(d) all of the above.

691. For power factor improvement static capacitors have the drawback(s) of
(a) short-service life.
(b) getting damaged by high voltage.
(c) not repairable. (d) all of the above.

692. A synchronous condenser is a
(a) synchronous generator. (b) paper condenser.
(c) synchronous motor. (d) none of these.
[West Bengal Univ. of Technology 2009-10]

693. A synchronous compensator absorbs inductive reactive power. It is
(a) over excited. (b) normally excited.
(c) under excited. (d) none of these.
[West Bengal Univ. of Technology 2009-10]

694. For power factor improvement synchronous condensers have the drawback(s) of
(a) comparatively higher maintenance and operating costs.
(b) requirement of auxiliary equipment for their starting.
(c) possibility of falling out of synchronism causing in interruption of supply.
(d) all of the above.

695. Advantages of the improved power factor are
(a) increase in operating efficiency of the power system.
(b) improvement in voltage regulation.
(c) reduction in overall cost per unit.
(d) better utilization of kW capacities of prime movers, transformers, switchgear and the lines.

696. The most suitable location for a power factor improvement device is
(a) near the electrical appliance which is responsible for the poor power factor.
(b) at the receiving end, in case of a transmission line.
(c) at the sending end.
(d) any where in the circuit.
(e) both (a) and (b).

697. An industrial installation has a power factor of 0.8 lagging. It would be economical to improve pf to
(a) unity. (b) about 0.8 leading.
(c) about 0.95 lagging. (d) about 0.95 leading.
[A.M.I.E. Sec. B. Power Systems, Summer 2000]

698. A 3-phase, 11 kV, 50 Hz, 200 kW load has a power factor of 0.8 lag. A delta-connected 3-phase capacitor is used to improve the pf to unity. The capacitance per phase of the capacitor in micro-farads is
(a) 3.948. (b) 1.316. (c) 0.439. (d) 11.844.
[GATE E.E. 1999]

699. The most economical limit of power factor correcting is governed by
(a) original power factor.
(b) relative costs of the supply and power factor correction equipment.
(c) both (a) and (b). (d) none of the above.

700. There is a limit beyond which it is not economical still further to improve the power factor in order to meet the increased demand on the generating station. The maximum value to which the pf can be economically raised entirely depends upon
(a) the cost of the power factor correction equipment.
(b) the cost of the generating plant.
(c) the relative costs of the generating plant and phase advancing plant.
(d) none of the above.

Answers: 685. (b), 686. (c), 687. (d), 688. (c), 689. (c), 690. (d), 691. (d), 692. (c), 693. (c), 694. (d), 695. (d), 696. (e), 697. (c), 698. (b), 699. (b), 700. (c).

ASSERTION (A)/REASON (R) TYPE QUESTIONS

Note : Each of the following questions consists of two statements, one labelled the 'Assertion (A)' and the other labelled the 'Reason (R)'. Examine the two statements carefully and decide if the Assertion (A) and Reason (R) are individually true and if so, whether the reason (R) is correct explanation of the Assertion (A). Select your answers to these questions using the codes given below:
Codes: (a) Both A and R are true and R is a correct explanation of A.
(b) Both A and B are true and R is not a correct explanation of A.
(c) A is true but R is false. (d) A is false but R is true.

1. (A): The prime mover and the transmission system are embraced by the common term drive.
(R): In a machine unit, the prime mover and the transmission system keep the working machine in motion.

2. (A): The modern trend is to use individual and multimotor drives.
(R): Individual and multi-motor drives are cheaper than group drives in initial cost.

3. (A): A dc series motor is most suitable for cranes and hoists.
(R): A dc series motor runs at high speed at light loads and low speed when load is heavy. Light loads are lifted quickly and heavy loads slowly.

Answers: 1. (a), 2. (c), 3. (a),

4. (A): Cumulative compound wound motor is most suitable for elevators, presses shears, compressors and reciprocating tools.
   (R): The speed-torque characteristic of a cumulatively compounded motor is intermediate between those of a series and a shunt motor. This motor is used when a series characteristic is desired and the load is likely to be almost entirely removed.
5. (A): For vacuum cleaners fractional kilowatt output ac series motors and universal motors are used.
   (R): AC series motors and universal motors operate at high speed and develop large starting torque.
6. (A): The capacitor-start, capacitor-run motor is well suited for driving fixed loads in hospitals, studios and other places where silence is very important.
   (R): The capacitor-start, capacitor-run motor behaves as a true 2-phase motor only when it operates on full load. The motor is then vibration free.
7. (A): The well suited motor for driving air compressors, large fans, pumps and high-inertia loads is capacitor-start induction motor.
   (R): The capacitor-start induction motor develops high starting torque.
8. (A): Solid-state fan regulators are preferred for speed control over resistance or inductance regulators.
   (R): Solid-state regulators are compact, energy efficient, more reliable, noise free and economical.
9. (A): AC motors are widely used in electric drive.
   (R): AC motors have excellent torque and speed operating characteristics in comparison with those of dc motors.
10. (A): Electrical connections to the motor are usually made by means of (a) flexible pipe rather than by a conduit pipe.
    (R): It avoids transmission of vibrations.
11. (A): The motor is usually mounted on a heavy concrete or cast-iron block or alternatively on a flexible support provided by rubber pads or spring.
    (R): It minimizes the transmission of noise from one part of the building to another.
12. (A): Electrical heating is economical in comparison to other heating systems.
    (R): Electric heating is clean, hygienic, efficient and safe in operation.
13. (A): Ni-Cr-Fe alloy is recommended to be used as an heating element for temperatures up to 1,150°C.
    (R): The addition of iron to the Ni - Cr alloy reduces the temperature at which oxidation takes place but the cost of the product is also reduced.
14. (A): The voltage across an oven can be controlled by varying the external resistance connected in series with the heating circuit.
    (R): The mehod of temperature control of an oven by external series resistance is not economical and so its use is limited to small furnaces.
15. (A): Salt bath furnaces are started with highest tap and as the temperature rises, the tap is gradually brought down.
    (R): The salt resistance increases with the increase in temperature of salt.
16. (A): The general trend is to use graphite electrode in arc furnaces.
    (R): Owing to low resistivity graphite is required half in size for the same current resulting in easy replacement and under normal conditions the consumption of graphite electrodes is about half of that of carbon electrodes.
17. (A): Alternating current is found most suitable for resistance welding.
    (R): AC can provide any combination of current and voltage with the help of a transformer.
18. (A): For flash-butt welding there is no need to level rough or irregular ends of the metal pieces to be welded and clean and pure weld is obtained.
    (R): In flash-butt welding all irregularities are burnt away during flashing period.
19. (A): Electric resistance spot welding is probably the best known and most widely used type of welding.
    (R): Electric resistance spot welding is cheap in cost, fast in action and dependable. It can be easily performed by even a semi-skilled operator.
20. (A): Coated electrodes are always to be stored in a dry place and if dampness is suspected they are to be dried in a warm cabinet for a few hours.
    (R): Dampness impaires the coated electrodes.
21. (A): An inductor is used in series with carbon arc for its stabilization in case of dc supply.
    (R): The carbon arc is unstable.
22. (A): In a sodium vapour lamp, the discharge envelope is usually bent into U-shape.
    (R): Long discharge paths are necessary in sodium vapour lamps.
23. (A): True colour rendition is not possible with mercury vapour lamps.
    (R): There is complete absence of red light from mercury vapour lamp radiations.
24. (A): For neon lamps used on ac the electrodes are of equal size but for dc the negative electrode is made larger in size.
    (R): When neon lamp is used on dc the gas glows near the positive electrode.
25. (A): A high series resistance of few thousands ohms is mounted in the cap of neon lamp.
    (R): In neon lamps, the current drawn may increase indefinitely owing to discharge of the gas between the electrodes in the form of an arc.
26. (A): Ammonia is preferred as a refrigerant in large commercial installations.
    (R): Ammonia is highly toxic and inflammable.
27. (A): Direct internal combustion engine drive is widely used for road transport.
    (R): Speed control and braking system employed in direct internal combustion drive is very simple.
28. (A): The trolley bus needs only a single driving axle while the tramway needs minimum, two driving axles.
    (R): The adhesion between a rubber-tyred wheel and the roadways is sufficiently high.
29. (A): Trolley buses are used for the inner suburbs with medium traffic density.
    (R): In trolley buses, the speed can be controlled by field weakening method.

Answers: 4. (a), 5. (a), 6. (a), 7. (a), 8. (a), 9. (c), 10. (a), 11. (a), 12. (b), 13. (d), 14. (b), 15. (c), 16. (a), 17. (a), 18. (a), 19. (a), 20. (a), 21. (d), 22. (a), 23. (a), 24. (c), 25. (a), 26. (c), 27. (b), 28. (a), 29. (b),

30. (A): The starting efficiency of ac locomotive is higher than that of a dc locomotive.
(R): In case of ac locomotive supply voltage for starting can be reduced by means of 'on load tap changer'.

31. (A): DC traction system is preferred for main line railway service.
(R): The dc locomotive and motor coach equipment is lighter in weight, cheaper in initial as well as in maintenance cost and more efficient.

32. (A): Traction motors are usually high speed motors.
(R): The weight of high speed motors is comparatively low and so the pay load capacity of the vehicle is increased.

33. (A): The traction motor should be capable to withstand temporary interruptions of supply without undue rush of current.
(R): Supply is interrupted at the instants of crossing over the cross-overs and section insulators.

34. (A): Traction motors should be of such speed-torque characteristics that when they are operated in parallel and mechanically coupled, they share the load almost equally.
(R): In traction work, usually more than one motor (two or four motors per motor car) are required.

35. (A): AC series motors are not very suitable for suburban services where stops are frequent.
(R): For given kW output, the single phase ac series motors are lighter in weight in comparison to dc series motors.

36. (A): The trolley collector is suitable only for comparative low speeds.
(R): There is a poor contact between the wheel and the trolley wire.

37. (A): The major power development programmes in Bihar, should be primarily based on conventional steam power plants.
(R): Bihar has abundant coal reserves.

38. (A): The operating cost of hydroelectric power plants are the lowest.
(R): There are no standby losses in hydroplants.

39. (A): The hydroplants have constant speed and constant frequency.
(R): Very acute governing is possible with water turbines.

40. (A): Availablity of storage may convert an uneconomical hydro-project into an economical one.
(R): Storage increases the firm capacity of the power plant and increase in firm capacity of hydroplant means dispensing with the alternative steam capacity.

41. (A): Pelton turbine is not suitable for heads below 200 m.
(R): The diameter of a pelton wheel runner, for the same output but low head becomes large and unwieldy.

42. (A): The draft tube in hydropower stations permits a negative of suction head to be established at the runner exit.
(R): The turbine can be installed above the tail-race level without loss of head. **[U.P.S.C. I.E.S. E.E.-II, 1995]**

43. (A): In a high or medium head hydroelectric power station having a long penstock, a surge tank is provided near the turbine.
(R): A surge tank stores additional water to be released during the peak load period. **[U.P.S.C. I.E.S. E.E., 2000]**

44. (A): Hydroelectric plants with large storage can be best used as base load plants with high load factor.
(R): For a plant to be used as a base load plant, the unit cost of energy generated by the plant should be low.
[U.P.S.C. I.E.S. E.E.-II, 2003]

45. (A): A pumped storage plant is very suitable for supplying peak loads. During off peak period, water is pumped back from tail race pond to head water pond.
(R): The starting time of pumped storage plant is very short.
[U.P.S.C. I.E.S. E.E.-II, 2005]

46. (A): Huge amount of coal is required for raising the steam (about 2,000 tonnes per day for a 200 MW station) in a steam power station.
(R): As per government policy only low grade coal with 30-40 percent ash content can be used in steam power plants for generation purposes.

47. (A): Fire tube boilers are commonly used in central steam power plants.
(R): Fire tube boilers are compact in size, have low initial cost and have the ability to raise rapidly large quantities of steam per unit area of fire grate.

48. (A): Regenerative air preheaters are widely used in steam power plants.
(R): The rotor elements, in regenerative air preheaters, provide large surface area to the flue gases.

49. (A): Commercial steam turbines use series combination of impulse and reaction types of turbines.
(R): Impulse-reaction turbines are cheaper in initial cost.

50. (A): For the optimum loading of a number generating units in a power plant, the units must operate at the same incremental fuel cost.
(R): When units operate at the same incremental fuel cost, the active and reactive load diversions are equal.
[U.P.S.C. I.E.S. E.E.-II, 1993]

51. (A): Nuclear power plants are not well suited for varying loads.
(R): The nuclear reactor does not respond to the load fluctuations efficiently.

52. (A): Diesel electric power plants are usually used as base load plants.
(R): Such plants can be located any where and there are no standby losses.

53. (A): Gas turbine power plants are generally used as base load power plants.
(R): The gas turbine power plants have somewhat higher fuel costs and low initial cost.

54. (A): Petroleum fuels such as kerosine, gas oil, diesel oil, residual oil have limited use, in gas turbines, as fuels.
(R): Petroleum fuels like gas oil, diesel oil, residual oil and kerosine have low calorific value.

55. (A): Intercooling in gas turbines reduces the size of the turbine and compressor for the same output.
(R): Intercooling removes the heat from the compressed air before it is sent to the combustion chamber.

56. (A): In almost all the fields open cycle turbine power plants are employed.
(R): Closed cycle gas turbine plants have the additional complexity of the plant including externally fired heater, gas cooler and various heat exchangers.

Answers: 30. (a), 31. (d), 32. (a), 33. (a), 34. (a), 35. (c), 36. (b), 37. (a), 38. (b), 39. (a), 40. (a), 41. (a), 42. (b), 43. (d), 44. (d), 45. (a), 46. (a), 47. (d), 48. (a), 49. (c), 50. (c), 51. (a), 52. (d), 53. (d), 54. (c), 55. (c), 56. (a),

57. (A): The seeding material ionizes more readily than the gas and increases the conductivity of the gas.
(R): Seeding material have comparatively higher ionization potential.

58. (A): Now great emphasis is being given for the development of new and alternative (non-conventional) methods of power generation.
(R): Non-conventional methods of power generation give low cost of production of electrical energy and also these methods are more reliable.

59. (A): Optimum loading of the power stations interconnected by transmission line is obtained when the incremental fuel cost of each station divided by its penalty factor is the same for all the stations in the system.
(R): The total load demand and transmission losses at any instant must be met by the total generation at that instant and the penalty factor is a function of the incremental transmission loss. **[U.P.S.C. I.E.S. E.E.-II, 1993]**

60. (A): During winter season the load on the system increases.
(R): There is an increase of load on the system during winter season due to heaters and geysers.

61. (A): As the load factor of the power plant increases, the cost per kWh of the energy generated decreases.
(R): Higher load factor means greater average load, resulting in larger number of units generated for a given maximum demand or capacity, and thus there is a distribution of standing charges over a large number of units supplied and therefore reduction in unit cost.

62. (A): Synchronous condensers are used for pf improvement only when kVAR requirement exceeds 5,000.
(R): Initial as well as operating costs of synchronous condensers are much higher than those of static capacitors and do not justify their installation for low ratings.

63. (A): The best location for the pf correction equipment to be installed is the generating end.
(R): Generators are relieved of carrying excessive current due to poor power factor.

64. (A): It is not economical to raise the power factor to unity or leading one.
(R) As the pf approaches unity, the capacity of pf correction device increases more rapidly.

Answers: 57. (c), 58. (c), 59. (d), 60. (a), 61. (a), 62. (a), 63. (d), 64. (a).

MATCHING TYPE QUESTIONS

Note: Each of the following questions consists of 2 lists. Match the items given in list 1 and those in list 2. Select your answers to these questions using the codes given in each of the questions.

1. **List 1 (Load)**
A. Hoist winches.
B. Blowers.
C. Boring machines.

List 2 (Characteristics)

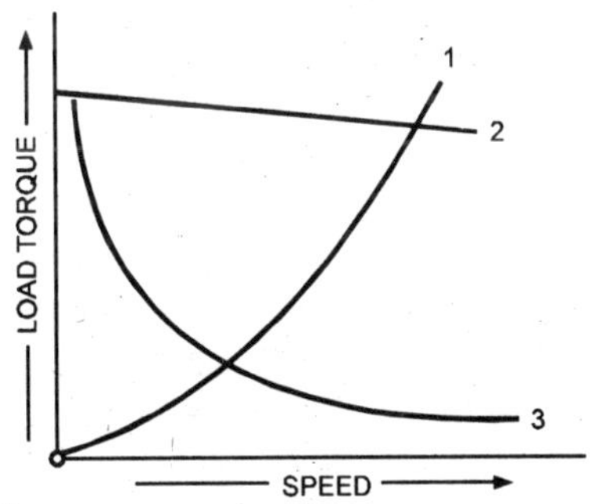

Codes:	A	B	C
(a)	2	1	3
(b)	1	2	3
(c)	3	1	2
(d)	3	2	1

2.

List 1 (Types of Electrical Loads)	List 2 (Torque Speed Characteristics)
A. Hoist	1. Torque ∝ $(\text{speed})^2$
B. Fans	2. Torque ∝ (speed)
C. Machine tools (lathe, milling machine etc.)	3. Constant torque
D. Loads with fluid friction	4. Torque ∝ 1/(speed)

Codes:	A	B	C	D
(a)	1	3	2	4
(b)	1	3	4	2
(c)	4	1	3	2
(d)	3	1	2	4

[U.P.S.C. I.E.S E.E.-II, 2002]

3.

List 1 (Motors)	List 2 (Applications)
A. DC series motor.	1. Shearing and pressing.
B. Squirrel-cage induction motor.	2. Haulage and hoisting.
C. DC shunt motor.	3. Rolling mill.

Codes:	A	B	C
(a)	1	2	3
(b)	2	3	1
(c)	3	1	2
(d)	3	2	1

[U.P.S.C. I.E.S. E.E.-II, 2001]

4.

List 1 (Machine)	List 2 (Performance)
A. Three phase induction motor.	1. Adjustable speed.
B. Synchronous motor.	2. High starting torque.
C. DC series motor.	3. Not self-starting.
D. DC shunt motor.	4. Self-starting.

Codes:	A	B	C	D
(a)	3	4	1	2
(b)	4	3	1	2
(c)	3	4	2	1
(d)	4	3	2	1

[U.P.S.C. I.E.S. E.E.-II, 2004]

Answers: 1. (a), 2. (c), 3. (b), 4. (d),

5.

List 1 (Types of Motors)	List 2 (Characteristics)
A. DC series motor.	1. Constant speed.
B. DC shunt motor.	2. High starting torque.
C. 3-phase induction motor.	3. Low starting torque.
D. Synchronous motor.	4. Poor stability.

Codes:	A	B	C	D
(a)	2	4	3	1
(b)	3	1	2	4
(c)	2	1	3	4
(d)	3	4	2	1

[U.P.S.C. I.E.S. E.E.-II, 2002]

6. The electromagnetic torque T_e of a drive and its connected load torque T_L are as shown below. Out of the operating points A, B, C and D, the stable ones are

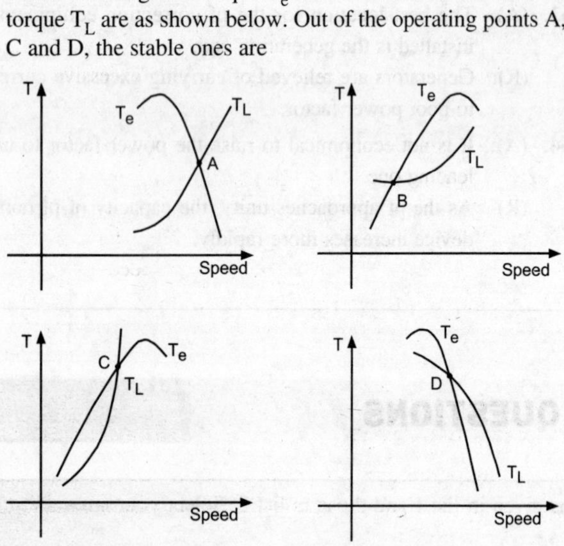

(a) A, C, D (b) B, C
(c) A, D (d) B, C, D **[GATE E.E. 2007]**

7.

List 1 (Type of Enclosure)	List 2 (Application)
A. Protected type.	1. Collieries.
B. Splash proof.	2. Saw mills.
C. Total enclosed fan cooled type.	3. Industrial purpose.
D. Flame proof type.	4. Rain.

Codes:	A	B	C	D
(a)	1	2	3	4
(b)	4	1	2	3
(c)	2	3	4	1
(d)	3	4	2	1

8.

List 1 (Applications)	List 2 (Type of Duty)
A. Blowers, fans.	1. Intermittent periodic duty.
B. Sirens.	2. Continuous duty.
C. Machine tools.	3. Short time duty.
D. Wire drawing machine.	4. Continuous duty with starting and braking.

Codes:	A	B	C	D
(a)	2	3	4	1
(b)	1	2	3	4
(c)	3	4	1	2
(d)	4	1	2	3

9.

List 1 (Type of Motor)	List 2 (Relative Cost)
A. Squirrel cage type motor.	1. 3–4
B. DC motor.	2. 2.0
C. Synchronous motor.	3. 1.0
D. Schrage motor.	4. 1.5–3.0

Codes:	A	B	C	D
(a)	4	1	2	3
(b)	2	3	4	1
(c)	3	4	2	1
(d)	1	2	3	4

10.

List 1 (Type of Motor)	List 2 (Application)
A. Synchronus motor.	1. Vacuum cleaner.
B. Multi-speed motor.	2. Haulage and hoist work.
C. 3-phase series motor.	3. Ventilating fans.
D. Single phase series motor.	4. Voltage control and pf improvement.

Codes:	A	B	C	D
(a)	4	3	2	1
(b)	1	2	3	4
(c)	2	3	4	1
(d)	3	4	1	2

11.

List 1 (Application)	List 2 (Type of Motor Suggested For Use)
A. Shapers and slotters.	1. Flame proof squirrel cage motor.
B. Very large refrigeration and air conditioning plants.	2. Synchronous motors.
C. Mines.	3. Constant speed squirrel cage induction motor.
D. Kiln.	4. Ward-Leonard controlled dc shunt motor.

Codes:	A	B	C	D
(a)	3	2	4	1
(b)	2	3	1	4
(c)	3	2	1	4
(d)	1	4	2	3

12.

List 1 (Type of Electric Heating)	List 2 (Applications)
A. Direct resistance heating.	1. Surface hardening.
B. Arc heating.	2. Commercial cooking.
C. Indirect resistance heating.	3. Electrode boiler.
D. High frequency eddy current heating.	4. Electric arc welding

Codes:	A	B	C	D
(a)	1	2	3	4
(b)	2	1	3	4
(c)	4	1	2	3
(d)	3	4	2	1

Answers: 5. (c), 6. (c), 7. (d), 8. (a), 9. (c), 10. (a), 11. (c), 12. (d),

13.

List 1 (Type of Furnace)	List 2 (Application)
A. Resistance furnaces.	1. Manufacture of ferro alloys such as ferrochrome and ferromanganese.
B. Direct arc furnaces.	2. Melting and refining of brass and other heavy non-ferrous metals.
C. Submerged arc furnaces.	3. Heat treatment of steel.
D. Vertical core type induction furnaces.	4. Melting and refining of steel.

Codes:	A	B	C	D
(a)	1	2	3	4
(b)	2	3	4	1
(c)	3	4	1	2
(d)	4	1	2	3

14.

List 1 (Type of Welding)	List 2 (Applications)
A. Laser beam welding.	1. Welding of pipes, conduits, tanks, gasoline tanks.
B. Submerged arc welding.	2. Welding of sheet steel, copper alloys, brass, bronze and aluminium.
C. Carbon arc welding.	3. Welding of low alloy high tensile steels, mild low carbon steels, nickel, monel and nonferrous metals such as copper.
D. Seam welding.	4. In aircraft and electronic industries for light gauge metals.

Codes:	A	B	C	D
(a)	1	2	3	4
(b)	2	3	1	4
(c)	3	4	1	2
(d)	4	3	2	1

15.

List 1 (Metal to be Refined)	List 2 (Energy Consumption in kWh/Tonne)
A. Copper.	1. 150-300
B. Nickel.	2. 300-350
C. Silver.	3. 400-425
D. Gold.	4. 2,500-4,000

Codes:	A	B	C	D
(a)	1	2	3	4
(b)	2	3	4	1
(c)	3	4	1	2
(d)	1	4	3	2

16.

List 1 (Service)	List 2 (Type of Traction Recommended)
A. Urban/suburban service.	1. Steam locomotive drive.
B. Main line railway service.	2. Diesel or electric locomotive drive.
C. Branch line railway service.	3. Direct-internal combustion engine drive.
D. Outer suburbs and country districts.	4. Tramways trolley bus and dc electric trains.

Codes:	A	B	C	D
(a)	4	2	1	3
(b)	1	2	4	3
(c)	1	2	3	4
(d)	4	2	3	1

17.

List 1 (Type of Traction Motor)	List 2 (Traction Service)
A. DC series motor.	1. None.
B. AC series motor.	2. Urban/suburban services.
C. Linear induction motor.	3. Main line railway services.
D. DC shunt motor.	4. Trolley cars for internal transport in workshops or as a propulsion unit in marshalling yards.

Codes:	A	B	C	D
(a)	1	2	3	4
(b)	2	3	1	4
(c)	2	3	4	1
(d)	3	4	1	2

18.

List 1 (Current Collector)	List 2 (Application)
A. Trolley collector.	1. Railways for operating speeds as high as 100-130 kmph.
B. Bow collector.	2. Tramways.
C. Pantograph.	3. Trolleys and tramways operating at comparative slow speeds.

Codes:	A	B	C
(a)	1	2	3
(b)	2	3	1
(c)	3	1	2
(d)	3	2	1

19.

List 1 (Appliance)	List 2 (Load)
A. Airconditioner.	1. Few watts.
B. Ceiling fan.	2. 1–2 kW.
C. Electric shaver.	3. 60–100 watts.
D. Fluorescent tube.	4. 40–60 watts.

Codes:	A	B	C	D
(a)	1	2	3	4
(b)	2	3	1	4
(c)	2	3	4	1
(d)	3	2	4	1

20.

List 1 (Type of Consumers)	List 2 (Demand Factor)
A. Domestic and commercial lighting upto 1 kW.	1. 0.5
B. Domestic power.	2. 0.5–0.8
C. Industrial power.	3. 0.5–0.7
D. Tube-well.	4. 0.9–0.95

Codes:	A	B	C	D
(a)	1	2	3	4
(b)	2	3	4	1
(c)	3	1	2	4
(d)	4	3	1	2

Answers: 13. (c), 14. (d), 15. (d), 16. (a), 17. (c), 18. (d), 19. (b), 20. (c).

TRUE(T)/FALSE(F) TYPE QUESTIONS

Note : Identify whether the following statements are TRUE (T) or FALSE (F).

1. The group drive is economical in consideration of the initial cost of the motors and the control gear.

2. For driving heavy machines such as for lifts, cranes, shapers, group drive is preferred.

3. For paper mills and textile industry individual drive is preferred.

4. Multimotor drive is essential in complicated metal-cutting machine tools, paper making machines and rolling mills.

5. The motor designed for high speed and incorporating a reduction gear in the motor frame is usually found cheaper than that of the corresponding low speed motor.

6. For the same output kW, a higher speed motor is costlier than a slow speed motor.

7. When the motor is loaded, the electrical power input to the motor automatically changes to match the load.

8. Intermittent rating motor is desired for traction loads.

9. Series-parallel control is widely used in electric traction.

10. The motor produces noise because of bearings, vibrations and magnetic pulsations.

11. For a reversing drive (such as for colliery winder), the Ward-Leonard control system is generally used.

12. Smaller size machine have their critical speeds above their synchronous speeds.

13. An amplidyne is basically a dc amplifier.

14. A body cools from 60ºC to 50ºC in 10 minutes. The body will take the same time (*i.e.* 10 minutes) in further cooling down from 50ºC to 40ºC, the external conditions remaining unchanged.

15. In electric press mica is used because it is bad conductor of heat and good conductor of electricity.

16. Molybdenum is not suitable for heating elements of vacuum furnaces because of its high vapour pressure.

17. Nickel-chrome-iron alloy is the most useful material as heating element in electric furnaces.

18. Graphite resistors are suitable for any temperature that can be used but need protection against oxidation about exceeding 600ºC.

19. In indirect resistance heating, an enclosure is required for transfer of heat by conduction.

20. The inside proportions of a heating chamber are made to suit the character of the charge and type of the furnace or oven.

21. In case of batch type chambers, the charge is heated as it moves through the chamber.

22. Direct arc furnaces are usually of 3-phase type where as indirect arc furnaces are usually of single phase type.

23. Stirring action by rocking the furnace is achieved in single phase arc furnaces and by electromagnetic force is achieved in 3-phase arc furnaces.

24. For direct core type induction furnaces high frequency supply is essential.

25. Ajax Wyatt vertical core type furnace is suitable only for continuous operation.

26. Gas welding, arc welding, thermit welding, electroslag welding fall under the category of plastic welding.

27. In order to avoid the surface distortion, the portion of the metal adjacent to the weld or joint should not be allowed to be over-heated.

28. For welding thick materials the resistance of the current path in the work is kept minimum.

29. For welding thick materials of low conductivity, the resistance of the current path is of comparatively greater value and the control of contact resistance is not necessary.

30. For welding thick materials of high conductivity either reduced pressure or high resistance electrodes having melting point higher than that of metal to be joined, can be used.

31. For welding two dissimilar metals having different conductivity, low conductivity electrode on high conductivity metal side and vice versa are used in order to prevent overheating on the high conductivity metal and to develop sufficient heat to melt light conductivity metal side.

32. Materials above 5 mm thickness should be shot-blasted before being taken to the resistance welding machine and there should be no delay between the shot- blasting and welding in order to avoid new corrosion which might eliminate the advantage gained by the former.

33. The lower cost of equipment makes spot welding economical for small quantities.

34. Spot welding is used for thick metal sheets.

35. The range of open-circuit voltage for an electric arc welding is usually 100–250 V.

36. Down hill welding is preferred for thick metals because it is faster than uphill welding.

37. Uphill welding is preferred for thin metals because it produces stronger welds.

38. Arc can be produced by either ac or dc.

39. The carbon arc is very stable and easy to maintain.

40. The length of the carbon arc can be varied over wide limits without causing the arc to go out.

41. There is a tendency for the electrode to freeze or stick as in the case of metallic electrode.

42. For carbon arc welding both dc and ac can be used.

43. In metal-arc welding, too high voltage results in too small beads.

44. In metal arc welding, too high welding speed causes excessive piling up of weld metal.

45. Post-welding cleansing is required in MIG welding.

46. The load taken by welding transformer is highly inductive.

47. The power factor of load using welding transformer is usually very low of the order of 0.4 lagging.

Answers: 1. (T), 2. (F), 3. (T), 4. (T), 5. (T), 6. (F), 7. (T), 8. (T), 9. (T), 10. (T), 11. (T), 12. (T), 13. (T), 14. (F), 15. (F), 16. (T), 17. (T), 18. (T), 19. (F), 20. (T), 21. (F), 22. (T), 23. (T), 24. (F), 25. (T), 26. (F), 27. (T), 28. (F), 29. (T), 30. (T), 31. (F), 32. (T), 33. (T), 34. (F), 35. (F), 36. (F), 37. (F), 38. (T), 39. (T), 40. (T), 41. (F), 42. (F), 43. (T), 44. (T), 45. (F), 46. (T), 47. (T),

48. The sensation of colour is due to the difference in the wavelengths of the light radiations.
49. Colour of light depends upon wavelength only.
50. Solid angle subtended at a point in space is 2π steradians.
51. While illumination of surface is independent of reflection factor, the value of brightness is not independent of it.
52. For the same power output low voltage lamps willl be more sturdy.
53. Filament lamps at starting will take current less than its normal operating current.
54. Incandescent filament lamps are generally of copper filaments.
55. Incandescent lamps should be used with luminaires.
56. Maintenance of discharge lamps is more cumbersome.
57. Halogen lamps are preferred for outdoor illumination of buildings.
58. In GLS lamps the cap is secured to the glass shell by cementing compound.
59. Sodium vapour lamp is only suitable for operation on dc.
60. A leak transformer used in sodium vapour lamp acts as a step-down transformer initially.
61. In a sodium vapour lamp, the neon gas serves as to start the discharge and to develop sufficient heat to vapourize the sodium.
62. In mercury vapour lamp the harmful ultraviolet rays are absorbed by outer bulb.
63. Sodium vapour lamps use metallic sodium sealed in translucent aluminium oxide tubes.
64. Power factor of high pressure mercury vapour lamp is nearly unity and so do not need capacitor for power factor correction.
65. In a colour-corrected high pressure mercury vapour lamp, the outer glass envelope is coated with fluorescent powder.
66. Mercury iodide lamps are not suitable for use for floodlighting.
67. The luminous efficiency of mercury iodided lamps ranges from 75 to 90 lumens per watt.
68. Starters used in fluorescent tubes are usually of thermal type.
69. In fluorescent tubes low voltage is required for starting.
70. Fluorescent tubes operating on dc are generally free from stroboscopic effect.
71. In a fluorescent tube starter is required for preheating the electrodes.
72. According to the Stefan Boltzmann's law intensity of emitted radians is proportional to T^4 where T is the absolute temperature.
73. For the same lumens output, the operating cost of fluorescent lamp is less than the filament lamp.
74. Low voltage generally facilitates starting of fluorescent tubes.
75. With use, walls of a fluorescent tube blacken a little and dark rings appear near the ends owing to the deposition of active material from the electrodes.
76. The lumen output of a fluorescent tube is reduced, after about 4,000 hours of use, by 15-20 per cent.
77. Power factor of fluorescent lamps is high.
78. Interior lighting is designed by lumen method.
79. Lumen method takes into account the interreflections of light inside a room.
80. Luminaires are generally categorized as industrial, commercial or residential.
81. Lenses with prismatic patterns refract light sources to disperse the rays or to direct them most effectively.
82. Greater the size of the room lesser will be the coefficient of utilization.
83. Maintenance factor and depreciation factor are two different names of same concept.
84. Space-height ratio is less with dispersive type reflectors than that with concentric type reflectors.
85. Smaller the space-height ratio lesser the number of lamps required by a given installation.
86. In a lighting installation selection of proper space-height ratio ensures uniformity of illumination.
87. Ratio of light flux reaching the working plane to that emitted by the lamp(s) is called the maintenance factor.
88. Depreciation factor is merely the inverse of utilization factor.
89. The load factor of chemical plants is about 50 per cent.
90. For copper plating, material used is copper sulphate and nitric acid.
91. The main drawback of thermoelectric refrigeration system is its initial prohibitive cost.
92. The main functions of a condenser in a refrigeration system is to remove the heat absorbed by the evaporator.
93. A room air-conditioner should, preferably, be fitted to a window which is not exposed to the direct Sun rays.
94. The use of air-conditioner at peak load periods should be the minimum.
95. Steam locomotives in India, use single acting reciprocating engines.
96. The overall efficiency of a steam locomotive is 10-15 per cent.
97. Maintenance and repair cost of steam locomotives is less in comparison to that of electric locomotives.
98. Steam locomotives cannot be used for underground railways.
99. The coefficient of adhesion is better in steam traction than that in electric traction.
100. In direct steam engine drive, the reciprocating steam engine is invariably used for getting the necessary motive power because of its inherent simplicity, operational dependability, simplified maintenance and easy speed control.
101. In direct IC engine drive, speed control is possible only by employing a gearbox.
102. A diesel engine has an overload capacity of 35%.
103. Steam electric drive (employing steam turbine for driving a generator used for supplying current to electric motors) is commonly used in railways.
104. In IC engine-electric drive, the reduction gear and gearbox are eliminated.
105. IC engine-electric drive has found considerable favour for railway work and locomotives of this type are becoming widely used.
106. Regenerative braking cannot be employed with IC engine-electric drive but there is no bar in using rheostatic braking.
107. Battery-electric drive is commonly used for urban and suburban services.

Answers: 48. (T), 49. (F), 50. (F), 51. (T), 52. (T), 53. (F), 54. (F), 55. (T), 56. (T), 57. (T), 58. (T), 59. (F), 60. (F), 61. (T), 62. (T), 63. (T), 64. (F), 65. (T), 66. (F), 67. (T), 68. (F), 69. (F), 70. (T), 71. (T), 72. (T), 73. (T), 74. (F), 75. (T), 76. (T), 77. (F), 78. (T), 79. (T), 80. (T), 81. (T), 82. (F), 83. (T), 84. (F), 85. (F), 86. (T), 87. (F), 88. (F), 89. (F), 90. (F), 91. (T), 92. (T), 93. (T), 94. (T), 95. (F), 96. (F), 97. (F), 98. (T), 99. (F), 100. (T), 101. (T), 102. (F), 103. (F), 104. (T), 105. (T), 106. (T), 107. (F),

108. Battery vehicles are started by series-parallel grouping of batteries in parallel for starting and running at the speed upto half maximum speed and in series for running at maximum speed.

109. In Kolkata city tramways are still a means of public transport.

110. Owing to better speed control in electric drive locomotion on steep gradients at slow speed is possible.

111. The electric drive is healthier from the hygienic point of view owing to complete absence of smoke and fumes.

112. The braking employed in case of tramways is regenerative one.

113. The equipment used in tramways is similar to that used in railways but of considerably smaller output (not exceeding 50 kW).

114. The trolley bus employs a dc compound motor of output of 50-100 kW.

115. The trolley bus makes use of series-parallel control method for its speed control.

116. Plain rheostatic braking is used in trolley buses.

117. Power for lighting in passenger coaches, in a long distance train, is provided through individual generator of bogie and batteries.

118. The overhead distribution system in the case of traction is identical with that of power system.

119. The cost involved in substations in case of dc traction system is less than that involved in ac system.

120. Cost of ac electrification of track is more than that of dc electrifcation.

121. In similar conditions the energy consumption in dc traction system is more as compared with ac traction system.

122. For exerting the same torque, current required in dc traction system is more than that in ac system.

123. The acceleration is constant during notching up.

124. During notching period, the tractive effort developed by the traction motor fluctuates between a maximum and minimum about the mean.

125. Distance between stops divided by the actual time of run is called the schedule speed.

126. In main line service free running and coasting periods are longer but acceleration and braking periods are smaller.

127. Weight transfer is a desirable quality.

128. Dead weight is smaller than accelerating weight.

129. The ratio of the dead weight to the maximum value of tractive effort developed without wheel slipping is called coefficient of adhesion.

130. The coefficient of adhesion reduces as the speed falls.

131. Increase of motor torque beyond adhesive limit does not increase the tractive effort but causes wheel skidding.

132. Increasing of braking torque beyond adhesive limit causes wheel slipping.

133. Gears are used in electric traction to increase the tractive effort at the driving wheel.

134. During starting the train resistance increases linearly with speed.

135. Power output from the driving axles is the product of tractive effort and speed.

136. Steep gradient will involve less energy consumption.

137. In a train, the energy output of the driving axles is used in accelerating the train, overcoming the gradient and train resistance.

138. The energy output of a train for a given speed-time curve is independent of the type of drive used, but energy consumption depends upon the type of drive used.

139. Specific energy output is independent of locomotive overall efficiency but the specific energy consumption depends upon the overall efficiency of the locomotive.

140. In a single phase ac traction system ac series motors are employed for getting necessary motive power.

141. Initial rush of current on temporary interruption of supply is more with dc series motors than that with dc shunt motors.

142. In a dc series motor operating on undulating dc supply, the undulating factor will be less than unity.

143. Single phase compensated series motors, which can be built in sizes up to several hundred kW output, are usually used for electric traction work.

144. The speed of the linear induction motor drive will be lesser than that of the travelling field.

145. Linear induction motor has superiority over conventional rotary induction motor for speeds exceeding 200 kmph.

146. Speed control of motors in ac traction is difficult and wasteful than that in dc traction.

147. Rheostatic control of 3-phase induction motors is used in light-locomotives and motor coaches where single economical speed is sufficient and energy consumption is of no importance.

148. In a buck and boost method of control, the number of machines used is 4.

149. Regenerative braking is more efficient and less complicated in ac system in comparison to that with dc system.

150. Negative boosters have to be used in dc traction to confine ground currents to rails.

151. In ac traction, booster transformers are necessary in order to confine the return current through the rails.

152. The rectifiers used in modern electric traction are based on the principle of silicon controlled rectifiers.

153. Staggar in the contact wire is provided to even out the wear on the contact strip of pantograph.

154. Single ended fairelay type pantograph is employed in dc traction whereas the diamond type pantograph is used in ac traction.

155. Neutral section in OHE is provided to avoid short circuiting of two live phases by the pantographs.

156. Interference with communication system is more with dc than that with ac.

157. Magnetic brakes are usually employed on trolley buses and cars.

158. Individual drive is employed for high speed locomotives and collective drive for low speed locomotives (goods locomotives).

159. Overhead construction is universal for all ac railways.

160. The substations for track electrification are unattended type and arranged for remote control operation.

161. A plant layout is the first step in installing a power plant. **[A.M.I.E. Sec. B. Winter 1996]**

162. The annual depreciation of a hydropower plant is about 5 per cent.

163. The power plant capacity can be determined by studying the load duration curve and anticipated future demands.

Answers: 108. (T), 109. (T), 110. (T), 111. (T), 112. (F), 113. (T), 114. (T), 115. (F), 116. (F), 117. (T), 118. (F), 119. (F), 120. (F), 121. (F), 122. (F), 123. (T), 124. (T), 125. (F), 126. (T), 127. (F), 128. (T), 129. (F), 130. (F), 131. (T), 132. (T), 133. (T), 134. (T), 135. (T), 136. (F), 137. (T), 138. (T), 139. (T), 140. (T), 141. (T), 142. (T), 143. (T), 144. (T), 145. (T), 146. (F), 147. (T), 148. (T), 149. (F), 150. (T), 151. (T), 152. (T), 153. (T), 154. (F), 155. (T), 156. (F), 157. (T), 158. (T), 159. (T), 160. (T), 161. (T), 162. (F), 163. (T),

164. Maximum demand on a power system is the instantaneous maximum value of kVA supplied during a given period.
165. Higher the maximum demand, better is the utilisation of total plant capacity.
166. The main purpose of design and operation of the power plant is to generate the electrical energy at minimum possible cost.
167. Transmission lines also form a part of capital cost of the power plant.
168. Depreciation is a fixed cost for a power plant.
169. The salaries of high officials is accounted for as running charges.
170. Fuel consumption depends upon the installed capacity of the power plant.
171. Cost of labour is considered as operating cost where as the depreciation is taken as fixed cost.
172. Obsolescence is because of technological advancements.
173. Salvage value of a plant is always zero.
174. Depreciation accounts for the wear and tear with use.
175. Operating cost of a nuclear plant is more than that of a hydroplant.
176. Operating cost of a nuclear plant is more than that of a steam power plant.
177. Operating cost of a diesel power plant is less than that of a gas turbine plant.
178. The peak of the load curve represents the maximum demand on the power station on the particular day.
179. Load curve is derived from the load duration curve.
180. Higher the value of load factor, lower is the cost of generation.
181. Higher the value of the diversity factor lesser will be the value of fixed charges per unit generated.
182. Domestic consumers have maximum demand during early night hours.
183. There is an inherent reserve in most electrical supply systems in that the load can be reduced by reducing the voltage.
184. A power plant having maximum demand more than installed capacity will have utilisation factor of lesser than unity.
185. While selecting units continuous rating of generators is taken into account.
186. In a generating station it is always desirable to have two or more generating units.
187. For a generator at a particular load, when incremental heat rate is equal to heat rate, it will operate at maximum efficiency.
188. In two part tariff, the consumer with low power factor will be charged more on account of demand charges.
189. In a two part tariff energy charges do not depend on power factor.
190. In a power factor tariff, charge per unit to be recovered from a consumer having high power factor is less.
191. Hopkinson demand rate is a three part tariff.
192. Doherty rate tariff is a two part tariff.
193. Industrial plants usually operate on leading power factor .
194. Power factor tariff is charged on the basis of maximum demand in kVA.
195. Low power factor leads to higher energy loss.
196. Low power factor results in poor voltage regulation.
197. It is advisable to improve power factor to unity.
198. Improvement in power factor of an installation results in increase in kVA demand of the installation.
199. Capacitor banks are installed for improvement of power factor.
200. For smaller kVAR requirements it is economical to use static capacitors rather than synchronous capacitors.

Answers: 164. (F), 165. (T), 166. (T), 167. (T), 168. (T), 169. (F), 170. (F), 171. (T), 172. (T), 173. (F), 174. (T), 175. (T), 176. (F), 177. (F), 178. (T), 179. (F), 180. (T), 181. (T), 182. (T), 183. (T), 184. (F), 185. (T), 186. (T), 187. (T), 188. (F), 189. (T), 190. (T), 191. (F), 192. (F), 193. (F), 194. (T), 195. (T), 196. (T), 197. (F), 198. (F), 199. (T), 200. (T).

FILL IN THE BLANKS TYPE QUESTIONS

Note : Fill in the blanks with appropriate word or words.

1. Hoist and cranes require _______ starting torque.
2. _______ is defined as a form of machine equipment designed to convert electrical energy into mechanical energy and provide electrical control of this process.
3. _______ rating is an output which a motor can deliver for a specified period.
4. Metadynes should _______ be operated on open circuit.
5. Amplidynes should never be operated on _______.
6. Thermal conductivity is measured in _______ .
7. The ratio of heat flow from two walls of same thickness having thermal conductivity in the ratio of 1 : 0.8 and for the same temperature difference will be _______ .
8. In case of immersion type water heater, the heat is transferred by _______ .
9. In electric press, mica is used because it is _______ conductor of heat but _______ conductor of electricity.
10. Heating element in an incandescent lamp is of _______.
11. Heating elements used in household appliances are made of _______.
12. An enclosure of a charge to be heated by radiation or convection or by the joint effect of these modes of heat transfer is called the _______.
13. In the indirect resistance heating method, maximum heat transfer takes place by _______ .
14. Heating chambers may be of _______ or _______ type.
15. Large furnace installations are often equipped with _______ charging.
16. Indirect furnaces are usually made in _______ sizes than direct arc furnaces.
17. Standard size of Ajax Wyatt vertical core type furnaces ranges from 60 to 300 kW, all _______ phase, 50 Hz for standard voltage upto _______ V.
18. Indirect core type induction furnace has comparatively _______ power factor in comparison to that of resistance ovens.

19. In the plastic or ______ welding, the pieces of metal to be joined are heated to plastic state and then forged together under external pressure.
20. In the fusion or ______ welding, the material at the joint is heated to a molten state and allowed to solidify.
21. Project welding uses the same equipment as the ______ welding.
22. ______ welding can be defined as series of continuous spot welds.
23. For spot welding the tips of the electrodes used are made of ______ alloy.
24. In metal arc welding, too excessive welding current results in ______.
25. A 10 SWG electrode has approximate diameter of ______.
26. The electrode that can be used for welding two 3 mm thick mild steel sheets is ______.
27. In atomic hydrogen welding, the electrodes used are made of ______.
28. In argon arc welding, the electrodes used are made of ______.
29. Argon is an ______ gas.
30. TIG welding is ______ welding.
31. MIG welding is ______ welding.
32. DC generator in a M-G set used in arc welding is of ______ characteristics.
33. The transformer used in a welding set is ______ transformer.
34. The dc generator used in M-G set employed in arc welding is of ______ type.
35. A rectifier used for welding has voltage-current characteristic as ______.
36. ______ supply is used in atomic hydrogen welding.
37. One Angstrom is equal to ______.
38. Human eye is most sensitive for a wavelength of ______ Å.
39. The amount of light entering the eye is controlled by ______.
40. Light is a form of ______ energy.
41. Light wave travels with a velocity of ______.
42. The colour having shortest wavelength is ______.
43. The colour having longest wavelength is ______.
44. Steradian is defined as the solid angle that subtends a surface on the sphere equivalent to the square of the ______
45. Light is expressed in ______.
46. ______ may be defined as the brightness within the field of vision of such a character as to cause annoyance, discomfort, interference with vision or eye fatigue.
47. The unit of illumination is ______.
48. Candela is the unit of ______.
49. 1 lumen/m^2 is equal to ______.
50. The luminous flux is measured in ______
51. Lumens/watt is the unit of ______.
52. A lamp has a total luminous flux of 1,570 lumens. Its *mscp* is ______.
53. ______ factor is defined as the ratio of reflected light to incident light and is always ______ than unity.
54. ______ lamp has least capacity to sustain voltage fluctuations.
55. ______ lamps emit only a faint glow and are used as indicating lamps on panels and appliances.
56. The diameter of tungsten filament for GLS lamps is of the order of ______
57. In GLS lamps, button incorporates ______ wires.
58. In GLS lamps, the blackening of lamps is on account of the deposits of ______.
59. Caps of GLS lamps are made of brass or ______.
60. Translucent materials are used as ______ materials for light sources.
61. Power factor of a sodium vapour lamp is nearly ______ lagging.
62. Leak transformer used in a sodium vapour lamp acts as a choke coil for limiting the ______ when the lamp starts conducting.
63. The light from a sodium vapour lamp is nearly ______.
64. Neon lamps are basically ______ lamps.
65. The usual operating voltage for neon tubes is ______ volts.
66. Neon tube consists of a length of glass tubing containing two electrodes, normally cylindrical in shape, of iron, steel or ______.
67. Fluorescent lamp operating on dc supply needs ______ in addition to a starter and a choke.
68. In comparison to typing rooms, drawing office needs ______ level of illumination.
69. To achieve uniform illumination the whole area to be illuminated is divided into a number of ______ as far as possible.
70. The SI unit of electrochemical equivalent is ______.
71. Rating of a domestic refrigerator is of the order of ______.
72. The refrigerating capacity of 165 domestic refrigerator is approximately equal to ______ tonne.
73. The capacity of a 1 tonne air conditioner is ______ kcal/hour.
74. One ton capacity of the refrigeration plant needs approximately ______ watt motor.
75. Various refrigerants used are SO_2, CO_2, methylchrolide, ammonia and ______.
76. Temperature inside the refrigerator is to be maintained between ______ and ______ °C.
77. Temperature inside the freezer of a refrigerator is ______ °C.
78. In air-conditioning system the quantity of free air required per person is usually ______ m^3/minute.
79. The air coming out of the outlet is known as ______ air.
80. As compared to men, women need ______ °C higher effective temperature.
81. Package units may be of ______ or ______ type.
82. A desert cooler operates on the principle of ______ cooling.
83. The temperature of water in a water cooler is controlled by a ______ switch.
84. In a steam locomotive, electric power is provided through a ______.
85. The overall thermal efficiency of steam locomotive is about ______ %.
86. Overload capacity of diesel engine is usually restricted to ______ %.
87. The efficiency of diesel locomotives is about ______ %.
88. ______ is cheaper for low traffic density areas.
89. For standard passenger locomotives the cruising range is around ______ km.
90. Because of low coefficient of adhesion power-weight ratio of steam locomotive is ______.
91. Owing to high centre of gravity of steam locomotive, the speed is ______.
92. The efficiency of IC engine at its normal speed is about ______ per cent.
93. Maintenance and operating costs of diesel-electric locomotives are ______.

94. For the same power diesel-electric locomotive is _______ than steam or electric locomotive.
95. _______ drive is ideally suitable for underground and tube railways.
96. An electric locomotive needs much _______ time for maintenance and repairs in comparison to a steam locomotive.
97. Operating voltage for dc traction is _______ volts.
98. The tramcar is provided with at least _______driving axles in order to achieve necessary adhesion and make control possible from either end.
99. Efficiency of dc substations is _______ in comparison to that of ac substations.
100. During free running _______ remains constant.
101. During coasting the speed of the train _______.
102. During coasting the power supply remains _______.
103. _______ precedes braking.
104. Coasting is followed by _______.
105. The maximum speed attained by the vehicle during the run is called the _______ speed.
106. Suburban railway service needs _______ acceleration and braking retardation and _______ stopping and starting.
107. The total weight of the locomotive and train to be pulled up is called the _______ weight.
108. In electric traction the torque exerted is _______ whereas in steam traction it is _______.
109. Slipping will take place when tractive effort is _______ than the maximum frictional force between the wheels and the track.
110. The coefficient of adhesion is _______ on wet rails in comparison to that of clean dry rails.
111. The effective force necessary to propel the train, at the wheels of the locomotive is called the _______.
112. The accelerating weight is taken _______ to _______ per cent more than dead weight.
113. The starting efficiency with four motors in series parallel is _______ whereas with rheostatic starting it is 50%.
114. Mechanical braking is required at a speed of _______ kmph to bring the locomotive to rest.
115. Generally regenerative braking is desirable and necessary for services having long gradients exceeding _______ per cent.
116. Electromechanical drum brakes are usually employed on _______.
117. _______ drive is not suitable for high speed locomotives owing to stresses and vibrations in the connecting rods at high speeds.
118. Power supply frequency for 25 kV single phase ac system in India is _______ Hz.
119. The motor for _______ drive needs not be paid so much attention in designing with regard to dimensions as much that for _______ drive.
120. The gearing ratio usually employed in transmission of drive in electric traction is usually _______.
121. _______ is usually employed for motor coaches and tramways.
122. _______is not used on large locomotives because of the relatively large unsprung weight.
123. _______ system is cheap and easy to repair and inspection and is suitable for heavy current collection (top contact system for voltages up to 750 V and side contact system up to 1,200 V).
124. The trolley collector is suitable for comparative low speeds, say _______ to _______ kmph and beyond this speed there is every possibility of its jumping.
125. In addition to main traction motors, the motor-generator set is installed on the electric locomotive to supply lighting, control system and other power circuits at low voltage from _______ to _______V.
126. The value of the plant at the end of its useful life is known as _______ value.
127. A load curve is the relationship between _______. [A.M.I.E. Sec. B. Summer 1994]
128. Load curve is a graph between _______ on X-axis and _______ on Y-axis. [A.M.I.E. Sec. B. Summer 1999]
129. _______ curve gives the number of hours for which a particular load lasts during the day.
130. The sum of continuous ratings of all the electrical equipment connected to the supply system is known as _______ load.
131. The greatest of all 'short time interval averaged' during a given period on the power station is called the _______.
132. Load factor is the ratio of _______ to _______. [A.M.I.E. Sec. B. Summer 1999]
133. Demand factor is the ratio of _______ to _______. [A.M.I.E. Sec. B. Summer 1999]
134. Diversity factor is the ratio of _______ to _______. [A.M.I.E. Sec. B. Summer 1999]
135. The peak load of a 40 MW capacity power station is 30 MW. The loads having maximum demands of 25 MW, 10 MW, 5 MW and 7 MW are connected to the system. The diversity factor of the plant is _______. [A.M.I.E. Sec. B. Summer 1993]
136. In Q. 135, if the load factor for the year is 50%, the energy supplied by the system is _______ kWh/year. [A.M.I.E. Sec. B. Summer 1993]
137. Plant capacity factor is the ratio of _______ to _______. [A.M.I.E. Sec. B. Summer 1999]
138. The utilisation factor is defined as the ratio of _______ to _______. [A.M.I.E. Sec. B. Summer 1999]
139. Firm power is the power intended to be available _______ even under _______condition. [A.M.I.E. Sec. B. Summer 1999]
140. Cold reserve is the reserve generating capacity which is available for _______ but is not in _______. [A.M.I.E. Sec. B. Summer 1999]
141. Hot reserve is the reserve generating capacity which is in _______ but not in _______. [A.M.I.E. Sec. B. Summer1999]
142. Spining reserve is that generating capacity which _______ and is ready to _______. [A.M.I.E. Sec. B. Summer 1999]
143. Most of the power generating units are conservatively rated. They can be easily overloaded by about _______ per cent for short durations.
144. _______ load includes domestic lights, power for domestic appliances and small motors for pumping water.
145. _______ load includes lighting for commercial establishments, advertising, and electrical appliances used in shops and restaurants.
146. _______ load consists of street lighting, power required for water supply and drainage purposes.
147. _______ load includes electrical power required for driving water pumps to supply water to fields.
148. _______are the different methods of charging the consumers for electrical energy consumed by them.
149. _______ is applied to medium industrial consumers.
150. _______ is applied to bulk supplies.

ANSWERS

1. high **2.** An electric drive **3.** Short-time **4.** never
5. short circuit **6.** MJ/m/°C/hour **7.** 1:0.8 **8.** convection
9. good; bad **10.** tungsten **11.** nichrome **12.** heating chamber
13. radiation **14.** batch; continuous **15.** side door **16.** smaller
17. single; 600 **18.** poor **19.** pressure **20.** non-pressure
21. spot **22.** Seam **23.** copper **24.** excessive spatter
25. 3.3 mm **26.** 10 SWG **27.** tungsten **28.** tungsten
29. inert **30.** tungsten inert gas **31.** metal inert gas **32.** drooping
33. step-down **34.** differential compound wound **35.** drooping
36. AC **37.** 10^{-10} m **38.** 5,500 **39.** pupil
40. radiant **41.** 3×10^8 m/s **42.** blue **43.** red
44. radius **45.** lumen-hours **46.** Glare **47.** lux
48. luminous intensity **49.** 1 lux **50.** lumens **51.** luminous efficiency
52. 125 **53.** Reflection; less **54.** Incandescent **55.** Glow
56. microns **57.** support **58.** tungsten vapour **59.** aluminium
60. screening **61.** 0.3 **62.** current **63.** monochromatic
64. gas discharge **65.** 6000 **66.** copper **67.** a resistor
68. higher **69.** squares **70.** kg C^{-1} **71.** 0.1–0.3 tonne
72. 0.1 **73.** 3,000 **74.** 750 **75.** freon
76. 5; 10 **77.** – 7 to –15 **78.** 1.2 **79.** primary
80. 0.5 **81.** window; console **82.** evaporative **83.** thermostatic
84. small turbo-generator **85.** 6–8 **86.** 10 **87.** 25
88. Steam-engine drive **89.** 160 **90.** low **91.** limited
92. 25 **93.** high **94.** costier **95.** Electric
96. less **97.** 1,500 **98.** two **99.** less
100. speed **101.** decreases **102.** off **103.** Coasting
104. braking **105.** crest **106.** rapid; frequent **107.** dead
108. continuous; pulsating **109.** more **110.** less **111.** tractive effort
112. 5; 10 **113.** 72.72 **114.** 6.5 **115.** 0.6
116. trolley buses and cars **117.** Collective **118.** 50 **119.** collective; individual
120. 3 to 5:1 **121.** Nose suspension **122.** Nose suspension **123.** Conductor rail
124. 24; 32 **125.** 30; 100 **126.** salvage
127. load on the power system with the time. **128.** time; kW **129.** Load duration
130. connected **131.** maximum demand **132.** average load; maximum demand
133. maximum demand; connected load
134. sum of individual maximum demand; maximum demand of power plant. **135.** 1.5667
136. 131.4×10^6 **137.** average load; rated capacity **138.** maximum demand; rated plant capacity
139. always; emergency **140.** service; operation **141.** operation; service **142.** is connected to bus; take load
143. 10–15 **144.** Domestic **145.** Commercial **146.** Municipal
147. Irrigation **148.** Tariffs **149.** Hopkinson demand rate **150.** Doherty rate.

□□□

Index

A

AC series motor, 340
AC track electrification, 383
— high voltage ac supply, 383
Advantages of power factor improvement, 443
Air conditioning, 290
Anodizing, 278
Applications of electric drive, 171
Arc furnaces, 193
— power supply and control, 195
— types of arc furnaces, 194
Arc welding with dc and ac, 229
Arc lamps, 242
— carbon-arc lamp, 242
— flame-arc lamp, 243
— magnetic-arc lamp, 243
Automatic regulating equipment, 112
— amplidyne (rotary amplifier), 114
— ignitron, 114
— magnetic amplifiers, 116
— thyratron, 112
Auxiliary equipment, 374

B

Basic principles of light control, 255
Bearings, 85
Block diagram of an ac electric locomotive, 386
Booster transformer, 391
Braking, 363
— plugging, 363
— regenerative braking, 364
— rheostatic or dynamic braking, 363
Braking; mechanical considerations and control equipment, 363
Brushless (or commutatorless) dc motor, 164
Buck and boost method of speed control, 357

C

Calculation of current required for depositing given amount of metal, 270
Calculation of rating of electrical equipment, 296
Capacity scheduling, 407
Causes of low power factor, 439
Central air conditioning systems, 294
Choice of drive, 87
Choice of equipment, 449
Choice of frequency, 207
Choice of rating of motors, 92
Choice of size and number of generating units, 408
Chopper control of dc motors, 141
— chopper control of separately excited dc motor drive, 146
— principle of operation of chopper, 142
— step-down chopper with R-L-E_b load, 147
— step-up choppers, 144
— step-up/down chopper, 145
Classification of costs, 399
Classification of power converters, 130
— ac/dc converters, 131
— ac/ac converters, 131
— dc/dc converters, 132
— dc/ac converters, 132
Classification of electric heating methods, 186
Comfort air conditioning, 291
Comparison between dc and ac systems of railway electrification from the point of view of main line and suburban line railway services, 310
Comparison between desert cooler and window type air conditioner, 294
Comparison between resistance and arc welding, 229
Comparison between tungsten filament lamps and fluorescent tubes, 253
Comparison of converter fed and chopper fed dc drives, 150
Compound or multi-stage vapour compression refrigeration system, 285
Contactor type controller, 354
Contactors, 125
— magnetic starters, 126
Control devices for industrial motors, 112
Control equipment, 372
Control of traction motors, 348
Cost analysis of power plants, 400
— fixed and operating costs of hydropower plants, 400
— fixed and operating costs of nuclear power plants, 401
— fixed and operating costs of steam power plants, 400
Crest speed, average speed and schedule speed, 314
Current collection systems, 377
— conductor rail system, 377
— overhead system, 378
Current collectors for overhead system, 378
Current efficiency, 272

D

DC motor control through converters, 133
DC series motor, 335
DC shunt motor, 337
DC track electrification, 381
— dc substations, 381
— feeding and distributing systems, 382
— location of dc substations, 382
— transmission system for traction substations, 381
Dead weight, accelerating weight and adhesive weight, 327
Deposition of alloys, 278
Desert cooler, 289
Design of lighting schemes, 256
Determination of specific energy output using simplified speed-time curve, 322
— factors affecting energy consumption, 323
Dielectric heating, 204
Different systems of traction, 301
Disadvantages of low power factor, 439

Domestic refrigerator, 285
— trouble shooting of refrigerator, 287
Drum controller, 353
Dynamics of braking, 82
Dynamics of motor-load combination, 12
Duty cycles 90
— maximum temperature attained with intermittent loading, 91

E

Economic considerations, 399
Economical comparison of the two methods of increasing the power supplied, 448
Economics of power factor improvement, 446
Economics of power generation, 404
Eddy current brakes, 370
Effective temperature, 291
Electric arc welding, 218
— atomic hydrogen arc welding, 219
— carbon arc welding, 218
— inert gas metal arc welding, 220
— metal arc welding, 219
— submerged arc welding, 220
Electric braking, 67
Electrical braking of polyphase induction motors, 76
— dynamic (or rheostatic) braking, 77
— plugging (or counter-current braking), 76
— regenerative braking, 80
Electric braking of rectifier controlled separately excited dc motor, 152
Electric braking of synchronous motors, 81
Electric braking with dc motors, 69
— dynamic (or rheostatic) braking, 70
— plugging or counter-current braking, 69
— regenerative braking, 71
Electric drives, 3
Electric heating, 185
Electric traction motors, 333
Electric welding equipment, 228
Electric welding, 211
— advantages and disadvantages of welding, 212
— classification of welding processes, 212
— importance of welding, 211
— personnel protection and safety rules, 212
Electrodeposition, 275
Electrodeposition of rubber, 278
Electrodes, 224
Electrofacing, 278
Electroforming, 278
Electrolysis—basic principle, 269
Electrolytic processes, 269
Electromagnetic relays, 126
Electromechanical drum brakes, 370
Electrometallisation, 278
Electron beam welding, 222
Electronic control in resistance welding, 226
— ac timer circuit, 226
— energy storage welding processes, 227
— heat control unit, 226
— ignitron contactor, 226
Electroplating, 276
Electropolishing, 278
Electrotyping, 278
Energy output from driving axles, 322
Energy efficiency, 273
Extraction of metals, 273

F

Factors affecting schedule speed, 314
Factors affecting specific energy consumption of an electric train operating on a given schedule speed, 323
Factors governing selection of electric motors, 4
Factory lighting, 257
— industrial lighting fittings, 258
— maintenance, 258
— types of lamps, 258
Faraday's laws of electrodeposition, 269
Floodlighting, 262
Four quadrant operation of electric motors, 7
Frame size, 99

G

Gaseous discharge lamps, 245
— compact fluorescent lamps (CFLs), 252
— fluorescent tubes, 248
— high pressure mercury vapour discharge lamp, 246
— mercury iodide lamps, 247
— neon lamp, 248
— neon tubes, 248
— sodium vapour discharge lamp, 246
General features of traction motors, 333
General configuration of a motor drive, 130

H

Heating of buildings, 295
— factors affecting the choice of heating systems, 295
— types of heating equipments used for space heating, 295
High frequency eddy current heating, 202
High frequency power supply sources, 203
Hydraulic braking, 370

I

Illumination engineering, 232
Incandescent lamps, 243
Induction heating, 198
— coreless induction furnace, 200
— core type furnaces, 198
Industrial air conditioning, 291
Insulating materials, 88
— classification of insulating materials, 88
Integrating sphere, 242
Interest and depreciation, 401
— life of generating plant, 401
Interference in telecommunication circuits, 390

J

Joint speed-torque characteristic of an electric motor and a driven unit, 11

L

Laser beam welding, 222
Laws of illumination, 235
Linear induction motor, 343
Load equalization, 100
Load forecasting, 424
Load sharing between base load and peak load plants, 407
Low frequency single phase system, 383

M

Machines for resistance welding, 225
Magnetic track brakes, 370
Major substation equipment, 386
Matching power electronic converter and motor, 132
Measurement of illumination, 242
Mechanical braking, 369

Mechanical considerations, 370
— drive, 371
— transmission of drive, 371
— wheel arrangement, 371
Mechanical regenerative braking, 369
Mechanics of train movement, 318
Metadyne control, 357
Methods of determination of depreciation, 402
Methods of lighting calculations, 258
Methods of power factor improvement, 440
— by use of static capacitors, 440
— by use of synchronous or high power factor machines, 441
— location of power factor correction equipment, 442
Methods of ventilation and cooling of machines, 98
Modes of transfer of heat, 185
Motor speed-torque characteristics, 11
Motors for particular services, 171
Multiple unit control, 373

N

Nature of electric supply, 4
Negative booster, 392
Noise, 87

O

Operating characteristics of dc motors, 334
Operating or running characteristics of electric motors, 15
dc motors, 16
— three-phase ac motors, 20
— single-phase ac motors, 27
Overhead construction for tramways and trolley buses and railways, 379
— railways, 380
— tramways and trolley buses, 379
Overhead equipment, 388
— automatic weight tension and temperature compensation, 388
— insulated overlap, 389
— neutral section, 389
— section insulators, 390
— uninsulated overlap, 389

P

Photometry, 240
Pilot devices and accessories, 123
Plain rheostatic starting (notching), 350
Polar curves, 239
Power factor, 438
Power output from the driving axles, 320
Power supply arrangement for ac track electrification, 384
— elementary section, 386
— feeding post, 385
— sectioning and paralleling post, 385
— sectioning arrangement for power feeding, 384
— subsectioning and paralleling post, 386
— substations, 384
Power supply for arc welding, 228
Power supply for electric traction, 376
Power supply for electrolytic processes, 278
Power supply for resistance welding, 225
Preparation of work, 224
Production of chemicals, 274
Protective devices and interlocks, 127
Protective devices, 394

R

Rating and ventilation, 345
Rating of machines, 92
— effect of altitude on rated output, 92
— overload capacity of induction motors, 92
Referred load torques and moments of inertia, 13
Referring forces and masses having translational motion to a rotating shaft, 14
Referring torques and translating masses when speed is variable, 15
Refining of metals, 273
Refrigerants, 285
Refrigeration and air conditioning, 282
Refrigeration cycle, 282
Refrigeration systems, 283
— thermoelectric refrigeration system, 284
— vapour absorption refrigeration system, 284
— vapour compression refrigeration system, 283
Remote control centre, 387
Requirements of good weld, 223
Resistance welding, 212
— butt welding, 214
— percussion welding, 217
— projection welding, 216
— seam welding, 217
— spot welding, 215
Resistance heating, 187
— causes of failure of heating elements, 190
— design of heating element, 189
— efficiency and losses, 189
— heating element materials, 188
— infrared or radiant heating, 191
— maximum operating voltage, 188
— salt-bath heating, 191
— special types of resistance furnaces, 191
— temperature control of resistance ovens/furnaces, 190
Room air conditioners, 293
— comparison between desert cooler and window type air conditioner, 294

S

Sag and tension calculation for a trolley wire, 380
Selection of electric drive, 8
— load torques depending on path or position of load during motion, 10
— load torque-time characteristics, 9
— load torques varying with shaft displacement angle, 10
— speed-torque characteristics of machines or mechanisms, 9
Series-parallel starting, 350
— practical requirements, 352
Signalling, 387
Significance of load factor and diversity factor, 407
Simplified speed-time curves, 315
Single quadrant operation of electric motors, 6
Solid state devices, 118
— diac, 122
— diode, 118
— gate turn-off (GTO) thyristors, 122
— power transistors, 123
— silicon controlled rectifier (SCR), 120
— triac, 121
— thyristor, 119

Sources of light, 242
Special features of thyristor drive motors, 150
— effects of thyristor power supply on the performance of a dc motor, 150
— special features of thyristor drive motors, 150
Speed control and starting of single phase series motors, 360
Speed control of dc motor using feedback loops, 151
Speed control of electric motors, 54
— speed control of 3-phase induction motors, 60
— speed control of 3-phase synchronous motors, 64
— speed control of ac commutator motors, 64
— speed control of dc motors, 54
— speed control of single-phase induction motors, 63
Speed control of single phase induction motor using triac (fan regulator), 163
Spot pricing, 435
Starting and speed control of 3-phase induction motors, 360
Starting and speed control of dc traction motors, 348
Starting of electric motors, 39
— acceleration time, 41
— energy consumption during starting, 42
— starting of 3-phase induction motors, 46
— starting characteristics of electric motors, 40
— starting of dc motors, 43
— starting methods of synchronous motors, 51
Street lighting, 261
Summer air conditioning system, 292
Switched reluctance motor, 164
Systems of electric traction, 305
— diesel electric traction, 306
— tramways, 305
— trolley buses, 305
Systems of track electrification, 308

T

Tariffs and power factor improvement, 432
Tariffs, 432
— objective of tariff, 432
— principal factors affecting framing of tariff, 432
Temperature rise in electrical machines, 89
Terms used in illumination, 233
Three phase induction motor, 343
Thyristor control of dc series motors, 152
— fully controlled rectifier, 152
— semicontrolled rectifier, 154
Thyristor control of electric motors, 129
Thyristor control of separately excited dc motor drive, 133
— single phase drives, 133
— three phase drives, 140
Thyristor control of three phase induction motors, 156
— ac regulators, 156
— cycloconverters, 160
— reduced voltage starting (soft start), 161
— rotor resistance control, 161
— slip-power recovery scheme, 161
— variable current variable frequency control, 160
— variable voltage and variable frequency control, 157
Thyristor control of traction motors, 359
Thyristor control of a 3-phase synchronous motor, 163
— load commutated inverter drive for synchronous motor, 163
Torque-speed and speed-torque characteristics, 8
Traction systems, 301
Tractive effort for propulsion of train, 319
Train movement and energy consumption, 313
Transition methods, 352
Transmission of drive, 85
Types of air conditioning systems, 292
Types of consumers and their tariffs, 435
— agricultural consumers, 435
— bulk consumers, 436
— commercial consumers, 435
— domestic consumers, 435
— industrial consumers, 436
— street lighting, 436
— water supply, 436
Types of drives, 5
Types of enclosures, 83
Types of loads, 6
Types of motors used in industrial drives, 171
Types of tariffs, 432
Types of lighting schemes, 255
Typical speed-time curves, 314

U

Ultrasonic welding, 221

V

Voltage, 272

W

Water cooler, 288
— capacity of water coolers, 289
— instantaneous type water coolers, 288
— storage type water cooler, 288
Winter air conditioning system, 292

Y

Year-round air conditioning system, 292

❑❑❑

Win Prizes

ATTENTION READERS

Win Prizes

We request for your frank assessment regarding various aspects of the book as given below :

UTILIZATION OF ELECTRIC POWER AND ELECTRIC TRACTION

By : J.B. GUPTA

Please Fill up the Sheet in Neat and Clean Writing

(i) What topics of your syllabus, important for your examination point of view, are not covered in this book ?

....

....

....

....

(ii) What are the chapters wherein the treatment of the subject matter is not systematic or organised or updated ?

....

....

....

....

(iii) Have you come across misprints/mistakes/factual inaccuracies in the book ? Please specify the chapters and the page numbers.

....

....

....

....

(iv) What other books on the same subject have you found/heard better than the present book ? Please specify in terms of price and quality.

....

....

....

....

(v) Further suggestions and comments for the improvement of this book.

....

....

....

....

(PS : If need be please attach a separate sheet to write your views)

OTHER DETAILS

(i) Who recommended you this book ? (Please tick (✓) in the box near the option relevant to you.

☐ Teacher ☐ Friend ☐ Bookseller

....

....

....

....

(ii) Name of the recommending teacher, his designation and address:

....

....

....

....

(iii) Name and address of the bookseller you purchased the book from :

....

....

....

(iv) Name and address of your institution (Please mention the University or Board as the case may be)

....

....

....

....

(v) Your name and complete address :

....

....

....

....

(vi) Course you are enrolled in :

....

....

....

....

The best assessment will be awarded each month : The award will be in the form of our publications as decided by the Editorial Board.

Please mail the filled up coupon at the vearliest to :

S.K. Kataria & Sons®
4885/109, Prakash Mahal, Dr. Subhash Bhargav Lane,
Opposite Delhi Medical Association, Daryaganj, New Delhi-110002
Phones: +91-11-43551243, Mobile: +91-9871775858
e-mail: katariabook@gmail.com, katariabooks@yahoo.com
Website: www.skkatariaandsons.com